Advances in Intelligent Systems and Computing

Volume 999

The series “Advances in Intelligent Systems and Computing” contains publications on theory, applications, and design methods of Intelligent Systems and Intelligent Computing. Virtually all disciplines such as engineering, natural sciences, computer and information science, ICT, economics, business, e-commerce, environment, healthcare, life science are covered. The list of topics spans all the areas of modern intelligent systems and computing such as: computational intelligence, soft computing including neural networks, fuzzy systems, evolutionary computing and the fusion of these paradigms, social intelligence, ambient intelligence, computational neuroscience, artificial life, virtual worlds and society, cognitive science and systems, Perception and Vision, DNA and immune based systems, self-organizing and adaptive systems, e-Learning and teaching, human-centered and human-centric computing, recommender systems, intelligent control, robotics and mechatronics including human-machine teaming, knowledge-based paradigms, learning paradigms, machine ethics, intelligent data analysis, knowledge management, intelligent agents, intelligent decision making and support, intelligent network security, trust management, interactive entertainment, Web intelligence and multimedia.

The publications within “Advances in Intelligent Systems and Computing” are primarily proceedings of important conferences, symposia and congresses. They cover significant recent developments in the field, both of a foundational and applicable character. An important characteristic feature of the series is the short publication time and world-wide distribution. This permits a rapid and broad dissemination of research results.

**** Indexing: The books of this series are submitted to ISI Proceedings, EI-Compendex, DBLP, SCOPUS, Google Scholar and Springerlink ****

More information about this series at http://www.springer.com/series/11156

Asit Kumar Das · Janmenjoy Nayak ·
Bighnaraj Naik · Soumen Kumar Pati ·
Danilo Pelusi
Editors

Computational Intelligence in Pattern Recognition

Proceedings of CIPR 2019

Editors
Asit Kumar Das
Department of Computer Science
and Technology
Indian Institute of Engineering Science
and Technology
Howrah, West Bengal, India

Janmenjoy Nayak
Department of Computer Science
and Engineering
Sri Sivani College of Engineering
Srikakulam, Andhra Pradesh, India

Bighnaraj Naik
Department of Computer Application
Veer Surendra Sai University of Technology
Burla, Sambalpur, Odisha, India

Soumen Kumar Pati
Department of Bioinformatics
Maulana Abul Kalam Azad University
of Technology
Kolkata, West Bengal, India

Danilo Pelusi
Faculty of Communication Sciences
University of Teramo
Teramo, Italy

ISSN 2194-5357 ISSN 2194-5365 (electronic)
Advances in Intelligent Systems and Computing
ISBN 978-981-13-9041-8 ISBN 978-981-13-9042-5 (eBook)
https://doi.org/10.1007/978-981-13-9042-5

This Springer imprint is published by the registered company Springer Nature Singapore Pte Ltd.
The registered company address is: 152 Beach Road, #21-01/04 Gateway East, Singapore 189721, Singapore

Organizing Committee

Chief Patron

Prof. Parthasarathi Chakrabarti, Director, Indian Institute of Engineering Science and Technology, Shibpur, Howrah, West Bengal, India

Patrons

Prof. Amit Kumar Das, Professor, Department of CST, IIEST, Shibpur, Howrah, West Bengal, India
Prof. Jaya Sil, Professor, Department of CST, IIEST, Shibpur, Howrah, West Bengal, India
Prof. Sipra Das Bit, Professor, Department of CST, IIEST, Shibpur, Howrah, West Bengal, India
Prof. Susanta Chakraborty, Professor, Department of CST, IIEST, Shibpur, Howrah, West Bengal, India

Honorary Advisory Chair

Prof. Michael Pecht, Chair Professor and Director, University of Maryland, College Park, USA
Prof. V. E. Balas, Senior Member IEEE, Aurel Vlaicu University of Arad, Romania

Honorary General Chair

Prof. Uma Bhattacharya, Professor, Department of CST, IIEST, Shibpur, Howrah, West Bengal, India
Prof. Sulata Mitra, Professor, Department of CST, IIEST, Shibpur, Howrah, West Bengal, India
Prof. Sekhar Mandal, Professor, Department of CST, IIEST, Shibpur, Howrah, West Bengal, India

General Chair

Prof. Ajith Abraham, Director, Machine Intelligence Research Lab, USA & Professor, Technical University of Ostrava, Czech Republic
Prof. Asit Kumar Das, Associate Professor, Department of CST, IIEST, Shibpur, Howrah, West Bengal, India
Prof. H. S. Behera, Associate Professor, Department of Information Technology, Veer Surendra Sai University of Technology, Burla, Odisha, India

Program Chair

Prof. D. P. Mohapatra, Professor, Department of Computer Science and Engineering, National Institute of Technology, Rourkela, Odisha, India
Prof. Janmenjoy Nayak, Associate Professor, Department of Computer Science and Engineering, Sri Sivani College of Engineering, Srikakulam, Andhra Pradesh, India
Prof. Bighnaraj Naik, Assistant Professor, Department of Computer Application, Veer Surendra Sai University of Technology, Burla, Odisha, India
Prof. Apurba Sarkar, Assistant Professor, Department of CST, IIEST, Shibpur, Howrah, West Bengal, India
Prof. G. T. Chandra Sekhar, Associate Professor, Department of EEE, Sri Sivani College of Engineering, Srikakulam, Andhra Pradesh, India

Organizing Chair

Prof. Dipak Kumar Kole, Associate Professor, Department of Computer Science and Engineering, The Jalpaiguri Government Engineering College, Jalpaiguri, West Bengal, India

Prof. Soumen Kumar Pati, Associate Professor, Department of Computer Science and Engineering, MAKOUT, West Bengal, India
Prof. Soumi Dutta, Assistant Professor, Institute of Engineering and Management, Kolkata, West Bengal, India

International Advisory Committee

Prof. A. Abraham, Machine Intelligence Research Labs, USA
Dr. Claude Delpha, HDR, Université Paris-Sud, France
Dr. Chi-Chang Chang, Chung-Shan Medical University, Taiwang
Dr. S. K. Panda, National University of Singapore, Singapore
Dr. M. S. Obaidat, Monmouth University, USA
Prof. Raouf Boutaba, University of Waterloo, Canada
Dr. Hisao Ishibuchi, SUSTech, China
Prof. M. Murugappan, University of Malaysia
Prof. Kenji Suzuki, University of Chicago
Dr. Raffaele Mascella, University of Teramo, Italy
Prof. Christos Douligeris, University of Piraeus
Prof. Shaikh A. Fattah, BUET, Bangladesh
Prof. Basabi Chakraborty, Iwate Prefectural University, Japan
Dr. Swapnoneel Roy, University of North Florida, USA

National Advisory Committee

Prof. Sushmita Mitra, Machine Intelligent Unit, ISI, Kolkata, India
Prof. Sipra Das Bit, Department of CST, IIEST, Shibpur, India
Dr. Arnab Bhattacharya, Department of CSE, IIT Kanpur
Prof. Susanta Chakraborty, Department of CST, IIEST Shibpur, India
Dr. Partha Bhowmik, Department of CSE, IIT Kharagpur
Dr. Swagatam Das, Indian Statistical Institute, Kolkata, India
Dr. B. K. Panigrahi, Indian Institute of Technology, Delhi, India
Prof. Sulata Mitra, IIEST Shibpur, Howrah, West Bengal, India
Prof. Debdatta Sinha, Department of CSE, University of Calcutta
Prof. Haffizur Rahman, Department of IT, IIEST Shibpur, West Bengal
Prof. Monojit Mitra, Department of ETC, IIEST Shibpur, West Bengal
Prof. Shila Ghosh, Principal, St. Thomas College of Engineering and Technology, Kolkata, India
Dr. Abhik Mukherjee, Department of CST, IIEST Shibpur, India
Prof. Santi Prasad Maity, Department of IT, IIEST Shibpur, India
Dr. Haris Karnick, Department of CSE, IIT Kanpur
Prof. Paramartha Dutta, Visva Bharati University, India

Dr. Arindam Biswas, Department of IT, IIEST Shibpur, West Bengal, India
Dr. Sekhar Mandal, Department of IT, IIEST Shibpur, West Bengal, India
Dr. Nabendu Chaki, Department of CSE, University of Calcutta
Prof. Amitava Chatterjee, Department of EE, Jadavpur University, West Bengal
Prof. Mita Nasipuri, Department of CSE, JU, India
Dr. Sandip Kumar Shukla, Department of CSE, IIT Kanpur
Dr. S. V. Rao, Department of CSE, IIT Guwahati
Dr. Arijit Sur, Department of CSE, IIT Guwahati
Prof. Chandan Kumar Chanda, Department of EE, IIEST Shibpur, West Bengal, India
Dr. J. K. Mandal, University of Kalyani, West Bengal, India

Technical Committee

Dr. Pabitra Mitra, Department of CSE, IIT, Kharagpur, India
Dr. Alok Kumar Dutta, Visva Bharati University, West Bengal, India
Dr. Rogers Mathew, Department of CSE, IIT, Kharagpur, India
Dr. Danilo Pelusi, University of Teramo, Italy
Dr. Tanmay De, NIT Durgapur, West Bengal
Dr. Saptarshi Ghosh, Department of CSE, IIT, Kharagpur, India
Dr. Somnath Bhattyacharya, Department of CSE, Assam University, India
Dr. Indrajit Banerjee, Department of IT, IIEST Shibpur, West Bengal, India
Dr. Santanu Phadikar, MAKAUT, West Bengal, India
Dr. Bibhudendra Acharya, Department of ETC, NIT Raipur, India
Dr. Satchidananda Dehuri, Fakir Mohan University, Balasore, India
Dr. G. T. Chandra Sekhar, Sri Sivani College of Engineering, Andhra Pradesh, India
Dr. Aditya Nigam, IIT Mandi, India
Dr. Malay K. Pakhira, Kalyani Government Engineering College, West Bengal, India
Dr. Vinay P. Namboodiri, Department of CSE, VSSUT, India
Dr. Ranjit Ghoshal, Department of IT, St. Thomas College of Engineering and Technology, West Bengal, India
Dr. Madhulina Sarkar, Department of CSE, Government College of Engineering and Textile Technology, West Bengal, India
Dr. Ram Sarkar, Department of CSE, Jadavpur University, West Bengal, India
Dr. Mousumi Dutt, Department of CSE, St. Thomas College of Engineering and Technology, West Bengal, India
Dr. Prasun Ghosal, Department of IT, IIEST Shibpur, West Bengal, India
Dr. Chandan Giri, Department of IT, IIEST Shibpur, West Bengal, India
Dr. Piyush Rai, Department of CSE, IIT Kanpur, West Bengal, India
Prof. Amitabha Das, Calcutta Institute of Engineering and Management, Tollygunge, West Bengal, India

Dr. Mamata Dalui, National Institute of Technology, Durgapur, India
Dr. Nibaran Das, Department of CSE, Jadavpur University, India
Dr. Arijit Ghosal, Department of IT, St. Thomas College of Engineering and Technology, West Bengal, India
Dr. J. C. Bansal, South Asian University, New Delhi, India
Dr. Surajeet Roy, Department of IT, IIEST Shibpur, India
Dr. Debasish Mittir, National Institute of Technology, Durgapur, India
Dr. Imon Mukherjee, Indian Institute of Information Technology, Kalyani, India
Dr. Amit Paul, Department of CSE, St. Thomas College of Engineering and Technology, West Bengal, India
Dr. T. P. R. Vittal, Sri Sivani College of Engineering, Srikakulam, Andhra Pradesh, India
Dr. Bhanu Prakash Kolla, K. L. University, Vijayawada, Andhra Pradesh, India
Dr. Santosh Kumar Majhi, Veer Surendra Sai University of Technology, Burla, Odisha, India

Publicity Chair

Dr. Samit Biswas, IIEST, Shibpur
Dr. Shampa Sengupta, MCKV Institute of Engineering, Howrah, West Bengal, India

Logistic Chair

Ashish Kumar Layek, IIEST, Shibpur
Tamal Pal, IIEST, Shibpur

Convenors

Dr. Asit Kumar Das, IIEST, Shibpur
Dr. Apurba Sarkar, IIEST, Shibpur

Publication Chair

Dr. Surajeet Ghosh, IIEST, Shibpur
Dr. Sunanda Das, Neotia Institute of Technology, Management and Science, Kolkata, West Bengal, India

Finance Chair

Dr. Asit Kumar Das, IIEST, Shibpur
Dr. Apurba Sarkar, IIEST, Shibpur
Malay Kule, IIEST, Shibpur

CIPR Reviewers

Dr. Asit Kumar Das, IIEST, Shibpur
Dr. Soumen Kumar Pati, St. Thomas College of Engineering, Kolkata
Dr. Santosh Ku Sahoo, CVR College of Engineering, Hyderabad
Dr. Kola Bhanu Prakash, K L University, Vijayawada
Dr. Sarat Ch. Nayak, CMR College of Engineering and Technology (Autonomous), Hyderabad, India
Dr. Sunanda Das, Neotia University, Kolkata
Dr. G. T. Chandra Sekhar, Sri Sivani College of Engineering, Srikakulam
Mrs. D. Priyanka, Sri Sivani College of Engineering, Srikakulam
Dr. P. Vittal, AITAM, Andhra Pradesh
Mr. P. Suresh Kumar, Dr. Lankapalli Bullayya College, Visakhapatnam
Dr. P. Pradeepa, Jain University, Bangalore
H. Das, Kalinga Institute of Industrial Technology (KIIT), Bhubaneswar
Dr. P. M. K. Prasad, GVP College of Engineering for Women Visakhapatnam, Andhra Pradesh, India
Rashmi Ranjan Sahoo, Parala Maharaja Engineering College (Government of Odisha), Berhampur, Odisha
D. K. Behera, Trident Academy of Technology, Bhubaneswar
Mr. Umashankar Ghugar, Berhampur University (Odisha)
Dr. Sourav Kumar Bhoi, Parala Maharaja Engineering College (Government of Odisha), Berhampur, India
Sibarama Panigrahi, Sambalpur University Institute of Information Technology, Sambalpur
Dr. Chittaranjan Pradhan, Kalinga Institute of Industrial Technology, Bhubaneswar
Dr. M. Marimuthu, Coimbatore Institute of Technology, Coimbatore
Aditya Hota, VSSUT, Burla
Dr. S. K. Majhi, VSSUT, Burla
Dr. Nageswara Rao, Velagapudi Ramakrishna Siddhartha Engineering College
Mr. Shyam Prasad Devulapalli, JNTU Hyderabad
Radhamohan Pattnayak, Godavari Institute of Engineering and Technology, Rajahmundry, Andhra Pradesh
Dr. K. Suvarna Vani, Velagapudi Ramakrishna Siddhartha Engineering College, Andhra Pradesh

Dr. P. Vidya Sagar, Velagapudi Ramakrishna Siddhartha Engineering College, Andhra Pradesh
Dr. Shampa Sengupta, MCKV Institute of Engineering, West Bengal
Dr. Santanu Phadikar, Maulana Abul Kalam Azad University of Technology, West Bengal
Arka Ghosh, IIEST, Shibpur
Dr. Samit Biswas, IIEST, Shibpur
Dr. Apurba Sarkar, IIEST, Shibpur
Sarita Mohapatra, SOA University, Bhubaneswar
Subhra Swetanisha, Trident Academy of Technology, Bhubaneswar
P. Raman, GMR Institute of Technology, Andhra Pradesh
Yasmin Ghazaala, St. Thomas College of Engineering, Kolkata
Soumi Dutta, Institute of Engineering and Management, Kolkata
Dr. Amit Paul, St. Thomas College of Engineering, Kolkata
Dr. Arijit Ghoshal, St. Thomas College of Engineering, Kolkata
Amiya Halder, St. Thomas College of Engineering, Kolkata
Dr. Ramesh Prusty, VSSUT, Burla
Ms. D. Mishra, CET, Bhubaneswar
Dr. Bighnaraj Naik, VSSUT, Burla
Dr. Janmenjoy Nayak, Sri Sivani College of Engineering, Srikakulam
H. Swapnarekha, Sri Sivani College of Engineering, Srikakulam
Abhishek Bhattacharya, Institute of Engineering and Management, Kolkata
Sujata Ghatak, Institute of Engineering and Management, Kolkata
Rupam Bhattacharya, Institute of Engineering and Management, Kolkata
Abhijit Sarkar, Institute of Engineering and Management, Kolkata
Ankita Mondal, Institute of Engineering and Management, Kolkata
Sayan Das, Bankura Unnayani Institute of Engineering, West Bengal
Akash Choudhyry, Lexmark, India
Gourav Dutta, Cognizant Technology Solutions, India
Soumya De, Tata Consultancy Services, India
Diptyasish Chakraborty, Cognizant Technology Solutions, India
Keya Kundu, Cognizant Technology Solutions, India
Aranya Bandopadhyay, Capgemini, India
Debopriya Das, Amazon, India
Mr. P. B. Dash, VSSUT, Burla

Preface

Since the last few decades, there has been an urgent need for intelligent computing in different fields of science and technology due to the flawless increment in data. Development in science and technology is authoritative to emerge the internal strength of the country. However, demonstrating the faults and methodologies of progress with an impartial lookout benefits to update the existing knowledge and technologies. In the current situation, all the areas of related disciplines of pattern recognition tools present adaptive systems which authorize the understanding of theory as well as data in some difficult and changing environments.

The first international conference entitled "Computational Intelligence in Pattern Recognition" (CIPR-2019) is organized by Indian Institute of Engineering Science and Technology (IIEST), Shibpur, West Bengal, India, on 19 and 20 January 2019. The conference is directed towards the knowledge and structure of positive research in different applications of pattern recognition which are leading and governing the technological domain. CIPR has attracted more than 200 submissions based on the thematic areas. With thorough rigorous peer-review process, the editors have selected only 88 high-quality papers by the sophisticatedly knowledgeable domain expertise evaluators preferred from country and abroad. The papers are reviewed regarding the contribution, technical content, clarity and originality of some latest findings and research. The entire system (includes the submission of paper, review and acceptance process) is done electronically. We had a great time in collaboration with advisory committee, programme committee and technical committee for call for papers, and review and confirm papers for the proceedings of CIPR.

The international conference CIPR aims at encircling new type of educators, technologists, researchers for global success. The conference features various keynote addresses in the scope of advanced information regarding CIPR topics. The sessions are largely categorized according to the significance and interdependency of the papers pertaining to the key concept of the conference, and they have established plentiful prospects for presentations.

The accepted papers (both research and review papers) have been estranged to highlight the modern spotlight of computational intelligence techniques in pattern recognition. We anticipate the author's individual study and estimations add value to it. The success of CIPR is the outcome of good quality research works on the current applications from the authors. We require an enormous time for putting together this proceeding. The CIPR conference is an acknowledgment to a vast collection of people, and everybody should feel proud of the outcome. We enlarge our deep sense of thankfulness to all those for this affectionate encouragement, motivation and support for building it attainable. Last but not least, the editorial members of Springer have helped us in an immense way to publish this proceeding in a graceful way. Hope all of us will be profusely thankful for the good assistance made and defend our hard work.

Howrah, India — Asit Kumar Das
Srikakulam, India — Janmenjoy Nayak
Burla, India — Bighnaraj Naik
Kolkata, India — Soumen Kumar Pati
Teramo, Italy — Danilo Pelusi

Acknowledgements

The key proposal and significance of the CIPR conference has been interested more than 200 academicians, professionals or researchers throughout the world facilitated as to prefer superior quality papers and offer to exhibit the reputation of the CIPR conference for original research findings, exchange of ideas and sharing of knowledge with precisely national and international communes in different aspects and fields of data analysis and pattern recognition. Heartfully thanks to all those who have offered in produces such an extensive conference proceedings of the CIPR.

The level of eagerness demonstrates by the CIPR conference committee from day one is admiring and commendable. Organizing committee have assurance that the members of CIPR preserve a outstanding paradigm as we have reassessed papers, afford opinion and build a strong body of obtainable work in this assembly of proceeding. The astonishing strength and determination as shown by organizing committee in all phase throughout the conference deserves profusely thanks from our heart.

We have been opportune adequately to work in with a luminous international as well as national advisory committee. The call for papers, reviewers in the papers and finalizing papers are accomplished by reviewers and technical committee consisting of eminent academicians to be included in the proceedings.

All the submitted papers have been double blind peer re-evaluated before finalization as the accepted papers for oral presentation at the conference venue. We would like to convey our heartfelt thanks and responsibility to the compassionate reviewers for sparing their valuable time and putting in effort to re-evaluate the papers in a stipulated time and providing their supportive ideas and appreciations in brainstorming the presentation, content and quality of proceeding. The valuable these accepted papers is an immense admiring comment not only to authors but also to reviewers who have directed towards excellence.

We want to profusely thank to Organizing committee members, International Advisory and Technical Committee members and all other Committee members for their effortless support and cooperation to formulate this conference successful in all sense.

Last but not the least, the editorial members of Springer publishing procure a special declare and our profound authentic gratitude to them not only for assembly our apparition come true in the form of proceeding but also for its luminous in-time and get-up publications in graceful, innovation intelligent scheme and technologies, Springer.

The CIPR conference and proceedings are recognitions to an immense assemblage of people and all should be advance of the outcome.

About the Conference

The term *Computational Intelligence* refers to the ability of a computer to learn a specific task from data or experimental observation. It is a set of nature-inspired computational methodologies and approaches to address complex real-world problems to which mathematical modeling may not be effective. Computational Intelligence provides solutions to problems for which precise mathematical model cannot be formulated due to presence of uncertainties or the process of modeling might be stochastic in nature. On the other hand *Pattern Recognition* is the process of automated recognition of patterns and regularities in data. The conference *Computational Intelligence in Pattern Recognition* (*CIPR*) is aimed at applying computational intelligence to find pattern in data and to develop useful application. The conference will provide a common platform for presentation of original research findings, exchange of ideas and dissemination of innovative, practical development experiences in different aspects and fields of data analysis and pattern recognition. The conference also aims at bringing together the researchers, scientists, engineers, industrial professionals and students in this fascinating area of research. The conference also addresses rapid dissemination of important results in the field of pattern recognition, soft computing technologies, clustering and classification algorithms, rough set and fuzzy set theory, evolutionary computations, neural science and neural network systems, image processing, combinatorial pattern matching, social network analysis, audio and video data analysis, data mining in dynamic environment, bioinformatics, hybrid computing, big data analytics, deep learning etc. It promotes novel and high quality research findings, innovative solutions to the challenging engineering problems.

At the end of the conference, we anticipate that the participants will enhance their knowledge by latest perceptions and examinations on existing research topics from leading academician, researchers and scientists around the globe, contribute their own thoughts on significant research topics mentioned above as well as to networks and conspire with their international compeer.

Contents

About the Editors

Asit Kumar Das is currently a Professor at the Department of Computer Science and Technology, Indian Institute of Engineering Science and Technology, Shibpur, India. His primary research interests are in the areas of data mining, bioinformatics, evolutionary algorithms, audio and video data analysis, text mining and social networks. He has published more than 50 research articles in respected peer-reviewed journals.

Janmenjoy Nayak is working as an Associate Professor at Sri Sivani College of Engineering, Srikakulam, India. He has published more than 50 research papers in various respected peer-reviewed journals and international conference proceedings, as well as a number of book chapters. He has received several awards, such as an INSPIRE Fellowship from DST, Govt. of India, and best researcher award by JNTU, Kakinada, AP. His areas of interest include data mining, nature-inspired algorithms and soft computing.

Bighnaraj Naik is an Assistant Professor at the Department of Computer Applications, VSSUT, Burla, Odisha, India. He has published more than 50 research papers in various respected peer- reviewed journals, referred journals and international conference proceedings, and has reviewed a number of books. He has over nine years of teaching experience in the field of computer science and information technology. His interests include data mining and soft computing.

Soumen Kumar Pati is currently an Associate Professor at the Department of Bio-Informatics, Maulana Abul Kalam Azad University of Technology, West Bengal, India. Dr. Pati received his Ph.D. degree in Engineering from the Department of Computer Science and Technology, Indian Institute of Engineering Science and Technology, Shibpur. His primary research interests are in the areas of data mining, artificial intelligence, soft computing, pattern recognition, social network, network security, machine learning, and he has published numerous articles in these fields.

Danilo Pelusi is an Associate Professor at the Faculty of Communication Sciences, University of Teramo. Associate Editor of IEEE Transactions on Emerging Topics in Computational Intelligence, IEEE Access and guest editor for Elsevier, Springer and Inderscience journals, he served as program member of many conferences and as editorial board member of many journals. His research interests include Fuzzy Logic, Neural Networks, Information Theory and Evolutionary Algorithms.

On the Synthesis of Unate Symmetric Function Using Memristor-Based Nano-Crossbar Circuit

Subhashree Basu and Malay Kule

Abstract VLSI technology can integrate a large number of electronic components into a single chip. But as per the Moore's prediction, this technology will soon hit a wall and no further miniaturization of VLSI chips will be possible. Because of this blockade that the technology would soon face, the scientists are turning their focus to emerging nanotechnology. Memristor can be such an alternative, which has the potential of very high speed, smaller size, and minimum power consumption compared to transistor-based technology. Memristor can be considered synonymous to "memory plus resistor" where its internal resistance acts as the data value stored in it, and it has the capability of retaining its previous resistance value even when the power supply is withdrawn. This paper utilizes this unique property of memristor in a new way to simulate unate symmetric functions. Implementation of 3- and 5-input unate symmetric functions are shown in the memristor-based memristor circuit.

Keywords VLSI · Nanotechnology · Memristor · Unate · Symmetric functions · Nano-crossbar

1 Introduction

Silicon industry has experienced an exponential growth in the past decades. But in the near future, this growth is bound to slow down. The chief technological limitations responsible for this slow down is interconnected problem and power dissipation. This slow down forced the scientists to search for an alternative device technology. Memristor can prove to be an interesting alternative. Initially, Chua [1] proposed this hypothetical device, which provided a relationship between flux and charge similar

S. Basu (✉)
ST. Thomas' College of Engineering and Technology, Kolkata, India
e-mail: basu.subhashree1984@gmail.com

M. Kule
Indian Institute of Engineering Science and Technology, Shibpur, Howrah, India
e-mail: malay.kule@gmail.com

A. K. Das et al. (eds.), *Computational Intelligence in Pattern Recognition*,
Advances in Intelligent Systems and Computing 999,
https://doi.org/10.1007/978-981-13-9042-5_1

Fig. 1 Memristor: symbolic representation

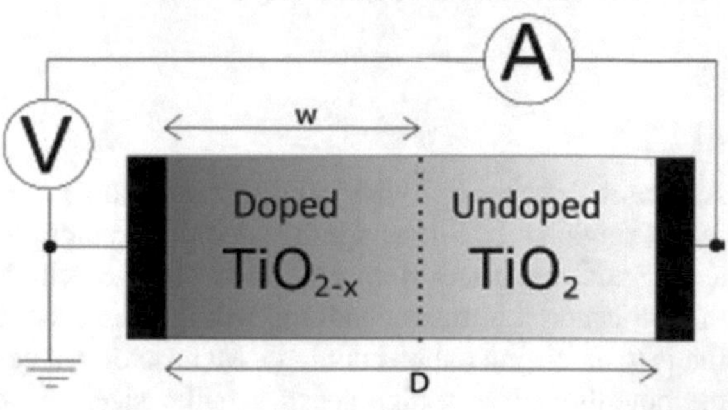

Fig. 2 Memristor: fabrication

to resistor which provides relation between voltage and current. This would mean that the device's resistance would vary according to the amount of charge passed through it and it would remember the resistance value even after the input current is turned off. Thus, memristor can be defined as a dynamical electronic circuit element that obeys the following relations:

$$V = R(w, i) * i \tag{1}$$

$$dw/dt = F(w, i) \tag{2}$$

The first equation is a quasi-static resistance equation, i.e., state-dependent Ohm's law. The second equation describes the temporal evolution of the state variable. In these equations, i is an independent input function, r is the resistance which is dependent on the physical state w which, in turn, imparts memory to the device. The symbolic representation of a memristor is shown in Fig. 1.

Titanium dioxide was the first material, which exhibited the memristance properties [2]. A thin layer of TiO_2 and $TiO_{2\text{-}x}$ are sandwiched between two platinum wires. TiO_2 is the undoped region and $TiO_{2\text{-}x}$ acts as the doped region. Positive current causes the oxygen ions to drift from $TiO_{2\text{-}x}$ to TiO_2 region, thus increasing the doped region and hence decreasing the memristance, which can be considered as the ON state. Applying negative current causes flow of ions in the reverse direction causing an increase in the memristance and hence, the memristor is considered to be in OFF state. If the voltage source is removed or ions reach saturation stage, i.e., no ions can flow either way, the memristor reaches a stable state and retains that state until further voltage is applied. Figure 2 depicts the conceptual diagrammatic representation of memristor fabrication.

Crossbar is an array of perpendicular wires where the junctions of two wires are connected by switches, which can be either opened or closed. Hence, crossbar arrays are basically a storage system where opened switch represents 0 and closed switch represents 1. The advantage of crossbar array is that it induces redundancy by allowing routing around the part of the circuit which is not working. Moreover, due to their simplicity, crossbar arrays have a much higher density of switches than a comparable integrated circuit based on transistors. Memristors are very much compatible with crossbar arrays with respect to size and fabrication. The operation time is in the range of picoseconds and power consumption in the range of picojoules. Hence, memristor-based crossbar circuit [3, 4] implementation has proved to be an efficient alternative to overcome the obstructions faced by the semiconductor industry. As a result, many types of circuits, both combinational and sequential [5–9] have been implemented using this architecture.

Let $f(x_1, x_2, \ldots, x_n)$ be a switching function of n Boolean variables. A vertex or minterm is a product of variables in which every variable occurs only once. The weight w of a vertex v is the number of uncomplemented variables that appear in v. A Boolean function is called negative (positive) unate, if each variable appears in complemented (uncomplemented) form but not in both in its minimized sum-of-products (SOP) expressions.

A function is called unate symmetric [10, 11] if it is both unate and symmetric. An unate symmetric function is always consecutive and can be expressed as S^n ($a_l - a_r$) where either $a_l = 0$ or $a_r = n$.

In our proposed work, we have synthesized the Unate Symmetric Function using Memristor-based Nano-crossbar Circuits in a new way. Simulation results support the correctness of our design.

This paper is divided into the following sections. Section 2 consists of the proposed method, which describes the design of 3- and 5-input unate symmetric functions. Section 3 discusses and analyses the results. Section 4 is the conclusion.

2 Proposed Method

The unate symmetric functions are designed so that the output lines U_j is high, i.e., they assume the value 1 if high voltage is applied to at least j inputs [10]. The output functions can be expressed as

$$U_1(n) = S^n(1, 2, \ldots n) = \sum x_i \text{ for } i = 1 \text{ to } n$$

$$U_2(n) = S^n(2, 3, \ldots n) = \sum x_i.x_j \text{ for } i, j = 2 \text{ to } n$$

$$U_n(n) = x_i.x_j \ldots x_n$$

For n = 3, there will be three output functions U_1, U_2, and U_3 where U_1 (3) is defined as

$$U_1(3) = S^n(1, 2, 3) = \sum x_i \text{ for } i = 1 \text{ to } 3$$

i.e., U_1 (3) output will be high when any one or more of the three inputs are high

$$U_2(3) = S^n(2, 3) = \sum x_2.x_3$$

i.e., U_2 (3) output will be high when any two or three of the inputs are high.

$$U_3(3) = x_1.x_2.x_3$$

i.e., U_3 (3) is high when all three of the inputs are high. Similarly for five inputs, i.e., n = 5, there will be five output functions U_1 (5), U_2 (5), U_3 (5), U_4 (5), and U_5 (5) and they can be derived following the above logic.

The implementation of 3- and 5-input unate symmetric functions are shown below. These are implemented using the concept of majority function. In majority function, whichever input has maximum occurrence will influence the output the most, i.e., if number of inputs n = 3, then if two or more inputs are equal to 1, then output will be 1 and if two or more inputs are equal to 0, then output will be equal to 0. To implement this majority function, we need a 5-input majority gate of which, three are the actual inputs and the other two are dummy inputs used to implement the majority function. This can be easily derived from the concept that if the number of input is x and the majority gate has g number of input lines, then by manipulating (g-x) lines, the unate symmetric functions can be designed. For example, in order to implement U_1 (3), two inputs of the five are made high. So, if at least any one of the other three inputs is high, according to the majority logic, the output will be high. For U_2 (3), one input is already made high, so only when any two or more of the inputs A, B, and C are high, U_2 (3) will be high satisfying the majority logic. For U_3 (3), two inputs of five majority gate are made 0. Hence, U_3 (3) will be high only when all the three inputs A, B, and C are high. Similarly, for the 5-input unate symmetric functions, we require a 9-input majority gate whereby manipulating four inputs, i.e., by making them either 1 or 0, the majority logic is determined and accordingly, we get the outputs U_1 (5)–U_5 (5), respectively.

When we implement the above logic in nano-crossbar circuit, then the input lines A, B, and C are connected to the positive end of the output memristors. Among the rest of the inputs D, E, F, and G, we connect those based on which output is required. Since the voltages across D, E, F, and G are constant, the voltages across A, B, and C will determine the majority function and will determine the outputs u1, u2, and u3 (the unate symmetric functions outputs), respectively, i.e., whichever voltage, i.e., high or low that will be predominant, will influence the output and the output will be equal to the predominant voltage. For example, in order to get the output u1, we connect D and E so that high voltage in any of the inputs A, B, or C will give a high value in u1. Similarly for u2, we connect D and G so that high voltage in any two of the inputs A, B, or C gives us the output u2 and so on. The same approach is followed while designing the 5-input circuit which requires a 9-input majority

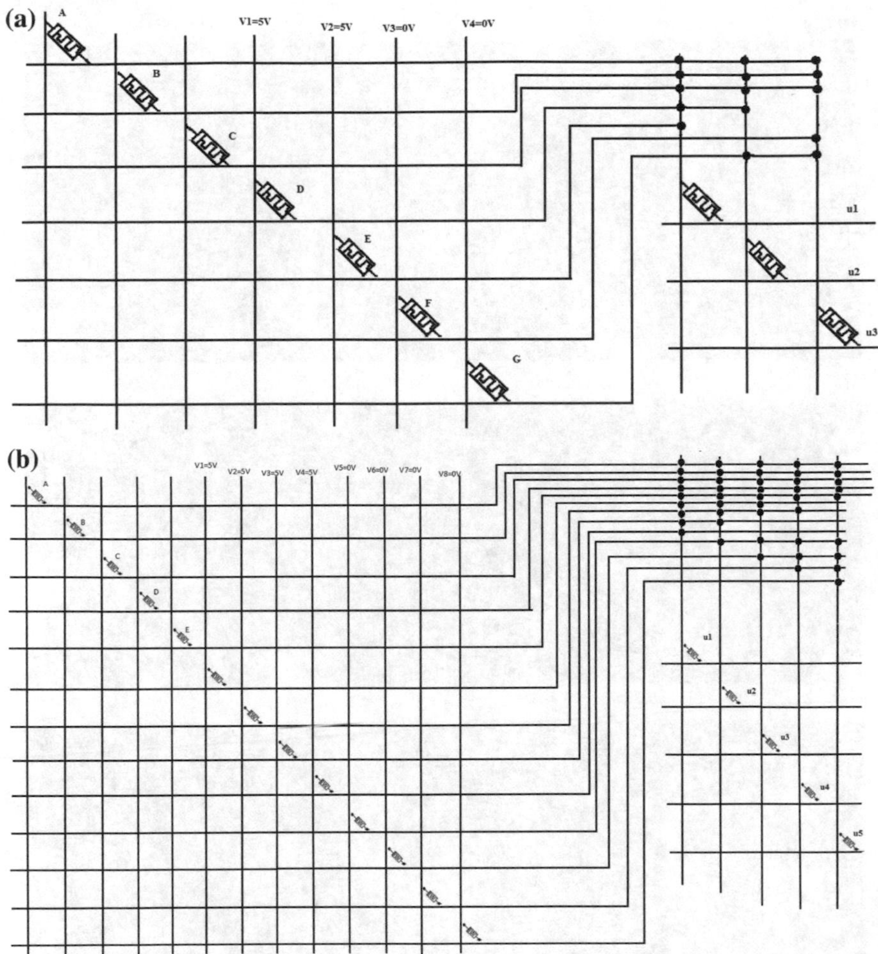

Fig. 3 **a** Nano-crossbar layout of 3-input unate symmetric function, **b** Nano-crossbar layout of 5-input unate symmetric function

gate. Figure 3a, b shows the crossbar layout for 3-input and 5-input unate symmetric functions, respectively.

3 Experimental Result and Analysis

The output waveforms of 3-input and 5-input unate symmetric functions show that the outputs are influenced by the number of inputs that are made high. For example, in Fig. 4a which shows the 5-input unate symmetric functions U_1 and U_2, the moment

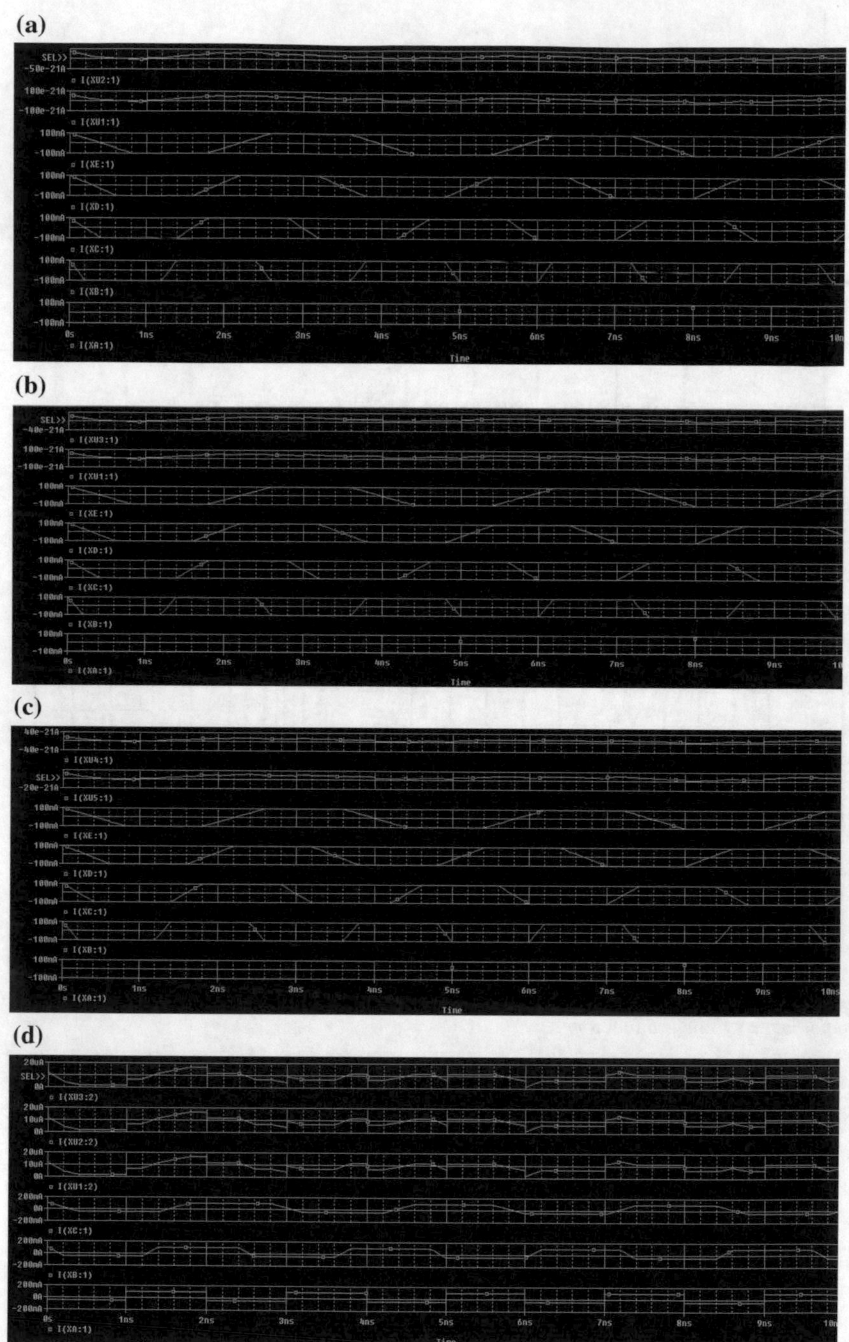

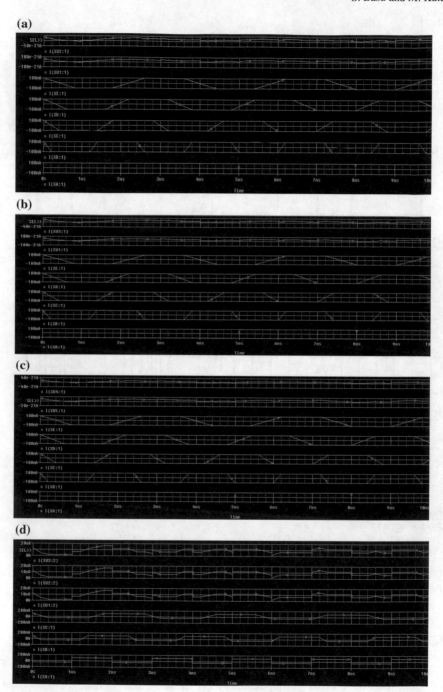

Fig. 4 **a** Output U_1 and U_2 of 5-input unate symmetric function, **b** Output U_3 of 5-input unate symmetric function, **c** Output U_4 and U_5 of 5-input unate symmetric function, **d** Output U_1, U_2, and U_3 of 3-input unate symmetric function

any one of the five inputs starts achieving the logic 1, the output XU1 also starts rising irrespective of whether the other inputs are high or low. The same phenomenon is seen in Fig. 4b, c which shows U_3, U4, and U_5, respectively. Figure 4d shows the output of unate functions for three inputs.

4 Conclusion

Symmetric functions have huge applications in the arithmetic circuit design and in many other applications. Hence, the design of the symmetric functions in nanoscale level can prove to be a very effective tool in designing nanocircuits in the future. Hence, designing of symmetric functions using memristor-based nano-crossbar circuit can be an optimal solution both from the time and space aspect. This paper has attempted to design nanoscale memristor-based crossbar circuit of unate symmetric functions for 3 and 5 numbers of inputs. A conscious effort has been made to avoid sneak path. If the memristors are connected in one single row, the possibility of sneak path arises although the size may be reduced. But we have given priority to elimination of sneak path against size as the size will anyway remain in nm range. Hence, in order to avoid sneak path problem, we have arranged the memristors in separate rows, so that the current from one row cannot sneak into the other rows as there are no common junction points to pass the current. Symmetric functions have already been implemented using other nanoscale devices like reversible gates [10, 12] and QCA [13, 14]. But this is, perhaps, the first attempt to implement symmetric functions using memristor based nano-crossbar circuits. This approach can also prove to be very efficient in solving all the limitations of the VLSI silicon technology and be cost worthy as well.

References

1. Chua, L.O.: Memristor-the missing circuit element. IEEE Trans. Circuit Theory, **CT–18**(5) (1971)
2. Haridas, M., Patil, S., Manjunath, T.C.: Recent ATC's in the design of memristors. Int. J. Comput. Electr. Eng. **2**(3) (2010)
3. Kule, M., Dutta, A., Rahaman, H., Bhattacharya, B.B.: High-speed decoder design using memristor-based nano-crossbar architecture. In: Sixth International Symposium on Embedded Computing and System Design (ISED) (2016)
4. Chen, Q., Wang, X., Wan, H., Yang, R.: A logic circuit design for perfecting memristor- based material implication. IEEE Trans. Comput.-Aided Des. Integr. Circuits Syst. **36**(2) (2017)
5. Owlian, H., Keshavarzi, P., Rezai, A.H.: A novel digital logic implementation approach on nanocrossbar arrays using memristor-based multiplexers. Microelectron. J. (2014)
6. Teimoory, M., Amirsoleimani, A., Shamsi, J., Ahmadi, A., Alirezaee, S., Ahmadi, M.: Optimized implementation of memristor-based full adder by material implication logic. In: 21st IEEE International Conference on Electronics, Circuits and Systems (ICECS) (2014)
7. Lavanya, A., Gopal, B.G.: Imply logic implementation of carry save adder using memristors. Int. J. Eng. Res. Appl. **5**(5), (Part -5), 105–109 (2015)

8. Chakraborty, A., Rahaman, H.: Implementation of combinational circuits via material implication using memristors. In: IEEE International Conference on Distributed Computing, VLSI, Electrical Circuits and Robotics (DISCOVER-2016/IEEE) (2016) https://doi.org/10.1109/discover.2016.7806227
9. Alamgir, Z., Beckmann, K., Cady, N., Velasquez, A., Jha, S.K.: Flow-based computing on nanoscale crossbars: design and implementation of full adders. In: IEEE International Symposium on Circuits and Systems (ISCAS) (2016)
10. Rahaman, H., Das, D.K., Bhattacharya, B.B.: Implementing symmetric functions with hierarchical modules for stuck-at and path-delay fault testability. J. Electron. Test. **22**(2), 125–142 (2006)
11. Lauradoux C., Videau, M.: Matriochka symmetric Boolean functions. IEEE ISIT, pp. 1631–1635 (2008)
12. Maslov, D.: Efficient reversible and quantum implementations of symmetric Boolean functions.Circuits, Devices and Systems. IEEE Proc. **153**(5), 467–472 (2006)
13. Rahaman, H., Sikdar, B.K., Das, D.K.: Synthesis of symmetric boolean functions using quantum cellular automata. In: International Conference on Design and Test of Integrated Systems in Nanoscale Technology (DTIS 06), pp. 119–124. Tunis, Tunisia
14. Bhattacharjee, P.K.: Use of Symmetric Functions Designed by QCA Gates for Next Generation IC. Int. J. Comput. Theory Eng. **2**, 1793–8201 (2010)

A Novel Enhancement and Segmentation of Color Retinal Image Based on Fuzzy Measure and Fuzzy Integral

Swarup Kr Ghosh, Biswajit Biswas and Anupam Ghosh

Abstract Image enhancement and segmentation are predominating methods in image processing and are widely used in ophthalmology for the diagnosis of various eye diseases such as diabetic retinopathy, glaucoma. Especially, retinal image segmentation is vastly required to extract certain features that can facilitate in diagnosis and treatment of eye. Due to the acquisition process, color retinal images can suffer from poor contrast, thus enhancement is essential in ophthalmology. In this work, a novel fuzzy color image enhancement and segmentation technique has been suggested to overcome the problem of low and variation of contrast, and segment in retinal images. This paper provides a special fuzzy computation on retinal image such as fuzzy measure which is used for enhancement and fuzzy integral is applied for segmentation. The proposed algorithm can accomplish excellent contrast and a segment of an image object that endows feasibility in diagnosis from retinal image.

Keywords Image enhancement · Image segmentation · Fuzzy partition · Fuzzy measure · Fuzzy integral

1 Introduction

Nowadays, a large number of retinal imaging apparatus such as flood imaging, optical scanning laser ophthalmology, and optical coherence tomography are used in several sophisticated image processing techniques (enhancement, segmentation,

S. K. Ghosh (✉)
Brainware University, Barasat, Kolkata 700124, India
e-mail: swarupg1@gmail.com

B. Biswas
University of Calcutta, Kolkata 700098, India
e-mail: biswajit.cu.08@gmail.com

A. Ghosh
Netaji Subhash Engineering College, Kolkata 700152, India
e-mail: anupam.ghosh@rediffmail.com

A. K. Das et al. (eds.), *Computational Intelligence in Pattern Recognition*,
Advances in Intelligent Systems and Computing 999,
https://doi.org/10.1007/978-981-13-9042-5_2

registration, etc.) for the diagnosis of various eye diseases such as diabetic retinopathy, macular degeneration, glaucoma [1, 2]. Image enhancement and segmentation which are application dependent, take a crucial role in retinal image processing and analysis [3]. A number of image enhancement algorithms have been implemented that cover a range of operations for deblurring, noise removal, gray level dynamic range modification, and so on [4]. Image contrast enhancement is an essential part of image preprocessing in medical imaging. Conventional contrast enhancement algorithm like histogram equalization or histogram specification [5], are not suitable for the ophthalmic images due to its nature such as variations of dynamic intensity and amplification of noise [2, 4]. Unsharp masking-based enhancement method is effectively used for handling regular illumination but is acquainted with artifacts due to amplification of noise [6, 7]. Matched filters-based enhancement method is partially developed for local contrast starching and it supports vessel segmentation but affects other textures in the retinal image [7, 8]. The normalized convolution has been used to reduce noise on retinal images [9]. Especially, histogram matching between the red and green planes of the retinal image has been utilized for vessel segmentation and it raises the contrast of shady features like vessels but shrinks the contrast of luminous objects and small shady substances [10]. Another image enhancement like contourlet transform is suggested, that suffers from poor contrast regions of the retinal image [8].

Image segmentation can be illustrated as the grouping of identical pixels in a parametric space, where they are related to each other in the same or distinctive images [3]. There are many conventional image segmentation algorithms such as thresholding methods which are quite straightforward, edge detection, region detection, etc[3, 11]. Edge-based methods and region detection methods are quite analogous to the contour detection but conscious to noise and fail to link together with the broken contours and are rather difficult [12]. There are several methods formulated for segmenting retinal images such as GVF active contours model and active shape models and drawback to need manual control point initialization [13, 14]. Recently, geometric active contours based on curve evolution and geometric flows with the level set have been designed for retinal image, which detects outer boundaries only [15]. Unsupervised learning-based algorithms like K-means clustering, Fuzzy C-means (FCM) clustering are becoming popular for color image segmentation [16]. Various probabilistic approaches such as Markov random field (MRF), Bayes method have been extensively used for image segmentation but they require prior information [17]. Graph theory-based image segmentation methods such as image partitioning by using normalized graph cut, and global partitioning on image region are also commonly used in medical imaging [18, 19].

In this work, the proposed method has been divided into two parts as automated enhancement and segmentation of retinal images by using fuzzy measure and integral. In enhancement phase, an image is partitioned into fuzzy sub-images by fuzzy extension process at which number of functional form is shaped by the dynamic range of image intensity sharing. Each fuzzy sub-image is contrasted by average weighted fuzzy computation of membership grades and hence, best membership grade values are achieved by mixing the computed membership degree of fuzzy partition sets [20], which can qualify for haziness and imprecision of the images in a more adaptive way. In the segmentation phase, it combines every one of appropriate membership grade values from fuzzy partition sets behind Sugeno measure and Sugeno fuzzy integration [21, 22]. In this paper, we have suggested a novel retinal image enhancing method, which uses fuzzy set partition approach and evaluation of the fittest fuzzy membership components from partition sets and color image segmentation, which merges best possible fuzzy membership components from partition sets in term of Sugeno measure and Sugeno integral. We have conducted the proposed model by several state-of-the-art methods.

The rest of the paper is framed as follows. Section 2 describes the proposed retinal image enhancing and segmentation algorithms along with mathematical preliminaries. Experimental results and discussions have been demonstrated in Sect. 3. Finally, conclusion along with future work is given in Sect. 4.

2 Methodology

In this section, different steps of the suggested methodology have been explained. We have emphasized the designed scheme as two phases, namely, the fuzzy image enhancement and image segmentation. The major steps of the methodology have been illustrated in Fig. 1.

2.1 Fuzzy Image Enhancement

Generally, retinal image is low contrast with dynamic intensity range. The primary objective of the suggested model is to find the optimal membership grade that starch contrast of retinal image actively. Fuzzy partition and computation of fuzzy membership grade with weighted progression are employed in fuzzy enhancement process [21, 22].

Fuzzification A retinal image **I** of dimension $\mathbf{M} \times \mathbf{N}$ along with intensity range 0 to $\mathbf{L} - \mathbf{1}$ is represented as a fuzzy singleton set as

$$\mathbf{I} = \bigcup_{i=1}^{\mathbf{M}} \bigcup_{j=1}^{\mathbf{N}} \frac{\mu_{ij}(\mathbf{x})}{\mathbf{x_{ij}}} \tag{1}$$

where $\mu_{ij}(\mathbf{x})$ represents the grade of membership function $\mathbf{x_{ij}} \in \mathbf{I}$, with dimension $\mathbf{M} \times \mathbf{N}$ of image **I**. The membership function $\mu_{\mathbf{I}}(\mathbf{x})$ for the image **I** is determined as

$$\mu_{ij}(\mathbf{x}) = \frac{\mathbf{x_{ij}} - \mathbf{x_{min}}}{\mathbf{x_{max}} - \mathbf{x_{min}}} \tag{2}$$

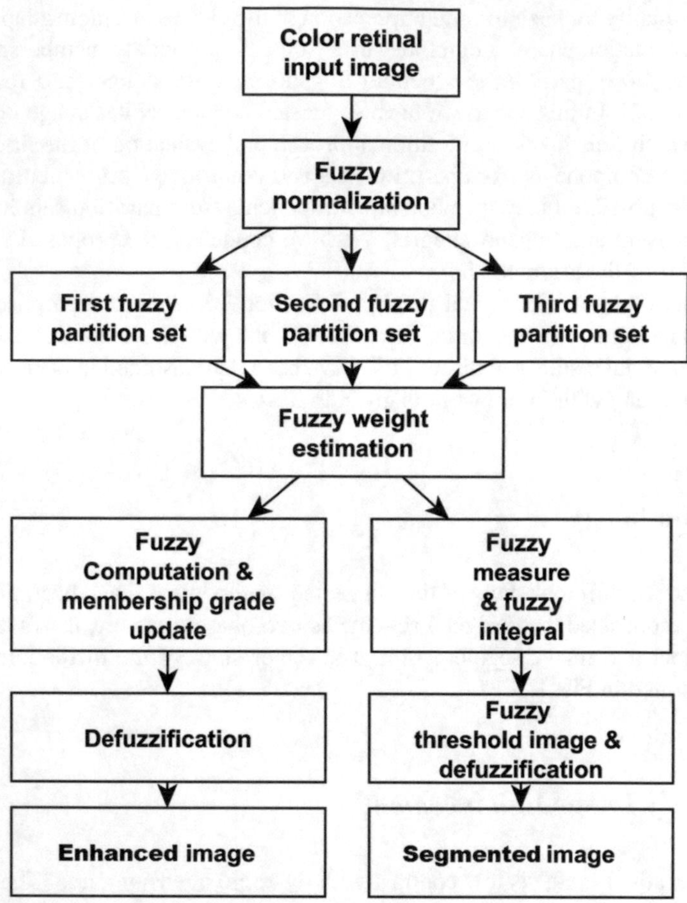

Fig. 1 Flowchart of the main steps of proposed methods

where $\mathbf{x_{ij}} \in \mathbf{I}$, $\mathbf{x}_{\max}$, and $\mathbf{x}_{\min}$ are ij^{th} pixel intensity, highest, and lowest intensity levels of the image **I**, respectively.

Fuzzy computation In this step, we have broken normalized intensity level set calculated from Eq. (2) of $\mu_{ij}(\mathbf{x}) \in \mathbf{I}$ into three fuzzy partition sets $\{\mathbf{I_0}, \mathbf{I_1}, \mathbf{I_2}\}$ as dark, middle range, and bright levels, respectively. The membership functions of partition sets $\mu_{\mathbf{I_0}}$, $\mu_{\mathbf{I_1}}$, and $\mu_{\mathbf{I_2}}$ are defined by using Eq. (2) as follows:

$$\mu_{\mathbf{I_0}}(\mathbf{x}) = \frac{(1-\mu(\mathbf{x}))^2}{2}, \forall \mu(\mathbf{x}) \in \mathbf{I} \tag{3}$$

$$\mu_{\mathbf{I_1}}(\mathbf{x}) = (1-\mu(\mathbf{x}))^2, \forall \mu(\mathbf{x}) \in \mathbf{I} \tag{4}$$

$$\mu_{\mathbf{I_2}}(\mathbf{x}) = \frac{(1+\mu(\mathbf{x}))^2}{2}, \forall \mu(\mathbf{x}) \in \mathbf{I} \tag{5}$$

As per definition of partition, the membership grades of the above three fuzzy sets defined by Eqs. (3), (4), and (5) must be sum to one. Thus, fuzzy membership degrees along with weighted values $\omega_{\mathbf{I}_p}(x)$ of intensity level among fuzzy partition sets of intensity levels $\mathbf{I}_p(x)$ for $p = \{0, 1, 2\}$ must hold the following condition:

$$\omega_{\mathbf{I}_0}(\mathbf{x}) + \omega_{\mathbf{I}_1}(\mathbf{x}) + \omega_{\mathbf{I}_2}(\mathbf{x}) = 1, \forall \mu(\mathbf{x}) \in \mathbf{I}$$

We consider a real positive regulation parameter $\theta \in [0, 1]$ to calculate fuzzy weighted normalization. However, three tunned weighted fuzzy partition sets using regulation parameter $\theta \in [0, 1]$ are represented as follows:

$$\omega_{\mathbf{I}_p}(\mathbf{x}) = \left(\frac{\mu_{\mathbf{I}_p}(\mathbf{x})}{\sum \mu_{\mathbf{I}_p}(\mathbf{x})} \right)^{\theta}, \forall \mu(\mathbf{x}) \in \mathbf{I}_p, p = \{0, 1, 2\} \tag{6}$$

After membership grade modification, we estimated membership upgradation for individual fuzzy partition sets $\mu_{\mathbf{I}_p}(x)$, $p = \{0, 1, 2\}$ by using Eq. (6) as:

$$\mu_{\mathbf{I}_p}(\mathbf{x}) = \prod_{p=0}^{2} \mu_{\mathbf{I}_p}(x) * \omega_{\mathbf{I}_p}(\mathbf{x}), p = \{0, 1, 2\} \tag{7}$$

Lastly, we have determined an aggregate value Λ from modified fuzzy partition sets by using Eqs. (6) and (7) as follows:

$$\Lambda(\mathbf{x}) = \sum_{p=0}^{2} \mu_{\mathbf{I}_p}(x) * \omega_{\mathbf{I}_p}(\mathbf{x}), p = \{0, 1, 2\} \tag{8}$$

$$\Lambda(\mathbf{x}) = \sum_{p=0}^{2} \mu_{\mathbf{I}_p}(x), p = \{0, 1, 2\} \tag{9}$$

Finally, the selection of new and modified membership grade from each three partition fuzzy set based on $\Lambda(\mathbf{x})$ is estimated from Eqs. (8) and (9) in the following expression:

$$\mu_{\mathbf{I}}(\mathbf{x}) = \begin{cases} \frac{1-\prod_{p=0}^{2} \mu_{\mathbf{I}_p}(x)}{1+\sum_{p=0}^{2} \mu_{\mathbf{I}_p}(x)}, \Lambda(\mathbf{x}) > 0.5 \\ \frac{1}{2} \log \left(\frac{1-\sum_{p=0}^{2} \mu_{\mathbf{I}_p}(x)}{1+\prod_{p=0}^{2} \mu_{\mathbf{I}_p}(x)} \right), \Lambda(\mathbf{x}) \leq 0.5 \end{cases} \tag{10}$$

At last, we found the best fitted membership grades for image $\mu_{\mathbf{I}_p}(x)$ by Eq. (10) from all mathematical computations that are used for defuzzification and finally get the excellent contrast enhancement retinal image.

Fuzzy manipulation and defuzzification Fuzzy manipulation function for the contrast enhancement of retinal image is achieved by using Eq. (10) in the order

$$\acute{\mu}_{\mathbf{I}}(\mathbf{x}) = \begin{cases} 2 * \mu_{\mathbf{I}}(\mathbf{x})^2, & \text{for } \mu_{\mathbf{I}}(\mathbf{x}) > 0.5; \\ 1 - 2 * (1 - \mu_{\mathbf{I}}(\mathbf{x}))^2, & \text{if } \mu_{\mathbf{I}}(\mathbf{x}) \leq 0.5; \end{cases} \tag{11}$$

Finally, we get enhanced retinal image $\mathbf{I_E}$ by defuzzification of intensification grade of membership $\acute{\mu}_{\mathbf{I}}(\mathbf{x})$ from Eq. (11) is furnished as follows:

$$\mathbf{I_E} = \mu_{\mathbf{I}}(\mathbf{x}) + \nabla\acute{\mu}_{\mathbf{I}}(\mathbf{x}) * (\mathbf{x}_{\max} - \mathbf{x}_{\min}) \tag{12}$$

where $\nabla\acute{\mu}_{\mathbf{I}}(\mathbf{x}) = \mu_{\mathbf{I}}(\mathbf{x}) - \acute{\mu}_{\mathbf{I}}(\mathbf{x})$ and $\mu_{\mathbf{I}}(\mathbf{x}) = \frac{\mathbf{x_{ij}} - \mathbf{x}_{\min}}{\mathbf{x}_{\max} - \mathbf{x}_{\min}}$ where $\mathbf{x}_{\max}$ and $\mathbf{x}_{\min}$ are highest, and lowest intensity levels of the image **I**, respectively. To get enhanced retinal color image, we have used Eq. (12) in image **I**.

2.2 Fuzzy Image Segmentation

A fuzzy integral based on fuzzy image region combination method merges with different fuzzy features in nonlinear process of the image (region, edge, etc.) has been used for segmentation. It is utilized to combine regions recursively as per the maximal fuzzy integral [21]. The objective of region merging is to fuse objective fuzzy evidence with fuzzy measure [20] that reflects in the regions of the image. The target regions can be achieved by using the process of maximal fuzzy integral recursively to integrate all detected regions. In this paper, we have used Sugeno measure and integral for image segmentation and have discussed the details in the following subsections.

Sugeno measure Let $\mathcal{X} = \{x_1, x_2, \ldots, x_n\}$ be collection of data sources, a fuzzy measure is a real valued function **g** defined as $\mathbf{g}\colon 2^{\mathcal{X}} \to [0, 1]$ on the family set of $\mathcal{X}$, which satisfies the following properties [22]:

$$\mathbf{g}(\phi) = 0 \text{ and } \mathbf{g}(\mathcal{X}) = 1$$
$$\mathbf{g}(\mathcal{A}) =\leq \mathbf{g}(\mathcal{B}) \text{ if } \mathcal{A} \subseteq \mathbf{g}$$
$$\text{if } \{\mathcal{A}_i\} \subseteq \mathcal{X} \text{ then } \lim_{i\to\infty}(\mathcal{A}_i) = \mathbf{g}\left(\bigcup_{i=1}^{\infty} \mathcal{A}_i\right)$$

However, a fuzzy measure **g** is known as Sugeno measure ($\mathbf{g}_\lambda$) if it satisfies the following conditions:

$$\forall \mathcal{A}, \mathcal{B} \subseteq \mathcal{X} \text{ with } \mathcal{A} \cap \mathcal{B} = \phi$$
$$\mathbf{g}_\lambda(\mathcal{A} \cup \mathcal{B}) = \mathbf{g}_\lambda(\mathcal{A}) + \mathbf{g}_\lambda(\mathcal{B}) + \lambda \mathbf{g}_\lambda(\mathcal{A})\mathbf{g}_\lambda(\mathcal{B}) \quad , \lambda > -1$$

where $\mathbf{g}_\lambda$ is fuzzy density in Sugeno measure. In Sugeno measure, the value of λ for set $\mathcal{X}$ is estimated by the following equation:

$$\lambda + 1 = \prod_{i=1}^{n} (1 + \lambda \mathbf{g}_\lambda), \lambda \in (-1, \infty) \tag{13}$$

Sugeno integral Let $(\mathcal{X}, \Omega)$ be a measurable space, and $\mathbf{h}: \mathcal{X} \rightarrow [0, 1]$ be a Ω measurable function. The fuzzy Sugeno integral over $\mathcal{A} \subseteq \mathcal{X}$ of the function $\mathbf{h}(\mathbf{x})$ with respect to a fuzzy measure $\mathbf{g}_\lambda$ is defined as follows [22]:

$$\int_A \mathbf{h}(\mathbf{x}) \circ \mathbf{g}_\lambda(\cdot) = \max_{i=1,\cdots,n} [\min\{\mathbf{h}(\mathbf{x}_i), \mathbf{g}_\lambda(\mathcal{A}_i)\}] \tag{14}$$

where $\mathcal{A}_i = \{\mathbf{x}_1, \mathbf{x}_2, \ldots, \mathbf{x}_n\}$, and for finite set, $\mathbf{h}(\mathbf{x})$ is a monotonous sequence, i.e., $\mathbf{h}(\mathbf{x}_1) \geq \mathbf{h}(\mathbf{x}_2) \geq \cdots, \mathbf{h}(\mathbf{x}_n)$ but $\mathbf{h}(\mathbf{x}_{n+1}) = 0$. During computation of fuzzy integral for image segmentation, $\mathcal{X}$ represents image features such as different intensities of varied regions. In this work, we used fuzzy measure and fuzzy integral to merge different image partition regions with similar level of intensity.

Fuzzy computation The estimated three fuzzy partition sets from Eq. (1) are $\mathbf{I}_p(x)$, $p = \{0, 1, 2\}$, and their resultant membership grade $\mu_{\mathbf{I}_p}(x)$, $p = \{0, 1, 2\}$ calculated from Eq. (7). These partition sets are contained different levels of intensity with membership grade. We use Sugeno measure and Sugeno integral on these three partition sets, and defined a target set that contained fused information. However, we estimate thrcc fuzzy density values from these three partition sets $\mu_{\mathbf{I}_p}(x)$, $p = \{0, 1, 2\}$ and consider variables of $\mathbf{a}_p$, $p = \{0, 1, 2\}$ as follows:

$$\mathbf{a}_0 = \mu_{\mathbf{I}_0}(x) \quad \mathbf{a}_1 = \mu_{\mathbf{I}_1}(x) \quad \mathbf{a}_2 = \mu_{\mathbf{I}_2}(x) \tag{15}$$

Three fuzzy density values have been estimated $\{\mathbf{g}_0, \mathbf{g}_1, \mathbf{g}_2\}$ from Eq. (15) which are performed as $\mathbf{g}_0 = \max(\mathbf{a}_0, \mathbf{a}_1)$, $\mathbf{g}_1 = \max(\mathbf{a}_1, \mathbf{a}_2)$ and $\mathbf{g}_2 = \max(\mathbf{a}_2, \mathbf{a}_0)$. Thus, we can estimate λ value from three defined fuzzy densities $\{\mathbf{g}_0, \mathbf{g}_1, \mathbf{g}_2\}$ by Sugeno measure. Then substitute these value of $\{\mathbf{a}_0, \mathbf{a}_1, \mathbf{a}_2\}$ into Eq. (13), and solve the quadratic equation formed as $a\lambda^2 + b\lambda + c = 0$, where λ is root and computed by $\lambda = \frac{b \pm \sqrt{b^2 - 4ac}}{2a}$. For positive λ, finally, we yield as follows:

$$\begin{aligned} a &= \mathbf{a}_0\mathbf{a}_1\mathbf{a}_2 \\ b &= \mathbf{a}_0\mathbf{a}_1 + \mathbf{a}_1\mathbf{a}_2 + \mathbf{a}_0\mathbf{a}_2 \\ c &= (\mathbf{a}_0 + \mathbf{a}_1 + \mathbf{a}_2 - 1) \end{aligned}$$

Thus, the estimated λ value should be used for fuzzy measure on sets $\mu_{\mathbf{I}_p}(x)$, $p = \{0, 1, 2\}$ in Eq. (7). Therefore, we have $\{\mathbf{g_0}, \mathbf{g_1}, \mathbf{g_2}\}$ and λ, so according to fuzzy measure, we can measure evidence on fuzzy densities set $\{\mathbf{g_0}, \mathbf{g_1}, \mathbf{g_2}\}$ with λ by the following computation:

$$\begin{aligned} \mathbf{g_{01}} = &\quad \{\mathbf{g_0} \cup \mathbf{g_1}\} = &\quad \mathbf{g_0} + \mathbf{g_1} + \lambda \mathbf{g_0} \mathbf{g_1} \\ \mathbf{g_{12}} = &\quad \{\mathbf{g_1} \cup \mathbf{g_2}\} = &\quad \mathbf{g_1} + \mathbf{g_2} + \lambda \mathbf{g_1} \mathbf{g_2} \\ \mathbf{g_{02}} = &\quad \{\mathbf{g_0} \cup \mathbf{g_2}\} = &\quad \mathbf{g_0} + \mathbf{g_2} + \lambda \mathbf{g_0} \mathbf{g_2} \end{aligned}$$

after fuzzy density measure on set $\{\mathbf{a_0}, \mathbf{a_1}, \mathbf{a_2}\}$, we defined, and compute measurable function $\mathbf{h}(\mathbf{x})$ on set $\{\mathbf{a_0}, \mathbf{a_1}, \mathbf{a_2}\}$ which is furnished as $\mathbf{h}(\mathbf{x_0}) = \min(\mathbf{a}_0, \mathbf{g}_0)$, $\mathbf{h}(\mathbf{x_1}) = \min(\mathbf{a}_1, \mathbf{g}_1)$ and $\mathbf{h}(\mathbf{x_2}) = \min(\mathbf{a}_2, \mathbf{g}_2)$ with satisfy $\mathbf{h}(\mathbf{x}_0) \geq \mathbf{h}(\mathbf{x}_1) \geq \mathbf{h}(\mathbf{x}_3)$. After all computation, finally we have a set of fuzzy density values $\{\mathbf{g_0}, \mathbf{g_1}, \mathbf{g_2}\}$ and measurable function $\mathbf{h}(\mathbf{x}_0) \geq \mathbf{h}(\mathbf{x}_1) \geq \mathbf{h}(\mathbf{x}_3)$ that should be used in fuzzy integral by using Eq. (14) on set $\{\mathbf{a}_0, \mathbf{a}_1, \mathbf{a}_2\}$, and yield

$$\acute{\mu}_{\mathbf{I_F}}(\mathbf{x}) = \max_{i=0,1,2} \left[\min(\mathbf{h}(\mathbf{x}_i), \mathbf{g}_{i,i+1}(\mathbf{x}))\right] \tag{16}$$

where $\acute{\mu}_{\mathbf{I_F}}(\mathbf{x})$ is fused fuzzy matrix that contain integrated membership grade from three partition fuzzy sets $\{\mathbf{I_0}, \mathbf{I}_1, \mathbf{I_2}\}$.

Threshold image construction In this step, we have formulated fuzzy membership grades classification, and generate threshold image from fused fuzzy matrix $\acute{\mu}_{\mathbf{I_F}}(\mathbf{x})$ by using Eq. (16). In order to classify a membership grade as an object into two class $\mathcal{C}_0 \in \mathcal{R}_0$, and $\mathcal{C}_1 \in \mathcal{R}_1$ where $\mathcal{R}_0, \mathcal{R}_1$ are image region. We assume that if any $\acute{\mu}_{\mathbf{I_F}}(\mathbf{x}) \in \mathcal{C}_0$ which is member of $\mathcal{R}_0$ then $\acute{\mu}_{\mathbf{I_F}}(\mathbf{x})$ assigned $\acute{\mu}_{\mathbf{I_F}}(\mathbf{x}) = 0$ else $\acute{\mu}_{\mathbf{I_F}}(\mathbf{x}) = 1$ for class $\mathcal{C}_1 \in \mathcal{R}_1$.

Thus, an object can belong to either class $\mathcal{C}_0 \in \mathcal{R}_0$ or $\mathcal{C}_1 \in \mathcal{R}_1$ depending on threshold. We can calculate threshold ζ from fused fuzzy matrix $\acute{\mu}_{\mathbf{I_F}}(\mathbf{x})$ on the following rules:

$$\begin{aligned} \mathcal{R}_0 \colon \acute{\mu}_{\mathbf{I_F}}(\mathbf{x}) < &\quad \zeta | \acute{\mu}_{\mathbf{I_F}}(\mathbf{x}) = 0 \in \mathcal{C}_0 \\ \mathcal{R}_1 \colon \acute{\mu}_{\mathbf{I_F}}(\mathbf{x}) < &\quad \zeta | \acute{\mu}_{\mathbf{I_F}}(\mathbf{x}) = 1 \in \mathcal{C}_1 \end{aligned}$$

where ζ is defined as $\zeta = \frac{1}{3(M \times N)} \sum \acute{\mu}_{\mathbf{I_F}}(\mathbf{x})$.

Defuzzification Finally, we get segmented retinal images $\mathbf{I_S}$ after defuzzification of membership grade $\acute{\mu}_{\mathbf{I}}(\mathbf{x})$ is furnished as follows:

$$\mathbf{I_S} = \acute{\mu}_{\mathbf{I_F}}(\mathbf{x}) * (\mathbf{x}_{\max} - \mathbf{x}_{\min}) \tag{17}$$

where $\mathbf{x}_{\max}$ and $\mathbf{x}_{\min}$ are largest and smallest intensity levels of the image $\mathbf{I}$, respectively. We get color retinal segmented image by using Eq. (17).

3 Result and Analysis

All the experiments for performance analysis of the proposed algorithms have been discussed in this section. We have used benchmark public retinal fundus image database "DRIVE" [23] for the effectiveness and comparison of the proposed algo-

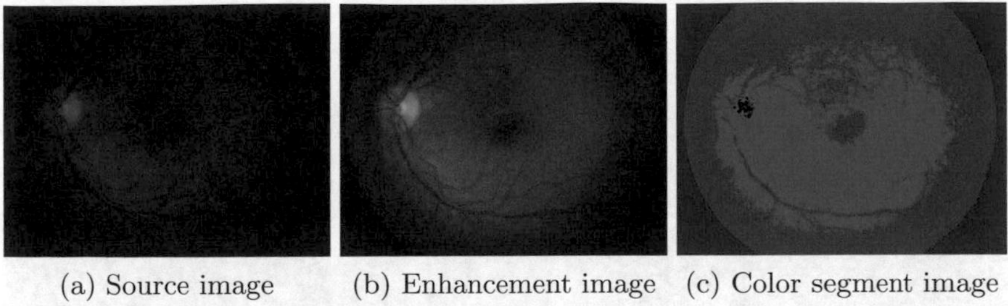

(a) Source image (b) Enhancement image (c) Color segment image

Fig. 2 Image data set 1

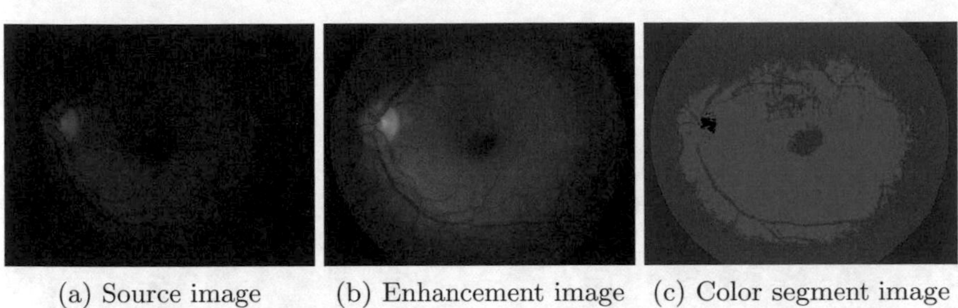

(a) Source image (b) Enhancement image (c) Color segment image

Fig. 3 Image data set 2

rithm. However, we have chosen baseline methods such as Histogram equalization [2, 22], Linear contrast starching [22], Color Retinal Image Enhancement based on Domain Knowledge [24], and Adaptive histogram method [2, 22] for validation purpose. Figures 2, 3, and 4 demonstrate the results that are achieved from the proposed enhancement and segment methods and Fig. 5 shows results obtained from different enhancement techniques. For all experimental results, the proposed method is superior with respect to better visual quality with uniform color distribution and preserves excellent contrast to retinal structures based on the comparison metrics. On the contrary, the existing enhancement techniques are unable to provide better clarity in enhancing the image and fail to maintain color distortion in the distribution of color and best retinal structures. For example, results of these enhancements methods shown in Fig. 5b, c, d, e are obviously unable to preserve improved color distribution and better retinal vasculature, macular pigment structures compare to proposed scheme in Fig. 5f.

To evaluate objective enhancement, we use the spatial frequency metric for performance analysis of the proposed method. Commonly, spatial frequency (**SF**) is estimated for the overall clarity level of an image, where a larger (**SF**) value implies a better visibility [25] and is determined as $\mathbf{SF} = \sqrt{\mathbf{RF}^2 + \mathbf{CF}^2}$, where **RF** is the row frequency, and **CF** is the column frequency defined by

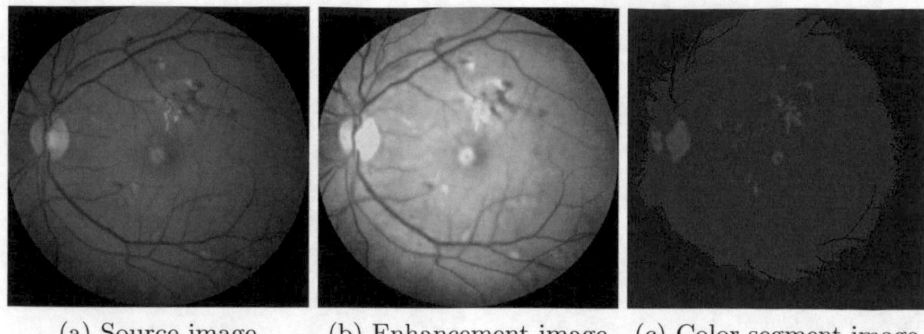

(a) Source image (b) Enhancement image (c) Color segment image

Fig. 4 Image data set 3

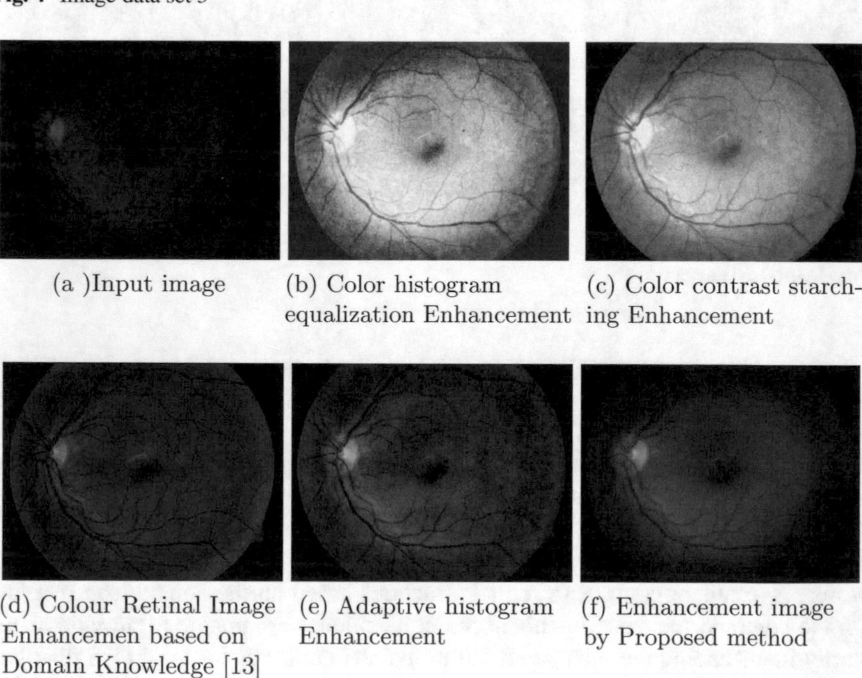

(a)Input image (b) Color histogram equalization Enhancement (c) Color contrast starching Enhancement

(d) Colour Retinal Image Enhancemen based on Domain Knowledge [13] (e) Adaptive histogram Enhancement (f) Enhancement image by Proposed method

Fig. 5 Compared with different image enhancement methods

$$\mathbf{RF} = \sqrt{\frac{1}{\mathbf{M}\,(\mathbf{N}-1)} \sum_{i=0}^{\mathbf{M}-1} \sum_{i=0}^{\mathbf{N}-2} \left(\mathbf{I_E}\,(i, j+1) - \mathbf{I_E}\,(i, j)\right)^2}$$

$$\mathbf{CF} = \sqrt{\frac{1}{\mathbf{N}\,(\mathbf{M}-1)} \sum_{i=0}^{\mathbf{M}-2} \sum_{i=0}^{\mathbf{N}-1} \left(\mathbf{I_E}\,(i+1, j) - \mathbf{I_E}\,(i, j)\right)^2}$$

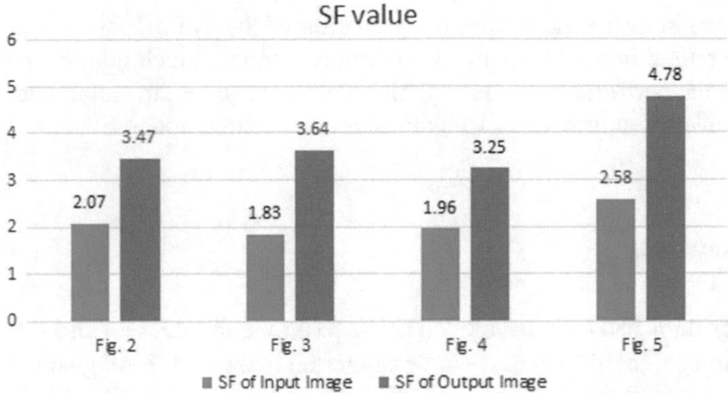

Fig. 6 Comparison with SF values

Table 1 Comparison analysis of enhancement results of different methods

Input image of Fig. 2a	Objective evaluation metric	
	SF value	Visibility improvement
Input image	2.0724	*
Color histogram equalization method [2, 22]	2.8496	37.51%
Color contrast starching method [22]	2.6971	30.14%
Image enhancement on Domain knowledge [24]	3.1745	53.18%
Adaptive histogram method [22]	2.9323	41.50%
Proposed method	**3.4695**	**67.24%**

All statistical data are shown in Fig. 6 and Table 1. Figure 6 represents **SF** of the proposed method, which shows that increase in the clarity of image has high **SF** score and percentage of clarity with respect to source image is 67.24, 98.83, 65.55, and 85.12%. Table 1 illustrates all comparative **SF** score and percentage of clarity against other enchantments methods. From the experimental results, it is clear that proposed enhancement technique maintains better-quality visual appearance without no artifact and color distortion. Other enhancement techniques such as Color Retinal Image Enhancement based on Domain Knowledge [24] maintains a modest enhancement and clarity but suffers from proper contrast level of small retinal vascular structures.

After that, we have performed image segmentation on all chosen benchmark color retinal images, and is shown in Figs. 2c, 3c, and 4c. It can be seen that the proposed method is able to segment different image regions with similar color patterns. For example, in Fig. 4c, all bright spots in fundus image are separated after segmentation.

Furthermore, in every segmented image region of the eye disc the macular darkly part of the retinal images is retained. Therefore, proposed techniques can be helpful to improve the performance of retinal blood vessel segment, macular detection, etc., and is flexible for automatic examination of color retinal images.

4 Conclusion

Many algorithms had been developed in the past for enhancement and segmentation of retinal image. In this paper, we have suggested fuzzy partitioning and computing-based color retinal image enhancement method to resolve the problem in a retinal image enhancement along with a novel retinal image segmentation method by using Sugeno fuzzy measure and integral. Experimental results demonstrate that the proposed scheme provides an effective and promising approach for retinal image enhancement and segmentation. Evidently, there is highest enhancing color content, and no new artifacts are established in resultant images.

We plan to improve the efficiency of our algorithms to assess on a large database of retinal images for robustness, stability in a noise background. In future work, deep denoising autoencoder for noise retinal image restoration and deep convolutional neural network (CNN) hybridization with fuzzy framework for retinal vessels abnormality detection will be elucidated.

References

1. Goldbaum, M., Katz, N., Chaudhuri, S., Nelson, M., Kube, P.: Digital image processing for ocular fundus images. Ophthalmol. Clin. N. Am. **3**(3), 447–466 (1990)
2. Zimmerman, J.B., Pizer, S.M.: An evaluation of the effectiveness of adaptive histogram equalization for contrast enhancement. IEEE Trans. Med. Imaging **7**(4), 304–312 (1988)
3. Burger, W., Burge, M.J.: Principles of Digital Image Processing Core Algorithms. Springer, London (2009)
4. Lin, T.S., Du, M.H., Xu, J.T.: The preprocessing of subtraction and the enhancement for biomedical image of retinal blood vessels. J. Biomed. Engg. **1**(20), 56–59 (2003)
5. Duan, J., Qiu., G.: Novel histogram processing for colour image enhancement. In: Proceedings of International Conference on Image and Graphics, pp. 55–58 (2004)
6. Foracchia, M., Grisan, E., Ruggeri, A.: Luminosity and contrast normalization in retinal images. Med. Image Anal. **3**(9), 179–190 (2005)
7. Sun, C.C., Ruan, S.J., Shie, M.C., Pai, T.W.: Dynamic contrast enhancement based on histogram specification. IEEE Trans. Consum. Electron. **51**(4), 1300–1305 (2005)
8. Feng, P., Pan, Y., Wei, B., Jin, W., Mi. D.: Enhancing retinal image by the contourlet transform. Pattern Recognit. Lett. **4**(28), 516–522 (2007)
9. Dai, P., Sheng, H., Zhang, J., Li, L., Wu, J., Fan, M.: Retinal fundus image enhancement using the normalized convolution and noise removing. Int. J. Biomed. Imaging **2016** (2016)
10. Starck, J.L., Murtagh, F., Candes, E.J., Donoho, D.L.: Gray and color image contrast enhancement by the curvelet transform. IEEE Trans. Image Process. **12**(6) (2003)
11. Vese, L.A., Chan, T.F.: A multiphase level set framework for image segmentation using the Mumford and Shah model. Int. J. Comput. Vis. **50**(3), 271–293 (2002)

12. Xu, W., Xia, S., et al.: A model based algorithm for mass segmentation in mammograms. In: IEEE Engineering in Medicine and Biology 27th Annual Conference, pp. 2543–2546 (2006)
13. Ghosh, S.K., Ghosh, A., Chakrabarti, A.: VEA: Vessels extraction algorithm and a novel Wavelet analyzer for diabetic retinopathy detection. Int. J. Image Graph. **18**(2) (2018)
14. Kass, M., Witkin, A., Terzopoulos, D.: Snakes: active contour models. Int. J. Comput. Vis. **1**(4), 321–331 (1988)
15. Brox, T., Weickert, J.: Level set based image segmentation with multiple regions. Pattern Recognit. Springer LNCS- **3175**, 415–423 (2004)
16. Chen, W., Giger, M.L., Bick, U.: A fuzzy c-means (FCM)-based approach for computerized segmentation of breast lesions in dynamic contrast-enhanced MR images. Acad. Radiol. **13**(1), 63–72 (2006)
17. Li, H.D., Kallergi, M., Clarke, L.P., Jain, V.K., Clark, R.A.: Markov random field for tumor detection in digital mammography. IEEE Trans. Med. Imag. **14**, 565–576 (1995)
18. Carballido-Gamio, J., Belongie, S.J., Majumdar, S.: Normalized cuts in 3-D for spinal MRI segmentation. IEEE Trans. Med. Imaging **23**(1), 36–44 (2004)
19. Shi, J., Malik, J.: Normalized cuts and image segmentation. IEEE Trans. Pattern Anal. Mach. Intell. **22**(8), 888–905 (2000)
20. Havens, T.C., Anderson, D.T., Wagner, C.: Data-informed fuzzy measures for fuzzy integration of intervals and fuzzy numbers. IEEE Trans. Fuzzy Syst. **23**(5) (2015)
21. Haubecker, H., Tizhoosh, H.R.: Fuzzy Image Processing, Computer Vision and Applications, vol. 2. Academic, New York (1999)
22. Klir, G.J., Yuan, B.: Fuzzy Set and Fuzzy Logic: Theory and Applications. Prentice Hall, USA (1995). ISBN 043-W1171-5
23. The database Available: http://www.isi.uu.nl/Research/Databases/DRIVE/
24. Joshi, G.D., Sivaswamy, J.: Colour retinal image enhancement based on domain knowledge. In: Sixth Indian Conference on Computer Vision, Graphics and Image Processing (2008)
25. Paulus, J., Meier, J., Bock, R., Hornegger, J., Michelson, G.: Automated quality assessment of retinal fundus photos. Int. J. Comput. Assist. Radiol. Surg. **5**(6), 557–564 (2010)

On the Cryptanalysis of S-DES Using Binary Cuckoo Search Algorithm

Ritwiz Kamal, Moynak Bag and Malay Kule

Abstract The cryptanalysis of Simplified Data Encryption Standard (S-DES) is an NP-Hard combinatorial problem. The goal of this paper is twofold. First, we study the cryptanalysis of S-DES via a metaheuristic algorithm named Binary Cuckoo Search Algorithm, which mimics the obligate brood parasitic behavior of some cuckoo species and resembles the Lévy flight behavior of certain birds. Second, we compare the efficiency of Cuckoo Search Algorithm with that of previous research works using Genetic Algorithm and Memetic Algorithm in regard to cryptanalysis of S-DES. Extensive tests and experimentations show that the proposed Cuckoo Search-based algorithm outperforms the previous works based on both genetic and memetic algorithms.

Keywords Cryptanalysis · Simplified data encryption standard · Cuckoo search algorithm · Genetic algorithm · Memetic algorithm · Lévy flight

1 Introduction

Cryptanalysis [1] is the study of methods for retrieving the information from the encrypted message, without any knowledge about the secret information. It is the process of identifying shortcomings in the design of ciphers [2] and attempting to overcome those or moving on to further improved cryptographic systems. Cryptanalysis of simplified data encryption standard [1] can be categorized as an NP-Hard combinatorial problem. It requires some effort, in terms of time and memory require-

R. Kamal (✉) · M. Bag · M. Kule
Department of Computer Science and Technology, Indian Institute of Engineering Science and Technology, Shibpur, Howrah 711103, West Bengal, India
e-mail: ritwizkamal@gmail.com

M. Bag
e-mail: moynakbagklvm@gmail.com

M. Kule
e-mail: malay.kule@gmail.com

A. K. Das et al. (eds.), *Computational Intelligence in Pattern Recognition*, Advances in Intelligent Systems and Computing 999,
https://doi.org/10.1007/978-981-13-9042-5_3

ments, to solve these problems thereby increasing the size of the problem. Traditional optimization techniques like brute force can prove to be very inefficient in this case. We, therefore, use a nontraditional optimization technique—binary cuckoo search algorithm [3] with an aim to find a sufficiently *good solution*.

A few recent studies have been made in the relevant area. In 2009, Xin-She Yang and Suash Deb [4] proposed a bio-inspired algorithm called Cuckoo Search Algorithm (CSA) for solving optimization problems. In 2012, Gherboudj et al. [5] proposed a discrete binary version of CSA for solving 0-1 knapsack problems. In 2013, Yang et al. [6] proposed a binary cuckoo search algorithm for feature selection. In both these papers, the authors used the sigmoid function to convert the continuous-valued outputs of CSA to binary values. In 2009, Garg [7] carried out interesting studies on the cryptanalysis of S-DES via evolutionary computation techniques, namely genetic algorithm, memetic algorithm, and simulated annealing. The results showed the superiority of memetic algorithm.

In our proposed work, we have used Binary Cuckoo Search algorithm to perform cryptanalysis of Simplified Data Encryption Standard algorithm. This paper is divided into five sections, the first one being an introduction to the work conducted. The second section gives an insight into the prerequisites or concepts related to the subject of the paper. The third section deals with the methodology proposed and used while the fourth section shows the results of the experimental works carried out. The fifth and final section concludes the study and suggests ideas for further research.

2 Preliminaries

This section gives an insight into concepts like the S-DES algorithm, the Cuckoo Search Algorithm, the binary version of Cuckoo Search, and the cost function used.

2.1 The S-DES Algorithm Description

In the following, we discuss briefly the S-DES algorithm [1, 7–9]. The inputs to the S-DES encryption algorithm are a 10-bit key and an 8-bit block of plaintext. The algorithm yields a ciphertext block also of 8 bits as the output. The algorithm for decryption is very similar to the encryption algorithm with the steps being carried out in reverse order. It must be noted that the key generation process [7, 9] involves a sequential application of permutation and substitution operations. Figure 1 [7] shows a schematic block diagram [7] of the complete S-DES algorithm.

The S-DES Encryption [7–9] involves the application of the following operations in a proper sequence. There is an initial permutation (IP = [2 6 3 1 4 8 5 7]) and a final permutation (IP^{-1}). A complex function f_k consists of a combination of permutation and substitution functions performed on the left and right 4 bits of the input, respectively. We have, f_k(L, R) = (L XOR f(R, key), R) where L and R are

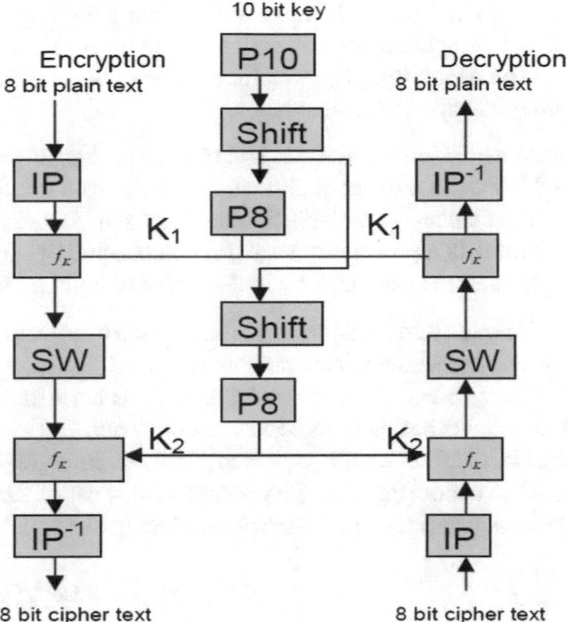

Fig. 1 The S-DES algorithm [7]

left and right halves of input, key is a subkey. Computation of f(R,key) is done using the expansion/permutation E/P = [4 1 2 3 2 3 4 1], the S-Boxes S_0 and S_1 and P4 = [2 4 3 1].

$$S_0 : \begin{pmatrix} 1\,0\,3\,2 \\ 3\,2\,1\,0 \\ 0\,2\,1\,3 \\ 3\,1\,3\,2 \end{pmatrix} \text{ and } S_1 : \begin{pmatrix} 0\,1\,2\,3 \\ 2\,0\,1\,3 \\ 3\,0\,1\,0 \\ 2\,1\,0\,3 \end{pmatrix}$$

The S-boxes work in the following manner: We consider the first and fourth input bits as an 2-bit number. This number specifies a particular row of the S-box. Similarly, the second and third input bits denote a column. The 2-bit output would be the entry in that row and column in base 2. The function f_k is applied first to the left 4 bits of input and then to the right 4 bits by using the switch function (SW). In this first instance, the key input is K1and it is K2 in the second instance.

2.2 The Cuckoo Search Algorithm

Cuckoo Search [4] is one of the most recently developed bio-inspired metaheuristic algorithms. It is based on the typical behavior of certain cuckoo species [3], which lay eggs in other birds' nests and also on Lévy flights [3], which is a typical flying pattern adopted by certain fruit flies and birds [3].

The Algorithm. For simplicity in describing cuckoo search, three idealized rules are used which are stated below:

- At a time, only one egg is laid by each cuckoo and is assigned to any random nest.
- Nests with a better quality of eggs are passed on to the subsequent generations.
- The number of available host nests is fixed. A cuckoo egg is discovered by the host bird with a probability Pa $\in$ [0,1]. Subsequently, this egg is either thrown away or the nest is abandoned by the host bird to build a brand new nest.

We can approximate the last assumption by replacing the fraction *Pa* of the *n* nests by new nests having new random solutions.

The generation of new solutions x_i^{t+1} is performed using Lévy flights [Eq. (1)]. Lévy flights basically serve as a random walk. The random step sizes are drawn from a Lévy distribution for large steps. Lévy distribution has an infinite variance and an infinite mean [Eq. (2)]. The consecutive steps of the cuckoo form a random walk process that obeys a power law step-length distribution with a heavy tail [10].

$$x_i^{t+1} = x_i^t + \alpha \oplus \text{Lévy}(\lambda) \tag{1}$$

$$\text{Lévy} \sim u = t^{-\lambda}, \quad (1 < \lambda \leq 3) \tag{2}$$

Here, x_i^{t+1} and x_i^t represents the solutions "i" at times t + 1 and t, respectively. Also, here, α (>0) is the step size, which depends on the scales of the concerned problem. We usually take $\alpha = O(1)$. The product $\oplus$ means entry-wise multiplication. The efficiency of a random walk via Lévy flight lies in its ability to explore the search space given its step size is much longer in the long run.

2.3 *Binary Cuckoo Search Algorithm*

The original CS algorithm operates in a continuous search space and is based on Lévy flights. Consequentially, the output of the algorithm is a set of real numbers. However, for binary optimization problems like cryptanalysis of S-DES, real solutions are not acceptable. Therefore, the real valued output of the algorithm must be converted into binary value. Gherboudj et al. [5] proposed the use of the Sigmoid function to constrain a real solution x_i to the interval [0, 1] as follows in Eq. (3):

$$S(x_i) = 1/\left(1 + e^{-x_i}\right) \tag{3}$$

where $S(x_i)$ is the flipping chance of bit x_i'. To get a binary solution, a random number is generated from the range [0, 1] and it is compared with the real solution as per Eq. (4).

$$x_i' = \begin{Bmatrix} 1\ if\ \gamma < S(x_i), & \gamma \epsilon [0, 1] \\ 0, & otherwise \end{Bmatrix} \tag{4}$$

Using this binary valued output, all relevant computations are made during the execution of the S-DES algorithm. A similar approach involving the use of sigmoid function was adopted by Yang et al. [6] in 2013.

2.4 *Cost Function*

The cost or fitness function is the most important factor of any optimization-based metaheuristic algorithm. The language characteristics almost entirely decide the choice for the fitness function. Nalini [11] used the following technique to evaluate the candidate key (k): comparing n-gram statistics of the decrypted message with those of the language (assumed to be known). Equation (5) is a general formula used for the same. Here, K refers to the known language statistics [12] (for English alphabets Ã = [A,….,Z]) and D refers to the decrypted message statistics. The unigram, bigram, and trigram statistics are denoted by *u/b/t* respectively. The values of α (0.2), β (0.4), and γ (0.4) [7, 9, 11] allow assigning of different weights to these n-gram types where $\alpha + \beta + \gamma = 1$.

$$C_k \approx \alpha . \sum_{i \epsilon \tilde{A}} \left| K_{(i)}^u - D_{(i)}^u \right| + \beta . \sum_{i,j \epsilon \tilde{A}} \left| K_{(i,j)}^b - D_{(i,j)}^b \right| + \gamma \cdot \sum_{i,j,k \epsilon \tilde{A}} \left| K_{(i,j,k)}^t - D_{(i,j,k)}^t \right| \tag{5}$$

When trigram statistics are used, Eq. (5) has a complexity of $O(P^3)$. Here, P corresponds to the alphabet size. Therefore, calculating the trigram statistics is an expensive task. We, therefore, use only the bigram for assessment function. We use Eq. (5) as the fitness function for cuckoo search algorithm attack on S-DES.

3 Proposed Method

In this section, we discuss the method proposed for cryptanalysis of S-DES using binary CSA. The attractiveness of CSA lies in the fact that it is a very simple algorithm compared to other population-based metaheuristic algorithms. CSA requires managing only two parameters, viz., number of nests *n* and the probability of a cuckoo egg being discovered *Pa*. This is a significant respite when compared to the likes of PSO [13] or GA [7], thereby making it easier to implement and experimented on. A pseudocode for the algorithm is given below.

```
Input: Pa and n
Output: The best key(last best solution)
Objective function: f(x), x=(x1,…,xd)T
Initialize a population of n host-nests xi(i=1,2,…,n)
Find the current best solution
while(t < MaxGeneration)
  Get a cuckoo randomly by Lévy flight
  Convert this to binary using sigmoid function
  Evaluate its fitness Fi
  Select a nest j randomly
  If(Fi > Fj)
   Replace solution j by new solution
  End
  Convert solutions to binary and find current best
  Abandon Pa fraction of worse nests
  Build new solutions via Lévy flight in their places
  Re-evaluate solutions to find current best k
  If (Fk= MaxFitness)
    break
  End
End while
Return the last best solution(key)
```

The proposed algorithm consists of the following basic steps. First, an initial population, i.e., candidate solution keys are generated randomly and each solution represents an egg in a nest. Next, the current best solution among the initial population is found. After that, in a loop, Lévy flight is performed to get a new and potentially better solution followed by the abandonment of a fraction of worse solutions. The loop terminates on reaching a terminating condition and the best solution is returned.

These basic steps involved in the algorithm can be explained with the help of an example. Let us say we have a ciphertext "cuckoo". First, *n* random candidate 10-bit keys are generated and the ciphertext is decrypted using each of these keys. Next, the fitness values corresponding to the "n" plaintexts are calculated to get the most fit solution. After that, a Lévy flight is performed to get a real solution say [2.39, 2.84, −1.20, 4.79, −1.71, 4.02, 2.29, −1.73, −4.04, −3.68], which is then converted into binary [1, 1, 0, 1, 0, 1, 0, 0, 0, 0] using the sigmoid function. This key is then used for decryption of "cuckoo" and its subsequent fitness calculation. If this fitness is more than that of a randomly chosen nest, it replaces the previously existing solution. This is followed by abandoning *Pa* fraction of the worse nests and the best solution is returned. The steps from Lévy flight till the abandonment are carried out in a loop that terminates when either the maximum number of generations is reached or the best solution key has a fitness equal to a predefined maximum value of fitness possible for our ciphertext "cuckoo".

4 Experiments and Results

This section describes the experiments that were carried out to outline the effectiveness of Binary Cuckoo Search Algorithm. These experiments also compare the performance of CSA with that of Genetic Algorithm (GA) [7, 9] and Memetic Algorithm (MA) [7] for the cryptanalysis of S-DES algorithm. For the purpose of comparison, the corresponding data for GA and MA, from Garg's [7] paper was used. Experimental results obtained for the algorithms were generated with a fixed number of runs per data point. For example, 5 messages were created and the algorithm was run 20 times for each message. The results were averaged to get the data point. It must be highlighted that all experimentations have been carried out using Octave 4.4.0 on a PC Intel® Core i5-7300HQ 2.50 GHz with 8 GB RAM running Windows 10 as operational system. For cryptanalysis, the initial number of nests n is taken to be 40 and the probability $Pa = 25\%$ [4, 14].

Table 1 depicts a tabulation of the performance of CSA in terms of the number of generations taken by the algorithm to arrive at the convergence versus the number of nests (population size) used by the algorithm. The population size was varied from 10 to 100. This table also gives a comparative study of the effectiveness of CSA in comparison with GA and MA. Basically, the table compares the average number of key bits (out of 10) recovered successfully by the algorithms and the execution time of the algorithm as the size of the ciphertext was varied from 100 to 1000.

It can be observed from Fig. 2 that the performance of CSA improves with the increase in the population size (number of nests) and gradually gains a near constant slope. This can be attributed to the fact that CSA is a random walk with step sizes drawn from a heavy-tailed Lévy distribution. More the number of nests, more is

Table 1 Performance of cuckoo search algorithm and comparison of CSA, GA, and MA

Population size		10	20	30	40	50	60	70	80	90	100
Generation of convergence		20	17	5	4	4	3	3	3	4	4
Cipher text size		100	200	300	400	500	600	700	800	900	1000
CSA [proposed]	Key bits matched	7	7	8	9	9	8	8	7	8	9
	Time (min)	2.04	2.94	3.23	4.19	5.18	4.63	3.89	4.32	4.61	5.91
GA [7]	Key bits matched	6	6	4	6	6	7	6	8	6	7
	Time (min)	2.62	4.50	2.13	2.35	2.52	2.07	4.07	2.40	2.53	2.17
MA [7]	Key bits matched	8	6	5	7	6	8	7	8	9	9
	Time (min)	5.10	14.0	15.3	12.5	10.0	5.50	3.05	2.85	2.24	2.14

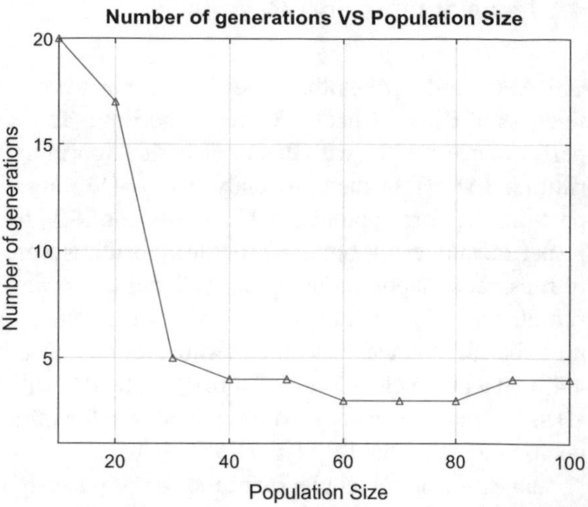

Fig. 2 Number of generations versus population size for CSA

the efficiency of exploring the search space. We can also conclude that the average number of generations taken for reaching convergence is 4. It must be noted that the values 20 and 17 of generation for population sizes 10 and 20, respectively, indicate that the algorithm performs well only when the size of the population is sufficiently big enough. Usually, a population size of 30 or more provides better results with the optimal results being in the range 50-70.

A number of extensive experiments were carried out to compare the performance of binary CSA with that of already established metaheuristics like Genetic and Memetic Algorithms [7]. Figures 3 and 4 show a graphical depiction of the results of the experiments. Figure 3 compares the accuracy of the above stated three algorithms in terms of the number of key bits recovered correctly. On the other hand, Fig. 4 gives an account of the efficiency of the three algorithms in terms of their run-time.

From Fig. 3, it can be observed that CSA outperforms GA and MA in terms of the accuracy. The number of key bits recovered by CSA is better than that by GA 90% of times. As with MA, CSA outperforms MA 50% of times and is on par 20% times. In simple words, the average number of key bits recovered by CSA is better than that by GA and MA. This highlights the fact that Cuckoo Search Algorithm is more efficient than Genetic Algorithm and Memetic Algorithm in terms of accuracy of achieving the goal which is efficient cryptanalysis in this case.

Figure 4 compares the time taken to execute cryptanalysis of S-DES by CSA, GA, and MA. It can be observed that CSA is quite efficient in terms of execution time of the algorithm. The average time taken by CSA is 4.09 min while that of GA and MA are 2.73 min and 7.27 min, respectively. Although GA has a lower average execution time than that of CSA, the difference is more than compensated by the better accuracy of CSA in retrieving the original key. On the other hand, in case of MA, the average execution time is evidently much greater than that of CSA.

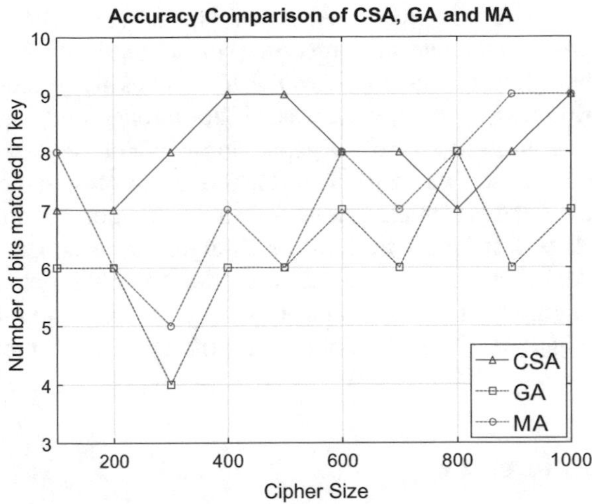

Fig. 3 Accuracy comparison of CSA [proposed], GA [7] and MA [7]

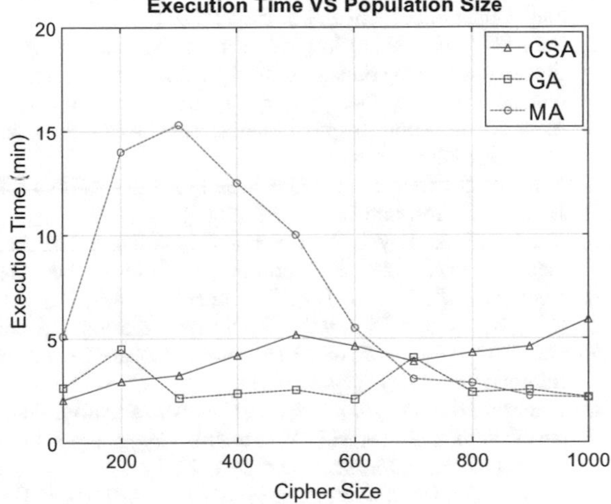

Fig. 4 Execution time versus cipher size for CSA [proposed], GA, [7] and MA [7]

Therefore, from Figs. 3 and 4, it is quite evident that CSA is the most efficient out of the three in terms of time and memory requirements for our problem concerning the cryptanalysis of simplified data encryption standard.

5 Conclusion and Future Works

In this paper, we have presented a new approach for the cryptanalysis of simplified data encryption standard (S-DES) algorithm—by using binary cuckoo search algorithm, which can be categorized as a nature-based metaheuristic algorithm. Our objective is twofold: one, to study the efficiency of binary CSA in solving the said problem and two, to compare the performance with that of already established nature-based evolutionary optimization techniques like genetic algorithm and memetic algorithm. The first objective was fulfilled through studying the dependence of generation of convergence over the population size. For the second objective, the performance comparison was done on the basis of the average number of key bits correctly recovered and execution time versus the amount of ciphertext. Our experimental results show that binary cuckoo search algorithm is superior than Genetic and Memetic Algorithms. There are, however, some possibilities of future work which include attempting to compare the efficiency of CSA with other metaheuristic techniques, which have been developed in recent times or that are being developed currently.

References

1. Stallings, W.: Cryptography and Network Security Principles and Practices, 4th edn. Prentice Hall, Upper Saddle River, NJ, USA (2005)
2. Palit, S., Sinha, SN., Molla, MA.,Khanra, A., Kule, M.: A cryptanalytic attack on the knapsack cryptosystem using binary firefly algorithm. In: 2nd International conference on computer and communication technology (ICCCT), IEEE, India, pp. 428–432 (2011)
3. Yang, X.-S.: Nature-Inspired Metaheuristic Algorithms, 2nd edn., Luniver Press, BA11 6TT, United Kingdom (2010)
4. Yang, X.-S., Deb, S.: Cuckoo Search via Lévy Flights. In: World Congress on Nature & Biologically Inspired Computing (2009)
5. Gherboudj, A., Layeb, A., Chickhi, S.: Solving 0–1 knapsack problems by a discrete binary version of cuckoo search algorithm. Int. J. Bio-Inspired Computation **4**(4), 229–236 (2012)
6. Rodrigues, D., Pereira, L.A.M., Almeida, T.M.S., Papa, J.P., Souza, A.N., Ramos, C.C.O., Yang, X.: BCS: A Binary Cuckoo Search Algorithm for Feature Selection, IEEE (2013)
7. Garg, P.: Cryptanalysis of S-DES via evolutionary computation techniques. Int. J. Comput. Sci. Informat. Secur. **1**(1) (2009)
8. Forouzan, B.A.: Cryptography and Network Security. Tata McGraw-Hill, New Delhi (2007)
9. Vimalathithan, R., Dr. M.L. Valarmathi.: Cryptanalysis of S-DES using genetic algorithm. Int. J. Comput. Sci. Informat. Secur. **2**(4) (2009)
10. Heavy-tailed Distribution: https://en.wikipedia.org/wiki/Heavy-tailed_distribution
11. Nalini: Cryptanalysis of data encryption standard via Optimization heuristics. Int. J. Comput. Sci. Informat. Secur. **6**(1B) (2006)
12. Known Language Statistics. http://practicalcryptography.com/cryptanalysis/letter-frequencies-various-languages/english-letter-frequencies/
13. Particle Swarm Optimisation: https://en.wikipedia.org/wiki/Particle_swarm_optimization
14. Yang, X.-S., Deb, S.: Engineering optimization by Cuckoo Search. Int. J. Math. Modell. Numeric. Optimisat. **1**(4), 330–343 (2010)

Feature Selection Using NSGA-II for Event Extraction on Genetic and Molecular Mechanisms Involved in Plant Seed Development

Amit Majumder, Asif Ekbal and Sudip Kumar Naskar

Abstract Molecular network structure to regulate plant seed development is very complex and to understand this from biomedical literature is a big challenge. Seed development is based on coordinated growth of different tissues, which are involved with complex genetics and environmental regulation. We develop a system for binary event extraction using statistical-, syntactic-, and dependency-based features. Experiments on the benchmark datasets of BioNLP-2016 SeeDev shared task show the recall, precision, and F-score values of 0.517, 0.399, and 0.451, respectively.

Keywords BioNLP-2016 SeeDev shared task · Binary event extraction · Dependency graph · Feature selection · Genetic algorithm

1 Introduction

Biomedical documents in electronic form are growing rapidly in the internet. Hence, the cost of retrieving relevant information and inferring scientific knowledge from this huge information is becoming a challenging task. Event expression consists of entities and relations. Entities represent proteins, genes, etc. in the biomedical literature. Relations represents an association between two or more entities. Events are represented by entities along with relation between entities. Research is going on some important NLP challenges conducted in BioNLP-2009, BioNLP-2011, and BioNLP-2013 [1–3]. Existing systems for extracting relations and events have taken

A. Majumder (✉)
Techno India University, Kolkata 700091, West Bengal, India
e-mail: jobamit48@yahoo.co.in

A. Ekbal
IIT Patna, Patna 801106, India
e-mail: asif@iitp.ac.in

S. K. Naskar
Jadavpur University, Kolkata 700032, India
e-mail: sudip.naskar@gmail.com

A. K. Das et al. (eds.), *Computational Intelligence in Pattern Recognition*, Advances in Intelligent Systems and Computing 999,
https://doi.org/10.1007/978-981-13-9042-5_4

two main approaches, which are rule-based and feature-based. We use feature-based approach for event extraction. In feature-based technique, the features which are in text format are converted into feature vector represented in numerical format. The data in feature vector format is supplied as input to classification algorithm like Support Vector Machine (SVM), which has been used by one of the popular event extraction systems TEES [4]. The SeeDev task of the BioNLP-2016 [5, 6] is concerned with relation extraction or event extraction. The task contributes to the problem of biomedical relation extraction. Some of the approaches that have been used by researchers on SeeDev binary relation extraction are convolution network approach [7], machine learning with a shallow linguistic kernel along with distant supervised approach [8], and SVM based on linear kernel [9]. We work on SeeDev task of binary event extraction using feature-based approach by Support Vector Machine (SVM). We have done feature selection by NSGA-II [10] optimization algorithm.

2 Motivation

Event extraction from biomedical data has started in the year 2009 (BioNLP-09) [2] and subsequently in BioNLP-11 [1] and BioNLP-13 [3]. These tasks are very challenging. Recently, BioNLP-16 [5] introduces extraction of the event on genetic and molecular mechanisms associated with the development of plant seed. This is a challenging task, as the experimental results achieved so far by the researchers are around 0.42 F-score, which indicates very low performance. So, we try to develop a system for binary event extraction using contextual, content-based, and dependency-based features and then apply feature selection to improve the performance of system. We also see that a system alone cannot provide a good result. So, we merge the results of several systems by selecting the best results from individual system.

3 Proposed Approach

We work on SeeDev task of BioNLP-2016 shared task. This task involves the extraction of 22 different types of binary events. Each event has entity type arguments. Total number of entity types are 16 in dataset. We define binary relation as an event consisting of two entity arguments. For each type of binary event, we create one classifier using supervised machine learning technique by Linear SVM (Support Vector Machine). We apply feature selection technique to find out the set of relevant features for each event type. Our approach is based on the following four steps:

1. For each event type,
   - Generate examples from training data and development data using features mentioned in Sect. 5.
   - Apply NSGA-II algorithm for feature selection and find the best set of features.

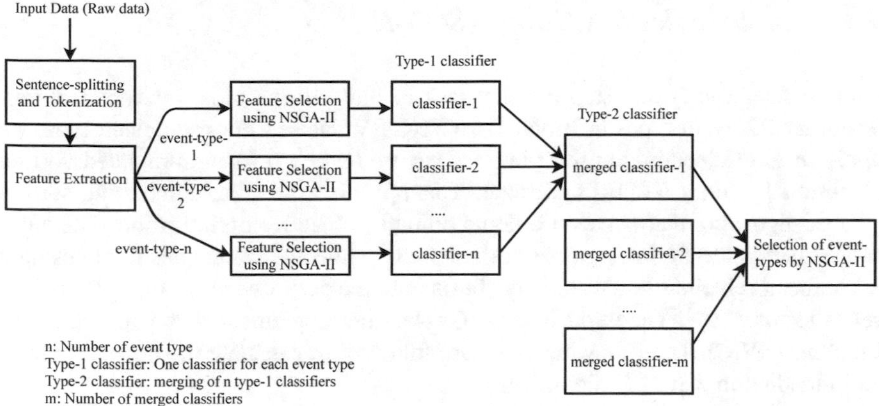

Fig. 1 Flow diagram of proposed method

- Use the feature set to generate classifier and apply the classifier on test data to find binary events.

2. Merge the results generated by the classifier of each event type.
3. Repeat step 1 and step 2 to generate several classifiers.
4. Selection of event types from several classifiers.

4 Major Steps for Event Extraction

In this section, we describe our major steps for binary event extraction. Figure 1 shows the flow diagram of our proposed method.

4.1 Segmentation and Tokenization

In this step, raw text data are segmented into sentences and each sentence is tokenized. For segmentation and tokenization, we use Stanford coreNLP tool.[1]

4.2 Feature Extraction

Different types of features is needed to identify binary events. These features include contextual, semantic, and syntactic features, which are extracted from the dataset. The features used in our experiment have been described in Sect. 5.

[1] https://stanfordnlp.github.io/CoreNLP/.

4.3 Feature Selection Using NSGA-II

Feature selection is one step in NLP task to find out the most relevant features. There are 22 event types in BioNLP-16 SeeDEV dataset. For each event type, we apply feature selection. For this purpose, we use NSGA-II (Non-dominated Sorting Genetic Algorithm II) [10] optimizer. This optimizer is a non-dominating sorting genetic algorithm that solves problems on multi-objective optimization. The algorithm attempts to perform global optimization, while enforcing constraints using a tournament selection-based strategy. To find out the performance of the system for a set of features, we have used LinearSVC[2] classifier implemented as Support Vector Machine (SVM). The following steps are followed to calculate the fitness value of an individual in NSGA-II algorithm.

For each event type:

1. Generate data in feature format from training and development and data.
2. Merge training and development data.
3. Create fivefolds of merged data (i.e., training plus development data).
4. Create five classifiers, where for each classifier, we use onefold as test data and remaining fourfolds as training data
5. Merge test result of each fold and calculate recall, precision, and F-score from this merged result.
6. Use recall and precision as two objective functions (for NSGA-II) which are to be maximized.

4.4 Generate Classifiers for Each Event Type

For each event type, we create classifier (say, *type-1 classifier*) using the optimal set of features as found in feature selection stage for that event type. We use the same classification algorithm as used in feature selection stage. We integrate the results of all event types to generate the performance of system.

4.5 Merge the Classifiers for Each Event Type

There are 22 different event types. As one classifier is generated for each event type, the total number of classifiers is 22 (i.e., 22 *type-1 classifiers*). Results of all these 22 classifiers are merged to generate the overall result (say *type-2 classifier*) of 22 event types and find out overall accuracy of the result in terms of F-score.

[2]http://scikit-learn.org/stable/modules/generated/sklearn.svm.LinearSVC.html.

4.6 Selection of Event Type from the Result of Classifiers

Each *type-2 classifier*'s result contains 22 different event types. We keep a record of recall, precision, and F-score per event type as shown in Table 2. We create 20 number of *type-2 classifiers*, where first 10 classifiers are formed using feature set with top 10 F-score values (i.e., top 10 fitness as found in feature selection by *NSGA-II*) and the rest 10 classifiers are by randomly selecting the feature set from the pool of feature sets to get diversity in the result.

5 Features

To develop a system for event extraction, we use statistical-, syntactic-, and dependency-based features along with some other features which are described below.

1. **Entity features**: Each binary event contains two entities. We consider words, POS tags of these two entities as entity feature. We also consider the distance between entities as one feature.
2. **BOW**: All words in a sentence are prefixed with positional information with respect to entity mentions. If a word appears before an entity, it is prefixed with *"before"*. If a word appears between two entities, it is prefixed with *"mid"* and if a word appears after an entity, it is prefixed with *"after"*.
3. **token_distance**: We find tokens of each entity and find the position of the first token in that entity, as an entity may have more than one word. Distance between the first word of each entity is considered as one feature value.
4. **Part of Speech (POS) feature**: POS tags are extracted for all the words that appear as the previous word, next word, and the words inside an entity as mentioned. These tags are concatenated separately for the three cases (before, after, and middle) and considered as three categories of POS feature. Each binary event has two arguments. POS tags of the tokens of these arguments are considered as another category of POS features.
5. **Verb feature**: We identify the verb tokens, which appear as previous token, next token, and tokens inside an entity mentioned in a sentence. The lemma of the token is prefixed with before or middle or after according to the position of the token in the sentence.
6. **Entity type**: Binary events consist of entities and relation among entities. Each entity belongs to one entity type. Some of the entity types are *RNA, Protein, Gene, Gene_Family, Protein_Family*, etc.
7. **Trigger pattern**: In text document, equivalent representation of a part of the text is represented using parenthesis. If two entities are expressed in this way, then we use this information as one feature.

Table 1 BioNLP-ST 2016 SeeDev binary event dataset

Attribute	Training data	Development data	Test data
Number of articles	39	19	29
Entities	3259	1607	2216
Events	1628	819	unknown

8. **Dependency path feature**: We create dependency graph for a sentence using Stanford CoreNLP.[3] From the graph, we find the shortest dependency path between the arguments of binary event. Edge labels, lemmas of tokens, and POS of tokens are included in dependency path features. From the dependency graph, we also find out input and output edges for the arguments. Dependency labels of these input and output edges are also included in dependency path features. Dependency path features are required to extract binary event, because, there are relations between pairs of token in the sentence and we find that arguments of binary event present in a sentence are highly related.

6 Experimental Results and Analysis

The purpose of the BioNLP-2016 SeeDev Task [5] is the extraction of binary events from scientific articles of the descriptions of genetic and molecular mechanisms involved in seed development of the model organism, *Arabidopsis thaliana*. We perform feature selection using NSGA-II to find out the most relevant features. We first describe dataset and then compare our experimental results with existing systems.

6.1 Dataset

The corpus used for BioNLP-2016 SeeDev task is composed of paragraphs collected from relevant scientific articles. The SeeDev dataset consists of 86 paragraphs from 20 articles on seed development of Arabidopsis thaliana. Statistics of BioNLP-16 dataset for SeeDev binary event extraction is shown in Table 1.

6.2 Result

We apply Genetic Algorithm (NSGA-II) to find out the most relevant features. Experimental setting is Population Size = 10, Number of Generations = 20, Crossover

[3]https://stanfordnlp.github.io/CoreNLP/.

Table 2 Feature selection of BioNLP-ST 2016 SeeDev binary event dataset. *F1–F19* are features. *F1*: entity1, *F2*: entity2, *F3*: before_after_mid, *F4*: token_distance, *F5*: postagseq_before, *F6*: postagseq_after, *F7*: postagseq_middle, *F8*: postagseq_arg1, *F9*: postagseq_arg2, *F10*: Verb feature, *F11*: entitytype_signature, *F12*: trigger, *F13*: deps_sdp, *F14*: deps_sdp_lemmas, *F15*: deps_sdp_POS, *F16*: eps_inEdges_arg1, *F17*: deps_inEdges_arg2, *F18*: deps_outEdges_arg1, *F19*: deps_outEdges_arg2. Value "1" in feature column indicates presence ("0" for absence) of the feature during training

Event type	F1	F2	F3	F4	F5	F6	F7	F8	F9	F10	F11	F12	F13	F14	F15	F16	F17	F18	F19	F-score
Binds_To	1	1	0	1	0	1	0	1	0	1	1	0	1	1	1	1	0	0	0	32.5582289056
Composes_Primary_Structure	0	1	0	1	1	0	0	0	0	0	1	1	1	1	1	1	0	0	0	40.0
Composes_Protein_Complex	1	0	0	0	1	1	0	1	0	1	1	1	0	1	1	1	0	1	1	0
Exists_At_Stage	1	0	0	0	0	0	0	0	0	1	1	1	1	1	0	1	0	1	0	38.4615384615
Exists_In_Genotype	0	0	0	1	0	1	0	1	1	1	1	0	0	1	0	1	0	1	0	36.0824742268
Has_Sequence_Identical_To	1	0	1	1	1	0	1	1	1	1	1	0	1	0	1	0	0	0	0	30.3246518909
Interacts_With	0	0	0	1	1	1	1	1	1	1	1	0	1	1	1	1	0	1	0	38.1780766615
Is_Functionally_Equivalent_To	0	0	1	1	1	0	0	1	1	1	1	0	1	1	1	1	0	0	0	33.2922148223
Is_Involved_In_Process	1	1	0	1	0	0	1	1	0	0	1	1	0	0	1	1	0	1	1	37.8368979308
Is_Linked_To	0	1	0	1	1	1	1	1	0	0	1	1	1	1	1	1	0	1	1	35.5275315254
Is_Localized_In	0	0	1	0	1	0	0	1	0	1	1	0	0	0	1	1	0	1	0	37.6780626781
Is_Member_Of_Family	1	0	1	0	0	1	0	1	1	0	1	0	1	0	1	0	1	1	0	43.6164995641
Is_Protein_Domain_Of	0	0	0	0	1	0	1	1	0	0	1	0	0	0	1	1	1	0	1	50.0
Occurs_During	1	1	0	1	0	0	0	1	0	0	0	0	1	1	0	1	1	0	0	62.5
Occurs_In_Genotype	0	0	0	0	1	1	0	1	0	0	0	1	1	1	0	1	1	1	1	100.0
Regulates_Accumulation	0	0	0	1	0	1	0	1	1	1	1	1	0	0	1	0	0	0	0	33.0946365092
Regulates_Development_Phase	0	1	0	1	0	0	1	1	1	1	0	1	1	0	0	1	0	0	0	52.7027027027
Regulates_Expression	0	1	1	0	0	0	1	0	1	0	1	1	1	1	1	1	0	1	0	46.4584504411
Regulates_Molecule_Activity	0	0	0	0	0	1	0	1	1	1	1	0	1	1	1	0	0	1	0	28.8979675896
Regulates_Process	1	0	0	1	1	0	0	1	0	0	1	0	1	0	1	1	0	1	0	47.628047628
Regulates tissue development	0	0	0	1	0	0	0	1	0	0	1	1	0	1	0	1	0	1	0	29.2727272727
Transcribes_Or_Translates_To	0	1	0	1	1	0	0	1	1	1	1	0	0	0	0	1	0	0	0	41.6666666667

Table 3 Detailed result of event extraction on development data, SeeDEV task, BioNLP-2016

Event type	Predictions	False negatives	False positives	Recall	Precision	F-score
Binds_To	21.0	17.0	14.0	0.2917	0.3333	0.3111
Composes_Primary_Structure	10.0	10.0	5.0	0.3333	0.5	0.4
Composes_Protein_Complex	2.0	0.0	2.0	0.0	0.0	0.0
Exists_At_Stage	3.0	8.0	3.0	0.0	0.0	0.0
Exists_In_Genotype	97.0	46.0	62.0	0.4321	0.3608	0.3933
Has_Sequence_Identical_To	33.0	5.0	8.0	0.8333	0.7576	0.7937
Interacts_With	19.0	22.0	9.0	0.3125	0.5263	0.3922
Is_Functionally_Equivalent_To	38.0	16.0	14.0	0.6	0.6316	0.6154
Is_Involved_In_Process	3.0	20.0	3.0	0.0	0.0	0.0
Is_Localized_In	52.0	24.0	29.0	0.4894	0.4423	0.4646
Is_Member_Of_Family	62.0	23.0	30.0	0.5818	0.5161	0.547
Is_Protein_Domain_Of	16.0	21.0	8.0	0.2759	0.5	0.3556
Occurs_During	2.0	8.0	0.0	0.2	1.0	0.3333
Occurs_In_Genotype	5.0	11.0	0.0	0.3125	1.0	0.4762
Regulates_Accumulation	16.0	18.0	5.0	0.3793	0.6875	0.4889
Regulates_Development_Phase	34.0	42.0	17.0	0.2881	0.5	0.3656
Regulates_Expression	107.0	68.0	64.0	0.3874	0.4019	0.3945
Regulates_Process	143.0	104.0	68.0	0.419	0.5245	0.4658
Regulates_Tissue_Development	1.0	9.0	1.0	0.0	0.0	0.0
Regulates_Molecule_Activity	2.0	0.0	2.0	0.0	0.0	0.0
Transcribes_Or_Translates_To	12.0	8.0	7.0	0.3846	0.4167	0.4
Is_Linked_To	1.0	22.0	0.0	0.04348	1.0	0.08333
All relations	668.0	502.0	351.0	0.3871	0.4746	0.4264

rate = 0.9, Mutation rate = 0.1, and selection strategy is tournament selection. Our experimental result on BioNLP-16 dataset for SeeDev binary event extraction is shown in Table 2.

Table 3 shows event extraction result on development dataset. This performance is low as compared to the results of existing system. To improve performance, we further apply one more stage, which basically, integrate the results of more than one system. In this regard, we create more systems and merge the results of the systems by selecting the best results from the individual system. By applying this technique, we find out a system whose result is 0.451 F-score (overall) as shown in Table 4.

6.3 Error Analysis

To analyze the error, we consider experimental result on the development data set as shown in Table 3, because, output label for each example in development data

Table 4 Event extraction on test data, SeeDEV task, BioNLP-2016

Event type	Recall	Precision	F-score
Binds_To	0.2188	0.25	0.2333
Composes primary structure	0.9375	0.625	0.75
Exists_In_Genotype	0.3937	0.5	0.4405
Composes protein complex	0.6667	0.5	0.5714
Exists at stage	0.1	1.0	0.1818
Has_Sequence_Identical_To	0.7679	0.6324	0.6935
Interacts_With	0.2407	0.65	0.3514
Is_Functionally_Equivalent_To	0.7037	0.5588	0.623
Is_Involved_In_Process	0.08333	1.0	0.1538
Is_Localized_In	0.5161	0.4103	0.4571
Is_Member_Of_Family	0.7791	0.2978	0.4309
Is_Protein_Domain_Of	0.5161	0.4	0.4507
Occurs_During	0.1667	0.6667	0.2667
Occurs_In_Genotype	0.2857	0.2857	0.2857
Regulates_Accumulation	0.375	0.5	0.4286
Regulates_Development_Phase	0.3896	0.411	0.4
Regulates_Expression	0.4348	0.3243	0.3715
Regulates_Process	0.7301	0.3981	0.5153
Regulates_Tissue_Development	0.0	0.0	0.0
Regulates_Molecule_Activity	0.1111	1.0	0.2
Transcribes_Or_Translates_To	0.125	0.5	0.2
Is_Linked_To	0.0	0.0	0.0
All relations	0.5177	0.3995	0.451

is known to us but in case of test data, it is unknown. From Table 3, we see that most of the error occurs in "*Composes Protein Complex*", *"Exists At Stage", "Is Involved In Process", "Regulates Tissue Development", and "Regulates Molecule Activity"* type events. Adding more efficient features may improve the performance for these event types. If we consider the *"Predictions", "False Negatives", and "False Positives"* values for these event type, we see that number of examples for these event types are very less as compared with other event types. Less number of examples of these event types in training data is also one of the reasons for low performance.

6.4 Comparison with Existing Systems

We have compared our experimental results with existing systems. Comparison is shown in Table 5. It is clear from the table that our experimental result outperforms

Table 5 Result of binary event extraction systems, SeeDev, BioNLP-2016

Teams	Recall	Precision	F-score
VERSE	0.458	0.273	0.342
UniMelb	0.386	0.345	0.364
LitWay	0.448	0.417	0.432
Our system	0.517	0.399	0.451

the best system (i.e., LitWay [11]). Our system achieves an increase of 0.02 (i.e., 2%) in F-score value.

7 Conclusion

In this paper, we propose an efficient technique for binary event extraction using feature selection technique by NSGA-II. Overall performance of the system is recall, precision, and F-score values of 0.517, 0.399, and 0.451, respectively, which outperforms the best event extraction system (F-score 0.432 by LitWay [11]) for SeeDev task of BioNLP-16. Overall evaluation results suggest that there is still room for further improvement. In our future work, we would also like to identify more efficient features to improve the performance. We also like to apply deep learning approach to check whether the performance improves or not.

References

1. Kim, J.-D., Pyysalo, S., Tomoko, O., Robert, B., Ngan, N., Jun'ichi, T.: Overview of BioNLP shared task 2011. In: Proceedings of the BioNLP Shared Task 2011 Workshop, pp. 1–6 (2011)
2. Kim J.D., Ohta, T., Pyysalo, S., Kano, Y., Tsujii, J.: Overview of BioNLP09 shared task on event extraction. In: BioNLP 09: Proceedings of the Workshop on BioNLP, pp. 1–9 (2009)
3. Lishuang, L., Wang, Y., Huang, D.: Improving feature-based biomedical event extraction system by integrating argument information. In: Proceedings of the BioNLP Shared Task 2013 Workshop, pp. 109–115 (2013)
4. Bjorne, J., Salakoski, T.: Generalizing biomedical event extraction. In: Proceedings of BioNLP Shared Task 2011 Workshop (2011)
5. Chaix, E., Dubreucq, B., Fatihi, A., Valsamou, D., Bossy, R., Ba, M., Deléger, L., Zweigenbaum, P., Bessières, P., Lepiniec, L., Nedellec, C.: Overview of the regulatory network of plant seed development (seedev) task at the BioNLP shared task 2016. In: BioNLP (Shared Task), pp. 1–11. Association for Computational Linguistics (2016)
6. Deléger, L., Bossy, R., Chaix, E., Ba, M., Ferré, A., Bessières, P., Nedellec, C.: Overview of the bacteria biotope task at BioNLP shared task 2016. In: Proceedings of the 4th BioNLP Shared Task Workshop, BioNLP 2016, Berlin, Germany, pp. 12–22 (2016). https://doi.org/10.18653/v1/W16-3002. Accessed 13 Aug 2016
7. Li, H., Zhang, J., Wang, J., Lin, H., Yang, Z.: DUTIR in BioNLP-st 2016: utilizing convolutional network and distributed representation to extract complicate relations. In: Proceedings of the

4th BioNLP Shared Task Workshop, BioNLP 2016, Berlin, Germany, pp. 93–100 (2016). https://doi.org/10.18653/v1/W16-3012. Accessed 13 Aug 2016
8. Lamurias, A., Rodrigues, M.J., Clarke, L.A., Couto, F.M.: Extraction of regulatory events using kernel-based classifiers and distant supervision. In: Proceedings of the 4th BioNLP Shared Task Workshop, BioNLP 2016, Berlin, Germany, pp. 88–92 (2016). https://doi.org/10.18653/v1/W16-3011. Accessed 13 Aug 2016
9. Chandrasekarasastry, N.P., Khirbat, G., Verspoor, K., Cohn, T., Ramamohanarao, K.: SeeDev binary event extraction using svms and a rich feature set. In: BioNLP (Shared Task), pp. 82–87. Association for Computational Linguistics, Berlin (2016)
10. Deb, K., Pratap, A., Agarwal, S., Meyarivan, T.: A fast and elitist multiobjective genetic algorithm: NSGA-II. IEEE Trans. Evol. Comput. **6**, 181–197 (2002). (Velingrad)
11. Li, C., Rao, Z., Zhang, X.: Litway, discriminative extraction for different bio-events. In: BioNLP (Shared Task), pp. 32–41. Association for Computational Linguistics, Berlin (2016)

Detection and Identification of Parasite Eggs from Microscopic Images of Fecal Samples

Kaushik Ray, Sukhen Shil, Sarat Saharia, Nityananda Sarma and Nagappa S. Karabasanavar

Abstract Detection and identification of parasite eggs is one of the important tasks in diagnosis of many diseases. Manual detection and identification process of parasite egg is time consuming and prone to error. Automation of this detection and identification process of various parasite eggs can save time and reduce the error in the diagnosis process. In this paper, we proposed a system that automatically detect the parasite eggs present in the microscopic fecal images of pig and identify Ascaris lumbricoides from the detected eggs. In our study, first, different objects are segmented including parasite egg and other non-egg artifacts present in the microscopic images. In the feature extraction stage, we extracted five different types of features from the segmented images. For classification, Artificial Neural Network (ANN) and Multi-class Support Vector Machine (MC-SVM) are used. The experimental result shows about 95% and 93% accuracy rate in identifying Ascaris eggs using MC-SVM and ANN, respectively.

Keywords Parasite · Moment · Artificial Nueral Network (ANN) · Support Vector Machine (SVM) · Ascaris · Non-ascaris

K. Ray (✉) · S. Shil · S. Saharia · N. Sarma
Tezpur University, Tezpur 784028, India
e-mail: ray.kaushik90@gmail.com

S. Shil
e-mail: sukhen.shil@gmail.com

S. Saharia
e-mail: sarat.saharia@gmail.com

N. Sarma
e-mail: niytasarma@gmail.com

N. S. Karabasanavar
Karnataka Veterinary, Animal and Fisheries Sciences University, Bidar 585401, India
e-mail: pub.nag@gmail.com

A. K. Das et al. (eds.), *Computational Intelligence in Pattern Recognition*,
Advances in Intelligent Systems and Computing 999,
https://doi.org/10.1007/978-981-13-9042-5_5

1 Introduction

Parasite egg detection and identification in fecal samples is one of the important tasks in diagnosis of many human and animal's diseases. Traditional approach of identification of parasite eggs from microscopic images is quite difficult for laboratory persons due to presence of various similar objects as parasite egg. This requires a trained personal to correctly identify various kinds of eggs which is tedious and prone to error. Therefore, an automated process of parasite egg identification is very much required in order to produces accurate result in a short time. Since the last decade, researchers are exploring digital image processing and pattern recognition techniques to automatically recognize the parasite eggs in microscopic images. The primary goals of these studies are to reduce the human error in detecting eggs manually and produce highly accurate result within short time without the help of experts.

Fecal samples are common substances that contain parasite eggs which makes easy to produce microscopic slides and analyze it. This is the reason why most of the researchers use microscopic images of fecal samples and use digital image processing techniques to eliminate fecal impurities and detect the parasite eggs present. In this study, we proposed a method to detect and identify common type of parasite eggs called Ascaris lumbricoides from microscopic images of fecal samples of pig which causes Ascariasis in pigs. Pigs of all ages are affected by ascariasis but it is more dreadful for young and growing pigs. An accurate detection and identification system of this ascaris egg is an essential need for diagnosis.

The rest of the paper is organized as follows: Sect. 2 presents review of some related previous works. Section 3 describes about the proposed system. Section shows the experimental result and in Sect. 5 we draw the conclusion and future work.

2 Literature Survey

Several studies were carried out in microscopic image analysis to detect parasite in the late 90s. These studies were able to differentiate several parasite eggs in animal using the image analysis tools and extracted morphometric features like length, shell smoothness from digital microscopic images [1]. In 2001, Yoon Seok Yang et al. [2] has used Artificial Neural Network (ANN) to automatically identify the species of human helminth eggs from the microscopic images of fecal samples. They have used a two-stage ANN architecture where ANN-1 removes non-egg artifacts and isolate only the parasite eggs and ANN-2 is used to identify the species of the eggs. Their experimental result shows an 86% detection rate and 90% identification rate.

Esin Dogantekin et al. [3] proposed a technique that is based on Hu's seven invariant moments and ANN classifier, which shows about 93% accuracy. In a similar work from Derya Avci et al. [4], also used Hu's seven invariant moments with a multi-class Support Vector machine (MC-SVM) for classification of 16 different types of human parasite eggs and achieved 97% accuracy rate in result.

The authors of [5] presented a work to automatically detect two types of human parasite worms: Ascaris Lumbricoides Ova (ALO) and Trichuris trichiura Ova (TTO). They have selected five features: Area, length, width, boundary length, and roundness based on three characteristics: shape, shell smoothness and size. For classification task, a logical classification method called Threshold with Logical Classification Method (TLCM) were used, where at each pass those objects are removed that doesn't fall in range of each feature type.

Suzuki et al. [6] have presented a method for automatic segmentation and identification of 15 common types of human intestinal parasite eggs. They have used eclipse matching technique to locate the objects and Image Foresting Transform (IFT) algorithm in segmentation. Simple object descriptor was used to represent different objects and their combinations are selected using genetic programming. At last, an optimum-path forest classifier was used to classify the different objects.

Quispe et al. [7] proposed an automatic identification method for eight different types of human parasite eggs. The first feature extraction was carried out using Multitexon Histogram (MTH) descriptor and the second, a multi-class support vector machine was used to classify the features that obtained overall identification rate of 94.78%.

Sengul [8] proposed an approach for classification of parasite eggs using Gray-level Co-occurrence Matrix (GLCM) and kNN classifier. GLCM is used to extract texture features at different gray levels for 14 different types of parasite eggs, which are then classified using kNN classifier.

Jimenez et al. [9] developed a system that identify and also quantify different species of helminth eggs in wastewater using digital image processing techniques. In order to classify the different species, they have used naive Bayesian classifier with shape features like area, perimeter, and eccentricity and texture features like energy, mean gray level, contrast, correlation, and homogeneity.

Abdalla and Seker [10] have described an automatic identification method of protozoan parasite where they have extracted features based on mean pixel intensity values of rows and columns of an image. Column Features (CF) set were taken from the column of the image matrix and Row Features (RF) set were taken from the rows. Another set of features CRF, obtained by merging the CF and RF [10]. These features were calculated for both RGB and Grayscale images and two classifiers: K-Nearest Neighbor (kNN) and Artificial Neural Network (ANN) were used in classification.

3 Proposed System

The proposed method consists of four stages, viz., (i) Preprocessing, (ii) Object segmentation, (iii) Feature extraction, and (iv) Classification. Each stage consists of a number of tasks. The complete architecture of our proposed system is shown in Fig. 1.

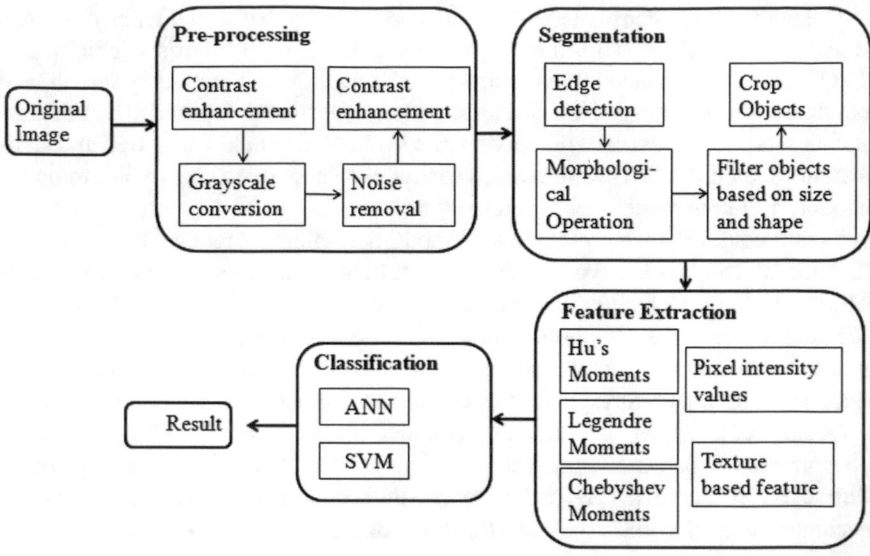

Fig. 1 Architecture of proposed system (small boxes inside a box define the various tasks of that stage)

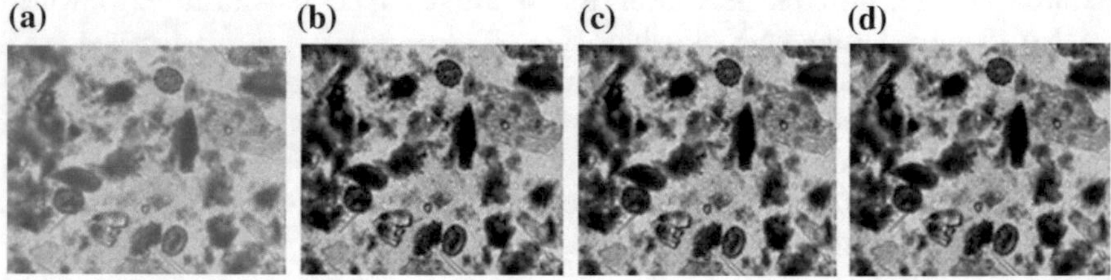

Fig. 2 **a** Original RGB image, **b** Contrast enhanced RGB image, **c** Grayscale image, **d** median filtered image

3.1 Preprocessing

At first, we enhanced the contrast of the original RGB images (Fig. 2a) and converted into grayscale to reduce the computation (Fig. 2b, c). After converting to grayscale, we used median filtering that remove noise and smoothen the image (Fig. 2d). In median filtering, the output pixel values are replaced by the median value of the neighboring pixels [3]. Smoothed images are further processed to increase the sharpness and to enhancement of the contrast. The contrast of an image can be increased by stretching the range of gray levels in the image [3]. Enhancement of contrast helps in detecting the edges of the different objects in the image and separates them from background.

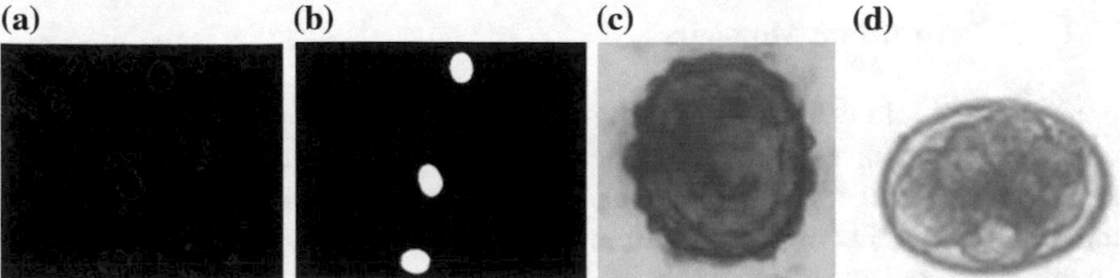

Fig. 3 **a** Edge detection image, **b** image after applying morphological operation and removed unwanted objects, **c** Segmented ascaris egg, **d** Segmented non-ascaris egg in grayscale

3.2 *Object Segmentation*

After preprocessing of the images, canny edge detector is used with Otsu's global threshold value [11] in order to detect the edges of different objects and separate them from background as shown in Fig. 3a. Morphological operations, namely dilation, erosion, and filling holes are performed to make the objects more clear for segmentation (Fig. 3b). Depending on the size and circularity, we remove some of the objects that are not relevant. From the binary image, coordinates of the detected objects including parasite eggs are taken and segmented out from gray scale images. Fig. 3c, d shows segmented ascaris and non-ascaris egg in grayscale.

3.3 *Feature Extraction*

From the segmented grayscale images, five different types of feature vectors are calculated. These feature vectors are (i) Hu's seven invariant moments, (ii) Legendre moments, (iii). Chebyshev moments, (iv) Pixel intensity-based feature: standard deviation, mean pixel intensity, maximum, and minimum pixel intensity, and (v) Shape- and texture-based features, viz., area, roundness, major axis length, minor axis length, eccentricity, entropy, contrast, and homogeneity. Moments are used as features to characterize the patterns in image processing applications [4]. Legendre and Chebyshev moments of nth order can be calculated as described below. For our experiment, we have calculated these two types of moments upto 10th order. For pixel intensity-based features, first, we resized the cropped images into 120×120 pixels and divided into 4×4 blocks of equal size. For each block, mean, standard deviation, maximum, and minimum pixel intensity values were calculated. A total of 64 features are extracted for each image and is used as feature vector. Area of an object is represented by the total number of pixels inside the region. Roundness of an object can be defined as $Roundness = 4Area/P^2$, where P represents perimeter of the object. To calculate texture features like eccentricity, homogeneity, and contrast, we have used Gray-level Co-occurrence Matrix (GLCM) [8].

3.3.1 Hu's Invariant Moments

In 1962, M. K. Hu describes seven functions of moments that are invariant under translation, rotation, and scale [12]. The moment invariants are proved to be efficient and adequate measure for objects pattern recognition regardless of its size, position in the image. Let an image with intensity distribution function f(x, y), the two dimensional moments of order (p + q) is defined in Eq. 1 [12]

$$m_{pq} = \int_{-\infty}^{\infty}\int_{-\infty}^{\infty} x^p y^q f(x, y)dxdy \tag{1}$$

where p, q = 0, 1, 2, 3…These moments are not invariant to translation, rotation and scaling. To be invariant one has to calculate central moments which is defined in Eq. 2 [12]

$$\mu_{pq} = \int_{-\infty}^{\infty}\int_{-\infty}^{\infty} (x-\overline{x})^p (y-\overline{y})^q f(x, y)dxdy \tag{2}$$

where $\overline{x} = m_{10}/m_{00}$ and $\overline{y} = m_{10}/m_{00}$.

The central moments are only invariant under translation. Scale invariance can be achieved by normalizing these central moments as defined in Eq. 3 [12]

$$\eta = \mu_{pq}/\mu_{00}^{\gamma}, \gamma = (p+q+2)/2, p+q = 2, 3, 4, \ldots \tag{3}$$

Hu calculated seven invariant moments based on these normalized central moments as expressed in Eqs. 4 [12]

$$\begin{aligned}
\phi_1 &= \eta_{20} + \eta_{02}\\
\phi_2 &= (\eta_{20} - \eta_{02}) + 4\eta_{11}^2\\
\phi_3 &= (\eta_{30} - 3\eta_{12})^2 + (3\eta_{21} - \eta_{03})^2\\
\phi_4 &= (\eta_{30} - \eta_{12})^2 + (\eta_{21} - \eta_{03})^2\\
\phi_5 &= (\eta_{30} - 3\eta_{12})(\eta_{30} + 3\eta_{12})[(\eta_{30} + 3\eta_{12})^2 - 3(\eta_{21} + \eta_{03})^2]\\
&\quad + (3\eta_{21} - \eta_{03})(3\eta_{21} + \eta_{03})[3(\eta_{30} + \eta_{12})^2 - (\eta_{21} + \eta_{03})^2]\\
\phi_6 &= (\eta_{20} - \eta_{02})[(\eta_{30} + \eta_{12})^2 - (\eta_{21} + \eta_{03})^2] + 4\eta_{11}(\eta_{30} + \eta_{12})(\eta_{21} + \eta_{03})\\
\phi_7 &= (3\eta_{21} - \eta_{03})(\eta_{30} + \eta_{12})[(\eta_{30} + \eta_{12})^2 - 3(\eta_{21} + \eta_{03})^2]\\
&\quad - (\eta_{30} - 3\eta_{12})(\eta_{21} + \eta_{03})[3(\eta_{30} + \eta_{12})^2 - (\eta_{21} + \eta_{30})^2]
\end{aligned} \tag{4}$$

3.3.2 Legendre Moments

Legendre Moments are used in image pattern recognition because they can represent information contained in an image with minimum amount of redundancy [13, 14]. Legendre moments are defined using Legendre polynomials [13, 14] as their kernels.

Legendre moment of (p + q)th order with intensity function f(x, y) can be defined by Eq. 5 [13]

$$L_{p,q} = \frac{(2p+1)(2q+1)}{4} \int_{-1}^{1} \int_{-1}^{1} P_p(x) P_q(y) f(x, y) dx dy \quad (5)$$

where $P_p(x)$ is the pth order Legendre polynomial defined in Eq. 6 [13]

$$P_p(x) = \sum_{k=0}^{p} C_{pk}[(1-x)^k + (-1)^p (1+x)^k] \quad (6)$$

Here,

$$C_{pk} = \frac{(-1)^k (p+k)!}{2^{k+1}(p-k)!(k!)^2}, \quad k = 0, 1, 2, \ldots \quad (7)$$

For a grayscale image of size $N \times N$, Legendre moment L_{pq} can be expressed by Eq. 8 [13]

$$L_{p,q} = \frac{(2p+1)(2q+1)}{(N-1)^2} \sum_{i=1}^{N} \sum_{j=1}^{N} P_p(x_i) P_q(y_j) f(x_i, y_j) \quad (8)$$

where, $x_i = \frac{(2i-N-1)}{(N-1)}$ and $y_j = \frac{(2j-N-1)}{(N-1)}$.

3.3.3 Chebyshev Moments

Chebyshev moments are discrete orthogonal moments based on discrete Chebyshev polynomials [15, 16]. The nth-order Chebyshev polynomial can be calculated using Eq. 9 [15]

$$t_n(x; N) = \sum_{k=0}^{N-1} a_{k,n} x^k = (1-N)_n {}_3F_2 \left(\begin{matrix} -n, & -x, & 1+n \\ 1, & 1-N & \end{matrix} \middle| 1 \right) \quad (9)$$

where n, x = 0, 1, 2,. . . , N–1, and

$${}_3F_2 \left(\begin{matrix} -n, & -x, & 1+n \\ 1, & 1-N & \end{matrix} \middle| 1 \right) = \sum_{k=0}^{\infty} \frac{(a_1)_k \ldots (a_p)_k z^k}{(b_1)_k \ldots (b_p)_k k!} \quad (10)$$

is the generalized hypergeomatric function with the Pochhammer symbol defined in Eq. 11 [15]

$$(a)_k = \frac{\tau(a+k)}{\tau a} = a(a+1)\ldots(a+k-1) \quad (11)$$

For an $N \times N$ image, set of Chebyshev polynomials tn(x;N) satisfy the orthogonality condition (Eq. 12) [15]

$$\sum_{x=0}^{N-1} t_m(x; N) t_n(x; N) = \rho(n; N)\delta_{mn} \tag{12}$$

where δ_{mn} is the Kronecker delta function [15] and

$$\rho(n, N) = \frac{N(N^2-1)(n^2-2^2)\ldots(N^2-n^2)}{2n+1} \tag{13}$$

The Chebyshev moment of order (m+n) for an image with intensity function f(x, y), x = 0, 1, 2,…,M-1 and y = 0, 1, 2,…,N-1 is defined in Eq. 14 [15, 16]

$$T_{mn} = \sum_{x=0}^{M-1}\sum_{y=0}^{N-1} \bar{t}_m(x; M)\bar{t}_n(y; N) f(x, y) \tag{14}$$

where $\bar{t}_m(x; M) = \frac{t_m(x;M)}{\sqrt{\rho(m;M)}}$ and $\bar{t}_n(y; N) = \frac{t_n(y;N)}{\sqrt{\rho(n;N)}}$.

3.4 *Classification*

For identification of parasite eggs types, we used two classifiers: Multi-class Support Vector Machine (MC-SVM) and Artificial Neural Network (ANN). ANNs are connected networks of input and output nodes with one or multiple layers of hidden nodes between input and output. In our experiment, each value of a feature vector represent an input node. For output nodes, we have used three class labels, viz., Ascaris, Non-Ascaris, and Non-Egg. We have used four hidden layers, which is chosen experimentally. An MC-SVM classifier draws a set of hyperplanes between the objects of different classes [4]. Hyperplanes are constructed depending on the feature values of objects from different categories. For our experiment, we have used MC-SVM classifier with Gaussian kernel and three output classes.

4 Result and Discussion

In our experiment, we have used 612 images of size 1080 × 1920 pixels where 336 images contain ascaris eggs, 137 images contain non-ascaris eggs, and 139 images that do not contain any parasite egg. From these images, 1372 objects are segmented, out of which 298 are ascaris eggs, 153 non-ascaris, and 921 non-egg artifacts. 70% of them are used in training and the rest of 30% images are used in testing purpose

Table 1 Classification result of identifying ascaris and non-ascaris eggs using MC-SVM

Total number of objects	Type of feature	Number of correctly classified eggs		Accuracy	
		Ascaris	Non-ascaris	Ascaris (%)	Non-ascaris (%)
Ascaris = 298 Non-Ascaris = 153	Hu's moments	261	101	87.58	66.01
	Legendre moments	273	98	91.61	64.05
	Chebyshev moments	278	109	93.28	71.24
	Intensity based feature	284	113	95.30	73.85
	Texture feature	270	110	90.60	71.89

Table 2 Classification result of identifying ascaris and non-ascaris eggs using ANN

Total number of objects	Type of feature	Number of correctly classified eggs		Accuracy	
		Ascaris	Non-ascaris	Ascaris (%)	Non-ascaris (%)
Ascaris = 298 Non-Ascaris = 153	Hu's moments	254	95	85.23	62.09
	Legendre moments	262	103	87.91	67.32
	Chebyshev moments	269	102	90.26	66.67
	Intensity based feature	278	108	93.28	70.58
	Texture feature	272	111	91.28	72.54

for both MC-SVM and ANN. In addition, we have also performed tenfold cross-validation process. Along with the classification task, our system also calculates the number of parasite eggs presented in the image. Classification process classifies the objects in three categories, viz., ascaris, non-ascaris, and non-egg artifacts. Tables 1 and 2 show the results for detecting and identifying two types of parasite eggs using MC-SVM and ANN, respectively.

From Tables 1 and 2, we observed that the rate of detecting ascaris eggs is higher than non-ascaris eggs. The reason for lower identification rate of non-ascaris eggs is due to fewer training sample than ascaris and we also observed that some non-ascaris eggs have similar texture as the impurities present in the images. The experimental result shows that performance of MC-SVM is better than ANN in identification of Ascaris eggs. Comparison of MC-SVM and ANN in identification of Ascaris and Non-Ascaris eggs are shown in Fig. 4.

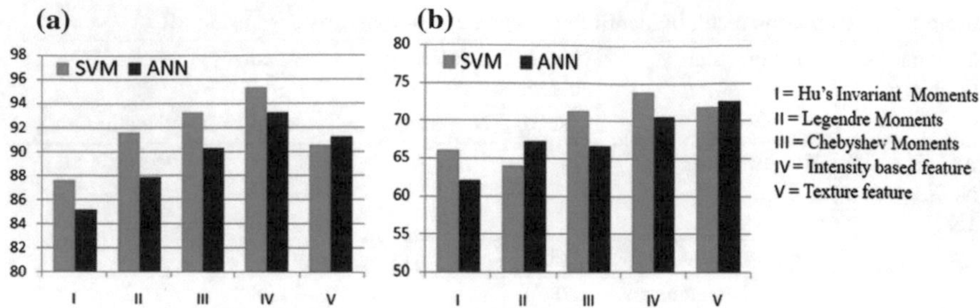

Fig. 4 Comparison of MC-SVM and ANN in identification of **a** Ascaris, **b** Non-Ascaris eggs

5 Conclusion and Future Work

In this study, we have proposed a system to automatically detect and calculate the number of parasite eggs from microscopic fecal images of pig. Different objects, including parasite eggs and non-egg artifacts are segmented from the images and extracted five different types of feature vectors. For classification of these objects, two classifiers: Artificial Neural Network (ANN) and Multi-class Support Vector Machine (MC-SVM) are used that classify the detected objects into one of the three classes: Ascaris, Non-Ascaris, and Non-Egg.

Here, in this study, we have mainly focused on identifying ascaris eggs from the detected parasite eggs. The classification accuracy rate of both MC-SVM and ANN are near about 95% and 93%, respectively, for Ascaris. We believe that these accuracy rates can be increased using moments of higher order and also using combination of different feature vectors. Future work of this study is to detect and identify more different types of parasite eggs using deep learning techniques.

Acknowledgements We are thankful to the ITRA, Digital India Corporation (formerly known as Media Lab Asia) for supporting and funding this research work.

References

1. Joachim, A., Dulmer, N., Daugschies, A.: Differentiation of two Oesophagostomum spp. from pigs, O. dentatum and O. quadrispinulatum, by computer-assisted image analysis of fourth-stage larvae. Parasitol. Int. l. **48**, 63–71 (1999)
2. Yang, Y.S., Park, D.K., Kim, H.C., Choi, M.-H., Chai, J.-Y.: Automatic identification of human helminth eggs on microscopic fecal specimens using digital image processing and an artificial neural network. IEEE Trans. Biomed. Eng. **48**(6), 718–730 (2001)
3. Dogantekin, E., Yilmaz, M., Dogantekin, A., Avci, E., Sengur, A.: A robust technique based on invariant moments ANFIS for recognition of human parasite eggs in microscopic images. Expert Syst. Appl. **35**(3), 728–738 (2008)
4. Avci, D., Varol, A.: An expert diagnosis system for classification of human parasite eggs based on multi-class SVM. Expert Syst. Appl. **36**(1), 43–48 (2009)

5. Hadi, R.S., Khalidin, Ir.Z., Ghazali, K.H., Zeehaida, M.: Human parasitic worm detection using image processing technique. In: Computer Applications and Industrial Electronics (ISCAIE). IEEE Symposium (2012)
6. Suzuki, C.T., Gomes, J.F., Falcao, A.X., Papa, J.P., Hoshino-Shimizu, S.: Automatic segmentation and classification of human intestinal parasites from microscopy images. IEEE Trans. Biomed. Eng. **60**(3), 803–812 (2013)
7. Flores-Quispe, R., Escarcina, R.E.P., Velazco-Paredes, Y., Castanon, C.A.B.: Automatic identification of human parasite eggs based on multitexton histogram retrieving the relationships between textons. In: 33rd International Conference on Chilean Computer Science Society (SCCC) (2014)
8. Sengul, G.: Classification of parasite egg cells using gray level co-occurrence matrix and kNN. Biomed. Res.-India **27**(3), 829–834 (2016)
9. Jimenez, B., Maya, C., Velasquez, G., Torner, F., Arambula, F., Barrios, J.A., Velasco, M.: Identification and quantification of pathogenic helminth eggs using a digital image system. Exp. Parasitol. **166**, 164–172 (2016)
10. Abdalla, M.A.E., Sekar, H.: Recognition of protozoan parasites from microscopic images: Eimeria species in chickens and rabbits as a case study. In: 39th Annual International Conference of the IEEE Engineering in Medicine and Biology Society (EMBC) (2017)
11. Otsu, N.: A threshold selection method from gray-level histograms. IEEE Trans. Syst. Man, Cybern. **9**(1), 62–66 (1979) 11; Hu, M.-K.: Visual pattern recognition by moment invariants. IEEE Trans. Inf. Theory **8**, 179–187 (1962)
12. Hu, M.-K.: Visual pattern recognition by moment invariants. IEEE Trans. Inf. Theory **8**(2), 179–187 (1962)
13. Shu, H., Luo, L., Bao, X., Yu, W.: An efficient method for computation of legendre moments. Graph. Model. **62**(4) (2000)
14. Saharia, S., Bora, P.K., Saikia, D.K.: A comparative study on discrete orthonormal chebyshev moments and legendre moments for representation of printed characters. In: Proceedings of 4th Indian Conference on Computer Vision. Graphics and Image Processing. Indian Statistical Institute. Kolkata, 16–18 Dec 2004
15. Yap, P.T., Raveendran, P.: Image focus measure based on Chebyshev moments. IEE Proc.-Vis. Image Signal Process. **151**(2) (2004)
16. Saharia, S., Bora, P.K., Saikia, D.K.: Improving character recogniton accuracy of tchebichef moments by splitting of images. In: Proceedings of National Conference on Communications (NCC-2009). IIT Guwahati, Guwahati, pp. 390–393. 16–18 Jan 2009

Identification of Some Transposable Elements of DNA Using "BP Suche" Algorithm

Rachita Ghoshhajra, Sanghamitra Chatterjee and Soma Barman (Mandal)

Abstract Deoxyribo Nucleic Acid (DNA) is the blueprint of all living organisms. In order to draw a clear picture of genome, the researchers concentrated on DNA sequence analysis. Coding region of DNA sequence, responsible for protein synthesis, has been studied extensively in the past decade but exploration of "Non-Coding" DNA (NCDNA) part, which is formerly known as "junk DNA" is gaining interest in recent years. The mystery of human evolution is hidden inside Transposon or Transposable Element (TE), a part of NCDNA. Recently, scientists aim to shed light on Transposable Element to uncover the evolutionary history. But the identification of Transposable Elements using computational method is yet to be explored. The authors in this paper address such problem of identifying some of the transposable elements by DNA sequence analysis. "BP Suche" algorithm is used here for the analysis of Transposon database of closely related species publicly available in the NCBI website.

Keywords DNA sequence · Transposon · ORF · LINE · SINE

1 Introduction

DNA is the building block of all living organisms. DNA molecules are double-stranded helices, consisting of two long biopolymers made of simpler units called nucleotides that contain genetic information. The year 2003 marks the 50th anniver-

R. Ghoshhajra
MCKV Institute of Engineering, Liluah, Howrah, India
e-mail: rghhajra@gmail.com

S. Chatterjee (✉)
Camellia Institute of Technology, Kolkata, India
e-mail: sangha3030@gmail.com

S. Barman (Mandal)
Institute of Radio Physics & Electronics, University of Calcutta, Kolkata, India
e-mail: barmanmandal@gmail.com

A. K. Das et al. (eds.), *Computational Intelligence in Pattern Recognition*, Advances in Intelligent Systems and Computing 999,
https://doi.org/10.1007/978-981-13-9042-5_6

sary of the discovery of DNA (Crick, Watson, Franklin, Wilkins 1953) [1] and also the completion of the Human Genome Project (HGP). DNA nucleobase contains four chemical bases: Adenine (A), Cytosine (C), Thymine (T), and Guanine (G) where each base pair with their complementary base, i.e., A always with T and C always with G, forming a unit called base pairs (BP). The human genome contains approximately 3 billion base pairs, which reside in 23 pairs of chromosomes. A segment of DNA called genes, wrapped in chromosomes, pass information from parents to offspring. The sequential arrangement of these base pairs are different in various species, and so the researchers look for a specific pattern among species and then make logical inferences from the information extracted. Three bases combined together to create a codon where 64 codons are created from the combination of A, C, T, and G. These codons are grouped into a set of 20 naturally occurring amino acids [2, 3]. In the field of biotechnology, researchers are also working on diseases by using evolutionary algorithms [4].

1.1 Coding Region of DNA

Classification, characterization, and structure of amino acids are thoroughly studied by different bio-scientists. Coding region inside any DNA sequence can be identified between Start Codon (ATG) and Stop Codon (TAA, TAG, and TGA) and this part of the sequence is responsible for protein synthesis. Researchers having technological background are trying to analyze the coding region using different computational tools like Automata Theory [5], Graph Theory [6], Randomized AXelerated Maximum Likelihood (RAXML) Program [7], etc. These analysis revealed the evolutionary relationship of various species according to the phylogenetic tree of evolution [5–7]. The comparison between various species has been studied extensively by different scientists from different domains [8, 9]. But according to Human Genome Project, only 2% of DNA sequences are protein coding part and rest are treated as "junk DNA". Recently, researchers show their interest in such part of DNA, which remained unexplored. Some portion of "junk DNA" recently explored as ORF and other as "Non–Coding" DNA.

1.2 Open Reading Frame (ORF)

The portion of DNA sequence that starts and ends with Stop Codons (TAA, TAG, and TGA) and separated by a whole number of nucleotide triplets are known as Open Reading Frames (ORF) [10]. ORFs are short DNA sequences with AT (Adenine + Thymine) levels >50% as Stop Codons are rich in A and T. Nucleotide composition regulates the frequency of Stop Codons. ORF plays an important role in classifying "Non–Coding" DNA sequence.

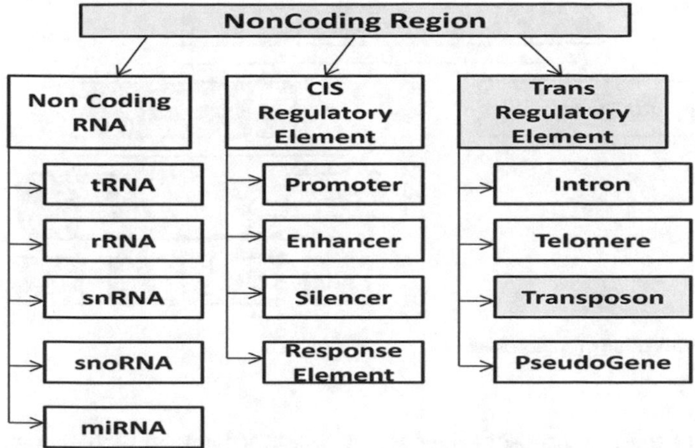

Fig. 1 Classification of noncoding DNA

1.3 Noncoding DNA

The functional region of DNA that does not code for a protein is a "Non-Coding" DNA region [11] spreading throughout the sequence. Over the years "Non-Coding" DNAs have emerged with new insights. These regions are responsible for genetic diversity, genetic disorders, and genetic diseases in human. Though initial classification and characterization of NCDNA were proposed by Patrushev et al. in 2014 [10], proper classification, structuring, and annotation of these segments help in understanding their specific functions. The detail classification of NCDNA has been proposed by Shanmugam et al. in 2017 [11], which are demonstrated in Fig. 1.

Among the classes of Trans-Regulatory Elements, Transposons play decisive role in evolution process [12–14]. As we want to study the evolution process of different species, we focus our research mostly in this region. Section 2 deals with a brief background of Transposon, Sect. 3 describes the methodology to analyze TE and find LINE and SINE separately. Results and discussions are presented under Sect. 4 and based on the analysis, concluding remarks have been made in Sect. 5.

2 Brief Background of Transposon

Transposons or Transposable Elements (TE) were defined by Haren et al. in 1999 [15] as "discrete segments of DNA capable of moving from one locus to another in their host genome or between different genomes" which is popularly known as Mobile DNA. Transposons have the ability to enhance genetic diversity and at the same time, genome has the capacity to reduce TE activity. This resulted in a balance that makes TE a crucial part of evolution and genome regulation in all living organisms. The

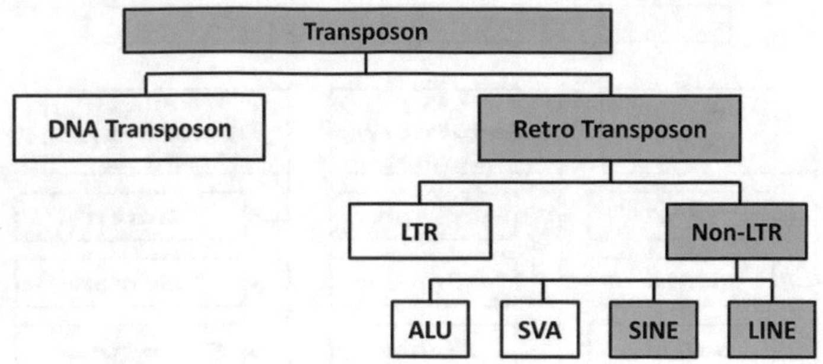

Fig. 2 Classification of transposon

behavior of these transposons depends on where it lands in the genome. Landing inside a gene may result in a mutation. TEs make up roughly 50% of the human genome and up to 90% of the plant genome. In 2016, Arensburger et al. [16] proposed that "TEs are discrete segments of DNA capable of moving within a host genome from one chromosome or plasmid location to another, or between hosts by using parasitic vectors that they use for lateral transfers".

TEs consists of two major classes: Class I or Retrotransposons and Class II or DNA transposons. DNA transposons are capable of moving and inserting into new genomic sites (cut–paste). Retrotransposons replicate by forming RNA intermediates (copy–paste), which are then reverse-transcribed to make DNA sequences and inserted into new genomic locations [12]. Based on the presence of Long Terminal Repeats (LTRs), retrotransposons are further classified into two groups: LTR and non-LTR transposons. Non-LTR retrotransposons include Long Interspersed Nuclear Element (LINE), Short Interspersed Nuclear Element (SINE), *Alu*, and SVA elements, collectively account for about one-third of the human genome [17]. Figure 2 depicts the most recent classification of TE. Studies have revealed that Non-LTRs are the only Transposon, which is currently active in human. Theoretical and experimental studies prove that Transposable Elements are useful for studying genome evolution and gene function [18–20].

The structure of SINE and LINE, proposed by Bowen in 2002 [12], is shown in Fig. 3. The figure shows that SINE is repetitive structures with a set of consecutive "A" found in the DNA sequence of ACTG.

On the other hand, Long Interspersed Nuclear Element (LINE), consisting of ORF, Reverse Transcriptase (RT), and a set of "A", can be found in a transposon sequence. In the year 2007, Wicker et al. [21] proposed the classification system of transposable elements present in all eukaryotes, which can be used by technologists in the field of bioinformatics. The structure proposed by Wicker is shown in Fig. 4. This structure of LINE and SINE in the DNA sequence is almost similar to the proposed structure by Bowen.

In 2018, scientists found that LINE-1 (L1) is an autonomous retrotransposon, which actively participated in the evolution of mammals and still contributes to human genotypic diversity and genetic diseases [22]. Full length or fragments of

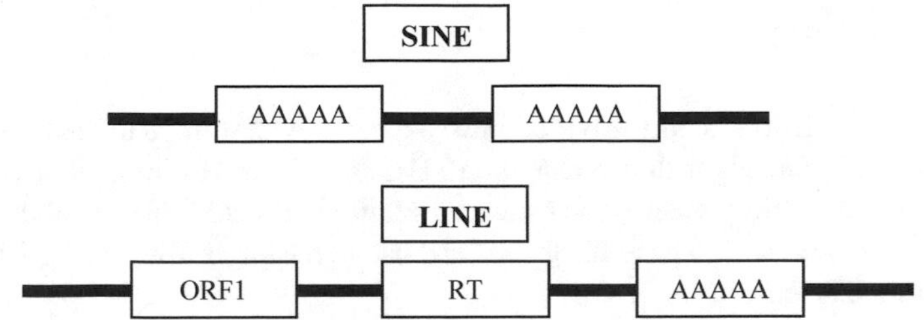

Fig. 3 Structure of LINE and SINE available in literature

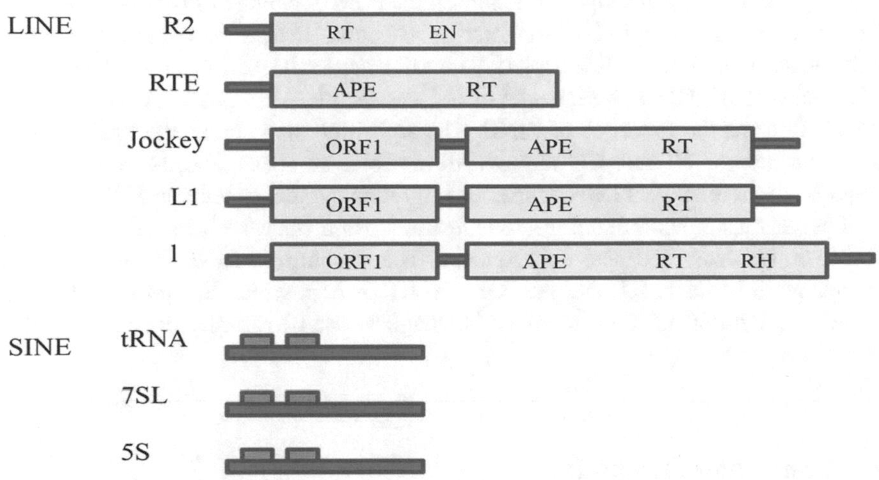

Fig. 4 Detailed structure of LINE and SINE

L1 occupies roughly 17% of the human genome sequence, whereas only 2% of the sequence is occupied by coding region. L1 not only helps itself to mobilize but also takes initiative to mobilize nonautonomous *Alu* and SVA elements. Though mobile DNA (TE) plays an important role in human evolution, there is no standard method available to locate LINE and SINE sequences automatically. We are trying to identify different features and location of LINE and SINE from Transposable Element sequence using "BP Suche" algorithm. A comparative study of LINE and SINE of three different species, e.g., Homo sapiens (Human), Pan troglodytes (Chimpanzee), and Macacamulatta (Rhesus Monkey) are also presented in this paper.

3 Methodology

"BP Suche" algorithm, mainly divided into two parts, is described in the flowchart as shown in Fig. 5. The algorithm is tested on TE databases of Homo sapiens (Human), Pan troglodytes (Chimpanzee), and Macacamulatta (Rhesus Monkey) collected from the NCBI website [23]. These databases are used as sample for finding LINE and SINE within TE region.

A set of five consecutive "A"s or sequence of A's is the key factor to identify both LINE and SINE. Our aim is to find out such key features within the TE sequence. This algorithm identifies Open Reading Frame (ORF), which lies between two Stop Codons (TAA, TAG, and TGA) (denoted as "SCodon" in flowchart) and separated by a whole number of nucleotide triplets. This ORF marks the starting point of an LINE. The presence of ORF and sequence of five consecutive A ("AAAAA") denoted as "A5" confirms the presence of LINE. The algorithm also counts the total number of occurrence of AT and CG along with the total number of occurrence of "A5" sequence, where at least two occurrences signifies the presence of an SINE.

The chromosomes containing LINE are used to find "Reverse Transcriptase" (RT), which lie between ORF and A5 sequence. The structure of LINE is the sequential arrangement of ORF, RT, and A5 (ORF + RT + A5), which is represented in the flowchart. Python 3.6 programming language is used to implement "BP Suche" algorithm.

4 Result and Discussion

The Transposable Element sequence database has been analyzed using "BP Suche" algorithm for three closely related species, i.e., Homo sapiens (Human), Pan Troglodytes (Chimpanzee), and MacacaMulatta (Rhesus Monkey).

The existence of LINE within a chromosome is shown in the following tables. The percentage of AT and CG present in TE and existence of LINE within TE for different Homo sapiens (Human) chromosomes is shown in Table 1. A similar result for Pan Troglodytes (Chimpanzee) and MacacaMulatta (Rhesus Monkey), respectively, are shown in Table 2. Results signify that all chromosome may or may not contain LINE.

The chromosome containing LINE are selected to compute "Reverse Transcriptase" (RT) length. RT is a very significant element in DNA sequence, which regulates cell activity of organism [24]. Percentage of RT inside TE sequence for Human, Chimpanzee, and Monkey, respectively, are displayed in Table 3.

Percentage of RT present in TE sequence for each chromosome is depicted in Fig. 6 for each species. It is clearly observed from the result that the length of RT varies with chromosome.

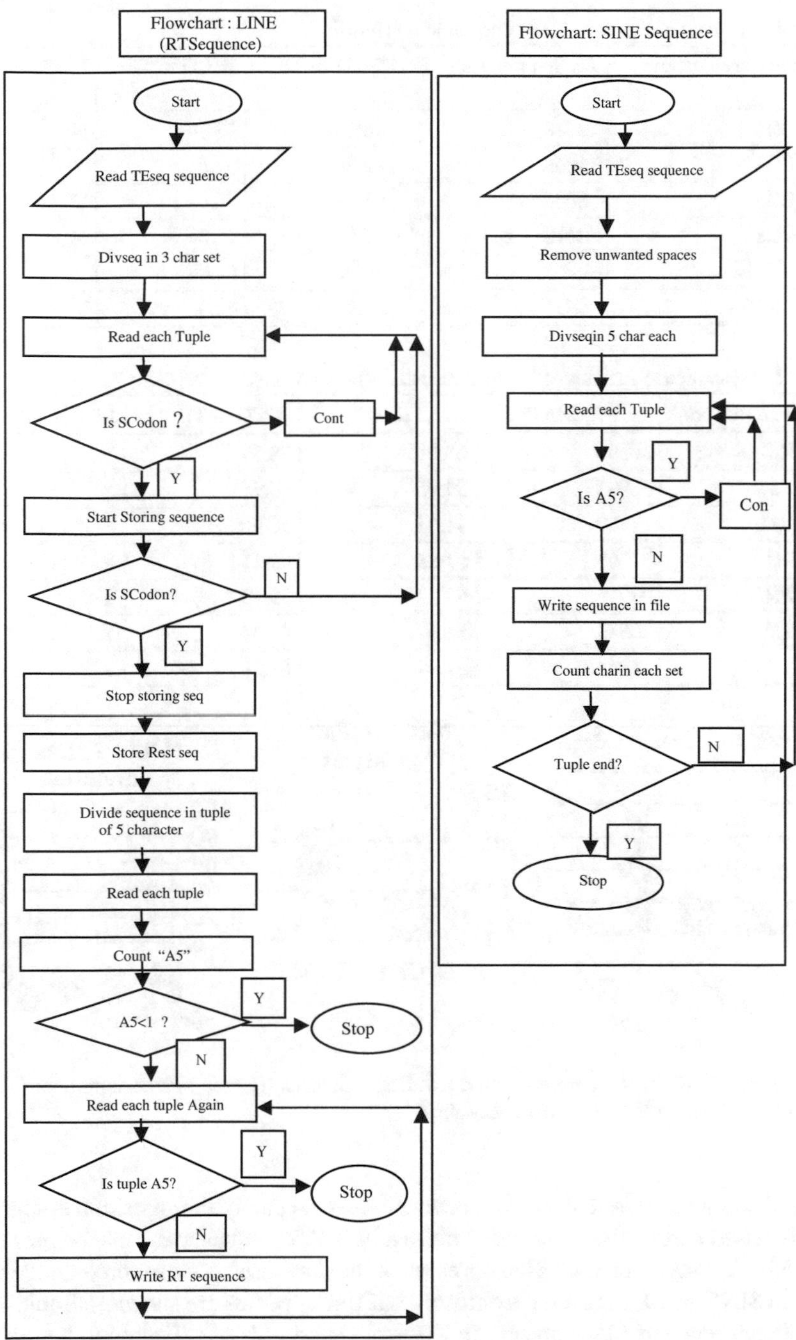

Fig. 5 Flowchart of "BP Suche" algorithm

Table 1 TE sequence analysis of Homo sapiens (Human)

Human Chromosome No.	TE Seq Length (BP)	% AT in TE	% CG in TE	LINE exist?
HS_CH2	2618	56.26	43.74	N
HS_CH4	11063	62.39	37.61	Y
HS_CH5	8045	53.93	46.07	Y
HS_CH8	2412	29.93	70.07	N
HS_CH11	2803	36.92	63.08	Y
HS_CH16	6632	57.1	42.9	Y

Table 2 TE sequence analysis of Chimpanzee and Rhesus Monkey

Chromosome No.	TE Seq Len (BP)	% AT in TE	% CG in TE	LINE exist?
C_CH4	10396	61.61	38.39	Y
C_CH5	8033	54.05	45.95	Y
C_CH11	4739	41.04	58.96	N
M_CH1	57609	57.83	42.17	Y
M_CH4	21949	60.32	39.68	Y
M_CH6	6310	52.66	47.34	Y
M_CH14	2661	37.39	62.61	Y

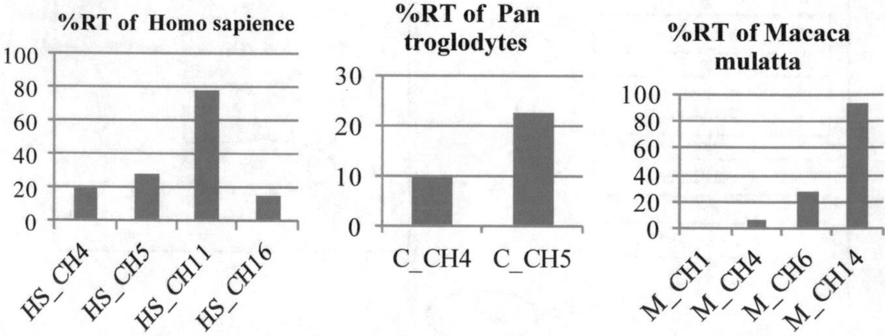

Fig. 6 Plot of %RT versus chromosome no. of Homo Sapiens (Human), Pan Troglodytes (Chimpanzee), and Macaca Mulatta (Rhesus Monkey)

For finding SINE from the DNA sequence, the next part of the algorithm is applied. As discussed earlier, the presence of at least two "A5" sequence means the presence of SINE. A total number of SINE present in the databases is considered and length of each SINE in BP for each chromosome of three species are shown in Table 4. A sample reference of SINE present in TE is depicted in Fig. 7. Each number within bracket indicates the number of bases (ACTG) that are present between two sets of "A5", indicating the presence of SINE.

Table 3 Identification LINE in DNA sequence of (I) Homo sapiens (Human), (II) Pan Troglodytes (Chimpanzee), and (III) Macaca Mulatta (Rhesus Monkey)

Human Chromosome No.	TE Seq Length (BP)	RT Len (BP)	% RT in TE	Chimp Chromosome No.	TE Seq Len (BP)	RT Len (BP)	% RT in TE	Monkey Chromosome No.	TE Seq Len (BP)	RT Len (BP)	% RT in TE
HS_CH4	11063	2200	19.9	C_CH4	10396	1010	9.72	M_CH1	57609	360	0.62
HS_CH5	8045	2275	28.3	C_CH5	8033	1800	0.22	M_CH4	21949	1505	6.86
HS_CH11	2803	2190	78.1					M_CH6	6310	1710	27.1
HS_CH16	6632	1015	15.3					M_CH14	2661	2480	93.2

Table 4 Analysis of SINE of Human, Chimpanzee, and Monkey

Human Ch. No.	No. of. SINE	SINE Length (BP)	Chimp Ch. No	No. of. SINE	SINE Length (BP)	Monkey Ch. No	No. of. SINE	SINE Length (BP)
HS_CH2	0	Not found	C_CH4	11	1975…535	M_CH1	123	20…620
HS_CH4	20	1145…280	C_CH5	08	1965…375	M_CH4	39	1230…610
HS_CH5	09	2440…430	C_CH11	0	Not found	M_CH6	05	165…490
HS_CH8	0	Not found				M_CH14	0	Not found
HS_CH11	01	Not found						
HS_CH16	14	1435…160						

GATTC ACGTA… …(1145)…….GACCT GTCAG **AAAAA** TTGGT CACTA……(150)….. TCCAT CTAGG CCCTA
GAAGT GAGAG GAGGA **AAAAA** TCTTA CGAAC ….(635)…..TCATT TTCTA GTTGC TTTGT **AAAAA** TCTTA
GTTGT TAATT ……..(345)……ATCTT GTTCT AGTAA TCACC **AAAAA** TTAGA CACTT ……(180)……. TGTTG
GGGAA ACGTA AGCGT GTGGT GTTGA CAATT **AAAAA** CTTGA GGAAG …..(280) .GCATC TCTTT **AAAAA**
CACTT TCCGT ……..GGTGT ACGGA ATTGG ……TGAAT CCACA GTTCG TGATA **AAAAA** GAACA AAGAA

Fig. 7 Presence of SINE in Homo sapiens (Human)

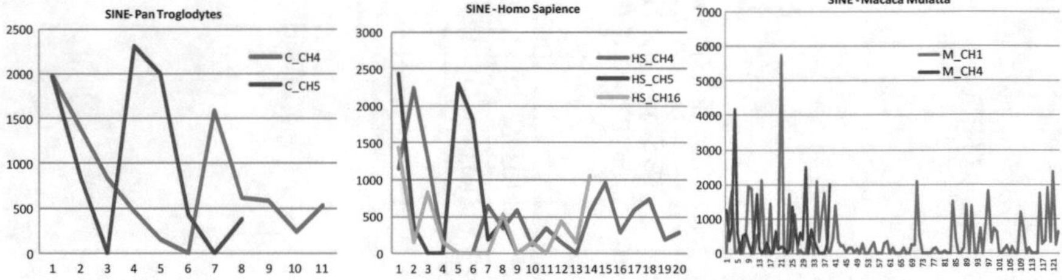

Fig. 8 SINE size distribution of Human, Chimpanzee, and Monkey

Figure 8 shows the size distribution, occurrence of SINE, and number of bases in each occurrence for all three species. It is observed from the graph that SINE size of human has range of 5–2250, chimpanzee 150–2305, and monkey 5–5720.

5 Conclusion

The signature of TE is significant in evolution process because LINE-1 (L1) is an autonomous retrotransposon, which actively participates in the evolution of mammals. In the human genome, L1 is the only autonomously active mobile genetic element. But there is no standard method available, which detects the presence of LINE automatically for chromosome. Identification of LINE and size of RT for any chromosome may be detected automatically by using this "BP Suche" algorithm. Also, this algorithm can be used to find the frequency of SINE and its size in TE. In this paper, a comparative study of related results on different closely related species has been conducted on a small set of freely available TE database from the NCBI site. Though huge study is required to uncover the history of evolution, this paper has tried to show a glimpse of evolutionary research that will definitely take shape when a set of algorithms will be tested on a large number of dataset.

References

1. Watson, J.D., Crick, F.H.C.: A structure for DNA. Nature (1953)
2. Shu, J.-J.: A new integrated symmetrical table for genetic codes. BioSystems **151**, 21–26 (2017)
3. Dutta, M., Barman, S.: Codon characterization based on electrical response. In: International Conference on Microelectronic Circuit and System (Micro-2015)
4. Das, H., Naik, B., Behera, H.S.: Classification of diabetes mellitus disease (DMD): a data mining (DM) approach. Progress in Computing, Analytics and Networking, pp. 539–549. Springer, Singapore (2018)
5. Lin, T.-Y., Shah, A.H.: Stochastic finite automata for the translation of DNA to protein. IEEE International Conference on Big Data (2014)
6. Sengupta, A., Das, J.K., Pal Choudhury, P.: Investigating evolutionary relationships between species through the light of graph theory based on the multiplet structure of the genetic code. In: 2017 IEEE 7th International Advance Computing Conference (2017)
7. Rokas, A.: Phylogenetic analysis of protein sequence data using the randomized axelerated maximum likelihood (RAXML) program. Current Protocols in Molecular Biology, pp. 19.11.1–19.11.14 (2011)
8. Pinheiro, H.P., Pinheiro, A.D.S., Abe, A.S., Reis, S.F.D.: Phylogenetic relationships and DNA sequence evolution among species of Pitvipers
9. Nater, A., Burri, R., Kawakami, T., Smeds, L., Ellegren, H.: Resolving evolutionary relationships in closely related species with whole-genome sequencing data. Syst. Biol. **64**(6), 1000–1017 (2015)
10. Patrushev, L.I., Kovalenko, T.F.: Functions of noncoding sequences in mammalian genomes. Biochemistry (Moscow) **79**(13), 1442–1469 (2014)
11. Shanmugam, A., Nagarajan, A., Pramanayagam, S.: Non-coding DNA–a brief review. J. Appl. Biol. Biotechnol. **5**(05), 42–47 (2017)
12. Bowen, N.J., King Jordan, I.: Transposable elements and the evolution of Eukaryotic complexity. Curr. Issues Mol. Biol. **4**, 65–76 (2002)
13. Britten, R.J.: Transposable element insertions have strongly affected human evolution. PNAS **107**(46), 19945–19948 (2010)
14. Ayarpadikannan, S.: The impact of transposable elements in genome evolution and genetic instability and their implications in various diseases (2014)
15. Haren, L., Ton-Hoang, B., Chandler, M.: Integrating DNA transposons and retroviral integrases. Annu. Rev. Microbiol. **53**, 245–281 (1999)
16. Arensburger, P., Piegu, B., ves Bigot, Y.: The future of transposable element annotation and their classification in the light of functional genomics. Mob. Genet. Elem. **6**(6) (2016)
17. Kojima, K.: Human transposable elements in Repbase: genomic footprints from fish to humans. Mob. DNA **9**, 2 (2018)
18. Ewing, A.D.: Transposable element detection from whole genome sequence data. Mob. DNA **6**, 24 (2015)
19. Mills, R.E., Andrew Bennett, E., Iskow, R.C., Luttig, C.T., Tsui, C., Stephen Pittard, W., Devine, S.E.: Recently mobilized transposons in the human and chimpanzee genomes. Am. J. Hum. Genet. **78**, 671–679 (2006)

20. Cordaux, R., Batzer, M.A.: The impact of retrotransposons on human genome evolution. Nat. Rev. Genet. **10**(10), 691–703 (2009). https://doi.org/10.1038/nrg2640
21. Wicker, T., et al.: A unified classification system for eukaryotic transposable elements. Nat. Rev. Genet. **8**, 973 (2007)
22. Khazina, E., Weichenrieder, O.: Human LINE-1 retrotransposition requires a metastable coiled coil and a positively charged N-terminus in L1ORF1p. eLife **7**, e34960 (2018). https://doi.org/10.7554/eLife.34960
23. https://www.ncbi.nlm.nih.gov
24. Sciamanna, I., Gualtieri, A., Piazza, P.F., Spadafora, C.: Regulatory roles of line-1-encoded reverse transcriptase in cancer onset and progression. Oncotarget **5**(18), 8039–8051 (2014)

An Approach for Sentiment Analysis of GST Tweets Using Words Popularity Versus Polarity Generation

Sourav Das, Dipankar Das and Anup Kumar Kolya

Abstract This paper represents an experimental approach for sentiment analysis of GST tweets, along with generating the most popular GST buzzwords with their respective popularity to polarity scores within a given range of tweets. In India, one of the most trending issues of 2017 was the implementation of Goods and Services Tax (GST) during June–July 2017. GST is a single tax-settled nation wise upon the supply chain of goods and services, primitively from the manufacturer to the consumer. GST was launched in India on the midnight of June 30, 2017, marked by a midnight (June 30–July 1, 2017) session of both the houses of parliament coalesced at the Central Hall of the Parliament. This new taxation system is governed by an GST Council formed by the Govt. of India and the chairman of this council is the Finance Minister of India, himself. As an immediate effect of the cumulation of these ongoing events, a lot of opinion contrast emerged on popular social networks such as Twitter, Facebook regarding this new taxation system. Inspired from this entire event, we propose a new Twitter-based polarity–popularity framework and words cluster for identifying most popular GST-exclusive word counts and sentiment analysis within a given range of tweets, which can be the good indicators of words occurrence probability specific to such an event regarding this topic. This work gives us a much more detailed aspect of the lexical-level analysis of tweets from several directions, along with the future improvement prospects.

Keywords Twitter · GST · Opinion mining · Lexicons · Sentiment analysis

---

S. Das (✉)
National Institute of Electronics and Information Technology, Kolkata 700032, India
e-mail: sourav.das17.91@gmail.com

D. Das
Department of Computer Science and Engg, Jadavpur University, Kolkata 700032, India
e-mail: ddas@cse.jdvu.ac.in

A. K. Kolya
Department of Computer Science and Engg, RCC Institute of Information Technology, Kolkata 700015, India
e-mail: anup.kolya@gmail.com

A. K. Das et al. (eds.), *Computational Intelligence in Pattern Recognition*, Advances in Intelligent Systems and Computing 999,
https://doi.org/10.1007/978-981-13-9042-5_7

1 Introduction

Sentiment analysis on social media not only helps to identify the mass opinion on certain topics but also indicates the ongoing as well as future trends of the social pathway, political future, even economy of a country. Twitter witnessed a large number of such recently happened cases, be it an anti-harassment movement like Me Too,[1] or the implementation of Goods and Services Tax (GST)[2] in India. People's curiosity and opinion about GST attained its peak and highly motivated us to gain the actual insights of this new taxation system for the world's largest democracy.

Here, we present a topic sentiment model inspired word-based popularity and polarity equation to analyze the word popularities and thereafter sentiment from tweets on GST over time. We developed a Twitter dataset consisting of 200 k tweets solely on GST issue, using the topic sentiment model. Implying various state-of-the-art sentiment lexicons, the Bag of Words (BOW) are generated. Next, we assigned sentiment ratings using NLTK-based Naïve Bayes model to our labeled data corpus, so that the system identifies the sentiment word clusters as well as sentiment polarity assigned tweets related to GST along with their sentiment ratings. Therefore, in order to track the most popular words with occurrence probability for such a topic, a popularity–polarity model has been implemented with the help of such Bag of Words and sentiment rated tweets to assign probability score to the most frequently occurred words in regarding the context of GST related tweets, and we further applied it to preserve each of the topic words exclusively for an event like this. We have also demonstrated the polarity mapping or relation of such words within a given miniature range of tweets. These words from the clustering carry respective probability scores, i.e., they reflect the dense or sparse possibilities to appear within the tweets about GST. Finally, we compared our results with four state-of-the-art lexicons to evaluate the performance of our proposed model.

The rest of the sections are organized as follows: In Sect. 2, we present a complete insight of our dataset, with our proposed mathematical approach as Feature I for tweet's Topic and Sentiment modeling while live streaming from Twitter. We present the data preprocessing in Sect. 3. In Sect. 4, we discuss about the popularity versus polarity approach as Feature II with results visualization. Furthermore, we state our analytical observations and comparisons with previous works in Sect. 5. With that, we pave the future prospects of our research and conclude in Sect. 6.

---

[1] http://time.com.

[2] Economictimes.indiatimes.com.

2 Preparing GST Tweets Corpus

India has a number of 26.7 million active Twitter users in 2017,[3] i.e., currently, the second highest in the world. Since GST was the clearly one of the largest taxation reforms in the history of independent India, Twitter witnessed a social opinion outburst on this topic during June–July 2017. In this context, we gathered tweets by employing a live Twitter streaming API in two phases. At the early stage, we collected tweets in synchronization with the implementation phase of GST in India during June–July 2017. GST was one of the top trending topics at that time, and people were tweeting about it frequently rather than any other topics. Since the time slot between 1 p.m. to 3 p.m. is considered to be the "peak-time" for Twitter activities,[4] we aimed primarily to gather tweets between the aforesaid time window. While Twitter API mostly allows its users to livestream only 1–2% of the total tweets on any keywords, we observed that we were able to collect tweets at a rate of 28,000 tweets per day. In this context, one thing is worth mentioning that up to October 2017, Twitter used to support a highest of 140 characters per tweet, whereas, from November 7, 2017, Twitter expanded its character limits to 280 characters per tweet. However, since we started collecting the GST tweets from June 2017, initially we were able to collect tweets with 140 characters only. One thing worth mentioning is that we are stating the total number of our collected tweets while considering all the impurities within tweets that were streamed by us.

It was observed that after the appliance of GST all over India, the dust was settled after a few months and it seemed that finally, people are not tweeting about it like before. But meanwhile, India's GST council held a meeting on November 10, 2017,[5] which eventually led the way for a rate shift of 177 products. This decision again slightly boosted up the topic and also the tweets about these reforming rates, which motivated us again to collect the tweets as the second phase of our data collection. Though during this phase, we could stream live tweets at a slower rate than before, at around 7,000–11,000 tweets per day. We randomly streamed tweets both from normal people as well as the Twitter handles of *@narendramodi*, *@arunjaitley*, *@FinMinIndia*, *@RBI*, *@GST_Council,* and so on. After combining two phases, we gathered a number of 1,99,864 tweets, or almost 200 k unprocessed and raw tweets containing the hash-tagged keywords such as *#gst*, *#gsttax*, *#gstlaunch*, *#gstrollout*, *#gsteffect*, *#onenationonetax,* etc. along with the main tweet bodies.

---

[3]www.statista.com.

[4]https://www.forbes.com.

[5]www.cbec.gov.in/resources.

2.1 Feature I: Tweet Model

Streaming live tweets from Twitter can be a tricky task. A lot of heterogenous tweets based on solely different topics can get collected as long as they consist of the same themes of sentiment or same types of hash-tagged words. To overcome this problem, we used a topic sentiment model to stream the live tweets. Our proposed model ensures the relevance of the tweets with GST topic, as well as it also determines if the tweets contain any sentiment or not. From the axiomatic perspective, it is worth mentioning that this model is not only restricted to GST topic, but can also be employed for crawling tweets only based on any particular topic of discussion.

2.1.1 Topic Modeling

In order to identify whether a tweet is relevant to our target/topic or not, we have considered a parameter κ as the keyword (here GST) of the tweet. Now, a keyword makes the most impact determining the relevance of the tweet while streaming it from Twitter. Moreover, as the tweeting person shifts the keyword position (any word containing the term "GST" or related to that), within the end of the tweet, i.e., closer towards the 140 characters limit, relevance of the keyword or its association with the tweet topic along with the context it is based on actually increases or decreases. Next, we developed the following equation(s) from our cognitive concept. More formally, if the keyword is found at the beginning of the tweet (Eq. 1):

$$t_{posi} = \kappa + \left(n - \text{text}_j\right) \tag{1}$$

where "pos_i" is the position of the parameter (keyword), and here, $pos_i = 1$, n is the total tweet (base) and $text_j$ is the remaining part of the tweet ($j = n\text{–}1$). Similarly, for (2), if the keyword is found in the middle of a tweet:

$$t_{posi} = \frac{n-(\kappa + \text{text}_j)}{2} \tag{2}$$

Finally, if the keyword is found at the last of a tweet:

$$t_{posi} = (n + \kappa) \tag{3}$$

Now, combining all the possibilities of searching a relevant tweet for our target/topic, we formulated the model as in (4):

$$\sum_{i=0...n}^{pos_i} t = \frac{na(\kappa + n)}{2} \tag{4}$$

where t is the entire body of the tweet, α is the odd coefficient unit of keyword position, and n is the remaining text position. Using this technique, the relevance of the tweet with GST and its associated words/phrases are achieved.

2.1.2 Sentiment Modeling

After detection of the matching keyword(s) that we are looking for, the probability of finding the polarity from the remaining text as matching to our topic is determined and expressed as

$$\mathrm{S_m} = \kappa + \mathrm{s} \tag{5}$$

where S_m is the sentiment model, κ is the relevant keyword, and s is the sentiment expression found related to that keyword. The sentiment is of any flavor, i.e., positive, negative, or neutral. But, only if the sentiment is found in a tweet along with the keyword, then it can be streamed. Finally, from Eqs. (4) and (5), we form Eq. (6):

$$\sum_{\mathrm{i=0...n}}^{\mathrm{pos_i}} \mathrm{t} = \mathrm{S_m} = \kappa + \mathrm{s} \tag{6}$$

where, if the tweet consists of both the keyword and sentiment, then only we streamed it.

3 Data Preprocessing

Since the tweets are livestreamed using Twitter streaming API, our first task was the preprocessing of noisy texts. We mostly aimed to manually remove Unicode, gibberish words, and URLs using simple Python snippets as they generally keep no impact on the extraction of underlying meaning or opinion of a natural English text. We also removed the tweets with only GST relevant keywords but no tweet bodies, as those tweets do not bear any types of implacable information. After preprocessing or cleaning the tweets, we tokenized the tweets into unigrams, bigrams, and trigrams with a frequency of 10,000 words for each type, while removing the stop words from unigrams, but not from bigrams and trigrams, as keeping the semantics of such phrases intact. With that, we kept a dense frequency distribution of the words. The purpose behind this is to catch as many as unigrams, bigrams or trigrams possible within a tweet. With that, we also checked continuously for duplicate tokens when the tokens are being collected, till it reaches the EOF.

Fig. 1 Some of the most matched words from our corpus with previous state-of-the-art lexicons

3.1 Coverage with State-of-the-Art Lexicons

After completion of the stemming for collecting frequently occurred grams, which we received previously, it was followed by part-of-speech (POS) tagging and this helped to shrink down our filtered and extracted lexicons further. Now, after obtaining the final list of POS tagged words from our dataset, we intended to match them with four state-of-the-art sentiment lexicons, such as Hu-Liu Positive_Negative Dataset [1], SenticNet 5.0 [2], SentiWordNet 3.0 [3], and finally Vader [4], to find out the coverage of our words in the standard lexicons. A number of covered words are presented as *wordcloud* representation in Fig. 1, where we represent a dense *wordcloud* diagram of some of the mostly matched words with previous gold standard lexicon corpora. We list these newly matched words in a separate file.

From the total number of matched words that we obtained, furthermore we approached for stemming and POS tagging of the words. These words, along with our previously extracted and POS-tagged grams, further served as the mixed bag of words for our Sentiment Model.

4 Popularity Versus Polarity

We retained two distinctly separate sets of extracted grams turned POS tagged words and tokens matched with state-of-the-art datasets for implying both of these lists as a bidimensional mixed bag of words for generating the sentiment scores for our labeled data corpus. For assigning the sentiment rating, deploying this mixed bag of words, we developed an *NLTK*-based *Naïve Bayes* sentiment analyzer to assign sentiment scores to the tweets which were previously labeled using the topic and sentiment to ensure their relevance with the subject matter. Since tweets are short in nature, for such short texts or textual fragments, Naïve Bayes tends to perform better than other baseline algorithms [5]. The scale of the sentiment scores provided by us was on a range from 1 to 5, such as *very negative* (1.0), *negative* (2.0), *neutral* (3.0), *positive* (4.0), and *verypositive* (5.0). Utilizing tweets with these scores, we created a popularity–polarity model, which consists a probability list containing the exclusive words related to GST event, with their respective probability score(s) to indicate the probability of their occurrence within a given range of tweets.

4.1 Feature II: Popularity–Polarity Modeling

From the sentiment rated tweets, one of our key aims was to identify the words, which occur repeatedly in our tweets but might not be found on any standard lexical or Twitter dataset before. Hence, we considered these words to be specifically related to the event of GST in Indian circumstances. Furthermore, we made a file containing these words and their respective scores in this type of economic event. We adopted two parameters, namely polarity and popularity to identify the scores of such crucial words.

$$\text{Score(word)} \approx |Polarity| \quad \text{and}$$
$$\text{Score(word)} \approx |Popularity|$$

where polarity defines the sentiment rating from our previously stated sentiment score and popularity defines the score (number) of occurrences for that word, say for 1,000–10,000 sample number of tweets from our entire dataset. Furthermore, if we denote score as δ, hence for the changing value of δ, the score of topics can be given as

$$\delta = \text{Score}(topic) \approx \delta_1.|Polarity| + \delta_2.|Popularity| \tag{7}$$

Based on this simple equation, we represent Table 1 to demonstrate the word polarity and popularity measure with respect to sentiment score and word occurrence.

Now, as we have already mentioned the compact relationship between the topic and sentiment in the tweets that we streamed, hence, the score of topic words are actually derived from the tweets consisting only GST topics and sentiments. More formally:

$$\sum_{i=0\ldots n}^{pos_i} t = S_m = \text{Score(topic)} \tag{8}$$

Table 1 Word polarity and popularity measure with respect to sentiment score and word occurrence

Polarity			Popularity		
δ_1	Tweet count	Score (word) ≈ |Polarity|	δ_2	Tweet count	Score (word) ≈ |Popularity|
1.0	1000	1000	556	1000	556000
2.0	2000	4000	500	2000	100000
3.0	4000	12000	353	4000	1412000
4.0	8000	32000	278	8000	2224000
5.0	10000	50000	130	10000	1300000

i.e., the polarity and popularity scores are derivable from each tweet with respective sentiment and topic.

Conclusively, the complete polarity and popularity model based on the topic sentiment dependent tweet streaming can be expressed by (9)

$$\sum_{i=0...n}^{pos_i} t = S_m = \delta_1.|Polarity| + \delta_2.|Popularity| \tag{9}$$

where δ1 is sentiment polarity score and δ2 is the word occurrence count within a given number of tweets. Applying the one to one δ_1 versus δ_2 combination, i.e., popularity versus polarity calculation for our labeled- and sentiment-rated tweets, we found out a list of the most unique and only GST-exclusive topic words within 10,000 tweets. A few sample topic words from a total list of 9,871 words in alphabetical order along with their respective popularity to probability scores are*Aadhar (556.0), Allaboutgst (248.0), Demonetization (117.66), Economy (69.5), Govttax (26.0), Iitians (140.0), Jewelery (128.58), Kollywoodvoice (89.72), Laws (102.36), Modi (227.92), PM (169.06),* etc.

4.2 Visualization

Once we obtained the complete list of words along with their respective probability scores, we calculated a visual representation in Fig. 2 of the mostly occurred 36 words within just 1,000 tweets.

To observe the relations between popularity versus the polarity of words, we plotted the following visual comparison in Fig. 3. This comparison graph represents the varieties of a sample number of words in the *x-axis* with their respective positive or negative threading parallelly along with a minimized scale of polarity scores ranging from 0 (*negative*) to 1 (*positive*) in the *y-axis*, with also the intermediate polarity scales in between. We analyzed this relation based on the most popular words that

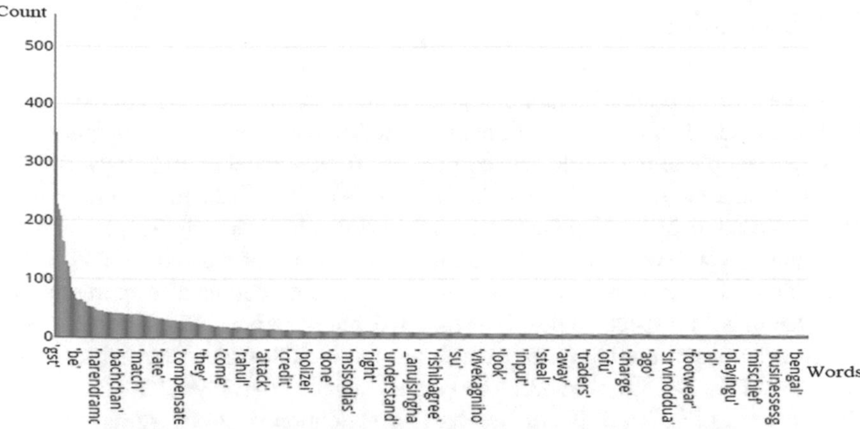

Fig. 2 Word popularity count within a sample number of 1,000 tweets

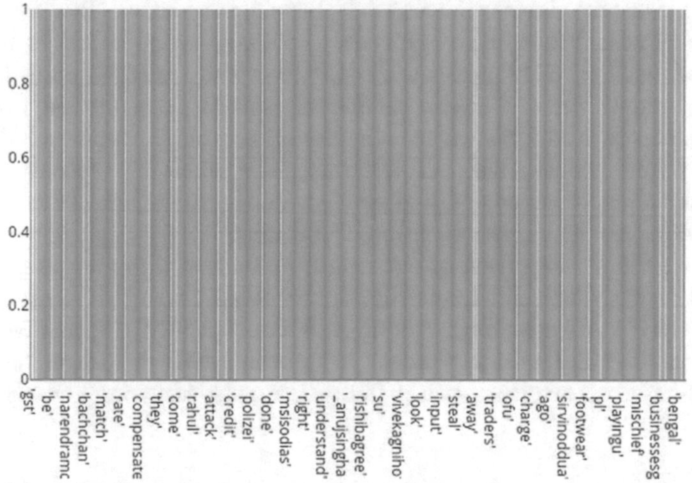

Fig. 3 Polarity mapping for some most frequent GST-exclusive words from a number of tweets

we obtained previously. In this graph, the horizontal green threads are the indicators of carrying the "*positive*" polarity tag for their respective words, while quite similarly, the horizontal white threads in between depict the "*negative*" polarity tagged words within a miniature group of (here a number of 1,000) tweets. Besides visualizing this graph for a handful number of tweets, this analytical deduction can also be deployed for the entire data corpus.

5 Observations

Using the aforesaid model and the popular topic words that we gained, we implied the Feature set II to produce sentiment prediction-based accuracy outcomes for a fractional sample of the whole data corpus, i.e., for a number of 10,000 tweets from our entire dataset. The main reason behind this is to reduce the time complexity as much as possible, by making the overall analysis much faster. Another reason is that a fractional overview of the results can be the showcase of the entire data corpus's characteristics. This analytical observation of our data corpus also represents the error rate in both respects to positive predicted and negative predicted tweets. With 10,000 tweets, our training (standard) set has total 9908 entries (tweets) with 5413 actual positive (54.63% of 9908), 4100 actual negative (41.38% of 9908), and 395 actual neutral (25.08% of 9908) tweets. Our validation set has a number of 4917 positive predicted tweets (90.83%), 3468 negative predicted tweets (84.58%), and 531 neutral predicted tweets (74.38%). We represent the predicted results in Table 2. Next, we present the classification report in Table 3, in which we show the statistical measures calculated from Table 6. Entries with 43.01% negative, 56.99% positive, and test set has total 5413 entries with 41.94% negative, 58.06% positive with an overall accuracy score of 83.44%. Validation result for 10,000 tweets reveals that the null accuracy is 54.72%, whereas the overall accuracy score is 83.85%, which is 29.13% more accurate than null accuracy. Here, Table 2 represents the actual and predicted prediction counts, and simultaneously, Table 3 shows the statistical classification parameters generated from that counts.

Next, we did a comparison of our results in Table 4 with four state-of-the-art sentiment analysis lexicons: Linguistic Inquiry Word Count (LIWC) [6], General Inquirer (GI) [7], Affective Norms for English Words (ANEW) [8], and Word-Sense Disambiguation (WSD) using WordNet [9], for analyzing the performance of our sentiment-based accuracy. As seen in Table 8, in most scenarios, our approach for

Table 2 Confusion matrix and classification report shows the differences between the predicted and actual tweets along with the prediction parameters

	Predicted		
Actual		Positive	Negative
	Positive	4917 (TP)	496 (FP)
	Negative	632 (FN)	3468 (TN)

Table 3 Confusion matrix and classification report shows the differences between the predicted and actual tweets along with the prediction parameters

Parameters	Results
Precision	0.9038
Recall	0.8861
Accuracy	0.8385
F1-Score	0.897

Table 4 Comparison of classification performance on social media posts

	Overall accuracy ground truth (of the respective corpora)	Classification accuracy metrics		
		Overall precision	Overall recall	Overall F1 score
Social media texts (10,000 Tweets)				
Our approach	**0.838**	**0.90**	**0.88**	**0.89**
GI	0.512	0.79	0.51	0.67
LIWC	0.606	0.91	0.49	0.60
ANEW	0.451	0.79	0.46	0.57
WSD	0.401	0.69	0.44	0.52

social media texts (here tweets) provides better overall classification parameters than the manually given human ratings, and outperforms the other previously well-established lexicons for sentiment analysis.

6 Conclusion and Future Work

In this paper, we represented a topic sentiment-based tweet crawling and thereafter, sentiment analysis from words popularity versus polarity generation for discovering and clustering the GST-exclusive topic words from GST tweets in India. For our further work, we want to keep the most occurred words from this event, and deploy them if any such event takes place, to predict the course and trend of that event. Also, our popularity–polarity model faced performance bottlenecks at times against code-mixed tweets. For our extended work, we would also like to develop a system to successfully evaluate bilingual or font-mixed tweets to enhance the accuracy of our experiment. And finally, since our work is one of the first on such a large-scale economic reformation, we are keen to publish our GST data corpus with all the frameworks and codes on an open-source data repository platform like Github,[6] so that other researchers feel free to experiment with our findings from this event, and they compare their achieved results with that of ours.

References

1. Hu, M., Liu, B.: Mining and summarizing customer reviews. In: Proceedings of the Tenth ACM SIGKDD International Conference on Knowledge Discovery and Data Mining. ACM (2004)
2. Cambria, E., et al.: SenticNet 5: discovering conceptual primitives for sentiment analysis by means of context embeddings. In: Proceedings of AAAI (2018)

[6]https://github.com/SouravDme.

3. Baccianella, S., Esuli, A., Sebastiani, F.: Sentiwordnet 3.0: an enhanced lexical resource for sentiment analysis and opinion mining. Lrec. vol. 10, No. 2010 (2010)
4. Gilbert, E., Hutto, C.J.: Vader: A parsimonious rule-based model for sentiment analysis of social media text. In: Eighth International Conference on Weblogs and Social Media (ICWSM-14). Available at (20/04/16) (2014). http://comp.social.gatech.edu/papers/icwsm14.vader.hutto.pdf
5. Wang, S., Manning. C.D.: Baselines and bigrams: Simple, good sentiment and topic classification. In: Proceedings of the 50th Annual Meeting of the Association for Computational Linguistics: Short Papers, vol. 2, pp. 90–94. Association for Computational Linguistics (2012)
6. Pennebaker, J.W., et al.: The development and psychometric properties of LIWC2015 (2015)
7. Wilson, T., Wiebe, J., Hoffmann, P.: Recognizing contextual polarity in phrase-level sentiment analysis. In: Proceedings of the Conference on Human Language Technology and Empirical Methods in Natural Language Processing. Association for Computational Linguistics (2005)
8. Bradley, M.M., Lang, P.J.: Affective norms for English words (ANEW): Instruction manual and affective ratings, vol. 30, No. 1. Technical report C-1, the center for research in psychophysiology, University of Florida (1999)
9. Gao, N., et al.: Word sense disambiguation using wordnet semantic knowledge. In: Knowledge Engineering and Management, pp. 147–156. Springer, Berlin (2014)

Smart Health Monitoring System for Temperature, Blood Oxygen Saturation, and Heart Rate Sensing with Embedded Processing and Transmission Using IoT Platform

Samik Basu, Sinjoy Saha, Soumya Pandit and Soma Barman (Mandal)

Abstract Traditional methods of healthcare demand the patient to visit the clinic on a regular basis. These conventional methods can be quite cumbersome for both the patient as well as the medical staff. The focus of this paper is to implement a smart IoT-based health monitoring system which uses noninvasive sensors to read different health parameters and displays it instantly on an LCD module. Additionally, the user has the option of sending the data to the cloud to store it securely. The stored data can be easily accessed from anywhere in the world. Low cost, portable, user-friendly interface and ease of data visualization are some of the key features of this system.

Keywords Smart health monitoring system · Arduino UNO · Internet of things · ESP8266 · DS18B20 · MAX30100 · ThingSpeak · REST API

1 Introduction

Human health condition is a major concern nowadays globally. In traditional methods of health checkup, a patient has to visit regularly to a clinic for his or her health checkup. This needs an involvement and availability of health care professionals for advising and diagnosis. A problem occurs when many patients are present at a time in the clinic and each individual does not get proper care which further aggravates their condition. Another major drawback of all traditional approaches is the lack

S. Basu (✉) · S. Saha · S. Pandit · S. Barman (Mandal)
Institute of Radio Physics & Electronics, University of Calcutta, 92, APC Road, Kolkata 700009, India
e-mail: samikbasu2010@gmail.com

S. Saha
e-mail: sinjoy@ieee.org

S. Pandit
e-mail: soumya.pandit.rpe@gmail.com

S. Barman (Mandal)
e-mail: barmanmandal@gmail.com

A. K. Das et al. (eds.), *Computational Intelligence in Pattern Recognition*, Advances in Intelligent Systems and Computing 999,
https://doi.org/10.1007/978-981-13-9042-5_8

of constant or time-based health monitoring. Hence, essentially, we need a system which is user-friendly and monitors the vital clinical parameters of a human being.

The Internet of Things (IoT) provides a solution to the problems faced by traditional methods. The IoT technology can be adapted to health care systems where dedicated sensors and processors continuously monitor an individual's physiological parameters and notify them of any health issue in advance. Such personalized embedded systems may send the data over the Internet via Wi-Fi or Ethernet cable to be viewed by a doctor or physician for further diagnosis. This type of system will address the demands in rural healthcare too.

With the advent of technology, health care systems have been constantly improving with time and many countries have achieved large gains in life expectancy [1]. Many discrete works have been done in the past to cater to the needs of the patient as well as the medical staff. In *WISH: A Wireless Mobile Multimedia Information System in Healthcare Using RFID,* a system was presented to use modern wireless technology, Radio Frequency Identification (RFID) tools and multimedia streaming to help physicians and medical staff [2]. Wearable devices are also becoming very popular today. A study showed the integration of Internet of Things (IoT) and eHealth solutions to effectively manage as well as monitor students' health [3]. Another system which integrated web services with multiple sensors controlled by Arduino UNO using Service Oriented Architecture (SOA) with Rich Internet Application (RIA) has been presented [4]. A similar project integrating the technologies of Wireless Sensor Networks (WSN) and public communication networks, measuring heartbeat and body temperature, aimed at senior citizens at home without interfering their daily activities was also done [5]. A similar type of work by Kaur and Jasuja on Arduino UNO used an Ethernet shield and the RJ45 cable to connect to the network [6]. Other works had also been done using the LPC 2129 ARM 7 Processor [7] and the Raspberry Pi [8].

Our proposed system uses the Arduino UNO microcontroller, ESP8266 Wi-Fi module compatible with the UNO and an LCD module. The UNO supports a wide array of sensor modules and is of low cost and low power. It can also be powered via battery, making the system portable. The system communicates over a secure network. It uses HTTP protocol and based on REST architecture and API key to transmit information over a private channel [9]. The user interface consists of an LCD display and two push buttons for accepting user's choice of sensor and whether to send data to cloud. This arrangement of our proposed IoT-based embedded health monitoring system provides security as well as accuracy and the system is user-friendly, low cost and low power. The comparative study and improvements over previous works have been given in Sect. 4.

The paper is organized as follows. Section 2 presents our System Architecture. In Sect. 3, we describe Experimental Setup. In Sect. 4, we present the results in graphical format and extension of to our webpage. Finally, we conclude the paper in Sect. 5.

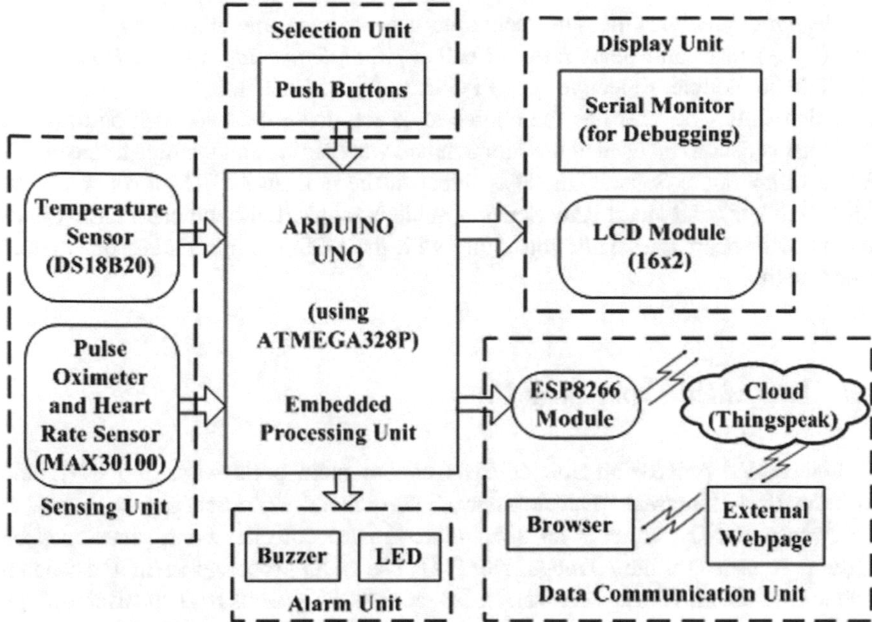

Fig. 1 General block diagram showing the different modules and subsystems used

2 System Architecture

A general block diagram of the health monitoring system is shown in Fig. 1. The system is divided into three main parts—Input Unit (sensing unit and selection unit), Embedded Processing Unit and Output Unit (display unit, alarm unit and data communication unit).

2.1 Input Unit

Sensing unit. Two Maxim Integrated sensors are used in this unit—one is DS18B20 used to measure body temperature and another is MAX30100 which measures both blood oxygen saturation and heart rate.

DS18B20. The *DS18B20* digital thermometer provides 9–12-bit Celsius temperature measurements and communicates over a 1-Wire bus that by definition requires only one data line for communication [10].

MAX30100. The *MAX30100* is an integrated pulse oximetry and heart rate monitor sensor solution which combines two LEDs, a photo detector, optimized optics and low-noise analog signal processing. The MAX30100 operates from 1.8 to 3.3 V power supplies [11].

Photoplethysmography. The technique used to measure blood oxygen saturation (SpO_2) level and heart rate are called *photoplethysmography (PPG)*. In the MAX30100 module, reflective type PPG is used [11, 12, 13].

Selection unit. The user has the choice to select, using the two push buttons, the particular sensor to be used at the moment and whether to send the data to the cloud. When Temperature Sensing mode is selected, the indicator LED glows and when Pulse Oximetry and Heart Rate Sensing mode is selected, the indicator LED blinks in sync with heart rate. This unit along with the LCD module makes our system user-friendly.

2.2 Embedded Processing Unit

The embedded processing unit consists of two main parts—Arduino UNO and Arduino IDE (Integrated Development Environment). The sensors send data to the Arduino UNO, which is an open-source microcontroller board developed by Arduino.cc based on the ATmega328P [14]. The UNO processes them and sends it over to the LCD and Serial Monitor. It also receives commands from the user through the push buttons. During transmission over Wi-Fi, it sets up the ESP8266 module and sends data using "AT" commands.

2.3 Output Unit

Display unit. The display unit consists of the Serial Monitor (used for debugging purposes) and the LCD module.

Serial monitor. The *Serial Monitor* is the "tether" between the computer and the Arduino—it lets one send and receive text messages, handy for debugging and also controlling the Arduino from a keyboard. It uses the built-in *Serial* library for USB to serial communication.

Liquid crystal display. The 16-pin 16 × 2 *Liquid Crystal Display (LCD)* module, compatible with the Hitachi HD44780 driver, is used. The *LiquidCrystal* library simplifies the low-level instructions for controlling the module. The Hitachi-compatible LCDs can be controlled in two modes: 4-bit or 8-bit. For this system, the LCD is used in the 4-bit mode [15].

Alarm unit. The alarm unit consists of a 5 V HYDZ piezoelectric buzzer and LED. When the sensing parameters are outside the range of typical values, the microcontroller starts the buzzer and turns on the LED. The typical range for human body temperature may be 97.8°F (36.5°C)–99°F (37.2°C) for a healthy adult. The heart rate of a healthy adult can range from 60 to 100 beats per minute [16]. Normal blood oxygen saturation levels range from 94 to 99% in adults [17].

Data communication unit. The data communication unit using the IoT platform consists of four main parts—ESP8266 Wi-Fi module, ThingSpeak.com (data storage), External Webpage and finally accessing it through a Web Browser.

ESP8266 Wi-Fi module. The *ESP8266* is a low-cost Wi-Fi microchip with full TCP/IP stack, microcontroller capability produced by Espressif Systems. It operates at 3.3 V. This module can act as a standalone microcontroller but in this system, it is used only as a Wi-Fi module.

ThingSpeak. It is an open-source *Internet of Things* (IoT) application and Application Program Interface (API) to store and retrieve data from things using the HTTP protocol over the Internet or via a Local Area Network [18]. The data is sent to ThingSpeak using the Representational State Transfer (REST) API. It uses HTTP requests to *GET, PUT, POST* and *DELETE* data, along with the API key provided by ThingSpeak for a secure communication [9, 18].

Webpage and browser. The contents of the channel can be viewed on *webpage* hosted on a personal website. ThingSpeak lets us embed the graphs on an external website. An HTTP request using RESTful architecture and corresponding read API key can be used [19]. The user finally gets to view the data through a *web browser* of any connected device like computer or smartphone across the world.

3 Experimental Setup and Methodology

Experimental setup for our system is shown in Fig. 2. The sensors, buttons and different modules are connected to the Arduino UNO board with the help of jumper wires and a breadboard. When the setup is turned on, the system initializes the Wi-Fi module and joins an access point. In this case, a mobile hotspot was used. It then initializes the LCD module as well as all the sensor modules and the system starts in the default mode, which is the temperature sensing mode. A push button can be used to toggle the different modes. Once a mode is selected, the setup receives the signal from the respective sensor and calculates the corresponding body parameter and displays it on the LCD module.

The Arduino sketch mainly consists of two parts—the *setup()* and the *loop()*. In the *setup()* function all the initializations of different modules are done. Once the system is initialized, it enters the *loop()* function which continuously iterates its contents until the system is turned off.

This complex system has limited program memory of 32 KB and dynamic memory of 2 KB (used for global variables) which is challenging. The *update()* function of the *PulseOximeter* object should be called as frequently as possible. Also, some parts of the program loop like reading input from a push button, displaying data and glowing LEDs need to be delayed after every iteration. A simple solution is using the *delay()* function but implementing this stops the entire program loop for the specified time, thus affecting the overall performance as calling the update() gets delayed. Thus, a more involved solution is necessary. In this system, the loop keeps on iterating and at specified intervals, kept track by individual global variables, the corresponding

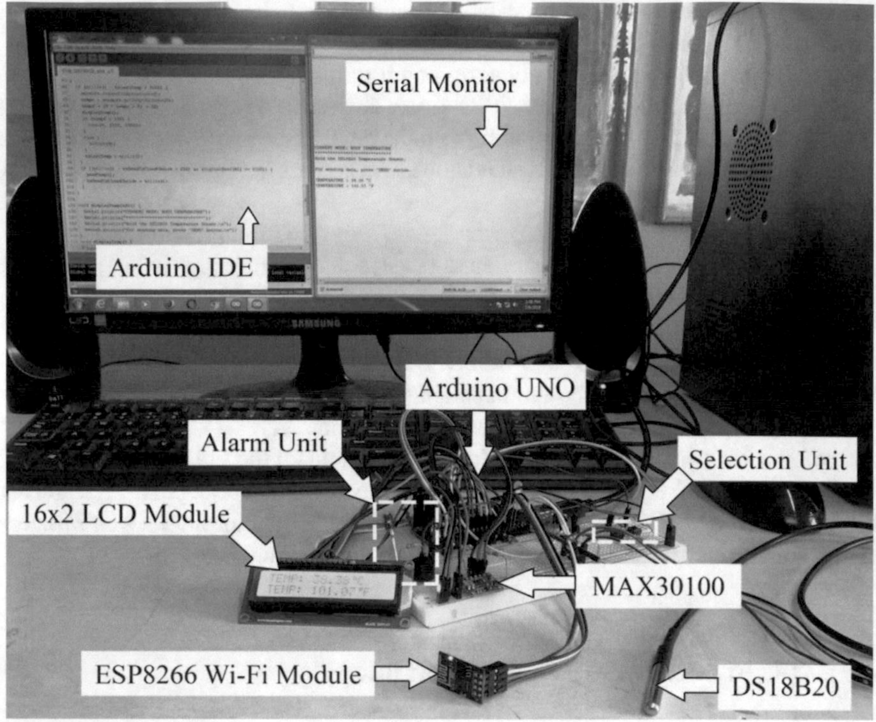

Fig. 2 Overall experimental setup showing the main components of the system as labeled

tasks are performed. Hence, we trade-off memory for performance and a balanced compromise between the two is achieved for a stable and successful implementation.

At first, in the *loop(),* the program checks for the toggling of the choice button. If it is toggled, the boolean variable *choice* is also toggled. This checking is done at an interval of 250 ms, kept track by the unsigned long variable *tsModeChoice*. This value depends on the user reaction time and is found by a trial and error method.

Then, the program enters into an if-else part for implementing either of the two sensors. If *choice* is true, temperature sensor value is displayed, else pulse oximeter value is displayed. Both the cases have their respective threshold values for turning on the buzzer and LED as discussed in Sect. 2.1 [16, 17].

For implementing the DS18B20 1-Wire digital thermometer, the *OneWire* and *DallasTemperature* libraries were used. For implementing the MAX30100 sensor, *MAX30100* library package is used. Since the Arduino operates at 5 V, 4.7 K Ω pull-up resistors are used on the SCL, SDA and INT pins of MAX30100 as its output ranges from 1.8 to 3.3 V [20]. The temperature values are read and displayed every three seconds while the heart rate and SpO_2 values are displayed every fifteen seconds. The SpO_2 provided by library function is used but the instantaneous heart rate value provided by the library function is found to be erroneous. To counter this, a beat detector function along with a counter variable is used. At every fifteen seconds, the current heart rate is given by Eq. (1), where "b_{pm}" denotes the heart beats per minute, "v_c" denotes the current counter value and "v_p" denotes the previous counter value. The multiplicative factor 4 is to change the value to per minute unit.

$$b_{pm} = (v_c - v_p) \times 4 \tag{1}$$

The user has the choice to send the displayed data to the IoT server by pressing another push button. Similar to the choice button, the program checks for toggling of the send button every 250 ms. Once this button is pressed, the Wi-Fi module starts a Transmission Control Protocol (TCP) connection with ThingSpeak and sends the data to the corresponding fields using the REST API. The data fields in ThingSpeak can be updated once every fifteen seconds for the free license. To counter this problem, the data transfer process is slowed down using the *delay()* function. Hence, after the data is sent, the Pulse Oximeter sensor is initialized again before the program control returns to its original state of measuring the heart rate and SpO_2 level. This is not necessary in case of the temperature sensor.

The ESP8266 can communicate with the Arduino UNO via the TX and RX pins using the *SoftwareSerial* library, which sets up two custom TX and RX pins on the UNO using the software. Before the data from Arduino's software TX pin reaches the RX pin of the ESP8266, the voltage is dropped down from 5 to 3.3 V using a 2N3904 NPN bipolar transistor. The data is sent in the form of "AT" commands. "AT" commands are commands which are used to control the modems where AT stands for ATtention. The command *"AT + CWJAP"* is used to join an access point with a Service Set Identifier (SSID) and Password. The command *"AT + CIPSTART"* is used to set up a TCP connection with the server and *"AT + CIPSEND"* is used to send data [21, 22, 23].

ThingSpeak provides eight fields for every channel. The three different sensors have three different fields in a single channel named "Person A". ThingSpeak makes it easy to visualize the data with the help of graph plots. The graphs have additional features of scaling the axes and setting maximum and minimum data values. The graphs are embedded in the HTML code of an external webpage, which can ultimately be viewed on a browser [19]. For maintaining a secure environment, the user has to sign into ThingSpeak to view the data of private channels.

4 Results

The temperature values are displayed on the LCD every three seconds and the heart rate and blood oxygen saturation values are displayed every fifteen seconds. The values can then be sent to ThingSpeak cloud depending on the user's choice. A set of values from the sensors sent to the ThingSpeak channel are displayed in a graphical format for ease of visualization. The data can be imported to an external webpage for the doctor or physician to view. In case of an abnormal change in any of the body parameters, the buzzer and LED starts to notify the individual as shown in Fig. 2.

In Fig. 3, the graphs show the variation of body temperature (in Fahrenheit) with time in Field 1. Here the values were sent at an interval of about thirty minutes, which can be altered via the selection unit. Figure 3a shows values from DS18B20 temperature sensor as displayed locally on LCD module. Figure 3b shows the graph provided by ThingSpeak after sending the temperature data. It is observed that the values of both the graphs are consistent with each other.

Similarly, in Figs. 4 and 5, the graphs show the variation of heart rate (in beats per minute) and variation of blood oxygen saturation (in percentage) with time in Field 2 and Field 3, respectively. Here also the consistency of the health parameter values can be observed. Hence, we can safely conclude that the data communication unit of the proposed system is extremely reliable and can be used by any individual for their self-health monitoring.

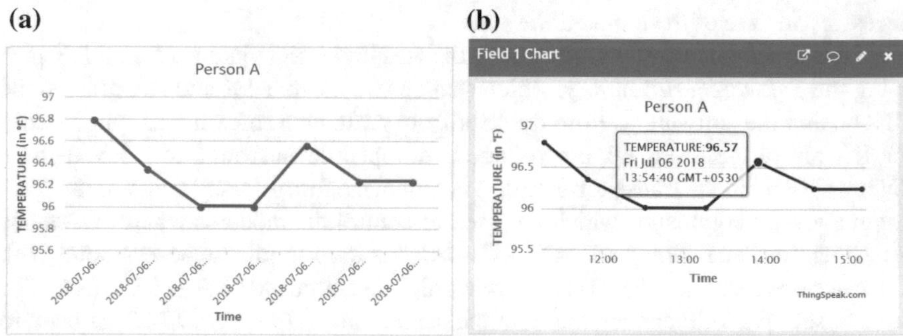

Fig. 3 The graphs show changes in body temperature with time in °F of Person A. **a** The graph shows values from DS18B20 temperature sensor as displayed on LCD module. **b** The graph is provided by ThingSpeak after sending the data

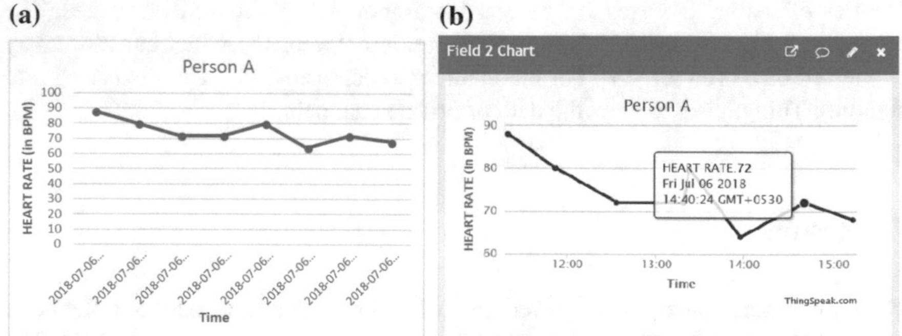

Fig. 4 The graphs show changes in heart rate with time in beats per minute of Person A. **a** The graph shows values from MAX30100 heart rate sensor as displayed on LCD module. **b** The graph is provided by ThingSpeak after sending the data

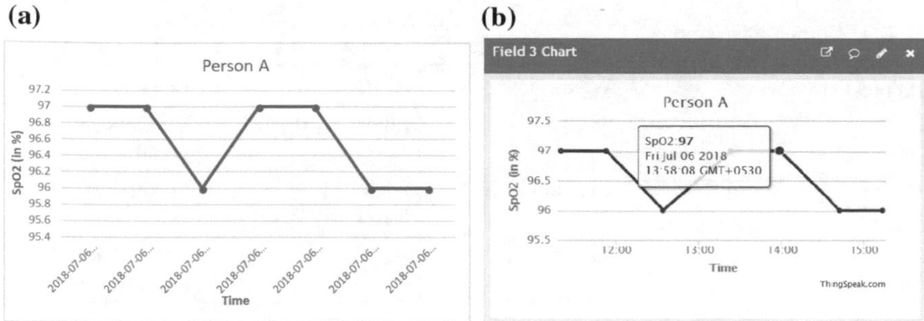

Fig. 5 The graphs show changes in blood oxygen saturation with time in % of Person A. **a** The graph shows values from MAX30100 pulse oximeter as displayed on LCD module. **b** The graph is provided by ThingSpeak after sending the data

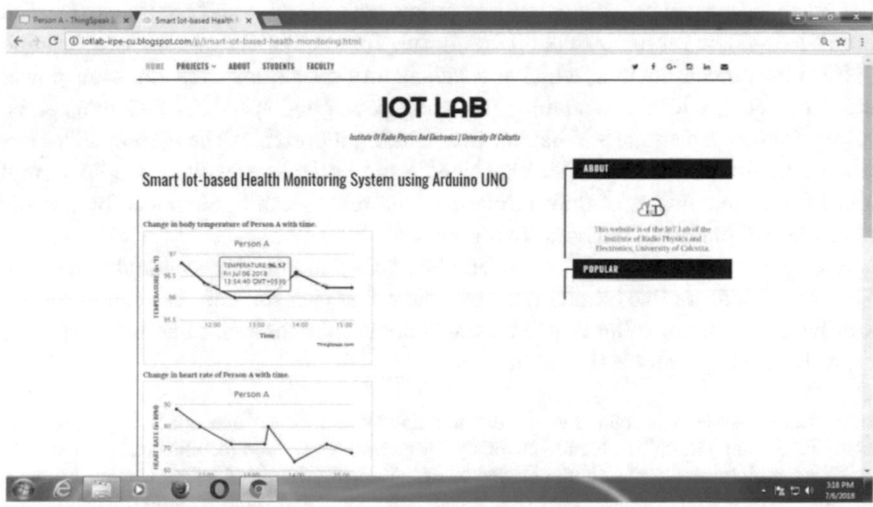

Fig. 6 The data can be imported to an external webpage as shown

Moreover, ThingSpeak allows us to embed the graphs from the ThingSpeak channel to an external webpage which is shown in Fig. 6 [19].

A comparative study between the proposed system and previous works have been given in Table 1. The first column specifies the component used in the proposed system, whereas the second column specifies the components used in different previous works. The third column indicates the improvements of this system over previous systems.

Table 1 Comparative study between the proposed system and previous works

Proposed system	Previous works	Improvement
Wi-Fi	RFID [2, 3]	Longer range
	Ethernet [6]	Wireless and portable
Arduino UNO	ARM [7], Raspberry Pi [8]	Lower cost
Alarm unit	–	Localized indication

5 Conclusion and Future Work

The proposed system provides a low power, low cost, portable and accurate IoT-based embedded system for remote health monitoring for people. The system uses Arduino UNO as the processing unit, which is a satisfactory choice as it consumes low power and is low cost, with a reasonable processing power and speed. The system can be powered through a single 9 V battery thus making it portable. The system addresses self-health monitoring of people, especially senior citizens as well as people in rural areas, from the comfort of their homes and the results can be reviewed by doctors using the IoT platform, thus improving their quality of life.

The system can be turned into a complete health monitoring system by incorporating more sensors like blood pressure sensor, ECG sensor, and respiration sensor. A different microcontroller or even a single board minicomputer like the Raspberry Pi can be used as a processing unit.

Acknowledgements The authors would like to thank UGC UPE II, "Modern Biology Group B: Signal Processing Group", Calcutta University for providing research facilities and also SMDP-C2SD project, University of Calcutta funded by MeitY, Govt. of India, for their technical support and West Bengal Higher Education, Science and Techonolgy and Biotechnology-funded project "Cytomorphic CMOS Circuit Modeling and Ultra-Low Power Design of P53 Protein Pathway for Synthetic Biology Applications".

References

1. Human Development Report 2010.: The Real Wealth of Nations: Pathways to Human Development. United Nations Development Programme. http://hdr.undp.org/sites/default/files/reports/270/hdr_2010_en_complete_reprint.pdf. Accessed 10 July 2018
2. Yu, W., Ray, P., Motoc, T.: WISH: a wireless mobile multimedia information system in healthcare using rfid. J. Telemed. E Health. **14**(4), 362–370 (2008)
3. Takpor, T.O., Atayero, A.A.: Integrating Internet of Things and EHealth solutions for students' healthcare. In: WCE (2015)
4. Hameed, R.T., Mohamad, O.A., Hamid, O.T., Ţăpuş, N.: Patient monitoring system based on e-health sensors and web services. In: 8th International Conference on Electronics, Computers and Artificial Intelligence (2016)

5. Huo, H., Xu, Y., Yan, H., Mubeen S., Zhang, H.: An elderly health care system using wireless sensor networks at home. In: 3rd International Conference on Sensor Technologies and Applications, pp. 158–163 (2009)
6. Kaur, A., Jasuja, A.: Cost effective remote health monitoring system based on IoT using arduino UNO. Adv. Comput. Sci. Inf. Technol. **4**(2), 80–84 (2017)
7. Pereira, M., Kamath, N.K.: A novel IoT based health monitoring system using LPC2129. In: IEEE International Conference on Recent Trends in Electronics Information & Communication Technology, pp. 564–568 (2017)
8. Kaur, A., Jasuja, A.: Health monitoring based on IoT using RASPBERRY PI. In: International Conference on Computing, Communication and Automation, pp.1335–1340 (2017)
9. Rouse, M.: What is RESTful API? https://searchmicroservices.techtarget.com/definition/RESTful-API. Accessed 24 June 2018
10. Maxim Integrated: DS18B20 Programmable Resolution 1-Wire Digital Thermometer, Rev 4. https://datasheets.maximintegrated.com/en/ds/DS18B20.pdf (2015)
11. Maxim Integrated: MAX30100 Pulse Oximeter and Heart-Rate Sensor IC for Wearable Health, Rev 0. https://datasheets.maximintegrated.com/en/ds/MAX30100.pdf (2014)
12. Dosinas, A., Vaitkūnas, M., Daunoras, J.: Measurement of human physiological parameters in the systems of active clothing and wearable technologies. Electron. Electr. Engineering. Med. Technol. (Research Journal) **71**(7), 77–81 (2006)
13. Oak, S.-S., Aroul, P.: How to Design Peripheral Oxygen Saturation (SpO2) and Optical Heart Rate Monitoring (OHRM) Systems Using the AFE4403. Texas Instruments, Application Report (2015)
14. Arduino: Arduino Uno Rev3. https://store.arduino.cc/usa/arduino-uno-rev3. Accessed 5 July 2018
15. Schmidt, M.: Arduino - HelloWorld. https://www.arduino.cc/en/Tutorial/HelloWorld (2015)
16. University of Rochester, Medical Center, Health Encyclopedia. https://www.urmc.rochester.edu/encyclopedia/content.aspx?ContentTypeID=85&ContentID=P00866
17. Amperor Direct.: How to Interpret Pulse Oximeter Readings. https://www.amperordirect.com/pc/help-pulse-oximeter/z-interpreting-results.html. Accessed 2 July 2018
18. ThingSpeak.: IoT Analytics. https://thingspeak.com/. Accessed 24 June 2018
19. IoT Lab.: Smart IoT-based Health Monitoring System using Arduino UNO. Institute of Radio Physics & Electronics. http://iotlab-irpe-cu.blogspot.com/p/smart-iot-based-health-monitoring.html. Accessed 6 July 2018
20. Roland.: How to use the MAX30100 as Arduino Heart Rate Sensor. https://www.teachmemicro.com/max30100-arduino-heart-rate-sensor/. Accessed 18 June 2018
21. Chen, D.: Wireless Communication with ESP8266. http://fab.cba.mit.edu/classes/865.15/people/dan.chen/esp8266/. Accessed 22 June 2018
22. Electronincsforu.com: All You Wanted to Know About AT and GSM Commands. https://electronicsforu.com/resources/cool-stuff-misc/wanted-know-gsm-commands. Accessed 23 June 2018
23. Darrah, K.: Wi-Fi Tutorials. https://www.youtube.com/user/kdarrah1234. Accessed 24 June 2018

Design of a Health Monitoring System for Heart Rate and Body Temperature Sensing Including Embedded Processing Using ARM Cortex M3

Mahasweta Ghosh, Samik Basu, Soumya Pandit and Soma Barman (Mandal)

Abstract A home-based self-health monitoring system is in huge demand in recent times. This can be achieved by a low-power, cost-effective, accurate, and easy-to-use system, which can measure, display, and warn the patient regarding their vital parameters. The proposed system uses non-penetrative sensors to accurately monitor the vital parameters like heart rate and body temperature and uses ARM Cortex M3-based processor LPC1768 to process, display, and alert the patient. The stability and accuracy of the proposed system have been verified by studying the vital signs of volunteers as measured by the system with reference to their actual data as per medical practitioners.

Keywords ARM Cortex M3 · LPC1768 · Heart rate measurement · Surface body temperature measurement

1 Introduction

One of the major challenges in today's world is to lead a healthy lifestyle. Monitoring of various vital health parameters not only acts as a prevention routine for many diseases by identifying their early symptoms but also helps us to keep the right track of our health. However, in our busy schedule or in the case of senior citizens and rural people, it is difficult to regularly visit a hospital or a physician's chamber. So, there

M. Ghosh · S. Basu (✉) · S. Pandit · S. Barman (Mandal)
Institute of Radio Physics and Electronics, University of Calcutta, 92, APC Road, Kolkata 700009, India
e-mail: samikbasu2010@gmail.com

M. Ghosh
e-mail: mahasweta94g@gmail.com

S. Pandit
e-mail: soumya.pandit.rpe@gmail.com

S. Barman (Mandal)
e-mail: barmanmandal@gmail.com

A. K. Das et al. (eds.), *Computational Intelligence in Pattern Recognition*, Advances in Intelligent Systems and Computing 999,
https://doi.org/10.1007/978-981-13-9042-5_9

is a dire need for a self-health monitoring system, which can indicate an abnormality in the vital signs of a person.

Though work has already been done in this field, they usually use generic sensors and Arduino-based microcontroller system for processing. Arduino despite being cheap and easy to use, has very low flash memory and slow processing speed. S. Das et al. proposes a robust real-time heart rate measurement using Arduino Uno processor [1]. B. Changela et al. propose a digital thermometer using LM35 [2]. P. N. V. Rajya Laxmi et al. in their paper discusses the development of a potential smartwatch for dementia patients [3]. A. Leone et al. in their paper propose the prototype of an NFC-based vital sign monitoring system using Temp100 sensor [4]. L. Yu et al. designs a health monitoring system for elderly people in Hong Kong [5].

In our proposed system, we measure two of the vital signs, heart rate and body temperature, which are essential to monitor health problems [6]. For the health monitoring system design, we use a low-power, cost-effective microcontroller LPC1768 based on ARM Cortex M3 core [7]. Our system has low power consumption, faster response, cost-effectiveness, and accuracy.

The accuracy of the system has been improved by proper adjustment of the threshold voltage in the comparator stage of our sensor module and the selectivity of the system has been increased by cascading the two filter stages. A case study of these health parameters based on gender has been performed and analyzed.

The paper is organized as follows: Sect. 2 presents the entire system architecture including our sensor module designs and interfacing of the embedded processor LPC1768 with the sensor and the other modules. Section 3 describes the experimental setup and methodology of our proposed system. In Sect. 4, the results obtained by monitoring the health parameters of the volunteers are analyzed. Finally, we conclude the paper in Sect. 5.

2 System Architecture and Design

The proposed health monitoring system can measure the heart rate and surface body temperature from the fingertip of the patient, display them on Hyper Terminal emulator and LCD screen and indicate abnormalities in both the health parameters by glowing a red LED and turning ON a buzzer. This system consists of four different modules: the *sensor module* containing both the heart rate and the body temperature sensors with their associated signal conditioning circuitry, the *embedded processor module* containing the ARM Cortex M3 microcontroller LPC1768 and the 12-bit successive approximation ADC as the input port of the microcontroller package, the *display module* containing both the personal computer (PC) terminal emulator, HyperTerminal screen, and a portable 16 × 2 LCD display and lastly the *indicator module* containing a red LED and a buzzer assembly. The entire system representation is given in Fig. 1.

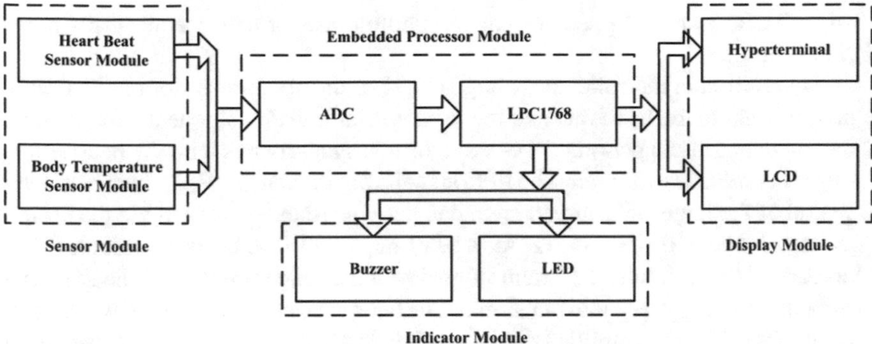

Fig. 1 System block diagram

2.1 Sensor Module

The sensor module is subdivided into the following two sections.

Sensor module for heartbeat measurement. This module consists of an optical sensing mechanism placed beneath the finger and signal conditioning stages to filter and amplify the *photoplethysmography* (PPG) signal to appropriate voltage levels, which is then processed by LPC1768 to display the heart rate [8]. The entire heart rate sensor module circuit is shown in Fig. 2. The sensor module comprising of an optical sensor and signal conditioning are explained as follows.

Optical sensor (TCRT5000). TCRT5000 is a reflective optical sensor comprising an IR LED and a phototransistor in a package that minimizes the ambient light incidence [9]. It is suitable for our purpose as the phototransistor in TCRT5000 has a high current transfer ratio of 1 mA collector current for a LED current of

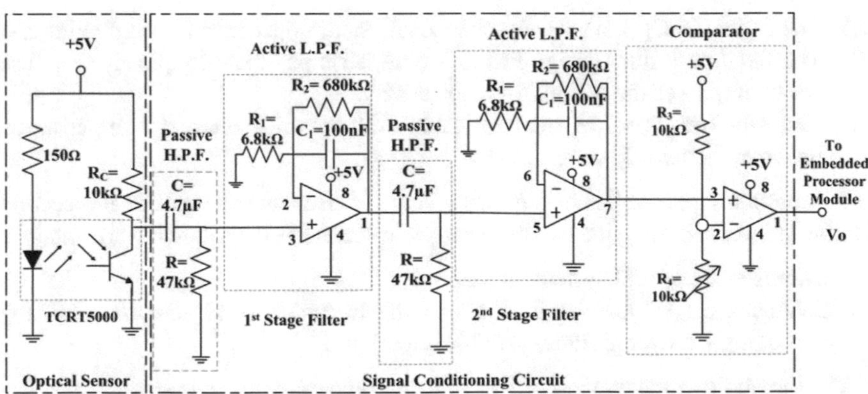

Fig. 2 Heartbeat sensor module

10 mA. TCRT5000 is biased so that the phototransistor can operate in its *active region* (Fig. 2).

This variation in the collector voltage (V_{sen}) of the phototransistor of TCRT5000 is proportional to the heart rate and the weak pulsatile AC component over the large DC component of the periodic PPG waveform is synchronous with the heartbeat.

Signal conditioning circuit. The PPG signal from the sensor (V_{sen}) must be filtered to get rid of the large DC component and to pass heartbeat pulses only in the desired frequency range of 0.72–2.34 Hz (for a heart rate of 43–140 beats per minute). This takes care of the unwanted AC mains signal (~50 Hz) interference and the irrelevant harmonics that may creep into our signal. The heartbeat signal being very weak needs to be amplified. These amplified pulses are then fed into a voltage comparator circuit and converted into a square wave signal enabling easy counting.

The filtering is achieved by using a band-pass filter assembly (Fig. 2) comprised of a passive high-pass filter and an active low-pass filter. The filters are separately designed so that the high-frequency noises are filtered out before the desired AC signal is amplified. The passive high-pass filter has a cut off frequency (f_{ch}) of 0.72 Hz. Taking $f_{ch} = 0.72$ Hz and $C = 4.7\ \mu F$, we get $R = 47\ k\Omega$ using the Eq. (1) [10].

$$f_{ch} = 1/(2\pi RC) \tag{1}$$

An active low-pass filter is used to provide a gain of 101 and having a cut off frequency (f_{cl}) of 2.34 Hz. Taking $f_{cl} = 2.34$ Hz and $C_1 = 100$ nF, we get $R_2 = 680\ k\Omega$ using the Eq. (2) and $R_1 = 6.8\ k\Omega$ from Eq. (3) [10].

$$f_{cl} = 1/(2\pi R_2 C_1) \tag{2}$$

$$A_V = 1 + (R_2/R_1) \tag{3}$$

Our circuit design has the following advantages:

(i) The capacitor C_1 prevents any changes in input impedance from affecting f_{cl}.
(ii) The band-pass filter design has an asymmetric passband frequency response which improves the overall filter selectivity.
(iii) The Op-Amp (LM358 dual Op-Amp IC) provides internal compensation between the two Op-Amps [11].

The output signal of the first filtering stage is used as the input of the second filtering stage. Cascading the two filter stages in series has the following advantages:

(i) Increases the overall system gain and
(ii) Enhances the stopband rejection and the transition band steepness thereby improving the overall filter system's selectivity.

The threshold voltage (V_{Th}) of the voltage comparator stage is set at 2.5 V and using Eq. (4), the potentiometer R_4 is set at 10 kΩ when $R_3 = 10\ k\Omega$ and $V_S = 5$ V.

$$V_{Th} = V_S R_4/(R_3 + R_4) \tag{4}$$

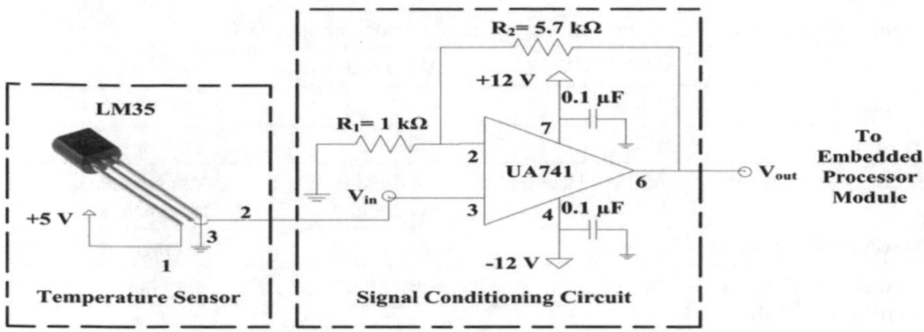

Fig. 3 Body temperature sensor module

The entire signal conditioning stage helps to filter, amplify and shape the PPG pulses so as to eliminate any noise signal from reaching the microcontroller stage, which produces error-free results.

Sensor module for body temperature measurement. The entire circuit for the body temperature measurement of the patient from the fingertip is given in Fig. 3. The different components in this module are discussed below:

Temperature sensor (LM35). LM35 is a precision temperature IC with an output voltage linearly proportional to the temperature in °C [12]. The sensor has low cost, low self-heating, operates within 4–30 V and has low output impedance which makes it suitable to connect it directly to the Op-Amp input in the signal conditioning stage.

Signal conditioning circuit. The signal conditioning stage of the temperature sensor comprises an operational amplifier UA741 [13].

The UA741 is used as a non-inverting amplifier to amplify theLM35 sensor output and is accompanied by noise eliminator components. Bypass capacitors provide low impedance power sources local to the analog circuitry and hence reduce the coupled noise which propagates into the circuit through the Op-Amp and the power pins.

2.2 *Embedded Processor Module*

The microcontroller used in our proposed system is ARM LPC1768 [7]. The embedded processor consists of the educational practice board EPBLPC1768, a stand-alone card allowing developers to evaluate the NXP LPC1768, ARM Cortex M3 based microcontroller as our hardware while we have used Eclipse Kepler 4.3 C/C++ IDE for compiling, debugging and executing our code written in Embedded C language [14, 15]. A comparison of the different microcontrollers is given in Table 1.

Table 1 Comparativestudy of the different types of microcontrollers [16]

	ARM LPC1768	AVR (Arduino)	8051
Buswidth	32-bit	8 or 32-bit	8-bit
ISA	RISC	RISC	CISC
Memory architecture	Modified Harvard architecture	Modified Harvard architecture	Von Neumann architecture
Power consumption	Low	Low	Average
Cost (as compared to features provided)	Low	Average	Very low
CPU frequency	Up to 100 MHz	16 MHz	12 MHz

2.3 *Display Module*

The results are displayed simultaneously on a *16X2 LCD screen* driven by Hitachi HD44780 LCD controller which receives the data from the microcontroller LPC1768 in 8-bit mode and on the *HyperTerminal window* of the personal computer (PC).

2.4 *Indicator Module*

When the results of the measured health parameters are out of the typical range the microcontroller LPC1768 sends a 5 V signal to the onboard buzzer terminal, where a piezo HYDZ buzzer and a red LED are connected, which will turn ON.

3 Experimental Setup and Methodology

The entire setup of our health monitoring system is shown in Fig. 4.

The patient is required to place his/her finger on the optical sensor TCRT5000 of the heart rate sensor module and reset the EPB LPC1768 to start taking the data. While the fingertip is placed over the sensor the volumetric pulsing of the blood inside the fingertip varies the intensity of the reflected beam and this variation of intensity is in accordance with the heartbeat. During ventricular systole, blood volume in the finger's artery increases. So, less light is reflected back on the phototransistor. This causes the phototransistor to conduct less, thereby decreasing the collector current and increasing the collector voltage. Similarly, for less blood volume in the artery, reflected light on the phototransistor is more. Hence, its collector current increases and collector voltage increases. The signal conditioning circuit conditions this signal and sends it to the ADC of the embedded processor module. LPC1768 processes this

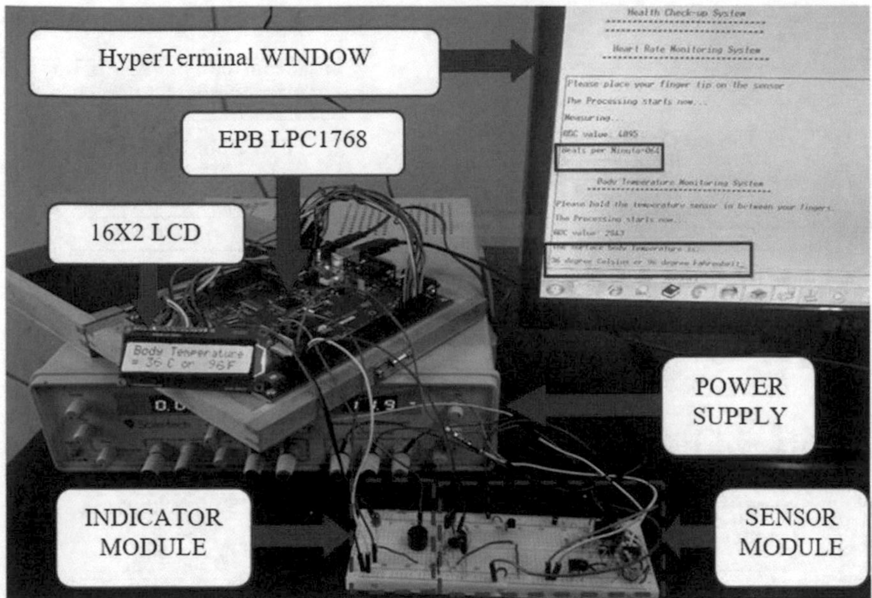

Fig. 4 An experimental setup of the entire health monitoring system

digitized data, counts the number of falling edges of the pulses (count) for 15 s and displays the heart rate as Eq. (5).

$$\text{Heart rate (in beats per minute)} = (\text{count} \times 4) \tag{5}$$

After the heart rate is displayed, the patient is asked to hold the LM35 sensor and the ADC input is connected to the output of the body temperature sensor module. The amplified sensor output is digitized by the ADC and then the surface body temperature of the patient is calibrated and displayed by the processor. If anyone or both of the health parameters are not within typical range, then buzzer and a red LED will turn ON. Figure 5 shows the general flowchart for the entire operation.

4 Results and Analysis

The data obtained by our system is verified by registered medical practitioners—the heart rate is measured from the radial artery manually and the surface body temperature at the fingertip is measured by an analog clinical thermometer. The data measured by the heart rate sensor module of our proposed system is plotted in comparison to the manually obtained data for 55 volunteers (20 female and 35 male) within the age group of 20–52 years, in Fig. 6a and b. Whereas the data measured by

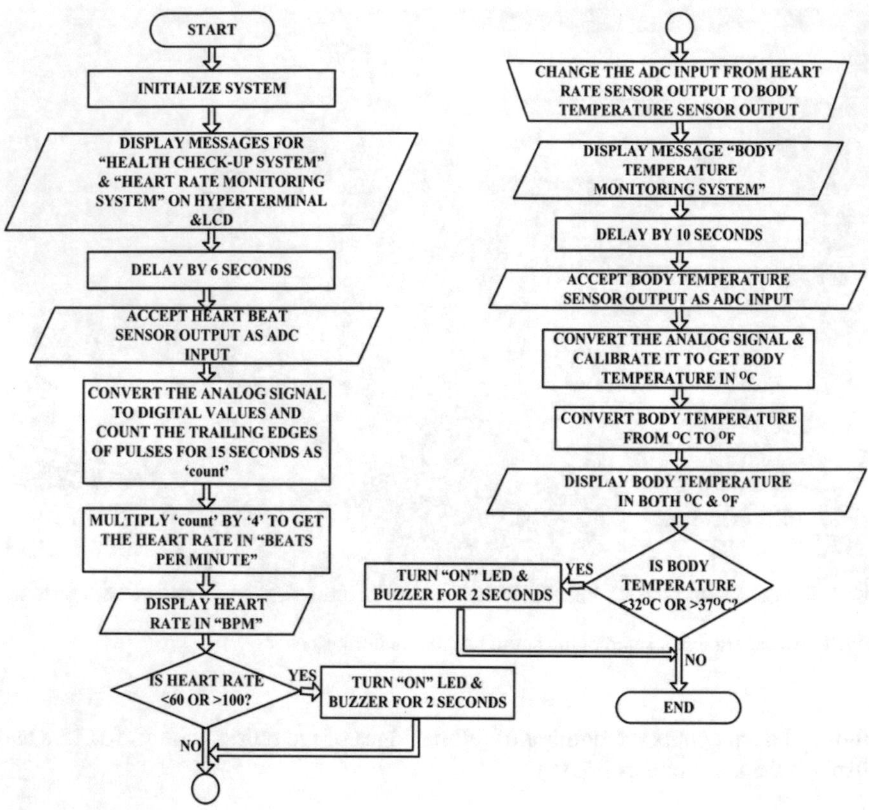

Fig. 5 Flowchart of the entire system operation

the body temperature sensor module of our system is plotted in comparison to the manually obtained data for 10 volunteers (3 female and 7 male) within the age group of 20–48 years, in Fig. 6d. During all the measurements the ambient temperature was in the range 32–34°C. Figure 6c shows the comparison of the average heart rate of male and female volunteers.

These data are validated with literature which state normal adult human heart rate to be within 60–100 beats per minute [17, 18]. Although it has been found to be lesser in athletes and other physically active persons, the condition of heart rate lower than 60 beats per minute is called *bradycardia* in people who are not so physically active. Bradycardia may occur due to the use of beta-blocker medications, like medication to reduce high blood pressure. It is a common symptom of diseases like atrioventricular heart block, bundle branch block, etc. [19]. Similarly, people having heart rate higher than 100 beats per minute are said to suffer from the condition *tachycardia*. Persons showing emotional stress, having a fever or an obese body, or are regular smokers tend to have tachycardia. Higher heart rate may also be observed due to high ambient temperature during the measurement. Tachycardia may include various diseases

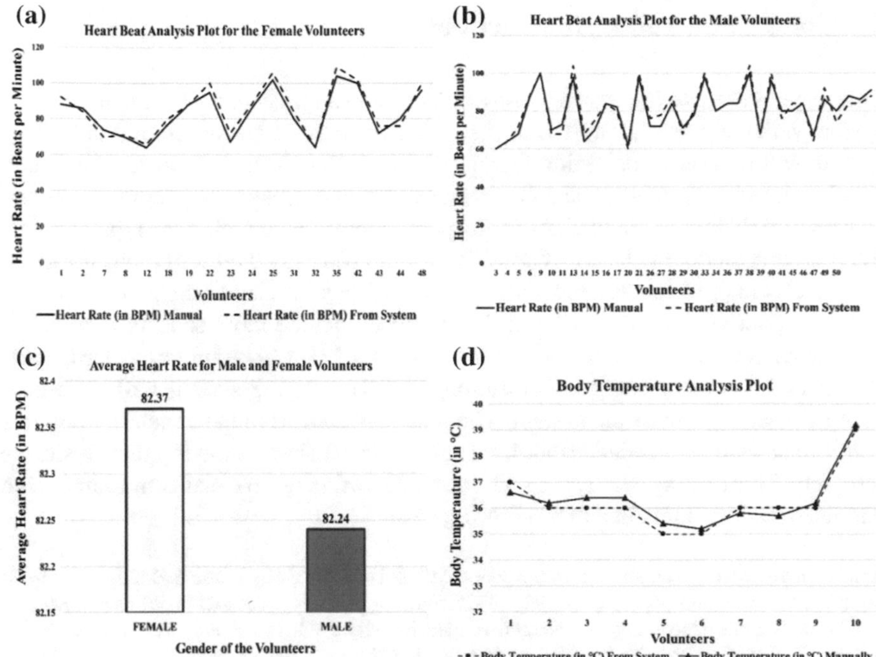

Fig. 6 Results of the health monitoring system. **a** Plot for heart rate of female volunteers; **b** Plot for heart rate of male volunteers; **c** Comparison between female and male heart rate; **d** Plot for body temperature of the volunteers (from a fingertip)

and symptoms like ventricular or supraventricular tachycardia, inappropriate sinus tachycardia, atrial flutter, atrial or ventricular fibrillation, etc. [20].

The heart rate of the female body is found to be *higher* than the male body (Fig. 6c). The probable cause of this may be the fact that the overall female body, as well as, the heart is smaller in size compared to their male counterparts. So, the heart needs to pump faster and circulate the blood more of times through the entire body [21]. However, the variation of the resting heart rate (RHR) with age is very less. They vary only when these persons are exercising or cooling off after exercising [22].

The typical temperature of the human body is 36°C or 98.6°F [23]. But the surface body temperature is found to vary for different body parts. The typical range is set at 32–37 °C [23, 24]. This is because the human skin temperature, when measured from the fingertip, is found to depend on the ambient temperature and this temperature during our measurement was in the range 32–34 °C.

A fever is the body's natural response to disease-specific stimuli. Our immunity system raises the body temperature to defend itself against diseases. Thus, body temperature can give us an insight into our state of health.

During our study, the indicator module has turned ON in six separate cases, five cases show that the volunteer has tachycardia and in one case, the volunteer has a fever.

5 Conclusion and Future Scope

This paper presents a low-power, cost-effective, and high-accuracy health monitoring system with the potential of PCB implementation to make the system portable. This system will be useful for senior citizens and rural people for whom regular health checkup is not accessible. The patients can self-monitor their body temperature and heart rate at their homes and if any of the parameters are not within the typical range, the indicator module will turn "ON". The patient can only then visit a physician for further checkup, which will prevent the hassle of regular visit to the physician.

This system can be improved in the future to include more sensors to measure other essential health parameters like blood pressure, ECG, respiration pattern, SpO_2 (Peripheral Capillary Oxygen Saturation), etc. Hence, the system will become a complete health monitoring system. The data collected from the patient-end by this system can also be uploaded to the cloud which can then be accessed by the physicians remotely. Thus, the system can be incorporated within an IoT platform to make it a patient-friendly remote health monitoring system.

Acknowledgements The authors would like to thank UGC UPE II, "Modern Biology Group B: Signal Processing Group", the University of Calcutta for providing research facilities and SMDP-C2SD project, the University of Calcutta, funded by MeitY, Govt. of India, for their technical support. The authors are also thankful to Antara Chaudhuri, student of Institute of Radiophysics and Electronics, for her kind help in this work and West Bengal Higher Education, Science and Technology and Biotechnology (Sci. & Tech.) funded project "Cytomorphic CMOS Circuit Modeling and Ultra-Low Power Design of P53 Protein Pathway for Synthetic Biology Applications " for partial support of infrastructure. The authors of the paper are thankful to all the student volunteers and the faculty members of the Institute of Radio Physics and Electronics, the University of Calcutta for their active support to conduct our research work. There is no conflict of interest.

References

1. Das, S., Pal, S., Mitra, M.: Real time heart rate detection from ppg signal in noisy environment. In: International Conference on Intelligent Control Power and Instrumentation, pp. 70–73 (2016)
2. Changela, B., Parmar, K., Daxini, B., Sanghani, K., Shah, K.: Digital thermometer: design & implementation using arduino UNO based microcontroller. Int. J. Sci. Res. Dev. **4**(03), 840–843 (2016)
3. Rajya Lakshmi, P.N.V., Venkatesan, P., Prasad, G.R.: Smart watch for healthcare monitoring. Int. J. Eng. Technol. Sci. Res. **5**(3), 1107–1111 (2018)
4. Leone, A., Rescio, G., Siciliano, P.: An open NFC-based platform for vital signs monitoring. In: XVIII AISEM Annual Conference (2015)
5. Yu, L., Chan, W.M., Zhao, Y., Tsui, K.L.: Personalized health monitoring system of elderly wellness at the community level in hong kong. IEEE Access https://doi.org/10.1109/access.2018.2848936
6. Signs, V.: Johns Hopkins Medicine: Health Library. https://www.hopkinsmedicine.org/healthlibrary/conditions/cardiovascular_diseases/vital_signs_body_temperature_pulse_rate_respiration_rate_blood_pressure_85,P00866. Accessed 1 July 2018

7. NXP Semiconductors.: LPC1769/68/67/66/65/64/63, Rev.9.7. https://www.nxp.com/docs/en/data-sheet/LPC1769_68_67_66_65_64_63.pdf (2017)
8. Hu, S., Zheng, J., Chouliaras, V., Summers, R.: Feasibility of imaging PPG. In: International Conference on BioMedical Engineering and Informatics, pp. 72–75 (2008)
9. Vishay Semiconductors: Reflective Optical Sensor with Transistor Output, Rev 1.1. https://www.vishay.com/docs/83760/tcrt5000.pdf (2009)
10. Sedra, A.S., Smith, K.C.: Microelectronic Circuits: Theory and Applications, 7th edn
11. Texas Instruments: LMx58-N Low-Power, Dual-Operational Amplifiers. https://www.ti.com/lit/ds/symlink/lm358a.pdf (2014)
12. Texas Instruments: LM35 Precision Centigrade Temperature Sensors, Rev H. https://www.ti.com/lit/ds/symlink/lm35.pdf (2017)
13. Texas Instruments: UA741 General-Purpose Operational Amplifiers datasheet, Rev. G. https://www.ti.com/lit/ds/symlink/ua741.pdf (2018)
14. Edutech Learning Solutions Pvt. Ltd.: Educational Practice Board LPC1768, Rev 1.0. https://www.edutechlearning.com/learningresoursedesc/electronics-engineering/kits-tools/educational-practice-board-for-arm-cortex-m3 (2013)
15. Miranda, T.: 15 years of open source embedded tools at eclipse. https://www.eclipse.org/community/eclipsenewsletter/2017/october/article2.php. Accessed 1 May 2018
16. Firmcodes.: Difference Between ARM and Other Microcontrollers. http://www.firmcodes.com/difference-arm-microcontrollers/. Accessed 7 Aug 2018
17. Mishra, T.K., Rath, P.K.: Pivotal role of heart rate in health and disease. J. Indian Acad. Clin. Med. **12**(4), 297–302 (2011)
18. American Heart Association: All About Heart Rate (Pulse). http://www.heart.org/HEARTORG/Conditions/HighBloodPressure/GettheFactsAboutHighBloodPressure/All-About-Heart-Rate-Pulse_UCM_438850_Article.jsp#.Wz811dIza00. Accessed 03 May 2018
19. Mayo Clinic: Bradycardia. https://www.mayoclinic.org/diseases-conditions/bradycardia/symptoms-causes/syc-20355474. Acccssed 5 May 2018
20. Mayo Clinic: Tachycardia. https://www.mayoclinic.org/diseases-conditions/tachycardia/symptoms-causes/syc-20355127. Accessed 05 May 2018
21. Nealen, P.M.: Exercise and lifestyle predictors of resting heart rate in healthy young adults. J. Hum. Sport. Exerc. **11**(3), 348–357 (2016)
22. Kostis, J.B., Moreyra, A.E., Amendo, M.T., Di Pietro, J., Cosgrove, N., Kuo, P.T.: The effect of age on heart rate in subjects free of heart disease. Circulation **65**(1), 141–145 (1982)
23. Denton, E.: Learn how to measure body temperature accurately and cost effectively. https://www.ti.com/lit/ml/slyw051/slyw051.pdf (2015)
24. Bierman, W.: The temperature of the skin surface. JAMA **106**(14), 1158–1162 (1936)

A Robust Multi-label Fruit Classification Based on Deep Convolution Neural Network

Biswajit Biswas, Swarup Kr. Ghosh and Anupam Ghosh

Abstract The problem of multi-label classification by machine learning algorithms such as support vector machine (SVM) and convolutional neural network (CNN) is that a data sample can be classified into a multi-label class cannot be categorized into any class. Automated multi-label classification and recognition of fruit images play a crucial role in decision-making scheme in agro-based applications. In this paper, we proposed a fruit recognition and classification scheme based on deep convolutional neural network (CNN). The CNN is applied to the tasks of fruit detection and recognition through parameter optimization. To validate the proposed scheme, four normal fruits are considered, i.e., apple, lemon, tomato, and plum in testing. In this experiment, we demonstrate that the accuracy is improved by the CNNs over the conventional SVM and other classification methods. The result of the proposed method has good accuracy in the classification and it is close to 98% for the four classes of 1200 fruit images, indicating that the proposed method could be used in agro-based applications.

Keywords Computer vision · Convolution neural network · Support vector machine · Multi-label fruit classification

B. Biswas
University of Calcutta, Kolkata 700098, India
e-mail: biswajit.cu.08@gmail.com

S. Kr. Ghosh (✉)
Brainware University, Barasat, Kolkata 700124, India
e-mail: swarupg1@gmail.com

A. Ghosh
Netaji Subhash Engineering College, Kolkata 700152, India
e-mail: anupam.ghosh@rediffmail.com

A. K. Das et al. (eds.), *Computational Intelligence in Pattern Recognition*,
Advances in Intelligent Systems and Computing 999,
https://doi.org/10.1007/978-981-13-9042-5_10

1 Introduction

Computer vision is a wide research field for several areas such as video surveillance systems, face recognition, fingerprint recognition, handwriting recognition, virtual reality systems, medical imaging, and several classification problems. Recently, fruit and vegetable classification is a challenging task on computer vision since fruit and vegetables have similar shapes, texture, and colors as features [3, 9]. It takes too much time to classify and count the fruit and vegetables for labeling before sale. Automatic classification of fruit is a multi-label classification problem which becomes difficult for traditional computer vision system [8, 12]. Machine learning (ML) becomes a popular technique in the field of automatic image recognition and classification for the past decade. Hence, an artificial neural network (ANN)-based machine learning has solved the problem of multi-label classification in computer vision [7, 13].

In recent years, many researchers have worked together in the field of fruit and vegetables recognition and classification. A multiple-feature-based algorithm for detecting the orientation of the fruit object suggested by Hetal et al. [9]. Ninawe et al. (2014) described a fruit classification method by using K-Nearest Neighbors (KNN) classifier with a texture of images as a new feature [8]. Li Wu (2012) explored a fruit classification method form on multi-class support vector machine [11]. Zhang et al. (2014) presented a methodology of fruit classification with good accuracy based on fitness-scaled chaotic bee colony algorithm and feedforward neural network to which they added more features in accounting such as colors, shape, texture, etc for good result [14] but it still suffers on computational cost. In the meanwhile, a random-forest-based food recognition and classification method is furnished by Lukas et al. [1]. They performed an experiment on the algorithm on a large food and non-food image database taking discriminative components in account of random forest but it increases computation time. Guoxiang Zeng (2017) developed a fruit and vegetable classification scheme by using image saliency and convolutional neural network in which he had used the famous VGG net for classification and achieved 95.6% of accuracy [13]. Recently, deep CNN architectures developed by Chen et al. [2] to extract the spectral, spatial, and spectral-and-spatial-based deep features from hyperspectral images and $L2$ regularization used to overcome the overfitting but the algorithm was developed only on HSI images.

As discussed in literature, all methods suffer from computational cost as well as accuracy since the multi-class problem for large data takes so much time to implement in computer. To overcome the problems such as classification accuracy, computation complexity, and high cost of hardware devices on traditional machine learning (ML), or the drawbacks of traditional computer vision algorithm, lots of research are flourishing on classification and recognition by using deep learning. Recently, deep convolutional neural network (CNN) [5] plays a crucial role in the field of classification because it offers a smaller amount of memory, decreases computation complexity, and provides high accuracy [15]. The journey of CNNs has been started since 2012 AlexNet and then CNNs makes excellent success in several models such as VGGNet, GoogLeNet, and ResNet [6].

In this paper, the basic method of using convolutional neural network (CNN) for multi-label fruit classification (four classes) such as apple, lemon, tomato, and plum followed by fruit recognition has been put forward. For comparison purpose, we have designed multi-class support vector machine (SVM) for fruit and vegetable classification.

The rest of this paper is framed as follows. Section 2 represents the proposed multi-label classification CNN architecture. Experimental results, description of databases, and discussions have demonstrated in Sect. 3. Finally, conclusion along with future work is given in Sect. 4.

2 Proposed Multi-class Convolutional Neural Network (CNN)

In this section, the proposed robust multi-label fruit classification scheme in the agro imagery by using convolutional neural network (CNN) in deep learning framework is illustrated in brief. The schematic diagram of the proposed method is shown in Fig. 1.

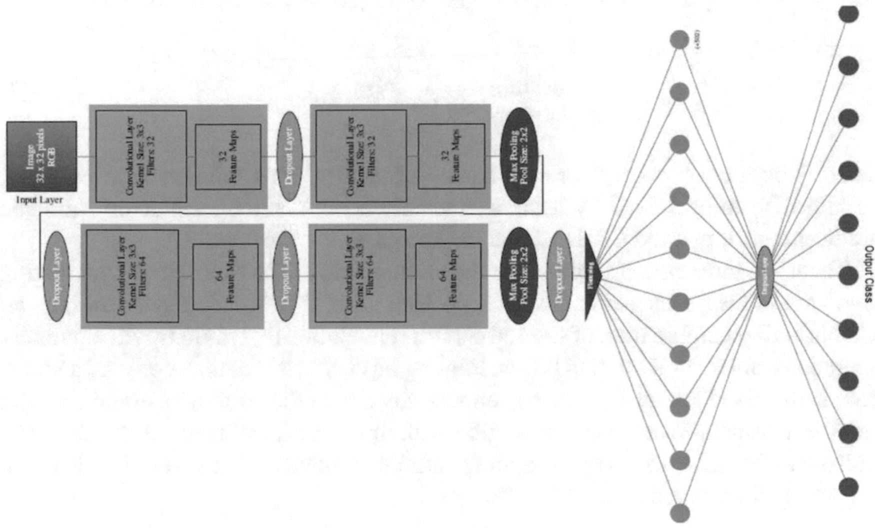

Fig. 1 Schematic diagram of the proposed model

2.1 Design of Convolution Neural Network

In practice, different multi-level classification tasks are often solved by various CNNs in literature [5]. Suppose, given an N training samples $\{(\mathbf{x}_i, \mathbf{y}_i)\}_{i=1}^{N}$ belong to the $\mathbf{K}$ number of classes, where $\mathbf{x}_i$ represents sample data point and $\mathbf{y}_i$ denotes the relevant feature labels [10]. Therefore, the CNN can measure a complex mapping in relation between input vector $\mathbf{x}_i$ and output vector $\mathbf{y}_i$ by a predefined model $\mathbf{F}$ in the training phases. The entire training process is completed under two learning phases, first is the forward propagation step and second is backpropagation step [15].

Next, the given train samples $(\mathbf{x}_i, \mathbf{y}_i)$ are fed as input to neural network then the processed sample $\mathbf{x}_i$ is transferred from input layer to final layer with back-to-back steps. Hence, for an expected output $\mathbf{o}_i$, the entire learning process of the network is mathematically expressed as follows [6]:

$$\mathbf{O}_i = \mathbf{F}_L\left(\mathbf{F}_{L-1}\left(\mathbf{w}_{L-1}^T\mathbf{x}_i + \mathbf{b}_{L-1}\right)\right) \circ \left(\mathbf{F}_1\left(\mathbf{w}_1^T\mathbf{x}_i + \mathbf{b}_1\right)\right) \tag{1}$$

where $\mathbf{L}$ denotes the number of neuron layers and $\mathbf{w}_i$ is the weighted vector at ith layer in $\mathbf{F}_i$. In training phases, the CNN performs several operations such as convolution with different kernel functions, max pooling, and activation in the series of layers in $\mathbf{F}_i$. After completion of these operations, the estimated weighted vector $\mathbf{w}_1, \mathbf{w}_2 \ldots \mathbf{w}_L$ will be solved by a minimization of objective functions [5] and it is defined as

$$\underset{\mathbf{w}_1, \mathbf{w}_2 \ldots \mathbf{w}_L}{\text{argmin}} \frac{1}{N} \sum_{i=1}^{N} \xi\left(o_i, y_i\right) \tag{2}$$

where ξ be a cross entropy or sum-squared function. This minimization problem, i.e., Eq. (2) can be solved by using a gradient decent method such as probabilistic gradient decent method (PGDM) and backpropagation process [10].

There are three primary blocks in the architecture of CNNs such as convolution layer, a nonlinear transformation, and pooling layer [5]. A deep CNN can be well established by combination of stacked certain convolution layers along with nonlinear operations and several pooling layers. Pooling layer consists of max pool and average pool with a fixed kernel size, we can choose any one of the pooling operation. CNNs have been structured as local connection with three blocks and shared weights. The CNNs are inclined to serve an appropriate performance. A convolutional neural network is formulated as [5, 15] follows:

$$\mathbf{x}_s^t = g\left(\sum_{r=1}^{\mathbf{N}} \mathbf{x}_r^{t-1} * \mathbf{k}_{rs}^t + \mathbf{b}_s^t\right) \tag{3}$$

where matrix $\mathbf{x}_r^{t-1}$ is the rth feature map of the preceding $(t-1)^{th}$ layer, the sth feature map of the current t^{th} layer is denoted as $\mathbf{x}_s^t$, and the number of input is $\mathbf{N}$. $\mathbf{k}_{rs}^t$ and $\mathbf{b}_s^t$ are weight and bias parameter which initialized with zero. Then, they are fine-tuned through backpropagation. $g(\cdot)$ and $*$ are represented as nonlinear function and the convolution operator, respectively.

2.2 Loss Function

In order to adjust the loss of trained model for multi-label fruit classification, we have used the pixel-wise cross-entropy loss $\Phi(x)$ that is defined as [10] follows:

$$\Phi(x) = \frac{-1}{N} \sum_{i=1}^{\mathbf{C}} \left(\sum_{x \in S_i} \log \left(\hat{p}_i \left(x \right) \right) \right) \tag{4}$$

where $\hat{p}_i(\cdot)$ for each pixel is the softmax class probabilities determined as

$$\hat{p}_i(\cdot) = \frac{e^{(-|x_i(\cdot)|)}}{\sum_{j=1}^{\mathbf{C}} e^{(-|x_j(\cdot)|)}} \tag{5}$$

$\mathbf{N}$ denotes the total number of pixels $x \in \mathbf{I}$, $\mathbf{S}_i$ represents the set of pixels within one class in $\mathbf{L}_n$, and i is the ground truth class label where $i \in [1, 2, \ldots, \mathbf{C}]$.

The loss function $\Phi(x)$ defined in Eq. (4) generates the real-valued output predictions and the input of the loss is the output of the preceding convolutional layer in the network as $\mathbf{x} \in [-1, +1]$. In this work, we have used the above loss function $\Phi(x)$ for all data sets in experiment.

2.3 Dropout, ReLU, and Batch Normalization

The overfitting is a common problem in deep neural network due to the huge dimension of dense layers along with large size batches and small size of training examples. Although the famous $L1$ and $L2$ regularization can handle the overfitting issue, a recent dropout method has been suggested to overcome overfitting [5]. The dropout is nothing but a predefined cutoff probability network in dense layer. The "ReLU" is a nonlinear activation function which accepts the output of a neuron if it is positive, else returns zero (0) for negative output. ReLU has some advantages than other conventional activation function which is more efficient in gradient propagation, as well as sparse activation, and low computation load.

Furthermore, to upgrade the performance of the network, we have adopted batch normalization (BN) technique in this work [10]. BN has been used in the network to mitigate the problem caused by inadmissible network initialization in the training phase. BN can also efficiently speed up the training procedure by prohibiting "gradient vanishing" [10].

To design the proposed network, we have used three convolutional layers, three ReLU layers, and three pooling layers. In the training phase, we have fixed 0.5 as a dropout ratio and the stride value set to $s = 2$. Finally, to produce the corresponding labels for four-class classification, all the nodes are furnished with fully connected layers (f_c) followed by a "softmax" classifier defined in Eq. (5).

2.4 Building the CNN for Classification Scheme

In this study, we have used Keras with backend Tensorflow library in the proposed scheme. The data set comprises 300 and 400 training examples, and 140 and 160 test examples of the fruits classification from databases "FIDS30" and "Fruits-360" [16, 17], formatted as 128×128-pixel color images. Full designed CNN architecture is defined in Eqs. (1) and (3). Let us build a model to classify the fruit images in the data set using the following multi-class CNN architecture:

1. Conv2D Layer #1: Uses 32, 3×3 filters, with ReLU activation function;
2. Pooling Layer #1: Executes max pooling with a 2×2 filter and stride $s = 2$;
3. Conv2D Layer #2: Uses 64, 3×3 filters, with ReLU activation function;
4. Pooling Layer #2: Executes max pooling with a 2×2 filter and stride $s = 2$;
5. Dense Layer #1: A total $1,024$ neurons, with dropout ratio set to 0.5;
6. Dense Layer #2 (FCL): A total 4 neurons, one for each pixel as target class.

3 Experimental Result and Discussion

The effectiveness of our algorithm, along with comparisons followed by description of databases, has been shown in this section. For simulation purpose, we have considered four-label fruit classification, viz., apple, lemon, tomato, and plum followed by recognition.

3.1 Dataset

We have used benchmark publicly available fruit image databases "FIDS30" and "Fruits-360" [16, 17] for the effectiveness and comparison of the proposed algorithm. The "FIDS30" database consists of 971 common fruit images which are classified into 30 different classes and each class contains 32 fruit images [17]. The fruit images are located in JPEG format with standard size and images are very diverse in nature. The number of fruit in a single image is variable, it may be single fruit or more fruit or fruit with leaves in an image. For the training and verification of the network in the proposed scheme, we have considered only four classes, viz., apple, lemon, tomato,

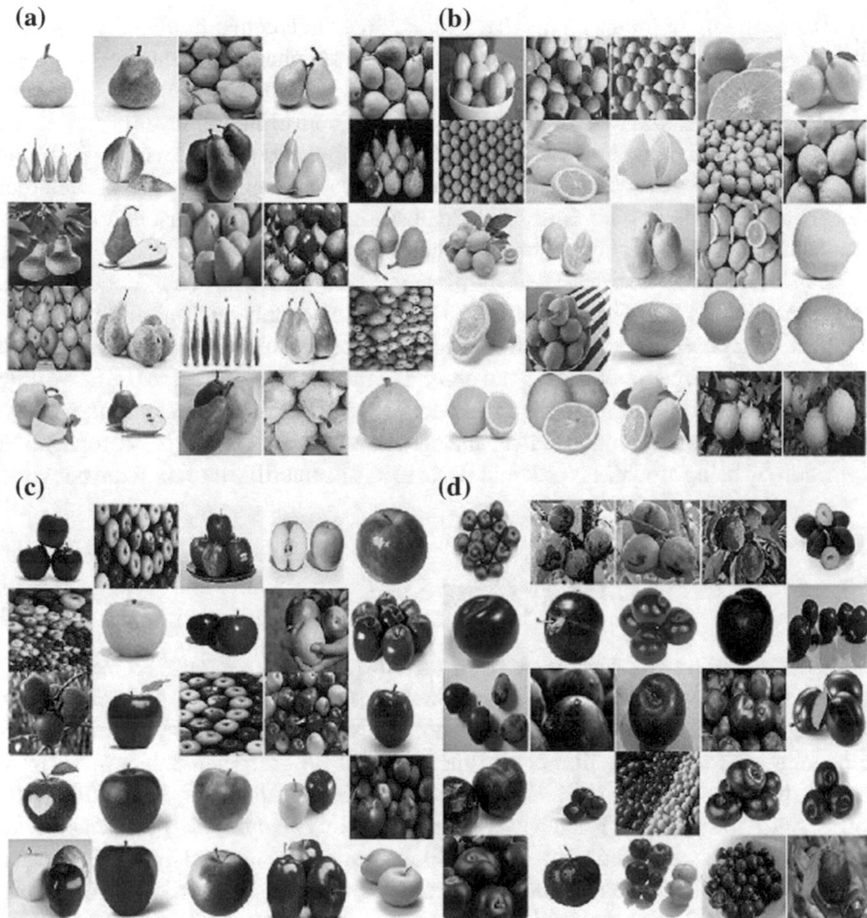

Fig. 2 Sample fruit images of four classes

and plum from the FIDS30 fruit database manually. A sample four-class fruit images such as apple, lemon, tomato, and plum are shown in Fig. 2.

The "Fruits-360" database consists of 55,244 common fruit images which are classified into 81 different classes and data set is divided into training and test images [16]. We have prepared manually four-class fruit database for the experimental purpose.

3.2 Implementation of CNN for Fruit Classification

In this paper, we have designed fruit classification CNN model with eight layers at which six is hidden layers and the last two are dense layers. The network comprises convolution, max pooling, and ReLU function except dense layer. The first input layer takes the input images with size 256 × 256 × 3. The first hidden layer is built on convolutional operation followed by an input layer which includes 96 feature maps of size 55 × 55. Total 96 kernels of size 11 × 11 × 3 have been utilized for each feature map in the first hidden layer. The first convolutional layer produces the output and second convolutional layer takes it as input and filters it with 256 kernels of size 5 × 5 × 48. The rest of the convolutional layers is connected with one another without any interceding pooling or normalization layers. The four-class labels are defined in the last dense layer which produces the final output.

The simulation of the proposed multi-class CNN model has been conducted on the NVIDIA GTX 950 GPU framework in python environment. To build the model in the training phase, we initialize the weights and biases of each layer with a random number. The hyperparameter has been chosen as the learning rate set to 0.01, learning rate decay to 0.005, the momentum is chosen 0.9, the weight decay is fixed with 0.0001, 50 is early stopping patience, and epochs is set to 1500. The backpropagation algorithm by using stochastic gradient descent has been utilized to train the network.

3.3 Fruits Recognition by Using SVM

Many classification problems have been solved by support vector machines (SVM) and hence it is used to solve multi-label classification problem [4, 11]. In this study, we have designed a four-class classification SVM structure along with one versus remaining approach and radial basis function (RBF). For implementation, we have used Python environment with Scikit package in GPU framework. Here, “FIDS30” and “Fruits-360” [16, 17] databases have utilized to analyze the performance. In FIDS30, 300 are employed to train and 140 for test data from total 440 fruit images. Similarly, out of 560 images, 400 are selected for training data and rest 160 as test data for Fruits-360 database. Table 1 demonstrates the performance of designed SVM classification and which are 84.38 and 82.62% for the mentioned two databases.

Table 1 SVM and CNN classification accuracy

Fruits data with SVM and CNN classification rates

Data set used	Total number of samples	Number of training sample	Number of test sample	Accuracy SVM (%)	Accuracy CNN (%)
Fruits data1 (FIDS30)	440	300	140	84.38	98.32
Fruits data2 (Fruits-360)	560	400	160	82.62	98.63

3.4 Fruits Recognition by Using CNNs

In this work, we have designed a multi-class CNN model for fruit recognition as well as multi-class classification is shown in Fig. 1. The proposed CNN scheme has recognized the fruits object first and then has classified fruits into four classes which is shown in Figs. 3 and 4. Thus, for object classification, we figure out the mean score of two different tasks. The results on fruit data sets are shown in Table 1.

After simulation, we can conclude that the proposed method gives better result than other methods especially patch-based methods for quite a lot. The designed CNN architecture even outperforms the classification during training, which shows the efficiency of CNNs in fruit classification. It is clear from Table 1, proposed scheme achieves 98.63 and 98.32% increment on fruit data sets FIDS30 and Fruits-360,

Fig. 3 Comparison of classification results by **a** SVM method **b** CNN method

Fig. 4 Comparison of classification results by **a** SVM method **b** CNN method

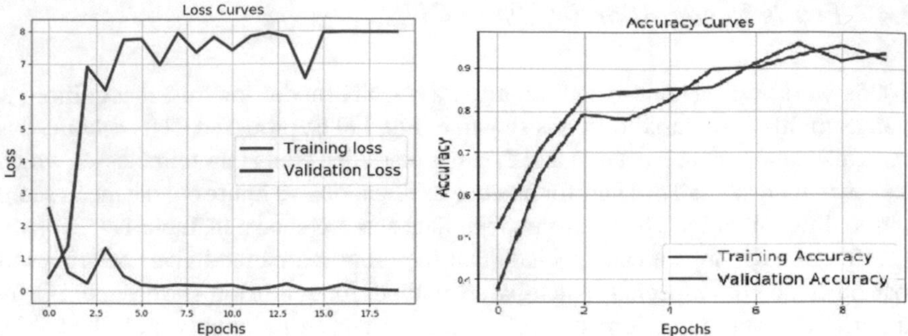

Fig. 5 Performance measurement

respectively. These results demonstrate that the proposed scheme can be achieved the state-of-the-art performance on object recognition as well as classification. On fruit data sets, the best result has been reported in literature which is the assemblage of CNNs than SVM, which achieves 84.38 and 82.62%, but it just simply averages the predicted score by two methods. The performance measure of the proposed CNN method has been shown in Fig. 5.

4 Conclusion

In this paper, an automated multi-class fruit classification and recognition system based on deep convolution neural network (CNN) has been illustrated. Here, we have considered multi-class SVM with proposed CNN-based classification scheme to perform the simulation and comparison process which is performed by using graphic processing unit. To simulate the designed classification and recognition scheme, huge data are required for training purpose and hence we have built an image database for processing since accuracy of deep learning is directly proportional to the size of training data. Finally, we have introduced a vision-based automated decision-making systems and control system in agro-based application. The proposed scheme provides high accuracy which is close to 98% for the four-class fruit classification and recognition. The main advantage of this method is the fact that it improves the computational cost for the multi-class problem and this indicates that it could be used for vision-subsystem-based control applications.

In future, we will represent an automated multi-label fruit classification model for all 30 and 81 classes for the given databases FIDS30 and Fruits-360, respectively, on GPU framework for the development of agro-industry. We will also predict the probable harvesting time and also predict whether the fruit will be going to rot prematurely.

References

1. Bossard, L., Guillaumin M., Gool, L.V.: Food-101 mining discriminative components with random forests. In: European conference on computer vision, pp. 446–461 (2014)
2. Chen, Y., Jiang, H., Li, C., Jia, X., Ghamisi, P.: Deep features extraction and classification of hyspectral images. IEEE Trans. Geosci. Remote Sens. **54**(10), 6232–6251 (2016)
3. Fisher, R.A.: Classification of fruits and vegetables. J. Food Compos. Anal. **22**, S23–S31 (2009)
4. Hsu, C.W., Lin, C.J.: A comparison of methods for multi-class support vector machines. IEEE Neural Netw. **13**, 415–425 (2002)
5. Krizhevsky, A. Sutskever, I., Hinton, G.E.: Imagenet classification with deep convolutional neural networks. In: Advances in Neural Information Processing Systems, pp. 1097–1105 (2012)
6. Liu, W., Wang, Z.: A survey of deep neural network architectures and their applications. Neurocomputing **234**, 11–26 (2017)
7. Muresan, H., Oltean, M.: Fruit recognition from images using deep learning. Acta Univ. Sapientiae Inform. **10**(1), 26–42 (2018)
8. Ninawe, P., Pandey, S.: A completion on fruit recognition system using k-nearest neighbors algorithm. Int. J. Adv. Res. Comput. Eng. Technol. **3**(7) (2014)
9. Patel, H.N., Jain, R.K., Joshi, M.V.: Fruit detection using improved multiple features based algorithm. Int. J. Comput. Appl. **13**(2), 1–5 (2011)
10. Srivastava, N., et al.: Dropout: a simple way to prevent neural networks from overfitting. J. Mach. Learn. Res. **15**(1), 1929–1958 (2014)
11. Wu, L.: Classification of fruits using computer vision and a multiclass support vector machine. Sensors **12**, 12489–12505 (2012)
12. Yang, C., Lee, W.S., Williamson, J.G.: Classification of blueberry fruit and leaves based on spectral signatures. Biosyst. Eng. **113**, 351–362 (2012)
13. Zeng, G.: Fruit and vegetables classification system using image saliency and convolutional neural network. In: IEEE 3rd Information Technology and Mechatronics Engineering Conference (ITOEC) (2017)
14. Zhang, Y., Wang, S., Ji, G., Phillips, P.: Fruit classification using computer vision and feedforward neural network. J. Food Eng. **2014**, 167–177 (2014)
15. Zhang, G., Chen L., Ding, Y.: A multi-label classification model using convolutional netural networks. In: 29th IEEE conference on Chinese Control And Decision Conference (2017)
16. Database available: https://www.kaggle.com/moltean/fruits
17. Database available: http://www.vicos.si/Downloads/FIDS30

Classification of Speech and Song Using Co-occurrence-Based Approach

Arijit Ghosal, Suchibrota Dutta and Debanjan Banerjee

Abstract Discriminating song and speech from audio is an exigent problem. This is a step toward self-executing categorization in case of audio signal. Foregoing attempts were mostly involved for discriminating speech with nonspeech but relatively not much works were involved for classifying speech and song from audio signal. Mainly perceptual and frequency depended features were associated with the foregoing attempts. Song, whether it is associated with instrument or not, reveals some type of periodicity, whereas this periodicity is absent in case of normal speech. For accurate study of these periodic nature textural features based on its co-occurrence matrix along with its mean and standard deviation are also considered. Support Vector Machine (SVM), Neural Network (NN), and k-Nearest Neighbor (k-NN) have been brought into play for the purpose of taxonomy of speech from song. Speech and song classification precision obtained in this work has been compared with that of some other previous works done to reveal effectiveness of the advised feature set.

Keywords Speech and song classification · Periodicity · Co-occurrence matrix

---

A. Ghosal (✉)
Department of Information Technology, St. Thomas' College of Engineering and Technology, Kolkata 700023, West Bengal, India
e-mail: ghosal.arijit@yahoo.com

S. Dutta (✉)
Department of Information Technology and Mathematics, Royal Thimphu College, Thimphu, Bhutan
e-mail: suchibrota@gmail.com

D. Banerjee
Department of MIS, Sarva Siksha Mission Kolkata, Kolkata 700042, West Bengal, India
e-mail: debanjanbanerjee2009@gmail.com

A. K. Das et al. (eds.), *Computational Intelligence in Pattern Recognition*, Advances in Intelligent Systems and Computing 999,
https://doi.org/10.1007/978-981-13-9042-5_11

1 Introduction

Quantity of multimedia data is gradually rising every day. Multimedia data includes audio, video, and image. As this repository is increasing, they need to be stored properly so that they can be retrieved properly in future. In this work, only audio data, specifically speech and song have been considered. For proper storing and retrieval of audio data, concept of classification is required based on certain unique features of speech and song. Speech and song discrimination finds its utility in some important purposes reminiscent of audio retrieval, audio indexing, instrument and music taxonomy, etc. Song and speech are two fundamental types of aural information. Hence, sorting of aural information for easy future retrieval may be initiated through classifying speech along with song. Speech and song differentiation is extremely motivating and admired research area to explore as it finds its usage in media services, search engines as well as in intellectual computer systems. Classification of speech and song may be judged as an essential as well as fundamental footstep toward implementation of self-activating speech identification scheme. Development in fields of data mining and processing of signal is also one of the causes of enrichment of investigating effort of aural information recovery. Though speech and song classification is a very potential research field, still a very few work have been carried out in this area till now. These few preceding works were carried out mostly using frequency and perceptual domain audio facets. Pitch, spectral flux, rhythm as well as spectral roll-off depended audio traits were involved in those efforts. As pitch-based features are very common for speech signal processing, this work aims to propose a low-dimensional and computationally simple audio feature set which do not use pitch depended traits for classifying speech from song.

In this work, brief description related to preceding works is illustrated in Sect. 2. Proposed methodology for classifying speech and song is described in Sect. 3. Experimental results in comparison with other works are explained in Sect. 4 followed by conclusion.

2 Related Works

Haralick [1] has proposed statistical features extraction method in the domain of image processing. These statistical features can be utilized in other domains also. Gerhard [2, 3, 4] has carried out some experiments in this domain. Pitch-based audio features have been proposed by Gerhard [2] to classify speech and song. Bugatti et al. [5] have performed a comparison study between neural and statistical approaches to discriminate speech and music. For fuzzy speech–song classification, Gerhard [3] has used perceptual features. His feature set was based on pitch. Effect of rhythm and silence for the purpose of analyzing speech and song has been studied by Gerhard [4]. Features of song-specific singing voice have been studied by Tzanetakis [6]. He has considered classification decision taken in every 20 ms. For classifying audio signals into some basic categories, a real-time-based classification scheme has been proposed by Lin and Chen [7]. Co-occurrence matrix based features have been conferred by Umbaugh [8]. Co-occurrence matrix based features are very helpful for minutely studying the characteristics of a certain feature.

Harmonic structure modeling has been used by Zhang and Zhang [9] to separate musical waves. Ruinskiy along with Lavner [10] have suggested one process to detect sounds of breath in speech as well as in song signals. An algorithm based on decision tree has been proposed by Lavner and Ruinskiy [11] to classify and segment music and speech. By applying features based on histogram equalization, audio signal is classified into music, speech, and song by Gallardo-Antolín and Montero [12]. Classification of song and speech using rhythmic and melodic patterns has been studied by Salselas and Herrera [13]. Spectral features have been used by Sonnleitner et al. [14] to identify speech from mixed audio signal. Bhavsar and Panchal [15] have done a review to find the applicability of Support Vector Machine (SVM) to classify data. Diverse schemes to differentiation of music and speech have been reviewed by Velayatipour and Mosleh [16]. Ramalingam and Dhanalakshmi [17] have proposed features calculated using wavelet to discriminate music and speech.

3 Proposed Methodology

Past research works reveal that mostly frequency-domain features are used while handling speech audio signal. Moreover, pitch is a perceptual domain feature which has been employed by several researchers during processing of speech signal. Too much usage of frequency-domain features and pitch-based features has created an opportunity to search for alternative acoustic features during processing of speech signal. In this work, the aim is set to propose a short dimensional aural trait independent of frequency-based traits and pitch-based traits to classify speech as well as song.

Songs can be sung with instrument as well as without instrument. If it is sung without instrument it appears like by speech but there are lots of differences between them. When a person sings, the person repeats some part of the song after a certain interval. This repetition is the main characteristics of song. Whatever way a song is sung this recurrence phenomenon is unavoidable. This repetition nature of song makes it different from speech. In speech, this repetition characteristics is absent. This surveillance has stimulated to consider periodicity as a trait. Moreover, there is a basic difference between articulation and song irrespective of whether melody is sung with the support of instrument or not. Song always maintains some rhythm which is absent in case of speech. This dissimilarity can be measured through skewness. Skewness has the capability to represent the asymmetry present between speech and song. Based on this observation skewness is also considered as a feature for discriminating speech and song. Computation of audio features and classification of audio signal into speech and song are explained in the following subsections. The steps of the proposed plan are explained in Fig. 1.

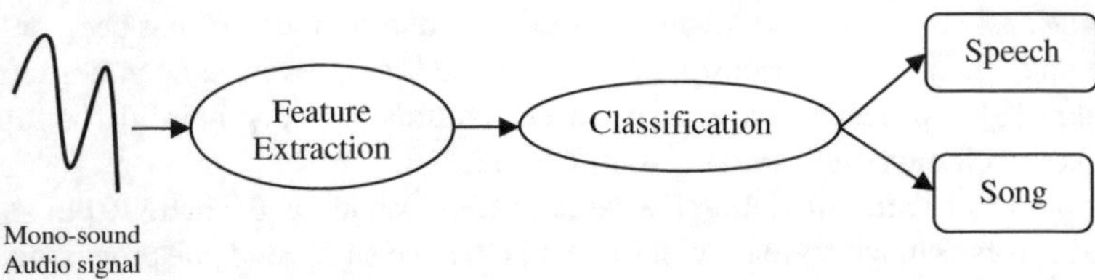

Fig. 1 Overview of propounded plan

3.1 Computation of Features

This is well known that a good feature set makes the classification task of a classifier easy. A good trait set maintains high interclass distance and at the same time low intraclass distance. To design a good feature set, the features are required to be chosen thoughtfully. Those features which exhibit high discriminating power can be chosen. As it is observed that speech and song mostly differ from the point of view of periodicity and energy distribution, they are chosen to discriminate speech and song. While designing a good trait set it is also to be taken care that it should be low in dimension otherwise it will be computationally expensive.

3.1.1 Features Based on Periodicity

Periodicity is by nature a perceptual feature. For computation of periodicity, initially the input audio signal, say X, is broken into frames. These frames need to be nonoverlapping in nature. There are 100 samples in each of the frames. After breaking the input signal into nonoverlapping frames, absolute Pearson's correlation coefficients [18] between two consecutive frames are estimated. For all the frames, same step has been repeated. An array of Pearson's correlation coefficients, $\{\mathrm{PCC}_i\}$, is formed for the corresponding input audio signal in this way. This correlation actually represents similarity between two consecutive frames. If some frames are repeated due to repetition nature of song then those frames will be highly correlated and other frames will not be.

It is well known that standard deviation and mean do not represent the overall concept of any array or distribution. Exact information related to it cannot be retrieved from that. For minutely study, the characteristics of correlation, co-occurrence matrix of Pearson's correlation coefficients [8] computed for all frames in successive nature is formed. Textural features are computed from that co-occurrence matrix and they are used as features. Positioning of different Pearson's correlation coefficient values enclosed by a region exhibits the characteristics of its corresponding audio signal. Consequently, a matrix CO_m having dimension $P \times P$ (where $P = \max\{\mathrm{PCC}_i\} + 1$) is formed.

A component in the matrix $\mathrm{CO}_m(x, y)$ represents the number of incidences of Pearson's correlation coefficient x and y in consecutive time instances. Co-occurrence

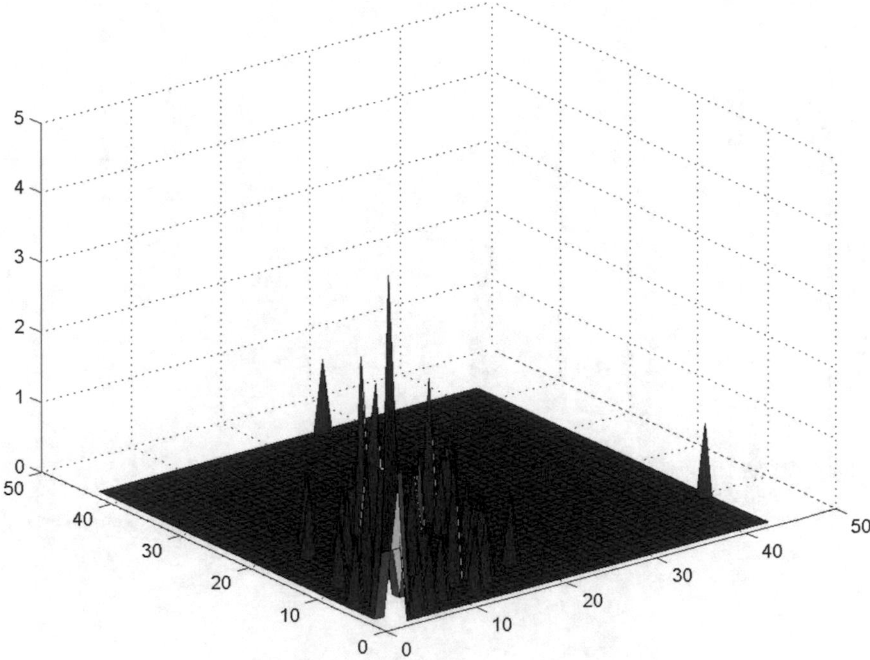

Fig. 2 Plot of Pearson's correlation coefficients co-occurrence matrix for speech

matrix based [1] statistical features like energy, entropy, contrast, correlation, and homogeneity are calculated.

Co-occurrence matrix plots of Pearson's correlation coefficients corresponding to speech as well as song are portrayed in Fig. 2 and Fig. 3, respectively. These plots reveal that repetitive nature of Pearson's correlation coefficients is different for speech and song.

3.1.2 Features Based on Skewness

Skewness is the third-order moment of the spectral circulation. Skewness is a statistical feature. Skewness measures the asymmetry or irregularity present for a certain normal distribution with respect to its mean position. Speech and song always maintain some amount of irregularity in their respective normal distribution. Skewness sk is measured in following Eq. (1):

$$sk = \left[E(b-\mu)^3\right]/\sigma^3 \tag{1}$$

where μ stands for the mean of sampled data b, standard deviation of b is represented by σ, and the expected value of the quantity y is denoted by $E(y)$.

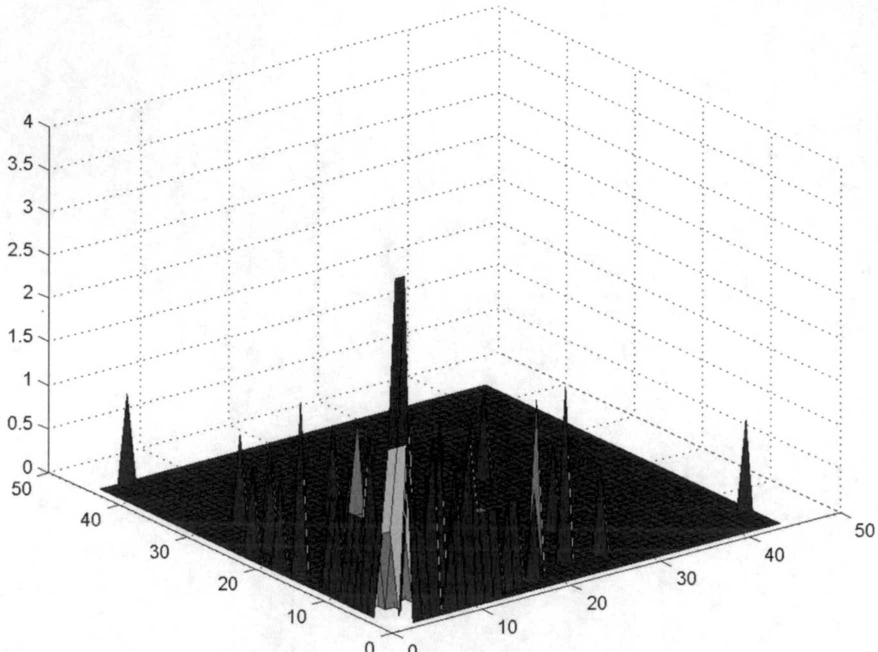

Fig. 3 Plot of Pearson's correlation coefficients co-occurrence matrix for song

3.2 Classification

The main goal of this work is to propose a small dimensional feature set which can discriminate speech and song very well. To test the differentiating power of the proposed periodicity and skewness-based feature set some popular as well as standard classifiers such as Neural Network (NN), Support Vector Machine (SVM) along with k-Nearest Neighbor or k-NN in short are applied. Multi-layer perceptron or MLP in short is pertained for in order to implement Neural Network (NN). An aural data pool is formed comprising 200 audio files consisting of speech and song category. Each category bears equal division in the custom audio dataset. For classification, the whole audio dataset was divided into two subsets—training data and testing data. To build training and testing dataset, half of the audio files from each category are taken into account. This means 50 files are taken into account from both speech and song category audio files resulting in a sum of 100 audio files both for training and testing. Support Vector Machine (SVM) has been implemented with kernel type as "Quadratic". For implementation of Multi-layer Perceptron (MLP)-type Neural Network (NN), six neurons have been considered in the input layer denoting six features values—five statistical features based on periodicity and one skewness-based feature, two neurons have been considered in the output layer indicating two breeds of audios—speech as well as song, four neurons have been considered in the masked level. k-NN classifier is used with $k = 3$, distance metric as "City Block" as well as nearest neighbor directive to break tie.

4 Experimental Results

To perform this investigation aiming to classify speech and song, a form of dataset has been organized which consists of 100 speech cases as well as 100 song cases. Every audio file belonging to this dataset are mono type and of 90 s duration. To provide wide variety in the dataset, diverse types of languages and diverse age groups have been considered for both the categories of audio data. These audio files are collected from CD recordings, audio recordings of different live programs, and also from the Internet. To evaluate actual performance of the proposed feature set in the real-life scenario, some of the audio files of the dataset are considered as noisy also for both the categories.

These audio files are chopped into frames. These frames are 50% extended beyond its previous frames so that boundary characteristics of any frame do not get missed out. 50% of dataset has been used as training data and rest 50% has been used as testing data for each of the classifier. Again same classification task is done once again by overturning training and testing data set. Mean of these two steps are considered as classification result and the same is tabularized in Table 1.

4.1 Relative Perusal

For establishing stratification strength of the proffered trait set, the same has been compared with few precedent efforts which have dealt with classification of speech as well as song. Identical audio data set that has been used in this effort has been employed to put into operation the schemes suggested by David Gerhard [2, 3] in his past two efforts. In the first work of Gerhard, pitch-based features have been used to attain the goal. Perceptual features like pitch, rhythm, etc. have been used by Gerhard in his second attempt. The second approach uses fuzzy classification of speech and song. But none of these two works use periodicity-based features, whereas periodicity is the main difference between speech and song. The relative

Table 1 Speech and song stratification accuracy

Stratification scheme	Classification precision (in %) for proffered effort	
	Speech data	Song data
Support vector machine	94.5	93.0
Neural network	92.5	91.5
k-NN	91.5	91.0

Table 2 Relative % accuracy presentation of proffered effort with former efforts

Past effort	Speech	Song
David Gerhard [2]	89.0	88.0
David Gerhard [3]	90.0	89.5
Proposed method (SVM based)	94.5	93.0

performance of proposed feature set and the feature set proffered by David Gerhard [2, 3] are tabulated in below table which is denoted as Table 2. From the relative presentation, the fact is clear that periodicity depended proffered trait set performs better than the feature set proffered by David Gerhard [2, 3].

5 Conclusion

Aiming to differentiate song from speech-, periodicity-, and skewness-based perceptual trait set has been suggested in this effort. This trait set is mainly based on statistical traits calculated from the co-occurrence matrix of Pearson's correlation coefficients computed for all frames of input audio signal in successive nature. Investigational consequences point to the information that proffered trait set do better compared to other existing trait set for classifying speech as well as song. The feature set which is suggested in this work is low dimensional as well as computationally simple and user-friendly. In future, advanced techniques like deep-learning-based approach may be adopted for this classification task.

References

1. Haralick, R.M., Shapiro, L. G.: Computer and Robot Vision, vol. I (1992)
2. Gerhard, D.: Pitch-based acoustic feature analysis for the discrimination of speech and monophonic singing. Can. Acoust. **30**(3), 152–153 (2002)
3. Gerhard, D.: Perceptual features for a fuzzy speech-song classification. In: IEEE International Conference on Acoustics Speech and Signal Processing, vol. 4, p. 4160 (2002)
4. Gerhard, D.: Silence as a cue to rhythm in the analysis of speech and song. Can. Acoust. **31**(3), 22–23 (2003)
5. Bugatti, A., Flammini, A., Migliorati, P.: Audio classification in speech and music: a comparison between a statistical and a neural approach. EURASIP J. Adv. Signal Process. **4** (2002)
6. Tzanetakis, G.: Song-specific bootstrapping of singing voice structure. In: 2004 IEEE International Conference on Multimedia and Expo, ICME 2004, vol. 3, pp. 2027–2030. IEEE (2004)
7. Lin, R.S., Chen, L.H.: A new approach for classification of generic audio data. Int. J. Pattern Recognit. Artif. Intell. **19**(01), 63–78 (2005)
8. Umbaugh, S.E.: Computer Imaging: Digital Image Analysis and Processing. CRC Press (2005)
9. Zhang, Y.G., Zhang, C.S.: Separation of music signals by harmonic structure modeling. In: Advances in Neural Information Processing Systems, pp. 1617–1624 (2006)
10. Ruinskiy, D., Lavner, Y.: An effective algorithm for automatic detection and exact demarcation of breath sounds in speech and song signals. IEEE Trans. Audio Speech Lang. Process. **15**(3), 838–850 (2007)

11. Lavner, Y., Ruinskiy, D.: A decision-tree-based algorithm for speech/music classification and segmentation. EURASIP J. Audio, Speech Music Process. **2009**(1) (2009)
12. Gallardo-Antolín, A., Montero, J.M.: Histogram equalization-based features for speech, music, and song discrimination. IEEE Signal Process. Lett. **17**(7), 659–662 (2010)
13. Salselas, I., Herrera, P.: Music and speech in early development: automatic analysis and classification of prosodic features from two Portuguese variants. J. Port. Linguist. **10**(1) (2011)
14. Sonnleitner, R., Niedermayer, B., Widmer, G., Schlüter, J.: A simple and effective spectral feature for speech detection in mixed audio signals. In: Proceedings of the 15th International Conference on Digital Audio Effects (2012)
15. Bhavsar, H., Panchal, M.H.: A review on support vector machine for data classification. Int. J. Adv. Res. Comput. Eng. Technol. (IJARCET) **1**(10), 185 (2012)
16. Velayatipour, M., Mosleh, M.: A review on speech-music discrimination methods. Int. J. Comput. Sci. Netw. Solut. **2**(2), 67–78 (2014)
17. Ramalingam, T., Dhanalakshmi, P.: Speech/music classification using wavelet based feature extraction techniques. J. Comput. Sci. **10**(1), 34 (2014)
18. Walk, M.J., Rupp, A.: Pearson product-moment correlation coefficient. Encycl. res. des. 1023–1027, (2010)

Intelligent Analysis for Personality Detection on Various Indicators by Clinical Reliable Psychological TTH and Stress Surveys

Rohit Rastogi, D. K. Chaturvedi, Santosh Satya, Navneet Arora, Piyush Trivedi, Akshay Kr. Singh, Amit Kr. Sharma and Ambuj Singh

Abstract Psychologists seek to measure personality to analyze the human behavior through a number of methods. As the personality of an individual affects all aspects of a person's performances, even how he reacts to situations in his social life, academics, job, or personal life. The purpose of this research article is to enlighten the use of personality detection test in an individual's personal, academics, career or social life, and also provide possible methods to perform personality detection test. One of the possible solutions to detect the personality study is based on the individual's sense of humor. Throughout the twentieth century, psychologists show an outgoing interest in the study of an individual's sense of humor. Since individual differences in humor and their relation to psychological well-being can be used to detect the particular personality traits. We have used machine learning used for personality

R. Rastogi (✉) · A. Kr. Singh · A. Kr. Sharma · A. Singh
Department of Computer Science & Engineering, ABESEC, Ghaziabad 201009, India
e-mail: rohit.rastogi@abes.ac.in

A. Kr. Singh
e-mail: akshay.17bcs1075@abes.ac.in

A. Kr. Sharma
e-mail: amit.17bcs1074@abes.ac.in

A. Singh
e-mail: ambuj.17bcs1078@abes.ac.in

D. K. Chaturvedi
Department of Electrical Engineering, DEI-Agra, Agra, India
e-mail: dkc.foe@gmail.com

S. Satya
Department of Rural Development, IIT-Delhi, Delhi, India
e-mail: ssatya@rdat.iitd.ernet.in

N. Arora
Department of Mechanical Engineering, IIT-Roorkee, Roorkee, India
e-mail: navneetroorkee@gmail.com

P. Trivedi
Center of Scientific Spirituality, DSVV- Hardwar, Hardwar, India
e-mail: piyush.trivedi@dsvv.sc.in

A. K. Das et al. (eds.), *Computational Intelligence in Pattern Recognition*,
Advances in Intelligent Systems and Computing 999,
https://doi.org/10.1007/978-981-13-9042-5_12

detection that involves the development and initial validation of questionnaire, which assesses four dimensions relating to individual differences in uses of humor. Which are Self-enhancing (humor use to enhance self), Affiliative (humor use to enhance the relationship with other), Aggressive (humor use to enhance the self at the expense of others), and Self-defeating (humor use to enhance relationships at the expense of self).

Keywords Sense of humor · Personality · Affiliative · Self-enhancing · Self-defeating · Aggressive · Python · Anaconda · Machine learning · Regression · Linear regression · Multiple linear regression · Decision tree · Random forest

1 Introduction

Machine learning is a study, which performs learning without being explicitly programmed. So, if we want our program to predict the personality, we can run different machine learning algorithms with data about some previously performed personality tests (Experience), and if learning is successfully done then, it will help in performing a better prediction. We have used one of the supervised learning algorithms called regression. Regression is a modeling and analyzing technique, which computes the relationship between the target (dependent variables) and predictors (independent variables). Regression allows comparing the effects of algorithms on variables of dataset [1]. To make predictions, various kinds of regression techniques are available:

1. Linear Regression—One of the most widely used techniques. It establishes the relationship between independent variables (one or more) and dependent variable using a best-fit regression line. It is represented by an equation [2]

$$Y = b + aX + e \tag{1}$$

   a. Where b is an intercept, a is slope of line, and e is error.

2. Multiple Linear Regression—It is the most common form of linear regression analysis. It helps to establish the relationship when the dependent variable, which is not only dependent on anyone of the independent variable but on multiple independent variables. It is represented by an equation

$$Y = a_0x_0 + a_1x_1 + a_2x_2 + \cdots + a_kx_k + e \tag{2}$$

   b. Where Y is response by k predictor variable $x_1, x_2, x_3 \ldots x_k$; $a_0, a_1, a_2 \ldots a_k$ are regression coefficients; and e is the error [3].
   c. It consists of five methods of building models ,namely Backward Elimination, Forward Selection, Bidirectional Elimination, and Score Comparison.

3. Decision Tree Regression—Decision tree algorithms are nonparametric supervised learning algorithms that are used for building regression and classification models, and through this algorithm, we try to train our model to predict values of target variable by learning decision rule inferred from data features. Decision trees are able to manage both numeric and categorical data and they can also handle multi-output problems [4]
4. Random Forest Regression—Random forest is like a forest of decision trees, i.e., it takes decision based on various decision trees. The final model of this technique is developed by the base models of various decision trees. It uses averaging for improving the predictive accuracy and controlling the overfitting. The equation used in this technique is

$$hx = k_0x + k_1x + k_2x + k_3x + k_4x + \cdots \quad (3)$$

Here, hx is final model and k_0x, k_1x, k_2x, k_3x, are base models.

2 Previous Work Study

In a survey of Personality Computing [5], according to Alessandro Vienciarelli, Gelareh Mohammadi researches, personality can be explained by a few measurable individual characteristics. We can use various technologies (i.e., which can deal with human personalities) for personality computing approaches, i.e., understanding, prediction, and synthesis of the human behavior. This paper is a survey of such approaches and it provides the three main problems indicated in the literature, i.e.,

(1) Automatic Recognition of personality,
(2) Automatic Perception of personality, and
(3) Automatic Synthesis of Personality [6].

In results of Personality Recognition on Social Media with Label Distribution Learning [7], the main aim of the experiment was to efficiently, reliably, and validly recognize an individual's personality. Di Xue, Zheng Hong, and Shize Guo have stated that the traditional ways of personality assessment are interviews, quiz prepared by the psychologists but these are expensive and not much practical in social media domain [8]. It proposes the method of big five personality recognitions (PR) with a new machine learning paradigm named label distribution learning. Out of 994 features, 113 features were extracted from active profiles and microblogs. Nine nontrivial conventional machine learning algorithms and 8 LDL algorithms are used to train the personality traits prediction models [8, 9]. Results show that two of the proposed LDL approaches outperform in predictive ability.

In the development of Agile Person Identification through Personality Test and k-NN Classification Technique [10]. Software methodology is a planning of the development of software; agile methodology is one of the most efficient software development methods. Agile method requires the cooperation and understanding between the employees, so this is important for the software project manager to assign the right work to the right people. Rintaspon Bhannarai, Chartchai Doungsard have revealed the fact that the researchers provide five personality traits to predict suitable people for the agile method [11]. A predicting method is done by using k-NN (*k*-nearest neighbor) classification technique. The *k*-NN techniques classify the unknown data from the known data by comparing the distance calculated between them [12]. The most common method to calculate the distance between data is using Euclidean Distance [13]. Using the participants which involve software developers, testers, managers, etc. the pilot study is used to explore this technique that *k*-NN technique can be explored to predict the correct people for agile method.

In a view to check individual differences in sense of humor and their relation with psychological well-being [14], the main aim of this survey was to identify the most common personality traits in males and females and which traits dominate in male and female. Here, the analysis was done by the quiz, which was developed on the basis of different senses of humor of a person. After collecting the data from candidates, the analysis was done on the basis of some particular personality traits [15, 16]. The objective of the experiment was to analyze the proportionality of these personality traits in male and female.

It anticipates the personality of an individual by giving the dominance of each personality trait (aggressive, affiliative, self-enhancing, and self-defeating) in the range of 0–5. If the particular personality trait is 0 that means it is least in the particular person and if it reaches near 5 that means it is a very dominant personality trait in an individual [17].

3 Motivations

Machines learning are getting popular day by day and enable the computers to perform computations without being explicitly programmed. Science has made our life simpler and proved to be boon for solving many complex problems. Regression techniques mostly differ based on the number of independent variables and the type of relationship between the independent and dependent variables. So, the motive of linear regression is used to minimize the error and predict the actual result [18]. The proposed work will be an effort to utilize the machine intelligent techniques like Python programming on Spyder framework. Human beings are complex to understand and their personality traits are changed according to the changes in them. So, we will try to learn more and more about them and we test the app also [19].

4 Study Plot

The dataset consists of the Humor Style Questionnaire [20]. The dimensions of data are 39 attributes and 1071 instances. There are no missing values in the dataset. Now, while exploring the dataset

Step 1: Apply data ETL operations
That is, apply data feature extraction, Data Transformation, and Loading of dataset.
[Loop Begins] Till all the dataset is completed.
Step 2: Estimation of summary of each attribute including max, min, mean, count value, and also some percentiles.
Step 3: Check for any null value.
Step 4: Check the data type of various attributes in dataset [End of Loop].

For data visualization, first, we start with univariate plots, that is, plot of each variable to get an idea of the distribution.

Figure 1 shows the range of affiliative personality among individuals, Fig. 2 shows the range of self-enhancing personality among individuals, Fig. 3 shows the range of aggressiveness personality among individuals, and Fig. 4 shows the range of self-defeating personality among individuals.

The next step is data preprocessing where we initially declare dependent and independent variables [4]. The dataset is split into Training set and Test set. Dataset is split in the ratio of 80:20. In other words, 80% of the data is kept in training set and 20% data is kept in test set. Now, we have a training data in "X_train" and "Y_train" and test data in "X_test" and "Y_test". We are building some ML models by establishing the correlation between dependent variable and independent variable and once the correlation estimated, then we will test the model using test

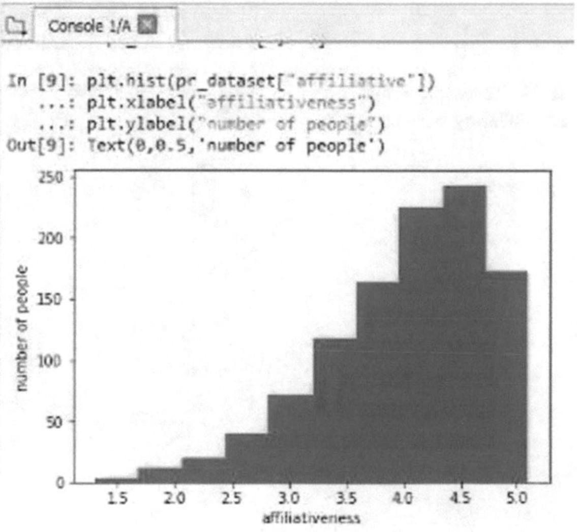

Fig. 1 Depicting range of affiliative personality among individuals

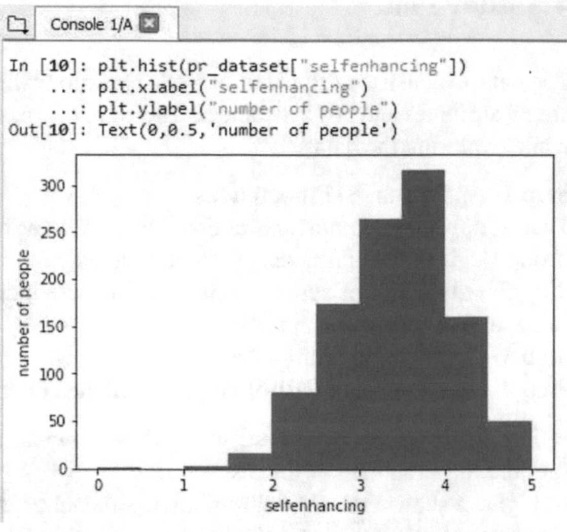

Fig. 2 Analyzing range of self-enhancing personality among individuals

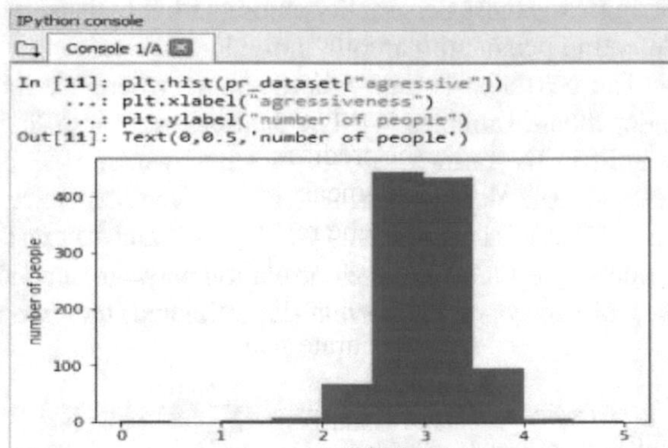

Fig. 3 Showing range of aggressive personality

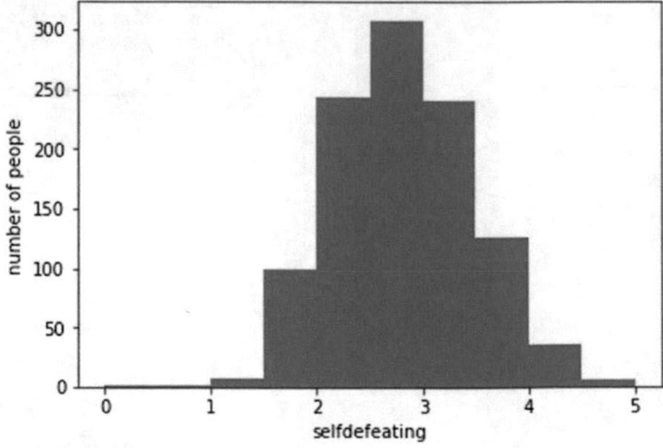

Fig. 4 Revealing range of self-defeating personality

data (“X_test” and “Y_test”) that whether it can understand the correlation between the training set on test set.

Prediction is an action performed on the test data to forecast the values that up to which extent model has learned from the trained data.

As different regression models like Linear Regression (by Eq. 1), Multiple Linear Regression (by Eq. 2), Decision Tree Regression, and Random Forest Regression (by Eq. 3). Machine learning regressors usually support a single-dependent variable. So, here, we create multi-output regressors for Multiple Linear Regression, Decision Tree Regression, and Random Forest Regression.

5 Algorithm(s) Implementation Results—Graphs and Tables

Upon experiments on the dataset available, we are getting interesting results with the help of LR, MLR, DTR, and RFR algorithms. They are shown with the help of graphs available.

Figure 5 graph shows the relationship between the tested and predicted values computed from the Linear Regression model (LR) using Eq. 1. As we can see in the graph, blue color dots are coinciding with each other and are not scattered in the whole plane because the predicted values are very close to the tested values. Accuracy achieved is 99.56% with mean square regression loss is 0.00156390272662.

Figure 6 graph shows the relationship between the tested and predicted values of Multiple Linear Regression 1 (MLR1) using Eq. 2 The dots are lying very close to each other and are also not scattered in the whole plane, so it shows that predicted values are very much accurate with accuracy 99.79%. Mean square regression loss is 0.00111408573375.

Figure 7 depicts the visualization of predicted and tested values of Multiple Linear Regression 2 (MLR2). This model also predicts the accurate results and this is clearly

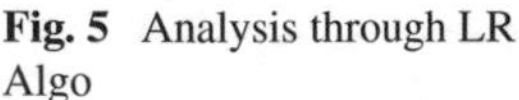
Fig. 5 Analysis through LR Algo

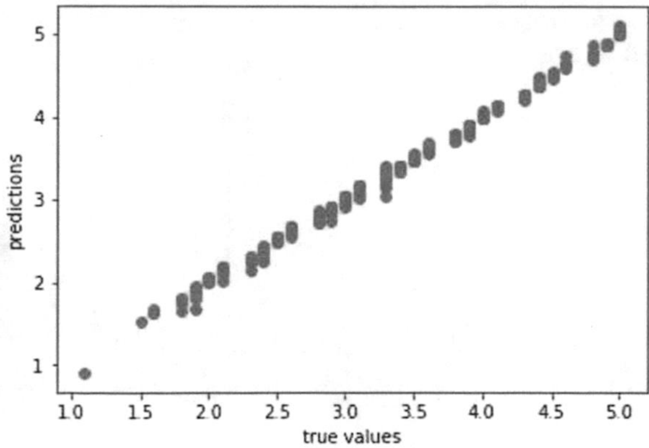

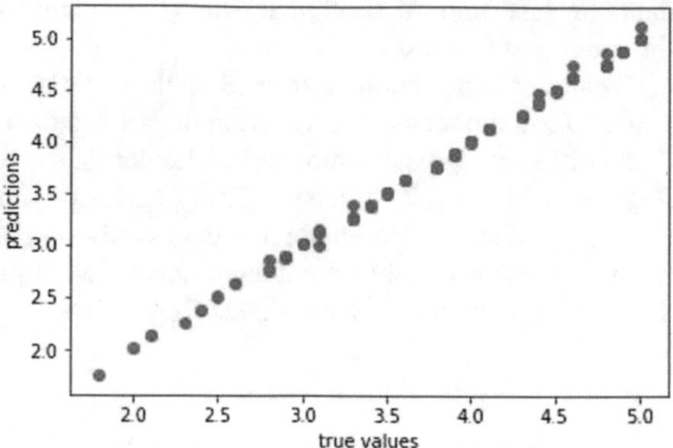

Fig. 6 Analysis through MLR1 Algo

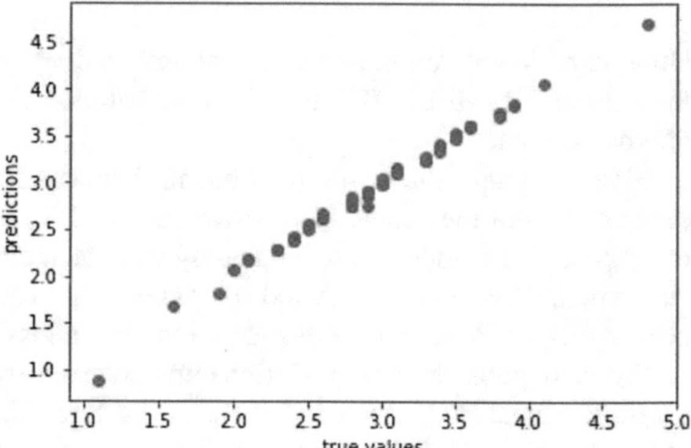

Fig. 7 Analysis through MLR2 Algo

visible through scattering of points linearly in a plane. Accuracy is 99.51% and mean square regression loss is 0.00169502757472.

Figure 8 performs the visualization of predicted and tested values of Multiple

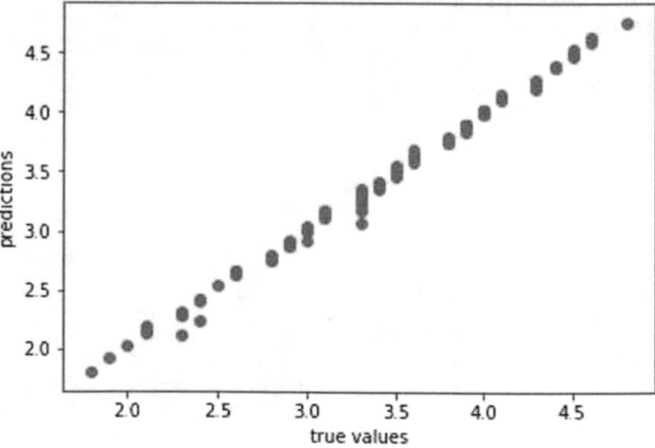

Fig. 8 Analysis through MLR3 Algo

Linear Regression 3 (MLR3). This model also predicts the accurate result but in comparison to Multiple Linear Regression 1 and Multiple Linear Regression 2m there is a minor difference in accuracy. Accuracy is 99.14% and mean square regression loss 0.00159577024357.

Figure 9 gives the visualization of predicted and tested values of Multiple Linear Regression 4 (MLR4). This model predicts the accurate results with an accuracy of 99.56% and mean square regression loss is 0.00159151039372.

Figure 10 shows the visualization of predicted and tested values of Decision Tree Regression 1 (DTR1). As we can see in the graph, all points are scattered in a plane because the prediction is not much accurate as in the case of Linear Regression and Multiple Linear Regression. Scattering shows that predicted and test values are not lying much close to each other. Accuracy is 87.52% and mean square regression loss 0.0675178855488.

Figure 11 exhibits the visualization of predicted and tested values of Decision Tree Regression 2 (DTR2). In this graph, points are much scattered initially that is lying at some distance. It is a good model but not accurate models in comparison to

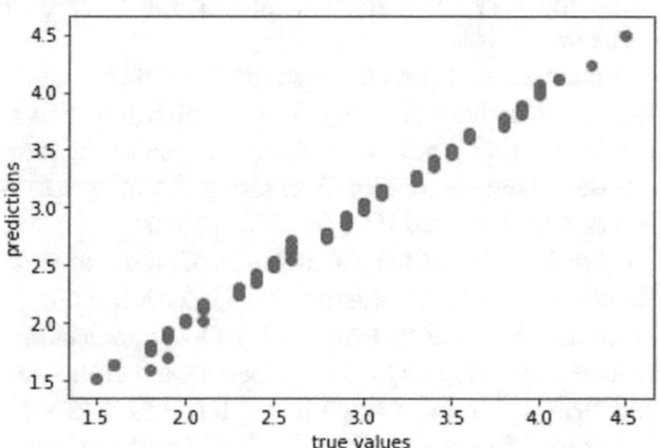

Fig. 9 Analysis through MLR4 Algo

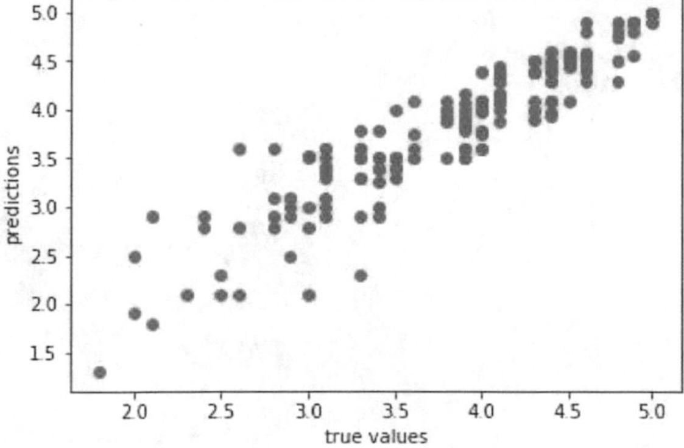

Fig. 10 Analysis through DTR1 Algo

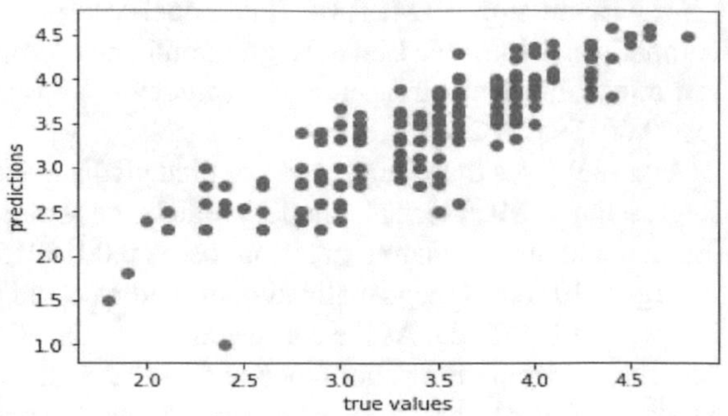

Fig. 11 Analysis through DTR 2 Algo

Decision Tree Regression 1. Accuracy is 71.48% and mean square regression loss is 0.0984915683583.

Figure 12 is the visualization of predicted and tested values of Decision Tree Regression 3 (DTR3). This model is least accurate as we can see in the graph that points are clustered at various places and are not scattered linearly. This model has very less accuracy. Accuracy is 28.75% and mean square regression loss 0.132380568167.

Figure 13 performs the visualization of predicted and tested values of Decision Tree Regression 4 (DTR4). This is far better than Decision Tree Regression 3 as we can see that points are scattered in the linear form. This is very much similar to Decision Tree Regression 2 in terms of prediction. Accuracy is 72.84% and mean square regression loss is 0.0982561775804.

Figure 14 reveals the visualization of predicted and tested values of Random Forest Regression 1 (RFR1) (using Eq. 3). This is an accurate model as in comparison to Decision Tree Regression where average accuracy is 65.1475%. All points are scattered linearly in a plane. So, better predictions are obtained. Accuracy is 94.16% and mean square regression loss is 0.0315995487491.

Figure 15 expresses the visualization of predicted and tested values of Random Forest Regression 2 (RFR2). This model predicts the results but with less accuracy

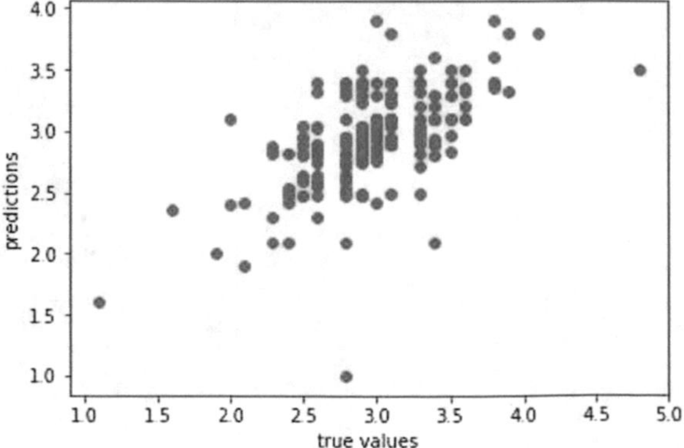

Fig. 12 Analysis through DTR3 Algo

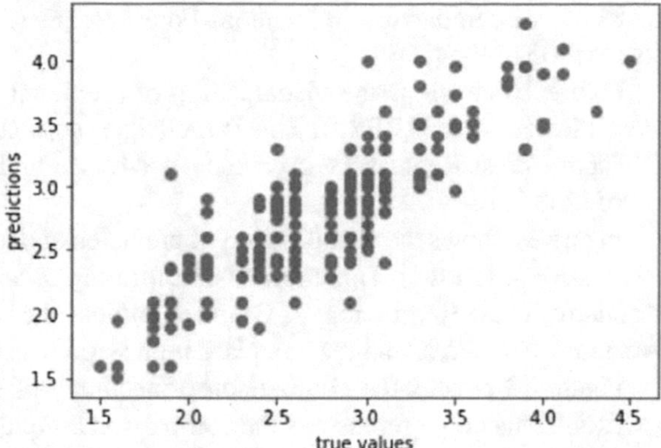

Fig. 13 Analysis through DTR4 Algo

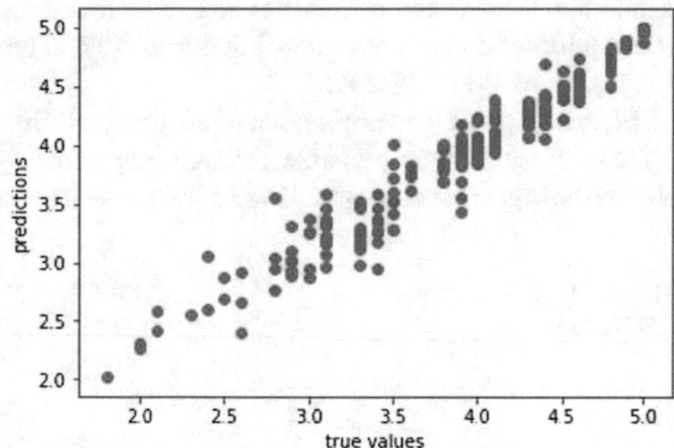

Fig. 14 Analysis through RFR1 Algo

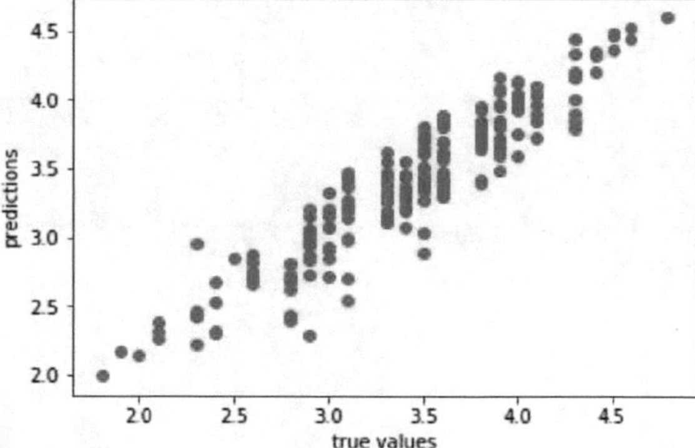

Fig. 15 Analysis through RFR 2 Algo

of 88.0% in comparison to Random Forest Regression 1. Mean square regression loss is 0.0315995487491.

Figure 16 develops the visualization of predicted and tested values of Random Forest Regression 3 (RFR3). This is model has comparatively less accuracy to other Random Forest Regressors. Accuracy is 64.84% and mean square regression loss 0.06532686210.

Figure 17 shows the visualization of predicted and tested values of Random Forest Regression 4 (RFR4). This model is performing predictions in the same manner as Random Forest Regression 2. Graph is almost the same for both the regressors. Accuracy is 89.35% and mean square regression loss 0.0385353877922.

Figure 18 creates the comparison of accuracy of different multiple regressions (MLRs). This graph represents the comparison among the accuracies of four regressors of Multiple Linear Regression. All the regressors are computing almost same results and there is not much difference in their accuracies. But the highest accuracy is of Multiple Linear Regression 1 that is 99.79%. The average accuracy of Multiple Linear Regression is 99.5%.

Figure 19 gives the comparison of accuracy of different decision tree regressions (DTRs). As it is clearly visible through graph that Decision Tree Regression 1 is having the highest accuracy that is 87.52% greatest among all the regressors. Decision

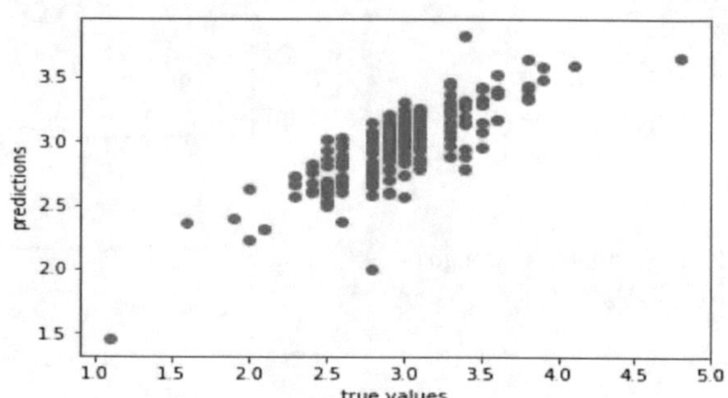

Fig. 16 Analysis through RFR3 Algo

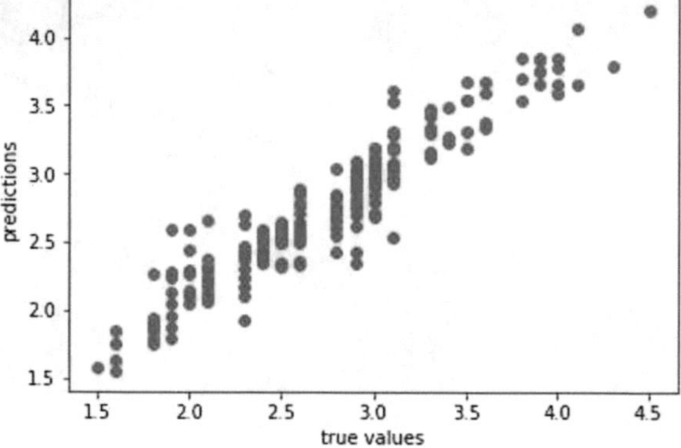

Fig. 17 Analysis through RFR4 Algo

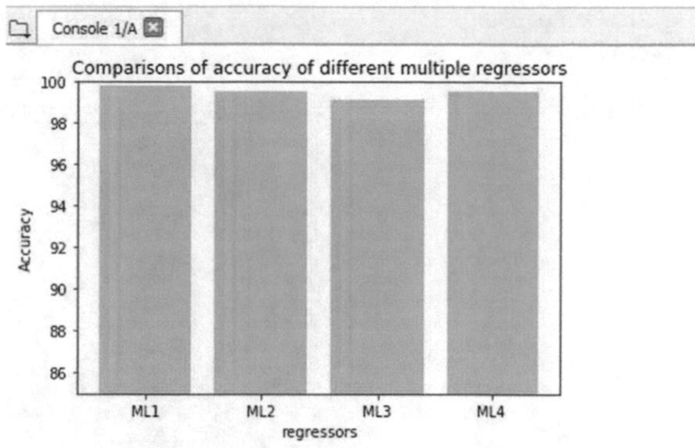

Fig. 18 Comparision between MLRs

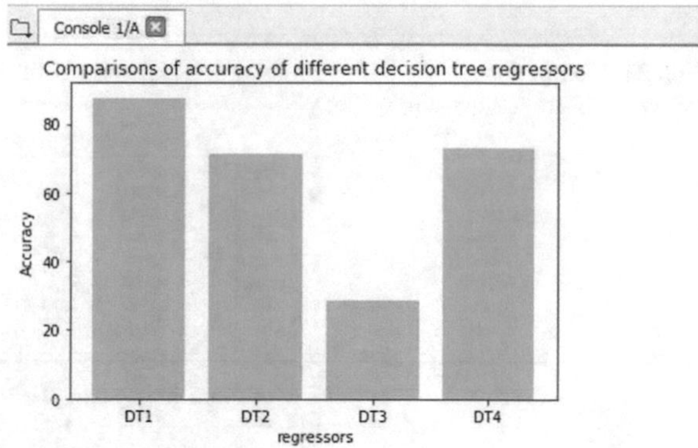

Fig. 19 Comparison between DTRs

Tree Regression 2 and Decision Tree Regression 4 achieves almost the same accuracy that is near to 70%. Least accurate is Decision Tree Regression 3 of 28.75% accuracy. The average accuracy of Decision Tree Regression is 64.175%.

Figure 20 makes the comparison of accuracy of different Random Forest regressions (RFRs). As it is clearly visible through graph that Random Forest Regression 1 is having the highest accuracy among all the regressors that is 94.16%. Random Forest Regression 2 and Random Forest Regression 4 achieves almost the same accuracy but less than of Random Forest Regression 1. Least accurate is Random Forest Regression 3 with accuracy 64.84%.

Figure 21 shows the comparison graph of accuracies of algorithms. This graph represents the performance of all the models in predicting the values in the form of accuracy. As graph depicts that Linear Regression and Multiple Linear Regression attained almost the same accuracy.

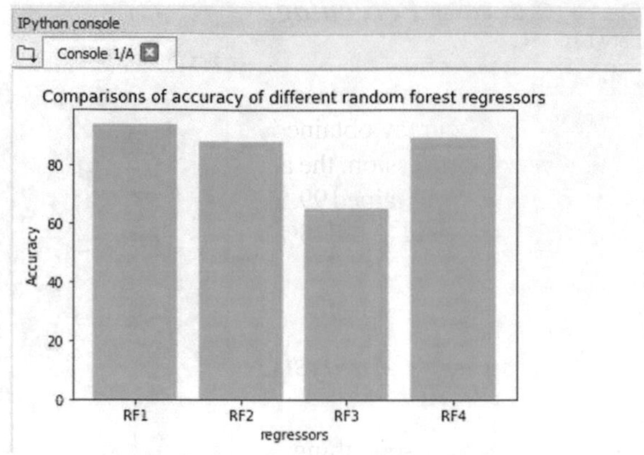

Fig. 20 Comparison between RFRs

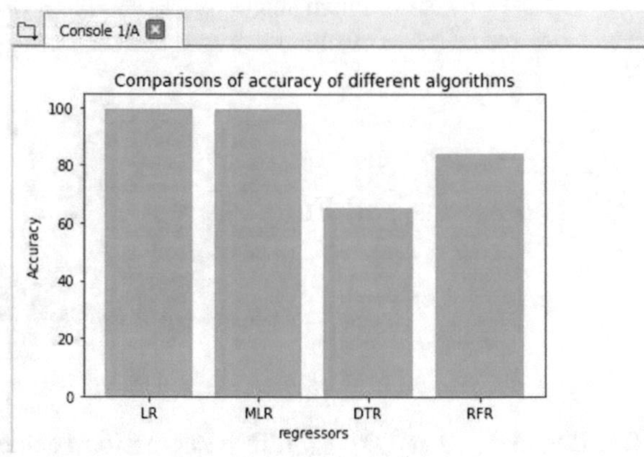

Fig. 21 Comparison between different regressions

6 Performance Evaluations

Among the different algorithms used in the experiment, Linear Regression and Multiple Linear Regression are proved to be the best. Both of them are the best algorithm that can be used to implement the prediction of the personality of individuals. On the basis of different datasets used, different algorithms are required to be implemented accordingly. This is because calculated root mean square error value was the minimum, i.e., 0.00156 and 0.0016. Here, we can see that decision tree regression and random forest regression are not giving good result.

6.1 Accuracy Percentage

On experiment over the datasets with four machine learning supervised/classification methods, the accuracy obtained is as below.

For Linear Regression, the accuracy is 99.56%, while for Multiple Linear Regression, the accuracy obtained 99.5%, the dataset is giving 64.175% accuracy in Decision Tree Regression and 84.0875% for Random Forest Regression.

6.2 Complexity Analysis

The personality is something, which describes your actions and reactions toward in any situations or how you will react in particular circumstances. Prediction of personality traits is not much accurate when performed manually and it is a complex process too [1]. Using machine learning, the subjective work can now be performed using mathematical models.

7 Conclusions and Future Directions

7.1 Conclusions

To predict the personality traits of an individual here, the authors have implemented a method that involves the supervised learning. Supervised learning has a subset of algorithms called regressions. These regressors have provided us with the models which are trained for learning. Models learned from experiences that is 80% of train data and the remaining 20% are used to test the accuracy of prediction. Among the four models, Linear Regression and Multiple Linear Regression have attained the best prediction accuracy of 95% about. Decision Tree Regression has prediction accuracy of 64.175% which is very less. So, this model can't be used for predicting the personality traits using dataset available with us. Random Forest Regression attained the accuracy of 84.0875%, which is better than Decision Tree Regression and it can also be used as a secondary model to predict the personality traits.

Since the personality of an individual can adversely affect the professional attitude of a person. All the MNCs and even our defense organization select the candidate for the job on the basis of their personality; therefore, they use some procedures to predict the personality of an individual, which can be done through machine learning. This experimental study can be further modified to be used in these organizations to get more efficient prediction.

The analysis is performed on the personality detection to get the domination of some particular personality traits in male or female [21]. The prediction of personality has been done through k-nearest neighbor classification technique [2]. Personality detection is less efficient and costly in terms of manual operations so authors have tried to perform it by machine learning. Regression techniques used for the prediction are linear regression, decision tree, and random forest. Individual can use the predicted result to develop their personality and improve their personal and professional life.

7.2 Future Directions

The method used by the authors is a new way for computing personality detection test. Machine learning is something, which makes the human work easier and gives more efficient output. Process of personality test is automated in order to reduce the human error. This is one of the reasons that this method is much accurate and preferable in order to predict the personality of an individual. These tests are better to understand the perception of an individual for self-reflection and understanding [22].

The model used for personality detection is Linear Regression. But Linear Regression is prone to overfitting such that regression begins to model noise in the data along with the relationship of the variables. This is pretty extrapolation and not necessarily accurate. Decision Tree Regression is not estimating the good results, so we can't use this for learning and prediction of personality traits.

One may find useful to explore oneself from a completely different perspective to find a successful career and improve social and personal life by understanding different personality traits [7].

Personality tests in interviews is needed in Govt. organizations and in corporate too, as the jobs offered in them demand different characteristics in the personality of an individual, i.e. the OLQ (officer like qualities) are needed in IAS officer, that can differ from those of an IPS officer, so in order to find right candidate the personality detection can be a great help.

Acknowledgements The authors are obliged for the guidance of ABES Engineering College staffs and faculties. Python mentor, Asst. Prof., Mr. Shubham Sidana, Asso. Prof., Mr. Abhishek Goyal, and evaluation team of experimental presentation to understand the concept well and for showing the path ahead. The acknowledgement to all those forces, which inspire us to work hard and to learn something and to make a difference in society.

References

1. Eysenek, H.J., Eysenck, H.J.: Dimensions of Personality, vol. 5. Transaction Publishers (1950)
2. Ruiz-Falcó, A.: /nstituto de Astrofísica de Andalucía, CSIC. Entiempo real Lenguajes de Descripción Hardware: Conceptos y Perspectivas **126**, 35 (1997)
3. Bandura, A., Walters, R. H., Riviere, A.: Aprendizaje social y desarrollo de la personalidad. Alianza Editorial Sa (2007)

4. De Juan, M.: Personalidad y Criminalidad, Apuntes de Psicología Criminológica. la asignatura Psicología Criminológica, Universidad Autónoma de Madrid (No publicado) (2005)
5. Francis, L.J., Brown, L.B., Philipchalk, R.: The development of an abbreviated form of the revised Eysenck personality questionnaire (EPQR-A): Its use among students in England, Canada, the USA and Australia. Personality Individ. Differ. **13**(4), 443–449 (1992)
6. Costa, P.T., McCrea, R.R.: Revised neo personality inventory (neo pi-r) and neo five-factor inventory (neo-ffi). Psychological Assessment Resources (1992)
7. Rossmo, D.K.: Geographic Profiling. CRC press. ISBN-0849381290 (1999)
8. Caprara, G.V., Barbaranelli, C., Borgogni, L., Perugini, M.: The "Big five questionnaire": A new questionnaire to assess the five-factor model. Personality Individ. Differ. **15**(3), 281–288 (1993)
9. Reddy, K.H.K., Das, H., Roy, D.S.: A data aware scheme for scheduling big-data applications with SAVANNA Hadoop. Futures of Network. CRC Press, USA (2017)
10. Mairesse, F., Walker, M.: Automatic recognition of personality in conversation. In: Proceedings of the Human Language Technology Conference of the NAACL, Companion Volume: Short Papers, pp. 85–88. Association for Computational Linguistics (2006)
11. Mishra, B.B., Dehuri, S., Panigrahi, B.K., Nayak, A.K., Mishra, B.S.P., Das, H.: Computational intelligence in sensor networks, vol. 776. Studies in computational intelligence. Springer (2018)
12. Das, H., Jena, A.K., Nayak, J., Naik, B., Behera, H.S.: A novel PSO based back propagation learning-MLP (PSO-BP-MLP) for classification. In: Computational intelligence in data mining, vol. 2, pp. 461–471. Springer, New Delhi (2015)
13. Das, H., Jena, A.K., Rath, P.K., Muduli, B., Das, S.R.: Grid computing-based performance analysis of power system: a graph theoretic approach. In: Intelligent computing, communication and devices, pp. 259–266. Springer, New Delhi (2015)
14. Goldberg, L.R.: Language and individual differences: The search for universals in personality lexicons. Review of personality and social psychology **2**(1), 141–165 (1981)
15. Kar, I., Parida, R.R., & Das, H.: Energy aware scheduling using genetic algorithm in cloud data centers. In: International Conference on Electrical, Electronics, and Optimization Techniques (ICEEOT), pp. 3545–3550. IEEE (2016)
16. Sarkhel, P., Das, H., Vashishtha, L.K.: Task-scheduling algorithms in cloud environment. In: Computational Intelligence in Data Mining, pp. 553–562. Springer, Singapore (2017)
17. Villena-Roman, J., Garcia-Morera, J., Moreno-Garcia, C. Ferrer-Urefia, L., Lana-Serrano, S., Carlos Gonzalez-Cristobal, J. Waterski. A. Martinez-Camara, E., umbreras, M.A. Martin-Valdivia, N1. T., Alfonso Urefia-Lopez, L.: TASS-Workshop on Sentiment Analysis at SEPLN, Workshop on Sentiment Analysis, Sociedad Espanola para el Procesamiento del Lenguaje (2012)
18. de Juan-Espinosa, M.: Personalidad Artificial: Hacia Una Simulación De Las Diferencias DePersonalidadEnSituaciones De Interacción. Universidad Autónoma de Madrid, Madrid (1997)
19. Polzehl, T., Möller, S., Metze, F.: Automatically assessing acoustic manifestations of personality in speech. In: Spoken Language Technology Workshop (SLT), pp. 7–12. IEEE (2010)
20. Francis, J.W.P.M.E., Booth, R.J.: Linguistic Inquiry and Word Count. Technical Report. Technical Report, Southern Methodist University, Dallas, TX (1993)
21. Ivanov, A.V., Riccardi, G., Sporka, A.J., Franc, J.: Recognition of personality traits from human spoken conversations. In: Twelfth Annual Conference of the International Speech Communication Association (2011)
22. Gill, A.J., & Oberlander, J.: Taking care of the linguistic features of extraversion. In: Proceedings of the Annual Meeting of the Cognitive Science Society, vol. 24, no. 24 (2002)

Automatic Multilingual System from Speech

Suchit Agarwal, Akshay Chatterjee and Ghazaala Yasmin

Abstract Language recognition is the way by which the language of a digital speech utterance is recognized automatically by a computer. Commenced Language Identification Systems sequentially transform the speech signal into discrete units, and then apply statistical methods on the resultant units to extract their language information. Today, a large number of audio retrieval features exists for automatic speech and language recognition. The proposed method has nominated an automatic system for well-known multi-languages. The identification has been done using a new set of audio features. The suitable feature has been adopted. This includes Zero-Crossing Rate, Spectral Flux, Pitch, Mel-frequency Cepstral Coefficients, Tempo, and Short-Time Energy. These features have been used exclusively for identifying the language along with the help of classifiers and feature selection algorithms.

Keywords Mel-frequency cepstral coefficients (MFCC) · Pitch · Tempo

1 Introduction

Communication is an elementary behavior that spans throughout the animal species. Human language is a one of a kind system, which segregates humans from the other animals. It is a process of sending and receiving information among people. The most common communication mode for humans is speech. Speech can be in varying languages depending on the culture the person belongs to. Humans can best identify language from speech. They can identify very fast whether the language is known

S. Agarwal · A. Chatterjee (✉) · G. Yasmin
Department of Computer Science and Engineering, St Thomas' College of Engineering and Technology, Kolkata 700023, India
e-mail: akshay.chatterjee2015vit@gmail.com

S. Agarwal
e-mail: suchitagarwal1995@gmail.com

G. Yasmin
e-mail: me.ghazaalayasmin@gmail.com

A. K. Das et al. (eds.), *Computational Intelligence in Pattern Recognition*,
Advances in Intelligent Systems and Computing 999,
https://doi.org/10.1007/978-981-13-9042-5_13

or not. The frequency, time, and perceptual domain characteristics of speech help in determining these. This has motivated us to introduce a methodology for automatic language identification.

Humans, as a part of their human intelligence, have the ability to differentiate between spoken languages. The quest to automate work has never stopped. Just like any other Artificial Intelligence technology, this spoken language identification work also aims to replicate the human ability to differentiate various spoken languages. The proposed work includes the identification of popularly spoken languages in Asia. During the training phase, the speech messages for each language are analyzed, resulting in a model for each type of language. During testing, a previously unknown test message is entered into the system, and the system outputs the language compared with the model that most closely matches with the test message. It is estimated that there exist several thousands of spoken languages in the world. In this paper, we will be dealing with only a few languages. Every country in Asia has its official languages spoken in the country. We aim at providing a way to identify the languages spoken in Asia based on the speech audio.

This is done with the help of classifiers. The extracted feature vectors are fed to a classifier that based on certain rules determines the class of the incoming vector. After that, we use feature selection algorithm to narrow the feature set that yields the desired output.

1.1 Related Work

The differences in perception between speech and other sounds have been a source of interest for many authors and researchers since the ages. Till date, researchers have found many features from speech, which helped in discrimination of speech. Many audits have gone through the issues and the solution related to language discrimination techniques and its applications [1–3]. Mahadev and Lakhani [4] have propounded the highlights of language classification by applying the notion of neural network has found important to incorporate feature Mel-Frequency Cepstral Coefficients (MFCC). In the proposition of Chandrasekhar, Sargin and Ross [5], Gaussian Mixture Models and Shifted Delta Cepstral Features for the language identification has been included. Itrat, Ali et al. [6] have taken into account the Pakistani language discrimination using MFCC and vector quantization. Adami et al. [7] have highlighted speech segmentation for the speaker as well as language recognition. Dehak et al. [8] and Yu et al. [9] both have given priority to I-vectors as an important factor for the classification of language. In the strategy of Gwon et al. [10] for language categorization, sparse coding has taken to be a prior notion. Behravan et al. [11] have propounded scheme of I-Vector as this notion has earned more attention in the domain of language categorization. Vatanen et al. [12] have identified different languages from Short Text Segments using N-gram models. Torres-Carrasquillo et al. [13] have presented an approach for Language Identification where they have used

Gaussian mixture model. Karpagavalli and Chandra [14] did a detailed study on speech recognition. Lartillot et al. [15] presented the MIRtoolbox for the extraction of various audio features.

2 Proposed Methodology

Audio can be of various types, namely, speech and nonspeech. Speech is an audible form of communication that humans use that is based upon the syntactical combination of items taken from their vocabulary. The proposed work uses certain audio features to differentiate Asian languages. The features are fed to the classifier to further identify the class of the language. A basic flowchart for the system is given in Fig. 1.

For determining a language from an audio speech, we need to first classify it. This is depicted in Fig. 2. Audio dataset has been subjected for feature extraction and then significant feature has been selected using feature selection algorithm. The selected feature has been served for classification for 28 different classes of languages in Asia.

2.1 *Feature Extraction*

2.1.1 Pitch

Pitch is defined as the degree of highness or lowness of speech. Sharp speech has a high pitch frequency and reverse for bass speech. So, the pitch becomes a significant feature for discrimination and that's why we have used this feature. The signal is broken into 88 frequency bands, which are further divided into short-term features. Short-term mean square power has been found out from these frames by averaging the value of STMSP for each band. Figure 3 shows the variation of the pitch with

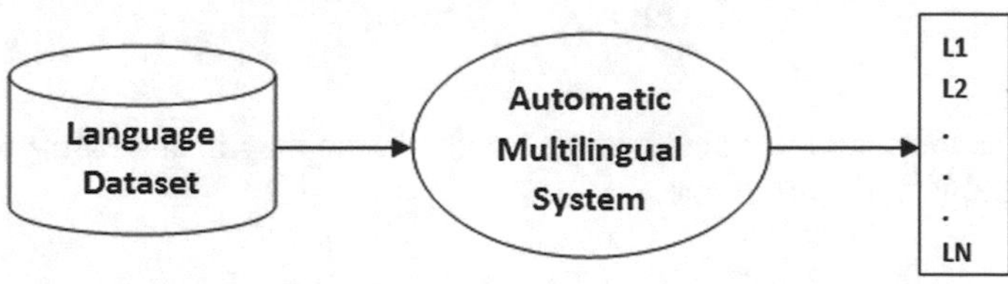

Fig. 1 Basic workflow of the system

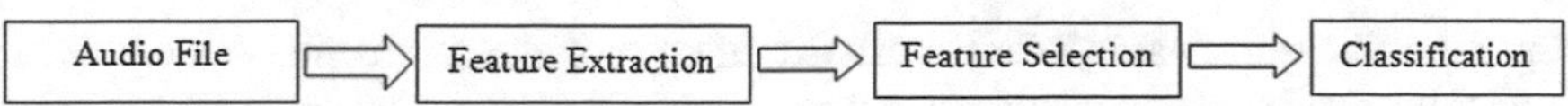

Fig. 2 Audio classification process

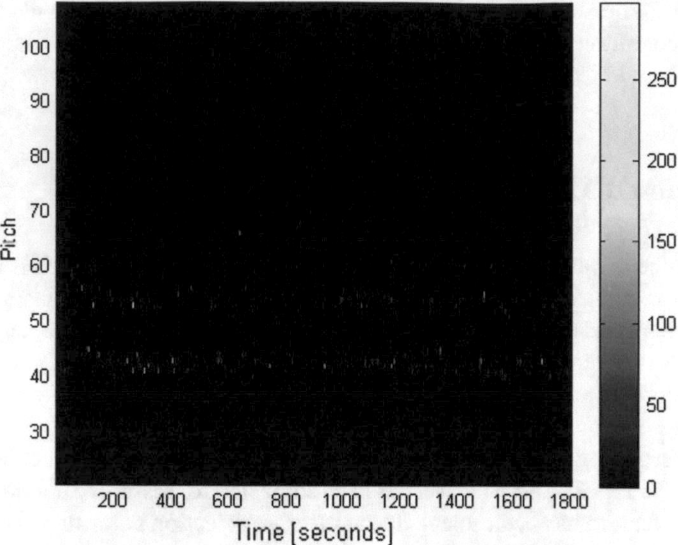

Fig. 3 Pitch plot of the sample language file

respect to time. As well as, it shows the variation of the pitch scale with respect to the energy bands.

2.1.2 Short-Time Energy (STE)

Short-Time Energy (STE) is a simple time domain feature, which is widely used by researchers. STE is suitable for discriminating speech and music. Speech consists of words interleaved by silence, which gives variation of STE which is quite different from the pattern obtained for music. The energy of the nth frame is defined in Eq. 1.

$$\text{En} = \frac{1}{r}\sum_{m=1}^{r}[xn(m)]^2 \quad (1)$$

where E_n denotes the nth frame energy, r is the frame length, and $x_n[m]$ represents the mth sample in the nth frame.

2.1.3 Zero-Crossing Rate (ZCR)

The Zero-Crossing Rate (ZCR) is an extremely popular short-time feature. ZCR depicts the concentration of energy in the spectrum. It can be measured as the number of times the signal crosses zero (i.e., changes sign) within the frame (Eq. 2).

$$\text{ZCRn} = \sum_{m=2}^{r} sign[\text{xn}(m-1) * \text{xn}(m)] \tag{2}$$

where r is the number of sample values in the nth frame and

$$\text{sign[v]} = \begin{cases} 1, & if\ v < 0 \\ 0, & otherwise \end{cases} \tag{2a}$$

The ZCR can be viewed as a measure of the dominant frequency. Since unvoiced speech typically has significantly hiked in the value ZCR values than voiced speech, ZCR can be used to differentiate between unvoiced and voiced speech. Figure 4 shows the variation of zero-crossing rate throughout the signal of a sample speech file of any sample language.

2.1.4 Mel-Frequency Cepstral Coefficients (MFCC)

Mel-Frequency Cepstral Coefficients (MFCCs) are coefficients that altogether makeup MFCC. It has been extracted from the Mel-scale cepstral representation of an audio clip. The key point in MFCC is the equal gap of the bands in frequency is done on the Mel scale, which approximates the human auditory setup more efficiently than the linearly spaced frequency bands that are applied in the usual cepstrum environment. Two properties of MFCCs, one being that the first coefficient directly

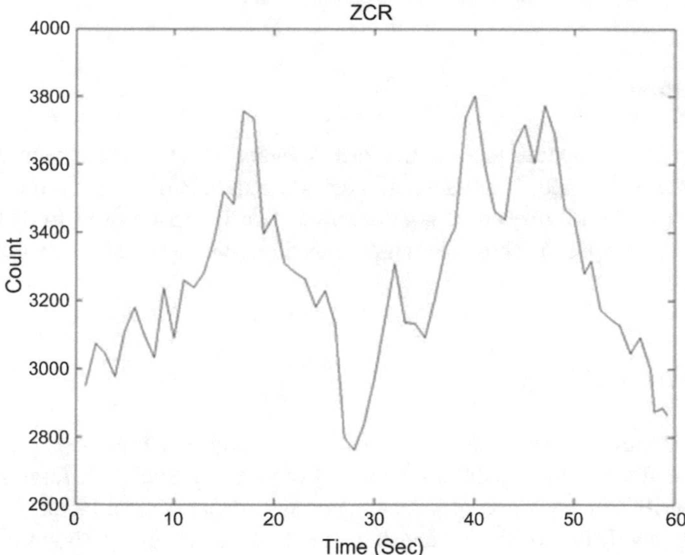

Fig. 4 Zero-crossing rate of the sample language file

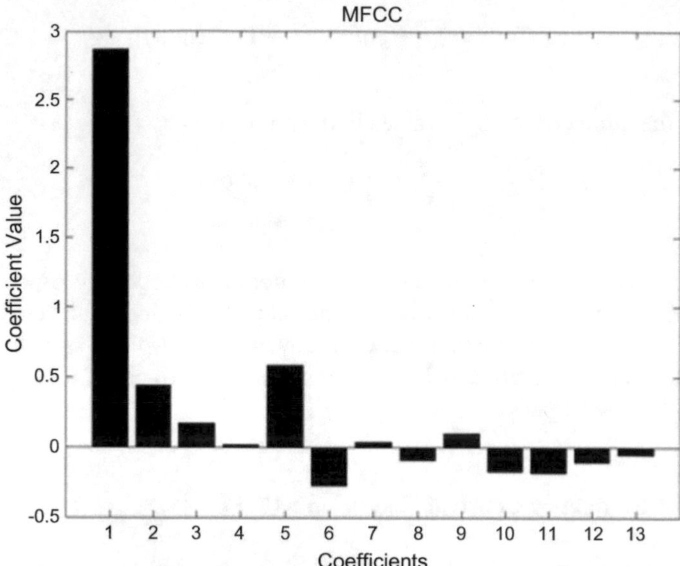

Fig. 5 MFCC plot of the sample language file

varies with the energy of audio and no correlation exists among different coefficients, this pattern made MFCCs much attractive in audio classification. Figure 5 shows the variation of a sample language speech file. The X-axis is depicting the 13 coefficients and Y-axis is depicting the value of each coefficient.

2.1.5 Tempo

Every sound has a specific speed in the form of beats. This speed can be measured by a property called tempo. It measures the beats per minute of the sound. It measures how the periodic occurrence of sound varies, which can be used to differentiate speech from nonspeech. Thus, the tempo-based feature has been plucked up for the proposed feature set.

2.1.6 Spectral Flux

It is the measure of spectral change between two adjacent frames (Eq. 3). It is an important feature for the separation of music from speech. Speech distinguishes itself between periods of transition of frequency or any other physical factor, while other speeches typically has a different constant rate of change. Figure 6 shows the plot for spectral flux for a sample language file. Here, the value of the flux has been notified against the real-time value in seconds. The figure shows that the flux value changes

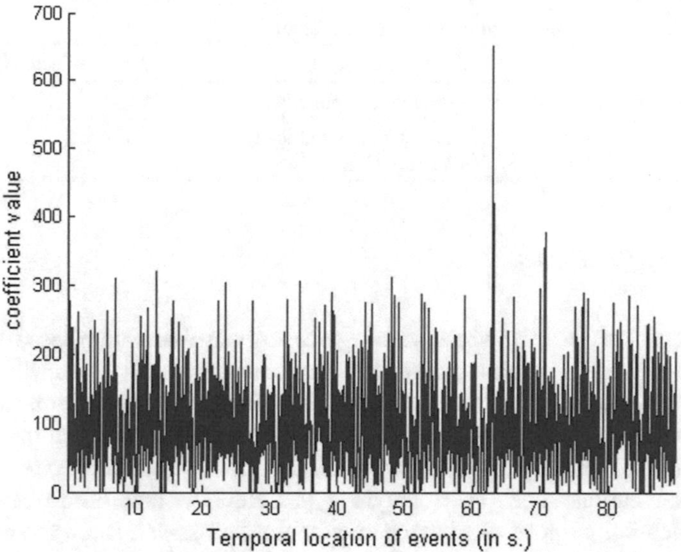

Fig. 6 Spectral flux of the sample language speech file

minutely with respect to the whole signal. So, it must give discriminative values with respect to each language file.

$$\mathrm{SFn} = \sqrt{\sum_{k=1}^{K} (|A(n+1,k)| - |A(n,k)|)^2} \tag{3}$$

2.2 *Dimensionality Reduction*

Feature selection or Dimensionality Reduction is the process of selecting a set of relevant features in model construction. It reduces the effort of classification. Here, dimensionality reduction has been achieved by reducing the size of the features by Principal Component Analysis PCA and Consistency-based selection algorithm CFS.167 features got depressed to the final set of features of size 21. Table 1 shows the selected feature after using feature selection algorithm. The selected feature with minimum number of attribute that is 18 has been subjected for classification. We have used Weka tool [16] for feature selection algorithm.

Table 1 Feature selection algorithm for extracted feature

System	Attribute evaluator	No. of selected attributes
Language identification system	Principal components	21
	Consistency-based feature selection CFS	18

2.3 *Classification*

Classification is the way by which we categorize a given observation based on the past observation. There is a training set of data that contains observations whose classes are known. These sets are used for the testing phase. Classification is considered under the domain of supervised learning. It is an approach in which the computer program learns from the input data and then uses this to authenticate new observations. Classification models are of varying types. Here, we have used Weka tools for the classification purpose. We have chosen tenfold cross-validation to analyze and compute predictive models by dividing the original sample into a training set that trains the setup, and a test set that evaluates it.

2.4 *Dataset Collection*

The experiment has been conducted on 600 audio files containing 31 official languages in different countries of Asia. It is very difficult to get benchmark audio data. The audio data has been collected from different sites of Internet. The data contains speeches, debates, and the recordings of public speakers.

3 Experimental Result

The feature selection algorithm was run on the entire feature extracted. The output of the algorithm was fed into a number of classifiers. Upon classification, we obtained an average accuracy of 81%. We were successfully able to identify 81% of the Asian languages correctly with the help of the overall system developed. Different classifiers have been audited to analyze the efficiency of the system. The accuracies have been summarized in Table 2. In Table 2, different types of classifiers has been taken into consideration. Furthermore, different classifiers of different types have been considered. The result has been calculated on tenfold cross-validation. Table 2 summarizes the result of classification.

Table 2 Classification accuracy (in %) for proposed work

Classifier type	Classifier name	Accuracy (%)
Bayes	Naïve Bayes	91.76
Function	SimpleLogistics	89.54
Function	SMD	89.6
Lazy	IBK	84.63
Lazy	Kstar	91.54
Meta	Bagging	91.15
Meta	Iterative Classifier Optimization	81.00
Meta	Random Committer	87.15
Meta	Randomizable Filtered Classifier	92.77
Rules	PART	90.31
Trees	J48	91.61
Trees	LMT	89.76
Trees	RandomForest	94.61
Trees	RandomTree	91.69
Trees	REPTree	91.07

Table 3 Comparison table for various methods

Method	Accuracy (%)
Mahadev and Lakhani [4]	83
Itrat et al. [6]	89
Gwon et al. [10]	85
Proposed method	91

3.1 Comparing Various Methods

The methods used in previous works has been used for comparing with our proposed methodology. We have done a comparative study with three papers and have achieved better results with the proposed work in Table 3. Mahadev and Lakhani [4] have used the MFCC approach for the classification of language. While working on their method, we achieved an accuracy of 75%, whereas in our proposed feature algorithm, we have attained an accuracy of 81%. Itrat et al. [6] have worked on differentiating Pakistani languages using MFCC and vector quantization. Their approach gave us an accuracy of 79%. Gwon et al. [10] have proposed the technique of sparse coding and their method gave an accuracy of 80%.

4 Conclusion

In the propounded work, a contrasting approach toward automatic language identification has been propounded. The application of the system has been pursued to the latest problem. The approach presented will abridge the use of classical methods while maintaining their reliability. The proposed system will be explored to another problem to analyze the performance of the system. The well-known Asian languages have been successfully identified from a set of various audio inputs with an accuracy of 81%. The future work on this project includes expansion of the language set on global level based on seven continents. Furthermore, dialect identification for global language will be identified.

Acknowledgements This chapter does not contain any studies with human participants or animals performed by any of the authors.

References

1. Garg, A., Gupta, V., Jindal, M.: A survey of language identification techniques and applications. J. Emerg. Technol. Web Intell. **6**(4), 388–400 (2014)
2. Karpagavalli, S., Chandra, E.: A review on automatic speech recognition architecture and approaches. Int. J. Signal Process. Image Process. Pattern Recogn. **9**(4), 393–404 (2016)
3. Grothe, L., De Luca, E.W., Nürnberger, A.: A Comparative Study on Language Identification Methods. *LREC* (2008)
4. Lakhani, V.A., Mahadev, R.: Multi-Language Identification Using Convolutional Recurrent Neural Network. arXiv preprint arXiv: 1611.04010 (2016)
5. Chandrasekhar, V., Sargin, M.E., Ross, D.A.: Automatic language identification in music videos with low level audio and visual features. In: 2011 IEEE International Conference on Acoustics, Speech and Signal Processing (ICASSP). IEEE (2011)
6. Itrat, M. et al.: Automatic Language Identification for Languages of Pakistan. Int. J. Comput. Sci. Netw. Secur. (IJCSNS) **17.2**
7. Adami, A.G., Hermansky, H.: Segmentation of speech for speaker and language recognition. In: Eighth European Conference on Speech Communication and Technology (2003)
8. Dehak, N., Torres-Carrasquillo, P.A., Reynolds, D., Dehak, R.: Language recognition via i-vectors and dimensionality reduction." In *Twelfth annual conference of the international speech communication association*. 2011
9. Yu, Chengzhu et al. "UTD-CRSS system for the NIST 2015 language recognition i-vector machine learning challenge. In: 2016 IEEE International Conference on Acoustics, Speech and Signal Processing (ICASSP). IEEE (2016)
10. Gwon, Y.L., et al.: Language recognition via sparse coding. *INTERSPEECH* (2016)
11. Behravan, H., Kinnunen, T., Hautamäki, V.: Out-of-set i-Vector selection for open-set language identification. Odyssey **2016**, 303–310 (2016)
12. Vatanen, T., Väyrynen, J.J., Virpioja, S.: Language identification of short text segments with N-gram models. *LREC* (2010)
13. Torres-Carrasquillo, P.A., Reynolds, D.A, Deller, J.R.: Language identification using Gaussian mixture model tokenization. In: 2002 IEEE International Conference on Acoustics, Speech, and Signal Processing (ICASSP), vol. 1. IEEE (2002)
14. Karpagavalli, S., Chandra, E.: A review on automatic speech recognition architecture and approaches. Int. J. Signal Process. Image Process. Pattern Recogn. **9**(4), 393–404 (2016)

15. Lartillot, O., Toiviainen, P.: A Matlab toolbox for musical feature extraction from audio. In: International Conference on Digital Audio Effects (2007)
16. Hall, M., et al.: The WEKA data mining software: an update. ACM SIGKDD Explor Newsl **11.1**(2009), 10–18

An Improved Data Hiding Scheme Through Image Interpolation

Manasi Jana and Biswapati Jana

Abstract Hidden data communication through image interpolation is a challenging research issue in the field of information hiding, which maintains the tradeoff between payload and quality. This paper presents an improved interpolation-based data hiding schemes. Tang et al. proposed a high-capacity steganographic scheme through multi-layer embedding (CRS) with average embedding capacity 1.79 bpp with nearly 34 dB PSNR. To improve the interpolation-based data hiding scheme, we proposed an advanced data hiding scheme which divides a fixed length sub-message into floor and ceiling values. Then one value is added and another is subtracted from two adjacent interpolated pixels to minimize the image distortion. Here, two interpolated pixels are calculated between each adjacent pixels to increase the embedding capacity. The experimental results show that the proposed scheme has better performance than other existing interpolation methods such as NMI, IMP, and CRS. Moreover, the suggested scheme has a low-time complexity and provide good visual quality.

Keywords Reversible data hiding · Image interpolation · Steganography

1 Introduction

As per the development of digital technology, multimedia communication has been increased day by day. People started hidden communication through multimedia documents such as Image, Audio, and Video. Many researchers [1–3] have proposed several data hiding techniques for hidden communication. Reversible Data Hiding (RDH) is one of the most useful and applicable in many human-centric applications

M. Jana
Department of Master of Application, Haldia Institute of Technology,
Haldia 721657, West Bengal, India
e-mail: manasi.das30@gmail.com

B. Jana (✉)
Department of Computer Science, Vidyasagar University, Midnapore 721102,
West Bengal, India
e-mail: biswapatijana@gmail.com

A. K. Das et al. (eds.), *Computational Intelligence in Pattern Recognition*,
Advances in Intelligent Systems and Computing 999,
https://doi.org/10.1007/978-981-13-9042-5_14

such as military and medical data processing which grows interest among researchers. For many applications such as medical image processing, remote sensing images application, military maps navigation, RDH is a good solution to restore original images.

Among several RDH methods, difference expansion-based RDH [4, 5] is one major class in which a payload bit is hidden into an image by expanding the difference between two pixels. Tian [6, 7] suggested a difference expansion- based data hiding approach to conceal the secret data into the difference of two consecutive pixels with high payload. In [8], Zhenfe-Zhao developed an RDH method using histogram modification. In the data embedding stage, a multilevel histogram modification strategy is employed to enhance hiding capacity. In [9], Hsien-wen Tseng, designed an RDH scheme based on prediction error expansion (PEE). The prediction value is calculated using several predictors. Interpolation-based data hiding scheme is reported by the papers [10–13]. Recently, some data hiding scheme [14–19] has been developed but have limited capacity and visual quality. Image interpolation is the new mechanism used in reversible data hiding methods. It is widely applied to medical imaging fields such as image generation in computed tomography (CT) or magnetic resonance imaging (MRI). In many areas, Image interpolation has been widely used. Such as image resizing, up or down sampling, zooming, scaling, magnification and resolution enhancement. To achieve a high capacity, a small digital image is amplified to the larger size. So, it is important to design an innovative interpolation scheme to improve the processing time, increase payload as well as maintain the high quality of reconstructed images.

The rest of the paper is organized as follows. Section 2 describes the proposed data hiding scheme in detail. Section 3 shows the experimental results and analysis. Finally, the conclusions are drawn in Sect. 4.

2 The Proposed Scheme

In this section, we propose a new data hiding scheme using image interpolation method. Unlike other methods, the proposed scheme generates two interpolated pixels between each pair of pivot pixels to enhance the embedding capacity. We use an equation $(7 \times a + b)$ derived from $((a + (a + (a + b)/2)/2)/2)$, where a is the nearest pivot pixel and b is the farthest pivot pixel to get an interpolated value which is very close to the nearest pivot pixel. Instead of one pixel, a secret sub-message is embedded into two interpolated pixels to get a high visual quality.

Consider an original image I of size $(R \times L)$. The proposed scheme is applied to produce a $(3 \times R - 2) \times (3 \times L - 2)$ cover image C using the given Eq. 1.

$C(i, j) = I(k, l)$ where $i = 1, 4, 7 \ldots 3 \times M - 2$, $j = 1, 4, 7, \ldots 3 \times N - 2$, $k = 1, 2, 3 \ldots M$, $l = 1, 2, 3, \ldots N$.

$C(i, j) = \lfloor((7 \times C(i, j - 1) + C(i, j + 2))/8)\rfloor$ where $i = 1, 4, 7 \ldots 3 \times M - 2$,

$$
\begin{aligned}
&j = 2, 5, 8 \ldots 3 \times N - 2\\
&C(i, j) = \lfloor (7 \times C(i, j+1) + C(i, j-2))/8) \rfloor \text{ where}\\
&i = 1, 4, 7 \ldots 3 \times M - 2,\\
&j = 3, 6, 9 \ldots 3 \times N - 2\\
&C(i, j) = \lfloor (7 \times C(i-1, j) + C(i+2, j))/8) \rfloor \text{ where } i = 2, 5, 8 \ldots 3 \times M - 2,\\
&j = 1, 4, 7, \ldots 3 \times N - 2\\
&C(i, j) = \lfloor (7 \times C(i+1, j) + C(i-2, j))/8) \rfloor \text{ where } i = 3, 6, 9 \ldots 3 \times M - 2,\\
&j = 1, 4, 7, \ldots 3 \times N - 2\\
&C(i, j) = \lfloor (C(i-1, j) + C(i, j-1) + C(i-1, j-1))/3 \rfloor \text{ where } i = 2, 5, 8 \ldots 3 \times M - 2,\\
&j = 2, 5, 8 \ldots 3 \times N - 2\\
&C(i, j) = \lfloor (C(i+1, j) + C(i+1, j+1) + C(i, j+1))/3 \rfloor \text{ where } i = 3, 6, 9 \ldots 3 \times M - 2,\\
&j = 3, 6, 9 \ldots 3 \times N - 2\\
&C(i, j) = \lfloor (C(i-1, j) + C(i-1, j+1) + C(i, j+1))/3 \rfloor \text{ where } i = 2, 5, 8 \ldots 3 \times M - 2,\\
&j = 3, 6, 9 \ldots 3 \times N - 2\\
&C(i, j) = \lfloor (C(i, j-1) + C(i+1, j-1) + C(i+1, j))/3 \rfloor \text{ where } i = 3, 6, 9 \ldots 3 \times M - 2,\\
&j = 2, 5, 8 \ldots 3 \times N - 2
\end{aligned} \tag{1}
$$

The pixel $C(i, j) = I(i, j), i = 1, 4, 7 \ldots 3 \times M - 2, j = 1, 4, 7 \ldots 3 \times N - 2$ is considered as the pivot pixel not changed during data embedding. The Embedding Procedure for proposed scheme is described in Algorithm 1.

2.1 *Algorithm for Embedding*

Figure 1 shows an example of the proposed scheme to generate cover image from original image. Let us consider the original image I (Fig. 1a). The pivot pixels in cover image C (Fig. 1b) can be found by the suggested scheme as follows: $C(1, 1) = I(1, 1) = 86$, $C(4, 1) = I(2, 1) = 120$, $C(1, 4) = I(1, 2) = 27$ and $C(4, 4) = I(2, 2) = 40$.

The interpolated values can be obtained as follows: $C(1, 2) = (7 \times 86 + 27)/8 = 78$, $C(1, 3) = (7 + 86)/8 = 34$, $C(2, 1) = (7 + 120)/8 = 90$, $C(3, 1) = (7 \times 120 + 86)/8 = 115$, $C(2, 2) = (86 + 78 + 90)/3 = 84$, $C(2, 3) = (34 + 27 + 28)/3 = 29$, $C(3, 2) = (115 + 120 + 110)/3 = 115$, $C(3, 3) = (38 + 40 + 50)/3 = 42$.

Check the value I_{max} among four pivot of each 4×4 overlapping block using Eq. 2.

$$
\begin{aligned}
&I_{max} = max[C(i, j), C(i+3, j), C(i, j+3), C(i+3, j+3)] \text{ where } i = 1, 4, 7 \ldots 3 \times M - 5,\\
&j = 1, 4, 7 \ldots 3 \times N - 5\\
&avg = \lfloor ((C(i, j) + C(i+3, j) + C(i, j+3) + C(i+3, j+3))/4) \rfloor \text{ where } i = 1, 4, 7 \ldots 3 \times \mathrm{M} - 5,\\
&\mathrm{j} = 1, 4, 7, \ldots 3 \times \mathrm{N} - 5\\
&\mathrm{d} = I_{max} - avg
\end{aligned} \tag{2}
$$

Algorithm 1 Algorithm for data embedding

Input: Cover image I sized R × L. I(i,j) is a pixel at (i,j)th position in the cover image I where i=1, 2, 3,…R, j=1, 2, 3…L
Output: A stego image S sized (3 × R-2)× (3×L-2) is produced.
Begin
Step 1: A cover image C is produced based on Eq. (1)
Step 2: Repeatedly select a 4 × 4 overlapping block until all data are embedded
  Step 2.1: Compute d using Eq. (2)
  Step 2.2 : Compute n using Eq. (3)
  Step 2.3: Compute b_k, f_k and c_k using Eq. (4)
  Step 2.4: Secret message is embedded using Eq. (5)
  Step 2.5: If all secret message is embedded completely then
    a stego image S is produced and proceed to **step 4**.
  Step 2.6: else
    goto **step 2**
  Step 2.7: end if
Step 3: End of step 2
Step 4: End(Algorithm)
End

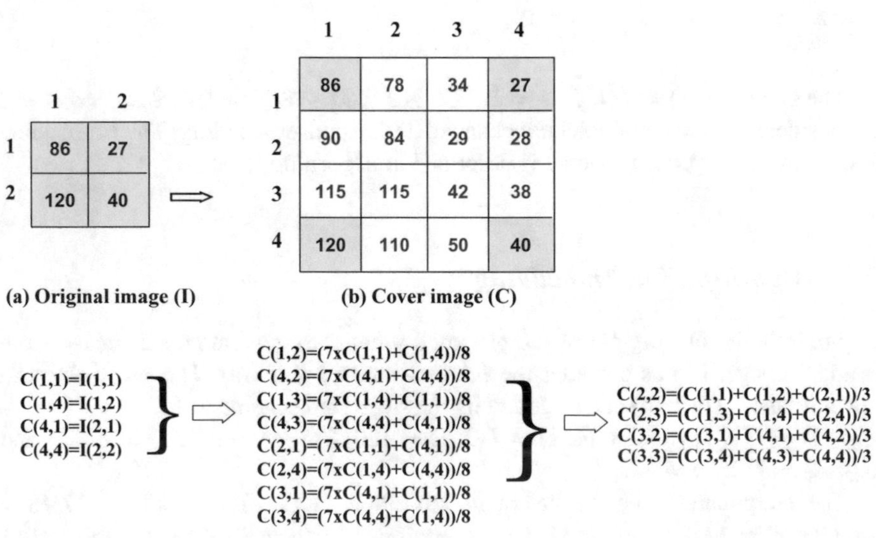

Fig. 1 Numerical example of developed image interpolation method

Then, the difference value d is calculated which decide the length of the hidden message. The number of bits can be represented as n which is defined in Eq. 3.

$$n = \log_2 d \tag{3}$$

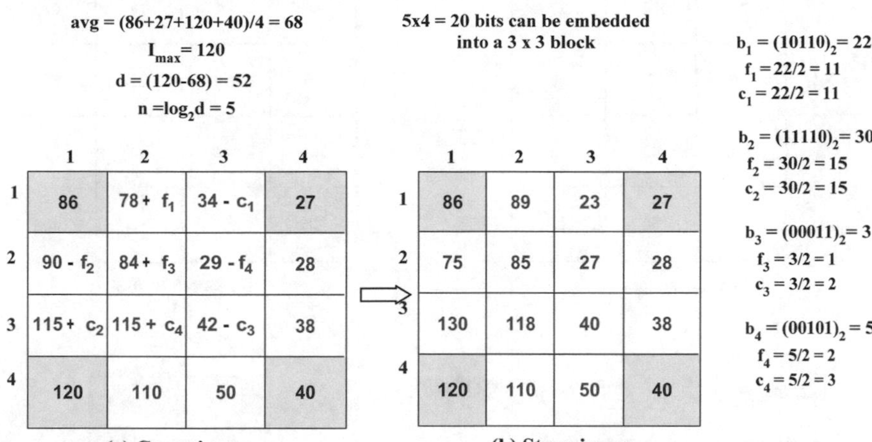

Fig. 2 Example of data embedding

Let b_k is the substring of secret message to be embedded. Then the floor and ceiling value of b_k are calculated using Eq. 4. Instead of one pixel these two values are embedded into two non-pivot pixels to get better image quality. $f_k = b_k/2$, $c_k = b_k/2$ where k = 1, 2, 3, 4 and f_k and c_k are the floor and ceiling value of b_k (Figs. 2, 3).

$$f_k = \lfloor b_k/2 \rfloor, c_k = \lceil b_k/2 \rceil \tag{4}$$

where k=1, 2, 3, 4

The secret information b_k, for k = 1, 2, 3 and 4 has been embedded using non-pivot pixels $C(i, j+1)$, $C(i, j+2)$, $C(i+1, j)$, $C(i+2, j)$, $C(i+1, j+1)$, $C(i+2, j+2)$, $C(i+1, j+2)$, $C(i+2, j+1)$.

Now, the stego pixels can be computed using Eq. 5.

$S(i, j) = C(i, j)$
if $C(i, j+1) \geq C(i, j)$ then
$S(i, j+1) = C(i, j+1) - f_1$
$S(i, j+2) = C(i, j+2) + c_1$
else
$S(i, j+1) = C(i, j+1) + f_1$
$S(i, j+2) = C(i, j+2) - c_1$
endif
if $C(i+1, j) \geq C(i, j)$ then
$S(i+1, j) = C(i+1, j) - f_2$

86	78	34	27
90	84	29	28
115	115	42	38
120	110	50	40

(a) Cover image

86	89	23	27
75	85	27	28
130	118	40	38
120	110	50	40

(b) Stego image

86	27
120	40

(c) Original Image

$f_1=(89-78)=11$, $c_1=(34-23)=11$ } $b_1=f_1+c_1=22=(10110)_2$

$f_2=(90-75)=15$, $c_2=(130-115)=15$ } $b_2=f_2+c_2=30=(11110)_2$

$f_3=(85-84)=1$, $c_3=(42-40)=2$ } $b_3=f_3+c_3=3=(00011)_2$

$f_4=(29-27)=2$, $c_4=(118-115)=3$ } $b_4=f_4+c_4=5=(00101)_2$

msg=10110 11110 00011 00101

Fig. 3 The original image and secret message generated from the stego image

Fig. 4 Four secret images

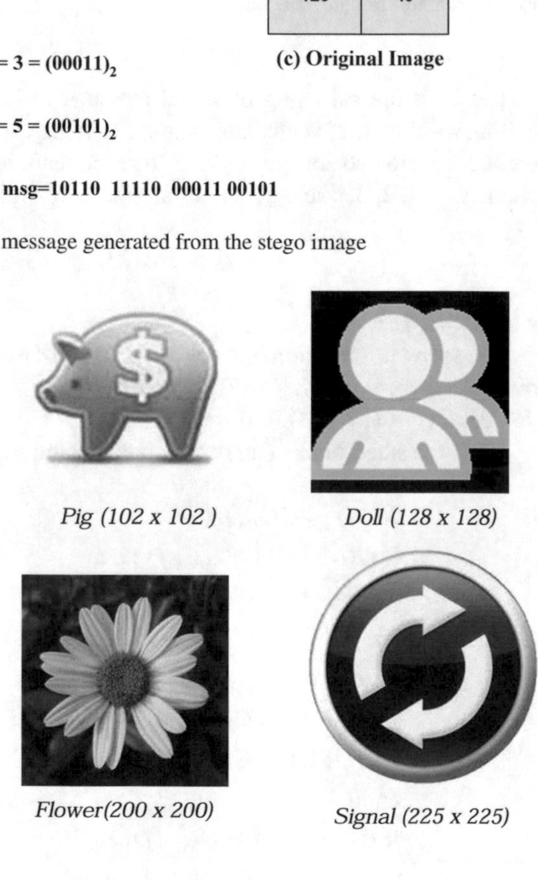

$$
\begin{aligned}
&S(i+2, j) = C(i+2, j) + c_2 \\
&\text{else} \\
&S(i+1, j) = C(i+1, j) + f_2 \\
&S(i+2, j) = C(i+2, j) - c_2 \\
&\text{endif} \\
&\text{if } C(i+1, j+1) \geq C(i, j) \text{ then} \\
&S(i+1, j+1) = C(i+1, j+1) - f_3 \\
&S(i+2, j+2) = C(i+2, j+2) + c_3 \\
&\text{else} \\
&S(i+1, j+1) = C(i+1, j+1) + f_3 \\
&S(i+2, j+2) = C(i+2, j+2) - c_3 \\
&\text{endif} \\
&\text{if } C(i+1, j+2) \geq C(i, j+3) \text{ then} \\
&S(i+1, j+2) = C(i+1, j+2) - f_4 \\
&S(i+2, j+1) = C(i+2, j+1) + c_4 \\
&\text{else} \\
&S(i+1, j+2) = C(i+1, j+2) + f_4 \\
&S(i+2, j+1) = C(i+2, j+1) - c_4 \\
&\text{endif} \\
&\text{where } i = 1, 4, 7, \ldots .3 \times M - 5, j = 1, 4, 7 \ldots 3 \times N - 5
\end{aligned}
\tag{5}
$$

2.2 Numerical Example

The original and cover image is used in this experiment has been depicted in Fig. 5. Here, four secret images are used as shown in Fig. 4 to compare the image quality with different schemes. Consider the secret information as binary bits as s_1 = “10110111100001100101”. Let i, j = 1 and find the maximum value.

$I_{max} = max\{C(1, 1), C(1, 4), C(4, 1), C(4, 4)\} = max\{86, 27, 120, 40\} = 120$ of the pixels C(i, j), C(i, j + 3), C(i + 3, j) and C(i + 3, j + 3) in a 4 × 4 block. Compute the value d using Eq. 3 to be sure about the number of secret bits to be embedded.

d $= I_{max} - avg = 120 - 68 = 52$ where avg $= \lfloor (C(i, j) + C(i, j+3) + C(i+3, j) + C(i+3, j+3))/4 \rfloor$

Calculate $n = log_2 d = 5$. Here we embed the n bits of secret message to each pair of non-pivot pixels. Here number of pairs of non pivot pixels is four as shown in Fig. 2. So we can embed 4 × 5 = 20 bits of secret message into a 4 × 4 overlapping

(a) Lena (b) Tiffany (c) Goldhill (e) Airplane

(f) Boat (g) Camera man (h) Chemical (d) Clock

(i) F16 (j) House (k) Lake (l) LivingRoom

(m) Moon (n) Peppers (o) Walk bridge (p) Woman

Fig. 5 Sixteen 256 × 256 grayscale test images

block. Since n = 5, the sub-messages are b1 = "10110" = $(22)_{10}$, b2 = "11110" = $(30)_{10}$, b3 = "00011" = $(3)_{10}$ and b4 = "00101" = $(5)_{10}$, respectively. Then after calculating the floor and ceiling value of each sub message we can embed these values into a pair of non-pivot pixels as shown in Fig. 2a using Eq. 5. and stego is generated shown in Fig. 2b.

The extracting process is a reversible process of the embedding process shown in Fig. 3. The maximum value I_{max} is calculated using four pivot pixels S(i, j), S(i + 3,j), S(i,j + 3), S(i + 3, j + 3) for each 4 × 4 overlapping block i.e., $I_{max} =$ max(S(i,j), S(i + 3,j), S(i, j + 3), S(i + 3, j + 3)). Then using Eqs. 2 and 3 calculate d and n. Since pixel values of S(1, 1), S(1, 4), S(4, 1) and S(4, 4) in the marked image are same with cover image C(1, 1), C(1, 4), C(4, 1) and C(4, 4) as shown in Fig. 3. Afterward, the cover image is generated by the proposed interpolation method using Eq. 1. Then the floor and ceiling values of each sub-message are calculated using cover and stego image. A secret sub-message is generated adding these floor and ceiling values. Let f_1 = (89–78) = 11 and c_1 = (34–23) = 11. Then $b_1 = f_1 + c_1 = (22)_{10} = (10110)_2$. Using same method, $b_2 =$ "11110", $b_3 =$ "00011" and $b_4 =$ "00101" are calculated respectively. At last, the secret message s1 = "10110"||"11110"||"00011"||"00101" = "1011011110000110 0101" is generated.

3 Experimental Result

In our experiment, sixteen 8-bit gray scale test images as shown in Fig. 5, each of which is 256 × 256, were used in the experiment as input source of the interpolation method. A 766 × 766 cover image is produced after applying the proposed interpolation algorithm. Table 1 shows the comparison with the proposed scheme. It is shown that our scheme achieves better than others. Jung and Yoo's method [10] could embed from 17, 201 bits to 81, 106 bits within image. Lee and Haung's method [11] could hide from 50, 080 bits to 1, 31, 586 bits. Tang et al. method [12] could hide the secret data from 1, 46, 352 to 2, 20, 779 bits. The proposed method could hide the secret data from 2, 57, 612 to 7, 75, 832 bits. The corresponding graph is shown in Fig. 6.

Table 2 presented a comparison among RDH scheme. It is observed that our scheme is better than other existing schemes. Here, four secret images are used as shown in Fig. 4 to compare the image quality with a different scheme. The PSNR value between cover and stego image of Jung and Yoo's method [10] varies from 33.01 dB to 40.42 dB. In Lee and Huang's method [11], the PSNR value varies from 25.93 dB to 33.54 dB. In Tang et al. method [12] the PSNR value varies from 20.80 dB to 28.29 dB. Whereas in the proposed method the PSNR value varies from 36.71 dB to 42.82 dB. The corresponding comparison graph is shown in Fig. 7.

The results show that the NMI method is better than INP and CRS schemes in terms of image quality. Whereas CRS scheme is better than NMI and INP in terms of capacity. Results show that our proposed scheme has good PSNR value with larger capacity then NMI, INP, and CRS schemes.

Table 1 A comparison of different reversible methods on pure payload

	Scheme	NMI	INP	CRS	Proposed
Test	Lena	**52,518**	97,934	2,15,033	5,54,284
	Tiffany	54,072	1,00,758	1,85,195	5,27,304
	Goldhill	59,267	1,02,140	2,07,838	5,46,192
	Airplane	17,201	50,080	1,46,352	2,57,612
	Boat	58,409	1,00,927	2,20,779	5,63,656
	Cameraman	40,652	75,211	2,00,307	4,03,392
	Chemical	77,328	1,23,882	1,99,806	6,76,252
	Clock	34,063	67,248	2,02,250	3,39,688
	F16	48,744	89,992	2,09,513	4,49,700
Images	House	36,694	67,764	2,04,153	3,43,428
	Lake	65,731	1,11,557	2,20,617	5,92,732
	Living room	59,665	1,08,250	2,08,770	6,03,632
	Moon	40,867	84,919	1,51,295	5,89,456
	Peppers	51,551	98,761	2,16,910	4,75,140
	Walk bridge	81,106	1,31,586	2,14,125	7,75,832
	Woman	38,377	75,910	2,17,558	3,57,880

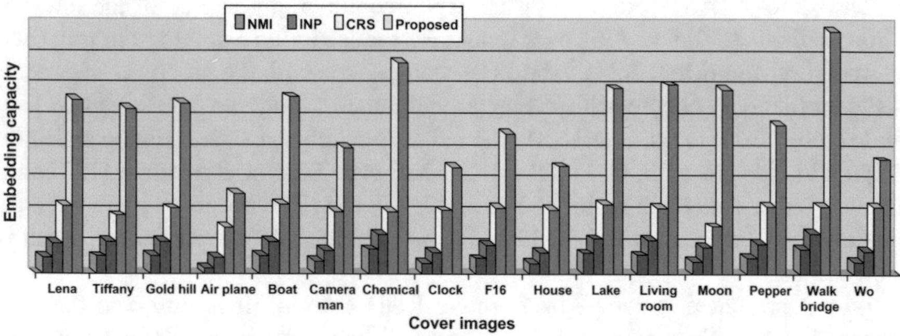

Fig. 6 Comparison of embedding capacity

4 Conclusions

In this investigation, a new interpolation approach has been proposed. Unlike other methods, the proposed method inserted two interpolated pixels between two pivot pixels to enhance the embedding capacity. The values of interpolated pixels are calculated in such a way that they are very much closer to the neighboring pivot pixel values. In stead of one pixel, a secret sub-message is embedded into two interpolated pixels to get a high visual quality. The proposed method is compared with other methods such as NMI, INP and CRS in terms of pure payload and PSNR value. Results demonstrated that the proposed method is better than the other three methods.

Table 2 A comparison of different reversible methods on PSNR

		Pig (83232 bits)				Doll (131072 bits)				Flower (320000 bits)				Signal (405000 bits)			
	Scheme	NMI	INP	CRS	Proposed	NMI	INP	CRS	Proposed	NMI	INP	CRS	Proposed	NMI	INP	CRS	Proposed
Test	Lena	36.27	29.78	25.92	48.41	40.51	32.46	27.85	48.79	39.94	32.63	24.97	42.52	35.00	28.66	22.48	38.77
	Tiffany	37.64	30.62	28.80	47.10	41.73	33.03	30.76	48.16	41.18	33.34	28.31	43.70	36.32	29.37	25.39	39.81
	Goldhill	38.17	31.44	24.88	50.45	41.69	33.95	29.15	50.81	41.67	34.06	25.55	45.46	36.89	30.37	22.11	41.06
	Airplane	38.84	31.81	29.34	43.08	49.84	36.53	32.62	49.94	43.10	35.96	31.15	43.70	38.01	30.73	27.44	40.49
	Boat	36.40	29.34	27.08	46.07	40.41	31.75	28.55	47.01	39.86	32.38	25.62	41.03	34.99	28.34	22.85	37.41
	Cameraman	34.93	28.18	27.63	43.25	41.55	31.98	29.07	43.23	39.04	31.65	26.36	39.82	33.64	26.83	23.78	36.96
	Chemical	35.39	30.05	28.57	46.01	38.41	31.20	30.90	47.50	38.82	31.10	27.19	43.29	34.59	27.66	24.56	39.51
	Clock	36.08	29.92	29.31	45.09	43.74	33.42	29.82	45.00	40.45	33.53	26.25	40.76	34.88	28.24	24.12	38.85
Images	F16	35.36	28.68	29.08	50.78	39.91	30.83	29.46	49.45	39.06	31.63	26.70	40.54	33.76	27.49	24.18	37.58
	House	36.52	30.03	25.72	45.49	43.18	33.96	27.80	46.99	40.53	33.86	25.69	43.91	35.25	29.07	22.73	40.98
	Lake	34.20	27.86	25.65	43.26	38.01	30.12	27.47	44.75	37.71	29.81	24.23	40.14	33.01	25.93	21.82	36.71
	Living room	36.84	30.38	27.10	47.90	40.31	32.25	28.55	48.21	40.40	32.57	26.42	42.43	35.43	28.85	23.78	38.38
	Moon	41.77	34.43	31.39	47.98	47.45	38.42	33.27	49.51	45.54	37.89	31.50	45.60	40.42	33.54	28.29	41.89
	Peppers	36.00	29.90	25.18	45.55	40.72	32.37	26.94	47.37	39.69	32.38	23.41	42.78	34.82	28.10	20.80	39.63
	Walk bridge	35.36	28.86	24.00	43.27	38.78	31.36	26.10	44.86	38.64	31.24	24.59	40.57	34.06	27.31	21.58	36.73
	Woman	40.08	31.47	26.73	48.87	45.03	34.71	28.98	51.94	43.54	34.95	25.87	46.38	38.50	30.83	23.11	42.82

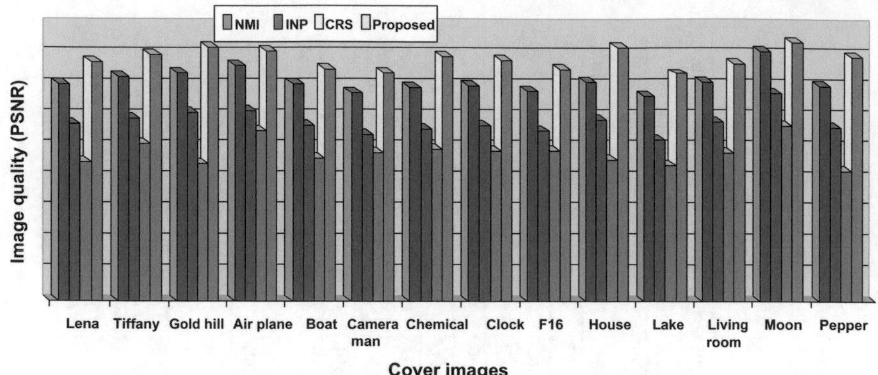

Fig. 7 Comparison of image quality after embedding 4,05,000 bits

References

1. Cheddad, A., Condell, J., Curran, K., Mc Kevitt, P.: Digital image steganography: survey and analysis of current methods. Signal Process. **90**(3), 727–752 (2010)
2. Luo, L., Chen, Z., Chen, M., Zeng, X., Xiong, Z.: Reversible image watermarking using interpolation technique. IEEE Trans. Inf. Forensics Secur. **5**(1), 187–193 (2010)
3. Jung, K.H.: A survey of interpolation-based reversible data hiding methods. Multimed. Tools Appl. **77**(7), 7795–7810 (2018)
4. Hu, Y., Lee, H.K., Chen, K., Li, J.: Difference expansion based reversible data hiding using two embedding directions. IEEE Trans. Multimed. **10**(8), 1500–1512 (2008)
5. Hu, Y., Lee, H.K., Li, J.: DE-based reversible data hiding with improved overflow location map. IEEE Trans. Circuits Syst. Video Technol. **19**(2), 250–260 (2009)
6. Tian, J.: Reversible data embedding using a difference expansion. IEEE Trans. Circuits Syst. Video Technol. **13**(8), 890–896 (2003)
7. Tian, J.: Reversible watermarking by difference expansion. In: Proceedings of Workshop on Multimedia and Security, vol. 19 (2002)
8. Zhao, Z., Luo, H., Lu, Z.M., Pan, J.S.: Reversible data hiding based on multilevel histogram modification and sequential recovery. AEU-Int. J. Electron. Commun. **65**(10), 814–826 (2011)
9. Tseng, H.W., Hsieh, C.P.: Prediction-based reversible data hiding. Inf. Sci. **179**(14), 2460–2469 (2009)
10. Jung, K.H., Yoo, K.Y.: Data hiding method using image interpolation. Comput. Stand. Interfaces **31**(2), 465–470 (2009)
11. Lee, C.F., Huang, Y.L.: An efficient image interpolation increasing payload in reversible data hiding. Expert. Syst. Appl. **39**(8), 6712–6719 (2012)
12. Tang, M., Hu, J., Song, W.: A high capacity image steganography using multi-layer embedding. Opt.-Int. J. Light. Electron Opt. **125**(15), 3972–3976 (2014)
13. Hu, J., Li, T.: Reversible steganography using extended image interpolation technique. Comput. Electr. Eng. **46**, 447–455 (2015)
14. Jana, B., Giri, D., Mondal, S.K.: Partial reversible data hiding scheme using (7, 4) hamming code. Multimed. Tools Appl. **76**(20), 21691–21706 (2017)
15. Jana, B., Giri, D., Mondal, S.K.: Dual image based reversible data hiding scheme using (7, 4) hamming code. Multimed. Tools Appl. 1–23 (2016)
16. Jana, B.: Dual image based reversible data hiding scheme using weighted matrix. Int. J. Electron. Inf. Eng. **5**(1), 6–19 (2016)

17. Jana, B., Giri, D., Mondal, S.K.: Dual-image based reversible data hiding scheme using pixel value difference expansion. IJ Netw. Secur. **18**(4), 633–643 (2016)
18. Jana, B.: High payload reversible data hiding scheme using weighted matrix. Opt.-Int. J. Light. Electron Opt. **127**(6), 3347–3358 (2016)
19. Jana, B.: Reversible data hiding scheme using sub-sampled image exploiting Lagrange's interpolating polynomial. Multimed. Tools Appl. 1–17 (2017)

Generalization of Multi-bit Encoding Function Based Data Hiding Scheme

Piyali Sanyal, Sharmistha Jana and Biswapati Jana

Abstract In this investigation, a generalize data embedding scheme with satisfactory quality has been suggested using multi-bit encoding procedure. Using a multi-bit encoding procedure, Kuo, Wen-Chung, et al. achieve high data hiding capacity with acceptable stego image quality, but security was another major issue which is possible to enhance. We have proposed a secure general form of multi-bit encoding function using an Adder-Element in which we achieve PSNR 42 (dB) while maintaining the same data hiding capacity with Kuo, Wen-Chung, et al. scheme. Two major achievements of our work are (1) it does not involve any complicated calculation to insert the secret message and (2) only Adder-Element is required to retrieve the hidden message. We have observed that the developed technique accomplishes high data hiding capacity while keep image quality better than other existing schemes.

Keywords Hidden data communication · Steganography · Original cover image · Exploiting modification direction (EMD) · Stego image

1 Introduction

Protection of personal information for online application is a challenging task. Information is communicated through public channel and hidden data communication is a challenge to protect them. Data hiding is an approach that hides secret data within a cover image and may be employed for ownership identification, tampered detection,

P. Sanyal · S. Jana
Department of Computer Science and Application, Midnapore College (Autonomous), Midnapore 721101, India
e-mail: piyasa1412@gmail.com

S. Jana
e-mail: sharmistha792010@gmail.com

B. Jana (✉)
Department of Computer Science, Vidyasagar University, Midnapore 721102, India
e-mail: biswapatijana@gmail.com

A. K. Das et al. (eds.), *Computational Intelligence in Pattern Recognition*, Advances in Intelligent Systems and Computing 999,
https://doi.org/10.1007/978-981-13-9042-5_15

and copyright protection. Two parameters, visual quality of marked image calculated by Peak Signal-to-Noise Ratio (PSNR (dB)) and data hiding capacity measured by bits per pixel (bpp) are used to measure the accomplishment of any proposed scheme. Stego image quality will be less when we embed at a high rate and vice versa.

A simple hidden data communication scheme is the Least Significant Bit—Replacement (LSB-R) has been introduced by Turner [18]. The LSB-R scheme is unbalanced because even valued pixel will never be decremented and odd valued pixel will never be incremented. This asymmetry is well observed by a few detectors [2]. To overcome this problem, Sharp [16] proposed LSB matching (LSB-M) scheme which does not replace LSB but randomly either increments or decrements *one* in LSB of cover image when no match is found with secret data bit. Embedded message within the scheme is also detected by the detector suggested by Ker [3]. To enhance the LSB-M scheme, Mielikainen [14] proposed the LSB matching revisited (LSB-M-R) where the payload was same as LSB-M but changes are fewer, which guarantees good quality stegos. Zang and Wang [19] assumed that the alteration direction of Mielikainen's scheme is not exploited fully; that is why they developed a data hiding scheme by Exploiting Modification Direction (EMD) which achieves maximum data hiding capacity through one bit per pixel (bpp).

Chang et al. [1] offered a new steganographic scheme using dual image through EMD method. They first established a mod function of a (256×256) magic matrix. Then convert the secret data bits into the numeral system of base-5 and embed two secret bits. Lee et al. [8] introduced a loss-less data hiding scheme. Later, Lee et al. [9] embedded private message applying center point direction of pixels to get the stego pixels. To protect the deterioration of the image quality, Lou et al. [12] proposed Reduced Difference Expansion (RDE) method. Lou's scheme is not only reversible but also meets low computational cost with high capacity data embedding scheme. Lee and Huang [7] suggested a RDH method. In their scheme, the average embedding rate is 1.07 (bpp). Qin et al. [15] presented a RDH using EMD. A LSB matching scheme has been designed by Lu et al. [13]. The stego images are obtained through the mod function. To achieve the reversibility in data hiding, the LSBs are checked via an averaging procedure then modification has been performed using a rule table. Chang et al. [1] embedded secret message bits by the mod function to accomplish a higher data hiding capability of 1.00 (bpp), but the visual quality of image was substandard to the method proposed by Lee et al. [11]. Zhang and Wang [20] suggested EMD techniques, that takes n pixels for encode ddata. Kieu and Chang [4] developed a new function for retrieval by altering the derivation function. Shen and Huang [17] developed a data hiding method through PVD and improved EMD that was irreversible. Qin et al. [15] presented only EMD as a the reversible data hiding technique. In 2016, Lee et al. [10] developed an efficient RDH using image interpolation. Kuo et al. [5] developed multi-bit encoding function for good hiding scheme with 3.2 bpp and 30 dB PSNR.

Thus, designing an innovative scheme is still an important issue which could maintain good quality image and increase data embedding capacity through dual image. In this paper, we proposed generalized data hiding schemes based on EMD which accomplish better qualities with improved capacity.

The paper is organized as mentioned below:
Section 2 presented the Literature survey. Section 3 contain the suggested scheme. Section 4 shows the results and comparison. Some steganalysis and steganographic attacks are listed in Sect. 5. Finally, Sect. 6 provides the summary of the technique.

Motivation and objective: In this article, a new data hiding technique using EMD has been proposed. The main motivation and objective of the proposed scheme are listed below:

- Data Hiding Capacity: Increase data hiding capacity or Payload.
- Security: Modification of embedding function and improve security.
- Imperceptibility: Secret data invisible and undecidable.

2 Literature Review

A data hiding method is called Exploiting Modification Direction (EMD) where a group of n pixels are used as an inserting unit and embed secret digit within $(2n + 1)$ notational system where $n \geq 2$. In 2006, Zhang and Wang [19] proposed EMD method where one pixel is increased or decreased by 1 during data embedding. In this method, a function $f()$ is used for embedding and extracting the secret message. For $n = 2$, there are four modification directions. In 2011, Kieu and Chang [4] proposed fully exploiting modification direction method by improving the $f()$ function. This method achieves higher results. In 2014, Qin et al. [15] suggested a reversible EMD method using two steganographic images. Kuo, Wen-Chung, et al. [5] proposed a multi-bit encoding function. Here, two EMD based steganographic method suggested by Zhang and Wang and Kuo, Wen-Chung, et al. are discussed.

2.1 Zhang and Wang's Scheme

Zhang and Wang's [19] developed a data hiding method based on Exploiting Modification Direction (EMD). Consider a cover image C of size $(X \times Y)$, where X is the height and Y is the breadth of the image. First the pixel values of cover image permuted using a key. Then partitioned into several blocks of n pixels $(x_1, x_2, \ldots, x_n)$, where $n \geq 2$. In this method, n cover pixels can hide secret digits in a $(2n + 1)$-ary notational system. Before embedding, the secret message is divided into several parts of k bits using the Eq. 1.

$$k = \lfloor m. \log_2(2n + 1) \rfloor \tag{1}$$

where m is the decimal value of k bits in the $(2n + 1)$-ary annotation system. There are $2n$ possible ways of modification which may happen for each block of n pixels.

During data embedding, only one pixel of each block is incremented or decremented by 1.

For data embedding, first calculate the function $f()$ using following Eq. 2.

$$f(x_1, x_2, \ldots, x_n) = (x_1 \times 1 + x_2 \times 2 + \cdots x_n \times n) \bmod (2n+1) \quad (2)$$

When the hidden digits $d = f()$, then pixel quantity remain same which means no modification is required. When $d \neq f()$, then compute s as $s = (d - f()) \bmod (2n+1)$. When $s < n$, then pixel amount of x_s is increased by one (1), differently, the pixel amount of x_{2n+1-s} is decreased by one (1). In the extraction process, it can easily retrieve hidden message by computing the function $f()$ from marked pixel block.

Example 1 Assume two pixels x_1 and x_2 are 113 and 120 respectively for $n = 2$. Secret message $D = (1011)_2$. Let $m = 1$. So, $k = \lfloor 1.\log_2(2 \times 2 + 1) \rfloor = 2$. Select two bits from secret message that is $(10)_2$ and convert it into 5-ary notational number that is $d = (2)_5$. Now, calculate f value using Eq. (2) as $f(113, 120) = (113 \times 1 + 120 \times 2) \bmod 5 = 3$. Since $2 \neq 3$, calculate s as $s = (2 - 3) \bmod 5 = 4$. Here, $4 \geq 2$, so the pixel value of $x_{5-4} = x_1$ that is 113 is decreased by 1 and we get 112. So, new pixel values are $x_1' = 112$ and $x_2' = 120$.

To extract data, calculate f value using Eq. (2). So, $f(112, 120) = (112 \times 1 + 120 \times 2) \bmod 5 = 2$. Then convert f value into 2 bits binary form that is 10. ■

2.2 *Kuo, Wen-Chung, et al.'s Scheme*

Kuo, Wen-Chung, et al. [5] proposed a high capacity data hiding scheme based on multi-bit encoding function. Before this scheme, none achieved payload greater than 2 bpp. To solve the problem, a novel multi-bit encoding function is proposed. The coefficients are produced by $a_1 = (2^1 - 1) = 2^0$, $a_2 = (2^2 - 1) = (2^0 + 2^1, \ldots, a_n) = (2^n - 1) = (2^0 + 2^1 + \cdots + 2^{n-1})$. In general form it is $a_i = 2^i - 1$. Therefore, they extend the idea of data embedding function with a suggested function and ratio $r = 2^k$ which are shown below in Eqs. 3 and 4:

$$f_m(x_i, x_2, \ldots, x_n) = \left[\sum_{i=1}^{n}(c_i.x_i)\right] mod\ 2^{nk+1} \quad (3)$$

and

$$c_i = \begin{cases} 1, & x = 1 \\ 2^k c_{i+1} + 1, & i \neq 1 \,\&\, i > 0 \end{cases} \quad (4)$$

3 Proposed Scheme

Here, we proposed a generalized EMD scheme of multi-bit encoding. We found that the suggested scheme enhanced the performance with the help of some extra variable, i.e., an Adder-Element (ae) as specified in the given Eq. 5.

$$f_{new}(x_1, x_2, \ldots, x_n) = \left[\sum_{i=1}^{n}(c_i \times (ae + x_i)\right] mod\ 2^{nk+1} \tag{5}$$

where x_i is the ith pixel in adjacent pixel and k = no of bits. The c_i value is calculated using Eq. 6.

$$c_i = \begin{cases} 1, & x = 1 \\ 2^k c_{i+1} + 1, & i \neq 1 \,\&\, i > 0 \end{cases} \tag{6}$$

and Adder-Element (ae) = (Length of secret data) mod 2^{α}, where $2 < \alpha \leq 8$. This Adder-Element also acts as a secret key without which it is hard to decode the secret message. It is possible to enhance the security using this function with the help of Adder-Element. We use generalized mathematical induction to prove the above equation as follows:

Step 1: For $i = 1$, compute $c_1 = (2^k)^0 = 1$.
Step 2: For $i = 2$, compute $c_2 = (2^k)^0 + (2^k)^1 = 2^k \times c_1 + c_1$.
Step 3: For $i = 3$, compute $c_3 = (2^k)^0 + (2^k)^1 + (2^k)^2 = 2^k \times [(2^k)^0 + (2^k)^1] + (2^k)^0 = 2^k \times c_2 + c_1$.
Step 4: For $i = n$, let $c_n = (2^k)^0 + 2^k \times [(2^k)^0 + (2^k)^1 + \cdots + (2^k)^{n-1}] = 2^k \times c_{n-1} + c_1$.
Step 5: For i = n + 1, compute $c_{n+1} = (2^k)^0 + 2^k \times [(2^k)^0 + (2^k)^1 + \cdots + (2^k)^{n-1} + (2^k)^n] = 2^k \times c_n + c_1$

To enhance the hiding capacity, we proposed encoding function which introduce an extra variable i.e. Adder-Element. We have used this Adder-Element as we are changing the value of the pixels at the very beginning and then we apply the algorithm that changes the values of pixels again to some extent. These changed values of pixels are the pixels of stego image. The advantage of using Adder-Element is that the values of new stego pixels are very close to the values that of the original pixels. The proposed algorithm is shown in Algorithm 1.

Algorithm 1: Embedding Scheme

Input: Original image I($M \times N$), Hidden message bits (S) = $\{(s_{nk}, s_{nk-1}, \ldots, s_0)_2\}$

Output: n adjacent marked-pixels $(x_1', x_2', \ldots, x_n')$

1: Compute function $t = f_{new}(x_1, x_2, \ldots, x_n)$;
2: Transform $(s_{nk}, s_{nk-1}, \ldots, s_0)$ to new coded value $(s = 2^{nk} \times s_{nk} + 2^{nk-1} \times s_{nk-1} + \ldots + 2^1 \times s_1 + 2^0 \times, s_0, where s_{nk}, s_{nk-1}, \ldots, s_0 \in 0, 1)$.
3: Compute the difference value D between s and t, i.e. $D = (s - t) \ mod \ 2^{nk+1}$
4: If $D = 0$, then not altered else if $D = 2^{nk}$, then $(x_n' = x_n + (2^k - 1))$ and $x_1' = x_1 + 1$. else if $D < 2^{nk}$, then transform $D = (d_{n-1}, d_{n-2}, \ldots, d_0)_2^k$ and $x_{i+1}' = x_{i+1} + d_i$ and $x_i' = x_i - d_i$ for $i = n-1, n-2, \ldots, 0$. else if $D > 2^{nk}$, then transform $D = 2^{nk+1} - D = (d_{n-1}, d_{n-2}, \ldots, d_0)_2^k$ and $x_{i+1}' = x_{i+1} - d_i$ and $x_i' = x_i + d_i fori = n-1, n-2, \ldots, 0$.
5: Therefore for each block follow these steps:
6: E1. Take n pixels $(x_1, x_2, \ldots, x_n)$ from the image I_c.
7: E2. Adjust the group pixels $(x_1, x_2, \ldots, x_n)$ using hidden message following Algorithm 1.
8: E3. Stego generated I_s.
9: end

Example 2 Consider four pixels = (10, 19, 5, 9), and hidden message $(0100100101001)_2$. Take 13 bits (n*k+1) from the message to hide within the above pixels. Here, n = 4, k = 3, The value of the Adder-Element(ae) = 12 mod 256 = 12. The resultant pixel will be = (11, 19, 8, 3) following Algorithm 1 given below:

Step 1: Compute t = $f_{new}(10, 19, 5, 9) = 13827 \bmod 2^{13}$
Step 2: Convert $(0100100101001)_2$ into 10-ary, that is s = $(2345)_{10}$
Step 3: Calculate D1 = $(2345 - 13827) \bmod 2^{13} = 4902$.
Step 4: D1 = 4902> 2^{12}. Afterward D1 = 2^{13}-D1 = 3290, Convert $(3290)_{10}$ into 2^3-ary = $(6332)_8$.
Step 5: Change (x_4, x_3, x_2, x_1) when $(d_3, d_2, d_1, d_0) = (6, 3, 3, 2)$; For $d_3 = 6$, compute $x_4 = 9 - 6 = 3$ and $x_3 = 5 + 6 = 11$; For $d_2 = 3$, compute $x_3 = 11 - 3 = 8$ and $x_2 = 19 + 3 = 22$; For $d_1 = 3$, compute $x_2 = 22 - 3 = 19$ and $x_1 = 10 + 3 = 13$; For $d_0 = 2$, compute $x_1 = 13 - 2 = 11$
Step 6: The resultant values are (11, 19, 8, 3)

The extraction has been performed using the following steps:

R 1: Consider n pixels $(x_1', x_2', \ldots, x_n')$ from I_s.
R 2: Compute the secret message $s = f_{new}(x_1', x_2', \ldots, x_n')$.
R 3: Convert s to binary form

Computer using the extraction function
$f_{new}(11, 19, 8, 9) = [1 * (12 + 11) + 9 * (12 + 19) + 73 * (12 + 8) + 585 * (12 + 9)] \bmod 2^{13}$ = 10537 mod 2^{13}= $(2345)_{10}$ that is $(0100100101001)_2$. Here Adder- Element = 12.

4 Implementation and Comparison

The images of size (512 × 512) shown in Fig. 1 has been used in our experiment. The Fig. 2 shows the image after embedding data. The Table 1 presented results of PSNR (dB). The payload is calculated as follows:

$$B = \frac{(H \times W) \times \frac{1}{4} \times r}{H \times W}, \tag{7}$$

where $H = 512$ and $W = 512$, r =13 bits. The payload $B = 3.25$ bpp.

The comparison is depicted in Table 2. The visual quality is lower than Shen and Huang [17] but higher than Lee et al. [6] and Zeng et al. [21]. This method is superior in terms of embedding capacity than the other existing methods. In terms of security, it is also better because the secret key is used to communicate the secret message through stego image. Again without secret key, it is hard to retrieve the secret message.

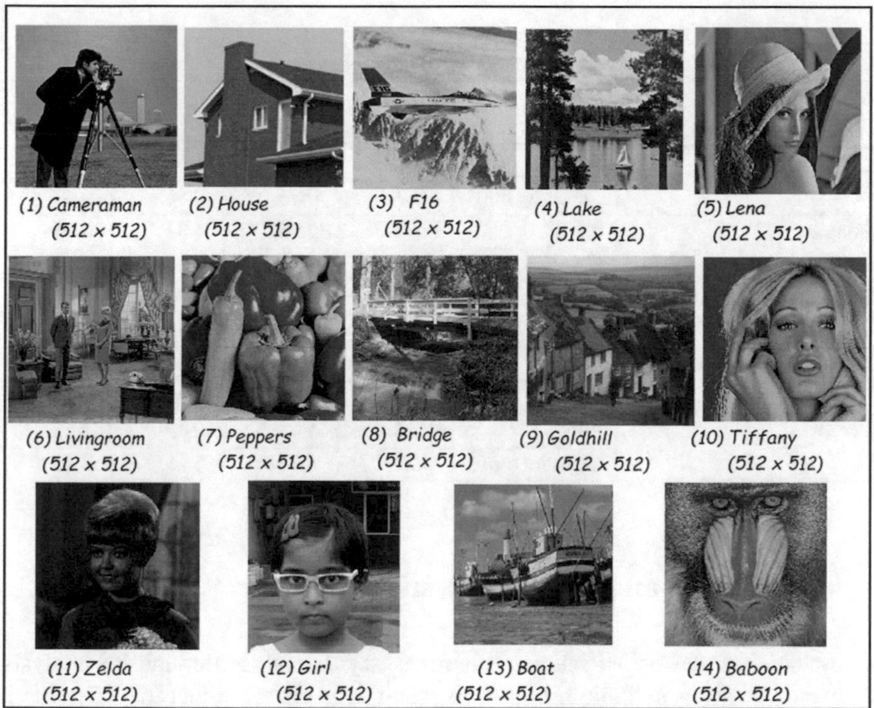

Fig. 1 Standard input images are used for experiment in our proposed scheme

Table 1 Experimental results with PSNR

Input image	Data (in bits)	PSNR (Stego versus Cover)
Boat (512 × 512)	327,680	52.87
	589,824	44.86
	851,968	37.31
House (512 × 512)	327,680	52.87
	589,824	44.87
	8,51,968	37.13
F16 (512 × 512)	327,680	52.87
	589,824	44.87
	851,968	37.13
Zelda (512 × 512)	327,680	52.88
	589,824	44.88
	851,968	37.13
Babbon (512 × 512)	327,680	52.87
	589,824	44.87
	851,968	37.13

Table 2 Comparison of proposed scheme with existing schemes

Methods	Measure	Images			
		Lena	Boat	Goldhill	Babbon
Lee et al. [6]	PSNR	34.38	33.12	32.08	30.03
	Capacity (bpp)	0.91	0.86	0.84	0.62
Zeng et al. [21]	PSNR	32.74	32.96	31.82	30.97
	Capacity (bpp)	1.04	1.04	0.80	0.51
Shen and Huang [17]	PSNR	42.46	41.60	41.80	38.88
	Capacity (bpp)	1.53	1.55	1.54	1.69
Kuo, Wen-Chung, et al. [5]	PSNR	37.46	37.60	37.80	37.88
	Capacity (bpp)	3.25	3.25	3.25	3.25
Proposed	PSNR	42.23	41.43	42.49	42.53
	Capacity (bpp)	3.25	3.25	3.25	3.25

5 Steganalysis and Steganographic Attacks

To evaluate the developed scheme it is necessary to analyze through RS analysis. The result is given in Table 3. From this result, it is observed that the RS value is nearer to zero which indicate good concealment.

Table 3 RS analysis of the stego image in our the proposed scheme

Image	Data (bits)	SM				
		R_M	R_{-M}	S_M	S_{-M}	RS value
Cameraman	160000	7118	7107	3551	3594	0.0051
	400000	6768	6851	3944	3895	0.0123
	600000	6304	5947	4943	5279	0.0616
Lena	160000	5617	5607	4067	4068	0.0011
	400000	5563	5476	4291	4337	0.0135
	600000	5636	5539	4517	4589	0.0166
Baboon	160000	5893	5815	4960	5105	0.0205
	400000	5897	5875	5076	5131	0.0070
	600000	6018	5813	5107	5313	0.0369

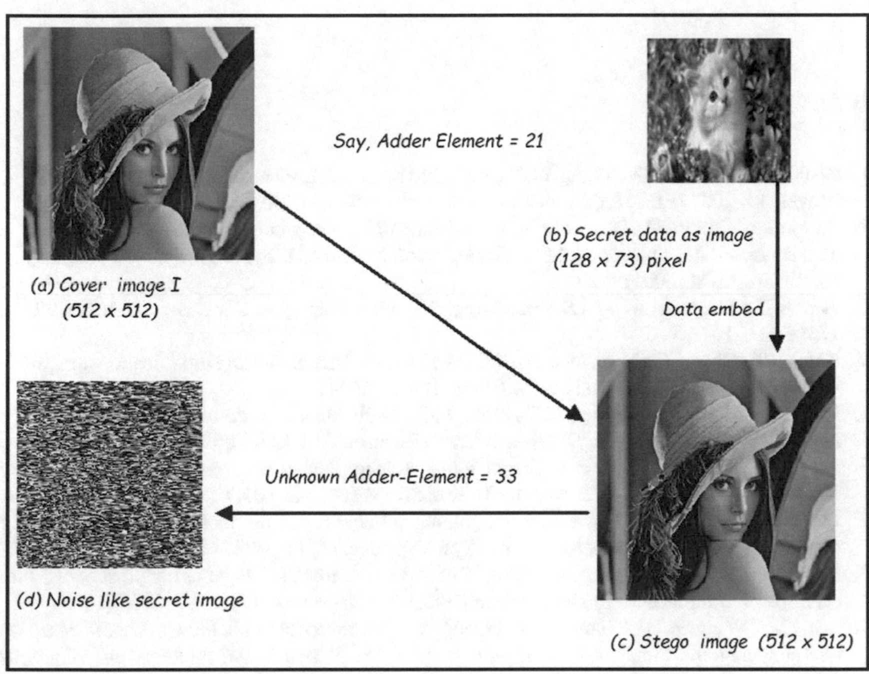

Fig. 2 Noise like secret data with unknown Adder-Element (ae) as secret key

5.1 *Brute Force Attack*

The brute force attack with unknown secret key is tested in this section.

The process prevents possible malicious attacks. The developed method constructs marked images that defend the actual message by embedding hidden data. Figure 2 depicted the results with wrong Adder-Element. The proposed scheme is robust against various attacks.

6 Conclusion

We proposed a generalized good embedding scheme using modified improved exploiting modification direction with shared secret key (Adder-Element). A secret key (Adder-Element) is employed that assures security. Also, the scheme still maintains good visual quality and it is superior than other existing scheme. In addition, the proposed scheme achieves payload 3.25 bpp that is better than previous scheme. Compared to our proposed scheme with existing schemes, it is superior in terms of hiding capacity and security. We have tested our scheme using RS analysis, Histogram attacks, statistical attack, and brute force attack. Also, the method maintains low relative entropy which ensures good concealment of hidden message. We have tested our proposed idea with standard image which assures the originality of the proposed work.

References

1. Chang, C.C., Kieu, T.D., Chou, Y.C.: Reversible data hiding scheme using two steganographic images. In: 2007 IEEE Region 10 Conference on TENCON 2007, pp. 1–4 (2007)
2. Fridrich, J., Goljan, M., Du, R.: Reliable detection of LSB steganography in color and grayscale images. In: Proceedings of the 2001 Workshop on Multimedia and Security: New Challenges, pp. 27–30. ACM (2001)
3. Ker, A.D.: Steganalysis of LSB matching in grayscale images. IEEE Signal Process. Lett. **12**(6), 441–444 (2005)
4. Kieu, T.D., Chang, C.C.: A steganographic scheme by fully exploiting modification directions. Expert. Syst. Appl. (Elsevier) **38**(8), 10648–10657 (2011)
5. Kuo, W.C., Kuo, S.H., Wang, C.C., Wuu, L.C.: High capacity data hiding scheme based on multi-bit encoding function. Opt. Int. J. Light. Electron Opt. **127**(4), 1762–1769 (2016)
6. Lee, C.F., Chen, H.L., Tso, H.K.: Embedding capacity raising in reversible data hiding based on prediction of difference expansion. J. Syst. Softw. (Elsevier) **83**(10), 1864–1872 (2010)
7. Lee, C.F., Huang, Y.L.: Reversible data hiding scheme based on dual stegano-images using orientation combinations. Telecommun. Syst. (Springer) **52**(4), 2237–2247 (2013)
8. Lee, C.C., Wu, H.C., Tsai, C.S., Chu, Y.P.: Adaptive lossless steganographic scheme with centralized difference expansion. Pattern Recognit. (Elsevier) **41**(6), 2097–2106 (2008)
9. Lee, C.F., Wang, K.H., Chang, C.C., Huang, Y.L.: A reversible data hiding scheme based on dual steganographic images. In: Proceedings of the 3rd International Conference on Ubiquitous Information Management and Communication, pp. 228–237. ACM (2009)
10. Lee, C.F., Weng, C.Y., Chen, K.C.: An efficient reversible data hiding with reduplicated exploiting modification direction using image interpolation and edge detection. Multimed. Tools Appl. (Springer), 1–24 (2016)
11. Lee, S.K., Suh, Y.H., Ho, Y.S.: Lossless data hiding based on histogram modification of difference images. Advances in Multimedia Information Processing-PCM 2004, pp. 340–347. Springer, Berlin (2004)
12. Lou, D.C., Hu, M.C., Liu, J.L.: Multiple layer data hiding scheme for medical images. Comput. Stand. Interfaces (Elsvier) **31**(2), 329–335 (2009)
13. Lu, T.C., Tseng, C.Y., Wu, J.H.: Dual imaging-based reversible hiding technique using LSB matching. Signal Process. (Elsevier) **108**, 77–89 (2015)
14. Mielikainen, J.: LSB matching revisited. Signal Process. (Elsevier) **13**(5), 285–287 (2006)

15. Qin, C., Chang, C.C., Hsu, T.J.: Reversible data hiding scheme based on exploiting modification direction with two steganographic images. Multimed. Tools Appl. (Springer) **74**(15), 5861–5872 (2015)
16. Sharp, T.: An implementation of key-based digital signal steganography, pp. 13–26. Information Hiding. Springer, Berlin (2001)
17. Shen, S.Y., Huang, L.H.: A data hiding scheme using pixel value differencing and improving exploiting modification directions. Comput. Secur. (Elsevier) **48**, 131–141 (2015)
18. Turner, L.F.: Digital data security system. Patent IPN wo **89**, 08915 (1989)
19. Zhang, X., Wang, S.: Efficient steganographic embedding by exploiting modification direction. IEEE Commun. Lett. **10**(11), 781–783 (2006)
20. Zhang, W., Wang, S., Zhang, X.: Improving embedding efficiency of covering codes for applications in steganography. IEEE Commun. Lett. **11**(8), 680–682 (2007)
21. Zeng, X.T., Li, Z.: Reversible data hiding scheme using reference pixel and multi-layer embedding. AEU-Int. J. Electron. Commun. **66**(7), 532–539 (2012)

A Machine Learning Approach to Comment Toxicity Classification

Navoneel Chakrabarty

Abstract Nowadays, derogatory comments are often made by one another, not only in offline environment but also immensely in online environments like social networking websites and online communities. So, an Identification combined with Prevention System in all social networking websites and applications, including all the communities, existing in the digital world is a necessity. In such a system, the Identification Block should identify any negative online behavior and should signal the Prevention Block to take action accordingly. This study aims to analyze any piece of text and detect different types of toxicity like obscenity, threats, insults and identity-based hatred. The labeled Wikipedia Comment Dataset prepared by Jigsaw is used for the purpose. A 6-headed Machine Learning tf–idf Model has been made and trained separately, yielding a Mean Validation Accuracy of 98.08% and Absolute Validation Accuracy of 91.64%. Such an Automated System should be deployed for enhancing the healthy online conversation.

Keywords Toxicity · Obscenity · Threats · Insults · Machine learning · tf-idf

1 Introduction

Over a decade, social networking and social media have been growing in leaps and bounds. Today, people are able to express their opinions and discuss different aspects via these platforms. In such a scenario, it is quite obvious that debates may arise due to differences in opinion. But often these debates take a dirty side and may result in fights over the social media during which offensive language termed as toxic comments may be used. These toxic comments may be threatening, obscene, insulting or identity-based hatred. So, these clearly pose the threat of abuse and harassment online. Consequently, some people stop giving their opinions or give up seeking different opinions, that leads to an unhealthy and unfair discussion. As a result,

N. Chakrabarty (✉)
Jalpaiguri Government Engineering College, Jalpaiguri, West Bengal, India
e-mail: nc2012@cse.jgec.ac.in

A. K. Das et al. (eds.), *Computational Intelligence in Pattern Recognition*, Advances in Intelligent Systems and Computing 999,
https://doi.org/10.1007/978-981-13-9042-5_16

different platforms and communities find it very difficult to facilitate fair conversation and are often forced to either limit user comments or get dissolved by shutting down user comments completely. The Conversation AI team, a research group founded by Jigsaw and Google have been working on tools and techniques for providing an environment for healthy communication [1]. They have also built publicly available models through the Perspective API on Comment Toxicity Detection [2]. But these models are sometimes prone to errors and do not provide the option to the users for choosing which type of toxicity, they are interested in finding. So, a more stable and versatile intelligent system is required for Toxic Comment Prevention in social communication. This model reads any piece of text (a text message or any comment appearing in social platform that can be toxic or nontoxic) and detects the type of toxicity it contains. The types of toxicity are simply toxic, severely toxic, obscene, threat, insult, and identity-based hate. This overcomes the drawback of the model developed using Perspective API, showing all the types of toxicity contained in the comment.

This paper has been structured as follows: Sect. 2 throws light on the existing works and approaches used by researchers as Literature Review, Sect. 3 describes the Proposed Methodology including the dataset used, data visualizations and model construction, Sect. 4 elucidates the Individual Training of the Pipelines, Sect. 5 mentions the Implementation Details, Sect. 6 deals with the Results, depicting the Model Performance. Finally, it is concluded with future scope or improvement in Sect. 7.

2 Literature Review

Many Machine Learning and Deep Learning Approaches have been attempted for detecting types of toxicity in user comments.

- Georgakopoulos et al. proposed a Deep Learning Approach involving Convolutional Neural Networks (CNNs) for text analytics in toxicity classification, obtaining a Mean Accuracy of 91.2% [3].
- Khieu et al. applied various Deep Learning approaches involving Long-Short Term Memory Networks (LSTMs) for the task of classifying toxicity in online comments, obtaining a Label Accuracy of 92.7% [4].
- Chu et al. implemented a Convolutional Neural Network (CNN) with character-level embedding for detecting types of toxicity in online comments, obtaining a Mean Accuracy of 94% [5].
- Kohli et al. proposed a Deep Learning Approach involving Recurrent Neural Network (RNN) Long-Short Term Memory with Custom Embeddings for comment toxicity classification, obtaining a Mean Accuracy of 97.78% [6].

3 Proposed Methodology

It consists of four subsections: Sect. 3.1 describes the Dataset used, Sect. 3.2 deals with Data Visualization, Sect. 3.3 deals with the Text Preprocessing and Sect. 3.4 illustrates the Pipelines.

3.1 The Dataset

The Wikipedia Talk Page Dataset prepared by Jigsaw and now publicly available at Kaggle is used [7]. The Dataset consists of total 159571 instances with comments and corresponding multiple binary labels: toxic, severe_toxic, obscene, threat, insult and identity_hate. Sample instances of the dataset are shown below in Fig. 1.

3.2 Data Visualization

Visualization is done in the form of Histogram showing the distribution of comment lengths over the whole corpus of the dataset in Fig. 2. The number of bins in the histogram is found using Freedman–Diaconis Formula given in Eq. (1).

$$h = 2 * IQR * n^{-1/3} \tag{1}$$

where

- IQR is the Inter Quartile Range,
- n is the Number of Data Points,
- h is the Bin-Width.

and formula for obtaining Number of Bins is given in Eq. (2).

$$No.\,Of\ Bins = (max - min)/h \tag{2}$$

where

- max is the maximum value of the observation (here Comment Length),
- min is the minimum value of the observation (here Comment Length).

Here, the number of bins comes out to be 659 approximately.

id	comment_text	toxic	severe_toxic	obscene	threat	insult	identity_hate
0044cf18cc2655b3	What page shoudld there be for important characters that DON'T reoccur?	0	0	0	0	0	0
00472b8e2d38d1ea	Void, Black Doom, Mephiles, etc	1	0	1	0	1	1

Fig. 1 Sample instances of the dataset

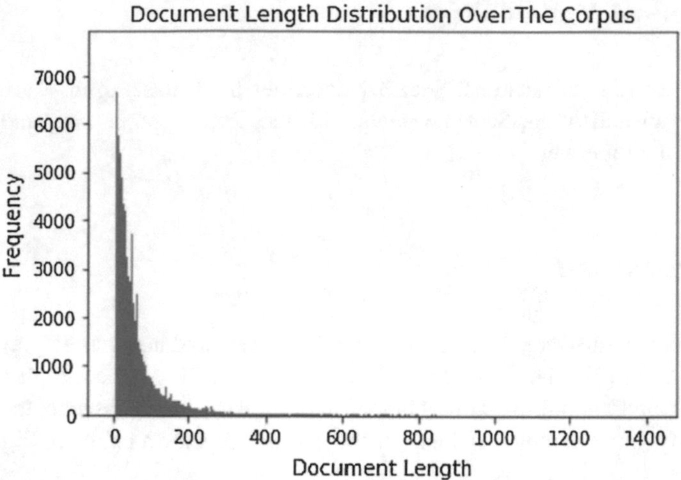

Fig. 2 Histogram showing distribution of comment length over the whole corpus

3.3 Text Preprocessing

The text preprocessing techniques followed before processing the text data are:

- **Removal of Punctuation**: All the punctuation marks in every comment are removed.
- **Lemmatization**: Inflected forms of words which may be different verb forms or singular/plural forms etc. are called lemma. For example, going and gone are inflected forms or lemma of the word, go. The process of grouping these lemma together is called Lemmatisation. So, lemmatization is performed for every comment.
- **Removal of Stopwords**: Frequently occurring common words like articles, prepositions, etc., are called stopwords. So, stopwords are removed for each comment.

3.4 The Pipelines

Six pipelines are used where each pipeline corresponds to each label. With the help of these pipelines, six models are instantiated and trained separately.

- The first, third and fifth Pipelines correspond to the labels toxic, obscene, and insult, respectively. The three stages of these pipelines are similar but these are trained separately. The stages of these pipelines are as follows:
  - **Bag-of-Words using Word Count Vectorizer**
    Bag-of-Words is a feature engineering technique in which a bag is maintained which contains all the different words present in the corpus. This bag is known

D1: He is a lazy boy. She is also lazy.

D2: Neeraj is a lazy person.

	He	She	lazy	boy	Neeraj	person
D1	1	1	2	1	0	0
D2	0	0	1	0	1	1

Fig. 3 Simpler example of bag-of-words

D1: He is a lazy boy. She is also lazy.

D2: Neeraj is a lazy person.

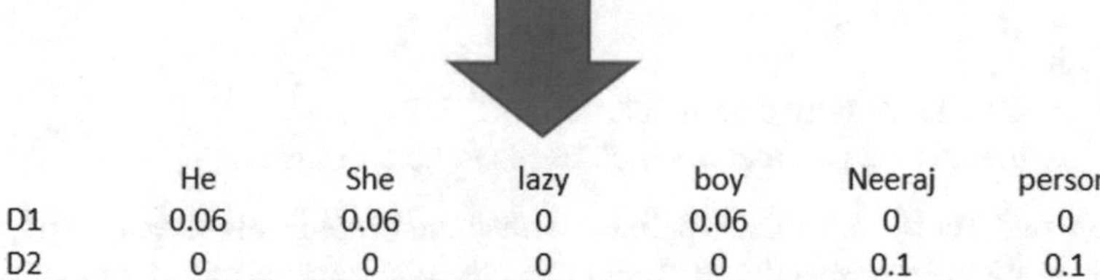

	He	She	lazy	boy	Neeraj	person
D1	0.06	0.06	0	0.06	0	0
D2	0	0	0	0	0.1	0.1

Fig. 4 Simpler example of tf-idf

as Vocabulary or Vocab. For each and every word present in the Vocabulary, counts of these words become the features for all the comments present in the corpus. A different and simpler example of Bag-of-Words is shown in Fig. 3.

– **tf-idf Transformer**
The featured Bag-of-Words Model or Matrix for the whole corpus is transformed into a matrix whose every element is a product of Term Frequency (TF) and Inverse Document Frequency (IDF), combined together as tf–idf.
Term Frequency (TF) is defined as the ratio of the number of times a word or a term appears in a comment to the total number of words in the comment.
The formula for Inverse Document Frequency (IDF) is given in Eq. (3)

$$IDF = log(N/n) \tag{3}$$

where,
N is the total number of comments,
n is the number of comments a word has appeared in.
A different simpler example of tf–idf is shown in Fig. 4.

- **Support Vector Machine Model with Linear Kernel**
  After tf-idf transformation, a complete numeric featured dataset is obtained. Now, a Support Vector Machine Model is instantiated with 100 as the maximum number of iterations and C (penalty parameter) as 1.0.
  Linear Support Vector Machine Algorithm:
  1. The p-dimensional training instances (with p features) are assumed to be plotted in space.
  2. A Hyperplane is predicted, which separates the different classes.
  3. The best hyperplane should be selected finally, which maximizes the margin between data classes. The data points, influencing the hyperplane are known as Support Vectors.
  4. The Large Margin Intuition for selection of best hyperplane for Linear SVM is given below:

$$\min_{\theta} C \sum_{i=1}^{m} \left[y^{(i)} \cos t_1 \, (\theta^T x^{(i)}) + (1 - y^{(i)}) \cos t_0 (\theta^T x^{(i)}) \right] + \frac{1}{2} \sum \theta_j^2$$

  where,
  * C is the penalty parameter,
  * theta is the parameter which needs to be optimized.

- The second, fourth and sixth Pipelines correspond to the labels severe_toxic, threat and identity_hate, respectively. Again the three components of these pipelines are similar but are trained separately. Also the first and second stages of these pipelines are similar to those in the first, third, and fifth Pipelines. Only the third stage is different and crucial.

  - **Bag-of-Words using Count Vectorizer**
  - **tf-idf Transformer**
  - **Decision Tree Classifier**
    After tf-idf transformation, a complete numeric featured dataset is obtained. Now, a Decision Tree Classifier is instantiated.
    Decision Tree Classifier Algorithm:
    1. The best feature of the dataset is selected on the basis of Gini-Impurity and placed at the root of the tree.
    2. The Training Samples are split into subsets such that each subset contains data with the same value for a feature.
    3. Steps 1 and 2 are repeated on all the subsets until leaf nodes are found in all the branches of the tree.

4 Individual Training of the Pipelines

The dataset is 80-20 random splitted into Training and Testing (Validation) Sets. Out of 159571 instances, 127656 instances are used for training the 6 pipelines individually and the remaining 31915 instances are used for the individual and combined

Validation and Performance Measure. So, the 6 pipelines are trained individually and tested.

5 Implementation Details

The text preprocessing, Bag-of-Words and tf–idf Transformer along with the training of the pipelines, i.e., the training of the Machine Learning Models are implemented using Python's NLTK (Natural Language Toolkit for Text Preprocessing) and are done on a machine with Intel(R) Core(TM) i5-8250U processor, CPU @ 1.60 GHz 1.80 GHz and 8 GB RAM, using Python's Scikit-Learn Machine Learning Toolbox.

6 Results

It consists of 2 subsections: Sect. 6.1 explains the Individual Results and Collective Results are shown in Sect. 6.2.

6.1 Individual Results

- Training Accuracy describes the accuracy achieved on the Training Set.
- Validation Accuracy describes the accuracy achieved on the test set.
- Precision is defined as the ratio of correctly predicted positive observations to the total predicted positive observations. The formula for Precision is shown in Eq. (4).

$$Precision = TP/TP + FP \tag{4}$$

- The Sensitivity or Recall is defined as the proportion of correctly identified positives. The formula for Recall is given in Eq. (5).

$$Recall = TP/TP + FN \tag{5}$$

- F1-Score is the Harmonic Mean of Precision and Recall.

After training, the pipelines are used for testing or validating the remaining 31915 samples. All the pipelines are allowed to give their predictions independently. But for the label severe_toxic, it is obvious that unless a comment is detected to be toxic, it has no chance of being severe_toxic. So based on the predictions made by the first Pipeline for the label toxic, those test instances which are not detected as toxic, are labeled 0, for the label severe_toxic, i.e., not detected as severe_toxic. Hence, only for the label severe_toxic, a second check is done with reference to the prediction

Table 1 Training accuracy, validation accuracy, precision, recall, and F1-score for all the pipelines/labels

Pipeline/Label	Training Accuracy	Validation Accuracy		Precision	Recall	F1-Score
1st Pipeline/Toxic	99.05%	96.01%		0.96	0.96	0.96
2nd Pipeline/Severe_Toxic	99.98%	98.81%	98.87%	0.99	0.99	0.99
3rd Pipeline/Obscene	99.58%	97.89%		0.98	0.98	0.98
4th Pipeline/Threat	99.99%	99.61%		1	1	1
5th Pipeline/Insult	99.36%	97.13%		0.97	0.97	0.97
6th Pipeline/Identity_Hate	99.99%	99.00%		0.99	0.99	0.99

made by 1st Pipeline for the label toxic, i.e., only those instances which are detected positive (Toxic) by first Pipeline are fed to the second Pipeline for predictions.
The Training Accuracy, Validation Accuracy (before and after second check for the label, severe_toxic), Weighted Precision, Weighted Recall, and F1-Score for all the pipelines/labels are tabulated in Table 1.

- Area Under Receiver Operator Characteristic Curve (AUROC)
  ROC Curve is the plot of True Positive Rate versus False Positive Rate. The Area under ROC Curve should be greater than 0.5 to be termed as an Acceptable Test. Here, the ROC Curves are plotted by taking the predictions given by the pipelines on the Validation Set as target scores.
  ROC Curves for all the pipelines/labels shown in Fig. 5.
- Confusion Matrix.
  Confusion Matrices for all the pipelines/labels are shown in Fig. 6.

6.2 Collective Results

- Mean Validation Accuracy is the average of the Validation Accuracies achieved by the 6 Pipeline Models. Hence, it is the Mean Validation Accuracy of the 6 Headed Model prepared.
  From this model, a Mean Validation Accuracy of 98.08% is achieved.
- Macro F1 is the average of the F1 Scores obtained by the 6 Pipeline Models. Hence, it is the Macro F1 of the 6 Headed Model prepared.
  From this model, a Macro F1 of 0.9817 is achieved.
- Macro AUC (Area Under ROC Curve) is the average of the AUCs achieved by the 6 Pipeline Models. Hence, it is the Macro AUC of the 6 Headed Model prepared.
  From this model, a Macro AUC of 0.73 is obtained.
- Absolute Validation Accuracy is the Validation Accuracy of the 6 Headed Model in which the accuracy is measured in such a way that, if all the predictions made by the 6 pipelines on the validation set, i.e., the predictions for all the 6 labels, matches exactly with the true values of the 6 labels for any instance/sample, then

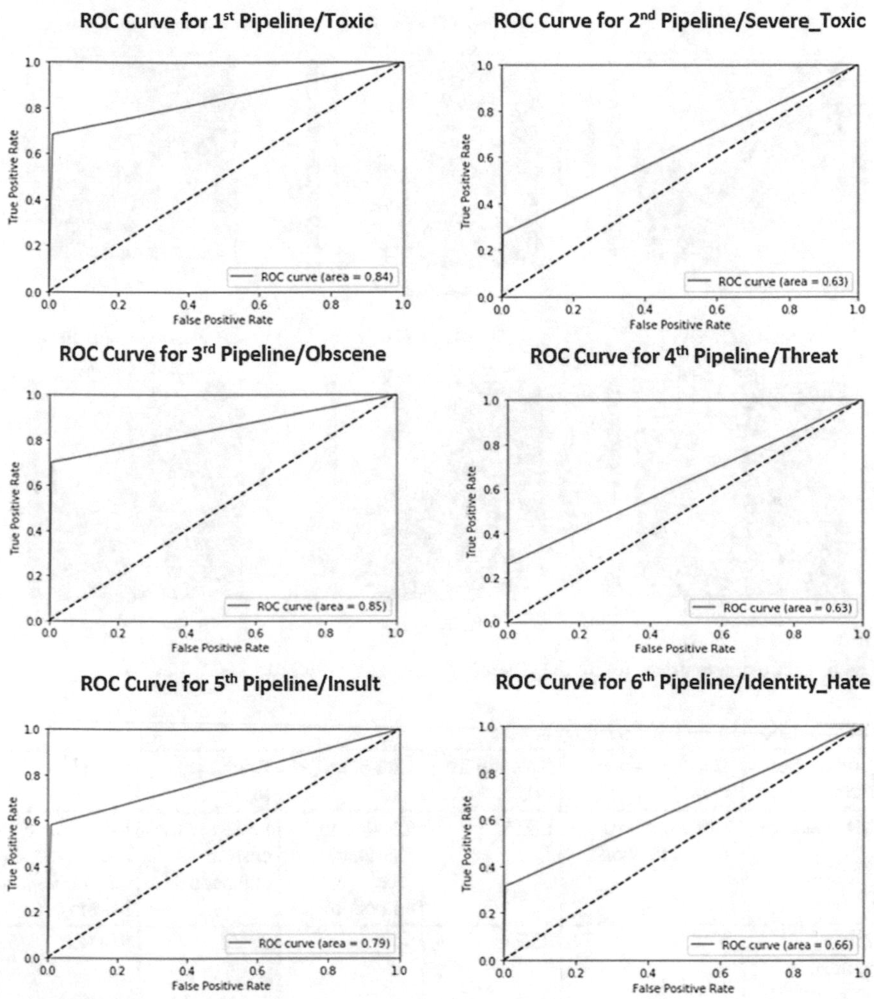

Fig. 5 ROC curves and area under ROC curves for all the pipelines/labels

the Model is said to be Absolutely Accurate.

From this model, an Absolute Validation Accuracy of 91.64% is achieved.

A Direct Comparison of this model with the existing models has been done on different parameters and shown in Table 2.

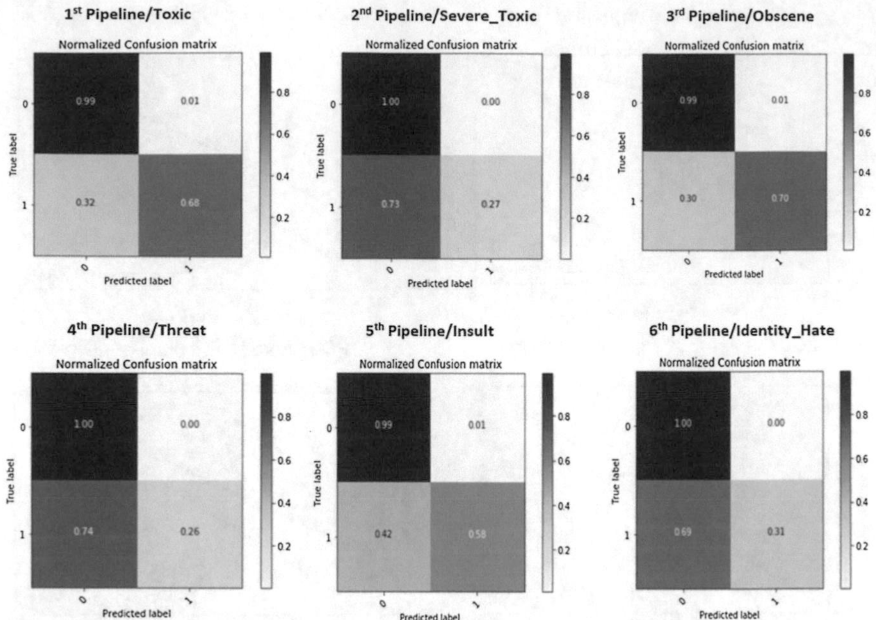

Fig. 6 Confusion matrices for all the pipelines/labels

Table 2 Comparison with existing models

Comparison Parameters	Georgeakopoulos et al. [3]	Khieu et al. [4]	Chu et al. [5]	Kohli et al. [6]	Model
Methodology	Convolutional neural network	LSTM	CNN with character-level embeddings	LSTM with custom embeddings	tf-idf with 6 headed machine learning
Mean validation accuracy (%)	91.2	92.7	94	97.78	98.08
Macro FI	–	0.706	–	–	0.9817
Macro AUC	–	–	–	0.7240	0.73

7 Conclusion

This paper proposed a Machine Learning Approach combined with Natural Language Processing for toxicity detection and its type identification in user comments. Finally, the Mean Validation Accuracy, so obtained, is 98.08% which is by far the highest ever numeric accuracy reached by any Comment Toxicity Detection Model. The research done in this paper is intended to enhance fair online talk and views sharing in social media.

A more robust model can be developed by applying Grid Search Algorithm on the same dataset over the Machine Learning Algorithms for every pipeline, being used in order to obtain more better results and accurate classifications.

References

1. Coversation AI Team. https://conversationai.github.io/
2. Perspective API. https://perspectiveapi.com/#/
3. Georgakopoulos, S.V., Tasoulis, S.K., Vrahatis, A.G., Plagianakos, V.P.: Convolutional neural networks for toxic comment classification. In: 10th Hellenic Conference on Artificial Intelligence (2018)
4. Khieu, K., Narwal, N.: Detecting and classifying toxic comments. https://web.stanford.edu/class/cs224n/reports/6837517.pdf
5. Chu, T., Jue, K., Wang, M.: "Comment abuse classification with deep learning. https://web.stanford.edu/class/cs224n/reports/2762092.pdf
6. Kohli, M., Kuehler, E., Palowitch, J.: Paying attention to toxic comments online. https://web.stanford.edu/class/cs224n/reports/6856482.pdf
7. https://www.kaggle.com/c/jigsaw-toxic-comment-classification-challenge/data

A Minutia Detection Approach from Direct Gray-Scale Fingerprint Image Using Hit-or-Miss Transformation

Debashis Das

Abstract Personal authentication through fingerprint matching merely depends on the proper identification of the minutia points of a fingerprint. In this paper, a minutia detection scheme is presented by employing gray scale hit-or-miss transformation. The work focuses exclusively on two widely used minutia points namely ridge bifurcations and ridge endings. To detect all the minutia points, a set of bifurcation shaped templates, oriented along various directions, are constructed. Ridge bifurcations are identified directly from original fingerprint through hit-or-miss transformation using the predefined templates. The ridge endings, on the other hand, are detected from the inverted image by using the same set of templates. The proposed method is implemented and tested on real fingerprint images. The experimental results show the efficiency and accuracy of the method. A comparative study is also provided between the proposed method and other relevant techniques which proves the efficacy of the method.

Keywords Minutia detection · Fingerprint image · Feature extraction · Hit-or-miss transformation

1 Introduction

Now-a-days, fingerprint based biometric system has gained immense popularity in various fields for person authentication. The automated system distinguishes an individual by comparing some specific topological or textural patterns of the fingerprint, commonly known as minutia. Therefore, detection of fingerprint minutia plays a crucial role in Automatic Fingerprint Identification System (AFIS). A fingerprint image may contain a number of minutia points among which two basic types of patterns namely—(a) ridge ending and (b) ridge bifurcation are widely accepted for matching

D. Das (✉)
Department of Information Technology, Institute of Engineering and Management, Kolkata 700091, India
e-mail: debashisitnsec@gmail.com

A. K. Das et al. (eds.), *Computational Intelligence in Pattern Recognition*, Advances in Intelligent Systems and Computing 999,
https://doi.org/10.1007/978-981-13-9042-5_17

purpose. Since past few decades, researchers have developed a number of minutia detection algorithms. Based on a compact literature survey, the existing methods can be classified into two orientations—(i) based on the mode of input images and (ii) based on the working principles. The existing algorithms have considered either gray-scale or skeletonized image as an input associated with two different types of working principles, one is template based filtering while the other focuses on ridge line tracing for detecting minutia points. It is experienced that, all the template based methods have been applied on skeletonized image while the ridge line tracing based approaches have been executed on both the gray-scale and skeletonized image. Here, a novel minutia detection technique is proposed that uses template based filtering on the gray-scale fingerprint image. In this context, some of the popular minutia detection algorithms have been discussed.

Ratha et al. [12] have proposed a ridge flow based feature extraction algorithm where ridge lines are extracted by analyzing the projection waveform of all the sub-image blocks which are subsequently skeletonized for locating the minutia points. A direct gray-scale based approach has been developed in [18] where the auxiliary ridge lines are constructed by following original ridge lines which are then tracked to find out the ridge ending and bifurcation points. Jiang et al. [25] have modified the method of Miao et al. by making it adaptive in ridge line tracing to make it computationally more efficient. On the other hand, Farina et al. [14] designed an algorithm for detecting minutia points that can be applied on skeletonized fingerprint image associated with a validation process to filter out the true minutia points. A chain code based scheme has been proposed by Shi and Govindaraju [2] in which the ridge contours are tracked through the chain code representation from which minutia points are determined by examining the turning direction of the contours. Similarly, Shin et al. [32] have adopted run length encoding to trace the ridge lines and extracting the minutia points accordingly. Another encoding based technique has been developed in [28] where principal curves are used to extract the ridge lines and minutia points are estimated based on some specific criteria. Fronthaler et al. [13] have suggested an unconventional technique which extracts minutia from direct gray-scale image by employing parabolic symmetry features along with the local symmetry features of the fingerprint. Besides, a robust feature extraction approach for low quality fingerprint images has been devised in [34]. Bansal et al. [3] have developed a template based feature extraction algorithm through morphological hit or miss transform on a thinned binary fingerprint where two sets of binary templates have been employed to treat the ridge endings and the ridge bifurcations separately. Another template based method has been proposed in [6] where a logical template is used along the ridge orientation to point out the minutia points from a binary fingerprint image. A Similar technique has also been devised in [8] with a modest difference in the selection procedure of templates which are learned from a training data set. A simple template based feature extraction algorithm has been proposed by Nallaperumal and Padmapriya [29] where a set of fixed templates have been used on thinned binary image regardless of the local ridge orientations. Besides, a number of minutia detection techniques have been proposed in the literature [1, 5, 7, 10, 11, 15, 16, 19, 21, 27, 33].

In this paper, a new approach for detecting minutia points from direct gray-scale image has been proposed by employing gray-scale hit-or-miss transformation using a pre-defined set of oriented templates. A set of bifurcation shaped templates have been constructed which can directly be applied on the input image to figure out the ridge bifurcation points. To the contrary, the ridge ending points are captured by executing hit-or-miss transformation on the inverted image using the same set of templates.

The rest of the paper is organized as follows. After this introductory section, an elaborative description of the proposed method is provided in Sect. 2. Section 3 illustrates the experimental results obtained by implementing the proposed algorithm. The performance analysis is presented in Sect. 4 and finally concluding remarks are drawn in Sect. 5.

2 Proposed Method

A scanned fingerprint may suffer from poor image quality. Hence, before extracting the minutia points, a pre-processing is needed to obtain an enhanced fingerprint impression. An essential enhancement step followed by the minutia detection mechanism and a post-processing are discussed in the following sections.

2.1 Image Enhancement

An enhancement method mentioned in [17] has been adopted for improving the input image clarity. The enhancement algorithm follows a number of steps which are discussed here in a nutshell. A raw fingerprint image is first normalized using a pre-defined mean and variance which helps to remove the noise without affecting the ridge and valley structures. After that ridge orientations and ridge frequencies are approximated from each local blocks of the normalized image. The region of interest is then separated by classifying each of the local blocks based on the waveform generated by the ridge-valley pattern. Finally, the segmented image is filtered out by employing a bank of Gabor filters tuned to the local ridge orientations to obtain the final enhanced fingerprint image. The enhancement result due to the Hong's method is illustrated in Fig. 1. Notably, it produces a gray-scale enhanced image in the output.

2.2 Minutia Detection

In this step, the minutia points are detected from the enhanced image so obtained by employing gray-scale hit-or-miss transformation. Here, the basic concept of the morphological hit-or-miss transformation—(i) the binary operation as well as (ii) the gray-scale extension is discussed.

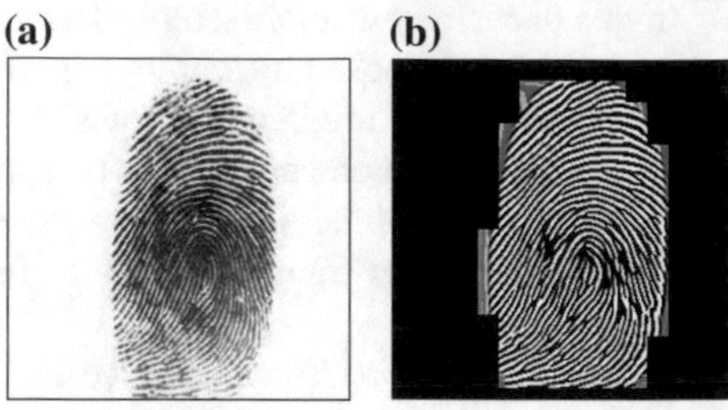

Fig. 1 Illustrates the fingerprint enhancement by Hong's method where **a** shows the input image and **b** shows the corresponding enhanced image

Binary hit-or-miss transformation (HMT). Morphological hit-or-miss transformation is a familiar tool used for shape detection of an object which has been introduced as binary operation [24]. The binary HMT is obtained by taking the intersection of two morphological erosions which can be expressed as provided in Eq. (1).

$$I_b \circledast (S_b^1, S_b^2) = (I_b \ominus S_b^1) \cap (I_b^c \ominus S_b^2) \tag{1}$$

where I_b is a binary image (set), I_b^c is the complement of the set I_b, S_b^1 is a binary structuring element (SE), S_b^2 is a background set containing S_b^1, '$\circledast$' symbolizes the hit-or-miss operator and '$\ominus$' specifies the morphological erosion operator.

Gray-scale hit-or-miss transformation (GHMT). The binary HMT can simply be extended for gray-scale image by considering gray-scale erosion. In the literature, a number of variations of GHMT has been developed [10, 20, 22, 26, 31]. In the proposed method, a modification on the concept mentioned in [31] is performed which is applied for template matching. The mathematical expression of the modified GHMT is given in Eq. (2).

$$I_g \circledast (S_g^1, S_g^2) = [\min_{p_1 \in S_g^1}{}^{2\text{nd}}(I_g + p_1)] - [\max_{p_2 \in S_g^2}{}^{2\text{nd}}(I_g - p_2)] \tag{2}$$

where I_g is a gray-scale image, S_g^1 is a foreground structuring element, S_g^2 is a background set containing S_g^1, min and max are the gray-level substitution of binary erosion and dilation operation respectively. Notably, $\min^{2\text{nd}}$ and $\max^{2\text{nd}}$ are respectively the second minimum and the second maximum value of a set.

Formation of templates. Here, a set of pre-defined oriented templates are considered for gray-scale hit-or-miss transformation. The templates that have been assumed, contain the shape of a ridge bifurcation of a fingerprint. A bifurcation shaped template consists of a foreground portion having pixels residing on the valley and a background set having pixels laying on the ridge line. The procedure of encoding and generation of a synthetic template is illustrated in Fig. 2 which imitates an original bifurcation shape of a real fingerprint. In Fig. 2b, the shape marked by '2' is assumed as the foreground SE whereas the shape marked by '1' is considered as the background SE.

Now, to treat all the oriented minutia points, a total of sixteen oriented templates have been selected. The template orientations can be considered as shown in Eq. (3).

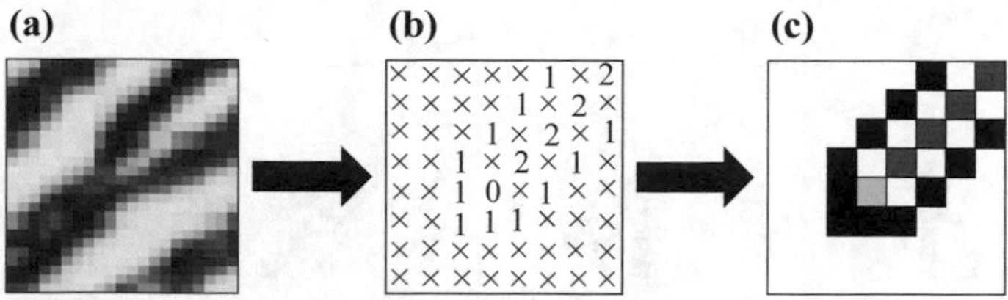

Fig. 2 Shows a bifurcation shaped template oriented along −45° where **a** shows a ridge bifurcation cropped from a real fingerprint image, **b** encoding of a bifurcation by marking the ridge pixels as '1', valley pixels as '2', center pixel as '0' and '×' symbolizes the don't care pixels, **c** shows the corresponding synthetic template

$$\theta_{S_g}^{k+1} = \theta_{S_g}^{k} + \theta_{S_g}^{0} \text{ where } \Theta_{S_g}^{0} = 360^\circ/n \tag{3}$$

In the above equation, θ^k is the orientation of a template S_g^k where $k \in \{0, 1, 2, \cdots, n-1\}$ and n is selected as 16.

It is mentioned that all the lines are drawn by employing *Bresenham* line drawing algorithm for constructing the templates [4, 23]. All the sixteen oriented masks, have been constructed for the proposed method are shown here in Fig. 3.

Detection of minutia points. Now, the minutia points are detected through gray-scale hit-or-miss transformation by using the set of structuring elements shown in Fig. 3. However, this set of templates are able to locate only the bifurcation points from a fingerprint. To detect the ridge ending points the original image is inverted. Mentioning that the minutia points are interchanged in the inverted image (Fig. 4). The image inversion has been implemented by employing the following equation Eq. (4).

$$I'(x, y) = P_{\max} - I(x, y) \tag{4}$$

where $I(x, y)$ and $I'(x, y)$ denote the intensity values of a pixel located at (x, y) of the original and inverted image respectively, $P_{\max}$ denotes the maximum pixel intensity of the original image.

To detect the minutia points, a pixel-wise hit-or-miss transformation is executed as expressed in Eq. (2) on both the original and inverted image. For each version of the input image (original and inverted), sixteen different filtered outputs are obtained for each of the oriented structuring elements. Equations (5) and (6) describe the mathematical expression of the mechanism.

$$G_{\text{org}}^{i} = I_{\text{org}} \circledast (S_1^{\theta_i}, S_2^{\theta_i}) \text{ where } i \in \{1, 2, \cdots, 16\} \tag{5}$$

$$G_{\text{inv}}^{i} = I_{\text{inv}} \circledast (S_1^{\theta_i}, S_2^{\theta_i}) \text{ where } i \in \{1, 2, \cdots, 16\} \tag{6}$$

In the above equations, I_{org} and I_{inv} are the original and inverted input images, G_{org}^{i} and G_{inv}^{i} are the original and inverted output images respectively obtained from the filtering process using a particular directional template oriented along θ_i, $S_1^{\theta_i}$ and $S_2^{\theta_i}$

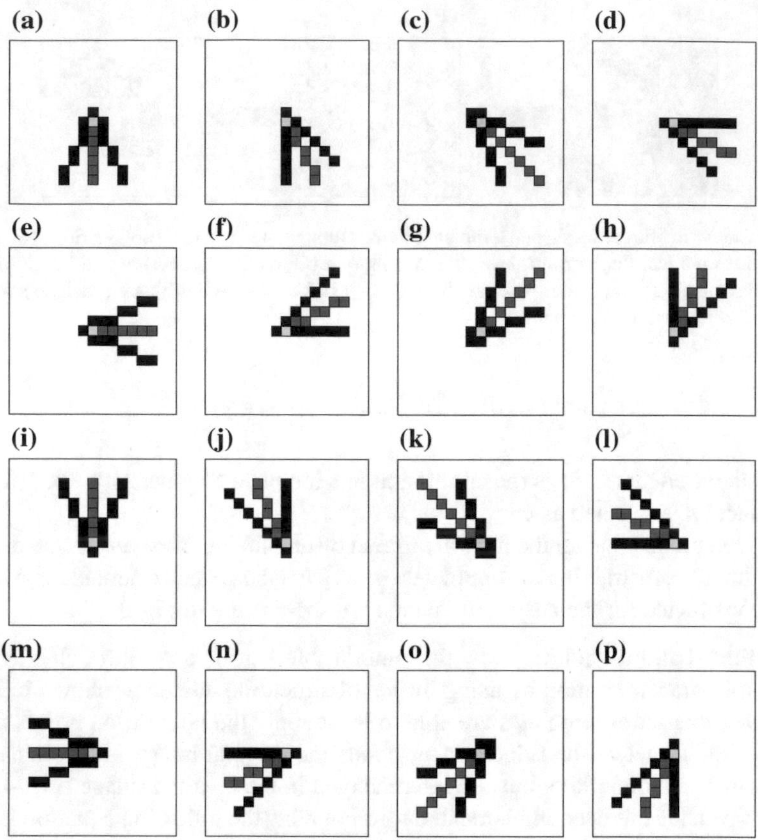

Fig. 3 Shows all the structuring elements constructed for minutia detection; here each component block representing a pixel

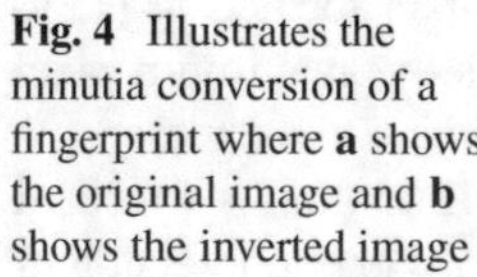

Fig. 4 Illustrates the minutia conversion of a fingerprint where **a** shows the original image and **b** shows the inverted image

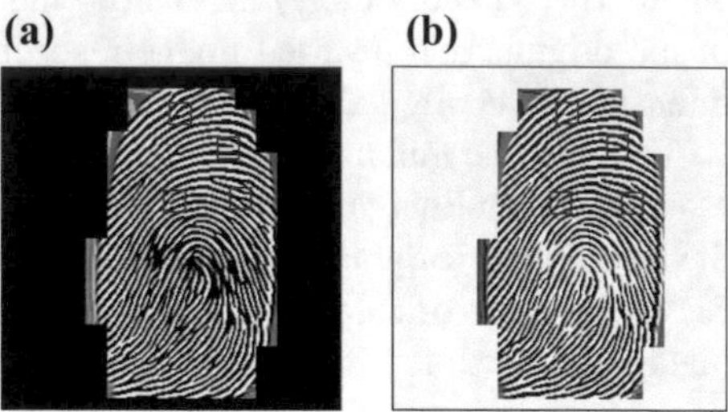

are the foreground and background structuring elements respectively along a specific orientation θ_i.

For each version of the image, the pixel-wise maximum value is found among all the sixteen filtered outputs and select a pixel position as the minutia point which secures a value higher than a specified threshold. The mathematical expression is provided in Eq. (7).

$$M = M \bigcup \{(x, y)\} \ if \max_{1 \le i \le 16} [G_{\text{org}}/\text{inv}^i(x, y)] > T \tag{7}$$

where M is the set of minutia points, $G_{\text{org}}/\text{inv}^i(x, y)$ denotes the intensity value of a pixel located at (x, y) of the ith filtered output obtained from either original or the inverted image and T is a threshold which is experimentally set as 20.

2.3 Post-processing

The set of minutia points so obtained, may contain some false minutia which can be detected due to one of the following reasons:

- two end points of each ridge line can be detected as minutia (referred as type-I spurious minutia)
- a set of surrounding points of a potential minutia location can also be detected as the gray-scale image is directly processed (referred as type-II spurious minutia)

Hence, a post-processing step is required to remove these false minutia points. The remedial measures are described next.

Removal of type-I spurious minutia. The entire image is divided into several 16×16 non-overlapping sub-image blocks. Now, each of the blocks are checked that are residing on the fingerprint contour. If any detected minutia point is found to be located in any of such blocks, the particular point is considered as the spurious one and hence excluded from the minutia set.

Removal of type-II spurious minutia. To remove the spurious minutia points of type-II, a block-wise processing is employed. A local block of size 5×5 is considered that is surrounded by a candidate minutia point. If there exits any other minutia points except the selected one, remove them from the minutia set.

3 Experimental Results and Discussions

The proposed algorithm has been implemented in *C* language on the *Linux* platform and tested on a set of real fingerprint images. For the experimental purpose, a set of real fingerprint images have been collected from a standard database namely FVC2004 [9]. Here, some of the experimental results have been shown as output in Figs. 5, 6 and 7. In each figures, (a) & (c) show the input fingerprint images where (b) & (d) show the corresponding outputs produced by the proposed algorithm.

The experimental results of the proposed method have been compared with ground truth (where the minutia points are manually marked by the human experts) as well as an existing method proposed in [3]. The comparative results are illustrated in Fig. 8. It is evident from Fig. 8 that the proposed algorithm is able to produce better results than the competing method.

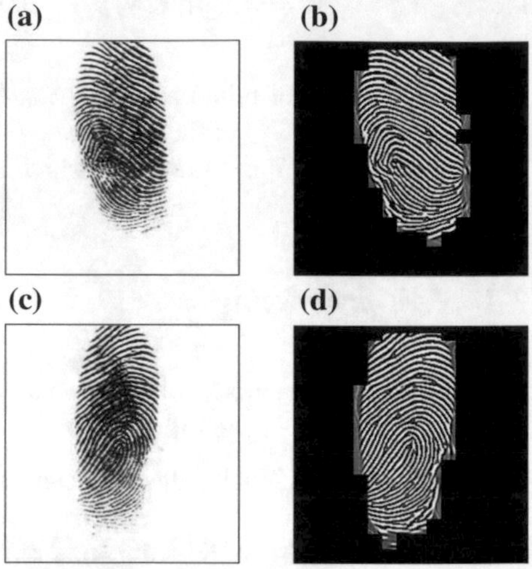

Fig. 5 Results on the images taken from *DB1_B* where **a** and **c** input images. **b** and **d** Output of proposed method

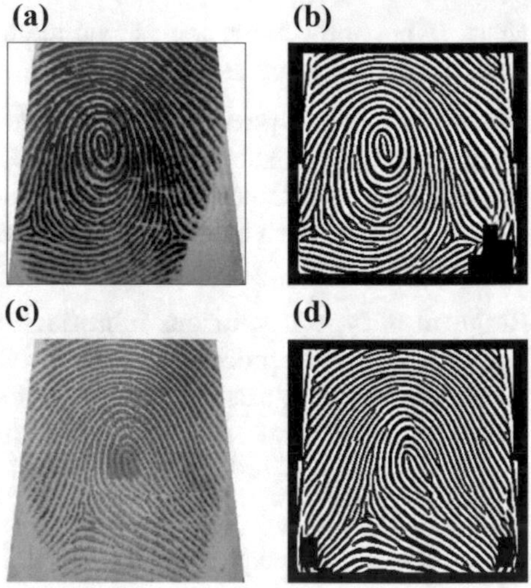

Fig. 6 Results on the images taken from *DB2_B* where **a** and **c** input images. **b** and **d** Output of proposed method

4 Performance Analysis

Beside visual inspections, a performance analysis has also been provided to establish the accuracy of the proposed method. The performance is assessed in measure of two parameters—(i) error rate and (ii) computational time. The minutia detection error is evaluated by measuring (a) number of dropped minutia (D_m), (b) number of false minutia (F_m) and (c) number of exchanged minutia (E_m). A comparative performance analysis between the proposed algorithm and the method mentioned in [3] are summarized in Table 1. It reveals that the proposed method provides reasonably good performance than the competing algorithm in terms of minutia detection accuracy.

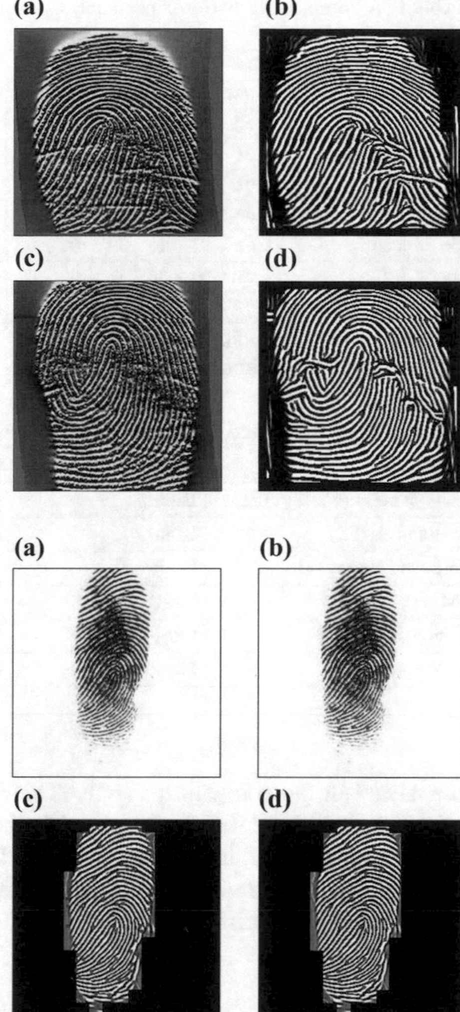

Fig. 7 Results on the images taken from *DB3_B* where **a** and **c** input images. **b** and **d** Outputs generated by the proposed method

Fig. 8 **a** Input image, **b** groundtruth output, **c** result due to the method [3] and **d** output of the proposed algorithm

It is observed that the number of dropped minutia and exchanged minutia for both the method is more or less same. On the other hand, Bansal et el.'s method produces large number of false minutia as the two end points of each ridge line get detected as the ridge ending points whereas the proposed method can efficiently remove them in the post-processing step. However, the execution of the proposed method takes marginally higher computational time for some of the images than the competitive method.

The proposed method has also been compared with few other existing methods on the basis of algorithmic accuracy and computational efficiency. For a fair comparison, both the binary image based method as well as gray-scale based methods have been

Table 1 A comparative performance analysis is provided in measure of minutia detection error and execution time

Images	Original minutia		Bansal et al.'s method							Proposed method						
			D_m		F_m		E_m		Time (s)	D_m		F_m		E_m		Time (s)
	R_e^*	$R_b^\#$	R_e^*	$R_b^\#$	R_e^*	$R_b^\#$	R_e^*	$R_b^\#$		R_e^*	$R_b^\#$	R_e^*	$R_b^\#$	R_e^*	$R_b^\#$	
5a	11	11	01	02	40	03	02	01	0.93	00	01	01	01	01	00	1.01
5c	11	12	02	02	42	03	01	01	1.13	01	01	01	02	01	01	1.09
6a	25	30	03	01	45	01	01	03	1.24	02	01	01	02	02	02	1.19
6c	20	16	01	03	48	01	02	02	1.06	01	02	00	01	01	02	1.17
7a	21	15	03	00	46	07	02	02	1.03	02	00	01	02	02	02	1.14
7c	17	11	01	00	38	05	04	02	1.08	01	00	02	02	02	01	1.26

$^*R_e$: number of ridge ending points
$^\#R_b$: number of ridge bifurcation points

Table 2 Comparison between the proposed and few existing methods

Method	Input image	Avg. error (%)	Execution time (s)
Shi and Govindaraju [2]	Binary	43.53	2.35
Shin et al. [32]	Binary	11.30	1.50
Nguyen et al. [34]	Binary	09.40	4.15
Maio et al. [9]	Gray	25.23	2.15
Jiang et al. [25]	Gray	15.07	0.89
Proposed	Gray	13.80	1.17

considered in the comparative study. The comparison is provided in Table 2. It is evident from Table 2 that the proposed method outperforms the most of the existing techniques. However, the method in [32] and [34] detects minutia points with lower error rate and the technique mentioned in [25] is computationally efficient compared to the proposed algorithm.

5 Conclusion

In this paper, a minutia detection algorithm has been designed by employing hit-or-miss transformation which executes on direct gray-scale image. To detect both ridge ending and bifurcation, a set of directional bifurcation shaped templates have been constructed. The bifurcation points are detected from the original image by using these pre-defined templates. On the other hand, ridge ending points are detected from the inverted image by using the same set of templates. The experimental results show the efficacy of the proposed method. A performance analysis has also been provided with due comparison against some existing methods which establish the accuracy and efficiency of the proposed algorithm.

However, it is experienced, during experiments, that the ridge-valley widths may differ for varying datasets. Therefore, the templates can be generated dynamically by analyzing the ridge-valley width of an input image for better performance which also reveals the new scope of the future work.

References

1. Arpit, D., Namboodiri, A.: Fingerprint feature extraction from gray scale images by ridge tracing. In: 2011 International Joint Conference on Biometrics (IJCB), pp. 1–8. IEEE (2011)
2. Bansal, R., Sehgal, P., Bedi, P.: Effective morphological extraction of true fingerprint minutiae based on the hit or miss transform. Int. J. Biom. Bioinform. (IJBB) **4**(2), 71–85 (2010)
3. Bansal, R., Sehgal, P., Bedi, P.: Effective morphological extraction of true fingerprint minutiae based on the hit or miss transform. Int. J. Biom. Bioinform. (IJBB) **4**(2), 71–85 (2010)
4. Barat, C., Ducottet, C., Jourlin, M.: Pattern matching using morphological probing. In: Proceedings 2003 International Conference on Image Processing, ICIP 2003, vol. 1, pp. 369–372. IEEE (2003)
5. Bhanu, B., Boshra, M., Tan, X.: Logical templates for feature extraction in fingerprint images. In: Proceedings of 15th International Conference on Pattern Recognition, vol. 2, pp. 846–850. IEEE (2000)
6. Bhanu, B., Tan, X.: Learned templates for feature extraction in fingerprint images. In: Proceedings of the 2001 IEEE Computer Society Conference on Computer Vision and Pattern Recognition, CVPR 2001, vol. 2, pp. 591–596 (2001)
7. Bresenham, J.E.: Algorithm for computer control of a digital plotter. IBM Syst. J. **4**(1), 25–30. IBM Corporation (1965)
8. Chikkerur, S., Wu, C., Govindaraju, V.: A systematic approach for feature extraction in fingerprint images. In: Biometric Authentication, pp. 344–350. Springer (2004)
9. Farina, A., Kovacs-Vajna, Z.M., Leone, A.: Fingerprint minutiae extraction from skeletonized binary images, Pattern Recogn. **32**(5), 877–889. Elsevier (1999)
10. Feng, J.: Combining minutiae descriptors for fingerprint matching. Pattern Recogn. **41**(1), 342–352. Elsevier (2008)
11. Fronthaler, H., Kollreider, K., Bigun, J.: Local feature extraction in fingerprints by complex filtering. In: Advances in Biometric Person Authentication, pp. 77–84. Springer (2005)
12. Fronthaler, H., Kollreider, K., Bigun, J.: Local features for enhancement and minutiae extraction in fingerprints. IEEE Trans. Image Process. **17**(3), 354–363 (2008)
13. Fronthaler, H., Kollreider, K., Bigun, J.: Local features for enhancement and minutiae extraction in fingerprints. IEEE Trans. Image Process. **17**(3), 354–363 (2008)
14. Gao, X., Chen, X., Cao, J., Deng, Z., Liu, C., Feng, J.: A novel method of fingerprint minutiae extraction based on Gabor phase. In: 2010 17th IEEE International Conference on Image Processing (ICIP), pp. 3077–3080 (2010)
15. Gao, Q., Moschytz, G.S.: Fingerprint feature extraction using CNNs. Eur. Conf. Circuit Theory Des. **1**, 97–100 (2001)
16. Hong, L., Wan, Y., Jain, A.K.: Fingerprint image enhancement: algorithm and performance evaluation. IEEE Trans. Pattern Anal. Mach. Intell. **20**(8), 777–789 (1998)
17. Hong, L., Wan, Y., Jain, A.K.: Fingerprint image enhancement: algorithm and performance evaluation. IEEE Trans. Pattern Anal. Mach. Intell. **20**(8), 777–789 (1998)
18. Jiang, X., Yau, W. Y., Ser, W.: Detecting the fingerprint minutiae by adaptive tracing the gray-level ridge. Pattern Recogn. **34**(5), 999–1013. Elsevier (2001)
19. Khosravi, M., Schafer, R.W.: Template matching based on a grayscale hit-or-miss transform. IEEE Trans. Image Process. **5**(6), 1060–1066 (1996)
20. Khosravi, M., Schafer, R.W.: Template matching based on a grayscale hit-or-miss transform. IEEE Trans. Image Process. **5**(6), 1060–1066 (1996)

21. Maio, D., Maltoni, D., Cappelli, R., Wayman, J.L., Jain, A.K.: FVC2004: Third Fingerprint Verification Competition. Biometric Authentication, pp. 1–7. Springer (2004)
22. Miao, D., Tang, Q., Fu, W.: Fingerprint minutiae extraction based on principal curves. Pattern Recogn. Lett. **28**(16), 2184–2189. Elsevier (2007)
23. Nallaperumal, K., Padmapriya, S.: A novel technique for fingerprint feature extraction using fixed size templates. In: 2005 Annual IEEE INDICON, pp. 371–374 (2005)
24. Nguyen, T.H., Wang, Y., Li, R.: An improved ridge features extraction algorithm for distorted fingerprints matching. J. Inf. Secur. Appl. **18**(4), 206–214. Elsevier (2013)
25. Ratha, N.K., Chen, S., Jain, A.K.: Adaptive flow orientation-based feature extraction in fingerprint images. Pattern Recogn. **28**(11), 1657–1672. Elsevier (1995)
26. Schaefer, R., Casasent, D.: Nonlinear optical hit—miss transform for detection. Appl. Opt. **34**(20), 3869–3882. Optical Society of America (1995)
27. Serra, J.: Image Analysis and Mathematical Morphology. Academic Press, Inc. (1983)
28. Shi, Z., Govindaraju, V.: A chaincode based scheme for fingerprint feature extraction. Pattern Recogn. Lett. **27**(5), 462–468. Elsevier (2006)
29. Shin, J.H., Hwang, H.Y., Chien, S.: Detecting fingerprint minutiae by run length encoding scheme. Pattern Recogn. **39**(6), 1140–1154. Elsevier (2006)
30. Short, N. J., Hsiao, M. S., Fox, E.: Robust feature extraction in fingerprint images using ridge model tracking. In: 2012 IEEE Fifth International Conference on Biometrics: Theory, Applications and Systems (BTAS), pp. 259–264. IEEE (2012)
31. Simon-Zorita, D., Ortega-Garcia, J., Cruz-Llanas, S., Gonzalez-Rodriguez, J.: Minutiae extraction scheme for fingerprint recognition systems. In: Proceedings 2001 International Conference on Image Processing, vol. 3, pp. 254–257. IEEE (2001)
32. Tiwari, S., Sharma, N.: Q-Learning approach for minutiae extraction from fingerprint image. Procedia Technol. (6), 82–89. Elsevier (2012)
33. Yang, J., Liu, L., Jiang, T.: Improved Method for Extraction of Fingerprint Features **552–558**, (2002)
34. Zhao, F., Tang, X.: Preprocessing and postprocessing for skeleton-based fingerprint minutiae extraction. Pattern Recogn. **40**(4), 1270–1281. Elsevier (2007)

Face Identification via Strategic Combination of Local Features

Rinku Datta Rakshit and Dakshina Ranjan Kisku

Abstract Face identification systems that use a single local descriptor often suffer from lack of well-structured, complementary and relevant facial descriptors. To achieve notable performance, a face identification system is presented, which makes use of multiple representations of a face by means of different local descriptors derived from Local Binary Pattern (LBP) and Local Graph Structure (LGS). Then, max, average and L2 pooling operations are applied on each representation of face image to scale down the features. Further, different representations are combined together and formed a strategically concatenated feature vector. Three classifiers including two correlation based and k-nearest neighbor (kNN) are used to produce matching proximities which are used to characterize a probe user with rank identity. The produced ranks are then fused together using rank level fusion techniques. The experimental results determined on the Extended Yale face B database and the Plastic Surgery database are encouraging and convincing.

Keywords Face identification · Local binary pattern · Local graph structure · K-NN classifier · Convolutional neural network · Pooling

1 Introduction

Automatic face recognition [8, 11] is a time demanding and routine activity in today's world to identify or verify a person, mainly where the identity is susceptible to attack. As the world is becoming a more vulnerable place, the need for face identification has been increasing dramatically in recent time. The automatic face identification system

R. D. Rakshit
Asansol Engineering College, Vivekananda Sarani, Kanyapur, Asansol 713305, West Bengal, India
e-mail: rakshit_rinku@rediffmail.com

D. R. Kisku (✉)
National Institute of Technology, Mahatma Gandhi Road, A-Zone, Durgapur 713209, West Bengal, India
e-mail: drkisku@cse.nitdgp.ac.in

A. K. Das et al. (eds.), *Computational Intelligence in Pattern Recognition*, Advances in Intelligent Systems and Computing 999,
https://doi.org/10.1007/978-981-13-9042-5_18

is achieved by the uprising of different state-of-the-art face recognition modalities [8, 11]. An automatic face identification system can be integrated into an application that offers secure access control. A face identification system [6] takes an unknown face as an input, and determines identity from a database of known individuals. Even though the current face identification systems have reached a certain level of maturity with higher accuracy rates, still there is a challenging task to correctly identify an unknown person in the unconstrained environment as well as when face deformation [21] is present. The recognition gets more challenging when faces are found to be degraded faces due to illumination and pose variations [17] or when plastic surgeries are done on distorted faces. As the interests are growing rapidly among researchers to solve face identification with higher identification rates across complex facial dynamics, novel ideas would be helpful.

To maximize the accuracy as well as security of a biometric identification process, the multimodal biometric process [19] is the best-suited solution, because the multimodal biometric system requires two or more than two biometric credentials for identification. Among various types of multimodal biometric systems, multiple representation system captures a single biometric sample using a single sensor and then using two or more feature extraction techniques to generate multiple representations of a captured image. It has been proved to be an essential multi-biometrics system where complementary information is generated in parallel. In a human identification system, rank level fusion [10, 14] is also an important component where it combines the decisions in terms of ranks obtained from different biometric modalities to raise the strengths and diminish the weakness of the separate measurements.

The main objective of this work is to exhibit a combined effort of a strategic combination of local features, pooling operations and rank level fusion in face identification system under various facial dynamics including post-surgery variations.

In this paper, a robust and efficient face identification system using multiple representations analogous to deep features are generated by a set of local descriptors which are variants of Local Binary Pattern (LBP) [3] and Local Graph Structure (LGS) [2]. The pictorial representation of analogous deep local features in shown in Fig. 1. Most of the variants are developed by the authors described in [6, 17]. The key experiments are performed on two challenging face databases, namely, the extended Yale face B database [22] and the Plastic Surgery face database [21], which present two different perspectives of face images captured in stimulating environments. Figure 2 illustrates how to combine multiple representations in the way filtering employed in a convolutional neural network with pooling strategy and rank level fusion in the proposed system. Multiple representations are created from multiple local descriptors, namely, V-LGS [17], VS-LGS [17], ZH-LGS [17], ZHM-LGS [17], ZV-LGS [17], ZVM-LGS [17], LE-LGS [17], LGS [2], SLGS [1], LBP [3], MS-LBP [12], LDP [23], and LC-LBP [6]. As most of the local feature extraction approaches are invariant to illumination variations, structural variations and local image distortions, hence multiple representations having complimentary information would be proved to be useful when they are combined and processed the way convolutional layer does. However, the convolution operation is not performed rather local changes are being processed and taken care of by local descriptors. In this process, three pooling

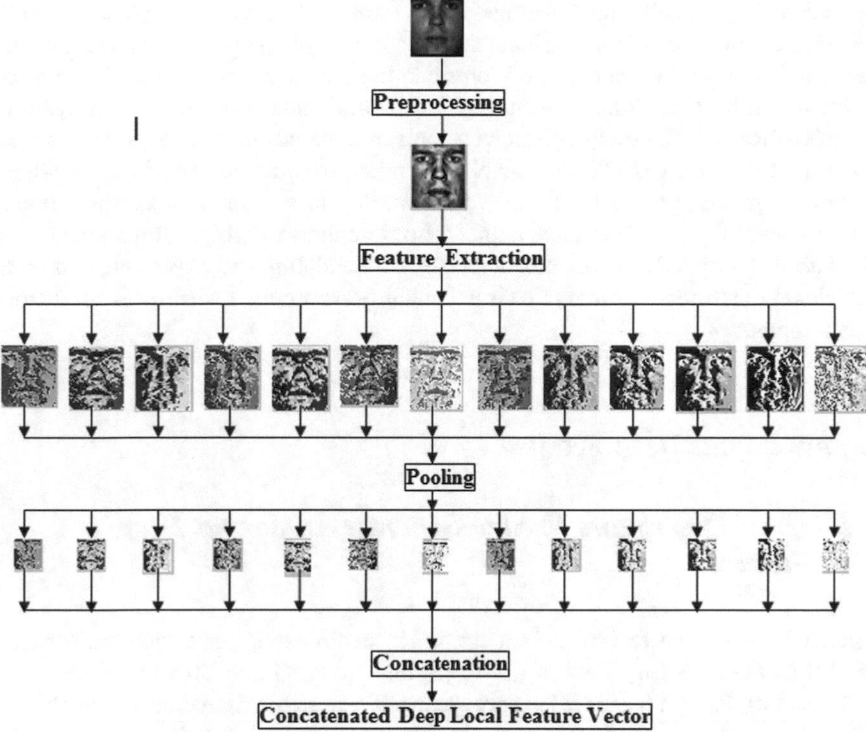

Fig. 1 Concatenated deep local feature at depth 1

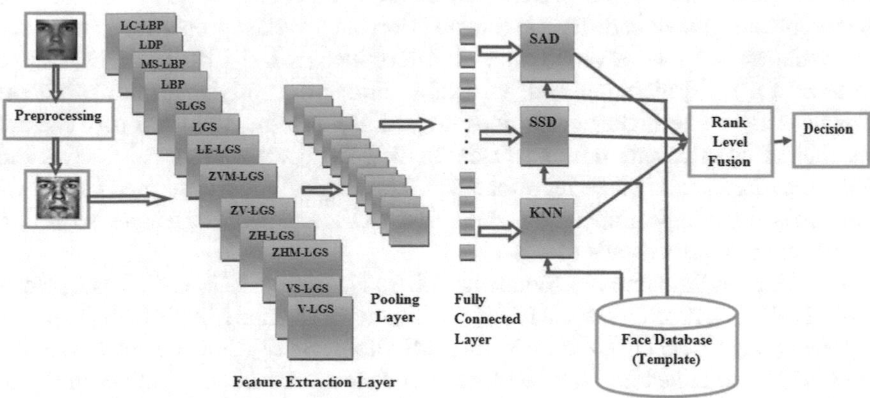

Fig. 2 Proposed face identification system

operations [8], namely, max, average, and L2 are applied to each representation to reduce the dimension of the original feature vector. Pooling operation is retaining the depth information by removing or consolidating less relevant features. Then these reduced feature vectors are concatenated to form an analogous deep local feature. For classification, three classifiers, viz., sum of squared difference (SSD), sum of absolute difference (SAD), and k-NN are applied to produce match scores, which in turn to generate ranks internally corresponding to a probe image. Some popular rank level fusion techniques, namely, Borda count, weighted Borda count, and Bucklin majority voting are considered for consolidating the ranks obtained from individual classifiers. To identify an unknown person rank 1 identity is used from this combination.

2 Analogous Deep Feature

2.1 Local Descriptors Used to Generate Analogous Deep Feature

The analogous deep feature uses a set of 13 local descriptors which are variants [6, 17] of Local Binary Pattern (LBP) [3] and Local Graph Structure (LGS) [2]. Local Binary Pattern (LBP) [3] is very facile and useful local descriptors which give binary labels to each pixel of an image by comparing the neighboring pixels with the value of center pixel of an image. LBP has become a popular local descriptor due to its discriminative ability and very less computational complexity. Multi-scale LBP [12] is one of the variants of LBP which encodes each pixel of an image by thresholding the neighboring pixels at different scales to bring out invariant properties from a face. The variation of a face at varied scales is determined by MS-LBP. Local Derivative Pattern (LDP) [23] elicits detailed facial information from outlying pixels of different directions by using higher order derivatives. LDP has the ability to derive more discriminating information from a face than the LBP. It seeks a more distinctive and well-structured local feature representation. The Logically Concatenated LBP (LC-LBP) uses three logical operations, viz., AND, OR, and XOR at feature extraction level to extract more details of a face.

Local Graph Structure [2], Symmetric LGS [1], Vertical LGS [17], Zigzag Horizontal LGS [17], Zigzag Vertical LGS [17], Zigzag Horizontal Middle LGS [17], Vertical Symmetric LGS [17], Zigzag Vertical Middle LGS [17], and Logically extended LGS [17] are directed graph based local descriptors for face recognition. In a face image, LGS [2] and its variants form a directed graph for a pixel with its neighboring pixels methodically to elicit the immutable and explicit spatial information. LGS and its variants are capable to bring out more discriminatory local texture information with compared to other existing local descriptor based techniques.

2.2 Analogous Form

Local descriptors are very useful for extracting discriminative spatial features which are an essential part of any object recognition. Applying a local descriptor on a face image is a process of mapping an image to an intermediate matrix, which produces a transformed image. Then a feature vector is derived from this intermediate matrix. These measures are very useful for classifiers to identify a person correctly. Alternatively, this transformed image can also be mapped to a histogram, from which a feature vector can be derived. The feature vector derived from the histogram can be used to classify a face image. Feature extraction from a face image using single local descriptor is not use effectively in all environments, because in conditions like illumination, pose and distortions due to plastic surgery, identifying a person is a difficult task. In these situations, combined and robust feature vectors having both relevant and complementary information would be an efficient approach for identification.

In this paper, multiple algorithms (multiple local descriptors) are applied to a face image separately. Multiple local descriptors are capable to represent various features of human faces, and reveal complementary facial property utilizing local image regions effectively by capturing illumination variations, image distortions, spots, flat areas, structural changes and edges in the face image. Despite the several advantages of multiple local feature representation, fusion at feature extraction level may lead to the problem of large dimensionality, which may require a strategic feature reduction approach. Every local descriptor that has been used in the proposed work, carrying some overlapping and disjoint characteristics. For example, LBP, LGS and SLGS are invariant to illumination changes, pose variations, spots, facial expression and various imaging modality. LDP is a high order pattern descriptor, can capture the changes of derivative directions and can alleviate the noise sensitivity problem. MS-LBP can extract feature from a face image at different scales. LGS and its variants are unvarying to any monotonic conversion such as logarithm, scaling, or shifting of pixel values. All these local descriptors are very efficient in terms of computational speed, complexity and memory requirement. The present approach uses analogous deep features which try to create the impact which is the same as that of the convolutional layer and pooling layer in CNN with less computational complexity. The proposed face identification framework can also be implemented in a dynamic or reinforcement way using deep learning. Already, there is a research work [13] in which the Local Binary Pattern (LBP) is combined with Deep Neural Network and designed LBPNet. Local binary comparisons and random projections are used jointly in LBPNet in place of conventional convolution operations only.

In this paper, multiple local descriptors (Sect. 2.1) are applied to generate multiple representations of a face image. Then, pooling operation is applied on a transformed face signature to reduce the size of the transformed pattern. Three different pooling techniques, namely, max pooling, average pooling, and L2-pooling are used to reduce the size of transformed face images as well as preserves discriminating features of face image.

3 Face Matching and Rank Level Fusion

3.1 Face Matching

Prior to matching, the face image is processed by 13 local descriptors independently and a set of patterns are generated. Afterward, pooling techniques [8] are applied on the transformed patterns which in turn reduced to a set of compact feature vectors either by removing (i.e., max pooling) or consolidating (i.e., average pooling) less significant spatial features. The pooling process not only retain depth features but also reducing the dimensionality of original transformed patterns in order to minimize the processing time as well as keep all relevant and complementary information together. Reduced feature sets are then concatenated together to form deep local feature set which is found discriminating in nature. This concatenated feature set is created for each face image in the dataset. During matching the template generated from a probe face image is compared to all face templates present in the training database and match scores are generated according to the type of classifiers used in the experiment. Then, a rank-order list is generated for a probe image according to match scores obtained from the classifiers. In these experiments, three different classifiers [8], namely, sum of absolute difference (SAD), the sum of squared difference (SSD), and k-nearest neighbor (K-NN) are applied to produce the match scores between probe template and all templates stored in face database. There are many sophisticated classifiers available for face recognition and Support Vector Machine (SVM) [9] is one such classifier which is used in many face recognition works and we have ample opportunity to apply the SVM in the proposed work. However, to alleviate or diminish the computational complicacy, simple distance-based metrics would be more useful. In the proposed work, three minimum distance-based metrics are used.

3.2 Rank Level Fusion

To improve the reliability in the face identification system, rank level fusion is found very obligate which put together decisions made by various classifiers to take favor of strengths of separate classifiers while restraining their failings. The proposed work uses rank level fusion [10, 14] approach which combines more than one rank-order lists produced by different classifiers to enhance the authenticity of a system. The rank level fusion is pertinent in an identification system where every classifier implicates a rank with each user those are already enrolled in the system. The proposed work uses three well-known rank level fusion approaches [8, 10, 14], viz. Borda Count, Weighted Borda Count, and Bucklin Majority Voting. The fused ranks are determined using these three fusion rules. To generate fused rank of every enrolled subject using Borda Count, the ranks out of every classifier are integrated. The enrolled subject's identity that makes the maximum rank is ascribed as the correct user identity. More details about these fusion rules can be found in [10, 14].

4 Proposed Face Identification Framework

Here, a complete framework of the proposed face identification system is presented. The framework comprises four phases—image enhancement, feature extraction, matching, and decision. In the image enhancement step, contrast-limited adaptive histogram equalization (CLAHE) technique [18] is used to improve the contrast in the face image. This technique improves the local contrast as well as enhances the edges in each region of the face image. In feature extraction, local features are extracted by using some effective multiple local descriptors and then concatenated deep feature vector is formed using pooling operations. During matching three different classifiers are used to calculate matching scores and then rank-order lists are generated from these matching scores. Further, rank level fusion rules are applied to generate a single combined rank-order list. Finally, at the decision phase, Rank1identity is taken as a final result of identification. The framework is shown in Fig. 2. The proposed framework which designed for face identification is very simple.

5 Evaluation

5.1 Databases

The experiment is conducted on two benchmark face databases such as the Extended Yale face B database [22] and plastic surgery database [21]. The databases themselves are considered very complex databases in nature and they are consisting of face images captured from different perspectives. The extended Yale face B database [22] comprises images of 38 persons. The images present in this database are captured in 9 different facial poses and under 64 illumination settings. The Plastic surgery face database [21] comprises 1800 pre-surgery and post-surgery face images concerning to 900 individuals. The database contains 519 image doublets resembling to local surgeries and 318 image doublets resembling to global surgeries. Specimen face images out of these two databases are exhibited in Fig. 3.

5.2 Experimental Results

This section presents experimental results obtained from extended Yale face B [22] and Plastic surgery [21] databases. The experiment exhibits the effectiveness of the proposed strategy. The evaluation is partitioned the databases into two disjoint sets: training set and probe set. The extended Yale face B database is divided into two parts where 380 face images are taken randomly from 38 individuals for query set and rest of 21,508 images are used as a learning set. In the Plastic surgery face database, it is very hard to discriminate between pre-surgery and post-surgery face images of

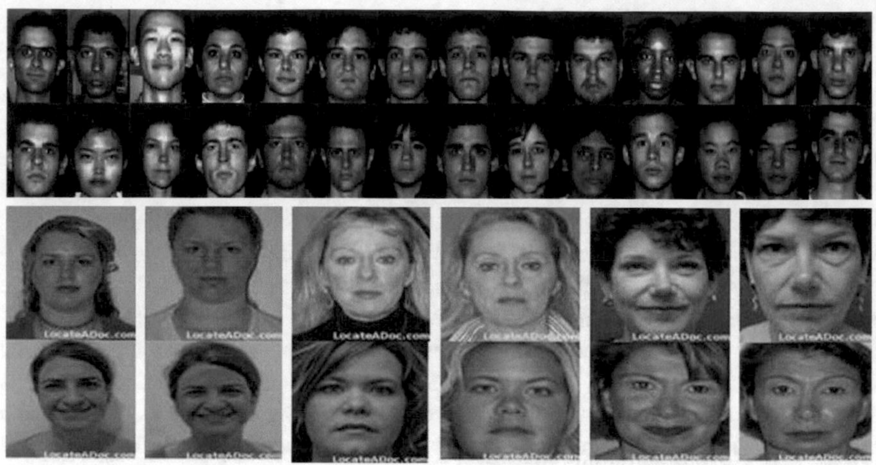

Fig. 3 Specimen face images from two databases are shown. Top-Row: Extended Yale face B database; Bottom-Row: Plastic Surgery database

some individuals as there is no isolate demarcation macula is provided among pre-surgery images and post-surgery images for several subjects. Therefore, prior to the experiment, the images are divided into different subsets as well as training and test subsets. Hence, to conduct the experiment, 53 post-surgery face images are taken for query set and rest 53 pre-surgery images are taken for the training set.

This work employs a well-known face identification protocol. A combined feature vector is extracted from one probe sample and then compared with every template stored in whole face database, and three different classifiers [8], namely, SAD, SSD, and KNN, are used to generate three different lists of match scores. Then from these match scores, three different rank-order lists are produced. Prior to taking the final decision, three rank-order lists are fused using three well-known rank level fusion techniques and then Rank 1 identification result is considered as a final decision. Experimental results achieved at Rank 1 on two databases using different rank level fusion approaches and varied pooling approaches are shown in Table 1.

Table 1 Identification accuracy at Rank 1 determined on the extended Yale face B and the Plastic Surgery databases

Rank level fusion	Identification accuracy (%) at Rank-1					
	Extended Yale face B database			Plastic Surgery database		
	Pooling approaches					
	Max	Average	L2	Max	Average	L2
Borda count	100	100	84.21	48	52	16
Weighted Borda count	100	100	86.84	64	80	16
Bucklin majority voting	100	100	86.21	68	80	20

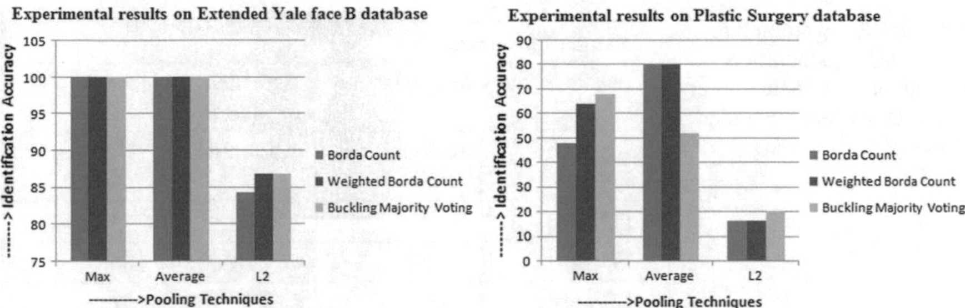

Fig. 4 Bar charts on identification accuracies determined on extended Yale face B and plastic surgery databases are shown

The extended Yale face B database is deliberated for the experiment to evaluate the effectiveness of the analogous concatenated deep local feature and rank level fusion in face identification when illumination and pose changes occur in face images captured in controlled environments. From the empirical outcomes shown in Table 1, it can be understood that at Rank 1, 100% accuracy is achieved for Borda count, Weighted Borda count and Bucklin majority voting rank level fusion approaches while max pooling and average pooling are applied at feature extraction phase. Results obtained by L2 pooling for three rank level fusions are also encouraging and convincing. Experimental results elicit that strategic combination of local descriptors with pooling operations and rank level fusions reveal that this approach is robust to pose changes and illumination variations in a degraded controlled environment.

The analogous deep features with rank level fusion strategies are evaluated in the proposed framework and are examined its efficacy on the Plastic surgery face database. The empirical outcomes reveal that identification precision is very encouraging and it is changing due to different pooling operations and rank level fusion strategies. Using average pooling at feature extraction level, Borda count and Weighted Borda count fusion rules achieve maximum 80% accuracy at Rank 1. Identification accuracies achieved on max pooling and L2 pooling with different rank level fusions are also encouraging. In real world it is very difficult to identify a person after plastic surgery is done on face and in automatic face identification system, it is very hard to capture local face variations, but experimental results achieved by the proposed framework is very encouraging. The identification accuracies determined on both the databases are shown in Fig. 4.

6 Performance Evaluation and Comparison

This section analyzes the effect of the single local descriptor in face recognition on the Plastic Surgery face database [21] and the Extended Yale face B database [22] and compares their performance with the proposed system. Only a few works on face identification is available in the literature. Plastic surgery can significantly deflect

Table 2 Recognition rate (%) of Face Recognition algorithms on Plastic Surgery face database and Extended Yale face B database

Name of face databases			
Plastic Surgery face database		Extended Yale face B database	
Algorithm	Recognition rate (%)	Algorithm	Recognition rate (%)
PCA [5]	29.1	LBP [3]	75
FDA [5]	32.5	G_LBP [23]	86.7
LFA [16]	38.6	LDP [23]	92
CLBP [4, 15]	47.8	G_LDP [23]	97.7
SURF [7]	50.9	Proposed method	100
GNN [20]	54.2		
Proposed method	80%		

the spatial regions of a face in terms of the facial feature as well as textures. The work presented in [21] accomplished 29.1% using PCA [5], 32.5% accuracy using FDA [5], 38.6% accuracy using LFA [16], 47.8% accuracy using CLBP [4, 15], 50.9% accuracy using SURF [7], 54.2% accuracy using GNN [20] at rank-1 on Plastic Surgery face database. The proposed approach achieves 80% identification accuracy at Rank 1 on Plastic Surgery face database and this outperforms existing results.

The work presented in [23] achieved 75.6% recognition rate using LBP [3], 86.7% recognition rate using G_LBP [23], 92.0% recognition rate using 3rd order LDP [23] and 97.7% recognition rate using G_LDP [23] on extended Yale face B database. The proposed system achieves 100% identification accuracy at rank-1 on extended Yale face B database. This results demonstrates outstanding performance over the existing works on the same database. The proposed work is compared with some state-of-the-art face recognition methods and is presented in Table 2.

7 Conclusion

Recently, several techniques are manifested to deal with face recognition using single local descriptor in feature extraction to extracts discriminating features from a face. However, in all circumstances, single local descriptor is not sufficient to extract discriminating features from a face and at the same time, it is necessary to cut the computational complexity of CNN. To achieve CNN like performance with less structural and computational complexity, an analogous deep representation of combined feature descriptors would be beneficial for obtaining higher accuracy. In this paper, an efficient and robust face identification system which uses concatenated local descriptors as analogous deep features for feature extraction and rank level fusion for individual identification. Through extensive experiments, it has been observed that the proposed method outperforms other state-of-the-art methods on extended Yale face B and plastic surgery databases.

References

1. Abdullah, M.F.A., Sayeed, M.S., Muthu, K.S., Bashier, H.K., Azman, A., Ibrahim, S.Z.: Face recognition with symmetric local graph structure (SLGS). Expert Syst. Appl. **41**(14), 6131–6137 (2014)
2. Abusham, E., Bashir, H.: Face recognition using local graph structure (LGS). Hum.-Comput. Interaction. Interact. Tech. Environ., 169–175 (2011)
3. Ahonen, T., Hadid, A., Pietikäinen, M.: Face recognition with local binary patterns. Comput. Vis.-ECCV **2004**, 469–481 (2004)
4. Ahonen, T., Hadid, A., Pietikainen, M.: Face description with local binary patterns: application to face recognition. IEEE Trans. Pattern Anal. Mach. Intell. **12**, 2037–2041 (2006)
5. Belhumeur, P.N., Hespanha, J.P., Kriegman, D.J.: Eigenfaces vs. Fisherfaces: recognition using class specific linear projection. Yale University, New Haven, USA (1997)
6. Datta Rakshit, R., Nath, S.C., Kisku, D.R.: An improved local pattern descriptor for biometrics face encoding: a LC–LBP approach toward face identification. J. Chin. Inst. Eng. **40**(1), 82–92 (2017)
7. Dreuw, P., Steingrube, P., Hanselmann, H., Ney, H., Aachen, G.: SURF-face: face recognition under viewpoint consistency constraints. In: BMVC, pp. 1–11 (2009)
8. Jain, A.K., Li, S.Z.: Handbook of face recognition. Springer, New York (2011)
9. Kisku, D.R., Mehrotra, H., Gupta, P., Sing, J.K.: Robust multi-camera view face recognition. Int. J. Comput. Appl. **33**(3), 211–219 (2011)
10. Kumar, A., Shekhar, S.: Personal identification using multibiometrics rank-level fusion. IEEE Trans. Syst., Man, Cybern., Part C (Appl. Rev.) **41**(5), 743–752 (2011)
11. Lenc, L., Pavel, K.: Automatic face recognition system (2013)
12. Liao, S., Zhu, X., Lei, Z., Zhang, L., Li, S.Z.: Learning multi-scale block local binary patterns for face recognition. In: International Conference on Biometrics, pp. 828–837. Springer, Berlin, Heidelberg (2007)
13. Lin, J.H., Yang, Y., Gupta, R., Tu, Z.: Local binary pattern networks (2018). arXiv preprint arXiv:1803.07125
14. Monwar, M.M., Gavrilova, M.L.: Multimodal biometric system using rank-level fusion approach. IEEE Trans. Syst., Man, Cybern., Part B (Cybern.) **39**(4), 867–878 (2009)
15. Ojala, T., Pietikainen, M., Maenpaa, T.: Multiresolution gray-scale and rotation invariant texture classification with local binary patterns. IEEE Trans. Pattern Anal. Mach. Intell. **24**(7), 971–987 (2002)
16. Penev, P.S., Atick, J.J.: Local feature analysis: a general statistical theory for object representation. Netw.: Comput. Neural Syst. **7**(3), 477–500 (1996)
17. Rakshit, R.D., Nath, S.C., Kisku, D.R.: Face identification using some novel local descriptors under the influence of facial complexities. Expert Syst. Appl. **92**, 82–94 (2018)
18. Reza, A.M.: Realization of the contrast limited adaptive histogram equalization (CLAHE) for real-time image enhancement. J. VLSI Signal Process. Syst. Signal, Image Video Technol. **38**(1), 35–44 (2004)
19. Ross, A., Jain, A.K.: Multimodal biometrics: an overview. In: 2004 12th European Signal Processing Conference, pp. 1221–1224. IEEE (2004)
20. Singh, R., Vatsa, M., Noore, A.: Face recognition with disguise and single gallery images. Image Vis. Comput. **27**(3), 245–257 (2009)
21. Singh, R., Vatsa, M., Bhatt, H.S., Bharadwaj, S., Noore, A., Nooreyezdan, S.S.: Plastic surgery: a new dimension to face recognition. IEEE Trans. Inf. Forensics Secur. **5**(3), 441–448 (2010)
22. UCSD Repository (2001). http://vision.ucsd.edu/~leekc/ExtYaleDatabase/ExtYaleB.html
23. Zhang, B., Gao, Y., Zhao, S., Liu, J.: Local derivative pattern versus local binary pattern: face recognition with high-order local pattern descriptor. IEEE Trans. Image Process. **19**(2), 533–544 (2010)

A Study on Reversible Image Watermarking Using Xilinx System Generator

Subhajit Das and Arun Kumar Sunaniya

Abstract The present work explains the design and implementation of the difference expansion method based reversible image watermarking using Xilinx System Generator and MATLAB. The method represents structural design for hiding information bits into the cover image by depending on the difference of neighborhood pixels of the cover image. The MATLAB-Simulink and Xilinx System Generator tools are a most preferable choice to improve the progress time and hardware resources by reducing the complication of image representation over FPGA environment. The reversible image watermarking using difference expansion is verified by taking a (4×4) sized test image and can be applicable for the larger size of cover images. The outcome of the results shows that the MATLAB-Simulink and Xilinx System Generator can be combined all the way through the graphical user interface by proposed structural design for image processing applications.

Keywords Xilinx system generator · Simulink · Difference expansion · Reversible watermarking

1 Introduction

Over the past few years, rapid change in development had been made in the field of image-based applications. Due to an increase in the activities of internet and digital devices, it very easy to access digital information as well as store them in electronic devices. This, in turn, gives rise to a new set of problems like distortion and illegal copying of digital data [1–3]. To overcome these problems, digital watermarking is most preferable to choose ever. When the distortion effect is less important, the reversible watermarking methods are used for copyright protection instead of digital watermarking [4]. There are mainly two type domains in which the reversible image

S. Das (✉) · A. K. Sunaniya
National Institute of Technology, Silchar 788010, Assam, India
e-mail: subhajitdas151@gmail.com

A. K. Sunaniya
e-mail: arun.sunaniya@gmail.com

A. K. Das et al. (eds.), *Computational Intelligence in Pattern Recognition*, Advances in Intelligent Systems and Computing 999,
https://doi.org/10.1007/978-981-13-9042-5_19

watermarking (RW) is mostly used. They are spatial domain and frequency domain. Spatial domain based RW algorithms are generally taken in account for hardware implementation as they provide less computational complexity over the frequency domain [5, 6]. Among numerous spatial domains based RW algorithms, the difference expansion (DE) based RW algorithm is the simplest and less complex algorithm [7–11]. The encoder and decoder are the main two parts of the RW. The embedded bits or watermark data are embedded in the cover image during the embedding process. The decoding is used to determine the watermark data present in the cover image and extract it to recover back both the watermark and the original cover image.

This paper mainly focuses to verify and provide an efficient model for hardware implementation of DE-based RW algorithm using Xilinx system generator tool. To archive the goal the MATLAB, Simulink, Xilinx system generator and an FPGA device or board have been used [12, 13]. The MATLAB software provides a model-based design environment which is known as Simulink. The Simulink has 18 block libraries, which are used for different application specific purposes. Xilinx system generator is the one of the block library provided by Simulink which is used to map digital signal processing (DSP) based MATLAB 2015a tool with field programmable gate array (FPGA) based Xilinx tool [7, 14–16].

2 The DE-Based RW Algorithm

The DE-based RW is first proposed by Tian [7]. This algorithm is based on the expansion of the difference between two neighborhood pixels. Let us consider a 4×4 sized 8-bit gray images which have 16 gray label intensity values as shown in Eq. (1). By taking two neighbor pixels there are total of 8 cases. In other words, case 1 has been formed by taking the first two neighbor pixels [183, 132]. Then case 2 is taking as [171, 168], case 3 as [109, 65] and so on. According to Tian [7], for case 1, the difference (D) and the average of those pixels (l) are calculated by Eq. (2) and Eq. (3), respectively.

$$f(x, y) = \begin{bmatrix} 183 & 132 & 171 & 168 \\ 109 & 65 & 126 & 156 \\ 72 & 46 & 114 & 136 \\ 79 & 78 & 133 & 116 \end{bmatrix} \quad (1)$$

$$D = x - y = 183 - 132 = 51 \quad (2)$$

and

$$l = \left\lfloor \frac{x + y}{2} \right\rfloor = \left\lfloor \frac{183 + 132}{2} \right\rfloor = 157 \quad (3)$$

The expanded new difference is calculated by appending a single embedded bit after the least significant bit (LSB) of the binary form of D. Let the embedded bit (b) is equal to 1 and the newly expanded difference is denoted as D'. This can be given by Eqs. (4) and (5).

$$D = (51)_{10} = (110011)_2 \tag{4}$$

$$D' = (110011b)_2 = (1100111)_2 = (103)_{10} \tag{5}$$

The corresponding watermarked pixels values for case 1 are calculated by Eqs. (6) and (7).

$$x' = l + \left\lfloor \frac{D'+1}{2} \right\rfloor = 157 + \left\lfloor \frac{103+1}{2} \right\rfloor = 157 + 52 = 209 \tag{6}$$

$$y' = l - \left\lfloor \frac{D'}{2} \right\rfloor = 157 + \left\lfloor \frac{103}{2} \right\rfloor = 157 - 51 = 106 \tag{7}$$

In a similar way, the watermarked pixels values for remaining cases have been calculated. Tian [7] also provides a boundary condition to overcome the underflow and overflow problems. The boundary conditions are explored from Eq. (8) to (11). The watermarked image is denoted as $w(x, y)$. The resultant watermarked gray-level intensity values are shown in Eq. (12)

$$0 \le l + \left\lfloor \frac{D+1}{2} \right\rfloor \le 255 \tag{8}$$

and

$$0 \le l - \left\lfloor \frac{D}{2} \right\rfloor \le 255 \tag{9}$$

Equations (7) and (8) can be rewritten as following ways.

$$|D| \le 2(255 - l) \quad \text{if} \quad 128 \le l \le 255 \tag{10}$$

$$|D| \le 2(l + 1) \quad \text{if} \quad 0 \le l \le 127 \tag{11}$$

$$w(x, y) = \begin{bmatrix} 209 & 106 & 173 & 166 \\ 132 & 43 & 172 & 111 \\ 86 & 33 & 148 & 103 \\ 80 & 77 & 142 & 107 \end{bmatrix} \tag{12}$$

For decoding processing, similarly with embedding process, eight numbers of cases have been considered by taking two neighborhood pixels from watermarked image. Then the average and difference have been calculated and denoted as l' and D' (Eq. 14), respectively. The actual difference D is then found out by extracting LSB of D'. For case 1, the calculations of l' and D are shown in Eq. (13) and Eq. (15), respectively.

$$l' = \left\lfloor \frac{x' + y'}{2} \right\rfloor = \left\lfloor \frac{209 + 106}{2} \right\rfloor = 157 \tag{13}$$

$$D' = x' - y' = 209 - 106 = 103 = (1100111)_2 \tag{14}$$

$$D = (110011)_2 = 51 \tag{15}$$

The original cover image value for case 1 is then calculated by following the inverse integer transform given in Eq. (16) and Eq. (17), respectively.

$$x = l' + \left\lfloor \frac{D + 1}{2} \right\rfloor = 157 + \left\lfloor \frac{51 + 1}{2} \right\rfloor = 157 + 26 = 183 \tag{16}$$

$$y = l' - \left\lfloor \frac{D}{2} \right\rfloor = 157 - \left\lfloor \frac{51}{2} \right\rfloor = 157 - 25 = 132 \tag{17}$$

This inverse integer transforms are applied for the remaining cases to find out the decoded image. It has been found that the gray level intensity values of the decoded image are exactly the same as the original cover image. Hence we can say that the DE-based RW algorithm preserves the reversibility properties by recover the original cover image without any loss of information about the original cover image.

3 The Block Model of the DE-Based RW Algorithm

In this section, the steps of implementation of the DE-based RW algorithm using Xilinx System Generator (XSG) has been explored. The purpose of this block model implementation is to verify the algorithm in FPGA environment. The basic block model diagram contains a function block named image from file, encoder, decoder, and video viewer. The first function is used to store the input cover image. The data type and size of the cover image are set to unit 8 and (4 × 4), respectively, as we are dealing with 8-bit grayscale (4 × 4) sized image. The basic block model diagram is shown in Fig. 1.

Both the encoder and decoder block have 3 sub-blocks which are shown in Fig. 2. The first sub-block of encoder and decoder is named as preprocessing of encoder and decoder, respectively, and they are shown in Figs. 3 and 4 in that order. Figures 3a and 4a shows the actual view of the preprocessing sub-block of encoder and decoder. The

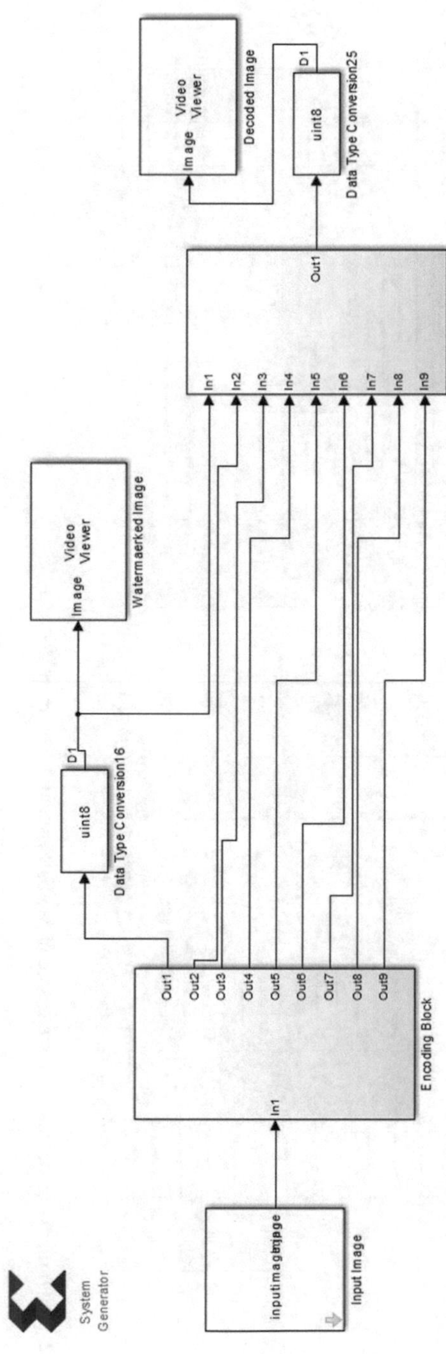

Fig. 1 The block model diagram DE-based RW using Xilinx system generator

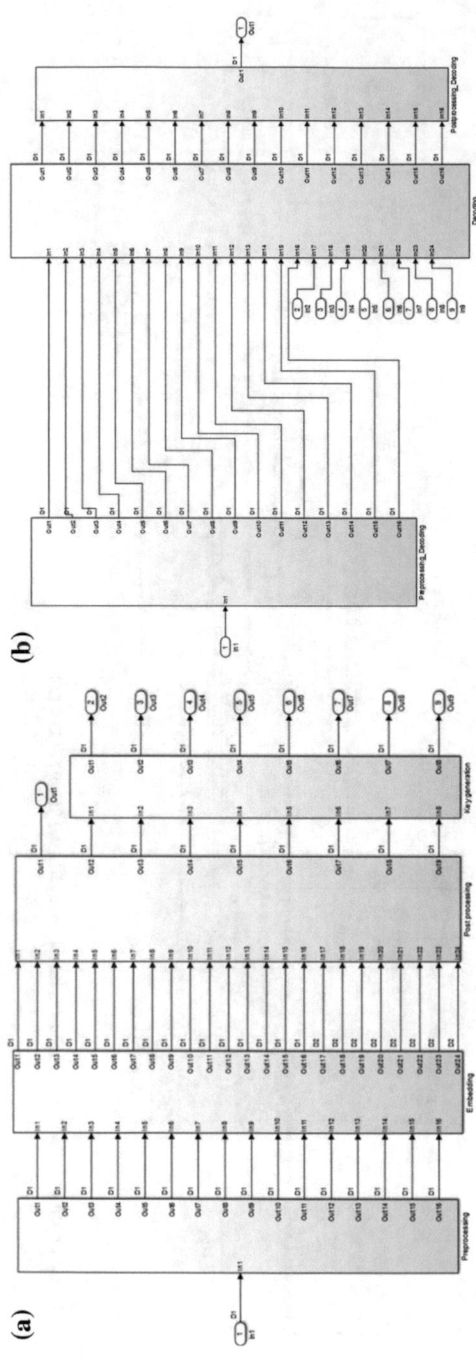

Fig. 2 **a** Sub-blocks model diagram of encoder. **b** Sub-blocks model diagram of decoder

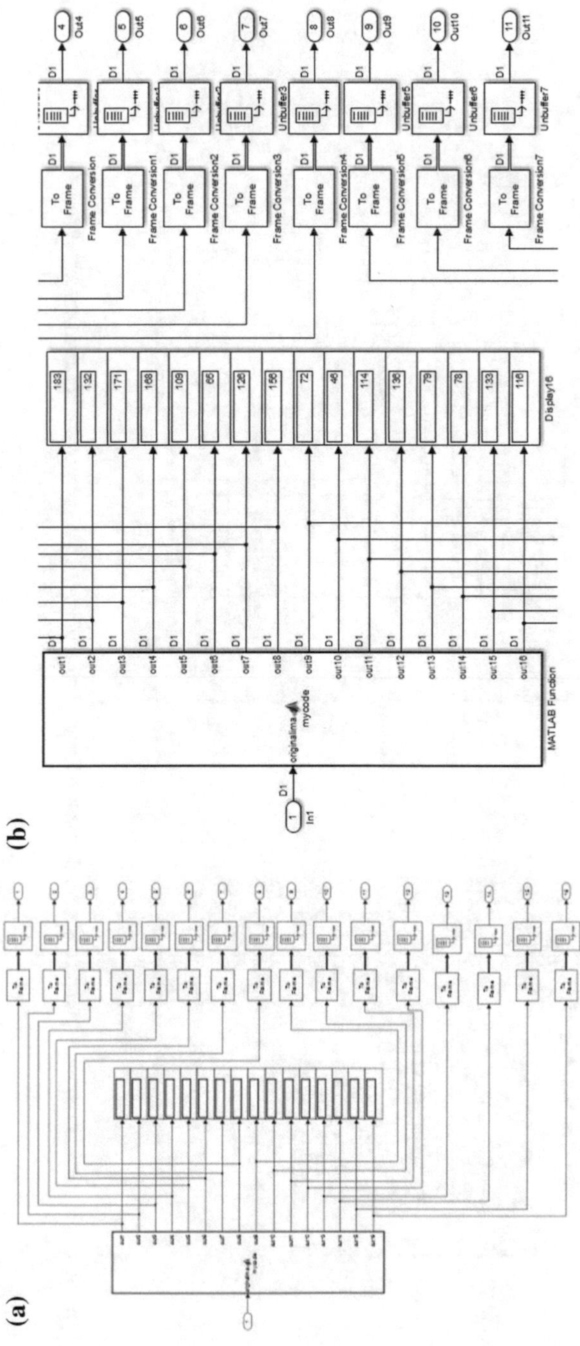

Fig. 3 **a** Sub-blocks model diagram of preprocessing of encoder. **b** Zoomed view of a partial portion of preprocessing block of encoder

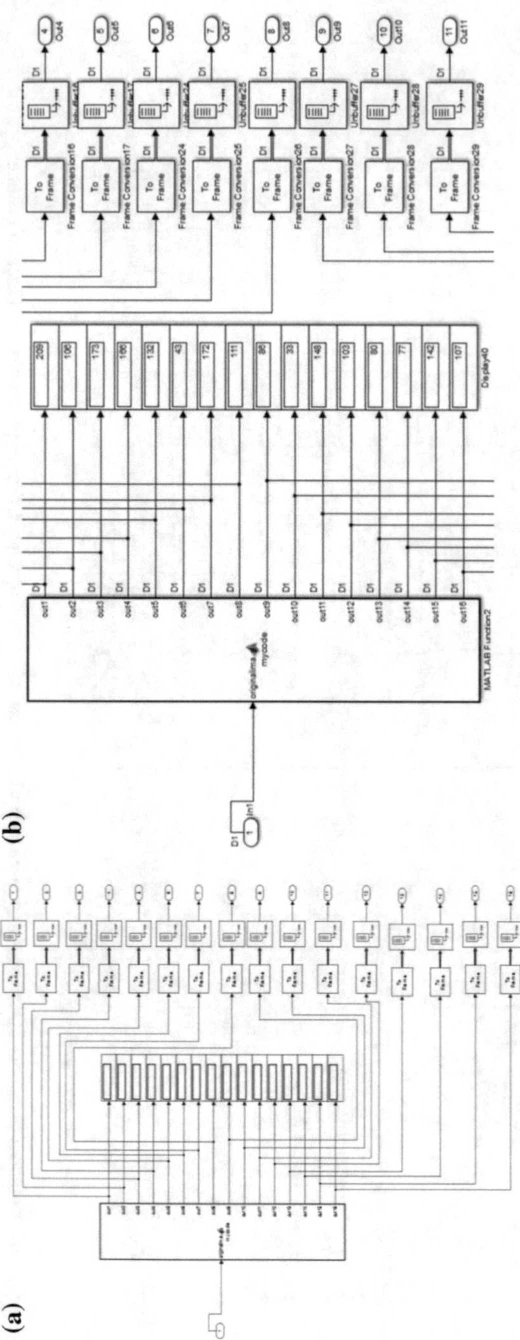

Fig. 4 **a** Sub-blocks model diagram of preprocessing of decoder. **b** Zoomed view of a partial portion of preprocessing block of decoder

zoomed view of partial portion of preprocessing block of encoder and decoder are shown in Fig. 3b and Fig. 4b, respectively. After reading the input image by the first function block, the preprocessing block helps to convert the image data which is in the form of m by n array into binary or intensity image. The output has a total (m × n) gray-level intensity values. Here both m and n are equal to 4. Then preprocessing block coverts the 2-D input data to 1-D format by 'convert 2-D–1-D' function block. The mode of the output is then set to sampling mode by 'frame conversion' function block. The last steps of the preprocessing block are to un-buffered the input data into row-wise which have been done by 'un-buffer' function block.

Initially, the preprocessing block for both encoder and decoder has been passed through an m-code block to show input data serially order. From Figs. 3b and 4b, it is observed that the original input data and decoded data are correctly read for further processing. Also, it noticed that both data are matched with desired data which are shown in Eqs. (1) and (12) of the previous section.

To implement the main two parts of the DE-based RW algorithm the XSG is used because our purpose is to process those parts of the algorithm through an FPGA board. In XSG, Hardware Description Language (HDL) codes are read through the function blocks named "black box" for processing the arithmetic operations for those parts. The embedding and decoding processing model blocks are shown in Fig. 5 and Fig. 7, respectively. Here two main function blocks named 'Gateway In' and 'Gateway Out' should be connected in between preprocessing and post-processing blocks of both encoding and decoding because those function blocks perform as input and output to Xilinx FPGA portion of the simulink design.

Figure 5b shows that the embedding process used four black boxes to perform four arithmetic operations. The operations are (1) difference calculation, (2) average calculation, (3) difference expansion, and (4) integer transform. The constant block set a constant bit value 1 which is act as an embedding bit. During embedding an 8 bit key is generated by depending on the boundary conditions which are given from Eq. (8) to (11).

The key generation process is shown in Fig. 6. They are used as an encryption key during decoding to recover back the original image without any loss. Like the embedding process, the decoding process also has 4 black boxes for four arithmetic operations which are shown in Fig. 7. Out of four operations, the first two operations are exactly the same as the encoding process. The last two arithmetic operations are used to recover the actual difference and reverse integer transform, respectively. The last sub-block is called post-processing for both embedding and decoding part. The working principle of the post-processing block is exactly opposite of the preprocessing sub-block. The post-processing sub-blocks for both embedding and decoding are shown in Figs. 8 and 9 in that order.

As it is difficult to read the text and values of post-processing block of encoder and decoder given in Fig. 8a and Fig. 9a, respectively, the zoomed view of a partial portion of those blocks are shown in Fig. 8b and 9b in that order. From Fig. 8b, we have noticed that the outcome intensity value of the watermarked image is exactly same as calculated intensity value of the watermarked image which has been shown in Eq. (12). Figure 9b shows the intensity value of decoded image after decoding

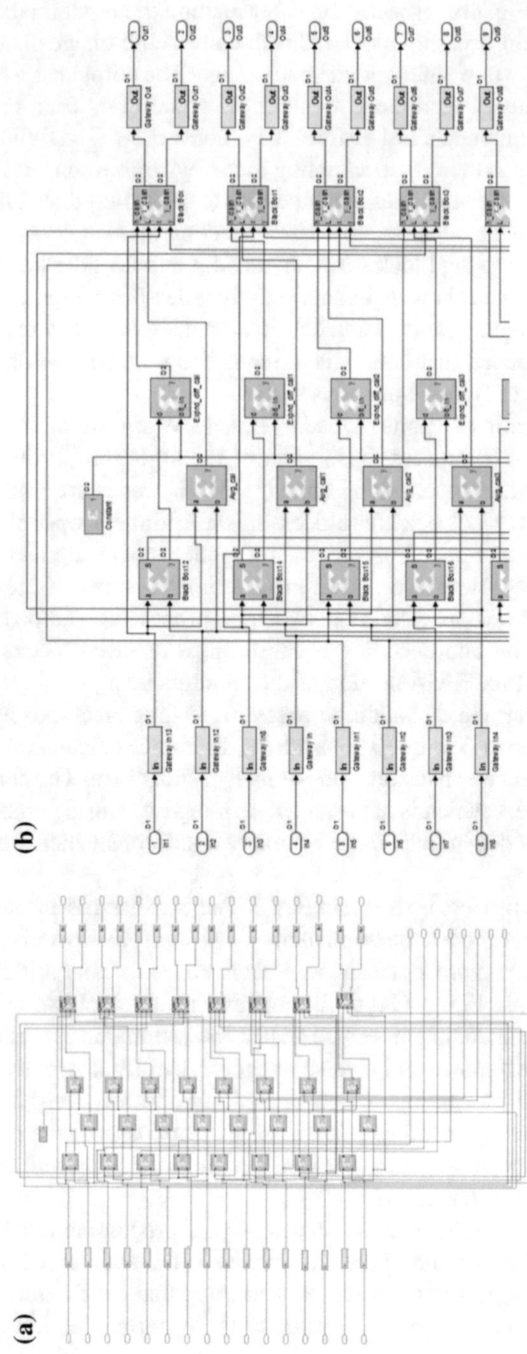

Fig. 5 **a** Sub-blocks model diagram of embedding algorithm. **b** Zoomed view of a partial portion of preprocessing block of embedding algorithm

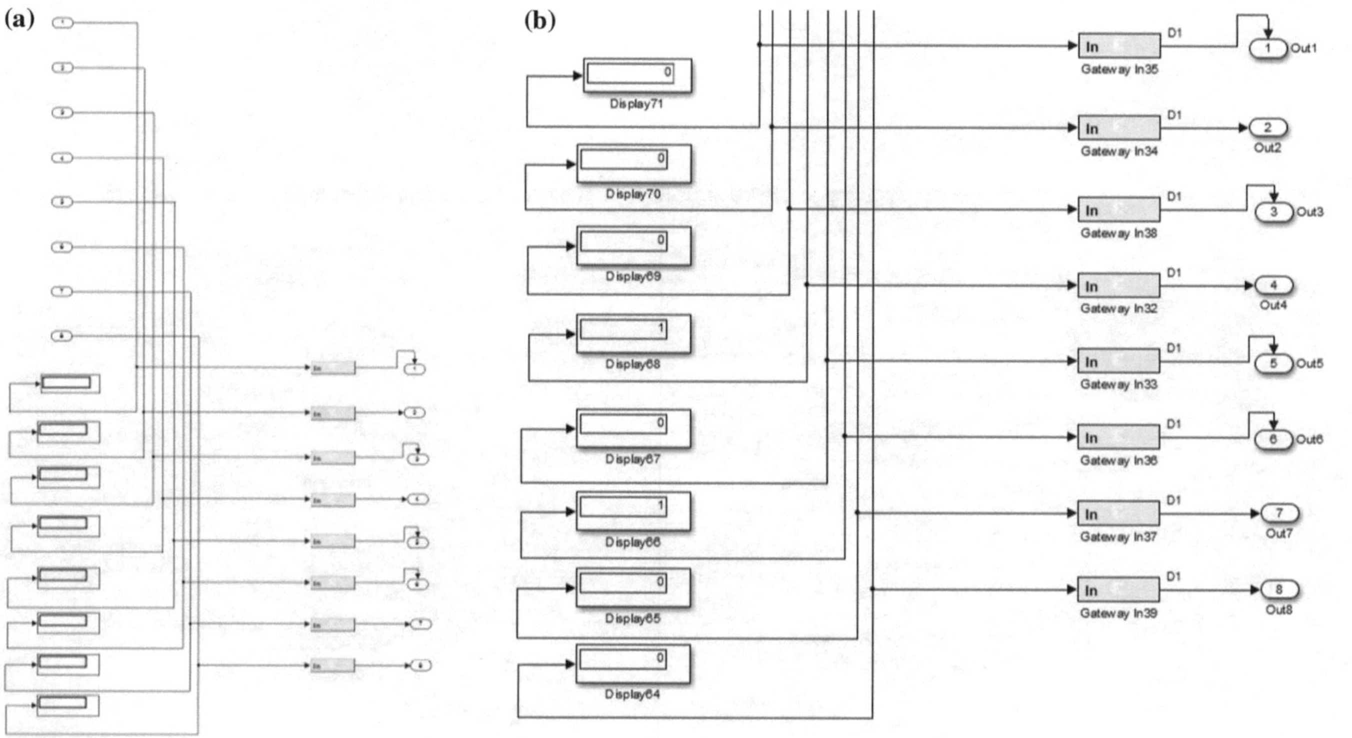

Fig. 6 **a** Sub-blocks model diagram of the key generation process. **b** Zoomed view of a partial portion of key generation process

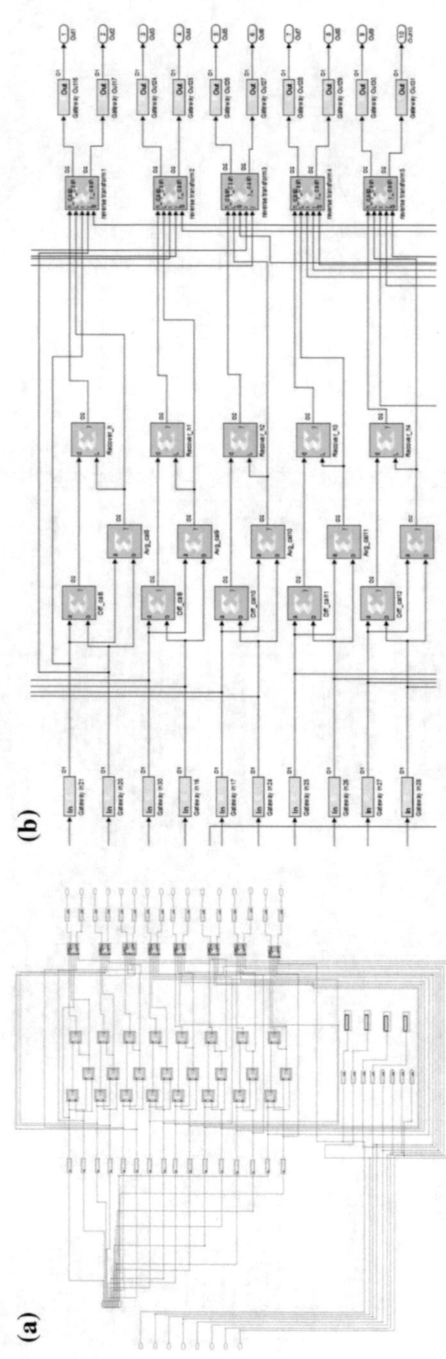

Fig. 7 **a** Sub-blocks model diagram of decoding process. **b** Zoomed view of a partial portion of decoding process

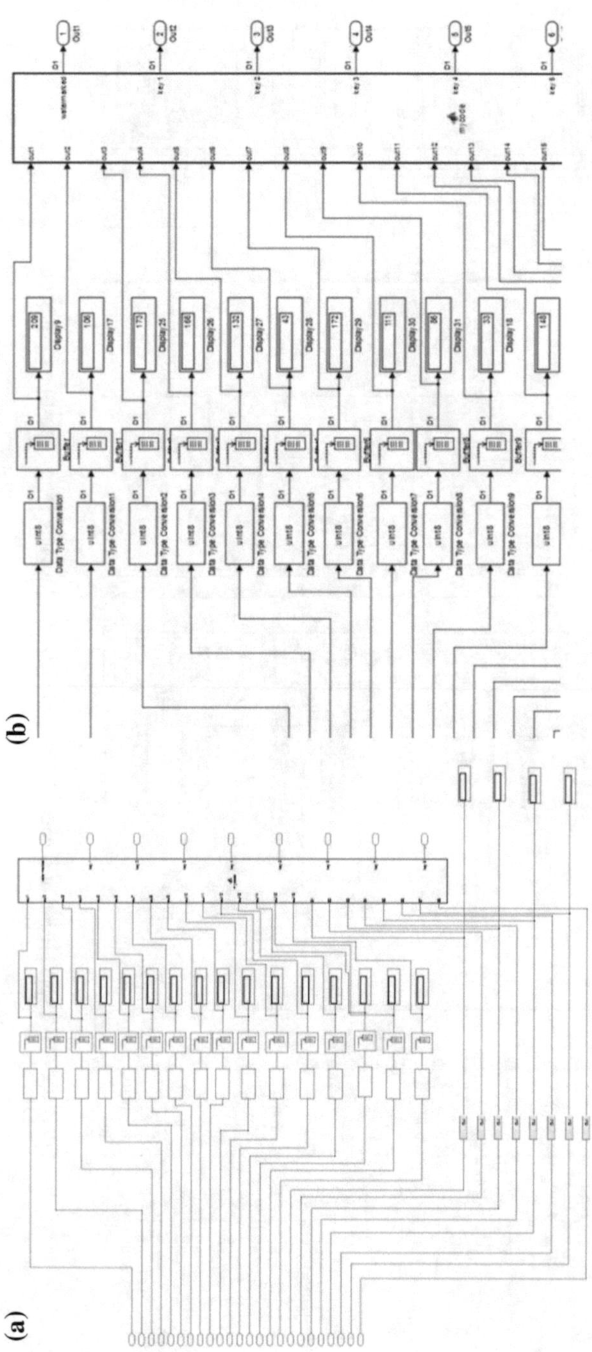

Fig. 8 **a** Sub-blocks model diagram of post-processing of encoder. **b** Zoomed view of a partial portion of post-processing block of encoder

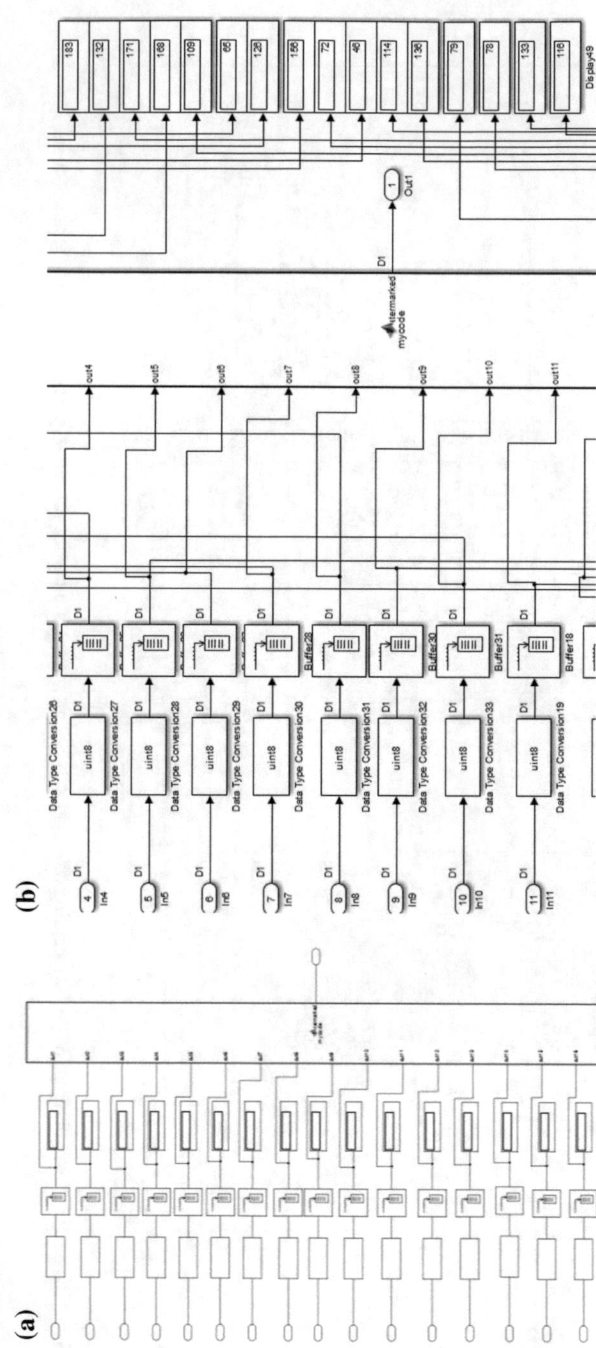

Fig. 9 **a** Sub-blocks model diagram of post-processing of decoder. **b** Zoomed view of a partial portion of post-processing block of decoder

process. We have noticed that it is exactly same as the intensity values of the original input cover image which has been shown in Eq. (1). So we can say that the algorithm preserve the properties of reversible watermarking.

4 Results and Analysis

The DE-based RW algorithm is first verified by taking 8-bit gray-level (4 × 4)-sized images whose gray level pixel intensity values are shown in Eq. (1). From Eq. (1) and Fig. 9(b), it is noticed that values of the original image and the decoded image are matched with each others. So we can say the reversibility property of the DE-based RW algorithm is verified by XSG. Two test images of (64 × 64) sized and their corresponding generated watermarked images using XSG are shown in Fig. 10. By using Zynq-7000 (Zed-board) technology XC7Z030 based target device, the device utilization summary for both encoder and decoder are shown in Table 1. Using the XSG, the algorithm is also verified by varying the image size from (4 × 4) to (256 × 256).

Fig. 10 **a** Test Image 1. **b** Test Image 2. **c** Watermarked Image 1 of Test Image 1. **d** Watermarked Image 2 of Test Image 2

Table 1 Device utilization summery

Process	Logic utilization	Used	Available
Encoder	Number of slices	12	1,57,200
	Number of slice flip flops	22	78,600
	Number of 4 input LUTs	22	78,600
	Number of bonded IOBs	18	469
	Number of GCLKs	1	24
Decoder	Number of slices	13	1,57,200
	Number of slice flip flops	28	78,600
	Number of 4 input LUTs	27	78,600
	Number of bonded IOBs	19	469
	Number of GCLKs	1	24

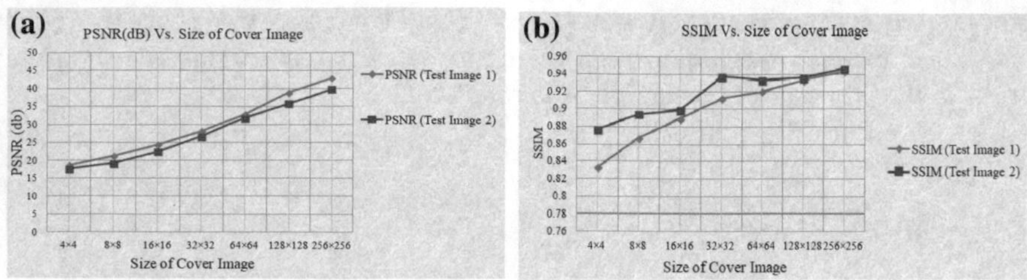

Fig. 11 Graphical analysis. **a** PSNR versus size of cover image. **b** SSIM versus size cover image

The maximum stimulation time using XSG for the DE-based RW algorithm is 147.31 ns. After getting the watermarked images using XSG for different size of cover images, the Peak signal to Noise Ratio (PSNR) and Structural Similarity Index Matrix (SSIM) between cover images and their corresponding watermarked images have been calculated. The PSNR and SSIM with respect to size of the two cover images are shown in Fig. 11a and b, respectively. It is observed that the PSNR and SSIM both increases with increase in the size of the cover images. It implies that we can embed more data by taking a larger size of the cover image. The blind watermarking properties are preserved for the larger size of cover images as the similarity between cover and watermarked is increased.

However, this model diagram is not suitable for JTAG co-simulation for real-time implementation of the algorithm using FPGA board like Zed-board because the four arithmetic operations are performed in a single simulation or pipeline cycle. We have to provide a suitable pipeline cycle for parallel processing of those arithmetic operations. This type of efficient implementation of block model using XSG for parallel processing is currently under process.

5 Conclusion

The step-by-step complete model diagram for DE-based RW algorithm has been presented and verified using XSG. The graphical analysis of the PSNR and SSIM between the various sizes of the cover and watermarked image are also presented and justified. The present block model implementation using XSG ensures that this process reduces the complexity and significantly increase the simulation time to handle image processing over parallel processing based software like Xilinx ISE.

Acknowledgements The authors would like to thank the TEQIP-III, NIT Silchar for providing financial assistance and VLSI Research Lab, Department of Electronics and Instrumentation Engineering, NIT Silchar to carry out the research work.

References

1. Fridrich, J., Goljan, M., Du, R.: Invertible authentication watermark for JPEG images. In: Proceedings International Conference on Information Technology: Coding and Computing, pp. 223–227. IEEE (2001)
2. Coltuc, D.: Low distortion transform for reversible watermarking. IEEE Trans. Image Process. **21**(1), 412–417 (2012)
3. Liu, Z., Zhang, F., Wang, J., Wang, H., Huang, J.: Authentication and recovery algorithm for speech signal based on digital watermarking. Signal Process. **123**, 157–166 (2016)
4. Ali, M., Ahn, C.W.: Comments on "Optimized gray-scale image watermarking using DWT-SVD and firefly algorithm". Expert Syst. Appl. **42**(5), 2392–2394 (2015)
5. Thongkor, K., Amornraksa, T., Delp, E.J.: Digital watermarking for camera-captured images based on just noticeable distortion and Wiener filtering. J. Vis. Commun. Image Represent. **53**, 146–160 (2018)
6. Pakdaman, Z., Saryazdi, S., Nezamabadi-Pour, H.: A prediction based reversible image watermarking in Hadamard domain. Multimed. Tools Appl. **76**(6), 8517–8545 (2017)
7. Tian, J.: Reversible data embedding using a difference expansion. IEEE Trans. Circuits Syst. Video Technol. **13**(8), 890–896 (2003)
8. Chen, C.C., Tsai, Y.H., Yeh, H.C.: Difference-expansion based reversible and visible image watermarking scheme. Multimed. Tools Appl. **76**(6), 8497–8516 (2017)
9. Arham, A., Nugroho, H.A., Adji, T.B.: Multiple layer data hiding scheme based on difference expansion of quad. Sig. Process. **137**, 52–62 (2017)
10. Chang, C.C., Huang, Y.H., Lu, T.C.: A difference expansion based reversible information hiding scheme with high stego image visual quality. Multimed. Tools Appl. **76**(10), 12659–12681 (2017)
11. Das, S., Maity, R., Maity, N.P.: VLSI-based pipeline architecture for reversible image watermarking by difference expansion with high-level synthesis approach. Cir. Syst. Sig. Process. **37**(4) 1575–1593 (2018)
12. Moreo, A.T., Lorente, P.N., Valles, F.S., Muro, J.S., Andres, C.F.: Experiences on developing computer vision hardware algorithms using Xilinx system generator. Microprocess. Microsyst. **29**(8–9), 411–419 (2005)
13. Selvamuthukumaran, R., Gupta, R.: Rapid prototyping of power electronics converters for photovoltaic system application using Xilinx System Generator. IET Power Electron. **7**(9), 2269–2278 (2014)
14. Elamaran, V., Praveen, A., Reddy, M.S., Aditya, L.V., Suman, K.: FPGA implementation of spatial image filters using Xilinx system generator. Procedia Eng. **38**, 2244–2249 (2012)
15. Krim, S., Gdaim, S., Mtibaa, A., Mimouni, M.F.: Implementation on the FPGA of DTC-SVM based proportional integral and sliding mode controllers of an induction motor: a comparative study. J. Circuits, Syst. Comput. **26**(03), 1750049 (2017)
16. Bahoura, M.: FPGA implementation of multi-band spectral subtraction method for speech enhancement. In: 2017 IEEE 60th International Midwest Symposium on Circuits and Systems (MWSCAS), pp. 1442–1445. IEEE, (2017)

SVM and MLP Based Segmentation and Recognition of Text from Scene Images Through an Effective Binarization Scheme

Ranjit Ghoshal and Ayan Banerjee

Abstract Text Binarization from scene images plays a crucial task for any text segmentation scheme and therefore in the OCR performance. So, an effective image binarization method is required for text segmentation and recognition tasks. This work describes an effective image binarization scheme for segmentation and recognition task of text from images. To binarize the image, Canny's edge information is incorporated into Otsu's method. It generates numerous components which are analyzed for segmentation of probable text components. Further, a few features are considered for classification of text and non-text. For this problem, SVM is considered. For training SVM classifier, information from ground-truth images of text and our own made non-text components are used. Finally, Multilayer Perceptron (MLP) is used for recognition of the text symbols. The MLP classifier is trained using 13496 samples of 39 classes. We tested our schemes on the publicly available ICDAR Born Digital data set. The outcomes are quite acceptable.

Keywords Segmentation of text · Text recognition · Binary image · MLP · SVM classifier

1 Introduction

With the increasing growth of technology nowadays, there are numerous camera-based applications are available. Everyone is capable to capture the text based scene images easily, but identification and recognition of scene text is a challenging project. So, scene text images have to be segmented correctly before recognized. To convert grayscale image into binary image is a big challenge, particularly where the outcome

R. Ghoshal
St. Thomas' College of Engineering and Technology, Kolkata 700023, India
e-mail: ranjit.ghoshal@rediffmail.com

A. Banerjee (✉)
Lexmark Research & Development Corporation, Kolkata, India
e-mail: ayanbanerjee.stcet@gmail.com

A. K. Das et al. (eds.), *Computational Intelligence in Pattern Recognition*, Advances in Intelligent Systems and Computing 999,
https://doi.org/10.1007/978-981-13-9042-5_20

of the binary image can enhance the OCR performance directly. Binarization is a key task of segmenting text. Many schemes are available for document image binarization but these schemes are not suitable for scene images. Scene images are more complicated than document images. The scene image is complex almost all the cases due to the background. Further, noise and shadows significantly increase the problem more complex. Therefore, to avoid this difficulty, new effective approaches are required. Generally, binarization schemes are categorized into two classes: *global* thresholding [1] and *local* thresholding [2, 3]. Recently several improvements over thresholding techniques are also proposed in document analysis and researchers are trying to extend these methods for scene text binarization. Gatos et al. [4] described a latest method for binarization of document images by applying a combination of a number of binarization schemes and adapted edge properties. Available schemes for text identification [5] can normally be grouped into two categories: window-based techniques and connected component based techniques.

CC-based methods filter out text components from scene images through analysis of the CC. CC-based techniques are widely used since the implementation is comparatively easy. The recognition problem also makes interest of many researchers. Although, all the attempts are still completed using scanned documents. Here, we present an effective new binarization scheme for segmentation and recognition of text.

2 Proposed Binarization Method

A text based color scene image, is transformed to a gray scale image which is the input of our scheme. Normally, edge finding schemes are used to detect the text boundary. Now, we intend to find edges in the gray image by applying Canny edge detector.

Proposed Algorithm:
Step 1: Obtain edges (E) from I_s (input image) applying Canny edge detection scheme and store this edge information in *edge1*.
Step 2: Initialize two matrices *img1* and *img2* (same size as I_s) by zeroes.
Step 3: Binarize I_s using Otsu's algorithm and store it in *img2*. Also, find the complement of *img2* and store it in *img1*.
Step 4: Run a $M \times M$ sliding window on I_s corresponding to the detected edge pixels. Binarize this window (BW) by Otsu's algorithm. Increment a pixel value by 1 in *img2* if its corresponding value is 1 in the window (BW). This matrix *img2* is used to count the number of ones in each pixel.
Step 5: Similarly to count the number of zeroes for each pixel, *img1* matrix is used. Complement the local window (i.e. BW) and increment a pixel value of *img1* by 1 if its corresponding value is zero, in the complement binarize window.
Step 6: Let $B(x, y)$ be the final binary image. It is calculated as follows:

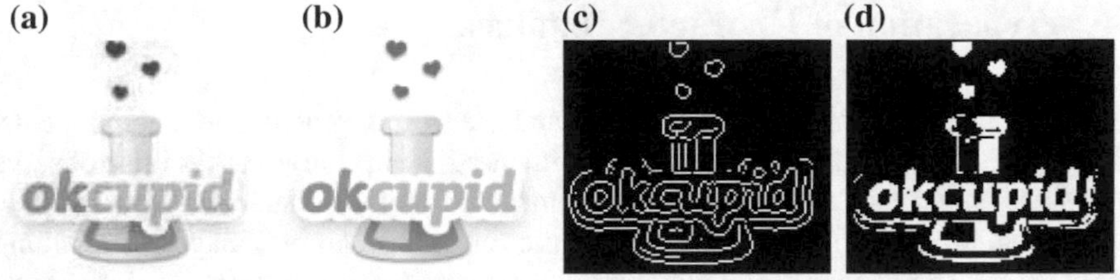

Fig. 1 **a** Input. **b** Grayscale image. **c** Canny edge **d** Binary image

$$B(x, y) = \begin{cases} 1\ ;\ \text{for any pixel } p_i,\ \text{if } img2(p_i) > img1(p_i) \\ 0\ ;\ \text{otherwise} \end{cases}$$

The main theoretical background of the steps 4 and 5 of the proposed binarization method is as follows. Let p_i be a pixel. Steps 4 and 5 mainly try to calculate the number of 1's and 0's of the pixel p_i. Based on these numbers, we assign the p_i label. Let, 1's count of p_i is larger. So, we assign p_i as 1. In Fig. 1, a few steps of the proposed scheme are illustrated. Figure 1a shows the input and Fig. 1b represents corresponding gray image. Canny edge maps are presented in Fig. 1c. Lastly, the binarization result is represented in Fig. 1c.

3 Feature Separation and SVM Based Segmentation of Text

Image binarization generates a number of connected components (CCs). To segment text components, we extract the following properties from each component.

S: The size ratio (S) is defined as: $S =$ (CC area/area of the input image).
A: It is (*Aspect ratio*) defined as $A = min\{(height/width), (width/height)\}$.
E: The nature of text components are normally elongated. To calculate elongatedness ratio (E), the measure using [6] is used.
O: It (O) is the ratio of the object pixels to background pixels [6].
AX: Axial ratio (AX) [5].
T: *Thickness* (T) of a CC is calculated using [6].
L: Length Ratio [5].
W: Width variance [5].

Joining these features, we create the vector for a component i.e. $\aleph = \{S, A, E, O, AX, T, L, W\}$.

We have applied an SVM classifier for classification of non-text and text CCs. The classifier is applied on the basis of the extracted feature vector from the non-text and text components. For each CC, the feature vector is constructed and fed to the SVM classier to classify whether the CC is text component or non-text component.

4 Preparation of Character Database

We have prepared an English character database from segmented text components. We examine that many characters touch with its adjacent characters. We also observe the same scenario in case of ground truth images. Touching characters directly affect the recognition accuracy. In Fig. 2 we represent a few sample images of touching characters. The separation of touching characters is a prime factor which affects the performance of recognition system. We examine the extracted and ground truth text components for separation of touching tendencies of the characters. We found 869 numbers of touching characters from our text components data set. This survey helps to make a statistics on which text characters have the tendency to touch its neighboring character. We observe that the characters ‘t’, ‘f’, ‘r’, ‘T’, ‘L’, ‘E’, ‘A’, ‘pg’, and ‘pc’ have the maximum tendency to touch with other characters. We also observe that when the character ‘p’ and ‘g’ are side by side to form a word, then there is a tendency to touch with each other and similarly for the characters ‘p’ and ‘c’. Figure 3a presents the characters which have a tendency to touch with other characters and their frequency. Further, in Fig. 3b we present a few sample touching characters which are formed by the abovementioned touching tendency characters. We design one simple but effective algorithm for touching character separation. Now, we have presented a touching characters separation technique. First, each text connected component (CC) is examined, whether it is touching or non-touching. We calculate the aspect ratio for each of the CC. On finding the aspect ratio, the CC’s are categorized into two classes namely, single character and touching character. We

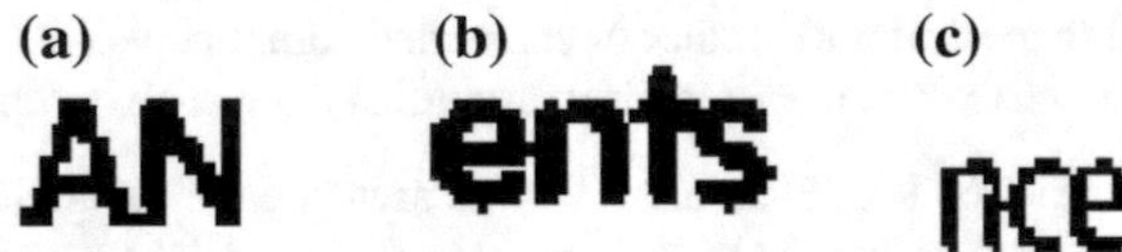

Fig. 2 A few images of touching characters

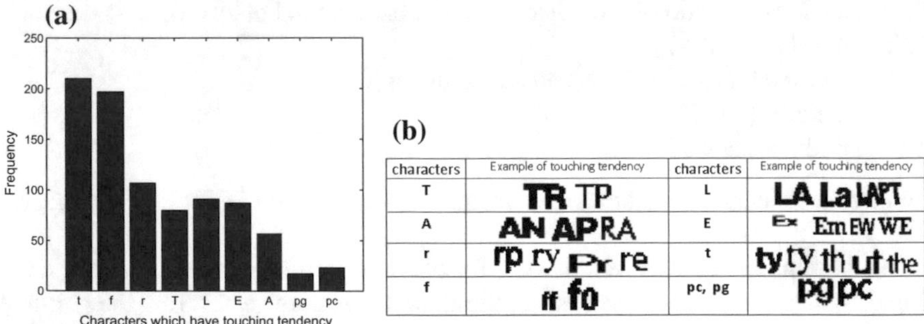

Fig. 3 **a** Touching characters which have a tendency to touch other characters and their frequency and **b** A few touching tendency characters and their examples

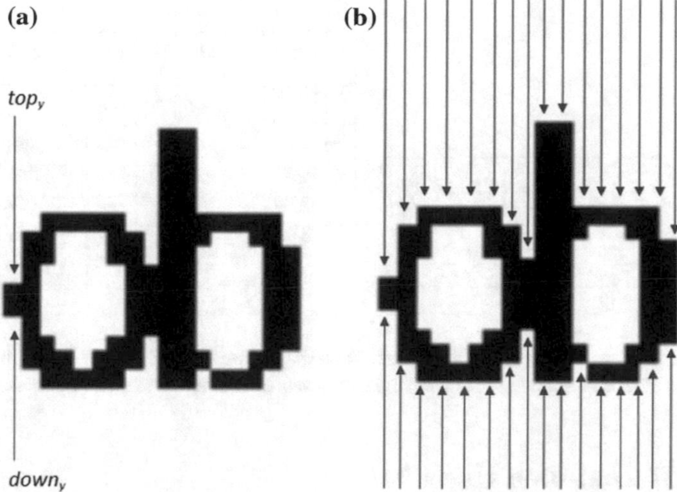

Fig. 4 **a** Two arrows in a touching component. **b** Arrow's to identify touching point

observe that a single character has maximum aspect ratio of 1.2 and minimum of 0.5. So, any CC lying in this range can be classified as a single character otherwise a touching character. Here, we consider two arrows to identify the separation point(s) of touching characters. Figure 4a shows that for each vertical strip of pixels, two arrows are shot, one from top and one from bottom, and their heads are marked as top_y and $down_y$, respectively. For each vertical strip of pixels, we traverse from top till we find a different-color pixel (background to foreground) and that pixel is marked as top_y, and similarly from bottom, the same is done and that pixel is marked as $down_y$. Now, the difference ($d_y = down_y - top_y$) of these are calculated for every strip as shown in Fig. 4b. To separate the touching character we proceed as follows. If the d_y at any point is below a predefined threshold then mark this pixel as a separation point and later perform segmentation on all these separation points for the entire CC. Now, we prepare a database of English characters which contains 13496 images of characters from the segmented text components. English alphabets contain 5 vowels and 21 consonants. Our concerned Born Digital dataset mainly contains upper and lower case text characters. However, the shapes of a few characters are the same in upper and lower cases. In Fig. 5 we present a few characters (V, X, C, W, O, K, P, Z, S, I, J and Y) whose lower and upper case shapes are almost similar. So, we consider (V, v) is a single class and similarly for (X, x), (C, c), (W, w), (O, o), (K, k), (P, p), (Z, z), (S, s), (I, i), (J, j) and (Y, y). Further upper and lower case of 'I' and lower case of 'L' are almost similar in terms of shape. So, we consider these three characters as a single class. Now we allocate all the English characters (13496) from our database over 39 classes. The allocation of samples in 39 classes is described in Fig. 5b by means of numbers of class (successive numbers). The number in bracket specify the total number of characters of the subsequent class of our data set.

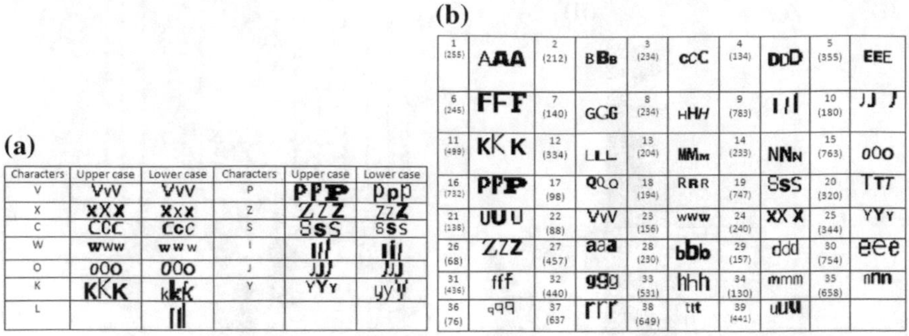

(a)

Characters	Upper case	Lower case	Characters	Upper case	Lower case
V	VvV	VvV	P	PPP	ppp
X	xXX	Xxx	Z	ZZZ	zzZ
C	CCc	CcC	S	SsS	sss
W	www	www	I	Il/	lil
O	oOo	oOo	J	JJJ	JJJ
K	KKK	kkk	Y	YYY	yyY
L		lll			

(b)

1 (255)	AAA	2 (212)	BBB	3 (234)	cCC	4 (134)	DDD	5 (355)	EEE
6 (245)	FFF	7 (140)	GGG	8 (234)	HHH	9 (783)	I/l	10 (180)	JJJ
11 (499)	KKK	12 (334)	LLL	13 (204)	MMM	14 (233)	NNN	15 (763)	oOo
16 (732)	PPP	17 (98)	QQQ	18 (194)	RRR	19 (747)	SsS	20 (320)	TTT
21 (136)	UUU	22 (88)	VvV	23 (156)	www	24 (240)	XXX	25 (344)	YYY
26 (68)	ZZZ	27 (457)	aaa	28 (230)	bbb	29 (157)	ddd	30 (754)	eee
31 (436)	fff	32 (440)	ggg	33 (531)	hhh	34 (130)	mmm	35 (658)	nnn
36 (76)	qqq	37 (637	rrr	38 (649)	ttt	39 (441)	uuu		

Fig. 5 **a** Characters whose lower and upper case shapes are similar except L and **b** Sample of examples for each character is presented in parentheses. Serial numbers point out class numbers

5 Text Recognition Using MLP

The outcome of any recognition scheme is based on the features being used for the classifier. Recognition system's performance can still be improved by considering good features. Here, we use one of the most famous and effective features, that is, the chain code feature [7] for character recognition. In the current work chain code histogram descriptions of a character component have been calculated.

MLP has been considered as the classifier of the current work of recognition of text from scene images. The famous backpropagation (BP) technique [8] is applied for the training of MLP classifier.

6 Experimental Outcomes

The outcomes are generated from Born Digital Data set [9] of ICDAR 2011. Our experiments are partitioned into three divisions which are described in the following subsections depends upon our objective of this study.

6.1 Outcomes of Image Binarization Algorithm

First, consider a few image binarization outcomes. A number of sample outcomes are shown in Fig. 6. We compare our algorithm with the text component of the binary image with the ground truth image for calculating the Precision, Recall and F-measure [10]. The values of these parameters our new binarization method obtained from the data set are respectively 0.93, 0.70 and 79.87. We now compare the performance of the proposed image binarization technique with Otsu's method. From the experiments presented in Table 1, our binarization scheme is better than the other methods.

Fig. 6 (i), (ii), (iii) Sample images and (iv), (v), (vi) the corresponding binarized image

Table 1 Comparison with other binarization scheme

	Recall	Precision	FM
Proposed scheme	**93.00**	**70.00**	**79.87**
Otsu [1]	88.98	65.36	65.05
Niblack [2]	87	36	38.17
Sauvola [3]	91	14	20.4
Bhattacharya et al. [11]	91.14	47.85	53.81
Kumar et al. [12]	85.56	47.09	46.81

6.2 Text Identification Results

This subsection presents the text identification outcomes acquired by our SVM classifier. Here, our dataset contains 20723 number of text samples and 10000 number of non-text samples. A few images and their corresponding identified text are shown in Table 2. It is clear that our scheme is good enough towards segmentation of text. A visual judgement is completely dependent upon user. To avoid this, we compare our algorithm by evaluation criteria. Regarding this, Clavelli et al. [13] proposed a few criterions to judge the text identification quality of each component of text described in the ground truth. From Clavelli et al., the text components are identified as *Well segmented*, *Merged* and *Lost*. The Recall, Precision and F-Measure measurements of our SVM based method generated based on our concerned data set images are 79.13%, 80.72% and 79.91% respectively.

Finally, we compare our technique with other well known schemes. In Robust Reading Competition of ICDAR 2011 published results of a number of schemes from numerous participants. Table 3 presents a number of these techniques. We compare

Table 2 A few images (1st rows), the corresponding segmented text (2nd rows)

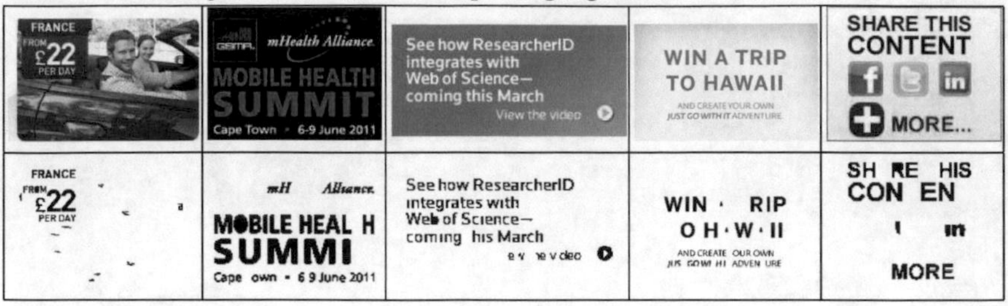

FRANCE FROM £22 PER DAY	mHealth Alliance MOBILE HEALTH SUMMIT Cape Town · 6-9 June 2011	See how ResearcherID integrates with Web of Science— coming this March View the video	WIN A TRIP TO HAWAII AND CREATE YOUR OWN JUST GO WITH IT ADVENTURE	SHARE THIS CONTENT MORE...
FRANCE FROM £22 PER DAY	mH Alliance MOBILE HEAL H SUMMI Cape own · 6 9 June 2011	See how ResearcherID integrates with Web of Science— coming his March	WIN · RIP O H·W·II	SH RE HIS CON EN MORE

Table 3 Comparison of our text extraction scheme with other techniques

Scheme	Well segmented	Merged	Lost	Recall	Precision	F-measure
Our scheme	**71.74**	11.39	**15.88**	**79.13**	**80.72**	**79.91**
Kumer et al. [12]	64.15	15.68	20.14	80.61	72.05	76.11
Adaptive EdgeDetection	66.55	9.23	24.20	78.23	70.97	74.42
Textorter	58.13	9.50	32.37	65.22	63.64	64.32
SASA	41.58	10.96	47.43	71.68	55.44	62.52

our proposed scheme with these methods. Our technique has attained precision and FM 80.72 and 79.91 respectively, which are higher than the other methods. Our scheme outperforms other techniques considering lost and well segment.

6.3 Recognition Results

Our English character data set contains 13496 isolated text symbols and these symbols are distributed over 39 classes. Next, these text symbol images are resized to a particular dimension. After that, median filter is used to overcome certain perturbations. Chain code histogram feature for a character component has been calculated of its contour description. Considering the histogram features, MLP classifier is applied for recognition.

The Table 4 represents the recognition results of the 39 classes of English text characters.

Table 4 Recognition results (RR) of English text components (in %) using MLP

Class No.	RR	Class No.	RR	Class No.	RR	Class No.	RR	Class No.	RR
1	88.3	9	89.7	17	79.5	25	77.2	33	88.1
2	83.3	10	85.9	18	93.8	26	94.7	34	84.4
3	79.2	11	91.3	19	86.3	27	87.4	35	73.6
4	75.3	12	94.4	20	94.6	28	91.1	36	87.7
5	83.7	13	89.7	21	86.6	29	95.4	37	70.3
6	92.2	14	85.1	22	82.7	30	90.3	38	71.2
7	98.1	15	97.3	23	88.6	31	81	39	86.7
8	95.1	16	78.2	24	93.6	32	67.5		

7 Conclusion and Future Scope

This paper gives an improved image binarization scheme for segmentation and recognition of text. Here, Canny's edge information is incorporated in Otsu's algorithm. Next, a number of features are described and an SVM classifier is applied for separation of text and non-text components. Further, MLP is applied for recognition of text. In future machine learning tools can be combined to progress the binarization algorithm.

References

1. Otsu, N.: A threshold selection method from gray-level histograms. IEEE Trans. Syst. Man Cybern. **9**(1), 377–393 (1979)
2. Niblack, W.: An Introduction to Digital Image Processing. Prentice Hall, Englewood Cliffs (1986)
3. Sauvola, J., Pietikinen, M.: Adaptive document image binarization. Pattern Recognit. **2**, 225–236 (2000)
4. Gatos, B., Pratikakis, I., Perantonis, S.J.: Document image binarization by using a combination of multiple binarization techniques and adapted edge information. In: Proceedings of the International Conference on Pattern Recognition (ICPR) (2008)
5. Ghoshal, R., Roy, A., Banerjee, A., Dhara, B., Parui, S.: A novel method for binarization of scene text images and its application in text identification. Pattern Anal. Appl. 1–15 (2018)
6. Ghoshal, R., Roy, A., Bhowmik, T.K., Parui, S.K.: Decision tree based recognition of Bangla text from outdoor scene images. In: Eighteen International Conference on Neural Information Processing (ICONIP), pp. 538–546 (2011)
7. Bhattacharya, U., Shridhar, M., Parui, S.K.: On recognition of handwritten Bangla characters. In: Proceedings of the Conference on Computer Vision, Graphics and Image Processing (ICVGIP), pp. 817–828 (2006)
8. Rumelhart, D.E., Hinton, G.E., Williams, R.J.: Learning internal representations by error propagation. Institute for cognitive science report 8506. San Diego: University of California (1985)
9. Karatzas, D., Robles Mestre, S., Mas, J., Nourbakhsh, F., Roy, P.P.: ICDAR 2011 robust reading competition-challenge 1: reading text in born-digital images (web and email). In: Proceedings of the 11th International Conference of Document Analysis and Recognition (ICDAR), pp. 1485–1490 (2011)

10. Dance, C.R., Seegar, M.: On the evaluation of document analysis components by recall, precision, and accuracy. In: Proceedings of the Fifth International Conference on Document Analysis and Recognition (ICDAR), pp. 713–716 (1999)
11. Bhattacharya, U., Parui, S.K., Mondal, S.: Devanagari and bangla text extraction from natural scene images. In: Proceedings of the International Conference on Document Analysis and Recognition (ICDAR), pp. 171–175 (2009)
12. Kumar, D., Ramakrishnan, A.G.: OTCYMIST:Otsu-Canny minimal spanning tree for born-digital images. In: Proceedings of the 10th IAPR International Workshop on Document Analysis Systems. DAS '12, pp. 389–393 (2012)
13. Clavelli, A., Karatzas, D., Lladós, J.: A framework for the assessment of text extraction algorithms on complex colour images. In: Proceedings of the 9th IAPR International Workshop on Document Analysis Systems. DAS '10, ACM, pp. 19–26 (2010)

User Identification and Authentication Through Voice Samples

Soubhik Rakshit

Abstract Voice authentication is a fundamental topic of research in today's technology. Reliable speech recognition is hard to achieve, but many approaches have been proposed in recent years to achieve such with an improved degree of accuracy. The following paper presents a novel approach through which users can be authenticated with reasonable accuracy using a small voice sample. The proposed method uses MFCCs, a well-known methodology for extracting features from the voice sample and finally uses Gaussian Mixture Models (GMM) for classification. An advantage of using MFCCs as the speech features is that the model is language independent. A model trained in one language can work equally well for a model trained in a different language.

Keywords MFCC · Mixture model · Voice authentication

1 Introduction

User recognition is currently a very well studied problem with the advent of numerous face recognition algorithms, behavior understanding through a range of sensors serving as the source of input, etc. We can unlock our smart phones using our face or with our fingerprints. We can use biometrics to authenticate ourselves at different locations. All of these can be used to identify and authenticate people. This paper tries to demonstrate a different way of recognizing users and authenticating them, where users need to submit a few seconds of their voice samples. First, a few voice samples are used to train the classifier model. Later, the user might submit another voice recording and the user corresponding to whom the classifier score is maximum is determined as the winner. In the testing phase, this model achieved an accuracy of 92% where it had to select the correct user from a pool of 1000 users. In voice

S. Rakshit (✉)
Pucho Technology and Department of Computer Science and Technology, Indian Institute of Engineering Science and Technology, Shibpur, Howrah 711103, India
e-mail: soubhik.dd2015@cs.iiests.ac.in

A. K. Das et al. (eds.), *Computational Intelligence in Pattern Recognition*, Advances in Intelligent Systems and Computing 999,
https://doi.org/10.1007/978-981-13-9042-5_21

authentication, where users first provide their identifier and proceeds to authenticate using their voice, 96% accuracy was achieved.

2 Background Research

There has been considerable research in this field for the past few years. In recent years, various methodologies have been proposed in this respect.

Campbell et al. in [1] have dealt with creating supervectors and have used the supervector in a SVM classifier to identify the voice. Two different kernels were based on distance metrics between GMM models have also been researched upon.

Paper [2] presents an automatic speech recognition on speaker independent connected digits using different robust features such as Revised perceptual linear prediction (RPLP), Bark Frequency Cepstral Coefficients (BFCC) and Mel frequency perceptual linear prediction (MF-PLP).

The method proposed in [3] designed a system for digit recognition mechanism in Hindi. In this proposed approach the authors have chosen HMM as the classifier and MFCC algorithm for features extraction. Here, voice samples having noise in them were used for classification.

Another methodology [4] works on a system for speech recognition for isolated English digit using the MFCC and DTW (Dynamic Time Warping) algorithm. Here, as the speaking speeds of different speakers is different, DTW is used. DTW is used for measuring similarity between two sequences, which may vary in time or speed.

In [5], Joint Factor Analysis was used as a feature extractor. This method makes the process faster and computationally less complex. The cosine distance kernel in the new total factor space is used to design two different systems, an SVM and a second one which directly uses this kernel to calculate the decision score.

3 Proposed Methodology

The current section demonstrates the work methodology elaborately. The objective of the present work is to identify a user from a pool of users using voice samples, and subsequently extending this to a fully voice authenticated system where the user first identifies themselves and then proceeds for authentication using voice. It is understood that if the first part is solved with a reasonable accuracy, it can be easily extended to find the solution for the second part of the objective (Fig. 1).

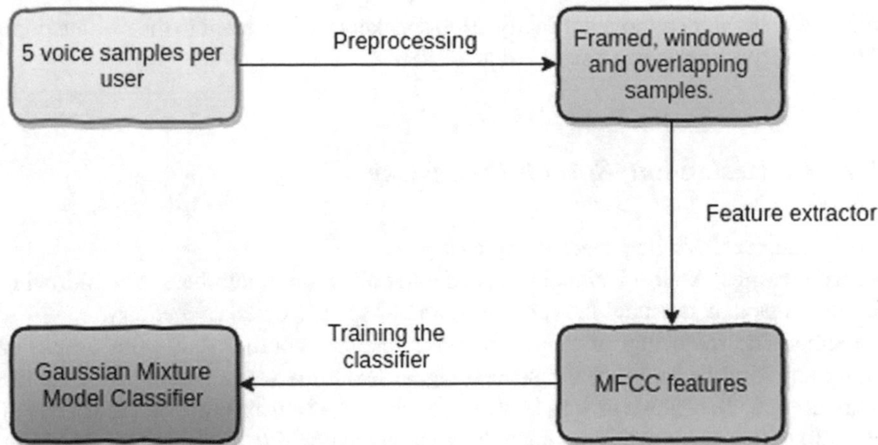

Fig. 1 Flowchart for the training process

3.1 Data Collection

Data is collected from [6]. VoxForge is a free speech corpus and acoustic model repository for open source speech recognition engines. The author is grateful to VoxForge for publishing the data repository online and providing it free of charge. The entire data for this work has been obtained from http://www.repository.voxforge1.org/downloads/SpeechCorpus/Trunk/Audio/Main/8kHz_16bit/ and multiple stages of preprocessing has been done.

Features	Description
20 MFCC features	These features are extracted directly from the audio sample using the feature extractor.
20 Delta features	These features are calculated from the MFCC features. These are the difference between two subsequent MFCC's.
Class	User Id

3.2 Preprocessing

The VoxForge URL as mentioned above has a list of compressed files, where each compressed file contains 10 different WAV sound files by the same speaker. In this project, 5 files were used for training the model and the other 5 to test the model. To prevent class imbalance, the compressed files which contained less than 10 different voice samples were removed. Finally 1000 speakers were present in the sample pool. All the voice samples were sampled at 16 kHz.

3.3 Understanding Speech Processing

Let us consider a single speech sample (Fig. 2).

Speech frames. A speech signal is an ordered collection of numbers. The following are the steps to understand the representation of speech signal in memory.

Framing. The frequency of a speech signal changes continuously with time. For doing any kind of analysis, the speech signal has to be made stationary for some point of time. To achieve this, the entire speech signal is broken into small frames of 20 to 30 milliseconds duration. It can be assumed that within this duration, the shape of our vocal tract remains constant. If longer frames were used, the above assumption might be risked as the signal may change too much within the frame. Frames shorter than this length will not have enough information to make the analysis fruitful.

Windowing. Windowing is performed before Fourier transform to avoid spectral leakage. This happens due to the discontinuities near the end-points of the frames. A special window function is multiplied with the waveform to generate the windowed frame (Fig. 3).

Overlapping. Due to windowing, some information is lost at the beginning and at the last part of the frames. This will have a negative effect by giving an incorrect

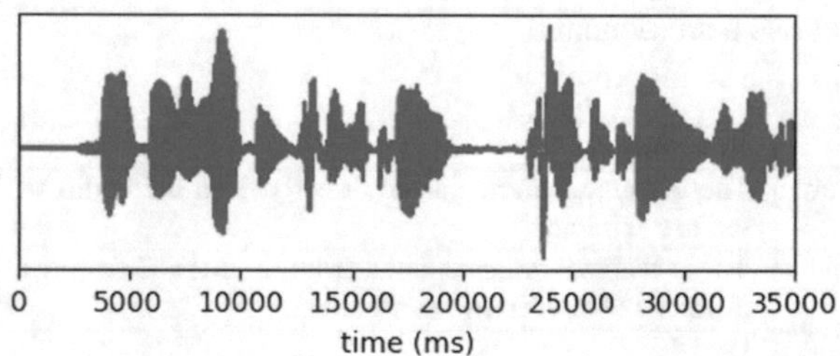

Fig. 2 Plot of the sound file

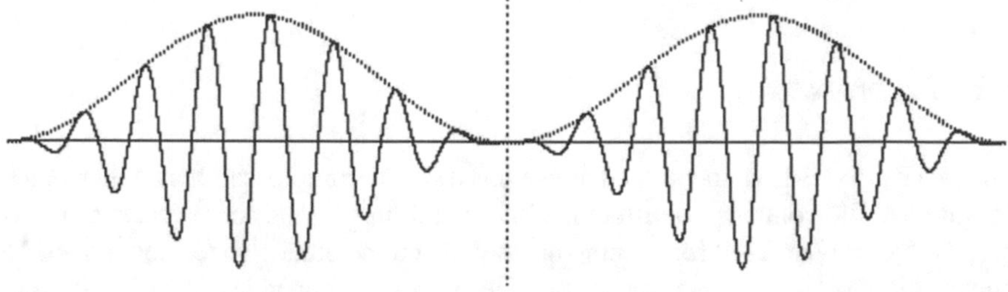

Fig. 3 The waveform is transformed to a windowed signal using a window function

frequency representation. To compensate for this, instead of using disjoint frames, overlapping frames are used. This means that the samples lost from the ith frame and the $(i + 1)$th frame will be fully included into the frame obtained from the overlap of these 2 frames.

3.4 Extracting MFCC Features

After obtaining the speech frames, the next step is to extract the Mel-Frequency Cepstrum Coefficients (MFCCs). MFCC's contain lots of information and is used widely for gender recognition, speech recognition, etc. Here it will be used for voice recognition. The voice recognition model will be built using a machine learning technique—Gaussian Mixture Models (GMM) (Fig. 4). A mixture model corresponds to the mixture distribution that represents the probability distribution of observations in the overall population. Here, MFCCs will be provided as the inputs to the GMM and the GMM will try to learn their representation, which will be representative of each user. During testing phase, when a new voice sample is provided, first the MFCCs of that voice sample will be calculated followed by calculating the scores of the features with the models of all the speakers. The model which contributes the highest score will be assigned as the speaker of the test sample.

According to the theory of speech production, speech is produced using a source of air and a filter. Our lungs act as the source and our vocal tract acts as the filter. The vocal tract gives shape to the spectrum of signal and it varies across speakers. The MFCCs best represent the shape of the vocal tract.

Here, the purpose is to simulate the vocal tract using MFCC features and remove the source part from the incoming audio signal. The steps are as follows:

1. First convert the speech signal which is in time domain to spectral domain using Fourier transform. Now source and filter parts are multiplied with each other.
2. Take log of the above values. Then the source and the filter are in addition with each other. Now, a linear filter can be used to separate the source from the filter in the log spectral signal.

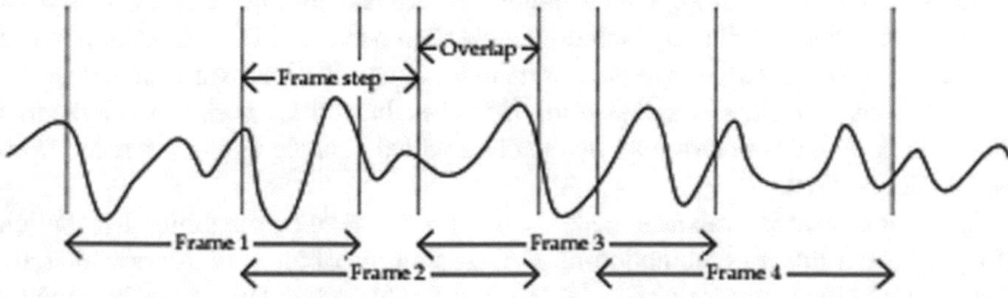

Fig. 4 Overlapping frames adopted from [7]

3. Discrete Cosine Transform (DCT) is applied to the log spectral signal to get MFCCs.
4. If Inverse-FFT is used to transform the log spectral signal to the time domain, log being a non-linear operation will create new frequencies called Quefrency. Now, the log spectral domain has been converted to cepstral domain.
5. Use the mel scale as humans are much better in discerning small changes in pitch at lower frequencies and not so much at higher frequencies. Incorporating this scale gives us the added benefit of extracting features that are as close to what humans actually hear.

From a single channel audio signal, get 20 different MFCCs are created. The derivatives of MFCCs that represent the change in values of MFCCs with time provides information regarding the dynamic nature of the signal. It is therefore better to incorporate the derivatives along with the original MFCCs which will subsequently be provided as input to the Gaussian Mixture Model (GMM). It turns out that calculating the MFCC trajectories and appending them to the original feature vector increases performance of the classifier.

The delta coefficients are calculated using the following formula:

$$d_t = \frac{\sum_{n=1}^{N} n(c_{t+n} - c_{t-n})}{2 * \sum_{n=1}^{N} n^2} \tag{1}$$

where, d_t is a delta coefficient of frame t computed in terms of static coefficients c_{t+N} to c_{t-N}. The value of N is generally taken to be 2. Many open source implementation of MFCCs is available in the internet.

3.5 Training the Gaussian Mixture Model

There are several reasons to use a Gaussian mixture model. GMM is very flexible in terms of cluster covariance. With GMM, each individual point is given an unconstrained covariance structure. The classifying planes can be rotated and/or elongated instead of just being spherical. Another advantage of GMM classifier is that it accommodates mixed membership. A single point can belong to multiple classes. This work might be extended in future by including more than one classifier to classify the data points. Here a weighted average of class memberships will successfully help in determining the specific class of a data point. However the current work is restricted to a single class, which is determined by the class which gets the maximum score from the GMM classifier.

Training a Gaussian mixture model is to approximate the probability distribution of a class by a linear combination of K Gaussian distributions/clusters which are also known as the components of a GMM. The likelihood of data points for a model is given by the following equation

$$P(X|\lambda) = \sum_{k=1}^{K} w_k P_k(X|\mu_k, \Sigma_k) \tag{2}$$

where, $P(X|\mu, \Sigma_k)$ is the Gaussian distribution

$$P(X|\mu, \Sigma_k) = \frac{1}{\sqrt{2\pi|\Sigma_k|}} e^{\frac{1}{2}(X-\mu_k)^T \Sigma_k^{-1}(X-\mu_k)} \tag{3}$$

The training set X_i belonging to class value λ are used to estimate the parameters mean μ, co-variance matrices Σ_k and weights w of these K components.

Python's *sklearn.mixture* package is used to learn a GMM from the feature matrix which contains the MFCCs along with the deltas.

3.6 Evaluation

A dataset of 1000 speakers, each speaker having 10 voice samples into 5 training and 5 test samples was first created. For test evaluation, each test sample was taken and 40 dimensional MFCCs and deltas were calculated. These MFCCs were compared with each of the trained models and it returns the speaker-id which gives us the maximum likelihood. This is repeated for all frames of the sample and the speaker model with the greatest likelihood is considered as the identified speaker.

4 Experimental Results

In the first experiment, the speakers were identified based on their voice samples from a pool of 1000 different speakers. Every speaker had 10 voice samples out of which 5 were used for training the model and 5 were used for testing. So, out of the 5000 test samples, this model achieved an accuracy of 92%.

In the speaker authentication task, the log likelihood of the speaker-id is calculated with the voice sample provided. After rigorous testing, the threshold value for the log likelihood value has been set to 0.9 which means that only scores greater than 0.9 is used to identify the speaker. This leads to an accuracy of 96%.

5 Conclusion

The best part of using MFCC's as features for the input is that the voice samples are language independent. A person may speak in any language. This method does not have a language barrier. However, even though speech recognition had evolved over the years, it is better to use multi-factor authentication like face recognition or biometric verification along with speech recognition. Since, a recorded voice of a person might as well be regarded as the original person, it is too dangerous to be used as the only form of authentication in high-security systems.

A solution might be to generate random text in a screen during authentication and ask the user to read the text. First, a voice to text conversion takes place and the text obtained is matched with the original text on the screen. If the pair of texts is same, only then the voice authentication system starts calculating the log likelihood function. This might be a temporary solution until even better artificial speech synthesis is developed. Thus, is definitely better to add more layers of authentication.

References

1. Campbell, W.M., Sturim, D.E., Reynolds, D.A.: Support vector machines using GMM supervectors for speaker verification. IEEE Signal Process. Lett. **13**(5), 308–311 (2006)
2. Mishra, A.N., Chandra, M., Biswas, A., Sharan, S.N.: Robust features for connected Hindi digits recognition. Int. J. Signal Process. Image Process. Pattern Recognit. **4**(2), 79–90 (2011)
3. Mishra, A.N., Astik, B., Chandra, M.: Isolated Hindi digits recognition: a comparative study. Int. J. Electron. Commun. Eng. India **3**(1), 229–238 (2010)
4. Dhingra, S.D., Nijhawan, G., Pandit, P.: Isolated speech recognition using MFCC and DTW. Int. J. Adv. Res. Electr. Electron. Instrum. Eng. **2**(8), 4085–4092 (2013)
5. Dehak, N., Dehak, R., Kenny, P., Brümmer, N., Ouellet, P., Dumouchel, P.: Support vector machines versus fast scoring in the low-dimensional total variability space for speaker verification. In: Tenth Annual Conference of the International Speech Communication Association (2009)
6. Voxforge. Free speech... recognition (linux, windows and mac) - http://www.voxforge.org. Accessed 17 June 2018
7. Overlapping frames image from https://appliedmachinelearning.files.wordpress.com/2017/06/overlap.png

Template Matching for Kinship Verification in the Wild

Aarti Goyal and T. Meenpal

Abstract Kinship verification is a new research problem in the field of computer vision. It has gained attention from different research communities due to the fact that parent/child facial resemblance could be utilized to perform kinship verification. This paper proposes kinship verification based on template matching. We present two different methods for template matching on specific facial parts of parent/child image pair. First method selects each facial part in parent image and computes similarity score with corresponding facial part in child image. It analyzes and compares role of each facial part in parent/child image pair for kinship verification. Second method performs fusion of all facial parts and selects the one with maximum similarity score. Experiments are performed on baseline KinFaceW-I/II datasets to show effectiveness of facial parts in kinship verification.

Keywords Kinship verification · Edge detection · Template matching · Facial parts · Parent/Child relation

1 Introduction

Kinship verification is an emerging research problem in the field of computer vision. Human understanding of kinship is an active research area and gained attention from different research communities [1, 2]. It is defined as the task to find similarity between parent/child image pair based on facial appearance. Recent evidences show that child resembles his/her parent more than other person [3, 4]. Anthrologists refer kinship as a network which consists of different relationships in society. While biologists refer kinship as degree of genetic similarities in individuals of same species

A. Goyal (✉) · T. Meenpal
Department of Electronics and Telecommunication Engineering, National Institute of Technology, Raipur 492010, Chhattisgarh, India
e-mail: agoyal.phd2016.etc@nitrr.ac.in

T. Meenpal
e-mail: tmeenpal.etc@nitrr.ac.in

A. K. Das et al. (eds.), *Computational Intelligence in Pattern Recognition*, Advances in Intelligent Systems and Computing 999,
https://doi.org/10.1007/978-981-13-9042-5_22

[5]. Kinship verification using computer visions techniques is termed as ability of machines to discriminate kinship relation from non-kinship relation for a given pair of facial images [6, 7].

Existing methods in kinship verification focus on full facial images for finding similarities between parent/child image pairs [8–11]. However, some literatures show that upper facial part has salient features than lower part [12]. While others show that specific facial part, such as, eye has the most promising role in verification [13, 14]. When we try to find similarity between parent and child, we see that complete full face of a child does not resemble his parent, rather some specific parts resemble. Some children have similar eye while others have similar nose. Hence we cannot generalize the region of resemblance.

This paper proposes kinship verification based on template matching for specific facial parts, left eye (LE), right eye (RE), nose (N) and lips (L) in parent/child image pair. We obtain a specific region covering only the required facial part and eliminate additional unrelated regions which may negatively affect kinship accuracy. Each facial part is selected separately in parent image and is compared with corresponding facial part in child image. For left eye of parent, we select left eye of child and for nose of parent, we select nose of child. Similarity between the selected facial part is computed by template matching.

Normalized cross correlation method (NCC) is used for template matching. There are other methods for template matching, such as, Sum of Absolute Differences, Sum of Squared Difference [15, 16], correlation. NCC is considered to be a better method because of its robustness [17, 18]. It overcomes the drawbacks of cross correlation by introducing normalization of the images to unit length and yielding the result in the range $(-1, 1)$.

We evaluate performance of the proposed method on four primary kinship relations of baseline KinFaceW-I/II datasets. We investigate the role of each selected facial part and compare their performance for kinship verification.

1.1 Paper Organization

The remainder of this paper is arranged as follows. In Sect. 2, we present a brief review of existing methods in kinship verification. Section 3 presents the proposed methodology. Sections 4 and 5 present the experimental results on kinship datasets and conclusion.

2 Literature Survey

The initial hypothesis that kinship verification could be performed using facial images is given by [19]. In this research, emphasis is to measure degree of genetic similarity between parent/child image pair. The pioneer attempt of kinship verification is originated from Bressnan et al. [20], in which the authors evaluated phenotype matching. These matchings focus on facial features and show how parent's visual resemblance could be correlated with his/her child. First idea to perform kinship verification using computational methods is given by Fang et al. [21] in 2010 in which features for facial images for a given image pair are extracted and compared for similarities. Existing methods in kinship verification are mainly categorized into feature-based methods and metric learning methods. Feature based methods extract discriminative prominent features to establish hand-crafted feature representation for facial images. While metric learning methods learn some semantic transformed subspace and increase the separability of interclass samples.

Some existing feature based methods for kinship verification are spatial pyramid based descriptor [22], DAISY descriptor [10], IFVF [23]. While metric learning methods for kinship verification are NRML [11], DMML [24], ESL in which intraclass samples i.e., images with kinship relation are pulled closer and interclass samples i.e., images without kinship relation are repulsed and pushed farther. Despite the aforementioned researches, there are still some shortcomings of the existing methods. Hence, this paper proposes novel template matching for kinship verification using salient facial parts.

3 Methodology

This paper aims to determine if kinship relation exists between parent/child image pair. Each kinship dataset has four subsets with different kinship relations such as, father-son (F-S), father-daughter (F-D), mother-son (M-S) and mother-daughter (M-D) relations. A sample example of this relation is shown in Fig. 1. Each of these relations is evaluated individually to verify if a given parent-child image pair shares kinship relation or not. This section describes training and testing stages of the proposed method separately for kinship verification.

3.1 Training Stage

Let $I = \{(I_i^p,\ I_i^c,\ y_i)\}_{i=1}^N$ denote training set of N pairs of parent/child full facial images with I_i^p and I_i^c as ith parent and child images respectively. Since the proposed method aims to find if a given parent/child image pair is kin or non-kin, the training samples are divided into two parts, positive training part where (i = j) with labeled output ($y_i = 1$) and negative training part where (i≠ j) and labeled output is ($y_i = 0$). Let P be positive training set represented as $P = \{(I_i{}^p, I_i{}^c)|y_i = 1, i = 1, 2, \ldots N\}$ and N be negative training set represented as $N = \{(I_i{}^p, I_j{}^c)|y_i = 0, i = 1, 2, \ldots N\}$. Training stage for parent/child full facial image pairs I_i involves three main steps: Facial part detection, template matching and similarity measurement. A detailed understanding is given in following subsections.

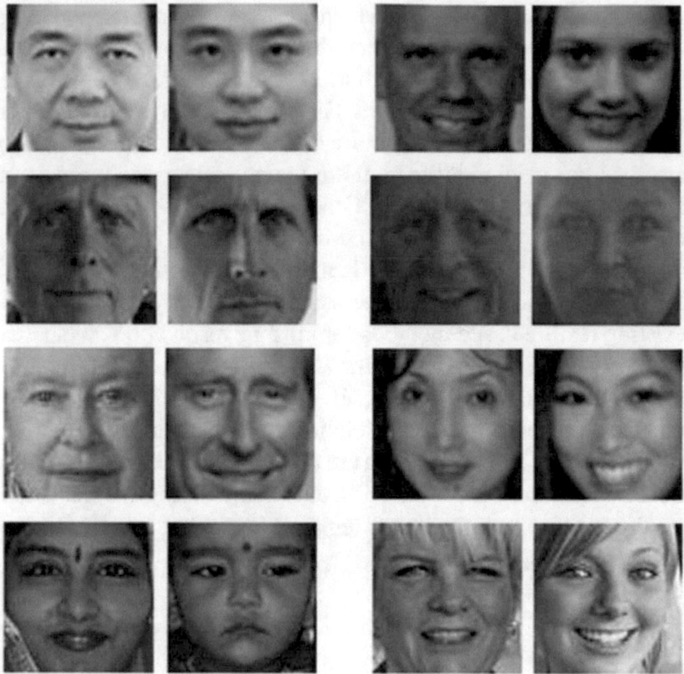

Fig. 1 Some examples of kinship relations in KinFaceW-I/II. First row is F-S kinship pair, second is F-D pair, third is M-S pair and fourth is M-D pair respectively

3.1.1 Facial Part Detection

We mainly consider four dominant facial parts, LE, RE, N, L for finding similarity between parent/child image pair. For the sake of brevity, let $I_{i,k} = \{(I^p_{i,k},\ I^c_{i,k}\ )\}^4_{k=1}$ represents set of k facial parts for ith parent/child image pair.

In this paper, facial parts are detected using Viola-Jones [25, 26]. Viola-Jones uses three different techniques for facial part detection. It detects rectangular boxes around facial parts and uses integral image for extraction of Haar like rectangular features. AdaBoost [25, 27] combines different small weak classifiers to form strong classifier using multiple iterations. Cascade classifiers [28] with strong classifiers are used to improve accuracy and speed. Output of Viola-Jones are rectangular cropped eyes area, nose area, and mouth area. These detected facial regions are then used to extract specific boundary regions. We perform edge detection followed by morphological operations to extract more specific facial regions.

Canny edge detection method [29] is used to detect edges of facial parts. It is an optimal operator for detecting edges in an image because of good detection, good localization and only one response to a single edge. The resultant image of edge detection is a binary image with pixel value '1' representing edge and pixel value '0' representing non-edge pixels. Morphological closing is performed to connect the

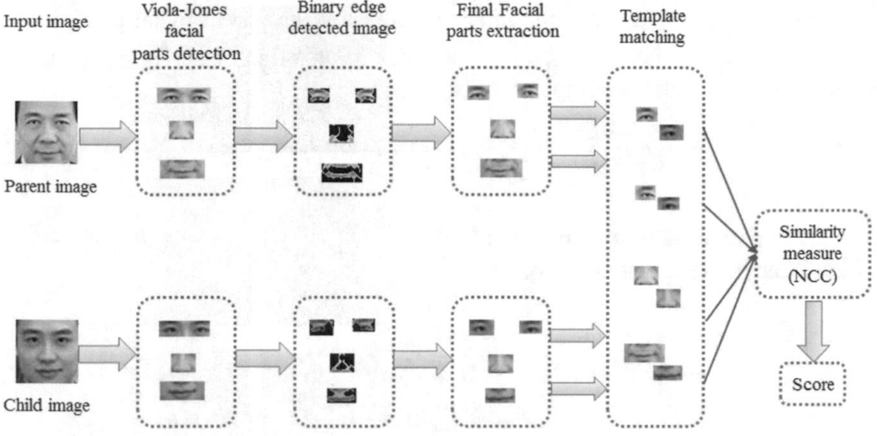

Fig. 2 Processing flow of training stage for proposed method by graphical representation

edges using structuring element and then each closed area is filled using Morphological dilation. Processing flow of training stage for proposed method is shown in Fig. 2.

3.1.2 Template Matching

We detect and extract four salient facial parts in each image of facial pairs in above section. Hence, reformulated training set of N samples is given in Eq. 1 as:

$$S_k = \{(I^p_{i,k}, I^c_{i,k}) | i = 1, \ldots, N, k = 1, \ldots, 4\} \quad (1)$$

where $I^p_{i,k}$ represents kth facial patch of ith parent image and $I^p_{i,k}$ represents kth facial patch of ith for child image.

The aim is to measure degree of similarity between a reference image and template image. We use parent facial part $I^p_{i,k}$ as reference image and child facial part $I^c_{i,k}$ as template image. NCC is used for template matching. It is independent of the size of images by normalizing the images to unit length. The size of the extracted facial patches depends on the extracted area and vary with images.

Let $c_{ik} = sim\{(I^p_{i,k},\ I^c_{i,k})\}$ represents normalized correlation coefficient [15] between kth facial part for ith pair as given in Eq. 2:

$$c_{i,k} = \frac{\Sigma_a \Sigma_b (I^p_{i,k}(a,b) - \overline{I}^p_{i,k_{(u,v)}})(I^c_{j,k}(a-u, b-v) - \overline{I}^c_{j,k})}{\sqrt{(\Sigma_a \Sigma_b (I^p_{i,k}(a,b) - \overline{I}^p_{i,k_{(u,v)}})^2)(\Sigma_a \Sigma_b (I^c_{j,k}(a-u, b-v) - \overline{I}^p_{j,k})^2)}} \quad (2)$$

In Eq. 2, $I^p_{i,k}(a, b)$ denotes the intensity values of the parent facial patch $I^p_{i,k}$ of size, say, $l \times m$ at pixel (a, b). $\overline{I}^p_{i,k_{(u,v)}}$ denotes mean value of $I^p_{i,k}$ in region under $I^c_{j,k}(a - u, b - v)$ and is computed in Eq. 3 as:

$$I^p_{i,k}(a, b) = \frac{1}{mn} \Sigma_{a=u}^{u+m-1} \Sigma_{b=v}^{v+n-1} (I^p_{i,k}(a, b)) \tag{3}$$

The simplified correlation part $\Sigma_b(I^p_{i,k}(a, b) - \overline{I}^p_{i,k_{(u,v)}})(I^c_{j,k}(a - u, b - v) - \overline{I}^c_{j,k})$ of Eq. 2 can be represented in Eq. 4 as:

$$\begin{aligned} \Sigma_a \Sigma_b(I^p_{i,k}(a, b) - \overline{I}^p_{i,k_{(u,v)}})(I^c_{j,k}(a - u, b - v) - \overline{I}^c_{j,k}) = \\ \Sigma_a \Sigma_b(I^p_{i,k}(a, b))(I^c_{j,k}(a - u, b - v) - \overline{I}^c_{j,k}) \\ - (\overline{I}^p_{i,k_{(u,v)}}) \Sigma_a \Sigma_b(I^c_{j,k}(a - u, b - v) - \overline{I}^c_{j,k}) \end{aligned} \tag{4}$$

Term $I^c_{j,k}(a - u, b - v) - \overline{I}^c_{j,k}$ in Eq. 4 represents the zero mean child facial patch.

The term $\Sigma_a \Sigma_b(I^p_{i,k}(a, b) - \overline{I}^p_{i,k_{(u,v)}})^2$ in Eq. 2 is further in Eq. 5 as:

$$\Sigma_a \Sigma_b(I^p_{i,k}(a, b) - \overline{I}^p_{i,k_{(u,v)}})^2 = \Sigma_a \Sigma_b(I^p_{i,k}(a, b))^2 + \Sigma_a \Sigma_b(\overline{I}^p_{i,k_{(u,v)}})^2 - 2\overline{I}^p_{i,k_{(u,v)}} \Sigma_a \Sigma_b I^p_{i,k}(a, b) \tag{5}$$

Equation 5 can be further simplified using Eq. 3, resulting in Eq. 6 as:

$$\Sigma_a \Sigma_b(I^p_{i,k}(a, b) - \overline{I}^p_{i,k_{(u,v)}})^2 = \Sigma_a \Sigma_b(I^p_{i,k}(a, b))^2 - \frac{1}{mn} \Sigma_a \Sigma_b(I^p_{i,k}(a, b))^2 \tag{6}$$

3.1.3 Similarity Score

Similarity score for parent/child image pair I_i is computed by two different methods. In first method, each facial part $I_{i,k}$ is selected and similarity score $S_{c_{i,k}}(I^p_{i,k}, I^c_{i,k})$ is computed. The role of each $I_{i,k}$ is analyzed which results in four different scores for four different facial parts (LE, RE, N, L). In second method, fusion of all facial parts is done and the one with maximum similarity scoreis selected. Similarity score $S_{c_{i,k}}(I^p_{i,k}, I^c_{i,k})$ for each facial part $I_{i,k}$ is computed similar to the first method. We then compare $S_{c_{i,k}}(I^p_{i,k}, I^c_{i,k})$ and select the maximum score $S^{max}_{c_{i,k}}(I_{i,k}) = S_{c_{i,k}}$ corresponds to $I_{i,k}$.

The proposed method is summarized in Algorithm 1.

Algorithm 1. Algorithm for kinship verification by template matching

---

Input: Training parent/child image pairs $I_i = \{(I_i^p,\ I_i^c)\}$
Output: Maximum similarity score $S_{c_{i,k}}^{max}(I_{i,k})$

Step 1: Facial part detection
for each pair I_i, do
Obtain $I_{i,k} = \{(I_{i,k}^p,\ I_{i,k}^c)\}$
end
Step 2: Template matching
for each pair I_i, do
for each facial part $I_{i,k}$, do
Compute normalized correlation $c_{i,k} = sim((I_{i,k}^p,\ I_{i,k}^c))$ using Eq. 2
end
end
Step 3: Similarity score
Initialize $\{a_k = 0\}_{k=1}^4$
for each pair I_i, do
Initialize $S_{c_{i,k}}^{max} = 0, postiion = 0$
for each facial part $I_{i,k}$, do
if $S_{c_{i,k}} > S_{c_{i,k}}^{max}$, do
$S_{c_{i,k}}^{max} = S_{c_{i,k}}, position = k,$
else
$S_{c_{i,k}}^{max} = S_{c_{i,k}}^{max}, position = position,$
end
end
Increment $a_k = a_k + 1$, for $position = k$

3.1.4 Testing Stage

For each image pair $(I_q^p,\ I_q^c)$, four facial parts $(I_{q,k}^p,\ I_{q,k}^c)$ is obtained and similarity score is computed. Kinship verification is a binary classification problem and SVM has performed excellent on binary problems [11]. We apply SVM RBF kernel for classifying positive and negative kinship. We perform fivefold cross validation for each subset in kinship datasets by randomly and independently dividing each subset to five folds. All five folds have approximately equal positive and negative pairs in which four are used for training and one for testing. Table 1 shows the range of folds for KinFaceW-I/II datasets.

4 Experimental Results

This section provides experiment methodology to evalaute performance of template matching method for kinship verification. We perform experiments on baseline KinFaceW-I/II datasets and describe the results in the following sections.

Table 1 Range of folds for different kinship datasets

Dataset		Fold				
		1	2	3	4	5
KinFaceW-I	F-S	1:32	33:64	65:96	97:128	129:156
	F-D	1:28	29:56	57:84	85:108	109:134
	M-S	1:24	25:48	49:72	73:96	97:116
	M-D	1:26	27:51	52:76	77:101	102:127
KinFaceW-II	All subsets	1:50	51:100	101:150	151:200	201:250

4.1 Results on KinFaceW-I/II Dataset

Mean verification accuracy of KinFaceW-I is shown in Table 2 and KinFaceW-II dataset is shown in Table 3.

The results are first shown for each subset and then for the mean of all the subsets in KinFaceW-I/II datasets. The comparison of mean verification accuracy (%) of KinFaceW-I/II datasets for proposed methods is graphically shown in Figs. 3 and 4 respectively for better illustration. Following observations are drawn from experimental results.

Table 2 Kinship verification accuracy (%) on subsets of KinFaceW-I dataset as a function of different facial parts. The best mean verification accuracy result is marked in bold

Facial part	KinFaceW-I (%)				
	F-S	F-D	M-S	M-D	Mean
Left eye	50.55	53.38	46.87	58.15	52.24
Right eye	51.19	55.55	48.17	50.77	51.42
Nose	47.33	52.94	55.98	49.92	51.54
Lips	47.65	53.82	44.70	61.19	51.84
Fusion	47.33	62.49	46.87	56.8	**53.38**

Table 3 Kinship verification accuracy (%) on subsets of KinFaceW-II dataset as a function of different facial parts. The best mean verification accuracy result is marked in bold

Facial part	KinFaceW-II (%)				
	F-S	F-D	M-S	M-D	Mean
Left eye	51.20	48.40	51.60	52.66	50.96
Right eye	53.10	53.60	49.80	49.00	51.37
Nose	49.00	46.00	48.00	45.00	47.00
Lips	55.10	50.00	54.00	56.20	53.82
Fusion	56.10	54.20	54.40	56.20	**55.23**

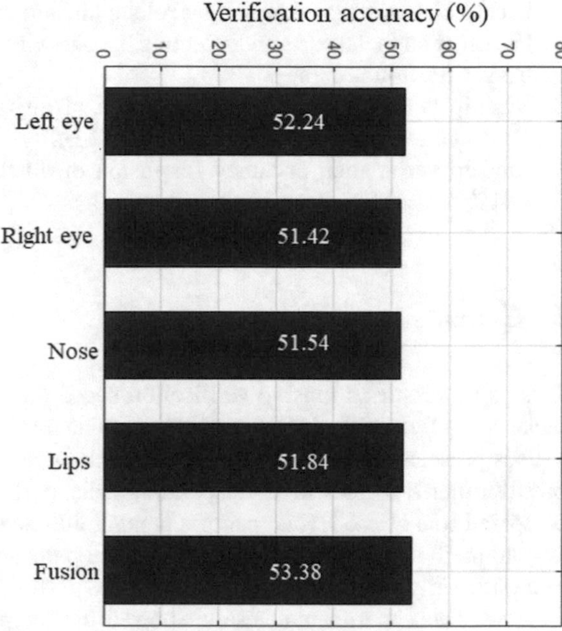

Fig. 3 Comparison of the proposed method for different facial parts on KinFaceW-I dataset graphically

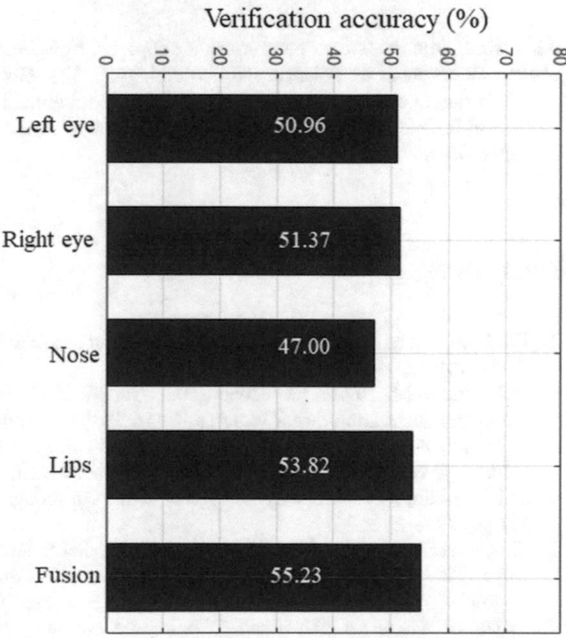

Fig. 4 Comparison of the proposed method for different facial parts on KinFaceW-II dataset graphically

1. Each facial part has an important role in kinship verification.
2. Fusion of all facial parts and selecting the one with maximum similarity improves the verification accuracy.
3. Kinship verification accuracy for fusion of all facial parts for KinFaceW-I is 1.14% better than second best accuracy (LE).
4. Kinship verification accuracy for fusion of all facial parts for KinFaceW-I is 1.41% better than second best accuracy (L).

5 Conclusion

This paper proposed kinship verification based on template matching for specific facial parts. Two methods for template matching on facial parts of parent/child image pairs is presented. First method selected each facial part in parent image and computed similarity score with corresponding facial part in child image. It analyzed and compared role of each facial part in parent/child image pair for kinship verification. Second method performed fusion of all facial parts and selected the one with maximum similarity score. Experiments have been performed on baseline KinFaceW-I/II datasets. The experimental results showed the template matching on facial parts effectively improved the verification accuracy for kinship verification.

Acknowledgements The research work is supported by Science and Engineering Research Board (SERB), Department of Science and Technology (DST), Government of India for the research grant. The sanctioned project title is "Design and development of an Automatic Kinship Verification system for Indian faces with possible integration of AADHAR Database." with reference no. ECR/2016/001659.

References

1. Dal Martello, M.F., Maloney, L.T.: Lateralization of kin recognition signals in the human face. J. Vis. **10**(8), 9–9 (2010)
2. DeBruine, L.M., Smith, F.G., Jones, B.C., Roberts, S.C., Petrie, M., Spector, T.D.: Kin recognition signals in adult faces. Vis. Res. **49**(1), 38–43 (2009)
3. Alvergne, A., Oda, R., Faurie, C., Matsumoto-Oda, A., Durand, V., Raymond, M.: Cross-cultural perceptions of facial resemblance between kin. J. Vis. **9**(6), 23–23 (2009)
4. Dal Martello, M.F., Maloney, L.T.: Where are kin recognition signals in the human face? J. Vis. **6**(12), 2–2 (2006)
5. Sills, D.L.: International Encyclopedia of the Social Sciences (1968)
6. Goyal, A., Meenpal, T.: Kinship verification from facial images using feature descriptors. In: Cognitive Informatics and Soft Computing, pp. 371–380. Springer, Singapore (2019)
7. Yadav, N., Goyal, A., Meenpal, T.: A feature averaging method for kinship verification. In: Cognitive Informatics and Soft Computing, pp. 381–391. Springer, Singapore (2019)
8. Yan, H., Lu, J., Zhou, X.: Prototype-based discriminative feature learning for kinship verification. IEEE Trans. Cybern. **45**(11), 2535–2545 (2015)
9. Zhou, X., Yan, H., Shang, Y.: Kinship verification from facial images by scalable similarity fusion. Neurocomputing **197**, 136–142 (2016)

10. Guo, G., Wang, X.: Kinship measurement on salient facial features. IEEE Trans. Instrum. Meas. **61**(8), 2322–2325 (2012)
11. Lu, J., Zhou, X., Tan, Y.P., Shang, Y., Zhou, J.: Neighborhood repulsed metric learning for kinship verification. IEEE Trans. Pattern Anal. Mach. Intell. **36**(2), 331–345 (2014)
12. Fisher, G., Cox, R.: Recognizing human faces. Appl. Ergon. **6**(2), 104–109 (1975)
13. Davies, G., Ellis, H., Shepherd, J.: Cue saliency in faces as assessed by the photofittechnique. Perception **6**(3), 263–269 (1977)
14. Heisele, B., Ho, P., Wu, J., Poggio, T.: Face recognition: codaly1982whommponent-based versus global approaches. Comput. Vis. Image Underst. **91**(1), 6–21 (2003)
15. Lewis, J.P.: Fast template matching. In: Vision Interface, vol. 95, pp. 15–19 (1995)
16. Di Stefano, L., Mattoccia, S.: Fast template matching using bounded partial correlation. Mach. Vis. Appl. **13**(4), 213–221 (2003)
17. Briechle, K., Hanebeck, U.D.: Template matching using fast normalized cross correlation. In: Optical Pattern Recognition XII, vol. 4387, pp. 95–103. International Society for Optics and Photonics (2001)
18. Luo, J., Konofagou, E.E.: A fast normalized cross-correlation calculation method for motion estimation. IEEE Trans. Ultrason., Ferroelectr., Freq. Control. **57**(6), 1347–1357 (2010)
19. Daly, M., Wilson, M.I.: Whom are newborn babies said to resemble? Ethol. Sociobiol. **3**(2), 69–78 (1982)
20. Bressan, P., Dal Martello, M.F.: Talis pater, talis filius: Perceived resemblance and the belief in genetic relatedness. Psychol. Sci. **13**(3), 213–218 (2002)
21. Fang, R., Tang, K.D., Snavely, N., Chen, T.: Towards computational models of kinship verification. In: 2010: 17th IEEE International Conference on Image Processing (ICIP), pp. 1577–1580. IEEE (2010)
22. Zhou, X., Hu, J., Lu, J., Shang, Y., Guan, Y.: Kinship verification from facial images under uncontrolled conditions. In: Proceedings of the 19th ACM international conference on Multimedia, pp. 953–956. ACM (2011)
23. Liu, Q., Puthenputhussery, A., Liu, C.: Inheritable fisher vector feature for kinship verification. In 2015: IEEE 7th International Conference on Biometrics Theory, Applications and Systems (BTAS), pp. 1–6. IEEE (2015)
24. Yan, H., Lu, J., Deng, W., Zhou, X.: Discriminative multimetric learning for kinship verification. IEEE Trans. Inf. Forensics Secur. **9**(7), 1169–1178 (2014)
25. Viola, P., Jones, M.J.: Robust real-time face detection. Int. J. Comput. Vis. **57**(2), 137–154 (2004)
26. Viola, P., Jones, M.: Rapid object detection using a boosted cascade of simple features. In: Proceedings of the 2001 IEEE Computer Society Conference on Computer Vision and Pattern Recognition. CVPR 2001, vol. 1, p. I. IEEE (2001)
27. Cristinacce, D., Cootes, T.F.: Facial feature detection using adaboost with shape constraints. In: BMVC, pp. 1–10. Citeseer (2003)
28. Wilson, P.I., Fernandez, J.: Facial feature detection using haar classifiers. J. Comput. Sci. Coll. **21**(4), 127–133 (2006)
29. Canny, J.: A computational approach to edge detection. In: Readings in Computer Vision, pp. 184–203. Elsevier (1987)

Attack Prevention Scheme for Privacy Preservation (APSP) Using K Anonymity in Location Based Services for IoT

Ayan Kumar Das, Ayesha Tabassum, Sayema Sadaf and Ditipriya Sinha

Abstract Internet of Things (IoT) is tagging our day-to-day usable objects across the internet. To make the life easy one should be more connective and communicative with the objects which are usable in daily life. In IoT, the privacy of users may be sacrificed while enjoying the conveniences and flexibilities provided by it. All the queries generated by the users, contain some relevant information about their location and identity. Location Based Server may uses this information maliciously and also reveal to third parties. In this paper, a new scheme, named APSP, has been proposed to preserve privacy of user's location and to make it difficult for the attacker to identify the real user's location. APSP implements an attack prevention algorithm on DLP for a given degree of anonymity. It selects number of dummy locations to ensure the maximum entropy with minimum variance. Simulation results show that, in APSP, there is least probability to reveal the identification of real user's location under attack and achieves better privacy preservation for selected dummy locations. APSP maximize the complexity for attacker by providing the optimal set of locations. The attacker gets confused with the same historical query probabilities of all selected locations and fails to identify the real user's location.

Keywords Degree of anonymity · Privacy preservation · Historical query probabilities · Internet of things · Dummy location · Location based services

---

A. K. Das (✉) · A. Tabassum · S. Sadaf
Department of CSE, Birla Institute of Technology, Mesra, Bihar 800014, India
e-mail: das.ayan777@gmail.com

A. Tabassum
e-mail: tabassum18at@gmail.com

S. Sadaf
e-mail: sadafsayema007@gmail.com

D. Sinha
National Institute of Technology Patna, Bihar 800005, India
e-mail: ditipriyasinha87@gmail.com

A. K. Das et al. (eds.), *Computational Intelligence in Pattern Recognition*,
Advances in Intelligent Systems and Computing 999,
https://doi.org/10.1007/978-981-13-9042-5_23

1 Introduction

In recent years, the tremendous growth of internet facilitates the user to increase the communication with the electronic devices and household devices as well to automate most of the systems. The word IoT (Internet of Things) [1] is made up of two words- 'Internet' as the backbone of connectivity and 'Things' as the objects/devices. The increase in IoT based systems demands security and privacy so that user can enjoy it conveniently [2]. Failures in IoT can lead to very serious consequences as it interacts with humans, machines and environments [3]. A study by HP [4] reveals that 70% of the most popular IoT devices contain serious vulnerabilities.

Location information is a large component of IoT information. User send queries to the Location Based Server (LBS) in order to retrieve the required information. LBS uses those queries containing their personal information to track users back. Maliciously, LBS can reveal the data to third parties. Thus, user privacy is of major concern.

Researchers are already being involved to address the privacy preservation in IoT and several techniques have been proposed. The authors of [5] use k-anonymity model for user's location privacy protection. It has widen the queried location spatially, so that server can communicate with $(k - 1)$ dummy locations via centralized location anonymizer [6, 7]. However, depending on Location Anonymizer is not always efficient, it may result in performance degradation. There are possibilities of the centralized location anonymizer to get failed and lose all the user information.

DLS algorithm [8] was proposed for providing location privacy in sending queries to LBS. It carefully selects dummy locations based on the entropy metric. Enhanced-DLS [8] has considered the Cloaking Region (CR) and entropy to expand the dummy locations selection area (i.e. cloak region gets bigger). An attack model on DLS, called ADLS [9], has been proposed to enhance the probability of successful identification of real user. This algorithm greedily selects $(k - 1)$ dummy locations on the basis of entropy metric for given k-anonymity. DLP [9] has been proposed to choose the optimal dummy locations among $k - 1$ dummy locations as per privacy requirement. In this paper, a novel scheme APSP is proposed to achieve better privacy-preservation. In APSP k-1 dummy locations are selected by DLS algorithm, where k is degree of anonymity. DLP is used to make those dummy locations optimal. Variance is calculated with the help of ADLS. In APSP greedy approach is used to select the locations, whereas for each location it also computes entropy and variance. Locations with minimum value of variance provide an optimal set. However, APSP minimizes the probability of identification of real user location and thus preserves the privacy of user location.

2 Review Work

The state of art study elaborates lots of existing literatures related to privacy preservation techniques in the location based services of IoT, which are discussed in this section.

Location privacy preservation for IoT based services is one of the major challenge. A large amount of data is communicating over the network, the security of which is another challenge. The authors of [10] proposed PAgIoT, which results in contribution to the aggregation of data based on the attribute-based queries and homomorphic Paillier cryptosystem. A trust information model, proposed in [11], prevents IoT servers or service providers from disclosing private information to the third parties, in the form of triggering conditions or functions. CPAL [12] is proposed to resists various security threats and provides more flexible privacy preservation. In [13], small- sized cryptographic key was used to provide IoT security, which reduced the computational cost when the resource of adversary is limited. Jin et al. [14] presents the smart city framework which uses information and communication technologies for betterment of city services through IoT.

K-anonymity [5] is introduced to measure the effectiveness of location privacy preservation. It hides the identification of real user's location and involves spatial temporal cloaking to anonymize messages. In [15] a model is proposed for QoS requirements, which leads to a good performance in terms of strength hiding of information and efficiency. The authors of [16] have developed a model using anonymization algorithms (k and pseudo) are used to preserve privacy of location due to unlimited usage of LBS. A k-anonymous cloaking technique [17] based on weighted graph has been proposed to protect the information of query senders, who are sending k-nearest neighbor query.

DLS [8] is proposed to achieve k-anonymity effectively and efficiently. This algorithm considers that the adversaries may reveal side information based on the entropy metrics. Privacy of DLS algorithm is enhanced by considering both entropy and CR (Cloaking Region) [8], in which computational cost is very high. To offer guaranteed large privacy region with reasonable query processing cost, PAD is proposed [18].

In the location based services of IoT, the adversary has targeted to obtain user's real location from its location information. ADLS [9] is proposed to use user's dummy location to identify their real location. Here (k − 1) dummy locations are selected based on entropy to obtain set of dummy locations by using greedy method.

DLP [9] selects optimal locations among k − 1 dummy locations. It chooses location on the basis of large entropy. In this approach there still exists the possibility for both active and passive adversaries to be compromised by the LBS server and to obtain all the details of it. However DLP can resist the colluding attack as well as inference attack.

3 Proposed APSP Scheme

The proposed scheme APSP is designed to increase the privacy preservation of DLP algorithm [9] under attack. In DLP, optimal dummy locations have been chosen for k-anonymity. Location privacy was introduced by DLS algorithm [8], having high computational cost. APSP uses DLS to select dummy locations based on entropy. The selection method (i.e. enumeration method) used for selecting k-1 dummy locations is the main reason for high computational cost as it makes the entropy larger. DLP reduces computational cost as it adopts greedy approach. The DLP provide the same level of privacy as that of DLS under the same scenario. Thus APSP is proposed to increase the privacy level of DLP with low computational cost. Obviously the description and analysis of DLS and DLP is required before the description of APSP scheme.

3.1 Notations

The key notations that are used in the description of APSP are described in Table 1.

Table 1 Key notations for proposed scheme

Notations	Description
k	Degree of anonymity
H	Historical query probability of all locations
i	Locations
E	Entropy
C	Set of optimal locations
σ	Variance

3.2 Algorithm for ADLP

Input:
P ← Historical query probability of all locations
R ← Information of user's location
k ← Degree of anonymity
Output:
Optimal set of locations, C with larger entropy E

```
Sort P and R
for (i=1; i < k; i++) do
    C_i ← Read any one location from R which was not read yet
    D_i ← Choose k-1 locations right after and k-1 left before real user location from the sorted list
    P_max ← max C_i;
    P_min ← min C_i;
    Calculate variance for each location of C_i,
    σ = Σ_{i=1}^{k} (r_i − x_i)^2,
    r_i ∈ R and x_i ∈ C_j
    return arg min σ
    From D_i, find P_min-max and P_max-min, such that,
    P_max-min = Minimum of probability set but greater than P_max;
    P_min-max = Maximum of probability set but smaller than P_min;
    H ← from sorted P, select locations having equal query probability with the real location of user
    if (size (H) ≥ k) then
        C ← randomly selects k-locations having same query probability as user's real location in H
    else if (k/4 < size (H) < k) then
        t ← size (H);
        C ← H;
        S ← Choose 2(k − t) locations having same query probability as real user's location
        Choose location from S and calculate entropy, E
        C ← C ∪ {l}, S ← S\{l};
    else
        S ← choose 4k-ω-ε candidate locations with the same query probabilities as of users real location;
        Choose location i from S randomly;
        C ← H ∪ {i};
    for (j = 1; j ≤ k − 2; j++) do
        Select a location h from S such that E(C, q) is maximum in the set S;
        C ← C ∪ {h}, S ← S\{h};
    end of for
end of if
return C (optimal set of dummy locations)
```

3.3 Description

APSP chooses k-1 dummy locations, where k is the total number of locations including the original one. Attacker may identify the real location based on the queries to LBS, either by choosing randomly or by making use of some attack algorithms. Dummy locations are selected in APSP based on the entropy metric by using greedy method. It obtains the set of dummy location and calculates the variance for different locations of dummy location set as in Eq. 1.

$$\sigma = \sum_{i=1}^{k} (r_i - x_i)^2 \tag{1}$$

where, $r_i \in R$ and $x_i \in C_j$ and C_j is the set where LBS server keeps user's real location ($j \in [1, k]$).

In order to achieve k, the given degree of anonymity for a user, the location of service area has to be grouped into $N \times N$ cells of same size. $k - 1$ dummy locations are selected from the service area by calculating entropy for each location. Location with larger entropy E is defined as in Eq. 2.

$$E = -\sum_{i=1}^{k} q_i \log_2 q_i \tag{2}$$

where, location i has the query probability (normalized) as $q_i = p_i / \sum_i^k p_i$.

And $\sum p_i = 1$.

Higher the value of k, the cost for selecting dummy locations also increases proportionally.

All the selected dummy locations should be optimal with larger entropy for a given degree of anonymity. In APSP, for given k, all locations of P and R are sorted in ascending order. P is a set for historical query probability of all locations and R is for user's real location. One location is selected from R for each location i ($1 \leq i \leq k$), and put it in C_i. Now, choose $2(k - 1)$ locations and place them with real user's location in sorted list. P_{max} and P_{min}(location having maximum and minimum number of same historical query probability) have been chosen from C_i, and find $P_{min\text{-}max}$ (location having minimum probability but greater than P_{max}) and $P_{max\text{-}min}$(location having minimum probability but smaller than P_{min}) from D_i(sorted candidate location set). Calculate entropy for each i. Choose $2(k - t)$ locations (where t is number of locations with same query probability) having same query probability as real user's location. Choose 4k-ω-ε candidate locations from sorted list for similar query probabilities. Calculate variance for each location with maximum entropy. Place all such locations in C (optimal set of dummy locations), which return the optimal set of dummy locations with larger entropy. The attacker tries to intercept the LBS server without any idea about the location of real user and keeps on trying. It can make some selection randomly, since it has to contact with k different dummy locations, the probability will be always 1/k. This helps to increase complexity for the attacker to identify the real user's location. Somehow, the attacker is successful in making some inference attack and can obtain the personal information of all dummy locations as LBS server knows all the query information. The attacks on the server will result in the confusion with the same historical query probabilities of all the dummy locations. Difference between number of locations with same historical query probability and k, have different impact on the attacking scenario for APSP. Reverse attacking is not possible in APSP as for each locations, entropy as well as variance is ensured in making selection of optimal location set. It will be more complex to obtain distinct information, when all the selected locations offer same information. The time complexity for making attacks on APSP will increase proportionally with number of tried attempts. Thus APSP preserves the privacy in better way. In order to achieve better privacy-preservation, APSP has larger computational cost for different k and number of locations where the query probability is same with that of real user. The algorithm of APSP uses greedy approach for selection of locations, but for each location it also has to computes entropy. Calculated entropy for current location (i) is to be compared with that of next location (i + 1). Variance is calculated for each location set with larger entropy. Thus APSP sacrifice computational cost to achieve better privacy-preservation. Maximum entropy can be calculated as in Eq. 3.

$$\mathrm{E_{max}} = \log_2 \mathrm{k} \tag{3}$$

Entropy comparison for chosen ith and (i + 1)th location can be defined as:

$$\mathrm{E_{i+1}} = -\sum_{\mathrm{i=1}}^{\mathrm{i+1}} \frac{1}{\sum_{\mathrm{q=1}}^{\mathrm{i+1}} \mathrm{P_q}} \log_2 \frac{1}{\sum_{\mathrm{q=1}}^{\mathrm{i+1}} \mathrm{P_q}} \tag{4}$$

where, l is historical query probability of user.

4 Performance Evaluation

Consider a location map of n × n cells, each cell having same size and query probability. It is assumed for some cells that they contents of equal historical query probability with compared to real user. Such number of locations is denoted as l. Various cases are analyzed for a given degree of anonymity k. Furthermore, the performance of proposed APSP is compared with DLS [8] and DLP [9] in terms of computational cost and privacy preservation under various anonymity degree.

4.1 *Case-I* (k < *l*)

In this case, user has contents of equal probability with the probability of real location of user. In Fig. 1, the increase of entropy of all the three algorithms remained same with the increase in anonymity degree. In Fig. 2, the computational time of DLP remains almost constant with the increasing value of k. In APSP and DLS computational time is rising with the increase in k. APSP uses greedy method to choose dummy location instead of sorting method as in DLS, which results in lower computational cost for APSP than DLS.

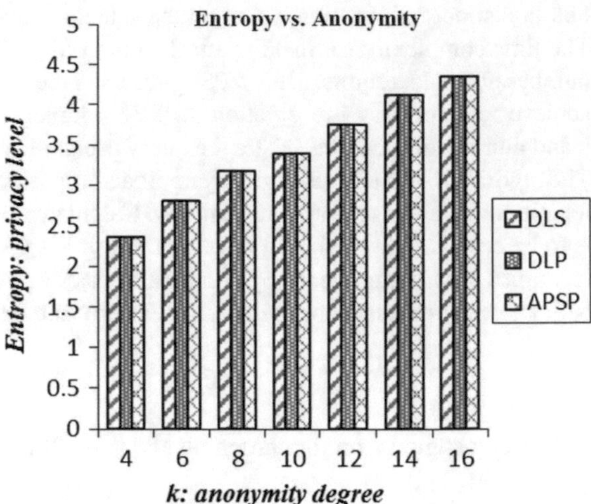

Fig. 1 Entropy for case-I

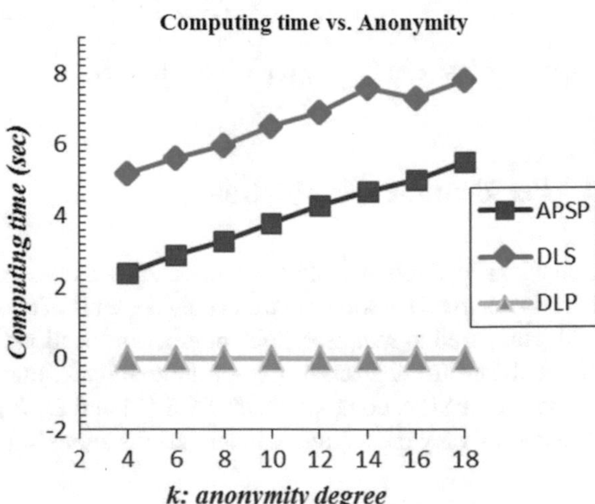

Fig. 2 Computing time for case-I

4.2 *Case-II* ($k/4 < l < k$)

In this case, user has enough locations that consist of equal probability with the probability of real location of user. In Fig. 3, the entropy of DLP slightly differs from that of APSP and DLS with the increasing k-anonymity. In DLS and APSP, privacy remains almost similar when the value of k increases. In Fig. 4, the computational time of DLP is almost constant, whereas, that of APSP and DLS is increasing proportionally with the increase in k. The computational cost of APSP is less than DLS.

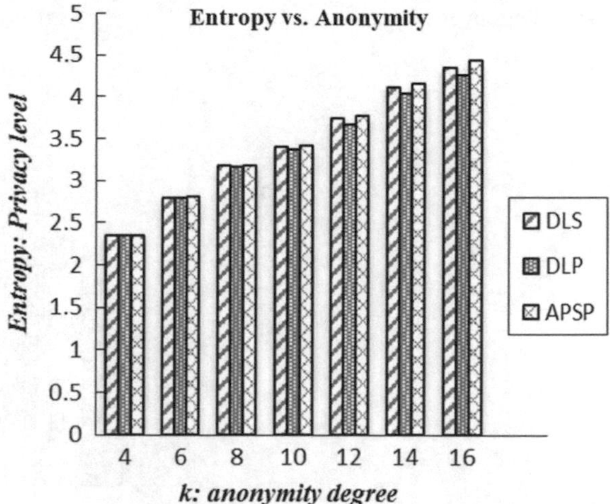

Fig. 3 Entropy for case-II

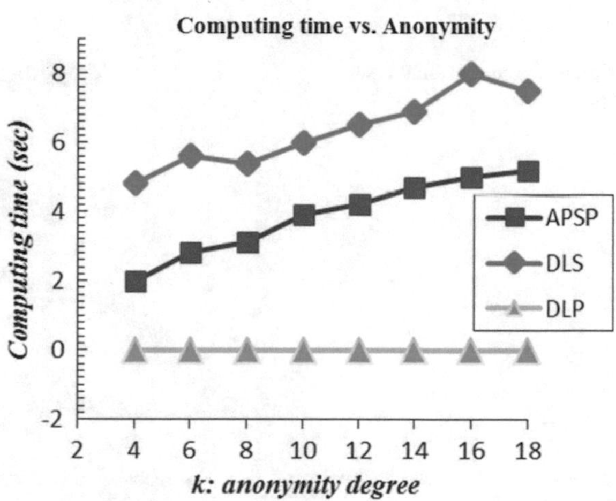

Fig. 4 Computational cost for case-II

4.3 *Case-III* $(1 < k/4)$

In this case, user has few locations that consist of equal probability with the probability of real location of user. In Fig. 5, the entropy of DLP slightly decreased from that of APSP and DLS with the increasing k-anonymity. In Fig. 6, the computational time of DLP remains constant, whereas the performance of APSP is better than DLS with increasing computational cost, which is proportional to k.

Hence, we can say that in case-I the entropy value of all the three algorithm remains same, but in case-II and case-III the entropy value of APSP is better than that of DLS and DLP. The computational cost of DLP remains best in all the given

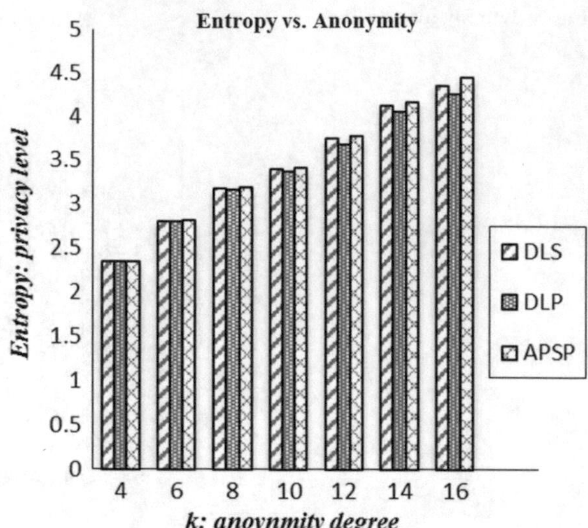

Fig. 5 Entropy for Case-III

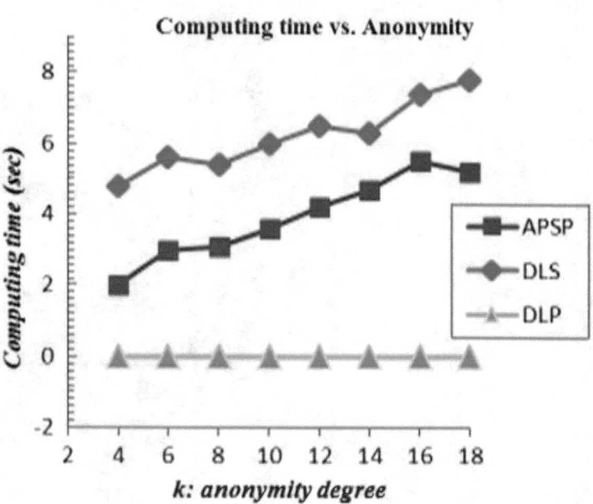

Fig. 6 Computational cost for case-III

three cases and in all the three cases, computational cost of APSP performs better than that of DLS.

5 Conclusion

APSP is designed and implemented to preserve the location privacy of users, who are using location based services in IoT. In APSP, dummy locations are selected by greedy method in order to provide better privacy and for each location, minimum variance is considered for a given value of k. The performance analysis of APSP show that it achieves better privacy level than DLS and DLP by minimizing the probability of identification of real user's location. However, APSP results in increased computational cost with compared to DLS and DLP.

References

1. Atzori, L., Iera, A., Morabito, G.: The Internet of Things: a survey. Comput. Netw. **54**(15), 787–2805 (2010)
2. Miorandi, D., Sicari, S., De Pellegrini, F., Chlamtac, I.: Internet of things: vision, applications and research challenges. Ad Hoc Netw. **10**(7), 1497–1516 (2012)
3. Kim, H., Wasicek, A., Mehne, B., Lee, E.A.: A secure network architecture for the internet of things based on local authorization entities. In: IEEE 4th International Conference on Future Internet of Things and Cloud (FiCloud) (2016)
4. Kovacs, E.: 70% of Internet of Things Devices Reveal Vulnerabilities. Hewlett Packard. http://www8.hp.com/us/en/hp-news/press-release.html?id=1744676#,V-e1I7Wa1fA
5. Sweeney, L.: k-anonymity: a model for protecting privacy. Int. J. Uncertain. Fuzziness Knowl.-Based Syst. **10**(5), 557–570 (2002)
6. Mokbel, M.F., Chow, C.-Y., Aref, W.G.: The new casper: query processing for location services without compromising privacy. In: ACM VLDB, pp. 763–774 (2006)
7. Chow, C.-Y., Mokbel, M.F., Aref, W.G.: Casper*: query processing for location services without compromising privacy. ACM Trans. Database Syst. **34**(4) (2009)
8. Niu, B., Li, Q., Zhu, X., Cao, G.: Achieving k-anonymity in privacy aware location-based service. IEEE INFOCOM, 754 –762 (2014)
9. Sun, G., Chang, V., Ramachandran, M., Sun, Z., Li, G., Yu, H., Liao, D.: Efficient location privacy algorithm for Internet of Things (IoT) services and application. J. Netw. Comput. Appl. YJNCA1738 (2016)
10. González-Manzano, L., de Fuentes, J.M., Pastrana, S.: PAgIoT-Privacy-preserving aggregation protocol for Internet of Things. J. Netw. Comput. Appl. **71**, 59–71 (2016)
11. Appavoo, P., Chan, M., Bhojan, A.: Efficient and privacy preserving access to sensor data for Internet of Things (IoT) based services. In: IEEE International Conference on Communication Systems and Networks, pp. 1–8 (2016)
12. Lai, C., Li, H., Liang, X.: CPAL A conditional privacy preserving authentication with access linkability for roaming service. IEEE Internet of Things J. **1**(1), 46–57 (2014)
13. Premnath, S., Haas, Z.: Security and privacy in the internet-of-things under time-and-budget-limited adversary mode. IEEE Wirel. Commun. Letters **4**(3), 277–280 (2015)
14. Jin, J., Gubbi, J., Marusic, S., et al.: An information framework for creating a smart city through internet of things. IEEE Internet Things J. **1**(2), 112–121 (2014)
15. Aryan, A., Singh, S.: Protecting location privacy in Augmented Reality using k-anonymization and pseudo-id. In: International Conference on Computer and Communication Technology, pp. 119–124 (2010)
16. Hossain, A., Jang, S., Chang, J.: Privacy-aware cloaking technique in location-based services. In: IEEE the First International Conference on Mobile Services, 978-0-7695-4754-1, pp. 9–16 (2012)
17. Niu, B., Li, Q., Zhu, X., et al.: Enhancing privacy through caching in location-based services. IEEE INFOCOM, 978-1-4799-8381-0,1017-1025 (2015)
18. Niu, B., Zhang, Z., Li, X., Li, H.: Privacy-area aware dummy generation algorithms for location-based services. IEEE ICC **957962**, 89–233 (2014)

Statistical Analysis of the UNSW-NB15 Dataset for Intrusion Detection

Vikash Kumar, Ayan Kumar Das and Ditipriya Sinha

Abstract Intrusion Detection System (IDS) has been developed to protect the resources in the network from different types of threats. Existing IDS methods can be classified as either anomaly based or misuse (signature) based or sometimes combination of both. This paper proposes a novel misuse-based intrusion detection system to defend our network from five categories such as Exploit, DOS, Probe, Generic, and Normal. Most of the related works on IDS are based on KDD99 or NSL-KDD 99 dataset. These datasets are considered obsolete to detect recent types of attacks and have no significance. In this paper, UNSW-NB15 (Moustafa and Slay, Military Communications and Information Systems Conference (2015) [1]) dataset is considered as the offline dataset to design intrusion detection model for detecting malicious activities in the network. The performance evaluation of proposed work with the UNSW-NB15 (benchmark dataset) shows higher accuracy and IDR compared to other existing approaches. Performance analysis proves that clustering technique is really useful in order to analyze similarity in behavior of different categories and hence helpful to improve the performance of IDS.

Keywords Intrusion detection system · Signature-based IDS · Clustering · Classification · Decision tree

V. Kumar (✉) · D. Sinha
Department of Computer Science and Engineering, National Institute of Technology, Patna 800005, India
e-mail: vika96snz@gmail.com

D. Sinha
e-mail: ditipriya.cse@nitp.ac.in

A. K. Das
Birla Institute of Technolgy, Patna 800014, India
e-mail: das.ayan777@gmail.com

A. K. Das et al. (eds.), *Computational Intelligence in Pattern Recognition*,
Advances in Intelligent Systems and Computing 999,
https://doi.org/10.1007/978-981-13-9042-5_24

1 Introduction

In today's world, almost every organization is providing online platforms for the convenience of its business and other activities. These online platforms need Internet connections for the purpose of communication over the globe, which makes it vulnerable toward several threats and attacks. Therefore, dependency of most of the population on the Internet makes security the most focused area of research. Most of the software designed for security are already in market but they are not competent to provide security for new attacks. Detecting attacks on the basis of their known signatures is used widely but it requires deep analysis of each attack by experts to define the signature. A new trend which is getting popularity is to use Intrusion Detection System (IDS) using machine learning [2] approach which has an advantage of getting updated by itself as the system comes across to any new type of attacks. Nowadays, Internet of Things (IOT) is getting very popular due to the property of devices to communicate with each other over the Internet without human intervention, in order to automate the systems around us. In IOT-based system, all the devices in the network are communicating wirelessly among themselves through the Internet which is vulnerable for different security threats. Here, computational capacity, power and storage capacity, hardware limitations, etc. [3] must be considered along with the security implementations. Since sensors have several constraints [3], the security must be imposed in the IOT-based system considering all the constraints. This paper has proposed an IOT-based framework shown in Fig. 1 in which the IDS is installed in the gateways of the local area network to protect internal nodes in IOT-based system from internal attacks as well as on the gateway through which local area network communicates with Internet. This paper analyzes the behavior of recent dataset using clustering technique and evaluates the performance on tree-based model which can be used to build IDS model to protects the IOT-based system from internal and external attacks. For proposing different IDS models, most of the papers have been worked with KDD99 [4], NSLKDD [5], etc. as a benchmark dataset. These datasets [5, 4] are very old and do not cover the modern attacks. Also, these datasets lack the latest normal activity traffic. So, this paper has considered UNSW-NB15 [1] dataset as benchmark, which covers several recent attacks as well as normal traffic compared to the other traditional dataset. Moustafa et al. have evaluated UNSW-NB15 [1] dataset with different machine learning techniques (Naïve Bayes, Decision Tree, Artificial Neural Network, Logistic Regression, and Expectation–Maximization Clustering) and analyzed their performance on different types of attacks (Analysis, Backdoor, DoS, Exploit, Fuzzers, Generic, Normal, Probe, Shellcode, Worm) in "The evaluation of Network Anomaly Detection Systems: Statistical analysis of the UNSW-NB15 dataset and the comparison with the KDD99 dataset (ENADS)" [6]. They have made a comparative analysis by evaluating the same models on KDD99 dataset. Though accuracy obtained on KDD99 was very high as compared to the accuracy shown using UNSW-NB15, dataset KDD99 did not cover recent attacks as well as normal traffic in the network. Their analysis has shown that the best result is obtained by decision tree model for both the datasets, i.e., with KDD99 the accuracy is 92.30%

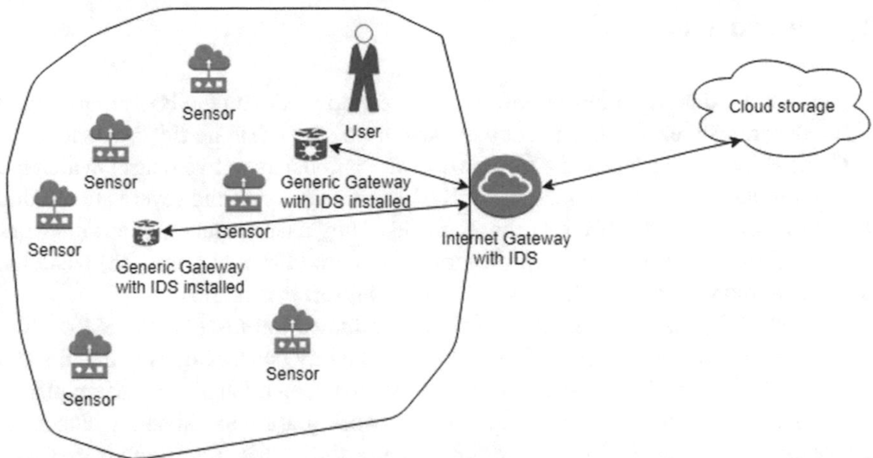

Fig. 1 Proposed IDS model in IOT environment

and False Alarm Rate (FAR) is 11.71%, while with UNSW-NB15 the accuracy is 85.56% and FAR is 15.78%. On the other hand, UNSW-NB15 dataset considers the most recent attacks.

Here, we analyzed several decision tree-based machine learning techniques (C5, CHAID, CART, QUEST) on upgraded dataset UNSW-NB15 [1]. Our proposed IDS model shows better error detection rate and generates lower FAR compared to ENADS [6]. Data preprocessing technique is applied in the proposed work to reduce the redundant data to make our learning model more efficient. This paper reduces the attack categories of UNSW-NB15 [1] dataset to 5 out of 10 due to overlapping nature of different attacks, which increases error detection rate and reduces FAR. In this paper, the proposed IDS model considers 13 features out of 47 features of UNSW-NB15 [1] dataset on the basis of information gain value. It is analyzed that if 13 features are considered instead of 47 features of UNSW-NB15 [1] dataset, FAR value does not increased significantly. Proposed IDS shows better intrusion detection rate on C5 model compared to other existing approaches.

The merit of proposed work is that it is enhancing the accuracy and detection rate while decreasing the FAR value compared to ENADS [6], and it covers most of the recent attacks in IOT-based environment.

Remaining of the paper is divided as follows: 1. Section 2 shows related works. 2. Section 3 describes proposed work in IOT-based environment. 3. Section 4 includes the result and analysis of different models using the processed dataset. Finally, we conclude our proposed work with future scope in Sect. 5.

2 Related Works

In this section, this paper has reviewed some related works on the IOT environment which describes their work, dataset they have used, and what are the drawbacks.

Ge et al. [7] have proposed a framework which consists of five phases which are used to model and assess the security of IOT. Here, they try to find severe attacks and their paths to nullify the effect of those attacks. They have evaluated the framework on three different areas of IOT application which are (1) smart home, (2) wearable health care monitoring, and (3) environment monitoring scenarios.

Raza et al. [8] have designed an IDS and evaluated the implementation on IOT and they named it as SVELTE. Here, they have mainly covered routing attacks and according to authors their model can be extended to detect other attacks. Main phases of intrusion detection in this model are (1) network graph inconsistency detection, (2) checking node availability, (3) routing graph validity, (4) end-to-end packet loss adaptation, and (5) Sybil and CloneID attack protection.

Mehare et al. [9] have proposed an IDS for IOT to detect DoS attack. It uses the location information and neighbor nodes information to find the attack and uses the signal strength to find out the attacker nodes. It uses UDP packet. They have built security manager by integrating DoS protection manager and the IDS with ebbits network manager [10].

Hodo et al. [11] have used an Artificial Neural Network (ANN) as an offline IDS to overcome with DDoS/DoS attacks by classifying the normal and threat patterns on an IOT network. They have validated the procedure through simulating an IOT network.

Koroniotis et al. [12] have proposed a network forensics mechanism to define botnets and investigate the location of botnets using four machine learning techniques which are (1) decision tree C4.5 (DT), (2) Association Rule Mining (ARM), (3) Artificial Neural Network (ANN), and (4) Naïve Bayes (NB) on UNSW-NB15 dataset. Investigation of location is done using the flow identifiers of packet flow. Four components of the proposed mechanism include (1) traffic collection, (2) network feature selection, (3) machine learning techniques, and (4) evaluation metrics.

3 Proposed Work

This section discussed the proposed work, i.e., the implementation of IDS in an IOT environment. IDS is installed on gateways which may be local to the IOT environment or could be the one through which devices communicate with Internet. In an IOT environment, addition of devices may be dynamic and hence any node could be malicious. To protect the IOT environment from being insider attack by such malicious devices, an IDS is installed on a generic gateway in LAN and from outsider attacks an IDS is placed at Internet gateway through which LAN connects to the Internet. Figure 1 shows the implementation of IDS in an IOT environment.

Table 1 Description of UNSW-NB15 dataset

No. of instances in training set	No. of instances in testing set	No. of attack categories
175,341	82,332	9

In Fig. 1, an IOT-based environment is demonstrated in which there are two internal gateways (Generic gateway with IDS installed) and several IOT devices are present. An Internet gateway with IDS is also present, through which the communication between IOT environment and Internet is achieved. Generic gateways monitor the traffic for any suspicious activity and also control traffic. If it finds any intrusion, it notifies the administrator. So these gateways work on internal intrusion. But the Internet gateway tries to detect intruder which are outside to the environment.

3.1 Dataset Description

In this section, the description of the dataset used has been provided. This paper has used UNSW-NB15 dataset which was generated using IXIA PerfectStorm tool and the analysis of the dataset is done by Moustafa and Slay [1]. This dataset includes nine recent attack categories and also includes normal traffic. There are a total of 49 features out of which 47 are related to IDS and other two are ID and attack categories. Most of the previous works on IDS show that KDD98 and NSLKDD are widely used dataset. But these datasets are old and do not cover the recent attacks as well as the normal traffic.

Features of UNSW-NB15 fall under the following categories: (a) Flow features, (b) basic features, (c) content features, (d) time features, and (d) additionally generated features. Dataset overview is shown in Tables 1 and 2. In Table 3, the definition of attacks is given.

3.2 Data Processing

This section is divided into two parts which are follows.

3.2.1 Dataset Reduction

This section has used clustering technique to reduce the UNSW-NB15 dataset where four different sizes of cluster are considered. Cluster quality is measured with the help of Silhouette measure. Table 4 shows the analysis for the selection of number of clusters which is used for further analysis. From Table 4, it is concluded that when the number of clusters is 15 the cluster quality is best.

Table 2 Features of the dataset [1]

#	Features	Description	Category
1	Srcip	Source IP address	Flow features
2	Sport	Source port address	
3	Dstip	Destination IP address	
4	Dsport	Destination port address	
5	Proto	Transaction protocol	
6	State	The state and its dependent protocol, e.g., Acc, clo, else	Basic features
7	Dur	Record total duration	
8	Sbytes	Source to destination bytes	
9	Dbytes	Destination to source bytes	
10	Sttl	Source to destination time to live	
11	Dttl	Destination to source time to live	
12	Sloss	Source packets retransmitted or dropped	
13	Dloss	Destination packets retransmitted or dropped	
14	Service	http, ftp, smtp	
15	Sload	Source bits transfer rate	
16	dload	Destination bits transfer rate	
17	spkts	Number of packets transferred from source to destination	
18	Dpkts	Number of packets transferred from destination to source	
19	Swin	Value of window advertisement of tcp at source	Content features
20	Dwin	Value of window advertisement of tcp at destination	
21	Stcpb	tcp sequence number at source	
22	Dtcpb	tcp sequence number at Destination	
23	Smeanz	Mean value of packet size transferred by source	
24	Dmeanz	Mean value of packet size transferred by destination	
25	Trans_depth	Pipelined depth of http request/response transaction in a connection	
26	Res_bdy_len	Size of data transferred by servers http service without compressing	
27	Sjit	Jitter produced at source	Time feature
28	Djit	Jitter produced at destination	
29	Stime	Start timestamp	

(continued)

Table 2 (continued)

#	Features	Description	Category
30	Ltime	Last timestamp	
31	Sinpkt	Time gap between two consecutive incoming packet at source	
32	Dintpkt	Time gap between two consecutive incoming packet at destination	
33	Tcprtt	Round trip time taken for tcp connection establishment	
34	Synack	Time gap between SYN and SYN_ACK	
35	Ackdat	Time gap between the syn_ack and ack packets	
36	Is_sm_ips_ports	If source and destination are same and port numbers are equal, then this feature assigns to 1 otherwise 0	Additional generated feature
37	Ct_state_ttl	No. of connection for every state based on the specific range of values of source to destination and destination to source time to live field	
38	Ct_ftw_http_mthd	No. of flows having methods such as Get and Post in http service	
39	Is_ftp_login	If the ftp session is accessed by user and password then this feature is set as 1 else 0	
40	Ct_ftp_cmd	No. of flows that has a command in ftp session	
41	Ct_srv_src	No. of records out of 100 records having same service and source based on the last time stamp	
42	Ct_srv_dst	No. of records out of 100 records, having same service and destination based on the last times stamp	
43	Ct_dst_ltm	No. of records out of 100 records, having same destination based on the last time stamp	
44	Ct_src_itm	No. of records out of 100 records having same source based on the last time stamp	
45	Ct_src_dsport_ltm	No. of records out of 100 records having same source and the destination port based on the last time stamp	
46	Ct_dst_sport_ltm	No. of records out of 100 records having same destination and the source port based on the last time stamp	
47	Ct_dst_src_ltm	No. of records out of 100 records having same destination and source based on the last time stamp	

Table 3 Depicts different types of attacks which act as the class labels in the training dataset

Attack	Description
Analysis	Used to penetrate web applications through emails using spam, web scripts, e.g., using HTML files, and port scans
Backdoor	A technique used to bypass authentication process of system which allow remote access and lead to an unauthorized access to a computer or device, which gives attacker an opportunity to issue commands remotely
DoS	An attack, through which attacker tries to bring down the services of a computer network (or server) or by making resources unavailable for authorized user requests
Exploit	Sequence of steps taken by an attacker in order to take advantage of any vulnerability, glitch, or bug present in a system or network
Fuzzers	Activity through which attacker tries to find security vulnerability in system, program, network, or operating system by flooding it with random data in order to crash it
Generic	It is a type of activity in which an attacker does not bother about the crypto-graphical implementation of any primitives and runs the attack. As an example, consider a cipher text with K-bit key, in the generic attack of brute force, attacker tries every combination possible using k bits, i.e., 2^k. combinations and try to decrypt the text
Probe	Process of gathering information related to computer system or network for evading security controls.
Shellcode	An attacker writes code and inject it to any application which triggers command shell in order to take control of compromised machine
Worm	Attackers try to replicate their functional copies and use system vulnerability or any social engineering techniques to enter the system

Table 4 Describes different number of clusters and their silhouette coefficient

Experiment no.	No. of clusters	Silhouette measure	Cluster quality
1	5	0.4	Fair
2	12	0.4	Fair
3	15	0.6	Good
4	20	0.4	Fair

Here, we have analyzed the categories with 15 number of clusters and this analysis shows that several categories are overlapping. So, in this paper some of the categories are removed because they are covered by others. Finally, after removal of some categories, only five categories (DoS, Exploit, Generic, Probe, Normal) are considered. Analysis for this is shown in Table 5.

From Table 5, it can be verified that maximum instances of DoS, Analysis, Backdoor, and Fuzzers are in cluster 1 and DoS dominates analysis and fuzzers attacks. On the other hand, maximum instances of Exploit and Shellcode attacks are in the same clusters and Exploit dominates Shellcode. Cluster 5 contains maximum instances of Generic attack. Clusters 2 and 9 contain with maximum instances of Probe attack.

Table 5 Describes similarity between different types of attacks

	C1	C2	C3	C4	C5	C6	C7	C8	C9	C10	C11	C12	C13	C14	C15
Analysis	1436					6			7			551			
Backdoor	1746		5						72			22			
DoS	10,473						1791								
Exploit			1615			5647						8677		17,454	
Fuzzers	6301		1		4045				1785		15	3680	2306		
Generic	503	1231			37,886					380					
Normal	6633			30,716		4123		166	225			5120			9012
Probe		8788							1703						
Shellcode														1133	
Worm	16	110										4			

Table 6 Reduced dataset

No. of instances in training set	No. of instances in testing set	No. of categories
97,172	54,593	5

Probe dominates worm. Same thing is true for Normal which is completely distinguishable compared to others. As a result, DoS, Exploit, Generic, Probe, and Normal are dominating on rest of the attacks. Now, Table 6 depicts the detail of our reduced dataset.

3.2.2 Feature Reduction

In this section, the process by which the number of features presents in the dataset UNSW-NB15 is reduced from 47 to 13. In this paper, the concept of *Information Gain (IG)* value is used, based on which individual feature is selected.

Information Gain

Information gain value of a feature is defined as its contribution for classifying the dataset. If D is the size of the given dataset and A is a feature, then information gain value for feature A is calculated as per Eq. (1).

$$Information\,Gain(A) = Entropy(D) - Entropy_A(D) \quad (1)$$

where
$Entropy(D)$ = Expected information needed to classify a tuple in D calculated as per Eq. (2).

$Entropy_A(D)$ = Extra needed expected information for exact classification when feature A is selected and it is calculated as per Eq. (3).

$$Entropy(D) = -\sum_{i}^{m} P_i \times \log_2(P_i) \quad (2)$$

$$Entropy_A(D) = \sum_{j=1}^{v} \frac{|D_j|}{|D|} \times Entropy\left(D_j\right) \quad (3)$$

where feature A has ‘v’ distinct values and D_j is number of tuples belonging to each distinct feature value of A and probability (P_i) that an arbitrary tuple in D belongs to class C_i is given by Eq. (4).

$$p_i = \frac{|C_i|}{|D|} \quad (4)$$

Table 7 Describes average information gain value of 47 features

Feature	Information gain
Srcip, Sport, Dstip, Dstport, State, Sload, dload, Swin, Dwin, Stcpb, Dtcpb, Trans_depth, Res_bdy_len, Djit, Stime, Ltime, Sinpkt, Tcprtt, Synack, Ackdat, Is_sm_ips_parts, Ct_ftw_http_mthd, Is_ftp_login, Ct_ftp_cmd, Ct_src_ltm	0
Proto	0.001325
Dur	0.000075
Sbytes	0.31735
Dbytes	0.02955
Sttl	0.19435
Dttl	0.000094
Sloss	0.000075
Dloss	0.004525
Service	0.240125
spkts	0.000075
Dpkts	0.002125
Smeanz	0.000231
Dmeanz	0.0003325
Sjit	0.000075
Dinpkt	0.00925
Ct_state_ttl	0.01805
Ct_srv_src	0.0063
Ct_srv_dst	0.02654
Ct_dst_ltm	0.002025
Ct_src_dport_ltm	0.00215
Ct_dst_sport_ltm	0.0718
Ct_dst_src_ltm	0.020675

Information gain value is taken average over several decision tree models (C5, CHAID, CART, QUEST) which is given in Eq. (5).

$$Information\ Gain_{AVG} = \frac{\sum_{for\ each\ model} Information\ gain}{Total\ number\ of\ model} \quad (5)$$

Table 7 shows the details of information gain. Among 47 features, average information gain value of 25 features in decision tree models is 0 and rest 22 features are greater than 0. We will consider only those 22 features.

	Features	C5			CHAID			CART			QUEST		
No. of features		Features Use	Importanc	Accurac	Features Used	Importan	Accuracy	Features Used	Importanc	Accurac	Features Used	Importanc	Accuracy
22	sttl spkts sbytes ct_srv_src service proto smean dpkts dinpkt dbytes sjit ct_srv_dst ct_dst_ltm ct_dst_src_ltm ct_dst_sport_ltm dur sloss dmean dloss	dpkt smean dmean ct_srv_dst dur ct_srv_src ct_dst_src_ltm dbytes sbytes sttl	0.0085 0.0109 0.0112 0.0113 0.0119 0.0184 0.0211 0.0382 0.3722 0.4588	89.86	dur ct_dst_ltm ct_srv_src dmean ct_srv_dst ct_state_ttl dttl ct_src_dport_ltm sttl smean	0.0028 0.003 0.0038 0.0134 0.0168 0.0309 0.0358 0.0864 0.0994 0.7015	83.68	sjit dinpkt spkts sloss dur ct_dst_src_ltm dbytes smean sttl ct_dst_sport_ltm	0.0003 0.0003 0.0003 0.0003 0.0003 0.0338 0.0495 0.0934 0.3599 0.4588	82.91	sbytes ct_dst_sport_ltm ct_src_dport_ltm ct_srv_dst ct_dst_src_ltm service	0.009 0.009 0.009 0.009 0.009 0.9548	57.59
13	sttl sbytes ct_srv_src service proto dbytes dinpkt ct_srv_dst ct_dst_ltm ct_dst_src_ltm ct_dst_sport_ltm ct_src_dport_ltm dloss	service dur dloss ct_srv_src ct_srv_dst dinpkt ct_dst_src_ltm dbytes sttl sbytes	0.0042 0.0044 0.0053 0.006 0.0074 0.008 0.0252 0.0713 0.3726 0.4682	89.76	dinpkt service ct_dst_ltm dur proto dbytes ct_dst_src_ltm ct_srv_dst sttl sbytes	0.0013 0.0016 0.0018 0.0021 0.003 0.0033 0.0167 0.0174 0.2009 0.752	81.76	dloss service dur proto dinpkt dbytes ct_dst_src_ltm sbytes sttl ct_dst_sport_ltm	0.0003 0.0003 0.0003 0.0003 0.0003 0.0448 0.0597 0.0618 0.329 0.4949	80.95	sbytes ct_dst_sport_ltm ct_src_dport_ltm ct_srv_dst ct_dst_src_ltm service	0.009 0.009 0.009 0.009 0.009 0.9548	57.59
6	sttl sbytes ct_srv_dst smean ct_dst_sport_ltm dbytes	ct_dst_sport_ltm ct_srv_dst smean dbytes sttl sbytes	0.0044 0.0213 0.0386 0.0503 0.4067 0.4786	84.35	ct_dst_sport_ltm dbytes sbytes ct_srv_dst sttl smean	0.002 0.0055 0.0302 0.691 0.0965 0.7967	76.23	sbytes ct_srv_dst dbytes smean sttl ct_dst_sport_ltm	0.0011 0.249 0.583 0.1179 0.3424 0.4554	75.6	sbytes smean dbytes ct_srv_dst ct_dst_sport_ltm sttl	0.0011 0.0011 0.0011 0.0768 0.4268 0.493	53.79

Fig. 2 Analysis of different classification models

3.2.3 Classification

This section discusses the classification models' accuracy with different sets of features and on different models. Figure 2 describes the performance of each classification model on different numbers of features, i.e., 22, 13, and 6 features, respectively. These three sets of features are selected randomly based on the information gain value in which the higher value feature get selected first and next is selected from the remaining features with higher information gain value and similarly for others.

It is observed from Fig. 2 that if a number of features are 22, accuracy of C5, CHAID, CRT, and QUEST are 89.86%, 83.68%, 82.91%, and 57.59%, respectively. On the other hand, when the number of features is 13, accuracy of C5, CHAID, CRT, and QUEST are 89.76%, 81.76%, 80.95%, and 57.59%, respectively. Similarly, when the number of features is 13, accuracy of C5, CHAID, CRT, and QUEST are 84.35%, 76.23%, 75.6%, and 53.79%, respectively. Hence, it is concluded that when the number of features is 13, its accuracy value is not deviated very much compared to 22 number of features So, these 13 numbers of features are selected for designing proposed integrated model.

4 Results and Analysis

In this section, this paper worked on the reduced dataset obtained in subsection of Sect. 3.2 with 13 number of features and 5 categories. It analyzes the performance of several classification models (C5, CHAID, CART, QUEST) and observed the accuracy, IDR, and FAR of each model using Eqs. (6), (7), and (10), respectively.

The proposed work is evaluated on the basis of accuracy, IDR, and FAR. Accuracy of an IDS system is defined as the number of data instances correctly classified to the total input data. IDR is defined as number of attacks correctly classified to the total number of attacks whether it is correctly or incorrectly classified and it is written as

$$Accuracy = \frac{TP + TN}{TP + TN + FP + FN} \tag{6}$$

$$Intrusion\, detection\, rate(IDR) = \frac{TP}{TP + FN} \tag{7}$$

where

TP—True Positive: number of attack instances correctly classified,
TN—True Negative: number of non-attack instances (i.e., Normal) correctly classified,
FP—False Positive: number of instances misclassified as attack, and
FN—False Negative: number of instances misclassified as normal or non-attack.
FAR is defined as the average value of False Positive Rate (FPR) and False Negative Rate (FNR). FPR and FNR are given by Eqs. (8) and (9).

$$FPR = \frac{FP}{FP + TN} \tag{8}$$

$$FNR = \frac{FN}{FN + TP} \tag{9}$$

So, the FAR is calculated as

$$FAR = \frac{FPR + FNR}{2} \tag{10}$$

It is shown from Fig. 3, for 13 features accuracy, is marginally decremented compared to 22 number of features. But when the numbers of features are 6 the accuracy is significantly less compared to others. Finally, 13 features are considered to design the proposed model. Figures 3, 4, and 5 depict accuracy of the dataset on different decision tree-based models with different features. Most of the previous works have compared their work with KDD99 or NSL-KDD. But in this paper, comparison work is done on most recent and complete dataset UNSW-NB15.

In Fig. 6, comparison between ENADS [6] and C5 model with processed dataset is shown. Here, the accuracy is higher and FAR is lower than that of ENADS [6]

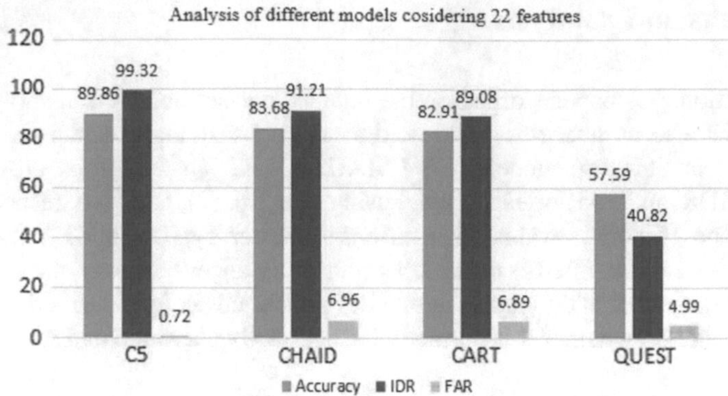

Fig. 3 Analysis with 22 features

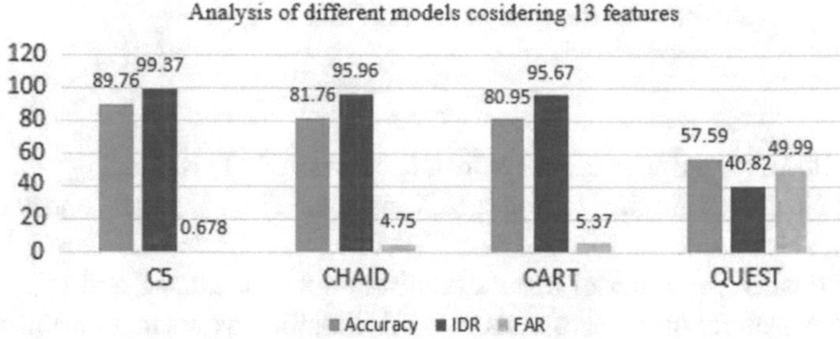

Fig. 4 Analysis with 13 features

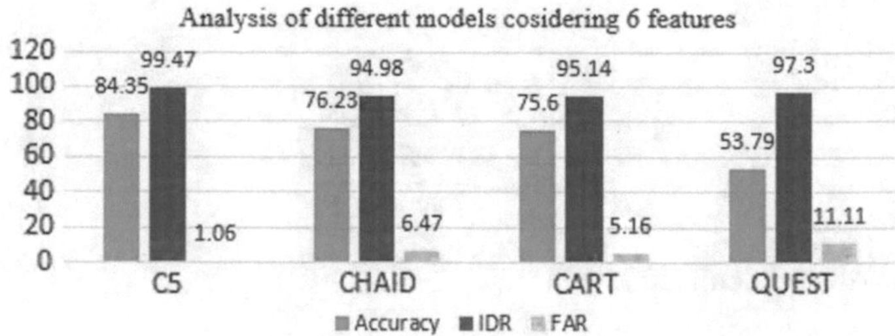

Fig. 5 Analysis with six features

because in this paper, the overlapping categories are removed using k-means clustering technique.

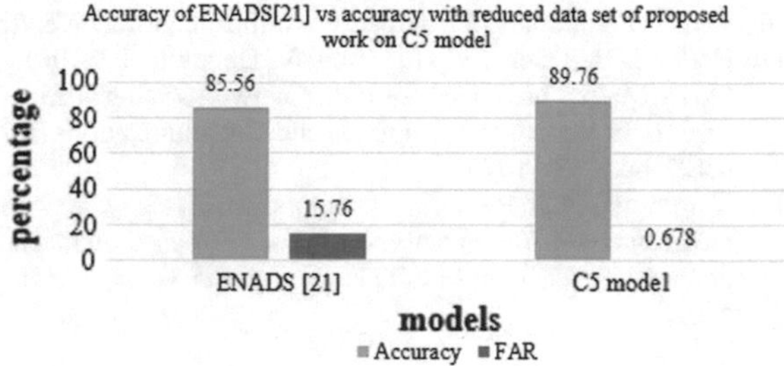

Fig. 6 Comparative analysis of ENADS [6] with accuracy on processed dataset using C5 model

5 Conclusion

This paper analyzes different tree-based models and evaluates its performance on traditional dataset. This dataset covers the most recent attacks (DoS, Exploit, Normal, Probe, and Generic) compared to KDD99 dataset. It is observed that IDR of the proposed C5 model is 99.37% which is higher compared to other existing traditional decision tree-based models. We can conclude that the proposed work provides a guideline to build efficient IDS which can prevent the system from internal and external malicious attacks.

Acknowledgements This research was supported by Information Security Education and Awareness (ISEA) Project II funded by Ministry of Electronics and Information Technology (MeitY), Govt. of India.

References

1. Moustafa, N., Slay, J.: UNSW-NB15: a comprehensive data set for network intrusion detection systems (UNSW-NB15 network data set). Military Communications and Information Systems Conference, pp. 1–6 (2015)
2. Malek, Z., Trivedi, B.: A Study of Anomaly Intrusion Detection Using Machine Learning Techniques, vol. 2, (1) (2013)
3. Haroon, A., Shah, M.A., Asim, Y., Naeem, W., Kamran, M., Javaid, Q.: Constraints in the IoT: the world in 2020 and beyond. Constraints J. **7**(11), (2016)
4. KDD 99 data set. http://kdd.ics.uci.edu/databases/kddcup99/kddcup99.html
5. Dhanabal, L., Shantharajah, S.P.: A study on NSL-KDD dataset for intrusion detection system based on classification algorithms. Int. J. Adv. Res. Comput. Commun. Eng. **4**(6), 446–452 (2015)
6. Moustafa, N., Slay., J.: The evaluation of network anomaly detection systems: statistical analysis of the UNSW-NB15 data set and the comparison with the KDD99 data set. Inf. Secur. J.: A Glob. Perspect. **25**(1–3), 18–31(2016)
7. Ge, M., Hong, J.B., Guttmann, W., Kim, D.S.: A framework for automating security analysis of the internet of things. J. Netw. Comput. Appl. **83**, 12–27 (2017)
8. Raza, S., Wallgren, L., Voigt, T.: SVELTE: Real-time intrusion detection in the Internet of Things. Ad Hoc Netw. **11**(8), 2661–2674 (2013)
9. Mehare, T., M., Bhosale, S.: Design and development of intrusion detection system for internet of things. Int. J. Innov. Res. Comput. Commun. Eng. 5(7), (2017)

10. Ebbits -Fraunhofer FIT. https://www.fit.fraunhofer.de/en/fb/ucc/projects/ebbits.html
11. Hodo, E., Bellekens, X., Hamilton, A., Dubouilh, P.L., Iorkyase, E., Tachtatzis, C., Atkinson, R.: (2016, May). Threat analysis of IoT networks using artificial neural network intrusion detection system. In Networks, Computers and Communications (ISNCC), International Symposium on IEEE, pp. 1–6 (2016)
12. Koroniotis, N., Moustafa, N., Sitnikova, E., Slay, J.: Towards developing network forensic mechanism for Botnet activities in the IoT based on machine learning techniques. In: International Conference on Mobile Networks and Management, pp. 30–44. Springer, Cham (2017)

Singer Identification Using MFCC and CRP Features with Support Vector Machines

Rajesh Sangeetha and N. J. Nalini

Abstract Singer identification is the process of identifying or recognizing the singers based on the uniqueness in their singing voice. It is a challenging task in music information retrieval because of the combined instrumental music with the singing voice. The work presented in this paper recognizes a singer using Mel Frequency Cepstral Coefficient (MFCC) features and Chroma-Reduced Pitch (CRP) features with Support Vector Machines (SVM). The proposed technique for singer identification has two phases: feature extraction and identification. During the feature extraction phase, MFCC and CRP features are extracted from the songs in a database of popular music. In the second phase, the extracted features are trained with the SVM classifier. To evaluate our work, a dataset of 50 music clips was tested against the trained models of various singers. An equal error rate of 8% and 56% is achieved with SVM using MFCC and CRP features, respectively. By combining MFCC and CRP features at score level, an EER of 6.0% is obtained which indicates a significant increase in identification rate.

Keywords Singer identification · Mel frequency cepstral coefficients · Chroma-reduced pitch · Support vector machines

1 Introduction

Music is an inevitable part of all ages of human life. With the rapid development in the digitization, the number of users and revenue generated from the digital music industry is growing amazingly. In the last few years, the music industry has shown a growth of 26% in revenue. The paid subscription of online music users has expected

R. Sangeetha (✉)
K. J. Somaiya Institute of Management Studies and Research, Mumbai 400077, India
e-mail: rajesh.sangeetha@gmail.com

N. J. Nalini
Annamalai University, Annamalai Nagar, Chidambaram 608002, India
e-mail: njnsce78@gmail.com

A. K. Das et al. (eds.), *Computational Intelligence in Pattern Recognition*, Advances in Intelligent Systems and Computing 999,
https://doi.org/10.1007/978-981-13-9042-5_25

an increase of 10–15% by 2020 [1]. With the introduction of music streaming, the digital music industry necessitates more efficient and effective music retrieval and indexing techniques. Music files can be retrieved by categories like artist, emotion, genre, etc. For recognizing the song, most people use the voice of the playback singer. The organization and retrieval of music data by singers is an eloquent way of retrieving music information. This paper designates a singer identification system using Mel Frequency Cepstral Coefficient (MFCC) features and Chroma Reduced Pitch (CRP) features with Support Vector Machines (SVM).

The identification of singer using songs is a challenging task due to the mixing of singing voice with the instrumental music which inversely affects the performance of the system. The effect of instrumental music can be reduced using various techniques available in audio processing tools. Another major issue is the lack of a standard dataset. The song dataset used in this work is collected from various music digital libraries and DVDs.

The objective of this paper is to propose a singer identification system. It entails two processes: (i) feature extraction and (ii) identification. Feature extraction converts the acoustic signal into feature vectors. The dataset collected is divided into testing and training sets. MFCC and CRP features are extracted from these music clips. The training data are used to train the models using the SVM classifier. The performance of the system is evaluated using the test data. An abstract model of the proposed singer identification system is revealed below (see Fig. 1).

The rest of the paper is systematized as follows: Sect. 2 is a review of the similar work done on singer identification. Feature extraction process is detailed in Sect. 3. Section 4 depicts the working of SVM model. Section 5 details the proposed singer identification model. Experiments and results are discussed in Sect. 6. Section 7 performs a comparison study with the existing literature. The paper is concluded by giving a summary of the work and the future scope.

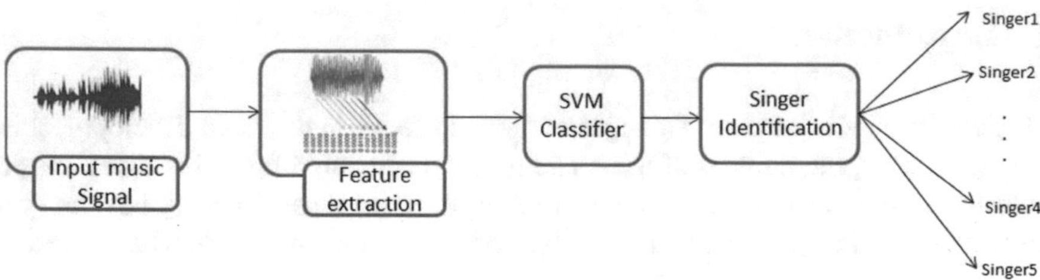

Fig. 1 Abstract model of the proposed singer identification model

2 Related Research: A Review

Singer identification is fundamentally a pattern recognition problem. The major tasks in this process are feature extraction and identification. In this section, features and classification models used for singer identification in various research literature are discussed.

Ratanpara et al. [2] presented an approach to identify the singer using the audio extracted from the Indian video songs. 13 MFCC coefficients and 13 Linear Predictive Coefficients (LPC) are extracted from the audio data. Dimensionality reduction is done using Principal Component Analysis (PCA). The audio is classified into different singer classes using Naïve Bayes classification model which has shown an accuracy of 71% and using the backpropagation neural network an accuracy of 78% is obtained. Tsai et al. [3] investigated the possibility of identifying the singer's voices using spoken data. The singing and speech voice is significantly different. SID using spoken data by the singers does not result in the optimum accuracy. Person's singing voice cannot be characterized by the person's speaking voice. The speech-derived model using Gaussian Mixture Model (GMM) is modified to find the singing voice characteristics. The speech and the singing voice relationships are used in transformation. The accuracy is improved using spoken data.

Regnier et al. [4] explored the use of TECC and INTO features to identify the singer of a given song. This work introduces two approaches using the combination of features in singer level and song level. GMM classifier is trained in both approaches using the combination of features. In singer level, an EER of 7.5% is obtained, whereas in song level EER is 9%. Patil et al. [5] explored MFCC and Cepstral Mean Subtracted MFCC (CMSMFCC) feature vectors for Singer Identification (SID). Second-order polynomial classifier is used to devise the model. Using MFCC, the feature classifier achieved an accuracy of 75.75%. CMSMFCC with the classifier model acquired accuracy of 84.5%. Score-level fusion of both results in an increase in the accuracy of 10.25%.

Dharini et al. [6] demonstrated the SID using Perceptual Linear Predictive (PLP) features. K-means clustering is used to build the model. The overall accuracy of 55.56% is achieved. Sarkar et al. [7] have explored the classification of song dataset using a voice signature inherent signal. Variation patterns of zero-crossing rates and short-time energy features are extracted which represents pitch and energy. Multilayer perceptron is trained using the features. 63% classification accuracy is achieved in the experiments. Seetharaman et al. [8] presented a cover song identification system using YouTube dataset.

3 Feature Extraction

Feature extraction is one of the major steps in any identification or classification task. The performance of the identification algorithm is centred on the features extracted from the input files. In this study, we have focused on MFCC and CRP features of the music clips. The lower MFCCs are closely related to timbre. Instead of extracting lower coefficients, only the upper coefficients are considered in chroma features. By discarding the timbre-related information, timbre invariance can be captured.

3.1 Mel Frequency Cepstral Coefficients (MFCC)

MFCCs [9] are the most used features in the speech and speaker recognition systems. It exploits the human auditory response better than any other frequency bands [10]. The MFCC is reliable for variations in speakers and recording conditions [11]. The feature extraction process of MFCC [12] from a music segment comprises the following processes.

3.1.1 Pre-emphasis

This stage compensates the high-frequency part by amplifying the high-frequency formants. The audio signal S(n) is sent to the high-pass filter as given in Eq. 1 [13, 14].

$$\mathrm{S}(n) = \mathrm{S}(n) - \alpha\mathrm{S}(n-1) \tag{1}$$

3.1.2 Frame Blocking

To reduce the variation within the segment, each segment is divided into 20–30 s frames as shown in Eq. 2.

$$x_l(n) = \hat{S}(Ml + n), \quad n = 0, 1, 2, \ldots, \mathrm{N} - 1, \quad 1 = 0, 1, 2, \ldots, \mathrm{L} - 1 \tag{2}$$

3.1.3 Windowing

It smoothens edge and reduces the effect of side lobe [15]. Hamming window is used in most of the music information retrieval work. The size of the window is 20–25 ms (ms) with 10 ms of overlap. The Hamming window can be represented as given in Eq. 3.

$$\mathrm{w(n)} = 0.54 - 0.46\cos\left(\frac{2\,\pi\,\mathrm{n}}{\mathrm{N}-1}\right), \quad 0 \leq \mathrm{n} \leq \mathrm{N} - 1 \tag{3}$$

where N is the total number of samples and n is the current sample.

3.1.4 Fast Fourier Transform

FFT is applied to perform the spectral analysis of the audio signal to obtain the magnitude frequency response of each frame as Eq. 4.

$$X_k = \sum_{n=0}^{N-1} x_n e^{-j2\pi kn/N}, \qquad k = 0, 1, 2, \ldots, N-1 \tag{4}$$

3.1.5 Mel Cepstral Coefficients

Mel-scale filter bank is a perceptually motivated scale. This scale is linear at lower frequencies and logarithmic at higher frequencies. The Mel-scale frequency can be computed as given in Eq. 5.

$$\text{Mel scaled frequency} = 2595 \log\left(1 + \frac{F_m}{700}\right) \tag{5}$$

3.1.6 MFCC

It is computed by applying DCT on the log of mel cepstral coefficients [11].

3.2 *Chroma Reduced Pitch (CRP) Features*

CRP features are designed in order to provide timbre-invariant features. The overview of computing the CRP features is shown [16–18] (see Fig. 2).

The process to compute CRP features is described below:

- Using a multi-rate filter bank, the music clip is divided into 120 frequency bands based on MIDI pitches.
- Convert it into pitch representation.
  Perform logarithm to the pitch representation, v = log (C.v + 1) with a constant C = 1000.

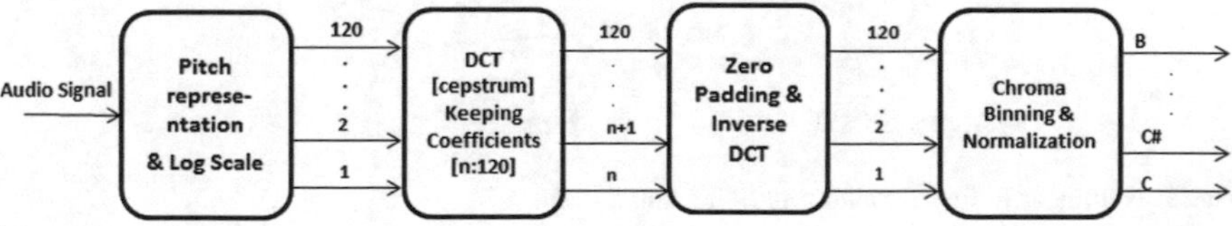

Fig. 2 Computation of Chroma Reduced Pitch (CRP) features

- DCT of size 120 is applied to the pitch vector.
- Discard the lower n − 1 coefficients and keep the upper coefficients.
- Apply inverse DCT and project it onto 12 chroma bins.
- All chroma vectors are normalized which gives CRP feature vector of the corresponding audio signal [19, 20].

4 Support Vector Machine (SVM) Classifier

The support vector machine is an efficient algorithm used in various areas for pattern recognition. SVM can be applied to linear and non-linearly separable data. In the latter case, using non-linear transformation, the input vectors are transformed into high-dimensional feature space (see Fig. 3). Then, the linear separation is done on the data. Using a different kernel function in the SVM algorithm, a variety of learning machines can be constructed [15, 21].

5 Singer Identification: Proposed Model

The major tasks in the proposed model are to extract the features from the music clips and identify the singers. The database used in this work contains 100 music clips of 10 different playback singers collected from various digital libraries. The collected data is divided into training and testing sets. The MFCC and CRP feature vectors are extracted from these music clips. The extracted feature vectors are trained using SVM classifier with Gaussian kernel function. Ten-fold cross-validation and 'OneVsAll' coding is used while training the model. The singer's voice is identified based on the features extracted from the test data of 5-s duration taken from different songs sung by the same singer. The proposed singer identification model is shown in Fig. 4.

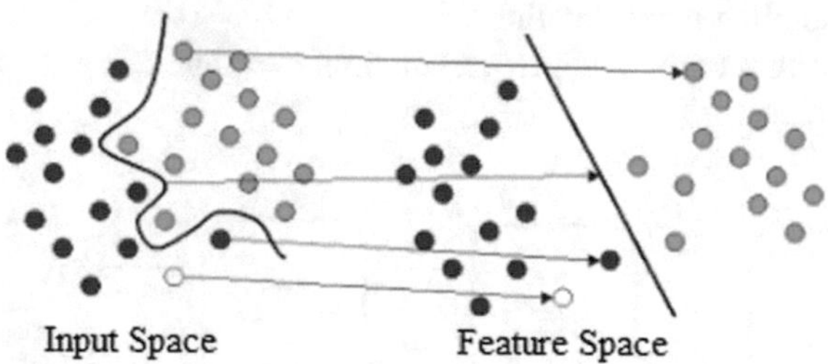

Fig. 3 Nonlinear to linear transformation using SVM

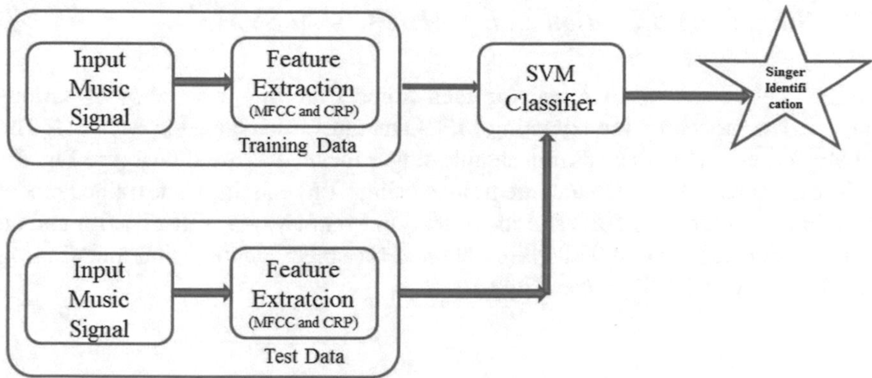

Fig. 4 Proposed singer identification model

6 Experiments and Results

This section describes the conducted experiments and analysis of results.

6.1 Music Corpus

The standard dataset is not available for Indian songs to perform various recognition tasks. So the necessary music data is collected from various online websites and DVDs. The database used in this work is a set of songs by ten playback singers with a sampling frequency of 44.1 kHz and 16-bit .wav format. 100 music clips of 10 s duration are collected which includes both male and female singers.

6.2 Performance Measures

The proposed system of singer identification is assessed using the dataset of popular music. The Equal Error Rate (EER) and identification accuracy are used to evaluate the performance of the system. EER is obtained by changing the threshold such that the False Acceptance Rate (FAR) and False Rejection Rate (FRR) are equal. FAR is the rate at which a singer identification model shows high confidence score when compared to the test singer identification model, whereas FRR is the rate at which the test model offers the low-confidence score when compared to other singer identification model. Accuracy is calculated using Eq. 6.

$$Accuracy = \frac{Number\ of\ samples\ correctly\ identified}{Total\ number\ of\ test\ cases} \tag{6}$$

6.3 Singer Identification Using MFCC with SVM

MFCC coefficients are extracted for each frame from the music clips of various singers. The procedure for extracting MFCC features is explained in Sect. 3.1. The 39 MFCC features extracted from sample singer music files are shown (see Fig. 5).

The extracted MFCC features are trained using SVM classifier. The trained model is evaluated in terms of EER and identification accuracy. An EER of 8.0% and an identification accuracy of 92.0% are obtained for singer identification model using MFCC features with SVM (see Fig. 6).

6.4 Singer Identification Using CRP with SVM

In this model, CRP features are extracted from the music clips. The process to extract CRP features is described in Sect. 3.2. The CRP features extracted from the sample singer music file are given (see Fig. 7).

In this model, the CRP features are trained using SVM classifier. The trained models have shown an identification accuracy of 44.0% and an EER of 56.0% (see Fig. 8).

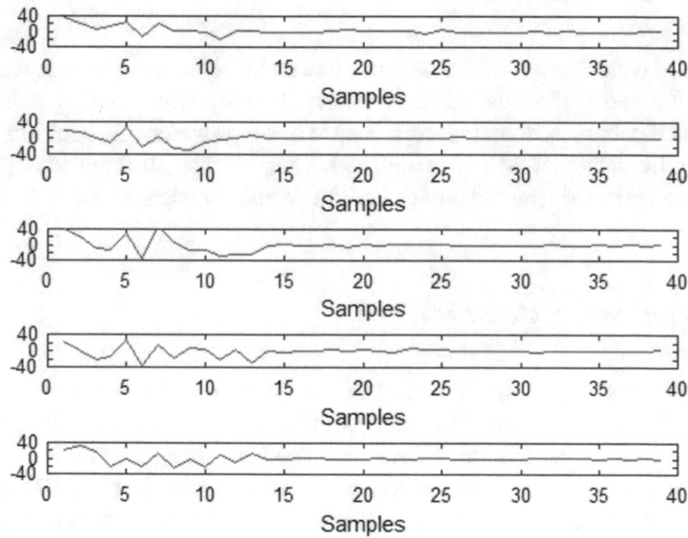

Fig. 5 MFCC features extracted from sample singer music files

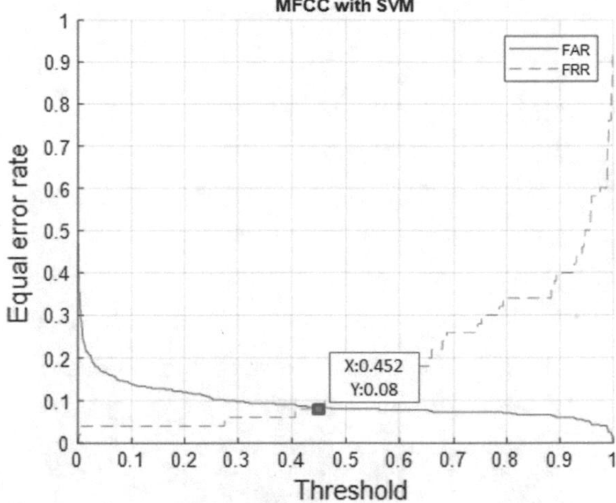

Fig. 6 FAR and FRR curves for singer identification performance using MFCC features

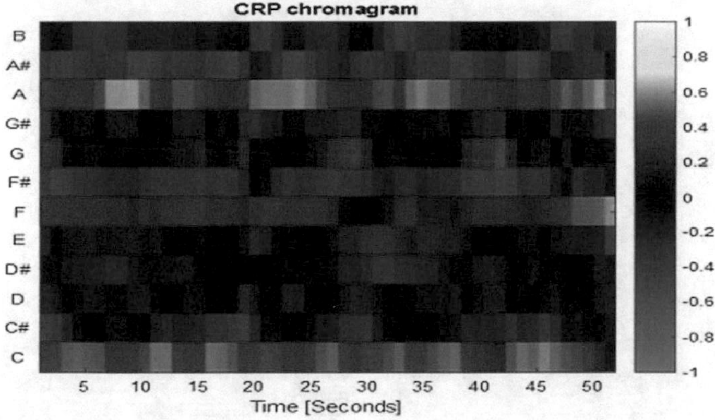

Fig. 7 CRP features extracted from the sample music file

6.5 Combining MFCC and CRP Features (Score-Level Fusion)

Both the models proposed above have limitations in identifying the singers. To improve the identification accuracy, we propose a model which combines the MFCC and CRP features. The complementary information in both the systems is combined at score level using the following equation:

$$Combined\ score = ws1 + (1 - w)s2 \tag{7}$$

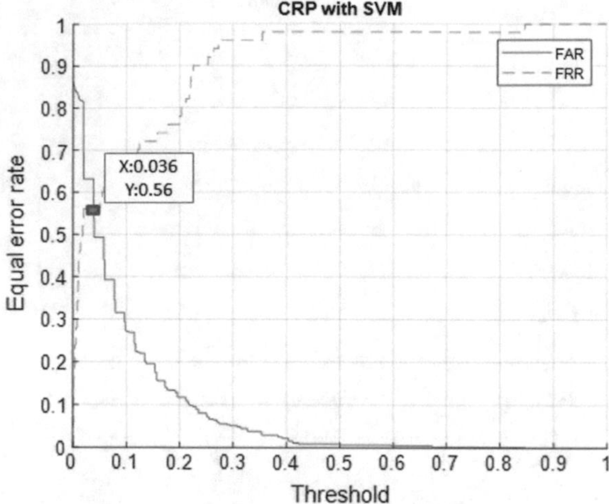

Fig. 8 FAR and FRR curves for singer identification performance using CRP with SVM

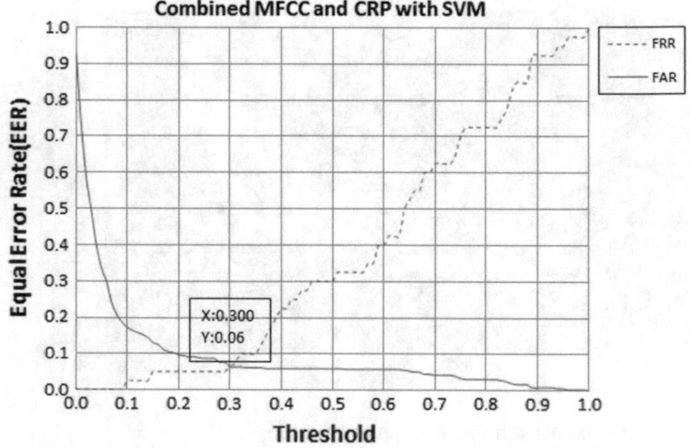

Fig. 9 FAR and FRR curves for singer identification using combined features

s1 and s2 are the confidence scores of MFCC and CRP features, respectively. The combined features at score level attain an EER of 6.0% and an identification accuracy of 94.0% (see Fig. 9).

7 Discussion

The singer identification technique using MFCC and CRP features with SVM enhances the performance compared to the state-of-the-art literature. The previous efforts on singer identification using MFCC and LPC with Naïve Bayesian classification and backpropagation neural networks show an accuracy of 71% and 78%, respectively [2]. The model proposed in [5] combined the evidences from MFCC and CMSMFCC and attained an accuracy of 86%. The proposed approaches, MFCC and CRP, are combined at the score level which gives an accuracy of 94%. As these features are complementary in nature, the combined accuracy shows increase in the performance of the model. The novelty of the work is the combination of MFCC and CRP with SVM.

8 Conclusion and Future Scope

In this paper, singer identification model is proposed using MFCC, CRP, and the combined features with SVM. Singer identification is done in two phases, feature extraction and identification. The experimental study is conducted using Indian songs database collected from various websites with sampling frequency of 44.1 kHz and bit rate of 16 bps. Initially, from the music clips, MFCC and CRP feature vectors are extracted. The feature vectors extracted are trained with SVM. The test files of the singers are identified using this trained model. A classification accuracy of 92.0% is achieved with SVM using MFCC and 44.0% with SVM using CRP features. MFCC and CRP features are combined at score level which gives superior performance than the individual feature for singer identification.

The work can be further extended using the neural network models for identification with different features. Identifying the singer's voice as male or female voice can be considered in the future work.

References

1. India Brand Equity Foundation. http://www.ibef.orglast. Accessed 25 Dec 2017
2. Ratanpara, T.: Singer identification using MFCC and LPCC from Indian video songs. In: Emerging the ICT for Bringing the Future, pp. 275–282 (2015)
3. Tsai, W.H., Lee, H.C.: Singer Identification based on spoken data in voice characterization. IEEE Trans. Acoust. Speech Signal Process. **20**(8), 2291–2300 (2012)
4. Regnier, L., Peters, G.: Singer verification: Singer model Vs. Song model. In: IEEE International Conference on Acoustics, Speech and Signal Processing, pp. 437–440. Japan (2012)
5. Patil, A, H., Radadia, G, P., Basu, T, K.: Combining evidences from Mel cepstral features and cepstral Mean Subtracted features for singer identification. In: International Conference on Asian Language Processing, pp. 145–148. Vietnam (2012)
6. Dharini, D., Revathy, A.: Singer identification using clustering algorithm. In: International Conference on Communication and Signal Processing, pp. 1927–1931. Bangkok (2014)
7. Sarkar, R., Saha, S.K.: Singer based classification of songs dataset using vocal signature inherent in signal. In: Fifth National Conference on Computer Vision, Pattern Recognition, Image Processing and Graphics. Patna (2015)
8. Seetharaman, P., Rafii, Z.: Cover song identification with 2D Fourier transform sequences, In: IEEE International Conference on Acoustics, Speech and Signal Processing (ICASSP), New Orleans (2017)

9. Davis, S.B., Mermelstein, P.: Comparison of parametric representations for monosyllabic word recognition in continuously spoken sentences. IEEE Trans. Acoust. Speech Signal Process. **28**, 357–366 (2009)
10. Youngmoo, E, Kim., Erik, M, Schmidt., Raymond, Migneco., Brandon, G, Morton.: Music emotion recognition: a state of the art review. In: 1st International Society for Music Information Retrieval Conference, pp. 255–266. Netherlands (2010)
11. Nalini, N.J., Palanivel, S., Balasubramanian, M.: Speech emotion recognition using residual phase and MFCC feature. Int. J. Eng. Technol. **5**(6), 4515–4527 (2014)
12. Dan, jurafsky.: Speech Recognition and Synthesis-Feature Extraction and Acoustic Modeling lecture notes (2007)
13. O'Shaughnessy, O.: Interacting with computers by voice: automatic speech recognition and synthesis. Proc. IEEE **91**(9), 1272–1305 (2003)
14. Peinado, A., Segura, J.: Speech Recognition over Digital Channels: Robustness and Standards, John Wiley and Sons (2006)
15. Vapnik, V.: Statistical learning theory. John Wiley and Sons, New York (1998)
16. Meinard, M., Sebastian, E., Sebastian, K.: Making chroma features more robust to timbre changes. In: Proceedings of IEEE International Conference on Acoustics, Speech, and Signal Processing (ICASSP), pp. 1869–1872, Taiwan (2009)
17. Müller, M., Kurth, F., Clausen, M.: Audio matchingvia chroma-based statistical features, In: International Music for Information Retrieval Conference, London (2005)
18. Ken, O'H., Sebastian, E., Johan, P., Mark, B., S.: Improved template based chord recognition using the CRP feature, In: IEEE International Conference on Acoustics, Speech and Signal Processing (ICASSP), New Orleans(2017)
19. O'Hanlon, K., Sandler, M.: A compositional approach to chroma estimation, In: Proceedings of the European Signal Processing Conference (EUSIPCO), Hungary (2016)
20. M¨uller, M., Ewert, S.: Towards timbre-invariant audiofeatures for harmony-based music. IEEE Trans. Audio Speech Lang Process. **18**(3), 649–662 (2010)
21. Dhanalakshm, P., Palanivel, S., Ramalingam,V.: Classification of audio signals using SVM andRBFNN, Expert systems with applications, 36(3.part 2), pp. 6069–6075 (2009)

Object Proposals Based on Variance Measure

Amit Verma, T. Meenpal and Bibhudendra Acharya

Abstract Object proposals have recently become an important part of the object recognition process. Current object proposals are mostly based on hierarchical grouping which generates too many proposals and consumes too much processing time. This paper presents a framework utilizing variance measure to produce object proposals. We divide complete image into small patches. Our algorithm identifies possible object patches and merged them together to form object proposals. We evaluate our algorithm on UT interaction dataset. Experimental results show that our method generates fewer but quality proposals. Our method also performs reasonably fast than the state-of-the-art approaches.

Keywords Object proposals · Object recognition · Variance · UT interaction dataset

1 Introduction

In recent years, object classification has become a major research area. It gave an opportunity to look at the traditional image segmentation approaches [1] from a different angle. In traditional segmentation, the goal was to assign every pixel in an image to one of N regions such that each object receives a unique region. But, unfortunately these methods fail to generate separate segments for different objects. Hence, the field of object proposal generation [2] got evolved. An object proposal algorithm generates image locations that may potentially contain an object. A variety of papers offer methods for generating region proposals [3–6].

A. Verma (✉) · T. Meenpal · B. Acharya
Department of Electronics and Telecommunication Engineering, National Institute of Technology, Raipur 492010, Chattisgarh, India
e-mail: averma.phd2016.etc@nitrr.ac.in

T. Meenpal
e-mail: tmeenpal.etc@nitrr.ac.in

B. Acharya
e-mail: bacharya.etc@nitrr.ac.in

A. K. Das et al. (eds.), *Computational Intelligence in Pattern Recognition*, Advances in Intelligent Systems and Computing 999,
https://doi.org/10.1007/978-981-13-9042-5_26

We propose an object proposal algorithm based on variance. We divide complete image into small patches, say 8 × 8. Active patches, i.e. patches having object entities, are segmented based upon a variance threshold. The central idea is that the background of an image is having almost similar pixel values, which limits the variance of a background patch to a very low value. In contrast to that, patches having object entity result in higher variance, especially in the boundaries of an object. Our method avoids the blind exhaustive search used in window-based approaches [4, 7–10] . In contrast to fixed size windows used in these approaches, our approach also adopts the bounding box size as per object. The main contribution of our approach is that it searches for an object possibility in active patches only, which makes our algorithm significantly faster.

Our method also avoids hierarchical grouping used in state-of-the-art segmentation-based approaches such as selective search [5, 6]. They start with number of proposals and finally converge to fewer ones. However, convergence will make this approach computationally expensive. In contrast to that, our approach merges neighbouring active patches and directly generates final proposals, which may also ease further processing such as object detection.

Our paper is organized as follows. Section 2 addresses previous studies related to object proposals. In Sect. 3, we explain the detailed procedures of the proposed method. Section 4 describes the experiments performed on UT interaction dataset [11] to verify the performance of the proposed method and evaluate the results. Finally, the conclusions are marked in the last section.

2 Related Works

One early approach to image segmentation is of splitting and merging regions [1] according to how well each region fits some uniform measures. Often such measures are aimed at finding either uniform intensity or uniform gradient regions. But still no uniformity criterion has been proposed to correctly segment images.

With the development of object detection algorithms, the field object proposal generation has also got a parallel growth. Object proposals can be broadly categorized into two categories: Window-based and segmentation-based. Window-based methods use image descriptors such as HOG [9], SIFT [12], and LBP [13] as features and sweep through the entire image to find regions having objectness. However, sweeping thorough entire image made it highly expensive with time. Also, size of the window fixed the size of the object.

In segmentation-based approaches, one of the earlier methods for object proposal was objectness [3]. They sample about 100,000 object boxes in each image with some objectness measures. These measures depend upon multiple characteristics such as edges, colour, etc. However, too many proposals limit the performance of this method. Another method [5] employs graph cuts with some initial seeds to detect multiple foreground/background regions. However, the method is quite slow (few minutes per image). One of the state-of-the-art algorithms proposed till date and

Fig. 1 Selective search algorithm for bounding box generation (Fig. source: [6])

currently being used in Fast RCNN [14] and Faster RCNN [15] for object detection is selective search [6]. In selective search, the authors carefully engineer a variety of features that greedily merge low-level superpixels. The method is reasonably faster than its earlier approaches. However, the hierarchical grouping generates multiple object boxes before being grouped to fewer ones as shown in Fig. 1. The number of proposals and processing time makes it computationally expensive.

In our approach, we first generated possible object patches based upon variance and then the close patches are merged together to form possible object locations. Our approach avoids blind exhaustive search and fixed aspect ratio used in window-based approaches [4, 7, 8] and cover objects in all scales. Our approach also avoids hierarchical grouping used in selective search [6]. Hence, it does not generate multiple object proposal boxes which makes the algorithm reasonably fast. In fact, our algorithm directly generates final bounding boxes.

3 Our Proposed Object Proposal Algorithm

In this section, we detail our object proposal algorithm. Our algorithm starts with two preprocessing steps.

3.1 Preprocessing

First, an RGB image is converted to grayscale as our algorithm requires a grayscale input. Followed by this conversion, the grayscale image is subjected to closing operation [16]. The closing operation fills small holes without affecting object regions and also eliminates unwanted objects as seen in Fig. 2.

Fig. 2 Grayscale input and output after closing operation

The closing of a set I by a structuring element s can be defined by Eq. 1:

$$I \bullet s = (I \oplus s) \ominus s \tag{1}$$

where $\oplus$ and $\ominus$ denote the dilation and erosion.

3.2 Elimination of Low-Variance Patches

As our algorithm requires patches to calculate variances, we have divided the whole image into set of 8×8 patches, i.e. Fig. 3. Hence, an image I may be represented in Eq. 2

$$: I \in \{I_1, I_2, I_3, I_4, I_5, \ldots, I_n\} \tag{2}$$

where $I_1, I_2, I_3, \ldots, I_n$ are set of 8×8 patches and n denotes number of such patches.

3.2.1 Calculating Variance

Variance of an image defines how each pixel value varies from its mean. In our case, the background of an image is having almost similar pixel values, which limits the variance of a background patch to a very low value. In contrast to that, patches having object entity result in higher variance, especially in the boundaries of an object. Variance of an image patch can be represented by Eq. 3:

$$Var_{I_k} = \frac{1}{M * N} \sum_{i=1}^{M} \sum_{j=1}^{N} (I_k(i, j) - \overline{I_k})^2 \tag{3}$$

Fig. 3 Image converted into small patches

where M and N denote size of an image patch, $I_k(i, j)$ denotes a pixel intensity at location (i, j) and $\overline{I_k}$ denotes mean pixel value.

We calculate variance of each image patch and form a variance vector Var for each image denoted by Eq. 4:

$$Var \in \{V_{I_1}, V_{I_2}, V_{I_3}, V_{I_4}, V_{I_5}, \ldots, V_{I_n}\} \tag{4}$$

3.2.2 Finding Low-Variance Patches

We decide the variance threshold after calculating variance for an image. All the patches with variance below threshold are considered as background patches. However, patches with variance above this threshold are object patches.

In Fig. 4, variance plots of different images with respective histograms are shown. The histogram shows that most of the image patches correspond to background (i.e. low-variance patches). Hence, number of low-variance patches is very high, i.e. close to 85%. To decide the threshold, we take bottom 85% values of the variance vector and find maximum of it, i.e. 85% of variance vector, i.e. Eq. 5.

$$Var_{threshold} = Var_{sorted}[0.85 * n] \tag{5}$$

where Var_{sorted} denotes variance vector sorted in ascending order and n denotes number of patches.

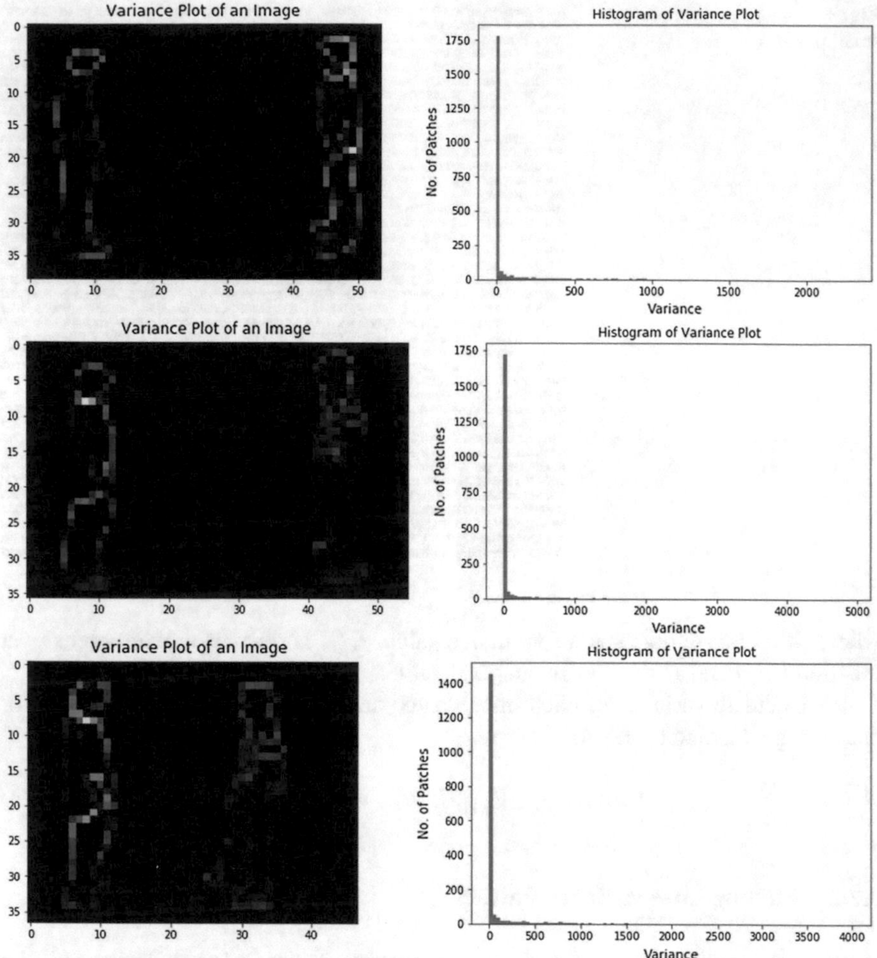

Fig. 4 Variance plots of images and corresponding histograms

3.2.3 Eliminating Low-Variance Patches

We identify low-variance patches based on variance threshold and eliminate them. Figure 5 shows the effect of eliminating low-variance patches. In Fig. 5, low-variance patches are replaced by null matrix.

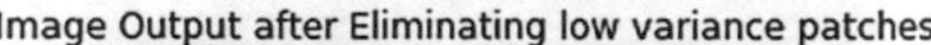

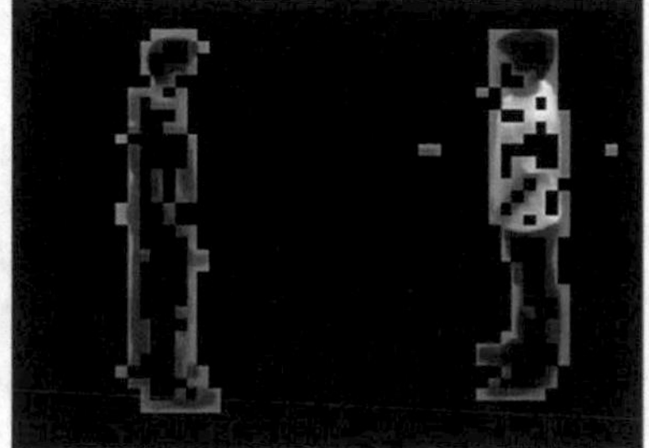

Fig. 5 Grayscale input and output after eliminating low-variance patches

Fig. 6 Active patches marked by green circle

3.3 *Bounding Box Generation*

After eliminating low-variance patches, we will have a set of patches with variance higher then threshold. We are naming them as active patches. The active patches can be defined as a vector A, i.e. Eq. 6:

$$A \in \{a_1, a_2, a_3, a_4, a_5, \ldots, a_m\} \tag{6}$$

where m denotes number of active patches in an image. Figure 6 shows active patches marked by green circle in an image.

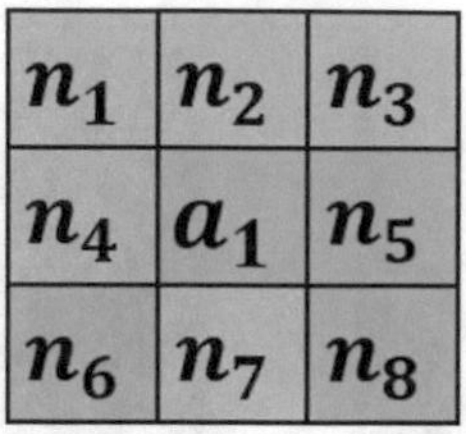

n_1	n_2	n_3
n_4	a_1	n_5
n_6	n_7	n_8

Fig. 7 Active (filled with green) and inactive or low-variance (filled with orange) neighbours for an active patch a_1

3.3.1 Active Neighbour Vector

For each active patch, we find active neighbouring patches and store them in an array. Figure 7 shows active (filled with green) and inactive or low-variance (filled with orange) neighbours for an active patch a_1. An active neighbour array for a_1 may be defined as in Eq. 7:

$$N_{a_1} \in \{a_1, n_5, n_6, n_8\} \tag{7}$$

For all active patches, an active neighbour vector may be denoted by Eq. 8:

$$N \in \{N_{a_1}, N_{a_2}, N_{a_3}, N_{a_4}, N_{a_5}, \ldots, N_{a_m}\} \tag{8}$$

where m denotes the number of active patches.

3.3.2 Grouping Connected Patches

Our grouping algorithm work as follows. We first compare each N_{a_i} with $N_{a_{i+1}}$. If they result in atleast one common patch, they get to merge and form a new one. Now this will be compared with next element as shown in Fig. 8. The process of grouping

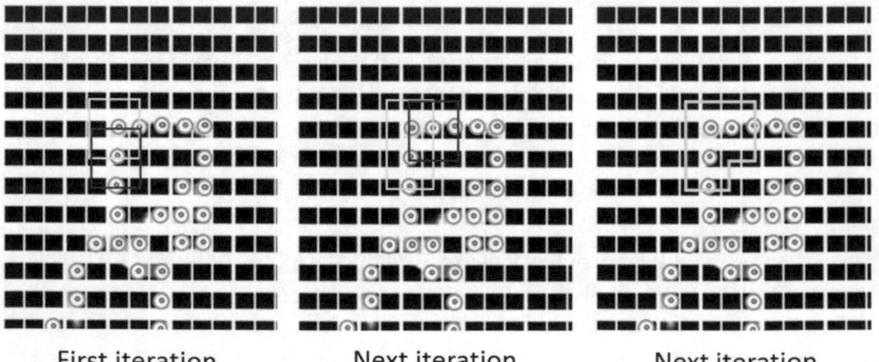

Fig. 8 Grouping procedure

is repeated until all N vector elements having common patches are grouped together. The grouping procedure is detailed in Algorithm 1.

Algorithm 1 Grouping Connected Patches

Input: Active neighbour vector N
Output: Object vector O

Initializing $X = N$, $O = [\ ]$, $NO = [\ ]$ and $j = 0$
while Until all connected N vector elements are grouped: $len(X) > 1$ **do**
    $O_j = X_1$
    for i = All X **do**
        Finding atleast one common element: $O_j \cap X_{i+1}$
        if True **then**
            Group elements: $O_j = O_j \cup X_{i+1}$
        else
            Group in next iteration: $NO.append(X_{i+1})$
    Ungrouped elements for next iteration: $X = NO$
    Increment object proposal counter: $j = j + 1$
end while

Algorithm 1 outputs one object proposal in each iteration of **for** loop. Object vector O consists all possible object proposals, Eq. 9:.

$$O \in \{O_1, O_2, \ldots, O_j\} \tag{9}$$

where $O_1, O_2, \ldots, O_j$ are set of connected N vector elements, each corresponding to one object and j denotes number of object proposals.

3.3.3 Plotting Bounding Boxes on Object Proposals

We have set of object proposals in vector O. We take each object proposal O_i and calculate the top-left (marked with red) and bottom-right (marked with yellow) patches, i.e. Fig. 9 and their respective pixel locations. Finally, we plot bounding box to each object proposal as shown in Fig. 10.

4 Evaluation

In this section, we evaluate the performance of our variance-based object proposal algorithm. We have divided our experiment into two parts. We compare our variance-based approach with state-of-the-art selective search algorithm [6] on UT interaction dataset [11]. The dataset consists of 50 videos that each contains multiple moving human objects. Later, we evaluate our object proposals on Intersection over Union (IoU) parameter.

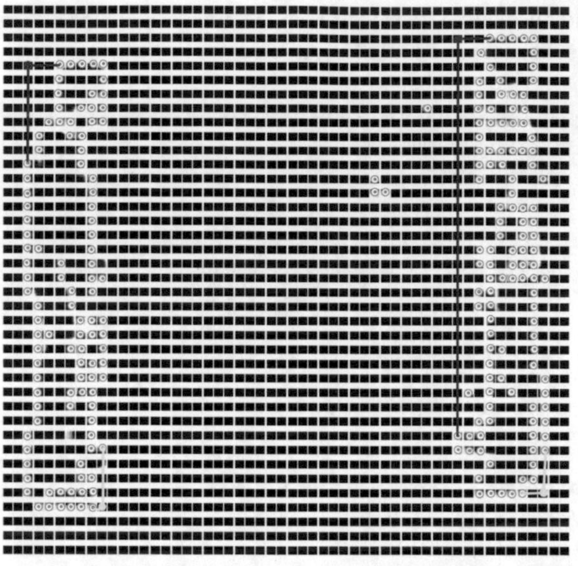

Fig. 9 Top-left (marked with red) and bottom-right (marked with yellow) patches

Fig. 10 Input image and final output with object proposals

4.1 Processing Time and Object Proposals

We evaluate our approach on 120 images from UT interaction dataset [11] and compare it with selective search algorithm outputs [6].

Processing time is the time difference calculated between providing image to the algorithm and getting output with object proposal. On computing the average processing time of our proposed algorithm on UT interaction dataset, it results in 0.184 sec., which completely outperforms selective search algorithm [6]. It concludes that the use of hierarchical grouping converges to fewer location but consumes a significant amount of time. Our method provides faster computation and may also be incorporated in real-time videos.

If we compare the number of proposals per object, our method outputs 1.12. However, for selective search [6], it comes out as 440.08. It suggests that segmenting with multiple criteria generates multiple object proposals but it does not result in good object proposals. The comparison is shown in Table 1. Few outputs are shown in Fig. 11.

Table 1 Comparison of average processing time and average proposal

S. no.	Method	Average processing time (in sec)	No. of proposal per object
1.	Selective search	1.344	440.08
2.	Our variance-based algorithm	0.184	1.12

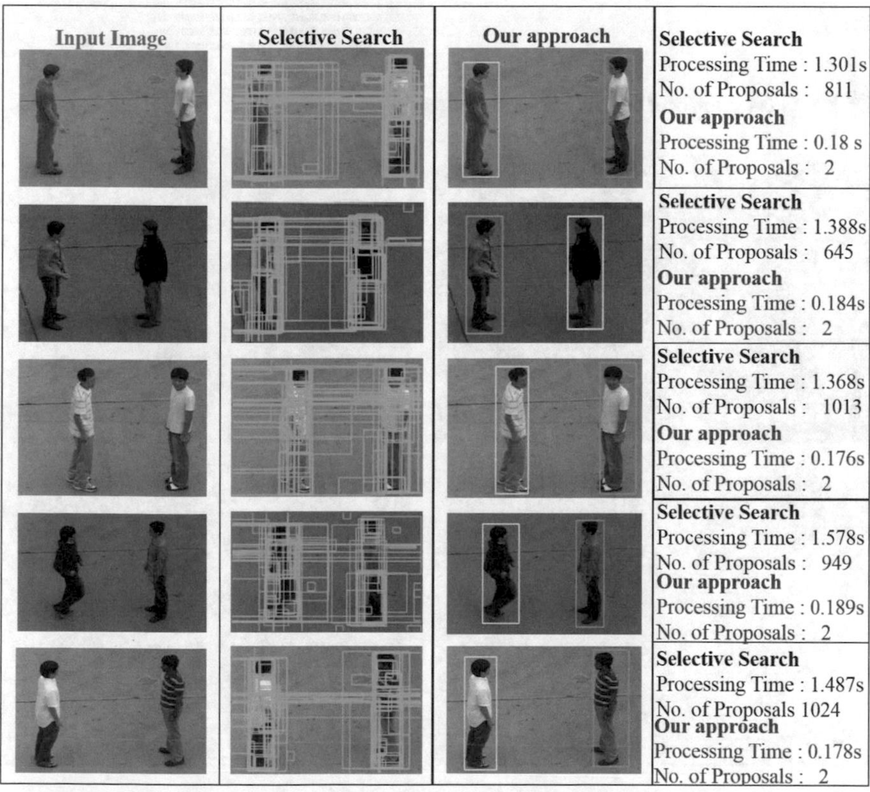

Fig. 11 Input image, selective search[] object proposals, object proposals generated by our method and their respective processing time and number of proposals

4.2 Quality of Proposals

We evaluate our variance-based algorithm in terms of Intersection over Union (IoU) to measure the quality of proposals . IoU is an evaluation matric used to measure overlap between two bounding boxes. The IoU calculation is denoted in Fig. 12. In our case, we compare bounding boxes, i.e. proposals generated by our method with the ground truth bounding boxes. Figure 13 shows the IoU scores calculated for few images. Ground truth object proposals are marked in red; however, proposals

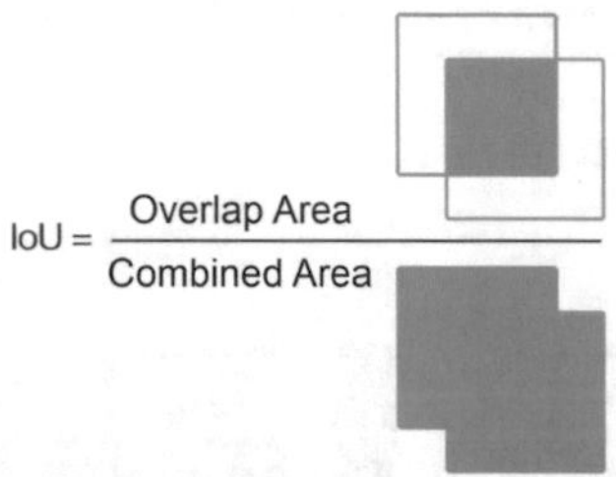

Fig. 12 Intersection over Union (IoU) calculation

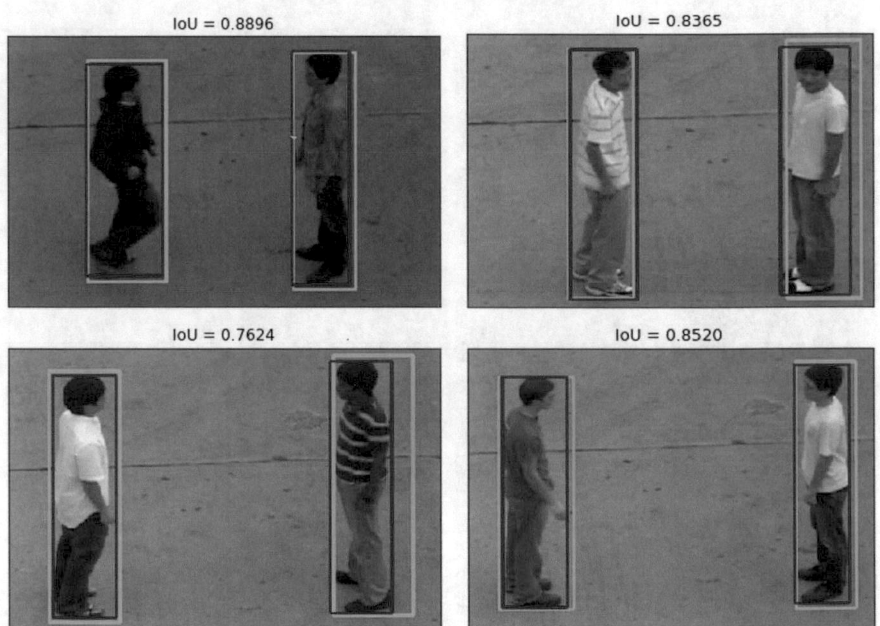

Fig. 13 Few examples with IoU scores. Ground truth bounding boxes are marked red and object proposals are marked green

generated by our method are green coloured. We achieve mean IoU score of **0.8561** for all 120 images. This illustrates that our variance-based algorithm yields high-quality object proposals.

5 Conclusion

We present a variance-based object proposal algorithm. We observe that previous approaches generate too many bounding boxes which makes them computationally expensive. Our method finds object patches and merges them together to form an object proposal. We tested our algorithm on UT interaction dataset [11]. Simulation results show that our method outperforms state-of-the-art selective search algorithm [6] both in computation time and number of proposals. Our method provides a fast computation and may also be extended to dynamic video sequences.

References

1. Ikonomatakis, N., Plataniotis, K., Zervakis, M., Venetsanopoulos, A.: Region Ggrowing and Region Merging Image Segmentation, vol. 08 (1997)
2. Hoiem, D., Efros, A.A., Hebert, M.: Geometric context from a single image. In: Tenth IEEE International Conference on Computer Vision (ICCV'05), vol. 1, pp. 654–661 (2005)
3. Alexe, B., Deselaers, T., Ferrari, V.: Measuring the objectness of image windows. IEEE Trans. Pattern Anal. Mach. Intell. **34**, 2189–2202 (2012). https://doi.org/10.1109/TPAMI.2012.28
4. Rahtu, E., Kannala, J., Blaschko, M.: Learning a category independent object detection cascade. In: 2011 International Conference on Computer Vision, pp. 1052–1059 (2011)
5. Carreira, J., Sminchisescu, C.: CPMC: automatic object segmentation using constrained parametric Min-Cuts. IEEE TPAMI **34**(7), 1312–1328 (2012)
6. Uijlings, J., van de Sande, K., Gevers, T., Smeulders, A.: Selective search for object recognition. Int. J. Comput. Vis. (2013). http://www.huppelen.nl/publications/selectiveSearchDraft.pdf
7. Zitnick, C.L., Dollár, P.: Edge boxes: Locating object proposals from edges. In: Fleet, D., Pajdla, T., Schiele, B., Tuytelaars, T. (eds.) Computer Vision - ECCV 2014, pp. 391–405. Springer International Publishing, Cham (2014)
8. Cheng, M., Zhang, Z., Lin, W., Torr, P.: Bing: Binarized normed gradients for objectness estimation at 300fps. In: 2014 IEEE Conference on Computer Vision and Pattern Recognition, pp. 3286–3293 (2014)
9. Dalal, N., Triggs, B.: Histograms of oriented gradients for human detection. In: 2005 IEEE Computer Society Conference on Computer Vision and Pattern Recognition (CVPR'05). vol. 1, pp. 886–893 (2005)
10. Harzallah, H., Jurie, F., Schmid, C.: Combining efficient object localization and image classification. In: 2009 IEEE 12th International Conference on Computer Vision, pp. 237–244 (2009)
11. Ryoo, M.S., Aggarwal, J.K.: UT-Interaction Dataset, ICPR contest on Semantic Description of Human Activities (SDHA). http://cvrc.ece.utexas.edu/SDHA2010/Human_Interaction.html (2010)
12. Lowe, D.G.: Distinctive image features from scale-invariant keypoints. Int. J. Comput. Vis. **60**(2), 91–110 (2004). https://doi.org/10.1023/B:VISI.0000029664.99615.94

13. Wang, X., Han, T.X., Yan, S.: An hog-lbp human detector with partial occlusion handling. In: 2009 IEEE 12th International Conference on Computer Vision. pp. 32–39 (2009)
14. Girshick, R.: Fast r-cnn. In: Proceedings of the 2015 IEEE International Conference on Computer Vision (ICCV), pp. 1440–1448. ICCV '15, IEEE Computer Society, Washington, DC, USA (2015). https://doi.org/10.1109/ICCV.2015.169
15. Ren, S., He, K., Girshick, R., Sun, J.: Faster r-cnn: Towards real-time object detection with region proposal networks. IEEE Trans. Pattern Anal. Mach. Intell. **39**(6), 1137–1149 (2017), https://doi.org/10.1109/TPAMI.2016.2577031
16. Soille, P.: Erosion and Dilation, pp. 63–103. Springer Berlin Heidelberg, Berlin, Heidelberg (2004). https://doi.org/10.1007/978-3-662-05088-0_3

Deep Learning Approach in Predicting Personal Traits Based on the Way User Type on Touchscreen

Soumen Roy, Utpal Roy and D. D. Sinha

Abstract The aim of this paper is to explore the possibility of revealing two common personal traits: age group (<18/18+) and gender (male/female) by measuring and analyzing the way of typing a simple daily-used text on the touchscreen. Deep learning method has been used to develop the model and LOUOCV (Leave-one-user-out cross-validation) has been used to evaluate the effectiveness of our model on the dataset created by 92 volunteers through a web application. Our method outperforms the entire model developed so far as per our knowledge. Accuracy more than 98% in identifying age group and more than 88% in identifying gender have been observed. Automatic age group and gender recognition could be used in a large number of possible application areas such as human-computer interaction, digital forensics, age-specific access mechanism, and targeted advertisement. The way user type on touchscreen concerns with the different issues in user identity verification has been studied well in literature but deep learning approach in revealing the age group and gender based on the typing pattern particularly on a touchscreen is an original idea. Analysis of data, results, and inferences here give a primary account toward achievability to the goal with ample scope for further work in this domain.

Keywords Biometrics · Keystroke dynamics · Deep learning · Personal traits · LOUOCV

S. Roy (✉) · D. D. Sinha
Department of Computer Science and Engineering, University of Calcutta, 92 APC Road, Calcutta 700 009, India
e-mail: soumen.roy_2007@yahoo.co.in

D. D. Sinha
e-mail: devadatta.sinha@gmail.com

U. Roy
Department of Computer and System Sciences, Visva-Bharati, Santiniketan 731235, India
e-mail: roy.utpal@gmail.com

A. K. Das et al. (eds.), *Computational Intelligence in Pattern Recognition*, Advances in Intelligent Systems and Computing 999,
https://doi.org/10.1007/978-981-13-9042-5_27

1 Introduction

Touch dynamics data are used to identify the user by the way the user touch the screen while typing, it is also possible to identify the traits of the user from that primary biometric data. The aim of this paper is to investigate the age group (child/adult) and gender (female/male) identification automatically based on monitoring and analyzing the way of typing a simple daily-used text on touchscreen primarily. Here, the child is considered as a user below age 18 and adult is considered as a user having age 18+, and gender is interchangeably used with sex. Machine learning method-deep learning was used to train and evaluate the effectiveness of our approach. More practical classification measurement was performed to verify the model performance. The experimental results on the dataset created by a study [1] from 92 users indicate that the typing pattern on a touchscreen can be used to predict such traits covertly instead of using face print, fingerprint, iris, hand geometry, and voice. It is the cheapest biometric solution and no extra effort is needed in data acquisition to predict such traits where users may not aware that their typing pattern is being measured and analyzed. It has many potential application areas such as human-computer interaction, digital forensics, surveillance system, social network, and targeted advertisement. Incorporation of predicting score can also be used to improve the performance of keystroke dynamics enables accurate and time efficient keystroke dynamics biometric systems.

Nowadays, automatic age group and gender identification is an active area of research in order to enhance the accuracy of face print, fingerprint, hand geometry, and iris recognition. Smartphone with all amenities are now available at low cost and becoming more common and popular electronics gadget. Therefore, it is more appropriate to identify the age group and gender through a smartphone. Some of the few researchers have put their effort in this area and most of the time typing patterns on a conventional keyboard were used to predict such traits.

Most popular and recently Kaggle (www.kaggle.com) machine learning competition dominated method was used to develop our model as well as to evaluate the efficiency of that model. Each model has been trained and evaluated 9200 times for each identifying trait. This study says that if we type "Kolkata" 7 times in one session then our approach can identify the age of more than 98% correctly where gender can be predicted 88% accurately. The popularity of deep learning technique is presented in Fig. 1. The "deepnet" package [2] is a very lightweight tool in R developed by Xiao Rong which was used to develop and evaluate the model.

Here machine learning is all about mapping inputs (touch dynamics pattern) to target (female/male, and child/adult), which is done by observing a large number of examples (labeled samples). Deep learning maps the inputs to targets via a deep sequence of data transformations (layers). Here each layer is initialized by some weights which correctly mapped the inputs and their associated targets. Finding the right values for each weight is the main job has been done during training.

Touch dynamics biometrics has added features compared to another type of biometrics characteristics because various sensors like gyroscope and accelerometer are available in recent mobile which gives an added opportunity to optimize the model

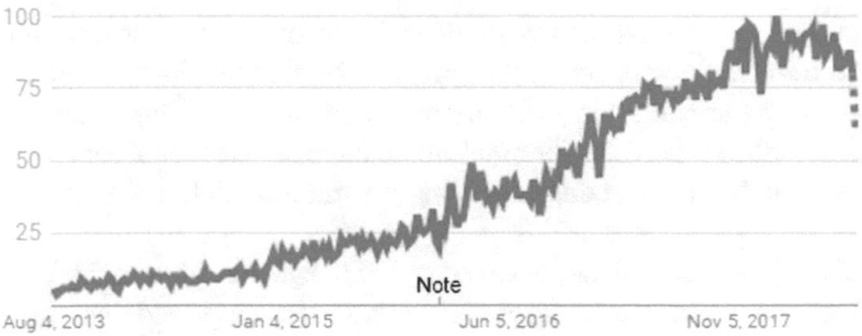

Fig. 1 Deep learning interest over time in Google trends

performance by feature-level fusion approach. In our experiment, only timing features were used to categorize age group and gender of the user. The way of typing of one category is identical and adequate dissimilar to other categories. Hand geometry, personality, familiarity level and involvement level on the touchscreen, neurophysiological, and neuropsychological parameters prominent and hidden tell the science behind it.

Identification of age group vis-a-vis gender automatically is very much imperative in today's fast world. This particular piece of work is an indicative pathway toward that goal based on the activities on a smartphone. This paper explains a solution for supporting the age group and gender identification automatically.

The major goal and contributions of the paper are as follows:

- Create a separate model to predict the age group and gender of the user based on the way they type on a touchscreen.
- Explain the feasibility of predicting personal trait by deep learning approach in keystroke dynamics domain.

The paper is combined as follows. Section 2 provides an overview of keystroke dynamics and the methodology used in literature to extract soft biometric trait based on the way of typing in general. The methodology of our study been discussed in Sect. 3. We present the results and statistics in Sect. 4. Finally, Sect. 5 explains the achievability to the goal with ample scopes for further work.

2 Identifying Traits in Literature and Analysis

Identifying gender and age group by measuring and analyzing gait, body, and facial images are common. Here we focused on keystroke dynamics pattern. In keystroke dynamics research, researchers spend more time in the data acquisition section rather than addressing challenging issues because data acquisition is the most fundamental and essential section of any behavioral biometric system. As a result, various datasets have been produced but none of this is considered to be balanced due to many factors ignored during the creation of dataset. Separate datasets have been created with different experimental setups in 37 years of on-going research and different evaluation criteria have been controlled on each. Researchers have developed datasets considering only their temporal requirement and method of application, this type of datasets are not so neither potential nor balanced. We need very versatile well balanced, reasonable large, containing many variations data set to make the study fruitful.

The study [3], proposed a model to recognize the gender identity based on the way a user type on a conventional keyboard. The study also concluded that this identifying trait significantly improves the keystroke dynamics user identification system. A similar study [4], has revealed the gender identity along with age group (≤30/30 + or ≤32/32+), hand(s) used, and handedness (left/right) of the user based on the dataset created by a conventional computer keyboard. But nowadays smartphone with various features are available at low cost and common and popular. Therefore, it is more significant to identify the traits based on the way a user type on the touchscreen.

Some of the few studies have worked on automatically gender identification using activities on a smartphone. As per the study [5], Gender can be reliably predicted by the way user strike on a touchscreen phone with the accuracy of 64.76% where 57.16% of accuracy is achieved on the swipe dataset collected from the same device. Keystroke dynamics data were collected from 42 users where swipe data were collected from 98. Random Forest (RF) classification algorithm was used to evaluate the model in different cross-validation methods. LOUOCV and 10-Fold Cross-Validation test options were used during the evaluation of the system proposed by them. Based on the limited dataset they have concluded that these biometric data can be used to predict the gender reliably. Obtained results are impressive but not acceptable in practice. It needs further study.

Another study [6], presented on approach for gender identification using behavioral biometrics pattern generated through smartphones. They have used an accelerometer and gyroscope sensor to capture the acceleration and gyroscope data while walking a user in three different modes: normal, fast and, slow speed. They have used the data collected from 42 subjects. They obtained the accuracy 72–75% maximum. It is possible to recognize the gender of the user base on the activities on the web browser as per the study [7]. They obtained 80% of accuracy in tenfold cross-validation test option. In cross-validation, the instances of one subject may be distributed among training and testing set. Overlapping the instances in both training and testing set increases the evaluation performance, which is not significant in practice.

A study [8], found that identifying a child is possible by the way of typing on a conventional keyboard as well as a touchscreen. Another study [9], extremely defined the possibilities of revealing many soft biometric features based on keystroke dynamics.

As per the literature is concerned, gender, age group, handedness, and hand(s) used can be explained through the typing pattern on a conventional keyboard but few of studies used LOUOCV method in model evaluation. But most of the researchers follow the cross-validation test option in system evaluation. This is not the effective evaluation process as per the study [5].

3 Research Methodology and Approach

The goal of this study is to develop a soft biometric keystroke dynamics benchmark dataset labeled by their subject identity, gender, and age group information. Therefore, we have developed a web-based application to collect the typing pattern of users in the uncontrolled environment. Then we have executed that application through a smartphone. Our second objective was to develop a suitable predicting model for each trait. Our third objective was to evaluate our model in a more appropriate system evaluation test option and compare with other machine learning techniques. The following series of steps have been followed in our experiment. The protocol used in our experiment are described in details.

3.1 Data Acquisition Protocol and Distribution of Classes

A web application was developed to acquire the data. For the present purpose, we have executed the application in a smartphone. The details of the dataset explained by the Authors in another paper [1].

All volunteers were advised to type a word "Kolkata" 7 repeated times in one session. The daily-used simple text was selected in order to get the habitual and consistent typing pattern. The samples from various categories are described in Fig. 2. It is clear from the chart that the samples of different classes are not properly distributed. However, variation in subject categories has been maintained. Key press and release time of all the entered characters were recorded during typing to get the timing sequences of key hold and latency time.

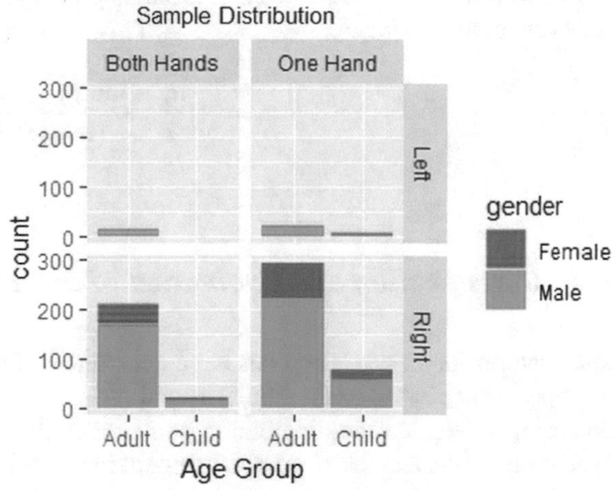

Fig. 2 Sample distribution of different classes in our dataset

3.2 Extraction and Selection of Feature

In keystroke dynamics, feature extraction method measures the data by transforming the typing behavior. Nowadays, various sensors are added to each smartphone which gives the added features like gyroscope and accelerometer. But, in this study, only the timing features were captured as key press and release time. We have used feature extraction method as per the suggestion suggested by the study [10]. The key hold duration time (KD), the time interval between two subsequent presses (DD), the time interval between two subsequent released (UU), the time interval between one key pressed and next key released (DU) and the time interval between one key release and the next key pressed (UD) were captured to be used in our experiment as feature subset.

Selection of universal features is also important to train the machine learning model faster and it reduces the complexity of the model. The selection of proper features also deals the overfitting problem. This part of the study is necessary when the dataset is dimensionally complex.

3.3 Pre-processing

Behavioral biometric characteristics like keystroke dynamics may change during one session or between two subsequent sessions. It depends on the mental and physical state of the person. Therefore, detecting the outliers and cleaning the data is necessary. Without this process, the performance of the model may vary significantly. We have detected the outliers (the value not in between the 1st quartile and 3rd quartile) and replace the outliers by the mean value of the multiple samples for each feature. Then normalization process is carried out. Equation 1 was used as a normalization process for faster computing.

$$x_i = \frac{x_i - \min(x)}{\max(\mathrm{x}) - \min(\mathrm{x})} \tag{1}$$

3.4 Classification and Evaluation Method

Once the pre-processing step has been completed, classification of age and gender were performed based on the similarity and variations among the dissimilar instances. Nowadays, deep learning method is common in the biometric pattern recognition system due to the fact that the performance of machine learning methods is impressive and operative in practice. Deep learning method has been applied to our created dataset. The performance was recorded with the default parameters of the deep learning technique which is available in "deepnet" package in R. Before developing the model for identifying each trait, we have divided the training and testing datasets subject wise. For each time, instances of one subject were used as the test dataset and remaining instances for the other subjects were treated as training dataset. This process was continued for each subject. To remove the imbalance in the training sample, we have selected the instances of all the minority class and randomly selected the majority class in order to balance the training set. The under sampling method is executed 100 times for each subject by taking a random selection of instances from the majority classes. The model was trained 9200 times for identifying each trait. The classifiers performances were recorded in different performance metrics. The average results were defined in this paper with a 95% Confidence Interval (CI).

LOUOCV method is used in the evaluation process, but there is a chance of getting an imbalanced training dataset and sometimes it influences the model performance. To overcome the issue, we have selected the instances from majority classes randomly in the same size as the number of instances presented in minority classes. The method is evaluated 100 times to get the variation in the obtained results.

3.5 *Performance Metrics*

Different metrics were used to evaluate the proposed model which shows how the model is effective against test samples which tells the suitability of the model in a different application. In this study, Accuracy, Sensitivity, Specificity, Area Under Curve (AUC), and Receiver Operating Curve (ROC) were used to measure evaluation performance of the proposed model. CI at 95% defined by Eq. 2 was used to define the vagueness linked with each metric.

$$CI = Mean\,value \pm 1.96\frac{\sigma}{\sqrt{N}} \tag{2}$$

Here, σ represents the variant and N represents the number of tries by changing training and testing samples.

4 Experimental Results

The obtained results of our proposed approaches are represented in Fig. 3 where all the performance evaluation metrics are presented for identifying each trait. Four important metrics were used to determine the performance of our approach to extract the personal identity of the user. The system was evaluated 100 times and all the results are depicted in the following graph. We have used violin chart to get the better observation in multiple evaluations. It is clear from the figure that the performance

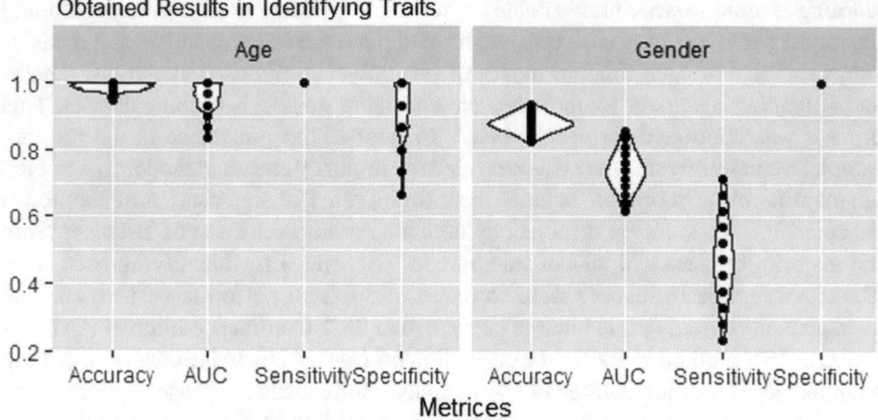

Fig. 3 Performance of our approaches in identifying traits

of identifying the age group and gender for a single predefined text is impressive and can be used in practice.

Figure 4 indicates the probability plot. Here two components are accuracy and the probability in percentage. In the first figure (a), the probability of predicting accuracy in age group identification is presented where figure (b) represents the probability

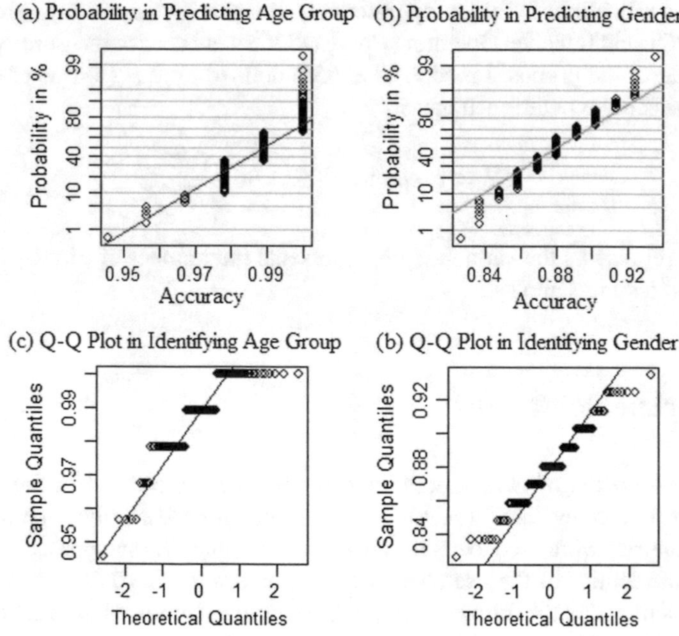

Fig. 4 Probability and Q–Q plot in predicting personal traits

Table 1 Obtained results in identifying soft biometric traits

Traits	Accuracy (%)	Sensitivity (%)	Specificity (%)	AUC (%)
Age group	98.74 ± 0.24	100 ± 0.00	92.27 ± 1.49	96.13 ± 0.74
Gender	88.08 ± 0.48	47.76 ± 2.10	100 ± 0.00	73.88 ± 1.05

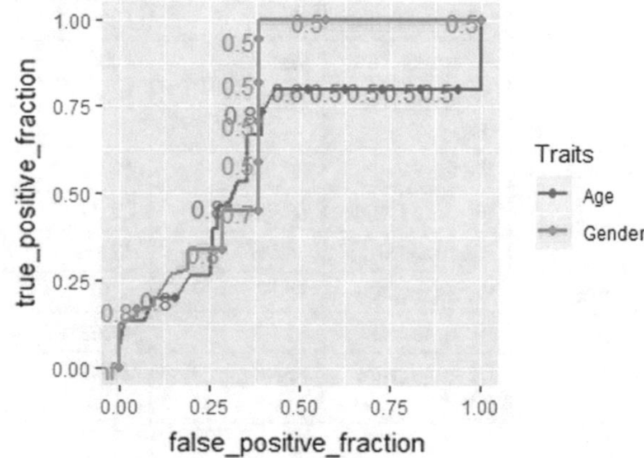

Fig. 5 ROC in identifying traits

of predicting accuracy in identifying gender, figure (c) represents the Q–Q plot in identifying age group where figure (d) represents the Q–Q plot in identifying gender.

Accuracy, Sensitivity, Specificity, and AUC were recorded with CI in identifying personal traits presented in Table 1. It is clear from the table that accuracy of identifying age group is 98.74% and identifying gender is 88.08%. The obtained results are impressive for considering a single short predefined text. The ROC of the obtained results is presented in Fig. 5.

Table 2 indicates that optimum results can be achieved by considering all the timing features instead of considering only key hold time or only flight time.

5 Conclusions and Future Work

Performance of the deep learning approach in predicting the probability of identifying age group and gender based on the way a user type on the touchscreen has been clearly described. As per the obtained results, typing pattern of the user below age 18 is not similar with the typing pattern of the adult user. Consequently, they can be separated by their typing style automatically. Physical characteristics such as the length of a finger have also a direct effect on the system. Performance of identifying gender is not acceptable in practice but impressive for a single short text. In a study [11], they found that user with longer thumps completes swipe gesture with shorter times,

Table 2 Timing features analysis in identifying each trait

Traits	Timing features	Accuracy (%)	Sensitivity (%)	Specificity (%)	AUC (%)
Age group	Vector	**98.74**	**100**	**92.27**	**96.13**
	PR time (KD)	90.22	100	40.00	70.00
	RR time (UU)	91.30	100	46.67	73.33
	PP Time (DD)	92.39	100	53.33	76.67
	RP time (UD)	92.39	100	53.33	76.67
	Di-graph time (DU)	88.04	100	26.67	63.33
Gender	Vector	**88.08**	**47.76**	**100**	**73.88**
	PR time (KD)	83.70	28.57	100	64.29
	RR time (UU)	86.96	42.86	100	71.43
	PP time (DD)	86.96	4.29	100	71.43
	RP time (UD)	88.04	47.62	100	73.81
	Di-graph time (DU)	83.40	28.57	100	64.29

higher speed, and with higher acceleration than a user with a shorter thump. This can be used in gender identification as they suggested.

In this paper, experimental setup and obtained results have been clearly described to extract age group and gender. There are many parameters prominent and hidden that pose a challenge for future redressing. Recently various sensors like gyroscope and accelerometer embedded in the smartphone give an added opportunity for further development in this domain. This particular piece of work is an indicative pathway toward age group and gender recognition based on most cheaply touch dynamics characteristics which could be used in a large number of applications.

References

1. Roy, S., Roy, U., Sinha, D.: Identifying Soft Biometric Traits Through Typing Pattern on Touchscreen Phone. In: Social Transformation—Digital Way Identifying, Kolkata: Springer Nature Singapore Pte Ltd., pp. 1–16 (2018)
2. Rong, X.: Package 'deepnet,' R Topics Documented (2015). https://cran.r-project.org/web/packages/deepnet/deepnet.pdf. Accessed: 01 Aug 2018
3. Giot, R., Rosenberger, C.: A new soft biometric approach for keystroke dynamics based on gender recognition. Int. J. Inf. Technol. Manag. Spec. Issue Adv. Trends Biometr **11**, 1–16 (2012)
4. Syed Idrus, S.Z., Cherrier, E., Rosenberger, C., Bours, P.: Soft biometrics for keystroke dynamics: Profiling individuals while typing passwords, Comput. Secur. **45** (2014)
5. Antal, M., Nemes, G.: Gender recognition from mobile biometric data. In: 11th IEEE International Symposium on Applied Computational Intelligence and Informatics, pp. 243–248 (2016)

6. Jain, A., Kanhangad, V.: Investigating gender recognition in smartphones using accelerometer and gyroscope sensor readings. In: Proceedings of International Conference on Computing Tech. Inf. Commun. Technol. ICCTICT 2016, pp. 597–602 (2016)
7. Kolakowska, A., Landowska, A., Jarmolkowicz, P., Jarmolkowicz, M., Sobota, K.: Automatic recognition of males and females among web browser users based on behavioural patterns of peripherals usage. Internet Res. **26**(5), 1093–1111 (2016)
8. Roy, S., Roy, U., Sinha, D.D.: ACO-Random forest approach to protect the kids from internet threats through keystroke. Int. J. Eng. Technol. 2–9 (2017) (Accepted)
9. Roy, S., Roy, U., Sinha, D.D.: User authentication: keystroke dynamics with soft biometric features. In: Internet of Things (IOT) Technologies, Applications, Challenges and Solutions, 1st ed., C. P. T. & F. Group, Ed. Boca Raton, FL 33487–2742, 2017, pp. 105–124
10. Roy, S., Roy, U., Sinha, D.D.: Comparative study of various features-mining-based classifiers in different keystroke dynamics datasets. In: Smart Innovation, Systems and Technologies (2016)
11. Bevan, C., Fraser, D.S.: Different strokes for different folks? Revealing the physical characteristics of smartphone users from their swipe gestures. Int. J. Hum Comput. Stud. **88**, 51–61 (2016)

Robust Non-blind Video Watermarking Using DWT and QR Decomposition

Chinmay Maiti and Bibhas Chandra Dhara

Abstract In this article, a robust and non-blind watermarking technique is presented for video data. In the embedding step, a gray-scale image is embedded into the key frames using DWT and QR decomposition. Before embedding, watermark is scrambled using Fibonacci-Lucas transform to enhance the security level of the watermark. The result of the experiment established that the proposed method is robust against different attacks like salt-and-pepper noise, rotation, Gaussian noise, blurring, cropping, frame averaging, compression, etc. The performance of the proposed technique is also compared with existing methods and it is established that current methods perform better than the existing methods.

Keywords Video watermarking · DWT · QR decomposition · Robustness

1 Introduction

With advancement of multimedia and communication technologies, large amount of digital data such as images, videos, etc. [1] are transferred easily from one end to other end through internet. As a result unauthorized access to those data are increasing day by day. Hence, it is necessary to restrict unauthorized access by the illegal users [2]. Over the last two decades, the film industries are facing one of the biggest problem known as camcorder theft. It is the single largest source of video piracy. More than 90% of the newly released films are pirated using the digital cam-cording device in a movie theater and then it is distributed or shared, through internet, to the people

C. Maiti (✉)
Department of Computer Science & Engineering, College of Engineering & Management, Kolaghat, Purba Medinipur 721171, India
e-mail: chinmay@cemk.ac.in

B. C. Dhara
Department of Information Technology, Jadavpur University, Kolkata 700098, India
e-mail: bcdhara@gmail.com

A. K. Das et al. (eds.), *Computational Intelligence in Pattern Recognition*, Advances in Intelligent Systems and Computing 999,
https://doi.org/10.1007/978-981-13-9042-5_28

without any copyright protection [3]. As a result, camcorder causes lost revenue for theater owners and producers.

Thus, digital video watermarking can be an effective solution to protect the video files from unauthorized access. In video watermarking technique, watermark information is embedded into the video frames by changing the content of video frames. Digital watermarking techniques are classified with respect to different view points. Firstly, the watermark techniques can be classified as spatial domain methods [4] and transform domain methods [5, 6]. Secondly, they can be classified as non-blind [7], semi-blind [8], and blind [9] methods depending on the information required during extraction of the watermark.

Video watermarking has certain factors which are not found in image watermarking. The data volume of video is huge and there is high correlation with successive frames and due to this videos are susceptible to piracy attacks, such as frame averaging, frame dropping, frame swapping, etc [10]. Most of the current video watermarking techniques are developed based on image watermarking. However, these methods are not good enough for copyright protection in video data. Watermarking of a video can be performed either in uncompressed video or in compressed video. Here, we have proposed a video watermarking technique in uncompressed video. Video watermarking in uncompressed video frames can be classified as spatial and transform domain based techniques.

The remaining part of this article is organized as follows: Sect. 2 describes the related techniques. Section 3 explores the proposed technique. The performance of the proposed technique is presented in Sect. 4. The conclusions are drawn in Sect. 5.

2 Related Technique

In this section we have described the techniques such as Fibonacci-Lucas transform, Discrete Wavelet Transform (DWT) and QR decomposition which are useful in our proposed method.

2.1 Fibonacci-Lucas Transformation

In this work, we have used Fibonacci-Lucas [11, 12] transformation to scramble the binary watermark image for enhancing the confidentiality of the proposed method. The Fibonacci-Lucas transform (FLT) is defined in Eq. 1.

$$\begin{pmatrix} x_{new} \\ y_{new} \end{pmatrix} = \begin{pmatrix} F_k & F_{k+1} \\ L_k & L_{k+1} \end{pmatrix} \begin{pmatrix} x \\ y \end{pmatrix} mod\ N \tag{1}$$

where image size is $N \times N$ and F_k and L_k represent the kth Fibonacci term and kth Lucas term, respectively.

The Fibonacci-Lucas transform (FLT) matrix is given by

$$FLT_k = \begin{pmatrix} F_k & F_{k+1} \\ L_k & L_{k+1} \end{pmatrix}$$

All of these transform will be periodic in nature with different periodicity and hence will produce different scrambling pattern. So, one can use Fibonacci-Lucas transform as more secured scrambling method.

2.2 Discrete Wavelet Transform

The discrete wavelet transform (DWT) is accepted in image watermarking since it has the capacity to decompose a signal in time/frequency domain, which match with the theoretical models of the human visual system [10]. Figure 1 shows the different sub bands of the image after applying 2D DWT twice. Where, LL represents the low-frequency sub bands, HH represents the high-frequency sub bands, HL and LH represent the middle-frequency sub bands.

Different researchers have used different sub bands depending on their requirements and applications. Many watermarking techniques [13] choose low- frequency sub band for embedding watermark because it shows more robust to different attacks. But, watermark in low frequency may degrade the image quality significantly. Some researchers have proposed watermarking techniques where high- frequency band is selected to embed the watermark since high frequency bands have edge and texture information and our vision system is less sensitive to these. Embedding watermark in the high frequency shows better image quality, but robustness is compromised. The selection of the sub band for embedding watermark depends on the requirement of the researchers with respect to the application.

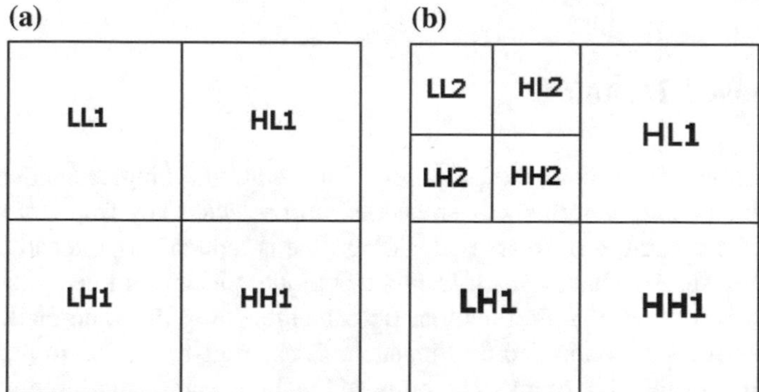

Fig. 1 2-level decomposition of the image using DWT

2.3 QR Decomposition

In linear algebra, a matrix $B_{m\times n}$ can be factored as $B = QR$, if B has linearly independent columns and in that case Q is an orthogonal matrix with size $m \times n$ and R is an upper triangular matrix of size $n \times n$. This factorization is known as QR decomposition. The matrix Q can be computed from B by applying Gram-Schmidt orthonormalization process. Suppose $B = [b_1, b_2, \ldots, b_n]$ and $Q = [q_1, q_2, \ldots, q_n]$, than q_i is computed as $q_i^{'} = b_i - \sum_{j=1}^{i-1} < b_i, q_j >$ and $q_i = \frac{q_i^{'}}{||q_i||}$ where, $< x, y >$ denotes the inner products of two vectors x and y.
The matrix R is computed using the Eq. 2.

$$R = \begin{bmatrix} < b_1, q_1 > & < b_2, q_1 > & \ldots & < b_n, q_1 > \\ 0 & < b_2, q_2 > & \ldots & < b_n, q_2 > \\ \vdots & \vdots & \ddots & \vdots \\ 0 & 0 & \ldots & < b_n, q_n > \end{bmatrix} \tag{2}$$

Here, the columns of Q form a new n-dimensional orthonormal basis and from Eq. 2 it obvious that R is the projection of B onto the basis Q. In this experiment, B is an image block. The matrix R has shown some interesting properties with respect to the matrix B. It can be proved that with a high probability, the absolute value of the elements of the first row of R is comparatively larger than other rows [14]. Therefore, due to a tiny change in first row, the change in image block B will be insignificant and cannot be visually tracked. When B is taken from a smooth region, the columns of B are more or less same and due to QR decomposition mostly insignificant (near to zero) values will be stored in 2nd row onwards in the matrix R. Therefore, except first row, other rows are not suitable for embedding watermark as visual quality of the image block will degrade significantly. The elements of the first row of R are used as trade-off between imperceptibility and robustness in designing effective watermarking techniques.

3 Proposed Technique

In this section, we discuss the proposed video watermarking technique. In this proposed technique, watermark is embedded into selected key frame of the video sequence. As a result, watermark embedding time is reduced significantly. The key frame is selected by chaotic map. In this technique, piecewise linear chaotic map (PWLCM) is used as chaotic function. Before embedding the watermark into key frame, watermark is scrambled by Fibonacci-Lucas transformation to increase the security level of the watermark. The proposed technique is a non-blind watermark technique as original video sequence is needed at the time of extraction. The proposed method is based on DWT and QR decomposition method. The PWLCM function is described in the following section followed by embedding and extraction process of the proposed technique.

3.1 The PWLCM Function

The piecewise linear chaotic map is used to generate the random numbers [15]. The PWLCM is a function that is described by multiple linear segments. The function is given in Eq. 3.

$$y_i = f(y_i, \beta) = \begin{cases} \frac{y_{i-1}}{\beta}, & 0 < y_{i-1} < \beta \\ \frac{y_{i-1}-\beta}{0.5-\beta}, & \beta \leq y_{i-1} < 0.5 \\ f(1-y_{i-1}, \beta), & 0.5 \leq y_{i-1} < 1 \end{cases} \tag{3}$$

where β is the control parameter and value of $\beta \in (0, 0.5)$. The value of $y_i \in (0, 1)$ and first value of it is the initial condition of the function. The PWLCM has good balance and is much closer to the uniform distribution [15]. This function is used to identify the key frame during embedding process.

3.2 Embedding Process

In the proposed technique, the watermark is embedded only into the key frame. The key frame is selected from the video sequence using PWLCM function and compute the frame difference respect to the key frame. The 2D Haar wavelet is applied on the key frame and four sub bands (LL1, LH1, HL1, HH1) are obtained, then LL1 sub band is chosen for the second level decomposition which gives the sub bands (LL2, LH2, HL2, HH2) (Fig. 1). The HH2 sub band is selected for further decomposition by QR decomposition method. At the same time, watermark image is also decomposed by 2D Haar wavelet up-to first level and then QR method is executed on the HH sub band. The elements of the first row of the matrix R of key frame is modified by the first of row of R matrix of watermark by using a linear function with a embedding strength α. The watermarked frame is obtained after inverse QR and inverse DWT process. Finally, all the frame differences are added to this watermarked frame to obtain watermarked video sequence. Block diagram of the embedding process is shown in Fig. 2 and the steps are discussed below.

Algorithm

1. Extract the frames from the original video one by one.
2. Identify the Key frame using PWLCM chaotic map and subtract the key frame from all other frames of the video.
3. Apply DWT on the key frame twice (Fig. 1).

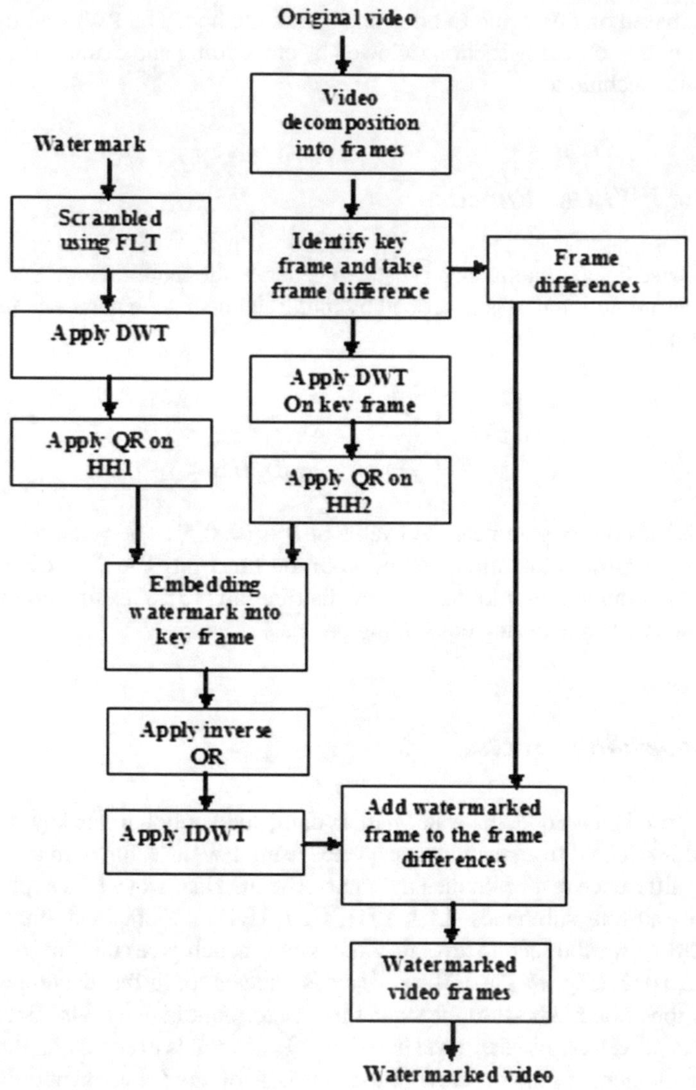

Fig. 2 Embedding process of the proposed video watermarking technique

4. Apply QR decomposition on HH2 sub band, i.e., $[Q_f\ R_f] = qr(HH2)$.
5. Apply DWT on the watermark image only once and QR decomposition is applied on high frequency sub band which gives us Q_w and R_w.
6. Embed the watermark following the logic as given in Eq. 4

$$R^{'} = R_f + \alpha \times R_w \tag{4}$$

7. Then compute inverse QR and inverse DWT to obtain watermarked frame.
8. Finally, add watermarked frame to all the frame differences to obtain watermarked video sequences.

3.3 Extraction Process

The watermark extraction is the reverse process of embedding. Since the current technique is non-blind, DWT is applied on watermarked video and original video as the step 3 of the embedding process and then QR decomposition method is applied to the high-frequency sub band of level-2 as step 4 of embedding process. The R matrix of watermark is computed from the R matrix of watermarked frame and original frame. The resultant R matrix is used to get the watermark.

4 Experimental Results

In the current method, we have performed the experiments on three video sequences namely "Bus", "Football", and "Foreman". The videos are in uncompressed CIF format with frame size 288 × 352. The frame rate of the videos used in this experiment is 30 frames per second. The number of frames of "Bus", "Football", and "Foreman" are 150, 90, and 300, respectively. In this experiment, Haar wavelet is used twice on the luminance component of the key frame to embed the watermark. A sample frame of the original videos is shown in Fig. 3a–c. The gray scale watermark image of size 144 × 176 is shown in Fig. 3d and the scrambled version of the watermark (with $k = 5$) is shown in Fig. 3e. The watermarked video frames of the proposed technique are shown in Fig. 3f–h.

In order to measure the imperceptibility property of the proposed technique, the metric PSNR and SSIM have used in this experiment. The PSNR values of the watermarked frames (shown in Fig. 3) are all above 65 dB. Therefore, we conclude that the proposed video watermarking technique shows very high imperceptibility. The SSIM values of the watermarked frames are all very close to "1" shown in Table 2. The robustness of the current method is tested considering different attacks such as salt-and-pepper noise, Poisson noise, Gaussian noise, cropping, flipping, frame rotation, sharpening, blurring, frame averaging, and JEPG compression. In Fig. 4a–j, different attacked watermarked video frames are shown and the extracted watermark from the video frames is displayed in Fig. 4k–t. The Table 1 shows correlation coefficient values of the original and watermarked videos.

The performance proposed technique is compared with the techniques of Sharma et al. [16] and Lai et al. [17]. These two existing techniques have embedded the watermark using DWT and SVD. The PSNR and SSIM values of the watermarked videos by the different techniques are shown in Table 2. It is found that the proposed technique shows better video quality than the other two techniques. The correlation

Fig. 3 **a** Original frame of Bus, **b** original frame of Football, **c** original frame of Foreman, **d** original watermark image, **e** scrambled watermark image, **f** watermarked frame of Bus (PSNR = 65.31), **g** watermarked frame of Football (PSNR = 65.39), **h** watermarked frame of Foreman (PSNR = 65.11)

coefficients of the attacked watermarked videos for different techniques are given in Table 3. It is observed that performance of the proposed method is same as Sharma et al. [16] method whereas our method performs better than Lai et al. [17].

5 Conclusion

The present technique is a non-blind watermarking scheme. The proposed video watermarking technique is based on DWT and QR decomposition. In this method, watermark is added only into a single frame which is selected by a chaotic sequence. Therefore, the embedding time is reduced remarkably. Also, the proposed method results in high quality watermarked video as the PSNR value for all test videos are above 65 dB and the SSIM value are very close to "1". This technique is also robust under various attacks. To enhance the security of the watermark, the proposed technique scrambled the original watermark by Fibonacci-Lucas transform before

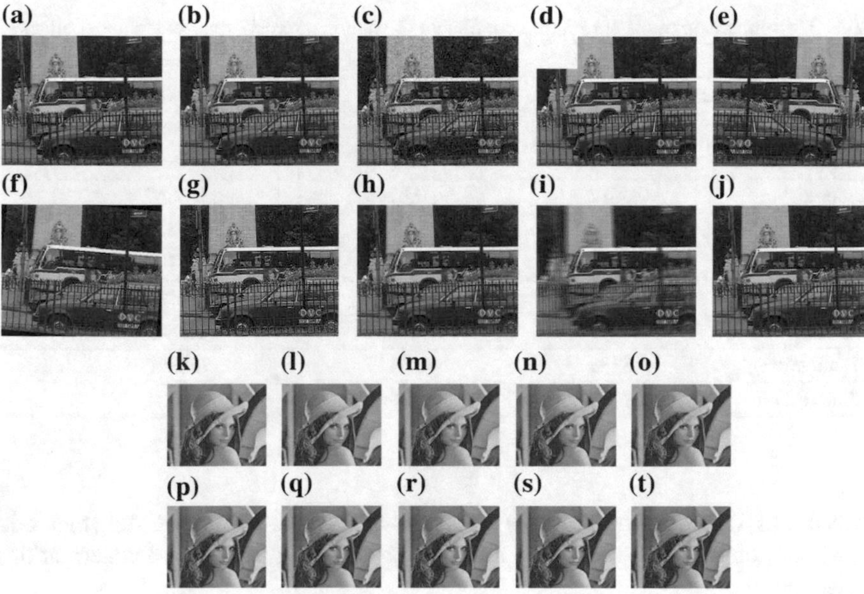

Fig. 4 **a–j** Attacked watermarked frames of Bus by different types of attack and **k–t** extracted watermark image from the corresponding attacked video frame

Table 1 Correlation coefficient values of watermarked videos under attacks: Bus, Football and Foreman

Attack name	Bus	Football	Foreman
Salt & pepper noise	0.9821	0.9650	0.9824
Poisson noise	0.9864	0.9577	0.9731
Gaussian noise	0.8945	0.7880	0.8921
Cropping	0.7487	0.6912	0.9406
Flipping	1.0000	1.0000	1.0000
Frame rotation	0.4905	0.4257	0.6533
Sharpening	0.9212	0.9374	0.9632
Blurring	0.9995	0.9997	0.9998
Frame averaging	0.9225	0.8384	0.9935
Compression	0.9931	0.9933	0.9962

Table 2 Comparison of PSNR and SSIM values of the watermarked videos by different techniques

Video	Proposed (PSNR, SSIM)	Sharma et al. [16] (PSNR, SSIM)	Lai et al. [17] (PSNR, SSIM)
Bus	65.31, 0.9999	60.72, 0.9665	37.68, 0.9139
Football	65.39, 0.9997	63.09, 0.9998	37.72, 0.9449
Foreman	65.11, 0.9999	62.38, 0.9888	37.69, 0.9289

Table 3 Comparison of the correlation coefficient of watermarked Football video under attacks

Attack name	Proposed	Sharma et al. [16]	Lai et al. [17]
Salt & pepper noise	0.9650	0.9615	0.9482
Poisson noise	0.9577	0.9574	0.9425
Gaussian noise	0.7880	0.7833	0.7803
Cropping	0.6912	0.6743	0.6862
Flipping	1.0000	1.0000	0.9823
Frame rotation	0.4257	0.4261	0.4192
Sharpening	0.9374	0.9087	0.8188
Blurring	0.9997	0.9862	0.9865
Frame averaging	0.8384	0.8117	0.8306
Compression	0.9933	0.9903	0.9821

embedding. The performance of the proposed method is better than existing methods. Our future objective is to enhance this method for designing a blind watermarking technique.

References

1. Ramkumar, G., Manikandan, M.: Uncompressed digital video watermarking using stationary wavelet transform. In: Proceedings of IEEE International Conference on Advanced Communication Control and Computing Technologies (ICACCCT), pp. 1252–1258 (2014)
2. Bhardwaj, A., Verma, V.S., Jha, R.K.: Robust video watermarking using significant frame selection based on coefficient difference of lifting wavelet transform. Multimed. Tools Appl. 1–20 (2017). https://doi.org/10.1007/s11042-017-5340-3
3. Asikuzzaman, M., Alam, M.J., Lambert, A.J., Pickering, M.R.: Imperceptible and robust blind video watermarking using chrominance embedding: a set of approaches in the DT CWT domain. IEEE Trans. Inf. Forenscis Secur. **9**(9), 1502–1517 (2014)
4. Yang, C.N., Lu, Z.M.: A blind image watermarking scheme utilizing BTC bitplanes. J. Digit. Crime Forensics **3**(4), 42–53 (2011)
5. Sharama, H., Kumar, A., Mandoria, H.L.: Study and comparison analysis of a video watermarking scheme for different attacks. Int. J. Emerg. Manag. Technol. **4**(9), 51–56 (2015)
6. Li, D., Qiao, L., Kim, J.: A video zero-watermarking algorithm based on LPM. Multimed. Tools Appl. **75**, 13093–13106 (2016)
7. Stutz, T., Autrusseau, F., Uhl, A.: Non-blind structure-preserving substitution watermarking of H.264/CAVLC inter-frames. IEEE Trans. Multimed. **16**(5), 1337–1349 (2014)
8. Thind, D.K., Jindal, S.: A semi blind DWT-SVD video watermarking. Procedia Comput. Sci. **46**, 1661–1667 (2015)
9. Karmakar, A., Phadikar, A., Phadikar, B.S., Maity, G.K.: A blind video watermarking scheme resistant to rotation and collision attacks. J. King Saud Univ. Comput. Inf. Sci. **28**, 199–210 (2016)
10. Agilandeeswari, L., Ganesan, K.: A robust color video watermarking scheme based on hybrid embedding technique. Multimed. Tools Appl. **75**, 8745–8780 (2016)
11. Mishra, M., Mishra, P., Adhikary, M.C., Kumar, S.: Image encryption using Fibonacci-Lucas transformation. Int. J. Cryptogr. Inf. Secur. **2**(3), 131–141 (2012)

12. Zou, J.C., Ward, R.K., Qi, D.X.: A new digital image scrambling method based on Fibonacci numbers. In: Proceedings of the International Symposium on Circuits and Systems, pp. 956–968 (2004)
13. Zhang, L., Li, A.: A study on video watermark based-on discrete wavelet transform and genetic algorithm. In: Proceedings of First International Workshop on Education Technology and Computer Science, pp. 374–377 (2009)
14. Naderahmadian, Y., Hosseini-Khayat, S.: Fast watermarking based on QR decomposition in wavelet domain. In: Proceedings of Sixth International Conference on Intelligent Information Hiding and Multimedia Signal Processing, pp. 127–130 (2010)
15. Xu, L., Li, Z., Hua, W.: A novel bit-level image encryption algorithm based on chaotic maps. Opt. Lasers Eng. **78**, 17–25 (2016)
16. Sharma, S.S., Tanay, C.P., Thapa, S.: A robust color video watermarking technique using DWT, SVD and frame difference. In: Proceedings of International Conference on Pattern Recognition and Machine Intelligence (PReMI 2017), pp. 148–154 (2017)
17. Lai, C.C., Tsai, C.C.: Digital image watermarking using discrete wavelet transform and singular value decomposition. IEEE Trans. Instrum. Meas. **59**(11), 3060–3063 (2010)

Detection and Classification of Earthquake Images from Online Social Media

Ashish Kumar Layek, Amreeta Chatterjee, Debanjan Chatterjee and Samit Biswas

Abstract Natural disasters present a huge challenge for Government and non-Governmental organizations, as the individuals and communities are heavily affected. Disasters like earthquakes are seismic events that result in shaking of the earth's surface causing large-scale death and destruction. Therefore, disaster management and control are of utmost importance. Nowadays individuals on site use social media platforms (like *Twitter, Facebook*, etc.) for sharing information and seeking remedy. If the extent of destruction can be estimated, one can assess the amount of relief needed. Thus, the data gathered from social media platforms can be used to evaluate the extremity of architectural damage induced by disasters like earthquakes. This paper proposes a CNN based segmentation of earthquake affected regions in images that are posted on online social media and classifying them into various classes according to the severity of damage. The approach is trained and tested using a dataset of images collected from social media posts during earthquakes and the results are encouraging.

Keywords Deep learning · CNN · Online social media · Twitter · Earthquake detection · Earthquake image classification

A. K. Layek (✉) · S. Biswas
Indian Institute of Engineering Science and Technology, Shibpur 711103, India
e-mail: ashish@cs.iiests.ac.in

S. Biswas
e-mail: samit@cs.iiests.ac.in

A. Chatterjee · D. Chatterjee
St. Thomas' College of Engineering & Technology, Kolkata 700023, India
e-mail: amritachatterjee96@gmail.com

D. Chatterjee
e-mail: debanjanchatterjee96@outlook.com

A. K. Das et al. (eds.), *Computational Intelligence in Pattern Recognition*, Advances in Intelligent Systems and Computing 999,
https://doi.org/10.1007/978-981-13-9042-5_29

1 Introduction

In modern times, popular Online Social Media (OSM) platforms like *Twitter* and *Facebook* are considered as sources of real time information, which maybe in the form of combination of texts, images, or videos. During some disaster, the social media platforms get flooded with images of the disaster-affected areas posted by the users as well as those who are affected. Apart from the text portion of the tweets, the images associated with it play an important role to understand the extent of damage in the affected areas. Sending relief to affected areas after a disaster is a concern for various organizations, as the extent of damage cannot be determined easily. This leads to inefficiency in response to a crisis. As natural disasters affect human life substantially, this becomes a very important issue that needs to be dealt with. The advanced technology in today's world helps the people at a disaster site to communicate using social media instantly. Pictures are posted from all over the world, which can play a tremendous role in deciding the relief required in an affected area. However, a significant number of posted images in the OSM during a particular disaster event actually doesn't convey any pictorial information about the extent of damages. As for example, half of the posted images during an earthquake might actually be relevant. From a survey of the images posted during the earthquake at Manipur in January 2016 (*MEQ2016* dataset [9]), it is observed that 63% images were not related to earthquake (see Table 1 for details), and the rest 37% of images portrayed significant damage or destruction, caused by the earthquake. It is to be noted that even though the tweets were collected using the search queries like "manipur earthquake", still several non-earthquake images appeared. This is expected because in a earthquake situation, several images of "emergency number", "epicenter location", "relief", "victim" etc. are posted. Examples of few true earthquake images and other kinds of images related to earthquake posted on *Twitter* during earthquake events are shown in Fig. 1.

Research on the utilization of information from social media for disaster control and management [5] has been going on for the past few years. However, most of the existing work focuses on extracting sensitive information from text-based social media content [6]. In [14], locations of *Twitter* users sending critical earthquake-related tweets, are used as sensors, thus the center of the earthquake event can be estimated using particle filtering.

Images can play a crucial role in real-time detection of crises events, and their significance has been demonstrated in [13]. During flooding in Saxony, 2013, posts from Instagram and Flickr were analyzed and relevant ones were filtered out, con-

Table 1 Survey on images posted on *Twitter* during Manipur Earthquake 2016

Dataset	Total images	Distinct images	Earthquake images *Images with debris*	Non-earthquake Images *Images without debris*
MEQ2016	2,342	780	291 (37.30%)	489 (62.70%)

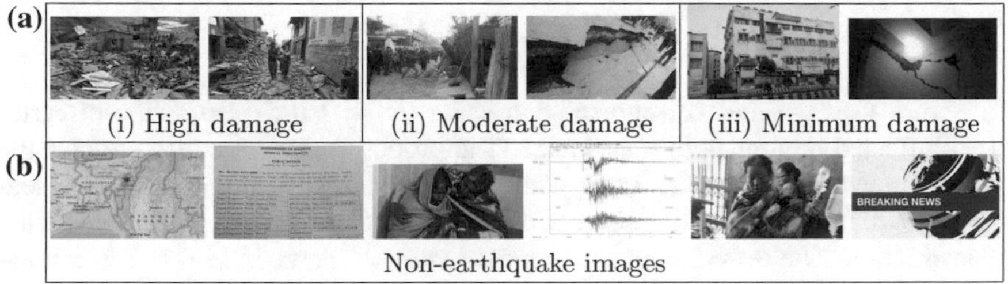

Fig. 1 Sample images posted on *Twitter* during disaster events: **a** with damage due to earthquake and **b** without damage but related to earthquake

cluding a theory that on-topic social media messages that also contain an image indicating a higher probability of relevant content. Work [1, 12] propose a real-time detection of disaster events and classification based on severity of damage. The proposed work focuses on one specific disaster category- earthquakes, unlike the two aforementioned papers which are not disaster specific.

A few other important works related to this domain are [11], which generates real-time crisis maps using a statistical approach. The work presented in [7] has a similar objective to the proposed work, but it uses remote sensing instead of social media images.

However, to the best of our knowledge, there has been no work on identifying the earthquake images out of several other images. In this paper, we have proposed a CNN-based model to detect earthquake-affected regions in an input image. The primary application of this model would be to segregate earthquake images from several other images posted on OSM during a disaster, to help various news media, as well as for various organizations to start acting toward disaster response. If an image is found to be an earthquake image, it is classified on the basis of the amount of affected portion within the image. Here our assumption is that the percentage of damage that appears in the image is directly proportional to the severity of destruction. This is just to get an idea of the magnitude of the earthquake across the image so that the rescue personnel can give a look, based on the severity of the damage. Figure 1a captures few earthquake-related images of different categories, i.e., high, moderate, and minimum damage. An image having a single or multiple completely destroyed buildings is considered to fall under the high damage category (Fig. 1a-i), whereas partially destroyed buildings or walls or road cracks fall in the moderate damage category (Fig. 1a-ii). If an image depicts slight infrastructural damage, e.g., buildings and roads having cracks (Fig. 1a-iii) then it is considered to fall under minimum damage category.

The rest of the paper is presented as follows: Sect. 2 explains in brief the proposed work, Sect. 3 presents the experimental results and evaluation, and finally the paper is concluded in Sect. 4.

2 Proposed Methodology

This work uses the idea of segmentation using CNN [10] for extraction of earthquake affected regions within an image. The proposed CNN architecture consists of some convolutional layers along with some *ReLU* and *Pooling* layers. The segmented earthquake affected regions within the image are used to identify the class of destruction. The overview of the proposed work is shown in Fig. 2. The following subsections brief the *earthquake-detector* CNN model (Fig. 2a) and Earthquake region detection and classification of destruction (Fig. 2b).

2.1 Earthquake-Detector: CNN Model and Training Dataset

Convolutional neural networks (CNNs) [8, 10] are a class of deep, feed-forward artificial neural networks that have successfully been applied to analyze visual imagery. This work proposes a CNN model architecture, that is used to segment regions within image consisting of debris or cracks due to earthquake. The following subsections explain CNN model for earthquake-detector and development of training dataset (see Fig. 2a).

2.1.1 CNN Model Architecture

This work uses the CNN model similar to the LeNet model [10]. It is a stacked structure of three blocks of layers, where each block consists of Convolution, ReLU (*Rectified Linear Unit*), and Pooling one after another. All the convolutional layers are followed by a rectification layer (ReLU) as suggested in [8]. These three blocks are followed by a couple of Fully Connected (FC) layers. The FC layer is equivalent to a convolutional layer, but the size of the filters are the same as the size of the input data. In the last layer, *softmax* [2] function is used for classification into two classes: *earthquake* and *non-earthquake* regions. The details of the layers of the CNN model are summarized in Table 2.

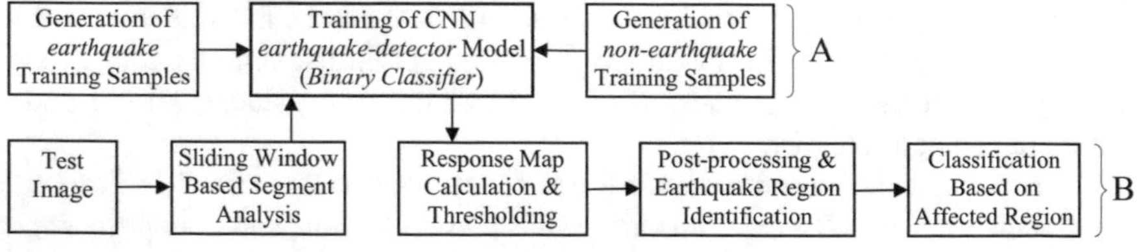

Fig. 2 Overview of the proposed work. **a** CNN based earthquake-detector model generation. **b** Affected region segmentation and classification work flow

Table 2 Layers of convolutional neural network architecture

S. No.	Layer type	Filter dimension	Number of filters	Output dimension
0	Input	–	–	48 × 48 × 3
1	Convolution + ReLU	9 × 9 × 3	96	42 × 42 × 96
2	Average pooling	3 × 3	–	14 × 14 × 96
3	Convolution + ReLU	3 × 3 × 96	128	12 × 12 × 128
4	Average pooling	3 × 3	–	4 × 4 × 128
5	Convolution + ReLU	4 × 4 × 128	256	3 × 3 × 256
6	Average pooling	2 × 2	–	2 × 2 × 256
7	Fully connected	–	–	1 × 1 × 256
8	Fully connected	–	–	1 × 1 × 2
9	Softmax (output)	–	–	1 × 1 × 2

It is to be noted that the parameters, such as number and spatial size of filters at various convolutional layer and the window size at pooling layers are set by performing a few rounds of experimentation and finalized the values of said parameters. For training the CNN model, the network is trained over 45 epochs. In each epoch, the network is trained over 500 batches (each of 100 samples), with the total of 50K training samples. After each training phase of the epoch, the network is validated using the 10K validation samples. The CNN model is implemented using the MATLAB MatConvNet library [15] 1.0 beta 23 version.

2.1.2 Generation of Training Dataset

The proposed CNN model is trained using two different kinds of image samples: (i) *earthquake* and (ii) *non-earthquake*. The training sample dimension is 48 × 48 × 3. A total of 60K samples are generated; keeping half of the samples from each class. We used 50K samples for training and 10K for validation of the CNN model. The training samples are generated from an input image; a window of size 48 × 48 is moved from left to right and from top to bottom and the images under the window are cropped to produce several image patches. Out of the huge number of generated samples, only 30K samples of each type are randomly selected. Note that contrast normalization [4] has been performed on samples before feeding them to the CNN.

Generation of "earthquake" Training Samples: The training samples for *earthquake* are generated from earthquake- related images, collected from the website *www.twitter.com* or *www.google.com* with the suitable keyword pattern:, i.e., "Nepal Earthquake", "Italy Earthquake" etc. We manually cropped only the part of the images where debris of collapsed buildings or road cracks appear. Then the samples are generated by sliding a window in a overlapping manner for the different manually cropped images segments.

However, the entire region of these cropped images do not always represent debris. Figure 3a shows example of one such cropped image of debris due to earthquake

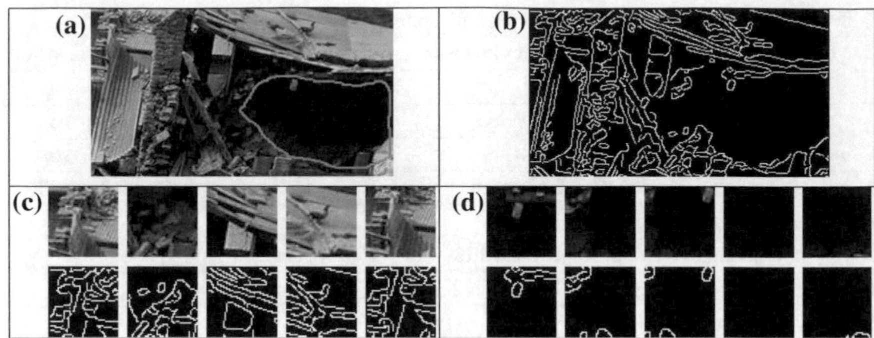

Fig. 3 Illustration of *earthquake* training samples generation process. **a**, **b** Cropped image segment of earthquake debris and equivalent edge detected image; few selected (**c**) and rejected (**d**) *earthquake* samples (along with edges)

(a) Example of *earthquake* samples (b) Example of *non-earthquake* samples

Fig. 4 Examples of few training samples (not in actual scale)

damage; it is seen that the portion marked with green of the image does not depict any characteristics of an earthquake. Hence, just running sliding window to generate training samples may lead to the generation of several patches with homogeneous background. Therefore, there is a need to exclude few samples which are mostly homogeneous in nature; this is done by applying *Canny edge detection* [3] on an input image. The portion of cracks or debris within an image are identified by the presence of edges (see Fig. 3b). While applying sliding window, those image patches are considered as training samples if a certain amount of edge-information is present in it (see Fig. 3a). The window moves from left to right and then from top to bottom in a overlapping fashion, where both the horizontal shift and vertical shift are considered as 16 pixels. A few of the earthquake samples are shown in Fig. 4a.

Generation of "non-earthquake" Training Samples: The *non earthquake* samples are generated by sliding a window in a non-overlapping manner for the different downloaded images of animals, cars, paintings, face, flowers, house, text, trees, etc. All the samples are extracted from a particular image, and in case two or more samples are too similar to each other, only one representative sample is included in the training dataset. This prevents biasing the CNN model in non-earthquake images with a solid color background. Here we have used a near-duplicate image detection technique, which is similar to the clustering technique proposed in [16]. A few of the earthquake samples are shown in Fig. 4b.

2.2 Earthquake Region Detection

In this section, we describe the process of locating regions having presence of debris or destruction due to earthquake in a given image. The process consists of response matrix generation and its analysis for identification of the area of interest. Then the class of destruction is computed based on the amount of affected area. Figure 2b summarizes the earthquake detection technique in brief.

2.2.1 Response Matrix Generation

We follow an overlapping sliding window approach with a shift (stride) of 8 pixels and the window size is 48×48. As a preprocessing step, the input image is padded with 48 pixels of value 0 along each side of the image, so that the responses at border region of the image are not victimized. These padded elements are discarded later to obtain the final response matrix of the actual image.

For each region selected under the window, we fed it to the *earthquake-detector* CNN model. At the *softmax* layer we received a 2×1 vector containing the probability values of the region being an earthquake and non-earthquake region. The region belongs to the class which has a greater probability. For given input image patch, it returns score (or confidence) for both *earthquake* and *non-earthquake* classes and they are denoted by S_{eq} and S_{ne} respectively. The value of S_{eq} (S_{ne}) lies in the range [0, 1] and $S_{eq} + S_{ne} = 1$. In a common practice [15], if the observed score S_{eq} (S_{ne}) is greater than the *threshold* $\mathcal{T}$, where $\mathcal{T} = 0.5$ in case of binary classification; then the input image patch is considered as earthquake (non-earthquake) region. Now the obtained *earthquake* score (S_{eq}) is added to the respective cell of an accumulator called *response matrix*. The size of the response matrix is same as input image (with zero padding) with initial value of each cell set to 0. On completion of the sliding window process, the calculated responses for each element of the matrix is normalized within the range [1..0] and we get normalized response matrix (M_{nr}). Figure 5b shows the normalized response matrix for the input image shown in Fig. 5a.

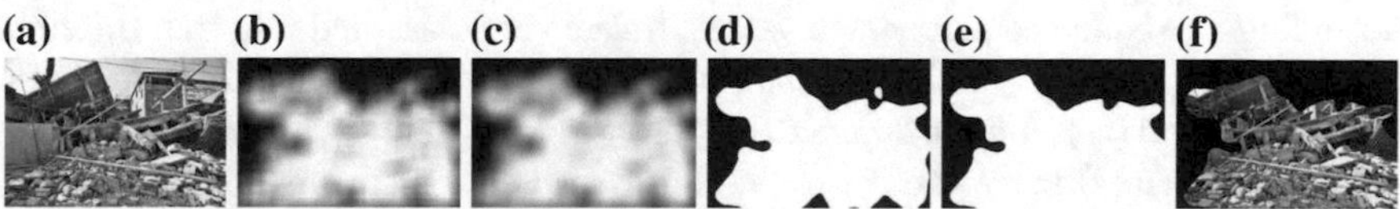

Fig. 5 Illustration of earthquake detection process: **a** Input image, **b** normalized response matrix (M_{nr}), **c** response matrix after Gaussian blurring ($\sigma = 7$), **d** region after thresholding ($C_t = 0.4$), **e** earthquake affected region after removal of small components, and **f** Detected earthquake region

2.2.2 Post-processing and Earthquake Region Identification

The normalized response matrix captures only the *earthquake* score (S_{eq}) for each pixel of an input image. Hence, M_{nr} represents the approximate probability of the presence of debris or damage at various pixel levels. It may be noted that the stride of the sliding window is 8 pixels, therefore, it may happen that in some earthquake regions, the value of the *earthquake* score is small. Hence, as a post-processing step, *Gaussian Blurring* is applied on M_{nr}, so that the values of the regions with lower scores are slightly increased. The kernel size of the Gaussian filter used is 29×29. Figure 5c shows a blurred response matrix after Gaussian blurring.

Next, we introduced a threshold value called *Confidence Threshold* C_t (where $0 < C_t < 0.5$) to determine whether each element of M_{nr} represents a part of the earthquake region or not. If the value of each element of M_{cr} is greater than C_t, then it returns 1, else 0 which yields a binary image shown in Fig. 5d. In this work, we considered $C_t = 0.4$, which is determined experimentally.

We also assumed that small isolated regions identified as debris may be a false alarm in reality. This is because in general earthquake debris won't be limited to a very small region. Similarly, if some small region is not identified as an earthquake, whereas the region surrounding that is an earthquake, then that small region should also be considered as an earthquake region. Hence, small component having area less than 3,136 pixels (56×56) is considered as noise in both foreground and background regions of binary image. Hence, those small components are removed from the binary image (Fig. 5e). The final output of the proposed earthquake-detector is shown in Fig. 5f.

2.2.3 Classification Based on Earthquake Affected Region

After the input images are processed, the earthquake affected areas are depicted by white regions and the black regions depict the unaffected areas. We calculate the percentage of the affected area in each image. Let's say that for tth image, $S_D(t)$ be the set of pixel ids of the detected earthquake region in the input image. For the tth input image, the percentage of affected region (AR_p) can be calculated as $AR_p(t) = \frac{100 \times |S_D(t)|}{Area(t)}$, where $Area(t)$ is the total number of pixels in the tth image. According to a range of percentage values, images are classified into four different classes: (i) *High*($AR_p \geq \alpha_3$), (ii) *Moderate*($\alpha_2 \leq AR_p < \alpha_3$), (iii) *Minimum*($\alpha_1 \leq AR_p < \alpha_2$), and (iv) *Non-earthquake* ($0 \leq AR_p < \alpha_1$). Here α_1, α_2 and α_3 are three thresholds, where $0 < \alpha_1 < \alpha_2 < \alpha_3 < 100$. Based on the study of several images of earthquake, we have considered $\alpha_1 = 5$, $\alpha_2 = 15$ and $\alpha_3 = 25$ for this work.

3 Experimental Results

To evaluate the performance of our proposed earthquake-detector, we have used 160 images randomly from the set of 780 distinct images from the CHF2015 [9] dataset as mentioned in Table 1. Human volunteers were asked to identify and separate these 160 distinct images into 4 classes like *High*, *Moderate*, *Minimum*, and *Non-earthquake* classes. While preparing these classes of images, volunteers visually observed the affected earthquake region in each image and put it under a class based on the approximate amount (%) of the affected region. An overview of the distribution of images in each class is shown in Table 3.

In our experiment, we have tested those 160 images by detecting the earthquake region and calculating affected region (AR_p) for each image. However, it is very unrealistic to categorize an image into one class, whose calculated AR_p falls in the border line of two classes. Hence, we considered a *hysteresis* (H_t) value and put an image under both classes if the calculated AR_p is in the range of $\alpha \pm H_t$. For example, for tth image, if $(\alpha_2 - H_t) < AR_p(t) < (\alpha_2 + H_t)$, then the tth image is put under both *Minimum* and *Moderate* classes. In this experiment, we have considered the value of $H_t = 2\%$.

The classification result is shown in Table 3. It can be seen that overall 80% classification accuracy is achieved by the proposed work. The Fig. 6 presents few images of MEQ2016 dataset where the classification of images are very accurate.

Table 3 Performance of classification based on 160 images of MEQ2016 dataset

Classes of damage	Damage amount (%)	Actual number of images	Number of images correctly classified	Classification accuracy (%)
High	25 or above	58	52	89.67
Moderate	15–25	27	19	70.37
Minimum	5–15	29	21	72.41
Non-earthquake	5 or less	46	37	80.43
Overall	–	160	129	80.63

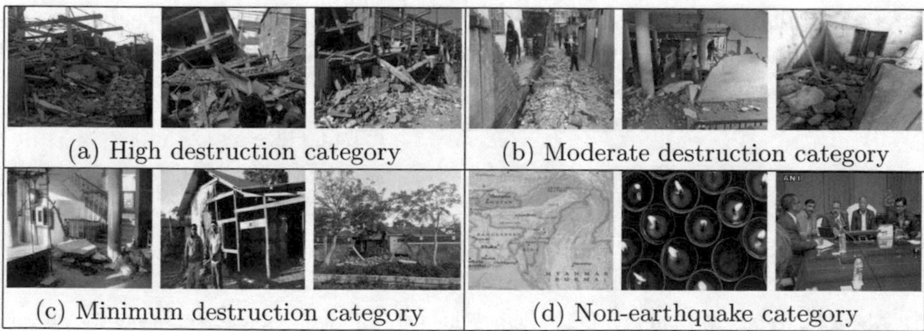

Fig. 6 Some images from the MEQ2016 dataset classified by our model. Aspect ratio of the images is altered to fit well in the figure

(a) Success scenarios (b) Failure scenarios

Fig. 7 Outputs for different images using the proposed earthquake-detector. Images in **a** and **b** show a few success and failure scenarios respectively

As we haven't seen any related work of earthquake region detection in the literature. Therefore, we couldn't compare the performance of the proposed method with any of the state-of-the-art work.

We have also presented a set of results in Fig. 7 for visual analysis to understand the performance and limitations of the earthquake-detector. The first row of the figure shows the input images and the second row represents the corresponding detected earthquake regions in the input images. Images in Fig. 7a show different cases where detected regions are accurate. Whereas, Fig. 7b captures a couple of cases where the detector produces false alarm. This typically happens in case of aerial view of a city or texture of wall, etc.

4 Conclusion

Detection of disaster-affected areas within images is still a challenging field for researchers. In this paper, we have presented a procedure of segmentation and classification of earthquake-affected areas in OSM images. The results presented in this work is acceptable and encouraging. This work used the concept of segmentation with the help of Convolutional Neural Network. A technique for classification of OSM images, into three earthquake classes (high, moderate, minimum) and a non-earthquake class, has also been put forward. We achieved the above objectives using the massive crowd-sourcing power of social media platforms. Therefore, a proposed system can be effectively used for damage assessment during an earthquake event which can help in achieving rapid disaster response.

References

1. Alam, F., Imran, M., Ofli, F.: Image4act: Online social media image processing for disaster response. In: Proceedings of the 2017 IEEE/ACM International Conference on ASONAM 2017, pp. 601–604. ACM (2017)
2. Bishop, C.M.: Pattern Recognition and Machine Learning. Springer (2006)

3. Canny, J.: A computational approach to edge detection. IEEE Trans. Pattern Anal. Mach. Intell. **8**(6), 679–698 (1986)
4. Coates, A., Lee, H., Ng, A.: An analysis of single-layer networks in unsupervised feature learning. In: Proceedings of the 14th International Conference on Artificial Intelligence and Statistics, pp. 215–223 (2011)
5. Hodgkinson, P.E., Stewart, M.: Coping with Catastrophe: A Handbook of Disaster Management. Taylor & Francis/Routledge (1991)
6. Imran, M., Elbassuoni, S., Castillo, C., Diaz, F., Meier, P.: Practical extraction of disaster-relevant information from social media. In: Proceedings of the 22nd International Conference on World Wide Web, pp. 1021–1024. ACM (2013)
7. Joyce, K.E., Belliss, S.E., Samsonov, S.V., McNeill, S.J., Glassey, P.J.: A review of the status of satellite remote sensing and image processing techniques for mapping natural hazards and disasters. Prog. Phys. Geogr. **33**(2), 183–207 (2009)
8. Krizhevsky, A., Sutskever, I., Hinton, G.E.: Imagenet classification with deep convolutional neural networks. In: Advances in Neural Information Processing Systems, vol. 25, pp. 1097–1105. Curran Associates, Inc. (2012)
9. Layek, A.K., Gupta, A., Ghosh, S., Mandal, S.: Fast near-duplicate detection from image streams on online social media during disaster events. In: 2016 IEEE Annual India Conference (INDICON), pp. 1–6, Dec 2016
10. Lecun, Y., Jackel, L.D., Eduard, H.A., Bottou, N., Cartes, C., Denker, J.S., Drucker, H., Sackinger, E., Simard, P., Vapnik, V.: Learning algorithms for classification: a comparison on handwritten digit recognition. In: Neural Networks: The Statistical Mechanics Perspective (1995)
11. Middleton, S.E., Middleton, L., Modafferi, S.: Real-time crisis mapping of natural disasters using social media. IEEE Intell. Syst. **29**(2), 9–17 (2014)
12. Nguyen, D.T., Alam, F., Ofli, F., Imran, M.: Automatic image filtering on social networks using deep learning and perceptual hashing during crises. CoRR arXiv:1704.02602 (2017)
13. Peters, R., de Albuquerque, J.P.: Investigating images as indicators for relevant social media messages in disaster management. In: ISCRAM (2015)
14. Sakaki, T., Okazaki, M., Matsuo, Y.: Tweet analysis for real-time event detection and earthquake reporting system development. IEEE Trans. Knowl. Data Eng. **25**(4), 919–931 (2013)
15. Vedaldi, A., Lenc, K.: Matconvnet: convolutional neural networks for MATLAB. In: ACM International Conference on Multimedia (2015)
16. Zhou, Z., Wu, Q.J., Huang, F., Sun, X.: Fast and accurate near-duplicate image elimination for visual sensor networks. Int. J. Distrib. Sens. Netw. **13**(2) (2017)

Blood Vessel Extraction from Retinal Images Using Modified Gaussian Filter and Bottom-Hat Transformation

Amiya Halder and Sneha Ghose

Abstract Extraction of Retinal blood vessels are important for computer-aided diagnosis of ophthalmologic diseases. This paper presents a novel method for the detection of blood vessels of the retina. In proposed algorithm, at first, the background of the retinal vessels is removed. Then we perform modified Gaussian filtering to remove the noise of the image. The resulting image is subjected to Bottom-hat transformation and thereafter a width-dependant threshold operation is carried out to extract the blood vessels effectively. The proposed method is very efficient and less computational time as compared to the previously implemented methods and is thus very useful.

Keywords Bottom-hat transformation · Gaussian filter · Retinal images · Contrast enhancement

1 Introduction

The retina is a thin layer of tissue that forms an inner lining of the posterior of the human eye close to the optic nerve. The lens of the eye focuses the light it receives onto the retina. The main purpose of the retina is to receive this light, convert it into signals, and send these neural signals to the brain for visual knowledge and understanding. The retina consists of photoreceptor cells, namely rods and cones. These cells are light-sensitive and help to perceive color and intensity of light. The information thus gathered from these photoreceptors are sent to the brain with the help of the optic nerve. The retina is supplied with necessary oxygen and nutrients via the central artery. The central artery reaches the eye and then branches and rebranches to form a mesh of very thin and fine blood vessels called capillaries. It is mainly through

A. Halder (✉) · S. Ghose
Department of CSE, St. Thomas' College of Engineering and Technology, 4 D. H. Road, Kidderpore, Kolkata 700023, India
e-mail: amiya.halder77@gmail.com

S. Ghose
e-mail: snhghose1313@gmail.com

A. K. Das et al. (eds.), *Computational Intelligence in Pattern Recognition*, Advances in Intelligent Systems and Computing 999,
https://doi.org/10.1007/978-981-13-9042-5_30

the capillaries that the nutrients, gases, and waste materials enter or leave the blood vessels. In case these blood vessels are somehow blocked or damaged it may cause starvation or any damage to the retina. Due to the major role played by the retina in the process of visual perception, any damage may result in difficulty in vision or even permanent blindness. It is thus very important to detect and analyze any discrepancy or disorder in the retinal blood vessels for proper treatment at an earliest.

Retinal blood vessel detection and analysis plays a vital role in the early detection and diagnosis of various cardiovascular disorders [1] and ocular diseases [2] such as vessel occlusions, diabetic retinopathy [3, 4] arteriosclerosis, aneurysm, hemorrhage, and stroke [5, 6]. Early detection of disorders in the retina is necessary for the timely treatment of the disease. The human retinal images are obtained through the fundus camera, which takes a photograph of the rear side of the eye which is known as the fundus. Specialized flash enabled cameras consisting of complex microscopes are used for the purpose of fundus photography. Such color fundus images of the retina serve as the basis for screening and diagnosis of retinal diseases. The manual analysis of such images is very tedious and time-consuming. It requires a huge amount of time, concentration, and observational skills. Neither is such a process feasible on a large scale. For this reason computer-aided detection techniques such as [7–12] are very essential so that the abnormalities in the retinal blood vessels can be detected easily and in a matter of seconds.

In this paper, we propose an efficient approach to detect the retinal blood vessels from fundus images by applying Gaussian noise removal and Bottom-hat transformation operations. Gaussian noise is statistical noise that follows the Gaussian distribution. The image is passed through a Gaussian filter for the removal of Gaussian noise [13]. In the proposed method, the green plane of the image is used, because the blood vessels can be distinguished more easily in the green plane than in the other planes.

2 Proposed Algorithm

In the proposed method initially the green plane of the fundus image is extracted for further processing. Thereafter we perform Contrast Limited Adaptive Histogram Equalization (CLAHE) on the image [14]. This is done to increase the contrast by applying Histogram Equalization on small data regions called tiles of the image rather than on the entire image. The resulting neighboring tiles are then joined back using bilinear interpolation to remove the boundaries. In this method, the amplification of image contrast is limited, thus diminishing the amplification of noise. The resulting image obtained is further subjected to morphological opening operations to remove the background. The detection of retinal blood vessels is based mainly on the contrast between the blood vessels and the immediate background. The removal of any kind of Gaussian noise [13] that may decrease this contrast is thus very necessary for further processing. Bottom-Hat transformation is then applied for further enhancement of contrast. The final image is obtained after binarization for proper analysis and comparison. The proposed algorithm is shown in Algorithm 1.

Algorithm 1: BLOOD VESSEL EXTRACTION FROM RETINAL IMAGE

Input: Color fundus image of the human retina $Img = f(\alpha, \beta, \gamma)$ of size $P \times Q$ intensity level $[0..L-1]$

Output: Extracted blood vessels

$I_g(\alpha, \beta) = f(\alpha, \beta, Greenplane)$ // Extract Green channel
$I_{inv}(\alpha, \beta) = 255 - I_g(\alpha, \beta)$
Segment the image into non-overlapping regions of size $\omega \times \omega$
Clip limit $\mu = \frac{P \times Q}{L}(1 + \frac{\varphi}{100}(\tau_{max} - 1))$
// $\varphi = clipfactor$, τ_{max} = maximum allowable slope
Total number of clipped pixels = μ_{clip} , where the number of pixels above μ have been clipped $\mu_{average} = \frac{\mu_{clip}}{L}$
for *each* $\omega \times \omega$ *region and* i^{th} *gray level* **do**
    Calculate histogram H_r of the region
    if $H_r(i) > \mu$ **then**
        $H_{r\_clip}(i) = \mu$
    else if $(H_r(i) + \mu_{average}) > \mu$ **then**
        $H_{r\_clip}(i) = \mu$
    else
        $H_{r\_clip}(i) = H_r(i) + \mu$
    // H_{r_clip} is the clipped histogram of the region
repeat
    $s = \frac{L}{\mu_{remain}}$ // μ_{remain} is the number of remaining clipped pixels, $s >= 1$
    for $i = graylevel_{min}........graylevel_{max}$ *with step* = s **do**
        if *number of pixels in i* $< \mu$ **then**
            distribute 1 pixel to the graylevel
until *all remaining pixels are distributed*;
Transform the clipped histogram to cumulative distribution for each region
Reunite the regions using bilinear interpolation to remove the boundaries giving the contrast limited adaptive histogram equalized image I_{clahe} [14]
$I_{open} = I_{clahe} \circ \varepsilon = (I_{clahe} \ominus \varepsilon) \oplus \varepsilon$
// ε is the structuring element
$I_{remove\_background} = I_{clahe} - I_{open}$
Remove Gaussian noise from image using [13] to form I_{gauss}
$I_{close} = I_{gauss} \bullet \sigma = (I_{gauss} \oplus \sigma) \ominus \sigma$
// σ is the structuring element
$I_{bottomhat} = I_{close} - I_{gauss}$
for $i = 1......P$ **do**
    for $j = 1......Q$ **do**
        if $I_{bottomhat}(i, j) < t$ **then**
            $I_{bottomhat}(i, j) = 0$
        else
            $I_{bottomhat}(i, j) = 255$
for *all elements in* $I_{bottomhat}$ **do**
    if *width of element* $< \Theta$ **then**
        discard element // Θ Predefined vessel width
    else
        do nothing

3 Result Analysis

The performance of the proposed method(PM) is compared with previously existing methods such as Prewitt, Canny, and Sobel edge detection methods, P1 [10], P2 [11], and P3 [12]. In result analysis, we have calculated the Accuracy (AC) and Overall Error (OE) by determining the Miss Alarm (MA), the False Alarm (FA) by comparing output images with the Ground Truth images. The results of the proposed method is evaluated on more than hundred retinal images. It is tested against the DRIVE [15] and STARE [16] databases which are publicly available. The comparative results are shown in Tables 1, 2, 3, 4 and 5 and vessel extracted images are shown in Fig. 1. It is observed from Tables 1, 2, 3, 4 and 5 that the proposed method gives better percentage accuracy as compared to the existing algorithms. It can also be observed that the values for MA, FA and OE show that the proposed method gives improved results than the previous methods, where Miss Alarm is a count of the number of pixels where the proposed algorithm fails to detect the vessels, False Alarm is the count of pixels where it falsely detects a vessel and Overall Error is the summation of MA and FA.

Table 1 Accuracy measure for Retinal image1

Methods	MA	FA	OE	Accuracy (%)
Canny	19170	20722	39892	87.91
Prewitt	23294	1973	25267	92.34
Sobel	23270	1977	25247	92.35
P1	11862	4882	16744	94.93
P2	5276	8726	14002	95.76
P3	7927	2869	10796	96.73
PM	7508	2764	10272	96.89

Table 2 Accuracy measure for Retinal image2

Methods	MA	FA	OE	Accuracy (%)
Canny	28357	21164	49521	84.99
Prewitt	36931	1764	38695	88.27
Sobel	36921	1763	38684	88.28
P1	19973	4050	24023	92.72
P2	12711	4077	16788	94.91
P3	16156	1531	17687	94.64
PM	13265	2950	16215	95.09

Table 3 Accuracy measure for Retinal image3

Methods	MA	FA	OE	Accuracy (%)
Canny	21902	33324	55226	83.26
Prewitt	25127	2862	27989	91.52
Sobel	25068	2880	27948	91.53
P1	10761	4504	15265	95.37
P2	6812	6267	13079	96.04
P3	9883	2826	12709	96.15
PM	9540	2610	12150	96.32

Table 4 Accuracy measure for Retinal image4

Methods	MA	FA	OE	Accuracy (%)
Canny	27618	21746	49364	85.04
Prewitt	32606	2303	34909	89.42
Sobel	32533	2314	34847	89.44
P1	16159	4480	20639	93.74
P2	11948	4763	16711	94.94
P3	13141	4070	17211	94.78
PM	12256	4234	16490	95.00

Table 5 Accuracy measure for Retinal image5

Methods	MA	FA	OE	Accuracy (%)
Canny	22910	30251	53161	83.89
Prewitt	27599	2032	29631	91.02
Sobel	27600	2042	29642	91.02
P1	12813	5309	18122	94.51
P2	6563	8096	14659	95.56
P3	8296	6275	14571	95.58
PM	9011	4826	13837	95.81

4 Conclusion

The paper proposes an efficient method for the detection of retinal blood vessels from fundus images. The experimental results obtained on comparison show that this method is more encouraging than different existing methods. This method only fails to detect very thin and fine vessels whose intensities are close to that of their immediate background. The easily implementable algorithm and its efficiency in the detection of retinal blood vessels, as well as the low computational complexity make this method convenient for use in the medical field.

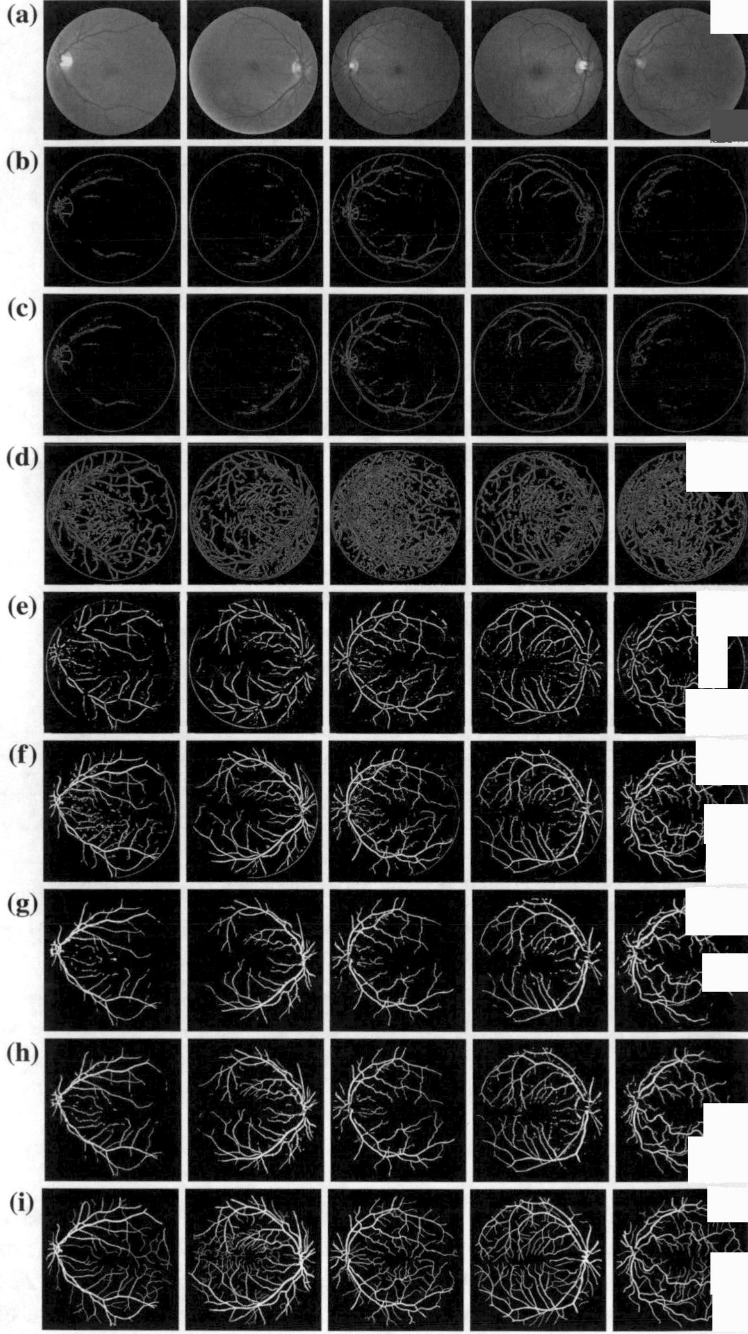

Fig. 1 Output images for retinal image1 to image5: **a** Original retinal images, **b** Prewitt, **c** Sobel, **d** Canny, **e** P1 approach, **f** P2 approach, **g** P3 approach, **h** proposed algorithm and **i** ground truth image

References

1. Martinez-Perez, M.E., Hughes, A.D., Stanton, A.V., Thom, S.A., Chapman, N., Bharath, A.A., Parker, K.H.: Retinal vascular tree morphology: a semiautomatic quantification. IEEE Trans. Biomed. Eng. **49**(8), 912–917 (2002)
2. Lowell, J., Hunter, A., Steel, D., Basu, A., Ryder, R., Kennedy, R.L.: Measurement of retinal vessel widths from fundus images based on 2-D modeling. IEEE Trans. Med. Imag. **23**(10), 1196–1204 (2004)
3. Teng, T., Lefley, M., Claremont, D.: Progress towards automated diabetic ocular screening: a review of image analysis and intelligent systems for diabetic retinopathy. Med. Biol. Eng. Comput. **40**, 2–13 (2002)
4. Jelinek, H.F., Cree, M.J., Leandro, J.J.G., Soares, J.V.B., Cesar, R.M., Luckie, A.: Automated segmentation of retinal blood vessels and identification of proliferative diabetic retinopathy. J. Opt. Soc. Am. A **24**, 1448–1456 (2007)
5. Ponto, K.A., Werner, D.J., Wiedemer, L., Laubert-Reh, D., Schuster, A.K., Nickels, S., Höhn, R., Schulz, A., Binder, H., Beutel, M., Lackner, K.J., Wild, P.S., Pfeiffer, N., Mirshahi, A.: Retinal vessel metrics: normative data and their use in systemic hypertension results from the Gutenberg Health Study. J. Hypertens. **35**, 1635–1645 (2017) (Wolters Kluwer Health Inc.)
6. Imani E., Javidi M., Pourreza H. R.: Improvement of retinal blood vessel detection using morphological component analysis. Comput. Methods Programs Biomed. **118**(3), 263–279 (2015). ISSN: 1872-7565
7. Wan Mustafa W.A.B., Yazid H., Bin Yaacob S., Bin Basah, S.N.: Blood vessel extraction using morphological operation for diabetic retinopathy. In: 2014 IEEE Region 10 Symposium (2014)
8. Hassan, G., El-Bendary, N., Hassanienc, A.E., Fahmy, A., Shoeb, A.M., Snasel, V.: Retinal blood vessel segmentation approach based on mathematical morphology. Procedia Comput. Sci. **625**, 612–622 (2015)
9. De, I., Das, S., Ghosh D.: Vessel extraction in retinal images using morphological filters. In: International Conference on Research in Computational Intelligence and Communication Networks (2015)
10. Halder, A., Bhattacharya, P.: An application of Bottom Hat transformation to extract blood vessel from retinal images, In: 2015 International Conference on Communications and Signal Processing, pp. 1791–1795 (2015)
11. Patil, V.P., Wankhede, P.R.: Pre-processing steps for segmentation of retinal blood vessels. Int. J. Comput. Appl. **94**(12), 34–37 (2014)
12. Halder, A., Sarkar, A., Ghose, S.: Adaptive histogram equalization and opening operation-based blood vessel extraction. In: Soft Computing in Data Analytics. Advances in Intelligent Systems and Computing, vol. 758, pp. 557–564 (2018)
13. Vijaykumar, V., Vanathi, P., Kanagasabapathy, P.: Fast and efficient algorithm to remove gaussian noise in digital images. IAENG Int. J. Comput. Sci. **37**(1), 78–84 (2010)
14. Ma, J., Fan, X., Yang, S.X., Zhang, X., Zhu, X.: Contrast Limited Adaptive Histogram Equalization Based Fusion for Underwater Image Enhancement. Preprints (2017)
15. Staal, J.J., Abramoff, M.D., Niemeijer, M., Viergever, M.A., Van Ginneken, B.: Ridge based vessel segmentation in color images of the retina. IEEE Trans. Med. Imaging **23**(4), 501–509 (2004)
16. Hoover, A., Kouznetsova, V., Goldbaum, M.: Locating blood vessels in retinal images by piecewise threshold probing of a matched filter response. IEEE Trans. Med. Imaging **19**, 203–210 (2000)

A Fast Approach for Text Region Detection from Images on Online Social Media

Ashish Kumar Layek, Sekhar Mandal and Saptarshi Ghosh

Abstract Microblogging sites are very popular sources of real-time information, which includes both textual information, images, and videos. Since individual posts on such sites are very small, they can convey only a small amount of information. Hence, in situations like an ongoing emergency or disaster, users wishing to share a large amount of information often resort to including the text in an image and then sharing the image. Utilizing such textual information within images requires a text-detection mechanism that not only needs to be accurate, but also very fast in order to process the hundreds of images posted on social media in real-time. In this work, we propose such a text-detection algorithm from images. Experiments over images posted on Twitter during a recent disaster event show that the proposed method achieves competitive accuracy with a state-of-the-art method, while being much faster.

Keywords Social Media · Twitter · Text detection · Convolutional neural network · CNN · Quadtree decomposition

1 Introduction

Online microblogging sites such as Twitter (http://twitter.com) and Sina Weibo (www.weibo.com) are presently being used by millions of users to exchange information on various events in real-time, i.e., as and when the event is happening. During

A. K. Layek (✉) · S. Mandal
Indian Institute of Engineering Science and Technology,
Shibpur 711103, India
e-mail: ashish@cs.iiests.ac.in

S. Mandal
e-mail: sekhar@cs.iiests.ac.in

S. Ghosh
Indian Institute of Technology, Kharagpur 721302, India
e-mail: saptarshi@cse.iitkgp.ernet.in

A. K. Das et al. (eds.), *Computational Intelligence in Pattern Recognition*,
Advances in Intelligent Systems and Computing 999,
https://doi.org/10.1007/978-981-13-9042-5_31

an event, various types of information, including textual information and images are posted by users in OSM in huge volumes, and at very rapid rates. While studying such images, we observe an interesting trend, which is as follows.

Individual posts in microblogging sites are very small—they can be at most 140 characters long. Hence the amount of information that can be provided in a post is very limited. There are certain situations where users wish to convey a large amount of information—for instance, during a disaster situation (e.g., an earthquake or flood), a user might want to give telephone numbers and contact information of hospitals in the region—which cannot be accommodated in a textual post. In such situations, users often take advantage of the fact that images are allowed to be attached to a textual post. Hence, they post the textual information as an image, e.g., taking a screenshot of the information using mobile phones. Apart from screenshots, several digitally created document images such as web-pages, notices, circulars, air/rail/bus time tables related to the events are also posted in Twitter. Figure 1 shows some examples of such images posted during a recent disaster event (the floods in the Indian city of Chennai in December 2015), containing varying amounts of text. From a survey of the images posted during this flood, it is observed that 52% images did not have any text (see Table 1 for details), and the rest 48% of images having text those portrayed significant information of damages due to the flood.

There is a large body of work on utilizing the textual content posted on OSM, e.g., gathering situational information during disaster situations [4, 15]. There have also been few recent attempts to utilize the images that are posted on OSM during disasters, e.g., to assess the damage [12]. However, to our knowledge, there has been no work on utilizing the *textual content within images*. We have observed that the textual content of images often contains very valuable situational information, which is not captured by the text-only methods. In this work, we propose a methodology

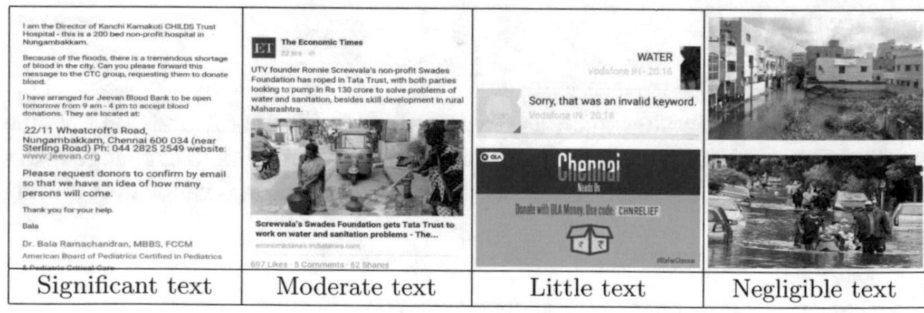

Fig. 1 Different kinds of images posted in Twitter while Chennai Flood 2015

Table 1 Survey on images posted on *Twitter* during Chennai in December 2015

Dataset	Total images	Unique images	Images with text	Images without text
CHF-2015	3,972	2,615	1,253 (48%)	1,362 (52%)

for identifying the textual content of such images posted on Online Social Media (OSM) sites like Twitter.

Text recognition is a very well known problem for several years in the field of machine learning and computer vision. The end-to-end text recognition is in general a two step process; (i) localization of text (text detection), and (ii) character/word recognition (text recognition).

The aim of any text detection method is to spot text first and then segment it from the given image. Detection of words in nonuniform background and cluttered images is a difficult task. The available approaches for text detection can be done either by *character regions approach* [3, 10] or *sliding window approach* [17]. In case of character regions approach, first each character is detected based on connected component analysis, then a set of neighboring characters forms a word. In [10], the text detection technique estimates the possible positions of strokes and detects characters based on known patterns of stroke orientations and their relative positions. The method proposed in [3] expands the use of Maximally Stable Extremal Regions (MSER) by using a CNN classifier to efficiently segregate the extremal regions resulting in less false-positive detections. However, performance of these techniques worsens under adverse conditions, such as disconnected strokes, blur, nonuniform illumination, and low resolution, etc. In case of sliding window approach, the text detection task is considered as object detection. The text detection technique presented in [16] uses a random ferns classifier trained on HOG features in a sliding window scenario to find characters in an image. Another method presented in [17] also uses the CNN to detect the text regions and identification of characters. In the detection phase, the input image is transformed into 13 different scales (as high as 150% and as low as 10%) and at each scale the response map is obtained. From these response maps, bounding boxes for candidate text lines in the original image is deduced. Then non-maximal suppression (NMS) is applied to obtain the final set of bounding boxes. The execution time of this method is quite high as well as the number of scales and scaling factor has to be determined manually.

In the literature, there are different methods of historical and handwriting document recognition. However, it is really challenging to recognize the text from highly variable foreground and background textures present in a scene image. This problem is attempted to handle either by character recognition or word recognition. The character recognition problem is considered as classification problem which is generally achieved with the use of CNN based strong classifiers [2, 17]. In general, CNN based character recognizers aim to recognize 62 classes for English script (upper-case, lower-case letters, and digits) which frequently appearing in the text window. The methods proposed in [2, 16, 17] produce satisfactory results in terms of detection and recognition, however these algorithms demand more time. On the other hand, in the word recognition approach, most of the works in this area rely upon segmentation-free lexicon dependent approaches. The lexicons are used to resolve the high confusion that exist in the text recognition problem. The works of [11, 17] are few of such category.

Present work: The problem of text detection and recognition has some unique features when the domain of images posted in OSMs during disasters is considered. First of all, this has to be very fast as thousands of images might be posted per minutes on Twitter during a major disaster. Most of the prior approaches of text detection are optimized for accuracy, but not the execution time required; hence, they are not likely to be practically usable for images posted in OSMs. These are the challenges that we address in this work. On the other hand, most of the text recognition system just recognizes abovesaid 62 English characters, which are not sufficient in true text recognition system. In OSM, the character like "#", "%", "@", "$", "?", "!", etc has very high importance. Apart from these, CNN based recognition systems are costly in terms of running time. Hence in the context of this problem, OCR based text recognition is still the only alternative as it is fast and quite accurate.

In this work, we propose a novel text-detector that is fast as well as accurate. The proposed text-detector consists of two new approaches—(i) a robust Convolutional Neural Network (CNN) based binary classifier that classifies text and non-text region of image, and (ii) a novel technique for text region detection that uses our proposed CNN model. In the text detection technique, we propose multi-transform based text detection using quadtree decomposition technique, which is very fast in comparing to existing technique like multi-scaling based text detection using sliding-window approach.

Rest of this paper is organized as follows. In the next section (Sect. 2), proposed method is described in detail. In Sect. 3, generation of ground-truths are presented. Experimental results are presented in Sect. 4. The concluding notes are put in Sect. 5.

2 Proposed Methodology

The proposed method of text region detection has two major tasks. At first, CNN model is trained using a large amount of training data. Then text detection technique is applied on the test images using the trained CNN model. The subsequent sections detail training of CNN model and text detection phases.

2.1 Text-Detector: CNN Model and Training Dataset

This work proposes a CNN model architecture, that is used to segment regions within image consisting of some text. The following subsections explain it.

2.1.1 CNN Model Architecture

The CNN architecture that we use is similar to the LeNet model [7]. The architecture consists of three convolution layers, two fully connected (FC) layers and one softmax

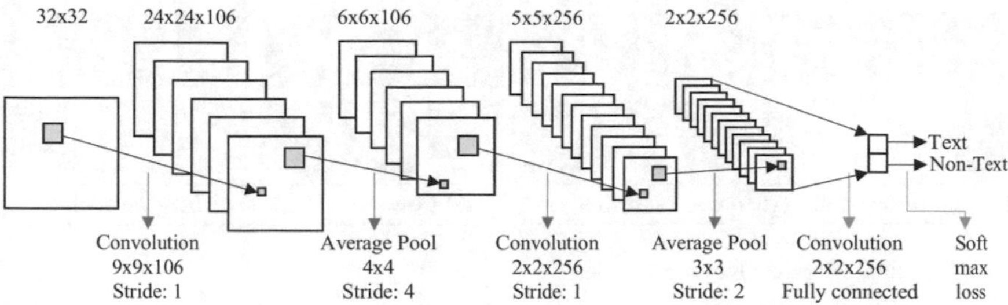

Fig. 2 Convolutional neural network architecture (ReLU layers not shown)

layer. Each convolution layer is followed by a rectification layer (ReLU) as suggested in [5]. The output of the ReLU is fed to the input of the average pooling layer and the output of the pooling layer is fed to the next convolution layer. The next two layers are the fully connected (FC) layer and finally softmax [1] function is used for classification. It was empirically verified that average pooling gives better result than the max(min) pooling. The detail of proposed network is summarized in Fig. 2.

2.1.2 Generation of Training Dataset

In this section, we briefly describe the dataset used to train text detector. To train the above-mentioned model, we use 60,000 image patches of size 32×32 out of which 30,000 training samples contain only text regions and other training samples have only non-text regions. The huge amount of training samples are generated from variety of Twitter posted and ICDAR-2003 training images [8].

Generation of *text* Training Samples: From ICDAR dataset, each word is considered and resized to $32 \times n$ without altering the aspect ratio of the word. The location of each word is provided in XML format along with the dataset. The heavily blurred words and words with large inter character gap are not considered for training purpose. On the other hand for Twitter posted images, we manually crop the text regions which will be used for training samples generation. The following points are kept in mind during selection of text regions: (i) consider only text portion which is clearly readable, (ii) text segments must contain only English letters and other special characters, (iii) text segments should cover varieties of fonts, sizes, colors, background texture and different lighting condition, (iv) if a text region contains a single word in that case the height of word is within the range 20–30 pixels, and (v) text regions may contain multiple lines where interline gap is very small (approximately 4–5 pixels). The cropped text regions are converted into gray-scale images and the actual training samples are created from these gray-scale images using sliding window approach. The size of window is 32×32 which moves from left to right and top to bottom. Both the horizontal shift and vertical shift is 8 pixels. Out of all the text training samples, approximately 20% samples are slightly blurred using *Gaussian Blurring* technique, where σ value is set to 1. A set of *text* training samples is shown in Fig. 3a.

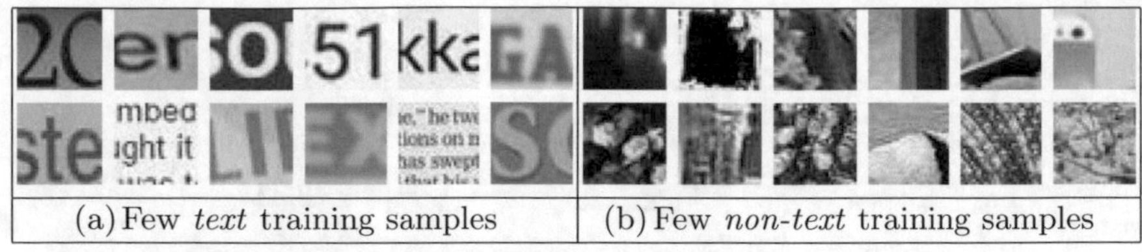
(a) Few *text* training samples (b) Few *non-text* training samples

Fig. 3 Example of a few training samples

Note that the CNN model proposed in several previous works [2, 17] used 32 × 32 image samples containing a well-centered character. In contrast, we also choose image samples containing multiple lines (several characters). This helps to detect text of smaller font size than 32 pixels using a bigger detection window.

Generation of *non-text* Training Sample: Here, a wide variety of images are considered for generating training samples to incorporate a wide range of non-text textures. A total of 313 images are selected which include images of animal, bird, human face, flower, house, tree, vehicle, and flood event, etc. These images do not have any text regions. We consciously haven't included images of pencil-sketch and cartoon because many segments of these kind of images have similarity with the texture of text regions, especially handwritten text.

Each selected image is converted into gray-scale image and a non-overlapping sliding window of size 32 × 32 is used to generate the training samples. It is observed that there are some training samples generated from the same image, may be near-duplicates. Hence, duplicate detection is necessary and we use a similar type of near-duplicate clustering technique proposed in [18] for near-duplicate detection. A set of *non-text* training samples is shown in Fig. 3b.

2.2 Text Region Detection

In this section, we describe the process of locating regions having presence of text in a given image. Figure 4 summarizes the text detection technique.

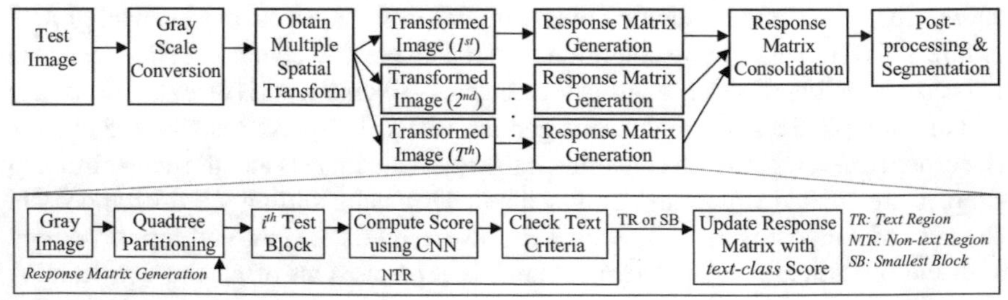

Fig. 4 Overview of the text detection work flow

2.2.1 Response Matrix Generation

The existing text detecting algorithms [2, 17] use multi-scale based *sliding-window* technique to locate the text regions of different font sizes present in a particular image. To reduce the run time of the text detection system, we avoid the multi-scaling. Instead, *quadtree decomposition* [13] along with dynamic thresholding technique is introduced to detect the text regions having different font size present in an image. The decomposition technique is elaborated below.

In this phase, a hierarchical decomposition algorithm is applied that recursively subdivides the input image into disjoint square blocks of varying size. As a precondition of square block quadrant of quadtree, the input image is padded with zeros to the nearest square size that is an integer power of 2. Let I_g and I_p be the original input image and image after zero padding respectively. Initially, the padded image I_p is subdivided into an arbitrary number of disjoint square regions (R_i). The size of each region is $D_{max} \times D_{max}$, where D_{max} (also integer power of 2) is maximum dimension of initial blocks that is determined adaptively based on the size of the input image. We take a reasonable assumption that average font size of texts in an image is proportional to the image resolution.

Now, each of the R_i is then tested whether it satisfies the predefined predicate function (Eq. 1) or not which is discussed later. If the region does not satisfy the predicated function, then the region under test is decomposed further in recursive manner. The decomposition process is terminated when the region satisfies the predicate function or the size of region is $D_{min} \times D_{min}$, where $D_{min} = 32$. Recall that image samples containing multiple text lines are also incorporated in our training dataset. Hence, it is sufficient that the minimum size of a decomposed region is same as the size of the training samples. After each iteration of the decomposition process, we capture the *text* class *score* (*Step 2*) of each decomposed region and store it in the response matrix (M_r). The size of the response matrix is same as I_p. The following steps detail the text detection method which is applied to each decomposed region.

Step 1: The current subblock R_i is resized to 32×32 using bicubic interpolation function and then contrast normalization is performed. The preprocessed image is denoted by $\hat{R}_i$.

Step 2: The preprocessed image patch is fed to the text-detector and response is observed at *softmax* layer. For any given input image patch, it returns score (or confidence) for both *text* and *non-text* classes and they are denoted by S_t and S_{nt} respectively. The value of S_t (S_{nt}) belongs within range [0, 1] and $S_t + S_{nt} = 1$. The elements of the matrix M_r corresponding to the pixels belonging to the test block R_i are updated with the value S_t.

Step 3: In a common practice [14], if the observed score S_t (S_{nt}) is greater than the *threshold* $\mathcal{T}$, where $\mathcal{T} = 0.5$ in case of binary classification; then the input image patch is considered as text (non-text) region. In contrary, we use dynamic thresholding which depends on the size of decomposed block (region) under test. For larger block size the value of $\mathcal{T}$ is higher which is determined empirically. If the value of S_t of a

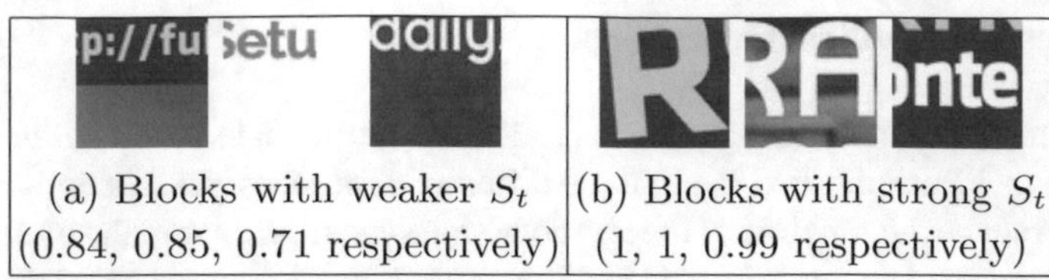

(a) Blocks with weaker S_t (0.84, 0.85, 0.71 respectively) (b) Blocks with strong S_t (1, 1, 0.99 respectively)

Fig. 5 Example of few 64 × 64 blocks and with their *text score*

test block is less the $\mathcal{T}$, then its decomposition depends on the following predicate function (Eq. 1).

$$P_s(\hat{R}_i) = \begin{cases} True, & \text{if } (D_b > D_{min})\ AND\ (S_t < \mathcal{T}_i) \\ & \text{where } D_b \text{ is the dimension of } \hat{R}_i \text{ and } 0.5 \leqslant \mathcal{T}_i \leqslant 1 \\ False, & \text{otherwise} \end{cases} \tag{1}$$

where $\mathcal{T}_i$ is threshold value for the ith test block and if the predicate function returns true then the test block will be decomposed further. The values of $\mathcal{T}$ that are determined empirically for different block size are as follows: $\mathcal{T}_{64\times64} = 0.90$, $\mathcal{T}_{128\times128} = 0.95$ and the block size higher than 128 × 128, the $\mathcal{T} = 1$. Figure 5 shows a set of decomposed blocks (regions) and their corresponding S_t values. It is also evident from Eq. 1 that the splitting process cannot stop if $D_b > 32$ even the value of *text* class score (S_t) of the test block is close to zero. This is because a large image block with some text may look like as non-text segment on resizing it to 32 × 32. Hence splitting large segments further and then inspecting small blocks may reveal some text.

Figure 6b illustrates the complete decomposed blocks (regions) for the input image shown in Fig. 6a. It is clear from Fig. 6b that decomposed blocks belonging to the background of the image are the smallest possible in size (i.e., 32 × 32), whereas there are some blocks which enclosed text regions are bigger in size (red squares). After completion of the decomposition process, the response matrix M_r (shown in Fig. 6c) is consulted to detect the text regions present in the test image. The decomposed block whose score (S_t) is greater than or equal to 0.5 is considered as a text region. The detected text regions are shown in Fig. 6d. It is evident from the text detected results that the performance of the text-detector at this point is far from satisfactory. Therefore some more sophistication is needed, which will be discussed next.

2.2.2 Multi-transform Based Text Localization

At this point we have obtained rough estimates of the different text regions present in the image. Recall that in existing work [2, 17], the detector response (score) is calculated at each pixel using sliding window approach and hence, the accurate results

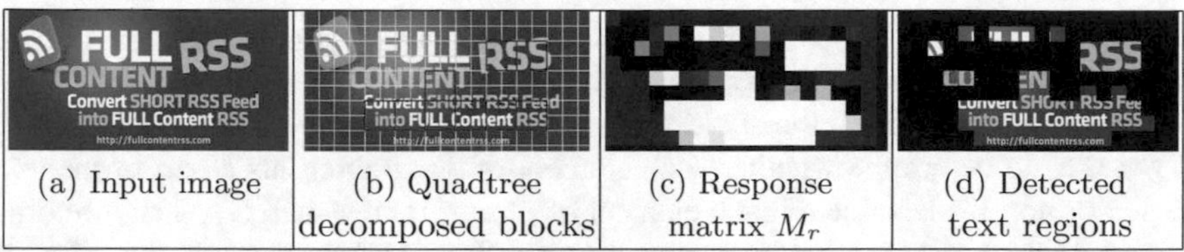

(a) Input image (b) Quadtree decomposed blocks (c) Response matrix M_r (d) Detected text regions

Fig. 6 Illustration of quadtree decomposition process

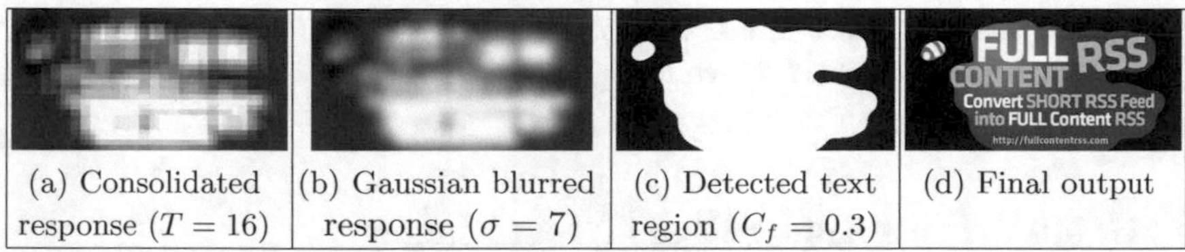

(a) Consolidated response ($T = 16$) (b) Gaussian blurred response ($\sigma = 7$) (c) Detected text region ($C_f = 0.3$) (d) Final output

Fig. 7 Illustration of multi-transform based text detection process

are obtained. Whereas the proposed work calculates the response at the disjoint block level. Hence, the performance of the text-detector is not satisfactory yet. Therefore, we propose a new technique to calculate the response at the block level for a set of different transformed version of the test image using the same CNN model and the technique discussed in previous subsection. Finally, the responses are consolidated into a single matrix which is used further.

Several transformed versions of an image are obtained by translation operation. Here circular shift operation is used to get the transformed image. The number of transformed images is $T = r^2$ where r is positive integer ($r > 1$). First, each pixel of input image is circularly shifted upward (downward) and the number of shifts is $\lfloor \frac{m}{r} \rfloor$, where m is the number of rows of the image. Consider I_s denotes this transformed image. Next, each such I_s is circularly shifted left (right) and the number of shifts is $\lfloor \frac{n}{r} \rfloor$, where n is the number of columns of the image. This operation is repeated r times. The upward (downward) circular shifting is also done r times. Note that setting of the value r is a trade-off between accuracy and speed of execution. To optimize both, the value of r is determined adaptively based on size of the input image, such that $r \propto (m \times n)$. For each transformed image, the response matrix (M_r) is generated as discussed earlier. Then the response matrix undergoes reverse circular shifts so that the transformed matrix is the equivalent response matrix (M_{rq}) of the original input image. Here, upward (downward) and/or left (right) shifts are required to get the M_{rq}. The consolidated response matrix is obtained using the following formula: $M_{cr} = \frac{1}{T} \sum_{i=1}^{T} {M_{rq}}^{i}$. The Fig. 7a shows consolidated response matrix of the input image shown in Fig. 6a.

The response matrix captures only the text score at disjoint block level of the input image, therefore the M_{cr} represents the approximate probability of presence of text at various pixel level. It is observed that in some text regions the value of text score is small. Hence, further processing on M_{cr} is required. The *Gaussian Blurring* (where $\sigma = 7$) is applied to blur the M_{cr} (refer Fig. 7b).

Next, we introduced a threshold value called *Confidence Factor* C_f (where $0 < C_f < 0.5$) to determine whether each element of M_{cr} represents a part of the text region or not. If the value of each element of M_{cr} is greater than C_f, then it returns 1, else 0 which yields a binary image shown in Fig. 7c. In this work, the value of C_f is set to 0.3 which is determined experimentally. In this work, we assume small isolated region may not be text. Hence, the small components having area less than 1024 pixels are considered as noise and removed from the binary image. The final output of the text-detector is shown in Fig. 7d.

3 Ground-Truth Generation of Text Images

This section briefly explains the ground-truth generation of the text images collected from Twitter during a natural disaster situation.

3.1 Dataset Overview

Using the Twitter search API, we have collected tweets related to a recent disaster event—the severe floods in the city of Chennai, India in December 2015. Specifically, the tweets were collected using the search terms "chennai flood". For brevity, we refer to this dataset as *CHF-2015*. We also collected all the images contained in the tweets. We start with a set of 3972 images and unique images are identified (i.e., duplicate images are removed from the set) using a semi-automated method. At first, near-duplicate image detection [6] is applied; the falsely detected images are corrected manually. Finally, we get 2,615 distinct images. Furthermore, three human volunteers were asked to identify and separate these images into two classes—(i) text class and (ii) non-text class. We classified an image as text class if and only if some part of an image has clearly readable text of any script. The text class contains 1,253 images and rest 1,362 images belong to non-text class. An overview of the dataset is given in Table 1.

3.2 Ground-Truths Generation

We have selected 132 images from the set of 1,253 text images of the above mentioned data set. While selecting this list, we tried to cover wide variety of images based on amount of text present in the image. Here the term "amount" implies the number of letters/characters appearing in the text, not the percentage of image area occupied

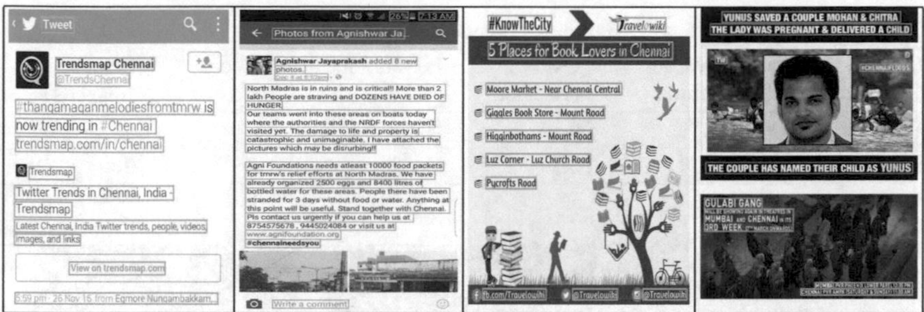

Fig. 8 Ground-truths of CHF-2015 dataset, manually observed text area marked

by the text. However, the estimation of number of characters/letters in the image is done by visual observation only.

We employ human annotators to generate the ground-truths of the bounding box of text segments within images. The information regarding the bounding boxes are captured in XML format as done in ICDAR-2003 dataset [8]. While capturing the bounding box, most of the time we coalesced multiple words of a line into a single segment, unless there is a sufficient gap (approx. 18 pixels) between consecutive words. At the same time we also coalesced multiple lines into a single segment if the interline gap is less than approximately 13 pixels and the width of lines is almost same. A set of marked images is shown in Fig. 8.

4 Experimental Results

We have implemented the proposed algorithm in MATLAB R2015a in a PC (Intel Core i7-6700 CPU @ 3.40GHz $\times$ 8, Memory 7.7GB) the OS being Ubuntu 16.04. We have tested the algorithm on one dataset. We compare the performance of proposed method with the state-of-the-art technique proposed by Wang et al. [17] (which we refer to as *Wang12*). In [17], authors proposed CNN based technique for both text detection and character recognition. However, we have only used the text detection algorithm.

For a quantitative comparison, we have considered the following metrics: (i) average accuracy, (ii) Overall Precision, (iii) Overall Recall, (iv) F_1 score, and (v) Average execution time (AET). The term weighted average precision (*Overall Precision*) and weighted average recall (*Overall Recall*) are calculated as defined in [9]. Note that while performing experimentation for CHF-2015 dataset, we have set the maximum block size $D_{max} = 128$ and number of transform $T = 16$ as most of the images of this dataset are of medium resolution.

We use 132 images of *CHF-2015* dataset for experimentation and a summary of these estimates is shown in Table 2. As evident from this summary, the proposed method competes quite well with the *Wang12* method in terms of accuracy, recall,

Table 2 Quantitative comparison of experimental results

CHF-2015 dataset (132 images)	Average accuracy (%)	Overall precision	Overall recall	F_1 score	AET per image (s)
Wang12	**92.00**	0.86	0.89	**0.87**	95
Proposed Method	88.34	0.79	**0.92**	0.85	**12**

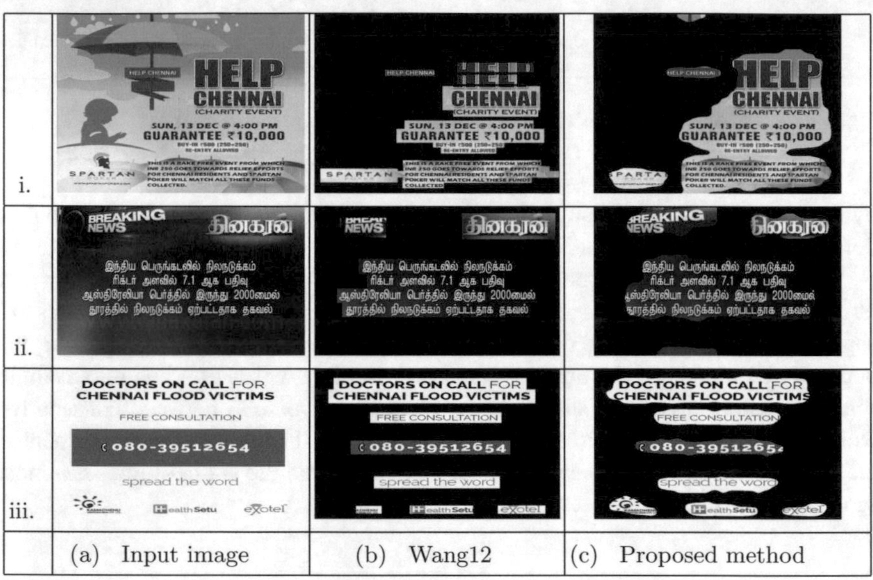

Fig. 9 Visual comparison of experimental results for few images of CHF-2015

and F_1 score, but the *Wang12* method outperforms in terms of precision. The reason behind it is that the proposed method computes the detector response at the block level instead of at each pixel, hence, the detected text regions encompass extra background region surrounding the actual text regions.

To substantiate the aforesaid statement, another set of results presented in Fig. 9 for visual comparison. Column (a) contains the input images. Column (b) and (c) contain the outputs produced by *Wang12* and proposed method respectively. It is clear from the Fig. 9 that the proposed method competes well with *Wang12* method. However, the proposed method is advantageous as it is much faster than the *Wang12* method. Hence the proposed method is applicable in practice over the huge number of images posted on OSM in near-real time.

Limitation: To discuss the limitation of the proposed method two types of results are presented in Fig. 10. Figure 10a, b shows some images where our text-detector is not able to detect text regions. This is due to the following reasons: (i) concentration of the text regions is very small and (ii) text regions are very blurred. Figure 10c, d illustrates some instances where our text-detector produces false alarm. Here, the

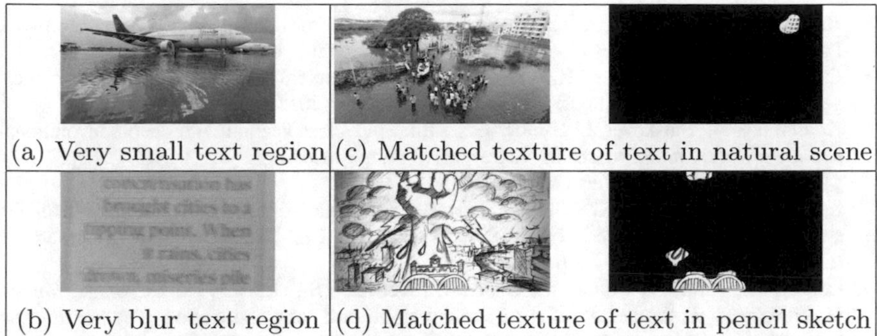
(a) Very small text region (c) Matched texture of text in natural scene
(b) Very blur text region (d) Matched texture of text in pencil sketch

Fig. 10 Few failure scenarios when performance of text-detector is wrong

texture of the detected regions is very much similar to the text region. These are very prominent in case of mesh structure of buildings (Fig. 10c), pencil sketch (Fig. 10d), cartoons, etc.

5 Concluding Discussion

In this work, we proposed a novel methodology for detecting text regions in an image. It achieves competitive performance with a state-of-the-art technique while taking much less time in processing an image. Hence, the proposed methodology is much more suitable for extracting textual information from images posted on OSM during disaster situations in real-time basis. The proposed methodology can be applied to find if an image has some text part. Then by using suitable text binarization techniques, the detected text segments can be binarized. Further the resultant binary image can be fed to the OCR to obtain the text in machine readable format, then text mining algorithms can be used.

Acknowledgements Corresponding author is thankful to Prof. Asit Kumar Das, Prof. Apurba Sarkar and Prof. Surajeet Ghosh from the Department of CST, IIEST Shibpur, Howrah, India for their recommendation letter Dated: 30/04/2019 for using publicly available Twitter data in this paper. The data collected does not have any conflict of interest and has been used ethically.

References

1. Bishop, C.M.: Pattern Recognition and Machine Learning. Springer (2006)
2. Coates, A., Carpenter, B., Case, C., Satheesh, S., Suresh, B., Wang, T., Wu, D.J., Ng, A.Y.: Text detection and character recognition in scene images with unsupervised feature learning (2011)

3. Huang, W., Qiao, Y., Tang, X.: Robust Scene Text Detection with Convolution Neural Network Induced MSER Trees, pp. 497–511. Springer International Publishing (2014)
4. Imran, M., Castillo, C., Diaz, F., Vieweg, S.: Processing social media messages in mass emergency: a survey. ACM Comput. Surv. **47**(4), 67:1–67:38 (2015)
5. Krizhevsky, A., Sutskever, I., Hinton, G.E.: Imagenet classification with deep convolutional neural networks. In: Advances in Neural Information Processing Systems, vol. 25, pp. 1097–1105. Curran Associates, Inc. (2012)
6. Layek, A.K., Gupta, A., Ghosh, S., Mandal, S.: Fast near-duplicate detection from image streams on online social media during disaster events. In: 2016 IEEE Annual India Conference (INDICON), pp. 1–6, Dec 2016
7. LeCun, Y., Jackel, L., Bottou, L., Cortes, C., Denker, J.S., Drucker, H., Guyon, I., Muller, U., Sackinger, E., Simard, P., et al.: Learning algorithms for classification: a comparison on handwritten digit recognition. Neural Netw. Stat. Mech. Perspect. **261**, 276 (1995)
8. Lucas, S.M., Panaretos, A., Sosa, L., Tang, A., Wong, S., Young, R.: ICDAR 2003 robust reading competitions. In: Proceedings of the Seventh International Conference on Document Analysis and Recognition, ICDAR '03, vol. 2 (2003)
9. Mariano, V.Y., Min, J., Park, J.H., Kasturi, R., Mihalcik, D., Li, H., Doermann, D., Drayer, T.: Performance evaluation of object detection algorithms. In: ICPR 2002, vol. 3 (2002)
10. Neumann, L., Matas, J.: Scene text localization and recognition with oriented stroke detection. In: Proceedings of the 2013 IEEE International Conference on Computer Vision, ICCV '13, pp. 97–104 (2013)
11. Neumann, L., Matas, J.: A method for text localization and recognition in real-world images. In: Proceedings of the 10th Asian Conference on Computer Vision, ACCV'10, vol. Part III, pp. 770–783 (2011)
12. Nguyen, D.T., Alam, F., Ofli, F., Imran, M.: Automatic image filtering on social networks using deep learning and perceptual hashing during crises. CoRR arXiv:1704.02602 (2017)
13. Spann, M., Wilson, R.: A quad-tree approach to image segmentation which combines statistical and spatial information. Pattern Recognit. **18**, 257–269 (1985)
14. Vedaldi, A., Lenc, K.: Matconvnet: convolutional neural networks for MATLAB. In: ACM International Conference on Multimedia (2015)
15. Vieweg, S., Hughes, A.L., Starbird, K., Palen, L.: Microblogging during two natural hazards events: what Twitter may contribute to situational awareness. In: Proceedings of the ACM SIGCHI (2010)
16. Wang, K., Babenko, B., Belongie, S.: End-to-end scene text recognition. In: Proceedings of the 2011 International Conference on Computer Vision, ICCV '11, pp. 1457–1464. IEEE Computer Society (2011)
17. Wang, T., Wu, D.J., Coates, A., Ng, A.Y.: End-to-end text recognition with convolutional neural networks, pp. 3304–3308, Nov 2012
18. Zhou, Z., Wu, Q.J., Huang, F., Sun, X.: Fast and accurate near-duplicate image elimination for visual sensor networks. Int. J. Distrib. Sens. Netw. **13**(2) (2017)

Numerical Integration Based Contrast Enhancement Using Simpson's Method

Amiya Halder, Pritam Bhattacharya and Nikita Shah

Abstract This paper is based on mathematical numerical integration method and provides an efficient algorithm for improving the contrast of an image. The aim of this unique method is to use neighboring pixel information and process them in accordance with mathematics (Simpson's $\frac{1}{3}$rd rule) to improve contrast of images. Simpson's rule is a numerical integration method for the accurate approximation of definite integrals. It gives the exact results for polynomials of degree three or less. We use the neighboring pixel information, interpolate among them, and produce the enhanced pixel information. Various parameters like Root Mean Square Error (RMSE), Peak-Signal-to-Noise-Ratio (PSNR) and Structural Similarity Index (SSIM) are used to measure the quality of the images. The experimental results of the proposed method are compared with some existing methods for further validation. Also, the computational time of the proposed method is much less as compared to the different existing methods.

Keywords Image enhancement · Contrast enhancement · Numerical integration · Simpson's $\frac{1}{3}$rd rule

1 Introduction

The perception of information in images for human viewers and to provide "better" input for other automated image processing techniques is termed as Image enhancement [9]. Reducing the noise, removing blurring effect, and increasing contrast of images are examples of enhancement operations. Contrast enhancement is one of

A. Halder (✉) · P. Bhattacharya · N. Shah
Department of CSE, St. Thomas' College of Engineering and Technology, 4 D. H. Road, Kidderpore, Kolkata 700023, India
e-mail: amiya.halder77@gmail.com

P. Bhattacharya
e-mail: prb2794@gmail.com

N. Shah
e-mail: nikitashah9714@gmail.com

A. K. Das et al. (eds.), *Computational Intelligence in Pattern Recognition*, Advances in Intelligent Systems and Computing 999,
https://doi.org/10.1007/978-981-13-9042-5_32

the most crucial problems in image enhancement. Actually, contrast refers to the striking difference in the pixel intensities or difference in luminance component of images. Vital information cannot be retrieved, if the images are too dark or low contrast images, in most of the cases. Thus, efficient algorithms are needed for the improvement of visual quality of images for further processing in computer vision, pattern recognition, the processing of digital images [6], etc.

The classical techniques of Histogram Equalization (HE) [16, 17] for image enhancement have undergone several modifications but still fail to produce desirable results when the contrast characteristics keep varying across the image. Histogram-based enhancement techniques does not succeed in preserving the brightness and lead to loss of information in case of medical images [2]. For improving the contrast in digital images, HE is the most commonly used method but it results in over-enhancement and noise amplification [4]. Also, several modifications have been made to HE such as Adaptive Histogram Equalization (AHE) [12, 18] which uses statistical information in the neighborhood pixel for equalization but the major problem with this is over enhancement, creating objects that are not visible in the original image. Sub image histogram equalization has also been implemented by dividing the input image into segments based on the mean and variance of each region [20]. In global histogram equalization method which is simple and fast, but its contrast-enhancement power is relatively low. On the other hand, the local histogram equalization, enhances contrast more effectively, but increases the complexity of computation due to its fully overlapped sub-blocks [8]. An adaptation of HE is Contrast Limited Adaptive Histogram Equalization (CLAHE) [15] which gives a more localized enhancement as it divides the input image into equal-sized blocks and performs HE locally, but it is computationally very intensive as compared to the classical method. Dynamic HE controls the effect of traditional HE and performs the enhancement without any loss of details in it but this is achieved by partitioning and repartitioning and it fails to give desired results when dominating portions appear in the image [1].

The Genetic Algorithm (GA) [14] approach makes use of several different parameters on the mating pool where convergence becomes very crucial. Image histogram adjustment (IM) [11] method is also based on histogram equalization where image histogram is adjusted automatically. In Adaptive Gamma Correction With Weighting Distribution (AGCWD) [5] method, where gamma correction and probability distribution for luminance pixels were used, is not desirable when the input image lacks bright pixels. The sigmoid function [10] method is done by a sigmoid function as contrast enhancer and then adaptive histogram equalization where unnatural intensity saturation occurs.

In proposed method, we try to develop a contrast enhancement method using mathematical function Simpson's $\frac{1}{3}$rd rule to improve the contrast of images. It makes use of the midpoint sum and the trapezoidal sum on each pixel intensity such that the midpoint sum is used twice. Simpson's $\frac{1}{3}$rd rule is a part of Newton-Cotes numerical integration method which is based on equally spaced arguments. The $\frac{1}{3}$rd rule is a three points formula in the interval specified, where the degree of precision is three. Moreover, the computational time of the proposed method is much less than the existing methods.

2 Proposed Method

To improve the contrast of images, different existing methods have been proposed. Most of these methods are based on gamma correction, sigmoid function, and several different approaches of histogram equalization. The proposed method uses mathematical operations to increase the contrast of different types of images. Simpson's $\frac{1}{3}$rd rule is used for integration calculation. To calculate the formula, $(p+1)$ interpolating points or data points $\Phi_j (j = 0, 1, 2, 3, .., p)$ are equi-spaced and the interval $h = \frac{b-a}{p}$, where $a = \Phi_0$, $\Phi_j = \Phi_0 + jh$ and $b = \Phi_p$ [3, 13]. From Lagrange's formula, we have the following Eq. 1 using Eq. 2:

$$\int_a^b \pi(\Phi)d\Phi \simeq I = \sum_{r=0}^{p} \pi(\Phi_r)H_r^{(p)} \tag{1}$$

where,

$$H_r^{(p)} = \int_a^b \frac{\Omega(\Phi)d\Phi}{(\Phi - \Phi_r)\Omega'(\Phi_r)} = \int_{\Phi_0}^{\Phi_0+ph} \frac{\Omega(\Phi)d\Phi}{(\Phi - \Phi_r)\Omega'(\Phi_r)} \tag{2}$$

To evaluate $H_r^{(p)}$, set $\Phi = \Phi_0 + ht$ and then the result is Eq. 3:

$$\Phi - \Phi_r = \Phi_0 + ht - (\Phi_0 + rh) = (t - r)h \tag{3}$$

and $d\Phi = hdt$. Further, obtained the following conditions as $\Phi = \Phi_0 = a, t = 0, \Phi = \Phi_p = b, t = p$ Thus, $\Omega(\Phi) = (\Phi - \Phi_0)(\Phi - \Phi_1)(\Phi - \Phi_2)...(\Phi - \Phi_p) = h^{(p+1)}t(t-1)(t-2)....(t-p)$ and $\Omega'(\Phi_r) = (\Phi_r - \Phi_0)(\Phi_r - \Phi_1)..(\Phi_r - \Phi_{r-1})$ $(\Phi_r - \Phi_{r+1})..(\Phi_r - \Phi_{p-1})(\Phi_r - \Phi_p) = h^p r!(p-r)!(-1)^{p-r}$.

Hence, obtained Eq. 4 using Eq. 5:

$$H_r^{(p)} = \int_0^p \frac{h^{p+1}t(t-1)(t-2)(t-3)..(t-p)hdt}{h^p.r!(n-r)!(-1)^{p-r}.h.(t-r)} = ph\mu_r^{(p)} \tag{4}$$

Where

$$\mu_r^{(p)} = \frac{(-1)^{p-r}}{p.r!(p-r)!} \int_0^p \frac{t(t-1)(t-2)..(t-p)dt}{(t-r)} \tag{5}$$

Since $\frac{(b-a)}{p} = h$, thus the following equation is obtained:

$$\int_a^b \pi(\Phi)d\Phi \simeq I = (b-a)\sum_{r=0}^{p} \pi(\Phi_r)\mu_r^{(p)} \tag{6}$$

Simpson's rule is a three points Newton-Cotes' formula in the interval $[a, b]$. Considering $p = 2$ and $h = \frac{(b-a)}{2}$ in Eq. 6, we get Eq. 7 using Eqs. 8, 9 and 10

$$I = (b-a)\sum_{r=0}^{2} \pi(\Phi_r)\mu_r^{(2)} = (b-a)[\pi(\Phi_0)\mu_0^{(2)} + \pi(\Phi_1)\mu_1^{(2)} + \pi(\Phi_2)\mu_2^{(2)}] \tag{7}$$

Where

$$\mu_0^{(2)} = \frac{1}{2.2!}\int_0^2 (t-1)(t-2)dt = \frac{1}{6} \tag{8}$$

$$\mu_1^{(2)} = \frac{1}{2.2.1!}\int_0^2 t(t-1)dt = \frac{2}{3} \tag{9}$$

$$\mu_2^{(2)} = \mu_{2-2}^{(2)} = \mu_0^{(2)} = \frac{1}{6} \tag{10}$$

Thus on substituting above conditions and Eq. 11, we reach Eq. 12

$$I = (b-a)[\frac{1}{6}\pi(\Phi_0) + \frac{2}{3}\pi(\Phi_1) + \frac{1}{6}\pi(\Phi_2)] \tag{11}$$

$$= (b-a)[\frac{1}{6}\alpha_0 + \frac{4}{6}\alpha_1 + \frac{1}{6}\alpha_2] \tag{12}$$

where $\alpha_0 = \pi(\phi_0)$, $\alpha_1 = \pi(\phi_1)$ and $\alpha_2 = \pi(\phi_2)$.

$$= \frac{b-a}{6}[\alpha_0 + 4\alpha_1 + \alpha_2] \tag{13}$$

$$= \frac{h}{3}[\alpha_0 + 4\alpha_1 + \alpha_2] \tag{14}$$

This mathematics is used for the foundation of proposed method. We consider the three neighboring pixels intensities $\alpha_0, \alpha_1, \alpha_2$ of the low contrast input image and apply Eq. 14 using Eq. 13 on them. This is done for each pixel position and the new approximated value is used to generate the contrast improved output image. The proposed technique is shown in Algorithm 1.

Algorithm 1: CONTRAST ENHANCEMENT USING SIMPSON'S RULE

Input: Low contrast image $\Omega(M, N)$
Output: High contrast image $\Theta(M, N)$

for $\theta = 1..M-1$ **do**
  for $\psi = 1..N-1$ **do**
    $\eta(\theta, \psi) = \frac{h}{3}[\Omega_{(\theta-1,\psi-1)} + 4 * \Omega_{(\theta,\psi)} + \Omega_{(\theta+1,\psi+1)}]$
    if $\eta(\theta, \psi) < 0$ **then**
      $\Theta(\theta, \psi) = 0$
    else if $\eta(\theta, \psi) > 255$ **then**
      $\Theta(\theta, \psi) = 255$
    else
      $\Theta(\theta, \psi) = \eta(\theta, \psi)$

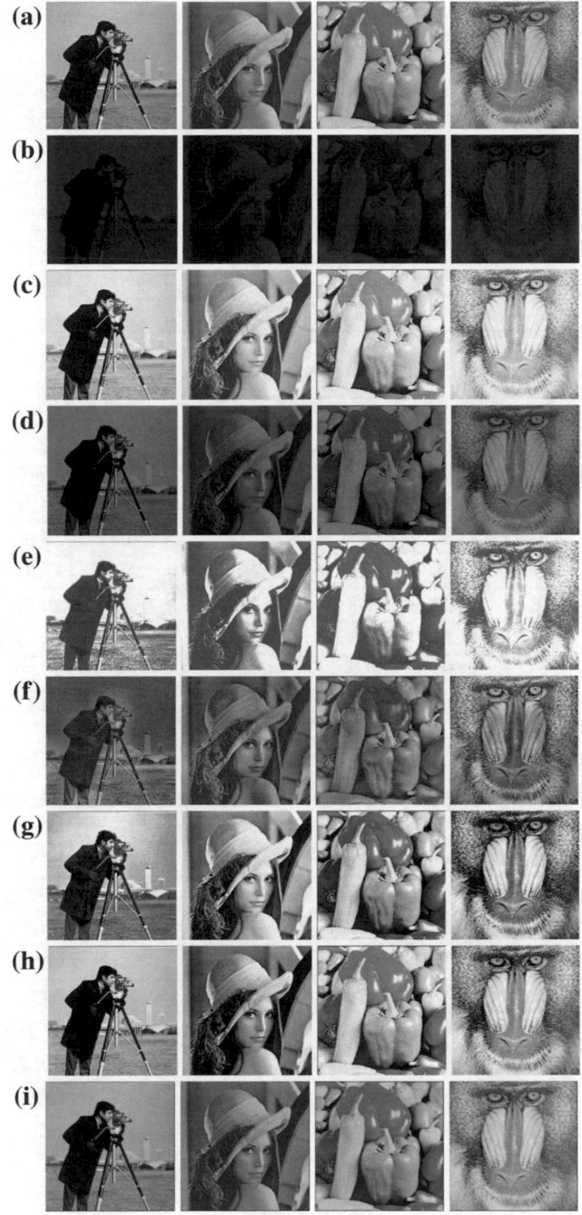

Fig. 1 Comparison of the output contrast images `Cameraman`, `Lena`, `Pepper` and `Mandril` using **c** AGCWD, **d** Sigmoid, **e** GA based, **f** AHE approach, **g** HE approach, **h** IM method and **i** PM, **a** is the original images and **b** is the low contrast images

Table 1 Compare RMSE values for different images using several existing methods and proposed method

Methods	Baloon	Barbara	Cameraman	Donald	Lena	Mandril	Pepper	Sania
Sigmoid	119.41	129.20	59.33	17.72	46.87	40.55	59.12	64.04
AGCWD	58.74	45.65	44.54	31.96	53.09	58.66	47.01	33.61
GA	55.76	54.91	65.84	63.84	54.31	74.31	56.23	36.40
AHE	44.18	32.81	64.42	47.83	24.09	31.40	39.23	64.23
HE	39.98	24.76	27.81	39.30	45.44	35.08	23.97	19.83
IM	32.63	16.21	21.45	34.81	23.80	27.75	23.62	12.58
PM	17.93	13.58	18.92	13.88	15.15	19.21	20.17	12.21

Table 2 Compare PSNR values for different images using several existing methods and proposed method

Methods	Baloon	Barbara	Cameraman	Donald	Lena	Mandril	Pepper	Sania
Sigmoid	6.58	5.90	12.66	23.15	14.71	15.96	12.69	12.00
AGCWD	12.75	14.94	15.15	18.03	13.63	12.76	14.68	17.60
GA	13.20	13.34	11.76	12.02	13.43	10.71	13.13	16.90
AHE	15.23	17.81	11.95	14.53	20.49	18.19	16.26	11.97
HE	16.09	20.25	19.25	16.24	14.98	17.23	20.54	22.18
IM	17.85	23.93	21.50	17.29	20.59	19.26	20.66	26.13
PM	23.05	25.46	22.59	25.27	24.51	22.45	22.03	26.39

Table 3 Compare SSIM values for different images using several existing methods and proposed method

Methods	Baloon	Barbara	Cameraman	Donald	Lena	Mandril	Pepper	Sania
Sigmoid	0.8017	0.7620	0.7688	0.9802	0.7766	0.8301	0.7761	0.7889
AGCWD	0.8353	0.8914	0.9298	0.9086	0.8371	0.8696	0.9007	0.9137
GA	0.7493	0.6684	0.6882	0.8212	0.6878	0.6007	0.7178	0.8007
AHE	0.8428	0.8428	0.7286	0.8110	0.8961	0.8852	0.8675	0.8255
HE	0.7118	0.8814	0.7729	0.5450	0.7789	0.7949	0.8579	0.8122
IM	0.7710	0.9129	0.9279	0.8149	0.8365	0.9085	0.9480	0.9601
PM	0.9695	0.9602	0.9412	0.9790	0.9497	0.8626	0.9614	0.9760

3 Experimental Results

For performance comparison, the proposed algorithm is compared with several existing methods such as Sigmoid function, AGCWD, GA, AHE, HE, and IM using three different parameters applied on several low contrast images namely `Baloon`, `Barbara`, `Cameraman`, `Donald`, `Lena`, `Mandril`, `Pepper` and, `Sania`, etc.

The qualitative parameters RMSE, PSNR [7] and SSIM [19] are used to measure and analyze the contrast of images. RMSE, PSNR, and SSIM values of the proposed method (PM) in comparison with different existing methods are shown in Table 1, Table 2, and Table 3, respectively. From the results, it shows that the proposed algorithm gives better outputs as compared to the other existing methods which is shown in Fig. 1. In proposed algorithm, the parameter h (the subinterval length) varies from 1.5 to 2.5 for better quality images. The parameters RMSE, PSNR, and SSIM are given in following Eq. 15, Eq. 16, and Eq. 17, respectively.

$$RMSE = \sqrt{\frac{1}{M \times N} \sum_{\theta=0}^{M-1} \sum_{\psi=0}^{N-1} |\Omega(\theta, \psi) - \Theta(\theta, \psi)|^2} \tag{15}$$

$$PSNR = 20 * log_{10} \frac{255}{RMSE} \tag{16}$$

Here $M \times N$ is the size of the images. $\Omega(i, j)$ represents the original input image whereas $\Theta(i, j)$ denotes the contrast improved output image.

$$SSIM(i, j) = \frac{(2\upsilon_i \upsilon_j + t_1)(2\beta_{ij} + t_2)}{(\upsilon_i{}^2 + \upsilon_j{}^2 + t_1)(\beta_i{}^2 + \beta_j{}^2 + t_2)} \tag{17}$$

where υ_i, υ_j are the mean values of the two windows i and j. β_i, β_j are the standard deviations of the two windows. β_{ij} is the covariance of i and j. t_1 and t_2 are the constants and the value of $t_1 = (0.01 \times 255)^2$ and $t_2 = (0.03 \times 255)^2$. From Fig. 1 and Tables 1, 2 and 3, it is noticed that visually as well as theoretically, the proposed algorithm gives the better results than other existing methods.

4 Conclusions

The proposed work presents a unique efficient method for enhancing the contrast of images based on numerical integration Simpson's $\frac{1}{3}$rd rule. Simpson's rule has distinct advantages over previously existing methods as it simply depends on the neighboring pixel intensities. This makes it useful and can be used for various other fields. However, further improvements are always possible for better contrast results.

References

1. Abdullah-Al-Wadud, M., Kabir, M.H., Dewan, M.A.A., Chae, O.: A dynamic histogram equalization for image contrast enhancement. IEEE Trans. Consum. Electron. **53**(2) (2007)
2. Agarwal, M., Mahajan, R.: Medical image contrast enhancement using range limited weighted histogram equalization. Procedia Comput. Sci. **125**, 149–156 (2018)

3. Datta, N., Jana, R.: Introductory Numerical Analysis. Shreedharprakashani (2010)
4. Gonzalez, R.C., Woods, R.E.: Digital Image processing. Prentice Hall (2002)
5. Huang, S.C., Cheng, F.C., Chiu, Y.S.: Efficient contrast enhancement using adaptive gamma correction with weighting distribution. IEEE Trans. Image Process. **22**(3), 1032–1041 (2013)
6. Ismail, N.H.B., Chen, S.D., Ng, L.S., Ramli, A.R.: An analysis of image quality assessment algorithm to detect the presence of unnatural contrast enhancement. J. Theor. Appl. Inf. Technol. **83**(3) (2016)
7. Kaur, R., Kaur, S.: Comparison of contrast enhancement techniques for medical image. In: Conference on Emerging Devices and Smart Systems (ICEDSS), pp. 155–159. IEEE (2016)
8. Kim, J.Y., Kim, L.S., Hwang, S.H.: An advanced contrast enhancement using partially overlapped sub-block histogram equalization. IEEE Trans. Circuits Syst. Video Technol. **11**(4), 475–484 (2001)
9. Kim, T.K., Paik, J.K., Kang, B.S.: Contrast enhancement system using spatially adaptive histogram equalization with temporal filtering. IEEE Trans. Consum. Electron. **44**(1), 82–87 (1998)
10. Lal, S., Chandra, M., et al.: Efficient algorithm for contrast enhancement of natural images. Int. Arab J. Inf. Technol. **11**(1), 95–102 (2014)
11. Mittal, N.: Automatic contrast enhancement of low contrast images using MATLAB. Int. J Adv. Res. Comput. Sci. **3**(1) (2012)
12. Moldvai, C.: Adaptive contrast enhancement. US Patent 7,760,961, 20 July 2010
13. Mollah, S.A.: Numerical Analysis and Computational Procedures. Publisher Arunabha Sen Books and Allied (P) Ltd. (2012)
14. Pal, S.K., Bhandari, D., Kundu, M.K.: Genetic algorithms for optimal image enhancement. Pattern Recognit. Lett. **15**(3), 261–271 (1994)
15. Reza, A.M.: Realization of the contrast limited adaptive histogram equalization (CLAHE) for real-time image enhancement. J. VLSI Signal Process. Syst. Signal Image Video Technol. **38**(1), 35–44 (2004)
16. Senthilkumaran, N., Thimmiaraja, J.: Histogram equalization for image enhancement using MRI brain images. In: 2014 World Congress on Computing and Communication Technologies (WCCCT), pp. 80–83. IEEE (2014)
17. Singh, R.P., Dixit, M.: Histogram equalization: a strong technique for image enhancement. Int. J. Signal Process. Image Process. Pattern Recognit. **8**(8), 345–352 (2015)
18. Stark, J.A.: Adaptive image contrast enhancement using generalizations of histogram equalization. IEEE Trans. Image Process. **9**(5), 889–896 (2000)
19. Wang, Z., Bovik, A.C., Sheikh, H.R., Simoncelli, E.P.: Image quality assessment: from error visibility to structural similarity. IEEE Trans. Image Process. **13**(4), 600–612 (2004)
20. Zhuang, L., Guan, Y.: Image enhancement via subimage histogram equalization based on mean and variance. Comput. Intell. Neurosci. **2017** (2017)

Crime Feature Selection Constructing Weighted Spanning Tree

Priyanka Das and Asit Kumar Das

Abstract The proposed work demonstrates a rough set based feature selection scheme for selecting crime features from online newspaper reports of crime performed against women in India. Only the verbs present in the crime reports are considered as the extracted features for crime analysis task. To select only the distinct verbs, all the words with common synonyms are identified and replaced by a single word. Most often the set of features contains the relevant as well as many irrelevant features. Hence, for any classification task, it is highly essential to select only the relevant features for accurate classification. In the proposed work, the rough set theory based relative indiscernibility relation is used to measure the similarity score between two features relative to the crime type. Then a weighted undirected graph has been generated that comprises the features as nodes and the inverse similarity score between two features as the weight of the corresponding edge. Prim's algorithm is applied to obtain a minimal spanning tree. Finally, a feature selection algorithm has been developed that selects the highest degree node and removes it from the tree iteratively until the modified graph becomes a null graph. The selected nodes are considered as the important features sufficient for crime reports categorization.

Keywords Feature selection · Rough set theory · Relative indiscernibility relation · Minimal spanning tree

1 Introduction

A text document is generally represented by vector of words. The words representing a document are known as conditional features and document type is the decision feature. There exist many irrelevant and noisy features providing less accuracy in

---

P. Das (✉)
Indian Institute of Engineering Science and Technology, Shibpur,
Howrah 711103, India
e-mail: priyankadas700@gmail.com

A. K. Das
e-mail: akdas@cs.iiests.ac.in

A. K. Das et al. (eds.), *Computational Intelligence in Pattern Recognition*,
Advances in Intelligent Systems and Computing 999,
https://doi.org/10.1007/978-981-13-9042-5_33

classification tasks. Therefore, in any classification scheme, it is highly essential to remove the irrelevant features and select optimal set of relevant features. The feature selection technique keeps only the relevant set of uncorrelated features having high similarity with the decision feature. Feature selection is almost similar to another data mining process called dimension reduction. Z. Pawlak was the first scientist to describe Rough Set theory (RST) [1], for reducing dimensions of massive datasets. Rough set theory helps in partitioning a data into some equivalent classes by measuring the importance of the data. Hu et al. [2] proposed a new rough set model with modified definitions of core features and reduct. Apart from dimension reduction, rough set theory is also used in feature selection [3]. A fast clustering [4] algorithm was proposed to select optimal set of features from high dimensional data. The efficiency of the fast clustering mechanism was validated with the efficiency of minimum spanning tree algorithm. A feature selection algorithm developed in [5] initially removed the irrelevant features from the data, measured the F-correlation score among the relevant features and constructed a fully connected weighted graph. Then minimal spanning tree was constructed using the Boruvka's algorithm. Alongwith the minimal spanning tree, fisher score was calculated for optimizing a feature selection technique [6]. Concepts of rough set and minimal spanning tree also have been applied in crime domain. Recently, [7] designed a forensic analysis system, where minimal spanning tree of a criminal network was generated and the leaders of the organization were identified. The strength pareto algorithm was chosen as a multi-objective optimization that provides non dominated pareto front for the optimal selection of crime features [8].

The proposed work describes a feature selection scheme based on crime data focusing on crime against women in India. The present work has tagged the content of the crime reports with their particular parts-of-speech and only the verbs present in the crime reports are considered as the conditional features. Relative indiscernibility relation measures how a conditional feature is related to the decision feature. A weighted undirected graph is constructed with features as nodes of the graph and the inverse of the similarity score measured by the relative indiscernibility has been assigned as the weight of the edges. Then Prim's algorithm has been applied to obtain the minimal spanning tree of the graph. Finally, the node with the highest degree is selected and removed from the tree. It may happen that two or more nodes are of the highest degree and in that case, the node having the maximum sum of weights of the adjacent edges gets selected. The process has been repeated until the tree becomes a null graph and the list of all removed features are considered as the optimal features. The proposed spanning tree based feature selection technique has been compared with some *state-of-the-art* feature selection methods in terms of number of features selected and different performance metrics like accuracy, precision, recall, and F-measure.

The rest of the paper is organized as follows: Sect. 2 describes the proposed feature selection technique in detail. Section 3 illustrates the experimental results and performance evaluation. Finally, Sect. 4 discusses the conclusion and future work.

2 The Proposed Methodology

The rough set based feature selection scheme from crime data is discussed in detail in the following subsections:

2.1 *Data Collection and Preparation*

The classified newspaper "The Times of India" has been chosen for collecting the crime data for the present work. Crime reports relating to the crime terms like "rape", "abduction", "molestation", etc. have been crawled from the web pages of the aforementioned newspapers. A total of 3647 crime reports (based on different crimes committed against women in India) have been extracted. The collected data comprises the crime reports for 29 states and 4 union territories of India for a time period of 2011–2016. As preprocessing, stopword removal and Porter Stemming [9] have been done. Then a part-of-speech (POS) tagger is used to tag each word in a sentence with their parts-of-speech. All these steps have been achieved by the Natural Language Tool Kit module available in Python [9].

2.2 *Collection of Verbs*

After the parts-of-speech tagging, we have chosen all the verbs present in the crime reports. The verbs in a sentence refer to some actions or events. So we have focused on the verbs for deriving useful crime related features. It may happen that a particular verb occurs more than once in the reports, so the verbs have been stored in a Python list, which has been later converted into a Python set of objects that removes all the duplicate verbs from the list. Now, there may also exist other synonyms of a verb. Hence, we stored all possible synonyms of the individual unique verbs obtained from the reports. We appended the "lemma_names from the "nltk.corpus.wordnet for the individual verbs in a list. Each individual verb in the reports has been replaced by their corresponding list of synonyms obtained previously. For example, a verb "v1" has synonyms "v2", "v3", "v4" and the verb "v2" has synonyms "v1", "v3", and "v5". Then while replacing the verbs with synonyms, if 'v2' appears first then it is replaced by "v1", but if afterwards "v1" appears, then it is not replaced. Hence, the synonym replacement operation is made idempotent in nature. Thus, some unique words (verbs) are generated that characterizes the crime reports. For each crime report, a context vector is created whose values have been obtained by calculating the term frequency and inverse document frequency (TF-IDF) of the unique terms. Let E_c be a collection of crime reports, c be a context word or feature, $R \in E_c$ be an individual crime report. The Term Frequency-Inverse Document Frequency (TF-IDF) of feature c in R is denoted by $f_R(c)$ and is defined by Eq. (1).

$$f_R(c) = f_{c,R} * log(\frac{|E_c|}{|\{R \in E_c : c \in R\}|}) \quad (1)$$

where, $f_{c,R}$ is the frequency of c in report R and $|\{R \in E_c : c \in R\}|$ is the number of reports in which the word c appears.

2.3 *Relative Indiscernibility and Feature Dependency*

Let DS be a decision system containing non-empty sets U, F and D usually represented as $DS = (U, F, D)$, where U defines the universal set of all crime reports and F is the set of features (verbs) and D defines the set of decision features, i.e., crime types. If any two crime reports are not discernible by their characteristics they are called equivalent crime reports and the equivalence relation between these two reports are known as indiscernibility relation. Let the universal set $U = \{N_1, N_2, \cdots, N_n\}$ comprises n reports and S is a subset of conditional features, i.e., $S \subseteq F$. The indiscernibility relation is defined as $IND(S) = \{(N, P) \in U^2 | \forall f \in S, f(N) = f(P)\}$. Here, N and P are two indiscernible crime reports and the relation $IND(S)$ partitions the set of reports into equivalence classes $[N]_S$. The equivalence class reports are indiscernible with respect to set S, whereas any two reports having different equivalence classes are discernible with respect to S. The present work considers the crime reports as the objects, the unique terms represent the conditional features and the crime type refers the decision feature. The relative indiscernibility relation yields the crime reports based on a particular term, relative to the crime class. Therefore, to obtain the indiscernible crime reports based on a term, the dataset has been partitioned based on that term as well as the crime class. All the partitions (induced by each term and the crime class) having similar set of crime reports are placed in the same class. Every conditional feature V_i of F determines a relative (relative to decision feature or crime class) indiscernibility relation (RIR) over the universal set U and is represented as $RI(V_i)$ denoted using Eq. (2).

$$RI(V_i) = \{(N, P) \in \Pi_{V_i}[N]_D \times \Pi_{V_i}[N]_D | f_{V_i}(N) = f_{V_i}(P) \forall [N]_D \in U/D\} \quad (2)$$

It is obvious that each equivalence class $\{[N]_{V_i/D}\}$ contains crime reports with same crime class which are indiscernible by the feature V_i.

2.4 *Feature Similarity Measurement*

In the context of classification, a feature V_i is said to have similarity with other feature V_j if they are placed in same equivalence classes of crime reports under their respective relative indiscernible relations. In general, a similarity factor is used to assess the similarity between two features. Here, the similarity between the feature V_i and feature V_j is measured using Eq. (3).

$$S(V_i, V_j) = \frac{1}{|U_D/V_i|} \sum_{[N]_{V_i/D} \in U_D/V_i} \frac{1}{[N]_{V_i/D}} \max[N]_{V_j/D} \in U_D/V_j([N]_{V_i/D} \cap [N]_{V_j/D}) \quad (3)$$

For each pair of conditional features (V_i, V_j), two different similarities $V_i \rightarrow V_j$ and $V_j \rightarrow V_i$ both have been computed. The higher value of $V_i \rightarrow V_j$ similarity score implies that the relative indiscernible relations $RI(V_i)$ and $RI(V_j)$ yields highly similar equivalence classes. This also implies similar classification power for both the features V_i and V_j. Since, two different similarity scores have been obtained, we have considered the average of these two similarity scores in the feature similarity set.

2.5 *Minimal Spanning Tree Generation*

Based on the feature similarity score, a complete weighted undirected graph, $G = (N, E, W)$ is formed, where N defines the set of nodes or set of features, E is the set of edges and W is the set of weights assigned to the edges in E. Now, as the present work focuses on minimal spanning tree generation, we have considered the inverse of the average similarity score as the weight. It is known that the traditional Prim's algorithm provides the optimal result for minimal spanning tree generation from undirected graphs. Therefore, once the graph G has been obtained, Prims algorithm has been used to yield the minimal spanning tree of the graph. A spanning tree of a graph G is a tree containing all the nodes of the original graph. The cost of the spanning tree is the sum of the weights of all the edges present in the tree. There may exist many spanning trees. Minimum spanning tree is considered to be the spanning tree where the cost is minimum among all the spanning trees. There also can exist many minimum spanning trees. The present work has taken the feature selection process one step closer by considering Prim's algorithm for generating only one minimal spanning tree of the feature similarity graph G. Figure 1a shows the original graph G along-with the minimal spanning tree obtained by Prim's algorithm in Fig. 1b. The advantage of using Prim's algorithm is that for our work, the graph was really dense with many number of edges than nodes. So, it performed significantly well for our proposed work.

2.6 *Feature Selection Based on the Minimal Spanning Tree*

This section reflects the most significant part of the proposed work by developing a graph-based feature selection technique that has been applied on the newly gener-

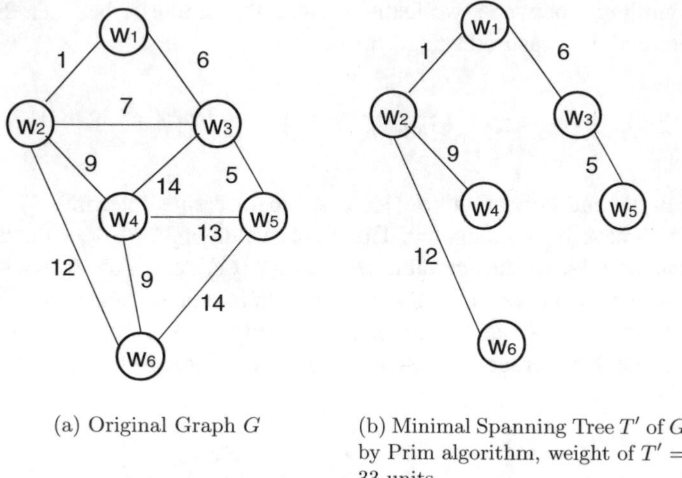

(a) Original Graph G

(b) Minimal Spanning Tree T' of G by Prim algorithm, weight of $T' =$ 33 units

Fig. 1 Minimal spanning tree of graph G by prim algorithm

ated minimal spanning tree of the original feature similarity graph G. The minimal spanning tree T contains all the vertices and some of the edges of the original graph. The proposed feature selection technique is an edge deletion based approach, where initially, the degree of each node present in the minimal spanning tree has been computed. The degree of a node is defined as the number of edges incident with that node. Now, the particular node, say n_{max} with the highest degree has been deleted from the graph. It may be noted that there may exist more than one node with the highest degree and in that case, importance has been given to the sum of weights of the adjacent edges of the nodes. Here, the node with the highest sum of the weights of adjacent edges has been deleted. Also, it may happen that multiple such nodes exist with similar weight of the adjacent edges and then a random deletion of a node has been done. The process of deletion and selection of nodes is repeated until the graph becomes a null graph provides a list of features. The detail of the feature selection technique is described in Algorithm 1.

3 Experimental Results

The experiments for proposed methodology have been performed in Python and tested on a crime dataset collected from online newspaper. The preprocessed dataset contains a total of 407 crime features and 3647 crime reports based on 11 different types of crime, namely “abduction”, “rape”, “abuse”, “murder”, “human trafficking”, “child sexual abuse”, “infanticide”, “domestic violence”, “sexual harassment at work”, “street harassment”, and “molestation”, performed against women in India

Algorithm 1: FSST(T)

```
Input: Minimal spanning tree T = (N, Ê, Ŵ), N= set of nodes in T, Ê = set of edges, Ŵ =
       weight of the edges = actual similarity scores between every pair of features;
Output: F_D, the selected features;
begin
    F_D = φ;
    PUSH (stack S, T) /* Insert T into stack S*/;
    repeat
        T ← POP(S) /* Remove top element from S */;
        for each node n_i ∈ N in T do
            Calculate the degree of the node n_i;
        end
        Let N_1 ⊆ N contains nodes with highest degree;
        if |N_1| = 1 then
            n_max ← node in N_1;
            else
                for each n_j ∈ N_1 do
                    compute w(n_j) = sum of weights of edges adjacent to n_j
                end
            end
        end
        Let w_max = max{w(n_j)∀n_j ∈ N_1};
        if w_max occurs for single node n_P ∈ N_1 then
            n_max ← n_P;
            else
                n_max ← any one of these nodes for which w_max occurs
            end
        end
        F_D = F_D ∪ {n_max};
        N = N − n_max /* delete node n_max from N */;
        Let number of deletion of n_max creates t number of components
        T_1, T_2, ··· , T_t /* each component has atleast one edge */;
        for i = 1 to t do
            PUSH (S, T_i)
        end
    until (S is empty);
    Return F_D;
end
```

for over a time period of 2011–2016. The proposed feature selection technique yields optimal subset of 74 crime features from the crime dataset. These features are very useful for discriminating different types of crime reports. The effectiveness of the proposed method is shown in the following subsections as follows:

3.1 Comparison with Graph Based Algorithms

The proposed graph based feature selection technique has been compared with some existing graph based methods like Infomap, Louvain, Fastgreedy, and Girvan-Newman. The proposed methodology has used these graph based algorithms from the Python i-graph package [10] implementation. These graph based algorithms divide the whole graph into several partitions, where each partition contains a collection of features. Now, the clustering coefficient has been computed using Eq. (4) for all the nodes (features) present in each partition.

$$c_{n_i} = \frac{1}{deg(n_i)(deg(n_i) - 1))} \sum_{n_i n_j} (\hat{w}_{ij} \hat{w}_{jk} \hat{w}_{ki})^{1/3} \quad (4)$$

where, c_{n_i} is the clustering coefficient of node n_i, summation is taken over all the triangles $\Delta n_i n_j n_k$ formed through the node n_i and the weight w_{ij} of edge (n_i, n_j) is normalized to $\hat{w}_{ij}$ by the maximum weight in the network, i.e, $\hat{w}_{ij} = w_{ij} / \max\{w_{ij} / w_{ij} \in W\}$. The value of c_{n_i} is assigned as 0 if $deg(n_i) < 2$.

Once the clustering coefficient is being calculated, the node with highest clustering coefficient has been selected. The process has been applied to all the partitions of the graph and a feature subset is obtained by each of the graph- based algorithms. This implies that number of features will vary with number of partitions. The methods Infomap Community Detection (IN), Fastgreedy (FS), Girvan-Newman (GR) and Louvain (LV) with this approach are named in experimental results as Infomap Community Detection with clustering coefficient (INCC), Fastgreedy with clustering coefficient (FSCC), Girvan-Newman with clustering coefficient (GRCC) and Louvain with clustering coefficient (LVCC) respectively. These four methods select 42, 47, 53, and 59 features respectively from the processed crime dataset of 407 features.

3.2 Evaluation of Feature Selection on Various Classifiers

The evaluation of the proposed and compared graph-based feature selection methods has been performed on various existing classifiers. K-Nearest Neighbor (KNN), Classification Tree (CT), Support Vector Machine (SVM), Random Forest (RF), Neural Network (NN), and Logistic Regression (LR) have been used for measuring the accuracy of the feature selection techniques. Apart from the classification accuracy (ACC), precision (PRN), recall (RCL), and F-measure (FMR) have also been computed. The precision value for a classifier with respect to a particular class is the ratio of the number of objects correctly predicted to the total number of objects of that class. The recall for a class is the ratio of the objects correctly predicted to the total number of objects actually belongs to the class. The precision and recall and F-measure are defined in Eqs. (5–7). The accuracy of the classifier is the proportion of total number of objects of all objects in the dataset that are correctly classified, defined in Eq. (8).

$$PRN = \frac{T_P}{T_P + F_P} \tag{5}$$

$$RCL = \frac{T_P}{T_P + F_N} \tag{6}$$

$$FMR = \frac{2(PRN)(RCL)}{PRN + RCL} \tag{7}$$

$$ACC = \frac{T_P + T_N}{T_P + F_P + F_N + T_N} \tag{8}$$

Table 1 Comparison of proposed FSST method with graph based methods using different classifiers

Classifier	Metric	FSST (74)	INCC (42)	FSCC (47)	GRCC (53)	LVCC (59)
KNN	ACC	**73.41**	62.12	70.21	73.14	62.77
	PRN	74.12	65.13	72.23	71.11	63.17
	RCL	75.32	64.17	70.14	70.22	62.25
	FMR	74.71	64.64	71.16	70.66	62.70
CT	ACC	**85.23**	71.25	81.21	77.23	68.42
	PRN	85.23	71.51	79.24	73.41	68.11
	RCL	85.23	74.42	81.23	75.52	69.23
	FMR	85.23	73.45	80.22	74.45	68.66
SVM	ACC	72.15	67.50	72.15	**73.51**	64.13
	PRN	71.13	67.22	72.00	70.14	65.24
	RCL	70.14	68.46	67.22	69.25	67.22
	FMR	70.63	67.83	69.52	69.69	66.21
RF	ACC	**79.21**	78.26	72.33	70.12	67.15
	PRN	77.32	78.34	71.34	72.11	70.23
	RCL	78.22	79.27	72.19	72.13	72.45
	FMR	77.76	78.80	71.76	72.11	71.32
NN	ACC	**83.11**	82.15	72.13	70.16	71.23
	PRN	84.25	80.29	73.14	70.25	71.14
	RCL	83.72	82.15	74.42	72.16	72.33
	FMR	83.98	81.20	73.77	71.19	71.73
LR	ACC	**84.26**	68.34	73.11	76.17	70.14
	PRN	85.70	67.23	71.22	77.48	72.24
	RCL	84.14	65.37	73.25	76.56	70.25
	FMR	84.91	66.28	72.22	77.01	71.23

where, terms T_P, F_P, T_N and F_N refers to true positive, false positive, true negative and false negative respectively.

The experimental results are achieved using 10 fold cross validation method and the final evaluation results as shown in Table 1 are obtained by averaging the results of the 10 evaluations. It is also observed from the Table 1 that KNN, CT, RF, NN, and LR provides the highest accuracy scores for the proposed approach, whereas SVM gives the highest accuracy for GRCC method.

4 Conclusion and Future Work

The proposed methodology demonstrated a minimal spanning tree based feature selection technique for crime data analysis. Each feature in a dataset contains some information about the crime report and the feature selection approach eliminates all the unwanted information from the dataset and helps in achieving optimal features. Here, the features have been taken from only one newspaper and the optimal set of features describes crime-related activities performed against women in India. The relative indiscernibility relation, the basic concept of rough set theory is used to make the graph weighted and finally the minimal spanning tree generation algorithm, namely PRIM's algorithm is used. It is also verified that Kruskal algorithm, another popular minimal spanning tree generation algorithm provides the same feature subset. The same task can be performed in future using other part-of-speech like nouns and adjectives together with directed spanning tree and different parametric measures of the nodes, instead of considering only the clustering coefficient. The experimental results may be compared using crime reports collected from different newspapers from different countries. Based on the selected features a novel classifier may be generated in future to analyze and identify the future crime trends of different districts and cities of our country. Not only for crime data, this proposed work will perform equally well for other datasets also.

References

1. Pawlak, Z.: Rough set theory and its applications to data analysis. Cybern. Syst. **29**(7), 661–688 (1998)
2. Hu, X.T., Lin, T.Y., Han, J.: A new rough sets model based on database systems. In: Rough Sets, Fuzzy Sets, Data Mining, and Granular Computing, pp. 114–121 (2003)
3. Zhang, M., Yao, J.T.: A rough sets based approach to feature selection. In: IEEE Annual Meeting of the Fuzzy Information, 2004. Processing NAFIPS '04, vol. 1, pp. 434–439 (2004)
4. Song, Q., Ni, J., Wang, G.: A fast clustering-based feature subset selection algorithm for high-dimensional data. IEEE Trans. Knowl. Data Eng. **25**(1), 1–14 (2013)
5. Yaswanth Kumar Alapati, K., Sindhu, S.S.: Relevant feature selection from high-dimensional data using mst based clustering. Int. J. Emerg. Trends Sci. Technol. **2**(3), 1997–2001 (2015)

6. Singh, B., Sankhwar, J.S., Vyas, O.P.: Optimization of feature selection method for high dimensional data using fisher score and minimum spanning tree. In: 2014 Annual IEEE India Conference (INDICON), pp. 1–6 (2014)
7. Taha, K., Yoo, P.D.: Using the spanning tree of a criminal network for identifying its leaders. IEEE Trans. Inf. Forensics Secur. **12**(2), 445–453 (2017)
8. Das, P., Das, A.K.: An application of strength pareto evolutionary algorithm for feature selection from crime data. In: 2017 8th International Conference on Computing, Communication and Networking Technologies (ICCCNT), pp. 1–6 (2017)
9. Loper, E., Bird, S.: NLTK: The natural language toolkit. In: Proceedings of the ACL-02 Workshop on Effective Tools and Methodologies for Teaching Natural Language Processing and Computational Linguistics. ETMTNLP '02, pp. 63–70 (2002)
10. Csardi, G., Nepusz, T.: The igraph software package for complex network research. InterJ. Complex Syst. **1695** (2006)

Empirical Analysis of Proximity Measures in Machine Learning

Nazrul Hoque, Hasin A. Ahmed and Dhruba Kumar Bhattacharyya

Abstract Availability of abundant and various types of proximity measures often projects a challenge in both supervised and unsupervised learning processes. There are various similarity and dissimilarity measures proposed in the literature of machine learning. These measures differ with respect to various issues imposed by different application domains such as ability to handle noise, ability to detect various types of correlation, and coping with large number of dimensions. In this work, we pick-up eighteen proximity measures and apply them on two well known distance-based learning frameworks. One framework uses a widely used supervised learning method, i.e., KNN classifier and the other uses an unsupervised learning method called k-means clustering.

Keywords Proximity measure · Distance-based machine learning · KNN classifier · K-means clustering

1 Introduction

Due to the tremendous growth of database industries, advanced data analysis techniques have been used to extract valuable knowledge from data. Different learning models or algorithms are used to extract hidden knowledge from data during analysis. A learning model takes a set of input data, analyzes it, extracts knowledge from the data and predicts about data. Feature selection methods are also used to select

N. Hoque (✉) · D. K. Bhattacharyya
Department of CSE, Tezpur University, Sonitpur 784028, Assam, India
e-mail: tonazrul@gmail.com

D. K. Bhattacharyya
e-mail: dkb@tezu.ernet.in

H. A. Ahmed
Department of Information and Computer Science, Assam Women's University, Jorhat 785004, Assam, India
e-mail: hasin@tezu.ernet.in

A. K. Das et al. (eds.), *Computational Intelligence in Pattern Recognition*, Advances in Intelligent Systems and Computing 999,
https://doi.org/10.1007/978-981-13-9042-5_34

informative features from a large feature space during learning [6, 11]. However, the type or pattern of input data plays a vital role in learning models. To analyze different types of data, a learning model is categorized into two classes, viz., supervised learning model and unsupervised learning model. A supervised learning model is used to analyze data where the data elements contain their class labels. On the other hand, unsupervised learning model uses unlabeled data during analysis.

A supervised learning technique is also known as a classification process. A classification process consists of two phases, viz., training phase and testing phase. In training phase, a learning model takes a set of input objects/data with their class labels and the model extracts valuable insights from the labeled objects. During testing, the model takes objects without any label and predicts labels for the unknown objects based on the knowledge extracted during training phase. The classification process uses different approaches, such as statistical measures and probabilistic measures to predict class labels of unknown objects. A common unsupervised learning technique is clustering. A clustering process groups similar types of objects into clusters. To group the similar types of objects, a clustering method uses proximity measure to determine inter object similarity.

1.1 Proximity Measures for Learning

A distance-based supervised or unsupervised learning method uses various proximity measures such as Euclidean distance, Manhattan distance, Mahalanobish distance, Pearson, Spearman, and Kendall correlation. The main advantage of a distance-based learning scheme is that the learning method can apply any proximity measure to find the nearest objects.

A proximity measure cannot be applied directly to all types of data during data analysis. Since, a data object may be of different types such as numeric, categorical, or mixed type and hence proximity measures are also applied based on the types of data. For numeric data analysis, people typically use Euclidean, Seuclidean, Manhattan, BrayCurtis, Canberra, Cosine, Pearson, Spearman, Kendall, or normalized Euclidean distance. However, for boolean data analysis, Hamming, Jaccard, or Dice distance are used. In this paper, we focus on measures suitable for numeric data types.

1.2 Motivation and Contributions

Inter-object distance computation in a multidimensional dataset, especially when dimensionality is high, is an issue. Researchers do not resort to a single point due to existence of various potential ways to compute inter-object distance in a multidimensional scenario. This conflict obviously is inherited to the machine learning techniques that adopt such computation as part of the learning process. At the expense of this dissidence, distance-based learning techniques offer flexibility to the users to pick a distance or proximity computation method that best suits the requirement of the problem in an application domain that is being addressed. Finding a proximity measure that fits to a problem is again a daunting task as it requires adequate knowledge on large number of measures on the problem and finally an evaluation system has to be adopted to decide which of the measures perform well in the learning process. As far as our knowledge is concerned, there is no appropriate article that tries to analyze effectiveness of proximity measures empirically both on supervised and unsupervised learning framework. Here in this article, we consider UCI machine Learning and Disease Diagnosis datasets to address the issue of selecting the best proximity measure during learning. Two major distance-based supervised and unsupervised techniques namely KNN classifier and K-means clustering techniques are used to evaluate effectiveness of eighteen proximity measures over the chosen datasets. A widely used evaluation system that can be generalized to both supervised and unsupervised set up is also adopted in the analysis.

2 Proposed Framework

We propose a framework to evaluate effectiveness of the measures in terms of ability to detect class distribution in a distance-based supervised and unsupervised learning setup. In our analysis, we use eighteen measures such as Euclidean distance [10], Squared Euclidean distance [20], Cityblock distance [10], Pearson's correlation coefficient [19], Spearman rank correlation [16], Kendall rank correlation coefficient [13], Cosine similarity [10], Minkowski distance [15], Chebyshev distance [5], Mutual information [14], Dice similarity index [9], Jaccard index [17], Canberra distance [12], BrayCurtis dissimilarity [4], Bhattacharyya distance [3], Angular separation [2], Normalized Mean Residue Similarity [18], Footrule distance [2], and SSSim [1]. The similarity measures are normalized and then the normalized value is subtracted from one to convert it to a dissimilarity score. These dissimilarity measures are then used in K-means clustering algorithm and KNN classification algorithm in an unsupervised and supervised learning setup, respectively, as shown in Fig. 1. One factor in K-means clustering algorithm is the value of K, which is set typically by user. Since K-means clustering algorithm is employed in a dataset to check it's ability to detect given class distribution for a measure, we decided to use total number of classes present in the dataset as the value of K. The clusters are then evaluated using $F\text{-}measure_{unsup}$. $F\text{-}measure_{unsup}$ is computed as geometric mean of $precision_{unsup}$ and $recall_{unsup}$. For a clustering solution C with clusters $c_1, c_2, ..., c_n$ and set of classes G with classes $g_1, g_2, ..., g_m$, $precision_{unsup}$ can be computed using Eq. 1 as follows [21].

$$Precision_{unsup} = \frac{|\{c_i | c_i \in C, \exists g_j \in G, Overlap\_score(c_i, g_j) \geq \alpha\}|}{|C|} \quad (1)$$

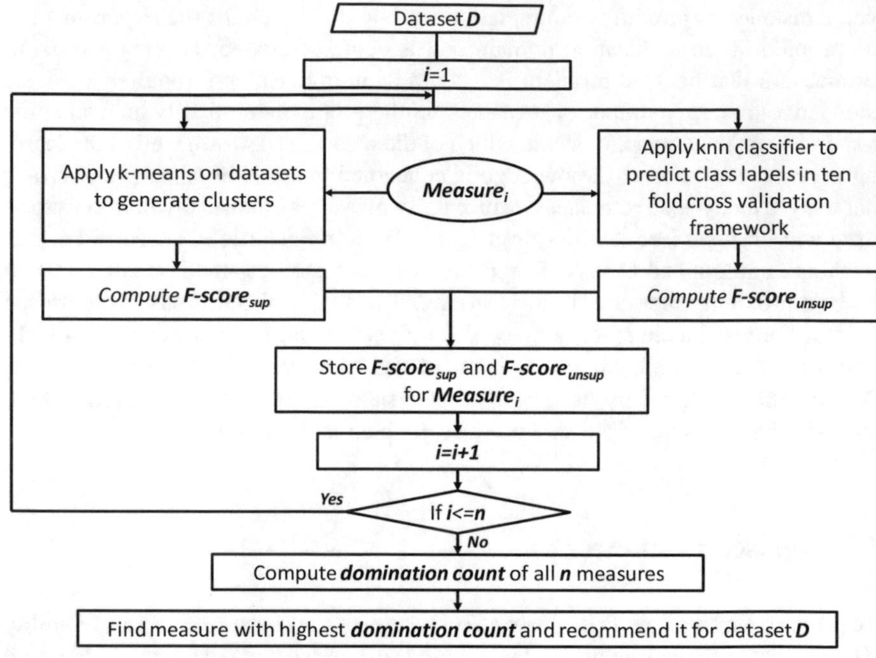

Fig. 1 Framework for evaluation of *n* proximity measures for a dataset in a distance based learning set-up

Here, $Overlap\_score(c_i, p_j)$ is the Jaccard index between cluster c_i and class g_j. The overlap score between c_i and g_j is computed using Eq. 2 as follows [21].

$$Overlap\_score(c_i, g_j) = \frac{|c_i \cap g_j|}{|c_i \cup g_j|} \tag{2}$$

Let, α is the overlap threshold value. Similarly, $recall_{unsup}$ can be computed using Eq. 3 as follows. $Recall_{unsup} =$

$$\frac{|\{g_i | g_i \in G, \exists c_j \in C, Overlap\_score(g_i, c_j) \geq \alpha\}|}{|G|} \tag{3}$$

For KNN classifier, we assign value 1 to the input parameter k. We evaluate the result of classification using F-measure$_{sup}$. F-measure$_{sup}$ is computed as geometric mean of precision$_{sup}$ and recall$_{sup}$.

Precision$_{sup}$ is computed using Eq. 4 as follows [7].

$$Precision_{sup} = \frac{TP}{TP + FP} \tag{4}$$

where TP (True Positive) is the number of correctly predicted objects, FP (False Positive) is the number of negative objects predicted as positive. Recall_{sup} is computed using Eq. 5 as follows [7].

$$Recall_{sup} = \frac{TP}{TP + FN} \tag{5}$$

where FN (False Negative) is the number of positive objects predicted incorrectly as negative.

As shown in Fig. 1, domination counts of measures in both the framework are computed. Domination count [8] of a measure in a learning framework is defined as the number of other measures whose $F - measure_{sup}/F - measure_{unsup}$ are less than that of the measure. For a dataset, measure with highest average domination count over the two frameworks is considered the best proximity measure.

3 Experimental Result Analysis

Experiments were carried out on a workstation with 12 GB main memory, 2.26 Intel (R) Xeon processor and 64-bit Windows 7 operating system. We implement our algorithm using MATLAB R2015 software. We choose some benchmark datasets of varied instances and dimensionality. We categorized them as $UCI_{general}$ and $UCI_{disease}$.

3.1 $UCI_{general}$ Datasets: Results and Analysis

From the empirical analysis of different proximity measures applied on $UCI_{general}$ datasets using supervised learning, it is observed that cityblock and bray-curtis give higher accuracy than other measures. As shown in Table 1, performance of Dice correlation measure is very poor in predicting the class label of unknown data object using KNN classifier. This is mostly because of the data variance of the training data objects while predicting the class labels of unknown objects. However, bray-curtis fails to perform well in case of D18, i.e., NIC dataset. The performance of proximity measures in terms of $F - measure_{unsup}$ on various UCI general datasets, we observed that Spearman and Pearson correlation give better accuracy that other measures. Table 2 shows that the average domination counts (ADC) of Spearman, Pearson, Kendall, Cityblock, Bray-curtis, and Footrule are greater than 4, i.e., greater than all other competing measures.

Table 1 Domination count of different measures and their average domination count (ADC) on $UCI_{general}$ datasets using $F - measure_{sup}$

	D1	D2	D3	D4	D5	D6	D7	D8	D9	D10	D11	D12	D13	D14	D15	D16	D17	D18	ADC
Euclidean	8	10	13	13	16	14	7	9	10	5	13	12	7	8	13	10	6	13	10.389
Cityblock	15	16	16	12	10	10	9	8	15	9	15	14	10	17	15	16	3	16	12.556
Pearson	11	6	4	5	11	15	11	13	12	1	12	6	3	12	7	15	10	17	9.5
Spearman	13	11	10	9	12	4	13	16	6	13	5	3	11	14	4	5	13	14	9.7778
Kendall	14	9	11	7	15	5	15	15	4	10	6	5	13	15	3	6	14	15	10.111
Cosine	12	15	14	4	17	13	12	6	13	2	11	11	5	9	8	17	9	11	10.5
Minkowski	9	7	12	17	14	12	5	7	9	8	14	9	6	10	12	9	4	12	9.7778
Seuclidean	4	2	2	1	2	2	3	0	0	12	4	0	12	3	2	3	16	5	4.0556
Chebychev	5	3	5	11	6	17	6	4	8	6	10	8	4	6	10	11	7	8	7.5
Dice	0	0	0	0	0	0	0	1	3	17	1	15	1	0	0	0	1	1	2.2222
Jaccard	1	1	1	2	1	0	1	2	5	16	0	16	0	1	1	0	0	2	2.7778
Canberra	6	17	17	14	8	9	17	11	17	0	9	2	15	7	11	12	15	6	10.722
Bray-Curtis	17	14	15	15	13	11	10	14	16	4	16	10	9	11	17	13	5	0	11.667
Bhattacharyya	3	5	7	16	3	7	4	17	2	7	2	1	2	4	9	7	2	3	5.6111
Angular	2	4	3	10	5	3	2	5	1	15	8	17	16	2	16	2	17	7	7.5
NMRS	10	8	9	8	7	16	8	10	14	3	17	13	8	16	14	14	8	9	10.667
Footrule	16	13	8	6	9	6	14	12	7	14	7	4	14	13	5	4	12	10	9.6667
SSSim	7	12	6	3	4	8	16	3	11	11	3	7	17	5	6	8	11	4	7.8889

D1 = Zoo, D2 = Accute1, D3 = Accute2, D4 = HayesRoath, D5 = House Voter, D6 = Iris, D7 = Wine
D8 = TicTacToe, D9 = Glass, D10 = Cleavland, D11 = Ecoli, D12 = CMC, D13 = Automobile, D14 = Sonar
D15 = Pima, D16 = Seed, D17 = German, D18 = NIC

Table 2 Domination count of different measures and their average domination count (ADC) on $UCI_{general}$ datasets using $F - measure_{unsup}$

	D1	D2	D3	D4	D5	D6	D7	D8	D9	D10	D11	D12	D13	D14	D15	D16	D17	D18	ADC
Euclidean	8	0	0	0	2	4	5	1	10	5	9	7	2	1	0	1	0	12	3.7222
Cityblock	16	0	0	7	2	4	5	1	14	0	16	1	5	1	0	1	0	8	4.5
Pearson	15	0	0	3	2	4	5	1	9	13	16	7	5	1	0	1	4	13	5.5
Spearman	12	0	0	2	2	4	5	1	17	13	13	7	5	1	0	0	4	17	5.7222
Kendall	8	0	0	7	2	0	1	1	5	5	14	7	16	1	0	1	4	14	4.7778
Cosine	13	0	0	3	2	4	5	1	10	5	6	0	5	1	0	1	4	8	3.7778
Minkowski	11	0	0	7	2	4	5	1	10	5	9	7	2	1	0	1	0	14	4.3889
Seuclidean	1	0	0	7	2	4	5	1	4	15	2	7	10	1	0	1	4	0	3.5556
Chebychev	5	0	0	7	0	4	5	0	8	5	9	7	1	1	0	1	0	0	2.9444
Dice	3	0	0	7	2	0	1	1	5	15	0	1	14	1	0	0	4	5	3.2778
Jaccard	4	0	0	7	2	0	1	1	0	0	5	7	9	1	0	1	4	6	2.6667
Canberra	6	0	0	7	2	4	5	1	2	4	6	7	10	1	0	1	4	0	3.3333
Bray-Curtis	13	0	0	7	2	4	5	1	16	0	6	1	10	1	0	1	4	10	4.5
Bhattacharyya	0	0	0	0	0	4	5	1	5	5	1	7	2	0	0	1	4	0	1.9444
Angular	1	0	0	7	2	4	0	1	0	15	2	7	10	1	0	1	4	0	3.0556
NMRS	8	0	0	3	2	4	5	1	14	0	14	1	0	1	0	1	4	11	3.8333
Footrule	16	0	0	3	2	4	1	1	3	5	9	1	14	1	0	1	4	14	4.3889
SSSim	7	0	0	7	2	0	5	1	10	5	4	1	17	1	0	1	4	7	4

D1 = Zoo, D2 = Accute1, D3 = Accute2, D4 = HayesRoath, D5 = House Voter, D6 = Iris, D7 = Wine
D8 = TicTacToe, D9 = Glass, D10 = Cleavland, D11 = Ecoli, D12 = CMC, D13 = Automobile
D14 = Sonar, D15 = Pima, D16 = NIC, D17 = Seed, D18 = German

Table 3 Domination count of different measures and their average domination count (ADC) on $UCI_{disease}$ datasets using $F - measure_{sup}$

	D1	D2	D3	D4	D5	D6	D7	D8	D9	D10	ADC
Euclidean	4	12	14	14	15	12	11	10	11	9	11.2
Cityblock	12	15	11	16	10	17	14	9	7	3	11.4
Pearson	8	11	15	11	16	7	16	13	9	5	11.1
Spearman	15	5	8	7	11	5	17	16	13	13	11
Kendall	16	4	12	6	12	3	13	15	3	12	9.6
Cosine	9	13	17	9	17	8	15	17	12	4	12.1
Minkowski	7	14	16	12	14	13	12	11	14	6	11.9
Seuclidean	0	3	4	3	4	2	4	7	16	11	5.4
Chebychev	3	8	6	10	9	11	8	8	8	10	8.1
Dice	2	0	0	1	1	0	0	0	4	16	2.4
Jaccard	5	1	1	0	0	1	0	1	2	15	2.6
Canberra	14	16	7	17	8	14	7	4	6	8	10.1
Bray-Curtis	10	10	13	15	13	16	2	3	5	2	8.9
Bhattacharyya	1	2	3	5	2	10	6	2	15	1	4.7
Angular	11	17	2	2	3	9	3	6	17	17	8.7
NMRS	6	9	10	13	7	15	10	14	10	0	9.4
Footrule	17	6	9	8	6	4	9	12	1	14	8.6
SSSim	13	7	5	4	5	6	5	5	0	7	5.7

D1 = StatHeartLog, D2 = Liver, D3 = Colon Cancer, D4 = Breast Cancer
D6 = Diabetes, D7 = Lung, D8 = Lymphoma, D9 = Leukemia, D10 = Wpbc

3.2 $UCI_{disease}$ Datasets: Results and Analysis

In terms of $F - measure_{sup}$, this study reveals that five measures give higher accuracy compared to other competent measures in classifying unknown objects of disease datasets using KNN classifier. As shown in Table 3, the average dominations count of Cosine, Minkowski, Cityblock, Euclidean, and Pearson measures are greater than 11. Among these five measures, Cosine is giving the highest accuracy.

In case of unsupervised learning, performances of Pearson and Spearman correlation measures are better than other measures using $F - measure_{unsup}$ on UCI disease datasets. Besides, Cityblock, Angular, and Seuclidean are also giving high accuracy, as shown in Table 4.

3.3 Overall Performance Analysis

In unsupervised framework using K-means clustering technique as shown in Fig. 2, Spearman rank correlation is the winner with average domination count 6.5 across

Table 4 Domination count of different measures and their average domination count (ADC) on $UCI_{disease}$ datasets using $F - measure_{unsup}$

	D1	D2	D3	D4	D5	D6	D7	D8	D9	D10	ADC
Euclidean	0	0	2	2	13	0	9	0	4	2	3.2
Cityblock	8	0	2	2	8	0	17	0	4	2	4.3
Pearson	8	10	2	2	13	0	15	3	4	2	5.9
Spearman	8	11	2	2	4	0	15	3	4	2	5.1
Kendall	0	11	2	2	4	0	9	3	4	2	3.7
Cosine	0	0	0	2	4	0	12	3	0	2	2.3
Minkowski	0	0	0	2	1	0	9	3	0	2	1.7
Seuclidean	8	11	2	2	8	0	3	3	4	2	4.3
Chebychev	8	0	2	2	13	0	2	3	4	2	3.6
Dice	8	0	2	2	8	0	12	3	4	2	4.1
Jaccard	0	0	2	2	8	0	12	3	4	2	3.3
Canberra	8	0	2	2	13	0	3	3	4	2	3.7
Bray-Curtis	8	11	2	2	2	0	1	3	4	2	3.5
Bhattacharyya	8	11	2	2	0	0	0	0	0	2	2.5
Angular	8	11	2	0	8	0	6	3	4	0	4.2
NMRS	0	0	2	2	13	0	7	3	0	2	2.9
Footrule	0	0	2	2	4	0	8	3	4	2	2.5
SSSim	0	11	2	0	3	0	5	3	4	0	2.8

D1 = StatHeartLog, D2 = Liver, D3 = Colon Cancer, D4 = Breast Cancer
D6 = Diabetes, D7 = Lung, D8 = Lymphoma, D9 = Leukemia, D10 = Wpbc

all datasets, while Bhattacharyya distance is the worst performer with domination count 2.03. Pearson and Kendall correlation followed Spearman correlation in the order of performance with average domination counts 6.17 and 4.83, respectively. In supervised framework using KNN classifier as shown in Fig. 3, Cityblock distance is the best performer with average domination count 12.13 across all the datasets, while Dice distance performs the worst with domination count 2.16. Cosine distance, Euclidean distance, and Minkowski distance follow Cityblock distance in the order with domination counts 11.06, 11.03, and 10.83, respectively. Average domination counts of measures over both the learning frameworks across all the datasets are reported in Fig. 4. Cityblock distance is ultimately the best performer with average domination count 8.33 over both supervised and unsupervised framework across all datasets. Spearman, Pearson correlation, and Cosine distance follow the order with domination counts 8.22, 8.16, and 7.35, respectively. Jaccard distance is the worst performer with average domination count 2.75.

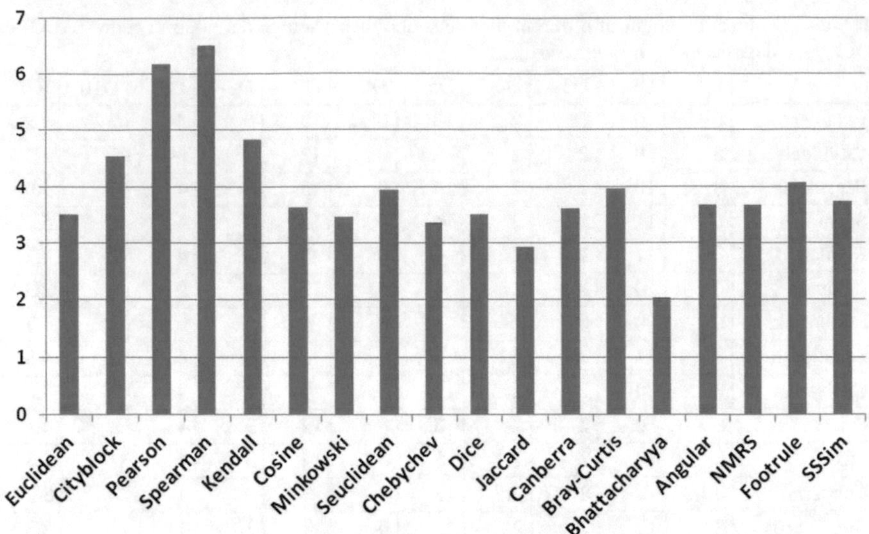

Fig. 2 Average domination count of proximity measures with respect to $F - measure_{unsup}$ over the datasets in unsupervised learning framework using K-means clustering

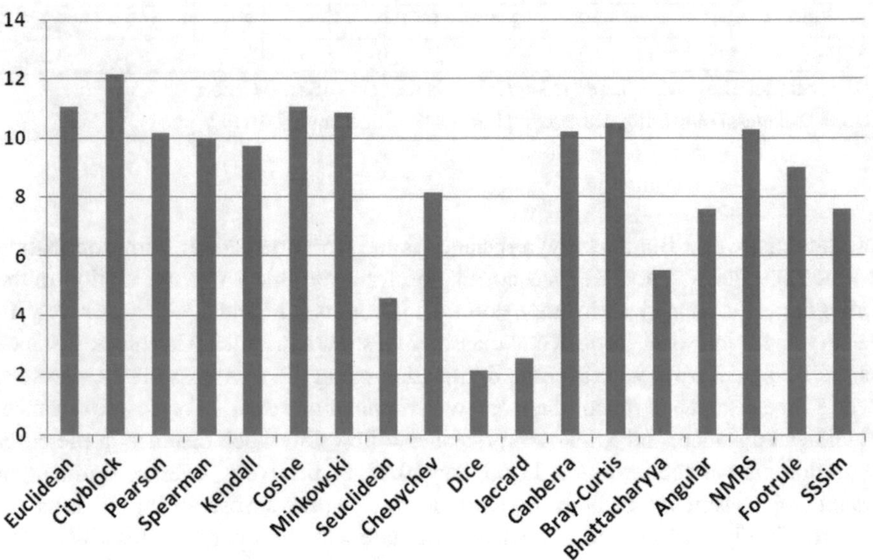

Fig. 3 Average domination count of proximity measures with respect to $F - measure_{sup}$ over the datasets in supervised learning framework using KNN classifier

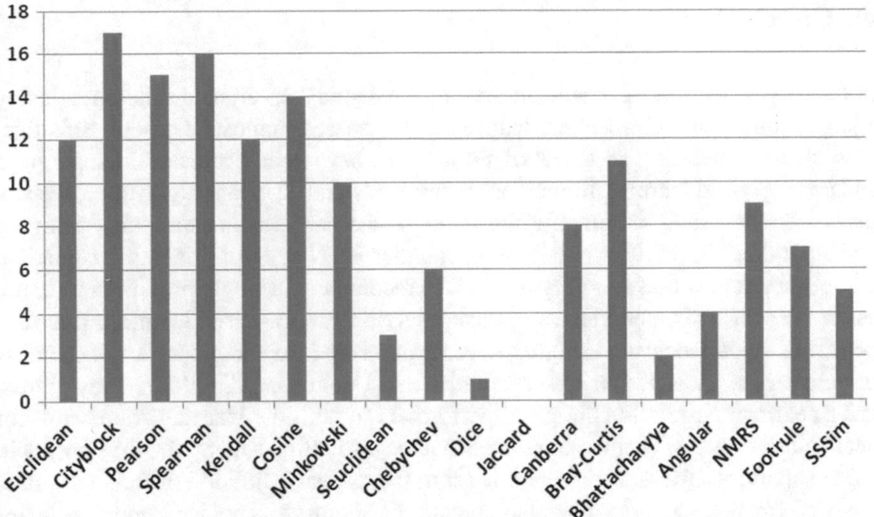

Fig. 4 Average domination count of proximity measures with respect to $F-measure$ over the datasets in both supervised and unsupervised learning framework

3.4 *Effectiveness*

To choose a measure that can be found effective across all the validity measures and for different learning techniques, is an extremely difficult task. Our experimental study also reveals that none of the measures show consistently best performance while used with supervised as well as unsupervised learning framework and evaluated in terms of F-measure and Domination Count. Although measures like Cityblock, Euclidean, Bray-curtis, Spearman, and Pearson show significantly good performance in both supervised and unsupervised framework in terms of domination count based on F-measure over benchmark datasets, however, in terms of shifting, scaling and shifting-and-scaling correlation handling, except Pearson and Spearman from this list, others fail to perform well over some synthetic datasets while Footrule, Kendall, NMRS, Cosine, and SSSim are found to yield good performance. In this empirical study we have analyzed the effectiveness of various proximity measures in terms of F-measure using KNN and K-means learning methods. From the experimental analysis, we found that cityblock distance gives better performance than other competing measures using KNN classifier. On the other hand, in case of unsupervised learning Spearman correlation performs well in comparison to other proximity measures.

4 Conclusion

In this paper, we have carried out an empirical study on eighteen measures using a large number of UCI machine learning and disease diagnosis datasets. Effectiveness of these measures in terms of F-measures have been evaluated in supervised and unsupervised learning framework using KNN and K-means algorithms, respectively. We use ADC to compute the ranks of the participating measures from the corresponding F-measures obtained from either KNN or K-means implementation. Our observation is that it is very difficult to recommend a single proximity measure as the best for both supervised and unsupervised distance-based learning. Because, performance of a measure is highly dependent on the intrinsic properties of the class instances of a dataset. Our observation is that Euclidean, Cityblock, Bray-Curtis, and Spearman correlation are good performers in both the learning framework for benchmark datasets. Further, our assessment is also influenced by the threshold α in the unsupervised framework which is set to 0.2 as per [22]. For a different α value, the performance scenario may also change. Furthermore, another important factor that might be the cause behind the degraded performance of some measures is the consideration of full space of features during distance computation. In future, we aim to carry out an analysis of these measures over optimal feature subspaces.

References

1. Ahmed, H.A., Mahanta, P., Bhattacharyya, D.K., Kalita, J.K.: Shifting-and-scaling correlation based biclustering algorithm. IEEE/ACM Trans. Comput. Biol. Bioinform. (TCBB) **11**(6), 1239–1252 (2014)
2. Bandyopadhyay, S., Saha, S.: Unsupervised Classification: Similarity Measures, Classical and Metaheuristic Approaches, and Applications. Springer, Berlin (2012)
3. Bhattachayya, A.: On a measure of divergence between two statistical population defined by their population distributions. Bull. Calcutta Math. Soc. **35**, 99–109 (1943)
4. Bray, J.R., Curtis, J.T.: An ordination of the upland forest communities of southern wisconsin. Ecol. Monogr. **27**(4), 325–349 (1957)
5. Cantrell, C.D.: Modern Mathematical Methods for Physicists and Engineers. Cambridge University Press, Cambridge (2000)
6. Chowdhury, H.A., Bhattacharyya, D.K.: mRMR+: An effective feature selection algorithm for classification. In: International Conference on Pattern Recognition and Machine Intelligence, pp. 424–430. Springer (2017)
7. Davis, J., Goadrich, M.: The relationship between precision-recall and roc curves. In: Proceedings of the 23rd International Conference on Machine Learning, pp. 233–240. ACM (2006)
8. Deb, K., Pratap, A., Agarwal, S., Meyarivan, T.: A fast and elitist multiobjective genetic algorithm: NSGA-II. IEEE Trans. Evol. Comput. **6**(2), 182–197 (2002)
9. Dice, L.R.: Measures of the amount of ecologic association between species. Ecology **26**(3), 297–302 (1945)
10. Han, J., Kamber, M., Pei, J.: Data Mining: Concepts and Techniques. Elsevier, New York (2011)
11. Hoque, N., Bhattacharyya, D., Kalita, J.K.: MIFS-ND: a mutual information-based feature selection method. Expert. Syst. Appl. **41**(14), 6371–6385 (2014)

12. Jurman, G., Riccadonna, S., Visintainer, R., Furlanello, C.: Canberra distance on ranked lists. In: Proceedings, Advances in Ranking–NIPS 09 Workshop, pp. 22–27 (2009)
13. Kendall, M.G.: A new measure of rank correlation. Biometrika **30**, 81–93 (1938)
14. Kraskov, A., Stögbauer, H., Andrzejak, R.G., Grassberger, P.: Hierarchical clustering using mutual information. EPL (Eur. Lett.) **70**(2), 278 (2005)
15. Kruskal, J.B.: Multidimensional scaling by optimizing goodness of fit to a nonmetric hypothesis. Psychometrika **29**(1), 1–27 (1964)
16. Lehman, A.: JMP for Basic Univariate and Multivariate Statistics: A Step-by-Step Guide. SAS Institute, Cary (2005)
17. Levandowsky, M., Winter, D.: Distance between sets. Nature **234**(5323), 34–35 (1971)
18. Mahanta, P., Ahmed, H.A., Bhattacharyya, D.K., Kalita, J.K.: An effective method for network module extraction from microarray data. BMC Bioinform. **13**(Suppl 13), S4 (2012)
19. Pearson, K.: Note on regression and inheritance in the case of two parents. In: Proceedings of the Royal Society of London, pp. 240–242 (1895)
20. Sherali, H.D., Tuncbilek, C.H.: A squared-euclidean distance location-allocation problem. Nav. Res. Logist. (NRL) **39**(4), 447–469 (1992)
21. Wu, H., Gao, L., Dong, J., Yang, X.: Detecting overlapping protein complexes by rough-fuzzy clustering in protein-protein interaction networks. PloS one **9**(3), e91856 (2014)
22. Wu, M., Li, X., Kwoh, C.K., Ng, S.K.: A core-attachment based method to detect protein complexes in PPI networks. BMC Bioinform. **10**(1), 1 (2009)

Memorizing and Retrieving of Text Using Recurrent Neural Network—A Case Study on Gitanjali Dataset

Rajat Subhra Bhowmick and Jaya Sil

Abstract This paper presents an application of Recurrent Neural Network for retrieving text from a given sequence of words of Gitanjali dataset. A Recurrent Neural Network (RNN) is trained to predict the poem corresponding to which the sequence of words are given. We demonstrate the experiment with two major RNN architectures and state the results to show which hyper-parameters like RNN size, sequence length, number of stacked layers affect the RNN most while completely memorizing the content of the poem. We also state the challenges to train the model in both forward and backward ways. We largely emphasis on the memorizing capability of RNN and put forward an application which depends on it.

Keywords Recurrent neural network · Memorizing capability · Long Short-Term Memory

1 Introduction

Recurrent neural networks (RNN) [3] are one of the main pillars in the field of deep learning. In many application associated to speech recognition [4], image captioning [7, 18], Question answering [19], Anomaly detection in time series [9] and language translation [21], breakthrough improvements have been observed using RNN compare to previously available state of the art systems. Furthermore, in last few years RNN is mostly involved with every natural language processing (NLP) task like paraphrase detection [14], word embedding extraction [10]and text normalization [15].

R. S. Bhowmick (✉) · S. Sil
Indian Institute of Engineering Science and Technology, Shibpur,
Howrah 711103, India
e-mail: rajatb.rs2017@cs.iiests.ac.in

S. Sil
e-mail: js@cs.iiests.ac.in

A. K. Das et al. (eds.), *Computational Intelligence in Pattern Recognition*,
Advances in Intelligent Systems and Computing 999,
https://doi.org/10.1007/978-981-13-9042-5_35

To our knowledge, there is no application which uses RNN for memorization and retrieving task. But there are lots of application apart from the major ones stated above which runs RNN under the hood like movie review sentiment analysis [13] for web platforms which recommend movie, application for real time detection of small natural earthquakes [20], Stock price pattern recognition [6]. RNN architecture is typically designed to find patterns in sequences and our dataset is nothing but sequence of words.

Human brain does both operations, learning (or memorizing) and generalization ensuring complete task such as speech recognition (that include learning long sequence of speech and understanding of speech). As biological neurons are affected by dropout or dying cells, may be that is why we only retain parts of any memorized text. So, if we could do duplicate the same memorizing phenomenon using our artificial network, it will be more productive as it is not susceptible to biological restriction and at the same time works just like our biological network. In the paper we use Long Short-Term Memory (LSTM) [5], a type RNN for memorizing the content of the poems. In contrast to most of recent applications of deep learning which primarily focuses on the generalization ability of the network, our application mainly depends on the ability of RNN to memorize every bit of our training information.

In our application the system predicts the full poem of Gitanjali dataset when the users only remember small part of the poem. Stacked LSTM layers are used to remember every sequence of the poem. Two types of network have been used to demonstrate the application, one is a stacked multiple layer in a single LSTM network and the other is seq2seq [16] modeling using encoder– decoder architecture [2]. We then compare the results of both the architecture.

2 Methodology

We applied two type of RNN architecture, one is single stacked LSTM network and the other consist of two LSTM network which is commonly known as Encoder–Decoder LSTM Network. The basis of using both of these architecture is the ability

Fig. 1 A LSTM cell

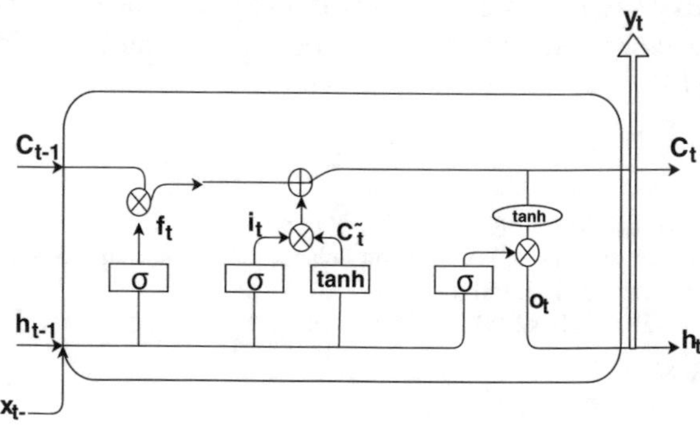

to produce a output sequence given a input sequence. The architecture of a single LSTM cell is shown in Fig. 1 [11].

$$f_{(t)} = \sigma(W_f[h_{t-1}, x_t] + b_f) \tag{1}$$

$$i_{(t)} = \sigma(W_i[h_{t-1}, x_t] + b_i) \tag{2}$$

$$C'_{(t)} = tanh(W_C[h_{t-1}, x_t] + b_C) \tag{3}$$

$$C_{(t)} = f_{(t)} * C_{(t)} + i_{(t)} * C'_{(t)} \tag{4}$$

$$o_{(t)} = \sigma(W_o[h_{t-1}, x_t] + b_o) \tag{5}$$

$$h_{(t)} = o_{(t)} * tanh(C_{(t)}) \tag{6}$$

Equation from 1 to Eq. 6 are used to update LSTM cell, where W_f, W_i, W_C, W_o are the weights corresponding to each operation, while respective biases are represented by b_f, b_i, b_C, b_o.

2.1 Single Stacked LSTM Network

The first network architecture we use is a simple LSTM network containing multiple stacked layers as shown in Fig. 2. The words are passed as input which then passed into a word embedding [1] layer, that converts the word into a vector. The word vectors are then passed to the first LSTM layer, which after processing produces outputs and the final state. The outputs are forwarded to the next stacked LSTM layer while the final state is retained and used for processing of next input. We use two to four stacked LSTM layers for experimentation. The final LSTM layer output is passed to the fully connected layer that converts the output of LSTM to the size of vocabulary for determination of next word in the sequence through one hot vector.

2.2 Encoder–Decoder LSTM Network

In the second network, we employ seq2seq architecture which is mainly used in language translation task. It basically has two networks in which the first network acts as an encoder by providing the summary information of the given input. The second network acts as a decoder and fetches next sequence of words using the encoded information supplied by the encoder, as shown in Fig. 3 . The outputs of the encoder are not used, only the cell-state has been preserved that is considered as encoded information. The decoder works just as Single Stacked LSTM Network mentioned above for the prediction of next sequence of words.

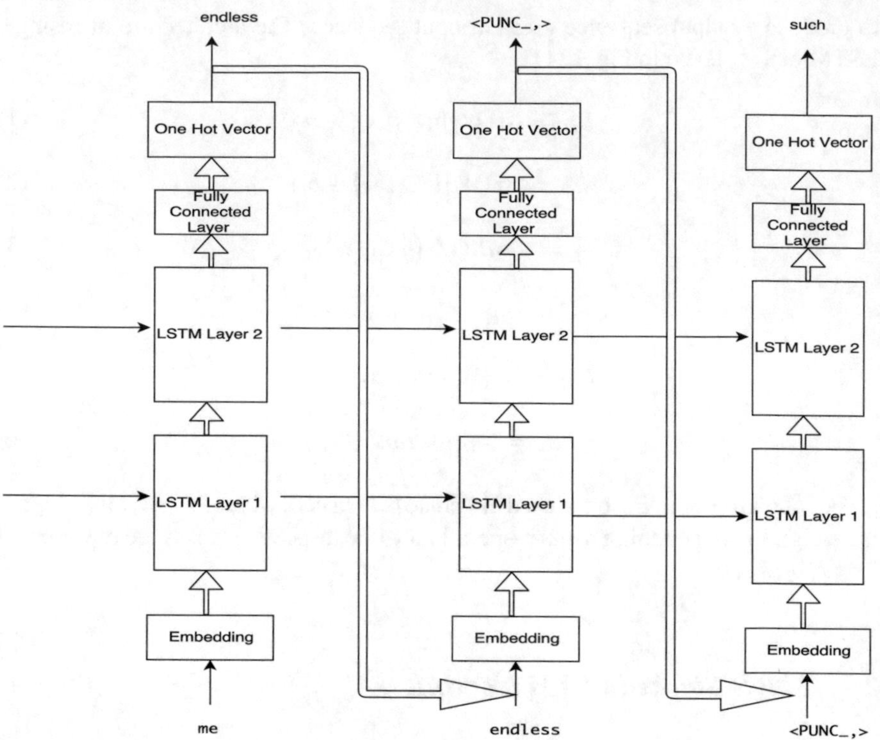

Fig. 2 Single Stacked LSTM Network

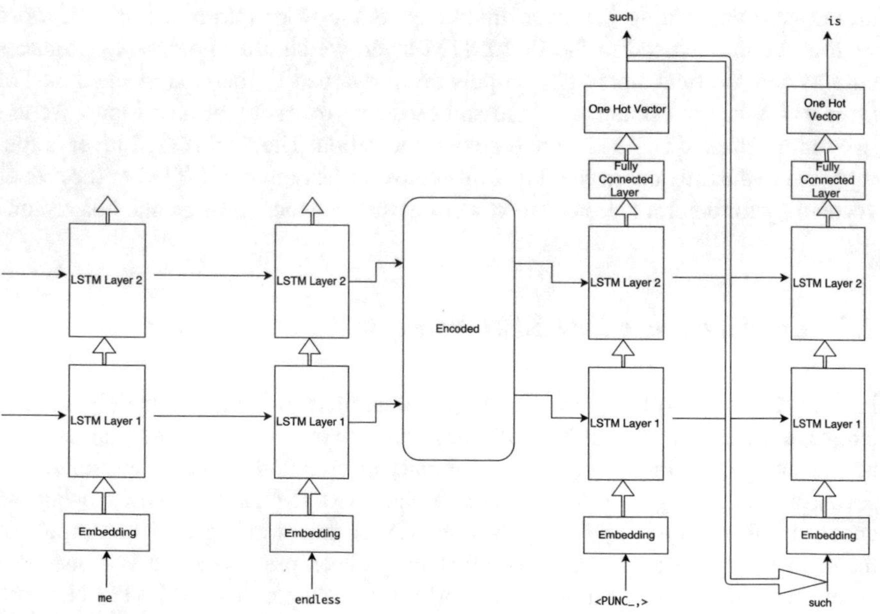

Fig. 3 Encoder–Decoder LSTM network

3 Experimental Setup

In our experiment we used a small self-made dataset. The preprocessing is one of the important stage of our experiment as the input to both the architecture is different from each other altogether. The Table 1 represent a sample sentence from the dataset.

3.1 Dataset

We select the poems from Gitanjali book [17], a collection of poems written by the Bengali poet Rabindranath Tagore. Tagore received the Nobel Prize in Literature for Gitanjali on 1931. One hundred three poems are included in the dataset for training. We consider this dataset to reduce training time complexity, and at the same time poem dataset provides just about the perfect sample to test other necessary criteria to retrieve the memorized information. The vocabulary of the set is 2206.

3.2 Preprocessing

The poem's content is concatenated with each other using <EOF> and <SOF> file token, which helps to identify the start and end of the poems. During prediction, these tokens assist in retrieving the whole poem working as flags.

All the punctuation letters are also converted into tokens. The tokens and words in the poem together make the vocabulary of the training set. The padding token <PAD> is used to complete the training set during batch creation, as we require every data in sequential order.

3.3 Training Set

There are two types of training sets for each of the proposed model. The first dataset is designed for Single Stacked LSTM Network in which the output of a given word is the next word in the sequence. The representation for the same is shown in Table 2.

The dataset of the encoder–decoder is different from that for the Single Stacked LSTM. Its input is a sequence of words whose length is equal to the number of

Table 1 Sample poem content

"Thou hast made me endless, such is thy pleasure. \n This frail vessel thou emptiest again and again"

Table 2 Training sample for Single Stacked LSTM Network (forward training)

	Input	Output
Original text	Thou hast made me endless, such is thy	hast made me endless, such is thy pleasure
Text with punctuation token	Thou hast made me endless <PUNC_,> such is thy	hast made me endless <PUNC_,> such is thy pleasure
Vocab index in vocabulary set	[163, 952, 1197, 1216, 689, 15, 1837, 1066, 1926]	[952, 1197, 1216, 689, 15, 1837, 1066, 1926, 1447]

Table 3 Training sample for Encoder–Decoder LSTM Network (forward training)

	Input	Output
Original text	Thou hast made me endless, such is thy	pleasure. \n This frail vessel thou emptiest again
Text with punctuation token	Thou hast made me endless <PUNC_,> such is thy	pleasure < PUNC_.> <TOKEN_NL> This frail vessel thou emptiest again
Vocab index in vocabulary set	[163, 952, 1197, 1216, 689, 15, 1837, 1066, 1926]	[1447, 17, 38, 161, 833, 2057, 1908, 680, 217]

sequence steps at the time of training. The output is another sequence starting from the next word after the input sequence. Table 3 shows the training set for Encoder–Decoder LSTM Network.

3.4 *Forward Training process*

During the forward training, the inputs and outputs are used to train the model so as to produce the next sequence of words until the <EOF> token is found. In the forward training of Single Stacked LSTM Network, the training sample remains the same as stated above. During training, each word of the input set is fed to the network and computed output logits are compared with the target output label vector, to calculate the softmax cross entropy loss. The word predicted by the network is not used during training.

For Encoder–Decoder LSTM Network the inputs are allowed to flow in the encoder network to produce encoded version of input which then fed to decoder. The decoder produces the set of outputs whose length is equal to number of steps. The output of the decoder is compared to the training output that produces sequence loss. Sequence loss is used to train the network.

Table 4 Training sample for Encoder–Decoder LSTM Network (backward training)

	Input	Output
Original text	again emptiest thou vessel frail This \n .pleasure	thy is such, endless me made hast Thou
Text with punctuation token	again emptiest thou vessel frail This <TOKEN_NL> <PUNC_.> pleasure	thy is such <PUNC_,> endless me made hast Thou
Vocab index in vocabulary set	[217, 680, 1908, 2057, 833, 161, 38, 17, 1447]	[1926, 1066, 1837, 15, 689, 1216, 1197, 952, 163]

3.5 Backward Training process

The training process of both the network remains the same as forward training. But, the dataset is processed in a way to retrieve the content of the poem until the <SOF> (start of file) token has been reached. The training sample to train Encoder–Decoder LSTM Network in backward direction is shown in Table 4. The training process with each model input and output are explained below.

3.6 Prediction

In order to retrieve the full poem, we need to run the forward and backward trained model and combined the output so as to obtain the entire poem. The forward trained model is used to predict until <EOF> (end of file) token is produced for the first Single Stacked LSTM Network. In the Encoder–Decoder LSTM Network, we fed the input sequence to produce an output sequence of same size. We again fed the produced output sequence as input until the output contains <EOF> token. There is an upper limit to the number of such runs. The backward trained model act similarly for both the network but the model needs to terminate as soon as <SOF> (start of file) token appears.

4 Results

We have executed through different values of the listed hyper-parameters in Table 5. For training we have used Adam Optimizer [8] and to mitigate the problem of exploding gradient we have used the gradients clipping [12]. The best results for both the network architecture is shown in Table 6. To test overall model efficiency in application level we have picked three to four sequences each size equal to step size from every poem as input. After executing both the forward and backward model on these inputs and combining their outputs, we have tested the accuracy of both the archi-

Table 5 The major hyper parameters upon which both the network architecture was tested given the available system configuration

Hyper-parameter	Description
num_layers	Number, of LSTM layers
num_steps	Number of sequence steps or words for a sample
lstm_size	Size of, LSTM
batch_size	Sequences, per batch during training
embed_length	Length, of embedded vector for each word or token

Table 6 The best results for the both the network architecture with varying hyper parameters

Network architecture	Num layers	Num steps	LSTM size	Batch size	Embed length	Mean loss after training (per epoch)	Application level accuracy
Single Stacked LSTM Network	3	7	256	128	64	0.0022	41.25%
Encoder–Decoder LSTM Network	4	11	256	1024	64	0.0026	96.05%
Encoder–Decoder LSTM Network	4	19	256	1024	64	0.0021	99.70%

tecture at application level. Table 6 shows what percent of the total input sequences is able to retrieve the whole poem from <SOF> to <EOF> token correctly.

The results confirm that Encoder–Decoder LSTM Network outperforms the Single Stacked LSTM network in terms of memorizing capability, by large margin. The most important hyper-parameter Encoder– Decoder LSTM Network is number of steps which is the sequence length of input to the network. Figure 4. represents the relation between sequence length and accuracy keeping the other parameters constant.

It is obvious that with increase of sequence length retrieving accuracy also increases, as more amount of data is available as input in the encoder level. The most of the incorrect prediction is due to overlapping of sequences in the poem.

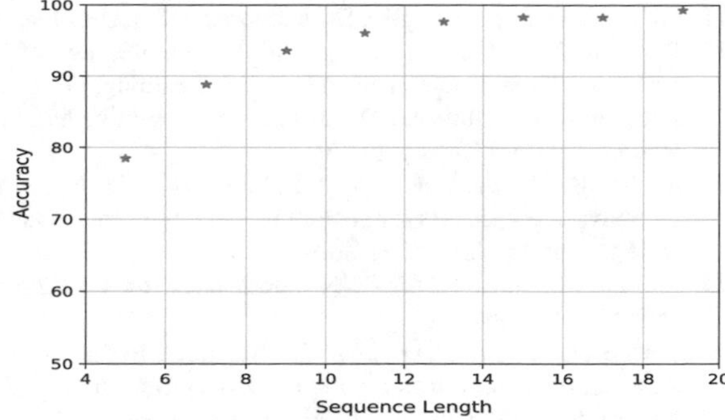

Fig. 4 Relation between sequence length and accuracy

5 Conclusion

The focus of our work is to explore the memorization ability of RNN. The dataset we used is small in size due to limited hardware resources. Through the results, we are able to show that with small dataset RNN can memorize the sequences and can retrieve the forward and backward sequences respect to it. In future it will be interesting to test the network on larger dataset and to explore the solution for overlapping sequences. Another aspect where work can be done is, to determine the number of stack layer required to remember a given amount of data.

References

1. Bengio, Y., Ducharme, R., Vincent, P., Jauvin, C.: A neural probabilistic language model. J. Mach. Learn. Res. **3**(Feb), 1137–1155 (2003)
2. Cho, K., Van Merriënboer, B., Gulcehre, C., Bahdanau, D., Bougares, F., Schwenk, H., Bengio, Y.: Learning phrase representations using RNN encoder-decoder for statistical machine translation (2014). arXiv:1406.1078
3. Elman, J.L.: Finding structure in time. Cogn. Sci. **14**(2), 179–211 (1990)
4. Graves, A., Mohamed, A.r., Hinton, G.: Speech recognition with deep recurrent neural networks. In: 2013 IEEE International Conference on Acoustics, Speech and Signal Processing (ICASSP), pp. 6645–6649. IEEE (2013)
5. Hochreiter, S., Schmidhuber, J.: Long short-term memory. Neural Comput. **9**(8), 1735–1780 (1997)
6. Kamijo, K.I., Tanigawa, T.: Stock price pattern recognition-a recurrent neural network approach. In: 1990 IJCNN International Joint Conference on Neural Networks, pp. 215–221. IEEE (1990)
7. Karpathy, A., Fei-Fei, L.: Deep visual-semantic alignments for generating image descriptions. In: Proceedings of the IEEE Conference on Computer Vision and Pattern Recognition, pp. 3128–3137 (2015)
8. Kingma, D.P., Ba, J.: Adam: a method for stochastic optimization (2014). arXiv:1412.6980
9. Malhotra, P., Vig, L., Shroff, G., Agarwal, P.: Long short term memory networks for anomaly detection in time series. In: Proceedings, p. 89. Presses universitaires de Louvain (2015)
10. Mikolov, T., Chen, K., Corrado, G., Dean, J.: Efficient estimation of word representations in vector space (2013). arXiv:1301.3781

11. Olah, C.: Understanding LSTM networks. GITHUB blog. Accessed 27 Aug 2015
12. Pascanu, R., Mikolov, T., Bengio, Y.: On the difficulty of training recurrent neural networks. In: International Conference on Machine Learning, pp. 1310–1318 (2013)
13. Pouransari, H., Ghili, S.: Deep learning for sentiment analysis of movie reviews. Technical Report, Stanford University (2014)
14. Socher, R., Huang, E.H., Pennin, J., Manning, C.D., Ng, A.Y.: Dynamic pooling and unfolding recursive autoencoders for paraphrase detection. In: Advances in Neural Information Processing Systems, pp. 801–809 (2011)
15. Sproat, R., Jaitly, N.: RNN approaches to text normalization: a challenge (2016). arXiv:1611.00068
16. Sutskever, I., Vinyals, O., Le, Q.V.: Sequence to sequence learning with neural networks. In: Advances in Neural Information Processing Systems, pp. 3104–3112 (2014)
17. Tagore, R., et al.: Gitanjali: Kītāñcali. Sura Books (2005)
18. Vinyals, O., Toshev, A., Bengio, S., Erhan, D.: Show and tell: a neural image caption generator. In: 2015 IEEE Conference on Computer Vision and Pattern Recognition (CVPR), pp. 3156–3164. IEEE (2015)
19. Wang, D., Nyberg, E.: A long short-term memory model for answer sentence selection in question answering. In: Proceedings of the 53rd Annual Meeting of the Association for Computational Linguistics and the 7th International Joint Conference on Natural Language Processing (vol. 2: short papers), vol. 2, pp. 707–712 (2015)
20. Wiszniowski, J., Plesiewicz, B., Trojanowski, J.: Application of real time recurrent neural network for detection of small natural earthquakes in poland. Acta Geophysica **62**(3), 469–485 (2014)
21. Wu, Y., Schuster, M., Chen, Z., Le, Q.V., Norouzi, M., Macherey, W., Krikun, M., Cao, Y., Gao, Q., Macherey, K., et al.: Google's neural machine translation system: Bridging the gap between human and machine translation (2016). arXiv:1609.08144

Shuffled Differential Evolution Algorithm Based Combined Heat and Power Emission Dispatch

S. Nagaraju, A. Srinivasreddy and K. Vaisakh

Abstract In this paper, a metaheuristic algorithm known as "Shuffled Differential Evolution (SDE)" had been applied to resolve the Combined heat and power emission dispatch (CHPEmD) issue. This SDE algorithm combines features of both the Differential evolution algorithm, shuffled frog-leaping algorithm by incorporating splitting into the partitions and shuffling. To verify the efficacy of this SDE algorithm and also to determine the exemplary solution for the CHPEmD problem, two caliber test systems are considered. The outcomes realized by this SDE algorithm are confronted with the optimization algorithms available in the previous literary works. The contrast of the results shows that SDE technique exhibits impressive performance in delivering the optimal results in terms of convergence and solutions.

Keywords Differential evolution · Shuffled frog-leaping algorithm · Shuffled differential evolution and combined heat and power emission dispatch (CHPEmD)

1 Introduction

Burning of the fossil fuels for the production of electric power in a conventional thermal unit is not an efficient process. Even the latest power plants having a collaborative cycle are operating efficiency around 60% only. The flue gases were released into the environment through cooling towers contains most of the heat energy. Combined heat and power (CHP) units also familiar as cogeneration unit, recovers heat and to make productive use of this heat, because of this net efficacy can be raised to

S. Nagaraju (✉)
AITAM(A), Tekkali 532201, India
e-mail: Subuddi_nagaraju@yahoo.com

A. Srinivasreddy
Sir.C.R.Reddy College of Engineering, Eluru 534007, India
e-mail: srinivasareddyalla@yahoo.co.in

K. Vaisakh
Andhra University College of Engineering(A), Visakhapatnam 530003, India
e-mail: vaisakh_k@yahoo.co.in

A. K. Das et al. (eds.), *Computational Intelligence in Pattern Recognition*,
Advances in Intelligent Systems and Computing 999,
https://doi.org/10.1007/978-981-13-9042-5_36

as much as 90% [1, 2]. In addition to this combined heat and power unit's reduces the amount of greenhouse gases (CO_2, NO_X, and SO_x emissions) by about 13–18. The power output of a cogeneration unit depends on its heat output and vice versa. This correlative relationship between power and heat introduces complexity while integrating cogeneration units with the other exclusive power generating units and exclusive heat generating units in the emission based electric power dispatch.

Combined heat and power emission dispatch (CHPEmD) plays a vital role for reducing the emissions of operation of power system. Because of nonlinear and non-convex attributes of CHPEmD problem, CHPEmD problem is a complex subject. The prime goal of the CHPEmD is to exploit the optimal values of the heat and power for all the units, such that emission levels are optimum. While solving CHPEmD problem, the system heat and power requirements and other system needs are taken care and at the same time, while the cogeneration units should operate in their feasible operating area enclosed by heat against power for a given unit. Therefore this complex CHPEmD problem needs a powerful optimization technique in attaining the quality solution for this problem. The usages of conventional methods such as gradient-based techniques are unsuccessful in achieving the caliber optimal results.

With the recent developments and advancements in the stochastic search algorithms over the past decades, these stochastic techniques are becoming more favored to solve the multi-objective problems [3]. Therefore, numerous clever algorithms turn up on the horizon, to solve the CHPEmD problem.

The population-based techniques such as real coded genetic algorithm [RCGA] [4, 5], Grey Wolf optimization [6], non-dominating sorting genetic algorithm II [NSGA II] [7] and Strength Pareto Evolutionary algorithm [8] have been exercised to resolve the CHPEmD problem. In this paper, a shuffled differential evolution (SDE) algorithm had been adapted to exploit the optimal solution for the CHPEmD problem.

The good attributes of both shuffled frog-leaping algorithm (SFLA) [9] and differential algorithm (DE) [10] are incorporated into SDE algorithm and this SDE algorithm standardized by solving large scale economic dispatch problem [11, 12]. In this paper, the SDE technique is used to solve the CHPEmD problem having power units with valve-point loading, heat only generating units and CHP units along with additional constraints of the system. To justify the performance of SDE, two caliber test systems are taken [7] and results are compared with other evolving techniques.

The outline of this paper:

Section 2 describes CHPEmD problem mathematical formulation.
Section 3 SDE application to CHPEmD problem.
Section 4 furnishes the resulted obtained by application of SDE algorithm to two test cases. Last Sect. 5 Paper's cessation.

2 Problem Formulation

The prime requisite of CHPEmD issue is to determine the ideal point of operation for each unit's power and heat values such that emissions are minimized, subjected to satisfying the system operating constraints. Figure 1 shows the bounded feasible territory of operation for a cogeneration (CHP) unit by PQRSTU curve.

The output power of power only unit and heat output of the heat only unit are confined by the respective unit's lower and upper boundaries. The system power demand is met by conventional thermal units along with CHP units, whereas system heat demand is met by CHP units along with heat only units. The prime intention of the CHPEmD problem is to exploit the operating values of CHP units, exclusive power units and exclusive heat units such that emission levels are minimized while system's demands and other constraints are taken care. The fitness function of CHPEmD can be expressed mathematically as

$$\text{Minimize emission cost (EC)} = \sum_{j=1}^{N_p} E_{Pj}(P_j) + \sum_{k=1}^{N_c} E_{Ck}(P_k, H_k) + \sum_{l=1}^{N_o} E_{hl}(H_l) \tag{1}$$

Subjected to

$$\text{The operating cost } FC_T = \sum_{j=1}^{N_p} C_{Pj}(P_j) + \sum_{k=1}^{N_c} C_{Ck}(P_k, H_k) + \sum_{l=1}^{N_o} C_{hl}(H_l) \tag{2}$$

where E_{pj}, E_{ck}, and E_{hl} are the emission costs and C_{pj}, C_{hl}, and C_{ck} are running costs of the exclusive power unit, exclusive heat unit, and CHP unit, respectively. N_p is the number of conventional thermal units, N_c is the number of CHP units, and N_o is the

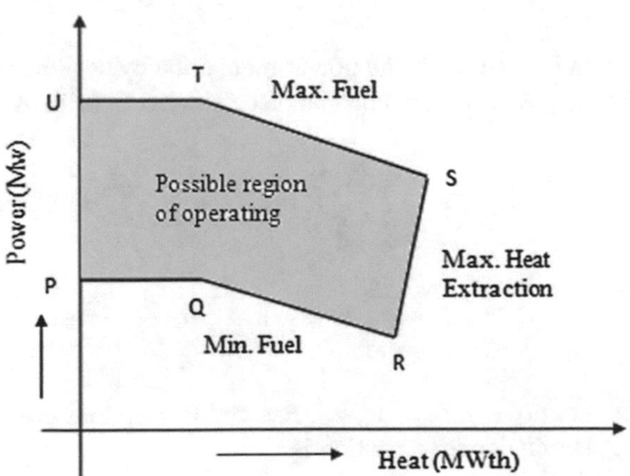

Fig. 1 CHP unit feasible operation zone

number of exclusive heat units, respectively. The indices j, k, and l are used for the exclusive power units, cogeneration unit, and exclusive heat units, respectively. FC_T is the total fuel cost; the power generation and heat generation levels of exclusive power unit and exclusive heat units are represented by P_j and H_l respectively; The heat and power outputs of kth CHP are indicated by H_k and P_k respectively.

The jth conventional thermal unit overhead cost of in Eq. (2) can be expressed as:

$$C_{pj}(P_j) = a_j * (P_j)^2 + b_j * P_j + c_j + \left|d_j * \sin\left(e_j * \left(P_j^{min} - P_j\right)\right)\right| (\$/\mathrm{h}) \quad (3)$$

where $C_{pj}(P_j)$ represents the cost of operating jth conventional thermal unit for generating P_j power; a_j, b_j, c_j, d_j, and e_j stand for cost coefficients.

The kth CHP unit's expense function in Eq. (2) is defined as

$$C_k(P_k, T_k) = a_k + b_k P_k + c_k (P_k)^2 + d_k H_k + e_k (H_k)^2 + f_k H_k P_k \left(\$/\mathrm{h}\right) \quad (4)$$

where $C_k(P_k, H_k)$ stands for the kth cogeneration unit operating cost function; a_k, b_k, c_k, d_k, e_k, and f_k are kth CHP unit's cost coefficients.

And the lth exclusive heat unit running expense function in Eq. (2) is defined as

$$C_l(H_l) = a_l + b_l H_l + c_l (H_l)^2 \left(\$/\mathrm{h}\right) \quad (5)$$

In Eq. (5), $C_l(H_l)$ stands for the of the lth exclusive heat generation unit's expense function for generating heat H_l; a_l, b_l, and c_l are cost coefficients of lth heat only unit.

The constraints given below are considered for solving the CHPEmD problem given in Eq. (1):

Power balance restriction is

$$\sum_{j=1}^{N_p} P_j + \sum_{k=1}^{N_c} P_k - P_{loss} = P_d \quad (6)$$

In Eq. (6), P_d is the power requisition by system and P_{loss} is transmission losses in the power system and can be calculated as follows:

$$P_{loss} = \sum_{i=1}^{N_p} \sum_{m=1}^{N_p} P_i B_{im} P_m + \sum_{i=1}^{N_p} \sum_{j=1}^{N_c} P_i B_{ij} P_j + \sum_{j=1}^{N_c} \sum_{l=1}^{N_c} P_j B_{jl} P_l + \sum_{j=1}^{N_p} B_{0j} P_j + \sum_{j=1}^{N_c} B_{oj} P_j + B_{00} \quad (7)$$

For Eq. (7), B_{im}, B_{ij}, B_{jl}, B_{0j}, and B_{00} are loss coefficients of the power system.

Heat balance restriction is

$$\sum_{j=1}^{N_C} H_j + \sum_{k=1}^{N_o} H_k = H_d \tag{8}$$

In Eq. (8), H_d is heat requirement by the system.

Conventional thermal unit in (2) has power generation restrictions are expressed as

$$P_j^{min} \leq P_j \leq P_j^{max} \quad \mathrm{i} = 1, 2, 3, \ldots \ldots \mathrm{N_p} \tag{9}$$

In Eq. (9), P_j^{min} and P_j^{max} are respectively lower and upper power boundaries of the jth exclusive power unit.

CHP unit's generation restrictions in Eq. (3) is

$$P_k^{min}(H_k) \leq P_k \leq P_k^{max}(H_k), \quad \mathrm{k} = 1, 2, 3, \ldots \mathrm{N_C} \tag{10}$$

$$H_k^{min}(P_k) \leq H_k \leq H_j^{max}(P_k), \quad \mathrm{k} = 1, 2, 3, \ldots \mathrm{N_C} \tag{11}$$

In Eq. (10), $P_k^{min}(H_k)$ and $P_k^{max}(H_k)$ are respectively the lower and upper power ceiling of the kth CHP unit. In Eq. (11), $H_k^{min}(P_k)$ and $H_k^{max}(P_k)$ are respectively the minimal and maximal heat values of the kth CHP unit.

Exclusive heat unit's heat generation limits in Eq. (4) is

$$H_l^{min} \leq H_l \leq H_l^{max}, \quad \mathrm{l} = 1, 2, 3, \ldots \mathrm{N_o} \tag{12}$$

In Eq. (12), H_l^{min} and H_l^{max} are respectively lth exclusive heat units, the lower and upper heat generation limits.

3 SDE Algorithm Implementation for Solving CHPEmD Problem

SDE algorithm was successfully applied to the Economic load dispatch problem [11, 12]. SDE algorithm is a hybrid method which integrates the good attributes of both SFLA and DE. SDE algorithm has overcome the congenital difficulties of SFLA and DE while providing solutions for complex problems like large area economic load dispatch having non-convex and non-differentiable cost function [11, 12].

The exploration process for the SDE technique is explained below.

1st Step: Initialize Parameters: To resolve the CHPEmD problem by using SDE algorithm, set the total number of units, mention the cost and emission coefficients of all the units (conventional, CHP unit, and heat only unit). Also set the algorithm parameters like initial population (P), quantity of memeplexes (m), memeplexes size (n) so that P = mxn.

2nd Step: Population Creation: Randomly generate the initial values of population (P) within each unit's operating region. This initial population must obey the equality constraints given by equations.
3rd Step: Assigning Ranks to the Frogs: Each frog's fitness is evaluated and they are arranged in the decreasing order of their fitness value. Also identify best frog (X_g) location among the entire population.
4th Step: Splitting into Memeplexes: The total frogs (P) are distributed among the m memeplexes each memeplexes having n frogs, so that P = mn. The distribution of the entire population is done based on the fitness of the frog, the 1st memeplex will have 1st highest fitness frog, the 2nd memeplex will have next highest fitness frog, and mth memeplex will have the frog m. The m + 1 frog will be assigned again the 1st memeplex and so on.
5th Step: Memetic Progression: Determine the best frog (Xb) and worst frog (Xw) in each memeplex. The next better position of the worst frog in each memeplex is determined using the following equation.

$$X_w \text{ New} = X_w \text{ old} + E_i \quad (-E_{i,\max} \leq E_i \leq E_{i,\max}) \tag{13}$$

$$E = Acc * rand() \text{ X } (X_b - X_w) \tag{14}$$

In Eq. (13) $E_{i,max}$ is the allowed maximum change in the position of the frog, in Eq. (14) Acc is the acceleration factor and rand() is random value between 0 and 1. If this adaption process yields an improved frog (better fitness), this new improved frog substitutes the older one. Else, X_b is substituted by X_g in Eq. (13) and the process is repeated. In the case of there is no betterment viable in the frog's fitness, a randomly generated frog substitutes the older frog.
6th Step: Local Search: To perform local search, repeat this process from 3rd step for a predefined set of iterations.
7th Step: Global Search: After completion of predefined internal evolutions (6th step) within the every memeplex, shuffling of the population is done by merging frogs from all the memeplexes into one group. This periodic shuffling assists in the information exchange between the frogs globally.
8th Step: Convergence checking: If the convergence conditions are not met, go to 3rd step. Else, Stop.

4 Results Simulation and Analysis

For validating SDE optimization ability, this SDE algorithm is exercised on two test systems. The outcomes of SDE algorithm are used to study the performance of the SDE with the past evolutionary methods. A personal computer with Intel core 2 duo processor of 1.66 GHz and 2 GB RAM having MATLAB 10 is used to for the execution of SDE algorithm.

4.1 Result Analysis

This SDE algorithm is applied for two standard test systems taken from [7]. 1st test system contains 7 units out of which there are 4 conventional thermal units, 2 CHP units, and 1 exclusive heat unit. 2nd test system involves one convention thermal unit, 3 CHP units, and 1 exclusive heat unit. The study of SDE algorithm on these test cases shows that SDE is productive in accomplishing quality solutions.

4.1.1 First Test System

SDE technique is exercised on this test system, whose Power requirement is 700 MW and heat requirement is 150 Mwth.

From Table 1, the emission's obtained by using SDE algorithm method is 14.2150 kg/hr which is less than 16.9208 kg/hr emission obtained by real coded genetic algorithm (RCGA) method and also the emission value 16.789 kg/hr achieved by DE.

From the convergence characteristics, shown in Fig. 2 it is observable that SDE is capable to attain its optimal values in 15 iterations as compared with DE, whereas DE took nearly 29 iterations to attain the final solution.

Table 1 Test system 1 optimal solutions

Methods	Emission dispatch		
	RCGA [7]	DE	SDE
P1 (MW)	73.3318	75.0000	61.3347
P2 (MW)	81.0489	84.9654	71.0628
P3 (MW)	93.4210	114.2879	87.8190
P4 (MW)	125.2112	86.0103	117.3100
P5 (MW)	214.9958	231.5578	247.0000
P6 (MW)	125.7907	118.5004	125.8000
H5 (MWth)	104.7715	33.2665	0.0000
H6 (MWth)	31.9272	29.5407	0.0000
H7 (MWth)	13.3013	87.1928	150.0000
Total Power	713.7994	710.3218	710.3265
Total Heat	150.0000	150.0000	150.0000
Cost ($/h)	17749.3100	17300.3315	18340.6802
Emission (kg/h)	16.9208	16.789	14.2150
CPU Time	22.7813	20.4250	18.9317

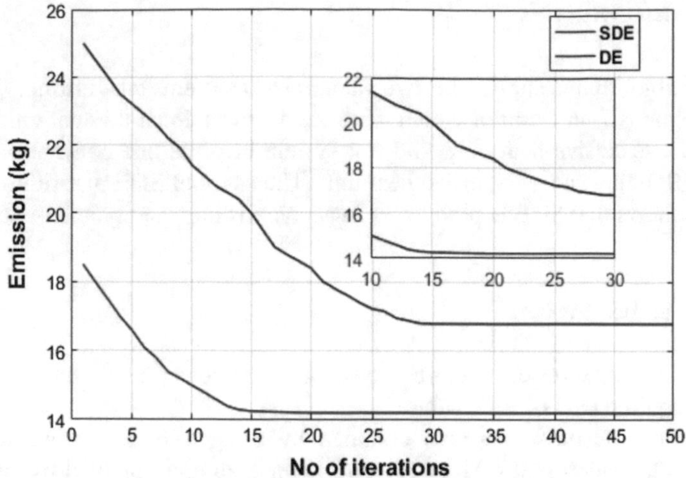

Fig. 2 1st test system-emission convergence characteristics

4.1.2 Second Test System

This test system contains 1 conventional thermal unit, 2 CHP units, and one heat only unit. The system power requirement is 300 MW and heat requirement is 150 Mwth.

Table 2 shows the emissions obtained by using SDE algorithm method is 1.18 kg/hr which is less than 1.446 kg/hr emission obtained by RCGA method and 1.19 kg/hr achieved by DE.

Table 2 Test system 2 optimal solutions

	Emission dispatch		
Methods	RCGA [7]	DE	SDE
P1 (MW)	39.2000	35.2000	35.0000
P2 (MW)	125.8000	114.0000	116.2558
P3 (MW)	45.0000	45.8000	43.7442
P4 (MW)	90.0000	105.0000	105.0000
H2 (MWth)	32.3998	105.0000	95.5382
H3 (MWth)	55.0000	45.0000	54.4618
H4 (MWth)	24.9999	0.0000	0.0000
H5 (MWth)	37.6002	0.0000	0.0000
Total Power	300.0000	300.0000	300.0000
Total Heat	149.9999	150.0000	150.0000
Cost ($)	17048.7500	17055.6102	17062.1460
Emission (kg)	1.446	1.1900	1.18
CPU time (s)	20.5417	18.5543	16.3251

5 Conclusion

In this paper, a novel SDE technique is applied to resolve the multi-objective CHPEmD issue. Results attained by the SDE algorithm are compared against the DE, RCGA published in the earlier literature. From the observation of results, the SDE method attained the desired goal of reduced emissions successfully and also has the ability to settle to the final value in less number of iterations. Hence SDE algorithm can be considered as a reliable means in providing a solution for the CHPEmD issue.

References

1. Karki, S., Kulkarni, M., Mann, M.D., Salehfar, H.: Efficiency improvements through combined heat and power for on-site distributed generation technologies. Cogener. Distrib. Gener. J. **22**(3), 19–34 (2007)
2. Vasebi, A., Fesanghary, M., Bathaee, S.: Combined heat and power economic dispatch by harmony search algorithm. Electr. Power Energy Syst. **29**, 713–719 (2007)
3. Spall, J.C.: Introduction to Stochastic Search and Optimization: Estimation, Simulation, and Control, vol. 65. Wiley, Hoboken (2005)
4. Herrera, F., Lozano, M., Verdegay, J.L.: Tracking real-coded genetic algorithms: operators and tools for behavioral analysis. Artif. Intell. Rev. **12**(4), 265–319 (1988)
5. Deb, K., Agarawal, R.B.: Simulated binary crossover for continuous search space. Complex Syst. **9**(2), 115–148 (1995)
6. Jayakumar, N., Subramanian, S., Ganesan, S., Elanchezhian, E.B.: Grey wolf optimization for combined heat and power dispatch with cogeneration systems. Electr. Power Energy Syst. **74**, 252–264 (2016)
7. Basu, M.: Combined heat and power economic emission dispatch using nondominated sorting genetic algorithm-II. Electr. Power Energy Syst. **53**, 135–141 (2013)
8. Abido, M.A.: Environmental/economic power dispatch using multi objective evolutionary algorithms. IEEE Trans. Power Syst. **18**(4), 1529–1537 (2003)
9. Eusuff, M., Lansey, K., Pasha, F.: Shuffled frog-leaping algorithm: a memetic metaheuristic for discrete optimization. Eng. Optim. **38**(2), 129–154 (2006)
10. Storn, R., Price, K.: Differential evolution—a simple and efficient heuristic for global optimization over continuous spaces. J. Global Optim. **11**, 341–359 (1997)
11. Srinivasa Reddy, A., Vaisakh, K.: Shuffled differential evolution for large scale economic dispatch. Electr. Power Syst. Res. **96**, 237–245 (2013)
12. Srinivasa Reddy, A., Vaisakh, K.: Shuffled differential evolution for economic dispatch with valve point loading effects. Electr. Power Syst. **46**, 342–352 (2013)

Classification of Ball Mill Acoustic for Predictive Grinding Using PCA

Sonali Sen, Arup Kumar Bhaumik and Jaya Sil

Abstract The aim of this paper is to design a classification model to predict the terminating condition of a ball mill in raw material specific application. The acoustics of a running ball mill in different phases of grinding is analyzed to derive the signatures of the signal with different raw materials. Here we classify the sound signatures of the mill, which helps us to predict the characteristics of different running conditions with different size distribution of the materials. Using stereophonic microphones acoustic of a running ball mill is captured. The acoustic signal is then fragmented based on one rotation of the mill and saved in an appropriate format for further analysis. The statistical moments of each fragment are taken as the parameters and applied Principal Component Analysis to select the important features. Finally, we classify the analyzed data to find the running state of the mill. The classification model is tuned in order to achieve an experimental result from the simulated result. Also in real time the model has been tested.

Keywords Principal component analysis · Classification · Statistical moments

1 Introduction

Particle grinding is an essential part of any mining industry, to obtain desired particle size distribution required for different purposes. But this important phase cannot be operated dynamically. Till now it follows a purely static manual process since the sensor installation is not feasible within this kind of mill, which produces a huge amount of dust and contains metallic balls for grinding operation. The manual involvement may cause wastage of raw materials. In the current procedure [1, 2],

S. Sen (✉)
Department of Computer Science, St. Xavier's College, Kolkata, India
e-mail: sonalisen_2004@yahoo.co.in

A. K. Bhaumik
Department of Computer Science, RCCIIT, Kolkata, India

J. Sil
Department of Computer Science, IIEST, Kolkata, India

A. K. Das et al. (eds.), *Computational Intelligence in Pattern Recognition*, Advances in Intelligent Systems and Computing 999,
https://doi.org/10.1007/978-981-13-9042-5_37

input mixes are prepared using dozers with different size materials. By controlling the speed of the dozer motor with a particular size ranged input mix the amount of dozing is controlled. Prepared input mix is then given to the mill vessel with grinding medias for grinding. Then the mill motor is started for crushing the material for a specific time (i.e., particle residence time) and after that it will be stopped manually. After removing the grinding media from the vessel the crushed materials are screened to separate different sized materials. If we get the desired particle size range, then that must be collected from the classifier otherwise we need to follow the same procedure for further dozing. This screening and crushing operation is a closed loop operation. It is purely an offline procedure where we cannot predict the exact breakage state for different materials in a different time-domain from outside. Researchers like Rosin-Rammler, Gaudin-Schuman, Benette, and Colleman [3, 4] worked on this field and proposed some mathematical relationships, which can forecast the breakage state to some extent for different particles with different size ranges, but none of them found were perfect to interpret the grinding state for fine grinding. Also, these equations are not suitable for any autonomous system where only using information transaction control is possible. We also published some works [5, 6] describing some procedures using acoustic signature to operate the mill.

Here, we have chosen the ball mill for our experiment among the other grinding mills like jaw mill, rod mill, pebble mill, etc. We propose a classification model to predict the terminating condition of the mill with different raw materials while operated under different load conditions. Mill acoustics is captured for continuous and discrete time for different crushing states of the mill. The recorded signal is then fragmented and analyzed to find the statistical moments of the fragments. Then using those characteristics principal component analysis is applied to find the principal component. Finally, the classification method is applied to distinguish the fragment on which the mills need to stop.

The paper has been divided into four sections. The proposed method and experimentation details are discussed in Sect. 2 while the result with discussions are provided in Sect. 3. Finally the conclusion is given in Sect. 4.

2 Predictive Grinding Model

In this work, at the beginning the running mill sound in varied load conditions and empty condition is captured. The proposed experimental procedure is described in Fig. 1.

2.1 Experimentation

We used zoom microphones to record the sound of the mill. There are two different sets of microphones for capturing the ball mill sounds in different raw material's

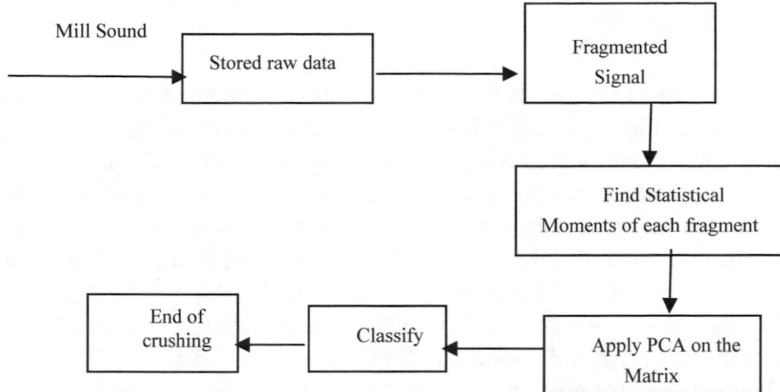

Fig. 1 Proposed predictive grinding model

load varying condition and empty condition. One microphone used to capture sound in discrete time and the other one for continuous time. Time for discrete recording is decided based on testing with different type and quantity of data, which produces sufficient amount of raw data for analysis. We sampled the acoustics signal and stored it in wav format. Continuous time-domain signal of the captured sound is given in Fig. 2.

We performed our experiment in a laboratory where we have tried to maintain a noise free environment. To make sure of any unwanted sound we also used a bandpass filter, which removes the frequencies outside the mill frequency band [5].

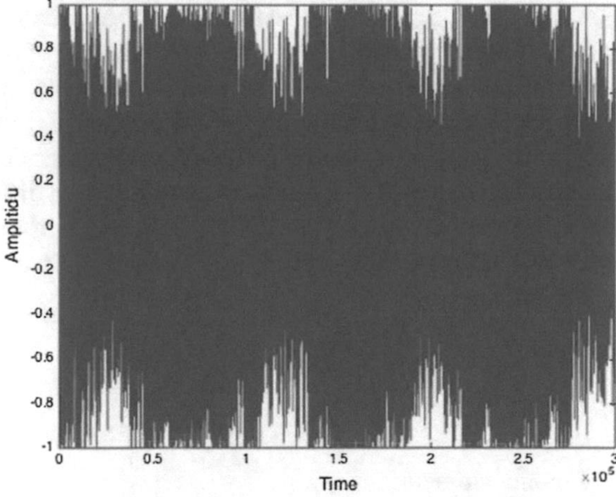

Fig. 2 Time domain plot of the running mill sound

2.2 Fragmentation

For this work we take coal and iron ore as our experimental crushing materials. During the experiment the captured acoustics produces a voluminous amount of raw data. In this proposed method our aim is to design an online method in which that huge amount of data cannot be analyzed in a few seconds. To provide the flexibility we proposed to fragment the sound based on one full rotation of the mill. In one of our previous work [4] we have noticed that the ball mill sound gradually changes over time when the material approaches toward the crushing state. At the time of fragmentation we have ensured that the data loss is not significant. Like load varying condition we also prepare the spectrum for the empty condition.

Different state of the coal crushing sound signal is shown in Fig. 3.

We have applied the following steps on the fragmented acoustic signals:

i. Finding the parameters of the signal using Statistical moments and form the matrix
ii. Perform PCA on the parameter matrix
iii. Classify the data set to predict the state of grinding.

2.3 Statistical Feature Extraction

In general the features of an acoustic signal are Amplitude, Frequency, and Phase, but these are not sufficient for signal analysis. For analysis we need some features, which can most efficiently and meaningfully represent the information integrated into the signal and can be used for classification. For extracting some important features from each of the fragmented data set we took the help of Statistical Moments. The *mean* and *variance* are the first two statistical moments, which provide the information about the location and variability of the data set. *Skewness* and *kurtosis* are the third and fourth moments that provide the information about the shape of the distribution. Based on these statistical moments we extract ten different features and form a 2-D feature matrix. Following are Eqs. 1, 2, 3, and 4 of mean-based first, second, third, and fourth moments respectively.

First Moment:

$$\sum_{i=1}^{n}(x_i - \bar{X})^1 \tag{1}$$

Second Moment:

$$\sum_{i=1}^{n}(x_i - \bar{X})^2 \tag{2}$$

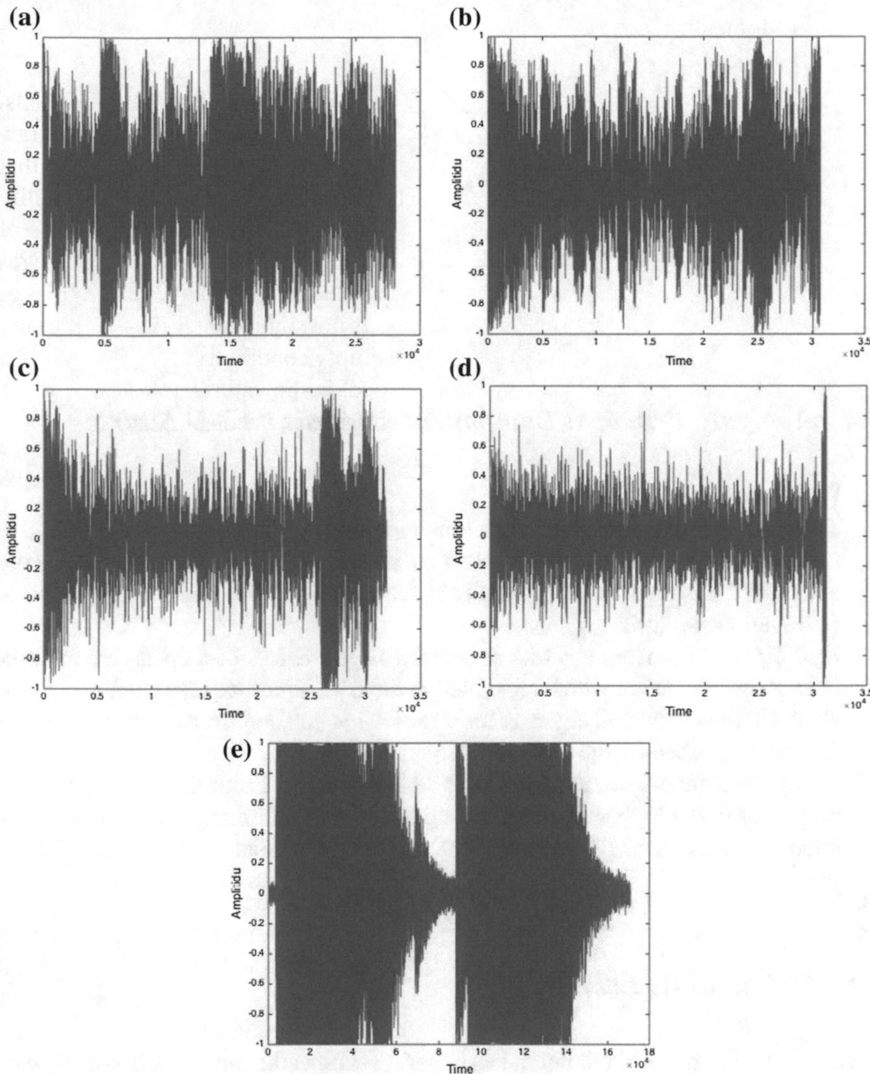

Fig. 3 **a** Crushing starts, **b** after 1 min of crushing, **c** 1 min before the crushing ends, **d** crushing ends, **e** empty vessel's sound signal with grinding media

Third Moment:

$$\frac{\sum_{i=1}^{n}\left(x_i - \bar{X}\right)^3}{ns^3} \tag{3}$$

Fourth Moment:

$$\frac{\sum_{i=1}^{n}\left(x_i - \bar{X}\right)^4}{ns^4} \tag{4}$$

2.4 Applying Principal Component Analysis on 2-D Matrix

1. We take a 98 × 10 matrix.
2. Then subtract the mean from each dimension.
3. Calculate the covariance Matrix: Next the covariance matrix is formed. Since the data is in 2-D format and the number of features is 10, this step produces a *10 × 10* covariance matrix.
4. Find eigenvector from covariance matrix: Eigenvectors of a covariance matrix extract the parameter, which characterize the data. Since we take a matrix of size 98 × 10, the calculated eigenvector size is 10 × 1. Then from that we prepare a rank matrix based on the eigenvalues.
5. Choosing components and forming a feature vector: From the rank matrix we have chosen the highest eigenvalue and based on that Principal component featured vector is formed. The size of the newly created matrix is also 98 × 10.

2.5 Define Class Labels

In this work, our aim is to predict the terminating state of the mill for different type of raw materials. In the experiment we have taken 98 data sets, which consist of initial, intermediate, and terminating conditions of the mill in the load varying condition. To define the class labels we have used *K-means* algorithm and store the corresponding data in a file.

2.6 Classify the PCA Output

After updating the data set with class labels, we have applied the classification algorithm on that data to classify the states of the mill. Data classification has two phases—in the first step the classifier algorithm is trained with 70% data set and in the second step the trained model is tested to analyze the performance with 30% data set.

Bayesian Classifiers [7] are the statistical classifiers, provides high accuracy and speed when applied to large databases. In this proposed work, Bayes theorem is applied to build the classifier to predict the class label of a 2-D matrix $m \times n$ without the class label. The Naive Bayesian classification predicts the output as $Pr(C|d)$ *represents* the probability that of a document d belongs to a class C. The Bayes algorithm is based on the following description:

- Let G be the training set with N tuples where each tuple is represented as a "k" dimensional attribute vector M, where $M = \{M_1, M_2, \ldots, M_k\}$
- Let there be "p" classes C_1, C_2, …, C_p. According to this classifier, a tuple T belongs to class C_x only when it has a higher conditional probability than any other class C_y, where $x \neq y$ and can be defined in Eq. 5.

$$P(C_x|T) > P(C_y|T) \quad \text{and} \quad P(C_x|T) = \left(P\left(T|C_x^* P(C_x)\right)/P(T)\right. \tag{5}$$

- Class conditional independence is assumed as described in Eq. 6,

$$P(M|C_x) = \prod_{i-1}^{k} P(M_k|C_x) = P(M_1)^* P(M_2) \ldots^* P(M_k) \tag{6}$$

- Class C_x is predicted as the output class when Eq. 7 follows,

$$\begin{aligned} &P(M|C_x)^* P(C_x) > P(M|C_y)^* P(C_y) \\ &\text{where } 1 \leq x, y \geq p \text{ and } x \neq y \end{aligned} \tag{7}$$

So we have applied Naive Bayes classification algorithm on the PCA output, i.e., *98* × *10* to classify the data set to predict the terminating condition of the mill. Ultimately, we got a matrix as *98* × *11*.

2.7 Performance Measures

Confusion matrix, True Positive (TP) rate, False Positive (FP) rate, Precision, Recall, F-Measure, and ROC are generally used to measure the performance of the classification output. The representation of a confusion matrix with two classes is given in Table 1.

Table 1 Example of a confusion matrix

		Predicted	
Actual		Class A	Class B
	Class A	TP	FN
	Class B	FP	TN

Meaning of the terms mentioned in the matrix:

- **True Positive (TP)**: Observed as **positive**, predicted as **positive**
- **False Negative (FN)**: Observed as **positive**, predicted as **negative**
- **True Negative (TN)**: Observed as **negative**, predicted as **negative**
- **False Positive (FP)**: Observed as **negative**, predicted as **positive**

The classifier's *accuracy* can be calculated as follows given in Eq. 8:

$$AC = \frac{a+d}{a+b+c+d} \tag{8}$$

It can be defined as the percentage of test set tuples that are correctly classified by the classifier on a data set.

The *recall* is calculated as in Eq. 9:

$$FP = \frac{b}{a+b} \tag{9}$$

It can be defined as the ratio of positive cases that were correctly identified.

The *true negative rate (TN)* can be defined, as the ratio of negative cases that were classified correctly is defined in Eq. 10:

$$TN = \frac{a}{a+b} \tag{10}$$

The f*alse negative rate (FN)* is defined as the ratio of positives cases that were incorrectly classified as negative which is defined in Eq. 11:

$$FN = \frac{c}{c+d} \tag{11}$$

Finally, *precision (P)* is defined as the ratio of the predicted positive cases that were correct that defined in Eq. 12:

$$P = \frac{d}{b+d} \tag{12}$$

The *error rate* or misclassification rate is the rate of incorrect classification. Also using *sensitivity* and *specificity* we can measure the accuracy of the classifiers. *Sensitivity* is referred to as the true positive rate, while *specificity* is the true negative rate. The process divides all the data into equal parts usually k = 7 and the model was trained for k−1 times and *k*th time to allow for both training and testing of each set.

3 Result and Discussion

In the proposed method, various experiments with different materials are performed. We take the PCA output represented, as a 2-D matrix 98 × 11 where the number of samples is considered 98 and the number of attributes are 11. Out of 11 attributes 11th attribute represents (column) the class label while the rest are conditional attributes. Here we take $k = 2$ in *k-means* algorithm to find the state as beginning and end. On that 2-D array we have applied Naïve Bayes classifier using sevenfold cross validation technique. The output of the classification method provides a confusion matrix, which is given in Table 2.

Our proposed classification method is applied on 98 tuples. Thirty-one positive tuples are correctly labeled as positive and 53 negative tuples are labeled as negative. Only 14 tuples are not correctly labeled. The other performance measures of classification output are shown in Table 3. The output is the average across the sevenfolds. The correctly classified instances of the experiment are approximately 83%. The ROC curve area is similar to life chart, used in signal detection to show trade of between hit rate and false alarm rate over noisy channel. The experiment provides the ROC area as 0.936 for all classes. The ROC graph is shown in Fig. 4, where Fig. 4a represents the ROC curve for class1 and Fig. 4b represents the ROC curve for class 2.

Table 2 Confusion matrix of the given data set

31 (TP)	12 (FN)
2 (FP)	53 (TN)

Table 3 Test mode: Naive Bayes classifier using sevenfold cross-validation

	TP rate	FP rate	Precision	Recall	F-Measure
Weighted Avg.	0.721	0.036	0.939	0.721	0.816
	0.964	0.279	0.815	0.964	0.883
	0.857	0.173	0.870	0.857	0.854

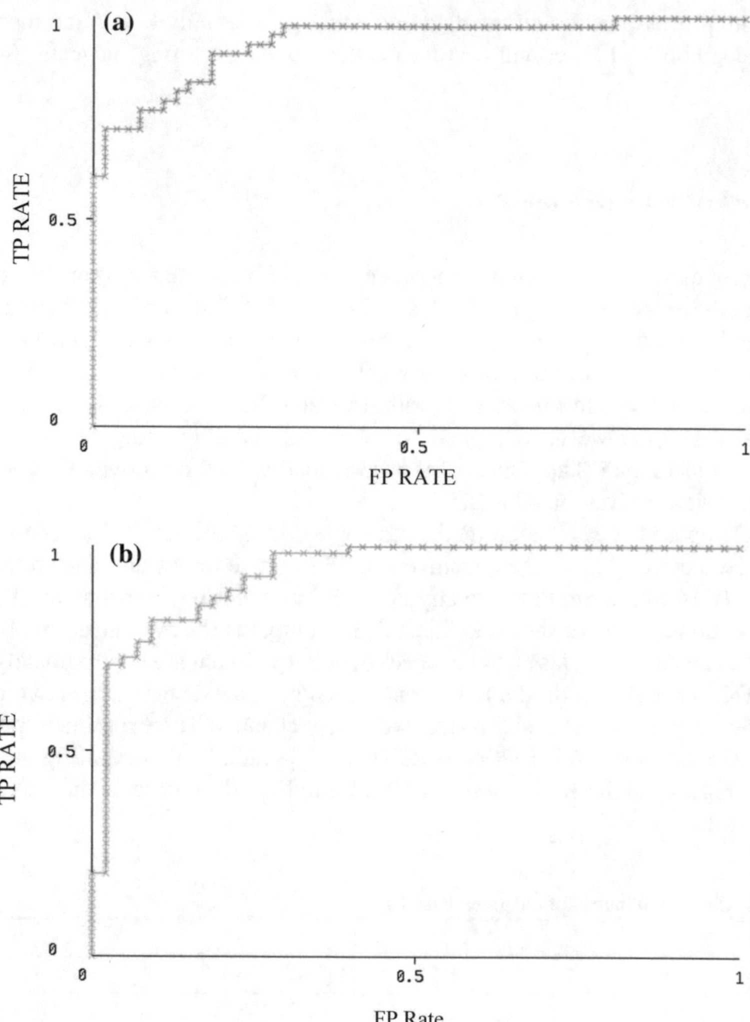

Fig. 4 **a** ROC for Class 1, **b** ROC for Class 2

4 Conclusions

The proposed method is designed to achieve the predictive grinding operation of a ball mill with different grinding materials in different load varying and empty condition. The simulated output of the proposed approach similarly behaves as the experimental result. We can also design the embedded system for this. For the prediction of the grinding states of the ball mill there are different works that have been proposed [8–10] but none are intelligent enough to come with a mathematical model. Here, we have done the work offline but this analysis is also possible in real time. For our work we have taken iron ore and coal as raw material and for each case we found that the simulated method satisfies the experimental output. Ultimately we can conclude that the proposed method can classify different states of the crushing procedure with different materials. We can predict the state from the outside of the mill without any manual intervention. In this work, we have concentrated on the acoustic signal analysis because we found that the distinction of different states can be easily identified from the outside world based on the sound. So we build a knowledge based on the proposed model. The same method can also be applied on different grinding mills like jaw mill, rod mill, pebble mill, etc. We achieved satisfactory outputs after performing the experiments under different training conditions.

References

1. Gills, W.A.: Mineral Processing. Springer, Germany (1991)
2. Taggert, H.: Hand Book of Mineral Processing. SME Publishers (1951)
3. Wills, B.A.: Mineral Processing Technology. Pergamon Press, UK (1986)
4. Bhaumik, A., Sil, J., Banerjee, S.: Hybrid control of tumbling mill using NN based schematics. Int. J. Inf. Comput. (IJICS) **2**(1), 21–28 (1999)
5. Bhaumik, A., Sil, J., Banerjee, S.: A new supervised training algorithm for generalized learning. In: Third International Conference on Computational Intelligence and Multimedia Applications, ICCIMA '99, pp. 489–493 (1999)
6. Sen, S., Bhaumik, A.K., Sil, J.: Prediction of the state of grinding materials in run time using genetic operator. In: Advances in Intelligent Systems and Computing Applications and Future Directions—Volume I, Advances in Intelligent Systems and Computing, vol. 798, pp 475–486. https://doi.org/10.1007/978-981-13-1132-1_37
7. Karthika, S., Sairam, N.: A Naïve Bayesian classifier for educational qualification. Indian J. Sci. Technol. **8**(16), 1–5.
8. Bhaumik, A., Sil, J., Banerjee, S.: Designing of intelligent hybrid control system for tumbling mill operation. In: Proceedings of the 6th International Conference on Information Technology (CIT2003), pp. 246–249. Allied Publishers Pvt. Ltd., Bhubaneswar (2003)
9. Bhaumik, A., Sil, J., Banerjee, S.: Designing of intelligent expert control system using petri net for grinding mill operation. WSEAS Trans. Inf. Sci. Appl. **2**(4), 360–365 (2005). ISSN 1790-0832
10. Sen, S., Bhaumik, A.: Design of intelligent control system using acoustic parameters for grinding mill operation. In: National Conference on Advancement of Computing in Engineering Research, pp. 261–268 (2013). https://doi.org/10.5121/csit.2013.3224

Particle Swarm-Based Approach for Storage Efficiency Optimization of Supercapacitors

Rudrendu Mahindar, Sahana Mukherjee, Sankhadeep Dutta, Tanusree Dutta and Rabindranath Ghosh

Abstract In the recent years, with the advent of artificial intelligence in multidisciplinary fields, more and more efficient mechanisms, to enhance the performance of various systems, are gaining prominence. This paper exploits the technique of computational intelligence to optimize the storage efficiency of supercapacitors. While a supercapacitor is a high-value capacitor, it can be used as an energy storage device like batteries. The storage efficiency of a supercapacitor depends directly on the capacitance value and inversely on the Equivalent Series Resistance (ESR) value. Supercapacitors with high storage efficiency can be deployed in numerous applications requiring high power-handling capacity. Different optimization methods have been used to maximize the storage efficiency of supercapacitors. In this paper, Particle Swarm Optimization (PSO) technique has been used in which the values of storage efficiency obtained are higher than those from other conventional algorithms.

Keywords Supercapacitor · Storage efficiency · ESR · Capacitance · PSO

R. Mahindar (✉) · S. Mukherjee · S. Dutta · T. Dutta · R. Ghosh
Electronics and Communication Engineering Department, St. Thomas' College of Engineering and Technology, Kolkata 700023, India
e-mail: rudrendumahindar@gmail.com

S. Mukherjee
e-mail: sahanamukherjee8@gmail.com

S. Dutta
e-mail: tpdutta998@gmail.com

T. Dutta
e-mail: momtanu@gmail.com

R. Ghosh
e-mail: rnghosh@gmail.com

A. K. Das et al. (eds.), *Computational Intelligence in Pattern Recognition*,
Advances in Intelligent Systems and Computing 999,
https://doi.org/10.1007/978-981-13-9042-5_38

1 Introduction

Demand for energy is increasing daily, and as supply of nonrenewable sources of energy is limited, there has been an urgent need to shift to renewable sources, which are inexhaustible, as well as nonpolluting [1]. Solar energy is one of the most abundant forms of renewable energy. Its supply is unlimited and at the same time it causes hardly any pollution to the environment. The technique to trap the solar energy and convert the solar power into electric power is by using the Photo Voltaic cells (PV cells). The efficiency of the PV cells is about 10–25% which can go even up to 35%. To obtain the maximum power from these cells, tracking of the incident solar radiation is necessary and among many methods, one efficient technique is the Maximum Power Point Tracking algorithm (MPPT) [2]. An MPPT is basically an electronic converter, which converts DC to DC that is used to step down a higher DC output voltage from the PV cells, down to a lower voltage which is required to charge the load. The load in most cases is a battery. Batteries, as they store energy electrochemically, continuous charging and discharging process can degrade the chemical compound inside it, over the course of time. Supercapacitors, on the other hand, do not involve chemical reactions and store energy on the surface electrostatically. As a result, they possess high power density and unlike batteries, they retain their storage capabilities over time, even up to millions of charging–discharging cycles. The storage efficiency of supercapacitors depends primarily on two factors, Equivalent series resistance (ESR) and capacitance values. Optimization technique allows us to achieve the highest performance under given system constraints by altering values of dependable parameters. In this system, the optimization technique has been used to maximize or enhance the storage efficiency of the supercapacitor by taking advantage of the best compromise between values of ESR and capacitance. The utmost utilization of resources can be done by using optimization techniques [3, 4] like Evolutionary computation, Gradient decent, Nonlinear optimization [5], Simulated annealing [6], Stochastic tunneling [7], Subset sum algorithm [8], and Linear programming [9]. The proposed Particle Swarm approach has shown a rise in storage efficiency of supercapacitor, when compared with results of other optimization techniques like Genetic Algorithm (GA) [10, 11] and Simulated Annealing (SA).

This paper has been organized in the following sections. In the next section (Sect. 2), a literature survey on the project has been discussed. Section 3 describes the mathematical and electrical models used in the simulation and also the various algorithms used in this work. Section 4 shows the variation of different parameters of the solar cell. Section 5 contains the tables of results obtained from our simulation, algorithm, and observation. Conclusion of this project has been drawn in Sect. 6.

2 Literature Survey

There has been a wide range of study on how to obtain the maximum energy output from the solar PV arrays or modules. The array output depends on the maximum power point and appropriate tracking of the maximum power point is essential, which in turn depends on the panel conditions as well as the irradiance hitting the solar array. Patel et al. [12] in their paper have analyzed the Perturb and Observe (P&O) Algorithm and discussed the benefits of the algorithm over other algorithms. The P&O algorithm is one the most common algorithms for MPPT. Femia et al. [13] in their work have further discussed the issue of MPPT. Their paper has proposed to limit the drawbacks of the P&Oalgorithm , which is the most common algorithm used for MPPT, by customizing the P&O and MPP factors or constraints according to the dynamic behavior of the particular converter adopted. Koutroulis et al. [14] have proposed to procure the maximum power from the PV arrays by using a DC to DC converter, which is a buck type one. Their proposed system is low cost and less complex. It provides higher efficiency and can be used to handle multiple sources of energy. Li et al. [15] have put forward the proposal of a scheme for charging supercapacitors, connected in series, and this method has given better utilization of electric power from the PV cells as well as shortened charging time. The research of Kim et al. [16] proposes the optimization technique for the determination of the optimal efficiency of the supercapacitor and also to obtain maximum energy transfer from the solar cells into the supercapacitors. Further works are found in literature, which are concerned with maximizing the power output from the solar arrays. Dondi et al. [17] in their proposal have put forward the analytical method for optimizing a solar harvester with maximum power point tracking for Wireless Sensor Network (WSN) nodes. Their proposal could reach a maximum efficiency of about 85% with discrete components. Iqbal et al. [18] present a general exploration of various optimization methods, and their research takes into account, approaches for application with respect to different end users, and is not limited to any specific issue or terrane factors. Kirkpatrick et al. [19] propose a different kind of optimization technique. Their work draws a detailed analogy between statistical mechanics and multivariate or combinational optimization. The technique of optimization by SA is discussed in detail in their literature. Furthermore, new researches have come up in the course of time, taking into consideration new techniques like those of SA and GA. Seth et al. [6] propose the method of SA to determine the maximum storage efficiency of the solar cells by considering the ESR values and capacitance values of the supercapacitor. The method of GA has been employed to maximize the storage efficiency of the supercapacitor of the solar array and such works are found in abundance in the literature [10, 11].

3 Mathematical Models and Algorithms

3.1 Model of Photo Voltaic Panel

A PV cell is essentially a silicon PN junction diode, made of extremely thin wafers or films. When sunlight is incident on the junction, a potential gets developed across the junction, which provides the momentum to the light stimulated electrons and flow of electrons will constitute electric current when the solar cell is connected to an external load. About 3–30% of incident solar energy is converted to DC electricity, and this conversion is dependent on various parameters like intensity of solar irradiation, temperature, and materials of the solar cell. The physical constructional features of a PV cell are depicted in Fig. 1.

The equivalent circuit of a solar cell consists of a current source which is connected in parallel with a diode. No solar cell is practically ideal in nature, and for this reason, a shunt resistance and a series resistance are added to the model. The equivalent model of the solar cell is shown in Fig. 2.

From the above circuit of the solar cell, it is observed that the current produced by the solar cell is equal to the current generated by the current source, from which the current drawn by the two elements, the diode, and the resistance in shunt are

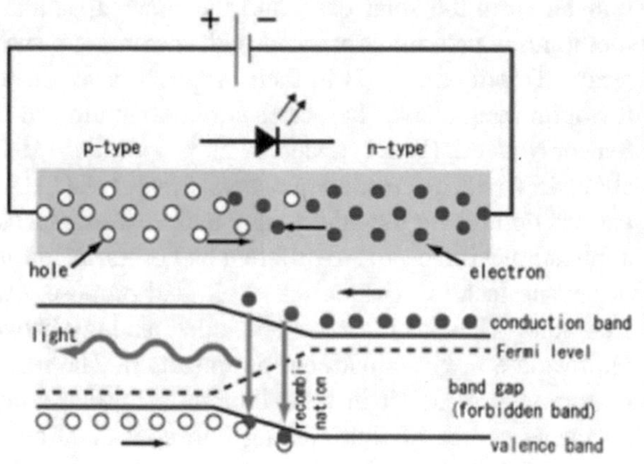

Fig. 1 Physical constructional features of solar cell

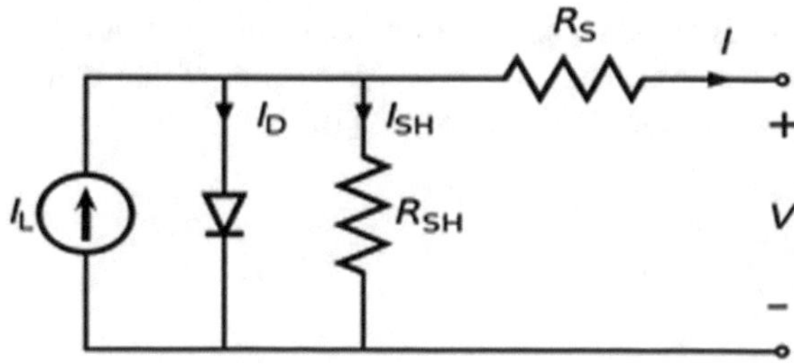

Fig. 2 Equivalent circuit of solar cell

needed to be subtracted, because they carry the current in the opposite direction. The equation of the current is reflected by Eq. 1,

$$I = I_L - I_D - I_{SH}. \tag{1}$$

where, I is the generated solar current, I_L is the photogenerated current, I_D is the diode current, and I_{SH} is the current in the branch where the shunt resistance is connected. The current through these elements is, controlled by the voltage across them, and is given by Eq. 2,

$$V_j = V + IR_S. \tag{2}$$

where, V_j is the voltage across both the diode and the shunt resistance $R_{SH,}$ V is the voltage over the output terminals, I is the output current and R_S is the series resistance. Equation 3, namely the Shockley diode equation gives the relationship between the current through the diode and other abovementioned parameters,

$$I_D = I_O(e^{V_j/nV_T} - 1). \tag{3}$$

where, I_O = reverse saturation current, n = diode ideality factor, q = elementary charge, k = Boltzmann's constant, T = absolute temperature the thermal voltage at 25 °C, V_T = 0.0259 V. The Ohm's Law gives the relationship between the shunt current and shunt resistance as shown in Eq. 4,

$$I_{SH} = V_j/R_{SH}. \tag{4}$$

where, R_{SH} = shunt resistance. Substituting these values in the first equation produces the characteristic equation of a solar cell, which is represented by Eq. 5,

$$I = I_L - I_O(e^{\frac{V+IR_s}{nV_T}} - 1) - \frac{V + IR_{SH}}{R_{SH}} \tag{5}$$

3.2 *Output Characteristics of Solar Cell*

The I-V characteristic of the solar cell is a graphical representation that tells about the relationship between current and voltage at different conditions of temperature and solar irradiance. The power delivered by the solar cell is the product of I and V, i.e., current and voltage (I × V). When point to point multiplication is done, starting from short circuit to open circuit conditions, the power graph is acquired for the particular intensity of irradiance. When the solar cell is operated at no load condition, i.e., it is open circuited, the voltage across the cell is maximum, whereas the current is at minimum. The Open Circuit Voltage is expressed by Eq. 6,

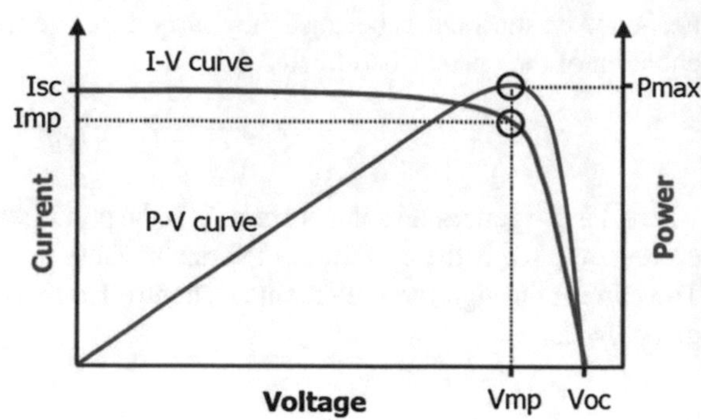

Fig. 3 I-V characteristics of solar cell

$$V_{OC} \approx nV_T \ln(I_L/I_O + 1). \tag{6}$$

When the solar cell is operated at short circuit condition, then the voltage is at the minimum value, and the current is at the maximum value. The short circuit current is illustrated by Eq. 7,

$$I_{SC} \approx I_L. \tag{7}$$

For one particular value of current and voltage, I_{MP} and V_{MP}, the power reaches the maximum value, and this point is referred to as the Maximum Power Point (MPP), which is located near the bend of the IV characteristics curve. The I-V curve for the solar cell is portrayed in Fig. 3, showing the important parameters.

3.3 Simulation Model of Supercapacitor

Supercapacitors have the ability to hold many times more electric charge than a standard conventional capacitor. They are most convenient in devices which require relatively low current and low voltage. They are usually double-layered capacitors or DLCs, which have intensive applications requiring impulsive power. They are widely used in driving motor, pump, and electric vehicles sometimes in combination with battery [13]. The small leakage current facilitates a long period of energy storage in the supercapacitors and efficiency could increase as high as up to 95%. This paper proposes two resistance models of the supercapacitor for software simulation as shown in Fig. 4 where C0 is a constant capacitor, R1 is the equivalent series resistance, and KV*V is a voltage-dependent capacitor and this first branch models the voltage dependencies of the supercapacitor. The second branch consisting of R2 and C2 are responsible for considering the charge redistribution occurring within the supercapacitor. The equivalent parallel resistor, R3 is used to model the leakage behavior of the capacitor. The storage efficiency of supercapacitors is dependent

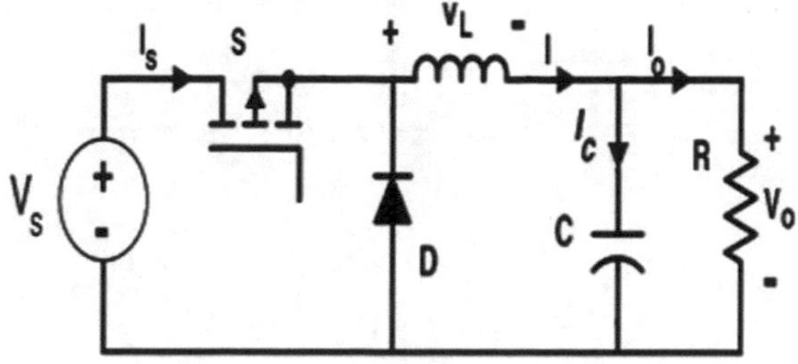

Fig. 4 Two resistance model of supercapacitor for software simulation

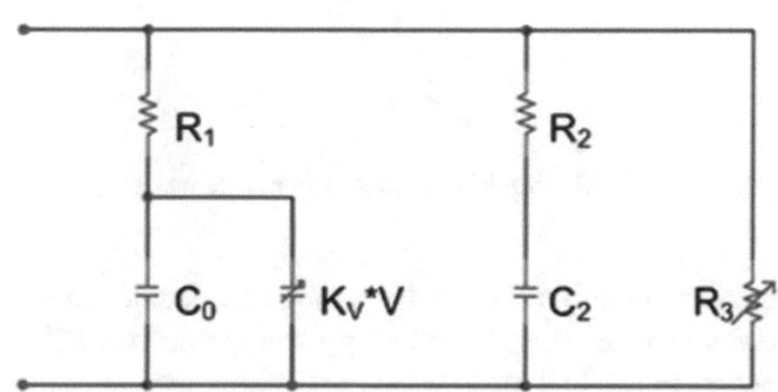

Fig. 5 Equivalent circuit of buck converter

on the values of both ESR and the capacitance value. The efficiency increases with increasing values of capacitance and decreases with increasing ESR value.

3.4 Simulation Model of Buck Converter

In this system, a buck converter is used, because we need to decrease the input voltage (usually greater than 12 V) from solar panel by proper voltage conversion to charge the supercapacitor, which requires a lower voltage level. The buck converter, a DC to DC step down converter, steps down voltage from the input to the output or the load. The buck converter can transfer energy from the input, to the output, in the form of energy packets, on the basis of frequency of the switching signal applied to the switch. Most of the time it is a MOSFET and the switching action is controlled by its gate. The equivalent circuit of the buck converter is shown in Fig. 5 and its corresponding MATLAB simulation model is shown in Fig. 6.

3.5 Maximum Power Point Tracking for Efficient Power Delivery to the Load

MPPT works on the principle of extracting the utmost power obtainable from the PV arrays, by operating them at the MPP. MPPT charge controller takes into consid-

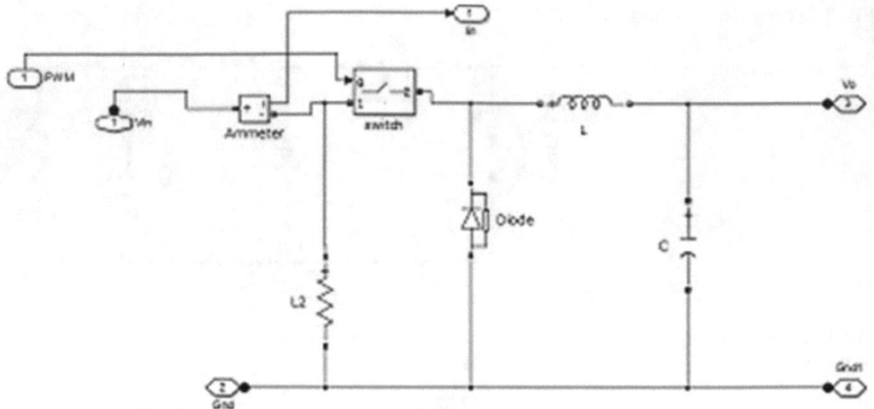

Fig. 6 MATLAB simulation model of buck converter

eration the output of the PV array, and decides the most optimum power the panel can deliver to charge the energy storage device. MPPT also facilitates in supplying power to the DC load, which is directly connected to the battery. In order to boost the power delivery to the supercapacitor from the PV panel, this paper utilizes the MPPT algorithm. Different MPPT algorithms are available which include the P&O Method, the Incremental Conductance Method, and the Current Sweep Method. Of these, this paper uses the P&O method, which is the most common algorithm due to its ease of implementation. The P&O algorithm, also known as the Hill Climbing Method, regulates the voltage by a small amount from the PV module, and computes the power. If the power is found to increase, then more adjustments are done until the power no longer increases. The flowchart for P&O algorithm is depicted in Fig. 7. The simulation model of the P&O algorithm is shown in Fig. 8.

3.6 Complete Simulation Model of the System

The simulation model of the system is shown in Fig. 9.

3.7 Choosing Particle Swarm Optimization (PSO) over Genetic Algorithm (GA) and Simulated Annealing (SA)

This paper has proposed the advantageous approach of PSO to cause a surge in storage efficiency of supercapacitor and eventually in Sects. 5 and 6, results of PSO have been compared with that of GA and SA and a suitable conclusion has been drawn,

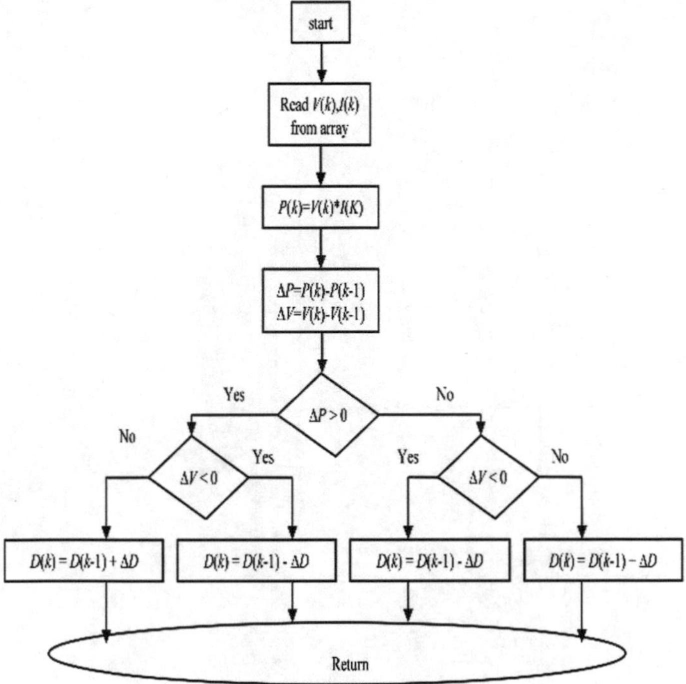

Fig. 7 Flowchart for P&O algorithm

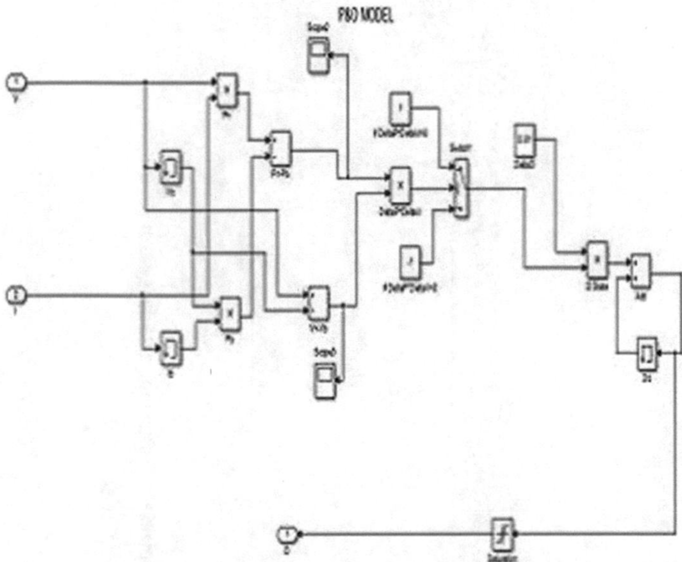

Fig. 8 Simulation model of P&O algorithm

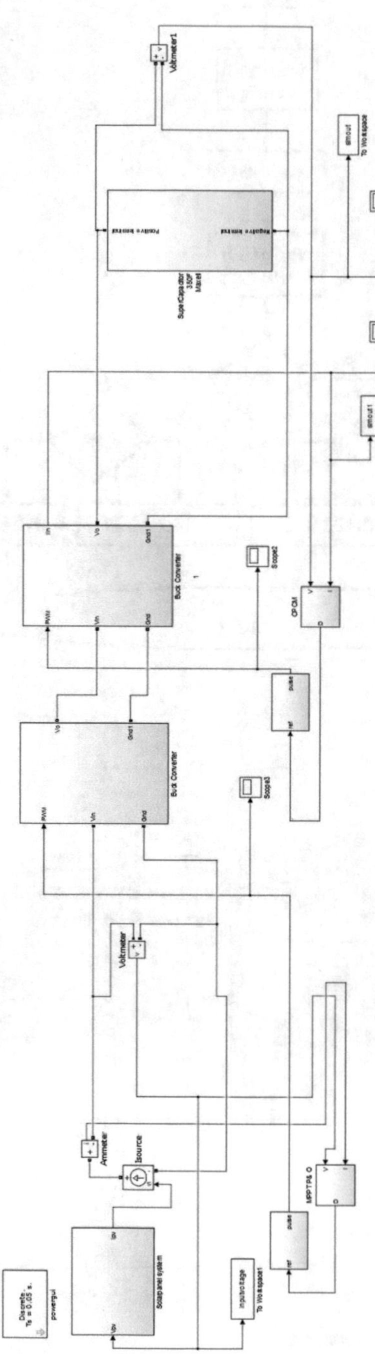

Fig. 9 Complete simulation model of the system

respectively. GA [10] searches artificially by the use of natural selection and natural genetics. Biological evolution is initiated by assigning present population of parents and then using them for producing future offspring. The increasing population finds the optimal solution. Unlike many algorithms like hill climber, SA [6] does not get caught at local optimums. It mimics the annealing process of metallurgy and uses temperature as a parameter to optimize the problem. Initially at high temperature there is high randomness of searching solutions when the algorithm may accept worse solutions which help to come out of local minima. With the lowering of temperature, the randomness reduces and a solution close to global minima is obtained by focusing on a specific area of search space. PSO algorithm [20] was inspired by the flocking and schooling patterns of birds and fish. This optimization technique shares many similarities with evolutionary computation techniques like GA. Unlike GA, PSO has no evolution operators such as crossover and mutation. Many computation techniques were inspired by biological systems. Social system, a type of biological system, shows the collective behaviors of simple individuals interacting with their environment and each other. Such a system was called swarm intelligence. PSO is a popular swarm inspired method in computational intelligence areas. In the process of optimization using PSO, the potential solutions to the optimization problem area swarm of particles, which flows through the parameter space defining trajectories which are driven by their own and the entire swarm's best-known position. At each time step of location update of swarm particles, the discovery of improved positions guides the movements of the swarm. The combination of exploration and exploitation of known good positions in the search space is repeated to obtain the satisfactory optimal solution. The flowchart for PSO algorithm is shown in Fig. 10.

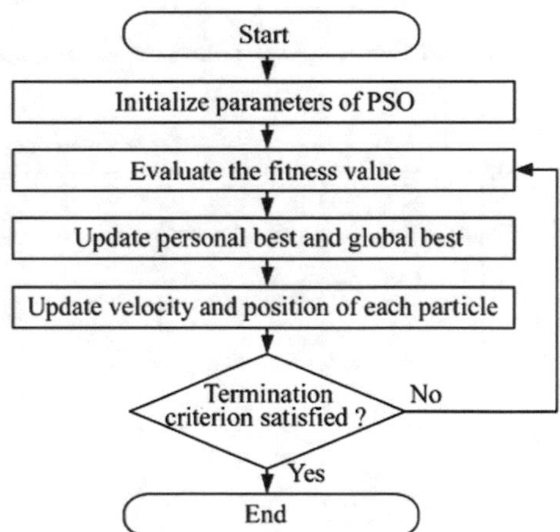

Fig. 10 Flowchart for PSO algorithm

4 Variation in Storage Efficiency with ESR and Capacitance

The model has been simulated and used to observe the variation in storage efficiency with variation in ESR and capacitance. From the model, the response obtained for storage efficiency vs. ESR is shown in Fig. 11 and the response of storage efficiency vs. capacitance is shown in Fig. 12.

The 3D plot with axes of storage efficiency, ESR, and capacitance are obtained after extrapolation of sample data of the simulation model and it is shown in Fig. 13. It can be observed from the figures that storage efficiency increases with a decrease in ESR and increase in capacitance of a supercapacitor.

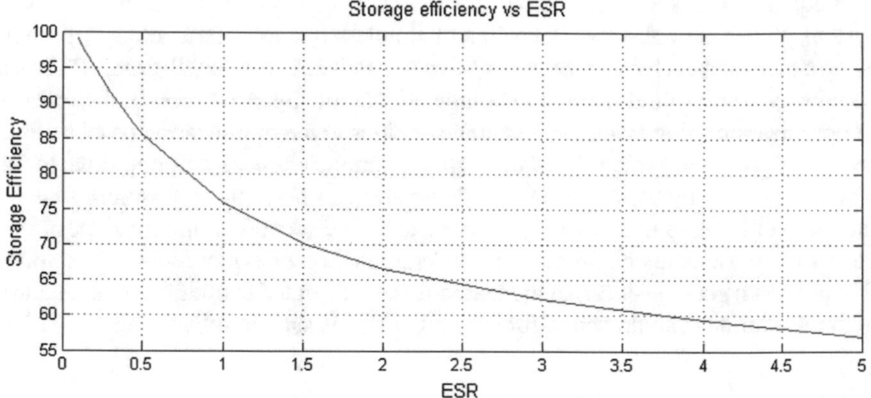

Fig. 11 Variation of storage efficiency with respect to ESR

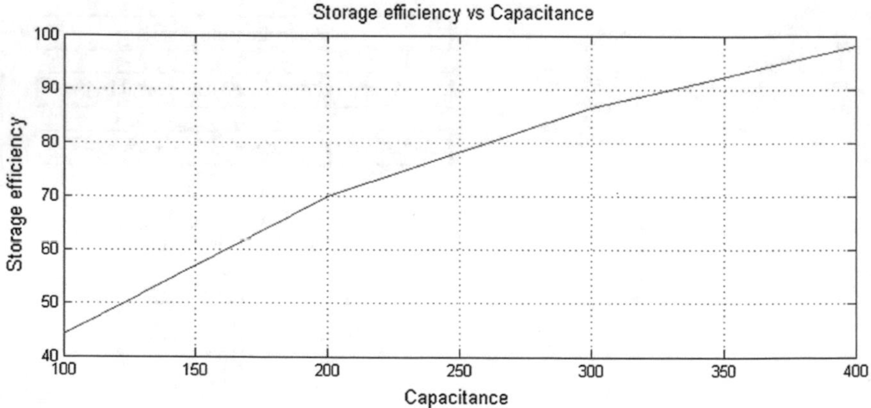

Fig. 12 Variation of storage efficiency with respect to capacitance

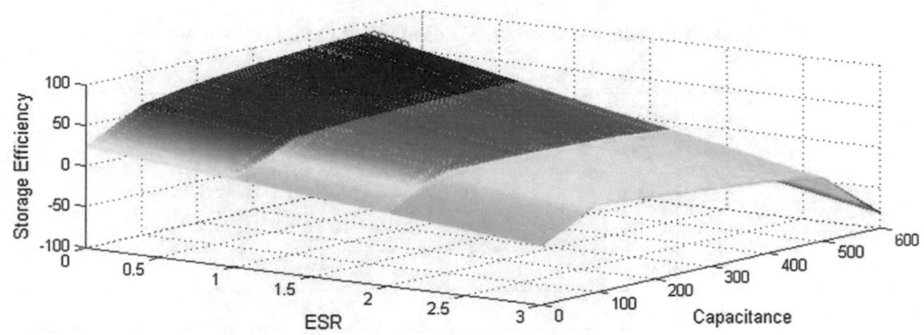

Fig. 13 3-D plot of storage efficiency, ESR, and capacitance

5 Experimental Results

The simulation of the model was done to get enough sample points for further processing by the PSO algorithm. Firstly, the model was used to determine the variation in efficiency with variation in ESR and capacitance individually and independently and the data points obtained from simulation model were plotted, as shown in the above graphs in Sect. 4. These data points were then used to find the polynomial function which will eventually be modified and used as an objective function in the PSO. The data points obtained from the model are given in Table 1. These data have been used to find the polynomial function using the curve fitting toolbox of MATLAB. The obtained polynomial is of degree 2 and is given as,

$$z(x, y) = p_{00} + p_{10}x + p_{01}y + p_{20}x^2 + p_{11}xy + p_{02}y^2$$

where, $p_{00} = 48.01$, $p_{10} = -22.45$, $p_{01} = 0.1337$, $p_{20} = 23.6$, $p_{11} = -0.04754$, $p_{02} = -0.0001107$, x is the ESR and y is the capacitance of the supercapacitor.

We have used the polynomial $z(x, y)$ in the PSO algorithm for maximizing the storage efficiency. The PSO algorithm is used to find the minima of the objective function $f(x, y)$.The minimization of $f(x, y)$ will lead us to maximization of $z(x, y)$, which is essentially the overall storage efficiency. Here, $f(x, y) = 1/z(x, y)$.

The efficiencies obtained for different values of ESR and capacitance by using particle swarm solver at different lower and upper bounds are compared with that of GA as shown in Table 2. Table 3 shows a comparison between efficiencies obtained by using PSO algorithm and SA. The efficiencies are compared in each of the tables for similar values of ESR and capacitance. Table 2 has been created from [10] at different stall generations for GA. Table 3 uses data from [6] at different initial temperatures of SA and Table 4 uses data from [6] at constant initial temperature 100 and different re-annealing intervals. The constrained optimization was performed by using PSO algorithm within specified upper and lower bounds satisfying the ranges of ESR and capacitance obtained from Table 1, which contains data from the simulation model. The result of the PSO algorithm has been recorded in Table 5.

Table 1 Sample data points collected from the simulation of the model shown in Fig. 9

ESR (ohm)	Capacitance (F)	Storage efficiency (%)
0.05	100	58.63
	200	69.26
	300	76.83
	400	81.37
	500	85.06
0.1	100	57.45
	200	67.88
	300	74.69
	400	79.45
	500	82.96
0.15	100	56.52
	200	66.58
	300	73.11
	400	77.66
	500	81
0.20	100	55.64
	200	65.37
	300	71.64
	400	75.99
	500	79.18
0.25	100	54.82
	200	64.22
	300	70.26
	400	74.43
	500	77.48

Table 2 Comparison drawn between the storage efficiencies of GA and PSO

Stall generation for GA	ESR (ohm)	Capacitance (F)	Storage efficiency (%) by using GA	Storage efficiency (%) by using PSO
20	1.00E−003	498	**86.86**	**87.0925**
21	3.00E−003	473	86.3	86.34
22	0	458.15	85.95	86.0286
23	2.30E−002	432.1	84.09	84.1366
24	1.20E−002	414.35	83.82	83.9006
25	1.60E−002	479.79	85.9	85.9568

Table 3 Comparison drawn between the storage efficiencies of SA (for different initial temperatures of SA) and PSO

Initial temperature	ESR (ohm)	Capacitance (F)	Storage efficiency (%) by using SA	Storage efficiency (%) by using PSO
10	1E−003	982.15	73.68	72.4711
20	0	964.79	74.81	73.9607
30	7E−003	963.77	76.4	73.5653
40	0	948.5	77.89	75.2329
50	0	833.05	82.71	82.5661
60	2E−003	818.18	83.34	83.1734
70	8E−003	779.37	84.59	84.4961
80	1E−003	884.6	79.77	79.5919
90	1E−003	701.82	**87**	**87.2621**
100	0	620.25	88.36	88.3500

Table 4 Comparison drawn between the storage efficiencies of SA (for constant initial temperature 100 and different re-annealing intervals of SA) and PSO

Re-annealing interval	ESR (ohm)	Capacitance (F)	Storage efficiency (%) by using SA	Storage efficiency (%) by using PSO
10	1E−003	597.99	88.34	88.3250
20	2E−003	667.01	87.86	87.8303
30	0	598.19	**88.36**	**88.3761**
40	0	665.64	88	87.9575
50	0	658.18	88.08	88.0533
60	2E−003	611.05	88.29	88.2711
70	0	636.49	88.29	88.2620
80	1E−003	609.23	88.35	88.3251

Table 5 Results of storage efficiency optimization using PSO algorithm

Iteration	Function value	Storage efficiency (%)
0	1.205938e−02	82.92
1	1.205694e−02	82.94
2	1.197423e−02	83.51
3	1.185877e−02	84.33
4	1.166719e−02	85.71
5	1.146241e−02	87.24
6	1.133751e−02	88.20
7	1.131684e−02	88.36
8	1.131484e−02	**88.38**

6 Conclusion

It has been observed from the graphs in Sect. 4 that lower values of ESR and higher values of capacitance are necessary to increase the storage efficiency of supercapacitors. From comparisons shown in the above tables it is evident that PSO provides better storage efficiency compared to all cases of GA. The results of optimization at different stall generations of GA showed highest storage efficiency of 86.86% at ESR value 1.00E−003 Ω and capacitance value 498 F. The PSO algorithm showed storage efficiency of 87.09% at the same values of ESR and capacitance. For optimization by using SA, the storage efficiency was initially determined at different initial temperatures, for different values of ESR and capacitance. For lower initial temperatures of SA, the storage efficiency determined from PSO was much lower than that of SA when capacitance was over 900 F and ESR was 0 Ω in many of those cases. With the gradual increase in initial temperature of SA, the storage efficiency obtained from PSO increased to a level almost equal to that of SA and even showed greater efficiency values in some cases. Then different re-annealing intervals, for different values of ESR and capacitance at constant initial temperature 100, resulted in highest storage efficiency of 88.36% from SA at ESR 0 Ω and capacitance 598.19 F. While the storage efficiency observed from PSO algorithm at ESR 0 Ω and capacitance 598.19 F was 88.38%. In addition, the use of PSO divulges the possibility of a soar in the storage efficiency of supercapacitor at a lower range capacitance compared to that of SA. The highest storage efficiency recorded in PSO algorithm (88.38%) is higher than the highest storage efficiencies of SA (88.36%) and GA (86.86%) which shows that particle swarm approach fulfilled our objective of finding ESR and capacitance values which would cause growth in storage efficiency figures. The best solution point was found corresponding to the best objective function value and from this location of the best (lowest) objective function value found in PSO, ESR value was noted as 0 Ω and capacitance value was noted as 603.88 F. From comparisons drawn about experimental results from different optimization algorithms in Sect. 5, it is evident that PSO provides us the desired parameter values for higher storage efficiency compared to GA and SA. So, PSO can be widely used as an effective optimization technique for practical applications concerned with the design of supercapacitors.

References

1. Panwar, N.L., Kaushik, S.C., Kothari, S.: Role of renewable energy sources in environmental protection: a review. Renew. Sustain. Energy Rev. **15**(3), 1513–1524 (2011)
2. Brito, D., Gomes, M., et al.: Evaluation of the main MPPT techniques for photovoltaic applications. IEEE Trans. Industr. Electron. **60**(3), 1156–1167 (2013)
3. Coello, C.A.C.: A comprehensive survey of evolutionary-based multiobjective optimization techniques. Knowl. Inf. Syst. **1**(3), 269–308 (1999)
4. Bottou, L.: Stochastic gradient descent tricks. In: Neural Networks: Tricks of the Trade, pp. 421–436. Springer, Berlin, Heidelberg (2012)

5. Schwefel, H.-P.: Evolution strategies: a family of non-linear optimization techniques based on imitating some principles of organic evolution. Ann. Oper. Res. **1**(2), 165–167 (1984)
6. Seth, S., Mukherjee, S., Dutta, T., Ghosh, R.: Storage efficiency optimization of a supercapacitor charged by a photovoltaic cell using stimulated annealing. Int. J. Eng. Res. Appl. **7**(8) (Part-2) (2017)
7. Wenzel, W., Hamacher, K.: Stochastic tunneling approach for global minimization of complex potential energy landscapes. Phys. Rev. Lett. **82**(15) (1999)
8. Coster, M.J., et al.: An improved low-density subset sum algorithm. In: Workshop on the Theory and Application of Cryptographic Techniques. Springer, Berlin, Heidelberg (1991)
9. Karmarkar, N.: A new polynomial-time algorithm for linear programming. In: Proceedings of the Sixteenth Annual ACM Symposium on Theory of Computing. ACM (1984)
10. Seth, S., Mukherjee, S., Dutta, T., Ghosh, R.: Storage efficiency optimization of a supercapacitor charged by a photovoltaic cell using genetic algorithm. IARJSET **4**(5) (2017)
11. Garg, R., Mittal, S.: Optimization by genetic algorithm. Int. J. Adv. Res. Comput. Sci. Softw. Eng. **4**(4) (2014)
12. Patel, G.R., Patel, D.B., Paghdal, K.M.: IASET analysis of P&O MPPT algorithm for PV system. Int. J. Electr. Electron. Eng. (IJEEE) **5**(6) (2016)
13. Femia, N., Petrone, G., Spagnulo, G., Vitelli, M.: Optimization of perturb and observe maximum power point tracking method. IEEE Trans. Power Electron. **20**(4) (2005)
14. Koutroulis, E., Kalaitzakis, K., Voulgaris, N.C.: Development of a microcontroller-based, photovoltaic maximum power point tracking control system. IEEE Trans. Power Electron. **16**(1) (2001)
15. Li, N., Zhang, J., Zhong, Y.: A novel charging control scheme for supercapacitor energy storage in photovoltaic generation system. In: Third International Conference on Electric Utility Deregulation and Restructuring and Power Technologies 2008, DRPT (2008)
16. Kim, Y., Cheng, N., Pedram, M.: Maximum power transfer tracking for a photovoltaic-supercapacitor energy system. In: ISLPED '10 Proceedings of the 16th ACM/IEEE International Symposium on Low Power Electronics and Design
17. Bondi, D., Bertachini, A., Brunelli, D., Larcher, L., Benini, L.: Modeling and optimization of a solar energy harvester system for self-powered wireless sensor networks. IEEE Trans. Industr. Electron. **55**(7) (2008)
18. Iqbal, M., Azam, M., Naeem, M., Khwaza, A.S., Anpalagan, A.: Optimization classification, algorithms and tools for renewable energy: a review. **39** (2014)
19. Kirkpatrick, S., Gelatt, C.D., Vecchi, M.P.: Optimization by simulated annealing. Sci. New Ser. **220**, 671–680
20. Eberhart, R., Kennedy, J.: A new optimizer using particle swarm theory. In: Proceedings of the Sixth International Symposium on Micro Machine and Human Science 1995, MHS'95. IEEE (1995)

Top-k Influential Nodes Identification Based on Activity Behaviors in Egocentric Online Social Networks

Soumyadeep Debnath, Dhrubasish Sarkar and Premananda Jana

Abstract The Online Social Networks (OSNs) are growing rapidly due to its mass popularity and easy accessibility. Nowadays, OSNs have a major impact on business, healthcare, agriculture, and society. Reaching the target audience efficiently through OSNs is most desirable to various organizations. Identifying influential nodes through online social network analysis enables to reach the target audience most effectively and thus has drawn significant attention from the researchers. In the recent past, several techniques have been proposed for finding influential nodes in OSNs. Though the activity behaviors of individual user play an important role in the influence maximization, most of the works in this area does not consider this. Our current work proposes a model to identify top-k influential nodes by evaluating Normalized Influence Factors of network members based on their activity behaviors in egocentric OSNs.

Keywords Influential nodes · Activity behavior · Egocentric online social network · Normalized influence factor

S. Debnath
Department of Computer Science and Engineering, Jalpaiguri Government Engineering College, Jalpaiguri 735102, India
e-mail: soumyadebnath13@gmail.com

D. Sarkar (✉)
Amity Institute of Information Technology (AIIT), Amity University, Kolkata 700135, India
e-mail: dhrubasish@inbox.com

P. Jana
Department of Computer Science & Engineering, MCKV Institute of Engineering, Howrah 711204, India
e-mail: prema_jana@yahoo.com

A. K. Das et al. (eds.), *Computational Intelligence in Pattern Recognition*, Advances in Intelligent Systems and Computing 999,
https://doi.org/10.1007/978-981-13-9042-5_39

1 Introduction

Online Social Networks (OSNs) are growing very fast and have become an important part of the daily life of millions of users around the globe. The OSN platforms provide the users with different tools for creating and sharing texts, photographs, audio-visual and web contents, as well as enable the users to share, like, and comment against the contents produced by other users. As the users of these OSNs spend a significant amount of their time and share information on these platforms, analyzing these OSNs helps the business and other organizations to understand the people, their choices, influences, etc. The connection network structure of the users, their activities and behaviors help to analyze these online social networks. There are two approaches of online social network analysis—sociocentric network analysis which considers the whole or complete network and egocentric network analysis which focus on individual (called ego) networks and involves quantification of interactions between an individual and all other connected persons (called alters) related to ego directly or indirectly.

In the present task, we are aimed to measure the influence to identify the top-k influential nodes by analyzing the activity behaviors of any particular user in its own social networks independently based on two sentiment types (positive and negative) for individual classified particular social fields.

We worked on three different social networks of the target user separately; Facebook,[1] Twitter,[2] LinkedIn[3] and only considered them from the egocentric network scenario because it is useful when the foci of the research are individuals in the network and capturing the complete network is less important.

2 Literature Review

Identifying influential nodes in OSNs has got attention due to its significant business applications. Different algorithms have been used to identify influential or top users of a network.

Domingous and Richardson [1] did the first study in this field. The social networks had been represented as Markov random field in their work. They developed three algorithms for determining influential users which to be used for viral marketing. Later, few other Models have been proposed to identify influential users.

Identifying influential nodes is to basically influence maximization problem. In [2], the authors analyzed various techniques used to identify influential nodes on OSNs. Kempe et al. [3] formulated influence maximization as a discrete optimization problem and proved that this optimization problem is NP-hard. The experiments show that their algorithm significantly outperforms the classic degree and centrality

[1] https://www.facebook.com/.

[2] https://twitter.com/.

[3] https://www.linkedin.com/.

based heuristics in influence maximization. In [4], the author has considered a data mining perspective for influence diffusion. However, some studies [5, 6] revealed that the diffusion of different topics is not same because of the different preferences of the users that will affect their role in the spread of a specific topic [7]. The approaches discussed above do not consider user preferences or choices instead uniform probability has been used. Later, Zhou et al. [8] designed a two-stage mining algorithm for mining top-K influential nodes in OSN based on user choices. Li et al. [9] defined information propagation for heterogeneous social networks and proposed to consider the individual behaviors of persons to model the influence propagation. In [10], the authors have discussed about sociocentric and egocentric approaches for social network analysis. Sociocentric and egocentric measures of network centrality have been discussed in [11].

In [12], the authors proposed a model for analyzing user activities using vector space model in OSNs. The Association Rule mining technique has been used in [13] to measure users' activity in OSNs. In [14], the users' activities were analyzed by lexicon and citation parameters for a particular Twitter network.

3 Dataset Preparation

In order to start the implementation of our methodological process, we needed to collect raw data within a particular time range from the distinguished online social networks individually; Twitter, Facebook, and LinkedIn. So, we initially selected a long time period of the past two years to crawl all the activity data of both the particular user and the particular network members associated with that user's activities. Thus, finally we prepared nine important classified datasets sequentially for each social network separately. The detailed steps are briefly described below (shown in Fig. 1).

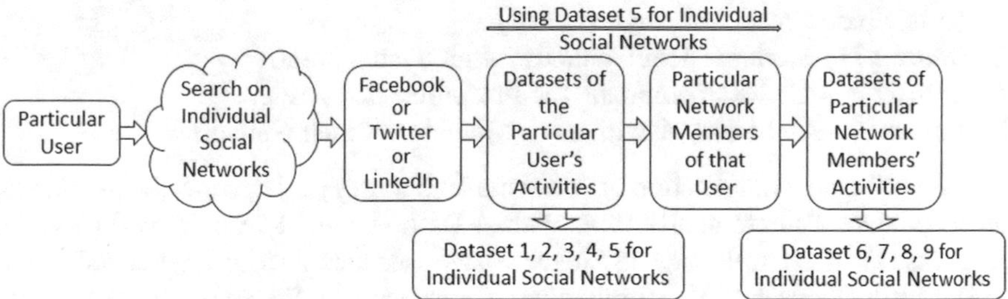

Fig. 1 Diagram of dataset preparation process

3.1 Stage 1: *Datasets of the Particular User's Activities*

Here, we first retrieved the list of all the activity details from the activity log portal within that time range of the particular user for which we need to identify the influential nodes from the individual social network. Using this list we crawled the data mentioned below;

1. *Dataset 1*—Text Contents of all Posts (Reacted posts, Commented posts, Shared posts, Tagged posts, and Own Tagging posts).
2. *Dataset 2*—Additional Text Contents for all Shared posts.
3. *Dataset 3*—All Text Comments for all Commented posts.
4. *Dataset 4*—All Messages with other Members of that user's network.
5. *Dataset 5*—(from that user's network)

    5.1. Owner Members of Reacted posts, Commented posts, Shared posts, and Tagged posts.
    5.2. Tagged Members for the Own Tagging posts.
    5.3. Associated Members who have interacted through Messages.

3.2 Stage 2: *Datasets of Particular Network Members' Activities of That User*

Similarly here we also retrieved the list of all the activity details from the activity log portals of all particular members (**Dataset 5**) of that user's network for individual social network who are associated with that user's activities. Using this list we crawled the data mentioned below;

6. *Dataset 6*—Text Contents of all Posts (Own posts, Reacted posts, Commented posts, Shared posts, and Tagged posts).
7. *Dataset 7*—Aditional Text Contents for all Shared posts.
8. *Dataset 8*—All Text Comments for all Commented posts.
9. *Dataset 9*—All Messages with other Members of their Networks.

Here we utilized the application of social media libraries [15] to crawl the previously described data. Therefore, all data excluding **Dataset 3** and **8** from individual social networks (Twitter, Facebook, and LinkedIn), we used their java libraries called 'Twitter4J'[4] [16], 'Facebook4J',[5] 'LinkedIn4J',[6] respectively. Somehow these libraries don't support any function to reach the reply/comment conversations of any particular post directly. So, to crawl the **Dataset 3** and **8** from an individual social network,

[4]http://twitter4j.org/en/index.html.

[5]https://facebook4j.github.io/en/index.html.

[6]https://github.com/Aristokrates/Freelancer-aggregator/tree/master/linkedin4j.

we used a very well-known process of web crawling, called "Web Scraping" using a java library called "jsoup"[7] [17] and parsed the HTML from an URL.

4 Dataset Preprocessing

Before implementation of our algorithmic process with our prepared datasets, we operated four preprocessing tasks on the data of all datasets (except **Dataset 5**) to make that more suitable as the inputs of our methodological process. The detailed steps are briefly discussed below (shown in Fig. 2).

4.1 Language Translation and Tweet's Tagged Usernames Retrieval

Here firstly, we tried to detect the languages of all the post contents and if it was a non-English language then we translated the languages of the post contents into English language. After that for only Twitter posts, we searched the '@' symbol in each tweet text to distinguish the next postfix word attached with that symbol as it is the username of the tagged member from that user's Twitter network and added all such tagged members in **Dataset 5** for only Twitter network.

4.2 Textual Noise Removal

This stage contains a set of four small substages which are performed programmatically and explained below sequentially;

1. Removal of all links or URLs.
2. Removal of all special symbols ("@","#","$","-","_" etc.) and set of special symbols ("….",",,,,","----","____" etc.) excluding the single comma (",") and single full stop (".").
3. Removal of all emoticons ("":D","":p","":)","%") etc.).
4. Removal of all unrecognized characters for the attached images and GIF files.

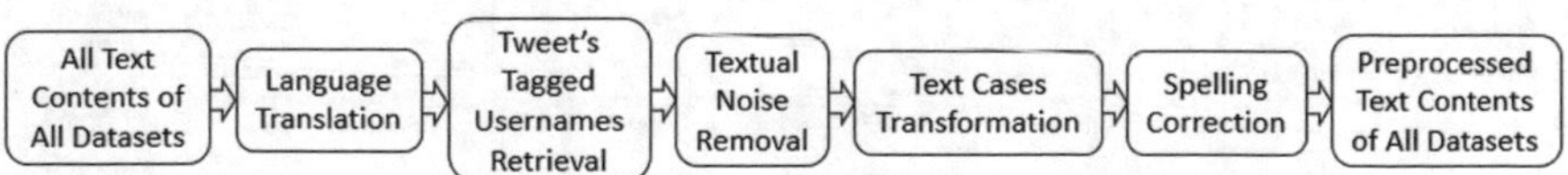

Fig. 2 Diagram of dataset preprocessing process

[7] https://jsoup.org/.

4.3 *Text Cases Transformation and Spelling Correction*

Here firstly, we transformed all upper case alphabets into lower case and then the major important task was identification and correction of noisy words. For that we checked the spelling of each word from the text contents and corrected the spelling.

Here we used "TextBlob"[8] [18], a python library for processing textual data to do the specified tasks of the "Language Translation" and "Spelling Correction'"stages.

5 Methodology

In order to evaluate the influence by analyzing the activities or working behaviors between the particular user and the particular network members (**Dataset 5**) in that particular time range, we generated a new algorithmic mathematical model as Influence Measurement Model. The sequential detailed process is briefly described below (shown in Fig. 3).

5.1 Stage 1: *Nodes and Communities Selection*

In this initial stage for each individual social network, we considered the particular user as *target node* for which we need to identify the *influential nodes* and all the network members mentioned in **Dataset 5** as the most interacted nodes which are associated with that target node's activities because the other network members' activities never influenced that particular user in that mentioned time period so inconsideration of them resulted in very less complexity for our algorithmic evaluation.

Here these most interacted nodes belong to the list of friends, following profiles, connections for each individual social network, respectively; Facebook, Twitter, and LinkedIn and also the list of following pages for all those social networks. They are assumed as separate influence community for each individual social network. We have explained this representation below (shown in Table 1).

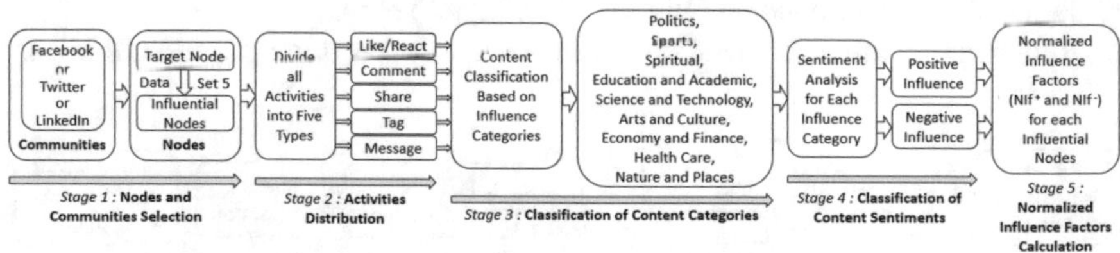

Fig. 3 Diagram of influence measurement model

[8] http://textblob.readthedocs.io/en/dev/quickstart.html.

Table 1 General representation of nodes and communities

Communities	Facebook	Twitter	LinkedIn
Nodes	Friends	Following profiles	Connections
	Facebook Pages	Twitter Pages	LinkedIn Pages

5.2 Stage 2: *Activities Distribution*

We categorized the activities from different datasets into five types for both that target node and those interacted nodes and each node can be considered for multiple actions. For normalization in future those were assigned different weight-factors according to significance in ascending order which are mentioned below in details;

1. *Likes/Reacts*—from Dataset 1—Weight Factor 1
2. *Comments*—from Dataset 3 and 8—Weight Factor 1.2
3. *Shared posts*—from Dataset 1 and 6—Weight Factor 1.4
4. *Tagged posts*—from Dataset 1 and 6—Weight Factor 1.6
5. *Messages*—from Dataset 4 and 9—Weight Factor 1.8

5.3 Stage 3: *Classification of Content Categories*

Here, initially we selected nine different classified social fields for influence categories, they are; Politics, Sports, Spiritual, Education and Academic, Science and Technology, Arts and Culture, Economy and Finance, Health Care, Nature, and Places to classify the text contents of all Posts and Messages from **Dataset 1, 4** (data of target node) and **6, 9** (data of interacted nodes). We did not classify other **Datasets** (**2, 3** and **7, 8**) because they are dependent on **Dataset 1** (data of target node) and **6** (data of interacted nodes) respectively.

In order to classify the activity contents into those social categories, so we used 'Magpie',[9] a deep learning python library for multi-label text classification. From experimental works it was observed that the accuracy result is always better for large sample training data, so in our case for small tweet samples we considered high subjectivity accuracy. Here, we manually labeled the training dataset of 1.27 GB, so the testing dataset can be labeled with multiple categories with the accuracy above 80% as threshold value.

[9]https://github.com/inspirehep/magpie.

5.4 Stage 4: *Classification of Content Sentiments*

Here, we analyzed the influence into two different types for each social category, they are; Positive and Negative to classify the text contents of all data (posts, additional texts, comments, and messages, respectively) from **Dataset 1, 2, 3, 4** (data of target node) and **6, 7, 8, 9** (data of interacted nodes). We again used the "TextBlob" python library here to utilize its own sentiment analysis algorithm on the above activity data in order to classify them into the positive or negative types based on polarity.

5.5 Stage 5: *Normalized Influence Factors Calculation*

Here, we measured the impact of those interacted nodes on the target node mathematically through the activities of the target node on their activity contents using the activity statistics (numbers of activities) mentioned in a table format below (shown in Table 2) for total five activities (total five tables).

Using these activity statistics data we calculated the Normalized Influence Factors (NIf) for both the Positive Influence (NIf^{+}) and Negative Influence (NIf^{-}) individually for each Interacted Node (N_n) with respect to Target Node (X) by our algorithm using the mathematical formulas mentioned below;

Algorithm 1:

Let: A_i, D_j, N_n are represented as each of the particular Activities, Influence Domains, Interacted Nodes respectively.

Let: The Total Number Counts of ith Activity (Ai),

In jth Domain (Dj) are $D_jA_i^{(Nn)}$ and $D_jA_i^{(X,Nn)}$ respectively for nth Interacted Node (N_n) and Target Node (X) with respect to nth Interacted Node (N_n).

In all Domains are $D_jA_i^{(Nn)}$ and $D_jA_i^{(X,Nn)}$ respectively for nth Interacted Node (N_n) and Target Node (X) with respect to nth Interacted Node (N_n).

Let: A_iW = Weight Factor (W) of ith Activity (Ai).

1: Select any particular Member from Dataset 5 of that target node's network.

2: Select each of the five activities one by one for that member.

3: **for** i = 1 to 5 **and** j = 1 to 9 **do**

4: Calculate, two types of Normalized Influence Factors (NIf) separately.

$$\mathrm{NIf}^{+} = \sum_{i=1}^{5}\left[\left(\frac{\left(\frac{D_jA_i{(X,N_n)}^{+}}{TA_i{(X,N_n)}^{+}}\right)*\left(\frac{D_jA_i{(N_n)}^{+}}{TA_i{(N_n)}^{+}}\right)}{\sum_{j=1}^{9}\left(\frac{D_jA_i{(X,N_n)}^{+}}{D_jA_i{(N_n)}^{+}}\right)}\right)*A_iW\right] \quad \mathrm{NIf}^{-} = \sum_{i=1}^{5}\left[\left(\frac{\left(\frac{D_jA_i{(X,N_n)}^{-}}{TA_i{(X,N_n)}^{-}}\right)*\left(\frac{D_jA_i{(N_n)}^{-}}{TA_i{(N_n)}^{-}}\right)}{\sum_{j=1}^{9}\left(\frac{D_jA_i{(X,N_n)}^{-}}{D_jA_i{(N_n)}^{-}}\right)}\right)*A_iW\right]$$

5: end for

6: Store, both the NIf^{+} and NIf values of that network member separately.

7: Repeat, the same process for all other Members (n-1) of that target node'snetwork.

8: Sort, both the NIf^{+} and NIf values separately in ascending order for all network members.

Table 2 Statistical representation format of a particular activity

Particular activity by particular node (in particular time period)	Total activity		Domain 1 (politics) activity		Domain 2 (sports) activity		Domain 3 (spiritual) activity		Domain 4 (education) activity		Domain 5 (science) activity		Domain 6 (culture) activity		Domain 7 (finance) activity		Domain 8 (health) activity		Domain 9 (nature) activity	
	+ve	−ve	+ve	−ve	+ve	−ve	+ve	−ve	+ve	−ve	+ve	−ve	+ve	−ve	+ve	−ve	+ve	−ve	+ve	−ve
By target node associated with an influential node N1 (X, N1)																				
By that influential node																				

5.6 *Top-k Influential Nodes Evaluation*

At the end of the last stage, we got two lists with the normalized influence factors of both the positive and negative influences separately for all Influential Nodes with respect to Target Node. Now, we operated the following operations on the result data to evaluate top-k influential nodes according to activity measures and network size.

Algorithm 2:
Step 1: For any value of k ($0<k\leq10$), select at most top 5*k number of influential nodes.
Step 2: The nodes having connections more than 10000 are removed from the list, considered as *Default Influential Nodes* (example, verified profiles and pages).
Step 3: Select top k number of influential nodes, considered as *top-k Influential Nodes.*

6 Result Analysis

In order to analyze the result of our methodological process, we selected only our facebook data (as dataset) on the positive type of the political category (as influence domain) for testing. The details of the result analysis process are elaborated below.

6.1 *Normalized Influence Factors Analysis and Top-k Influential Nodes Evaluation*

Here we plotted a graph (shown in Fig. 4a) considering k as 10 using our result data, where the values of X-axis denotes individual influential node or the influential network members and the values of Y-axis denotes the normalized influence factors of those influential nodes on the target node.

In the above graph (Fig. 4a), the less numbers of separated blue dots at the top portion represents the highly popular *default influential nodes* which are always highly influential members for any target node for that particular influence domain in general. In order to analyze their popularity we also plotted another graph (shown in Fig. 4b) by changing the values of Y-axis with the number of connections/followers of those influential nodes and after removing the *default influential node* (using Algorithm 2), we plotted the same graph (Fig. 4c) as like Fig. 4a with the *top-k influential nodes*.

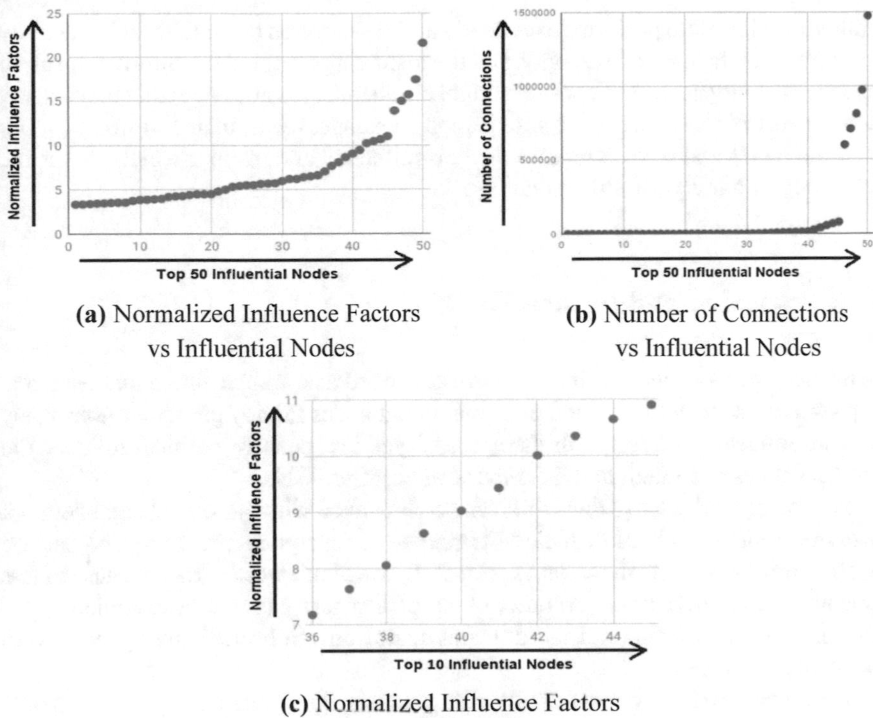

(a) Normalized Influence Factors vs Influential Nodes

(b) Number of Connections vs Influential Nodes

(c) Normalized Influence Factors vs Influential Nodes

Fig. 4 Scatter chart on influence factors analysis

6.2 *Testing Result of Top-k Influential Nodes' Activities*

In order to analyze the result accuracy in real life scenarios, we observed the activity data of those top-k influential nodes individually for next six months' time interval and measured the involvements through different activities of target node on them which are represented by the values of Y-axis in the graph (shown in Fig. 5).

Here individual color represents the real result of individual activity, so from the above graphical representation, it is already observed that our top influential nodes

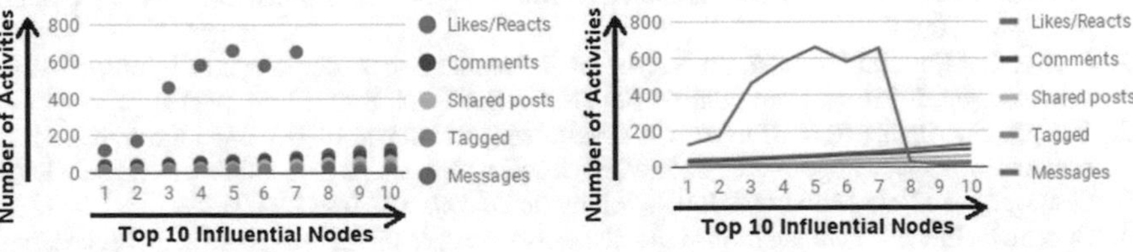

Fig. 5 Scatter and line chart on real activity analysis

evaluation algorithm performs excellently and is supported by all the real life activity results except the message activity. But the justification behind the informal result of message activity is, the top nodes are highly influential (as any page) that they mostly don't involve in any direct communication (messages) with their followers (in our case the target node), so it represents their influence in a manner which proves the efficiency of our algorithmic model.

7 Conclusion and Future Work

In the end, we can conclude that we have generated a new algorithmic mathematical approach to identify top-k number of influential nodes for any particular user in any specific influence domain with the specific type (ex, positive political influence in our result) based on their activities in the egocentric OSNs.

For future work, firstly this work can be elaborated with our other social network datasets; Twitter, LinkedIn which will improve our analytical results by comparison but in spite of retrieving those data successfully we didn't work with it because of less amount of user interactions. Another of our present task can also be extended with a large number of domains and sub-domains apart from our limited nine domains with two sentiment types.

Also apart from only text activity data analysis, our system can also be capable for image activity data analysis by using the Optical Character Recognition (OCR) for text data retrieval from image [19] and the Convolutional Neural Network (CNN) for image category classification [20] (according to the influence domains). And also here we only considered the activities for measuring influence without analyzing the network structures of the nodes, so structural behaviors can be incorporated.

Lastly, our research work can be highly applicable to influence diffusion node analysis effectively based on the influence domains with types for any particular incident like demonetization, GST, etc.

References

1. Doming, P., Richard, M.: Mining the network value of customers. In: Proceedings of the 7th ACM SIGKDD International Conference on Knowledge Discovery and Data Mining (KDD), pp. 57–66 (2001)
2. Sarkar, D., Kole, D.K., Jana, P.: Survey of influential nodes identification in online social networks. Int. J. Virtual Communities Soc. Netw. (IJVCSN) **8**(4), 57–69 (2016)
3. Kempe, D., Kleinberg, J., Tardos, E.: Maximizing the spread of influence through a social network. In: Proceedings of the Ninth ACM SIGKDD International Conference on Knowledge Discovery and Data Mining (KDD), Washington DC, pp. 137–146 (2003)
4. Bonchi, F.: Influence propagation in social networks: a data mining perspective. In: Proceedings of the 2011 IEEE/WIC/ACM International Conferences on Web Intelligence and Intelligent Agent Technology, vol. 01 (2011)

5. Saito, K., Kimura, M., Ohara, K., Motoda, H.: Learning continuous-time information diffusion model for social behavioral data analysis. In: Proceedings of the 1st Asian Conference on Machine Learning: Advances in Machine Learning (ACML), Nanjing, China, pp. 322–337 (2009)
6. Saito, K., Kimura, M., Ohara, K., Motoda, H.: Behavioral analyses of information diffusion models by observed data of social network. In: Proceedings of the 2010 International Conference on Social Computing, Behavioral Modeling, Advances in Social Computing Prediction (SBP10), USA, pp. 149–158 (2010)
7. Tang, J., Sun, J., Wang, C., Yang, Z.: Social influence analysis in large-scale networks. In: Proceedings of the 15th ACM SIGKDD International Conference on Knowledge Discovery and Data Mining (KDD), Paris, France, pp. 807–816 (2009)
8. Zhou, J., Zhang, Y., Cheng, J.: Preference-based mining of Top-K influential nodes in social networks. Futur. Gener. Comput. Syst. **31**, 40–47 (2014)
9. Li, C., Lin, S., Shan, M.: Influence propagation and maximization for heterogeneous social networks. In: Proceedings of the 21st International Conference Companion on World Wide Web, Lyon, France, pp. 559–560 (2012)
10. Chung, K.S., Hossain, L., Davis, J.: Exploring sociocentric and egocentric approaches for social network analysis. In: Proceedings of the 2nd International Conference on Knowledge Management in Asia Pacific, pp. 1–8 (2005)
11. Marsden, P.V.: Egocentric and sociocentric measures of network centrality. Soc. Netw. **24**, 407–422 (2002)
12. Sarkar, D., Jana, P.: Analyzing user activities using vector space model in online social networks. In: Proceedings of the National Conference on Recent Trends in Information Technology & Management (RTITM 2017), India, pp. 155–158 (2017)
13. Sarkar, D., Kole, D.K., Jana, P., Chakraborty, A.: Users activity measure in online social networks using association rule mining. In: Proceedings of the IEMCON 2014: 5th International Conference on Electronics Engineering and Computer Science (Elsevier Science & Technology), Kolkata, India, pp. 172–178 (2014)
14. Debnath, S., Das, D., Das, B.: Identifying terrorist index (T+) for ranking homogeneous Twitter users and groups by employing citation parameters and vulnerability lexicon. In: International Conference on Mining Intelligence and Knowledge Exploration, pp. 391–401. Springer, Cham (2017)
15. Jonsén, F., Stolpe, A.: The feasibility and practicality of a generic social media library (2017)
16. Loria, Yamamoto, Y.: Twitter4j-a java library for the twitter api (2014)
17. Hedley, J.: jsoup Java HTML Parser, with best of DOM, CSS, and jquery. In: Jsoup Java HTML Parser: with best of DOM, CSS, and jquery (2009)
18. Loria, S., Keen, P., Honnibal, M., Yankovsky, R., Karesh, D., Dempsey, E.: Textblob: simplified text processing. In: Secondary TextBlob: Simplified Text Processing (2014)
19. Patel, C., Patel, A., Patel, D.: Optical character recognition by open source OCR tool tesseract: a case study. Int. J. Comput. Appl. **55**(10) (2012)
20. Rakshit, S., Debnath, S., Mondal, D.: Identifying land patterns from satellite imagery in Amazon Rainforest using deep learning (2018). arXiv:1809.00340

Music Tagging and Similarity Analysis for Recommendation System

Laxman Kumar, Anirban Mitra, Mamta Mittal, Vatsal Sanghvi, Sudipta Roy and Sanjit Kumar Setua

Abstract Many of the websites follow the system of retrieving and recommending music based on the metadata. Metadata is generally a text file that attached to the music file has title and genre. Without attached metadata, it is very difficult for such websites to recommend or retrieve music. A regularly utilized rundown of the fundamental components incorporates pitch, timber, surface, volume, span, and frame. In the proposed methodology to process such a vast amount of data, the distributed storage and data processing systems like Hadoop and Spark has been used. Hadoop Distributed File System has been used for storing the music files and extracting feature information. Kafka queues has been used for asynchronous feature extraction in the background and finally Spark has been used for feature analysis using

L. Kumar · V. Sanghvi
Department of Computer Science and Engineering, U.V. Patel College of Engineering, Ganpat University, Kherva, Mehsana 384012, India
e-mail: laxman123prajapati@icloud.com

V. Sanghvi
e-mail: vatsal.bda1505d@ict.gnu.ac.in

A. Mitra (✉)
Department of Computer Science and Engineering, Academy of Technology, Adisaptagram 712121, West Bengal, India
e-mail: anirban.mitra.cse@gmail.com

A. Mitra · S. K. Setua
Department of Computer Science and Engineering, Calcutta University Technology Campus, JD-2, Sector-III, Salt Lake, Kolkata 700098, India
e-mail: sksetua@gmail.com

M. Mittal
Department of Computer Science & Engineering, G. B. Pant Government Engineering College, Okhla, New Delhi 110020, India
e-mail: mittalmamta79@gmail.com

S. Roy
Mallinckrodt Institute of Radiology Department, Washington University in St. Louis, 510 South Kingshighway Boulevard, St. Louis, MO 63110-1076, USA
e-mail: sudiptaroy01@yahoo.com

A. K. Das et al. (eds.), *Computational Intelligence in Pattern Recognition*, Advances in Intelligent Systems and Computing 999,
https://doi.org/10.1007/978-981-13-9042-5_40

machine learning algorithms. This Proposed automated system for assigning genres for music provides very promising accuracy with a high true positive value.

Keywords Music tagging · Plagiarism detection · Genre classification · Bigdata technology · Machine learning

1 Introduction

These days different kinds of music exist in the market, organizing them without metadata or incomplete metadata can be a tiring process. Also, music does not come with mood-related metadata, it is very subjective and a bit depends on the listener, e.g., one might say a song is aggressive and loud another might find the same song energizing. The use of different effects in songs and in instruments being played make this even harder. The data are rapidly growing in the last two decades; one terabyte could fit approximately 200,000 songs or 17,000 h of music. So for almost two continuous years without repeating you can listen to the songs with just one TB of data. In present days more and more music streaming services are coming up with recommending music to its users. Our main focus is to implement a distributed system that classifies genre. In [1], the author had clarified that advanced music of good quality around 1 MB; the extraction of highlights from the entire music can be restrictive because of the required preparing time. For these reasons, highlights are separated from three thirty-seconds music fragments. One from the earliest starting point, one from the center, and one from the end. Some portion of a music piece is chosen and highlight vectors are separated from each section. It helps in searching and indexing. In [2], author had predicted the mood of music from song lyrics using machine learning naïve Bayes model. The predicted mood was classified into happy and sad. In [3], mood prediction-based music system was implemented. The prime focus was to categorize the audio into different moods. Songs with a similar pattern or their similar audio feature range had grouped together to yield a particular mood. So for every mood there was a different playlist and there is also facility for the user to view all the songs. In [4], the author had assessed music likenesses that don't depend on the sound flag but instead takes different parts of the setting in which a music substance happens into thought. In [5], the authors had utilized Gaussian blend models and Gaussian classifiers. They had clarified a segment on include extraction and given a melodic class pecking order. In any case, with ten kinds, characterization precision does not achieve enough exactness. In [6], the author had created a graphics user interface bases system with more than two million songs that uses machine learning algorithm on audio files and natural language processing algorithm on lyrics to celebrated create new metadata, build on SPARQL endpoint with Not Only Structure Query Language(popularly known as NOSQL) databases. Those above examined approach takes after specific deficiencies including information inadequately and fame inclination, since a bigger number of information is accessible for famous specialists than for lesser known ones, which regularly twists inferred likeness measures.

To overcome those shortcomings authors have implemented a distributed system and prediction algorithm that is used to classify genre.

The rest of this article is organized as follows. Section 2 contains a discussion on the proposed solution. Within this section there are three subsections as: Sect. 2.1 will contain the methodology for implementation of automated classification, Sect. 2.2 will contain the description of methodology for feature extraction of music, and Sect. 2.3 will contain the description of the genre classification technique. In Sect. 3 results of the proposed methodology are illustrated with graphical and tabular representation. Finally, Sect. 4 concludes the article with a brief summary and challenges ahead.

2 Proposed Solution

One of the biggest challenges of music genre identification is the isolation of noise from the music as noise is often an integral part of the music. So, removing noise might lead to data loss in preprocessing. Identifying the mood of music can be a confusing process as there are songs which incorporate multiple moods. Separating vocals from music is also not a straightforward process in western music as vocals might not be center panned at all. The idea is to create a feature representation of music which will help in genre and moods and find similar tracks. These can be put into a search engine and implementing all this in a way which is scalable and uses Big Data technologies. Preprocessing is an important step before implementing any algorithm in order to remove noise, data loss. There may be the case that this music is in different languages, and may have multiple moods which are required for separating vocals of the music.

Conversion of MP3 file To Wav: The file size of the MP3 is very less whereas the Wav file is much larger. When compared to MP3 it also has good sound quality and is good for processing the music file. As the MP3 file is in compressed format so the conversion of Mp3 to Wav is necessary for audio analysis.

Channel merging, stereo (L&R) to mono: Having one channel to process is simpler and easier so Left and Right channels are merged to one channel.

Standard frequency 44,100 Hz: Human ear can hear only till 20,000 Hz, so to fully store these frequencies range from +20,000 Hz to −20,000 Hz is selected and it is also an industry standard.

2.1 Implementation

For implementation of automated classification system, authors have used Spark, RabbitMQ, and Python technologies. Spark is quick and its processing engine is good with Hadoop information. It can keep running in Hadoop cluster through YARN or Spark's stand-alone mode, and it can process information in HDFS, HBase, Cassan-

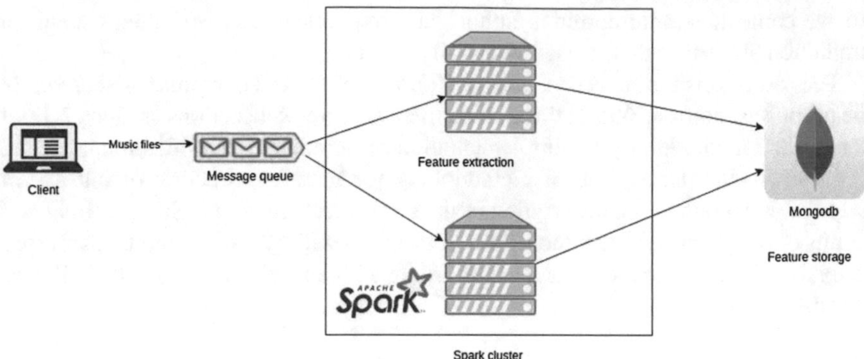

Fig. 1 Proposed system architecture

dra, Hive, and any Hadoop Input Format. It is intended to perform both batch processing (like MapReduce) and new workloads like spilling, intelligent inquiries, and machine learning. RabbitMQ is lightweight and simple to convey on-premises and in the cloud. It supports multiple different types of protocol for messaging. RabbitMQ can be conveyed in disseminated and combined arrangements to meet high-scale, high-accessibility necessities. Python is a widely used high-level programming language for general-purpose programming. The proposed Big Data architecture has been shown in Fig. 1.

In this architecture, Client can be any machine which will publish the file to a message queue topic. Then the message queue will deliver the file to its listeners and it will be consumed by the feature extracting service and then extracted features will be stored in mongo dB for later usage. Authors can then use those features to train our dataset continuously. For classification we will use the model generated by the proposed algorithm. This is usually a CSV file with model values for different genres. With the help of such architecture we can process a large number of music files and extract features, as well as provide the classification of music in real time.

2.2 Feature Extraction

Music can be dissected by thinking about an assortment of its components, or parts (angles, qualities, highlights), exclusively or together. A usually utilized rundown of the fundamental components incorporates pitch, timbre, texture, volume, length, and frame. Musical features are calculated over chunks of small frames of 0.5 s from the song. So, the final outcome is an array of values. The list of features that have to be considered are as follows:

Zero Crossing Rate: The zero-intersection rate is the rate of sign-changes along a flag, i.e., the rate at which the flag changes from positive to negative or back. This component has been utilized vigorously in both discourse acknowledgment and music data recovery is a key element to order percussive sounds. Statistical Features: Statistical features are calculated based on musical features.

Mean: It recognizes the central area of the information, now and then alluded to in English as the normal.

Standard deviation: The standard deviation is the most well-known measure of inconstancy, estimating the spread of the informational index and the relationship of the intend to whatever is left of the information.

2.3 Genre Classification

SVMs are used in text and hypertext arrangement as their application can fundamentally decrease the requirement for named preparing examples in both the standard inductive and transductive settings. Classification of music can likewise be performed utilizing SVMs. The trial comes about to demonstrate that SVMs accomplish essentially higher inquiry precision than customary question refinement conspires after only three to four rounds of pertinence input. Utilizations a subset of preparing focuses on the choice capacity (called bolster vectors), so it is additional memory proficient. Distinctive Kernel capacities can be indicated for the choice capacity. Regular portions are given, yet it is additionally conceivable to determine custom bits. The part traps capacities take low dimensional information space and change it to a higher dimensional space, i.e., it changes over the not detachable issue to distinct issue, these capacities are called kernels.

The kernel functions are considering as: Linear (x, x′), Polynomial: $(\gamma(x, x') + r)^d$ where d is specified by key word degree, γ by coefficient zero, and RBF: $\exp\left(-\gamma \, ||x - x'||^2\right)$ where γ is specified by keyword gamma, must be greater than 0. Sigmoid $(\tanh(\gamma \, (x, x') + r)$, where r is specified by coefficient zero.

The RBF portion is a prominent bit work utilized as a part of different kernelized learning calculations. Specifically, it is ordinarily utilized as a part of help vector machine arrangement. It takes gamma and C esteem as contribution with highlight set.

The RBF portion on two distinctive example set x and x′, spoke to as highlight vector in some information space can be characterized by K(x, x′) as the following equation Eq. (1):

$$K(x, x') = \exp\left(-\gamma \, ||x - x'||^2\right) \tag{1}$$

where $\gamma > 0$.

Result of SVM(RBF): Confusion matrix for SVM(RBF kernel) given in Table 1

Table 1 Table showing confusion matrix for SVM-RBF kernel

	C	PRE	REC	F1
Classical	0.001	33.3	0.0	0.0
	0.010	33.3	0.0	0.0
	0.500	92.6	88.0	90.3
	1.000	94.0	85.5	89.5
	5.000	97.1	99.0	98.0
	10.000	94.0	94.5	94.3
	20.000	94.9	92.5	93.7
Jazz	0.001	33.3	0.0	0.0
	0.010	33.3	0.0	0.0
	0.500	84.6	89.7	87.1
	1.000	85.0	88.7	86.8
	5.000	93.3	93.3	93.3
	10.000	87.9	90.0	89.0
	20.000	86.4	91.0	88.6
Rock	0.001	37.5	100.0	54.5
	0.010	37.5	100.0	54.5
	0.500	94.2	91.7	92.9
	1.000	92.5	94.0	93.2
	5.000	95.3	94.0	94.6
	10.000	93.8	91.3	92.6
	20.000	94.1	94.7	92.4

has precision, recall and f1-score value for all the three-genre classical, jazz, and rock. The overall accuracy we had attained in this model is 95.0 (using Eq. 7) for c = 5.000 and overall f1-score is 95.3 (using Eq. 6) for c = 5.000. T

3 Results and Discussion

We have tested the proposed method on 50 music files and analyzed the performance by some performance evaluation metric. The upside of cross-approval is that the extent of the preparation/approval split isn't subject to the quantity of iteration (folds). It generates a confusion matrix of true positives and false positives, which helps assess the performance of the algorithm and helps to determine which algorithm is the best suited for the problem. It is used to estimate the C value. The C parameter exchanges off misclassification of preparing cases against the straightforwardness of the choice surface. A low C settles on the choice surface smooth, while a high C goes for grouping all preparation cases effectively by giving the model opportunity to choose more examples as help vectors. Different algorithms will have different sets of C values to be tested. In performance metric [7, 8] precision is abbreviated as PRE, recall is abbreviated as REC, f1-score is abbreviated as F1 and accuracy is abbreviated as ACC. The F1 score otherwise called F-Score or F-measure is a measure of a test's exactness. Both exactness p and review r is utilized to compute the score. P is the quantity of the number of right positive qualities separated by the quantity of all positives that come about given by the classifier and r is the quantity of right positive qualities partitioned by the quantity of all relevant. Precision is the fraction of correctly identified cases belonging to class c among all the cases which classifier claims that they belong to class c as Eq. (2).

$$\text{Precision} = \frac{\text{tp}}{\text{tp} + \text{fp}} \tag{2}$$

where, tp = no of True positive, fp = no of false positive. Review otherwise called affectability is the part of accurately recognized cases having a place with class c among every one of the cases which really have a place with class c as Eq. (3).

$$\text{Recall} = \frac{\text{tp}}{\text{tp} + \text{fn}} \tag{3}$$

where, tp = no of True positive, fn = no of false negative. In the following equation Eq. (4), the F1-Score is the consonant normal of the accuracy and review, where best estimation of f1 is 1 and the worst is 0.

$$\text{F-score} = \frac{(1 + \beta^2)(\text{pre} \cdot \text{rec})}{(\beta^2 \cdot \text{pre} + \text{rec})} \tag{4}$$

where, β is commonly 0.5, 1 or 2. Accuracy can be calculated as total no True Positive and True Negative Prediction among the total no of dataset positive and negative which will be described at Eq. (5).

$$\text{Accuracy} = \frac{(\text{TP} + \text{TN})}{(\text{P} + \text{N})} \tag{5}$$

Performance results of the proposed method for plagiarism and music tagging has been shown in Table 1.

Figure 2 shows the overall accuracy using the combination of SVM-RBF. We have calculated accuracy and SVM-RBF proves to be best for our problem. It gives the highest average true positive value. Highest true positive value helps us to determine the best solution. From these values correct prediction is possible. These are the cases in which we have predicted that the clip is of certain genre and it is actually that genre. The number of cases of True Positive result are more as the column chart representation of true positive values has been shown in Fig. 3.

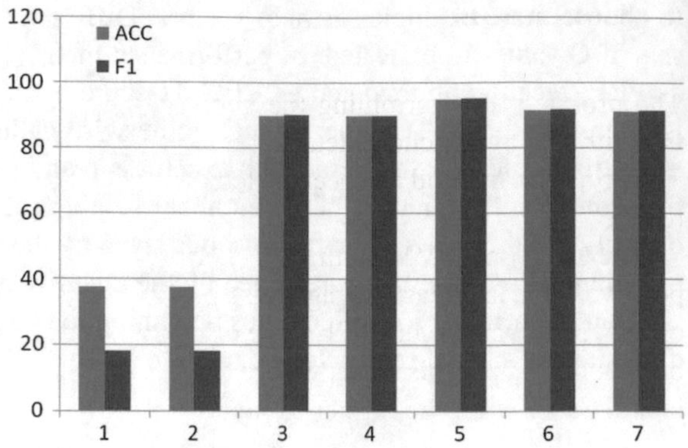

Fig. 2 Overall accuracy of the proposed method

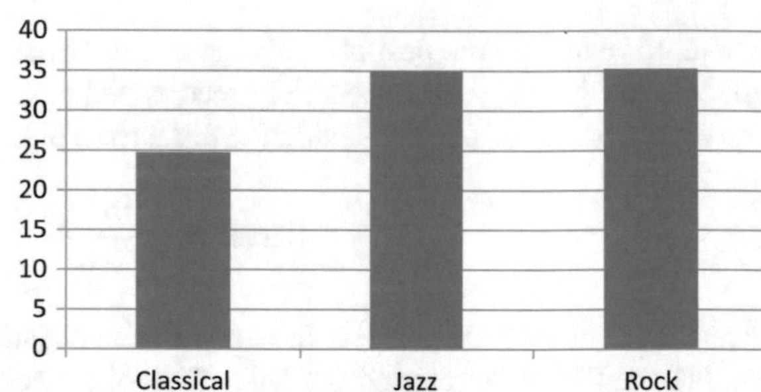

Fig. 3 True positive value of the proposed method

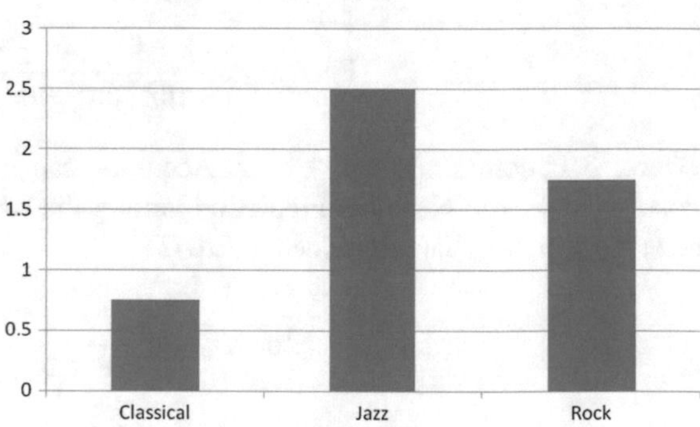

Fig. 4 False positive value of the proposed method

False Positive [7, 8] is the cases in which authors have predicted that the clip is of certain genre and it is actually not of that particular genre. Figure 4 contains the false positive [7, 8] values for the entire implemented algorithm. It is clear from Fig. 4 that the false positive has least average misclassification in the proposed method.

The proposed method works well on the huge amount of music as Big Data technologies for real-time processing have been used by the authors. That increases the parallel processing and helps in reduced processing time. The SVM-RBF strategy gives exceptionally encouraging outcomes in setting to exactness and mistake.

4 Conclusion

The problem of determining the genre of songs by analyzing the raw input song file without any attached Metadata is correctly determined by the proposed method. The proposed method has been tested on real-time music and gives a very promising result.

The proposed SVM with RBF kernel performs best giving high average true positive value and least average false positive value. In this method, overall accuracy of above 90% and overall F1-score of above 90% using SVM with RBF kernel have been achieved. At this stage, we are able to find only three-genre classical, rock, and jazz. The work can be further expanded for other genres.

Acknowledgements The authors are extremely thankful to Raj Geriya, Ganpat University for his support to work on this research.

References

1. Silla, Jr., C.N., Kaestner, C.A.A., Koerich, A.L.: Automatic music genre classification using ensemble of classifiers. In: 2012 Fourth International Conference on Computational Intelligence, Modelling and Simulation, Proceedings of IEEE, pp. 237–243 (2012)
2. Raschka, S.: Music Mood: Predicting the Mood of Music from Song Lyrics Using Machine Learning. Paper from the Michigan State University November 2014
3. Saswadkar, A., Kodale, S., Lalge, S., Somavanshi, M., Thakre, P.U.: Mood Prediction Based Music System, vol. 4, no. 3 (2017)
4. Schedl, M., Knees, P.: Context-Based Music Similarity Estimation. Paper from Johannes Kepler University Linz
5. Tzanetakis, G., Cook, P.: Musical genre classification of audiosignals. IEEE Trans. Speech Audio Process. **10**(5), 293–302 (2002)
6. Meseguer-Brocal, G., Peeters, G., Pellerin, G., Buffa, M., Cabrio, E., Zucker, C.F., Giboin, A., Mirbel, I., Hennequin, R., Moussallam, M.: A Two Million Song Database Project with Audio and Cultural Metadata plus WebAudio enhanced Client Applications, 19 September 2017
7. Roy, S., Bhattacharyya, D., Bandyopadhyay, S.K., Kim, T.-H.: An iterative implementation of level set for precise segmentation of brain tissues and abnormality detection from MR images. IETE J. Res. (Taylor & Francis) **63**(6), 769–783 (2017)
8. Roy, S., Bhattacharyya, D., Bandyopadhyay, S.K., Kim, T.-H.: An effective method for computerized prediction and segmentation of multiple sclerosis lesions in brain MRI. Comput. Methods Programs Biomed. **140**(C), 307–320 (2017). ELSEVIER

Elastic Window for Multiple Face Detection and Tracking from Video

Aniruddha Dey, Satadal Chakraborty, Debaditya Kunduand and Manas Ghosh

Abstract This paper deals with an efficient method for the detection and tracking of multiple moving faces from a video sequence. Appropriate detection of multiple faces from a video sequence is a challenging task due to the different combination of noise, illuminations, pose, and locations of the human face which is likely to differ from one frame to another. This paper presents a unique technique for multiple face detection from a video sequence. In this study, our major objective is to detect and track locations of multiple faces from video using elastic window. Additionally, the face tracking system includes the tracking of face motion. Firstly, for each pixel, local entropy is calculated by considering a 3×3 window for detecting the face edges. Subsequently, Gaussian filtering technique is used to eliminate the undesired edges. In this context, it may be noted that a video frame passes through a number of preprocessing steps in order to eliminate the background noise to realize the thin binary image consisting of face boundaries. The human face from video sequences can be tracked by calculating the scalar and vector distances of four corner points between two adjacent frames. The movement of corner points represents the position and location change of the face in the upcoming frame. The presented method has been tasted on several video database and obtained efficient detection and tracking of multiple faces from the video sequences.

A. Dey (✉)
Bankura Unnayani Institute of Engineering, Pohabagan, Bankura 722146, West Bengal, India
e-mail: anidey007@gmail.com

S. Chakraborty
Department of Computer Science and Engineering, SIT, Siliguri 734009, India
e-mail: satadal.chakraborty@gmail.com

D. Kunduand
Department of Information Technology, SIT, Siliguri 734009, India
e-mail: debaditya.kundu@gmail.com

M. Ghosh
Department of Computer Application, RCCIIT, Kolkata 700015, India
e-mail: manas.ghosh@rcciit.org

A. K. Das et al. (eds.), *Computational Intelligence in Pattern Recognition*,
Advances in Intelligent Systems and Computing 999,
https://doi.org/10.1007/978-981-13-9042-5_41

Keywords Local entropy · Multiple face detection · Multiple face tracking · Thinning · Face contour

1 Introduction

Face detection and tracking is one of the significant aspects in the arena of computer vision and artificial intelligence. The major applications of face detection and tracking include surveillance, industry, medicine, and military. Face tracking is a formidable task due to several factors like movement of the camera, partial and full object occlusions, object fading, clutter, illumination change, and complex object shapes/motion. Furthermore, the change in background results in variation in the appearance, position, and object scale, which causes information loss. Tracking algorithms therefore needs to be adaptive under such complex environments [1, 2]. In the case of video the main challenge is the unrestricted nature of faces and their surroundings. Furthermore, the scale of the faces, occlusion, and poses may vary in case of video sequences which hinders the detection and tracking process [3]. Most of the algorithms are computationally expensive. Therefore, efficient algorithms for processing a large amount of information are needed [3, 4]. The face recognition systems start with detecting the faces. A wide range of researches presented in the literature in the biometric area are mostly works on the still images [5] which fail to work in case of video sequences. Several researchers introduced a neural network (NN) and statistical methods for face tracking [6]. The facial features can efficiently be detected and tracked by pixel counting method. In this context, it may be noted that the face tracking can broadly be classified as facial feature tracking, head tracking, and combination of both [8, 9]. The face tracking includes connection and boundary matching as its key steps [10]. Another important aspect of face recognition in a video clip is face detection. A significant number of works for frontal face detection are presented in the literature. Among this, the support vector machine classifier and discriminating feature analysis, neural network-based face detection [8], fuzzy-based face detection for color images [9], etc. In case of face detection, facial color information is significant information [10].

Our main objective is to present a new technique for detecting multiple faces from the video sequences. It is much more challenging to locate and track the multiple faces from the video sequences. The presented method is validated over several video sequences to validate the claim.

The remaining part is organized as follows: Sect. 2 defines proposed the idea and described the proposed method. The experimental results on the face database Sect. 3. Finally, we have summarized the entire work in Sect. 4.

2 Proposed Method

Detection of the faces from a video sequence is the primary step of a face recognition system. The face recognition from a video is a formidable task as the human faces may have a wide variety of positions and orientations. In addition to this, the faces may even degrade by means of background noise and different illumination conditions. If the person is the only moving object in a video scene, the detection of the outline region is relatively easy on the basis of frame differencing.

2.1 *Face Edge Detection Using Local Entropy*

Let $g(x, y)$ represents the gray value of a pixel (x, y) such that, $g(x, y) > 0$, $M \times N$ is the height and width of the image, the local entropy H_f can be expressed as follows in Eq. (1):

$$E_l = -\sum_{i=1}^{M}\sum_{j=1}^{N} S_{ij}\, ln\, S_{ij} \tag{1}$$

where, $g(x, y)/\sum_{i=1}^{M}\sum_{j=1}^{N} S_{ij}\, ln\, S_{ij}$ is gray distribution; E_l the local entropy; M and N is the height and width of the local window.

The dispersion of the image gray levels can be realized by local entropy [1]. The entropy at the edge of an object is always high. The detection of the edges can be done by applying the idea of big dispersion and gray mutation.

Due to $S_{ij} \ll 1$, in Eq. (2) we have used the Taylor series expansion is used. The equation is expressed as follows:

$$E_l \approx -\sum_{i=1}^{M}\sum_{j=1}^{N} S_{ij}(S_{ij} - 1) = 1 - \sum_{(i,j)\in(M,N)} S_{ij}^2 \tag{2}$$

Local entropy generally reflects the image gray dispersion in the local window. Therefore, if a squared neighborhood contains contour pixel, the respective entropy will be very small. On the contrary, if there are no edge pixels in the neighborhood, their local entropies will be very high. Therefore, by setting a threshold we can identify a center pixel as the edge pixel if its local entropy is smaller than the aforementioned threshold. Figure 1 shows the output result of local entropy based contour detected face images. The algorithm is summarized as follows:

Step 1: The squared window is scanned throughout the image and calculated the local entropies of every pixel $x_{i,j}$. Now, a threshold (*Th*) is selected. The local entropy $E_l < Th$, denotes that the pixels are having large dispersion inside the local window and $x_{i,j}$, is an edge pixel. Conversely, $E_l > Th$

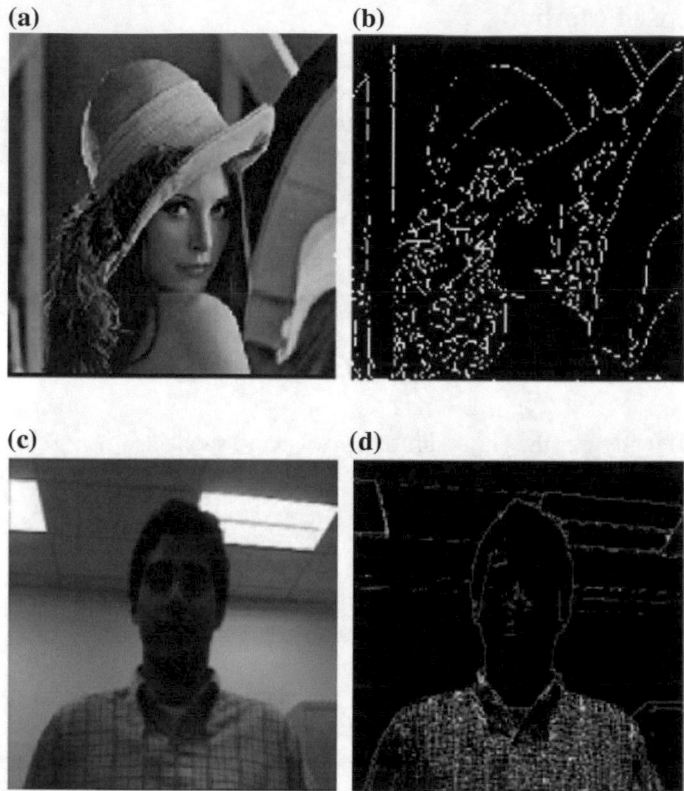

Fig. 1 **a, c** Original face images; **b, d** qualitative outputs of the local entropy based contour detected face images

indicates the pixels in local window that are having small dispersion, so $x_{i,j}$, isn't an edge pixel [1].

Step 2: After calculating the local entropy for every pixel in the image, we obtained the gray values for edge (255) and non-edge (0) pixels. This can be expressed as follows as in Eq. (3):

$$D_{i(x,y)} = \begin{cases} 0; & if\ D_{i(x,y)} \geq Th \\ 255; & Otherwise \end{cases} \tag{3}$$

where, *Th* is a threshold value of pixel intensity which is fixed empirically.

Step 3: The thinning algorithm [5] is performed on every binary images to obtain the thin binary images having only dots and arcs.

Step 4: The inessential noises caused by the scattered arcs, dots, and lines are eliminated from each thin images by scanning the entire image with several window sizes. i.e., $l \times l$ windows ($l = 3, 5\ and\ 7$). Thinning face and noise removing is clarified in Fig. 2.

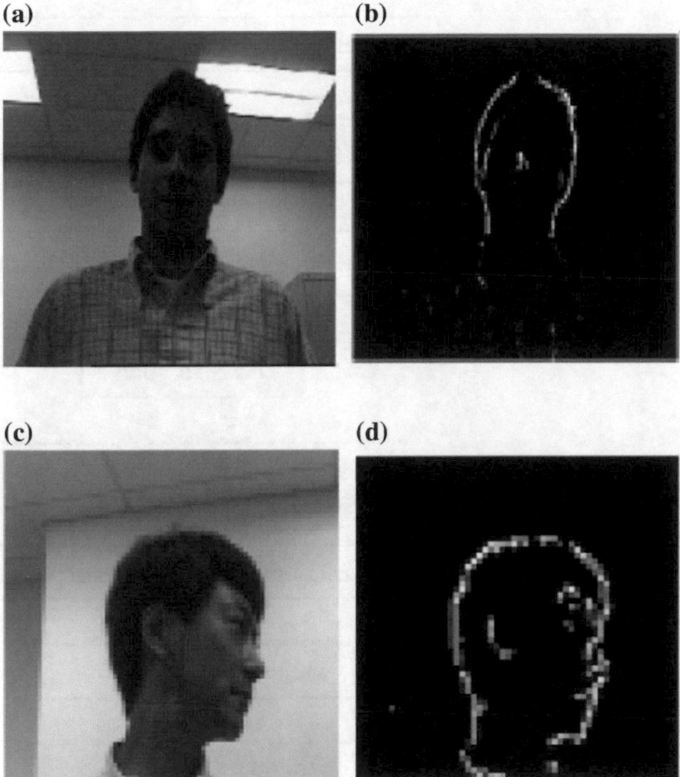

Fig. 2 **a**, **c** Two original video frames whose frame; **b**, **d** corresponding frame thinning and noise removing

The entire workflow of the presented method for face detection and tracking is illustrated in Fig. 3.

Let F_v is the number of video frames of resolution $M \times N$. Our key objective is to devise a technique which is more robust to noise. The lines and arcs form the face edge and then firstly track single face image and then track multiple face images from video sequences.

2.2 *Face Detection Using Elastic Window*

The key idea of the presented method is to find the lines and arcs which represent the left and right side of the face boundary [5]. Then these are convolved together to obtain the location of the face in the image frame. The extreme lines representing the entire face contour is found by a window which is elastically expanded according to the local image gradients. The elastic window $\boldsymbol{EW}_{AB}$ can be represented as follows

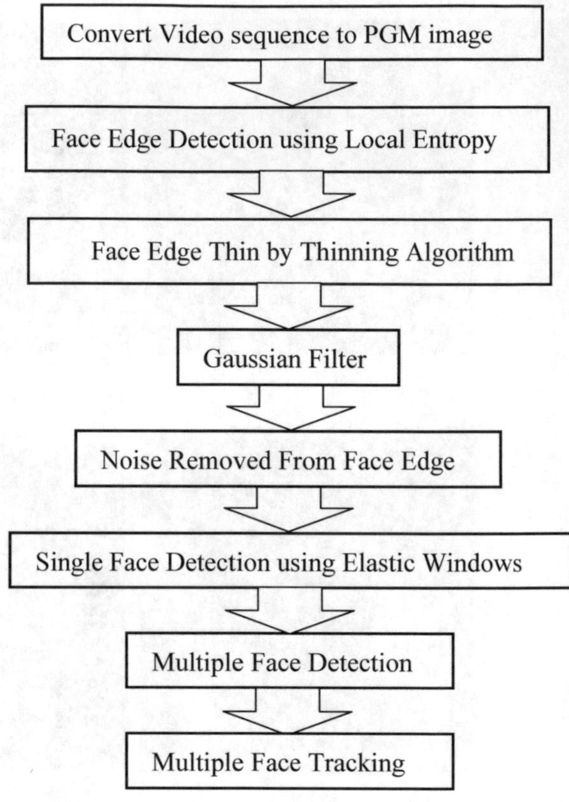

Fig. 3 Workflow of the proposed technique

in Eq. (4):

$$\boldsymbol{EW_{AB}} = \{\boldsymbol{A}, \boldsymbol{B}\} \tag{4}$$

where, A and B denote the top left most corner and bottom right most corner, respectively.

More specifically, a small window with size $l \times l$ ($l = 3\, or\, 5$) is scanned from top to bottom to find the leftmost line which represents the face boundary. The $\boldsymbol{EW_{AB}}$ is updated by expanding the window driven by the local pixel gradients which include all the pixels that constitute the line [5]. The $\boldsymbol{EW_{AB}}$ can be expressed as follows in Eq. (5):

$$\boldsymbol{EW_{AB}} = \{A + \delta A,\ B + \delta B\} \tag{5}$$

where, δA and δB denote the displacement magnitude of the points A and B, respectively. The window is expanded in upward and downward directions if the elastic window size is less than the threshold. Then, the two elastic windows are convolved together to locate the entire face image.

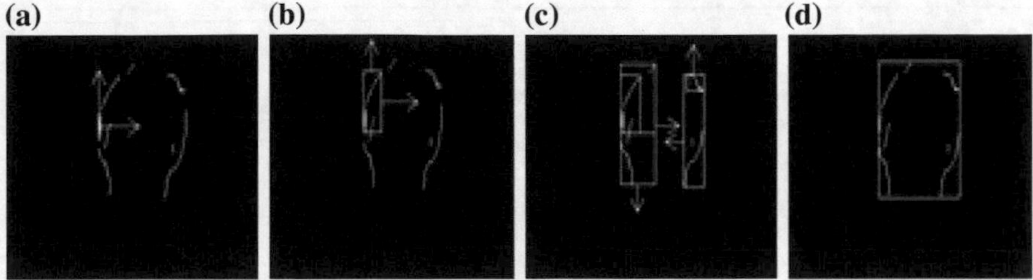

Fig. 4 **a** Starting to generate elastic window containing at first on the left side of face contour; **b** generate elastic window covering the left side of face contour; **c** generate elastic window covering the right side of face contour; **d** after convolution right and left elastic windows

Figure 4 indicates the qualitative outputs obtained from elastic window algorithm. This algorithm can be performed to detect the face region in the difference frame. Figure 4a–d are the outcomes of the intermediate steps of the elastic windows. It may be also noted that a single face edge is properly identified by convolving elastic windows (two) as shown in Fig. 4d.

2.3 Multiple Face Detection Using Elastic Window

The idea described in Sect. 2.2 is performed iteratively to locate the boundary of multiple faces from video sequences. Equation (6) for the elastic window for detecting multiple faces is described below:

$$\sum_{i=1}^{N} \boldsymbol{W}_{\boldsymbol{AB}} = \{\boldsymbol{A}, \boldsymbol{B}\} \tag{6}$$

N is the number of images in a single frame.

Figure 5 includes the qualitative outputs obtained from our face detection technique from video sequences. The pictorial representation clearly demonstrates that the proposed method correctly detected the face locations. The detection of faces is a challenging task due to background noise, illumination, and wide range of orientations. In this context, it may be noted that the moving face detection can easily be done by means of frame differencing [2].

2.4 Multiple Face Tracking

The boundary rectangles are drawn for each face in the video sequence by means of four corner points. The Euclidean (scalar) and angular (vector) distance between two adjacent frames are computed afterward. Now, the angular distance between the

Fig. 5 1st row: original color video sequence; 2nd row: converted PGM video sequence; 3rd row: multiple face locations detected using elastic window; 4th row: face locations detected from whole frames from video sequences

points (x, y) can be computed as follows in Eq. (7):

$$cos\,\theta = (x \cdot y) \div (\|x\|\|y\|) \tag{7}$$

The displacement of the corner points between two consecutive frames indicates the movement of the faces. Therefore, the face movement can be detected between two sequential frames.

Fig. 6 Face tracking results of video sequences

3 Empirical Results

The performance of the proposed method is validated by means of video sequences with challenging video criterion. In this context, it may be noted that for single face the calculations are made using standard Honda/UCSD [9] video database. In case of multiple face detection, a video of the press conference is taken with 15 frames per second. The resolution of each frame is 640×480. The proposed method can efficiently track multiple moving faces from the video sequences.

Figure 6 clearly demonstrates the results of multiple faces detection and tracking from video sequences achieved by our algorithm. The efficiency of the presented method is validated by performing the same on several video databases which can also handle the situations like size variation of face, occlusions, and out of plane rotation, etc. The periodical face detection is done to ensure recovery from errors. Detection of multiple faces can easily handle by using elastic window.

4 Conclusion

We have presented a new method to redress the problems of tracking and detection of multiple faces in video sequences involving occlusion and pose variance, by incorporating the constraints of inter-frame temporal smoothness and within-frame identity. The algorithm uses an elastic window for multiple face detection and tracking from video sequences. Moreover, scalar and vector distance is estimated to track the multiple faces from video database. Hence, the face tracking from the video sequence is accomplished. The elastic windows expand the left to the right side end of the frame and whenever the left or right arcs are found, then convolve and this process iteratively goes on. To validate our claim the proposed method is performed on several video sequences. The experimental results clearly demonstrate the supremacy of the proposed technique.

Acknowledgements Authors are indebted to Mr. Sayan Kahali for his valuable inputs for improving the quality of the manuscript.

References

1. Dai, W., Wang, K.: An image edge detection algorithm based on local entropy. In: Proceeding of the IEEE International Conference on Integration Technology 2007, pp. 418–420 (2007)
2. Dey, A.: A contour based procedure for face detection and tracking from video. In: Proceeding of the RAIT 2016, pp. 252–256 (2016)
3. Sappa, A., Dornaika, F.: An edge-based approach to motion detection. In: Proceeding of the ICCS 2006, pp. 563–570 (2006)
4. Jabri, S., Duric, Z., Wechsler, H., Rosenfeld A.: Detection and location of people in video images using adaptive fusion of color and edge information. In: Proceeding of the International Conference on Pattern Recognition 2000, pp. 627–630 (2000)
5. Chowdhury, S., Dey, A., Sing. J.K., Basu, D.K., Nasipuri, M.: A novel elastic window for face detection and recognition from video. in: Proceeding of the ICCICN 2014, pp. 252–256 (2014)
6. Sheikh, Y., Shah, M.: Bayesian object detection in dynamic scenes. In: Proceeding of the CVPR 2005, pp. 252–256 (2005)
7. Radke, R., Andra, S., Al-Kofahi, O., Roysam, B.: Image change detection algorithms: a systematic survey. IEEE Trans. Image Process. **14**, 294–307 (2005)
8. Farin, D., de With, P., Effelsberg, W.: Robust background estimation for complex video sequences. In: Proceeding of the ICIP 2003, pp. 145–148 (2003)
9. Lee, K., Ho, J., Yang, M., Kriegman, D.: Visual tracking and recognition using probabilistic appearance manifolds. Comput. Vis. Image Underst. **99**, 303–331 (2005)
10. Suganya Devi, K., Malmurugan, N., Manikandan, M.: Object motion detection in video frames using background frame matching. Int. J. Comput. Trends Technol. **6**, 1928–1931 (2013)

Assessment of Reading Material with Flow of Eyegaze Using Low-Cost Eye Tracker

Aniruddha Sinha, Sanjoy Kumar Saha and Anupam Basu

Abstract Reading text is a complex process involving visual, language, and motor functions. The movement characteristics of eye depend on the understanding of a content. In this paper, we present the methodology of quantifying the difficulty level of a textual content for a silent reading scenario using the eyegaze data captured from low-cost eye-tracking device, EyeTribe. Broadly, two types of contents are chosen such that one is easy to understand and the other is difficult. Standard readability indices are used to benchmark the easiness level of the contents. Experiment is performed with 15 individuals, and eyegaze data is used to quantify the smoothness in the flow of the reading. Statistical features are derived from the adjacency matrix obtained from the successive fixations within and adjacent lines. Results demonstrate statistically significant difference between two types of texts, enabling a set of materials to be relatively graded. A scoring metric based on a mixture of partial sigmoid functions is proposed to quantify the easiness of the text experienced by an individual. The scores reflect that even within the broad category of easy and difficult content, the level of easiness varies between individuals.

Keywords Textual content · Eyegaze · Adjacency matrix · Sigmoid

---

A. Sinha (✉)
TCS Research and Innovation, Tata Consultancy Services,
Ecospace 1B, Kolkata 700156, India
e-mail: aniruddha.s@tcs.com

S. Kumar Saha
Computer Science and Engineering, Jadavpur University,
Kolkata 700032, India
e-mail: sks_ju@yahoo.co.in

A. Basu
Computer Science and Engineering, Indian Institute of Technology,
Kharagpur 721302, India
e-mail: anupambas@gmail.com

A. K. Das et al. (eds.), *Computational Intelligence in Pattern Recognition*,
Advances in Intelligent Systems and Computing 999,
https://doi.org/10.1007/978-981-13-9042-5_42

1 Introduction

Understanding of a learning material heavily depends on the cognitive and psychological factors [1] of an individual. It is important to maintain a balance between the skill of an individual and challenge of the task. It helps an individual to be in a flow state [2] and performs optimally. Thus, to enhance the learning performance, it is important to deliver a material that matches the state of the subject. With the popularity of distant learning [3] scenario, there is a need for personalization in order to match the educational content and learner's profile. For this purpose, it is essential to categorize the difficulty level of tasks or educational contents in advance and also understand the characteristics of reading for individuals. The mental workload or cognitive load [1] experienced during tasks can be applied to study individual learning behavior. Various electrophysiological sensing [4] mechanisms, used for measuring the cognitive load, include photoplethysmogram, galvanic skin response, electrocardiogram, skin temperature, electroencephalogram, etc. However, all these methods require individuals to wear some sensors to assess the cognitive load. On contrary, eye movement is a direct reflection of cognitive load, attention, and engagement in educational task [5], where the sensor is placed at a distance from the individual.

Promising results are obtained for content analysis using superior Infrared (IR) eye trackers. During text comprehension, the cognitive processes are evaluated [6] by analyzing the fixations in eyegaze using *EyeLink-1000* device. In the area of medical, during silent reading, eye tracking is used to evaluate glaucoma patients [7]. In children, the effect of malnutrition is studied by analyzing the eye movement using the *Tobii-X2-60* eye tracker [8]. In a reading type of task, the eyegaze characterizes the flow of reading [9] in a detailed manner. During a newspaper reading, the entry points and reading paths are studied for semiotic research using a costly head-mounted eye tracker [10]. The eye movement is analyzed during reading of second language using *Tobii-X2-60* [11]. Thus, all these studies mainly focus on very costly devices and are not practical for mass deployment in an e-learning scenario. A very limited amount of study is done on characterization of silent reading using low-cost eye tracker devices. The major challenge for such device lies in handling the eyegaze with high noise characteristics [12] and then derive useful features from the same. Recently, studies are done with commercially available eye trackers on noise cleaning [13] and to quantify the cognitive load using pupil dilation during mental addition and anagram tasks [14]. We are interested to evaluate the difficulty level of a collection of content for a set of readers as well as characterize a reader for a known content.

In this paper, we use a low-cost eye-tracking device, EyeTribe [13], and analyze the flow of the eyegaze during reading of relatively easy and difficult texts. In order to justify the choice of the texts, the difficulty levels of two types of reading contents are benchmarked using the standard readability indices [15]. The contributions of this paper are as follows:

1. An adjacency matrix is formed based on the fixations belonging to the same line or adjacent lines during the reading of a line in the text. Statistical features are derived from the adjacency matrix to demonstrate the capability of being able to grossly differentiate between two types of contents.
2. We have proposed a novel metric to provide a relative score of the easiness experienced by an individual while reading a textual content. This metric is derived using a mixture of partial sigmoid functions which maps the statistical findings to a normalized easiness score which an individual experiences. The score enables us to compare individual traits of the eyegaze flow during reading of textual contents with various difficulty levels.

The paper is organized as follows. Section 2 gives the methodology of analyzing the fixation data and generation of easiness score. Section 3 describes the experimental setup, the textual content, participants, and protocol of the data capture. Section 4 provides the results followed by conclusion in the final section.

2 Methodology

Subject matter experts usually design the educational contents. However, the perspective of the target audience is also important to be considered in quantifying the level of complexity. This is usually done through feedback using a standard Likert scale [16]. However, the feedback fails to provide detailed insights on how an individual experiences the content relative to others. It only captures a broad opinion about what the individual thinks about the content.

When we read a textual content, we usually search for the next word or segment of the text. Hence, the primary components of eyegaze, during textual reading, are saccades and fixations. A sudden jerky movement characterized by a shift in 2–10° of visual angle in a relatively shorter duration (25–100 ms) having a high rotational velocity (500–900°/s) [17] is called a saccade. On the other hand, when the eye remains in a localized region (<2° of visual angle) for 250–500 ms, then it is called a fixation [17]. During a reading task, alternate fixation and saccades enable us to process the visual information.

In the current work, the eyegaze of an individual is analyzed with a focus on two aspects, namely, (i) categorizing a content based on the collective analysis of all participants and (ii) scoring the easiness of reading for an individual. This helps to assess an individual's reading capability. The overall flowchart of the processing is shown in Fig. 1.

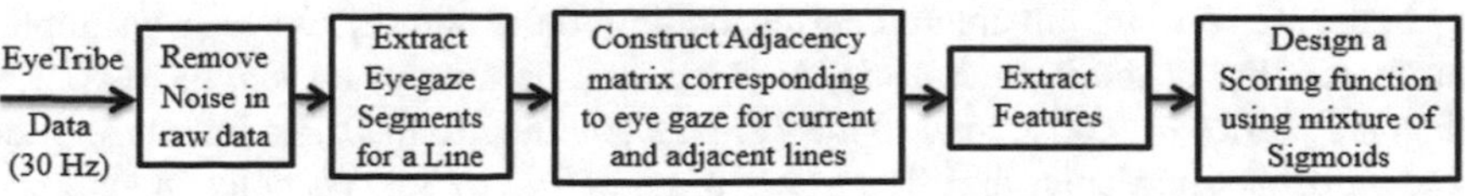

Fig. 1 Flowchart for the analysis of textual reading

2.1 Removal of Noise in Eyegaze Data

The horizontal and vertical movements of the eye are captured as a time series data of XY coordinates using an eye-tracking device, EyeTribe. This low-cost eye-tracking device is associated with large variable and systematic noise due to the inherent characteristics of the infrared (IR) receiver of the hardware. These noises are removed using Kalman and graph signal processing-based filtering methods as proposed by Gavas et al. [13]. Additionally, the segments of eyegaze where the eye is intermittently not detectable, due to head movements or eye blinks, are identified using the metadata of EyeTribe. In such scenarios, an interpolation is done with a standard bicubic interpolation using the adjacent data. This interpolated data is then used for subsequent processing.

2.2 Line Segments in Eyegaze

While reading a text, the eyegaze usually moves from left to right reflecting in a gradual increase in the X-coordinate, and it suddenly decreases when the gaze moves from the end of a line to the beginning of the next line. A trough detection [18] is performed to extract the time stamps (using the sample points) belonging to the start of a line. Within the time segment corresponding to each line, the Y-coordinate of the eyegaze provides information of the shift in the vertical direction. The combined information of X- and Y-coordinates indicates whether the gaze is confinement within a line or there are momentary shifts to adjacent lines.

2.3 Construction of Adjacency Matrix and Feature Extraction

Once a line is completely read, then the eyegaze moves to the beginning of the next line. However, due to the difficulty in understanding a portion in the content, one may occasionally move to previous line and again come back to the current position. On the other extreme, for easy texts one may perform predictive reading [19] and sometimes move forward to next lines for verification. In such scenarios, it is desirable to analyze the direction of the eyegaze scanning to previous or next lines during reading of a particular line.

In order to do that, an adjacency matrix A of size $N * N$ is formed with entries a_{ij} in the ith row and jth column corresponding to the number of eyegaze samples during reading of the ith line. Here, N is the total number of lines in the text. The diagonal entries a_{ii} correspond to the eyegaze confined to the same line during the reading of that particular line. The upper triangle $(a_{ij}, \forall i < j)$ denotes the amount of forward movement and is indicative of the flow of the reading. The lower triangle

$(a_{ij}, \forall j < i)$ denotes the amount of backtracing while reading a line. These upper and lower triangles contain information on the easiness and difficulty level experienced during reading of the text by an individual.

In order to capture the normal flow and the backtracing, we derive two features. One is the normalized mean (F_1) of the upper triangle as shown in (1). The intuition behind this feature is that for easy text most of the time readers would experience an enhanced normal flow of reading, and hence F_1 is expected to be high for an easy text compared to a difficult one. The other feature is the normalized standard deviation (F_2) of the lower triangle as shown in (1). In case of difficult text, it is expected that participants would experience backtracing in a rather uniform manner, and hence the standard deviation (STD) of the entries in the lower triangle would be lower. On the other hand for an easy content, there may be occasional difficult segments which would lead to a large variation in the lower triangle. Hence, the F_2 is expected to be higher for an easy content compared to a difficult one. Analysis of Variance (ANOVA) [20] is performed on F_1 and F_2 obtained from all the participants to demonstrate whether the difference between the features of easy and difficult texts is statistically significant, where $p < 0.05$ and F indicating the extent of difference.

$$F_1 = \frac{\sum_{i=1}^{N}\sum_{j=i+1}^{N} a_{ij}}{\sum_{i=1}^{N}\sum_{j=1}^{N} a_{ij}}, \quad F_2 = \frac{STD(a_{mn})}{\sum_{i=1}^{N}\sum_{j=1}^{N} a_{ij}}, \forall 1 < m \le N, 1 \le n < m \tag{1}$$

2.4 *Easiness Score Using Mixture of Partial Sigmoids*

In order to generate a score from the features F_1 and F_2, it is necessary to study the distribution of those features for easy and difficult textual contents and then derive a mapping function f to map the features (positive real values including 0 denoted by R^+) to a score lying between 0 and 1 as given by (2).

$$f : \mathbb{R}^+ \rightarrow [0, 1] \tag{2}$$

A sample distribution of box plot for the F_1 feature is shown in Fig. 2a. The median values for the features corresponding to easy and difficult contents are indicated by E_M and D_M, respectively. The 25th and 75th percentiles of the distributions are denoted by E_{25}, D_{25}, E_{75}, D_{75}. In order to map the F_1 to a score, it is required to choose a function that satisfies the relationship as shown in (2). The design of the function is given in the subsequent part of this section. Finally, in order to generate the easiness score, a weightxed average is performed on the mapping functions corresponding to two features F_2 and F_2.

Design of the Mixture of Partial Sigmoids: We have considered popularly used sigmoid type of functions which is monotonically increasing, limited within 0 and 1,

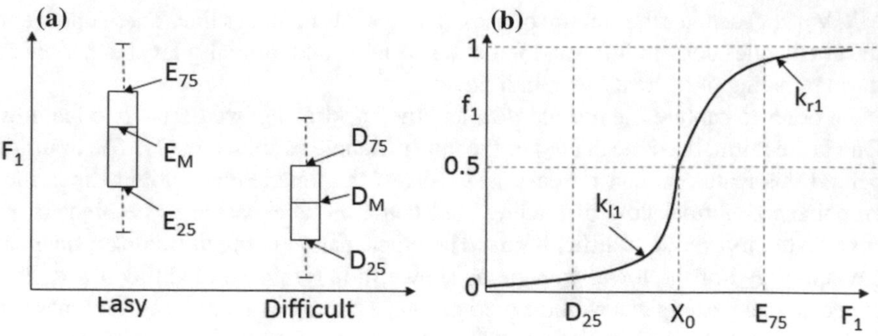

Fig. 2 **a** An example boxplot of F1 feature for easy and difficult content, **b** corresponding example of mixture of partial sigmoid functions

has an inflexion point in the center where the slope of the function is momentarily constant. At any point less than the inflexion point, the slope rises monotonically and after the inflexion point the slope falls monotonically. The family of logistic or sigmoid functions is given in (3) where the point x_0 is the inflexion point and the value of k determines the steepness of the slope at any point x.

$$f(x) = \frac{1}{1 + e^{-k(x - x_0)}} \tag{3}$$

The desired function (2) is constructed with a mixture of two partial sigmoid functions using (4) with one lying in the range $0 \le x < x_0$ and the other in $x > x_0$. One such example function is shown in Fig. 2b. The value of the two partial sigmoids is same (0.5) at $x = x_0$; hence, it is a continuous function, though not differentiable at $x = x_0$, which is taken care of later in (5). The slope parameters are different, which are k_l *and* k_r for two sigmoids, respectively.

$$\begin{gathered} f(x) = \frac{1}{1 + e^{-k_l(x - x_0)}} \quad \forall\, 0 \le x < x_0, \\ f(x) = \frac{1}{1 + e^{-k_r(x - x_0)}} \;\forall\, x > x_0, \quad f(x) = 0.5 \;\; at \;\; x = x_0 \end{gathered} \tag{4}$$

Design of the Inflexion Point and Slope parameters: It is desired that the mapping function (4) is able to separate the feature values (e.g., F_1) corresponding to easy and difficult texts into the high and low output values while retaining the limit between 0 and 1. Thus, the inflexion point x_0 is chosen to be the average of the two median values (Fig. 2) as given by (5).

Next, the choices of the k_l and k_r are done by tuning the slope of the sigmoids such that there is a balance between the dynamic range of the input features and the separation of easy and difficult content. The slope of a sigmoid is given by (5). At any given point x_p as the k is reduced from a very high value, the slope S_k increases and the dynamic range also increases. However, beyond a point with further increase in k the value of S_k starts reducing which worsen the discriminating power of the mapping function. Thus, the values of k_l and k_r are chosen such that the derivative of the slope S is maximized (6) at $x = D_{25}$ and $x = E_{75}$, respectively.

$$S = \frac{df}{dx} = \frac{k * e^{-k(x-x_0)}}{(1 + e^{-k(x-x_0)})^2}, \; x \neq x_0 \; where \; x_0 = \frac{E_M + D_M}{2} \tag{5}$$

$$\frac{dS}{dk_l} = 0 \;\; at \;\; x = D_{25} \;\; and \;\; \frac{dS}{dk_r} = 0 \;\; at \;\; x = E_{75} \tag{6}$$

Generation of Weighted Score: Let us assume that solving (6), we obtain the sigmoid parameter pairs as k_{l1}, k_{r1} and k_{l2}, k_{r2} for two feature F_1 and F_2, respectively. Moreover, let us assume that using (4) the functions derived for the respective features are f_1 and f_2. Consider the F values (higher the values better is the discriminating power of the feature) obtained from ANOVA analysis of the two features F_1 and F_2 are F_{F_1} and F_{F_2}, respectively. Then, the final easiness score is given by (7) where α is the weighing parameter. The feature generating higher F value for the ANOVA analysis gets a proportionately higher weight in the computation of the easiness score.

$$Score = \alpha * f_1 + (1 - \alpha) * f_2, \;\; where \;\; \alpha = \frac{F_{F_1}}{(F_{F_1} + F_{F_2})} \tag{7}$$

The score indicates relative differences between the easiness experienced by the participants while reading a particular text. The range of the score is based on the stimulus or textual contents used in the experiment. It is to be noted that there is no concept of absolute easy or difficult content; hence, the proposed analysis is meant for a sample space comprising a given collection of textual content.

3 Experimental Paradigm

In this section, we describe the experimental setup, the sensors used for eyegaze tracking, description of the textual contents used for the reading task, and the details on the participants.

3.1 Setup

A 21-inch display screen, with resolution of 1600 × 1200 pixels, is placed at an approximate distance of 60 cm from the participant. The eye-tracking device (EyeTribe), having sampling rate as 30 Hz, is placed below the screen facing the subject as

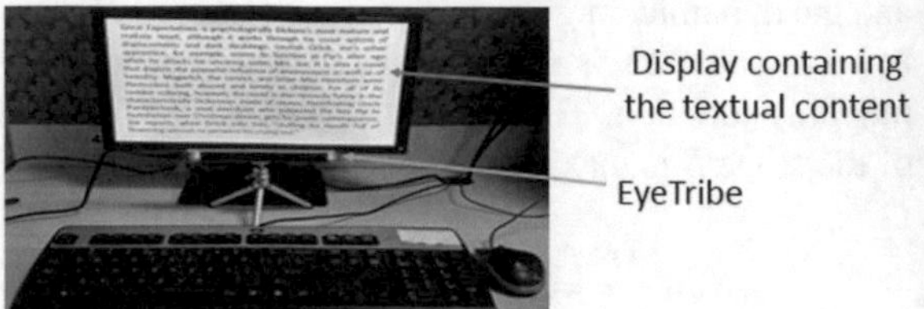

Fig. 3 Experimental setup for textual reading

shown in Fig. 3. The experiment is performed in a quiet room with constant lighting condition. Participants are instructed to minimize the extent of head movement in order to capture a reliable eyegaze data.

3.2 Description of the Stimulus for the Reading Task

Three paragraphs of easy texts (T1) are created using materials[1] used for teaching English in primary schools. For difficult texts (T2), three paragraphs are created from notes[2,3] on novels, demonetization, and nationalism. Each paragraph consists of 12–14 lines and 132–164 words. Standard readability indices like Flesch Kincaid Grade Level, Flesch Kincaid Reading Ease, SMOG Index, and Coleman Liau Index [15] are used to justify the difference in the difficulty levels of T1 and T2. As an example, the Flesch Kincaid Grade Level for T1 is within 3 and 6 and T2 within 13 and 16.

3.3 Participant Details

The eyegaze data is collected from 15 engineering students (8 males, 7 females, age: 21–44 years) having normal or corrected to normal vision. For this experiment, the data were anonymized, participants voluntarily signed the informed consent forms and we followed the Helsinki Human Research guidelines.[4]

---

[1]http://www.preservearticles.com/2011080510137/12-short-paragraphs-in-english-language-for-school-kids-free-to-read.html.

[2]https://www.enotes.com/topics/great-expectations.

[3]https://www.quora.com/Why-is-there-so-much-criticism-for-demonetization-in-India-when-it-is-undoubtedly-a-correct-step.

[4]https://www.helsinki.fi/en/research/research-environment/research-ethics.

3.4 Experiment Protocol

Initially, a calibration is performed using the SDK of EyeTribe where the participants are asked to follow a ball on the screen. Then, the participants need to fixate on a '+' for 15 s, shown in the screen. After that, the text paragraphs are displayed on the screen with 32 Calibri font in black text with white background. Textual content of T1 (3 easy) and T2 (3 difficult) are shown one after another. Participants need to press a button for the next content. The sequences of T1 and T2 are randomized to get rid of any bias. At the end of the experiment, feedback on the difficulty of the contents and their reading habits are taken on a 5-point Likert scale [16].

4 Results

Initially, we present the statistical analysis of the eyegaze features for easy and difficult types of texts. Later, an easiness score is generated which reflects the ease of the reading experienced by individuals.

The one-way ANOVA analysis was done on the features, namely, F_1 and F_2 obtained from 15 subjects with three easy and three difficult types of texts. Both the features indicate significant difference between the easy and difficult contents with p=0.023 and F=5.35 (=F_{F1}, which is the F value for the F_1 feature) and p=0.0051 and F=8.24 (=F_{F2}, which is the F value for the F_2 feature).

In order to derive the easiness score, the parameters for the mixture of partial sigmoid functions (f_1 and f_2) are derived using (5) and (6). The parameters for f_1 are $X_0 = 7.077$, $k_{l1} = 0.78$, $k_{r1} = 0.38$ and for f_2 are $X_0 = 21.3$, $k_{l2} = 0.2$, $k_{r2} = 0.18$. These values are derived using the median and interquantile values obtained from the ANOVA analysis of both the features. The α used for generating the easiness score is 0.394 as obtained from (7). The average scores obtained from 15 individuals for easy and difficult contents using (7) are shown in Table 1. The higher the value of the score, the easy is the reading experience. The average easiness score is 0.58 considering all the easy contents, whereas it is 0.42 for difficult contents.

Table 1 Comparison of the average scores derived from all participants

Content type	Individual content score	Average score
Easy 1	0.62	0.58
Easy 2	0.6	
Easy 3	0.53	
Diff 1	0.42	0.42
Diff 2	0.4	
Diff 3	0.43	

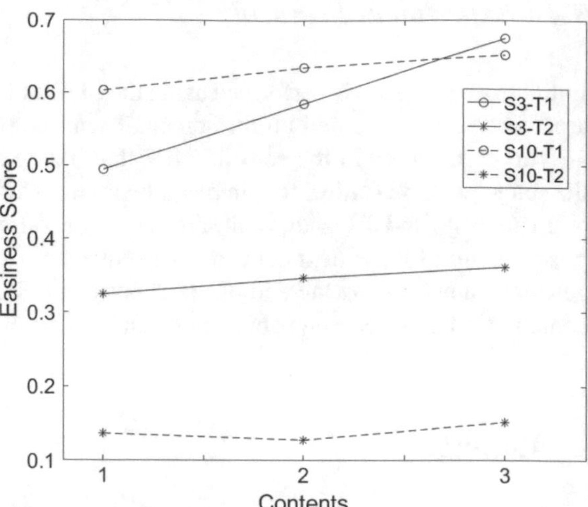

Fig. 4 Comparison of easiness scores of subjects S3 and S10 for T1 and T2

We further investigate the relative effect of the contents on the individuals. We choose two subjects, one (S3) who often reads English novels and one (S10) who rarely. Figure 4 shows the scores for both the subjects obtained from three easy (T1) and three difficult (T2) contents. Though there is gross separation of easiness score between easy and difficult contents for each subject, the relative individual experiences vary a lot. The easiness experienced by S3 for difficult content is much higher than S10 which also match with the feedback on reading habits.

5 Conclusion

The movement of the eyegaze during reading of a text provides information on the easiness of the content experienced by an individual. For easy content, the forward flow of the eyegaze is more compared to difficult ones. An adjacency matrix is constructed from the eyegaze data containing information on the amount of movement to adjacent lines during reading. The mean and standard deviation features extracted from the upper and lower triangle of the adjacency matrix quantify the forward flow and backtracing. ANOVA analysis establishes that these features are able to significantly differentiate between two broad types of contents. Additionally, an easiness score is generated from the features using a novel mixture of partial sigmoid-based mapping function. This score is able to provide further insights into the difference in the easiness experienced by individuals for the same content. In future, we would like to further investigate on the distribution of the spatial and temporal spread of fixations and its conditional dependencies on the previous and future saccades.[5]

[5]https://www.helsinki.fi/en/research/research-environment/research-ethics.

Acknowledgements The clearance on ethical issues for handling and analysis of the data collected has been acquired on October 24, 2017 from Institutional Review Board (IRB) of Tata Consultancy services Ltd. Authors are thankful to the IRB members for conducting the review, monitoring, and approval for the eyegaze-related study. We have followed the Helsinki Human Research guidelines for the data collection.

References

1. Sweller, J.: Cognitive load during problem solving: effects on learning. Cogn. Sci. **12**(2), 257–285 (1988)
2. Czisikszentmihalyi, M.: Flow-the psychology of optimal experience (1990)
3. Mehlenbacher, B., Bennett, L., Bird, T., Ivey, M., Lucas, J., Morton, J., Whitman, L.: Usable e-learning: a conceptual model for evaluation and design. In: Proceedings of HCI International, vol. 2005. Citeseer (2005)
4. Haapalainen, E., Kim, S., Forlizzi, J.F., Dey, A.K.: Psycho-physiological measures for assessing cognitive load. In: Proceedings of the 12th ACM International Conference on Ubiquitous Computing, ACM, pp. 301–310 (2010)
5. Paas, F., Tuovinen, J.E., Tabbers, H., Van Gerven, P.W.: Cognitive load measurement as a means to advance cognitive load theory. Educ. Psychol. **38**(1), 63–71 (2003)
6. Raney, G.E., Campbell, S.J., Bovee, J.C.: Using eye movements to evaluate the cognitive processes involved in text comprehension. J. Vis. Exp. JoVE (83) (2014)
7. Murata, N., Miyamoto, D., Togano, T., Fukuchi, T.: Evaluating silent reading performance with an eye tracking system in patients with glaucoma. PloS one **12**(1), e0170230 (2017)
8. Forssman, L., Ashorn, P., Ashorn, U., Maleta, K., Matchado, A., Kortekangas, E., Leppänen, J.M.: Eye-tracking-based assessment of cognitive function in low-resource settings. Archives of Disease in Childhood (2016) archdischild—2016
9. Just, M.A., Carpenter, P.A.: A theory of reading: from eye fixations to comprehension. Psychol. Rev. **87**(4), 329 (1980)
10. Holsanova, J., Rahm, H., Holmqvist, K.: Entry points and reading paths on newspaper spreads: comparing a semiotic analysis with eye-tracking measurements. Vis. Commun. **5**(1), 65–93 (2006)
11. Indrarathne, B., Kormos, J.: The role of working memory in processing l2 input: insights from eye-tracking. Biling. Lang. Cogn. **21**(2), 355–374 (2018)
12. Johansen, S.A., San Agustin, J., Skovsgaard, H., Hansen, J.P., Tall, M.: Low cost vs. high-end eye tracking for usability testing. In: CHI'11 Extended Abstracts on Human Factors in Computing Systems, pp. 1177–1182. ACM (2011)
13. Gavas, R.D., Roy, S., Chatterjee, D., Tripathy, S.R., Chakravarty, K., Sinha, A.: Enhancing the usability of low-cost eye trackers for rehabilitation applications. PloS one **13**(6), e0196348 (2018)
14. Gavas, R.D., Tripathy, S.R., Chatterjee, D., Sinha, A.: Cognitive load and metacognitive confidence extraction from pupillary response. Cogn. Syst. Res. **52**, 325–334 (2018)
15. Kincaid, J.P., Fishburne Jr., R.P., Rogers, R.L., Chissom, B.S.: Derivation of new readability formulas (automated readability index, fog count and flesch reading ease formula) for navy enlisted personnel. Technical report, Naval Technical Training Command Millington TN Research Branch (1975)
16. Hinkin, T.R.: A brief tutorial on the development of measures for use in survey questionnaires. Organ. Res. Methods **1**(1), 104–121 (1998)
17. Salvucci, D.D., Goldberg, J.H.: Identifying fixations and saccades in eye-tracking protocols. In: Proceedings of the 2000 symposium on Eye Tracking Research & Applications, pp. 71–78. ACM (2000)

18. Jacobson, A.: Auto-threshold peak detection in physiological signals. In: Engineering in Medicine and Biology Society, 2001. Proceedings of the 23rd Annual International Conference of the IEEE, vol. 3, pp. 2194–2195. IEEE (2001)
19. Garrison, J.W., Hoskisson, K.: Confirmation bias in predictive reading. Read. Teach. **42**(7), 482–486 (1989)
20. Keselman, H., Huberty, C.J., Lix, L.M., Olejnik, S., Cribbie, R.A., Donahue, B., Kowalchuk, R.K., Lowman, L.L., Petoskey, M.D., Keselman, J.C., et al.: Statistical practices of educational researchers: an analysis of their anova, manova, and ancova analyses. Rev. Educ. Res. **68**(3), 350–386 (1998)

NCBI: A Novel Correlation Based Imputing Technique Using Biclustering

Hussain A. Chowdhury, Hasin A. Ahmed, Dhruba Kumar Bhattacharyya and Jugal K. Kalita

Abstract Presence of missing values (MV) in gene expression data is commonplace. It significantly affects the performance of statistical analysis and machine learning algorithms. Discarding objects or attributes with missing values and inappropriate estimation of MVs lead to high information loss and misleading results. So, it is necessary to have an accurate technique for missing value imputation. In this paper, we present a novel correlation based missing value imputation technique for gene expression datasets. We refer to our method as NCBI. We compare the estimation accuracy of our technique with two widely used methods such as KNNI and KMI, on four benchmark datasets by randomly knocking out data values as missing. Our technique can estimate missing values almost 20–25% more accurately than KNNI and KMI in all datasets.

Keywords Imputation · Missing value · Microarrays · Shifting-and-scaling · Machine learning · Biclustering

---

H. A. Chowdhury · D. Bhattacharyya (✉)
Computer Science and Engineering, Tezpur University,
Sonitpur 784028, Assam, India
e-mail: hussain@tezu.ernet.in

D. Bhattacharyya
e-mail: dkb@tezu.ernet.in

H. A. Ahmed
Department of Information and Computer Science, Assam Women's University,
Jorhat 785004, Assam, India
e-mail: hasin@tezu.ernet.in

K. Kalita
Computer Science, University of Colorado,
Colorado Springs, CO 80933-7150, USA
e-mail: jkalita@uccs.edu

A. K. Das et al. (eds.), *Computational Intelligence in Pattern Recognition*,
Advances in Intelligent Systems and Computing 999,
https://doi.org/10.1007/978-981-13-9042-5_43

1 Introduction

There are a variety of reasons for missing data values in different application domains. For example, in bioinformatics, microarrays generate missing data due to the presence of scratches or dust in microarray slides. Such missing values may lead to failure in the underlying gene expression measurements. Incomplete dataset generation is considered a random process that can be decomposed into two such processes [4]: (i) a complete data process that generates complete datasets, and (ii) a missing data process that determines which elements of the complete dataset are missing.

We can try to explain the missing data process by a theory in terms of its impact on a machine learning algorithm, in particular, the inference and prediction of an algorithm. This theory should be able to differentiate missing data into three categories [9]: (i) data values missing at random (MAR), (ii) not missing at random (NMAR), and (iii) missing completely at random (MCAR). When the MAR condition holds in the data, we can simply discard examples with missing data values and inference can be made on fully observed data examples only. However, if NMAR condition holds in the data, then ignorance of missing data values leads to the biasing of results of a machine learning algorithm.

The success of any machine learning algorithm depends on the quality of data provided to the algorithm. Often, some preprocessing of data is required before analysis. Some widely used preprocessing techniques include normalization, discretization, and missing value imputation.

Most missing value imputation techniques are local. Here, for imputation of missing values for an object, a subset of highly correlated objects with that object are used. Some widely used imputation algorithms are K nearest-neighbor imputation (KNNI) [11] and K-means clustering imputation (KMI) [7]. Most methods use all features when computing the correlation between a pair of objects to identify highly correlated objects. However, there may be some irrelevant features which provide low amount of information during correlation computation, and leading to the inappropriate selection of highly correlated objects. Thus, removal of irrelevant features before calculation of correlation of a pair of objects is important [6].

In this work, we propose a novel correlation based missing value imputation technique for gene expression datasets. The paper is organized as follows. Sections 2 and 3 briefly describe background and related work. The proposed method and experimental results are presented in Sects. 4 and 5, respectively. Finally, Sect. 6 concludes this paper.

2 Background

Some important preliminaries such as treatment of missing values, different types of patterns present in gene expression datasets and validation of imputation methods are discussed below.

2.1 Missing Values and Their Treatment

Missing values are common occurrences, and one needs to have a strategy for treating them. Some well-known reasons for missing values are lack of response, inconsistencies among recorded data measurements not made by human, and machine error.

Generally, in machine learning, missing values are treated in the following ways: (i) discard objects or features with MVs, (ii) acquire missing values by incurring some cost, and (iii) impute missing values from available cases. Missing value imputation approaches are categorized into four different classes based on the type of information needed in the algorithm. These approaches are [8]: (i) global, (ii) local, (iii) hybrid, and (iv) knowledge assisted. In global approaches, the algorithms perform MV imputation based on the correlation structure derived from the whole data, whereas in local approaches, only local similarity structure present in the data is used for MV imputation. Knowledge-assisted approaches integrate external information or domain knowledge into the imputation process. Hybrid approaches use both local and global correlation structures of the correlation during the imputation process. They can also use external information or domain knowledge in the imputation process.

2.2 Patterns in Gene Expression Datasets

Various types of correlations exist among examples or objects in genes in gene expression datasets. This phenomenon poses challenge in developing effective techniques for gene expression data analysis. The existence of patterns in a gene expression dataset helps us identify most correlated genes. The different types of correlations that may exist in such datasets are given below. These correlations can be captured by existing proximity measures such as (a) Cosine distance [1], which can capture scaling patterns, (b) NMRS [10], which can capture shifting patterns, and (c) SSSim [1], which can capture both shifting and scaling patterns like Pearson's correlation coefficient usually does [3].

Definition 1 Correlation: Two objects are said to be correlated if one object can derived from the other by shifting or scaling, or by shifting and scaling.

Definition 2 Shifting correlation: Two objects $O_1 = [a_1, a_2, \ldots, a_n]$ and $O_2 = [b_1, b_2, \ldots, b_n]$ are said to exhibit shifting correlation if $a_i = b_i + p, i = 1, 2, \ldots, n$ where p is an additive constant.

Definition 3 Scaling correlation: Two objects $O_1 = [a_1, a_2, \ldots, a_n]$ and $O_2 = [b_1, b_2, \ldots, b_n]$ are said to exhibit scaling correlation if $a_i = b_i \times p, i = 1, 2, \ldots, n$ where p is a multiplicative constant.

Definition 4 Shifting-and-scaling correlation: Two objects $O_1 = [a_1, a_2, \ldots, a_n]$ and $O_2 = [b_1, b_2, \ldots, b_n]$ are said to exhibit shifting-and-scaling correlation if $a_i = q + p \times b_i, i = 1, 2, \ldots, n$ where q is an additive constant and p is a multiplicative constant.

2.3 *Biclustering*

Biclustering is an unsupervised machine learning technique to group subsets of objects, considering subsets of features which show high correlation among them. It aims to find a set of co-expressed objects across a number of experimental conditions. So, a bicluster generally contains high amount of local structure information in the data. The objects present in a bicluster show high shifting, scaling or shifting-and-scaling correlations. Thus, biclustering can help in imputing missing values by considering relevant correlated values.

2.4 *Validation of Imputation Results*

Validation of imputation results helps estimate the effectiveness of an MV imputation algorithm. A common way of validation is by comparing imputed values with the original values. To obtain the ground truth, first, we ignore objects containing missing values in a dataset and later the resulting reduced dataset is used in our experiments as a complete dataset. Next, missing datasets are generated by randomly knocking out some values as missing from reduced dataset. A widely used metric to check the accuracy of an imputation method is the root mean squared error (RMSE) between imputed values and original values [5]. The equation of RMSE is given as

$$RMSE = \sqrt{\frac{\sum_{i=1}^{m} \sum_{k=1}^{n} (g_{ik} - \tilde{g_{ik}})^2}{m \times n}} \tag{1}$$

where g_{ik} is the value in the kth experiment for gene g_i, and g and $\tilde{g}$ represent true value and the imputed value respectively. m and n are the number of objects and number of features in the dataset, respectively. Missing values in the datasets are randomly generated for validation and lower RMSE means more accurate imputation.

3 Related Work

The missing values in an object are imputed based on the knowledge of other similar objects. There are several methods for this purpose. We discuss two well-known missing value imputation techniques based on the local correlation structure present in the data.

KNN imputation is a K-nearest neighbor-based imputation technique (also known as KNNI) [2]. It is considered an established MV imputation technique because of two main reasons: (i) It is based on a correlation function between two objects and has high adaptability, and (ii) It is able to deal with attributes with both numeric and nominal values. Another major advantage of KNNI is that it does not require one to establish a model for the data. So, it is widely used as a baseline for comparison when establishing a new imputation technique. However, selection of the appropriate K-value is difficult, and is usually estimated empirically.

Clustering is a widely used data mining technique used to group a subset of objects which are highly similar to each other. The main aim of clustering is to minimize intra-cluster dissimilarity and to maximize the inter-cluster dissimilarity. In K-means clustering, the intra-cluster dissimilarity is measured by the summation of Euclidean distances between the objects and the centroid of the cluster they are assigned to. The centroid is represented by the mean values of the objects present in the cluster.

MV imputation technique based on K-means clustering works in three steps. First, K complete objects are selected randomly as K centroids. Second, we minimize the sum of distances for each object iteratively until this sum becomes less than a user defined threshold $\in$ from the centroid of the cluster to which that object assigned to. Finally, we perform MV imputation of an incomplete object by considering other objects present in that cluster as reference objects. Data objects that belong to the same cluster are considered as reference objects to each other for MV imputation.

4 Proposed Work

To understand our proposed work, we present a few definitions and lemmas.

Definition 5 Reference Object: An object O_i is a reference object for O_j, iff $O_j, O_j \in C_i$ i.e., ith cluster. In other words, two objects are references to each other if they belong to the same cluster.

Definition 6 Seed pair: A pair of objects/genes $\{O_i, O_j\}$ is considered the seed pair for formation of a bicluster, if O_i and O_j are strongly correlated.

Definition 7 Bicluster: It is a group of objects/genes $\{O_i, \ldots, O_m\} \in O$ under a group of features/conditions $\{F_i, \ldots, F_m\} \in F$ which show high correlation among them.

Definition 8 Rank of an object: It is a quantitative index which tells how much an object O_j is correlated to the reference object O_i than other objects in $O - \{O_i, O_j\} \in O$.

Lemma 1 *Estimation of missing attribute values of an object/gene is dependent on/influenced by the non-missing attribute values in its correlated objects.*

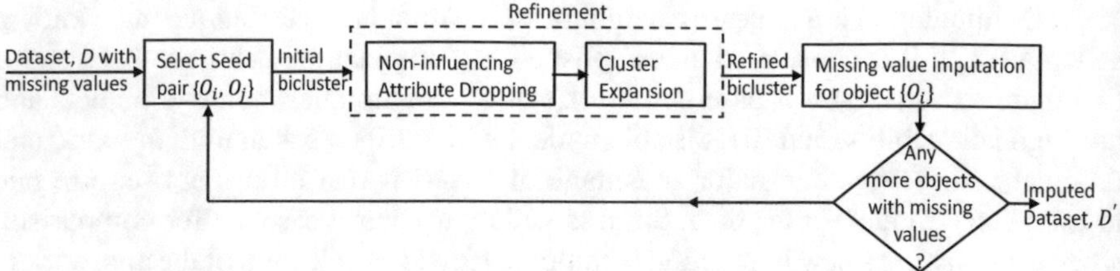

Fig. 1 Framework of the proposed method

Lemma 2 *Estimation of a missing attribute value of an object O_i/gene g_i is not dependent on non-missing attribute values of the same object O_i or g_i.*

Lemma 3 *Estimation of a missing attribute value of an object or gene is not dependent on non-missing attribute values of the correlated objects.*

4.1 Proposed Framework

Generally, a bicluster provides several many much local information about the data, which can be used for better missing value imputation. Our technique is based on the framework given in Fig. 1. For each missing valued object O_i, we create a bicluster to select highly correlated objects. First, we select a pair of strongly correlated objects $\{O_i, O_j\}$ and consider it as the seed pair for formation of a bicluster. We explore the effectiveness of three correlation measures, viz; SSSim [1], NMRS [10], and Cosine similarity [1] when finding the seed pair. Next, in the refinement step, we drop attributes which have lowest influence in correlating O_i and O_j and then add other nonselected objects which are correlated with $\{O_i, O_j\}$ by considering average correlation between objects included in the bicluster and the nonselected objects. Now, we have a bicluster containing most local information about the object O_i. Finally, ranking of the selected objects is performed based on correlation values with object O_i and then all missing values of O_i are imputed using a weighted mean function.

4.2 Proposed Algorithm

The proposed NCBI method operates on gene expression data to impute missing values based on four input parameters, namely dimensionalities of a bicluster, i.e., K_1 and K_2, the correlation measure η, and the dataset D with missing values. Different symbols used in the proposed Algorithm 1 with description is provided in Table 1. Assume that an object O_i contains some missing values. First, we identify a pair of strongly correlated object pairs $\{O_i, O_j\}$ to form of an initial bicluster. The correlation between objects O_i and O_j is calculated by only including attributes where both

Table 1 Symbol table for the Algorithm 1

Symbols	Descriptions
D	Dataset with missing values
D'	Imputed dataset
K_1	Number of objects in a bicluster
K_2	Number of attributes in a bicluster
O	Set of objects
F	Set of attributes in the dataset
O_i	ith object in O
ρ	Partial bicluster
$\rho_{\cdot objects}$	Set of objects in ρ
$\rho_{\cdot features}$	Set of features in ρ
η	Used correlation measure
X	Set of objects in $\rho_{\cdot objects}$
$\eta_{\rho_{\cdot features}}(O_j, O_i)$	Correlation between objects O_i and O_j over features$\rho_{\cdot features}$
O_s	Set of selected objects in descending order of correlation
$meanImpute(O_i)$	Replaces all the missing values in O_i with mean
$LinearModel(O_i, O_j)$	Solves a linear equation and returns the slope α and intercept β
$O_{s_j}^{F-(\mu\cup\mu')}$	Object j form O_s with features $F-(\mu\cup\mu')$
$O_i^{f_p}$	Value of O_i at feature f_p

objects have non-missing values. Further, in this study we experiment with three measures, viz; SSSim [1], NMRS [10], and Cosine similarity [1] when computing correlated pairs, and finally recommend the best one which shows consistent performance. Then, we drop one at a time those attributes which have low influence in correlating objects O_i and O_j. We use backward search for the selection of K_2 best attributes in the bicluster. Now, we select $K_1 - 2$ most correlated objects $\rho_{objects}$ from nonselected objects by considering average correlation with the objects present in the bicluster. When finding correlation we consider attributes present in the bicluster.

Next, we rank all the objects present in $\rho_{objects}$ in descending order of correlation with object O_i and use them as selected objects O_s for missing value imputation for object O_i.

We assume that the target object O_i and the reference object O_j are related by a linear regression model [5] $O_i = \alpha + \beta O_j + \in$. To impute MVs in the object O_i, we obtain regressed objects $\hat{O_{i1}} \ldots \hat{O_{iK_1}}$ from each of the objects present in O_s. For each $\hat{O_i}$, the parameters α_i and β_i are based only on attributes where neither O_i nor O_s have missing values.

Finally, we compute a weighted average of all $\hat{O_i}$s and replace the MVs of the object O_i. The weighting is designed to help identify the objects most correlated with O_i with large weights, and they are expected to give best estimates of the missing value.

```
Input: Dataset, K_1, K_2, η
Output: Imputed dataset
for i=1,2,3.... |O| do
    μ=Set of all missing valued attributes of O_i;
    if |μ| > 0 then
        F' = F − μ;
        ρ.objects ← {O_i};
        ρ.features ← {F'};
        Find O_j ∈ O − O_i so that there is no O_m ∈ O − O_i such that η_{ρ.features}(O_m, O_i) > η_{ρ.features}(O_j, O_i) ;
        ρ.objects ← ρ.objects ∪ O_j;
        while |ρ.features| > K_2 do
            Find f ∈ ρ.features so that there is no f' ∈ ρ.features such that
            η_{ρ.features−f'}(O_i, O_j) > η_{ρ.features−f}(O_i, O_j) ;
            ρ.features = ρ.features − f
        end
        for l=1,2.... K_1 − 2 do
            Find O_j ∈ O − ρ.objects so that there is no O_m ∈ O − ρ.objects such that
            η_{ρ.features}(O_m, ρ.objects) > η_{ρ.features}(O_j, ρ.objects) ;
            ρ.objects ← ρ.objects ∪ O_j;
        end
        X = ρ.objects − O_i;
        X_i ← X;
        for j=1,2,3.... |X| do
            Find O_j ∈ X_i so that there is no O_m ∈ X_i − O_j such that η_{ρ.features}(O_m, O_i) > η_{ρ.features}(O_j, O_i) ;
            X_i ← X_i − O_j;
            O_s ← O_s ∪ O_j;
        end
        for each f_p ∈ F' do
            for j=1,2,3.... |O_s| do
                μ'=Set of all missing valued attributes of O_{s_j};
                if |μ ∪ μ'| > |F| × 0.4 then
                    O_i ← meanImpute(O_i);
                    O_{s_j} ← meanImpute(O_{s_j});
                    (α, β) = LinearModel(O_{s_j}, O_i);
                else
                    (α, β) = LinearModel(ranked_{O_j}^{F−(μ∪μ')}, O_i^{F−(μ∪μ')});
                end
                if O_{s_j}^{fp} is not empty then
                    weight ← (|O_s| − j);
                    S_j ← (α + β × O_{s_j}^{fp}) × weight;
                    totalWeight ← totalWeight + weight;
                end
            end
            O_i^{fp} ← sum(S)/totalWeight;
        end
    end
end
```

Algorithm 1: Proposed Algorithm for Missing Value Imputation

4.3 Complexity Analysis

Overall time complexity of the method to impute all missing values in a gene expression dataset is $O(N(((nl + n\ log\ n)k + (m^2 l + m^2\ log\ m)) + km)) \approx O(N(nkl +$

$m^2 l)) = O(Nl(nk + m^2))$ assuming $l > log\ n$. Where l is the complexity of correlation measure η, N is the number of objects with missing values, n is the total number of objects, m is the total number of attributes and k is the number of objects required in a bicluster. If we can assume $nk \gg m^2$, we can write the complexity simply as $O(Nlnk)$.

5 Experimental Results

We carry out this study on a machine with 2 GB main memory, Intel (R) core(TM) i3-2120 processor and 64-bit Windows 7 operating system. We use MATLAB R2015a and WEKA 3.7 to perform our experiments.

We select four benchmark gene expression data sets. The reason for using the Arrhythmia dataset is to distinguish between the presence and absence of cardiac arrhythmia and to classify an example into one of the 16 groups. The colon tumor dataset contains 62 attributes collected from colon cancer patients. Among them, 40 tumor biopsies are from tumors and 22 normal biopsies are from healthy parts of the colon for the same patients. The leukemia dataset contains 38 bone marrow samples over 7129 probes from 6817 human genes. The lymphoma dataset includes information about 4026 objects and 96 attributes. Except diabetes and colon tumor the other datasets contain MVs. We cannot directly use these mentioned datasets in our experimentation because during RMSE computation we need complete and imputed datasets. So, we drop objects containing missing values from the original datasets and later we use WEKA 3.7 to randomly knock out different percentage of values as missing. After removing objects with missing values these mentioned datasets contain number of objects as follows: Arrhythmia 68, Leukemia 3051, and Lymphoma 854.

In our study, we find correlation between gene expression profiles using three measures, i.e., SSSim, NMRS and Cosine and analyze their performance when imputing MVs. We compare our method with KNNI and KMI, and report results in Table 2 and Fig. 2. To evaluate effectiveness, we inject missing values into the dataset randomly and later impute them. For validation, we compute RMSE values using Eq. (1) between the original dataset and the imputed dataset. We report the performance of our method for three different measures while comparing with KNNI and KMI in Table 2. Our method was implemented with these three measures and are referred to as NCBI_SSSim (when SSSim is used), NCBI_NMRS (when NMRS is used), and NCBI_Cosine (when Cosine measure is used), respectively. From the results in Table 2 and Fig. 2, we can observe that all the variants of NCBI show significantly improved performance over KNNI/KMI for all the four datasets. However, among these three measures, our observation is that NMRS is the best performing measure. Further, our experimental study reveals that for datasets with a small number of attributes, NCBI may not perform well (results are not reported in this paper). Work is in progress to extend the current version of NCBI toward more robustness for handling datasets of any number of features (≤ 10) as well as any number of instances.

Table 2 Average additional accurate imputation of missing values by NCBI over KNNI and KMI for upto 25% of missing values. The nearest round-off values are reported

Mesure used	Performance in terms of RMSE in %							
	Arrythmia		Colon		Lymphoma		Leukamia	
	KNNI	KMI	KNNI	KMI	KNNI	KMI	KNNI	KMI
SSSim	7	20	23	17	21	17	40	23
NMRS	8	21	21	14	28	24	43	27
Cosine	5	18	23	16	29	26	41	26

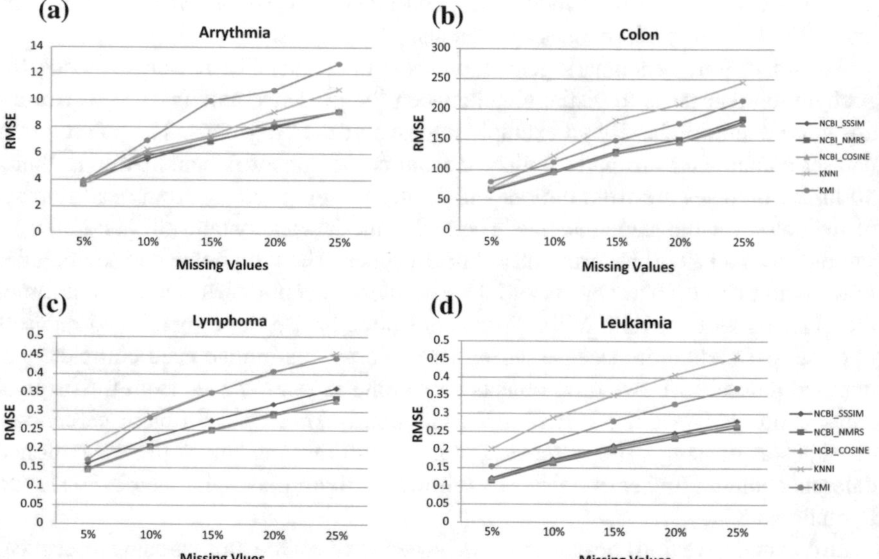

Fig. 2 Comparison of NCBI variants with KNNI and KMI in terms of RMSE values

6 Conclusion

This paper has presented an effective missing value imputation technique using a biclustering approach. The technique is capable of imputing up to 25% of missing values in real life datasets of moderate or higher dimensionality. The technique is independent of proximity measure because it has been validated through implementation of multiple proximity measures in the NCBI framework. The technique (considering all variants of NCBI) has been found to clearly outperform well-known KNNI and KMI approach in terms of RMSE.

References

1. Ahmed, H.A., Mahanta, P., Bhattacharyya, D.K., Kalita, J.K.: Shifting-and-scaling correlation based biclustering algorithm. IEEE/ACM Trans. Comput. Biol. Bioinform. (TCBB) **11**(6), 1239–1252 (2014)
2. Batista, G.E., Monard, M.C.: An analysis of four missing data treatment methods for supervised learning. Appl. Artif. Intell. **17**(5–6), 519–533 (2003)
3. Benesty, J., Chen, J., Huang, Y., et al.: Pearson correlation coefficient. In: Noise Reduction in Speech Processing, pp. 1–4. Springer (2009)
4. Bennett, D.A.: How can I deal with missing data in my study? Aust. N. Z. J. Public Health **25**(5), 464–469 (2001)
5. Bø, T.H., Dysvik, B., Jonassen, I.: LSimpute: accurate estimation of missing values in microarray data with least squares methods. Nucl. Acids Res. **32**(3), e34–e34 (2004)
6. Chowdhury, H.A., Bhattacharyya, D.K.: mRMR+: an effective feature selection algorithm for classification. In: International Conference on Pattern Recognition and Machine Intelligence, pp. 424–430. Springer (2017)
7. Li, D., Deogun, J., Spaulding, W., et al.: Towards missing data imputation: a study of fuzzy k-means clustering method. In: Rough Sets and Current Trends in Computing, pp. 573–579. Springer (2004)
8. Liew, A.W.C., Law, N.F., Yan, H.: Missing value imputation for gene expression data: computational techniques to recover missing data from available information. Brief. Bioinform. **12**(5), 498–513 (2011)
9. Little, R.J., Rubin, D.B.: Statistical Analysis with Missing Data. Wiley (2014)
10. Mahanta, P., Ahmed, H.A., Bhattacharyya, D.K., Kalita, J.K.: An effective method for network module extraction from microarray data. BMC Bioinform. **13**(13), S4 (2012)
11. Troyanskaya, O., Cantor, M., Sherlock, G., et al.: Missing value estimation methods for DNA microarrays. Bioinformatics **17**(6), 520–525 (2001)

ACO Based Optimization of DC Micro Grid Under Island Mode

B. Sateesh Babu, J. Vijaychandra, B. Sesha Sai, P. Jagannadh, Ch. Jaganmohana Rao and N. Naga Srinivas

Abstract Nature inspired algorithms are generally used in finding the solution for different optimization problems. In this paper, we have modeled and simulated the dc micro grid by using MATLAB/SIMULINK Software. A hybrid energy system has been taken i.e., Solar and wind energy systems which are connected to grid from their end. A solar MPPT technique is used to obtain maximum solar power. Optimization of solar MPPT is done by Nature inspired Ant Colony Optimization (ACO). This energy from solar is inverted to alternating power and is then connected to grid, while the power obtained from wind is connected in a beeline to the grid. The Grid is assumed to be under islanding mode.

Keywords Microgrid · ACO · Islanding mode

B. Sateesh Babu (✉) · J. Vijaychandra · B. Sesha Sai · P. Jagannadh · N. Naga Srinivas
Sri Sivani College of Engineering, Chilakapalem, Srikakulam 532402, India
e-mail: b.sateeesh@gmail.com

J. Vijaychandra
e-mail: vijaychandrajvc@gmail.com

B. Sesha Sai
e-mail: bseshasai211@gmail.com

P. Jagannadh
e-mail: jagannadh92@gmail.com

N. Naga Srinivas
e-mail: n.nagasrinivas@yahoo.com

Ch. Jaganmohana Rao
VSSUT, Burla, Sambalpur 768018, Odisha, India
e-mail: ch.jagan211@gmail.com

A. K. Das et al. (eds.), *Computational Intelligence in Pattern Recognition*,
Advances in Intelligent Systems and Computing 999,
https://doi.org/10.1007/978-981-13-9042-5_44

1 Introduction

Future of imminent generation is highly reliant on electrical energy due to its importance in the advancement of day to day technology. In this paper, it is shown that the generation of electrical power is done by solar and wind energy together with storage systems. The one cause that we go for Renewable energy is, the vast reduction of fossil fuels and the other is global warming which is mainly due to the burning of fossil fuels. The advent of the new technologies, new power devices, new topologies, and novel control approaches gives rise to the favorable outcome of the renewable energy generation technologies. The generated hybrid electrical power is connected to Micro grid and can be isolated further if needed, which presents islanding mode of operation. The Distributed generation which is located in close proximity to loads will diminish flows in transmission as well as distribution circuits with two major repercussions, those are loss reduction and caliber to the potential replacement of network resources. On top of that, the presence of generation neighboring to demand could enlarge service grade that is seen by the end consumers.

2 Modeling of a PV Cell

A PV array is formed after arranging the solar cells in series and in parallel fashion. It is done so to improve the efficiency and performance of a solar system. Series connected Solar cells impart more output voltage whereas the parallel connected solar cells provide increased current [1]. Figure 1 shows the Circuit diagram of one PV cell.

Photo-current of the module:

$$I_{ph} = [I_{scr} + k_i(T - 298)] * \lambda/1000 \quad (1)$$

Reverse saturation current of the module:

$$I_{rs} = I_{scr}/\left[\exp(qV_{oc}/nkT) - 1\right] \quad (2)$$

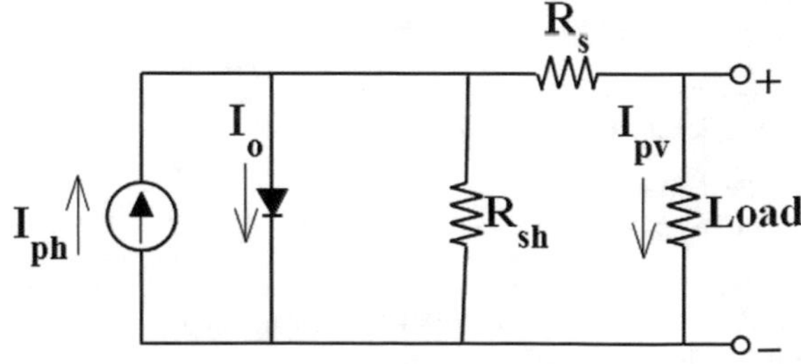

Fig. 1 Circuit diagram of one PV cell

Saturation current:

$$I_0 = Irs[T/T_r]^3 \exp\left[(qE_{g0}/BK)(1/\mathrm{T_r} - 1/\mathrm{T})\right] \tag{3}$$

The current output of PV module:

$$I_{PV} = N_p * I_{ph} - N_p * I_0\left[\exp\{(q * \mathrm{V_{pv}} + \mathrm{I_{pv}R_s})/\mathrm{N_sA_{kt}} - 1\right] \tag{4}$$

Equations (1)–(4) represent the modeling equations of PV system.

2.1 The Consequence of Different Solar Irradiation

The Solar irradiation mainly accomplishes the outcome of Voltage to Power characteristics and Voltage to Current characteristics of a solar PV cell. There is a different case of change in environment due to which distinct maximum point of power with change in the extent of solar irradiation. So an MPPT algorithm is used to sustain the maximum power to an everlasting value, even though there is a change in the solar irradiation level [2].

If the level of solar irradiation is higher, there will be an additional input to the input to the solar cell which results in supplemental magnitude of the power with the identical value of voltage. Figure 2 shows the Variation of P-V Characteristics at distinct Irradiation levels of a PV Module.

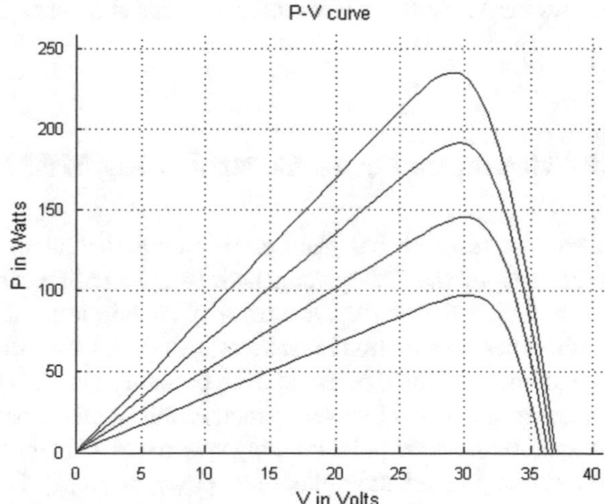

Fig. 2 Variation of P-V characteristics at distinct irradiation levels of a PV module

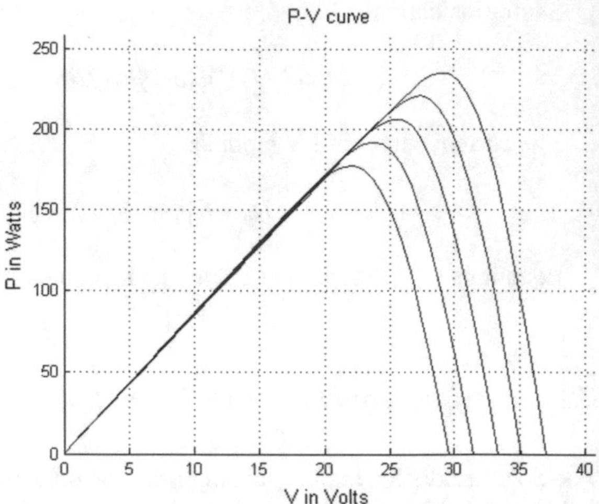

Fig. 3 Variation of P-V characteristics at distinct temperatures of a PV module

2.2 *Temperature Effect on the PV Module*

The amount of output of PV cell alters even with the dissimilitude of temperature also. The Potentiality of power generation will change with the change in temperature of a Solar cell, which is an unenviable character of a solar system. The temperature rise of a solar cell gives rise to the decrement of open circuit voltage. This causes increase in the band gap and hence an additional amount of energy is required to cross the barrier. As a result, the Overall efficiency of the solar cell decreases. Figure 3 shows the Variation of P-V Characteristics at distinct Temperatures of a PV Module.

3 Maximum Power Control Using MPPT

Plenty of research has been carried outearlier to improve the power quality and efficiency of the PV system. Due to the nonlinear and time-varying P-V and I-V characteristics of PV systems, they contain less energy transformation efficiency with respect to variation in temperature and solar irradiance. As a solar panel absorbs maximum amount of power at its maximum point, it is necessary to identify the point of maximum power for every panel. At this point the panel provides higher efficiency. A maximum power point tracking system (MPPT) is conventionally used in the solar PV system to track their maximum power point [3]. The power generation efficiency of the PV system is ameliorated with the help of MPPT system. Hence, MPPT is pondered as the most constitutive element in a solar PV system. There are several approaches to track the point of maximum power. Few of them are listed below:

(1) Hill climbing method (Perturb and Observe)
(2) Fractional short circuit current
(3) Incremental Conductance method
(4) Fuzzy logic
(5) Fractional open circuit voltage
(6) Neural networks.

The options of the algorithm rely on the time entanglement the algorithm takes to track the Maximum power point, cost of execution, and the ease of accomplishment.

3.1 Incremental Conductance

Incremental conductance mechanism employs two voltage and current sensors to identify the achieved voltages and currents of the solar PV array.

At MPP the slope of the PV curve is 0.

$$(dP/dV)MPP = d(VI)/\mathrm{d}V \tag{5}$$

$$0 = I + V dI/dV MPP \tag{6}$$

$$dI/dV MPP = -I/V \tag{7}$$

Equations (5)–(7), represents the modeling equations used in incremental conductance mechanism.

3.2 Proposed MPPT Technique

This paper presents a PI controller which is going to be tuned with an ACO algorithm technique. In order to track the maximum power from the PV system, the ACO based PI-MPPT controller was enhanced by one of the MPPT techniques available i.e., Fractional O.C. method. To drive the optimized PI controller, this technique amalgamates or incorporates with the Incremental conductance method with the Fractional OC method. This entire technique was based on the ant colony optimization algorithm [4]. Figure 4 shows the I-V and P-V characteristics of the studied PV module.

The Error signal is denoted by **e**, which is defined as the ratio of change in the power to the change in the voltage i.e., dp/dv. This error should always be zero in order to get the maximum power.

$$\text{Error,} \quad e = \frac{P(k) - P(k-1)}{V(k) - V(k-1)} \tag{8}$$

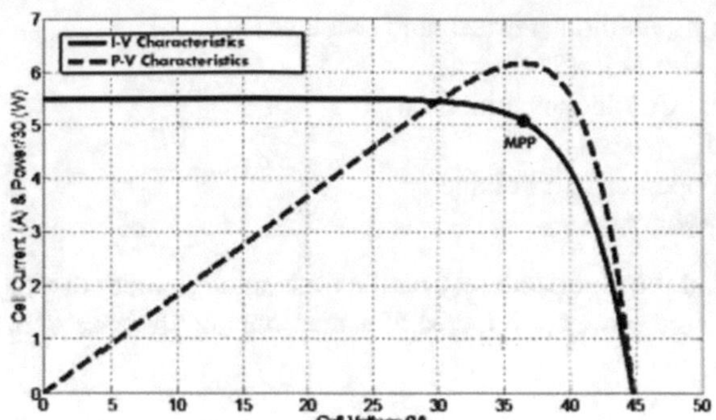

Fig. 4 I-V and P-V characteristics of the studied PV module

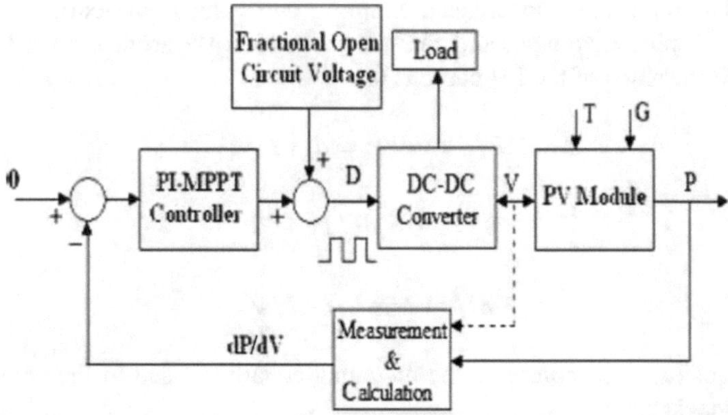

Fig. 5 PV MPPT system

Equation (8) represents the error signal which is zero is considered to be taken as the feedback variable and is supposed to be compared with the signal of zero reference and finally the error is passed to the aco PI controller. The output from this controller is fed to the duty cycle of the DC-DC converter. Figure 5 represents PV MPPT system.

4 Modeling of Wind Turbine

The wind turbine [5, 6] is modeled based on the following governing equations,

$$P = \frac{1}{2}C_p(\beta, \lambda)\rho\pi \mathrm{R}^2\mathrm{V}_\mathrm{w} \tag{9}$$

$$Cp = (0.44 - 0.0167\beta)\sin\frac{\pi(\lambda - 2)}{13 - 0.3\beta} - 0.00184(\lambda - 2)\beta \tag{10}$$

The aerodynamic torque (T_m)

$$T_m = \frac{C_p(\beta, \lambda)\rho\pi R_5}{2\lambda_3}\omega_m^2 \tag{11}$$

Equations 9–11 represent the modeling equations of wind turbine.

5 Simulation Results

Figures 6, 7, 8, and 9 represent Voltage of PV system, ACO MPPT of PV, Power of Hybrid system in KW, and P and Q of Hybrid systems respectively.

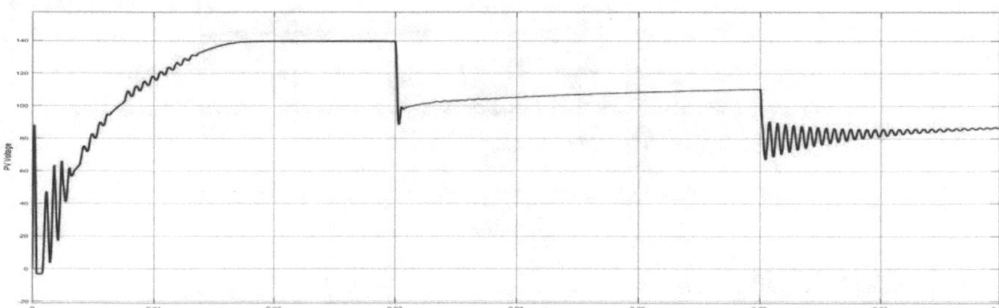

Fig. 6 Voltage of PV system

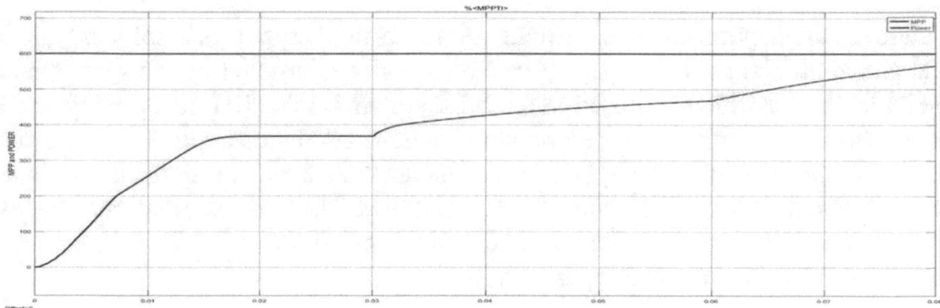

Fig. 7 ACO MPPT of PV

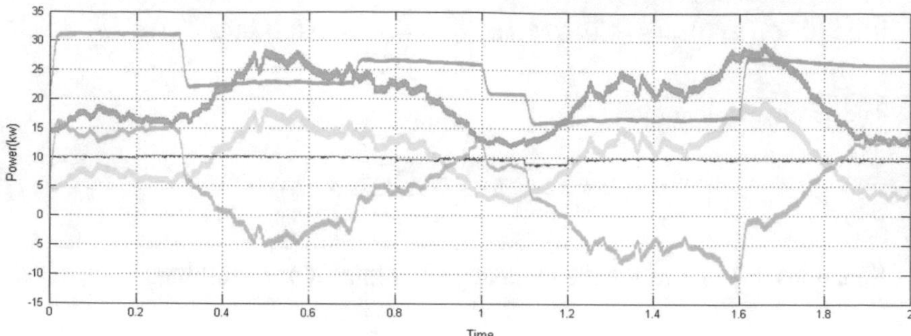

Fig. 8 Power of hybrid system in KW

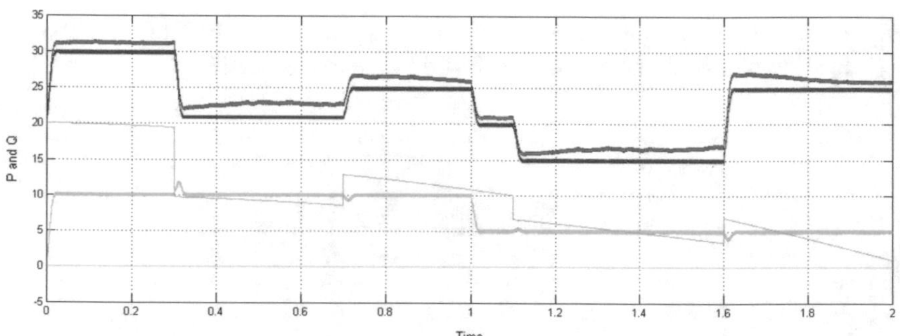

Fig. 9 P and Q of hybrid system

6 Conclusion

Nature inspired optimization algorithms are widely used nowadays in solving typical and global problems. In this paper, we have taken a solar-wind hybrid system and the PI controller which was used in the MPP system was optimized by using ACO technique. The proposed controller had given good results under drastic variations in the Irradiance levels. The hybrid energy model considered are simulated and the transient conditions under islanding mode are studied. The frequency and the voltages of the system are almost identical to the normal values which we considered at various loads under Islanded mode of operation.

References

1. Hansen, A.D., Sørensen, P., Hansen, L.H., Bindner, H.: Models for a Stand-Alone PV System. Risø National Laboratory, Roskidle, Technical Report Risø-R-1219 (EN)/SEC-R-12, December 2000
2. Prapanavarat, C., Barnes, M., Jenkins, N.: Investigation of the performance of a photovoltaic ac module. IEE Proc.-Gener. Transm. Distrib. **149**, 472–478 (2002)
3. Koizumi, H., Kurokawa, K.: A novel maximum power point tracking method for PV module integrated converter. In: 36th Annual IEEE Power Electron. SpeCialists Con!, pp. 2081–2086 (2005)
4. Hussein, K.H., Muta, I., Hoshino, T., Osakada, M.: Maximum photovoltaic power tracking: an algorithm for rapidly changing atmospheric condition. Proc. Inst. Electr. Eng. Gen. Transmiss. Distrib. **142**(1), 59–64 (1995)
5. Heier, S.: Grid Integration of Wind Energy Conversion Systems. Wiley, West Sussex (1998)
6. Manwell, J.F., McGowan, J.G., Rogers, A.L.: Wind Energy Explained Theory, Design and Application. Wiley, England (2002)

Non-invasively Grading of Brain Tumor Through Noise Robust Textural and Intensity Based Features

Prasun Chandra Tripathi and Soumen Bag

Abstract Identifying the tumor grade has an important role in surgery planning. A medical practitioner assesses the tumor grade through a biopsy. However, a biopsy usually leads to discomfort and post-biopsy complications. In this work, we propose a noninvasive method to detect the tumor grade using Magnetic Resonance Imaging (MRI). The grading of brain tumor is done through four stages: Skull stripping, brain MRI segmentation, feature extraction, and classification. We have used spatial fuzzy C-means for the segmentation of brain. For textural features extraction, we have used noise robust Local Frequency Descriptor (LFD). The classification is done using random forest, Support Vector Machine (SVM), and decision tree classifiers. The experimental results on a real brain MRI dataset confirm the effectiveness of the proposed method.

Keywords Brain MRI · Feature extraction · Segmentation · Tumor grading

1 Introduction

Excessive growth of cells in brain develops brain tumor. A brain tumor typically has high mortality rate [1]. It is essential to detect and characterize tumor in time for the effective treatment. The malignancy of a brain tumor is examined through brain tumor grading. However, all types of tumors are not malignant but effective treatment is essential in all cases. Once the tumor is detected, a physician usually suggests for a biopsy. A biopsy is an invasive method where a probe is inserted into brain to extract very small sample of tumor. The sample is tested on microscope to determine the tumor's grade. Even though, minimal invasive biopsy methods are

P. C. Tripathi (✉) · S. Bag
Department of Computer Science and Engineering, Indian Institute of Technology (Indian School of Mines) Dhanbad, Dhanbad 826004, Jharkhand, India
e-mail: prasunchandratripathi@gmail.com

S. Bag
e-mail: bagsoumen@gmail.com

A. K. Das et al. (eds.), *Computational Intelligence in Pattern Recognition*, Advances in Intelligent Systems and Computing 999,
https://doi.org/10.1007/978-981-13-9042-5_45

available, but still it may have certain complications. Imaging examinations have proven to be effective in the treatment of brain abnormalities. In particular, MRI is a very effective paradigm to treat brain tumors. It captures the images of organs with significantly distinguishable gray levels.

In literature, the problem of characterizing brain tumors has been solved using several image processing and pattern recognition methods. In [2], authors have proposed a system to classify the brain tumor into two types: Gliomas and Meningioma. Probabilistic neural network is trained using textural features to classify brain tumors. In [3], an interactive CAD (Computer-Aided Diagnosis) system is proposed for brain tumor treatment. Texture and histogram features have been used to train the model in this work. The CAD system can classify six types of brain tumors. In [4], an expert system has been developed to diagnosis astrocytoma tumors. Researchers have developed fuzzy type-II approximate reasoning for tumor diagnosis. In [5], a tumor diagnosis system has been proposed where the brain tumor is classified in two classes: low grade and high grade. Researchers have used rough set firefly based feature selection in this work. In [6], the authors have developed brain tumor classification system using bag-of-words (BoW) model. We have observed that the feature set, used in the above described methods, is not robust to noise. However, the brain MRI typically has inherently noise artifacts due to imperfection in magnetic coils or patient movement. As a result, these methods suffer in discrimination performance. Thus, we have used noise robust and rotation invariant texture feature LFD [7] in the proposed method. The proposed method can grade three types of brain tumors: meningioma, glioma, and pituitary tumor. The spatial fuzzy C-means clustering (sFCM) [8] is used for the segmentation of brain MRI. Furthermore, the combination of intensity and texture features is used to enhance the classification performance.

The remaining write-up has been classified as follows: Sect. 2 describes the methodology of the work, experimental results are presented in Sect. 3, and the concluding remarks have been pointed out in Sect. 4.

2 Methodology

2.1 Skull Stripping

A brain MRI scan also includes non-cerebral tissues such as skull, dura, meninges, and scalp. These tissues typically deteriorate the segmentation performance. Thus, the non-cerebral tissues in a brain MRI scan are eliminated before any further processing [9]. In this work, we have used intensity thresholding and morphological operations to extract the brain region from a brain MRI. The steps are:

1. Input image (Fig. 1a) is binarized with a global threshold T as shown in Fig. 1b. The threshold T is taken in the range [0.12, 0.23]. It has to be noted that the gray levels of input image are in the range [0, 1].

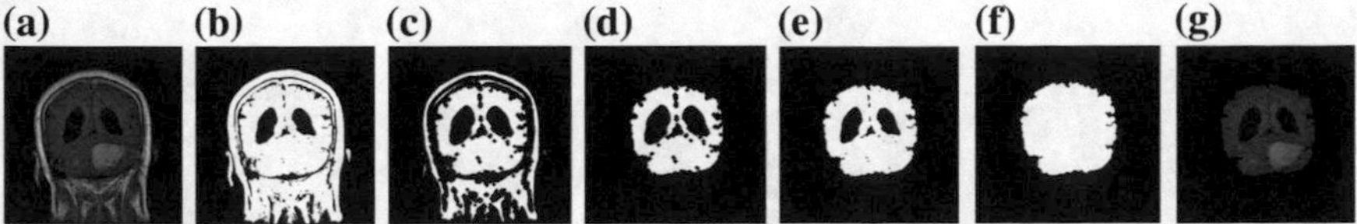

Fig. 1 Skull stripping of a brain MRI

2. Morphological erosion is applied on the binary image to separate brain with skull part (Fig. 1c) and largest region is selected as shown in Fig. 1d.
3. Morphological dilation is applied on output of Step 2 as shown in Fig. 1e and region filling algorithm is executed to obtain brain mask (Fig. 1f). The brain region is extracted using the mask as shown in Fig. 1g.

2.2 *Brain MRI Segmentation*

The segmentation of brain MRI is to partition the brain image into prominent tissues and abnormality (if any). The important regions of brain MRI include Gray Matter (GM), White Matter (WM), and Cerebrospinal fluid (CSF). The segmentation of brain MRI is not trivial task due to presence of intensity inhomogeneity artifact and complex anatomy of the human brain. We have used spatial fuzzy C-means (sFCM) [8] for the segmentation of brain MRI. This algorithm includes the local information in the clustering process. Due to that, it provides robustness in the segmentation in presence of intensity inhomogeneity artifact.

Suppose $Y = \{y_1, y_2, ..., y_i, ..., y_N\}$ is a brain MRI of size N, y_i denotes gray level of ith pixel, the sFCM algorithm decomposes the image Y into C tissue classes $(1 < C < N)$ using the following iterative equations (Eqs. 1 and 2):

$$\mu_{ik} = \frac{1}{\sum_{j=1}^{C} \left(\frac{||y_i - c_k||^2 + \frac{\alpha}{N_R} \sum_{r \in N_i} ||y_r - c_k||^2}{||y_i - c_j||^2 + \frac{\alpha}{N_R} \sum_{r \in N_i} ||y_r - c_j||^2} \right)^{\frac{1}{m-1}}} \tag{1}$$

$$c_k = \frac{\sum_{i=1}^{N} \mu_{ik}^m \left(y_i + \frac{\alpha}{N_R} \sum_{r \in N_i} y_r \right)}{(1 + \alpha) \sum_{i=1}^{N} \mu_{ik}^m} \tag{2}$$

where, c_k represents cluster prototype of kth cluster, μ_{ik} represents fuzzy membership in kth cluster of pixel i ($\mu_{ik} \in [0, 1]$), m is fuzzy weighting exponent ($m > 1$), N_i is a set of neighbors of pixel i, α is trade-off parameter, N_R is the size neighborhood,

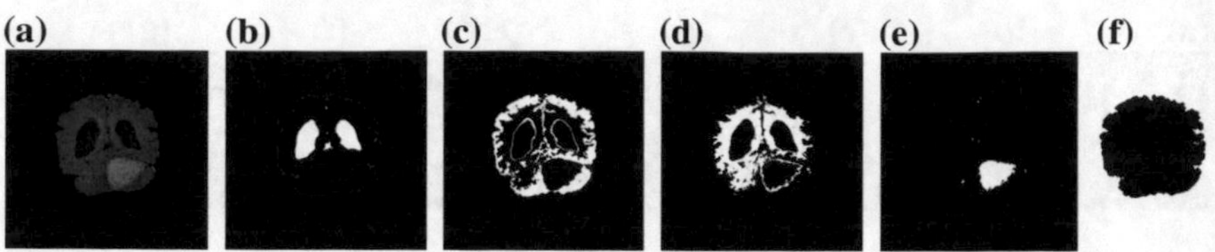

Fig. 2 Segmentation of a brain MRI: **a** input image, **b** CSF, **c** GM, **d** WM, **e** tumor, and **f** background

and ||•|| denotes L2-norm. In this work, we have taken number of tissues classes $C = 5$ corresponding to CSF, GM, WM, tumor, and background as shown in Fig. 2.

2.3 *Feature Extraction*

The extraction of suitable feature set is critical in computer-aided diagnosis of brain tumor using MRI. As it can improve or deteriorate the classification performance. In this work, we have used combination of intensity and textural features for the classification of brain tumors.

2.3.1 Intensity Features

The intensity-based features work on gray level of pixels individually. It captures the image characteristics based on the statistical properties. In this work, the following five intensity features are used: (a) mean, (b) standard deviation, (c) skewness, (d) kurtosis, and (e) entropy.

2.3.2 Texture Features

Local Binary Pattern (LBP) [10] encodes the neighboring pixels based on the differences between gray level of central pixel and neighbors in a local window. The encoding is done using a binary value. The histogram of the binary patterns is used for texture description. However, LBP and its variants [11, 12] face two problems: non resistant to noise and drastically increment of patterns with respect to number of samples. The brain MRI has noise artifact due to the inability of feature set to provide enough classification performance. This reason motivated us to use local frequency descriptor (LFD) [7] which is resistant to noise. LFD uses local frequencies despite of original gray value in the image. Suppose, N_s is the total number of neighbors, denoted by $t_u (u = 0, 1, ..., N_s - 1)$, around a center pixel (p, q) with radius ρ. Then, the coordinates of t_u can be written as $(p - \rho . \sin 2\pi u/N_s)$ and

$(q + \rho.\cos 2\pi u/N_s)$. The local frequency f_n is obtained by applying 1D DFT using Eq. 3:

$$f_n = \sum_{u=0}^{N_s-1} t_u e^{\frac{-i2\pi nu}{N_s}} \tag{3}$$

This gives $(\frac{N_s}{2} + 1)$ unique frequency channels. It is observed that low frequency channels contain most of the textural information, whereas high frequency channels contain mostly noise. The local frequency terms contain local texture characteristics only. So, to correlate local texture properties, we calculate 2D DFT CH_n of each frequency channel as follows in Eq. 4:

$$CH_n(w, x) = \sum_{p=0}^{P-1} \sum_{q=0}^{Q-1} |f_n(p, q)|.e^{-2\pi i(\frac{pw}{P} + \frac{qx}{Q})} \tag{4}$$

where, $|f_n(p, q)|$ is the magnitude of nth local frequency term at (p, q). The final feature vector is obtained by applying circular and directional filters on each frequency channel $CH_n(w, x)$.

2.4 Tumor Classification

The proposed method works for three class classification of brain tumors. The supervised approach of classification is opted in this work. We have used following three classifiers in this method for the classification purpose: (a) Decision Tree, (b) SVM, and (c) Random Forest.

3 Experimental Results

The experimental results indicate the performance of the proposed method in identifying different tumor grades. The method can classify the brain tumor into three types: meningioma, glioma, and pituitary tumor.

3.1 Evaluation Framework

We have evaluated the performance of the proposed method in classifying the tumor grade on a real brain MRI dataset. The performance of proposed system is evaluated using the tumor classification accuracy ξ which is given using Eq. 5.

$$\xi = \frac{TP + TN}{TP + TN + FP + FN} \tag{5}$$

where, TP is true positive rate, TN is true negative rate, FP is false positive rate, and FN is false negative rate. All implementation has been done on the Matlab R2016a with system configuration intel core $i5$, 2.20 GHz, 8 GB RAM, and 2 GB AMD graphics processing unit (GPU) running Windows 10 OS.

In the experimental setup for tumor segmentation, the neighborhood window is taken as a square window of size 3×3. We set the fuzzy weighting exponent $m = 2$, number of cluster $C = 5$, and parameter $\alpha = 1$. For textural features LBP and LFD, we have taken number of samples $N_s = 8$ with radius $\rho = 1$.

3.2 Real Brain Tumor Dataset

The brain tumor dataset has been acquired from the work [6]. The data set consists of three types of lesions: meningioma, glioma, and pituitary tumor of 233 patients. The images are in T1-weighted contrast-enhanced MRI. Each image is in the resolution of 512×512 with slice thickness of 6 mm.

3.3 Result Analysis

The results of first two stage are illustrated in Fig. 3. The different regions in the segmented image are shown using five colors: Red color shows the tumor region, Green color shows the gray matter, Cyan color shows the white matter, Blue color shows the CSF, and Black color shows the background.

We have extracted 59 features using LBP and 46 features using LFD. In case of LFD, we have extracted features from first two frequency channels only. In this work, three classifiers have been used they are decision tree, SVM, and random forest. We have used k-fold cross validation in the classification process. In this method, the data samples are partitioned into k parts. $k - 1$ parts are used for training the classifier and one part is used for testing. The mean accuracy of the k passes gives the overall performance of the model. $k = 5$ has been used in this work.

We have performed the comparative analysis of features into two phases. In first phase, we have compared the performance of LFD and LBP using three classifiers. Table 1 shows the comparison between LBP and LFD using three classifiers. It is noticed that LFD performs better for all classifiers. This gain in performance is achieved due to its noise handling ability. In second phase, we have added five intensity features to the LFD set and compared the performance of this set with LFD set. We have achieved significant gain in accuracy using this feature set. The maximum accuracy is achieved using Random Forest classifier as shown in Table 2.

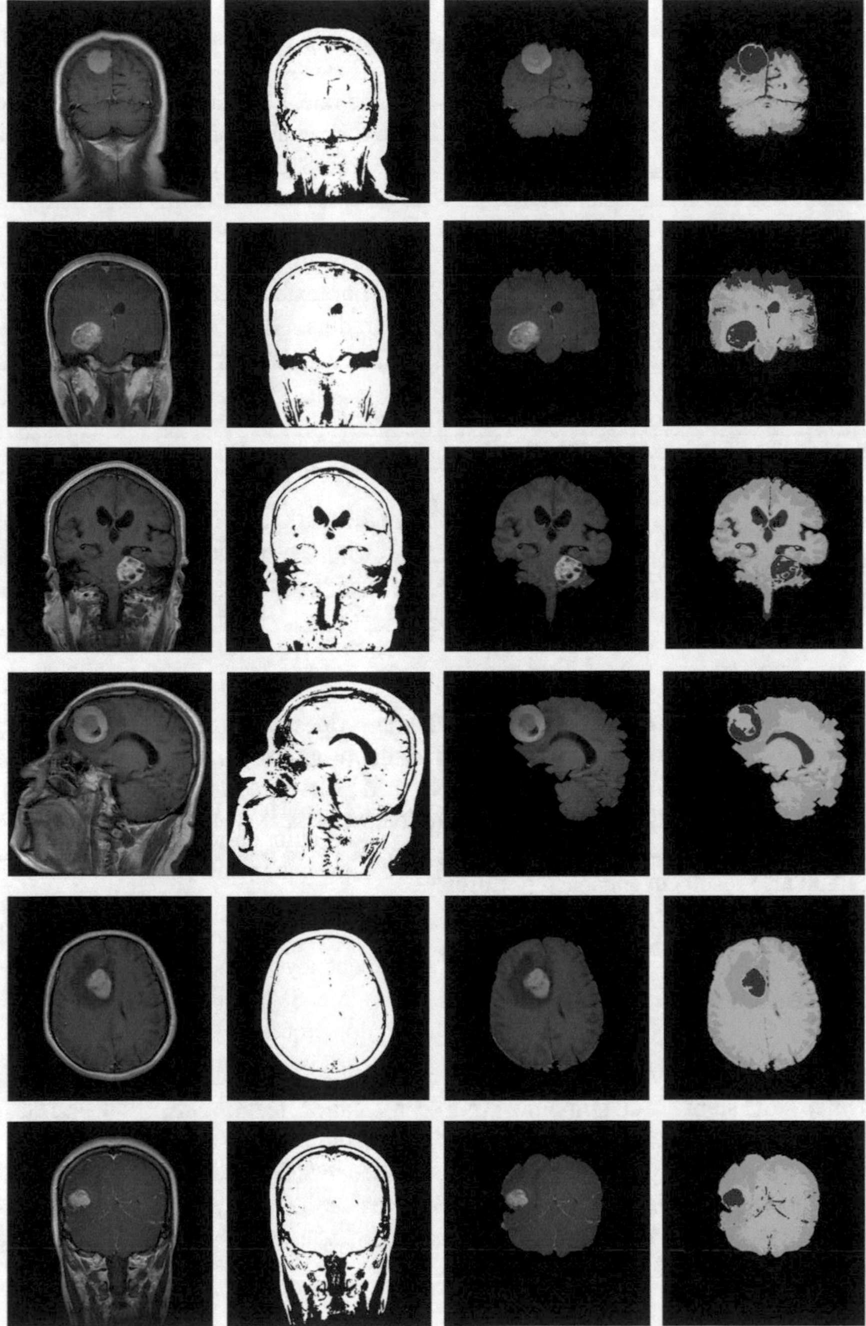

Fig. 3 Segmentation results of brain MRI for dataset (Cheng et al. [6]). First column: input brain scans, second column: binarized images. Third column: extracted brain regions after morphological operations on second column images. Fourth column: segmentation of brain (colors are used for distinguishing the regions)

Table 1 Accuracy using LBP and LFD features

Classifier	Accuracy (%) using	
	LBP	LFD
Decision tree	57.52	64.57
SVM	68.85	71.06
Random forest	69.03	73.83

Table 2 Accuracy using LFD and LFD+intensity features

Classifier	Accuracy (%) using	
	LFD	LFD+intensity
Decision tree	64.57	91.86
SVM	71.06	94.63
Random forest	73.83	94.64

Table 3 Comparison of the proposed method with an existing method

Method	Accuracy (%)
Cheng et al. [6]	91.14
Proposed method	94.64

We have compared the performance of the proposed method with an existing method of Cheng et al. [6]. It is shown that the proposed method performs better than Cheng's method as reported in Table 3. It is happened because the existing method does not use noise resistant texture features whereas our method uses noise robust texture feature. Moreover, the integration of two types of features in proposed method also results in enhanced performance.

4 Conclusion

In this paper, a noninvasive method for brain tumor classification is proposed. It involves four steps: Skull stripping, brain segmentation, feature extraction, and classification. The tumor segmentation is performed using spatial Fuzzy C-means clustering. Fifty one intensity and texture features are used for classification. It is noticed that noise robust texture features (LFD) enhance the classification performance. Experiments carried out on a tumor dataset provide effectiveness of the method comparing with an existing method.

References

1. Brain tumor research. https://www.braintumourresearch.org/. Accessed 06 Aug 2018
2. Georgiadis, P., Cavourous, D., Kalatzis, I., Daskalakis, A., Kagadis, G.C., Sifaki, K., Malamas, M., Nikiforidis, G., Solomou, E.: Improving brain tumor characterization on MRI by probabilistic neural networks and non-linear transformation of textural features. Comput. Methods Programs Biomed. **89**(1), 24–32 (2008)
3. Sachdeva, J., Kumar, V., Gupta, I., Khandelwal, N., Ahuja, C.K.: Segmentation, feature extraction, and multiclass brain tumor classification. J. Digit. Imaging **26**(6), 1141–1150 (2013)
4. Zarandi, M.F., Zarinbal, M., Izadi, M.: Systematic image processing for diagnosing brain tumors: a Type-II fuzzy expert system approach. Appl. Soft Comput. **11**(1), 285–294 (2011)
5. Jyothi, G., Inbrani, H.: Hybrid tolerance rough set-firefly based supervised feature selection for MRI brain tumor image classification. Appl. Soft Comput. **46**, 639–651 (2016)
6. Chung, J., Huang, W., Cao, S., Yang, R., Yang, W., Yun, Z., Wang, Z., Feng, Q.: Enhanced performance of brain tumor classification via tumor region augmentation and partition. PloS ONE **10**(10), e0140381 (2015)
7. Mani, R., Kalra, S., Yang, Y.H.: Noise robust rotation invariant features for texture classification. Pattern Recognit. **46**(8), 2103–2116 (2013)
8. Ahmed, M.N., Yamany, S.M., Mohmed, N., Farag, A.A., Moriarty, T.: A modified fuzzy c-means algorithm for bias field estimation and segmentation of MRI data. IEEE Trans. Med. Imaging **21**(3), 193–199 (2002)
9. Laha, M., Tripathi, P.C., Bag, S.: A skull stripping from brain MRI using adaptive iterative thresholding and mathematical morphology. In: Proceedings of the International Conference on Recent Advances in Information Technology, pp. 1–6 (2018)
10. Ojala, T., Pietikainen, M., Maenpaa, T.: Multiresolution gray-scale and rotation invariant texture classification with local binary patterns. IEEE Trans. Pattern Anal. Mach. Intell. **24**(7), 971–987 (2002)
11. Huang, X., Li, S.Z., Wang, Y.: Shape localization based on statistical method using extended local binary pattern. In: Proceedings of the International Conference on Image and Graphics, pp. 184–187 (2004)
12. Liao, S., Law, M.W.K., Chung, A.C.S.: Dominant local binary patterns for texture classification. IEEE Trans. Image Process. **18**(5), 1107–1118 (2009)

Line Encoding Based Method to Analyze Performance of Indian States in Road Safety

Samya Muhuri, Susanta Chakraborty and Debasree Das

Abstract Road accidents are becoming a global crisis which does not only cost human life but also affect the economic condition of a nation. Therefore it is gaining the interest of several researchers to analyze the influential factors behind road accidents and suggest the preventive measures. A large country like India is also suffering from a substantial amount of road accidents and try to incorporate improve road safety measures. In this paper, we evaluate the performances of different Indian states in road safety measures based on the recorded road accident data. The current analysis is based on a novel line encoding method which categorizes Indian states into three categories, i.e., best performer, average performer, and stragglers. Results have efficiently identified the states that reduce road accidents and the potential states which need to improve on road safety to a great extent. The outcome after ranking the states based on our approach shows there is a high linear correlation between accident rate with population and number of vehicles whereas accident and literacy rate shares an inverse correlation. The satisfactory results based on our approach definitely enlight the interdisciplinary research domains.

Keywords Line encoding · Pearson correlation · Road safety

1 Introduction

The vehicle density has been swiftly growing over few decades globally [1]. India is not an exception and tries to build its overall performance in different aspects worldwide by expanding road network and transportation system. The highly connected

S. Muhuri (✉) · S. Chakraborty · D. Das
Indian Institute of Engineering Science and Technology, Shibpur, Howrah 711103, India
e-mail: samyamuhuri.rs2015@cs.iiests.ac.in

S. Chakraborty
e-mail: sc@cs.iiests.ac.in

D. Das
e-mail: debasreedas1994@gmail.com

A. K. Das et al. (eds.), *Computational Intelligence in Pattern Recognition*, Advances in Intelligent Systems and Computing 999,
https://doi.org/10.1007/978-981-13-9042-5_46

road and highway network in India is the backbone of the continuous development of Indian social economics. Besides the overall growth progresses, road accidents in India are also accelerating. The frequent severe road accidents not only result in life threats but also damage the infrastructure and aggregates massive economic loss. According to the Ministry of Road Transport and Highways [2], the number of accidental deaths have increased by 31% from 2007 to 2017 and take a lot of attention. Indian Government and different state authorities already implemented different safety measures and strict traffic rules.

Road safety is one of the primary parameters for the growth of a country combining other factors such as population, literacy, etc. To address the problem of insufficient road safety, it requires the dedicated action of several ministries, most notably law, planning, transport, education, public information, and health. Road safety assessment in periodical manner has played a major role in the theory and practice of transport management systems. The states are maintaining lots of statistical data which can be used to extract the progress of individual states in road safety awareness. It can be noted that every state has different population density, road infrastructure, and geographical location and number of road accidents in different states are not comparable. But the alarming situation demands an alternative data mining tool to measure the performance of the states in road safety aspects and relate them to influential parameters like population, number of vehicles, and literacy rate. The correlation among these attributes with road accidents may reveal the actual performance of the different Indian states in reducing traffic mishaps. In this paper, we develop a novel line encoding method to demonstrate the progress of any entity from time-varying data. The method is very easy to implement and disclose the proper pattern of positive or negative growth rate as the time progresses. In the current analysis, we categorize Indian states into three categories, i.e., best performer, average performer, and stragglers. Our method is unique and the results are satisfactory which can open up a new framework for performance analysis.

The remaining paper is structured as follows. We elaborate some previous works in Sect. 2. In Sect. 3, our approaches for performance analysis are proposed. We illustrate our method with an example in Sect. 4. Experimental analysis and results are shown in Sect. 5. Then in Sect. 6, we conclude.

2 Literature Review

Different computational work are already employed to analyze road network [3–5] and efficient transportation system [6–8]. Assuming the road network as a graph with road junction as vertices and highways between them as edges, different topological properties are analyzed in [9]. Topological properties are studied and found road network possesses the properties of small world assortative network. Based on betweenness centrality, most important junction points are identified. Most important junction points are concluded as potential congestion points. Wenxue et al. [10] proposed an improved method to analyze hazardous material road transportation accident rate. To analyze and divide regions of a state based on the pattern of accidents, an agglomerative hierarchical clustering algorithm is introduced [11]. To evaluate best distance measure, the Cophenetic correlation coefficient is used for clustering. The approach is applied to analyze hourly road accident data of twenty-six districts of Gujarat. Tracking vehicles and monitoring driving habits is proposed in [12] to reduce the number of road accidents. Telematics system is used to track each vehicles position and report the errors for further analysis. A modified method [13] of principal component analysis is proposed to analyze the relationship between traffic accident evaluating index and the different causes of traffic accident which includes number of people and vehicles, type of road, and corresponding environment. Data mining tools are used to predict the probability of accidents on specific roads like State highways (SHs) or Ordinary district roads (ODRs) by estimating the severity of accidents [14]. There are no such works which have been proposed for performance analysis of the states from the recorded road accident statistics. Our proposed approach is an attempt to examine past accidental data to ensure the progress of different Indian states based on the road safety measures.

3 Our Approach for Performance Analysis

In this section, we propose a novel line encoding method to compare progress of the entities ($e_1, e_2, \ldots, e_n \in E$) from time-varying data. Progress means the performance variation of different entities with respect to time. An index like progressive score will serve as the comparison base among the entities and signifies how much progress each entity has made throughout a given time range. For a entity ($e_i \in E$), a set of random variables $V\{v_1, v_2, v_3, \ldots, v_n\}$ at different time instant $T\{t_1, t_2, t_3, \ldots, t_n\}$ is given. X-axis denotes time and Y-axis represents the real-valued attributes of different entities. From previous time instant (t_n) to successive time instant (t_{n+1}), value of a entity e_i can increase, decrease, or remain moreover same. The change is denoted as (Δv). $\Delta v = v_{n+1} - v_n$. This change can be interpreted as a ray I_α where the slope of the ray is α. The line can be oriented at any direction out of n positions. Here n is in the range from positive Y-axis to negative Y-axis in clock-wise direction. The aggregate value of all the I_α, i.e., $(I_{\alpha 1} + I_{\alpha 2} + I_{\alpha 3} + \cdots + I_{\alpha n-1})$ generate the summative score for a single entity e_i.

Encoded values for each $e_i \in E$ is calculated using the following formula:

$$\text{Progressive Score } (e_i) = \sum_{1}^{n-1} I_\alpha \tag{1}$$

$$I_\alpha = \begin{cases} 0 \text{ if } v_n \simeq v_{n+1} \\ (\frac{\Delta v}{|v_{max}|}) \text{ otherwise} \end{cases} \tag{2}$$

where $|v_{max}| = max\{v_1, v_2, v_3, \ldots, v_n\}$.

4 Illustration with Example

Figure 1, illustrates the performance of four entities denoted as L_1, L_2, L_3 and L_4. Let, up to the time instant t_1, all of there progressive score is 0. At time instant t_2, progress of each entity is calculated. $I_{L_1} = \frac{30-0}{30} = 1$, $I_{L_2} = \frac{10-0}{30} = 0.33$, $I_{L_3} = \frac{-20-0}{30} = -0.67$ and $I_{L_4} = \frac{-30-0}{30} = -1$. In our experiment, the statistical performance data in lap of a year is assumed as a time-varying set of values. Each state are defined as different entities. Progressive score is translated differently for individual cases. For road accident rate and literacy, the states with low (L_4) and high (L_1) progressive score respectively are denoted as the best performer.

Let us take a set of 5 variables as illustrated in Table 1. Each variable is having a set of values at different time instant. The difference of values between two consecutive time instant is shown Table 2. Now using the Eq. 1 we get the results in Table 3. Here, we get $V1$ and $V2$ as the lowest and highest valued entities respectively. To prevent the accident, $V1$ performs better than other entities.

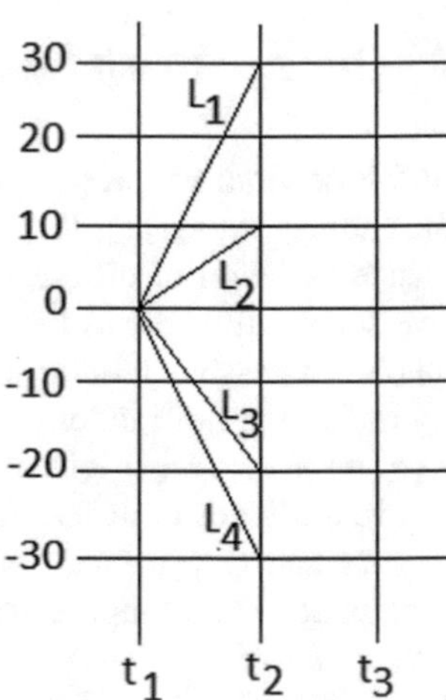

Fig. 1 Illustration of progressive score for different entities

Table 1 Example dataset of five variables

Variable name	T1	T2	T3	T4	T5
V1	5	10	2	0	−2
V2	4	4	2	10	5
V3	5	6	1	1	0
V4	6	5	4	5	5
V5	5	6	10	6	2

Table 2 Difference between two consecutive time instant

V1	5	−8	−2	−2
V2	0	−2	8	−5
V3	1	−5	0	−1
V4	−1	−1	1	0
V5	1	4	−4	−4

Table 3 Progressive score calculation

V1	0.5	−0.8	−0.2	−0.2	$\sum = -0.7$
V2	0	−0.2	0.8	−0.5	$\sum = 0.1$
V3	0.1	−0.5	0	−0.1	$\sum = -0.5$
V4	−0.1	−0.1	0.1	0	$\sum = -0.1$
V5	0.1	0.4	−0.4	−0.4	$\sum = -0.3$

5 Result and Analysis

In this section, we analyze the actual data based on the proposed method. We also validate the result over the benchmark dataset.

5.1 Data Collection

We have collected different statistical data maintained by the Government of India to compare the performance of different states and union territories (UT). The data includes road accident data from 2001 to 2014, literacy rate from 1951 to 2011, total population data from 1951 to 2011, total registered motor vehicles from 2001 to 2012 which are gathered from NCRB (National Crime Records Bureau) [15, 16], Ministry of Home Affairs Govt. of India [17], Census of India [18], Ministry of Highways and Road Transports, Govt. of India [2] respectively to assess the performance of our proposed approach.

5.2 Result over Different Attributes

Figure 2 depicts the results obtained from road accident data based on our proposed method. It shows West Bengal is the most progressive state in preventing road accident throughout 2001 to 2014 whereas Kerala performed the worst. Here, steep slope signifies the higher increase in accident which is costly for progress whereas gradual angle signifies lesser increase in number of accidents. If the angle is negative then it

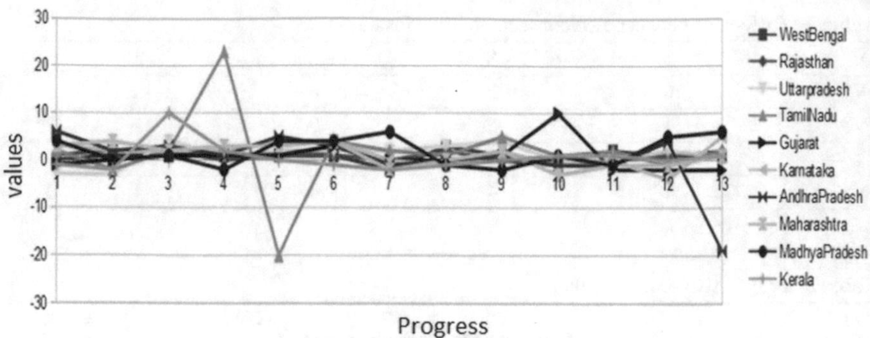

Fig. 2 Progress of the states based on road accident data

Table 4 Classification of the States and Union Territories

	Best performer	Average performer	Stragglers
States	Andhra Pradesh	Goa	
	Assam	Madhya Pradesh	Arunachal Pradesh
	Bihar	Maharashtra	Jammu & Kashmir
	Chhattisgarh	Odisha	Jharkhand
	Gujarat	Punjab	Kerala
	Haryana	Rajasthan	Manipur
	Himachal Pradesh	Sikkim	Meghalaya
	Karnataka	Tamil Nadu	Mizoram
	Uttarakhand	Tripura	Nagaland
	West Bengal	Uttar Pradesh	
Union Territories	Delhi	Chandigarh	
	Lakshadweep	Daman & Diu	A & N Islands
	Puducherry	D & N Haveli	

signifies the reduction in accident which is economic whereas a horizontal straight line signifies no change between two successive points and thus, non-increasing. Tamil Nadu, Rajasthan, Uttar Pradesh have more gradual angles than steeper angles turning out to be average progress.

Classification of the states and Union Territories is shown in Table 4 based on their performance in alphabetical order. Most of the north-east states are classified as straggler which is a real concern. The geographical condition and environmental hazards might be the reason for their unsatisfactory performance. Some newly born states like Chhattisgarh, Uttarakhand are grouped as the best performer which is very much encouraging.

Next, we examine the literacy progress of different states. Figure 3 depicts Delhi, Kerala and Andaman & Nicobar Islands have done the best progress in literacy. Here, steeper angle signifies an increase in literacy which turns out to be good progress,

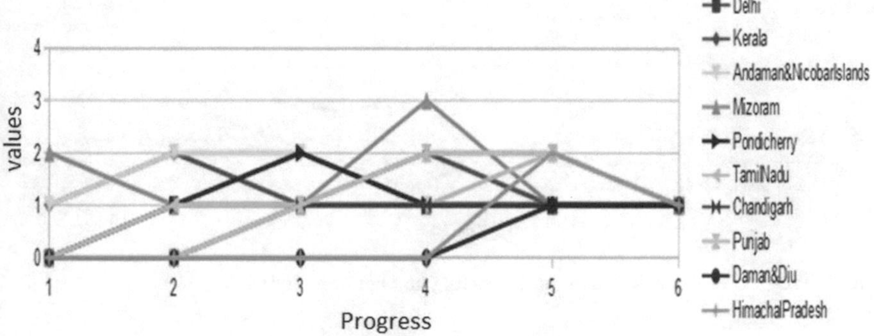

Fig. 3 Progress of the states based on literacy

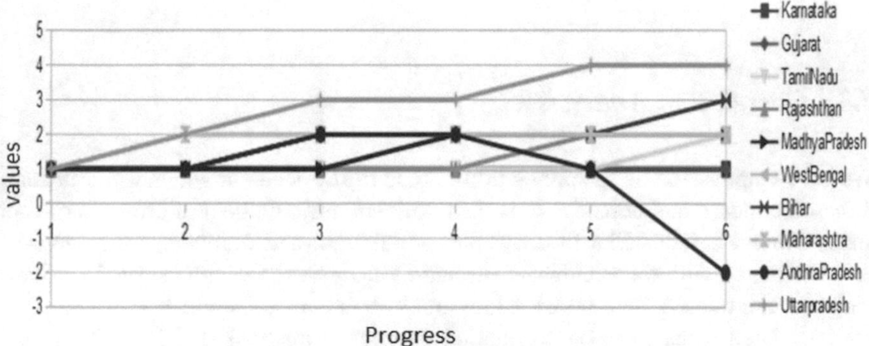

Fig. 4 Progress of the states based on preventing population

whereas gradual angle turns out to be bad progress and horizontal straight line signifies non-increasing literacy rate. Mizoram, Pondicherry, and Tamil Nadu having only a few numbers of steeper angle but no negative angle denote they performed well in literacy. Himachal Pradesh has the most number of horizontal straight lines thus making it average performer.

Population control is another important attribute of progress for the states as a whole. Figure 4 depicts Karnataka and Gujarat have done the best progress in preventing increasing population. Here, steeper angle signifies an increase in population which turns out to be bad progress, whereas gradual angles turn out to be good progress and horizontal straight lines signify retaining the same population as of previous year. Most of the states except Karnataka and Gujarat have gradual angles and thus making them less progressive in controlling population.

We analyze the number of motor vehicles registered on the roads in different states. Figure 5 depicts Gujarat, Madhya Pradesh, Rajasthan, and Punjab have done the best progress in controlling the number of motor vehicles. Only best performer states have all the angles as gradual thus making it more progressive than others

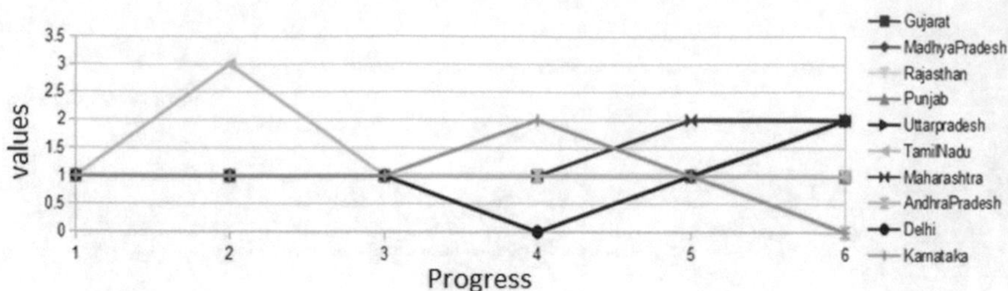

Fig. 5 Progress of the states based on preventing number of motor vehicles

whereas Delhi and Karnataka have made bad progress which signifies them as the worse performer.

5.3 *Correlation Analysis*

We have emphasized on the prevention of road traffic accident which is our primary research concern and consider it as the dominant attribute behind progress of each state. Other attributes like literacy rate, population, and number of transport are also correlated with the accident. Correlation analysis shows how strongly pairs of variables are related. Two set of values are $X = (x_1, x_2, \dots)$ and $Y = (y_1, y_2, \dots)$ are correlated using Pearson correlation coefficient analysis (PCCA). The PCCA constant [19] r is defined as

$$r(x, y) = \frac{cov(x, y)}{S_x.S_y} \tag{3}$$

S_x and S_y are the standard deviation of X and Y respectively. $cov(x, y)$ is the covariation within x and y.

$$cov(x, y) = \frac{1}{n-1} \sum (x_i - \overline{x})(y_i - \overline{y}) \tag{4}$$

here n is the number of paired data. $\overline{x}, \overline{y}$ is the mean and x_i and y_i is the ith instance value of set X and Y respectively. Pearson correlation coefficient between two variables always lie in between $+1$ and -1 where they denote observations have identical rank and dissimilar rank between themselves respectively. We analyze different attributes with road traffic accident to examine their relationship. Result has shown road accident is directly proportional with population and number transport increment with $r = 0.92$ and 0.84 respectively and inverse proportional with literacy rate with $r = -0.96$.

Table 5 PCCA similarity measure of Adenoma attributes

	A1	A2	A3	A4	A5	A6	A7	A8
A1	1	0.914	0.878	0.969	0.642	0.475	0.575	0.587
A2	0.914	1	0.978	0.953	0.790	0.641	0.729	0.768
A3	0.877	0.978	1	0.932	0.887	0.752	0.817	0.849
A4	0.642	0.790	0.888	0.709	1	0.935	0.947	0.952
A5	0.969	0.954	0.932	1	0.709	0.556	0.611	0.632
A6	0.475	0.641	0.752	0.556	0.935	1	0.950	0.918
A7	0.575	0.729	0.817	0.611	0.947	0.950	1	0.981
A8	0.587	0.768	0.849	0.632	0.952	0.918	0.981	1

Table 6 Strength of association

A1	A2	A3	A4			
A2	A1	A3	A4			
A3	A1	A2	A4	A5	A7	A8
A4	A3	A6	A7	A8		
A5	A1	A2	A3			
A6	A5	A7	A8			
A7	A3	A5	A6	A8		
A8	A3	A5	A6	A7		

5.4 Validation of the Proposed Approach

We have validated our result on Adenoma attributes ($A1, A2, \ldots, A8$) as shown in sample dataset [20], to find its strength of association. Table 5 shows PCCA similarity measure of Adenoma attributes based on our approach. Strength of association among different attributes are shown in Table 6. According to our experiment, primary attributes are A3 and A4 which is moreover similar to the biological conclusion. Therefore, we infer our method works well in finding primary attributes and their strength of association.

6 Conclusions

Road accident does not happen only due to the improper infrastructure of the roads, rather it is influenced by other factors as well like literacy rate, number of vehicles, the population of individual states, etc. Proper road safety management results in overall development of the nation. Road safety has been made one of the most important dimensions out of several development goals set for a country like India. In this paper, we develop a line encoding method to analyze the progress of different Indian states in traffic awareness. Our result concludes that traffic control measures are not the only solution. The states have to control population and number of vehicles and make their resident more literate to control road traffic accidents. For last few years, the focus of the Indian government is to develop a holistic traffic system in a hope that it will accelerate the growth of the country. In future, we may consider the geographical position of the states, weather condition and influence of different highway network to reveal other factors for road traffic accidents.

References

1. Li, J., Cheng, H., Guo, H., Qiu, S.: Survey on artificial intelligence for vehicles. Automot. Innov. **1**(1), 2–14 (2018)
2. Ministry of Road Transport and Highways, Government of India. http://morth.nic.in/
3. Chen, Y., Wu, C., Yao, M.: Dynamic topology construction for road network. In: 2008 Fourth International Conference on Networked Computing and Advanced Information Management, vol. 1, pp. 359–364 (2008). https://doi.org/10.1109/NCM.2008.45
4. Mambou, E.N., Nlend, S., Liu, H.: Study of the us road network based on social network analysis. In: 2017 IEEE AFRICON, pp. 974–978 (2017). https://doi.org/10.1109/AFRCON.2017.8095614
5. Mo, Y., Liu, J., Lv, S.: GIS-based analysis of fractal features of the urban road network. In: 2015 6th IEEE International Conference on Software Engineering and Service Science (ICSESS), pp. 845–848 (2015). https://doi.org/10.1109/ICSESS.2015.7339187
6. Chen, P., Li, K., Sun, J.: A method of traffic flow forecast and management. In: 2008 International Conference on Information Management, Innovation Management and Industrial Engineering, vol. 2, pp. 24–28 (2008). https://doi.org/10.1109/ICIII.2008.131
7. Muhuri, S., Das, D., Chakraborty, S.: An automated game theoretic approach for cooperative road traffic management in disaster. In: 2017 IEEE International Symposium on Nanoelectronic and Information Systems (iNIS), pp. 145–150 (2017). https://doi.org/10.1109/iNIS.2017.38
8. Spirou, S., Stiller, B.: Economic traffic management. In: 2009 IFIP/IEEE International Symposium on Integrated Network Management-Workshops, pp. 144–146 (2009). https://doi.org/10.1109/INMW.2009.5195951
9. Mukherjee, S.: Statistical analysis of the road network of India. Pramana **79**(3), 483–491 (2012)
10. Wenxue, C., Hengpeng, W., Shangjiang, S., Lilu, A.: An improved hazardous material road transportation accident rate analysis model. In: 2010 International Conference on Logistics Systems and Intelligent Management (ICLSIM), vol. 2, pp. 1081–1085 (2010). https://doi.org/10.1109/ICLSIM.2010.5461123
11. Kumar, S., Toshniwal, D.: Analysis of hourly road accident counts using hierarchical clustering and cophenetic correlation coefficient (CPCC). J. Big Data **3**(1), 13 (2016)
12. Khalil, O.K.: A study on road accidents in Abu Dhabi implementing a vehicle telematics system to reduce cost, risk and improve safety. In: 2017 10th International Conference on Developments in eSystems Engineering (DeSE), pp. 195–200. IEEE (2017)
13. Ming, H., Xiucheng, G.: The application of BP neural network principal component analysis in the forecasting the road traffic accident. In: 2009 Second International Conference on Intelligent Computation Technology and Automation, vol. 1, pp. 107–111 (2009). https://doi.org/10.1109/ICICTA.2009.35
14. Kaur, G., Kaur, E.H.: Prediction of the cause of accident and accident prone location on roads using data mining techniques. In: 2017 8th International Conference on Computing, Communication and Networking Technologies (ICCCNT), pp. 1–7 (2017). https://doi.org/10.1109/ICCCNT.2017.8204001

15. National Crime Records Bureau. http://ncrb.gov.in/
16. Persons Injured in Road Accidents. https://data.gov.in/catalog/persons-injured-road-accidents
17. Ministry of Home Affairs, Government of India. https://mha.gov.in/
18. Census of India website. http://censusindia.gov.in/
19. Wu, W.J., Xu, Y.: Correlation analysis of visual verbs' subcategorization based on pearson's correlation coefficient. In: 2010 International Conference on Machine Learning and Cybernetics (ICMLC), vol. 4, pp. 2042–2046. IEEE (2010)
20. Mujahid, A.K., Thirumalai, C.: Pearson correlation coefficient analysis (PCCA) on adenoma carcinoma cancer. In: 2017 International Conference on Trends in Electronics and Informatics (ICEI), pp. 492–495. IEEE (2017)

Efficient Energy Management in Microgrids Using Flower Pollination Algorithm

Meenakshi De, Gourab Das and K. K. Mandal

Abstract This paper presents energy management in microgrids using flower pollination algorithm. Here energy management in microgrid is done utilizing renewable energy such as wind and solar power. In this work, implementation of demand response (DR) schedules is carried out as incentive-based payment i.e., on offered price packages. In the microgrid system, several power components are utilized viz., wind turbines, photovoltaic cell, fuel cells, micro turbine, and battery. Case studies are conducted for obtaining minimum operating costs without demand response participation and with demand response participation respectively. The results obtained represent the superiority of the proposed approach.

Keywords Microgrids · Energy management · Flower pollination algorithm · Demand response

1 Introduction

The microgrid concept of power production is relatively new aimed at reducing harmful emissions from fossil-fueled power plants, at the same time enhancing the utilization of new technologies and concepts in the form of inclusion of renewable energy resources in the existing power networks. The disadvantage in renewable sources lies with their intermittent nature in power production. Thus, to maintain steady operation, power system operators provide certain system reserve to overcome uncertainties of power production. Some studies conducted previously investigated the inclusion of distributed generations (DG) such as wind and solar [1–3]. But these solutions suffer from drawbacks such as increase in costs and problems in

M. De (✉) · G. Das · K. K. Mandal
Department of Power Engineering, Jadavpur University, Kolkata 700106, India
e-mail: meenakshide.ju@gmail.com

G. Das
e-mail: gourabdas.ju@gmail.com

K. K. Mandal
e-mail: kkm567@yahoo.co.in

A. K. Das et al. (eds.), *Computational Intelligence in Pattern Recognition*, Advances in Intelligent Systems and Computing 999,
https://doi.org/10.1007/978-981-13-9042-5_47

commitment of power units, etc. Another take on this problem includes increasing the total energy reserve [4] to maintain system security. So a demand side reserve or demand response (DR) can be included by network operators by way of decreasing energy consumption during energy shortage period [5].

So, demand response (DR) is the change in consumption pattern of electricity in customer's side in response to changing electricity prices over a specified duration of time. Thus it increases incentive-based payment modes with respect to utilization of electricity. These incentive-based payments encourage lower use of electricity in times of high market price and higher use when prices are within threshold value. The incentive-based demand response management in microgrids provide diverse advantages such as it covers the uncertainty associated with solar and wind power production, and at the same time the customers will have a wide variety of choices from the offered packages to suit their needs and their budget. Some studies focused on microgrid (MG) operation and management involving a stochastic model in order to achieve optimized set points of operation [6, 7] while others stressed on investigating MG operation by heuristic algorithm [8]. Dynamic economic dispatch of a microgrid: Mathematical models and solution algorithm is proposed in [9].

2 Problem Formulation

The microgrid system considered in present research work is based on planning of existing units to supply demand by wind and solar power generation with inherent stochastic behavior and the way these types of power generating elements is covered by incentive-based demand response (DR) programs.

2.1 Specification of Demand Response Participants

A variety of power customers having different types of utilization patterns is analyzed for present work. In this research work, electric power customers are considered as residential, commercial, industrial types. Mathematically, Eqs. (1), (2) and (3) describe the respective contributions of residential, commercial and industrial consumers as follows:

$$R^{p}(r,t) = R^{c}(r,t) \times \pi_{r,t};\quad R^{c}(r,t) \le R_{t}^{c\max} \tag{1}$$

$$C^{p}(c,t) = C^{c}(c,t) \times \pi_{c,t};\quad C^{c}(c,t) \le C_{t}^{c\max} \tag{2}$$

$$I^{p}(i,t) = I^{c}(i,t) \times \pi_{i,t};\quad I^{c}(i,t) \le I_{t}^{c\max} \tag{3}$$

In the preceding equations r, c, i present three types of customers; R^p is the indicator of cost due to load reduction applied to residential customers. C^p is the indicator of expenses due to load reduction applied to commercial customers. Similarly I^p is the indicator of cost due to load reduction applied to industrial customers. $R_t^{c,\max}$, $C_t^{c,\max}$, $I_t^{c,\max}$ denotes maximum load reduction of each customer in time t. R^c, C^c and I^c represents quantity of load reduction for each type of customers; $\pi_{r,t}$ $\pi_{c,t}$, $\pi_{i,t}$ is incentive-based recompense toward residential, commercial, and industrial customers respectively.

2.2 Objective Function

The total operation cost of microgrid are storage expenses, startup and shutdown expenses in the generating units, costs for power from/to main grid, costs in demand response participation is stated mathematically in Eq. (4) as follows:

$$Min\ f_1(X) = \sum_{t=1}^{T} \text{Cost} = \sum_{t=1}^{T} \left\{ \sum_{i=1}^{Ng} [u_i(t)\, P_{DGi}(t) B_{Gi}(t) + \right.$$

$$S_{Gi}\, |u_i(t) - u_i(t-1)|] + \sum_{j=1}^{Ns} \left[u_j(t)\, P_{sj}(t) B_{sj}(t) + \right.$$

$$\left. S_{sj} \left| u_j(t) - u_j(t-1) \right| \right] + P_{Grid}(t) B_{Grid}(t) + P_{DR}(t) B_{DR}(t) \Big\} \quad (4)$$

$P_{DGi}(t)$, $P_{sj}(t)$ present active power in ith generator, jth storage in time t, where $B_{Gi}(t)$, $B_{sj}(t)$ present bids in generation and storage units respectively, S_{Gi}, S_{sj} present starting and shutting costs of ith generation unit, jth storage unit. u represents state vector that indicates ON/OFF states of all units in period t, $P_{Grid}(t)$ is the active power exchanged with the utility in time t, $B_{Grid}(t)$ present bids bought and sold from/to main grid for time t. $P_{DR}(t)$, $B_{DR}(t)$ present real power, bid costs in participation of demand response programs. T is total time duration.

2.3 Problem Boundaries

The problem boundaries of the objective function are illustrated below.

2.3.1 Power Balance

Cumulative power generated by distributed units should be equal to the total demand of power in microgrid (MG) for maintaining a steady supply of power which is represented in Eq. (5).

$$\sum_{i=1}^{Ng} P_{Gi}(t) + \sum_{j=1}^{Ns} P_{sj}(t) + P_{Grid}(t) = \sum_{k=1}^{Nk} P_{lk}(t) \tag{5}$$

Here, P_{lk} denotes quantity in kth load level, N_k presents total load levels, N_g represents generating units and N_s denotes count of storage units.

2.3.2 Active Power Production

Efficient and reliable MG system requires power of DG unit bounded within limits as presented by Eq. (6).

$$\begin{aligned} &P_{Gi,\min}(t) \le P_{Gi}(t) \le P_{Gi,\max}(t) \\ &P_{sj,\min}(t) \le P_{sj}(t) \le P_{sj,\max}(t) \\ &P_{Grid,\min}(t) \le P_{Grid}(t) \le P_{Grid,\max}(t) \end{aligned} \tag{6}$$

$P_{G,\min}(t)$, $P_{S,\min}(t)$, $P_{Grid,\min}(t)$ lower real powers in DG, storage and grid. $P_{G,\max}(t)$, $P_{S,\max}(t)$, $P_{Grid,\max}(t)$ upper real powers in DG, storage and grid.

2.3.3 Battery Limits

During each time period, boundaries on battery conditions are given by Eq. (7),

$$\begin{aligned} &\phi_j(t) = \phi_j(t-1) - \frac{1}{\lambda_{Dj}} u_{Dj}(t) P_{Dsj}(t) + \lambda_{Cj}(t) u_{Cj}(t) P_{Csj}(t) \\ &\phi_{\min} \le \phi_j(t) \le \phi_{\max} \end{aligned} \tag{7}$$

where $\phi_j(t)$ and $\phi_j(t-1)$ represent reserved energy in present time and at the previous time count. $P_{Dsj}(t)$ and $P_{Csj}(t)$ is allowed rate of discharge and charge during a definite time interval. $\lambda_{Dj}(t)$ and $\lambda_{Cj}(t)$ present battery efficiency at times of discharge and charge.

2.4 *Microgrid Model*

Microgrid comprising of several generation and storage units like micro turbine (MT), fuel cell (FC), battery, and renewable energy resources utilizing solar photovoltaic (PV), wind turbines (WT) is considered for the present study.

3 Flower Pollination Algorithm

Xin She Yang first proposed the flower pollination algorithm in 2012 [10] and since then it has found wide applications in solving various optimization problems. There are two major forms of pollination: abiotic and biotic [11].

The flower pollination algorithm consists of the following steps:

(1) Biotic and cross pollination are global pollination where pollen bearing pollinators perform L´evy flights.
(2) Abiotic, self-pollination are local pollination.
(3) Flower constancy is activated by insects, at a par with reproduction probability proportional of similarity in 2 flowers involved.
(4) Local, global pollination is controlled by switch probability p.

We represent the fittest as g_*. The global pollination, i.e., first rule with flower constancy is represented mathematically in following Eq. (8):

$$x_i^{t+1} = x_i^t + L(x_i^t - g_*) \tag{8}$$

where x_i^t represents pollen i/vector x_i in iteration t, g_* presents best solve. L represents pollination strength that represents step size and can be formulated mathematically as in Eq. (9).

$$L \approx \frac{\lambda \Gamma(\lambda) \sin(\pi\lambda/2)}{\pi} \frac{1}{s^{1+\lambda}}, (s\rangle\rangle s_0\rangle 0) \tag{9}$$

$\Gamma(\lambda)$ is the standard gamma function. Distribution remains valid in large steps $s > 0$. The chosen value of $\lambda = 1.5$. Second rule with flower constancy is given mathematically in following Eq. (10).

$$x_i^{t+1} = x_i^t + \in \left(x_j^t - x_k^t\right) \tag{10}$$

x_j^t, x_k^t represent pollens of different flowers in same plant species which mimics flower constancy within specified neighborhood. x_j^t, x_k^t comes from same species/same population, this forms local random walk. $\in$ is selected from uniform distribution within [0, 1]. Utilization of switch probability (rule 4) p switches between global and local pollination.

3.1 Pseudo Code of Flower Pollination Algorithm

The pseudo code of flower pollination algorithm (FPA) can be given as follows:

Objective min $f(x), x = (x_1, x_2, ..., x_d)$

Initializing a population of N flowers/ pollen gametes using random solutions

Find best solution g_* in initial population

Define: Switch-probability $p \in [0,1]$

While (t<max-generation)

for $i = 1 : N$ (all N flowers in Population)

if rand<p,

Define a (d-dimensional) step vector L which follows L′evy distribution

Global- Pollination; $x_i^{t+1} = x_i^t + L(g_* - x_i^t)$

else:

Define $\in$ from uniform distribution within [0, 1]

Randomly choose j & k within all Solutions;

Local Pollination; $x_i^{t+1} = x_i^t + \in (x_j^t - x_k^t)$

end if:

Compute New Solutions

If new solutions are excellent; Update them within population;

end for

Find present Best Solution: g_*

End: while.

4 Results and Discussion

Case 1: Operating cost without Demand Response

During the case study, it is considered that all the generating units are switched on and participate in microgrid network with their unique electrical characteristics. The excess or deficit of power supply in microgrid flows from/to main grid [12]. Total operating cost for case 1 when the demand response scheme is not taken into consideration is found out to be 221.4469 €ct using FPA depicted in Fig. 1.

Tables 1 and 2 show economic power dispatch for case 1 and 2 respectively. Table 3 shows detailed comparative analysis of Flower Pollination Algorithm (FPA) with Particle Swarm Optimization (PSO).

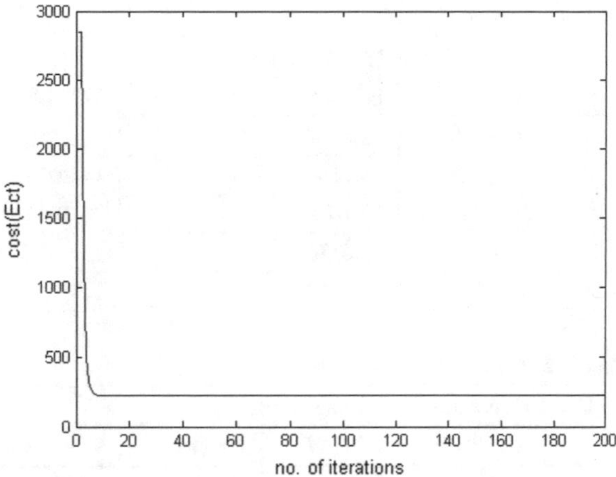

Fig. 1 Case 1 (without DR) using FPA

Table 1 Case 1: without DR

Hr	Generation and storage units					
	MT (kW)	FC (kW)	PV (kW)	WT (kW)	Batt (kW)	Utility (kW)
1	26.052	3.949	0	4.9555	30.00	−5.05
2	30.0	20.8305	0	1.7263	4.7588	−13.39
3	28.0003	13.2002	0	4.2251	29.9437	8.672
4	9.0488	27.4530	0	10.8748	14.8304	3.755
5	14.6290	14.8533	0	1.0546	8.6518	−9.55
6	12.4528	15.9629	0	13.5319	−11.0344	−9.59
7	30.0	12.7750	0	1.8484	−17.1430	28.31
8	28.5380	23.1285	5.9421	2.6494	−14.7158	−18.78
9	17.2052	7.4910	4.8579	9.7346	−26.0704	−18.70
10	10.0177	17.1384	14.7763	0.4031	−19.6657	11.29
11	12.6424	15.8113	18.5812	4.2894	−28.2435	6.925
12	11.3058	8.1003	20.4541	10.3934	5.4941	−26.65
13	13.0507	3.0	0.9692	3.1425	−2.2109	12.73
14	8.1291	21.4577	6.7837	6.1939	−4.2989	−7.37
15	15.2815	3.0	3.8031	4.6636	1.4512	21.67
16	9.0326	3.9632	16.2600	2.0136	−12.6248	1.576
17	7.4280	3.0	0.3344	1.7721	13.7202	15.70
18	8.6514	10.6587	0	2.8241	6.1007	−4.572
19	11.4057	15.5134	0	15.0	−24.0291	19.13

(continued)

Table 1 (continued)

Hr	Generation and storage units					
	MT (kW)	FC (kW)	PV (kW)	WT (kW)	Batt (kW)	Utility (kW)
20	24.8049	13.6214	0	5.5158	14.4560	−26.92
21	18.7812	5.7186	0	14.7056	−17.6824	−27.14
22	29.4257	29.7679	0	6.3843	−14.3629	30.0
23	28.6948	25.6576	0	15.0	30.0	26.50
24	21.1682	23.3117	0	11.0469	−5.9562	30.0

Table 2 Cost with demand response: (case 2)

Hr	Generation and storage units					
	MT (kW)	FC (kW)	PV (kW)	WT (kW)	Batt (kW)	Utility (kW)
1	11.9304	12.4719	0	5.2738	−18.5137	21.06
2	6.0	20.2558	0	1.6660	3.5228	−30.0
3	20.2969	18.0385	0	13.7369	26.2926	4.345
4	23.5876	30.0	0	8.4812	30.0	21.55
5	30.0	6.7268	0	5.2813	−22.7233	−4.65
6	26.6892	26.537	0	6.8510	−19.5711	−15.4
7	16.3850	24.6109	0	11.4211	9.8900	16.93
8	6.0	23.0291	24.1146	5.4707	11.2084	−25.7
9	17.0566	21.7885	13.6893	4.5467	−15.1567	6.145
10	6.0	10.2823	6.3390	3.1488	−30.0	6.001
11	9.9770	19.707	20.1251	3.9460	−30.0	8.930
12	14.5353	20.7005	3.9236	7.7339	−2.8571	19.33
13	14.0316	6.4775	7.4716	7.1788	−8.0159	−20.0
14	21.5082	23.6109	3.8884	11.1322	−29.6159	−30.0
15	24.4758	11.2317	3.3516	0.2906	−30.0	−2.17
16	13.4739	5.7531	0.7991	2.6208	−10.8637	−30.0
17	25.8828	24.8601	15.0733	8.5317	−11.9009	−3.85
18	10.4855	11.1107	0	2.3148	15.5461	−1.56
19	12.0688	11.0574	0	11.3104	30.0	−9.40
20	6.0	22.0579	0	13.5546	−1.0657	9.341
21	28.5089	14.7889	0	6.6767	0.3287	8.018
22	28.0748	14.3575	0	7.8322	26.3482	3.100
23	9.9638	6.0488	0	2.3145	−3.5479	−20.0
24	7.3629	22.4384	0	2.7245	17.4967	30.0

Table 3 Comparative analysis

Method	Parameter	FPA	PSO [12]
Without DR	Operating Cost (€ct)	221.44	241.3
With DR	Operating Cost (€ct)	211.22	231.3

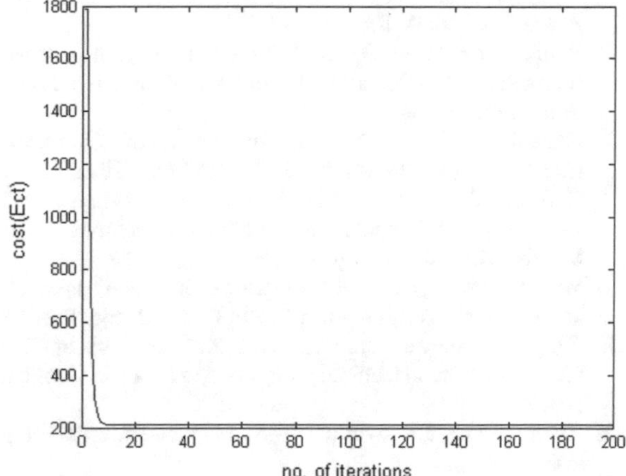

Fig. 2 Case 2 (with DR) using flower pollination algorithm

Case 2: Operating cost with Demand Response: The total operating cost for case 2 when the DR is taken into consideration is found out to be 211.2230 €ct using FPA depicted in Fig. 2.

Table 3 shows comparative study of FPA with particle swarm optimization (PSO).

5 Conclusion

In this paper, a microgrid system consisting of wind and PV resources are considered. The results are obtained and analyzed to point out that operating costs reduced by DR programs using FPA which prove superiority of proposed approach.

Acknowledgements The authors express gratitude to The Department of Power Engineering, Jadavpur University, Kolkata, India for providing support and assistance to this research work.

References

1. Ahn, S.J., Nam, S.R., Choi, J.H., Moon, S.: Power scheduling of distributed generators for economic and stable operation of a microgrid. IEEE Trans. Smart Grid **4**(1) (2013)
2. Ravichandran, A., Malysz, P., Sirouspour, S., Emadi, A.: The Critical Role of Microgrids in Transition to a Smarter Grid: A Technical Review. IEEE (2013)

3. Lasseter, R.H., Paigi, P.: Microgrid: a conceptual solution. In: 35th Annual IEEE Power Electronics Specialists Conference (2004)
4. Meiqin M., Meihong, J., Wei, D., Chang, L.: Multi-objective economic dispatch model for a microgrid considering reliability. In: 2nd IEEE International Symposium on Power Electronics for Distributed Generation Systems, pp. 993–998 (2010)
5. Srinivasa Rao, R., Lakshmi Narasimham, S., Raju, M.R., Srinivasa Rao, A.: Optimal network reconfiguration of large-scale distribution system using harmony search algorithm. IEEE Trans. Power Syst. **26**(3), 1080–1088 (2011)
6. Ramalakshmi, S.S.: Optimal siting and sizing of distributed generation using fuzzy-EP. In: International Conference on Recent Advancements in Electrical, Electronics and Control Engineering (2011), pp. 470–477
7. Ahmadi, A., Pedrasa, M.A.A.: Optimal design of hybrid renewable energy system for electrification of isolated grids. In: IEEE TENCON (2012)
8. Colson, C.M., Nehrir, M.H., Pourmousavi, S.A.: Towards real-time microgrid power management using computational intelligence methods. In: Power and Energy Society General Meeting IEEE (2010), pp. 1–8
9. Wu, H., Liu, X., Ding, M.: Dynamic economic dispatch of a microgrid: Mathematical models and solution algorithm. Int. J Electr. Power Energy Sys. **63**, 336–346 (2014)
10. Yang, X.S.: Flower pollination algorithm for global optimization. In: Unconventional Computation and Natural Computation. Lecture Notes in Computer Science, vol. 7445, pp. 240–249 (2012)
11. Glover, B.: Understanding Flowers and Flowering: An Integrated Approach. Oxford University Press (2007)
12. Aghajani, G.R., Shayanfar, H.A., Shayeghi, H.: Presenting a multi-objective generation scheduling model for pricing demand response rate in micro-grid energy management. Int. J. Energy Convers. Manag. **106**, 308–321 (2015)

Applications of Computational Intelligence Techniques for Automatic Generation Control Problem—A Short Review from 2010 to 2018

Tulasichandra Sekhar Gorripotu, Simma Gopi, Halini Samalla, A. V. L. Prasanna and B. Samira

Abstract This work provides a brief review report on applications of computational intelligence techniques for automatic generation control (AGC) problem. Various optimization techniques employed to the conventional multi-area power system and restructured power system are highlighted. With that, the review also highlights the types of controllers used for various types of systems.

Keywords AGC · Controllers · Intelligence techniques · Restructured power system

1 Introduction

The present day, power system is expanding day by day due to the expansion of power generation, development in renewable energy generation, and changing in time to time load variation. The motto of the power system engineers is to generate, transmit, and distribute the quality electric power to the consumers without any disturbances. When the generation is equal to the load demand plus losses of the system then it said to be stable system. But, due to lack of long-term planning and increase in load demand the system becomes unstable and the frequency and voltage

T. S. Gorripotu (✉) · S. Gopi · H. Samalla · A. V. L. Prasanna · B. Samira
Department of Electrical & Electronics Engineering, Sri Sivani College of Engineering, Srikakulam 532410, Andhra Pradesh, India
e-mail: gtchsekhar@gmail.com

S. Gopi
e-mail: simmagopi53@gmail.com

H. Samalla
e-mail: halini.samalla@gmail.com

A. V. L. Prasanna
e-mail: avlprasanna23@gmail.com

B. Samira
e-mail: ssameera.barla@gmail.com

A. K. Das et al. (eds.), *Computational Intelligence in Pattern Recognition*, Advances in Intelligent Systems and Computing 999,
https://doi.org/10.1007/978-981-13-9042-5_48

of the system will deviate from their nominal values. Hence, it is necessary to take some measures to keep them at nominal values with allowable tolerance bands. This paper essentially will focus on the various techniques, controllers used for frequency control of interconnected power system [1–3].

2 Preliminaries

This is well-known that, for controlling a frequency in an automatic generation control problem primary controller, i.e., concept of moment of inertia is not enough to do so and therefore, it is essential to have a secondary controller such as sub-optimal controllers [4], optimal controllers [5], artificial neural networks [6], classical controllers [7–10], fuzzy controllers [11–13], two-degree of freedom controllers [14, 15], fractional controllers [16–18], cascade controllers [19–21], sliding mode controllers [22–24] and, etc., in the system. The optimal values of the above mentioned controllers can be finalized by means of computational intelligence techniques [5, 25–106] such as Chemical Reaction Optimization (CRO), Artificial Bee Colony (ABC), Differential Evolution (DE), Evolutionary Algorithms (EA), Particle Swarm Optimization (PSO), Genetic Algorithm (GA) Firefly Algorithm (FA), Teacher Learning Based Optimization (TLBO), Pattern Search (PS), Moth-Flame Optimization (MFO), Bacterial Forging Optimization Algorithm (BFOA), Imperialist competitive algorithm (ICA), Bat algorithm, Whale optimization algorithm (WOA), Grey Wolf Optimization (GWO), Grasshopper Optimization Algorithm, Ant Lion Optimizer (ALO) algorithm, Quasi-Oppositional Harmony Search (QOHS), and etc. The problem of AGC can be categorized as conventional system and deregulated or restructured power system with a subcategory of linear system and nonlinear system which includes boiler dynamics, governor dead band (GDB), generation rate constraint (GRC) and communication or time or transport delay (TD) [107–113]. By keeping all these views in mind, this work is presented.

3 Literature Review

Since few decades researchers over the world are investigating the concept of AGC by employing with Various control strategies and optimization techniques. In this section, some of the recent advancements took place in the AGC concept are pointed out during this decade, i.e., from 2010 to 2018 [5, 25–113].

In [114], authors have elaborated the concept of AGC for two area nonlinearity power system and implemented it with a multi-objective evolutionary algorithm. The authors have also proposed HVDC link and Superconducting Magnetic Energy Storage (SMES) for the improvement of transient response of the system. In [115], the authors have proposed Hybrid particle swarm optimization with constriction factor approach (HPSOCFA) optimized PID controller for a four area deregulated power

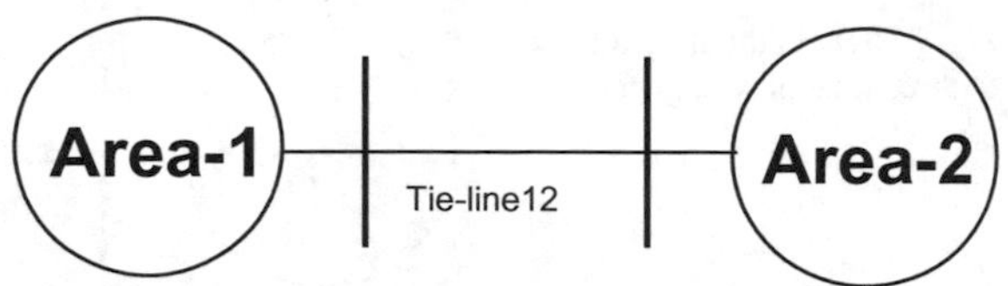

Fig. 1 Schematic diagram of two area power system

system. The authors have also shown the supremacy of proposed technique with binary coded genetic algorithm (BGA) and real coded genetic algorithm (RGA). In [116], authors suggested IMC mode two degree of freedom and unified PID controller for multi-area power system. Rasolomampionona et al. [117] have projected the usage of FACT devices for the improvement in frequency regulation. The authors have considered eigen values of the system for two area and three area power systems to evaluate the stability of the system. Golpira et al. [118] have designed a realistic three area power system with nonlinearities such as GDB, GRC, and TD for AGC problem. The authors have tuned the SIMULINK model by considering GA technique having PID controller as a secondary controller. Lalit et al. [119] have proposed five area reheat thermal system with classical controllers namely I, PI, ID, PID, and IDD. The final values of the controllers are attained with BFOA technique and results have proved that IDD is performing better than other controllers. The authors have also shown the effect of IDD controller for the variation of regulation parameter and other parameters. Ali and Abd [120] have considered a simple two area thermal power system and shown the ability of BFOA technique by comparing with GA technique and Conventional technique. The corresponding schematic diagram is shown in Fig. 1. Sudha et al. [121] have anticipated the performance of type-2 fuzzy controller for a two area power system and the authors have also proven that type-2 fuzzy performs better than conventional PI and type-1 fuzzy systems. In [122], the authors analyzed the effect of FACT devices (SSSC & TCPS) with the energy storage devices for ALFC problem of two area hydro-hydro power system. Haluk et al. [123] proposed ABC technique for a 2-area reheat thermal units. The authors showed the ability of proposed one by analyzing the system at various load conditions and also did comparative analysis with PSO technique. The schematic model of three area and four area are shown in Figs. 2 and 3 respectively. In Fig. 4 the sample model of FACT device incorporated system in deregulated environment is shown. In Table 1, some the conclusion points are mentioned against the computational technique.

4 Analytical Analysis

After careful review, the following points can be noted down:

a. Still, there is a scope of computational intelligence techniques usage for the better analysis of the AGC system.

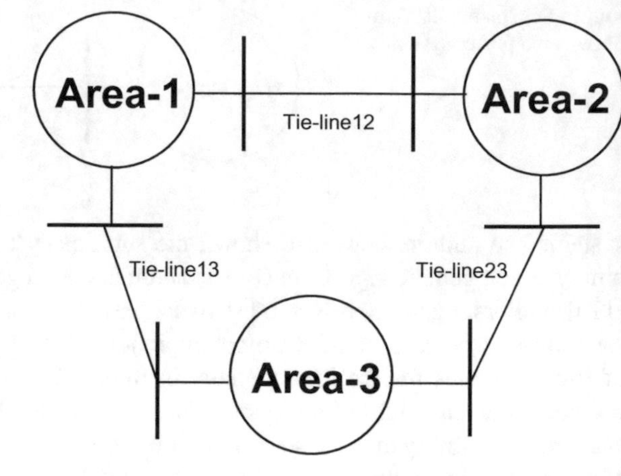

Fig. 2 Schematic diagram of three area power system

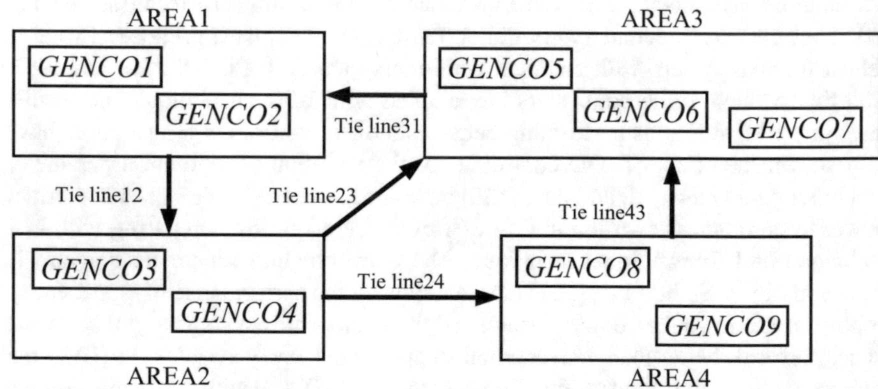

Fig. 3 Schematic diagram of four area power system

b. Lot of controller structures are coming into picture day by day and all those are competitive with one another.
c. Most of the research is confined to conventional rather than deregulated system
d. Inclusion of nonlinearities is required to have a realistic system
e. FACT devices and energy storage device incorporated systems will provide better results.
f. It is unfortunate that, no researcher can stay on a single computational intelligence technique but good part is that a lot of research is going on these techniques to solve AGC problem within very short time.

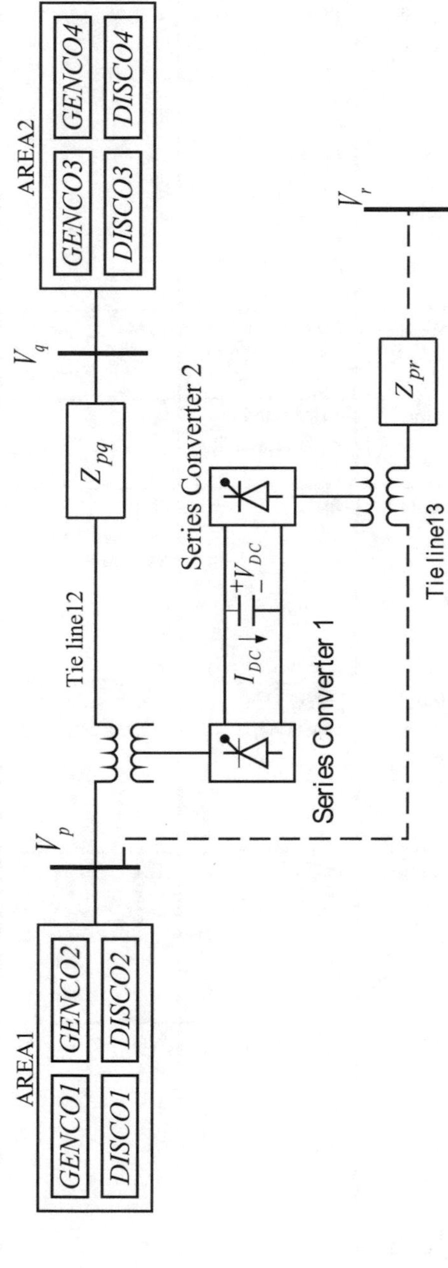

Fig. 4 Schematic diagram of two area deregulated power system with FACTS device

Table 1 Brief review on controllers and techniques

Authors	Type of the System	Proposed Controller	Technique used	Conclusion
Abdelaziz et al. [25]	Conventional	PI	Cuckoo search	Superior Performance than GA, PSO, NN, FL & ANFIS
Arya et al. [26]	Conventional and Deregulation	Fuzzy gain scheduling	–	Provides better performance than optimal, BFOA:PID, GSA:PID, hBFOA-PSO:PI
Bhateshvar et al. [27]	Deregulation	Fuzzy	GA	GA optimized FLC makes the system stable
Boonchuay [28]	Deregulation	Critical load level approach	–	CLL info will help to improve the regulation
Cai et al. [29]	Conventional	PID	Port-Hamiltonian system	Proves the validation of proposed concept for a four area thermal power system
Chandrakala et al. [30]	Conventional	Fuzzy gain scheduling	Fuzzy gain scheduling	Acknowledged the advantage of fuzzy gain scheduling over Z-N and GA techniques
Chathoth et al. [31]	Deregulation	FOPID	GA	Presents good results with proposed controller when compared GA: PI, PID and FLC
Dahiya et al. [32]	Conventional	Optimal	State space	The authors got better results when compared to techniques such as TLBO, DE, GA, PSO, FA, KHA, BAT, SFS, hBFOA-PSO, MFO & GSO

(continued)

Table 1 (continued)

Authors	Type of the System	Proposed Controller	Technique used	Conclusion
Debnath et al. [33]	Conventional	PD-Fuzzy PID	GWO-TLBO	Provides better results compared to PID and Fuzzy PID. The transient responses are improved when the system is incorporated with HVDC & UPFC
Demiroren et al. [34]	Conventional	ANN		Proposed system gives better results than conventional technique
Doolla et al. [35]	Conventional	Multi-pipe scheme		Authors shows the ability of multi pipe scheme in LFC concept
Golpîra et al. [36]	Conventional	Economic load dispatch	GA	Authors shown the ability of economic load dispatch based GA technique for LFC problem
Hasan et al. [37]	Conventional	LQR	State space	Solved the AGC problem with state cost weighted matrix 'Q'
Hasan et al. [38]	Conventional	ABT	–	The authors proposed ABT designed LFC for an interconnected power system
Hosseini et al. [39]	Deregulated	PID	ICA	Authors proposed ICA based PID controller for three area diverse source system

(continued)

Table 1 (continued)

Authors	Type of the System	Proposed Controller	Technique used	Conclusion
Huang et al. [40]	Conventional	Robust	LMI	Proposed technique is good enough compared to PI controller
Ibraheem et al. [41]	Conventional	Sub-optimal	State space	Sub-optimal controller provides better results than optimal controllers
Khezri et al. [42]	Conventional	Fuzzy	–	Provides better results compared to hybrid GA and LMI
Krishnan et al. [43]	Deregulated	PI	Lyapunov	Gives stable operating conditions
Kumar et al. [44]	Deregulated	PID	ICA	Proposed ICA-PID controller for 75-bus power system under deregulated scenario and also shown the supremacy of ICA-PID compared to GA-PID
Nasiruddin et al. [45]	Conventional	PID	BFOA	Authors executed diverse source power system and shown the superiority of BFOA over GA technique
Nguyen et al. [46]	Conventional	PID	NNFL	Implemented NNFL-PID with SMES to 5 area system
Novella-Rodríguez et al. [47]	Conventional	PD-PID		Applied frequency domain approach for delayed systems

(continued)

Table 1 (continued)

Authors	Type of the System	Proposed Controller	Technique used	Conclusion
Ojaghi et al. [48]	Conventional	Predictive	LMI	Implemented PLFC and NPLFC
Orihara et al. [49]	Conventional			Authors proposed frequency regulation by using a battery during shortage of power
Padhan et al. [50]	Conventional	PI, PID	FA	Shown the supremacy of FA technique over ZN, GA, DE, BFOA, PS and hBFOA-PSO techniques
Prakash et al. [51]	Conventional	Fuzzy	ANN	Proposed Neuro-fuzzy controller for three area power system
Shiva et al. [52]	Conventional	I, PI, PID	QOHS	Authors did comparative analysis of classical controllers for proposed system
Sönmez et al. [53]	Conventional	IMC	–	Proposed freedom of IMC method and it is better than Tan's and Anwar method
Tan et al. [54]	Conventional	PI	–	Authors shown the graphical approach to determine the stability
Vrdoljak et al. [55]	Conventional	LADRC	–	The authors shown the effect of LADRC in presence of GDB

5 Conclusion

In this present review, the advanced approaches in the problem of AGC for conventional and deregulated power systems are extensively reviewed. A detailed survey has been done and presented here in this manuscript. The recent controller techniques, computational techniques studied extensively. Moreover, the review on FACT devices and energy storage devices, nonlinearities is also presented.

References

1. Elgerd, O.I.: Electric Energy Systems Theory–An Introduction. Tata McGraw Hill, New Delhi (2000)
2. Bevrani, H.: Robust Power System Frequency Control. Springer (2009)
3. Bervani, H., Hiyama, T.: Intelligent Automatic Generation control. CRC Press (2011)
4. Hasan, N., Kumar, P.: Sub-optimal automatic generation control of interconnected power system using constrained feedback control strategy. Int. J. Electr. Power Energy Syst. **43**(1), 295–303 (2012)
5. Parmar, K.S., Majhi, S., Kothari, D.P.: Load frequency control of a realistic power system with multi-source power generation. Int. J. Electr. Power Energy Syst. **42**(1), 426–433 (2012)
6. Gorripotu, T.S., Sahu, R.K., Panda, S.: Comparative performance analysis of classical controllers in LFC using FA technique. In: International Conference on Electrical, Electronics, Signals, Communication and Optimization (EESCO), pp. 1–5. IEEE (2015)
7. Dash, P., Saikia, L.C., Sinha, N.: Automatic generation control of multi area thermal system using Bat algorithm optimized PD–PID cascade controller. Int. J. Electr. Power Energy Syst. **68**, 364–372 (2015)
8. Tammam, M.A., Aboelela, M.A., Moustafa, M.A., Seif, A.E.A.: A multi-objective genetic algorithm based PID controller for load frequency control of power systems. Int. J. Emerg. Technol. Adv. Eng. **3**(12), 463–467 (2013)
9. Sahu, R.K., Panda, S., Padhan, S.: Optimal gravitational search algorithm for automatic generation control of interconnected power systems. Ain Shams Eng. J. **5**(3), 721–733 (2014)
10. Gorripotu, T.S., Kumar, D.V., Boddepalli, M.K., Pilla, R.: Design and analysis of BFOA optimized PID controller with derivative filter for frequency regulation in Distributed Generation System. Int. J. Autom. Control. INDERSCIENCE **12**(2), 291–323 (2018)
11. Sun, Z., Wang, N., Srinivasan, D., Bi, Y.: Optimal tuning of type-2 fuzzy logic power system stabilizer based on differential evolution algorithm. Int. J. Electr. Power Energy Syst. **62**, 19–28 (2014)
12. Sahu, R.K., Panda, S., Yagireddy, N.K.: A novel hybrid DEPS optimized fuzzy PI/PID controller for load frequency control of multi-area interconnected power systems. J. Process Control **24**(10), 1596–1608 (2014)
13. Sahu, R.K., Panda, S., Sekhar, G.C.: A novel hybrid PSO-PS optimized fuzzy PI controller for AGC in multi area interconnected power systems. Int. J. Electr. Power Energy Syst. ELSEVIER **64,** 880–893 (2015)
14. Sahu, R.K., Panda, S., Rout, U.K.: DE optimized parallel 2-DOF PID controller for load frequency control of power system with governor dead-band nonlinearity. Int. J. Electr. Power Energy Syst. **49**, 19–33 (2013)
15. Dash, P., Saikia, L.C., Sinha, N.: Comparison of performances of several Cuckoo search algorithm based 2DOF controllers in AGC of multi-area thermal system. Int. J. Electr. Power Energy Syst. **55**, 429–436 (2014)

16. Taher, S.A., Fini, M.H., Aliabadi, S.F.: Fractional order PID controller design for LFC in electric power systems using imperialist competitive algorithm. Ain Shams Eng. J. **5**(1), 121–135 (2014)
17. Das, S., Pan, I., Das, S., Gupta, A.: A novel fractional order fuzzy PID controller and its optimal time domain tuning based on integral performance indices. Eng. Appl. Artif. Intell. **25**(2), 430–442 (2012)
18. Debbarma, S., Saikia, L.C., Sinha, N.: Automatic generation control using two degree of freedom fractional order PID controller. Int. J. Electr. Power Energy Syst. **58**, 120–129 (2014)
19. Dash, P., Saikia, L.C., Sinha, N.: Flower pollination algorithm optimized PI-PD cascade controller in automatic generation control of a multi-area power system. Int. J. Electr. Power Energy Syst. **82,** 19–28 (2016)
20. Padhy, S., Panda, S.: A hybrid stochastic fractal search and pattern search technique based cascade PI-PD controller for automatic generation control of multi-source power systems in presence of plug in electric vehicles. CAAI Trans. Intell. Technol. **2**(1), 12–25 (2017)
21. Padhy, S., Panda, S., Mahapatra, S.: A modified GWO technique based cascade PI-PD controller for AGC of power systems in presence of plug in electric vehicles. Eng. Sci. Technol. Int. J. **20**(2), 427–442 (2017)
22. Liao, K., Yan, X.: A robust load frequency control scheme for power systems based on second-order sliding mode and extended disturbance observer. IEEE Trans. Ind. Inf. **14**(7), 3076–3086 (2018)
23. Ma, M., Zhang, C., Liu, X., Chen, H.: Distributed model predictive load frequency control of the multi-area power system after deregulation. IEEE Trans. Ind. Electron. **64**(6), 5129–5139 (2017)
24. Mu, C., Tang, Y., He, H.: Improved sliding mode design for load frequency control of power system integrated an adaptive learning strategy. IEEE Trans. Ind. Electron. **64**(8), 6742–6751 (2017)
25. Abdelaziz, A.Y., Ali, E.S.: Load frequency controller design via artificial cuckoo search algorithm. Electr. Power Compon. Syst. **44**(1), 90–98 (2016)
26. Arya, Y., Kumar, N.: Fuzzy gain scheduling controllers for automatic generation control of two-area interconnected electrical power systems. Electr. Power Compon. Syst. **44**(7), 737–751 (2016)
27. Bhateshvar, Y.K., Mathur, H.D., Siguerdidjane, H., Bhanot, S.: Frequency stabilization for multi-area thermal–hydro power system using genetic algorithm-optimized fuzzy logic controller in deregulated environment. Electr. Power Compon. Syst. **43**(2), 146–156 (2015)
28. Boonchuay, C.: Improving regulation service based on adaptive load frequency control in LMP energy market. IEEE Trans. Power Syst. **29**(2), 988–989 (2014)
29. Cai, L., He, Z., Haitao, H.: A new load frequency control method of multi-area power system via the viewpoints of port-hamiltonian system and cascade system. IEEE Trans. Power Syst. **32**(3), 1689–1700 (2017)
30. Chandrakala, V., Sukumar, B., Sankaranarayanan, K.: Load frequency control of multi-source multi-area hydro thermal system using flexible alternating current transmission system devices. Electr. Power Compon. Syst. **42**(9), 927–934 (2014)
31. Chathoth, I., Ramdas, S.K., Krishnan, S.T.: Fractional-order proportional-integral-derivative-based automatic generation control in deregulated power systems. Electr. Power Compon. Syst. **43**(17), 1931–1945 (2015)
32. Dahiya, P., Mukhija, P., Saxena, A.R., Arya, Y.: Comparative performance investigation of optimal controller for AGC of electric power generating systems. Automatika. **57**(4), 902–921 (2016)
33. Debnath, M.K., Jena, T., Mallick, R.K.: Optimal design of PD-Fuzzy-PID cascaded controller for automatic generation control. Cogent Eng. **4**(1), 1416535 (2017)
34. Demiroren, A.: Automatic generation control using ANN technique for multi-area power system with SMES units. Electr. Power Compon. Syst. **32**(2), 193–213 (2004)
35. Doolla, S., Bhatti, T.S., Bansal, R.C.: Load frequency control of an isolated small hydro power plant using multi-pipe scheme. Electr. Power Compon. Syst. **39**(1), 46–63 (2011)

36. Golpîra, H., Bevrani, H.: A framework for economic load frequency control design using modified multi-objective genetic algorithm. Electr. Power Compon. Syst. **42**(8), 788–797 (2014)
37. Hasan, N., Ibraheem, Kumar, P.: Optimal automatic generation control of interconnected power system considering new structures of matrix Q. Electr. Power Compon. Syst. **41**(2), 136–156 (2013)
38. Hasan, N., Ibraheem, Ahmad, S.: ABT based load frequency control of interconnected power system. Electr. Power Compon. Syst. **44**(8), 853–863 (2016)
39. Hosseini, H., Tousi, B., Razmjooy, N., Khalilpour, M.: Design robust controller for automatic generation control in restructured power system by imperialist competitive algorithm. IETE J. Res. **59**(6), 745–752 (2013)
40. Huang, C., Zhang, K., Dai, X., Zang, Q.: Robust load frequency controller design based on a new strict model. Electr. Power Compon. Syst. **41**(11), 1075–1099 (2013)
41. Ibraheem, K.P., Hasan, N., Nizamuddin: Sub-optimal automatic generation control of interconnected power system using output vector feedback control strategy. Electr. Power Compon. Syst. **40**(9), 977–994 (2012)
42. Khezri, R., Golshannavaz, S., Shokoohi, S., Bevrani, H.: Fuzzy logic based fine-tuning approach for robust load frequency control in a multi-area power system. Electr. Power Compon. Syst. **44**(18), 2073–2083 (2016)
43. Krishnan, R., Pragatheeswaran, J.K., Ray, G.: Robust stability of networked load frequency control systems with time-varying delays. Electr. Power Compon. Syst. **45**(3), 302–314 (2017)
44. Kumar, N., Tyagi, B., Kumar, V.: Multi-area deregulated automatic generation control scheme of power system using imperialist competitive algorithm based robust controller. IETE J. Res. **64**(4), 528–537 (2018)
45. Nasiruddin, I., Bhatti, T.S., Hakimuddin, N.: Automatic generation control in an interconnected power system incorporating diverse source power plants using bacteria foraging optimization technique. Electr. Power Compon. Syst. **43**(2), 189–199 (2015)
46. Nguyen, N.-K., Huang, Q., Dao, T.-M.: A novel two-stage NNFL strategy for load-frequency control using SMES. IETE J. Res. **61**(4), 392–401 (2015)
47. Novella-Rodríguez, D.F., Cuéllar, B.D.M., Márquez-Rubio, J.F., Hernández-Pérez, M.Á., Velasco-Villa, M.: PD–PID controller for delayed systems with two unstable poles: a frequency domain approach. Int. J. Control. 1–13 (2017)
48. Ojaghi, P., Rahmani, M.: LMI-based robust predictive load frequency control for power systems with communication delays. IEEE Trans. Power Syst. **32**(5), 4091–4100 (2017)
49. Orihara, D., Saitoh, H.: Improvement of frequency stability by using battery to compensate rate shortage of LFC reserve. J. Int. Counc. Electr. Eng. **6**(1), 146–152 (2016)
50. Padhan, S., Sahu, R.K., Panda, S.: Application of firefly algorithm for load frequency control of multi-area interconnected power system. Electr. Power Compon. Syst. **42**(13), 1419–1430 (2014)
51. Prakash, S., Sinha, S.K.: Load frequency control of multi-area power systems using neuro-fuzzy hybrid intelligent controllers. IETE J. Res. **61**(5), 526–532 (2015)
52. Shiva, C.K., Mukherjee, V.: Automatic generation control of hydropower systems using a novel quasi-oppositional harmony search algorithm. Electr. Power Compon. Syst. **44**(13), 1478–1491 (2016)
53. Sönmez, S., Ayasun, S.: Stability region in the parameter space of PI controller for a single-area load frequency control system with time delay. IEEE Trans. Power Syst. **31**(1), 829–830 (2016)
54. Tan, W., Chang, S., Zhou, R.: Load frequency control of power systems with governor dead-band (GDB) non-linearity. Electr. Power Compon. Syst. **45**(12), 1305–1314 (2017)
55. Vrdoljak, K., Perić, N., Petrović, I.: Applying optimal sliding mode based load-frequency control in power systems with controllable hydro power plants. Automatika **51**(1), 3–18 (2010)
56. Kumar, N.V., Ansari, M. M.T.: A new design of dual mode Type-II fuzzy logic load frequency controller for interconnected power systems with parallel AC–DC tie-lines and capacitor energy storage unit. Int. J. Electr. Power & Energy Syst. **82**, 579–598 (2016)

57. Prasad, S., Purwar, S., Kishor, N.: Load frequency regulation using observer based non-linear sliding mode control. Int. J. Electr. Power Energy Syst. **104**, 178–193 (2019)
58. Lu, K., Zhou, W., Zeng, G., Zheng, Y.: Constrained population extremal optimization-based robust load frequency control of multi-area interconnected power system. Int. J. Electr. Power Energy Syst. **105**, 249–271 (2019)
59. Chen, M.R., Zeng, G.Q., Xie, X.Q.: Population extremal optimization-based extended distributed model predictive load frequency control of multi-area interconnected power systems. J. Frankl. Inst. (2018)
60. Saxena, S.: Load frequency control strategy via fractional-order controller and reduced-order modeling. Int. J. Electr. Power Energy Syst. **104**, 603–614 (2019)
61. Dhundhara, S., Verma, Y. P.: Capacitive energy storage with optimized controller for frequency regulation in realistic multisource deregulated power system. Energy **147**, 1108–1128 (2018)
62. Morsali, J., Zare, K., Hagh, M.T.: A novel dynamic model and control approach for SSSC to contribute effectively in AGC of a deregulated power system. Int. J. Electr. Power Energy Syst. **95**, 239–253 (2018)
63. Xiong, L., Li, H., Wang, J.: LMI based robust load frequency control for time delayed power system via delay margin estimation. Int. J. Electr. Power Energy Syst. **100**, 91–103 (2018)
64. Ahmadi, A., Aldeen, M.: An LMI approach to the design of robust delay-dependent overlapping load frequency control of uncertain power systems. Int. J. Electr. Power Energy Syst. **81**, 48–63 (2016)
65. Balamurugan, S., Lekshmi, R.R.: Control strategy development for multi source multi area restructured system based on GENCO and TRANSCO reserve. Int. J. Electr. Power Energy Syst. **75**, 320–327 (2016)
66. Baral, K.K., Barisal, A.K., Mohanty, B.: Load frequency controller design via GSO algorithm for nonlinear interconnected power system. In: International Conference on Signal Processing, Communication, Power and Embedded System (SCOPES), pp. 662–668. IEEE, (2016)
67. Dhillon, S.S., Lather, J.S., Marwaha, S.: Multi objective load frequency control using hybrid bacterial foraging and particle swarm optimized PI controller. Int. J. Electr. Power & Energy Syst. **79**, 196–209 (2016)
68. Elsisi, M., Soliman, M., Aboelela, M.A.S., Mansour, W.: Bat inspired algorithm based optimal design of model predictive load frequency control. Int. J. Electr. Power Energy Syst. **83**, 426–433 (2016)
69. Madasu, S.D., Kumar, M.S., Singh, A.K.: Comparable investigation of backtracking search algorithm in automatic generation control for two area reheat interconnected thermal power system. Appl. Soft Comput. **55**, 197–210 (2017)
70. Abdelaziz, A.Y., Ali, E.S.: Cuckoo search algorithm based load frequency controller design for nonlinear interconnected power system. Int. J. Electr. Power Energy Syst. **73**, 632–643 (2015)
71. Ali, E.S., Abd-Elazim, S.M.: BFOA based design of PID controller for two area Load Frequency Control with nonlinearities. Int. J. Electr. Power Energy Syst. **51**, 224–231 (2013)
72. Anwar, M.N., Pan, S.: A new PID load frequency controller design method in frequency domain through direct synthesis approach. Int. J. Electr. Power Energy Syst. **67**, 560–569 (2015)
73. Barisal, A.K.: Comparative performance analysis of teaching learning based optimization for automatic load frequency control of multi-source power systems. Int. J. Electr. Power Energy Syst. **66**, 67–77 (2015)
74. Bernard, M.Z., Mohamed, T.H., Qudaih, Y.S., Mitani, Y.: Decentralized load frequency control in an interconnected power system using coefficient diagram method. Int. J. Electr. Power & Energy Syst. **63**, 165–172 (2014)
75. Bhatt, P., Ghoshal, S.P., Roy, R.: Load frequency stabilization by coordinated control of thyristor controlled phase shifters and superconducting magnetic energy storage for three types of interconnected two-area power systems. Int. J. Electr. Power Energy Syst. **32**(10), 1111–1124 (2010)

76. Chaturvedi, D.K., Umrao, R., Malik, O.P.: Adaptive polar fuzzy logic based load frequency controller. Int. J. Electr. Power Energy Syst. **66**, 154–159 (2015)
77. Daneshfar, F., Bevrani, H.: Multiobjective design of load frequency control using genetic algorithms. Int. J. Electr. Power Energy Syst. **42**(1), 257–263 (2012)
78. Dey, R., Ghosh, S., Ray, G., Rakshit, A.: H∞ load frequency control of interconnected power systems with communication delays. Int. J. Electr. Power Energy Syst. **42**(1), 672–684 (2012)
79. Francis, R., Chidambaram, I.A.: Optimized PI + load–frequency controller using BWNN approach for an interconnected reheat power system with RFB and hydrogen electrolyser units. Int. J. Electr. Power Energy Syst. **67**, 381–392 (2015)
80. Khodabakhshian, A., Pour, M.E., Hooshmand, R.: Design of a robust load frequency control using sequential quadratic programming technique. Int. J. Electr. Power Energy Syst. **40**(1), 1–8 (2012)
81. Khooban, Mohammad Hassan, and Taher Niknam. A new intelligent online fuzzy tuning approach for multi-area load frequency control: Self Adaptive Modified Bat Algorithm. Int. J. Electr. Power Energy Syst. **71**, 254–261 (2015)
82. Khuntia, S.R., Panda, S.: Simulation study for automatic generation control of a multi-area power system by ANFIS approach. Appl. Soft Comput. **12**(1), 333–341 (2012)
83. Maher, R.A., Mohammed, I.A., Ibraheem, I.K.: Polynomial based H∞ robust governor for load frequency control in steam turbine power systems. Int. J. Electr. Power Energy Syst **57**, 311–317 (2014)
84. Mohanty, B., Panda, S., Hota, P.K.: Controller parameters tuning of differential evolution algorithm and its application to load frequency control of multi-source power system. Int. J. Electr. Power Energy Syst. **54**, 77–85 (2014)
85. Naidu, K., Mokhlis, H., Bakar, A.A.: Multiobjective optimization using weighted sum artificial bee colony algorithm for load frequency control. Int. J. Electr. Power Energy Syst. **55**, 657–667 (2014)
86. Nikmanesh, E., Hariri, O., Shams, H., Fasihozaman, M.: Pareto design of load frequency control for interconnected power systems based on multi-objective uniform diversity genetic algorithm (muga). Int. J. Electr. Power Energy Syst. **80**, 333–346 (2016)
87. Oliveira, E.J., Honório, L.M., Anzai, A.H., Oliveira, L.W., Costa, E.B.: Optimal transient droop compensator and PID tuning for load frequency control in hydro power systems. Int. J. Electr. Power Energy Syst. **68**, 345–355 (2015)
88. Panda, S., Mohanty, B., Hota, P.K.: Hybrid BFOA–PSO algorithm for automatic generation control of linear and nonlinear interconnected power systems. Appl. Soft Comput. **13**(12), 4718–4730, (2013)
89. Pappachen, A., Fathima, A.P.: Load frequency control in deregulated power system integrated with SMES–TCPS combination using ANFIS controller. Int. J. Electr. Power Energy Syst. **82**, 519–534 (2016)
90. Ponnusamy, M., Banakara, B., Dash, S.S., Veerasamy, M.: Design of integral controller for load frequency control of static synchronous series compensator and capacitive energy source based multi area system consisting of diverse sources of generation employing imperialistic competition algorithm. Int. J. Electr. Power Energy Syst. **73**, 863–871 (2015)
91. Rahmani, M., Sadati, N.: Two-level optimal load–frequency control for multi-area power systems. Int. J. Electr. Power Energy Syst. **53**, 540–547 (2013)
92. Rakhshani, E., Remon, D., Rodriguez, P.: Effects of PLL and frequency measurements on LFC problem in multi-area HVDC interconnected systems. Int. J. Electr. Power Energy Syst. **81**, 140–152 (2016)
93. Roy, R., Bhatt, P., Ghoshal, S.P.: Evolutionary computation based three-area automatic generation control. Expert. Syst. Appl. **37**(8), 5913–5924 (2010)
94. Sahu, B.K., Pati, S., Mohanty, P.K., Panda, S.: Teaching–learning based optimization algorithm based fuzzy-PID controller for automatic generation control of multi-area power system. Appl. Soft Comput. **27**, 240–249 (2015)
95. Sahu, B.K., Pati, T.K., Nayak, J.R., Panda, S., Kar, S.K.: A novel hybrid LUS–TLBO optimized fuzzy-PID controller for load frequency control of multi-source power system. Int. J. Electr. Power Energy Syst. **74**, 58–69 (2016)

96. Sathya, M.R., Ansari, M.M.T.: Load frequency control using Bat inspired algorithm based dual mode gain scheduling of PI controllers for interconnected power system. Int. J. Electr. Power & Energy Syst. **64**, 365–374 (2015)
97. Shankar, R., Chatterjee, K., Bhushan, R.: Impact of energy storage system on load frequency control for diverse sources of interconnected power system in deregulated power environment. Int. J. Electr. Power Energy Syst. **79**, 11–26 (2016)
98. Shiroei, M., Toulabi, M.R., Ranjbar, A.M.: Robust multivariable predictive based load frequency control considering generation rate constraint. Int. J. Electr. Power Energy Syst **46**, 405–413 (2013)
99. Shiva, C.K., Mukherjee, V.: Automatic generation control of interconnected power system for robust decentralized random load disturbances using a novel quasi-oppositional harmony search algorithm. Int. J. Electr. Power Energy Syst. **73**, 991–1001 (2015)
100. Sondhi, S., Hote, Y.V.: Fractional order PID controller for perturbed load frequency control using Kharitonov's theorem. Int. J. Electr. Power Energy Syst. **78**, 884–896 (2016)
101. Sudha, K.R., Santhi, R.V.: Load frequency control of an interconnected reheat thermal system using type-2 fuzzy system including SMES units. Int. J. Electr. Power Energy Syst. **43**(1), 1383–1392 (2012)
102. Tan, W., Zhou, H.: Robust analysis of decentralized load frequency control for multi-area power systems. Int. J. Electr. Power & Energy Syst. **43**(1), 996–1005 (2012)
103. Toulabi, M.R., Shiroei, M., Ranjbar, A.M.: Robust analysis and design of power system load frequency control using the Kharitonov's theorem. Int. J. Electr. Power Energy Syst. **55**, 51–58 (2014)
104. Tsay, T.S.: Load–frequency control of interconnected power system with governor backlash nonlinearities. Int. J. Electr. Power Energy Syst. **33**(9), 1542–1549 (2011)
105. Yazdizadeh, A., Ramezani, M.H., Hamedrahmat, E.: Decentralized load frequency control using a new robust optimal MISO PID controller. Int. J. Electr. Power Energy Syst. **35**(1), 57–65, (2012)
106. Yousef, H.: Adaptive fuzzy logic load frequency control of multi-area power system. Int. J. Electr. Power Energy Syst. **68**, 384–395 (2015)
107. Sahu, R.K., Sekhar, G.C., Panda, S.: DE optimized fuzzy PID controller with derivative filter for LFC of multi source power system in deregulated environment. Ain Shams Eng. J. ELSEVIER **6**, 511–530 (2015)
108. Sahu, R.K., Gorripotu, T.S., Panda, S.: A hybrid DE–PS algorithm for load frequency control under deregulated power system with UPFC and RFB. Ain Shams Eng. J. ELSEVIER **6**, 893–911 (2015)
109. Gorripotu, T.S., Sahu, R.K., Panda, S.: AGC of a multi-area power system under deregulated environment using redox flow batteries and interline power flow controller. Eng. Sci. Technol. Int. J. ELSEVIER **18**, 555–578 (2015)
110. Sekhar, G.C., Sahu, R.K., Panda, S.: Load frequency control of power system under deregulated environment using optimal firefly algorithm. Int. J. Electr. Power Energy Syst. ELSEVIER **74**, 195–211 (2016)
111. Sahu, R.K., Sekhar, G.C., Panda, S.: Automatic generation control of multi-area power systems with diverse energy sources using Teaching Learning Based Optimization algorithm. Eng. Sci. Technol., Int. J. ELSEVIER **19** 113–134 (2016)
112. Sahu, R.K., Panda, S., Biswal, A., Sekhar, G.C.: Design and analysis of tilt integral derivative controller with filter for load frequency control of multi-area interconnected power systems. ISA Trans. ELSEVIER **61**, 251–2644 (2016)
113. Sekhar, G.C., Sahu, R.K., Panda, S.: Firefly algorithm optimised PID controller for automatic generation control with redox flow battery. Int. J. Comput. Syst. Eng. INDERSCIENCE **3**(1/2), 48–57, (2017)
114. Ganapathy, S., Velusami, S.: MOEA based design of decentralized controllers for LFC of interconnected power systems with nonlinearities, AC–DC parallel tie-lines and SMES units. Energy Convers. Manag. **51**(5), 873–880 (2010)

115. Bhatt, P., Roy, R., Ghoshal, S.P.: Optimized multi area AGC simulation in restructured power systems. Int. J. Electr. Power Energy Syst. **32**(4), 311–322 (2010)
116. Tan, W.: Unified tuning of PID load frequency controller for power systems via IMC. IEEE Trans. Power Syst. **25**(1), 341–350 (2010)
117. Rasolomampionona, D., Anwar, S.: Interaction between phase shifting transformers installed in the tie-lines of interconnected power systems and automatic frequency controllers. Int. J. Electr. Power Energy Syst. **33**(8), 1351–1360 (2011)
118. Golpira, H., Bevrani, H.: Application of GA optimization for automatic generation control design in an interconnected power system. Energy Convers. Manag. **52**(5), 2247–2255 (2011)
119. Saikia, L.C., Nanda, J., Mishra, S.: Performance comparison of several classical controllers in AGC for multi-area interconnected thermal system. Int. J. Electr. Power Energy Syst. **33**(3), 394–401 (2011)
120. Ali, E.S., Abd-Elazim, S.M.: Bacteria foraging optimization algorithm based load frequency controller for interconnected power system. Int. J. Electr. Power Energy Syst. **33**(3), 633–638 (2011)
121. Sudha, K.R., Santhi, R.V.: Robust decentralized load frequency control of interconnected power system with generation rate constraint using type-2 fuzzy approach. Int. J. Electr. Power Energy Syst.: **33**(3), 699–707 (2011)
122. Bhatt, P., Roy, R., Ghoshal, S.P.: Comparative performance evaluation of SMES–SMES, TCPS–SMES and SSSC–SMES controllers in automatic generation control for a two-area hydro–hydro system. Int. J. Electr. Power Energy Syst. **33**(10), 1585–1597 (2011)
123. Gozde, H., Taplamacioglu, M.C., Kocaarslan, I.: Comparative performance analysis of Artificial Bee Colony algorithm in automatic generation control for interconnected reheat thermal power system. Int. J. Electr. Power Energy Syst. **42**(1), 167–178 (2012)

How Effective is the Salp Swarm Algorithm in Data Classification

Nibedan Panda and Santosh Kumar Majhi

Abstract In this paper, the Salp Swam Algorithm (SSA) is deployed in training the Multilayer Perceptron (MLP) for the task of data classification. The UCI machine learning repository standard datasets are used for evaluation of the proposed SSA based MLP. The performance of the proposed method is verified by considering various standard classification measures over the considered benchmark datasets. The result obtained by the proposed method is compared with other evolutionary algorithms such as Genetic Algorithm (GA), Particle Swarm Optimization (PSO), Differential Evolution (DE), Grey Wolf Optimization (GWO), etc. From the simulation study, it has been verified that the proposed method shows supremacy in results as compared with other evolutionary algorithm based MLP.

Keywords Classification · Salp swarm algorithm · PSO · DE · GWO · GA

1 Introduction

In everyday life, we are facing many situations where we have to take decision whether we have to accept or reject based on our past experience. In data mining, classification is a supervised learning technique which divides things into corresponding groups and assigns analogous class labels to unidentified patterns. Its objective is to create a model from the training dataset. It is a two-step approach problem: the training phase (or learning step) and the classification phase. The goal of classification is to predict the exact class or label. In the current scenario the application of classification task varies to different dimensions in the field of engineering and science such as text classification [1], document classification [2], biomedical data classification and gene expression classification [3], fault classification [4], sentiment classification

N. Panda (✉) · S. K. Majhi
Department of Computer Science and Engineering, Veer Surendra Sai University of Technology, Burla 768018, Odisha, India
e-mail: nibedan.panda@gmail.com

S. K. Majhi
e-mail: smajhi_cse@vssut.ac.in

A. K. Das et al. (eds.), *Computational Intelligence in Pattern Recognition*, Advances in Intelligent Systems and Computing 999,
https://doi.org/10.1007/978-981-13-9042-5_49

[5], image classification [6], video classification [7], voice classification [8], internet traffic intercept classification [9], and others. Major classification methods are Naive Bayes [10], Support Vector Machine [11], Artificial Neural Network [12], Decision Tree [13], and *K*-Nearest Neighbor [14]. Nowadays, researches and engineers are applying the Artificial Neural Network (ANN) optimized by evolutionary algorithms for solving the classification task with high accuracy.

Artificial Neural Network (ANN) is an information processing model, inspired by biological nervous systems. This is one of the highest achievement in the area of Artificial Intelligence. It is composed of a large number of highly interconnected processing elements called neurons. There are three layers such as input, hidden, and output. ANN allows for nonlinear classification. It works similar to human brain and learns by example meant for specific application: data classification, pattern recognition, and regression analysis. In 1943, McCulloch and Pitts first gave the idea of neural network as computing machine [15]. Currently a variety of different ANNs proposed in the literature: Feedforward neural network [16], Radial basis function neural network, Kohonen self-organizing neural network, Recurrent neural network [1], Convolutional neural network, Modular neural network. Some higher order ANNs are FLANN (Functional link artificial neural network) [17], PSNN (Pi-sigma neural network) [18] and JPSNN (Jordan pi-sigma neural network) [18].

Though there are lot of differences among ANNs, the learning process for all is common. It will learn from own experience. Multilayer perceptron (MLP) neural networks are the most wide spread and popular among the ANNs. They are automatically configured to take a given set of inputs. The learning processes for ANNs are classified into two categories: supervised and unsupervised. In case of supervised learning the ANNs accustomed itself to the feedbacks from the external source and for unsupervised learning it is accustomed to getting input without support of external feedback. Actually the effectiveness of ANNs is completely dependent on this learning nature. Again supervised learning method learns by two ways: gradient-based learning and heuristic approach for learning. One of the best example of gradient-based technique is back-propagation method. The advantages of back-propagation algorithm are its versatility and simplicity. But apart from these advantages there are few disadvantages of back-propagation method such as: very slow, requires small learning rate for stable learning, trapped during the process of local minimum finding, and finally the result is highly dependent on the primary inputs given.

So, to avoid such demerits of gradient-based method metaheuristic search algorithms is a best alternative. Quite good number of literature is available for optimization of the MLP using the heuristic search based algorithm. The merits of heuristic algorithm are that it can give faster and relatively inexpensive feedback to the trainer by randomly assigning the initial essential parameters. Allocating the correct heuristic approach may lead to best corrective measure. Also it will not suffer the drawback from local optima trap. But as compared to gradient-based technique it is much slower. Nowadays, a number of population-based optimization techniques are used to train the MLP structure. The population-based metaheuristic algorithms give more accurate result for finding local optimal solution. Evolutionary and swarm-based algorithms are the most commonly used metaheuristic method. The popular population-based approach available in the literature are: Genetic Algorithm (GA) [19], Differential Evolution (DE) [20], Particle Swarm Optimization (PSO) [21], Ant Colony Optimization (ACO) [22], Artificial Bee Colony(ABC) [23]. In this paper we have used a recently developed optimization technique named as Salp Swarm Algorithm for training the MLP to solve classification problem. The SSA is selected due to its proven advantages such as avoidance of local minima by balancing the process of exploitation and exploration.

The next part of the paper is organized as follows: In Sect. 2 the materials and methods are described. Section 3 represents overview of result and discussion. Finally the summary of findings of this work is concluded in Sect. 4.

2 Materials and Methods

2.1 Overview of MLP

MLP belongs to the class of feed-forward neural network which consists of three layers such as: input layer, output layer, and hidden layer. The first layer, i.e., input layer receives the input signal and maps it to the network. The final layer is the output layer which is responsible for making decision or prediction about the inputs. The hidden layer which may be selected as arbitrary that is present in between input and output layer and it is responsible for the true computation. All the layers communicate by the help of a set of processing elements called neurons. The connecting path through the neurons goes in one direction only, means the connection is a directed graph among the layers. A basic MLP structure is given in Fig. 1.

MLP is most suitable for solving tasks which require supervised learning approach. The training process contains a set of input–output pairs and examines what are the dependencies among these pairs. The training process involves adjust-

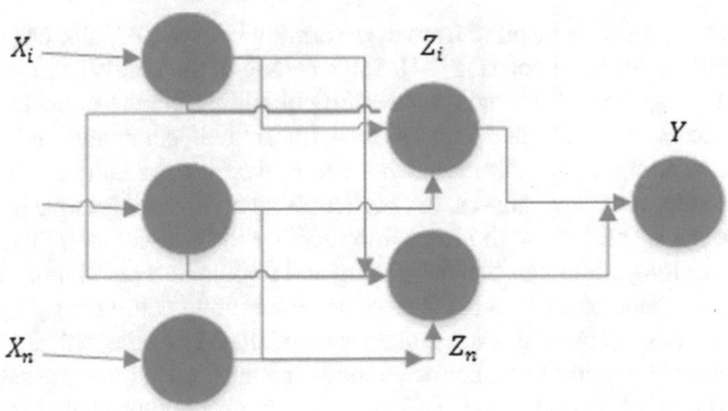

Fig. 1 Basic structure of MLP

ing the parameters, i.e., weights and biases of the network to optimize the error. The calculated error can be represented in variety of ways, popularly root mean square error (RMSE).

The Fig. 1 shows typical representation of MLP which contains one hidden layer. The desired output in case of MLP can be formulated in multiple steps:

Step1: The summation of inputs is evaluated by Eq. (1).

$$p_j = \sum_{i=1}^{n} w_{ij}x_j + b_j, \qquad j = 1, 2, \ldots m \tag{1}$$

Step2: Activation function is required to get the output of that layer and feed it as input to the next layer. Generally in the literature sigmoid is the most preferred activation function. The output from the hidden layer can be evaluated as given in Eq. (2).

$$P_j = \frac{1}{(1 + \exp(-p_j))}, \qquad j = 1, 2, \ldots m \tag{2}$$

Step3: The network final output can be evaluated as given in Eq. (3).

$$y_k = \sum_{i=1}^{m} w_{ih,p_j} + b_h, \qquad h = 1,2, \ldots m \tag{3}$$

$$Y_k = \frac{1}{1 + \exp(-y_k)}, \qquad h = 1, 2 \ldots m \tag{4}$$

2.2 *Salp Swarm Algorithm*

Salp swarm algorithm is inspired from the swarming behavior of salps based on their foraging and navigation in oceans [24]. Salps belong to the family salpidae of order salpida. They are one of the most successful colonizing marine animals. Salp is a planktonic tunicate which has a barrel shaped, transparent, gelatinous body. The salp contracts and pumps water through its body to move forward. During the life cycle of an adult salp, it has two phases. In the first phase a salp swims alone and lives in solitary. In the second phase the salps join together to form colonies. The colonies of salps form long chains for better foraging and improved locomotion. The chains may be in the shape of a wheel or a line or any other architectural design.

The swarming behavior of salps and the swarm chain is mathematically designed to solve optimization problems. For better modeling the population in the salp chain is categorized into leader and followers. The salp at the front of the chain is considered as leader and the rest of the salps in the chain are categorized as followers. The leader guides the chain and the followers follow the leader and also each other.

For mathematical modeling the position of all salps are taken in a two dimensional matrix x and there is another variable F, which is considered as the food source in the search space. The food source is the swarm's target. The position of the leader is updated using Eq. (5).

$$x_j^1 = \begin{cases} F_j + C_1((ub_j - lb_j)c_2 + lb_j)c_3 \geq 0 \\ F_j - C_1((ub_j - lb_j)c_2 + lb_j) \quad C_3 < 0 \end{cases} \tag{5}$$

where

x_j^1 Position of the leader in jth dimension
F_j Position of the food source in jth dimension
ub_j Upper bound of jth dimension
lb_j Lower bound of jth dimension

C_1 is a factor which is responsible for exploration and exploitation. It can be formulated as given in Eq. (5).

$$C_1 = 2e^{-\left(\frac{4l}{L}\right)^2} \tag{6}$$

Here l is the current iteration and L is the maximum number of iteration.

C_2, C_3 are random number in the interval [0, 1].

The position of followers is updated using Eq. (7).

$$x_j^i = \frac{1}{2}\left(x_j^i + x_j^{i-1}\right) \tag{7}$$

where $i \geq 2$

x_j^1 Position of the ith follower salp in jth dimension.
x_j^{i-1} Position of the *(i-1) th* follower salp in jth dimension

2.3 Proposed SSA Optimized MLP

In the current section, we elaborate the proposed approach based on Salp swarm algorithm for training MLP network, termed as SSA-MLP. The MLP network contains only one hidden layer for training process. For getting optimum classification accuracy using MLP, the most important aspect is the parameter setting, i.e., weights and biases. As discussed in the introduction, the initial parameters must be chosen randomly for metaheuristic problems, the trainer should provide a set of initial values as weights and biases. So the important parameters for SSA algorithm are weights, biases, and the objective function. The weights and biases are represented for SSA in Eqs. (8) and (9):

$$W = W_{1,1}, W_{1,2}, \ldots, \ldots . W_{i,j} \tag{8}$$

$$B_j = b_1, b_2, \ldots \ldots, b_j \tag{9}$$

where, i represents the number of input nodes, j represents number of hidden nodes, W_{ij} represents the corresponding weights from input to hidden node and B_j represents corresponding biases of the hidden node.

To achieve optimum accuracy which is the objective of training MLP a popular metric is used in the literature, i.e., root mean square error (RMSE). The RMSE is evaluated by calculating the difference between the output we received from the MLP and the target output. The RMSE is evaluated by the Eq. (10)

$$RMSE = \sqrt{\frac{\sum_{i=1}^{n} \left(Y_i^k - T_i^k\right)^2}{n}} \tag{10}$$

where n is the number of outputs, T_i is the target output and Y_i is the actual output. After finding the RMSE the effectiveness of the MLP can be evaluated by calculating the average of RMSE. By providing the initial weights and biases to the MLP, SSA receives the RMSE value for all the inputs. But to reduce the RMSE error rate SSA algorithm randomly changes the weights and biases repeatedly. Though the weights and biases are randomly chosen and iteratively SSA supplies the values so there is a chance of getting better result after each iteration. There is no guarantee that SSA always gives optimum result for a dataset due to the heuristic nature of the algorithm. However after each iteration the RMSE will be reduced. Finally it will give a better result as compared to the initial random values assigned.

3 Result and Discussion

The proposed SSA-based MLP is evaluated using six standard classification datasets available in the UCI machine learning repository such as: balloon, iris, cancer, heart, liver, and diabetes in terms of classification accuracy. The detailed description of the datasets is given in Table 1. There are some standard parameters to evaluate the effectiveness of the said model. Some important parameters such as: error rate, accuracy, sensitivity, specificity, and prevalence. We have evaluated all the above standard parameters for DE-MLP, GA-MLP, PSO-MLP, GWO-MLP [25], and compared with the proposed SSA-MLP.

Table 1 Structure of MLP for different datasets

Classification of datasets	Number of attributes	MLP structure
Balloon	4	4-9-1
Cancer	9	9-9-1
Diabetes	8	8-9-1
Heart	13	13-9-1
Iris	4	4-9-1
Liver	6	6-9-1

3.1 Experimental Setup

The proposed method and other considered algorithms are implemented in MATLAB R2014 environment. The datasets are divided into training and testing sets in the ratio 70:30. The train and test samples for the datasets are specified in Table 2. All the algorithms are evaluated for 20 numbers of generations. The results of each algorithm for all the considered datasets are reported in Tables 3 and 4.

a. Results

The results are reported in terms of minimum, average, and standard deviation of RMSE, sensitivity, specificity, prevalence, accuracy, and error rate. It is clear from Table 3 that the proposed SSA trained MLP gives better performance than other algorithms. The SSA algorithm reported better accuracy rate when compared with GA, PSO, and DE. However, SSA-MLP gives challenging result in comparison with GWO-MLP for the datasets such as cancer, balloon, and iris. The SSA-MLP avoids local minima and provides global solution as it balances the process of exploitation and exploration. The training problem of MLP needs local avoidance to perform better. The SSA trained MLP is a potential method for classification as it avoids local optima.

Table 2 Description of used standard datasets

Classification datasets	Number of attributes	Number of training samples	Number of test samples	Number of class
Balloon	4	20	16	2
Cancer	9	683	120	2
Diabetes	8	768	150	2
Heart	13	270	60	2
Iris	4	150	150	3
Liver	6	345	70	2

Table 3 Experimental results from datasets

Dataset\algorithm		DE	GA	GWO	PSO	SSA
Balloon	MIN RMSE	0.4233	0.4929	0.3829	0.4264	0.4555
	AVG	0.2144	0.4960	0.4272	0.4574	0.4985
	STD	0.0158	0.0021	0.0345	0.0148	0.0151
	SPECIFICITY	1	1	1	1	1
	SENSITIVITY	1	1	1	1	1
	PREVALENCE	50	50	50	50	50
Cancer	MIN RMSE	0.2551	0.2846	0.3384	0.3419	0.3733
	AVG	0.2852	0.2886	0.4087	0.3944	0.4030
	STD	0.0141	0.0034	0.0512	0.0250	0.0160
	SPECIFICITY	0.9662	0.9662	0.9662	0.9662	0.9662
	SENSITIVITY	1	1	1	1	1
	PREVALENCE	50	50	50	50	50
Diabetes	MIN RMSE	0.4418	0.4528	0.4601	0.4564	0.4632
	AVG	0.4512	0.4581	0.4729	0.4700	0.4678
	STD	0.0026	0.0043	0.0077	0.0063	0.0030
	SPECIFICITY	0.6712	0.6283	0.6422	0.5713	0.6856
	SENSITIVITY	0.8141	0.8422	0.8283	0.8714	0.8286
	PREVALENCE	40.71	42.14	41.42	43.57	41.42
Heart	MIN RMSE	0.4735	0.4556	0.3658	0.4890	0.4674
	AVG	0.4889	0.4629	0.3812	0.4956	0.4799
	STD	0.0072	0.0038	0.0099	0.0028	0.0085
	SPECIFICITY	0.7400	0.7777	0.7400	0.6666	0.8888
	SENSITIVITY	0.8141	0.8141	0.8511	0.8888	0.7777
	PREVALENCE	40.74	40.74	42.59	44.44	38.88
Iris	MIN RMSE	0.1921	0.2056	0.1764	0.2053	0.2206
	AVG	0.2144	0.2120	0.1983	0.2251	0.2807
	STD	0.0158	0.0050	0.0116	0.0165	0.0372
Liver	MIN RMSE	0.4930	0.4880	0.4918	0.4928	0.4922
	AVG	0.2144	0.2120	0.1983	0.2251	0.2807
	STD	0.0158	0.0050	0.0116	0.0165	0.0372
	SPECIFICITY	0.5000	0.5295	0.4413	0.4413	0.5882
	SENSITIVITY	0.6767	0.6175	0.8232	0.8232	0.6763
	PREVALENCE	33.82	30.88	41.17	41.17	33.82

Table 4 Accuracy results of data sets

Dataset\Algorithm		DE	GA	GWO	PSO	SSA
Balloon	Error rate (%)	0	0	0	0	0
	Accuracy (%)	100	100	100	100	100
Cancer	Error rate (%)	1.6	1.6	1.6	1.6	1.6
	Accuracy (%)	98.4	98.4	98.4	98.4	98.4
Diabetes	Error rate (%)	25.7	26.4	26.4	27.8	24.2
	Accuracy (%)	74.3	73.6	73.6	72.2	75.8
Heart	Error rate (%)	22.2	20.3	20.3	22.2	16.6
	Accuracy (%)	77.8	79.7	79.7	77.8	83.4
Iris	Error rate (%)	23.3	13.3	3.3	10	3.3
	Accuracy (%)	76.7	86.7	96.7	90	96.7
Liver	Error rate (%)	41.1	42.6	36.7	36.7	36.7
	Accuracy (%)	58.9	57.4	63.3	63.3	63.3

4 Conclusion

In this paper classification accuracy of six standard datasets are evaluated by using SSA trained MLP. The initial set of values, i.e., weights and biases are chosen randomly and after successive iteration the initial parameters reach its optimality after iterative training process. The proposed SSA-MLP is compared with other four population-based heuristic algorithms such as: DE-MLP, GA-MLP, PSO-MLP, and GWO-MLP. Result shows that the SSA-MLP exhibits better accuracy for classification and also gives optimal set of weights and biases. The SSA successfully avoids local optima problem as it balances the exploitation and exploration.

References

1. Yogatama, D., Dyer, C., Ling, W., Blunsom, P.: Generative and discriminative text classification with recurrent neural networks (2017). arXiv:1703.01898
2. Parlak, B. Uysal, A.K.: On feature weighting and selection for medical document classification. In: Developments and Advances in Intelligent Systems and Applications, pp. 269–282. Springer, Cham (2018)
3. Ludwig, S.A., Picek, S. Jakobovic, D.: Classification of cancer data: analyzing gene expression data using a fuzzy decision tree algorithm. In: Operations Research Applications in Health Care Management, pp. 327–347. Springer, Cham. (2018)
4. Abdelgayed, T., Morsi, W. Sidhu, T.: A new approach for fault classification in microgrids using optimal wavelet functions matching pursuit. IEEE Trans. Smart Grid (2017)
5. Arulmurugan, R., Sabarmathi, K.R. Anandakumar, H.: Classification of sentence level sentiment analysis using cloud machine learning techniques. Clust. Comput. 1–11 (2017)
6. Maggiori, E., Tarabalka, Y., Charpiat, G., Alliez, P.: Convolutional neural networks for large-scale remote-sensing image classification. IEEE Trans. Geosci. Remote Sens. **55**(2), 645–657 (2017)

7. Adigun, O. Kosko, B.: Using noise to speed up video classification with recurrent backpropagation. In 2017: International Joint Conference on Neural Networks (IJCNN), pp. 108–115. IEEE (2017, May)
8. Verde, L., De Pietro, G., Sannino, G.: A methodology for voice classification based on the personalized fundamental frequency estimation. Biomed. Signal Process. Control. **42**, 134–144 (2018)
9. Achunala, D., Sathiyanarayanan, M. Abubakar, B.: Traffic classification analysis using omnet ++. In: Progress in Intelligent Computing Techniques: Theory, Practice, and Applications, pp. 417–422. Springer, Singapore (2018)
10. Rish, I.: An empirical study of the naive Bayes classifier. In: IJCAI 2001 Workshop on Empirical Methods in Artificial Intelligence, vol. 3(22), pp. 41–46. IBM, New York (2001, August)
11. Romero, E., Toppo, D.: Comparing support vector machines and feedforward neural networks with similar hidden-layer weights. IEEE Trans. Neural Netw. **18**(3), 959–963 (2007)
12. Hopfield, J.J.: Artificial neural networks. IEEE Circuits Devices Mag. **4**(5), 3–10 (1988)
13. Lajnef, T., Chaibi, S., Ruby, P., Aguera, P.E., Eichenlaub, J.B., Samet, M., Kachouri, A., Jerbi, K.: Learning machines and sleeping brains: automatic sleep stage classification using decision-tree multi-class support vector machines. J. Neurosci. Methods. **250**, 94–105 (2015)
14. Zhang, S., Li, X., Zong, M., Zhu, X., Wang, R.: Efficient knn classification with different numbers of nearest neighbors. IEEE Trans. Neural Netw. Learn. Syst. **29**(5), 1774–1785 (2018)
15. McCulloch, W.S., Pitts, W.: A logical calculus of the ideas immanent in nervous activity. Bull. Math. Biophys. **5**(4), 115–133 (1943)
16. Majhi, S.K.: An efficient feed foreword network model with sine cosine algorithm for breast cancer classification. Int. J. Syst. Dyn. Appl. (IJSDA). **7**(2), 1–14 (2018)
17. Naik, B., Nayak, J., Behera, H.S. Abraham, A.: A harmony search based gradient descent learning-FLANN (HS-GDL-FLANN) for classification. In: Computational Intelligence in Data Mining, vol. 2, pp. 525–539. Springer, New Delhi. (2015)
18. Nayak, J., Naik, B., Behera, H.S. Abraham, A.: Particle swarm optimization based higher order neural network for classification. In: Computational Intelligence in Data Mining, vol. 1, pp. 401–414. Springer, New Delhi (2015)
19. Goldberg, D.E., Holland, J.H.: Genetic algorithms and machine learning. Mach. Learn. **3**(2), 95–99 (1988)
20. Qin, A.K., Huang, V.L., Suganthan, P.N.: Differential evolution algorithm with strategy adaptation for global numerical optimization. IEEE Trans. Evol. Comput. **13**(2), 398–417 (2009)
21. Kennedy, J.: Particle swarm optimization. In: Encyclopedia of machine learning, pp. 760–766. Springer, Boston, MA (2011)
22. Mavrovouniotis, M., Yang, S.: Training neural networks with ant colony optimization algorithms for pattern classification. Soft. Comput. **19**(6), 1511–1522 (2015)
23. Xue, Y., Jiang, J., Zhao, B., Ma, T.: A self-adaptive artificial bee colony algorithm based on global best for global optimization. Soft Comput. 1–18 (2017)
24. Mirjalili, S., Gandomi, A.H., Mirjalili, S.Z., Saremi, S., Faris, H., Mirjalili, S.M.: Salp Swarm Algorithm: A bio-inspired optimizer for engineering design problems. Adv. Eng. Softw. **114**, 163–191 (2017)
25. Mirjalili, S.: How effective is the Grey Wolf optimizer in training multi-layer perceptrons. Appl. Intell. **43**(1), 150–161 (2015)

Feature Selection Using Graph-Based Clustering for Rice Disease Prediction

Ankur Das, Ratnaprava Dutta, Sunanda Das and Shampa Sengupta

Abstract Rice is one of the main crop cultivated all over the world. It is attacked by several diseases and pests that reduce quantity and degrades quality of the product. There are different types of rice diseases like Leaf Blast, Brown Spot, Stem Rot, Bakanae, yellow Dwarf which damage various parts of the plant. Disease identification at the early stage and taking precautions in time help the farmers to sustain both the quality and quantity of the product. The present work extracts different types of features from the disease portions (i.e., images) of the plants and identifies the most valuable features that can distinguish the disease types. To identify the most valuable features, initially, a weighted graph is constructed with extracted features as nodes and similarity between every pair of features as the weight of the corresponding edge. Based on the weights assigned to the edges, importance of each node of the graph is calculated. Finally, a graph-based clustering algorithm namely, Infomap clustering algorithm is applied on the graph to partition it into a set of connected subgraphs. Then the most influential node from each subgraph of the partition is selected and this subset of nodes is considered as important feature subset useful for rice disease prediction. The experimental result shows the effectiveness of the proposed method.

A. Das · R. Dutta
Department of Computer Science and Engineering, Calcutta Institute of Engineering and Management, Tollygunge, Kolkata, West Bengal, India
e-mail: ankurdas8017@gmail.com

R. Dutta
e-mail: rdutta97.rd@gmail.com

S. Das (✉)
Department of Computer Science and Engineering, The Neotia University, Diamond Harbour Road, South 24 Pgs, Sarisha, West Bengal, India
e-mail: das.sunanda2012@gmail.com

S. Sengupta
Department of Information Technology, MCKV Institute of Engineering, Liluah, Howrah, West Bengal, India
e-mail: shampa2512@yahoo.co.in

A. K. Das et al. (eds.), *Computational Intelligence in Pattern Recognition*, Advances in Intelligent Systems and Computing 999,
https://doi.org/10.1007/978-981-13-9042-5_50

Keywords Feature extraction · Feature selection · Graph based clustering · Infomap algorithm · Rice plant disease · Classification

1 Introduction

Oryza sativa (also known as rice) is a principal food for major portion of the world's population. According to the reports of Thai Rice Exporters Association, India is the largest exporter of rice in the world. In India, rice production plays a vital role in the national economy. Various diseases caused by bacteria, viruses, or fungi significantly reduce rice grain production and quality. It is a big problem in rice farming due to farmers lack of knowledge in different rice diseases, and difficulty to identify the disease. To overcome from this problem, farmers need correct identification and good understanding of the mechanisms to deal with diseases. But for proper identification of the diseases and correct timing for pesticides, farmers either have to depend on guide books or on their own experiences which results in a long process. An automated system based on image processing and data mining techniques, can help for disease classification of rice plant. The system can give alert to the farmers regarding the possibilities of attacks on different organs starting from the very initial stage and the farmers can take the necessary precautions. Thus the automated system will be very useful to the farmers for crop cultivation. Image collection, image preprocessing, image segmentation, feature extraction, and feature selection are the basic tasks of an automated system. The rice plant images are captured and processed to predict diseases that occur at different parts of the plant. Finally, after feature selection, image diseased dataset is prepared to build classification models. These classification models are too efficient to discriminate distinct diseases and to predict future data trends. To create a database of rice disease, acquisition of rice images from farm is a necessary task for which a good quality and high-resolution digital camera like Nikon D810, Nikon Coolpix A100, Canon 450D, Canon PowerShot A1300 is mainly used nowadays. Still the original image may be degraded due to noise, shadow, etc. This degradation of image leads to erroneous pixel extraction which results in the wrong feature selection. So the main and primary task is to reduce the noises and enhance the quality of the captured images. For this reason, the diseased portions of the plants are separated from each other and from the background images. Then that images have to be resized and cropped to reduce the processing time. The presence of dust, dewdrops, insect's excrements on the plant may degrade the quality of the images which are considered as image noise. Moreover, water drop effects and shadow effect on captured images can create issues in segmentation of image and extraction of features from the images. In image restoration, several noise removal filters help to remove distortion. The captured image with shadow effect and low contrast which causes wrong feature extraction, can be ameliorated by removing blurring effect with the help of various filtering schemes and increasing contrast using various contrast enhancement algorithms. Image resizing and cropping are crucial task for processing of the images taken with high definition cameras because these images are very big in size. Not only that but also image reduction means reduction of the computing memory. The reliability of an optical inspection can significantly be increased by image preprocessing. Some image preprocessing techniques are explained below which are mainly used for improving the image quality before image database preparation.

A. Noise Removal: Noise Removal is a technique to remove the noise from the captured image in such a way that the original image is distinguishable. The captured images may get certain noises from the environment. There are various noise removal techniques [1] by which noises can be removed from the acquired image. To handle with Gaussian noise, the average filter method is used. To reduce or eliminate salt and pepper noise and subsequently sharpen the edges, median filtering is a very useful digital filtering scheme. Moreover, Bilateral filter is another method for noise removal and smoothing the images where, intensity value at each pixel of an image is replaced by a weighted average of intensity values from nearby pixels.

B. Shadow Removal: It [2] is a crucial preprocessing task for feature extraction. The captured image often may get shadow due to occluding of light from light source by some objects. The aim of the task is to segregate the shadow and improve performance of computer vision applications. Vegetative indices help to extract background pixels from the image to measure hydrological variables like leaf area index. Canny edge detection algorithm is one kind of algorithm for strip detection and shadow pixel removal.

C. Image Enhancement: Image enhancement methods [3] helps to enhance the quality of the image by contrast stretching, histogram equalization, gamma correction, log transform, etc. It improves the local contrast and enhances the edges in each region of an image. For this purpose, adaptive histogram is used in order to carry out contrast enhancement by equalizing the pixel intensity value.

D. Image Segmentation: Image segmentation is an important and challenging process which divides an image into multiple segments with similar features and properties. The objective of segmentation is to make an image more easier to analyze. It helps to separate infected regions from noninfected regions. Several segmentation methods are used in different types of applications which are mentioned in the literature [4]. Fermi energy based segmentation, K-means clustering based segmentation, Otsu's Method, Region-based segmentation, Boundary Segmentation, etc., are some frequently used image segmentation techniques which are also used for rice plant diseased image segmentation purpose. Generally, image segmentation performs in two different ways: (i) based on discontinuities and (ii) based on similarities. In the first case, an image is partitioned using certain changes in intensity values, e.g., edge detection. And in the second case, images are partitioned based on the specific threshold values, e.g., Otsu's method.

E. Feature Extraction: The features of image is capturing visual properties of an image which may be either global for the entire image or local for a small group of pixels. The relevant features extraction [5] are one of the most important task of the researchers for disease prediction. Greenish leaf color indicates healthy paddy leaf and it's a sign of good plant growth. But when Fungal and bacterial diseases infect paddy leaves, the color of the leaves become dark brown or grayish green in color and the different colored spots of different shapes become visible on leaves. So, both the color and shape of the diseased part are considered as features to identify the diseases; area, axis, angle, etc., belong to the shape based features.

The rest of the paper is organized as follows: Sect. 2 describes different types of rice plant diseases. Section 3 discusses different image preprocessing techniques applied on the diseased images and different features extracted from the processed image. Graph-based clustering algorithm for partitioning the graph into subgraphs to select important features for disease prediction is described in Sect. 4. Section 5 presents the experimental result using various classifiers to express the effectiveness of the proposed method and finally, the paper is concluded in Sect. 6.

2 Rice Plant Diseases

The diseases may occur at different parts of the rice plant at any time during its entire life. In India, Leaf Blast, Sheath Rot, Brown Spot, and Bacterial Blight are the diseases that occur frequently in rice plant. At initial stage, a sign of white to gray-green spots with dark green borders is visible for Leaf Blast disease. Older spots of infected leaves are either of elliptical or of spindle shaped and also whitish to gray centers surrounded by red to brownish or necrotic border. Some spots are of diamond shape where the center is wide and pointed toward either ends. With the time, the spots can enlarge day by day, which results to kill the entire leaves. Brown spot is one kind of fungal disease that infects leaf as well as leaf sheath. If any seedling gets infected, then a small, circular, yellow brown or brown spot may appear on leaves. At the beginning of tillering stage, spots of small, circular, and dark brown to purple-brown spots can be observed on the leaves. Fungi produce the toxin which makes the spots of circular to oval with a light brown to gray center, surrounded by a reddish brown margin. Stem Rot is a fungus infected disease in the stem with tiny white and black sclerotia and mycelium. It makes the infected culms with unfilled panicles and chalky grain. With the time, the infected portion grows rapidly from initial small, irregular black spots to large spots on the outer leaf sheath near water level. Xanthomonas oryzae pv. oryzaeca uses bacterial disease which makes the leaves yellowing and drying. At early stage, the leaves when infected turn into grayish green and then with the times, the leaves turn into yellow to straw-colored, which leads the whole seedlings to dry up and die. The images of the abovementioned diseases are shown in Fig. 1.

On older plant, Yellow-orange stripes on lesions can appear on leaf blades or leaf tips. Initially, lesion has a wavy margin and then it progresses toward the leaf base. At early stage, lesions look like milky dew drop and later dries up and becomes yellowish beads underneath the leaf. On last stage, lesions turn yellow to grayish white with black dots due to the growth of various saprophytic fungi.

Fig. 1 Diseased rice plant: **a** leaf blast **b** brown spot **c** stem rot **d** bacterial blight

3 Preprocessing on Diseased Images

The proposed method is applied on simulated rice disease image dataset [6] for the prediction of different rice diseases. The images acquired often contain various noises from environment and instruments collecting the images. These noises are eliminated or reduced by various noise removal methods to increase the quality of the image. In the paper, bilateral filter [7] is used for noise removal and smoothing the images while preserving edges where the intensity value at each pixel in an image is replaced by a weighted average of intensity values from nearby pixels. For strip detection and shadow pixel removal, Canny edge detection algorithm [8] is applied on the image. Adaptive histogram [9] is used in order to carry out contrast enhancement by equalizing the pixel intensity value. After equalization, the image edges become more prominent compared to the original image. The images are segmented using Otsu's method [10]. After preprocessing, Rice disease dataset is prepared from 500 infected rice plant images having 31 features with four disease classes. The four disease classes are Rice blast caused by fungus named *Pyricularia grisea*, Sheath Rot caused by Pathogen named *Sarocladium oryzae*, Leaf brown spot caused by fungus Bipolaris oryza, and Bacterial Blight caused by bacteria *Xanthomonas oryzae*. The features extracted are the color features and shape features [6], listed in Table 1.

4 Graph-Based Clustering Algorithm for Feature Selection

All the extracted features from diseased images may not be equally important for image separation. This is because some features may be irrelevant and some other may be redundant. So only relevant features are desired to develop a system for

Table 1 Color and shape features extracted from diseased portion of rice plants

Color features: mean of diseased pixels in red plane, mean of diseased pixels in green plane, mean of diseased pixels in blue plane, mean of background pixels in red plane, mean of background pixels in green plane, mean of background pixels in blue plane, mean of boundary region pixels of disease in red plane, mean of boundary region pixels of disease in green plane, mean of boundary region pixels of disease in blue plane, mean of core region pixels of disease in red plane, mean of core region pixels of disease in green plane, mean of core region pixels of disease in blue plane, standard deviation of diseased pixels in red plane, standard deviation of diseased pixels in green plane, standard deviation of diseased pixels in blue plane, standard deviation of background pixels in red plane, standard deviation of background pixels in green plane, standard deviation of background pixels in blue plane, standard deviation of boundary region pixels of disease in red plane, standard deviation of boundary region pixels of disease in green plane, standard deviation of boundary region pixels of disease in blue plane, standard deviation of core region pixels of disease in red plane, standard deviation of core region pixels of disease in green plane, standard deviation of core region pixels of disease in blue plane
Shape features: length, width, area and perimeter of the disease portion, best matched primitive shape of the disease portion, area-discrepancy, aspect-ratio

rice disease prediction [11]. This is one of the main objective of feature selection techniques. The sequence of steps used for proposed feature selection algorithm is given below:

(i) Let U is the set of m diseased images (called objects) and each object a in U is characterized by a set $F = \{F_1, F_2, \ldots, F_n\}$ of n extracted features. Initially, all the feature-values are normalized into the range [0, 1] using Min-max normalization algorithm [12] and then discretized into ten discrete values using smoothing by bin medians [12] so that the values in the range (x, x + 0.1] represent a unique discrete value, for x = 0.0, 0.1, 0.2, 0.3, 0.4, 0.5, 0.6, 0.7, 0.8, 0.9. Here, by convention, open parenthesis '(' indicates exclusion of value of x and closed parenthesis ']' indicates its inclusion. Only the first interval is considered as [0.0, 0.1].

(ii) Partition of the images of the dataset U into ten different groups $P_{Fi} = \{P_{i_1}, P_{i_2}, \ldots, P_{i_{10}}\}$ where, all the images in a group have same value for feature F_i and two images of two different groups have distinct feature values. Thus we have n different partitions $P_{F1}, P_{F2}, \ldots, P_{Fn}$ of images, each for one feature separately.

(iii) Let the two partitions obtained using features F_i and F_j are $P_{Fi} = \{P_{i_1}, P_{i_2}, \ldots, P_{i_{10}}\}$ and $P_{Fi} = \{P_{j_1}, P_{j_2}, \ldots, P_{j_{10}}\}$ respectively. Similarity of F_i to $F_j$is computed using Eq. (1).

$$S_{ij} = \frac{1}{10} \sum_{k=1}^{10} \max_{1 \leq l \leq 10} \left\{ \frac{|P_{i_k} \cap P_{j_l}|}{|P_{i_k} \cup P_{j_l}|} \right\} \tag{1}$$

Similarly, similarity of F_j to F_i is computed from Eq. (1), interchanging i, j and k, l. Thus, between every pair of features, similarities are computed based on the partitions of images.

(iv) Now a graph $G = (F, E, W)$ is constructed with F, the set of features considered as nodes, E, the set of edges and W, the set of weights assigned to the edges of the graph. The weight assigned to the edge between nodes F_i and F_j is set as $w_{ij} = (S_{ij} + S_{ji})/2$. Thus we get a complete weighted and undirected graph where weight of an edge represents the average similarity between two end nodes of the edge.

(v) The generated graph provides a structure of diseased dataset consisting of features and the relationships among the features. It is very important to develop a powerful method to partition the graph into subgraphs so that the original relationships are preserved. One such method is known as Infomap clustering algorithm [13] that identifies strongly intra-connected components of the graph, i.e., the nodes in each component are densely connected to each other. This graph-based partitioning algorithm helps in analyzing graph with higher degree of nodes. Infomap is one of the popular graph-based clustering algorithms that optimizes the map equation which balances the number of nodes in the subgraphs or component to detect important structures within the subgraphs. Initially, each node in the graph is assigned to its own module. Then each node is moved to the neighboring module in a random sequential order that results in the largest decrease of the map equation. If no move results in a decrease of the map equation then the node resides in its original module. The process is continued iteratively in a new random sequential order, until no move results in a decrease of the map equation further. Finally, the graph is reconstructed with the modules of the level forming the nodes in this level and exactly as at the previous level nodes are joined into modules. This hierarchical reconstruction of the subgraph is continued until the map equation cannot be reduced further. Thus we have a partition of graph into a set of connected subgraphs.

(vi) From each generated subgraph, a node is selected as a representative of all nodes. The node is representative of other nodes if it can give the information that is obtained from others. Each subgraph contains some nodes and weighted edges of the graph. To find the representative node, we first compute the importance of the nodes in the subgraph. The importance of a node n_i in a subgraph is computed as the sum of the weights of the edges adjacent with n_i in the subgraph. Since the subgraphs are modules where all nodes in a module are strongly intra-connected, so any one node from the module can be considered as representative node. But in Infomap algorithm, as we have balanced the number of nodes in the subgraphs to detect important structures within the subgraphs, so the node with the highest importance in a subgraph is selected as the representative node. This representative node is considered as the most influential node in the subgraph. Thus collection of most influential nodes from all the subgraphs are considered as the important feature subset used for disease prediction.

5 Experimental Results

The proposed method is applied on rice disease image data for predicting different rice diseases. The experiments are carried out on Computer Model: ACER emachines D725; CPU: Pentium(R) Dual-Core CPU T4400 @ 2.20 GHz 2; Memory: 1 GB; OS: Ubuntu 12.04 LTS—32 bit. Python programming Language is used for implementation of the work. The simulated rice disease dataset consists of 500 infected rice plant images having 31 features with four disease classes such as Rice Blast (RB), Sheath Rot (SR), Leaf Brown Spot (BS), and Bacterial Blight (BL). In our work, color features, and shape features, listed in Table 1, are considered for feature selection algorithm. The proposed graph-based feature selection algorithm (GFS) selects 7 features whereas the popular existing feature selection algorithms like "Correlation-based Feature selection" (CFS) [14] and "Consistency based feature selection" (CON) [15] select 10 and 5 features respectively, as shown in Table 2.

The classification accuracies of different classifiers on whole dataset and reduced datasets are listed in Table 3. The feature selection algorithms CFS and CON and the classifiers are run using weka3.9.1 toolbox [16], where 10-fold cross validation are carried out during classification accuracy measurement. The proposed GFS algorithm provides better accuracies for all most all classifiers except "Multilayer Perceptron" (MLP) and the ensemble based classifiers like Boosting and Bagging. The results show the importance of the proposed feature selection method.

Table 2 Selected features by different feature selection algorithms

CFS	Mean of diseased pixels in green plane, mean of boundary region pixels of disease in green plane, mean of core region pixels of disease in green plane, standard deviation of background pixels in red plane, standard deviation of core region pixels of disease in blue plane, standard deviation of diseased pixels in red plane, standard deviation of boundary region pixels of disease in green plane, standard deviation of boundary region pixels of disease in blue plane, area, aspect-ratio
CON	Mean of boundary region pixels of disease in green plane, mean of core region pixels of disease in blue plane, standard deviation of background pixels in green plane, standard deviation of boundary region pixels of disease in green plane, area
GFS	Mean of boundary region pixels of disease in green plane, mean of core region pixels of disease in red plane, mean of core region pixels of disease in green plane, standard deviation of background pixels in red plane, standard deviation of boundary region pixels of disease in green plane, best matched primitive shape of the disease portion, aspect-ratio

Table 3 Accuracy of classifiers on whole and reduced datasets

Classifier	Whole dataset	Dataset reduced by CFS	Dataset reduced by CON	Dataset reduced by GFS
Naïve Bayes	80.11	81.07	74.31	**83.07**
KSTAR	80.17	82.35	71.77	**83.63**
Ada-Boost	**53.71**	**53.71**	43.29	51.11
Bagging	78.31	**80.14**	75.54	77.12
PART	80.13	82.35	79.96	**83.19**
J48	80.79	83.13	79.06	**84.38**
MLP	**79.65**	77.04	73.07	75.42
SVM	78.27	**80.04**	77.74	**80.04**
Random forest	80.06	82.29	78.17	**83.97**
Logistic	80.17	80.17	80.17	**81.35**

Bold signifies the comparatively better results

6 Conclusion

Often a large quantity of crops are being damaged due to disease occur in rice plant. Rice disease prediction is an emerging research area in the field of Information Science and Agriculture Department. In this paper, a graph-based clustering method is applied to partition the graph into subgraphs and the most influential node of each subgraph is selected. The selected nodes are considered as the feature subset important for rice disease prediction. The experimental results in terms of number of features selected and classification accuracies show the importance of the proposed method. The method is compared with some existing feature selection methods to demonstrate its usefulness and acceptability by the research community.

References

1. Kaur, S.: Noise types and various removal techniques. Int. J. Adv. Res. Electron. Commun. Eng. (IJARECE) **4**(2), 226–230 (2015)
2. Guo, R., Dai, Q., Hoiem, D.: Single-image shadow detection and removal using paired regions. In: CVPR, p. 175 (2011)
3. Maini, R., Aggarwal, H.: A comprehensive review of image enhancement techniques. J. Comput. **2**(3), 8–13 (2010)
4. Pal, N.R., Pal, S.K.: A review on image segmentation techniques. Pattern Recogn. **26**(9), 1277–1294 (1993)
5. Kumar, G., Bhatia, P.K.: A detailed review of feature extraction in image processing systems. In: International Conference on Advanced Computing & Communication Technologies (2014). https://doi.org/10.1109/acct.2014.74
6. Phadikar, S., Sil, J., Das, A.K.: Rice diseases classification using feature selection and rule generation techniques. Comput. Electron. Agric. **90**, 76–85 (2013). Elsevier

7. Tomasi, C., Manduchi, R.: Bilateral filtering for gray and color images. In: IEEE International Conference on Computer Vision (1998)
8. Canny, J.: A computational approach to edge detection. IEEE Trans. Pattern Anal. Mach. Intell. **8**(6), 679–698 (1986)
9. Sund, T., Moystad, A.: Sliding window adaptive histogram equalization of intra-oral radiographs: effect on diagnostic quality. Dentomaxillofacial Radiol **35**(3), 133–138 (2006)
10. Otsu, N.: A threshold selection method from gray-level histograms. IEEE Trans. Syst. Man Cybern. **9**(1), 62–66 (1979)
11. Anitha, A.: A predictive modeling approach for improving paddy crop productivity using data mining techniques. Turkish J. Electr. Eng. Comput. Sci. **25**, 4777–4787 (2017)
12. Han, J., Kamber, M.: Data Mining: Concepts and Techniques. Morgan Kaufmann, San Francisco (2001)
13. Rosvall, M., Bergstrom, C.: Maps of random walks on complex networks reveal community structure. Proc. Natl. Acad. Sci. U. S. A. **105**, 1118 (2008)
14. Hall, M.: Correlation based feature selection for machine learning. Doctoral dissertation, Department of Computer Science, University of Waikato (1999)
15. Dash M., Liu H., Motoda, H.: Consistency based feature selection. In: Pacific-Asia Conference on Knowledge Discovery and Data Mining (PAKDD), pp. 98–109 (2000)
16. WEKA,http://www.cs.waikato.ac.nz/ml/weka

Shifted Peterson Network: A New Network for Network-on-Chip

M. M. Hafizur Rahman, M. N. M. Ali, Osinusi A. Olamide and D. K. Behera

Abstract The advancements in technology in this contemporary time require special computing devices to able to manage the increase of the computational power. However, the limitation of the sequential computing systems stands as an obstacle in fulfilling these requirements. Thus, the research community looking for a viable solution able to solve the grand computing problems. A parallel computing system is a collection of processing elements that cooperate and communicate to provide a fast solution for a problem. A new symmetric network named Shifted Peterson Network (SPN) has been proposed in this paper. The basic module (BM) of the suggested network is composed of ten nodes, and the BMs of SPN are connected through external nodes to construct an advanced level of this network. Furthermore, the connectivity of SPN is based on a shifting mechanism of the binary digits of each node. The architecture of SPN discussed in this paper, in addition, thc static network performance parameters of this network has been evaluated and compared to popular conventional interconnection networks. The networks evaluated in terms of node degree, diameter, average distance, arc connectivity, bisection width, cost, wiring complexity, packing density, and message traffic density. SPN showed good results in almost all aspects compared to these networks. Therefore, SPN is a suitable choice for the massively parallel computer (MPC) systems.

M. M. Hafizur Rahman
Department of Communications and Networks, CCSIT, King Faisal University, Al-Ahsa 31982, Kingdom of Saudi Arabia
e-mail: mhrahman@kfu.edu.sa

M. N. M. Ali
Department of Computer Science, KICT, IIUM, 50728 Kuala Lumpur, Malaysia
e-mail: moh.ali.exe@gmail.com

O. A. Olamide
Department of Computer Science, Fountain University Osogbo, Osogbo, Nigeria
e-mail: osinusi1993@yahoo.com

D. K. Behera (✉)
Indira Gandhi Institute of Technology, Sarang, Odisha, India
e-mail: dkb_igit@rediffmail.com

A. K. Das et al. (eds.), *Computational Intelligence in Pattern Recognition*, Advances in Intelligent Systems and Computing 999,
https://doi.org/10.1007/978-981-13-9042-5_51

Keywords Hierarchical interconnection networks · SPN · Static network performance · Packing density · Message traffic density

1 Introduction

The design of the interconnection network topology has a significant impact on the actual performance of the MPC system [1, 2]. The processing elements (PEs) within these networks are placed in a particular design, and the relationship between them is a subject to the architecture of the entire network. The interconnection network built based on a particular network topology; this topology has wide influence in controlling the data flow within these networks, and it has a crucial effect on the network performance by either growing or degrading the latency and the network congestion [3, 4]. Therefore, choosing a good design for the network topology became a strong factor in guarantee the effectivity of these networks. As a result, many topologies have been proposed to construct the interconnection networks of the multiprocessor systems [5]. Currently, the scalability of the interconnection networks is important to provide parallel computer systems with a high number of processing elements to increase the computing power of these systems in order to process a complex problem in a short period of time. These systems composed of programming tools, performance evaluation tools, debuggers, and optimization libraries [6]. This will participate in solving a high number of computational operations concurrently by giving a fast solution which will help in decreasing the network power consumption [7–9]. The earlier proposed topologies for the MPC systems proved poor performance with network scalability [10, 11]. Thus, the hierarchical interconnection networks (HINs) have been presented to be an alternative solution of the conventional ones [12]. Thereby, many HIN topologies have been proposed looking for an ideal hierarchical interconnection network [13, 14].

The network-on-chip (NoC) is a new paradigm of the interconnection networks to connect many cores in a chip [15]. In addition, it is considered the best scalable communication infrastructure for multiprocessor systems [16]. These networks used to enhance the connection between the processing elements (PEs) inside the chip by replacing the traditional on-chip interconnect schemes [17]. Currently, the advancement in technology has a great impact on improving the structure of these networks leading to a promising future in building the multiprocessor systems [18]. Moreover, the NoC has been invented to take the advantages of increasing the number of cores to employ it in creating efficient and scalable systems. These networks are a potential solution to respond to the high demand of computational power [19].

A Shifted Petersen Network (SPN) is a new symmetric hierarchical interconnection network. Besides, the architecture of SPN is a novel design of a hierarchical topology for MPC systems; the connectivity of this network in the higher levels proposed based on a shifting mechanism. In this paper, we presented SPN as a new HIN network. Besides, the static network performance parameters of SPN has been assessed. SPN shows afford low network diameter and low wiring complexity. These two parameters are important in developing the performance and degrading the price of building a new parallel computer system. SPN network composed of ten nodes connected to each other in a particular design. Multiple basic modules of SPN are connected by global links to form the higher levels network. In this paper, Sect. 2 is the description of the architecture design of SPN, Sect. 3 is the evaluation of the static network performance of the SPN, and finally, Sect. 4 is the conclusion.

2 Structural Design of SPN

SPN is a new design of a HIN topology; this topology composed of several modules interconnected hierarchically to build a complete parallel computer system.

The basic module (BM) of SPN contains ten nodes; every single node is connected directly to three adjacent nodes through electrical wires. Therefore, the node degree of the internal nodes of SPN is 3. On the other hand, the node degree of the external nodes is 4. The increasing of the node degree of the external nodes than the internal nodes due to use the first one as gate nodes in connecting the BMs of SPN to build the higher level networks. Figure 1 depicted the architecture of the basic module of SPN, in addition, it reveals the communication mechanism of these nodes. SPN has 5 internal nodes with node degree 3, and 5 external nodes with node degree 4. Increasing the node degree of a network will participate in improving the network performance by mitigating the congestion.

In order to expand the size of the SPN network, the external node in each BM of SPN will be connected to another external node from another BM. The BM of SPN has 5 external nodes and is represented as $(N_1, N_2, N_3, N_4, N_5)$. However, the

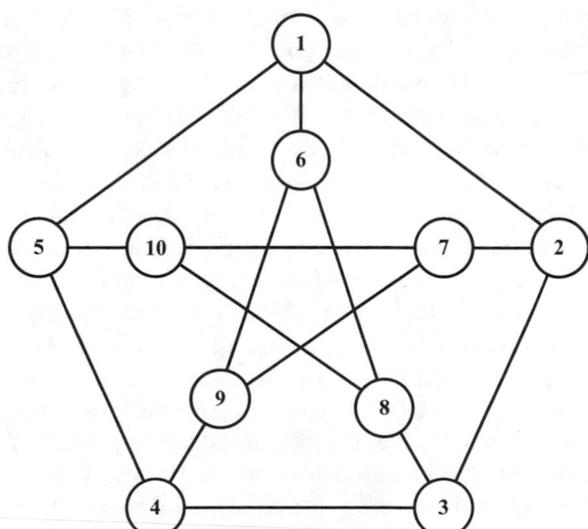

Fig. 1 Basic module of a shifted Peterson network

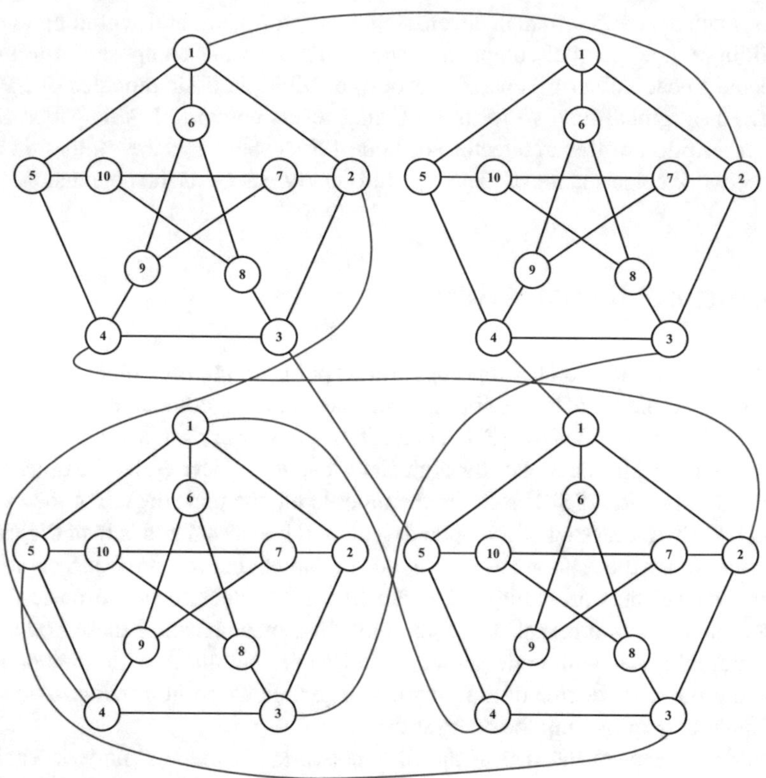

Fig. 2 Higher level network of a shifted Peterson network

remaining internal 5 nodes represented as (N_6, N_7, N_8, N_9, N_{10}). Figure 2 illustrated the advance level network of SPN; this level is composed of four groups of nodes. These groups are connected globally through either electrical or optical fiber wires. The external nodes of the BM of SPN are connected to other external nodes of another BM based on a particular mechanism; this mechanism is applying by relocating the binary digits of the node based on the distance between the alphabetic sequences of the groups in the higher level network. To deal easily with these groups we will donate each group by using a different alphabetic number from A to D. Thus we will have Group-A, Group-B, Group-C, and Group-D. The number of nodes within each group is 10, thus we address each node by using the group name as: the nodes in group A are (A_1, A_2, A_3, A_4, A_5, A_6, A_7, A_8, A_9, A_{10}), the nodes in group B are (B_1, B_2, B_3, B_4, B_5, B_6, B_7, B_8, B_9, B_{10}), and so on. The connectivity of the higher level of SPN is based on converting the decimal index of each node to the equivalent digits in binary, after that, we shift the binary digits of the destination node x moves to the left, where the left is the default shifting direction, and $1 \leq x \leq 3$. The shifting mechanism is related to the sequential distances between the groups. For instance, groups A and B have a single distance, groups A and C have a double distance,

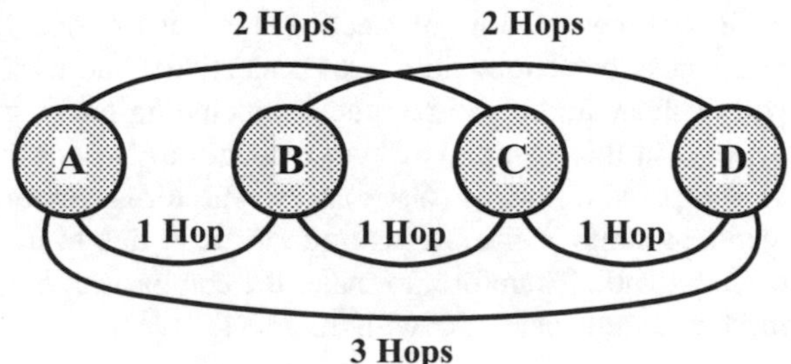

Fig. 3 Distance between groups of a shifted Peterson network

and groups A and D have a triple distance. The distances between the groups of the advanced network of SPN represented in Fig. 3.

For example, connecting two nodes from groups A as a source node (N_A) and B as a destination node (N_B) required several steps. Firstly converting the decimal index of (N_A) to the equivalent binary. Secondly, converting the indexes of all nodes in group B to the binary digits. Thirdly, shifting the binary digits of any (N_B) node one step in the left direction. Fourthly, comparing the binary of (N_A) node to that of each (N_B) node. Fifthly, (N_A) will be connected to that node which has the same arrangements of the binary digits. Likewise, we connect any particular (N_A) node to any (N_C) or (N_D) nodes from the groups C and D respectively. However, the distances between these groups will be considered based on Fig. 3. From Fig. 2, we can say that to connect node 2 in group A to any (N_B) node, the index number of the node (2) will be converted to its corresponding in binary (010), after that we will convert each decimal number in group B to the binary. Then the binary of each (N_B) node will be moved one step in the left direction and it will be compared to the binary of node 2 from A. In case a node from group B shows same arrangements of the binary digits to that of node 2 in group A; this means both nodes can be connected. To connect two nodes by using a bidirectional link, we will assume the left as the default direction for shifting, and the right one for the connection in the opposite direction. Finally, we need to highlight what we have mentioned earlier in this paper the connection between the BMs is applicable only between the external nodes of SPN.

3 Static Network Performance of SPN

The performance of the interconnection network affected widely by the architecture design of the network topology and each network has its unique design [21]. Thus, the topological properties of these networks have a crucial effect during the evaluation process, and in comparing the network to other networks. Evaluating and comparing the static network performance parameters of the hierarchical interconnection networks are important to prove the superiority of the network over the conventional interconnection networks. Therefore, the evaluation of the static network parameters of SPN is presented in this section in terms of node degree, diameter, average distance, cost, bisection width, arc connectivity, and wiring complexity. Moreover, we will evaluate additional parameters including packing density, and message traffic density. In this section, we will evaluate SPN with 40 nodes, and the results will be compared to that of (6×6) mesh and torus networks. Using topologies with an identical number of nodes is quite difficult due to the differences in the design of each network. Therefore, to make the comparison almost fair we will assume both mesh and torus networks with 36 nodes.

3.1 Degree

The maximum number of wires originating from a node of a network is known as the node degree [20], and it is denoted as D_g. The BM of SPN has ten nodes; whereby five interior nodes and five exterior nodes. The interior nodes using only three links to be connected internally. However, the exterior nodes are connected globally to other nodes from different BMs to compose the higher level of SPN. Therefore, the node degree of these nodes is 4. This implies that the maximum number of links emanation from a node of SPN is 4. Table 1 tabulated the results of the network degree comparison between SPN, mesh and torus networks. For a fair comparison, we have considered all the networks who have the node degree of 4.

3.2 Diameter

The diameter is defined as the maximum length in hop among the lengths of the shortest paths between all the possible pairs of nodes in the network [20]. Let us denote diameter as D and it is the worst-case distance between two nodes along the shortest path. The diameter of a network has a critical impact on the performance of this network. Short diameter is desirable because it implies a good performance and it provides an idea of the maximum possible latency of a network, which depends upon the traffic scenario in a network. The diameter comparison between SPN, 2D-mesh, and 2D-torus networks stated in Table 1. It is interesting that the diameter of SPN is smaller than that of the other networks. This implies the performance of SPN is better than that of mesh and torus networks. Hence, SPN is a good choice over these networks for constructing the future MPC systems.

3.3 Average Distance

Most of the time the packets traverse paths less than the network diameter. Therefore, the network diameter predicts the upper-bound of latency and the average distance

Table 1 Comparison of static network performance of various networks

Network	Node degree	Diameter	Average distance	Cost	Bisection width	Arc connectivity	Wiring complexity	Packing density	Message traffic density
2D-Mesh	4	10	4.0	40	6	2	60	0.9	2.4
2D-Torus	4	6	3.0	24	12	4	72	1.5	1.5
SPN	4	5	3.4	20	6	3	68	2.0	2.0

predicts the actual latency of a network. Thus, to calculate the actual network performance we need to know the average distance the packet has been traveled. The average distance is the mean distance between all distinct pairs of nodes in a network. Furthermore, the average distance has a significant effect on the interconnection network performance. Small average distance implies a less packet latency, and it means less number of links and buffers leading to mitigate the communication contention.

The average distance of SPN measured and thus tabulated in Table 1, the result compared to that results of the popular mesh and torus networks. It is shown that the average distance of SPN is lower than that of a 2D-mesh network and a little bit higher than that of a 2D-torus network. Therefore, SPN will enrich the performance of the future generation of parallel computer systems.

3.4 Cost

The cost of a parallel computer system depends upon the number of nodes used to build the parallel computer, number of connecting links used to connect those nodes, and the router used in each node. The router cost depends upon the node degree. High node degree incurs the high cost of a router and hence the cost of a node. On the other hand, the diameter is increasing with the increase in network size. Therefore, the product of the node degree (D_g) and diameter (D) is a good measure to statically assess and study an interconnection network topology [20, 21]. It is called the "Cost" (denoted as C) as represented in the Eq. 1.

$$C = D_g \times D \tag{1}$$

The cost of different networks along with the SPN is tabulated in Table 1. It is shown that the cost of SPN is lower than that of mesh and torus networks.

3.5 Bisection Width

The bisection width of an interconnection network is an important topological attribute and it has a critical effect on the cost of the Very-Large-Scale-Integration (VLSI) layout and the performance of a network. The bisection width is the minimum number of wires that have to be removed to partition the network into two equal parts. This will help widely in solving many problems, the data in each half will manipulate and the results will be merged to give the last result. Low bisection width provides low bandwidth and high congestion in the middle of the network. In contrast, high bisection width affects the design of VLSI. Therefore, the moderate bisection width is desirable. The bisection width comparison shown in Table 1 between SPN, (6×6) mesh and (6×6) torus networks, and it shows that the bisection width of SPN is equal to that of 2D-mesh and far lower than that of 2D-torus. This implies that SPN provides a moderate BW which is desirable in building the parallel computer systems.

3.6 Arc Connectivity

The arc connectivity of an interconnection network has a high influence on measuring the network robustness. In addition, it used to measure the diversity of paths between the nodes. Besides, it defines the minimum number of links that must be removed from the network to make this network disjoint into two parts. Higher arc connectivity implies a high-performance network with less path congestion and more fault tolerance. The ratio between the arc connectivity and the node degree provide the network static fault tolerance performance. If the result is 1, that means the network is reliable and robust. Table 1 tabulated a comparison between the arc connectivity of SPN, (6×6) mesh, and (6×6) torus networks, and it shows that the arc connectivity of SPN is higher than that of 2D-mesh and a little bit lower than that of 2D-torus. Thus, SPN has the preeminence over 2D-mesh and a proper alternative solution instead of 2D-torus network.

3.7 Wiring Complexity

The entire number of links/wires (denoted as L) that is required to connect every single node to the other nodes within the interconnection network known as the wiring complexity (WC) of this network. In addition, it has a crucial impact on the cost of the overall system. The wiring complexity of (6×6) 2D-mesh and 2D-torus is $\{N_x \times (N_y - 1) + (N_x - 1)N_y\} = 6 \times (6 - 1) + (6 - 1) \times 6 = 60$ and $(2 \times N_x \times N_y = 2 \times 6 \times 6) = 72$ links, respectively. N_i represents the number of nodes in the ith dimension. On the other hand, the wiring complexity of SPN has been calculated in Table 1. Due to hierarchical nature, the WC of SPN is a little bit greater than that of mesh and torus networks.

3.8 Packing Density

The ratio of the total number of nodes of a network (N) to its cost (C) is known as the network packing density. This implies that the network cost is an important factor in determining the network packing density. In VLSI, a small chip area and high packing density is desirable and required for VLSI layout. Equation 2 is using to calculate the packing density of the interconnection network.

$$Packing\ Density = \frac{N}{C} \tag{2}$$

Here, N is the number of nodes and C is the cost. The packing density of various networks stated in Table 1 and it is depicted that the packing density of SPN is far higher than that of 2D-mesh and 2D-torus networks. As a result, SPN is more suitable than these networks for creating the new generations of the MPC systems.

3.9 Message Traffic Density

As mentioned earlier, the number of communication links increased with the increase of nodes. Thus, the ratio between nodes (N) and links (L) indicates the average number of paths to travel a packet from a node to its neighboring nodes. Average distance (d) is the mean distance between all distinct pairs of nodes using shortest path algorithm. It affects the message traffic density. The product of the average distance and the ratio between the total number of nodes and the total number of wires is called the message traffic density (MTD) as shown depicted in Eq. 3. It represents the efficiency of traffic distribution in a network. The low message traffic density participates in reducing the network traffic congestion and providing wide bandwidth.

$$MTD = d \times \frac{N}{L} \tag{3}$$

The comparison between SPN, (6 × 6) mesh, and (6 × 6) torus networks stated in Table 1. SPN provides lower results compared to 2D-mesh, and a little bit higher results compared to 2D-torus. Hence, SPN qualified as a good option over these networks in creating the future MPC systems.

4 Conclusions

In this paper, we have presented a new design of an interconnection network. This network called Shifted Peterson Network, the basic module of this network composed of 10 nodes are connected through internal wires, these nodes divided to 5 internal nodes and 5 external nodes. To expand the network size, extra links added to some of the external nodes to be used in connecting the BM of SPN to other BMs to create the advanced level network. The architecture of SPN discussed in details in this paper. In addition, we have assessed the static network performance parameters of this network. The outcome results compared to that of (6×6) mesh and (6×6) torus networks. SPN showed perfect features compared to these networks, which implies that SPN is a good choice over these networks to construct the future NoC systems.

The architecture and static network performance is studied in detail in this paper. For the further exploration of SPN we would like to continue the research on (1) dynamic communication performance of an SPN using a deadlock-free routing algorithm [22, 23] (2) on-chip and off-chip power analysis of SPN network [24, 25].

References

1. Rahman, M.M.H., Nor, R.M., Awal, M.R., Sembok, T.M.T., Miura, Y.: Long wire length of midimew-connected mesh network. In: Distributed Computing and Internet Technology, pp. 97–102 (2016)
2. Fu, H., Liao, J., Yang, J., Wang, L., Song, Z., Huang, X., Yang, C., Xue, W., Liu, F., Qiao, F., Zhao, W., Yin, X., Hou, C., Zhang, C., Ge, W., Zhang, J., Wang, Y., Zhou, C., Yang, G.: The sunway TaihuLight supercomputer: system and applications. Sci. China Inf. Sci. **59** (2016)
3. Rahman, M.M.H., Nor, R.M., Akhand, M.A.H., Sembok, T.M.T.: Cost effective factor of a midimew connected mesh network. Asian J. Sci. Res. **10**(4) (2017)
4. Yunus, N.A.M., Othman, M., Mohd Hanapi, Z., Lun, K.Y.: Reliability review of interconnection networks. IETE Tech. Rev. **3**(6), 596–606 (2016)
5. Ali, M.N.M., Rahman, M.M.H., Nor, R.M., Sembok, T.M.T.: A high radix hierarchical interconnection network for network-on-chip. In: 12th International Conference on Computing and Information Technology (IC2IT), Bangkok, Thailand (2016)
6. Mishra, B.S.P., Dehuri, S.: Parallel computing environments: a review. IETE Tech. Rev. **28**(3), 240–247 (2011)
7. Sanchez, D., Michelogiannakis, G., Kozyrakis, C.: An analysis of on-chip interconnection networks for large-scale chip multiprocessors. ACM Trans. Archit. Code Optim. (TACO) **7**(1), 4 (2010)
8. Hag, A.A.Y., Rahman, M.M.H., Nor, R.M., Sembok, T.M.T.: Dynamic communication performance of a horizontal midimew connected mesh network. Int. J. Adv. Comput. Technol. **8**(1), 31 (2016)
9. Khan, Z.A., Siddiqui, J., Samad, A.: Linear crossed cube (LCQ): a new interconnection network topology for massively parallel system. Int. J. Comput. Netw. Inf. Secur. (IJCNIS) **7**(3), 18 (2015)
10. Rahman, M.M.H., Inoguchi, Y., Sato, Y., Miura, Y., Horiguchi, S.: Dynamic communication performance of a TESH network under the nonuniform traffic patterns Authors. In: 11th IEEE International Conference on Computer and Information Technology (ICCIT), pp. 365–370 (2008)

11. Rahman, M.M.H., Horiguchi, S.: Dynamic communication performance of a hierarchical torus network under non-uniform traffic patterns. IEICE Trans. Inf. Syst. **87**(7), 1887–1896 (2004)
12. Faisal, F.A., Rahman, M.M.H., Inoguchi, Y.: A new power efficient high performance interconnection network for many-core processors. J. Parallel Distrib. Comput. **101**, 92–102 (2017)
13. Ali, M.N.M., Rahman, M.M.H., Behera, D.K., Inoguchi, Y.: Static cost-effective analysis of a shifted completely connected network. In: Computational Intelligence in Data Mining, pp. 165–175. Springer, Singapore (2019)
14. Adhikari, N., Tripathy, C.R.: The folded crossed cube: a new interconnection network for parallel systems. Int. J. Comput. Appl. **4**(3), 43–50 (2010)
15. Kim, J., Balfour, J., Dally, W.: Flattened butterfly topology for on-chip networks. In: 40th Annual IEEE/ACM International Symposium on Micro-architecture (Micro-40), pp. 172–182 (2007)
16. Valinataj, M., Mohammadi, S., Safari, S.: Fault-aware and reconfigurable routing algorithms for networks-on-chip. IETE J. Res. **57**(3), 215–223 (2011)
17. Kim, J., Dally, W., Towles, B., Gupta, A.: Microarchitecture of a high-radix router. In: 32nd Annual International Symposium on Computer Architecture, vol. 33, pp. 420–431 (2005)
18. Amano, H.: Tutorial: introduction to interconnection networks from system area network to network on chips. In: 1st International Symposium on Computing and Networking, pp. 15–16 (2013)
19. Ahmed, A.B., Abdallah, A.B.: Graceful deadlock-free fault-tolerant routing algorithm for 3D network-on-chip architectures. J. Parallel Distrib. Comput. **74**(4), 2229–2240 (2014)
20. Rahman, M.M.H., Shah, A., Fukushi, M., Inoguchi, Y.: HTM: a new hierarchical interconnection network for future generation parallel computers. IETE Tech. Rev. **33**(2), 93–104 (2016)
21. Kumar, J.M., Patnaik, L.M.: Extended hypercube: a hierarchical interconnection network of hypercubes. IEEE Trans. Parallel Distrib. Syst. **3**(1), 45–57 (1992)
22. Moudi, M., Othman, M., Lun, K.Y., Rahiman, A.R.A.: x-folded TM: an efficient topology for interconnection networks. J. Netw. Comput. Appl. Elsevier **73**, 27–34 (2016)
23. Prasad, N., Mukherjee, P., Chattopadhyay, S., Chakrabarti, I.: Design and evaluation of ZMesh topology for on-chip interconnection networks. J. Parallel Distrib. Comput. Elsevier **113**, 17–36 (2018)
24. Candel, F., Petit, S., Sahuquillo, J., Duato, J.: Accurately modeling the on-chip and off-chip GPU memory subsystem. Future Gener. Comput. Syst. **82**, 510–519 (2018)
25. Andujar, F.J., Villar, J.A., Sanchez, J.L., Alfaro, F.J., Duato, J.: N-dimensional twin torus topology. IEEE Trans. Comput. **64**(10), 2847–2861 (2015)

Gathering Identification Using Image Metrics for Intelligent Situation Awareness System in Real Time Scenarios

Nandita Gautam, Debdoot Das, Sunirmal Khatua and Banani Saha

Abstract The advancement of image processing in the field of Artificial Intelligence has created various research prospects in the area of object detection, pattern recognition, etc. Face detection technology is applied for biometric authentication systems for face recognition and verification. This work aims at developing a tool that can be used in an Intelligent Situation Awareness System for a given Region of Interest. This has been done by identifying human face as objects from digital video frames through real time video streaming by using Recurrent Convolution Neural Network (RCNN) classifiers. The key frames from a video are identified and machine learning algorithm is being applied on it for performing the object identification. After the facial objects are being identified in a given frame, this can be utilized for performing semantic analysis in a given spatiotemporal scenario. The metrics identified in this work include object count, hand gestures, relative distance as well as density of objects for developing a robust system that could function in real time.

Keywords Face detection · Convolution neural networks

N. Gautam (✉) · D. Das · S. Khatua · B. Saha
Computer Science Department, University of Calcutta, Kolkata 700098, India
e-mail: nanditagautam43@gmail.com

D. Das
e-mail: dasdebdoot@gmail.com

S. Khatua
e-mail: enggnimu_ju@yahoo.com

B. Saha
e-mail: bsaha_29@yahoo.com

A. K. Das et al. (eds.), *Computational Intelligence in Pattern Recognition*,
Advances in Intelligent Systems and Computing 999,
https://doi.org/10.1007/978-981-13-9042-5_52

1 Introduction

The design and development of an Intelligent Situation Awareness System is an emerging area of research for monitoring a Region of Interest (RoI) covered by spatiotemporal real time video streams from a set of CCTV Cameras. This system can be used for perceiving environmental elements in the form of mob or gathering. These elements are useful for analyzing spatiotemporal data both syntactically and semantically in order to detect the unusual behavior of the mob. This requires feature extraction from the identified objects. Object detection is one of the emerging research areas in the field of pattern recognition, image classification, semantic segmentation, etc. Face detection is a fundamental problem in computer vision tasks that is mainly used for applications such as biometric authentication. Face detection can be utilized for several other tasks such as face verification, face recognition, and face clustering. Face detection technology is helpful for identifying or verifying a person from a digital image or a frame obtained from a video. Facial expression recognition (FER) systems can be developed by analyzing the features obtained from a face detection system. Several machine learning algorithms have been developed for training effective classifiers in order to obtain a face detection tool. It can be viewed as a subset of object detection tasks in computer vision. Classifiers have been built up using Convolution Neural Networks (CNN) for detecting a face in a given digital image. The facial objects that are obtained by applying these classifiers, can be used for performing syntactical and semantic analysis is spatiotemporal scenarios. For performing scene analysis in a given scenario, it becomes important to identify features that give some useful information about the occurring event.

This work focuses on identifying these metrics by performing object identification for developing such situation awareness systems. The metrics identified in this work deal with the head count, relative distance among objects, and density function. For monitoring a Region of Interest, it becomes important to identify the parameters on which the analysis must be performed in order to obtain efficient results. This can be understood when a large dataset of videos is compared and high accuracy rate is obtained. After the process of metric identification is complete, the results can be utilized for further semantic analysis in order to generate the necessary classification. These classifications are required for the decision-making process of the awareness system.

A threshold for the count and the relative distance has been selected for generating the metrics. The accuracy of the threshold has been established by obtaining and comparing results from large number of video inputs. In this work, the input video has been streamed in real time and the frames are fed to RCNN for obtaining the desired metrics. The metrics are then deployed over a large number of videos and comparative analysis is performed in order to obtain the rate of accuracy. The object here is identified as the human face which has been obtained by the help of trained machine learning classifiers like the HaarCascade classifier. It has classifiers such as frontalface, leftface, rightface which give the front view and side views respectively. The hand movements have been taken as another metric parameter for feature extraction.

The density cluster formation has been depicted by taking the relative distances among the objects. The related work in this context has been discussed in the next section. The entire problem has been detailed out in a comprehensive manner in Sect. 3. Methodology and workflow for generating the outcome has been discussed in Sect. 4. Further, the results have been obtained by taking different video data as the raw input. Based on the input, the task for identification of gathering has been performed. The cluster for a gathering has been highlighted in this work, so that it can be deployed for further classification in order to perform the semantic analysis in real time scenarios.

2 Related Work

With the development of software tools there has been an enormous development in the way data is processed nowadays. Digital image processing has been a vast area of research in the field of science and technology. Video processing is an important technique that helps in efficient content retrieval for a video stream. A video stream is understood to be a sequence of frames.

Human face detection is an area of extensive research in the literature of computer vision. In [1], a new face detection scheme has been developed for showing an improvement on the RCNN framework. This has been done by performing feature concatenation, multi-scale training, and calibration of key parameters. An extreme learning machine technique has been applied using Big data for performing the task of face tagging in social networks by Vinay Shekar [2]. This is a cognitive technique which takes the facial structures such as the jawline, eye positioning as the key features for identifying the face.

A work for heterogeneous face recognition has been proposed in [3]. A single layer hidden network has been used along with visual infrared database for conducting the experiments. In this regard, a pose-based image analysis technique has been proposed in [4]. In this pose-based images have been identified for recognizing the facial expression. Depth images have been used for implementing the machine learning algorithm. The concept of Local Directional Pattern (LDA) has been improvised to obtain a Modified Local Directional pattern (MLDA) in which the positive and negative edge pixels have been identified. For better face feature extraction, the further processing has been done through Generalized Discriminant Analysis (GDA). A learning algorithm for two-stage learning has been developed in [5]. It deals with recognition of face in two steps. Further works show the development of learning algorithms for building systems over the cloud platform. It uses the Local Binary Pattern (LBP) for designing the extreme learning machine classifier using Support Vector Machine [6].

3 Problem Statement

A video is a sequence of frames through which content is retrieved and information is extracted. Feature extraction is a fundamental task that needs to be executed accurately for effective facial recognition. The facial structures such as jawline, positioning of the parts can be derived from geometric feature extraction or appearance-based feature extraction [7]. The former category involves feature vectors such as lines, angles, etc. while the latter deals with applying filters on the whole image. In this work, appearance-based technique has been used by applying CNN filters for identifying the object. The idea is to extract features based on certain metrics in order to develop a robust system. The features identified in this work are head count for identified objects (facial objects), relative distance among the objects, and density of the objects.

The head count deals with the number of identified facial objects in a given frame. This ascertains that total how many objects are being dealt with in the feature extraction process. The hand gestures [8] are nonverbal cues for identifying human behavior in a given scenario. The hand movements of the objects is taken as a parameter as it is useful for further behavioral classification in case of human beings [9, 10]. The movement of the hands at various angles is captured for further analysis.

In order to determine the sparsity or density of a given number of objects, it is important to measure the relative distance between them. The relative distance needs to be represented in a structured fashion in order to perform detailed analysis. This gives the classification of objects as gathering. If the relative distance among the objects is less than a certain threshold then it is classified as a gathering or a cluster. The case of a single such gathering is being dealt in this work.

The density of the cluster is identified on the basis of the number of objects identified in a particular gathering. The count and distance metrics are required for determining the density of a gathering or cluster of objects. These metrics identified in this work contribute as elements for developing a robust situation awareness system.

4 Proposed Methodology

In order to perceive the environmental elements in an intelligent situation awareness system for a given RoI, the features of the objects must be accurately identified. The digital video frame or a digital image is taken as input for performing the analysis. In order to determine the head count the frame as fed as input to the RCNN which uses cascading classifiers for facial object detection [11] from various angles. These facial objects are then separately identified. The next feature extraction is done based on the hand movements of the identified objects. The hand gestures are an important feature for understanding behavioral patterns of humans. It helps in behavioral classification for performing semantic analysis. This can be understood by finding the change in position of hands in consecutive frames.

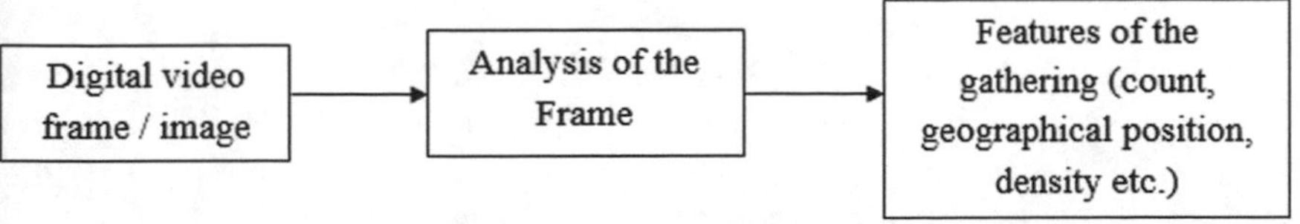

Fig. 1 Workflow for gathering identification resulting into a set of features related to the gathering as the output

The relative distance among the objects is identified by pointing the centroid position for each of the objects and then calculating the distance between the centroids. The distances are stored in the form of multidimensional matrix. In order to classify the objects as a gathering, a threshold value for the count and relative distance is selected. If the threshold value for count is greater than a certain value, then that group of objects is identified as a gathering. Similarly, if the distance between the objects exceeds the threshold value, then the gathering is identified as a fixed gathering. The number of objects inside the gathering form the density. The workflow for performing the feature extraction from a given frame has been depicted in Fig. 1.

5 Experimental Results

5.1 Experimental Setup

In this subsection, the experimental setup has been described. The input video has been taken in through real time streaming [12] from a CCTV camera. Initially, the number of heads are counted in each frame. A certain value of threshold is selected with which it is compared. In this work, a threshold value of 10 has been taken for the head count. This value of the threshold is subject to the training that needs to be done in the development of a robust situation awareness system using a machine learning algorithm. If the head count exceeds the threshold value, the objects are identified as a gathering. In this work, the set of distances between each possible pair of heads in a frame, has been calculated [13]. Then the change in the distance between the consecutive frames has been compared. If the average change in distance between the pairs crosses a threshold in consecutive frames, then it is classified as a moving crowd. If the relative distance between consecutive frames remains constant then it is classified as a standing crowd. In this work a fixed value of head count has been chosen as 10 for gathering identification. However, the data such as count of heads, their distance or change in relative distance over consecutive frames and other metrics give a better analysis upon classification. This is done by feeding the metrics into a classifier and applying a classification algorithm to classify it as a gathering. This can be further classified as a peaceful gathering or a mob.

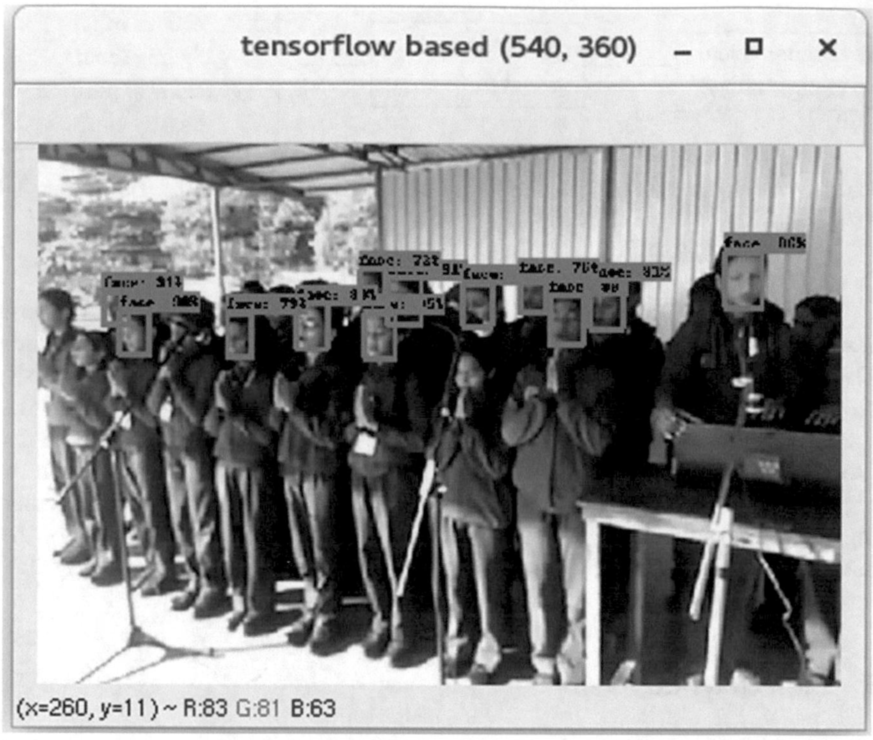

Fig. 2 Head count identification for a peaceful gathering

Initially, we stream a digital video in real time through a CCTV camera. Analysis is performed on the frames. A praying video that we have taken from YouTube as shown in Fig. 2 is identified as a peaceful gathering, on the basis of distance and head count. This has been done by using RCNN classifiers in Tensorflow Support Vector Machine [14]. The head count and relative distance has been calculated for the same. Further, a video that we have taken from 24 Ghanta news channel archive on YouTube as shown in Fig. 3 has been identified as a mob by using RCNN [15]. The CPU utilization has been shown for a particular frame in the next section.

5.2 CPU Utilization

This module uses about 5 GB of combined GPU memory and uses about 40% of about total 3900 CUDA cores to process a 720p60 video. CPU usage is about 8% per core, which is the computation required for handling the data transfer and assignment to the GPU. The frame shown in Fig. 3 has been considered for depicting CPU utilization.

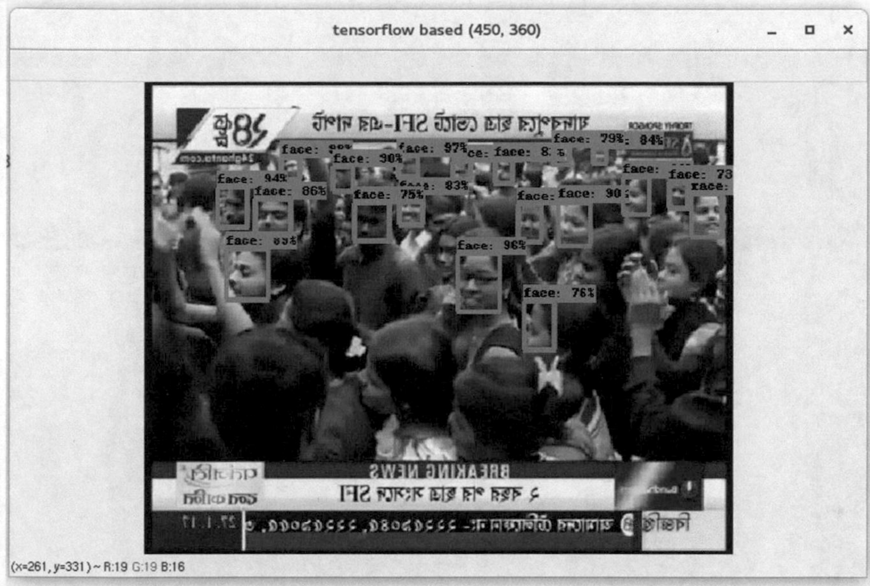

Fig. 3 Head count identification for a mob

The core distribution and percent GPU utilization for that frame have been depicted in Figs. 4 and 5 respectively.

6 Conclusion

In this paper, a methodology has been described and implemented for identifying the environmental elements in a Region of Interest for developing an Intelligent Situation Awareness System. The Tensorflow library has been used which has fast RCNN classifiers for performing the object identification based on the mentioned metrics. These metrics need to be classified for further analysis. The neural network used in this work can be further developed by training a classification algorithm by using datasets consisting of these metrics and the environmental elements which generated those metrics.

Acknowledgements This publication is an outcome of the Research and Development work undertaken project entitled “Object Identification through Syntactic as well as Semantic Interpretation from given SpatioTemporal Scenarios” under DRDO (ERIP/ER/1404742/M/01/1661) as well as the Visvesvaraya Ph.D. Scheme of Ministry of Electronics & Information Technology, Government of India, being implemented by Digital India Corporation.
We would like to express our sincere gratitude to all the members for this opportunity.

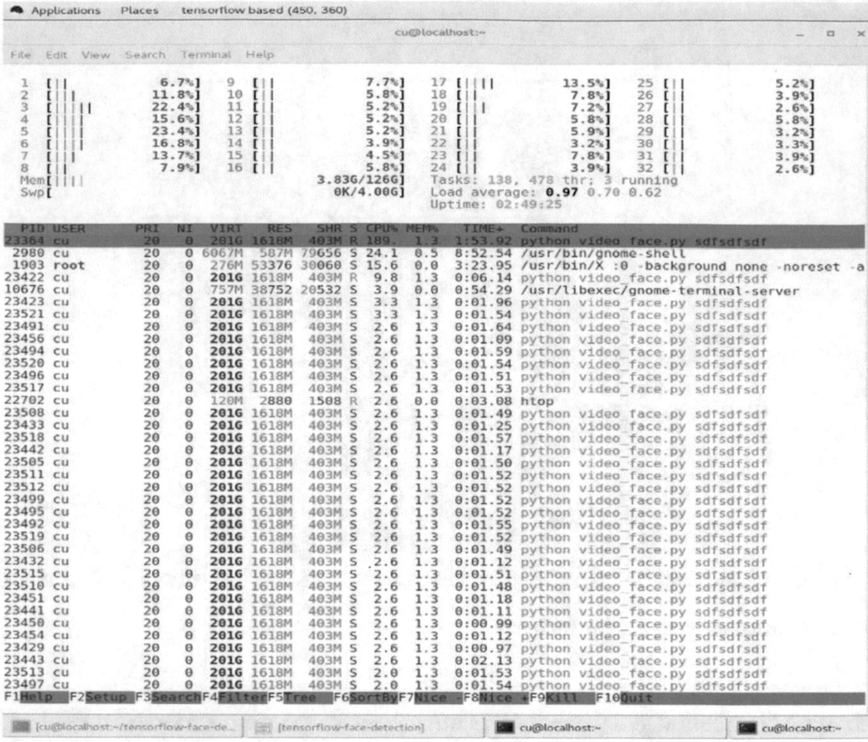

Fig. 4 CPU usage per core

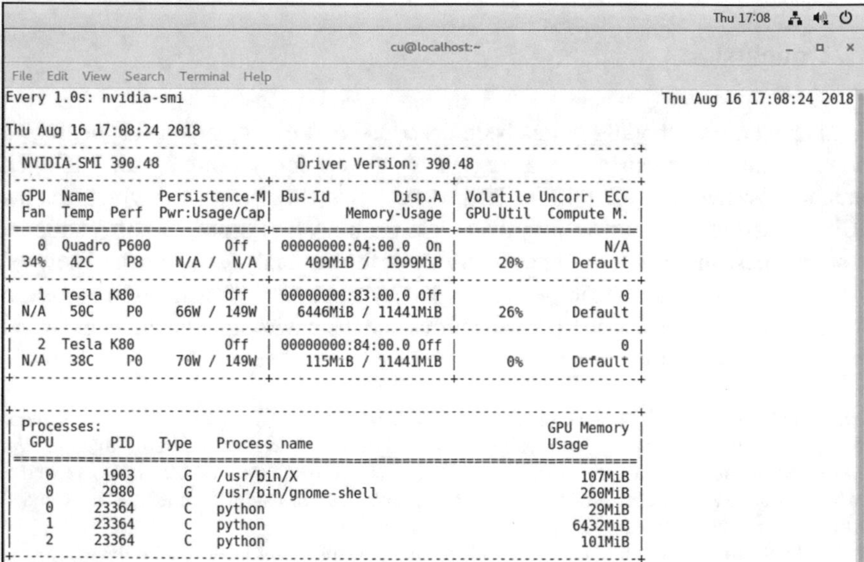

Fig. 5 GPU percent utilisation

References

1. Baxter, P., Trafton, J.G.: Cognitive architectures for human-robot interaction. In: Proceedings of the 2014 ACM/IEEE International Conference on Human-Robot Interaction, ACM Press, pp. 504–505 (2014)
2. Zhao, W., Krishnaswamy, A., Chellappa, R.: Discriminant analysis of principal components for face recognition. In: Proceedings of the Third IEEE International Conference on Automatic Face and Gesture Recognition, Japan, April 14–16, pp. 336–341 (1998)
3. Huang, G.B., Zhu, Q., Siew, C.K.: Extreme learning machine: theory and applications. Neurocomputing **70**(1), 489–501 (2006)
4. Kim, D.S., Jeon, I.J., Lee, S.Y., Rhee, P.K., Chung, D.J.: Embedded face recognition based on fast genetic algorithm for intelligent digital photography. IEEE Trans. Consum. Electr. **52**(3), 726–734 (2006)
5. Padgett, C., Cottrell, G.: Representation Face Images for Emotion Classification, Advances in Neural Information Processing Systems, 9. MIT Press, Cambridge, MA (1997)
6. Zhao, W., Chellappa, R.: Face Recognition-A literature survey. ACM Computing Surveys (2003)
7. Li, W., Zhang, Z., Liu, Z.: Expandable data-driven graphical modeling of human actions based on salient postures. IEEE Trans. Circ. Syst. Video Technol. **18**(11), 1499–1510 (2008)
8. Mitra, S., Acharya, T.: Gesture recognition: a survey. IEEE Trans. Syst. Man. Cybern. Part C **37**(3), 311–324 (2007)
9. Liu, X., Fujimura, K.: Hand gesture recognition using depth data. In: Proceedings of International Conference on Automatic Face and Gesture Recognition, pp. 529–534 (2004)
10. Mo, Z., Neumann, U.: Real-time hand pose recognition using low-resolution depth images. In: Proceedings of IEEE Conference on Computer Vision and Pattern Recognition, pp. 1499–1505 (2006)
11. Uddin, M.Z., Hassan, M.M., Almogren, A.: A facial expression recognition system using robust face features from depth videos and deep learning. Comput. Electr. Eng. **63**(1), 114–125 (2017)
12. Breitenstein MD., Kuettel D., Weise T., Van Gool L., Pfister H.: Real-time face pose estimation from single range images, In: Proceedings of IEEE conference on computer vision and pattern recognition, pp. 1–8 (2008)
13. Cootes, T.F., Edwards, G.J., Taylor, C.J.: Active appearance models. IEEE Trans. Pattern Anal. Mach. Intell. **23**(6), 681–685 (1998)
14. Phillips, P.J.: Support vector machines applied to face recognition. In: Proceedings of Advances in Neural Information Processing Systems II, pp. 803–809 (1999)
15. Hamed, A., Malekzadeh, M., Sanei, S.: A new neural network approach for face recognition based on conjugate gradient algorithms and principal component analysis. J. Math. Comput. Sci. 166–175 (2013)

Secure: An Effective Smartphone Safety Solution

Sampreet Kalita and Dhruba Kumar Bhattacharyya

Abstract With rapid advancements in smartphone and wearable technologies in the past decade, the world has virtually become a smaller place, facilitating connectivity between opposite ends of the globe just by the click of a button. However, certain emergency situations might not provide the opportunity or the window for that click as well. Hence, the need for a semi-automatically triggered distress signal activation scheme arises. In this paper, we introduce such a smartphone-based solution for all-round personal security that takes into account multiple modes of SOS broadcast activation, such as manual, Internet of Things (IoT) device data based and phone state data based. Also, we develop a couple of wearable device prototypes and test them for realtime detection and activation of SOS broadcast using a smartphone application.

Keywords SOS broadcast · Emergency SMS · Smartphone application · IoT · Wearable · Personal safety · Webserver

1 Introduction

In this connectivity-perked age, even though people might be separated by very long distances, they are always just a phone call away. This gift of technology has reduced a substantial amount of concern for the ones we care for, knowing that we are able to reach them as and when we wish to. However, given the recent surge in unprecedented incidents all over the world, a rising concern for one's personal security can be seen. As such, a scheme to relay an SOS broadcast instantaneously whenever one meets up with a difficult situation is of utmost necessity. Such a design should not only consider the activation of a manual distress signal but also take into

S. Kalita (✉) · D. K. Bhattacharyya
Tezpur University, Napaam, Sonitpur 784028, Assam, India
e-mail: sampreetkalita@gmail.com

D. K. Bhattacharyya
e-mail: dkb@tezu.ernet.in

A. K. Das et al. (eds.), *Computational Intelligence in Pattern Recognition*,
Advances in Intelligent Systems and Computing 999,
https://doi.org/10.1007/978-981-13-9042-5_53

account semi-automatic measures for all-round safeguarding. And what better an option to implement the same than on the smartphones one carries all along!

Given the computational potential and hardware features of a smartphone, it can be of good use during emergencies provided the proper application is installed and running. In fact, a number of solutions already exist for the same, featuring SMS broadcasts through manual activation [1, 2], or via other semi-automatic approaches like fall detection using accelerometer data [3], integrated omni-camera [4], etc. Solutions for safety beacon activation during large-scale disasters have also been developed [5, 6]. A few of the manual activation solutions have also implemented IoT device-based activation [7]. Further, wearable devices designed for the purpose of monitoring personal health are already being developed commercially [7–10]. Such devices also rely on a companion smartphone application to display the statistics and results obtained by them.

In this paper, an effective smartphone solution for personal safety with multiple modes of SOS broadcast activation is reported. Its workflow, designed to provide relevant security features is discussed in Sect. 2. The same is implemented and tested by developing an Android-based mobile application, an overview of which is given in Sect. 3. Also, the design of a wearable device compatible with the application is introduced. This, along with the server framework is discussed briefly in Sect. 4 with future research directions in Sect. 5. Finally, a conclusion is drawn from the work and is presented in Sect. 6.

2 Workflow Proposition

To help maintain all-round personal security, the process of activation of an SOS broadcast is divided into three modes, such as manual, IoT device data based and phone state data based. A foreground service monitors each of these modes and activation through any one of them triggers the required broadcast. A simplified workflow diagram is given in Fig. 1.

In this section, the various modes of activation are discussed followed by the set of instructions that are executed upon activation.

2.1 Manual Activation

The user manually activates the broadcast through the hardware keys of the smartphone. The pattern of the key-presses specific to the mode of activation is set up in the settings of the smartphone application. As such, the foreground service just keeps the hardware key-press events in check and triggers an SOS broadcast when the user-defined pattern is detected.

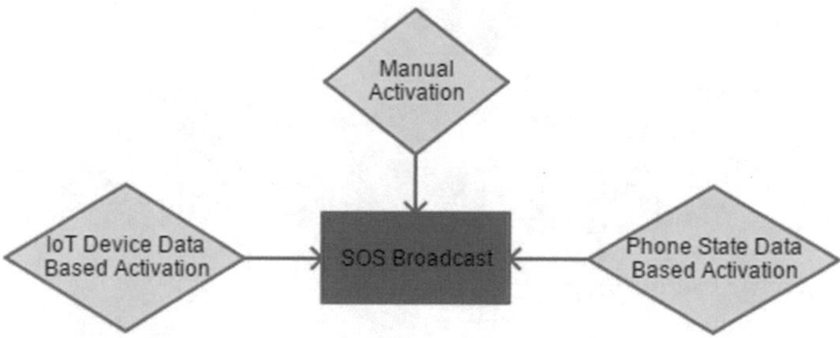

Fig. 1 A simplified workflow of the proposed solution

2.2 *IoT Device Data Based Activation*

A compatible IoT device (discussed in Sect. 4) triggers an SOS broadcast via the smartphone application whenever,

- the health measurement readings of the device exceed specific thresholds, or,
- the user manually presses the device's hardware keys in a particular pattern.

The foreground service keeps the IoT device connected and receives its data using a persistent Bluetooth Low Energy (BLE) connection.

2.3 *Phone State Data Based Activation*

Assuming that the smartphone application is trained to differentiate between the normal movements of the phone and shocks during a security threat, and that, in the latter case, the smartphone does not break down completely, an SOS broadcast is triggered as soon as sudden movement of the second type is detected. Here, the foreground service monitors the phone state at regular intervals and tests the data with its phone-specific model for detection of shocks.

Upon activation through any of the discussed modes, the following two tasks are executed.

1. The instantaneous location data obtained from the smartphone as well as additional information supplied by the user during the application's setup are sent as SMS to the contacts added by the user as trusted ones. The data is also uploaded to a primary web server (Sect. 4.1) over the Internet.
2. An audio recording starts in the background to capture the sound environment and is stored locally. The smartphone continues the recording until the service is stopped or another instance of the broadcast is activated. Once the recording is complete, it is uploaded to the primary server.

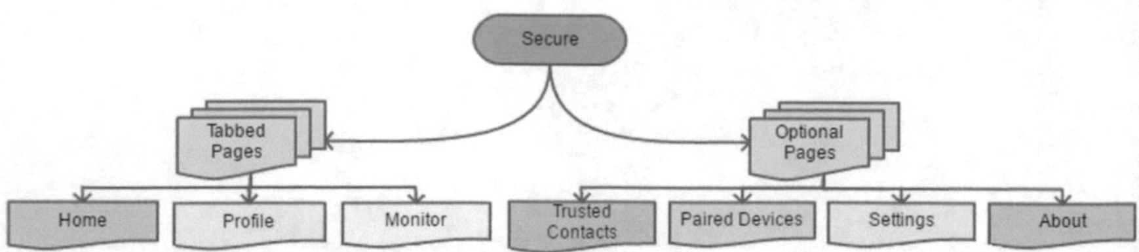

Fig. 2 An overview of *Secure*'s layout

Also, as a fall-back to poor Internet connectivity, the smartphone searches for local servers (Sect. 4.2) to upload the location data and the recorded audio file via Bluetooth. These servers then forward the same to the primary server.

3 Secure: An Overview

Secure is a smartphone application that monitors the state of the phone and has a few tricks up its sleeve that come handy at times of emergency. Its main features include:

- *Instant Geolocation Update* to the user-specified contacts and the Internet server.
- *Persistent Bluetooth Connections* with compatible IoT devices.
- *Phone Activity Monitoring* to detect sudden movements of the smartphone.

In this section, the layout, workflow and development of the application is discussed along with its comparison with available counterparts.

3.1 Layout of the Application

Secure's user interface is arranged in the form of pages that can be visited inside the application with ease. These pages are divided into Primary/Tabbed pages and Secondary/Optional pages based on their accessibility.

Tabbed pages are visible as soon as the application is opened. They can be accessed right from the first view and the user can navigate to them through swipes or just a tap on their tabs on the top bar.

Optional pages can be accessed through the secondary menu of the application. These pages contain additional settings that make the application whole.

The layout depicting the classification of the pages is illustrated in Fig. 2.

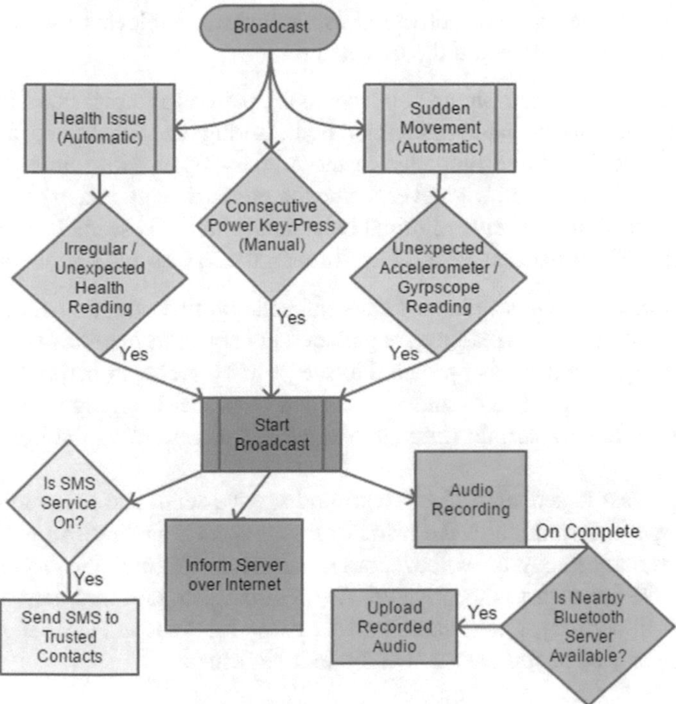

Fig. 3 An overview of *Secure*'s workflow

3.2 *Systematic Workflow*

The workflow of the application is designed following the one proposed in Sect. 2. The user sets up the preferences in the *Settings* page and after selecting the required service options, starts the service(s) from the *Home* page. A foreground service ensures that its required components are constantly monitoring the smartphone and are keeping up with their expected functionality. The services keep running even when the application is minimized. These services can be stopped manually from the *Home* page or by exiting the application completely.

The activation of the broadcast can be done through the modes discussed earlier. However, the manual mode is a stock feature of the application and the user defines the preferred number of presses as well as the maximum delay between successive key-presses for activation in the application settings. An outline of the workflow of the application is shown in Fig. 3. The workflow of each of the services, when opted for from the *Home* page, are discussed below:

Relay Messages Upon activation of the SOS broadcast via any of the modes, the foreground service sends an SMS containing the geolocation data of the smartphone to each of the trusted contacts added by the user. The format of the SMS is such that it also mentions the mode of activation as well as some relevant user information like the smartphone model and the user name.

Sync Wearable The smartphone connects to the paired IoT device and using BLE, keeps on listening to the data broadcasted by it, storing the same to a local database. This data is processed and uploaded to the primary server from time to time. An SOS broadcast is triggered whenever the health measurement data exceed their corresponding thresholds or any abnormal behaviour in them is seen. The service also checks if the SOS broadcast is triggered through the IoT device's panic button.

Monitor Motion The smartphone's state information like orientation, accelerometer reading and gyroscope reading are gathered in very short intervals and processed simultaneously. The same is uploaded to the primary server to be processed using Machine Learning algorithms and is classified as probable emergencies if sudden movements following certain traits are detected, hence sending the broadcasts remotely.

If a broadcast is activated, the foreground service sends the location as well as necessary user information to the primary server as well as via SMS to the trusted contacts. Instantaneously, the voice recorder starts recording until the service is turned off or the SOS broadcast is reactivated. As a fall-back to poor Internet connectivity or failed SMS delivery, it also searches for nearby local servers and sends both the information as well as the recorded audio to it via Bluetooth.

3.3 Application Development and the Codebase

To make the application scalable across platforms, we use the *Xamarin.Forms* [11] framework to implement the common functionality, and at the same time implement native platform-specific controls through inheritable interfaces. The APIs written for the application are fully documented and a GIT-based version control is used to keep track of periodic modifications to the application.

3.4 Comparison with Available Applications

Personal security applications that send text messages in emergency situations are not uncommon [1–7]. And Android being the leading platform for smartphone users, the list of applications developed on the platform that feature SOS broadcasts is unending. A wide range of them also support a login experience through the Internet, emails to selected contacts, emergency calling, a map integration, etc. for better security. Some of them even feature live location sharing with friends and relatives while travelling through potentially unsafe areas. But very few of them provide a smooth interface and a good user experience. And even fewer provide the scope to sync with wearable health monitoring devices. A detailed comparison among some of the applications is reported in Table 1.

Table 1 Comparison with available android applications (as on 15.08.2018)

Application name	KP[a]	Map	LI[a]	Call	AR[a]	HM[a]	DM[a]	MA[a]	Last updated	Additional features	Developer (package)[b]
Proposed	✓	✓	✓	✓	✓	✓	✓	3		Tweakable settings, IoT device	
My safetipin		✓	✓	✓				1	16.05.2018 [v1.0.14]	Safety score, friends	Safetipin (https://play.google.com/store/apps/details?id=com.safetipin.mysafetipin)
Raksha	✓	✓	✓	✓				1	11.08.2018 [v2.4]	Nearby police, organ donation	Portal Perfect.com (https://play.google.com/store/apps/details?id=com.portalperfect.sosapp)
Himmat plus			✓		✓			1	25.04.2018 [v1.1.1]	Groups, report journey	IT centre police head quarter (https://play.google.com/store/apps/details?id=com.dp.himmat)
Shake2Safety	✓			✓			✓	2	21.12.2017 [v2.0]	Shake, set sensitivity	photonapps (https://play.google.com/store/apps/details?id=com.photon.shake2safety)

(continued)

Table 1 (continued)

Application name	KP[a]	Map	LI[a]	Call	AR[a]	HM[a]	DM[a]	MA[a]	Last updated	Additional features	Developer (package)[b]
Personal safety app		✓	✓					1	14.08.2018 [v5.7.9]	Fake call, track contact, friends	Smart24 (https://play.google.com/store/apps/details?id=smart.emergencyservice)
bSafe		✓	✓	✓	✓			1	06.08.2018 [v3.6.13]	Follow me, timer alarm, fake call, guardians	Mobile software AS (https://play.google.com/store/apps/details?id=com.bipper.app.bsafe)
Telltail	✓		✓				✓	1	28.04.2015 [v2.1]	Track vehicle	DIMTS Ltd. (https://play.google.com/store/apps/details?id=com.dimts.ui)
Wearsafe personal safety	✓	✓	✓	✓				1	07.09.2017 [v1.10.1]	Wearsafe tag wearable, android wear	Wearsafe labs (https://play.google.com/store/apps/details?id=com.wearsafe.wearsafeapp)
Personal safety	✓	✓	✓	✓	✓			1	31.07.2018 [v2.2.4]	Detect scream, record video, track	Saferway mobile (https://play.google.com/store/apps/details?id=com.cocoa.saferway)

[a]KP = Key-press, LI = Log In, AR = Audio Recording, HM = Health Monitor, DM = Device Monitor, MA = Mode of Activation
[b]The corresponding URLs can be obtained by using https://play.google.com/store/apps/details?id={package} [12]

It can be seen from the table that compared to the existing applications, the proposed solution includes more features and functionality and is found to be more effective in view of the following points.

1. The three different modes of broadcast activation allow it to reach out to a much wider variety of situations, making it more versatile.
2. The design of the application is minimalistic and simple taps and swipes govern the flow between pages, most of which are single level apart.
3. Its cross-platform integrability stands out among the crowd of similar applications, giving it a universal mode of implementation and administration.
4. The layered web servers and their time to time upgradation to newer learning techinques help increase the accuracy of threat detection.
5. With the support for usage-specific updates to the various detection models, it resembles a far more personalized security application.

4 The Servers and the IoT Devices

To back *Secure* in its working, two types of web servers are designed to receive data from the application and process the same. Further, two wearables are designed to record heart rate and temperature of the user. This section gives a brief outline of these assisting elements of *Secure*.

4.1 The Primary Sever

The primary Internet server is built on *Node.js* [13] and uses the JavaScript Object Notation (JSON) format to exchange data and MongoDB [14] for the storage of information. This server is responsible for, but not limited to,

- obtaining the required information for application registration and sending back a registration token for subsequent correspondence.
- displaying the broadcast to the administrating team by pointing to the location on a map whenever an SOS broadcast is detected.
- learning and updating the detection models by processing the data received from the application and sending a distress signal remotely if any anomality in the data is discovered.

The primary server is structured into numerous modules that concentrate on different processes, creating a log of all events as they work.

4.2 The Local Servers

A local server is set up on a *Raspberry Pi 3 Model B* [15]. In the view of a smart city containing such a device in the vicinity with the server running, the application relays the information from the user and the audio recording to the server through a Bluetooth socket connection. This local server maintains a log of all the file and data transactions and automatically relays the information obtained to the primary server.

4.3 The Waist Bag: WearGen

WearGen, a *Arduino/Genuino 101*-based [16] wearable device is programmed to gather the health information using a Heartbeat sensor and a Temperature sensor interfaced to it. Figure 4 shows the proposed design of the waist bag containing a side-compartment for batteries and a front-container for the sensors. A prototype of the same was designed and tested, an image of which is shown in Fig. 5

The device can be powered using rechargeable cells plugged into the battery compartment. The openable compartment can contain an LCD display along with the sensors. The waist bag, with the empty space inside, can be used to store handy stuff when carried. The device syncs the health readings with the smartphone application through the BLE connection when the sensor compartment is opened.

4.4 The Wrist Watch: SecureWatch

SecureWatch, developed on the *ESP-32* chip [17], mimics a wristwatch that collects and relays health information of its user to the smartphone application *Secure*. The same device, when used as an independent health measurement device can connect upto four sensor inputs. The proposed design of this modular wearable is shown in Fig. 6.

It features a native Heartrate sensor and an OLED screen to display the sensor readings. The Wi-Fi module allows it to sync time online and receives updates from time to time. The BLE module advertises and notifies the paired smartphone of the health data changes in real time. An image of the designed prototype for *SecureWatch* can be seen Fig. 7.

5 Future Prospects

As a future direction to the work, we aim to explore the development of a flexible learning approach to create better shock detection models based on the smartphone

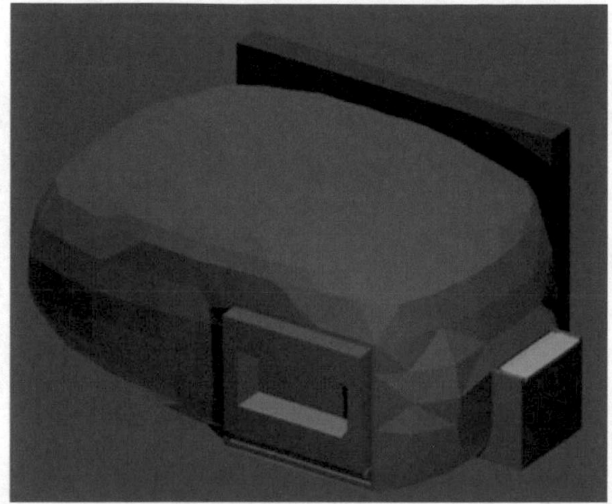

Fig. 4 *WearGen*: the design

Fig. 5 The prototype of *WearGen*

Fig. 6 *SecureWatch*: the design

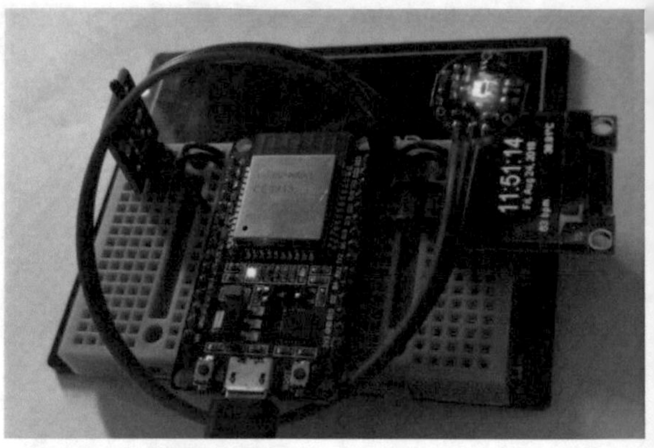

Fig. 7 The prototype of *SecureWatch*

brand. Also, health issue detection algorithms featuring machine intelligence will be implemented to avail medical assists at early stages. Furthermore, the designs of the wearable devices will also be revisited for ergonomic improvements and then implemented and tested in real life.

6 Conclusion

Although the Internet is flooded with applications that mimic an SOS broadcast, it can be seen that most of them lack a lot of functionality. Even the ones that are feature-packed seem to have missed a relevant idea or two in their implementation. The solution proposed in this paper tries to overcome the imperfections present in the already available ones. The application leaves a benchmark in the sea of personal safety applications by implementing not only the key features that justify the name, but also exploring the present advancements in the field of IoT devices and machine learning research, altogether building an application backed by a well-managed server and a reliable wearable interface.

References

1. Chand, D., Nayak, S., Bhat, K.S., Parikh, S., Singh Y., Kamath, A.A.: A mobile application for women's safety: WoSApp. In: 2015 IEEE Region 10 Conference on TENCON 2015, pp. 1–5, Macao (2015)
2. Yarrabothu, R.S., Thota, B.: Abhaya: an android app for the safety of women. In: Annual IEEE India Conference (INDICON). pp. 1–4, New Delhi (2015)
3. Sposaro, F., Tyson, G.: iFall: an android application for fall monitoring and response. In: Annual International Conference of the IEEE Engineering in Medicine and Biology Society, pp. 6119–6122, Minneapolis (2009)

4. Miaou, S.-G., Sung, P.-H., Huang, C.-Y.: A customized human fall detection system using omni-camera images and personal information. In: 1st Transdisciplinary Conference on Distributed Diagnosis and Home Healthcare, pp. 39–42, Arlington (2006)
5. Chen, Y., Lin, C., Wang, L.: A personal emergency communication service for smartphones using FM transmitters. In: IEEE 24th Annual International Symposium on Personal, Indoor, and Mobile Radio Communications (PIMRC), pp. 3450–3455, London (2013)
6. Suzuki, N., Zamora, J.L.F., Kashihara S., Yamaguchi, S.: SOSCast: location estimation of immobilized persons through SOS message propagation. In: Fourth International Conference on Intelligent Networking and Collaborative Systems, pp. 428–435, Bucharest (2012)
7. Wearsafe Homepage. www.wearsafe.com. Last Accessed 15 Aug 2018
8. Pantelopoulos, A., Bourbakis, N.G.: A survey on wearable sensor-based systems for health monitoring and prognosis. IEEE Trans. Syst. Man Cybern. Part C (Appl. Rev.) **40**(1), 1–12 (2010)
9. Hao, Y., Foster, R.: Wireless body sensor networks for health-monitoring applications. Inst. Phys. Eng. Med. Physiol. Meas. **29**(11), 27–56 (2008)
10. Fitbit Versa Homepage. www.fitbit.com/in/versa. Last Accessed 15 Aug 2018
11. Xamarin.Forms Homepage. https://docs.microsoft.com/en-us/xamarin/xamarin-forms/. Last Accessed 14 Sept 2018
12. Google Play Store Homepage. www.play.google.com/store. Last Accessed 15 Aug 2018
13. Node.js Homepage. https://nodejs.org/en/. Last Accessed 14 Sept 2018
14. MongoDB Homepage. https://www.mongodb.com/. Last Accessed 14 Sept 2018
15. Raspberry Pi Homepage. https://www.raspberrypi.org/. Last Accessed 14 Sept 2018
16. Arduino 101 Guide. https://www.arduino.cc/en/Guide/Arduino101. Last Accessed 14 Sept 2018
17. ESP32 Overview. https://www.espressif.com/en/products/hardware/esp32/overview. Last Accessed 14 Sept 2018

A Gamma-Levy Hybrid MetaHeuristic for HyperParameter Tuning of Deep Q Network

Abhijit Banerjee, Dipendranath Ghosh and Suvrojit Das

Abstract In this study we propose a novel metaheuristic algorithm; namely "*Gamma-Levy Hybrid Metaheuristic with Conditional Evolution (GLHM-CE)*". The proposed algorithm is evaluated over 28 Blackbox Problems of CEC-2013, Special Session on Real-Parameter Optimization and compared with modern metaheuristic and evolutionary algorithms like SHADE, Co-DE, and JADE. The statistical results show that GLHM-CE successfully circumvents local minimas on high dimensional blackbox functions and has a fast convergence. GLHM-CE is then used to optimize the hyperparameters of a static Deep Q Neural Network evaluated on OpenAI Gym Cartpole problem. The results evaluated over a total episodal run of 5000 shows a better stability of the DQN when the hyperparameters are optimized by GLHM-CE.

Keywords GLHM-CE · MetaHeuristics · Evolutionary algorithms · DQN · Hyper parameters · OpenAI

1 Introduction

Artificial Neural Networks with multiple hidden layers; thus qualifying as deep Neural Networks for DNN in their current state are principally trained through gradient-based approaches like Backpropagation [8] which uses Stochastic Gradient Descent (SGD) or any of its modern variants optimizing the weights of the entire network

A. Banerjee (✉)
Department of Electronics and Communication, Dr. B C Roy Enginering College, Durgapur 713206, India
e-mail: abhijit.banerjee@bcrec.ac.in

D. Ghosh
Department of Computer Science, Dr. B C Roy Enginering College, Durgapur 713206, India
e-mail: dipen.ghosh@bcrec.ac.in

S. Das
Department of Computer Science and Engineering, National Institute of Technology, Durgapur 713205, India
e-mail: suvrojit.das@gmail.com

A. K. Das et al. (eds.), *Computational Intelligence in Pattern Recognition*, Advances in Intelligent Systems and Computing 999,
https://doi.org/10.1007/978-981-13-9042-5_54

to reduce the overall error or loss function. With the advent of cheap, reliable, and massive computing capabilities in the form multicore CPU's and GPU's, large Deep Neural Networks (DNN's) with millions of weights can finally be trained. The effectiveness of DNN's when properly trained is documented in various publications [9, 11, 14]. SGD and its variants can lead to global optimal solution and not get stuck in local optima as there are many paths to the relative optima in a high dimensional ANN. If this applies to SGD, so does it also applies to Evolutionary Algorithms. After all EA is nothing but approximation of the gradient. But it has to be considered that EAs approximate gradients while SGD calculates exact gradient. Furthermore with the introduction of rectified linear units (ReLUs); gradient propagation among layer is improved and better than sigmoidal activation functions [7]. EA used to train weights in a DNN is somehow daunting if not unfeasible. For example, MNIST image classification dataset [4] has 60 k images. If a population-based EA is applied to train this dataset, for 50 population size; a total of 3 million objective function has to be evaluated for a single epoch. SGD does the same in one go. So, the use of population-based EA for weight training of DNN is still a massively challenging task.

SGD on the other hand has its hyperparameters like learning rate and batchsize to be hand tuned or pruned through a graph-based approach. This tuning of hyperparameters in a fixed sized ANN becomes an obligatory task which is where EAs can step in. Deep Q Network(DQN) [5] is a reinforcement learning technique in lines of Markov Decision Process (MDP) which entwines Bellman equation [1] to learn a sequence of steps to complete a task. The primary hyperparameters of DQN are:

- Learning rate η
- Reward Decay/Discount Factor λ
- Batch Size
- Node Size in a fixed layer ANN.

In subsequent sections, we first propose an evolutionary strategy, "*Gamma-Levy Hybrid Metaheuristic with Conditional Evolution (GLHM-CE)*" which converges faster than most other EAs and evades local optimas. We evaluate its performance on CEC-2013 [15] Real PArameter Benchmark functions and compare the results with modern Evolutionary Algorithms like SHADE, JADE and CoDE [10, 12, 16]. Once, its effectiveness as a fast converging metaheuristic is established, we use GLHM-CE to tune hyperparameters of a fixed architecture DQN and use to learn the CartPole problem from OpenAI Gym environment [2].

2 Gamma-Levy Hybrid Metaheuristic with Conditional Evolution (GLHM-CE)

The ideal equilibrium between exploration and exploitation in a population-based optimization method is very crucial. Population- based algorithms tend to remain stagnant in and around local minimas in higher dimension multimodal functions resulting in '*Curse of Dimensions*'. GLHM-CE addresses these issues by incorporating a dual GAmma-Levy stochastic model for *exploitation* and an Evolutionary Strategy for *exploration*. The switching between these two modules is randomly drawn from the uniform distribution.

2.1 Gamma-Levy: Exploitation (GLHM)

Let fitness value of the objective function be $I_i | i \in \Re$ of an agent A_i in population size N. The population is first sorted with respect to the fitness value I_i after evaluation.

In this module we loop from the median agent to the last agent (the worst performing agents) A_i with fitness value I_i. Then we pick a random agent from the better performing agents A_j with fitness value I_j. The stepsize α with which the worst agent will move toward the better agent is drawn from Levy's Distribution; derived from publication by Yang and Deb [13] is as follows:

$$\sigma = \left(\Gamma(1+\beta) * sin(\pi\beta/2) \Big/ (\Gamma((1+\beta)/2)) * 2\beta^{(\beta-1)/2}\right)^{1/\beta}$$
$$u = rand_{int}(dim) * \sigma, \quad v = rand_{int}(dim), \quad step = \big(u/abs(v)\big)^{1/\beta}$$
$$\alpha = 0.001 * (step * (A_i[dim_d] - A_j[dim_d])); \quad d \;\; \forall \;\; Dimension$$

where, $rand_{int}(dim)$ is a random integer between 1 and highest dimension dim.
where, σ is the standard deviation of the distribution.
where, u and v are vectors from normal distribution and Γ is the gamma function.
β is Power law Index $(1 < \beta < 2)$, usually kept at 1.

(1)

Next, we calculate the stochastic multiplier G_d randomly drawn from the two parameter Gamma Distribution.

$$shape = I_i \Big/ I_j, \quad scale = I_j \Big/ I_i$$
$$G_d = \Gamma(shape, scale) \tag{2}$$

Finally, the movement equation of A_i to A_j is given as:

$$A_i[1:Dim] = A_i[1:Dim] * (1 - G_d) + A_j[1:Dim] * G_d + (\alpha * \sqrt{|U_b - L_b|})$$

Where; Dim $=$ Dimension of the problem
Dimension Upper Bound $:= U_b$
Dimension Lower Bound $:= L_b$

(3)

2.2 Conditional Evolution: Exploration (CE)

An iteration is run from the best agent to the median agent in a sorted population. The next best agent is chosen and a crossover is performed between these parents at a randomly chosen dimension to produce a child. The child thus produced is then mutated on a single randomly chosen dimension. The $N/2$ number of child population thus created is directly copied to the $N/2$ number of worst performing agents.

Input: Fitness value of ith agent be denoted as I_i
Let $gt = int(N/2)$ denote the median agent index.
Let i denote the current iteration index.
$i = gt$
while $(i < N)$ **do**
  j=$randint[gt : N]$
  Calculate: Stepsize from Levy's Distribution :α (Eq. 1)
  Calculate: Sample random float from Gamma Distribution :G_d (Eq. 2)
  Move agent i toward j in all dimensions(Apply Eq. 3);
  i++
end

Algorithm 1: Pseudocode of GLHM.

Input: Encode ith agent gene as $A_i[D]$ vector,
(where D=Dimension or Gene length.)
Data: Let F_{ub}=max($A_0[0, D]$),
Let L_{lb}=min($A_0[0, D]$),
Let CrossoverPoint=randint[1:D]
Initialize child population (*ChildPop*) of $N/2$ vectors of dimension D.
$gt = int(N/2)$, $i = 1$
while $(i < gt)$ **do**
  Generate child popilation/offspring;
  $ChildPop_i = CrossOver(A_i[CrossoverPoint], A_{i+1}[CrossoverPoint])$.
  $ChildPop_i[randint[1 : D]] = random.uniform[0, 1] * (F_{ub} - L_{lb}) + L_{lb}$. i++
end
Copy offsprings to worst populations: $ChildPop => A_{[gt:N]}$

Algorithm 2: Pseudocode of Conditional Evolution (CE).

2.3 GLHM-CE Pseudocode

```
Input: Initialize population of size N with Agents A[1:N].
i = 1
while (i < MaxGeneration) do
    Evaluate population for function value as I[1:N]
    and sort population r=rand([0 : 1])
    if r < 0.05 then
        Process Conditional Evolution (CE) module [Algorithm 2
        ] else
            Process Gamma-Levy Hybrid Metaheuristic (GLHM) module
            [Algorithm 1]
        end
    end
    i++
end
Post Process and Visualize.
```

Algorithm 3: Pseudocode of combined Gamma-Levy Hybrid Metaheuristic with Conditional Evolution (GLHM-CE).

2.4 GLHM-CE Evaluation on CEC-2013 Benchmark Functions

Evaluation results of GLHM-CE on CEC2013 [15] and its performance analysis with SHADE [10], CoDE [12] and Jade [16] is presented in Table 1. The time complexity of the algorithm is shown in Table 2. The statistical data of SHADE, CoDE, and Jade is referenced from the publication by Tanabe and Fukunaga [10]. The Wilcoxon Signed Rank Test and t-test (both one-tailed and two-tailed) is also provided in Table 1.

Though GLHM-CE is not considerably significant over JADE,SHADE, and CoDE it is observed from the results that GLHM-CE holds rank 1 in most test cases. For rotated functions; Rotated Schaffers F7, Rotated Ackleys, Rotated Weierstrass, Rotated Griewanks Function GLHM-CE performs worse than all the rest algorithms under consideration. For all other functions, GLHM-CE shows superiority.

3 Deep Q Networks

A Deep Q network (DQN) is a MLP(Multi Layered Perceptron) which produces an output action values vector $Q(s, \theta)$, θ being the network parameters including hyperparameters. In standard Q learning, the parameter update equation when an

Table 1 Average error: CEC2013 real parameter benchmark functions f1 − f2, with their Wilcoxon Signed Rank Test (2-tailed). Function evaluation (FES) = 3E5, Population size = 100, Dimension = 30

F	GLHM-CE		JADE		CoDE		SHADE		GLHM-CE-BEST	Rank
	Mean	Std	Mean	Std	Mean	Std	Mean	Std		
1	0.00E+000	0.00E+000	0.00E+000	0.00E+000	0.00E+000	0.00E+000	0.00E+000	0.00E+000	0.00E+000	1
2	4.16E+003	4.13E+003	9.00E+003	7.47E+003	9.778E+004	4.81E+004	7.67E+003	5.66E+003	0.00E+000	1
3	8.05E+005	8.25E+005	4.02E+001	2.13E+002	1.078E+006	3.03E+006	4.71E+005	2.35E+006	7.33E+005	3
4	1.08E−014	1.47E−013	1.92E−004	3.01E−004	8.178E−002	1.09E−001	6.09E+003	1.33E+004	0.00E+000	1
5	0.00E+000	0.00E+000	0.00E+000	0.00E+000	0.00E+000	0.00E+000	0.00E+000	0.00E+000	0.00E+000	1
6	1.34E−001	1.32E−001	5.96E−001	3.73E+000	4.155E+000	9.00E+000	2.07E+000	7.17E+000	0.00E+000	1
7	1.31E+002	1.31E+002	4.60E+000	5.39E+000	9.318E+000	6.34E+000	3.16E+000	4.13E+000	1.28E+002	4
8	2.09E+001	2.09E+001	2.07E+001	1.76E−001	2.079E+001	1.18E−001	2.09E+001	4.93E−002	1.50E−001	3
9	2.83E+001	2.88E+001	2.75E+001	1.77E+000	1.445E+001	2.90E+000	2.65E+001	1.96E+000	1.38E+001	4
10	5.35E−001	5.35E−001	7.69E−002	3.58E−002	2.709E−002	1.50E−002	4.04E−002	2.37E−002	5.10E−001	4
11	0.00E+000	0.00E+000	0.00E+000	−0.00E+000	0.00E+000	0.00E+000	0.00E+000	0.00E+000	0.00E+000	1
12	1.98E+001	2.00E+002	2.30E+001	3.73E+000	3.978E+001	1.21E+001	2.29E+001	5.45E+000	0.00E+000	1
13	2.41E+001	2.41E+002	5.03E+001	1.34E+001	8.036E+001	2.74E+001	4.67E+001	1.37E+001	0.00E+000	1
14	0.00E+000	0.00E+000	3.18E−002	2.33E−002	3.656E+000	4.09E+000	2.86E−002	2.53E−002	0.00E+000	1
15	4.03E+003	4.03E+003	3.22E+003	2.64E+002	3.356E+003	5.31E+002	3.24E+003	3.17E+002	8.59E+002	4
16	1.30E+000	1.30E+000	9.13E−001	1.85E−001	3.379E−001	2.03E−001	1.84E+000	6.27E−001	9.70E−001	2
17	3.04E+001	3.04E+001	3.04E+001	3.83E−014	3.038E+001	1.17E−002	3.04E+001	1.95E−014	3.00E−002	2
18	1.77E+002	1.77E+002	7.25E+001	5.58E+000	6.688E+001	9.23E+000	7.76E+001	5.91E+000	1.10E+002	4
19	1.62E+000	1.62E+000	1.36E+000	1.20E−001	1.607E+000	3.58E−001	1.44E+000	8.71E−002	2.90E−001	4
20	1.04E+001	1.09E+001	1.05E+001	6.04E−001	1.056E+001	6.69E−001	1.04E+001	5.82E−001	0.00E+000	1
21	3.00E+002	3.00E+002	3.09E+002	5.65E+001	3.018E+002	9.02E+001	3.04E+002	6.68E+001	2.00E+000	2
22	4.99E−100	6.26E−009	9.81E+001	2.52E+001	1.168E+002	9.96E+000	9.39E+001	3.08E+001	0.00E+000	1

(continued)

Table 1 (continued)

F	GLHM-CE		JADE		CoDE		SHADE		GLHM-CE-BEST	Rank
	Mean	Std	Mean	Std	Mean	Std	Mean	Std		
23	3.84E+003	3.84E+003	3.51E+003	4.11E+002	3.556E+003	6.12E+002	3.36E+003	4.01E+002	4.81E+002	4
24	2.83E+002	2.84E+002	2.05E+002	5.29E+000	2.208E+002	9.28E+000	2.17E+002	1.57E+001	7.78E+001	4
25	2.03E+002	3.03E+002	2.59E+002	1.96E+001	2.567E+002	6.55E+000	2.74E+002	1.06E+001	0.00E+000	1
26	2.00E+002	2.00E+002	2.02E+002	1.48E+001	2.176E+002	4.48E+001	2.15E+002	4.11E+001	0.00E+000	1
27	1.23E+002	1.23E+003	3.88E+002	1.09E+002	6.188E+002	1.01E+002	6.70E+002	2.40E+002	0.00E+000	1
28	3.00E+002	3.00E+002	3.00E+002	0.00E+000	3.000E+002	0.00E+000	3.00E+002	0.00E+000	0.00E+000	1
			Wilcoxon Signed Rank Test (two-tailed)							
	Comparision			*R+*		*R−*		*P-value*		
1	GLHM-CE versus SHADE			145		131		0.83366		
2	GLHM-CE versus CoDE			114		162		0.4654		
3	GLHM-CE versus JADE			105		126		0.71884		
			t-Test (one tailed and two tailed)							
	Comparision			*t Stat*	*P (T = t) (one-tail)*	*t Critical (one-tail)*	*P (T = t) (two-tail)*	*t Critical (two-tail)*		
1	GLHM-CE versus SHADE			9.95E−001	1.64E−001	1.70E+000	3.29E−001	2.05E+000		
2	GLHM-CE versus CoDE			−1.28E+000	1.05E−001	1.70E+000	2.11E−001	2.05E+000		
3	GLHM-CE versus JADE			9.73E−001	1.70E−001	1.70E+000	3.39E−001	2.05E+000		

Table 2 Time complexity of GLHM-CE (miliseconds). T0, T1, T2 can be referenced from CEC-2013 [15]

Dimension	TO	T1	T2	(T2 − T1)/TO
10	220	2847	4963	9.618
30	220	4815	8491	16.709
50	225	28451	37158	38.697

action A_t is taken when in state S_t resulting in a reward of R_{t+1} in next state S_{t+1} is given as:

$$\theta_{t+1} = \theta_t + \alpha(Y_t^Q - Q(S_t, \theta_t))\nabla_\theta Q(S_t, \theta_t) \tag{4}$$

α is the stepsize and Y_t^Q target is given by the Bellman Equation:

$$Y_t^Q = R_{t+1} + \gamma \arg\max Q(S_{t+1}, \theta_t) \tag{5}$$

Mnih et al. [6] proposed a Target Network and Experience Replay for DQN. The target used by DQN is then given by:

$$Y_t^{DQN} = R_{t+1} + \gamma \arg\max Q(S_{t+1}, \theta_t^-) \tag{6}$$

where, θ_t^- is the parameters of the target network which is updated/copied from the online network every τ steps. We have used GLHM-CE to optimize these hyper-parameters in the reinforcement learning environment of CartPole v1 from OpenAI gym [2]. The cartpole problem is stated as "*A pole is attached by an un-actuated joint to a cart, which moves along a frictionless track. The system is controlled by applying a force of* $+1$ *or* -1 *to the cart. The pendulum starts upright, and the goal is to prevent it from falling over. A reward of* $+1$ *is provided for every timestep that the pole remains upright. The episode ends when the pole is more than* 15° *from vertical, or the cart moves more than 2.4 units from the center.*" [3].

3.1 Proposed Strategy for Hyperparameter Optimization

DQN is trained using SGD, ADAM, and RMSPROP optimizers while the hyperparameters are solely optimized by GLHM-CE. A population of 50 agents in GLHM-CE is used to create 50 different networks. All networks are evaluated for "score" in the game. The first 50% agent with best fitness value is sent to GLHM-CE for hyperparameter optimization. The GLHM-CE optimized networks are then directly copied to the last 50% of the population. Total epoch was taken as 5000 for all experiments. The plots of "Average score of last 100 epochs" is plotted for both vanilla DQN as shown in Fig. 1 and GLHM-CE optimized DQN. The vanilla DQN for baseline uses

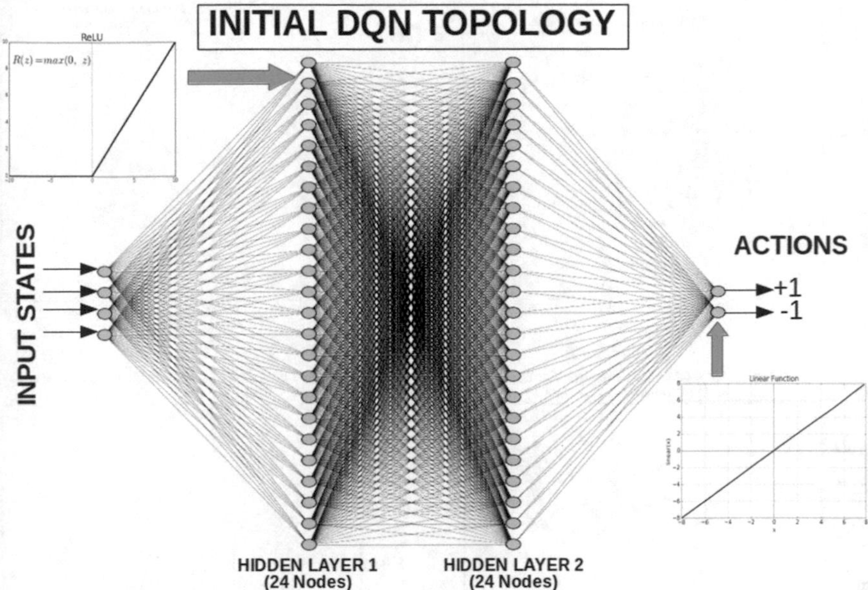

Fig. 1 Initial topology of the DQN

1 Input Layer with 4 nodes for 4 state values, 2 Hidden layer with 24 nodes each and Activation Function as ReLu.1 Output Layer with 2 nodes for 2 actions (+1, −1) is used with Activation Function = Linear.

Hyperparameters: learning rate = 0.0001, learning rate decay = 0.995, reward discount = 0.95, replay batchsize = 32.

3.2 *Results*

The network is deemed successful only when the average score per 100 epoch is greater than 495. Figure 2d shows GLHM-CE hyperparameter tuning with ADAM weight optimizer, being the only DQN network with the score stable after 2000 epochs. All other networks suffered from stability issues. DQN with no hyperparameter tuning resulted in '*Catastrophic Forgetting*' of learned weights.

The optimized parameters for this evolved network is: learning rate $\alpha = 0.0042$, reward decay $\lambda = 0.71$, batch size = 64, node length = 14 in both hidden layer.

Figure 3 shows the effects on average score on cartpole, when the hyperparameters of the DQN is optimized by Genetic Algorithm and Neuroevolution of augmenting topologies (NEAT). All score values are rounded off to the nearest highest integer in multiples of 100. Both vanilla GA and NEAT is observed to be prone to Catastrophic

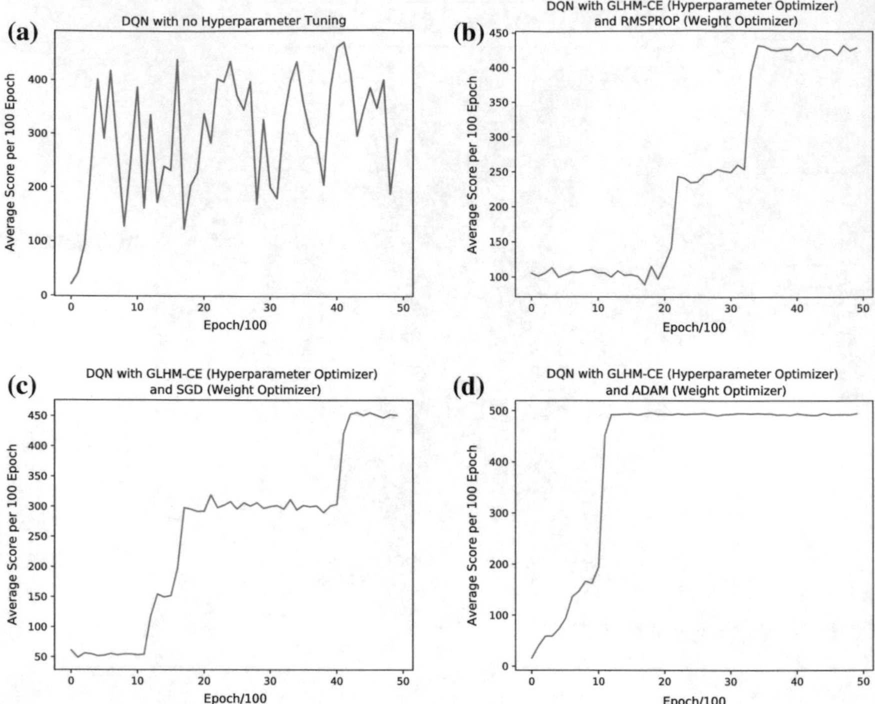

Fig. 2 Score per 100 Epoch for **a** DQN with no hyperparameter tuning. **b** GLHM-CE hyperparameter tuning with RMSPROP weight optimizer. **c** GLHM-CE hyperparameter tuning with SGD weight optimizer. **d** GLHM-CE hyperparameter tuning with ADAM weight optimizer

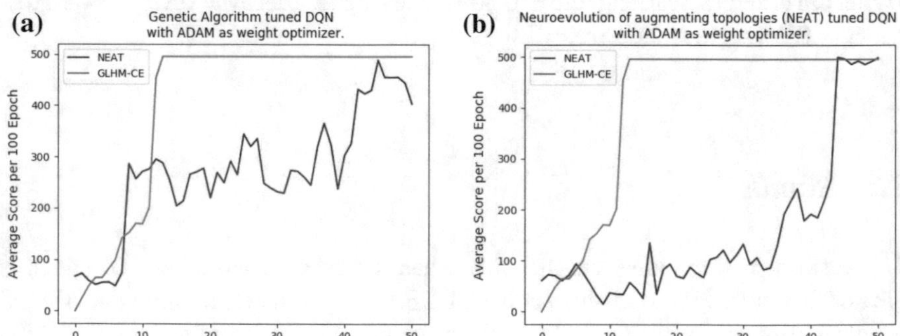

Fig. 3 Score per 100 Epoch for **a** Genetic algorithm hyperparameter tuning with ADAM weight optimizer. **b** Neuroevolution of augmenting topologies (NEAT) hyperparameter tuning with ADAM weight optimizer

Forgetting, though NEAT showed a consistent rise in average score, occasionally forgetting previously learned parameters. Though it is also observed from Fig. 3b that NEAT is slower in convergence to GA. GLHM-CE on the other hand is both fast in convergence and its memory retention capability with global broadcast feature evades high dimensional mode traps to reach global minima.

4 Conclusion

In this short study we proposed a metaheuristic algorithm (GLHM-CE) for blackbox optimization with fast convergence and compared the results of evaluating on CEC-2013 benchmarks with JADE, CoDE, and SHADE algorithms. GLHM-CE was used to optimize the hyperparameters of a DQN to learn Cartpole game. It is observed from experiments that GLHM-CE has successfully evolved a suitable DQN network with optimum parameters to win the CartPole game with less number of nodes in hidden layers. GLHM-CE will further be investigated on complicated game environments and its viability as a weight optimizer for DQN will be tested in future studies. It is concluded from the result that, GLHM-CE can successfully evolve small DQN in less epochs.

References

1. Bellman, R.: Dynamic Programming. Dover Publications (1957)
2. Brockman, G., Cheung, V., Pettersson, L., Schneider, J., Schulman, J., Tang, J., Zaremba, W.: Openai gym (2016). arxiv:1606.01540
3. cartpole v1. https://gym.openai.com/envs/CartPole-v1/
4. LECUN, Y.: The mnist database of handwritten digits. http://yann.lecun.com/exdb/mnist/, https://ci.nii.ac.jp/naid/10027939599/en/
5. Mnih, V., Kavukcuoglu, K., Silver, D., Graves, A., Antonoglou, I., Wierstra, D., Riedmiller, M.A.: Playing atari with deep reinforcement learning. (2013). CoRR arXiv:1312.5602
6. Mnih, V., Kavukcuoglu, K., Silver, D., Rusu, A.A., Veness, J., Bellemare, M.G., Graves, A., Riedmiller, M., Fidjeland, A.K., Ostrovski, G., Petersen, S., Beattie, C., Sadik, A., Antonoglou, I., King, H., Kumaran, D., Wierstra, D., Legg, S., Hassabis, D.: Human-level control through deep reinforcement learning. Nature **518**, 529 EP (2015). https://doi.org/10.1038/nature14236
7. Nair, V., Hinton, G.E.: Rectified linear units improve restricted boltzmann machines. In: Proceedings of the 27th International Conference on International Conference on Machine Learning, ICML'10, pp. 807–814. Omnipress, USA. http://dl.acm.org/citation.cfm?id=3104322.3104425 (2010)
8. Rumelhart, D.E., Hinton, G.E., Williams, R.J.: Parallel distributed processing: explorations in the microstructure of cognition. Learning Internal Representations by Error Propagation, vol. 1, pp. 318–362. MIT Press, Cambridge. http://dl.acm.org/citation.cfm?id=104279.104293 (1986)
9. Shi, X., Tian, S., Yu, L., Li, L., Gao, S.: Prediction of soil adsorption coefficient based on deep recursive neural network. Autom. Control. Comput. Sci. **51**(5), 321–330 (2017). https://doi.org/10.3103/S0146411617050066

10. Tanabe, R., Fukunaga, A.: Success-history based parameter adaptation for differential evolution. In: 2013 IEEE Congress on Evolutionary Computation, pp. 71–78 (2013). https://doi.org/10.1109/CEC.2013.6557555
11. Wang, W., Yang, J., Xiao, J., Li, S., Zhou, D.: Face recognition based on deep learning. In: Zu, Q., Hu, B., Gu, N., Seng, S. (eds.) Human Centered Computing, pp. 812–820. Springer International Publishing, Cham (2015)
12. Wang, Y., Cai, Z., Zhang, Q.: Differential evolution with composite trial vector generation strategies and control parameters. IEEE Trans. Evol. Comput. **15**(1), 55–66 (2011). https://doi.org/10.1109/TEVC.2010.2087271
13. Yang, X.S., Deb, S.: Multiobjective cuckoo search for design optimization. Comput. Oper. Res. **40**(6), 1616–1624 (2013). https://doi.org/10.1016/j.cor.2011.09.026
14. Yuan, Y., Mou, L., Lu, X.: Scene recognition by manifold regularized deep learning architecture. IEEE Trans. Neural Netw. Learn. Syst. **26**(10), 2222–2233 (2015). https://doi.org/10.1109/TNNLS.2014.2359471
15. Zambrano-Bigiarini, M., Gonzalez-Fernandez, Y.: cec2013: benchmark functions for the special session and competition on real-parameter single objective optimization at CEC-2013 (2015). http://CRAN.R-project.org/package=cec2013. R package version 0.1-5
16. Zhang, J., Sanderson, A.C.: Jade: adaptive differential evolution with optional external archive. IEEE Trans. Evol. Comput. **13**(5), 945–958 (2009). https://doi.org/10.1109/TEVC.2009.2014613

Dewarping of Single-Folded Camera Captured *Bangla* Document Images

Arpan Garai and Samit Biswas

Abstract The document images captured through mobile camera get warped due to the curved surfaces like an open thick book page, notices posted on a cylindrical lamppost or a paper hanging from a notice board attached at top-middle. The very recent state of the art techniques for dewarping, work mostly for document images containing alphabetic script like English; these cannot handle all the variety of warping. This paper proposed a method for dewarping of single folded warped document images containing alpha-syllabary script like *Bangla*. Here the information of top line and baseline increase the robustness of the algorithm. Unlike other related methods, this work has not used any additional information like focal length, distance from the camera or view angle. Experimental evaluation with the collected dataset shows promising results.

Keywords Single folded document · Dewarping · Bangla document

1 Introduction

The peoples especially students prefer to capture images using their digital/mobile camera phones. They capture not only natural scene images but also document page images like a scanner for future use. These documents may be book pages or open notices. Mobile cameras mostly used in various situations where conventional scanners cannot. These cameras used for capturing all types of documents which are hanging from notice board or posted on a lamppost; usually generates a distorted image which feed to the OCR like [5] and gives result with less accuracy. The quality of the captured document image needs to improve by removing this type of distor-

---

A. Garai (✉) · S. Biswas
Computer Science and Technology, Indian Institute of Engineering Science and Technology, Shibpur 711103, India
e-mail: ag.rs2016@cs.iiests.ac.in

S. Biswas
e-mail: samit@cs.iiests.ac.in

A. K. Das et al. (eds.), *Computational Intelligence in Pattern Recognition*, Advances in Intelligent Systems and Computing 999,
https://doi.org/10.1007/978-981-13-9042-5_55

tion. Various kinds of distortion may occur in camera captured document images like blurry during capturing with a shaky hand, poor luminance or warping. This work focuses on dewarping of a single folded document images containing Alpha-syllabary script like *Bengali/Devanagari*.

Generally, the dewarping methods as of now consist of two parts: Firstly, the estimation of the warped surface or the document page itself and secondly, correction of the distortion with the help of the approximated surface. The estimation of surface and correction/restoration followed three dimensional ($3D$) or two dimensional ($2D$) methods.

There were several $3D$ models like *cylindrical surface model* by Cao et al. [4], *shape-from-shading* model by Zhang et al. [20, 21], *general cylindrical surface (GCS) model* by Meng et al. [14]. All these $3D$ models were used to find the shape of surface of a document page. Liang et al. [10] approximated $3D$ shape using the information like texture flow and rectified the image of a warped document page. It first detected the region of text and the texture flow fields. The planar or curved regions within the image localized using the linearity property of the estimated texture. He et al. [8] proposed a *book dewarping system*; here, the boundary of a book used to approximate 3D surface for restoration. Kim et al. [9] read the information of text-flow from the discrete representation of text line. You et al. [19] used data and smoothness terms in the image to estimate the geometric property of the document surface to develop surface dewarping had done.

The restoration did in case of $2D$ methods based on the information of text-flow of lines and direction of the erection of characters. Lu et al. [12, 13] proposed a method in which the image divided into some grids. Authors used vertical stroke boundary (VSB) to compute the direction of text and estimated $X - line$ and base-line to approximate horizontal and vertical curvature. Ulges et al. [18] rectified both perspective and page curl distortion. Here, the geometric shape of the warped document approximated by the local distances among adjacent baselines. Gatos et al. [7] segmented words with lines. They calculated and corrected slant at the word level. Here, the dewarping performed by the neighbourhood relationship of each word. Bukhari et al. [2] proposed a line segmentation method warped document image by using coupled-snakes. Stamatopoulos et al. [17] proposed a two-step goal based technique to rectify the warped document image; these were coarse rectification from the text border and a fine rectification from the word baseline. Liu et al. [11] extracted the shape of each base-lines with the slant angle of each character. Finally, thin plate spline used for dewarping.

It is clear from the recent related works, dewarping uses the information like shades, luminance or page borders; these may be unavailable for all types of document images. Estimation of headline/baseline is important for dewarping of document images. The methods, which used the flow of text line often guided by lower baseline has following limitations for alpha-syllabary Indian scripts like *Bangla/Devanagari*: (a) Symbols in Alphabetic scripts like *English* are aligned with its lower portion on an invisible baseline, whereas symbols of a word in Alpha-syllabary scripts are aligned with the upper part and connected through a headline. The baseline cannot yield a better solution for alpha-syllabary scripts. (b) Most characters in a word of alphabetic

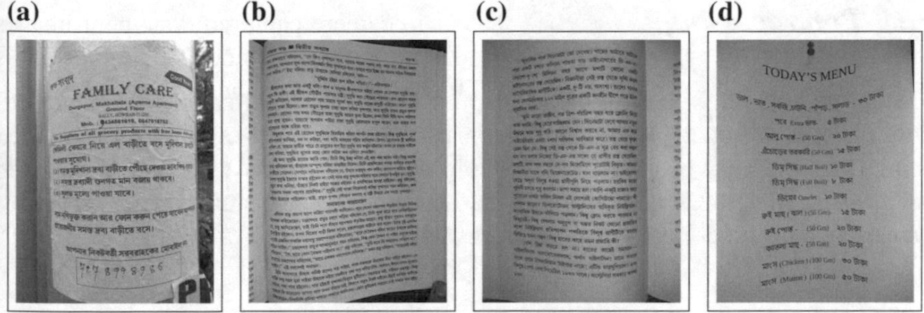

Fig. 1 Different types of warping: **a** Different fonts mixed with bold type; **b** Multiple fonts with Bold and/or Italic type (marked in green colour); **c** Text mixed *Bangla* and *English* with superscript; **d** Highly warped image with larger font between text

scripts like *English* are non-touching. The baselines estimated using the bottom points of these characters. In the case of Alpha-syllabary scripts, most characters of a word are connected. So the baseline estimation methods for Alphabetic scripts like *English* are not directly applicable to estimate headline for alpha-syllabary scripts like *Bangla*. (c) Components in warped document images are often shared. The distortion due to shear in each component is negligible due to the non-connecting nature of *English* characters. In the case of *Bangla* document images, the distortion is prominent and a correction is needed.

Most of the characters in *Bengali/Devanagari* script are connected at the top using a headline called *'matra'*. Different forms of warping (See Fig. 1) can present in camera-captured born-digital document images; this is due to the document surface and camera view angle. Here, dewarping performs with the help of the surface of the document page; this surface is estimated using the information of text lines. The next sections show the proposed method in brief, experimental results and concluding remarks.

2 Proposed Approach

The dewarping a document by the proposed approach mostly depends upon the 'matra' calculated for each text line. Here, we have considered only the single folded documents. In single folded documents the number of 'loop' or 'hump' in the flow of text line remains one. This method generally consists of the following steps: (a) Preprocessing, (b) Text line segmentation, (c) Bottom-line extraction, (d) Head-line extraction and (e) Rectify through relocation of components.

2.1 Preprocessing

Firstly, the image is binarized using adaptive document binarization technique proposed by Sauvola et al. [16]. After binarization, various types of noises are removed using the previously proposed well-established techniques. At the end of this step we have the image that contains only the text region of the entire image. That preprocessed image is forwarded to the next step.

2.2 Text Line Segmentation

The traditional projection profile-based methods of line identification does not work for warped documents. Line segmentation in warped document is a quite popular research domain recently. Among the already have developed approaches [1, 3, 15] we have used the technique proposed by Bukhari et al. [3] as it worked sufficiently well for *Bangla* documents. We proceed in the next step with the segmented lines.

2.3 Bottom Line Extraction

Although, the bottom line does not give the exact text orientation for top aligned script like *Bengala*, it gives a rough impression about the text line flow. That helps us to find the exact headline. So, we have approximated bottom line. In the *Bangla/Devanagari* scripts most of the characters are connected with each other by a horizontal line, called as 'matra'. We have separated the characters by erasing that 'matra'. Then from the partly erased characters found the bottom points. To separate the characters we have erased an amount of $(\xi^i \times 3)$ from the first occurrence of the text in each column. Here, ξ^i is the stroke width for ith text line. As a result, we have got a set of separated, partly erased characters. Next, we have done a connected component analysis and taken bottom points(B_j^i) of the sufficiently large components. Next, we have simulated a polynomial equation $(B_y = \sum_{i=1}^{N} a^i x^i)$ from these set of points (B_j^i). Here N is the degree of the polynomial. The process is described in Fig. 2.

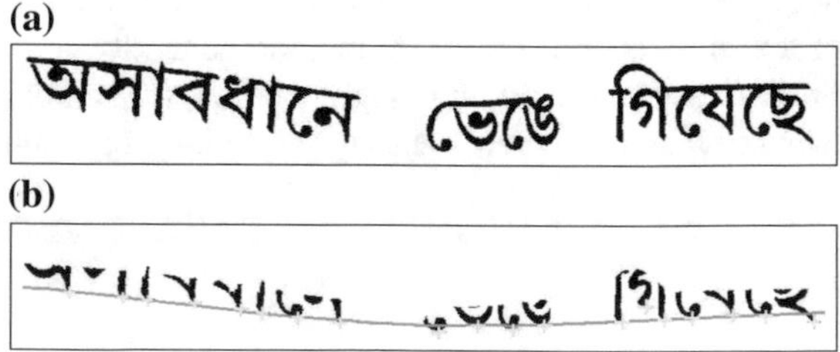

Fig. 2 **a** A warped text line; **b** Drawn baseline using partly erased text line

From the property of polynomial equations we know that for nth degree of polynomial the maximum number of *'hump's* or turning points present in their graphical representation is $(n - 1)$. We have observed that in single fold warped documents the number of *'hump's* or turning points is '1'. Sometimes, towards the edge of the documents like captured image of a thick book it tends to raise. So, for better result we have to consider the degree of the polynomial (N) as '3'.

2.4 Headline Extraction

The headline guides the text line flow accurately for the scripts like *Bangla/ Devanagari*. It is approximately parallel to 'bottom line' estimated earlier. Firstly, we have taken samples of top profile of the text line at fixed interval. In our experiment this interval is set empirically as $(\xi^i \times 2)$ for the ith line. We have clustered the sample points (S_j) into three sets based on the Euclidean distance from the bottom line (D_{B_y,S_j}). These sets are, 1. Accepted points (A_{pt}); 2. Rejected points (R_{pt}); 3. Others (O_{pt}). These points are defined as following:

1. $A_{pt} = \{S_j | (D_{B_y,S_j}) \geq m \times \xi^i, (D_{B_y,S_j}) < n \times \xi^i\};$
2. $R_{pt} = \{S_j | (D_{B_y,S_j}) < m \times \xi^i\};$
3. $O_{pt} = \{S_j | (D_{B_y,S_j}) > n \times \xi^i\};$

We have empirically set the value of m and n as 5 and 8 respectively. Here, we consider a new set (ς) where the 'accepted points' are added directly added and 'rejected points' are discarded. For the '*type 3*' points we scan downwards and take the first background to text transition point. Next, the perpendicular distance is measured from the point to the bottom line. This new point is again checked with the property of the 'Accepted points'. If it matches, the points are marked as corrected points C^r_{pt} otherwise this correction process is continued until the point satisfies the property of 'rejected point'. Finally, the set of corrected points are added with the set ς. In other words, $\varsigma = A_{pt} \bigcup C^r_{pt}$. Using the points in the set ς we have drawn a smooth curve using cubic spline. An example of that process is shown in Fig. 3a. The point marked in the green circle is corrected to the point marked as red '+'. The calculated 'matra' is shown using the dashed line.

The curve looks like the calculated 'matra'. But in some cases it does not give the accurate result. Sometimes, it fails to go along with the top horizontal line. It is so because some points are not corrected. So, we have followed an iterative approach to perform another level of correction to increase the accuracy of the estimated headline.

Firstly, we have considered the sufficiently large components for which the simulated 'headline' does not overlap with the text portion in more than 80% of its width. We have checked the nearest text pixel at each point of the portion of 'headline' where it did not overlap with a text. We have counted the number of nearest text pixel towards its upper side (C_u) as well as at lower side (C_l). If $C_l > C_u$ then we come to the conclusion that the 'headline' should be dragged down. Form the previously chosen set of points (ς) the point residing inside ROI of the connected component under consideration and above the 'headline' are rechosen. It is the first background

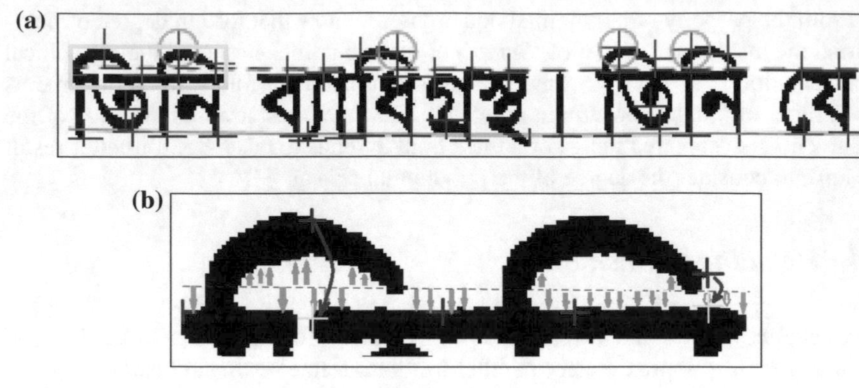

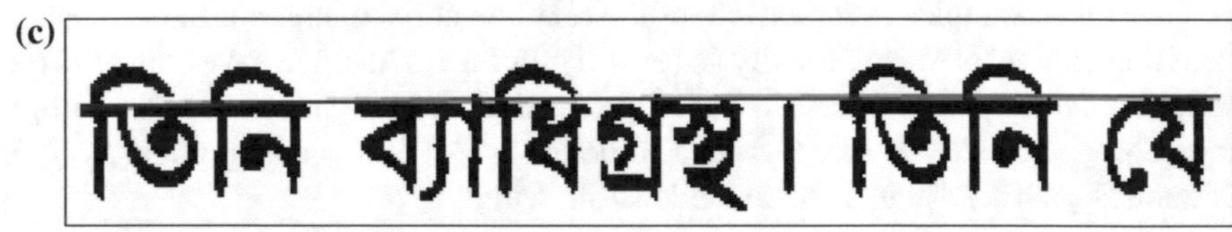

Fig. 3 **a** Estimated 'matra': Dashed blue Line; Estimated 'Bottom line': Continuous Orange line **b** Rectification of 'matra' (Only portion inside yellow box in Fig. 3a is shown); **c** Rectified 'matra (Continuous Line)

to foreground transition below the 'headline'. With the updated set of points we have again calculated the 'headline' using smooth cubic spline. This process is described in Fig. 3b. The final 'matra' is shown in Fig. 3c.

2.5 *Repositioning of the Components*

As we have found a smooth curve as headline, we have shifted the components pixe wise. We have estimated imaginary horizontal lines for each line as discussed by Arpan Garai et al. [6]. Next, we have shifted each of the pixel of the text line such that the calculated headline coincides the imaginary horizontal line. Rest of the tex pixels remains with the same distance from the headline.

3 Experimental Results

To develop and check the outcomes of our approach a suitable dataset is very essential As far as we have studied, three datasets containing warped images are available. Bu two (DFKI 2007 dataset and IUPR dataset) of them contain images that have only *English* script. Also, they do not have varieties like 'lamp post document image (Fig. 1a), 'notice board document image' (Fig. 1d). So, we have used the datase Warped Document Image dataset (WDID) where all the image is in *Bangla* and sometimes mixed with *English*. It also contains the previously mentioned variety

The dataset also consists of some multiple folded warped document images. We have not considered those images for the time being as our approach works only for known number of folds. Nevertheless, our approach can further be upgraded to dewarp the multiple folded document images.

The performance of these methods can be evaluated in two ways; they are direct and indirect. Here, we have followed an indirect way which is followed by many researchers. It is nothing but using a OCR. The output of the dewarping methods is fed to OCR. Let, the total number of word present in the corresponding document image is N_0 and the correctly recognized words by the OCR in the dewarped image is N_1. Now, the OCR accuracy rate after dewarping is defined as $\frac{N_1}{N_0} \times 100\%$. The accuracy is measured for input image, dewarped using Kim method [9] and proposed method. It is shown in Table 1. Here, we have used the *Google Doc OCR*.

The restoration accuracy of italic type texts is measured with the help of average slant angle. In *English* or *Bangla/Devanagari* script some characters consist of vertical strokes like for *English* 'H', 'I', 'L', etc. The average slant angle of italic type characters both in ground truth image (θ_g) and dewarped image (θ_d) is calculated using these vertical strokes as shown in Fig. 4.

Similarly, the correctness of restored bold texts is measured using stroke widths. For ground truth images, the stroke width (W_b) of the bold characters and stroke width (W_n) of the normal characters with the same font-size are measured. $B_g = \frac{W_b}{W_n}$ is obtained. The similar procedure is adopted for the dewarped image to obtain B_d.

Table 1 Performance of proposed dewarping method

Born digital database	Curve type	Language	OCR rate		
			Kim et al. [9]	Garai et. al [6]	Proposed approach
Set 1	Left	Bangla	52.03	97.6	98.6
Set 2	Right	Bangla	54.07	97.52	98.4
Set 3	Other	Bangla	53.04	96.41	98.23
Set 4	–	Mixed	53.05	95.7	97.72

Table 2 Restoration accuracy for *Bold* and *Italic* type characters

Text type	Restoration accuracy-Bold (β_i)		
	Kim et al. [9]	Garai et al. [6]	Proposed approach
Bold	88.1	91.1	91.8
Bold + Italic	80.5	92.0	92.5
	Restoration accuracy- Italic (α_i)		
Italic	61.1	92.48	92.3
Bold + Italic	60.5	91.36	92.5

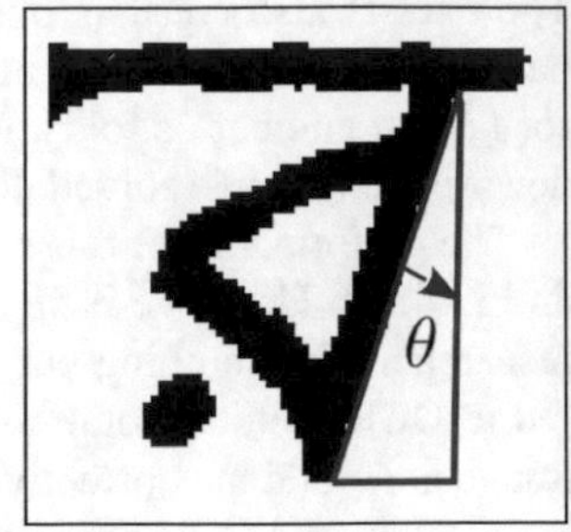

Fig. 4 Calculation of slant angle for italic type texts

The restoration accuracy for the bold character (β_i) and the italic character (α_i) are computed using the following formulas and shown in Table 2. $\beta_i = \frac{B_d}{B_g} \times 100\%$, $\alpha_i = \frac{\theta_d}{\theta_g} \times 100\%$.

A set of experimental results is shown in Fig. 5. It is evident from Fig. 5 that the proposed method performs well for warped documents having different font size, bold and italic characters, text with mixed scripts, highly warped image and multi-column text.

Proposed approach is compared with Kim et al. [9], where dewarping is dependent on focal length (FL) which may not be available every time with the image file. We also have compared with Garai et al. [6] which is developed for *Bangla* script. For visual comparison an example is shown in Fig. 5. It is evident from the Fig. 5 that the proposed method is performed better than the method proposed in [6, 9]. Moreover, the proposed approach does not need any additional information regarding image like focal length.

4 Conclusion

The proposed simple dewarping method successfully dewarps the warped documents having different font size, bold and italic characters, text with mixed scripts, highly warped image and multi-column text. The cubic spline function is used to generate a smooth curve and continuous headline (baseline) for the entire text line which is used for proper placement of components after inverse shear transform. The efficacy and efficiency of the proposed method are also endorsed by testing on a variety of document images available in our database with different font sizes and font styles.

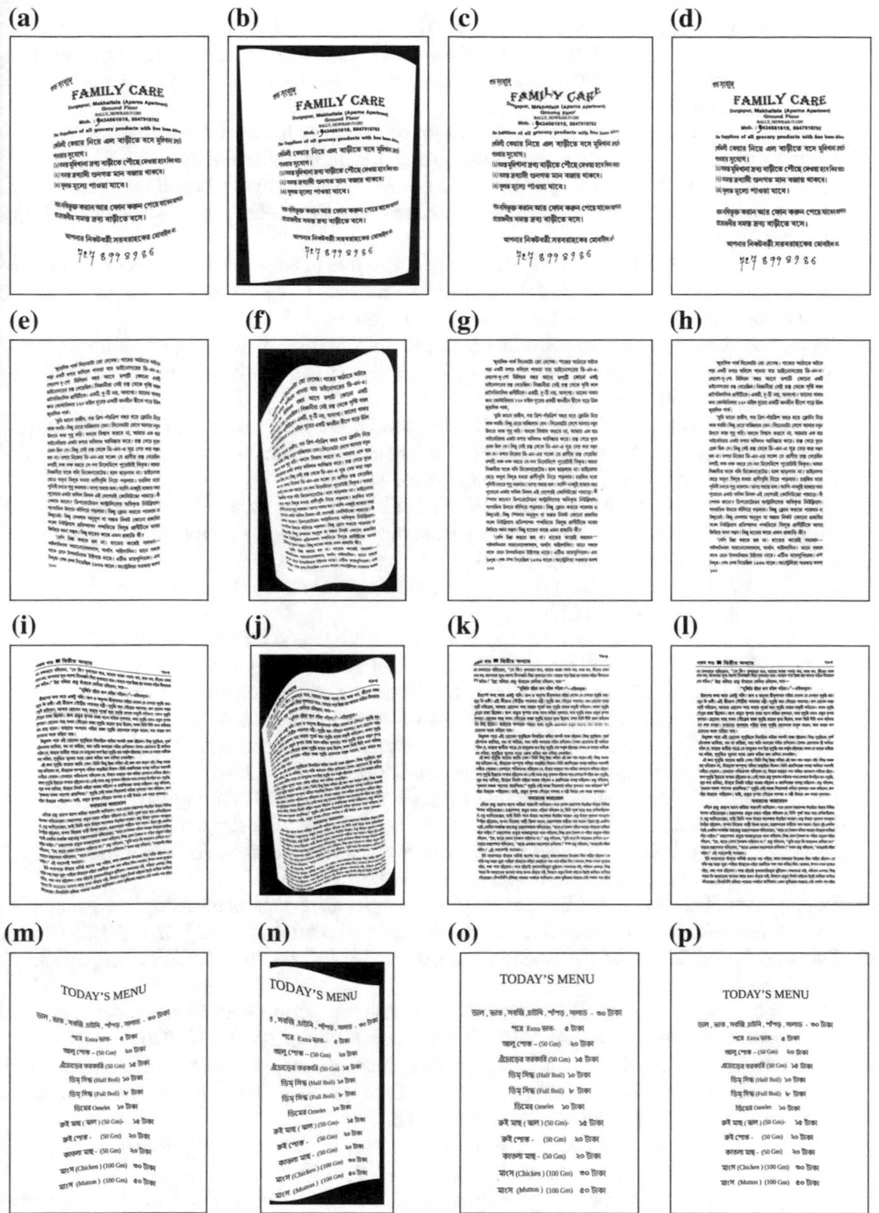

Fig. 5 **a**, **e**, **i**, **m** Preprocessed image of images shown in Fig. 1; Results of **b**, **f**, **j**, **n** Kim et al. [9] method; **c**, **g**, **k**, **o** Garai et al. [6] method **d**, **h**, **l**, **p** Proposed method

References

1. Bukhari, S.S., Shafait, F., Breuel, T.M.: Text-line extraction using a convolution of isotropic gaussian filter with a set of line filters. In: 2011 International Conference on Document Analysis

and Recognition, pp. 579–583, September 2011

2. Bukhari, S.S., Shafait, F., Breuel, T.M.: Dewarping of document images using coupled-snakes. In: Proceedings of Third International Workshop on Camera-Based Document Analysis and Recognition, pp. 34–41 (2009)
3. Bukhari, S.S., Shafait, F., Breuel, T.M.: Coupled snakelets for curled text-line segmentation from warped document images. Int. J. Doc. Anal. Recognit. (IJDAR) **16**(1), 33–53 (2013)
4. Cao, H., Ding, X., Liu, C.: A cylindrical surface model to rectify the bound document image. In: Proceedings Ninth IEEE International Conference on Computer Vision, vol. 1, pp. 228–233, October 2003
5. Chaudhuri, B.B., Pal, U.: An ocr system to read two indian language scripts: Bangla and devnagari (hindi). In: Proceedings of the Fourth International Conference on Document Analysis and Recognition, vol. 2, pp. 1011–1015, August 1997
6. Garai, A., Mandal, S., Biswas, S., Chaudhuri, B.B.: Automatic dewarping of camera captured born-digital bangla document images. In: Ninth International Conference on Advances in Pattern Recognition (ICAPR 2017), December 2017
7. Gatos, B., Pratikakis, I., Ntirogiannis, K.: Segmentation based recovery of arbitrarily warped document images. In: Ninth International Conference on Document Analysis and Recognition (ICDAR 2007), vol. 2, pp. 989–993, September 2007
8. He, Y., Pan, P., Xie, S., Sun, J., Naoi, S.: A book dewarping system by boundary-based 3d surface reconstruction. In: 2013 12th International Conference on Document Analysis and Recognition, pp. 403–407, August 2013
9. Kim, B.S., Koo, H.I., Cho, N.I.: Document dewarping via text-line based optimization. Pattern Recognit. **48**(11), 3600–3614 (2015)
10. Liang, J., DeMenthon, D., Doermann, D.: Geometric rectification of camera-captured document images. IEEE Trans. Pattern Anal. Mach. Intell. **30**(4), 591–605 (2008). April
11. Changsong Liu, Y., Zhang, B.W., Ding, X.: Restoring camera-captured distorted document images. Int. J. Doc. Anal. Recognit. (IJDAR) **18**(2), 111–124 (2015)
12. Lu, S., Chen, B.M., Ko, C.C.: Perspective rectification of document images using fuzzy set and morphological operations. Image Vis. Comput. **23**(5), 541–553 (2005)
13. Lu, S., Tan, C.L.: Document flattening through grid modeling and regularization. In: 18th International Conference on Pattern Recognition (ICPR'06), vol. 1, pp. 971–974 (2006)
14. Meng, G., Pan, C., Xiang, S., Duan, J.: Metric rectification of curved document images. IEEE Trans. Pattern Anal. Mach. Intell. **34**(4), 707–722 (2012). April
15. Roy, P.P., Pal, U., Lladós, J.: Text line extraction in graphical documents using background and foreground information. Int. J. Doc. Anal. Recognit. (IJDAR) **15**(3), 227–241 (2012)
16. Sauvola, J., Pietikinen, M.: Adaptive document image binarization. Pattern Recognit. **33**(2), 225–236 (2000)
17. Stamatopoulos, N., Gatos, B., Pratikakis, I., Perantonis, S.J.: Goal-oriented rectification of camera-based document images. IEEE Trans. Image Process. **20**(4), 910–920 (2011). April
18. Ulges, A., Lampert, C.H., Breuel, T.M.: Document image dewarping using robust estimation of curled text lines. In: Eighth International Conference on Document Analysis and Recognition (ICDAR'05), vol. 2, pp. 1001–1005, August 2005
19. You, S., Matsushita, Y., Sinha, S., Bou, Y., Ikeuchi, K.: Multiview rectification of folded documents. IEEE Trans. Pattern Anal. Mach. Intell. **PP**(99), 1–1 (2017)
20. Zhang, L., Tan, C.L.: Warped image restoration with applications to digital libraries. In: Eighth International Conference on Document Analysis and Recognition (ICDAR'05), vol. 1, pp. 192–196, August 2005
21. Zhang, L., Tan, C.L.: Restoringwarped document images using shape-from-shading and surface interpolation. In: 18th International Conference on Pattern Recognition (ICPR'06), vol. 1, pp. 642–645 (2006)

Deep Learning Techniques—R-CNN to Mask R-CNN: A Survey

Puja Bharati and Ankita Pramanik

Abstract With the advances in the field of machine learning, statistics, and computer vision, the advanced deep learning techniques have attracted increasing research interests over the last decade. This is because of their inherent capabilities of overcoming the drawback of traditional techniques. The main contribution of this work is to provide a comprehensive description of region-based convolutional neural network (R-CNN) and its recent improvement like fast R-CNN, faster R-CNN, region-based fully convolutional networks, single shot detector, deconvolutional single shot detector, R-CNN minus R, you only look once (YOLO), mask R-CNN, etc., with brief details. This survey paper presents an overview of the last update in this field and their practical applications and its classification for ease of understanding. The performances and challenges of these techniques in terms of speed, accuracy, or simplicity are also compared. In general, the speed performance of YOLO is approximately 21 ~ 155 fps which is the fastest and the average precision of Mask R-CNN is ~47.3 which outperforms all other techniques.

Keywords Deep learning techniques · Mask R-CNN · Object detection · R-CNN · YOLO

1 Introduction

Convolutional neural network (CNN) is a biologically inspired process. It is an artificial neural network which is feed forward and is most commonly used in computer vision and machine learning to examine visual imagery. The architecture of CNN consists of four layers: convolutional layers which perform convolution, rectified

P. Bharati (✉) · A. Pramanik
Department of Electronics and Telecommunication Engineering, IIEST, Shibpur 711103, West Bengal, India
e-mail: pujaciem@gmail.com

A. Pramanik
e-mail: ankita@telecom.iiests.ac.in

A. K. Das et al. (eds.), *Computational Intelligence in Pattern Recognition*,
Advances in Intelligent Systems and Computing 999,
https://doi.org/10.1007/978-981-13-9042-5_56

linear unit (ReLU) layers which apply element wise activation function, pooling layers which perform down sampling operation, and fully connected layers (FC) which compute scores of a class. Lecun et al. Network (LeNet) [1] is the first CNN model. After this, different CNN architectures are evolved which are deeper such as AlexNet [2], Zeiler and Fergus Network (ZF Net) [3], Google LeNet (GoogLeNet) [4], Visual Geometry Group Network (VGGNet) [5], Residual Network (ResNet) [6], etc.

Recently, advanced deep learning techniques, especially region-based CNN (R-CNN), have attained remarkable successes in a diversity of tasks in machine learning, statistics, and computer vision, for example, object detection, image categorization, image segmentation, etc. This is because CNN only identifies the object's class, not the location of object in an image. Especially when multiple objects are in the image then CNN cannot work well due to interference. R-CNN identifies region of interest (RoI) using selective search [7] and using a CNN it categories them as foreground and background. Nowadays, R-CNN [8] is the best CNN-based detectors as it can identify the location of the object and also work well for an image containing multiple objects.

Due to expensive training and slow object detection, R-CNN is modified and fast R-CNN [9], faster R-CNN [10], region-based fully convolutional networks (R-FCN) [11], single shot detector (SSD) [12], you only look once (YOLO) [13], grid-CNN (G-CNN) [14], spatial pyramid pooling network (SPP-Net) [15], R-CNN minus R [16], propose network (Pro-Net) [17], mask R-CNN [18], etc., are evolved one after another. These newer algorithms make the object detection and its classification more accurate and faster. For applying these techniques in real world applications such as face detection [19], people detection [20], traffic sign detection, and traffic surveillance [21], etc., the problem is to choose the suitable model or technique for their application. A trade-off among speed, precision, and simplicity of these techniques is essential to get the desired result. The main and basic applications of all these techniques are object detection and image classification [1, 7, 10–12, 18, 22, 23]. These models also help in the application of action recognition [24], human pose estimation [18], instance segmentation [18], fault detection, text recognition, etc. These techniques also help in medical diagnosis such as tumor detection, nucleus segmentation [25], and much more. For object detection, many datasets such as PASCAL VOC datasets [26], Microsoft COCO datasets [27], ImageNet datasets, [28] etc. are available.

New approach and new techniques in the area of deep learning are evolving each and every day. So, a survey of all the existing techniques is essential for its appropriate application and further development in the field of CNN. The main contributions of this paper are as follows: (1) to give a comprehensive description of R-CNN (2) proposing object detection inspirational tree for ease of understanding (2) to investigate how different techniques or descendants are evolved from one implementation to other and (3) to summarize performance comparisons of the different techniques.

The remaining part of the paper is organized as follows: In the succeeding section, different types of base networks are discussed. The basic principle of R-CNN and the various improvements on R-CNN are presented in Sect. 3. The different techniques are compared in Sect. 4 and this survey paper is finally concluded in Sect. 5.

2 Base Network

To understand the working of R-CNN, it is imperative to understand the base network. The performance of the various deep networks is dependent on the base network. The base networks are the base architecture of R-CNN. There are many different types of architecture evolved almost every year such as LeNet [1], AlexNet [2], ZF Net [3], GoogLeNet [4], VGG Net [5], ResNet [6], etc. Some common architecture is described below.

(i) **LeNet** [1]: the uses of convolutional networks were exaggerated in 1990s. Of these, the outstanding known is the LeNet construction that was secondhand to peruse handwritten digits, zip codes, and so on.
(ii) **AlexNet** [2]: The main work that advanced convolutional networks in computer vision was the AlexNet. The AlexNet participated in the imagenet large scale visual recognition challenge (ILSVRC) 2012. ILSVRC [29] challenge examine the techniques for object detection and image classification and also measure the progress of computer vision at large scale. AlexNet outflanked the first sprinter up (classification error rate of 15.3%).
(iii) **ZF Net** [3]: The ILSVRC 2013 [29] champion was a convolutional network known as ZF Net. It was a change on AlexNet by tweaking the design hyper parameters, specifically by growing the span of the center convolutional layers and influencing the stride and filter size on the primary layer lesser.
(iv) **GoogLeNet** [4]: The GoogLeNet convolutional network won the ILSVRC 2014 challenge. Its fundamental commitment was the advancement of an inception module that drastically diminished the quantity of parameters in the system (4 million, contrasted with AlexNet with 60 million). Also, this method utilizes average pooling rather than fully connected layers at the highest point of the convolutional network, dispensing with a lot of parameters that don't appear to make a difference much.
(v) **VGGNet** [5]: The ILSVRC 2014 [29] sprinter up was the system that ended up called the VGGNet. Its principle commitment was in demonstrating that the profundity of the system is a basic part for good execution. Their last best system accommodates sixteen convolutional or fully connected. There are some drawbacks of VGGNet. It is very costly to assess and it also uses significantly more memory and more parameters (140 million).

(vi) **ResNet** [6]: Residual Network was the victor of ILSVRC 2015. It highlights extraordinary skip associations and a substantial utilization of batch normalization. The design is additionally lacking completely associated layers toward the finish of the system. ResNets are as of now by a long shot best in class convolutional neural network models and are the default decision for utilizing.

3 R-CNN

The R-CNN [8] architecture includes three steps as shown in Fig. 1. First step is to take the input image and scan for probable objects using an algorithm called selective search [7] which generates ~2000 region proposals. In second step, on the top of each region proposals a CNN is run. Third step is to take the each CNN output and provide this output into (i) a support vector machine (SVM) and (ii) a linear regressor. SVM classifies the region of object in an image and linear regressor tightens the bounding box of the detected object.

R-CNN has some notable drawbacks. (i) To classify ~2000 region proposals per image, it takes huge amount of time to train the network, (ii) Object detection in real time is not possible as it takes approximately 47s for each test image, (iii) R-CNN should train 3 different models separately—CNN that generate image feature, classifier which predict the class, and the regressor model which help to tight the bounding boxes. Hence, the pipeline is really tough to train. Because of these limitations, R-CNN needs improvement.

Improvement on R-CNN

Taking the inspiration from R-CNN, different models are evolved year after year. The summary is shown in Fig. 2. The present work proposes this tree for understanding the evolution of the deep networks in a lucid fashion.

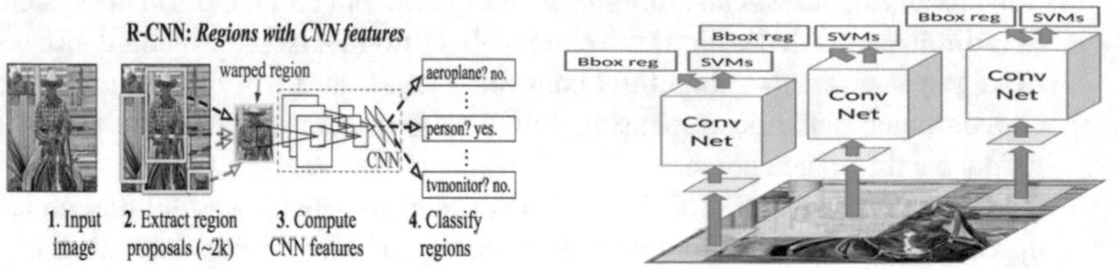

Fig. 1 R-CNN architecture

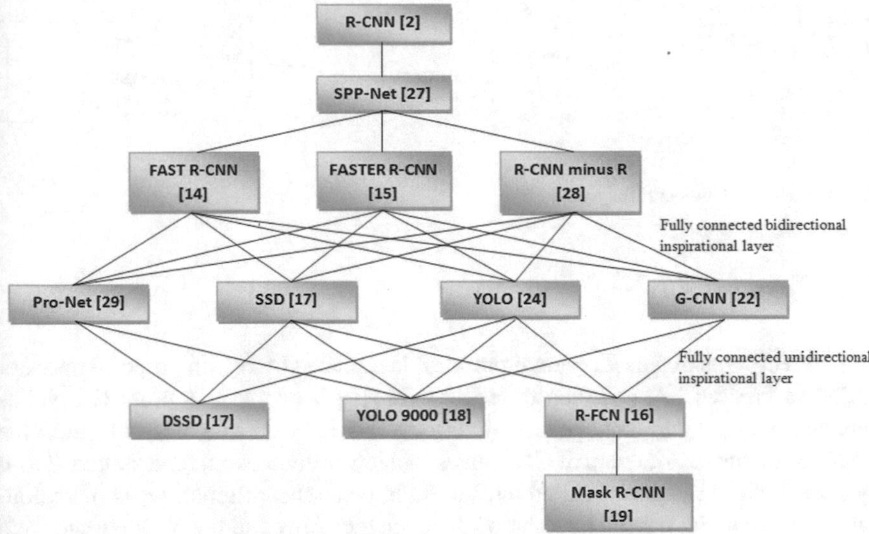

Fig. 2 Object detection inspirational tree

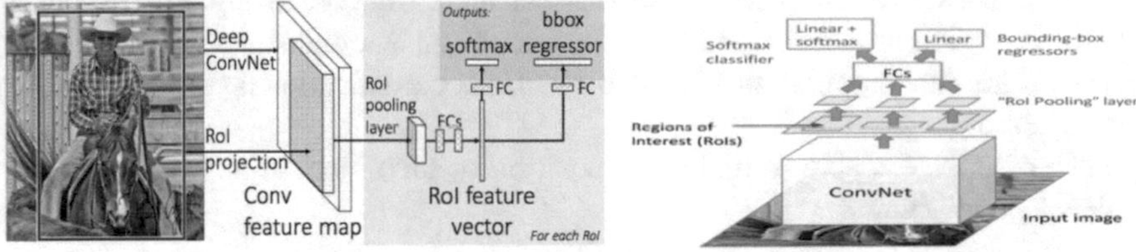

Fig. 3 Fast R-CNN architecture

3.1 *Fast R-CNN*

Fast R-CNN [9] approach is similar to that of the R-CNN. The difference is instead of providing the region proposals to the CNN; the input image is directly fed to the CNN for generating convolutional feature map. From this, the region proposals are identified and warped them into squares. To feed into fully connected layer, RoI pooling layer reshape them into a fixed size. To predict the class of the proposed region softmax layer is used instead of SVM as shown in Fig. 3.

This algorithm is faster than R-CNN because ~2 K regions proposals are not required to be fed to the CNN each and every time. It is fed only once to generate the feature map. The training of CNN, classifier and bounding box regressor of fast R-CNN are done simultaneously in a single model.

Instead of having all these advantages, region proposal is the main problem in this technique. These proposals were done using selective search which was a slow process and became bottleneck of the overall process.

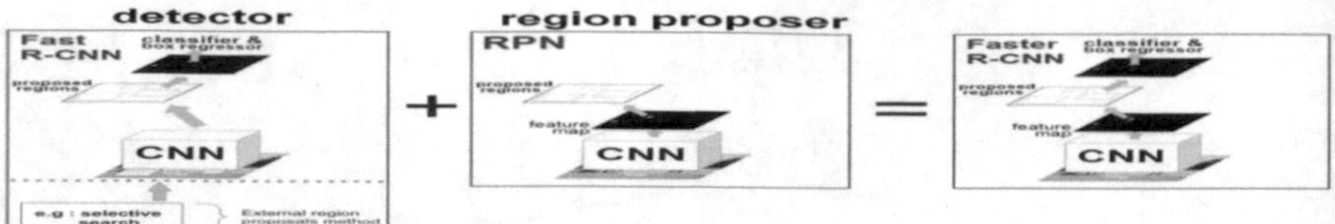

Fig. 4 Faster R-CNN architecture

3.2 *Faster R-CNN*

Slow selective search as a regional proposal is replaced by region proposal network (RPN) in fast R-CNN and then faster R-CNN [10] is evolved. In faster R-CNN as depicted in Fig. 4, input image is fed to the CNN. RPN works on the last layer of the CNN. Here, the feature map of CNN maps to a lower dimension, for example, 256-d by a 3×3 sliding window, called anchor. RPN generates different types of region-based anchor boxes having fixed ratio k for each location of sliding window such as a large box, a wide box, a tall box, etc. For each anchor box, if intersection-over-union (IoU) is more than 0.7 then there is an object or if IoU is less than 0.3 then there is no object, likewise an "objectness" score is calculated and object region is proposed. After this, region proposal is fed into a RoI pooling layer and some fully connected layers. At last, the output of this is fed to a softmax classification layer and bounding box regressor.

Faster R-CNN is able to find exact position of the object in an image and also achieves good accuracy.

3.3 *R-FCN*

Fast R-CNN share a single CNN computation to all region proposals for improving object detection speed. The motivational concept behind R-FCN [11] is maximum sharing of the computation to increase the speed. Here, input image is feed to the CNN. A $k \times k \times (C + 1)$ score bank of the "position-sensitive score maps" is created by a fully convolutional layer using the output of CNN, where $k \times k$ represents the number of relative location to partition an object and $(C + 1)$ represents the total number of classes including background. Also to generate RoI, a fully convolutional RPN is used. Each RoI is divided into $k \times k$ subregions. For each subregion, the score bank is checked to know if that subregion corresponds to the same position of some object. For each class, this process is repeated. Once if "object match" value for k^2 subregions of each class is calculated then average the value of subregions to get single object match score for each class. Then, the RoI is classified with a softmax for the remaining $C + 1$ dimensional vector. R-FCN architecture is approximately twenty times speedier than Faster R-CNN and achieved almost same precision as that of Faster R-CNN.

3.4 SSD

In all R-CNN and R-FCN, two different models carry regional proposal by using RPN to generate regions of interest and region classifications either by using fully connected layers (FCL) or by using position sensitive convolutional layers for classifying those regions. SSD perform these two tasks in only one shot. As the image processes, SSD [12] predict the bounding box and the class simultaneously.

According to SSD technique, first image is fed to a series of convolutional layers to get different sets of feature maps. Different feature map has different scales like 10×10 then 5×5 then 3×3 etc. For each feature maps, a 3×3 convolutional filter is used at each location for evaluating a small set of default bounding boxes which is equivalent to anchor boxes of faster R-CNN. The bounding box offset and the class probabilities are predicted for each box simultaneously. For each predicted boxes, IoU is calculated with the help of ground truth box. Among those predicted boxes, one box is marked as "positive" if intersection over union (IoU) exceeds 0.5 and greater than all other predicted boxes are selected and the rest boxes is discarded.

SSD is not entirely different from R-CNN and R-FCN. It draws different bounding boxes having different shapes and scales at every position of the image and classify simultaneously instead of "region proposal". SSD does this in a single shot; hence it is the fastest of the previous models with comparable accuracy.

3.5 YOLO

YOLO is different from region-based algorithm. Region-based algorithm takes the help of regions to know the position of object within an image. YOLO [22] looks parts of the image having high probabilities of containing the object. Here, a single CNN decides the bounding boxes with their class probabilities.

YOLO first take an image then split the image into $S \times S$ grid. For each and every grid, it draws M bounding boxes. A class probability and offset values are calculated for each bounding boxes. A bounding box is selected if it has a class probability greater than both the class probability of other bounding boxes and a threshold value of class probabilities; remaining all other bounding boxes is discarded. The selected bounding box corresponds to the object within an image.

YOLO is faster than other algorithm (45 frames per second). Instead of being faster, YOLO cannot recognize a group of small objects or irregularly shaped objects within an image.

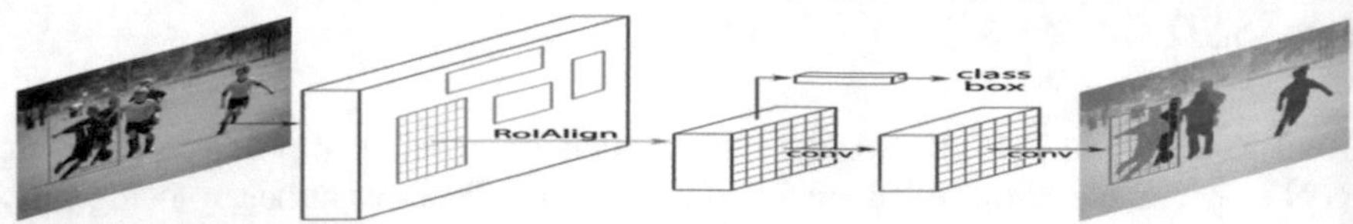

Fig. 5 Framework of mask R-CNN

3.6 *Mask R-CNN*

Faster R-CNN classifies the objects but it cannot find which pixel is a part of an object in an image. Hence, mask R-CNN [18] is evolved which is used for instance segmentation. Instance segmentation has two parts that are object detection and semantic segmentation, also known as pixel level segmentation. In mask R-CNN as shown in Fig. 5, faster R-CNN is used for object detection and fully connected network (FCN) applied on each RoI is used for semantic segmentation.

Mask R-CNN use RoI align (RoIAlign) layer (improved version of RoI pooling layer). This is because pixel level segmentation needs much more fine-grained alignment which is not necessary in bounding boxes. In RoI pooling, misalignment of location occurs due to quantization. For avoiding quantization and adjusting RoI pooling to more precise, bilinear interpolation is used and this method is called RoIAlign.

4 Comparative Study

The comparisons among these models are difficult. It cannot be determined easily that which model is best. There should be a trade-off between speed and accuracy for real life applications. Performance of these models or methods are highly dependent on feature extractors (base network), input image resolution, non-max suppression, IoU threshold, positive versus negative anchor ratio, number of proposals, bounding box encoding, data augmentation, training datasets, localization of loss function, training configurations, etc.

The accuracy of the object detection techniques is measured by mean average precision (mAP) which is the average of maximum precisions at different recall values. Precision measures prediction accuracy and recall measures true positives from all true positive and false negative cases. The mAP performance using PASCAL VOC 2012 dataset and AP and bounding box AP (AP^{bb}) performance using MS COCO datasets comparisons of the works discussed above are given in Tables 1, 2 and 3 respectively.

The accuracy comparison of the different base networks is given in Fig. 6. From different tables and Fig. 6 it can be inferred that SSD and R-FCN methods are faster but cannot beat the faster R-CNN in terms of precision or accuracy. In R-FCN and

Table 1 mAP (mean Average Precision) using VOC 2012 dataset [22]

Method	Data	mAP
Fast R-CNN	07++12	68.4
Faster R-CNN	07++12	70.4
YOLO	07++12	57.9
SSD300	07++12	72.4
SSD512	07++12	74.9
ResNet	07++12	73.8
YOLOv2 544	07++12	73.4

Table 2 AP (Average Precision) using MS COCO dataset [22]

	Backbone	AP	AP_{50}	AP_{75}	AP_s	AP_M	AP_L
Two-stage methods							
Faster R-CNN+++	ResNet-101-C4	34.9	55.7	37.4	15.6	38.7	50.9
Faster R-CNN w FPN	ResNet-101-FPN	36.2	59.1	39.0	18.2	39.0	48.2
Faster R-CNN by G-RMI	Inception-ResNet-v2	34.7	55.5	36.7	13.5	38.1	52.0
Faster R-CNN w TDM	Inception-ResNet-v2-TDM	36.8	57.7	39.2	16.2	39.8	52.1
One-stage methods							
YOLOv2	DarkNet-19	21.6	44.0	19.2	05.0	22.4	35.5
SSD513	ResNet-101-SSD	31.2	50.4	33.3	10.2	34.5	49.8
DSSD513	ResNet-101-DSSD	33.2	53.3	35.2	13.0	35.4	51.1
YOLOv3 608 × 608	Darknet-53	33.0	57.9	34.4	18.3	35.4	41.9

Table 3 Average Precision (AP) using MS COCO dataset [18]

	Backbone	AP^{bb}	AP^{bb}_{50}	AP^{bb}_{75}	AP^{bb}_{S}	AP^{bb}_{M}	AP^{bb}_{L}
Faster R-CNN+++	ResNet-101-C4	34.9	55.7	37.4	15.6	38.7	50.9
Faster R-CNN w FPN	ResNet-101 -FPN	36.2	59.1	39.0	18.2	39.0	48.2
Faster R-CNN by G-RMI	Inception-ResNet-v2	34.7	55.5	36.7	13.5	38.1	52.0
Faster R-CNN w TDM	Inception-Re sNet-v2-TDM	36.8	57.7	39.2	16.2	39.8	**52.1**
Faster R-CNN, RoIAlign	ResNet-101-FPN	37.3	59.6	40.3	19.8	40.2	48.8
Mask R-CNN	ResNet-101-FPN	38.2	60.3	41.7	20.1	41.1	50.2
Mask R-CNN	ResNeXt-101-FPN	**39.8**	**62.3**	**43.4**	**22.1**	**43.2**	51.2

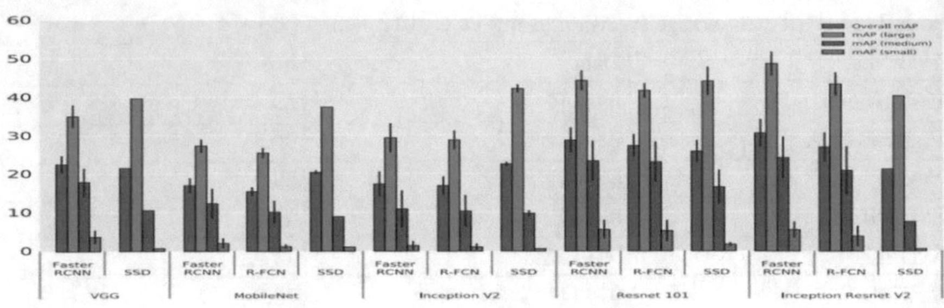

Fig. 6 Accuracy for different base network [23]

faster R-CNN, detection accuracy is affected by the choice of feature extractors. However, SSD is unaffected by such choices. SSD can outperform R-FCN and faster R-CNN in accuracy for large objects but performs worse for smaller objects. Hence, faster R-CNN balances between accuracy and speed. It matches the speed of R-FCN and SSD at ~32 mAP, for number of proposals = 50. Recently, mask R-CNN outperforms all other methods in accuracy and has an average precision of ~47.3. YOLO is the fastest single stage architecture having speed of approx. 21 ~ 155 frames per second.

5 Conclusions

A survey of the various models or techniques evolved year after year from R-CNN to its modified version is presented here. This work also carries out detailed performance comparisons of the existing works. It can be concluded that with the help of a good base network, YOLO performs faster and manage to have 21 ~ 155 fps. In terms of accuracy, mask R-CNN have an average precision ~47.3 with the help of ResNet architecture. So, based on the application area, the deep learning network should be appropriately chosen.

References

1. Lecun, Y., Bottou, L., Bengio, Y., Haffner, P.: Gradient-based learning applied to document recognition. Proc. IEEE **86**(11), 2278–2324 (1998)
2. Krizhevsky, A., Sutskever, I., Hinton, G.E.: ImageNet classification with deep convolutional neural networks. In: NIPS, pp. 1106–1114 (2012)
3. Zeiler, M.D., Fergus, R.: Visualizing and understanding convolutional neural networks. In: ECCV (2014)
4. Szegedy, C., Liu, W., Jia, Y., Sermanet, P., Reed, S., Anguelov, D., Erhan, D., Vanhoucke, V., Rabinovich, A.: A going deeper with convolutions. In: Computer Vision and Pattern Recognition, pp. 1–9 (2015)

5. Simonyan, K., Zisserman, A.: Very deep convolutional networks for large-scale image recognition. arXiv:1409.1556 (2014)
6. He, K., Zhang, X., Ren, S., Sun, J.: Deep residual learning for image recognition. In: Computer Vision and Pattern Recognition, pp. 770–778 (2015)
7. Uijlings, J., van de Sande, K., Gevers, T., Smeulders, A.: Selective search for object recognition. IJCV (2013)
8. Girshick, R.B., Donahue, J., Darrell, T., Malik, J.: Rich feature hierarchies for accurate object detection and semantic segmentation. In: Proceedings of CVPR (2014)
9. Girshick, R.: Fast R-CNN. In: Proceedings of the IEEE International Conference on Computer Vision 2015, pp. 1440–1448 (2015)
10. Ren, S., et al.: Faster R-CNN: towards real-time object detection with region proposal networks. In: Advances in Neural Information Processing Systems, pp. 91–99 (2015)
11. Dai, J., Li, Y., He, K., Sun, J.: R-FCN: object detection via region-based fully convolutional networks. In: Advances in Neural Information Processing Systems 2016, pp. 379–387 (2016)
12. Fu, C.Y., Liu, W., Ranga, A., Tyagi, A., Berg, A.C.: DSSD: deconvolutional single shot detector. arXiv:1701.06659, 23 Jan 2017
13. Redmon, J., Divvala, S., Girshick, R., Farhadi, A.: You only look once: unified, real-time object detection. In: Proceedings of the IEEE Conference on Computer Vision and Pattern Recognition 2016, pp. 779–788 (2016)
14. Najibi, M., Rastegari, M., Davis, L.S.: G-CNN: an iterative grid based object detector. In: Proceedings of the IEEE Conference on Computer Vision and Pattern Recognition 2016, pp. 2369–2377 (2016)
15. He, K., Zhang, X., Ren, S., Sun, J.: Spatial pyramid pooling in deep convolutional networks for visual recognition. In: European Conference on Computer Vision, 6 September 2014, pp. 346–361. Springer, Cham (2014)
16. Lenc, K., Vedaldi, A.: R-CNN minus R. In: British Machine Vision Conference (BMVC) (2015)
17. Sun, C., Paluri, M., Collobert, R., Nevatia, R., Bourdev, L.: ProNet: learning to propose object specific boxes for cascaded neural networks. In: Proceedings of the IEEE Conference on Computer Vision and Pattern Recognition 2016, pp. 3485–3493 (2016)
18. He, K., Gkioxari, G., Dollár, P., Girshick, R.: Mask R-CNN. In: 2017 IEEE International Conference on Computer Vision (ICCV), pp. 2980–2988. IEEE (2017)
19. Jiang, H., Learned-Miller, E.: Face detection with the faster R-CNN. In: 2017 12th IEEE International Conference on Automatic Face & Gesture Recognition (FG 2017), 30 May 2017, pp. 650–657. IEEE (2017)
20. Stewart, R., Andriluka, M., Ng, A.Y.: End-to-end people detection in crowded scenes. In: Proceedings of the IEEE Conference on Computer Vision and Pattern Recognition 2016, pp. 2325–2333 (2016)
21. Mao, T., Zhang, W., He, H., Lin, Y., Kale, V., Stein, A., Kostic, Z.: AIC2018 report: traffic surveillance research. In: CVPR Workshop (CVPRW) on the AI City Challenge (2018)
22. Redmon, J., Farhadi, A.: YOLO9000: better, faster, stronger. arXiv preprint, 17 Jul 2017
23. Huang, J., Rathod, V., Sun, C., Zhu, M., Korattikara, A., Fathi, A., Fischer, I., Wojna, Z., Song, Y., Guadarrama, S., Murphy, K.: Speed/accuracy trade-offs for modern convolutional object detectors. In: IEEE CVPR 2017, 1 July 2017, vol. 4 (2017)
24. Peng, X., Schmid, C.: Multi-region two-stream R-CNN for action detection. In: European Conference on Computer Vision, 8 October 2016, pp. 744–759. Springer, Cham (2016)
25. Johnson, J.W.: Adapting mask-RCNN for automatic nucleus segmentation. arXiv preprint arXiv:1805.00500, 1 May 2018
26. Everingham, M., Van Gool, L., Williams, C.K., Winn, J., Zisserman, A.: The pascal visual object classes (voc) challenge. IJCV **88**(2), 303–338 (2010)

27. Lin, T.-Y., Maire, M., Belongie, S., Hays, J., Perona, P., Ramanan, D., Dollár, P., Zitnick, C.L.: Microsoft COCO: common objects in context. In: ECCV (2014)
28. Deng, J., Dong, W., Socher, R., Li, L.-J., Li, K., Fei-Fei, L.: ImageNet: a large-scale hierarchical image database. In: IEEE Computer Vision and Pattern Recognition (CVPR) (2009)
29. Russakovsky, O., Deng, J., Su, H., Krause, J., Satheesh, S., Ma, S., Huang, Z., Karpathy, A., Khosla, A., Bernstein, M., Berg, A.C.: Imagenet large scale visual recognition challenge. Int. J. Comput. Vis. **115**(3), 211–252 (2015)

Comparison of Four Nature Inspired Clustering Algorithms: PSO, GSA, BH and IWD

Shankho Subhra Pal, Rupkatha Hira and Somnath Pal

Abstract Nature uses an optimized approach to do all type of jobs. Thus many natural phenomena are mathematically modelled as meta-heuristic optimization algorithms. These meta-heuristic optimization algorithms are used in different fields to find an optimal minimum or maximum solution where deterministic algorithm can't be used. Clustering is one such field where different meta-heuristic are used. Although various such algorithms claim superiority over others, in this work, four different meta-heuristic algorithm based clustering are compared to determine which one is superior.

Keywords Clustering · k-Means algorithm · Nature-inspired algorithm · Meta-heuristic

1 Introduction

All natural phenomena occur in an optimized way. Some examples are movement of grey wolves [7], movement of whales [9], evolution due to genetics [3] etc. Researchers are inspired to mathematically model such natural optimization phenomenon. Some natural phenomenon are mathematically modelled in PSO [15], GSA [12], BH [11] and IWD [13] which are modelled after movement of flocks of birds, movement of celestial bodies in the universe, movement of celestial bodies due to presence of black hole and movement of water droplets respectively.

S. Subhra Pal · R. Hira (✉) · S. Pal
Indian Institute of Engineering Science and Technology, Shibpur, Howrah 711103, India
e-mail: rupkatha.hira@gmail.com

S. Subhra Pal
e-mail: shankho.subhra.pal@gmail.com

S. Pal
e-mail: sp@cs.iiests.ac.in

A. K. Das et al. (eds.), *Computational Intelligence in Pattern Recognition*,
Advances in Intelligent Systems and Computing 999,
https://doi.org/10.1007/978-981-13-9042-5_57

These meta-heuristic algorithms are used for optimizing different functions used mainly in engineering. Some of the applications include training neural network [1, 8, 10], optimizing electrical problems [4, 16] and many more.

Clustering is the process of grouping similar samples together such that intra-cluster distance is minimized. As a minimization is needed, so many meta-heuristic optimization algorithms are modified so that they can be used for clustering. Since many such mets-heuristic algorithms for clustering claim to be superior to others, a comparison among them is called for. In this work PSO [2], GSA [6], BH [5] and IWD [14] based k-Means Clustering are compared.

In Sect. 2, Literature Review is discussed followed by Experimental Setup in Sect. 3. Experimental Results and Discussion appears in Sect. 4. Section 5 summarizes the Conclusion.

2 Literature Review

2.1 PSO Based k-Means Clustering

PSO stands for Particle Swarm Optimization which is a meta-heuristic optimization algorithm. It has been used in many engineering fields. Clustering has also been done using PSO.

In PSO, the position and velocity of the particles are randomly initialized and in each iteration they are updated stochastically depending on the fitness at the current position, global best position, current best position and current velocity. Global best position is the position of fittest particle found till now. Next the current best position and global best position is updated. For PSO based k-Means clustering [2], each particle has dimension equal to the number of features times the number of clusters. The fitness function is sum of intra-cluster distances.

2.2 GSA Based k-Means Clustering

GSA stands for Gravitational Search Algorithm which is a meta-heuristic optimization algorithm which has been used in many engineering applications including clustering. In GSA, the position of the particles are randomly initialized and velocity of each particle is initialized to zero. In each iteration the fitness of each particle is used to cluster the mass of the particle which is normalized. Then the force on each particle due to kBest other particles is calculated and is used to find the acceleration of each particle. Here force is found using Newton's Law of Gravitation. Next the velocity and position of each particle is updated using Newton's Laws of motion.

For GSA based k-Means clustering [6], each particle has dimension equal to the number of features times the number of clusters. The fitness function is sum of intra-cluster distances. To have a good quality of initial population, the position of one of the particles is initialized to the result of k-Means, the positions of one each is initialized to the minimum, maximum and mean of the data-set and the rest of the positions are initialized randomly.

2.3 BH Based k-Means Clustering

BH stands for Black Hole Optimization which is also applied for clustering problem. In BH, the position of the particles are randomly initialized and in each iteration they are updated stochastically depending on the fitness at the current position and the fittest particle. The fittest particle is the black hole and it is updated in each iteration. To ensure that there is no immature convergence, if a particle enters event horizon, i.e. comes too near to fittest particle, it is randomly re-initialized. For BH based k-Means clustering [5], each particle has dimension equal to the number of features times the number of clusters. The fitness function is sum of intra-cluster distances.

2.4 IWD Based k-Means Clustering

The Intelligent Water Drops (IWD) algorithm attempts to model the behaviour of natural water drops flowing in rivers in order to optimize an objective function of a particular problem. The problem is modelled as a weighted multigraph $G = (V, E)$ where V is the set of vertices (nodes) and E is its set of edges.

IWDs are placed randomly on the nodes of the graph which then travel along the edges of the graph from one node to another, altering the weight (*soil*) of the edges and constructing a solution to the problem. As time elapses, the edges of better solutions contain less *soil* than other edges of the graph. Moreover, at the end of each iteration of the algorithm, the total-best solution (TBS), which is the best solution found so far, is updated by the solutions obtained by all the IWDs. This process is repeated a number of times for new water drops. IWD algorithm combined with the k-Means algorithm [14] is used to cluster input data vectors.

3 Experimental Setup

This experiment was performed using python3 on Intel i7–6700 processor with 8 GB RAM in Ubuntu 18.04 LTS operating system. Twelve data-sets were used and they are compared. The data-set description is provided in Table 1.

Table 1 Data-set description

Dataset	No. of samples	No. of features	No. of clusters
Diabetes	768	8	2
Wine	178	13	3
Glass	214	9	6
Iris	150	4	3
Dataset_spine	310	12	3
Indian_liver_patient	583	10	2
Affairs	601	9	5
Vehicle	94	18	4
Ecoli	336	7	8
SPECTF_heart	267	44	2
Machine	209	7	30
Tae	151	5	3

4 Experimental Results and Discussion

The experimental results are presented in Table 2. Although the results of Table 2 are interesting themselves, they are summarized in Table 3. As it can be seen from Table 3, GSA is better than PSO in seven instances, equal in three instances and worse in two instances. GSA is better than BH in four instances, equal in six instances and worse in two instances. GSA is better than IWD in eleven instances, equal in one instance and worse in zero instance. BH is better than PSO in six instances, equal in three instances and worse in three instances. BH is better than IWD in ten instances,

Table 2 Comparison of the four methods

Dataset	No. of clusters	PSO	GSA	BH	IWD
Diabetes	2	121.751	121.271	121.271	143.484
Wine	3	49.112	48.954	48.954	71.669
Glass	6	19.166	18.436	19.366	40.652
Iris	3	7.060	6.998	6.998	8.910
Dataset_spine	2	171.602	171.656	170.858	180.712
Indian_liver_patient	2	90.969	90.969	90.969	90.969
Affairs	5	224.778	222.553	238.245	238.231
Vehicle	4	29.482	29.482	29.482	36.522
Ecoli	8	19.827	19.337	19.656	37.103
SPECTF_heart	2	214.274	214.352	214.161	217.117
Machine	30	4.447	4.226	5.935	8.256
Tae	3	33.672	33.672	33.672	56.162

Table 3 W-D-L

	PSO	GSA	BH	IWD
PSO	–	2-3-7	3-3-6	11-1-0
GSA	7-3-2	–	4-6-2	11-1-0
BH	6-3-3	2-6-4	–	10-1-1
IWD	0-1-11	0-1-11	1-1-10	–

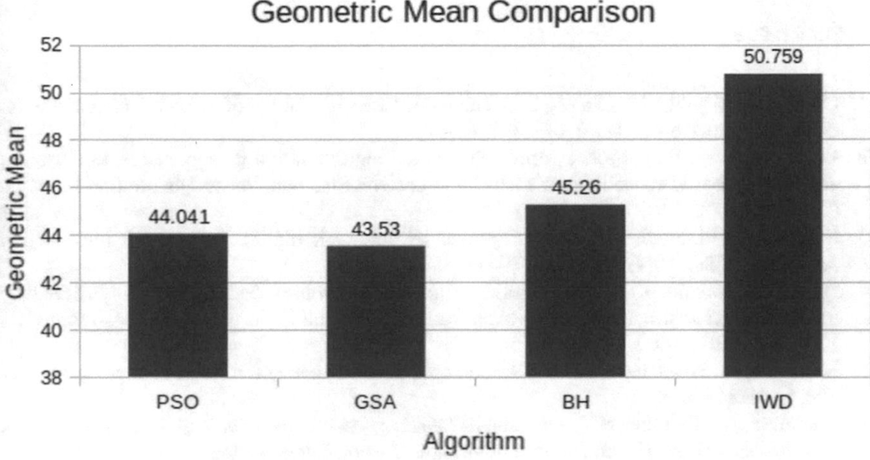

Fig. 1 Geometric mean comparison

equal in one instance and worse in one instance. PSO is better than IWD in eleven instances, equal in one instance and worse in zero instance.

Since the data-sets are from different domains, the geometric mean is more relevant compared to arithmetic mean. Figure 1 presents the geometric mean of all data-sets for each algorithm. From Fig. 1 we can see that GSA has best geometric mean followed by PSO, BH and IWD.

5 Conclusions

Since various nature-inspired algorithms have been developed, each claiming superiority over others, the present work attempts to evaluate comparative study among four state of the art algorithms for clustering problem. The algorithms considered in this work are Particle Swarm Optimization (PSO), Gravitational Search Algorithm (GSA), Black Hole Optimization (BH) and Intelligent Water Drops (IWD) for clustering. The meta-heuristic algorithms for clustering are evaluated based on sum of intra-cluster distances (which should be minimum), for twelve diverse domains.

From the Table 2 considering the draws at top position as good result, it can be seen that GSA is giving good results ten data-sets, BH in seven data-sets, PSO in four data-sets and IWD in one data-set. From Tables 2, 3 and Fig. 1 it can be concluded that GSA gives the best performance. However, a detailed study including more such algorithm and on various parameters to judge the clustered result can be considered as future work.

References

1. Chau, K.W.: Application of a pso-based neural network in analysis of outcomes of construction claims. Autom. Constr. **16**(5), 642–646 (2007)
2. Chen, C.-Y., Ye, F.: Particle swarm optimization algorithm and its application to clustering analysis. In: 2012 Proceedings of 17th Conference on Electrical Power Distribution Networks (EPDC), pp. 789–794. IEEE (2012)
3. Fraser, A.S.: Simulation of genetic systems by automatic digital computers I. Introduction. Aust. J. Biol. Sci. **10**(4), 484–491 (1957)
4. Haida, T., Akimoto, Y.: Genetic algorithms approach to voltage optimization. In: Proceedings of the First International Forum on Applications of Neural Networks to Power Systems, pp. 139–143. IEEE (1991)
5. Hatamlou, A.: Black hole: a new heuristic optimization approach for data clustering. Inf. Sci. **222**, 175–184 (2013)
6. Hatamlou, A., Abdullah, S., Nezamabadi-Pour, H.: A combined approach for clustering based on k-means and gravitational search algorithms. Swarm Evol. Comput. **6**, 47–52 (2012)
7. Long, W., Xu, S.: A novel grey wolf optimizer for global optimization problems. In: 2016 IEEE Advanced Information Management, Communicates, Electronic and Automation Control Conference (IMCEC), pp. 1266–1270. IEEE (2016)
8. Mirjalili, S.: How effective is the grey wolf optimizer in training multi-layer perceptrons. Appl. Intell. **43**(1), 150–161 (2015)
9. Mirjalili, S., Lewis, A.: The whale optimization algorithm. Adv. Eng. Softw. **95**, 51–67 (2016)
10. Pal, S.S.: Grey wolf optimization trained feed foreword neural network for breast cancer classification. Int. J. Appl. Ind. Eng. (IJAIE) **5**(2), 21–29 (2018)
11. Piotrowski, A.P., Napiorkowski, J.J., Rowinski, P.M.: How novel is the "novel" black hole optimization approach? Inf. Sci. **267**, 191–200 (2014)
12. Rashedi, E., Nezamabadi-Pour, H., Saryazdi, S.: Gsa: a gravitational search algorithm. Inf. Sci. **179**(13), 2232–2248 (2009)
13. Shah-Hosseini, H.: Intelligent water drops algorithm: a new optimization method for solving the multiple knapsack problem. Int. J. Intell. Comput. Cybern. **1**(2), 193–212 (2008)
14. Shah-Hosseini, H.: Improving k-means clustering algorithm with the intelligent water drops (iwd) algorithm. Int. J. Data Min. Model. Manag. **5**(4), 301–317 (2013)
15. Shi, Y., Eberhart, R.C.: Empirical study of particle swarm optimization. In: Proceedings of the 1999 Congress on Evolutionary Computation, CEC 99, vol. 3, pp. 1945–1950. IEEE (1999)
16. Yoshida, H., Kawata, K., Fukuyama, Y., Takayama, S., Nakanishi, Y.: A particle swarm optimization for reactive power and voltage control considering voltage security assessment. IEEE Trans. Power Syst. **15**(4), 1232–1239 (2000)

Detection of Diseases in Potato Leaves Using Transfer Learning

Soumik Ranjan Dasgupta, Somnath Rakshit, Dhiman Mondal and Dipak Kumar Kole

Abstract The problem of food shortage has grown rampant in the recent times in developing countries. In a tropical country like India, potato is one of the major staple food that is eaten throughout the year. Recently the production of potato is falling short due to various diseases like Early Blight and Late Blight which cause a huge loss of the cropped plants. This also leads to a major loss in the national economy as well. The emergence of deep learning has affected many fields of machine learning research. Since it is not required in deep learning to develop hand-crafted features, it has found widespread adoption in the scientific community. To tackle the need for a huge amount of data for deep learning, another heavily implemented technique is used, namely, transfer learning, to make the training process faster and more accurate with a relatively small dataset at hand. The performance of the model is demonstrated both quantitatively by computing the accuracy metric as well as visually. The model is lightweight and robust and thus can be added to an application in a handheld device like smartphone so that crop growers could spot the disease affected crops on the go and save them from getting ruined.

Keywords Deep learning · Computer vision · Transfer learning · Image classification

---

S. R. Dasgupta · D. Mondal · D. K. Kole
Jalpaiguri Government Engineering College, Jalpaiguri 735102, India
e-mail: srd1908@cse.jgec.ac.in

D. Mondal
e-mail: dhiman.mondal@cse.jgec.ac.in

D. K. Kole
e-mail: dipak.kole@cse.jgec.ac.in

S. Rakshit (✉)
Cyware Labs, Bangalore 560102, India
e-mail: somnath@cse.jgec.ac.in

A. K. Das et al. (eds.), *Computational Intelligence in Pattern Recognition*,
Advances in Intelligent Systems and Computing 999,
https://doi.org/10.1007/978-981-13-9042-5_58

1 Introduction

Potato is one of the most essential and important vegetables in a tropical country like India. This is one such vegetable which is accepted worldwide and taken by most households on a daily basis.[1] Potatoes are a major variety of economical food; they provide a source of low-cost energy to the human diet. They are a rich source of starch, vitamins especially C and B1 and minerals. West Bengal is a major potato producing state and accounts for about 33% of national output [1]. But several fungal and bacterial diseases create major obstacles in the process of potato production. The diseases can prove to be fatal which leads to a loss in production and eventually cause a loss in the national economy. *Alternaria Solani* or Early Blight and *Phytophthora Infestans* or Late Blight are two among such fatal diseases [2].

The growing popularity of smartphones among crop growers with an expected 5 billion smartphones by 2020 offers the potential to turn the smartphone into a handy tool for identification of crop diseases among the diverse communities growing food around the world. A light and robust model is required to identify the disease affected plants on the go so that prompt action could be taken and loss may be alleviated. Moreover, the crop growers can become more aware and can communicate with agricultural experts in case of a widespread fall-out of the diseases.

The dataset for this task is a subset of the PlantVillage Dataset [3]. In this dataset three classes of leaf images of potato are taken, namely *Alternaria solani*, *Phytophthora infestans* and healthy leaf images. In total there are 856 leaf images out of which 412 leaf images of *Alternaria solani*, 380 leaf images of *Phytophthora infestans* and 64 images of healthy leaf. The images are provided without any preprocessing. The dataset is split into an 80-20 ratio after randomly shuffling the images and used to train the model. The performance of this model is then evaluated by quantitatively by computing the accuracy metric as well as visually.

There are lots of work on plant disease classification using leaf images in the scientific literature. Fujita et al. [4] used a dataset of cucumber leaf images and used feedforward neural nets to obtain an accuracy of 82.3%. Alvaro et al. [5] effectively detected class and location of diseases in tomato leaves. Amara et al. [6] used a deep-learning based approach for banana leaf diseases classification and achieved a highest of 97% accuracy. DeChant et al. [7] used a CNN to automatically detect infected maize plants in fields from leaf images. A method is presented in this paper that is easy to use with the implementation done in Fastai[2] and is able to identify the diseases in leaf images. The performance of various easy to use pretrained models like ResNet [8], inception modules [9], DenseNets [10] and VGG networks [11] are compared.

Transfer learning is a well-known technique in deep learning where pre-trained models used on other datasets are used as a starting point to train and evaluate the data on hand. This is because deep learning architectures need huge time and computing resources and a very large dataset for training and evaluation with high accuracy.

[1] www.agricoop.gov.in.

[2] https://www.fast.ai.

Since this is not always feasible, the method of transfer learning has increasingly gained popularity. Multiple classification tasks have been done using transfer learning by researchers such as Quattoni et al. [12], Do et al. [13], Dai et al. [14] and many more.

2 Background

Alternaria solani [15] is a fungal pathogen, that produces a disease in potato plants called Early Blight. It is deuteromycete[3] that has a polycyclic life cycle, which means they can produce more than one infection cycle in a single crop cycle. The spores need free water to germinate and typically take two hours to do so on a wide range of temperatures. The pathogen sporulates best at around 26.6 °C in presence of abundant moisture. Primary damage caused by Early Blight is the premature defoliation of potato plants. In the initial stage concentric dark brown spot occurs at the centre of the leaf. In the later stage the leaf turns yellow and either dries out or falls off the stem.

Phytophthora infestans [16] is an oomycete or water mold, a microorganism which causes the serious potato disease known as late blight or potato blight. In case of Late Blight primary symptom is the appearance of dark blotches on leaf tips and plant stems. White mold appears under the leaves in humid conditions eventually and the whole plant may quickly collapse. Figure 1a–c are some images of disease infected leaves by Early Blight and Late Blight.

Several widely popular architectures are used to train the dataset. ResNet, short for Residual Networks, developed by He et al. [8] is a novel architecture with "skip connections" and heavy batch normalization. The network is designed in such a way that instead of mapping the input to the regular hypotheses function $H(x)$, it maps the input to a residual function $H(x) - x$, due to which it is named so. Therefore, the original function becomes $F(x) + x$. This additional x comes from the output of a previous layer and is same as the input to that layer as well, hence it is also called identity connection. The advantage of residual networks is that where regular deep networks show decreasing accuracy as the network gets deeper, a residual network performs remarkably well. This dataset is tested on ResNet of various depths, viz. 18, 34, 50, 101, 152. Figure 2 shows the architecture for a single ResNet block.

Another architecture that is used on the dataset is the Inception architecture. In this kind of architecture, convolutions of several filter sizes are applied on the input and stacked together in a single layer. This is done because the size of the convolutional filter to apply on the input is often confusing, so all the possible filter sizes are applied at once. Additionally, a 1×1 convolution is also applied before each operation in order to perform a dimensionality reduction and lower the number of computations to be performed. A single inception module is illustrated in Fig. 3. The advantage of this architecture is that the smaller convolutions can pick up local features and the

[3]http://202.127.145.151/agroprojects/dictionary/diseaseDictSci.htm.

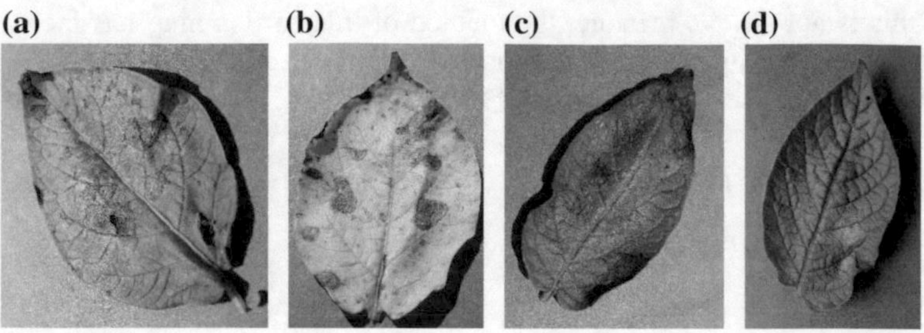

Fig. 1 Images of leaves affected with the following diseases: **a** Early Blight. **b** Early Blight. **c** Late Blight. **d** Late Blight

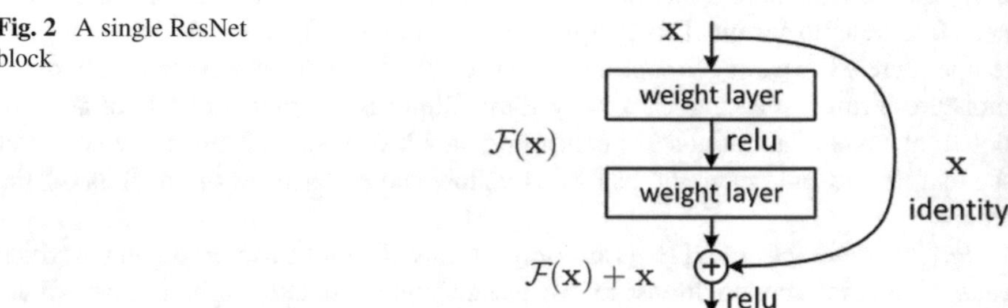

Fig. 2 A single ResNet block

larger ones can detect high-level abstract features. The dataset is used on Inception v4, the most recent version of the model.

DenseNet is also applied on the dataset. The idea behind Densenet is that the feature maps obtained by applying convolutions in each layer are concatenated to the feature maps of the subsequent layer, unlike ResNet, which performs the summation of previous layers and feeds it to the current one. The dataset is used on DenseNets

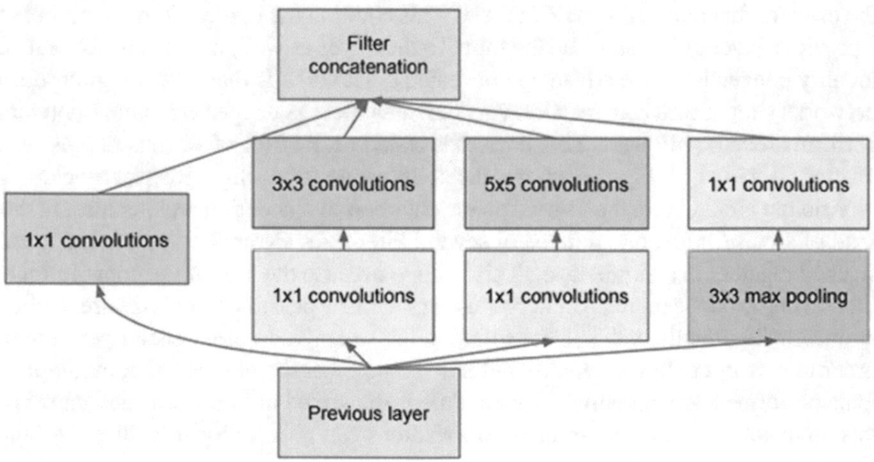

Fig. 3 Inception module

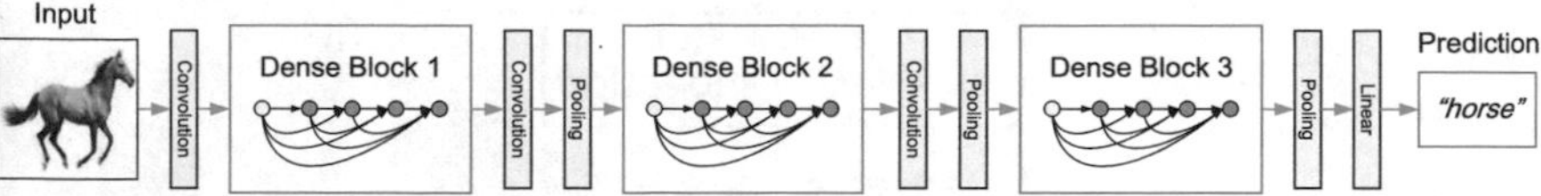

Fig. 4 A deep DenseNet with three dense blocks

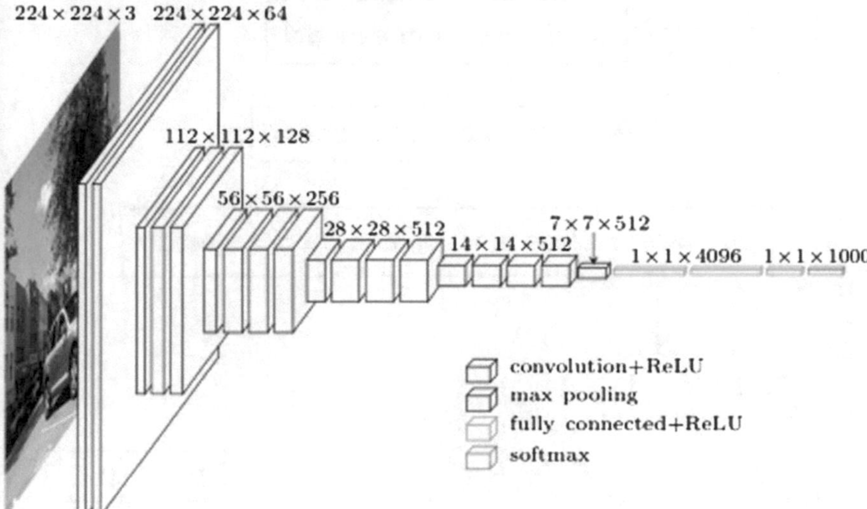

Fig. 5 VGG16 architecture

having various number of layers, viz. 121, 161, 169, 201. A deep DenseNet architecture with three dense blocks is illustrated in Fig. 4.

VGGNetwork is yet another novel architecture that replaces large 11 × 11 and 5 × 5 kernel-sized filters with multiple small 3 × 3 kernel-sized filters. This increases the depth of the network and enables it to learn more complex features at a lower cost. Figure 5 is an illustration of the VGG16 architecture.

3 Method

The method used to train and evaluate the models is described here. The execution pipeline is illustrated in Fig. 6.

Step 1. The image is first taken as an input.
Step 2. The image resolution is changed to an absolute 250 × 250 pixels.
Step 3. The image is trained using various pre-trained models, viz. Resnet18, Resnet34, Resnet50, Resnet101, Resnet152, Inception4, Densenet121, Densenet161, Densenet169, Densenet201, VGG16 and VGG19.
Step 4. The accuracy of the models is then calculated using the validation dataset.

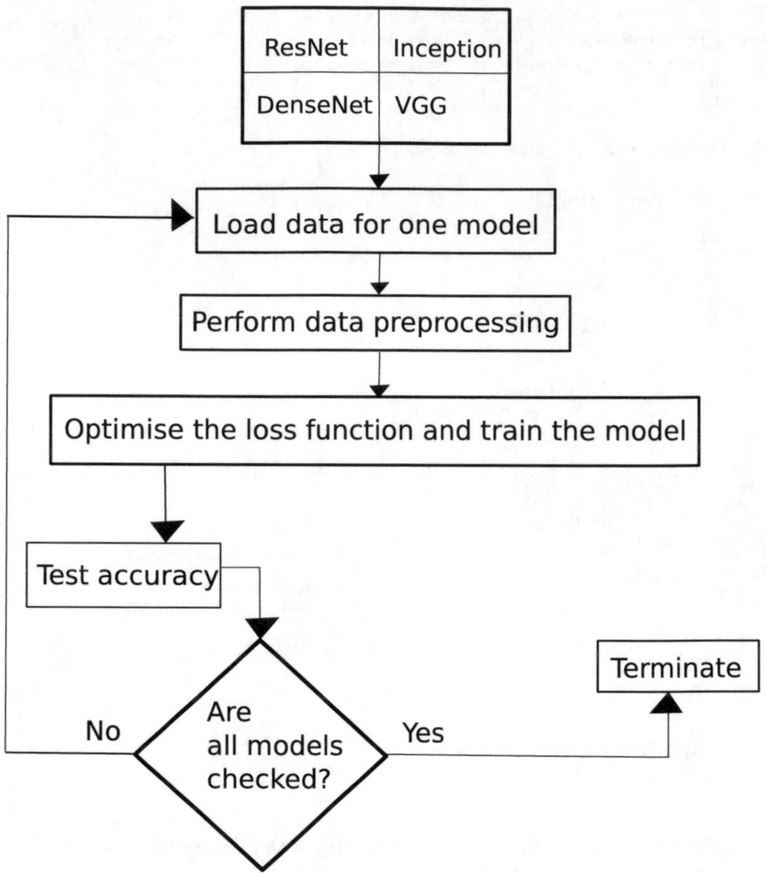

Fig. 6 Pipeline of execution

The optimizer used during the training procedure is Stochastic Gradient Descent (SGD) with a momentum value of 0.9. The equations of SGD with momentum update are given by Eqs. 1 and 2, where Eq. 1 calculates the derivatives and Eq. 2 updates the parameters.

$$\nu = \gamma\nu + \alpha\Delta_\theta J(\theta; x^{(i)}, y^{(i)}) \tag{1}$$

$$\theta = \theta - \nu \tag{2}$$

Here α is the learning rate which is set to 0.1 and γ is the momentum which has a value of 0.9. The function J(θ; $x^{(i)}$, $y^{(i)}$) represents the cost function of parameter θ and $x^{(i)}$, $y^{(i)}$ which represent a single instance of the dataset. The cost function is represented by the Eq. 3.

$$MSE = \frac{1}{n}\sum_{i=1}^{n}(Y_i - \hat{Y}_i)^2. \tag{3}$$

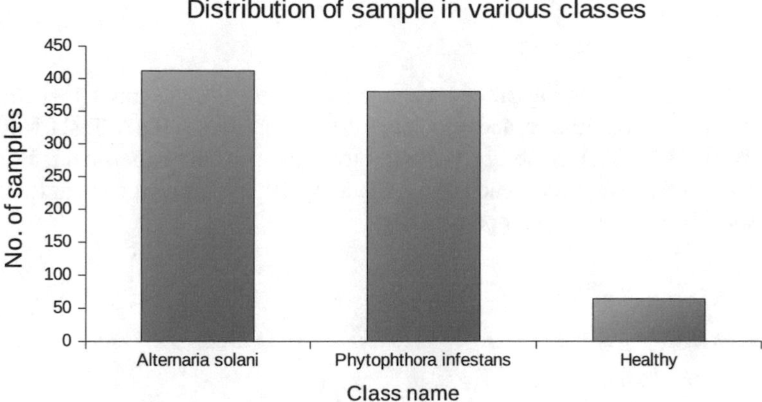

Fig. 7 Dataset distribution

The abbreviation *MSE* stands for Mean Squared Error which is equivalent to the summation of squares of the differences between the ground truth Y_i and the expected output $\hat{Y}_i$ divided by the total number of instances in the dataset.

4 Experimental Result

4.1 Dataset

The dataset in use is the PlantVillage dataset.[4] PlantVillage is a research and development unit by Pennsylvania State University that focuses on empowering the farmers by providing them with easy-to-use affordable technology and democratizing access to knowledge that can help them grow more crops. The dataset originally consists of 50,000 images of disease infected leaf images of 38 variety of plants. The leaf images of potato are taken separately and then split into an 80-20 ratio to obtain the training and validation set. The split is made such that the percentage of each class is equal in the test and train sets. Figure 7 shows the distribution of the dataset as a bar graph. There are three classes in the dataset classified on the basis of diseases affecting potato production, viz. *Alternalia solaris*, *Pytopthora infestans* and unaffected leaves. Since the resolution of the images is not uniform, the images are converted to sizes of dimensions 250 × 250 pixels before being fed to the model.

[4]https://plantvillage.psu.edu/.

4.2 Testbed

Both the training and testing procedure are executed on a single core hyper-threaded Xeon Processor at 2.30 Ghz and accelerated by a single NVIDIA Tesla K80 GPU with 12 GB RAM. FastAI library was used to implement the experiment. With this setup, 10 epochs were run for each of the model to achieve convergence of loss. Each epoch took 10 s to complete approximately.

4.3 Results

The performance of the various models used in the experiment is measured both quantitatively as well as visually. Figure 9 compares the accuracies from various models visually. It is observed that when Densenet121, Densenet169, and VGG19 are used for transfer learning, the accuracy is equal to 100%. However, in other cases also, the accuracy is more than 98% apart from Inception4 model where the accuracy goes to 94.94%. Table 1 shows the various implementations done by researchers for disease classification from leaf images and the accuracy obtained. Table 2 shows

Table 1 Various leaf disease classification techniques and their accuracies

Sl. No.	Authors	Accuracies obtained (%)
1	Fuentes et al. [5]	83
2	Lu et al. [17]	95
3	Oppenheim and Shani [18]	96
4	DeChant et al. [7]	97
5	Our best model	100

Table 2 Accuracies obtained from various networks

Sl. No.	Network model	Accuracy obtained
1	Resnet18	0.9887
2	Resnet34	0.9831
3	Resnet50	0.9943
4	Resnet101	0.9943
5	Resnet152	0.9831
6	Inception4	0.9494
7	Densenet121	1.0
8	Densenet161	0.9888
9	Densenet169	1.0
10	Densenet201	0.9943
11	VGG16	0.9887
12	VGG19	1.0

Fig. 8 Examples of correct classifications

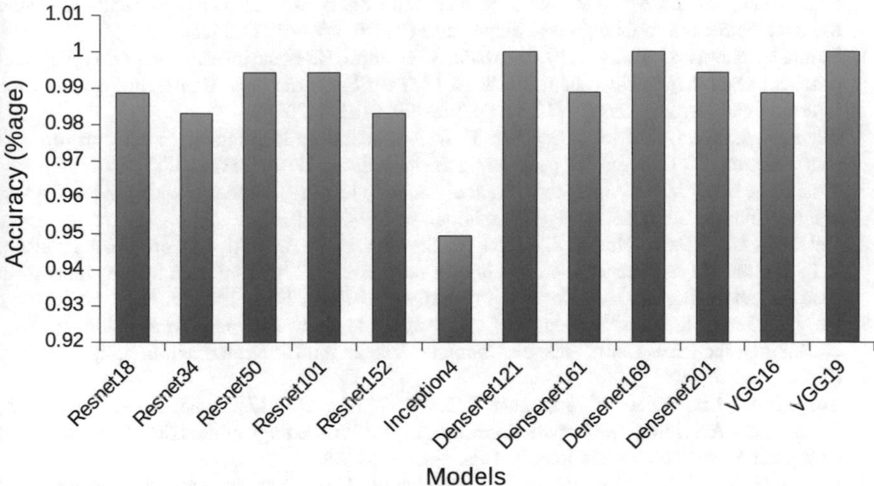

Fig. 9 Comparison of accuracies from various models

the accuracy obtained from various models from our experiment. The models were evaluated on 178 leaf images from various classes.

A comparative illustration of the accuracies obtained from various models is illustrated in Fig. 9. Apart from this, one example of the images classified correctly by our best performing model is shown side by side in Fig. 8a–c.

5 Conclusion

The task of leaf disease classification manually can be tiresome and error prone. The same when automated can lead to a significant boost in production of the crops and therefore, a gross increase in the national output. In this work, the way transfer learning can be applied in the agricultural domain and how various pretrained models can perform extremely well on an unseen dataset is shown. As a future scope, this method can be integrated into a smartphone application and uploaded on a public repository which can then be used by numerous crop growers all over the world.

References

1. Mitra, S., Sarkar, A.: Relative profitability from production and trade: a study of selected potato markets in West Bengal. Econ. Polit. Wkly. **38**, 4694–4699 (2003)
2. Henfling, J.W.: Late Blight of Potato, vol. 4. International Potato Center (1987)
3. Hughes, D., Salathé, M., et al.: An open access repository of images on plant health to enable the development of mobile disease diagnostics (2015). arXiv:1511.08060
4. Fujita, E., Kawasaki, Y., Uga, H., Kagiwada, S., Iyatomi, H.: Basic investigation on a robust and practical plant diagnostic system. In: 2016 15th IEEE International Conference on Machine Learning and Applications (ICMLA), pp. 989–992. IEEE (2016)
5. Fuentes, A., Yoon, S., Kim, S.C., Park, D.S.: A robust deep-learning-based detector for real-time tomato plant diseases and pests recognition. Sensors **17**(9), 2022 (2017)
6. Amara, J., Bouaziz, B., Algergawy, A., et al.: A deep learning-based approach for banana leaf diseases classification. In: BTW (Workshops), pp. 79–88 (2017)
7. DeChant, C., Wiesner-Hanks, T., Chen, S., Stewart, E.L., Yosinski, J., Gore, M.A., Nelson, R.J., Lipson, H.: Automated identification of northern leaf blight-infected maize plants from field imagery using deep learning. Phytopathology **107**(11), 1426–1432 (2017)
8. He, K., Zhang, X., Ren, S., Sun, J.: Deep residual learning for image recognition. In: Proceedings of the IEEE Conference on Computer Vision and Pattern Recognition, pp. 770–778 (2016)
9. Szegedy, C., Liu, W., Jia, Y., Sermanet, P., Reed, S., Anguelov, D., Erhan, D., Vanhoucke, V., Rabinovich, A.: Going deeper with convolutions. In: Proceedings of the IEEE Conference on Computer Vision and Pattern Recognition, pp. 1–9 (2015)
10. Huang, G., Liu, Z., Van Der Maaten, L., Weinberger, K.Q.: Densely connected convolutional networks (2017)
11. Simonyan, K., Zisserman, A.: Very deep convolutional networks for large-scale image recognition (2014). arXiv:1409.1556
12. Quattoni, A., Collins, M., Darrell, T.: Transfer learning for image classification with sparse prototype representations. In: IEEE Conference on Computer Vision and Pattern Recognition, CVPR 2008, pp. 1–8. IEEE (2008)
13. Do, C.B., Ng, A.Y.: Transfer learning for text classification. In: Advances in Neural Information Processing Systems, pp. 299–306 (2006)
14. Dai, W., Chen, Y., Xue, G.R., Yang, Q., Yu, Y.: Translated learning: transfer learning across different feature spaces. In: Advances in Neural Information Processing Systems, pp. 353–360 (2009)
15. Horsfield, A., Wicks, T., Davies, K., Wilson, D., Paton, S.: Effect of fungicide use strategies on the control of early blight (alternaria solani) and potato yield. Australas. Plant Pathol. **39**(4), 368–375 (2010)
16. Dyer, A., Matusralt, M., Drenth, A., Cohen, A., Splelman, L.: Historical and recent migrations of phytophthora infestans: chronology, pathways, and implications. Plant Dis. **77**, 653–661 (1993)
17. Yang, L., Yi, S., Zeng, N., Liu, Y., Zhang, Y.: Identification of rice diseases using deep convolutional neural networks. Neurocomputing **267**, 378–384 (2017)
18. Oppenheim, D., Shani, G.: Potato disease classification using convolution neural networks. Adv. Anim. Biosci. **8**(2), 244–249 (2017)

Self Driving Car

Ankit Kumar, Mayukh Mukherjee and Preetam Mukhopadhyay

Abstract The self-driving car is capable of sensing its environment and navigating without human input which it applies the technology of autonomous vehicle that allows the car to experiment driverless experience. Software designed to drive from point A to point B without requiring any human intervention is implemented.

Keywords Convolutional neural network · Self driving car · Max pooling · ReLU · Flattening · Full connection · DropOut layers

1 Introduction

The world is on the verge of a huge change in transportation. Throughout history to get to point A to point B a person had to be in control of vehicles movement. From horse-drawn carts to motorcars traveling has required time and concentration. This human operation is soon to fade and then computer will take its place.

Autonomous technologies have existed in many forms and are still spreading in today's market. In 1860 Robert Whitehead, an English engineer created the self-propelled torpedo and aeroplanes had a primitive version of autopilot just after 10 years of its invention. Cars have had a greater struggle with navigation than any other mode of transportation. They need to understand the urban world, the city streets, traffic control, and many more.

A. Kumar
IIEST, Hostel 13, Shibpur 711103, India
e-mail: ankittaxak5713@gmail.com

M. Mukherjee (✉) · P. Mukhopadhyay
IIEST, Hostel 15, Shibpur 711103, India
e-mail: mayukhranaghat2012@gmail.com

P. Mukhopadhyay
e-mail: ompreetam317@gmail.com

A. K. Das et al. (eds.), *Computational Intelligence in Pattern Recognition*,
Advances in Intelligent Systems and Computing 999,
https://doi.org/10.1007/978-981-13-9042-5_59

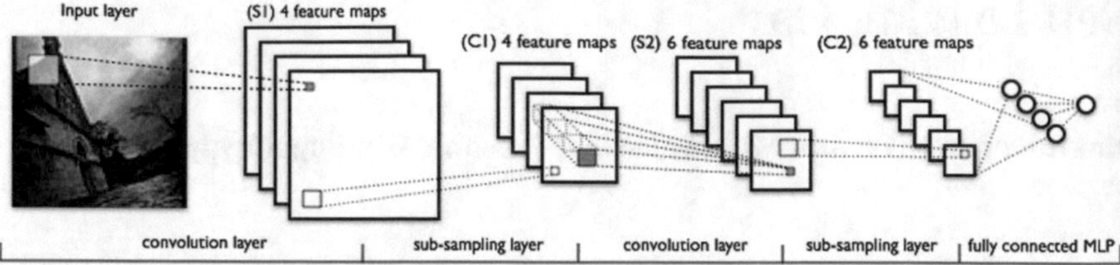

Fig. 1 CNN architecture

2 Steps of CNN

Convolutional neural networks are deep artificial neural networks that are used primarily to classify images (e.g., name what they see), cluster them by similarity (photo search), and perform object recognition within scenes.

A classic CNN architecture has the following steps `input->conv->ReLU->conv->ReLU->pool->ReLU->conv->ReLU->pool->fully connected`

The convolution is described as:

$$(f * g)[n] = \sum_{m=-\infty}^{\infty} f[m]g[n-m]$$

Then the resulting operation is a element-wise multiplication and addition of the terms (Fig. 1).

2.1 Feature Map and Feature Detector

A filter or feature detector acts in the receptive field. Now this filter contains an array of weights or parameters. Depth of the input dimensions should be same as the depth of the filter array. Subsequently we have to choose stride and padding.

As we keep applying conv layers, the spatial dimensions decrease. As we want to preserve most of the input size so that we can extract those low-level features. Zero padding pads the input volume with zeros around the border.

If you have a stride of 1 and if you set the size of zero padding to

$$zero\_padding = \frac{K-1}{2}$$

where K is the filter size, then the input and output volume will always have the same spatial dimensions. The formula for calculating the output size for any given conv layer is

$$O = \frac{W - K + 2P}{S} + 1$$

where O is the output height/length, W is the input height/length, K is the filter size, P is the padding, and S is the stride. As the filter slides or rather we can say it convolves over the image and computes element-wise multiplications then multiplications are all summed up and a value is obtained. We repeat this above process on each and every pixel of image and produces a number for each instance.

2.2 *Max Pooling*

In the first layer, the input was just the original image, then in the second convolutional layer, the input is the activation map(s) from the first layer and so on. Each layer of the input describes the location of certain low-level features. As we go down the network and come across with more convolutional layers, we get activation maps with more and more complex and higher level features. The filters have a larger receptive field, means that they are able to grasp data from a larger area of the original input image. After some ReLU layers, pooling layers (down-sampling layer) may be applied. This takes a filter (normally having the size of 2×2) and a stride of the same length. Subsequently, one applies it to the input and outputs the maximum number in every subregion that the filter convolves around [1].

Other options for pooling layers are average pooling and L2-norm pooling. Once we know that a specific feature is in the original input volume (occupying high activation value), its relative location becomes more important comparing to the other features. This layer drastically reduces the spatial dimension of the input volume serving two main purposes. Primarily, the amount of parameters or weights is reduced by 75%, thus lessening the computation cost. The second is that it will control overfitting [1].

2.3 *ReLU*

After each conv layer, it is convention to apply a nonlinear layer (or activation layer) immediately afterward. The purpose of this layer is to introduce nonlinearity to a system that basically has just been computing linear operations during the conv layers (just element-wise multiplications and summations). In the past, nonlinear functions like tanh and sigmoid were used, but researchers found out that ReLU layers work far better because the network is able to train a lot faster without making a significant difference to the accuracy. It also helps to alleviate the vanishing gradient problem, which is the issue where the lower layers of the network train very slowly because the gradient decreases exponentially through the layers. The ReLU layer applies the function $f(x) = \max(0, x)$ to all of the values in the input volume. This layer increases the nonlinear properties of the model and the overall network without affecting the receptive fields of the conv layer [1, 2].

2.4 Flattening

Just like any other regressor, an ANN regressor needs individual features as input which is also known as feature vector. So, whatever be the output of Convolutional part of CNN is, we just need to convert it into a 1D feature vector to feed the input of the ANN classifier. This operation is named as flattening. After getting the output of the convolutional layers of CNN, it flattens all its structure to create a single long feature vector for the usage of Dense Neural network [3, 4].

2.5 Full Connection

At the end of a Convolutional Neural Network we can attach a fully connected layer. This layer takes an input volume (which could be the output of a convolutional or ReLU or maxPool layer) and outputs a N dimensional 1D vector that signifies the number of classes that we can produce from the input volume. This fully connected layer implements softmax function and cross-entropy. It takes the output of the previous layer which basically represents the activation maps of high level features from the layers precede the fully connected layer [5, 6].

2.6 DropOut Layers

When we have deep neural network, the biggest problem to it is overfitting. Deep neural networks contain many hidden layers so we need to train many weights. So we need to avoid overfitting at any cost. To avoid overfitting we can use L1 and L2 regularization. Dropout is the technique to avoid overfitting in deep neural networks. As in case of random forest, we used randomization for regularization. This is the core idea in random forest to avoid overfitting? Can we take the same concept in case of deep neural networks to avoid overfitting. At any point whenever we do forward propagation and backward propagation. In any iteration in any given layer randomly make subset of given neurons from all neurons and remove connections of remaining neurons. So we randomly remove neurons from layer so all incoming and outgoing connections to that neuron get disconnected. At every iteration different set of neurons are dropped out. So in case of dropouts in layer subset of activation units or neurons becomes deactivated in that layer. If we know that there are high chances of overfitting, larger number of weights as compare to data points then we should use keep probability value small [7].

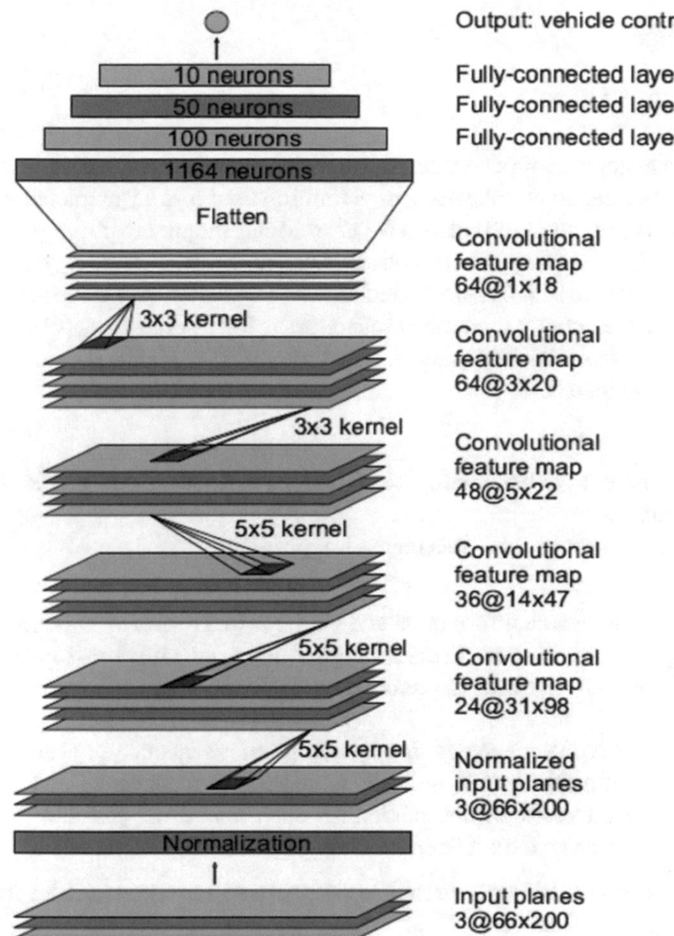

Fig. 2 End-to-end model

3 Dataset

We get the data collected by Sully, what he did was pretty interesting he added a front dash camera in front of his car and he had his OBD port in car that captured steering angle while driving the car. So front dash camera became the input and he drove his personal car and output the steering angle.

It's a 25 min dataset. Now we converted 25 min video into frames using Opencv. For every image, there is corresponding steering angel label associated with it. We have used 70% of total dataset as training dataset and 30% of dataset as testing dataset (Fig. 2).

3.1 Model

3.1.1 Baseline Model

We plotted a histogram of occurrence of steering angle than we find that zero radian is most found value. So we take mean of yi training used it as a y predicted for testing data. As this is a regression problem we used Mean square error as a performance metric. So we find mean square error on test data using mean_training_y as predicted y for test data was 0.191127. As zero radian was most often find value then we used zero radian as predicted y than mean square error for test data was 0.190891. This model we used it as a baseline model, now any model we build should have mean square error less than 0.19.

END- TO- END MODEL

Basically, in dataset we have sequence of images and we have to predict sequence of steering angles.

We have two ways to solve this regression problem

Case 1:
We just discard sequence information and we pose this problem as given an image we have to predict steering angle discarding sequence information. So we can build convolutional neural networks to predict steering for an image.
Case2:
We want to consider sequence of images and predict sequence of steering angles. There are class of models CNN + RNN, we take individual images we pass them through CNN get a vector representation for each image and pass this vector as a input to LSTM and as a output from LSTM we get steering angle.

Due to less computation resources, we simply used end-to-end model and results we get were phenomenon.

In non end to end model we need to create different models to detect road, lanes, other vehicles, pedestrians, so for each of these tasks we need an individual model. So we will build end to end model to solve this regression problem. Our model consists of five convolutional layers, three fully connected layers, one normalization layer. We take input as a tensor of depth 3, each value in tensor is RGB value, tensor has dimensions of 3@66×200. In first layer we applied normalization operation over tensor, now each value of tensor will lie between 0 and 1. After normalization layer we have five convolutional layers. In first convolutional layer we performed convolutional operation with 3@5×5 kernel, stride is equal to 1, padding is valid after first convolutional layer now our tensor shape becomes 24@31×98. After five convolutional layers shape of tensor becomes 64@1×18. After fifth convolutional layer we performed flattening operation and we obtain 1164 neurons, after flattening layer model has three fully connected layers. In all the three fully connected layers we have used dropouts with keep probability equal to 0.8. First, fully connected layer has 100 neurons, second, FC layer has 50 neurons, third, FC layer has 10 neurons. In last layer, we applied atan (tan inverse) function to get steering angle.

3.1.2 Testing

For checking the accuracy of model we have used mean square error as a metric. We had divided our dataset into two parts 70% of dataset is used for training our model and rest 30% is used for testing purpose.

$$MSE = \sum(|predicted\ value - actual\ value|)^2$$

3.1.3 Scoring

We have calculated loss using mean square error, lesser the mean square error better the performance of model. In baseline model mean square error of model was 0.1908 and accuracy is approximately 81%. In end to end model mean square error is 0.13 and accuracy is approximately 87%.

4 Conclusion and Future Scope

We will run RNN on the output given by CNN, as it will increase its accuracy to considerable amount.

References

1. https://arxiv.org/pdf/1609.04112.pdf
2. https://en.wikipedia.org/wiki/Kernel_(image_processing)
3. http://yann.lecun.com/exdb/publis/pdf/lecun-01a.pdf
4. https://peterroelants.github.io/posts/neural_network_implementation_intermezzo02/
5. https://cs.nyu.edu/~fergus/tutorials/deep_learning_cvpr12/
6. http://andrew.gibiansky.com/blog/machine-learning/convolutional-neural-networks/
7. http://ais.uni-bonn.de/papers/icann2010_maxpool.pdf
8. https://rdipietro.github.io/friendly-intro-to-cross-entropy-loss/

A Local-to-Global Approach for Document Image Binarization

Deepika Gupta and Soumen Bag

Abstract Document image binarization is a preprocessing step in Optical Character Recognition (OCR). This paper presents a novel hybrid approach for degraded document image binarization. An adaptive or local approach is applied to enhance the degraded image that separates the foreground and background considerably. After that, a new global approach for binarization is proposed. A postprocessing technique is applied after to get the final output. This transition from local-to-global approach results in document image binarization. Exhaustive experimental analysis has been carried out on a series of benchmark dataset having variations of degraded document images. Comparative analysis of the proposed method with other existing methods promises the effectiveness of the approach.

Keywords Binarization · DIBCO · Global thresholding · Hybrid algorithm · Local thresholding

1 Introduction

Document image binarization is performed as a preprocessing step in various document analysis algorithms including OCR. Main objective of binarization is to separate the foreground from the background. An accurate and efficient binarization algorithm is important as quality of binarized output affects the accuracy of OCR. Binarization has been remained an interesting area for researchers but the problem is still open for research. Various algorithms have been proposed till now. However, binarization is still a challenging problem due to uneven illumination, variable contrast, complex noisy background. Various document degradations such as shadow, smudge, smear,

D. Gupta (✉) · S. Bag
Department of Computer Science and Engineering, Indian Institute of Technology (ISM) Dhanbad, Dhanbad 826004, India
e-mail: deepika.guptaa19@gmail.com

S. Bag
e-mail: bagsoumen@gmail.com

A. K. Das et al. (eds.), *Computational Intelligence in Pattern Recognition*, Advances in Intelligent Systems and Computing 999,
https://doi.org/10.1007/978-981-13-9042-5_60

and ink bleed through make it even more difficult to design an efficient binarization approach.

The simplest approach for binarization is through the thresholding which acts like a filter. Intensity values above/below the threshold are categorized as foreground/background. Thresholding algorithm can be categorized the main into two groups: *global* and *local*. Global thresholding methods [11] use a single threshold value for whole image to separate foreground from background. A comparative study of global thresholding algorithms is given in [6]. Global thresholding algorithms are fast but are sensitive to the illumination in images. It works well for good-quality images but fails to segment foreground from background in badly illuminated images. To overcome the problem of variations in illumination, local or adaptive thresholding algorithms are proposed. The main difference between the two approaches is that in local method different thresholding values are computed for each pixel on the basis of the local neighborhood of the pixel. This approach is more robust to the illumination variation. A good deal of work based on adaptive thresholding is described in literature [1, 9, 18]. These algorithms work good for changing backgrounds but are computationally expensive. Also, determination of the appropriate size of the neighborhood window is a limitation.

In recent times, a combination of the above two thresholding algorithms hybrid algorithms has been proposed. These algorithms use both the approaches and benefited with advantages of both the methods and overcome the limitations of both the methods. Some of the hybrid approaches are reported in publications. Some of the research on combination of above two methods is reported in [2, 7]. In [7], hybrid binarization algorithm is proposed. This method uses Otsu's global thresholding method [11] and Sauvola's [18] adaptive thresholding method.

In this paper, we propose a novel hybrid binarization approach for degraded document images with complex background. The basic idea is divided into three parts. At first, we use the local approach for estimating the background of the image, and then we eliminate this background from the document image. This enhances the document image and separates the background from the text considerably. After that, we apply a new histogram-based global algorithm. At last, we apply a postprocessing scheme to remove small noisy regions which results in final binarized output. This progression from local preprocessing to global thresholding method results in binarization of document images.

Rest of the paper is organized as follows: Sect. 2 describes the proposed method. Experimental analysis and analysis of results have been discussed in Sect. 3. Concluding remarks have been given in Sect. 4.

2 Proposed Method

The proposed method takes the advantages of both adaptive and global methods. It consists of three phases: Preprocessing, global thresholding, and postprocessing.

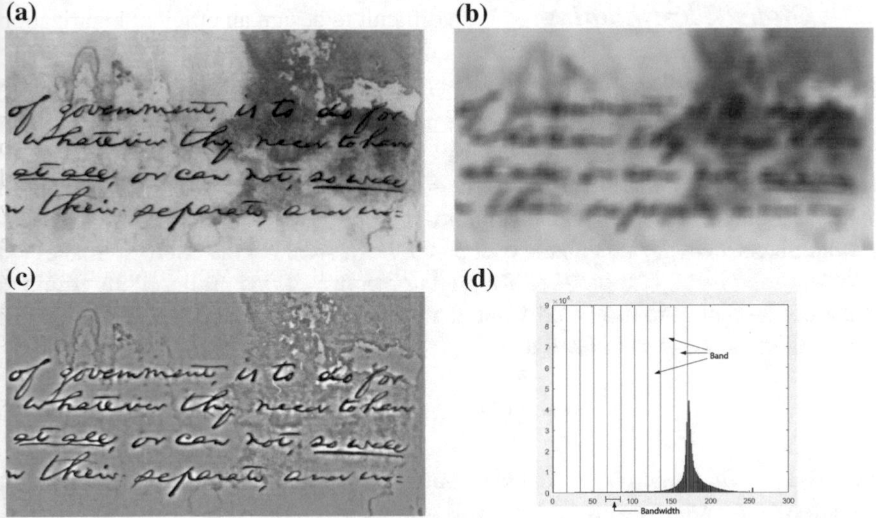

Fig. 1 Output of preprocessing on sample image of DIBCO-2009. **a** Input grayscale image; **b** adapted background of input image; **c** contrast image; and **d** histogram of the contrast image

2.1 Preprocessing

The first preprocessing step we propose is to eliminate as much degradation as possible that creates most difficulty for global thresholding. In this step, we locally calculate the background of the gray image I using a large-sized averaging filter. The basic assumption is that the text areas are less in comparison to background. After applying a larger sized average filter, output image I_{BG} (Fig. 1b) resembles a smooth adaptation of background. Since threshold value calculation does not depend on the neighborhood, selection of the optimum window size is not a limitation here as in case for adaptive thresholding algorithm. We just need to get the estimation of background so any large window size is enough to cover the stroke width of character. We have used a window size of [31 × 31].

Once the background I_{BG} is calculated, image contrast $\hat{I}$ is measured using the method described in [8] as

$$\hat{I} = \gamma \times \left(\frac{I}{I_{BG}} \right) \tag{1}$$

where γ is the constant and value of γ is set to the mean intensity of the image. After this step, background is separated from the text as shown in Fig. 1c.

2.2 Global Thresholding

Image pixels are classified into three main categories: background pixel, noise pixel, and text pixel. This proposed global algorithm works on the concept that major part of the image is background. After the background is removed from the image, foreground and backgrounds are separated considerably. The histogram of this enhanced image is plotted (Fig. 3d). In the histogram, most of the part is occupied by the background pixels and very less pixels occupied by text area. All the contrast images ($\hat{I}$) follow similar histogram pattern. The main objective of this proposed algorithm is to choose an optimum threshold value that removes background and noise. Global threshold value T_G is calculated as

$$T_G = \eta(\hat{I}) - (\lambda - 1) \times \sigma_G^2 \tag{2}$$

where η is the maximum occurring pixel value in image contrast $\hat{I}$, σ_G^2 is the standard deviation of $\hat{I}$, and λ is a constant whose value is given by

$$\lambda = \ell \times \frac{Band}{Bandwidth} \tag{3}$$

where $Band$ is the fixed number of intervals in which histogram $\hbar(\hat{I})$ for range $[0, \eta]$ is divided (on experimental basis, the value of $Band$ is set to 10) and $Bandwidth$ is the size of $Band$ and is given as

$$Bandwidth = \left(\frac{\eta}{Band}\right) \tag{4}$$

and ℓ is calculated as

$$\ell = \underset{i}{\operatorname{argmax}} \left| A_i - A_{i-1} \right| ; \quad 2 \le i \le 10 \tag{5}$$

in which A_i is the area of $\hbar$ in $Band$ B_i of $\hbar(\hat{I})$. The key idea behind the calculation of ℓ is that region occupied by background class is large in comparison to region occupied by text class.

With threshold value T_G calculated using Eq. 2, binary image B is obtained as

$$B(x, y) = \begin{cases} 1, & \text{if } \hat{I}(x, y) \le T_G \\ 0, & \text{otherwise} \end{cases} \tag{6}$$

where 1 represents foreground and 0 represents background.

Since noise pixels are mixed with both the background and the text pixels, the selected threshold value removes noise pixels along with the background. But when noise pixels are amalgamated too much with the text, it is more difficult to separate noise pixels from text. So, the binary output contains small regions of noise which are removed in postprocessing scheme.

2.3 *Postprocessing*

We have observed that since the distribution of noise and text is not considered in the images, some of noise is also classified as text in the resulting binarized result. This noise severely affects the binarized output. To reduce the noise, we apply a postprocessing method on binarized output which increases the efficiency of the proposed method. The following steps explain the postprocessing scheme.

1. Generate image I_d by applying dilation with structuring element of "disk" of radius 3 (value is chosen on the basis of experimental analysis) on binarized image B. This step ensures that the smaller size components which are part of text should not be removed (e.g., the dot of character "i" or "j" in English alphabet).
2. Detect all the connected components $C = \{c_1, c_2, \ldots c_n\}$ of the dilated image I_d.
3. Calculate the area $a_i \forall c_i \in C$ and then calculate average area $A_{avg} = \frac{1}{n}\sum_{i=1}^{n} a_i$.
4. The size of the noisy components is small in comparison to the text. Based on this, we eliminate the component c_i if $a_i < \frac{1}{3}A_{avg}$ in I_d.
5. Multiply B to I_d to get the final binary image.

3 Experimental Results and Analysis

3.1 *Experimental Dataset*

Datasets of DIBCO series (DIBCO-2009 [5], H-DIBCO-2010 [12], DIBCO-2011 [13], H-DIBCO-12 [14], DIBCO-13 [15], H-DIBCO-14 [10], H-DIBCO-16 [16], and DIBCO-2017 [17]) have been used to test the proposed method. These datasets are benchmark datasets which are used in document image binarization contest held in ICDAR 2009, ICFHR 2010, ICDAR 2011, ICFHR 2012, ICDAR 2013, ICFHR 2014, ICFHR 2016, and in ICDAR 2017, respectively. We have mentioned these datasets as DS-9, DS-10, DS-11, DS-12, DS-13, DS-14, DS-16, and DS-17, respectively.

3.2 *Binarization Results*

The implementation of the proposed method has been done on MATLAB. Method described above performs the binarization of the degraded documents, document with complex background, documents having faint characters, and documents with ink bleed problem. Binarization results on sample images from each dataset are shown in Figs. 2 and 3.

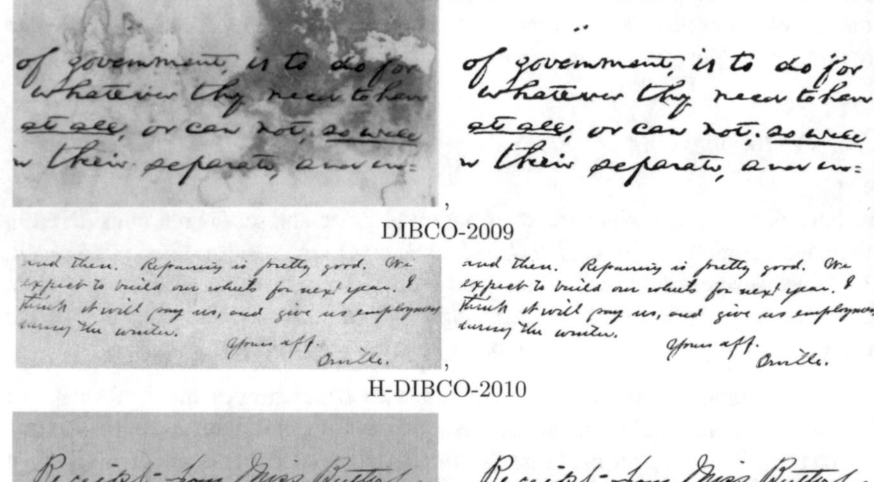

DIBCO-2009

H-DIBCO-2010

DIBCO-11

H-DIBCO-12

Fig. 2 Image results of the proposed method on a specimen image each from DIBCO datasets (2009–2012). Input images are shown in left and their corresponding binarized output is shown in right

3.3 Performance Evaluation

The performance of proposed method has been evaluated using different evaluation measures. These measures are F-measure and Peak Signal-to-Noise Ratio (PSNR), Negative Rate Metric (NRM), and Distance Reciprocal Distortion Metric (DRD) which are widely used as evaluation models in various binarization contests. F-measure gives the accuracy of algorithm. PSNR measures how close is the binarized output to the ground truth image. NRM calculates the mismatches between the ground truth and the binarized output, and DRD gives the visual distortion in binary image. Performance of the proposed method is evaluated on the basis of the above four parameters and results are given in Table 1.

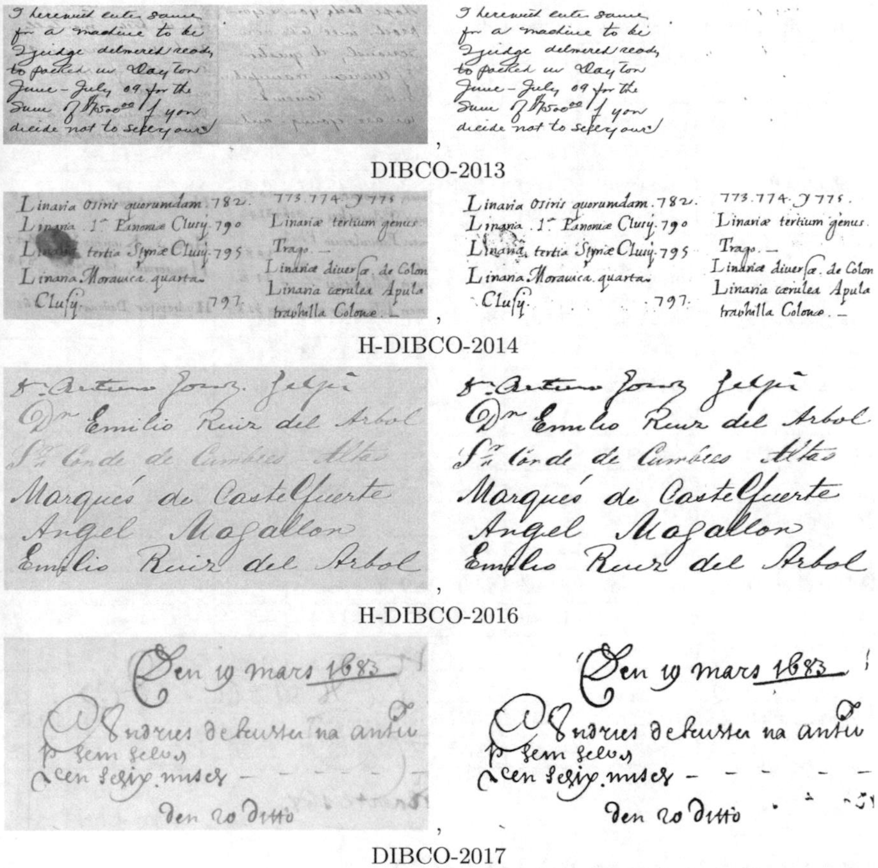

DIBCO-2013

H-DIBCO-2014

H-DIBCO-2016

DIBCO-2017

Fig. 3 Image results of the proposed method on a specimen image each from DIBCO datasets (2013–2017). Input images are shown in left and their corresponding binarized output is shown in right

Table 1 Evaluation measurements for proposed method on DIBCO series dataset

Parameters	DS-9	DS-10	DS-11	DS-12	DS-13	DS-14	DS-16	DS-17
F-Measure	87.37	87.82	88.22	88.64	88.65	91.46	89.59	88.02
PSNR	17.51	18.16	18.42	18.30	17.75	17.73	18.30	16.92
NRM	0.072	0.068	0.070	0.058	0.067	0.041	0.039	0.058
DRD	4.67	3.41	5.62	4.44	5.97	4.72	4.36	6.17

Table 2 Evaluation measurements among different binarization methods on DIBCO series datase

Method	Parameters	DS-9	DS-10	DS-11	DS-12	DS-13	DS-14	DS-16	DS-17
Otsu's [11]	F-Measure	76.82	77.96	68.87	77.24	72.70	80.94	74.16	75.03
	PSNR	13.98	15.47	12.24	15.14	13.94	13.92	13.84	13.01
Niblack's [9]	F-Measure	78.60	85.43	82.10	75.07	80.03	88.62	86.58	77.73
	PSNR	15.30	17.51	15.71	15.03	16.62	16.71	17.70	13.84
Sauvola's [18]	F-Measure	83.87	79.44	85.58	81.92	83.78	86.48	87.77	81.75
	PSNR	16.05	16.32	16.61	17.13	17.75	17.32	17.94	15.58
Liang et al. [7]	F-Measure	**88.35**	86.08	86.47	87.72	84.67	–	–	–
	PSNR	17.47	17.83	16.98	18.23	18.05	–	–	–
Biswas et al. [2]	F-Measure	83.66	–	–	–	–	–	–	–
	PSNR	15.60	–	–	–	–	–	–	–
Chen et al. [4]	F-Measure	–	–	81.72	73.09	81.57	–	–	–
	PSNR	–	–	15.92	16.05	16.83	–	–	–
Boudraa et al. [3]	F-Measure	–	–	–	–	87.23	**92.40**	85.08	–
	PSNR	–	–	–	–	15.87	17.59	17.47	–
Sehad et al. [19]	F-Measure	85.53	82.07	80.78	84.35	–	–	–	–
	PSNR	16.40	15.59	17.06	17.04	–	–	–	–
Garret et al. [20]	F-Measure	–	–	88.25	–	–	–	87.26	–
	PSNR	–	–	17.84	–	–	–	17.53	–
Proposed method	F-Measure	87.37	**87.82**	**88.22**	**88.64**	**88.65**	91.46	**89.59**	**88.02**
	PSNR	**17.51**	**18.16**	**18.42**	**18.30**	**17.75**	**17.73**	**18.30**	**16.92**

3.4 Comparison with Other Method

We compared our proposed method with other well-known methods, namely, Otsu's [11], Niblack's [9], Sauvola's [18], Liang et al. [7], Biswas et al. [2], Chen et al. [4], Boudraa et al. [3], Sehad et al. [19], and Garret et al. [20]. The quantitative evaluation results for above methods are given in Table 2. The comparison is carried out on the basis of two most common evaluation measures (F-Measure and PSNR) used by all the compared methods. The proposed method achieves better accuracy with the abovementioned methods except for few datasets which may be observed from the table.

4 Conclusion and Future Work

In this work, a hybrid binarization method is proposed for degraded document images with complex background. The proposed method first estimates the background using adaptive method. The background is then eliminated resulting in a less noisy image.

Then, a novel global thresholding approach is applied to the resultant image resulting in the binarized output. The adaptive method for estimating the background handles different types of noises and makes global binarization algorithm to work efficiently. The proposed method has been tested on DIBCO series and shows comparable result with other methods.

References

1. Bag, S., Bhowmick, P.: Adaptive-interpolative binarization with stroke preservation for restoration of faint characters in degraded documents. J. Visual Commun. Image Represent. **31**, 266–281 (2015)
2. Biswas, B., Bhattacharya, U., Chaudhuri, B.B.: A global-to-local approach to binarization of degraded document images. In: International Conference on Pattern Recognition, pp. 3008–3013 (2014)
3. Boudraa, O., Hidouci, W.K., Michelucci, D.: A robust multi stage technique for image binarization of degraded historical documents. In: International Conference on Electrical Engineering-Boumerdes, pp. 1–6 (2017)
4. Chen, X., Lin, L., Gao, Y.: Parallel nonparametric binarization for degraded document images. Neurocomputing **189**, 43–52 (2016)
5. Gatos, B., Ntirogiannis, K., Pratikakis, I.: ICDAR 2009 document image binarization contest (DIBCO 2009). In: International Conference on Document Analysis and Recognition, pp. 1375–1382 (2009)
6. Lee, S.U., Chung, S.Y., Park, R.H.: A comparative performance study of several global thresholding techniques for segmentation. Comput. Vis. Graph. Image Process. **52**(2), 171–190 (1990)
7. Liang, Y., Lin, Z., Sun, L., Cao, J.: Document image binarization via optimized hybrid thresholding. In: International Symposium on Circuits and Systems, pp. 1–4 (2017)
8. Lu, S., Su, B., Tan, C.L.: Document image binarization using background estimation and stroke edges. Int. J. Doc. Anal. Recognit. **13**(4), 303–314 (2010)
9. Niblack, W.: An Introduction to Digital Image Processing, vol. 34 (1986)
10. Ntirogiannis, K., Gatos, B., Pratikakis, I.: ICFHR2014 competition on handwritten document image binarization (H-DIBCO 2014). In: International Conference on Frontiers in Handwriting Recognition, pp. 809–813 (2014)
11. Otsu, N.: A threshold selection method from gray-level histograms. IEEE Trans. Syst. Man Cybern. **9**(1), 62–66 (1979)
12. Pratikakis, I., Gatos, B., Ntirogiannis, K.: H-DIBCO 2010-handwritten document image binarization competition. In: International Conference on Frontiers in Handwriting Recognition, pp. 727–732 (2010)
13. Pratikakis, I., Gatos, B., Ntirogiannis, K.: ICDAR 2011 document image binarization contest (DIBCO 2011). In: 2011 International Conference on Document Analysis and Recognition, pp. 1506–1510 (2011)
14. Pratikakis, I., Gatos, B., Ntirogiannis, K.: ICFHR 2012 competition on handwritten document image binarization (H-DIBCO 2012). In: International Conference on Frontiers in Handwriting Recognition, pp. 817–822 (2012)
15. Pratikakis, I., Gatos, B., Ntirogiannis, K.: ICDAR 2013 document image binarization contest (DIBCO 2013). In: International Conference on Document Analysis and Recognition, pp. 1471–1476 (2013)
16. Pratikakis, I., Zagoris, K., Barlas, G., Gatos, B.: ICFHR2016 handwritten document image binarization contest (H-DIBCO 2016). In: International Conference on Frontiers in Handwriting Recognition, pp. 619–623 (2016)

17. Pratikakis, I., Zagoris, K., Barlas, G., Gatos, B.: ICDAR2017 competition on document image binarization (DIBCO 2017). In: International Conference on Document Analysis and Recognition, pp. 1395–1403 (2017)
18. Sauvola, J., Pietikäinen, M.: Adaptive document image binarization. Pattern Recognit. **33**(2), 225–236 (2000)
19. Sehad, A., Chibani, Y., Hedjam, R., Cheriet, M.: LBP-based degraded document image binarization. In: International Conference on Image Processing Theory, Tools and Applications, pp. 213–217 (2015)
20. Vo, G.D., Park, C.: Robust regression for image binarization under heavy noise and nonuniform background. Pattern Recognit. **81**, 224–239 (2018)

Feature Extraction and Matching of River Dam Images in Odisha Using a Novel Feature Detector

Anchal Kumawat and Sucheta Panda

Abstract In the field of image processing, a piece of information can be defined as a feature, which can be utilised to determine the computational task with references to some applications. Features may refer to specific structures in the images such as points, edges or objects. Feature detection and feature description are the initial steps of image registration process. There are many algorithms of feature detection like BRISK, FAST, SURF, etc. In our earlier paper [15], a hybrid algorithm for feature detection and feature description has been proposed. The next step of image registration is feature matching. This paper deals with feature matching algorithm using BRISK [4], FAST [3] and hybrid [15]. The proposed hybrid algorithm performs better results than the above two existing algorithms in terms of elapsed time, CPU time and performance measuring time. This can be clearly concluded that the feature matching result of hybrid in terms of tables and figures performs well. For testing the algorithm, river dam images of Odisha are chosen.

Keywords Feature matching · Hybrid feature detector · Image registration · Feature extraction · Performance measure

1 Introduction

Image registration plays an important role in the task of image analysis. From the combination of various data sources like in image fusion, image restoration and Change detection, the final information can be obtained [2]. Registration is very much essential in remote sensing applications. Applications have been proposed in [2]. Advantage of the image registration are in the field of various areas like Medical

A. Kumawat (✉) · S. Panda
Department of Computer Applications, Veer Surendra Sai University of Technology (VSSUT), Burla, Sambalpur 768018, Odisha, India
e-mail: anchal.kumawat@gmail.com

S. Panda
e-mail: suchetapanda_mca@vssut.ac.in

A. K. Das et al. (eds.), *Computational Intelligence in Pattern Recognition*, Advances in Intelligent Systems and Computing 999,
https://doi.org/10.1007/978-981-13-9042-5_61

Application, Cartography and Computer Vision [1, 2]. The data which can be applied for image registration can be gather from different acquisition resources like times view points or sensors [1, 2].

In the past few years, image acquisition devices have been developed rapidly and automatic image registration is a major active research area. A detailed description of image registration methods has been introduced by Brown in 1992 [1]. The registration method geometrically aligned two images, i.e. the input and targeted images. In that paper, according to their nature, approaches of image registration are classified. One is based on area and another is based on feature. The basic steps of image registration procedures contain four steps, i.e. feature detection, feature matching, mapping function designed and image transformation and resampling.

In image processing, feature extraction and matching starts from an initial set of measured data and builds derived values (features) intended to be informative and non-redundant, facilitating the subsequent learning and generalisation steps, and in some cases leading to better human interpretations.

In this paper, two basic steps of image registration have been discussed. They are feature extraction and feature matching. For these, two Odisha River Dam images have been considered for extraction and matching. Two original river dam images have been taken and some portion of these two images must be same; otherwise there will be no matching. Feature extraction and matching have been performed based on two famous existing feature detection algorithms, i.e. BRISK [4] and FAST [3]. BRISK is invariant to rotation and scaling, but it is computationally expensive and FAST takes fewer time to extract and match the interest points and feature points, but not suitable for scale-invariant. To solve the drawbacks of BRISK and FAST feature extraction and matching algorithm, hybrid feature extraction and matching algorithm have been proposed. The main advantage of hybrid algorithm is that it consumes minor time in comparison to BRISK and FAST, to extract and match feature key point. This algorithm has been applied to Odisha River Dam images. All the above three algorithms compared in terms of computational time which are presented in a tabular form.

Organisation of the present paper is as follows: Sect. 2 describes related work, Sect. 3 presents the formulation of feature extraction and matching algorithm, Sect. 4 deals with simulation and results. Section 5 provide conclusion and Sect. 6 offers future work.

2 Related Work

In many situations, the comparison has to be done among the images, taken at different times, by different sensors or from different viewpoints [2]. These images are needed to align with each other so that one can detect the difference. For this purpose, the image registration method is the best approach. Taking two different images, having some part in common, we can get a single registered image by extracting and matching the feature points. So in the process of image registration, feature extraction and image matching play a crucial role.

The concept of machine learning can be utilised to derive a feature detector, having less computational time [3]. They have proved that in comparison to Harris detector and SIFT detector, the proposed detector is having low computational cost. Also in addition to this, they have proposed a corner detector based on the criteria applied to three-dimensional scenes. Leutenegger et al. [4] proposed an efficient method for feature points detection, description and matching, i.e. BRISK. It achieves comparable quality of matching with less computational cost. The relevance of feature-based image registration method has been described. A summary of three robust feature detection and matching methods has been presented in [5]. They are Scale-Invariant Feature Transform (SIFT), Principal Component Analysis (PCA)—SIFT and Speed Up Robust Feature (SURF). SIFT finds its interest points using Difference of Gaussian (DoG). PCA-SIFT shows its advantages in rotation and illumination change, and it is faster than SIFT. SURF is fastest one with the good performance. Madhuri [6] has presented a survey on different image registration algorithms. This survey establishes the fact that for panoramic image generation and creation, image registration is very crucial. Another review on image registration methods can be found in [7]. They have devided the concept of image registration into two types: (i) Image-to-image registration and (ii) image-to-map registration. For real-time face tracking applications, performance analysis of various feature detectors and descriptors can be found in [8]. A review of different techniques in image registration can be found in [9].

Hassaballah et al. [10] proposed a mathematical formulation for describing as well as detecting features. I present a complete analysis of various existing description and detection methods. It also demonstrates some concepts of feature matching and makes a performance evaluation of feature matching algorithms. A novel feature matching method has been proposed [11]. Their method measures the dissimilarity of the features through the path based on eigenvector properties. Here, the dissimilarity in the signal frequency along the path between two features is measured and they are accumulated in a 2D dissimilarity space, which allows accurate corresponding features to be extracted based on the cumulative space using a voting strategy. A comparative study of various feature detection and matching algorithms has been defined in [12].

Salahat et al. [13] studied many algorithms of feature detection and description proposed in literature. An exhaustive comparison of SIFT, SURF, KAZE, AKAZE, ORB and BRISK feature detectors is carried out in [14]. It can be concluded from the above paper that the performance of SIFT and BRISK is better than all other algorithms, and it can be concluded that SIFT is the most accurate algorithm.

These are also used in feature matching applications. Some of the feature detectors are STAR, BRIEF, FAST, FREAL and ORB. Using various parameters like average number of detected key points, average detection time of key point, frame per second and number of matches, they have measured the speed of tracking and accuracy of the above feature detector.

3 Feature Extraction and Feature Matching

Feature extractions are mainly used to extract the feature from the given two images, i.e. reference and sensed image. It helps in recognising and classifying the images. It is used in various image processing applications like character recognition, letter identification, etc. Basically, it is used to complete the task quickly for image matching. It also differentiates between the images. To extract the features, grayscale image is given as input. For the fast computation, the grayscale image input is again converted in floating point representation, i.e. double. Finally, it produces the array of extracted key points as an output. The obtained key points contain various fields, like coordinates X and Y, orientation, i.e. true or false, scale vectors and Laplacian value.

Feature matching plays an important role in image registration for establishing the correspondence between two images of same fields with different acquisition resources. For matching the features, it consists of a set of key points which are associated with each image descriptor. Once the features are extracted from two given images, the very next step is to perform matching. The main goal is to find the better correspondence from reference image to sensed image with their interest points, by comparing each feature key point of first sensed image from another reference image and measure the distance between these two points of different images with using BRISK, FAST and hybrid descriptor. The following distance equation is used between two key points (x_1, y_1) and (x_2, y_2) of referenced and sensed image to perform matching and then show the matched points with their orientation 'true' and montage/blend mode which gives the facility to perform reference and sensed image next to each other on the same image.

$$distance = \sqrt{(x_2 - x_1)^2 + (y_2 - y_1)^2} \quad (1)$$

4 Simulation and Results

In this paper, our proposed algorithm for feature extraction and matching has been successfully tested with using MATLAB 2017a on Windows 8 Operating System. Here, two river dam images have been used for the simulation. First river dam is related to Hirakud Dam, which is the longest dam in the world. Another one refers to Chiplima Dam, which is known as natural fall for generating electricity. It is the second hydro-electric project of the longest dam, i.e. Hirakud Dam. On every image, BRISK, FAST and hybrid algorithms are used for feature detection, extraction and matching. Here, two original images of river dam have been taken for image registration steps where some portion is common with another river dam image to perform matching.

Figure 1a shows the figure of Hirakud Dam image, Fig. 1b shows the same dam image but taken from some other viewpoints and these two figures are having some part common. Figure 1c, d shows the corresponding grayscale images of Fig. 1a, b, respectively. Figure 1e shows the BRISK feature detector applied to Fig. 1c, f that shows the BRISK feature detector applied to Fig. 1d. Then, Fig. 1g, h shows the feature extraction applied to Fig. 1c, d, respectively. Finally, Fig. 1i shows the feature matching of the two above figures with BRISK.

Figure 2 shows the same sequence of images as above but here we have applied FAST feature detection algorithm. Figure 3 shows the above sequence with the proposed hybrid feature detector. Figure 4 shows the BRISK, FAST and hybrid feature detection algorithms applied to the original and rotated Chiplima Dam images with angle of rotation 20°. Figure 4e shows feature matching of the above two Chiplima Dam images using BRISK. Figure 4f, g shows feature matching of the above two Chiplima Dam images using FAST and hybrid feature detectors, respectively. From the above results, it is clear that in comparison to existing two feature detectors, the proposed feature extraction with matching of hybrid feature detector gives the better result in terms of elapsed time, computation time and performance measuring time. This can be easily seen in Tables 1, 2 and 3. Elapsed time is calculated with the help of tic and toc function of MATLAB where tic starts the timer and toc find out the elapsed time. Equation 2 is used to find out the elapsed time which is shown in Table 1. It can be shown from Table 1 that, in terms of elapsed time, the proposed hybrid feature detector takes less time, i.e. 4.23391 seconds for Hirakud Dam and 6.97726 seconds in case of Chiplima Dam image, which is less than BRISK and FAST feature matching time.

$$Elapsed\_time = [Finish\_time - Initial\_time] \tag{2}$$

CPU time which is shown in Table 2 is calculated with the help of Eq. 3.

$$CPU\_time = \left[\frac{busy\_time}{busy\_time + idle\_time}\right] \tag{3}$$

Performance Measuring Time (PMT) is the time to measure the performance of our algorithm with the help of timeit function of MATLAB that follows Eq. 4 and the results are shown in Table 3, where timeit takes two arguments. First argument ip refers to number of input or it may be empty argument list and another argument op introduces to the desired number of output. This proves that PMT produces the less time for the case of hybrid in feature matching as compared to BRISK and FAST detectors.

$$PMT = timeit(ip, op) \tag{4}$$

From Table 3, it can be concluded that in case of new proposed feature detector, PMT is low, i.e. 4.232 in case of Hirakud Dam image and 6.98 in case of Chiplima Dam image in comparison to BRISK and FAST feature detector.

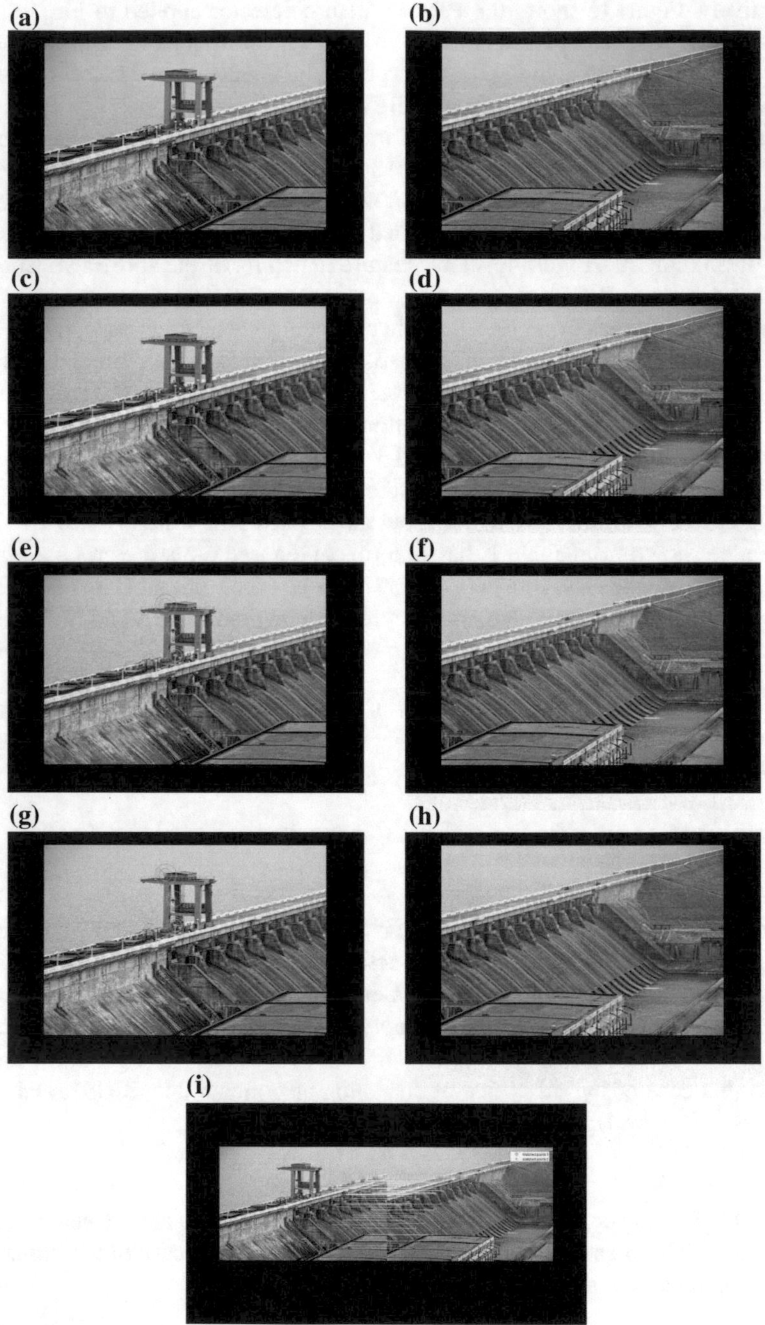

Fig. 1 **a** Hirakud Dam original one; **b** Hirakud Dam original two; **c** Hirakud Dam grayscale one; **d** Hirakud Dam grayscale two; **e** Hirakud Dam BRISK feature detection one; **f** Hirakud Dam BRISK feature detection two; **g** Hirakud Dam BRISK feature extraction one; **h** Hirakud Dam BRISK feature extraction two; **i** Hirakud Dam BRISK feature matching one and two

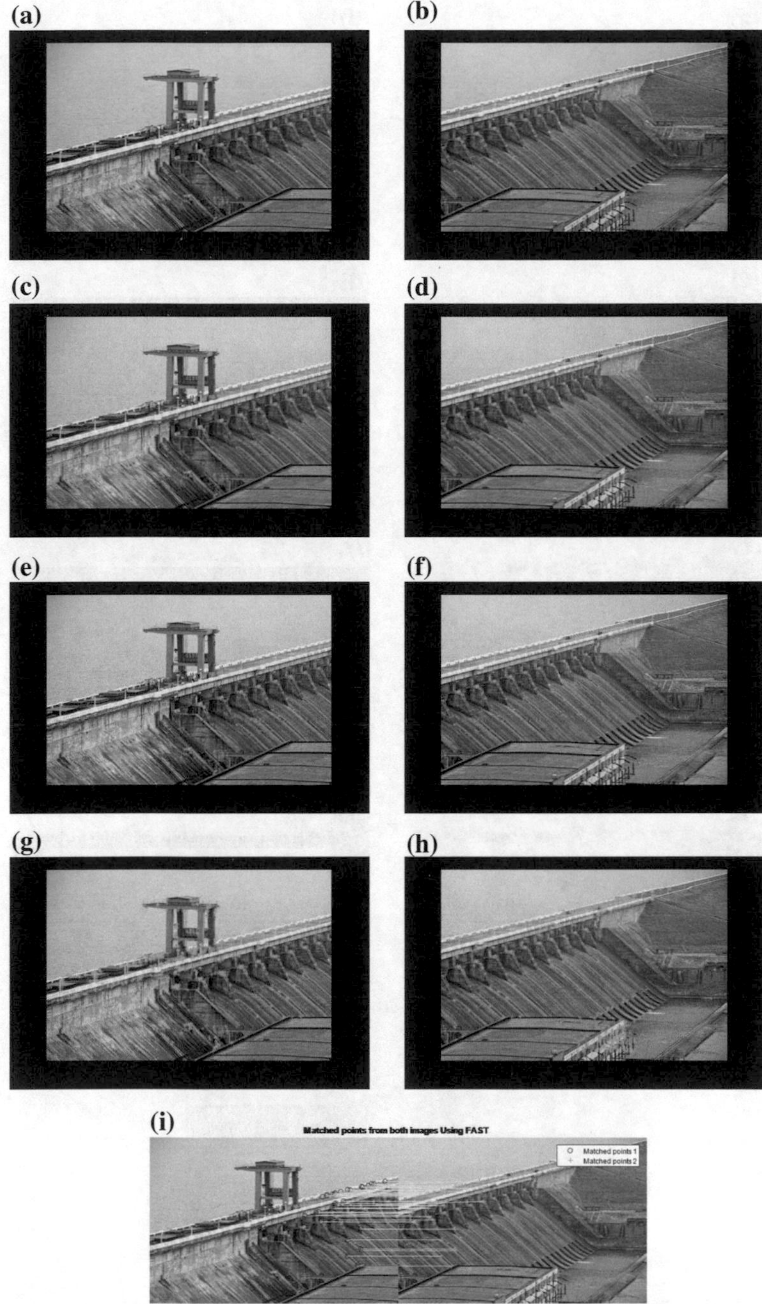

Fig. 2 **a** Hirakud Dam original one; **b** Hirakud Dam original two; **c** Hirakud Dam grayscale one; **d** Hirakud Dam grayscale two; **e** Hirakud Dam FAST feature detection one; **f** Hirakud Dam FAST feature detection two; **g** Hirakud Dam FAST feature extraction one; **h** Hirakud Dam FAST feature extraction two; **i** Hirakud Dam FAST feature matching one and two

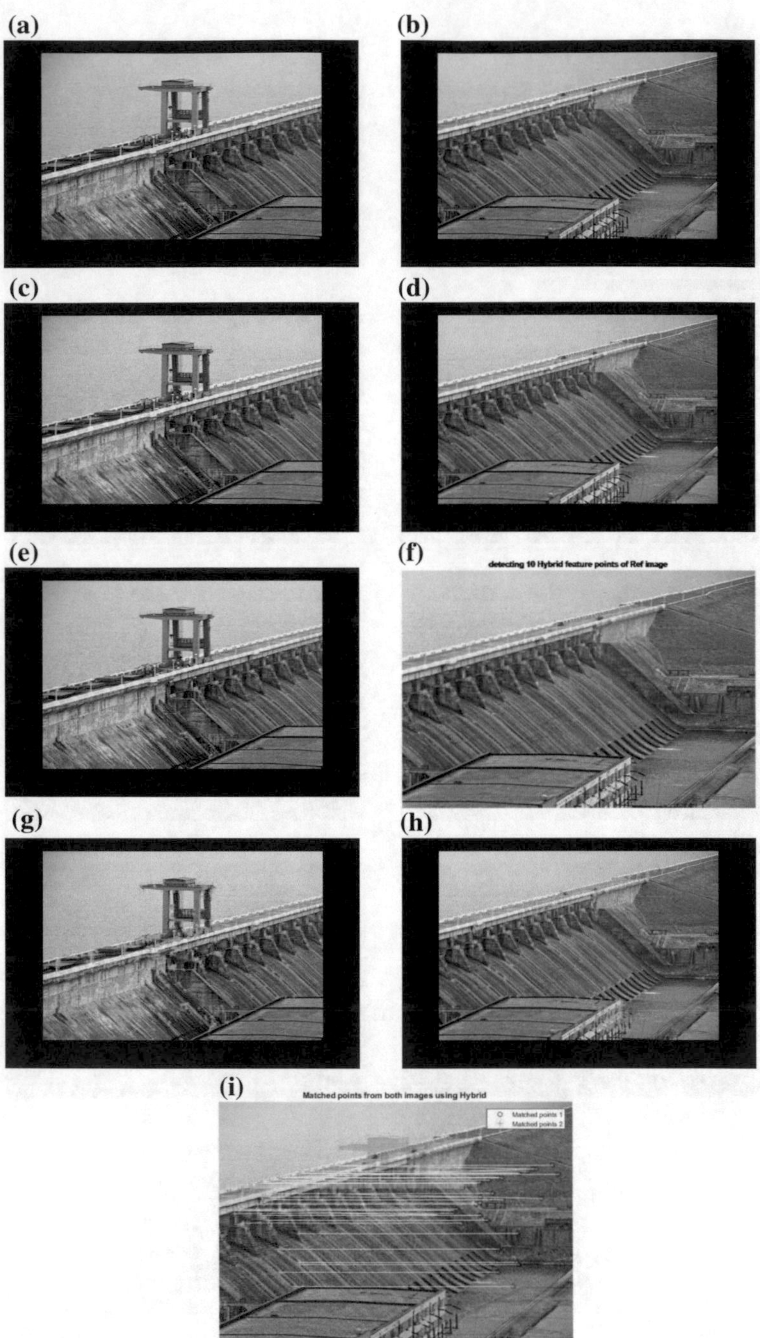

Fig. 3 **a** Hirakud Dam original one; **b** Hirakud Dam original two; **c** Hirakud Dam grayscale one; **d** Hirakud Dam grayscale two; **e** Hirakud Dam hybrid feature detection one; **f** Hirakud Dam hybrid feature detection two; **g** Hirakud Dam hybrid feature extraction one; **h** Hirakud Dam hybrid feature extraction two; **i** Hirakud Dam hybrid feature matching one and two

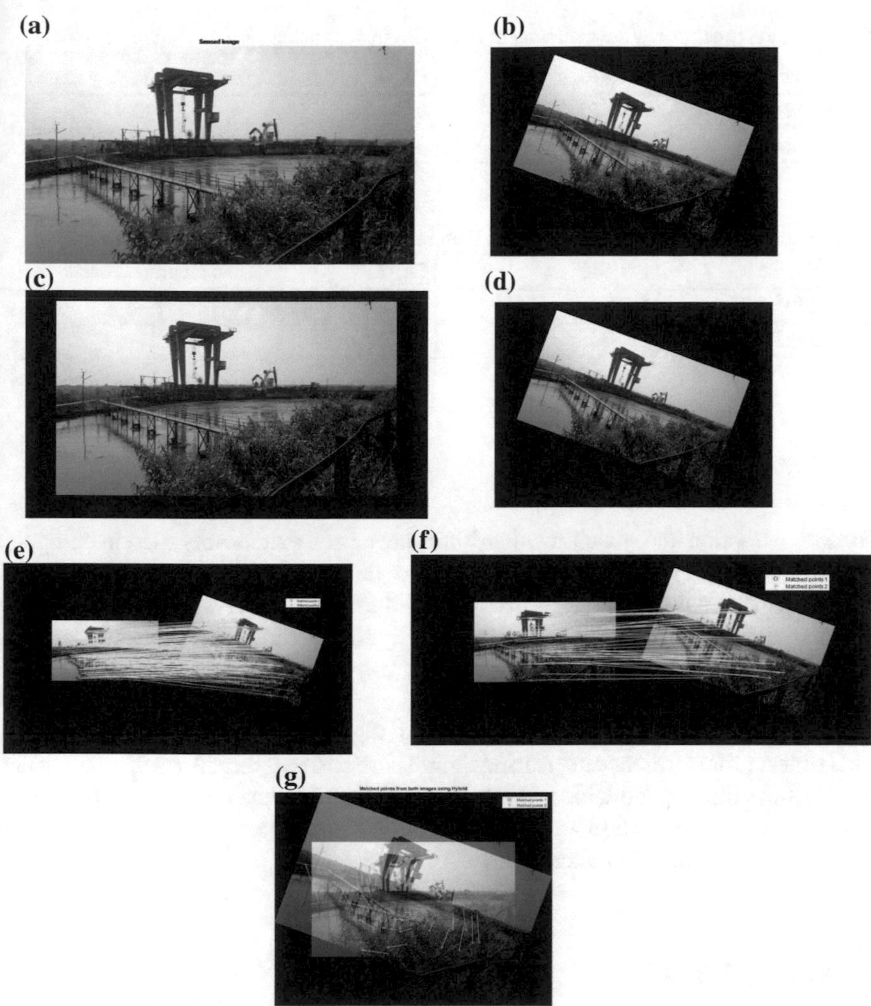

Fig. 4 **a** Chiplima Dam original one; **b** Chiplima Dam original two; **c** Chiplima Dam grayscale one; **d** Chiplima Dam grayscale two; **e** Chiplima Dam feature matching using BRISK of grayscale image of one and two; **f** Chiplima Dam feature matching using FAST of grayscale image of one and two; **g** Chiplima Dam feature matching using hybrid of grayscale image of one and two

Table 1 Elapsed time for feature matching of river dam images using various feature detectors

Figure name	BRISK	FAST	Hybrid
Hirakud Dam	10.9238	9.49545	4.23391
Chiplima Dam	9.57357	8.34241	6.97726

Table 2 CPU time for feature matching of river dam images using various feature detectors

Figure name	BRISK	FAST	Hybrid
Hirakud Dam	8.84375	8.28125	3.73438
Chiplima Dam	9.85938	8.4375	7.453

Table 3 PMT for feature matching of river dam images using various feature detectors

Figure name	BRISK	FAST	Hybrid
Hirakud Dam	10.906	9.488	4.232
Chiplima Dam	9.551	8.343	6.98

5 Conclusion

Feature extraction and matching algorithms have been extensively used in computer vision-based applications. And automated key points extraction and matching algorithms are nowadays used by the machines and robots for analysing and interpreting the surrounding world. In this paper, a novel hybrid feature extraction and matching algorithm has been proposed. This algorithm has been compared with two standard existing algorithms, i.e. BRISK and FAST. The proposed hybrid algorithm performs well in comparison to the above two existing algorithms in terms of elapsed time, CPU time and performance measuring time. These are represented in a tabular format. From all the figures and tables, it can be concluded that the proposed hybrid algorithm for feature extraction and matching outperforms the two existing algorithms of feature extraction and matching.

6 Future Work

For registering two different images, there are four steps, i.e. Feature detection and description [15], feature matching, outlier rejection, derivation of transformation function and image reconstruction and stitching [14]. Edges of an image represent the sharp transition in grey-level values. So, before feature detection and description step, edges can be detected from an image. Then, the output of edge detection step can be utilised for feature detection and description. Edge detected image can be used for feature detection, description and matching. This method can be explored in future work. Also, future work includes to carry out other remaining steps for image registration.

References

1. Brown Gottesfeld, L.: A survey of image registeration techniques. ACM Comput. Surv. **24**, 325–376 (1992)
2. Zitova, B., Flusser, J., Sroubek, F.: Image registration: a survey and recent advances. In; Institue of Information Theory and Automation Academy of Sciences of Czech Republic Pod Uarenskou vezi4, 18208 Prague and Czech Republic (ICIP Tutorial) (2005)
3. Rosten, E., Drummond, T.: Machine learning for high speed corner detection. In: European Conference on Computer Vision, vol. 9, no. 1, pp. 430–443 (2006)
4. Leutenegger, S., Chli, M., Siegwart, R.: BRISK: Binary Robust Invariant Scalable Keypoints. ETH Library (2011)
5. Shah, U., Mistry, D., Banerjee, A.: Image registration of multiview satellite images using best features points detection and matching methods from SURF, SIFT and PCA-SIFT. J. Emerg. Technol. Innov. Res. (JETIR) **1**(1) (2014)
6. Madhuri, S.: Classification of image registration technique and algorithm in digital image processing. Res. Surv. IJCTT **15**, 2 (2014)
7. Phogat, R.S., Damecha, H., Pandya, M., Choudhary, B., Potdar, M.: Different image registration methods. Overv. Int. J. Sci. Eng. Res. (IJSER) **5**(12), 44–49 (2014)
8. Patel, A., Kasat, D.R., Jain, S., Thakare, V.M.: Performance analysis of various feature detector and descriptor for real-time video based face tracking. Int. J. Comput. Appl. **93**(1), 0975–8887 (2014)
9. Parita, P., Vaghasiya, P.K.: Gautam. Image registration techniques: a review. Int. J. CS Eng. **2**(4), 10489–10492 (2015)
10. Hassaballah, M., Abdelmgeid, A.A., Hammam, A.: Image Features Detection, Description and Matching. Springer International Publishing Switzerland, Alshazly (2016)
11. Kahaki, S.M.M., Nordin, M.J., Ashtari, A.H., Zahra, S.J., Malo, J.: Invariant feature matching for image registration application based on new dissimilarity of spatial features (2016)
12. Babri, U.M., Tanvir, M., Khurshid, K.: Feature based correspondence: a comparative study on image matching algorithms. Int. J. Adv. Comput. Sci. Appl. (IJACSA) **7**(3) (2016)
13. Salahat, E., Qasaimeh, M.: Recent advances in features extraction an description algorithms: a comprehensive survey. In: Member, IEEE (2017)
14. Tareen, S.A.K., Saleem, Z.: A comparative analysis of SIFT, SURF, KAZE, AKAZE, ORB, and BRISK. In: International Conference on Computing, Mathematics and Engineering Technologies—(iCoMET). IEEE (2018)
15. Kumawat, A., Panda, S.: Feature detection and description in remote sensing images using a hybrid feature detector. In: International Conference on Computational Intelligence and Data Science (ICCIDS 2018), Procedia Computer Science, vol. 132, pp. 277–287 (2018)

Multimodal Segmentation of Brain Tumours in Volumetric MRI Scans of the Brain Using Time-Distributed U-Net

Jeet Dutta, Debajyoti Chakraborty and Debanjan Mondal

Abstract Brain tumour segmentation poses a challenging task even in the eyes of a trained medical practitioner. Traditional machine learning algorithms require hand-coding features from images before they can learn to identify the regions. Deep learning can solve the problem of detecting tumours with precision and even segment it. Neural networks can learn a hierarchical representation of features from the data by itself. We use a time-distributed architecture for U-Net based deep convolutional neural networks (TD-UNET). We tested our network against the MICCAI BRATS 2015 dataset that comprised 220 high-graded gliomas (HGG) and 54 low-graded gliomas (LGG) and yielded a test case accuracy of 58.3%.

Keywords Deep learning · Medical image segmentation · Neural networks · U-Net

1 Introduction

Cancer constitutes the phenomenon of abnormal division of cells leading to the formation of masses of tissues that affect the regulation of functions in the human body. These masses, called *tumours*, have the ability to travel to other parts of the body, called *metastasis*, latch themselves to vital organs and inhibit proper functions. This causes organ failure and ultimately, death of the patient. Some tumours may be benign, in which case they are mostly localized and poses no fatal threats to the person. In a study published in 2016 [1], 2% of malignant tumours are formed in the

J. Dutta (✉) · D. Chakraborty · D. Mondal
Department of Computer Science and Engineering, Netaji Subhash Engineering College, Garia, Kolkata 700152, India
e-mail: jtdutta1@gmail.com

D. Chakraborty
e-mail: debajyoti.chakraborty2u@gmail.com

D. Mondal
e-mail: debanjan.mondal8@gmail.com

A. K. Das et al. (eds.), *Computational Intelligence in Pattern Recognition*, Advances in Intelligent Systems and Computing 999,
https://doi.org/10.1007/978-981-13-9042-5_62

brain, mostly in the glial cells. The tumours are classified as *gliomas*. 59.5% of these gliomas are high graded, as in cancerous. Gliomas can be fatal even though they are benign, causing abnormalities in proper brain functions.

Detection of gliomas can be difficult even in the eyes of a trained medical practitioner. The blood–brain barriers (BBB) of the tumours are not easily distinguishable and may be mistaken for a normal lobe of the brain. Modern imaging techniques like MRIs can manipulate the images to enhance certain types of signals, highlighting the BBB of the tumours effectively.

In the past decade, *machine learning* and i*mage processing* have advanced exponentially leading to an explosion in the efficiency of computers classifying images, recognizing objects, speech and text recognition. But this efficiency stagnated with traditional algorithms. With the increase in computational power and data, deep learning fuelled the efficiency. Deep learning encompasses several layers of Artificial Neural Networks (ANNs). Convolutional Neural Networks (CNNs) [2–5] are a branch of ANNs that aid in learning and classifying unstructured data, like images and sequences.

In this paper, we introduce a novel architecture that can work with volumetric (3D) images and is computationally efficient in segmenting images. The paper is divided into various sections. Section 2 talks about some related works on which we built our model. Section 3 introduces the architecture of the model and talks about its inception. In Sect. 4, we highlight various problems related to our work and some of their solutions finally concluding this paper with Sect. 5. We also briefly explore some future scopes to improve upon this architecture.

2 Related Works

Deep learning has stripped off the necessity to hand-code features from a dataset. Neural networks have the ability to learn hierarchical representation of features [6, 7]. CNNs, developed by LeCun [2, 3], had the ability to classify a handwritten digit with a few layers of convolutions and subsampling (later renamed as max-pooling). The network, however, performed poorly when the availability of training data was lower than the number of hyperparameter, and the computational power to train them at the time was low. This architecture was later improved upon by Krizhevsky et al. [4] and used to classify thousands of images from the ILSVRC-2012 dataset. The AlexNet performed efficiently with a loss of 15.3%. VGG Nets [8] introduced a more general purpose learning hierarchy using multiple 3×3 kernels increasing exponentially on every convolutional layer. But deeper networks suffered from problems like exploding and vanishing gradients. ResNets [5] were able to eliminate this using residual learning.

Image segmentation involves identifying an object(s) in a single image and separating it from the rest of the image. HOG [9] and SIFT [10] were the state-of-the-art tools for detecting objects for a long time, but they involved hand coding a lot of the features. R-CNN [11] showed how CNNs can be used to effectively identify

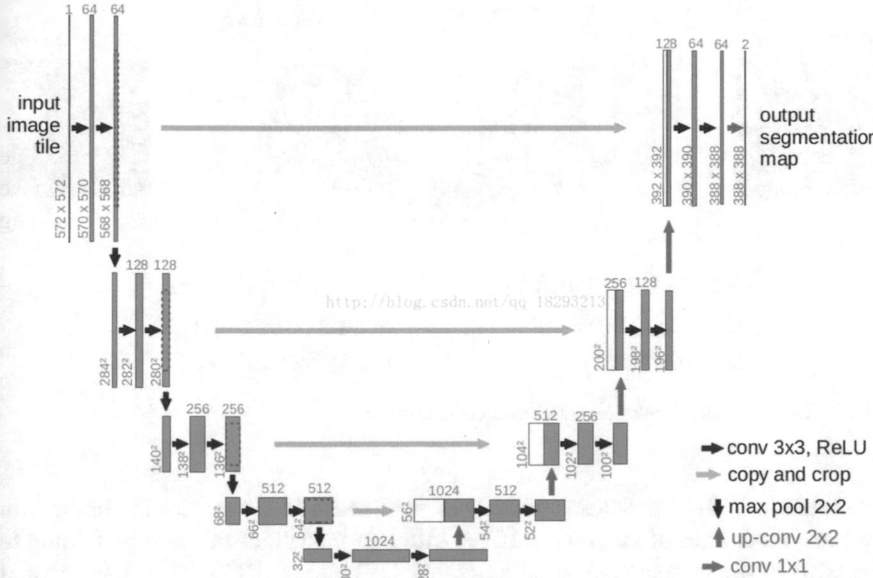

Fig. 1 U-Net architecture given by Ronneberger et al. [13]

objects in an image. It involved a selective search to find bounding boxes for possible objects, and then a CNN was applied on these segments of the image to classify them. You Only Look Once (YOLO) [12] is another object detection network that is much faster than the R-CNN with an increased precision in the detection of objects. Medical image segmentations, however, depend heavily on end-to-end learning.

U-Net [13] (see Fig. 1) is the de facto standard in medical image segmentation based on the end-to-end learning paradigm. It consists of an architecture similar to auto-encoders [7] but instead of fully connected layers, it incorporates convolutional and max-pooling layers [2, 3]. Utilizing fully convolutional layers improved the performance of the U-Net [14–16]. But these models utilized 2D images instead of 3D.

3 Proposed Method

In this section, we introduce our novel architecture based on the U-Net. We tried to recreate the works of Erden et al. [17] but due to low resources we were not able to use 3D convolutional layers. Instead, we used a novel approach to make use of 3D images by time distributing the slices. This reduced the amount of resource needed by the model and required a lower time period for training.

We use a U-Net, where every layer is time-distributed, for segmenting the tumours (see Fig. 2). We call this network the TD-UNet. We train our network on the BraTS

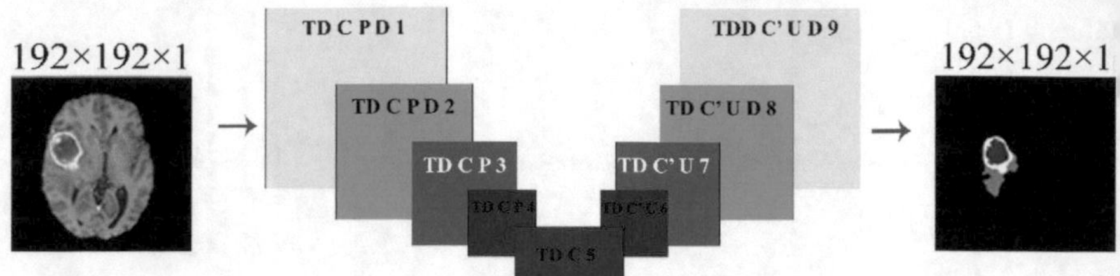

Fig. 2 The diagram provides a visualization of our model

2015 dataset [18]. The dataset consists of volumetric MRI scans of the brain with multiple modalities of the tumour (three different modalities of scans pertaining to the BBB of the tumour). We used a dropout [19] rate of 0.25 for the first two and the last two layers to prevent overfitting and trained the model for 13 h using a batch size of 1, with an Adam [20]-based gradient descent optimizer. The TD-UNet is well adapted to low-cost hardware but is difficult to train using standard metrics.

We considered many other approaches for our network, like using a recurrent ConvNet [21]. We also considered pixel-wise classification [22], an application of Multiple Instance Learning (MIL) and considered the possibility of using multidimensional LSTMs [23, 24], but we could not, due to limited resources.

We used max-pooling [2, 3] layers instead of fully convolutional layers to prevent increasing the amount of hyperparameters (due to limited resources). Table 1 provides the details of each layer of the network. Although convolutions and max-pooling layers are different, we consider them grouped as one layer for easier analysis. We even consider dropout as a different layer in addition to the convolution and the max-pooling layer.

4 Result and Discussions

The model was the culmination of intensive experimentations and analysis of various tasks. We divide them into various subsections, highlighting problems and their respective solutions that occurred throughout the workflow. The first subsection records the details of the dataset we used, its shortcomings and how we overcame them. The next subsection introduces the inception of the evaluation metric against which we optimized our network. The final subsection highlights the issues we faced during the training stage of the pipeline.

Table 1 Detailed descriptions of the layers with their hyperparameters and the output shape

	Layer name	Output shape	Parameters
1	Convolution	(None, 60, 192, 192, 32)	320
	Max-pooling	(None, 60, 96, 96, 32)	0
	Dropout	(None, 60, 96, 96, 32)	0
2	Convolution	(None, 60, 96, 96, 64)	18,496
	Max-pooling	(None, 60, 48, 48, 64)	0
	Dropout	(None, 60, 48, 48, 64)	0
3	Convolution	(None, 60, 48, 48, 128)	73,856
	Max-pooling	(None, 60, 24, 24, 128)	0
4	Convolution	(None, 60, 24, 24, 256)	295,168
	Max-pooling	(None, 60, 12, 12, 256)	0
5	Convolution	(None, 60, 12, 12, 512)	1,180,160
6	Deconvolution	(None, 60, 12, 12, 256)	1,179,904
	Up sampling	(None, 60, 24, 24, 256)	0
7	Deconvolution	(None, 60, 24, 24, 128)	295,040
	Up sampling	(None, 60, 48, 48, 128)	0
8	Deconvolution	(None, 60, 48, 48, 64)	73,792
	Up sampling	(None, 60, 96, 96, 64)	0
	Dropout	(None, 60, 96, 96, 64)	0
9	Deconvolution	(None, 60, 96, 96, 32)	18,464
	Up sampling	(None, 60, 192, 192, 32)	0
	Dropout	(None, 60, 192, 192, 32)	0
10	Deconvolution	(None, 60, 192, 192, 1)	289
Total number of parameters			3,135,489

4.1 Dataset

We used the MICCAI BraTS 2015 [18] dataset for this project. It utilizes multi-institutional preoperative MRI scans and focuses on the segmentation of intrinsically heterogeneous (appearance, shape and histology) brain tumours. Furthermore, it also focuses on the prediction of patient's overall survival, helping in determining false results. The dataset consists of 220 HGG and 54 LGG volumetric brain scans along with segmented and labelled output with the different modalities. Four types of MRI scans (see Fig. 3) were used, each bringing a unique variation to the resulting image.

- *Flair*. Fluid-Attenuated Inversion Recovery. It suppresses fluid imagery by setting an inversion time. This helps in suppressing cerebrospinal fluid and exposes hyper lesions.
- *T1 weighted*. It can measure the spin–lattice relaxation using short echo time (TE) and short relaxation time (TR). It is best for lowering signal from high watery places and boosts signals from fatty regions, boosting quality of tumours in the scan.
- *T2 weighted*. It can measure spin–spin relaxation using long Relaxation Time (TR) and long Echo Time (TE). This inverses what T1 can observe. So, we get more signal for watery areas and lower signal for fatty regions.
- *T1 weighted with Contrast Enhancements (T1c)*. T1-weighted MRI scan is subjected to contrast agent. This enhances signals from tumours substantially.

We used the T1c MRI scans as it showed the BBB of the tumours distinctly. The data is already skull-stripped and stored in the raw format. There are certain

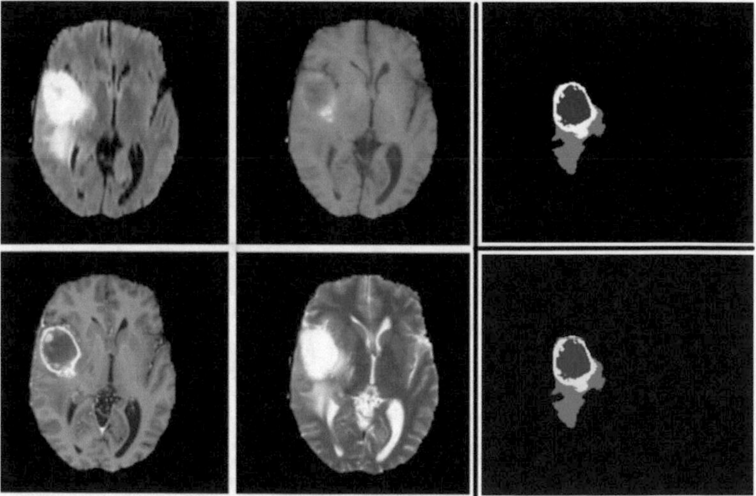

Fig. 3 (Left) From top left to bottom left clockwise, Flair, T1 weighted, T2 weighted and T1 weighted with contrast enhancements. (Right) This is the ground truth for the given set of scans

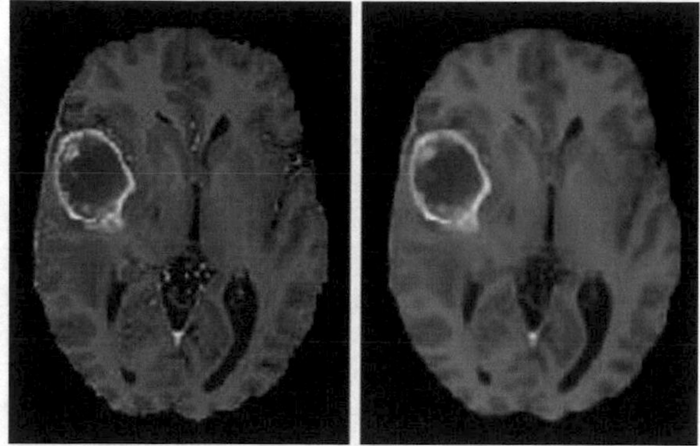

Fig. 4 The "T1c" MR image (left) and after preprocessing (right) using the curvature flow function, resulting in efficient image denoising as well as edge preservation

abnormalities that arose while reading the dataset (like high irregular pixel value variation) which had to be fixed manually and prepared for training the network. We used a curvature flow filter (see Fig. 4) to remove noise.

The ground truth of the scans consists of three different regions labelled as different modalities:

- *Necrosis*. It is a region at the core of the tumours consisting of dead tissue.
- *Advancing Tumour*. As the name suggests, this is the region where the tumour cells actively divide.
- *Edema*. This region mostly consists of swollen tissue filled with fluids.

4.2 Performance Evaluation Metric

We used the Sorensen–Dice index (in Eq. 1) performance evaluation metric, more commonly known as dice score metric. It calculates the ratio of number of true positives to the sum of the number of actual positives and the number of false positives.

$$\text{Dice Score} = 2\frac{|x \cap y|}{|x| + |y|} \tag{1}$$

where x and y represent the variables whose similarity is to be measured. A modified function is used for differentiability (in Eq. 2).

$$S = 2\frac{|GT \cdot FP|}{|GT|^2 + |FP|^2} \tag{2}$$

where GT = Ground Truth and FP = False Positive.

We now calculate this score for each time slice of the predicted data and the ground truth and sum all the values to get the global evaluation. The loss of the network can be calculated by subtracting this metric from 1.

4.3 Experimentations

The current model was the result of vigorous experimentations with almost all the factors affecting the end results. This begins from augmenting the data and works its way to tuning hyperparameters in the network.

The training time was irregular and long for non-preprocessed data. This is because of many irregularities present in the data itself. The optimizer has a higher probability of encountering plateaus and local minima before reaching the global minima. This is removed by normalizing the image, running smoothening functions (like Gaussian filters). We chose the curvature flow filter as it does a better job at preserving edges and does not blur out much of the features (see Fig. 4).

The dimensions of the image were reduced to $60 \times 192 \times 192$ from $155 \times 240 \times 240$ to remove excess portion of unwanted background data. The initial network was mostly of the same architecture, except that we have reduced the number of dropout layers to 4 instead of 10. The initial network failed to segment the tumours properly, due to the excess number of dropout layers. Also, the absence of the dropout layers made the network converge faster during training, but failed to segment the test cases correctly in a condition called overfitting.

All the convolution layers have Rectified Linear Unit (ReLU) activations. ReLU introduces non-linearity to the data and works very well non-normalized, randomly distributed data, even though we normalized the data for faster training. It also removes the problem of vanishing gradients because unlike functions like tanh and sigmoid, it has a range of $[0, \infty)$. ReLU is one of the most popular activation functions (in Eq. 3).

$$\mathrm{f(x)}=\begin{cases} 0, & \text{when } x \leq 0 \\ x, & \text{when } x > 0 \end{cases} \tag{3}$$

Stochastic gradient descent is the most commonly used optimizer to reduce the cost function performance evaluator. We tested the network against Adam-[20], AdaGrad-[25] and RMSProp [26]-based gradient descent and found that the network converges faster using Adam. All of them were tested with a batch size of 1, beyond which the system crashed.

4.4 Performance Evaluation

The TD-UNet was trained on an NVidia GTX 1050Ti GPU with 4 GB of VRAM, 100 GB of DRAM and an Intel i5-8600 K CPU overclocked to 5.10 GHz on all cores. The training period lasted for about 13 h, before it had to be stopped due to overfitting. Figure 5 represents the input, ground truth and output of our network, while Fig. 6 represents the loss evaluation and accuracy graphs on the training and test cases.

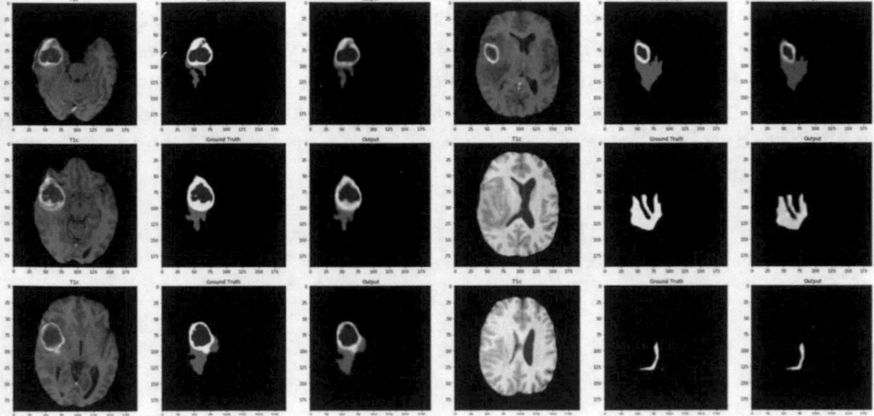

Fig. 5 (From left, set of 3) Input, ground truth and predicted output for some test data

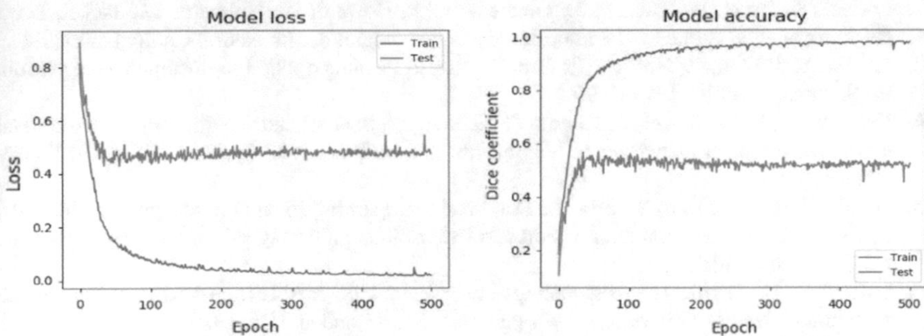

Fig. 6 (Left) Loss evaluation graph. (Right) accuracy graph

The model underwent 500 epochs of training five different times, and this was the best possible result among all of them. We had to consider other possibilities like hyperparameter tuning. Manually tuning the hyperparameters resulted in little to no variations to results during training but it affected the test cases ever so slightly.

5 Conclusion and Future Scope

The TD-UNet is a promising architecture that works on low-cost hardware. But on working with 3D image data, more experimentation is needed to improve the performance of the network, as it is currently not fit for real-world usage. More resource is needed to find the network's scalability for real-world usage. Also, we did not make use of all the training samples, and our model also suffers from needing to overfit over the training samples to correctly segment the test cases, which remains as one of the major scopes of the future development of this work. The network also needs a great deal of generalization to improve test case performance which is another area we will look into for the foreseeable future. Due to these limitations, we did not compare it with current models performing the same task. We believe that this architecture may be efficient computationally and in the future with more research, the performance will surely increase.

Acknowledgements We would like to show our gratitude towards MICCAI BraTS for providing their 2015 dataset as open source.

References

1. Dasgupta, A., Gupta, T., Jalali, R.: Indian data on central nervous tumours: a summary of published works. South Asian J. Cancer (2016)
2. LeCun, Y., Boser, B., Denker, J., Henderson, D., Howard, R., Hubbard, W., Jackel, L.D.: Backpropagation applied to handwritten zip code recognition. In: Neural Computation (1989)
3. LeCun, Y., Bottou, Y., Haffner, P.: Gradient-based learning applied to document recognition. In: Proceedings of the IEEE (1998)
4. Krizhevsky, A., Sutskever, I., Hinton, G.E.: ImageNet classification with deep convolutional neural networks. In: Advances in Neural Information Processing Systems (NIPS 2012), vol. 25 (2012)
5. He, K., Zhang, X., Ren, S., Sun, J.: Deep residual learning for image recognition. In: 2016 IEEE Conference on Computer Vision and Pattern Recognition (CVPR), Las Vegas, Nevada, US, 27–30 June 2016
6. Luttrell, S.: Hierarchical self-organising networks. In: 1989 First IEE International Conference on Artificial Neural Networks, (Conf. Publ. No. 313), London, UK (1989)
7. Hinton, G.E., Salakhutdinov, R.R.: Reducing the dimensionality of data with neural networks. Science **313**(5786), 504–507, 28 Jul 2006
8. Simonyan, K., Zisserman, A.: Very deep convolutional networks for large-scale image recognition. In: ICLR (2015)
9. Dalal, N., Triggs, B.: Histograms of oriented gradients for human detection. In: 2005 IEEE Computer Society Conference on Computer Vision and Pattern Recognition (CVPR'05), San Diego, CA, USA (2005)
10. Lowe, D.G.: Object recognition from local scale-invariant features. In: International Conference on Computer Vision (1999)
11. Girshick, R., Donahue, J., Darrell, T., Malik, J.: Rich feature hierarchies for accurate object detection and semantic segmentation. arXiv preprint, arXiv:1311.2524v5 [cs.CV], 22 Oct 2014
12. Redmon, J., Divvala, S., Girshick, R., Farhadi, A.: You only look once: Unified, real-time object detection. arXiv preprint arXiv:1506.02640v5 [cs.CV], 9 May 2016
13. Ronneberger, O., Fischer, P., Brox, T.: U-Net: convolutional networks for biomedical image segmentation. arXiv preprint arXiv:1505.04597v1 [cs.CV], 18 May 2015
14. Long, J., Shelhamer, E., Darrell, T.: Fully convolutional networks for semantic segmentation. arXiv preprint arXiv:1411.4038v2 [cs.CV], 8 Mar 2015
15. Pan, T., Wang, B., Ding, G., Yong, J.-H.: Fully convolutional neural networks with full-scale-features for semantic segmentation. In: Proceedings of the Thirty-First AAAI Conference on Artificial Intelligence (AAAI-17) (2017)
16. Dong, H., Yang, G., Liu, F., Mo, Y., Guo, Y.: Automatic brain tumour detection and segmentation using U-Net based fully convolutional networks. arXiv preprint arXiv:1705.03820v3 [cs.CV], 3 Jun 2017

17. Erden, B., Gamboa, N., Wood, S.: 3D Convolutional Neural Network for Brain Tumor Segmentation, Student Report in cs231n 526. Stanford University (2017)
18. MICCAI BRATS 2015 dataset. http://www.braintumorsegmentation.org/
19. Srivastava, N., Hinton, G., Krizhevsky, A., Sutskever, I., Salakhutdinov, R.: Dropout: a simple way to prevent neural networks from overfitting. J. Mach. Learn. Res. **15**, 1928–1958 (2014)
20. Kingma, D.P., Ba, J.L.: Adam: a method for stochastic gradient descent. arXiv preprint arXiv: 1412.6980v9 [cs.LG], 30 Jan 2017
21. Pinheiro, P.H.O., Collobert, R.: Recurrent convolutional neural networks for scene parsing. arXiv preprint arXiv:1306.2795v1 [cs.CV], 12 Jun 2013
22. Pinheiro, P.H.O., Collobert, R.: From image-level to pixel-level labeling with convolutional networks. arXiv preprint arXiv:1411.6228v3 [cs.CV], 24 Apr 2015
23. Graves, A., Fernandez, S., Schmidhuber, J.: Multi-dimensional recurrent neural networks. arXiv preprint arXiv:0705.2011v1 [cs.AI], 14 May 2007
24. Stollenga, M.F., Byeon, W., Liwicki, M., Schmidhuber, J.: Parallel multi-dimensional LSTM, with application to fast biomedical volumetric image segmentation. arXiv preprint arXiv:1506.07452v1 [cs.CV], 24 Jun 2015
25. Duchi, J., Hazan, E., Singer, Y.: Adaptive subgradient methods for online learning and stochastic optimization. J. Mach. Learn. Res. **12**, 2121–2159 (2011)
26. RMSProp. https://www.cs.toronto.edu/~tijmen/csc321/slides/lecture_slides_lec6.pdf

A Cooperative Co-evolutionary Genetic Algorithm for Multi-Robot Path Planning Having Multiple Targets

Ritam Sarkar, Debaditya Barman and Nirmalya Chowdhury

Abstract A cooperative co-evolutionary genetic algorithm based method has been proposed in this work to plan multiple robots' path. A domain knowledge-based operator named "Deletion Operator" and a method to find out the elite set of chromosomes have been developed for our proposed algorithm. A number of works have been carried out on multi-robot path planning, but most of them are suitable for robots having a single target only. Our proposed method can deal with the situation where each of the robots can have multiple targets without any predefined order. Hence, our objective is to get the shortest path from source to goal nodes considering that the goal nodes can appear in any order and the robots should not collide with any obstacles or other robots. The start nodes can also be the same as well as different for each of the robots. We have deployed our proposed method on a number of simulated environments to measure the efficacy of our proposed method. The experimental results reveal the effectiveness of the proposed method.

Keywords Genetic algorithm · Multi-robot path planning · Domain knowledge-based operator · Cooperative co-evolutionary genetic algorithm · Elitist strategy · Multiple targets

---

R. Sarkar · N. Chowdhury (✉)
Department of Computer Science and Engineering, Jadavpur University, Kolkata 700032, India
e-mail: nirmalya63@gmail.com

R. Sarkar
e-mail: ritamsrkr.cse@gmail.com

D. Barman
Department of Computer and System Sciences, Visva-Bharati, Santiniketan 731235, India
e-mail: debadityabarman@gmail.com

A. K. Das et al. (eds.), *Computational Intelligence in Pattern Recognition*,
Advances in Intelligent Systems and Computing 999,
https://doi.org/10.1007/978-981-13-9042-5_63

1 Introduction

Multi-robot path planning has been one of the important research topics of the Multi-Robot System (MRS) for the last few years. Generally, every solution in the multi-robot path planning problem is a set of paths (i.e., each path for each of the robots), where each path has a start node, a goal node (or a set of goal nodes), and some intermediate nodes. We need to take care of a number of criteria while planning a path for such multiple robots. These are as follows: (1) There must not be any collision between a robot and an obstacle, (2) There must not be any collision among the robots themselves, and (3) The set of paths as a whole needs to be optimal, whereas the path of any specific robot may or may not be optimal. Moreover, we need to impose additional criteria that will decide the appropriate order of goal nodes in case the robots are having multiple goal nodes.

There exist two traditional approaches for multi-robot path planning. These are centralized or coupled [1] and decentralized or decoupled multi-robot path planning [2–4]. In coupled path planning, multiple robots are considered as a single combined robot and the planning is executed on a combined configuration space (i.e., by combining every robot's configuration space). On the other hand in decoupled planning, the problem is divided into a number of independent subproblems, where each of the subproblems is associated with a single robot. Both approaches have merits and demerits. The decoupled planning technique requires less computational cost but it is incomplete in the sense that the planning technique does not ensure to find a path even though there may exist one. On the other hand, coupled path planning is complete but it has a large computational cost, since it searches the solution in the configuration space $C^n = C \times C \times C \ldots \times C$, where C is the original configuration space and n is the number of robots [1]. So, decoupled path planning approach can be preferred over coupled path planning approach, since, although it is not complete, it requires much less computational cost. Alternatively, we can say that the lack of completeness of the decoupled approach can be compensated by its less computational cost.

Sanchez and Latombe [5] made a comparison between the couple and decoupled techniques using a Probabilistic Roadmap (PRM) planner. The decoupled path planning can also be thought of as a cooperative co-evolutionary approach where each and every robot can be handled by separate evolutionary algorithms that are cooperative in nature. Cooperative co-evolutionary genetic algorithm based robot path planning have been proposed in [2, 4]. Svestka and Overmars [6] proposed a coordinated path planning which is probabilistically complete. Schneider and Wildermuth [7] developed a direct potential field-based approach. Ulusoy et al. [8] proposed a temporal logic constraints based method. Luna and Bekris [1] developed graph decomposition based approaches for multi-robot path planning. Liu et al. [9] proposed an ant colony and distributed navigation based algorithm. There also exist priority-based robot path planning where the paths are planned consecutively depending on some optimized priorities which are assigned to each of the robots [10, 11].

The aforesaid approaches are applicable for multi-robot path planning problem where every robot has only a single goal node. In the literature, we have not come across any work that deal with multiple robots having multiple targets. In this work, we have proposed a Cooperative Co-evolutionary Genetic Algorithm (CCGA) based planning for multiple robots where each one of them can have multiple predefined goal nodes. The rest of the paper has been divided into four sections. Section 2 presents the proposed methodology, the empirical results are discussed in Sect. 3, and the concluding remarks are stated in Sect. 4.

2 The Method of Using Genetic Algorithm

As mentioned in the introductory section, here in this work a Genetic Algorithm (GA)-based approach has been developed to find out the overall shortest path for multiple mobile robots. GA is a multidimensional searching procedure where genetic operators are applied to a population of chromosomes to generate a new population of chromosomes for the next generations until some termination criteria are fulfilled. These genetic operators are selection, crossover, and mutation. There is a fitness function that is used to evaluate the goodness of the chromosomes in the population, and finally, the Elitist strategy is used to copy the elite chromosome of the previous generation into the current generation, in order to preserve the best chromosome.

According to cooperative co-evolutionary approach [12], a large problem is divided into a number of subproblems and to solve each subproblem, separate Evolutionary Algorithms (EAs) are used. As a consequence, all the subproblems are taken care of by separate EAs concurrently and independently. Thus, the solution of each EA cooperates among themselves in order to generate the solution for the whole large problem.

Since path planning for multiple robots is more complex than that for the single robot and the cooperation among these robots plays a major role in multi-robot path planning which is not required for a single robot, regular GA does not fit well for this problem. For this purpose, CCGA has been proposed for this multi-robot path planning problem. In CCGA, the problem has been divided into a number of subproblems where each subproblem is associated with a single distinct robot and for every robot, there are separate subpopulations which are taken care of by separate GAs. So, here multiple GAs can evolve parallelly and independently. There is also an interaction phase at the end of each iteration to find out the elite set of chromosomes as the solutions generated by the separate GAs. These solutions are the paths for different mobile robots. So, in order to get the complete solution (i.e., for the entire multi-robot system), the interaction phase plays a vital role. Moreover, the set of goal nodes for each of the robots is pre-allocated irrespective of their order. So, the objective is to get a optimal set of paths having proper ordering of the goal nodes and intermediate nodes in between them. Our proposed method is based on the following assumptions.

1. All the environments are static in nature.
2. The velocity of the robots is fixed.

3. The complete information regarding the environment is known in advance.
4. The robot has been considered as a point, as the obstacle boundary consists o the actual boundary of the obstacles and minimum safety distance of the robo [13].

2.1 Representations of Environments

Grid-based environment has been considered as the environment which can be decomposed by a collection of orderly numbered grids which are equal in shape and size. The grids are rectangular in shape, and these are of two types as shown in Fig. 1a which are colored grids and blank grids. Colored grids represent obstacles and the blank grids represent free spaces through which a robot can move freely. The environment contains a predefined start node S and a set of target nodes $\{T_1, T_2, T_3\}$.

2.2 Representation of Paths

In this robot path planning problem, every chromosome is a path, represented by the decimal-coded representation where the first node is the start node, and then we have a set of goal nodes including the last node, a number of intermediate nodes between the start node and the first goal node as well as between two consecutive goal nodes. Chromosome's length has been considered as variable rather than fixed. If any of the segments of a path collides with any obstacles, then it will be an infeasible path, otherwise a feasible one. As a result of that, the path [1, 6, 36] is an infeasible path whereas another path, i.e., [1, 10, 14, 36], is a feasible path as shown in Fig. 1b.

(a)

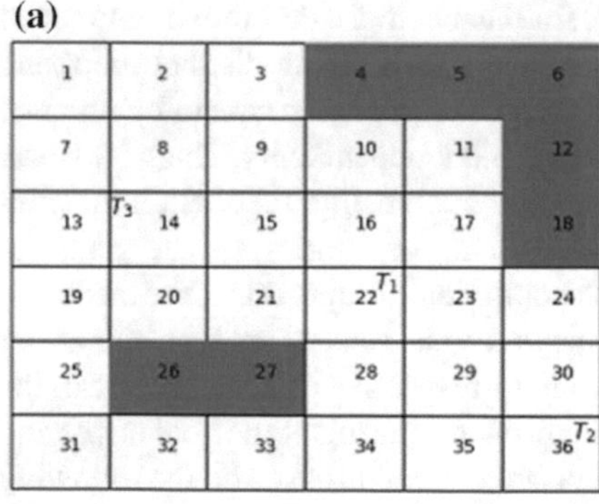

(b)

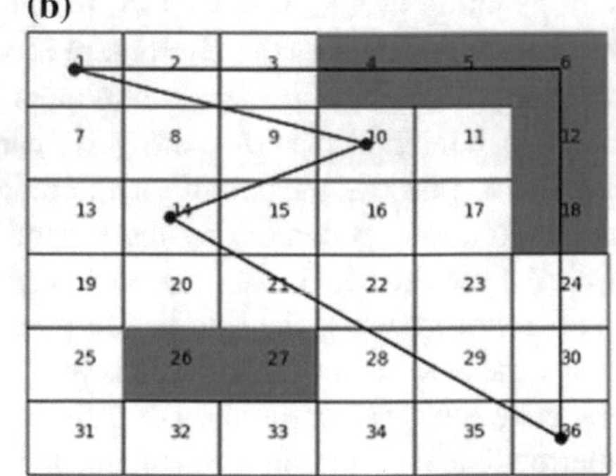

Fig. 1 **a** Robot's environment, **b** Robot's path

2.3 Initial Population

In this CCGA, for each and every subproblem, there is a separate subpopulation that has been generated randomly. It may be noted that only feasible solutions have been considered as the member of a subpopulation, because infeasible chromosome or path may reduce the convergence rate. The number of chromosomes for each subpopulation has been taken as same. So, the total size of the population is equal to the number of chromosomes in each subpopulation multiplied by the number of robots as the number of robots is equal to the number of subpopulations.

2.4 Fitness

The fitness function is used to evaluate the goodness of a chromosome with respect to the other chromosomes in the population. This fitness function may be a minimization or maximization function based on the problem scenario. Since our objective is to get the shortest path, the fitness function (f) happens to be a minimization function. Along with that, the path needs to be collision free. For this reason, a penalty value has been added to the fitness function (f) to include a positive penalty value if there is any collision between a path and obstacles. The fitness function (f) for a path P having n numbers of nodes has been described below:

$$f = \sum_{i=1}^{n} \sqrt{(x_{i+1} - x_i)^2 + (y_{i+1} - y_i)^2} + penalty \tag{1}$$

$$penalty = \begin{cases} \textit{longest path segment possible according to the environment}, & \textit{if path is infeasible} \\ 0, & \textit{if path is feasible} \end{cases} . \tag{2}$$

In Eq. 1, x_i and x_{i+1} are the x-coordinates of ith and $(i + 1)$th node of the path P, respectively. Similarly, y_i and y_{i+1} are the y-coordinates of ith and $(i + 1)$th of the path P. The penalty value has been described in Eq. 2.

2.5 Genetic Operators

There are three genetic operators. These are selection, crossover, and mutation. These three genetic operators have been described below.

Selection

The purpose of the selection operator is to select the fittest chromosomes based on their fitness values from a population of chromosomes to generate a mating pool, upon which the rest of the operators are applied to generate a new population of chromosomes. The selection probability of a chromosome is either directly or inversely proportional to its fitness value based on the fact that the problem is either maximization or minimization. Our problem is a minimization problem since the objective is to obtain the shortest path. As a consequence, the selection probability has been considered as inversely proportional to the fitness value of the corresponding chromosome.

Crossover

Since the target nodes are predefined and they can appear in any order, each and every path has been divided into a number of path segments before performing the crossover operation. For instance, if there is a path say P_1 $(S, \ldots, T_{11}, \ldots, T_{21}, \ldots, T_{i1}, \ldots, T_{n1})$ contains n targets where T_{i1} is the ith target of the path P_1 and this ith target of P_1 may not be same as that of ith target of another path say P_2 (i.e., T_{i2}). Here, dotted line indicates some intermediate nodes. So, the path segments for the path P_1 will be—$(S, \ldots, T_{11})$, $(T_{11}, \ldots, T_{21})$, ..., $\left(T_{(i-1)1}, \ldots, T_{i1}\right)$, $\left(T_{i1}, \ldots, T_{(i+1)1}\right)$, ..., $\left(T_{(n-1)1}, \ldots, T_{n1}\right)$. Therefore, the number of path segments would be equal to n if there exists n number of targets. After that, one-point crossover has been performed between the path segments of two different paths (i.e., P_1 *and* P_2). Crossover point has been chosen randomly except the start node and the target nodes of the path segment having shorter length. Under these circumstances, two path segments of two different paths will participate in the crossover operation if they follow one of the following three cases.

Case 1:

If the first path segments of paths P_1 and P_2 are $(S, \ldots, T_{11})$ and $(S, \ldots, T_{12})$, respectively, then one-point crossover will be performed between them if and only if targets T_{11} and T_{12} are same (i.e.,$T_{11} = T_{12}$).

Case 2:

Let ith segment of P_1 be $\left(T_{i1}, \ldots, T_{(i+1)1}\right)$ and jth segment of path P_2 be $\left(T_{j2}, \ldots, T_{(j+1)2}\right)$; now one-point crossover operation will be performed between $\left(T_{i1}, \ldots, T_{(i+1)1}\right)$ and $\left(T_{j2}, \ldots, T_{(j+1)2}\right)$ if and only if $T_{i1} = T_{j2}$ and $T_{(i+1)1} = T_{(j+1)2}$.

Case 3:

Consider the ith segment of path P_1 as $\left(T_{i1}, \ldots, T_{(i+1)1}\right)$ and the jth segment of P_2 as $\left(T_{j2}, \ldots, T_{(j+1)2}\right)$, now these path segments will participate in the crossover operation if and only if $T_{i1} = T_{(j+1)2}$ and $T_{(i+1)1} = T_{j2}$. In this type of situation, the path segments of P_2 need to be reversed before performing crossover operation.

At the end, all the path segments of P_1 need to be reordered and then merged according to their original order (i.e., the parents' path or chromosome). The same procedure has to be followed for P_2 also.

Mutation

The mutation operator used in this work is based on the mutation operator proposed in Tuncer et al. [3] where mutation operation has been performed once on a randomly chosen node (except the start node and goal node) out of the entire path, whereas,

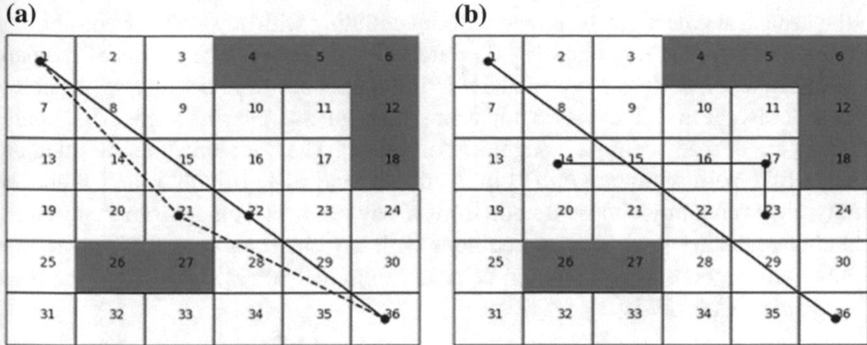

Fig. 2 **a** Example of mutation, **b** example of interaction cost

in this work, all the nodes except the start node and the goal node get a chance to participate in the mutation operation. This approach speeds up the solution space exploration process or in other words, takes less number of iterations to explore the solution space. According to this mutation operator, for a path *P* [1, 21, 36], shown by dotted line in Fig. 2a, 21 will be selected for mutation (since it is the only node except start node 1 and the goal node, i.e., 36). Now, neighbors of 21 are needed to be found out. These neighbors are 14, 15, 16, 22, 28, and 20 (26 and 27 have been discarded since they are obstacles). Finally, 21 will be replaced by one of the neighbors that generates the least fitness value including the original path, i.e., the path containing 21 itself. As a consequence, 21 has been replaced by 22 as shown in Fig. 2a.

2.6 Domain Knowledge-Based Operator

The introduction of some local search strategies may enhance the capability of conventional GA by not letting them converge into the local optima in case the obstacles as well as the environments are complex in nature. Domain knowledge-based operators are often developed using some local search strategies. Generally, these operators are based on some heuristics which are only applicable to the domain on which it is applied. One such operator named "Deletion Operator" has been proposed in this work.

Deletion Operator

The deletion operator is based on the principle that the length of one side of a triangle is always less than the summation of the length of the other two sides. It is also known as triangle inequality theorem. This deletion operator consists of two cases which are performed with 50% probability. According to case 1, it selects two adjacent path segments and checks whether the starting node of the first segment and the last node of the second segment can be joined without colliding with any obstacles or not. For example, consider $[P_i, P_j]$ and $[P_j, P_k]$ are two adjacent path segments of the path P, and then the node P_j can be deleted if $[P_i, P_k]$ does not collide with any obstacles. In case 2 also, it first selects two adjacent path segments. For instance, $[P_i, P_j]$ and $[P_j, P_k]$ are two adjacent path segments of path P. Then, it stores the intermediate nodes of the path segment $[P_i, P_j]$ in X and $[P_j, P_k]$ in Y. If both X and Y are not empty, then randomly choose a node from X say $rand_i$ as well as from Y say $rand_j$ and check whether $[rand_i, rand_j]$ collides with any obstacles or not. If they do not collide with any obstacles and then delete P_j from the path P and insert $rand_i$ after P_i and $rand_j$ after $rand_i$.

2.7 Elitist Strategy

The purpose of the elitist strategy is to keep the elite chromosome of the previous generation in such a way that, the elite chromosome is not lost by the application of other operators on the current generation. The elitist strategy used in this work is based on [14].

2.8 Interaction Cost

All the interaction positions between any two paths are required to compute the interaction cost between these two paths where each one of them belongs to separate robots. Thus, the interaction cost between two paths is the number of interaction positions for which the distance between that interaction position and the starting nodes of those robots are same, multiplied by some positive constant (k). For example, as shown in Fig. 2b, the number of interaction positions between the path of robot say r_1 [1, 36] and the path of robot say r_2 [14, 17, 23] is 1 (i.e., 15) and interaction cost is 0 as the distances between the interaction position (i.e., 15) and the starting nodes of each of the robots (i.e., 1 for r_1 and 14 for r_2) are not same.

2.9 Elite Set of Chromosome

Unlike single robot path planning, here in this multi-robot path planning, our objective is to find out an optimal set of paths in terms of shortest length where each of the individual paths may or may not be optimal but as a set they are optimal. For this purpose, we need an additional elitist strategy at the end of each generation to preserve an elite set of chromosomes where each chromosome may or may not be individually elite (i.e., optimal).

For an example, consider there are three robots (r_1, r_2, r_3) and the size of each subpopulation is 2. Let the subpopulations for r_1, r_2, r_3 are $[[ch_{r11}, ch_{r12}], [ch_{r21}, ch_{r22}], [ch_{r31}, ch_{r32}]]$. In order to calculate an elite set of chromosomes for one of the permutations of these three robots say (r_1, r_2, r_3), at first, we need to find out all the possible overall costs between the individuals of the populations of every two robots. The overall cost for a set of n chromosomes (i.e. $[ch_1, \ldots, ch_n]$) where each one of them belongs to n different robots has been stated in the following Eq. 3.

$$overall\_cost(n) = \sum_{i=1}^{n} fitness\ value\ of\ ch_i + \sum_{i=1}^{n-1}\sum_{j=i+1}^{n} interaction\ cost\ between\left(ch_i, ch_j\right) \tag{3}$$

Thus, all the possible overall cost between the paths of r_1 and r_2 have been shown in Eqs. 4–7.

$$[Fitness\ value\ of\ ch_{r11} + Fitness\ value\ of\ ch_{r21} + Interaction\ cost\ between\ ch_{r11}\ and\ ch_{r21}] \tag{4}$$

$$[Fitness\ value\ of\ ch_{r11} + Fitness\ value\ of\ ch_{r22} + Interaction\ cost\ between\ ch_{r11}\ and\ ch_{r22}] \tag{5}$$

$$[Fitness\ value\ of\ ch_{r12} + Fitness\ value\ of\ ch_{r21} + Interaction\ cost\ between\ ch_{r12}\ and\ ch_{r21}] \tag{6}$$

$$[Fitness\ value\ of\ ch_{r12} + Fitness\ value\ of\ ch_{r22} + Interaction\ cost\ between\ ch_{r12}\ and\ ch_{r22}] \tag{7}$$

Suppose we get the least overall cost value for $[ch_{r11}, ch_{r22}]$. Next, we need to find out the overall cost between ch_{r22} and each of the chromosomes of r_3. Suppose minimum value has been achieved for $[ch_{r22}, ch_{r33}]$. Finally, we get $[ch_{r11}, ch_{r22}, ch_{r33}]$ for the permutation (r_1, r_2, r_3) as one of the possible elite sets of chromosomes and the overall cost for this set has been shown in Eq. 8. The same calculation needs to be carried out for rest of the permutations of those three robots (i.e., $[r_1, r_3, r_2]$, $[r_2, r_1, r_3]$, $[r_2, r_3, r_1]$, $[r_3, r_1, r_2]$, $[r_3, r_2, r_1]$). There exist another possible elite set of chromosomes which is chromosome of r_1 having least fitness value, chromosome of r_2 having least fitness value, and chromosome of r_3 having least fitness value. Thus, the elite set of chromosome will be one of the total numbers of possible elite sets of chromosomes which is $(3! + 1)$ where 3 is the total number of robots. The set with the least overall cost value has been considered as the elite set of chromosomes.

Table 1 Experimental results for environments 1, 2, 3, and 4

Environment No.	Population size	Best fitness value	Average best fitness value	Average number of generations
Environment 1	40	95.6958	95.7153	41
Environment 2	40	100.1096	100.1256	38
Environment 3	50	85.2720	85.2875	39
Environment 4	40	98.4002	98.4162	21

$$\begin{aligned}&\big[Fitness\ value\ of\ ch_{r11} + Fitness\ value\ of\ ch_{r22}\\&\quad + Fitness\ value\ of\ ch_{r33} + Interaction\ cost\ between[ch_{r11}, ch_{r22}]\\&\quad + Interaction\ cost\ between\ [ch_{r22}, ch_{r33}]\\&\quad + Interaction\ cost\ between\ [ch_{r11}, ch_{r33}]\big]\end{aligned} \tag{8}$$

2.10 Termination Condition

Though there does not exist any universal stopping criterion for GA [14], the algorithm has been terminated in this work when either one of the following three conditions is satisfied:

1. The total number of generation exceeds 100.
2. The algorithm converges to the optimal solution within 100 iterations or generations.
3. The fitness value of the best chromosome remains unchanged for 50 consecutive generations.

2.11 Flowchart

The flowchart of the proposed method has been depicted in Fig. 3.

3 Experimental Results

We have conducted our experiment on four different simulated environments. The values of different operators have been taken as—crossover probability is 1, mutation probability is 0.25, and deletion probability for case 1 is 0.30 and for case 2 is 0.45. The experimental results have been reported in Table 1 which consist of environment number, population size, best fitness value, average best fitness value, and average number of generations in 25 experimental runs.

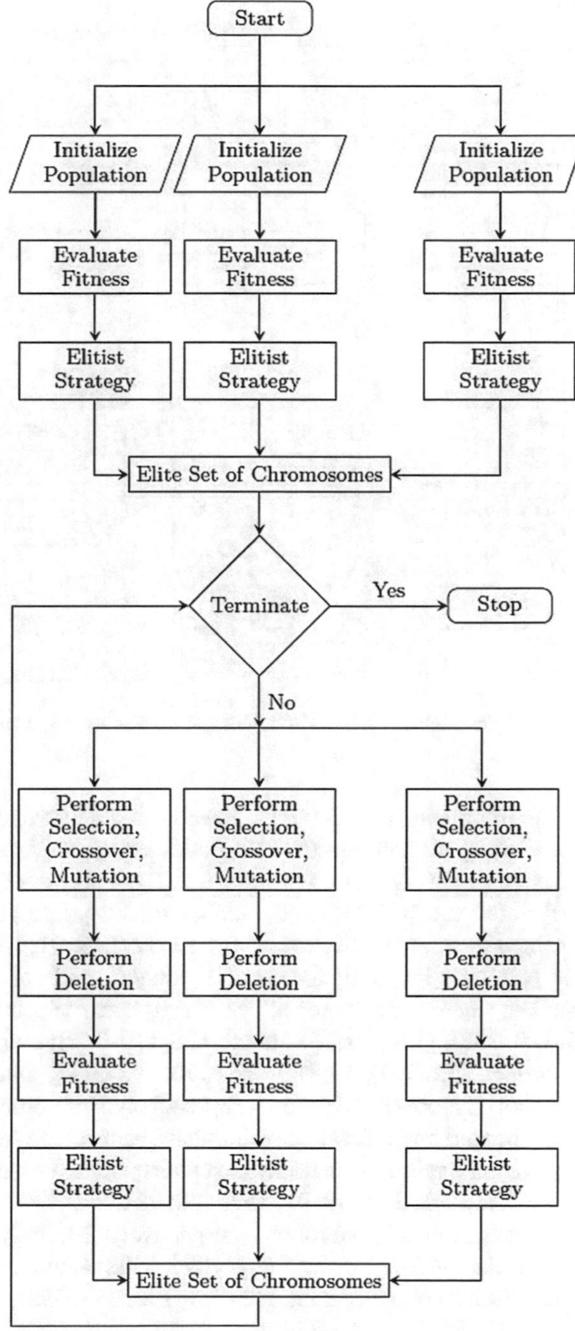

Fig. 3 Flowchart of the proposed algorithm

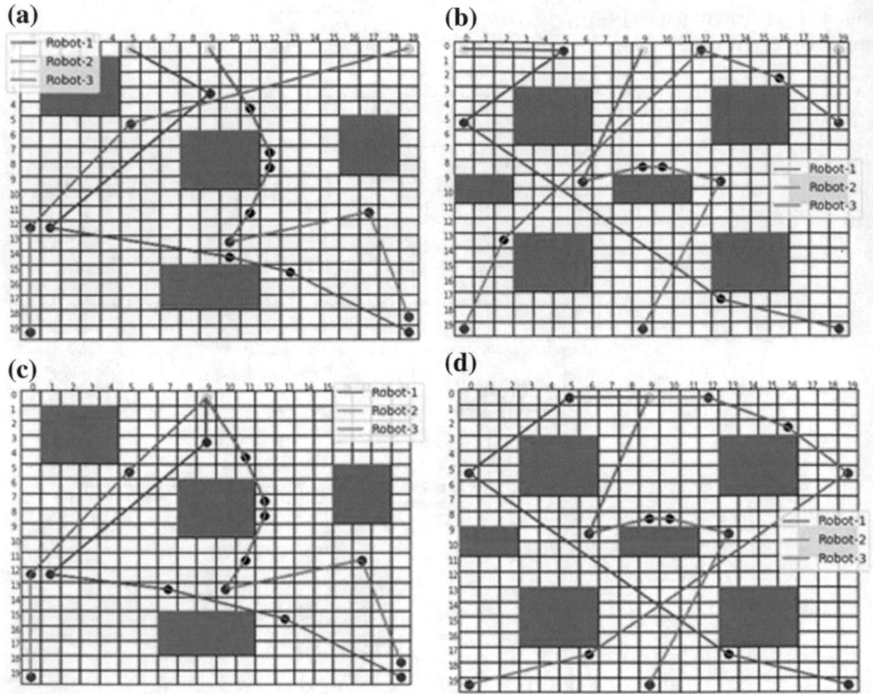

Fig. 4 **a** Environment 1, **b** environment 2, **c** environment 3, **d** environment 4

Environments 1, 2, 3, and 4 have been represented in Fig. 4a–d, respectively. The start, goal, and intermediate nodes are represented by yellow, red, and black color, respectively. For environment 1, the start nodes are 5, 9, and 19; the set of goal nodes are [69, 241, 399], [237, 270, 379], and [105, 240, 380]; and the best obtained paths are [5, 69, 241, 290, 313, 399], [9, 91, 152, 172, 231, 270, 237, 379], and [19, 105, 240, 380] for robot-1, robot-2, and robot-3, respectively. For environment 2, the start nodes are 0, 9, and 19; the sets of goal nodes are [5, 100, 399], [186, 193, 389], and [12, 119, 380]; and the best obtained paths are [0, 5, 100, 353, 399], [9, 186, 169, 170, 193, 389], and [19, 119, 56, 12, 262, 380] for robot-1, robot-2, and robot-3, respectively. Environment 3 is same as environment 1, but the start nodes for all the robots have been taken identical (i.e., 9) and the rest are same as for environment 1. The best paths which we have got for environment 3 are [9, 69, 241, 291, 399], [9, 91, 152, 172, 231, 270, 237, 379], and [9, 105, 240, 380] for robot-1, robot-2, and robot-3, respectively. Similarly, environment 4 is also same as environment 2 expect one thing which is the same start nodes for all the robots (i.e., 9) and the best obtained paths are [9, 5, 100, 353, 399], [9, 193, 170, 169, 186, 389], and [9, 12, 56, 119, 346, 380] for robot-1, robot-2, and robot-3, respectively.

4 Conclusion and Scope for Future Work

Most of the previous works on this field dealt with multi-robot path planning having a single goal node for each robot. In this work, we have proposed a GA-based approach named CCGA which can handle multiple robots where each one of them can have multiple goal nodes. Please note that the nature of the problem for robots with multiple targets is more complex since here one has to find the appropriate sequence of goal nodes in order to find the optimal set of paths. To judge the efficacy of the proposed method, we have applied it on four different environments considering both identical start node and different start nodes for each of the robots. It may be noted that a domain knowledge-based operator named "deletion operator" and a technique to find out the elite set of chromosome have been introduced in this method. Since we have not come across any work in the literature that deals with the multiple robots having multiple targets in the robot path planning problem, we could not provide any comparison of experimental results.

In this work, our objective is to find the shortest path. In other words, it is a single objective optimization problem. But one can also consider other parameters to optimize along with the shortest path, two such parameters may be: (1) distance between the path segments and the obstacles and (2) smoothness of the paths. Thus, multi-robot path planning using multi-objective optimization technique can be one of the possible directions of future work in this field.

References

1. Luna, R., Bekris, K.E.: Efficient and complete centralized multi-robot path planning. In: 2011 IEEE/RSJ International Conference on 2011 Intelligent Robots and Systems (IROS), Sept 25, pp. 3268–3275. IEEE
2. Kala, R.: Multi-robot path planning using co-evolutionary genetic programming. Expert Syst. Appl. **39**(3), 3817–3831 (2012)
3. Tuncer, A., Yildirim, M.: Dynamic path planning of mobile robots with improved genetic algorithm. Comput. Electr. Eng. **38**(6), 1564–1572 (2012)
4. Cai, Z., Peng, Z.: Cooperative co evolutionary adaptive genetic algorithm in path planning of cooperative multi-mobile robot systems. J. Intell. Rob. Syst. **33**(1), 61–71 (2002)
5. Sanchez, G., Latombe, J.C.: Using a PRM planner to compare centralized and decoupled planning for multi-robot systems. In: Proceedings—IEEE International Conference on Robotics and Automation, 11 May 2002, vol. 2, pp. 2112–2119 (2002)
6. Svestka, P., Overmars, M.H.: Coordinated Path Planning for Multiple Robots. Information and Computing Sciences, Utrecht University (1996)
7. Schneider, F.E., Wildermuth, D.: A potential field based approach to multi robot formation navigation. In: 2003 IEEE International Conference on Robotics, Intelligent Systems and Signal Processing. Proceedings. 8 Oct 2003, vol. 1, pp. 680–685. IEEE (2003)
8. Ulusoy, A., Smith, S.L., Ding, X.C., Belta, C., Rus, D.: Optimality and robustness in multi-robot path planning with temporal logic constraints. Int. J. Robot. Res. **32**(8), 889–911 (2013)
9. Liu, S., Mao, L., Yu, J.: Path planning based on ant colony algorithm and distributed local navigation for multi-robot systems. In: Proceedings of the 2006 IEEE International Conference on Mechatronics and Automation, 25 Jun 2006, pp. 1733–1738. IEEE (2006)

10. Bennewitz, M., Burgard, W., Thrun, S.: Finding and optimizing solvable priority schemes fo decoupled path planning techniques for teams of mobile robots. Robot. Auton. Syst. **41**(2–3) 89–99 (2002)
11. Bennewitz, M., Burgard, W., Thrun, S.: Optimizing schedules for prioritized path plannin of multi-robot systems. In: IEEE International Conference on Robotics and Automation. Pro ceedings 2001 ICRA, vol. 1, pp. 271–276. IEEE (2001)
12. Potter, M.A., De Jong, K.A.: A cooperative co evolutionary approach to function optimization In: International Conference on Parallel Problem Solving from Nature, 9 Oct 1994, pp. 249–257 Springer, Berlin, Heidelberg (1994)
13. Li, Q., Zhang, W., Yin, Y., Wang, Z., Liu, G.: An improved genetic algorithm of optimum path planning for mobile robots. In: Sixth International Conference on Intelligent Systems Design and Applications, 2006. ISDA'06. Oct 16, vol 2, pp 637–642. IEEE (2006)
14. Murthy, C.A., Chowdhury, N.: In search of optimal clusters using genetic algorithms. Pattern Recogn. Lett. **17**(8), 825–832. (1996)

A Shape Retrieval Technique Using Isothetic Cover of Digital Objects

Abdul Aziz Al Aman, Apurba Sarkar and Arindam Biswas

Abstract This work presents a novel approach to classify digital objects based on outer isothetic cover (OIC). The descriptor vectors of the isothetic cover are used as the shape descriptors of a particular digital object. Two objects may be of similar kind of shape if their shape descriptors are analogous. To find a match between two descriptors, common substrings are found out and based on their length of the common substring a decision is made. Experimental results show encouraging results.

Keywords Isothetic covers · Shape analysis · Shape descriptor · Longest common substring

1 Introduction

Shape analysis of digital object is the field of image processing, but more deeply, shape analysis is nothing but analysis of geometric properties of an object like length, area, volume, segments, etc. As world is moving rapidly toward digital world, shape analysis of digital objects plays an important role to identify similar objects. Shape extraction of a digital object is a very ground-level research work for its classification or retrieval. Shape analysis can be done in several ways. A detailed study of digital geometry methods may be seen in [9–11]. Many methods to measure complexity

A. A. A. Aman (✉) · A. Sarkar
Department of Computer Science and Technology, Indian Institute of Engineering Science and Technology, Shibpur, Howrah 711103, India
e-mail: mdaman.iiest@gmail.com

A. Sarkar
e-mail: as.besu@gmail.com

A. Biswas
Department of Information Technology, Indian Institute of Engineering Science and Technology, Shibpur, Howrah 711103, India
e-mail: barindam@gmail.com

A. K. Das et al. (eds.), *Computational Intelligence in Pattern Recognition*, Advances in Intelligent Systems and Computing 999,
https://doi.org/10.1007/978-981-13-9042-5_64

of shapes have been proposed. Guido et al. presented a bit rate-distortion technique to encode polygonal boundary [15]. Biswas et al. [2] proposed a shape analysis method where isothetic polygonal envelope is used to measure shape complexity of object. In another approach [5], they construct chain code using triangular cover and then find out the Levenshtein distance to compare similarities between objects. In yet another approach [3], they formulated shape code using edge length between consecutive vertices on the shape descriptor of the objects. Freeman [7] proposed quantization technique to encode boundary of digital curve. Thomas et al. proposed an algorithm to measure complexity of polygonal objects in [6]. Gdalyahu et al. [8] designed an algorithm to match curves under substantial deformations and arbitrary large scaling and rigid transformations. Jaume et al. [12] analyzed shape of an object based on integral geometry and information theory tools where they evaluate degree of structure from inside of the object and compute degree of interaction with the circumscribing sphere from outside. Although there are a number of techniques available in the literature to analyze shape of digital object, there is no reported work so far for shape analysis using isothetic cover. This paper uses isothetic cover of digital objects as the shape descriptor and uses common substrings technique to retrieve objects which are similar in shape of the given query image.

The rest of the paper is arranged as follows. Section 2 presents the way to construct the Outer Isothetic Cover (OIC) of the object based on the method proposed in [1, 4] on which the present work is based upon. The important idea of shape matching used in this work is based on longest common substring, which is briefly described in Sect. 3.1. The major steps of algorithm are explained in Sec. 3.2. Experimental results and analysis are presented in Sect. 4. Finally, Sect. 5 summarizes the proposed method along with possible future directions.

2 Construction of Outer Isothetic Cover (OIC)

As the initial step toward the present work, the outer isothetic cover of the digital object is found out. The construction of OIC is based on the combinatorial algorithm by Biswas et al. [1, 4]. The idea is to find tightest outer cover, A_{out}, of an object imposed on an isothetic grid. The isothetic grid consists of sets of horizontal and vertical parallel lines. A grid vertex is defined as an intersection of horizontal and vertical grid lines and the line segment between adjacent grid vertices is a grid edge, and the length of the segment between two consecutive vertices along horizontal/vertical line is termed as g. A smallest grid square defined by four grid edges is termed as unit grid block (UGB). A grid vertex is adjacent to four such unit grid block (UGB) and based on object occupancy of each adjacent UGB, vertices are classified into four types. A vertex in A_{out} is classified depending on the number of fully or partially occupied UGBs at that vertex. If the number of fully or partially occupied UGBs by the object at a point v is i, where $i \in [0, \dots, 4]$, then v falls in the class of C_i. The classification of vertex is shown in Fig. 1.

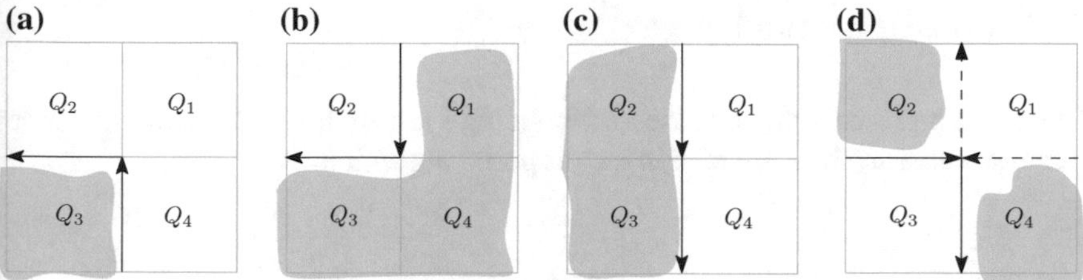

Fig. 1 Types of vertices present in the outer isothetic cover

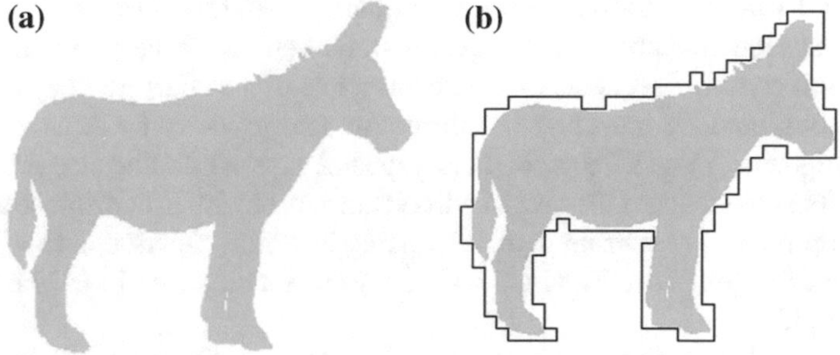

Fig. 2 **a** A digital object, **b** its outer isothetic cover

In summary, the classification of vertices is as follows. (i) C_0: v is not a vertex, since none of four neighbors has object containment; (ii) C_1: v will be a Type-1 vertex of A_{out} if only one adjacent UGB contains object shown in (Fig. 1a); (iii) C_2: If two adjacent UGBs are occupied by the object, then v is Type-2 vertex (Fig. 1c); (iv) C_3: v is classified as a Type-3 vertex if three UGB has object containment (Fig. 1b) or diagonally opposite UGB contain object (Fig. 1d); and another kind of vertex where four UGB's occupied object is not considered in case of OIC.

To construct the outer cover of a digital object, a start point has to be fixed and in this case it is chosen to be the top left vertex of the digital object and it can be obtained by row-wise scan of the grid vertex. It is to be noted that the start vertex chosen this way should definitely be 90° or Type-1 vertex. An anticlockwise traversal is then made starting from the start vertex. Let v_{i-1} and v_i be the two contiguous vertices in A_{out} and d_{i-1} and d_i be the direction of the vertices, respectively, also let t_i be the type of v_i. During the traversal, the direction from one vertex to the next vertex is obtained by the formula $d_i = (d_{i-1} + t_i)\ mod\ 4$, where $d_i \in [0, 3]$. The direction code 0, 1, 2, and 3 indicate the direction of movement along right, top, left, and downward, respectively. The traversal procedure continues and stops when it reaches the start point. Figure 2a shows a digital object and Fig. 2b shows object along with its outer isothetic cover.

3 Proposed Method

This section presents the proposed methodology in detail, i.e., deriving the longest common substring and use of it in the proposed algorithm.

3.1 Deriving Longest Common Substring (LCS)

The Longest Common Substring (LCS) problem is slightly different from longest common subsequence problem. Longest common subsequence problem generally gives longest common sequence, where occurrence of the character may or may not be contiguous, but LCS must preserve the contiguous property, i.e., character should occur contiguously. The LCS problem is a special case of edit distance problem and LCS problem also reduces the cost of edit distance problem. Edit distance algorithm basically transforms one string to another using insertion, deletion, and substitution operations with some cost. The LCS problem was first proposed in 1974 by Wagner and Fisher [13].

Let us consider a string, $S = s_1 s_2 s_3, \ldots, s_m$, where s_i denote the ith character in S. Let $S_{i \ldots j}$, $j \geq i$ be the substring $s_i s_{i+1} s_{i+2}, \ldots, s_j$. Let $W = w_1 w_2 w_3, \ldots, w_n$ be another string. So, $S_{i \ldots j}$ will be a common substring between S and W if $S_{i \ldots j} = W_{k \ldots l}$ such that $l - k = j - i$ for some i, j, k and l. $S_{i \ldots j}$ will be the longest common substring if it the largest common substring among all possible common substrings. In the proposed method, all possible common substrings are found out and the length of all common substrings is added if their length exceeds a particular threshold. The string that results after summing up is named as "Longest Common Substring Sum ($lcss$)". The fraction of $lcss$ present in source and target string is then calculated and if they are greater than certain percentage (70% in the proposed method), they are treated as similar.

3.2 Proposed Algorithm

The outer isothetic cover of a digital object is determined using the algorithm referred in Sect. 2. The directions of the vertices on the outer cover of the digital object are stored in an array and used as the descriptor for that particular object. This array of directions is termed as shape descriptor vector. Shape descriptor vector of a digital object consists of only direction of Type-1 and Type-3 vertexes. The shape database contains shape descriptor vectors for different images. To retrieve similar images of a given input image, its shape descriptor is constructed and the algorithm finds the set of images with which the input image has long enough (a threshold found out experimentally) common substring ($lcss$). The outline of the algorithm is presented in Fig. 3.

Algorithm Outline
INPUT : **Input Image**
OUTPUT : **Similarity between images**
1. $L = \text{FindOIC}(Q)$
2. $D_q = \text{FindDescriptor}(L)$
3. For $i = 1$ to N
4. $\quad lcss = 0$
5. $\quad D_{d_i} = \text{Retrieve\_Descriptor}[i]$
6. $\quad$ Do
7. $\qquad cs = \text{FindLCS}(D_{d_i}, D_q)$
8. $\qquad$ If $|cs| \geq cs\_threshold$
9. $\qquad\quad lcss = lcss + |cs|$
10. $\quad$ While $(|cs| \geq cs\_threshold)$
11. $\quad Q_f = \frac{|lcss|}{|D_q|} \times 100\%$
12. $\quad D_{i_f} = \frac{|lcss|}{|D_{d_i}|} \times 100\%$
13. $\quad$ If $Q_f \geq match\_threshold$ and $D_{i_f} \geq match\_threshold$
14. $\qquad$ Match found
15. End For

Fig. 3 Outline of the algorithm

In Step 1, function *FindOIC* computes *OIC* of a given query image (Q). The result of *FindOIC* is the list L of vertices in the outer cover. Step 2 converts the vertex list obtained in Step 1 into shape descriptor vector using function *FindDescriptor*. The function actually takes the outgoing direction of each vertex in L. The database (D) consists of shape descriptor vectors of several images (around 200). In Step 4, shape descriptor vector (D_{d_i}) from the database is chosen. The $Do - While$ loop (steps 5 through 10) continues until length of the longest common substring between Q and D_{d_i} is greater than $cs\_threshold$. Step 6 and 7, compute the common substrings between Q and D_i, respectively and if it is greater than $cs\_threshold$, they are added up in Step 8 and 9. Once $Do - While$ completes, $lcss$ contains the sum of all common substrings that are greater than $cs\_threshold$. In Steps 11 and 12, fractional presence of $lcss$ in query and database images are, respectively, calculated. Step 13 compares Q_f and D_{i_f} with some predefined threshold ($match\_threshold$ which is 70% in this case) and if they both are greater than it is concluded that the query image is similar to database image. The *For Loop* (Steps 3 through 15) is repeated for all the images in the database to retrieve all the similar images to the input image.

4 Experimental Results and Analysis

The proposed technique for shape retrieval of the digital object in a $2D$ integer plane has been tested on a database which consists of several objects of various shapes and sizes. The algorithm is also tested for different grid sizes (g) and different lengths of common substrings ($lcss$). Two images are considered to be of similar shape if percentage presence of common substring in both the images is greater than some threshold ($match\_threshold$) value. The threshold value is calculated depending on grid size (g) and threshold value of $cs\_threshold$ or sometimes depending on the application (the purpose where it will be used). The proposed method is tested with grid size 4, 8, 10 and with $cs\_threshold$ of 3, 7, 11. The value of $match\_threshold$ for different combinations of g and $cs\_threshold$ is chosen to be 70%. This value is selected after few sample runs of the algorithm. Table 1 shows the retrieval results for $g = 4$ and $cs\_threshold = 3$. The object shown in the first column of the table is the query object, and remaining columns in a particular row are the retrieved objects. The numeric figures below each object show the percentage presence of the common substring in the query object and the retrieved object, respectively. Table 2 shows similar results with $g = 4$ and $cs\_threshold = 7$. The proposed technique is also tested with other combinations of g and $cs\_threshold$ and produced similar kind of results. An important observation of the result shown in the tables and the results that have been performed is that if the grid size increases then the percentage of matching decreases and similarly if $cs\_threshold$ is increased then the percentage of matching decreases.

5 Conclusions

In this work, an efficient retrieval technique is designed and implemented to demonstrate the versatility of an outer isothetic cover. This technique may also be applied to any type of isothetic polygon to measure the similarities. The importance of the algorithm is that it can be used to retrieve objects which are roughly similar in shape with a less complex descriptor of objects. The complexity of the descriptor can also be tuned by tuning the grid size (g). The proposed technique can also be used for classification of digital object by tuning it suitably. In future, other combinatorial algorithms like the inner isothetic cover [4] or the Sandwich Cover [14] can be used to increase accuracy of the algorithm, which needs further study.

Table 1 Retrieved objects w.r.t. query objects with $g = 4$ and $cs\_threshold > 3$. Numeric figures below each object represent fraction of the common substring in the query and the retrieved images, respectively

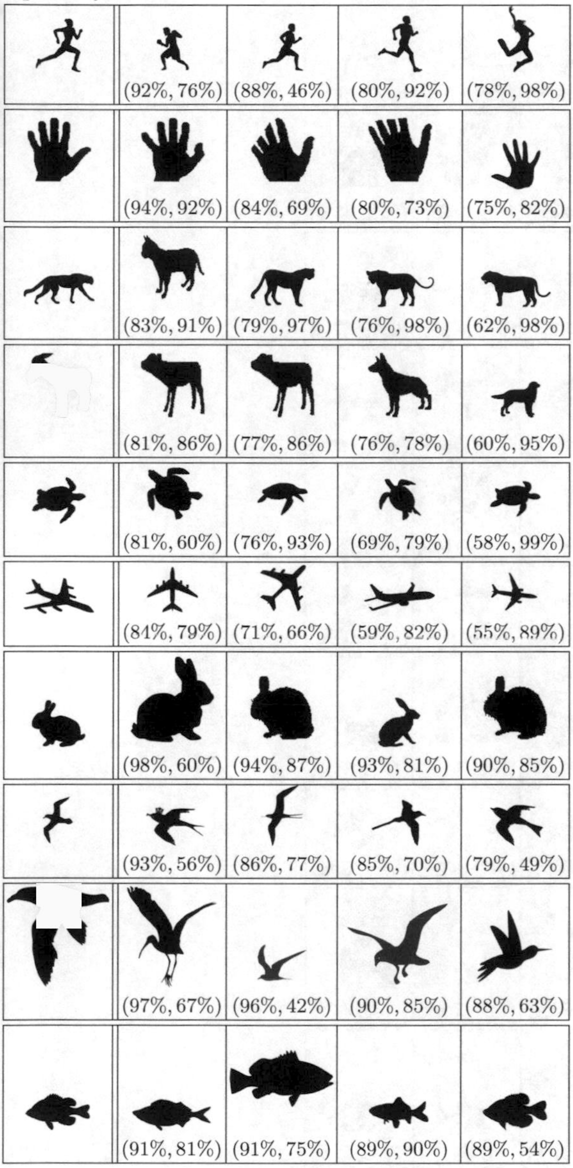

Query				
	(92%, 76%)	(88%, 46%)	(80%, 92%)	(78%, 98%)
	(94%, 92%)	(84%, 69%)	(80%, 73%)	(75%, 82%)
	(83%, 91%)	(79%, 97%)	(76%, 98%)	(62%, 98%)
	(81%, 86%)	(77%, 86%)	(76%, 78%)	(60%, 95%)
	(81%, 60%)	(76%, 93%)	(69%, 79%)	(58%, 99%)
	(84%, 79%)	(71%, 66%)	(59%, 82%)	(55%, 89%)
	(98%, 60%)	(94%, 87%)	(93%, 81%)	(90%, 85%)
	(93%, 56%)	(86%, 77%)	(85%, 70%)	(79%, 49%)
	(97%, 67%)	(96%, 42%)	(90%, 85%)	(88%, 63%)
	(91%, 81%)	(91%, 75%)	(89%, 90%)	(89%, 54%)

Table 2 Retrieved objects w.r.t. query objects with $g = 4$ and $cs\_threshold > 7$. Numeric figure below each object represent fraction of the common substring in the query and the retrieved image respectively

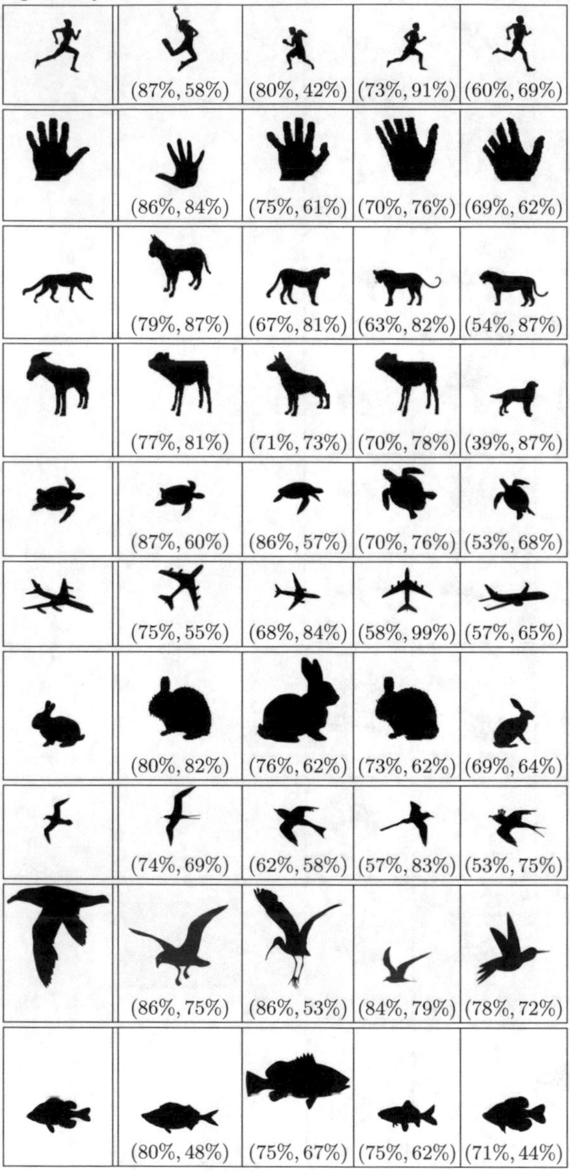

Query	Retrieved 1	Retrieved 2	Retrieved 3	Retrieved 4
	(87%, 58%)	(80%, 42%)	(73%, 91%)	(60%, 69%)
	(86%, 84%)	(75%, 61%)	(70%, 76%)	(69%, 62%)
	(79%, 87%)	(67%, 81%)	(63%, 82%)	(54%, 87%)
	(77%, 81%)	(71%, 73%)	(70%, 78%)	(39%, 87%)
	(87%, 60%)	(86%, 57%)	(70%, 76%)	(53%, 68%)
	(75%, 55%)	(68%, 84%)	(58%, 99%)	(57%, 65%)
	(80%, 82%)	(76%, 62%)	(73%, 62%)	(69%, 64%)
	(74%, 69%)	(62%, 58%)	(57%, 83%)	(53%, 75%)
	(86%, 75%)	(86%, 53%)	(84%, 79%)	(78%, 72%)
	(80%, 48%)	(75%, 67%)	(75%, 62%)	(71%, 44%)

References

1. Biswas, A., Bhowmick, P., Bhattacharya, B.B.: TIPS: on finding a tightisothetic polygonal shape covering a 2D object. In: 14th Scandinavian Conference on Image Analysis (SCIA). vol. 3540, pp. 930–939. Springer, Berlin (2005)
2. Biswas, A., Bhowmick, P., Bhattacharya, B.: Scope: shape complexity of objects using isothetic polygonal envelope, pp. 356–360 (2007)
3. Biswas, A., Bhowmick, P., Bhattacharya, B.B.: Musc: multigrid shape codes and their applications to image retrieval. In: Hao, Y., Liu, J., Wang, Y., Cheung, Y.M., Yin, H., Jiao, L., Ma, J., Jiao, Y.C. (eds.) Computational Intelligence and Security, pp. 1057–1063. Springer, Berlin, Heidelberg (2005)
4. Biswas, A., Bhowmick, P., Bhattacharya, B.B.: Construction of isothetic covers of a digital object: a combinatorial approach. J. Vis. Comun. Image Represent. **21**(4), 295–310 (2010)
5. Biswas, A., Bhowmick, P., Bhattacharya, B.B., Das, B., Dutt, M., Sarkar, A.: Triangular covers of a digital object. J. Appl. Math. Comput. (2017). https://doi.org/10.1007/s12190-017-1162-8
6. Brinkhoff, T., Kriegel, H.P., Schneider, R., Braun, A.: Measuring the complexity of polygonal objects (1995)
7. Freeman, H.: On the encoding of arbitrary geometric configurations. IRE Trans. Electron. Comput. EC **10**(2), 260–268 (1961)
8. Gdalyahu, Y., Weinshall, D.: Flexible syntactic matching of curves and its application to automatic hierarchical classification of silhouettes. IEEE Trans. Pattern Anal. Mach. Intell. **21**(12), 1312–1328 (1999)
9. Khan, J.F., Bhuiyan, S.M.A., Adhami, R.R.: Image segmentation and shape analysis for road-sign detection. IEEE Trans. Intell. Transp. Syst. **12**(1), 83–96 (2011)
10. Klette, R., Rosenfeld, A.: Digital Geometry: geometric Methods for Digital Picture Analysis. Morgan Kaufmann, San Francisco (2004)
11. Naser, M.A., Hasnat, M., Latif, T., Nizamuddin, S., Islam, T.: Analysis and representation of character images for extracting shape based features towards building an OCR for Bangla script. In: International Conference on Digital Image Processing, pp. 330–334 (2009)
12. Rigau, J., Feixas, M., Sbert, M.: Shape complexity based on mutual information. In: International Conference on Shape Modeling and Applications 2005 (SMI' 05), pp. 355–360 (2005)
13. Robert, F., Wagner, M.J.F.: The string-to-string correction problem. J. ACM (JACM) **21**(1), 168–173 (1974)
14. Sarkar, A., Dutt, M.: Construction of sandwich cover of digital objects. In: Combinatorial Image Analysis—17th International Workshop, IWCIA. pp. 172–184 (2015). https://doi.org/10.1007/978-3-319-26145-4_13
15. Schuster, G.M., Katsaggelos, A.K.: An optimal polygonal boundary encoding scheme in the rate distortion sense. IEEE Trans. Image Process. **7**(1), 13–26 (1998)

Complete Statistical Analysis to Weather Forecasting

Anisha Datta, Shukrity Si and Sanket Biswas

Abstract The primary objective of the model applied in this work is to predict the weather of a city named Austin in Texas using supervised machine learning algorithms. In this case, artificial neural networks and gradient boosting classifier were implemented to build models to predict weather and comparisons between these two models are also made for this dataset. Here, average temperature, average dew point, average pressure of sea level, average percentage of humidity, etc. are the parameters taken into consideration which influence the weather of the place. Using these parameters, the trained models performed a classification to predict whether the weather is rainy (thunderstorm or not), not rainy, snowy, or foggy.

Keywords Supervised learning · Neural networks · Gradient boosting classifier · Comparison

1 Introduction

Machine learning is a branch of data science that basically plays with the data in a statistical manner. If we have input and output data, then it builds models from the data and the models are used for prediction or solving the given tasks. We can break it into two parts: Learner and reasoner. With the help of the experience of the data and background knowledge, learner builds models and with the help of models,

A. Datta · S. Si (✉)
Department of Computer Science and Engineering, Jalpaiguri Government Engineering College, Jalpaiguri 735102, India
e-mail: sukriti.si98@gmail.com

A. Datta
e-mail: dattaanishadatta@gmail.com

S. Biswas
Computer Vision and Pattern Recognition Unit, Indian Statistical Institute, Kolkata 700108, India
e-mail: sanketbiswas1995@gmail.com

A. K. Das et al. (eds.), *Computational Intelligence in Pattern Recognition*, Advances in Intelligent Systems and Computing 999,
https://doi.org/10.1007/978-981-13-9042-5_65

reasoner solves tasks. Here, machine learning algorithms were applied to implemen this work.

Weather forecasting is the work of predicting the condition of atmosphere on th basis of some given data like temperature, pressure, wind speed, humidity, etc. at a specified area. Meteorological approach is very popular for predicting weather. Many meteorological instruments like thermometer, hydrometer, barometer, rain gauges, etc. are used for prediction. But it is very complex to merge all the information together and predict the state of atmosphere. But predicting the state of atmosphere is very important for our daily life especially for the farmers and for any outdoor work. If weather is predicted correctly, farmers can know when to cultivate crops, surfers can know when large waves are expected, people can plan accordingly when and where to go on their leisure time, and aircraft and shipping are also dependent on the state of atmosphere. The road accidents due to fog and alarm for thunderstorms and Tsunami are the main reasons why correct weather prediction is so important. So, we have introduced machine learning models to predict weather. Here is our project on weather forecasting of Austin, the capital of the U.S. state of Texas. So far, we are concerned about weather prediction and comparisons between the models used here.

Two models are used: artificial neural networks and gradient boosting classifier algorithm. A weather dataset for Austin, Texas was obtained and used to train this algorithm. The dataset contains some features of atmospheric condition as inputs and the outputs are generated based on the results whether the particular day will be rainy, snowy, or something else and the efficiency of these two models is also compared in our study for this dataset.

2 Related Work

We examined that some papers on machine learning for weather prediction, linear regression, and functional regression [1] were used in one paper and on the other hand, naive Bayes and C4.5 decision tree [2], multiclass support vector machines [3, 4], and ARIMA model [5] were used in the other papers and a good amount of literature is also witnessed based on artificial neural network [4, 6, 7]. In some papers, data mining technique [7–9] and normal equation methods [10] were also used. But these approaches did not become successful to identify the abnormal pattern of the weather. Between the two regression models, linear regression model was unstable to outliers due to its high variance and low bias and functional regression model was also not relevant due to its low variance and high bias for that paper [1]. The problem could be solved by collection of more data.

In other paper, the performance of C4.5 decision tree was very good but the performance of naive Bayes was very poor. With the increase of attributes in the dataset, the performance of naive Bayes drastically affected but the decision tree handled the problem in suitable manner [2]. But the time taken to build the models was less in case of naive Bayes. In the case of multiclass support vector machines and neural network, due to complex data systems, categorical and continuous pattern of the weather, noisy and high-dimensional data effective forecasting analysis could not be achieved [3, 4, 6]. So, effective models to predict weather are to be analyzed. In this paper, we have strived to build efficient models to predict weather. Neural networks and gradient boosting classifier are used in our study.

3 Preprocessing and Features Analysis

Here, average temperature, average dew point, average humidity, average pressure of sea level, average visibility, average wind speed, and average wind gust are used as features in the datasets to predict weather. The information is retrieved from mid-2014 to 2018 from the database of Austin airport. There are seven classes in this datasets: rainy, not rainy, snowy, foggy, thunderstorm, rainy with thunderstorm, and fog.

Now, we come to the point of how the selected features affect our weather. From the knowledge of fluid mechanics, we know that the air particles do not stay stable; rather they tend to move rapidly in free space. The more the space it gets, the more it spreads thus lowering the pressure and tries to cool off. As the earth gets heat during daytime, the air above the ground becomes more hot than the upper part of the atmosphere due to radiation. The hot air having low pressure gradually goes up to the space by convection heat transfer and spreads rapidly as the height increases. The vacuum created on the ground gets filled by the cold air of surrounding. If the temperature goes too high, then the vacuum is created more rapidly and the cold air tries to occupy it with high speed giving rise to powerful wind gust or storm. Now if there is some water body, the hot air carries moisture and rises up. Up there it gets cold and cold air cannot carry that much moisture. The temperature and amount of moisture it gets saturated with (when it cannot carry further) are called dew point and humidity, respectively. So it then gives rise to rain if the water droplets are big, otherwise fog in tropical region and snow in polar region. The low the dew point and humidity, the more the weather remains cool. The visibility through the air, depending on the wind speed or rain, can thus predict the weather also.

Here, we have calculated the central tendency of the datasets. 1305 sample features are used in the datasets. So, it can be useful to represent the entire data set with a single value that describes the average value of the entire set. In statistics, this value is called the central tendency. There are three types of central tendency: mean, median, and mode. The values are shown in the table. We are also interested in ones that describe the spread or variability of the data values. So, standard deviation is calculated. The values are shown in Table 1. Here SD stands for standard deviation.

To understand the distribution of the data better, Kernel density estimation is done. Kernel density estimation (KDE) is a nonparametric way of estimating the probability density function (pdf) of any distribution given a finite number of its samples. The pdf of a random variable X given finite samples, as per KDE formula, is given by

Table 1 Statistics of the features

Features	Mean	Median	Mode	Standard deviation
Average temp	70.557854	73	84	14.009579
Average dew point	56.636782	61	71	14.862556
Average humidity	66.662835	67	64	12.503302
Average pressure	30.022835	30	29.95	0.171879
Average visibility	9.162452	10	10	1.459463
Average wind speed	5.009195	5	4	2.081891
Average wind gust	21.383908	21	20	5.887797

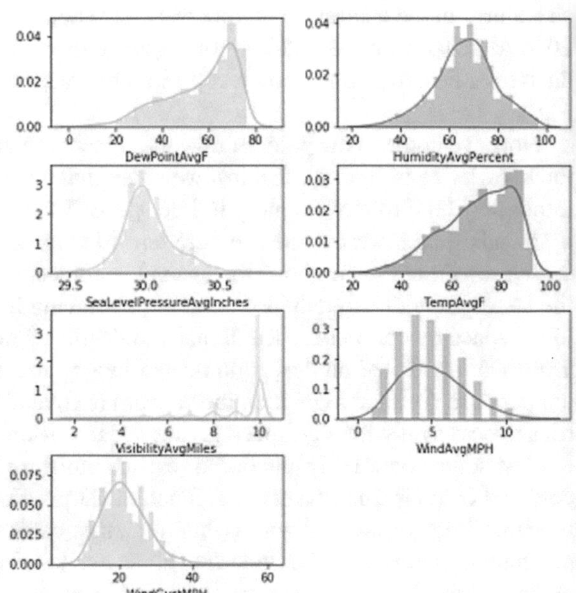

Fig. 1 Histogram plot

$$\overline{f_h}(x) = \frac{1}{n}\sum_{i=1}^{n} K_h(x - x_i) = \frac{1}{nh}\sum_{i=1}^{n} K\left(\frac{x - x_i}{h}\right) \quad (1)$$

x_i = independent and identically distributed sample, K = kernel (a nonnegative function that integrates to one), h(>0) = smoothing parameter (Bandwidth).
Graphs of the histograms of the dataset are shown in Fig. 1.

Now, we have found the correlation between the variables through correlation matrix. Correlation matrix is great way to explore new data. It can measure correlation between every combination of the variables. It does not really matter if there is an outcome at this point or not, but it will compare everything against everything. Correlation coefficient is a measure of similarity between two vectors of numbers.

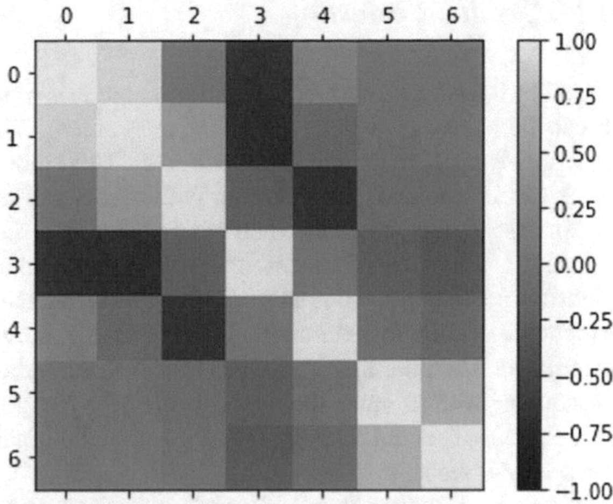

Fig. 2 Correlation matrix

The value can range between 1 and −1 where 1 is perfectly correlated, −1 is inversely correlated, and 0 is not correlated. And our correlation matrix is given in Fig. 2.

Observing the distribution and correlation of the data, we have used standardization for preprocessing of the dataset. Standardization is the process of putting different variables on the same scale, and a standardized variable is a variable that has been rescaled to have a mean of zero and a standard deviation of one. If we start with a variable x and generate a variable x*, then the process is

$$x^* = \frac{(x - m)}{sd} \tag{2}$$

where m is the mean of x and sd is the standard deviation of x.

Normalization is also used here for preprocessing of the dataset. Normalization scales all numeric variables in the range [0, 1]. If we start with a variable x and generate a variable x, and if, minimum value is called min and maximum value is called max, then the process is

$$x' = \frac{(x - min)}{(max - min)} \tag{3}$$

4 Proposed Methodology

Two algorithms are used for this prediction: Neural networks and gradient boosting classifier.

4.1 Gradient Boosting

Gradient boosting is a greedy algorithm, and it is a powerful model for prediction. It can be stated as hypothesis boosting problem. This algorithm consists of loss function, weak learner, and additive model. This algorithm works on the belief that a weak learner can be a better learner with the help of several regularization methods.

As base learner, decision trees of fixed size are used with gradient boosting algorithm. It is a method of ensemble learning that implements the sequential boosting algorithm. Basically, the goal of gradient boosting machine (GBM) is to reduce the expectation of the loss function. To achieve this, the residual of the initial model is calculated. Then the base (or, weak) learner is fitted to the residual by the gradient descent algorithm. Thus, the model is updated by adding the weighted base learner to the previous model. Finally, the target model is obtained by iteratively conducting the previous steps.

Now assuming that the number of leaves for each tree is "J", the space of the mth tree can be divided into "J" disjoint subspaces (or, leaves) such as $R_{1m}, R_{2m}, \ldots, R_{Jm}$ and the predicted value for subspace R_{jm} is the constant b_{jm}. The regression tree for input x_i can be expressed as

$$g_m(x_i) = \sum_{j=1}^{J} b_{jm}.I(x_i \epsilon R_{Jm}) \tag{4}$$

$$\begin{aligned} where \quad & I(x_i \epsilon R_{Jm}) = 1, if x_i \epsilon R_{Jm} \\ otherwise \quad & = 0 \end{aligned} \tag{5}$$

To minimize the loss function, we use the steepest descent method. We take the approximate solution as follows.

On summation,

$$F(x_i) = \sum_{m=0}^{M} f_m(x_i) \tag{6}$$

M denotes the index of the tree, f_m is the learning model (or, function) for each input x_i.

Here, we can write

$$f_m(x_i) = -\rho_m g_m(x_i) \tag{7}$$

where gradient $g_m(x_i)$ (pseudo-residual for a given tree) is

$$g_m(x_i) = \left[\frac{\partial L(y_i, f(x_i))}{\partial f(x_i)}\right], \text{with} f(x_i) = f_{m-1}(x_i) \tag{8}$$

And the multiplier ρ_m (a constant for optimisation) is

$$\rho_m = argmin_p \sum_{i=1} nL(y_i, f_{m-1}(x_i) + \rho_m g_m(x_i)) \tag{9}$$

The updated model is expressed as

$$F_m(x_i) = F_{m-1}(x_i) + \rho_m g_m(x_i) \tag{10}$$

this is the most important step (i.e., iteration) of this algorithm.

4.2 *Artificial Neural Networks*

In our brain, the neurons act as a unit cell which processes information through a connecting bridge called the synopses. These neurons communicate with other neurons through chemical signals and process the data. Like these neurons in biological system, the computer scientists have created a model named Artificial Neural Networks (ANNs) to process data in a big data system in order to predict future outcomes by observing the previously recorded data pattern. This model is the foundation of Artificial Intelligence (AI) which is nothing but a machine that works much faster than a human brain. This learns from the human activity pattern upon training and gives the future analysis based on the learnt algorithms. The more collective data it gets, the more it performs well.

The ANN model has an activation function which calculates the weighted sum of the inputs, i.e., given featured data. Here, the weighted inputs create a multilayered network. By backpropagation method (i.e., based on gradient descent algorithm for error calculation), we can learn the pattern and reduce the error with multiple iterations.

Here is the activation function (x, w in vector form)

$$A_j(\overline{x}, \overline{w}) = \sum_{i=0}^{n} x_i w_{ji} \tag{11}$$

where w_{ji} is the weight connecting neuron j to neuron i (each neuron of previous layer). We get the actual output per neuron j as

$$O_j(\overline{x}, \overline{w}) = \frac{1}{1 + e^{A_j(\overline{x}, \overline{w})}} \tag{12}$$

The error is the difference between the actual (O_j) and the desired output (d_j). We need to modify the weights in order to minimize the error. We can define the squared error function for the output of each neuron j as

$$E_j(\overline{x}, \overline{w}, d) = (O_j(\overline{x}, \overline{w}) - d_j)^2 \tag{13}$$

To adjust the weights of each neuron, we use backpropagation algorithm which iteratively adds some random weight to the previous weight. The additional weight depends on the learning rate and gradient of error E w.r.t each given weight as

$$\Delta w_{ji} = -\eta \frac{\delta E}{\delta w_{ji}} \tag{14}$$

where η is the learning rate ("+" constant),

Now, we calculate how much the error depends on the output:

$$\frac{\delta E}{\delta O_j} = 2(O_j - d_j) \tag{15}$$

The output can be expressed in terms of the inputs as

$$\frac{\delta O_j}{\delta w_{ji}} = \frac{\delta O_j}{\delta A_j} . \frac{\delta A_j}{\delta w_{ji}} = O_j(1 - O_j)x_i \tag{16}$$

Thus adjustment to each weight will be

$$\Delta w_{ji} = -2\eta(O_j - d_j)O_j(1 - O_j)x_i \tag{17}$$

So, the weights are modified as

$$w_{ji}(t + 1) = w_{ji}(t) + \delta w_{ji} \tag{18}$$

This modified weight (weight w_{ji} at time t + 1) helps in reducing the error generated in each layer and thus helps the model learn better from the training dataset and gives better prediction.

5 Results

In our study, we have observed accuracy, precision, recall, and F1 score of the used model and made a comparison between these models based on the values. We got these values by preprocessing the dataset with two methods—standardization and normalization and one with no preprocessing. We can see in each case the values are increased after preprocessing the dataset. It is also seen that the algorithm gradient boosting has worked well over ANN giving better results in most of the cases. Now we come to the confusion matrix from which we can calculate those things.

Table 2 Confusion matrix of binary classification

Predicted (cols)/True (rows)	Positive	Negative
Positive	TP	FN
Negative	FP	TN

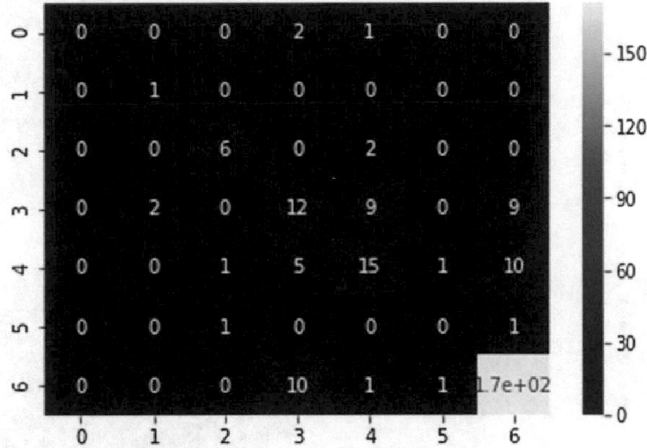

Fig. 3 GBM (no preprocessing)

The confusion matrix is a way to summarize the performance of a classifier for classification tasks. The square matrix consists of columns and rows that list the number of instances relative or absolute true class versus predicted class ratios.

For the binary classification, the two outputs are positive and negative and its confusion matrix is shown in Table 2. Here, TP—True Positive (predicted true and actually true), FN—False Negative (predicted false but actually true), FP—False Positive (predicted true but actually false), TN—True Negative (predicted false and actually false).

But, in our study, we have multiclass classification. For multiclass classification, if we have N classes then we will get N * N matrix. The heatmap representations of the confusion matrices of our study using gradient boosting classifier and artificial neural network are given in Figs. 3, 4, 5, 6, 7, and 8 with no preprocessing, standardization, and normalization, respectively.

We have calculated accuracy, recall, precision, and F score for multiclass classification using this N * N matrix (M) shown in Tables 3, 4, 5, and 6. The formulae are also given.

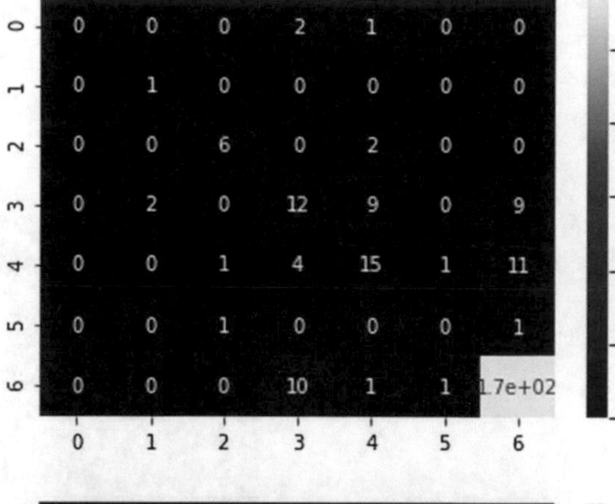

Fig. 4 GBM (standardized)

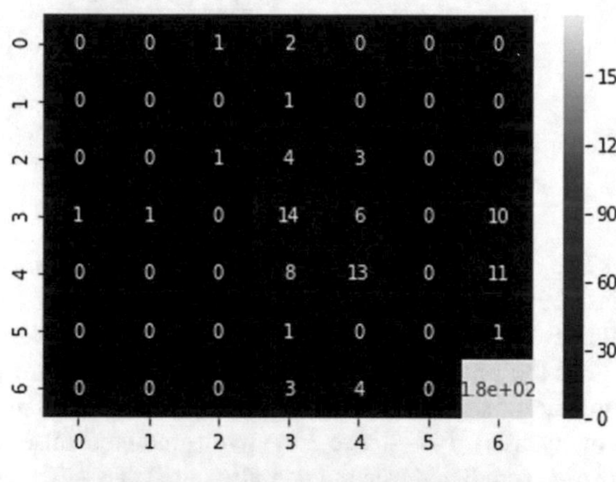

Fig. 5 GBM (normalized)

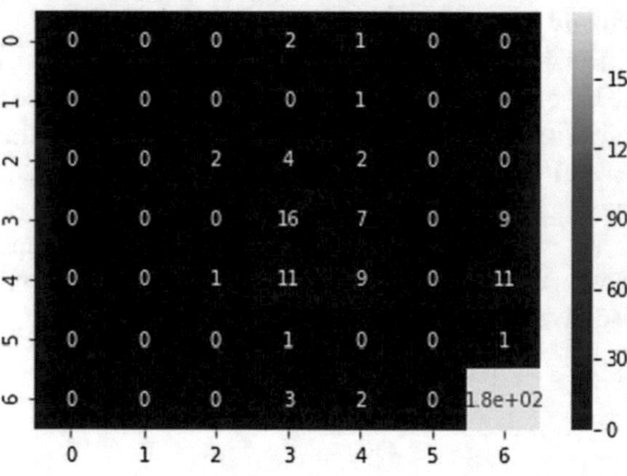

Fig. 6 ANN (no preprocessing)

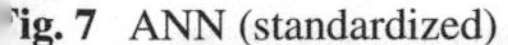

Fig. 7 ANN (standardized)

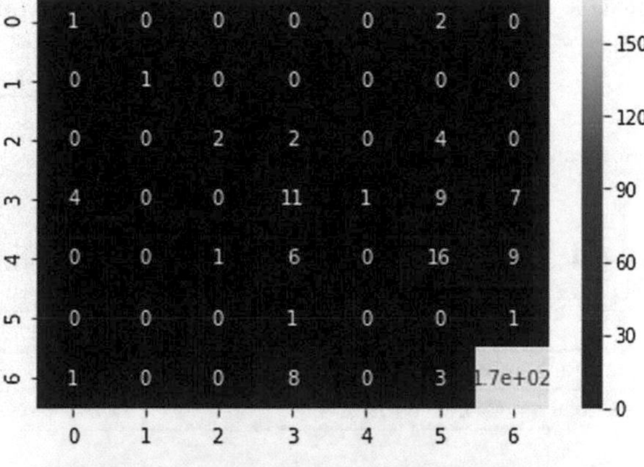

Fig. 8 ANN (normalized)

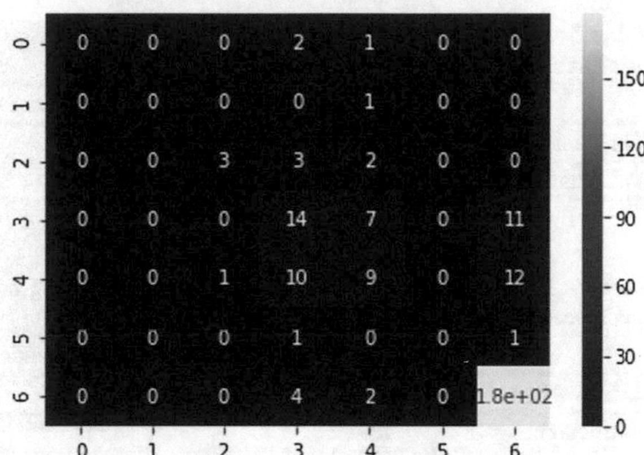

$$1.\ Accuracy = \frac{true\ \ prediction\ \ of'+'\ ve\ \ and\ \ '-'\ ve\ \ classes}{Population\ \ of\ \ all\ \ classes} \tag{19}$$

$$2.\ Precision_i = \frac{M_{ii}}{\sum_j M_{ij}} \tag{20}$$

$$3.\ Recall_i = \frac{M_{ii}}{P} = \frac{M_{ii}}{\sum_j M_{ji}} \tag{21}$$

$$4.\ F1 - score = \frac{2}{\frac{1}{recall} + \frac{1}{precision}} \tag{22}$$

(i.e., the harmonic mean of recall and precision).

M_{ii} = diagonal values of the matrix which represents true prediction (TP), M_{ij} = Every Case predicted as true but actually may be true or false (TP + FP), M_{ji} = total number of true cases where predictions were sometimes true and sometimes not (TP + FN).

Table 3 Accuracy prediction by these two algorithms using no preprocessing one time and two preprocessing methods standardization and normalization

	No preprocessing	Standardization	Normalization
Gradient boosting	78.16	78.92	78.54
Artificial network	78.16	78.54	77.78

Table 4 Precision prediction by these two algorithms using no preprocessing one time and two preprocessing methods standardization and normalization

	No preprocessing	Standardization	Normalization
Gradient boosting	76.37	77.023	75.641
Artificial network	75.134	77.291	74.270

Table 5 Recall prediction by these two algorithms using no preprocessing one time and two preprocessing methods standardization and normalization

	No preprocessing	Standardization	Normalization
Gradient boosting	78.16	78.93	78.54
Artificial network	78.16	78.544	77.778

Table 6 F1 score prediction by these two algorithms using no preprocessing one time and two preprocessing methods standardization and normalization

	No preprocessing	Standardization	Normalization
Gradient boosting	77.05	77.80	76.56
Artificial network	76.056	77.557	75.55

The more the value of diagonal elements with less value for the non-diagonal ones, the more correct predictions the model will give in distinguishing between the classes.

Here, we have made a comparison between gradient boosting and neural network algorithm on the basis of accuracy, precision, recall, and F1 score. The results are pretty much close but gradient boosting classifier with standardization model is better than the other models in this case. For this dataset, preprocessing has played a key role and standardization gives better result. But the result can be made better for neural network with both standardization and normalization if the study is extended.

6 Conclusion

In our study, we use machine learning algorithms for weather prediction and these models yield good results and can be considered as an alternative to traditional metrological approaches. The study explains the effectiveness of machine learning algorithms for predicting various weather phenomena like rain, thunderstorm, snow, fog, etc. Here, we use gradient boosting classifier and artificial neural network algorithms and after observing the comparison of the results between two models, we can say that they are well-suited models for this kind of application. It also concludes that the backpropagation algorithm is also capable of predicting weather and can also be applied to this kind of weather forecasting data. And further improvement can be made to the results of these models by doing proper preprocessing to the dataset at early stage.

References

1. Holmstrom, M., Liu, D., Vo, C.: Machine Learning Applied to Weather Forecasting. Stanford University, 15 Dec 2016
2. Sheikh, F., Karthick, S., Malathi, D., Sudarsan, J.S., Arun, C.: Analysis of data mining techniques for weather prediction. Indian J. Sci. Technol. **9**(38) (2016). https://doi.org/10.17485/ijst/2016/v9i38/101962
3. Zhang, W., Zhang, H., Liu, J., Li, K., Yang, D., Tian, H.: Weather prediction with multiclass support vector machines in the fault detection of photovoltaic system. IEEE/CAA J. Autom. Sinic **4**(3) (2017)
4. Janani, B., Priyanka, S.: Analysis on the weather forecasting and techniques. Int. J. Adv. Res. Comput. Eng. Technol. (IJARCET) **3**(1) (2014)
5. Krishna, G.V.: An integrated approach for weather forecasting based on data mining and forecasting analysis. Int. J. Comput. Appl. **120**(11) (2015)
6. Sawaitul, S.D., Wagh, K.P., Chatur, P.N.: Classification and prediction of future weather by using back propagation algorithm-an approach. Int. J. Emerg. Technol. Adv. Eng. **2**(1), (2012). ISSN 2250-2459
7. Samya, R., Rathipriya, R.: Predictive analysis for weather prediction using data mining with ANN: a study. Int. J. Comput. Intell. Inform. **6**(2) (2016)
8. Chauhan, D., Thakur, J.: Data mining techniques for weather prediction: a review. Data Min. Tech. Weather Predict. Rev. **2**(8)
9. Biradar, P., Ansari, S., Paradkar, Y., Lohiya, S.: Weather prediction using data mining. IJEDR **5**(2) (2017). ISSN: 2321-9939
10. Gupta, S., Indumathy, K., Singhal, G.: Weather prediction using normal equation method and linear regression techniques. Gupta, S., et al. (ed) (IJCSIT) Int. J. Comput. Sci. Inf. Technolo. **7**(3), 1490–1493 (2016). www.ijcsit.com 1490

An Improved Non-local Means Denoising Technique for Brain MRI

Saumyadipta Sarkar, Prasun Chandra Tripathi and Soumen Bag

Abstract Noise artifacts introduced in Magnetic Resonance Imaging (MRI) due to imperfection in radio-frequency coils typically deteriorate the performance of automated analysis of a brain MRI. Thus, it is necessary to eliminate the noise for effective analysis of MRI. The denoising techniques that work on pixels vicinity (e.g., Averaging filter and Gaussian filter) suffer in denoising due to loss of salient information in brain MRI. As a result, these techniques gain poor Peak Signal-to-Noise Ratio (PSNR) in denoising of an MRI. However, the methods that denoise an image based on non-local correlation of patches show significant improvement in the performance. In this work, we proposed an improved denoising method for brain MRI based on non-local correlation of patches. The input image is decomposed into two components: periodic component and smooth component using Fast Fourier Transform (FFT). These two components of the image are denoised separately using non-local based averaging. The filtered image is obtained by reconstructing these two components. Quantitative and qualitative results on a T1-w simulated brain MRI dataset and a local brain MRI dataset indicate the effectiveness of the proposed method.

Keywords Brain MRI · Denoising · FFT · Non-local means · Periodic component

S. Sarkar (✉) · P. C. Tripathi · S. Bag
Department of Computer Science and Engineering, Indian Institute of Technology (Indian School of Mines) Dhanbad, Dhanbad 826004, Jharkhand, India
e-mail: saumyadiptasarkar49@gmail.com

P. C. Tripathi
e-mail: prasunchandratripathi@gmail.com

S. Bag
e-mail: bagsoumen@gmail.com

A. K. Das et al. (eds.), *Computational Intelligence in Pattern Recognition*, Advances in Intelligent Systems and Computing 999,
https://doi.org/10.1007/978-981-13-9042-5_66

1 Introduction

Medical images are playing a crucial role in the treatment of several diseases. The images are taken to visualize the internal organs of the human body. The images typically visualize the inflammation in tissue with distinguishable gray levels that help the physicians to detect a disease. Computed Tomography (CT), X-ray, MRI and Ultrasound scans are few medical imaging modalities. In particular, MRI [1] is widely accepted in the medical field due to its ability to produce images without any harmful radiation. This technology uses strong magnetic fields, radio waves, and magnetic field gradients to delineate the tissues in a human body. However, the imperfect magnetic coils and patients movement lead to MRI scans contaminated with noise. The noise artifacts in MRI can hamper the performance of various stages of automated detection of a disease [2]. So, noise reduction is essential before any further processing of an MRI.

There are two types of methods to handle noise in MRI. In the first type, a noise-free image is obtained by taking multiple scans of an object and averaging them [3]. However, this method requires a significant amount of acquisition time, which is not possible in a typical clinical setting. In the second type, an MRI is acquired with the one-time scan, and later noise reduction methods are applied to achieve the enhanced image. The Gaussian smoothing that works on pixel's local neighborhood has also been used in some studies [4]. This method cannot preserve the edges of an MRI image. This problem has been tackled by denoising in the orthogonal directions of edges in Anisotropic Diffusion Filter (ADF) [5]. However, this filter struggles in preserving small details in an MRI. Recently, non-local means filter (NLM) [6] has shown its effectiveness in denoising of an image. In this method, the redundant information present in the images is used for effective filtering. This filter has also been applied for MRI denoising [7]. However, due to complex structure of human brain, it has insufficient number of similar patches. As a result, NLM produces low denoising performance. In [8], a collaborative non-local means (CNLM) denoising scheme is presented. In this method, an MRI is denoised by utilizing redundant patches across co-denoising images. The co-denoising images are obtained from different patients. The method is tested on simulated images. However, it is difficult to search patches across the images in real settings due to variations in brain images of different patients. In this work, we present a novel method for denoising of brain MRI. The method works on the transformed domain rather than spatial domain of the image. The image is decomposed into two components that are denoised using NLM filter. Finally, these two components are reconstructed to obtain noise-free image.

The rest of the paper is categorized as follows. Section 2 describes the proposed method, Sect. 3 presents the experimental results, and Sect. 4 concludes the paper.

2 Proposed Method

The proposed method works in three steps. In the first step, the image is decomposed into periodic and smooth components using FFT. In second step, these two components undergo non-local means filtering. Finally, these two components are reconstructed to obtain filtered image.

2.1 Decomposition of Brain MRI

The input brain MRI I of size $M \times N$ is decomposed [9] into two components as follows:

$$I = I_P + I_S \tag{1}$$

where I_P denotes periodic component of input MRI and I_S denotes smooth component of input MRI. The periodic component is visually likely with the actual image, and smooth component has very significant smooth variations over the image domain. These two components are estimated using the following equations:

$$I_S(i, j) = \frac{X(i, j)}{2\cos(\frac{2\pi i}{N}) + 2\cos(\frac{2\pi j}{M}) - 4} \tag{2}$$

$$I_P(i, j) = I(i, j) - \frac{X(i, j)}{2\cos(\frac{2\pi i}{N}) + 2\cos(\frac{2\pi j}{M}) - 4} \tag{3}$$

where, $X(i, j) = x_1(i, j) + x_2(i, j)$.
$x_1(i, j)$ and $x_2(i, j)$ are estimated as follows:

$$x_1(i, j) = \begin{Bmatrix} I(M-1-i, j) - I(i, j), \text{ if } i = 0 \text{ or } i = N-1 \\ 0 \qquad\qquad\qquad\qquad\qquad\qquad otherwise \end{Bmatrix}$$

$$x_2(i, j) = \begin{Bmatrix} I(i, N-1-j) - I(i, j), \text{ if } j = 0 \text{ or } j = M-1 \\ 0 \qquad\qquad\qquad\qquad\qquad\qquad otherwise \end{Bmatrix}$$

Figure 1 shows the decomposition of a T1-w axial brain MRI. The input brain MRI is shown in Fig. 1a which is decomposed into periodic and smooth components as shown in Fig. 1b, c, respectively.

2.2 Non-local Means Filtering

Let $NLM(P)(p_i)$ be the recomputed value of given voxel at location $p_i \in \mathbb{R}^2$, The computation within the search window for $W_s(p_i)$ is carried out as follows:

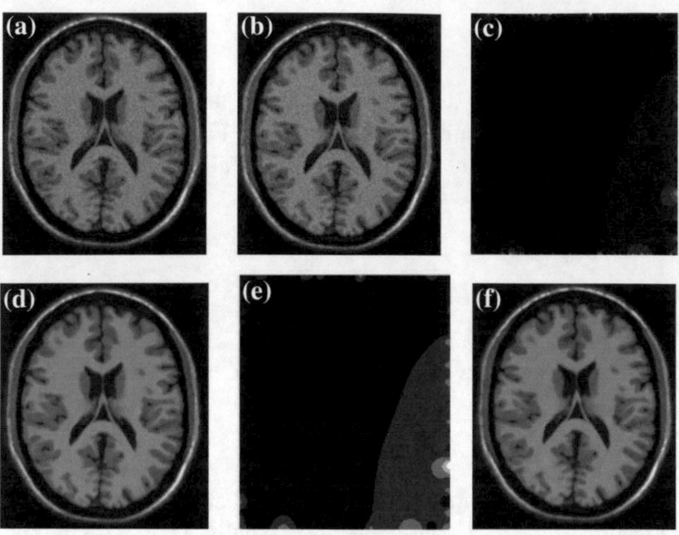

Fig. 1 Denoising of an axial brain MRI by proposed method. **a** A noisy brain MRI, **b** periodic component, **c** smoothing component, **d** NLM filtered image of periodic component, **e** NLM filtered image of smoothing component, and **f** Denoised MRI

$$NLM(P)(p_i) = \sum_{p_j \in W_s(p_i)} w(p_i, p_j)P(p_j) \tag{4}$$

where p_i represents coordinates of ith pixel, $W_s(p_i)$ is the search window of pixel p_i of certain radius, $P(p_j)$ is the intensity value of pixel p_j, and $w(p_i, p_j)$ is the weight. The weight $w(p_i, p_j)$ is estimated using the Euclidean distance between the vector $P(N(p_i))$ and $P(N(p_j))$,

$$w(p_i, p_j) = \frac{1}{Z_i} \exp^{-} \frac{||P(N(p_i)) - P(N(p_j))||^2}{h_i^2} \tag{5}$$

where $N(p_i)$ represents the neighbors of pixel p_i, h_i controls the decay of experimental function, and Z_i is a normalization constant to ensure that sum of $w(p_i, p_j)$ is one, and weights always follow the following condition:

$$Z_i = \sum_{p_j \in W_s(p_i)} w(p_i, p_j) \tag{6}$$

$$0 \leq w(p_i, p_j) \leq 1 \tag{7}$$

if p_i and p_j are same, then weight becomes very large. Hence, $w(p_i, p_j)$ is set according to

$$w(p_i, p_j) = max(w(p_i, p_j)), \forall i \neq j \tag{8}$$

2.3 Reconstruction of MRI Components

After applying non-local means algorithm in second step on MRI components, we obtain filtered MRI components ($NLM(I_P)$ and $NLM(I_S)$) as shown in Fig. 1d, e. These components are reconstructed to obtain noise-free image $\hat{I}$ (as shown in Fig. 1f).

$$\hat{I} = NLM(I_P) + NLM(I_S) \tag{9}$$

3 Experimental Results

3.1 Evaluation Framework

The performance of the proposed denoising method has been tested quantitatively and qualitatively. The quantitative results have been tested using PSNR which is given as follows:

$$PSNR = 10\log_{10}\left(\frac{255^2}{MSE}\right) \tag{10}$$

where MSE is Mean Squared Error which is estimated using the following equation:

$$MSE = \frac{1}{NM}\sum_{i=0}^{N-1}\sum_{i=0}^{M-1}(I(i, j) - \hat{I}(i, j)) \tag{11}$$

where $I(i, j)$ denotes reference image and $\hat{I}(i, j)$ denotes denoised image. The proposed method is implemented in MATLAB 7.0. All codes have been executed on a system having Intel Pentium 1.60 Ghz N3710 processor with 4 GB RAM. The parameter h has been taken in the range $\hat{\sigma} \leq h \leq 1.2\hat{\sigma}$ where $\hat{\sigma}$ is the estimated noise level. This value has been set experimentally. The squared windows of sizes 5×5 and 3×3 are used as a search window and neighborhood window, respectively.

3.2 Dataset Description

We have tested the performance of the proposed method on one volume of simulated BrainWeb dataset[1] and a local brain MRI dataset. The local dataset has been acquired from Asian Dwaraka Jalan Hospital, Dhanbad. The dataset has been acquired with 1.5 Tesla Siemens MRI scanner. These brain MR images are T2-w images of size 204×256.

We have taken one volume of T1-w brain MRI from BrainWeb. The volume has 0% noise level with slice thickness 1 mm. The size of each slice is 181×217. For BrainWeb data, we have added different levels of Rician noise as follows.

By adding $r\%$ Rician noise, we mean that the noise over complex domain has normal distribution $\aleph(0, P(r/100))$, where P is the intensity value of the brightest tissue in the image. Given a noise-free image, we generated Rician corrupted image for each pixel p_i by the following formula :

$$S = \sqrt{\left(\frac{x}{\sqrt{2}} + n_r\right)^2 + \left(\frac{x}{\sqrt{2}} + n_i\right)^2} \quad (12)$$

where n_r and $n_i \sim \aleph(0, \sigma^2)$. S is the realization of random variable with Rician probability density function with parameters x and σ.

3.3 Results Analysis

The performance of the proposed method is compared with three other methods: Averaging filter, NLM [7], and CNLM [8]. Table 1 shows the quantitative results in terms of PSNR. It is observed from the table that proposed method performs better for noise levels 3, 5, and 7%. However, CNLM method performs better for noise level 9%. CNLM method has one limitation. It requires a set of co-denoising images. These co-denoising images typically obtained from different patients. The denoising of an image is carried out by searching similar patches in the set of co-denoising images. This process takes significant amount of computation as compared to our method. We have also improved the computation time in the proposed method using parallel programming. We have achieved significant speed-up while running the proposed method on multiple cores as shown in Table 2.

We have also compared the denoising results of the proposed method qualitatively. Figure 2 shows the denoising results of proposed method on denoising an axial slice of brain MRI of BrainWeb dataset at different noise levels. It is observed from the figure that the proposed method can denoise effectively without loss of any information. Figure 3 shows the denoising results of three methods on local MRI dataset. It is

[1] http://www.bic.mni.mcgill.ca/brainweb/.

Table 1 Average PSNR of different methods on BrainWeb dataset

Noise level	3%	5%	7%	9%
Averaging filter	25.90	24.25	23.70	23.02
NLM [7]	32.53	31.19	30.09	29.17
CNLM [8]	37.18	33.88	33.01	31.58
Proposed method	39.68	35.81	33.03	30.96

Table 2 Computation time (in seconds) for denoising an MRI using proposed method

Physical core	Logical threads	Time
1	1	89
4	1	42.6
4	2	30.4
4	4	23.7
4	8	15.45

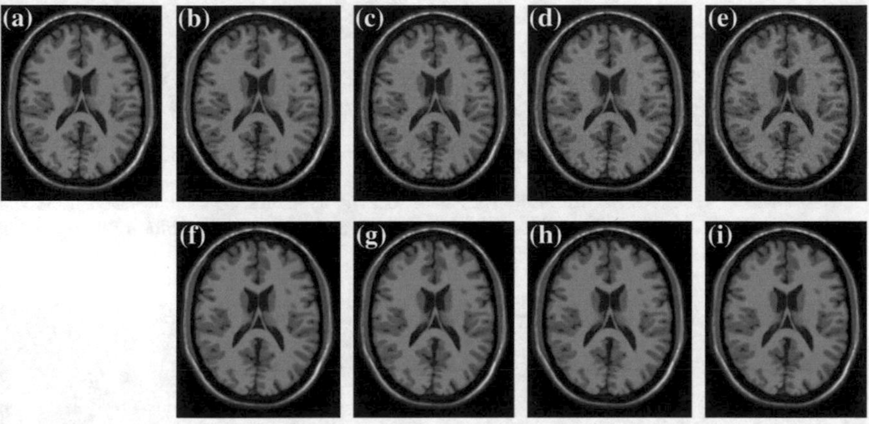

Fig. 2 Denoising of an axial brain MRI by proposed method at different noise levels. **a** An axial brain MRI, **b–e** noisy images (at 3, 5, 7, 9%), **f–i** corresponding denoised images

observed from this figure that our method can denoise the real brain MRI images effectively.

Figure 4 shows the visual results of three methods: proposed method, averaging filter, and NLM [7]. It is observed from the figure that the result of averaging filter does not retain the significant detail in the image. However, the results of proposed method and NLM retain the image detail.

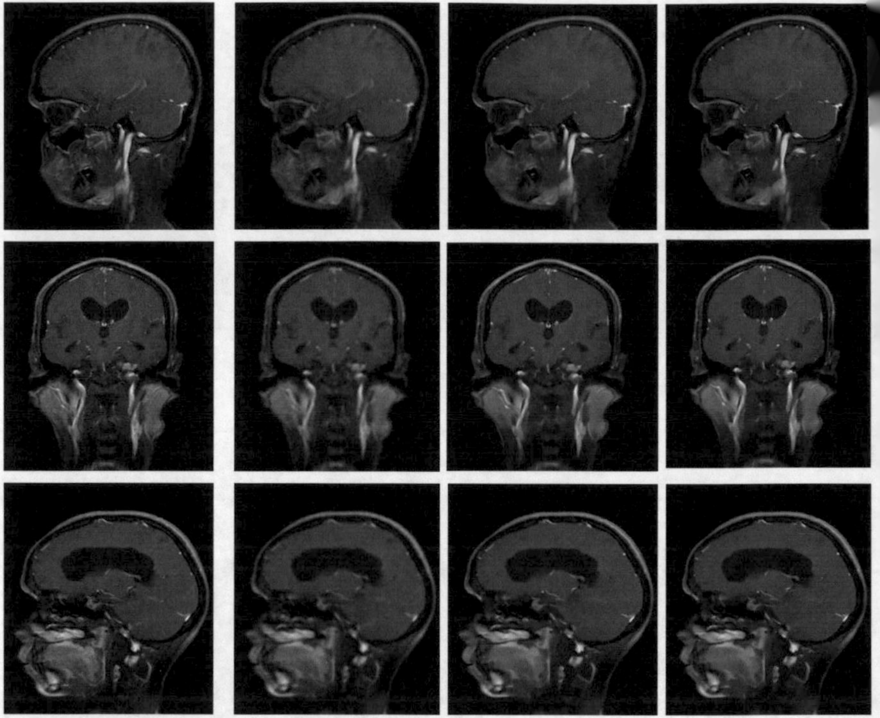

Fig. 3 Denoising results of the different methods on local dataset. First column: input MRI, Second column: result of averaging filter, Third column: result of NLM, Fourth column: result of the proposed method

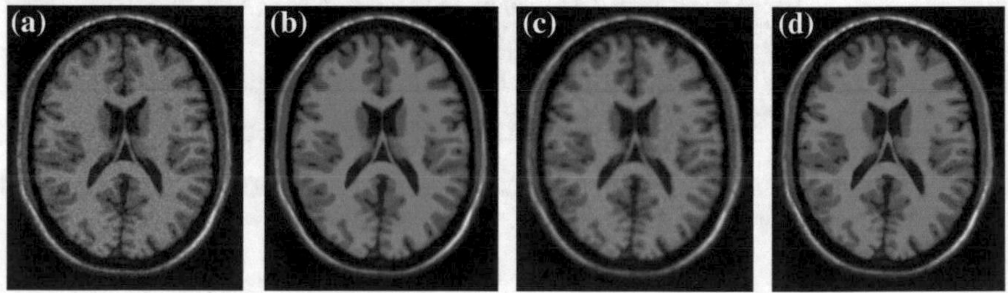

Fig. 4 Denoising results of different methods on BrainWeb slice number 90 at 9% noise. **a** An axial brain MRI, denoising results of **b** proposed method, **c** averaging filter, and **d** NLM

4 Conclusion

In this paper, a novel method for brain MRI denoising is proposed. It decomposes an MRI into two components using FFT. The non-local means filtering is carried out at component level. The filtered image is achieved by reconstructing these two components. The processing time of the proposed method is optimized through parallel programming. The experiments performed on simulated brain MRI dataset and a local brain MRI dataset give the usefulness of the proposed method.

Acknowledgements The authors would like to acknowledge Asian Dwaraka Jalan hospital, Dhanbad for providing MRI dataset to accomplish this work.

References

1. Magnetic resonance imaging. https://www.radiologyinfo.org/, Accessed 18 Aug 2018
2. Laha, M., Tripathi, P.C., Bag, S.: A skull stripping from brain MRI using adaptive iterative thresholding and mathematical morphology. In: International Conference on Recent Advances in Information Technology, pp. 1–6 (2018)
3. Gerig, G., Kubler, O., Kikinis, R., Jolesz, F.A.: Nonlinear anisotropic filtering of MRI data. IEEE Trans. Med. Imaging **11**(2), 221–232 (1992)
4. Ashburner, J., Friston, K.J.: Voxel-based morphometry the methods. Neuroimage **11**(6), 805–821 (2000)
5. Samsonov, A.A., Johnson, C.R.: Noise-adaptive nonlinear diffusion filtering of MR images with spatially varying noise levels. Magn. Reson. Med. Off. J. Int. Soc. Magn. Reson. Med. **52**(4), 798–806 (2004)
6. Buades, A., Coll, B., Morel, J.M.: A non-local algorithm for image denoising. In: Computer Vision and Pattern Recognition, pp. 60–65 (2005)
7. Manjón, J.V., Carbonell-Caballero, J., Lull, J.J., García-Martí, G., Martí-Bonmatí, L., Robles, M.: MRI denoising using non-local means. Med. Image Anal. **12**(4), 514–523 (2008)
8. Chen, G., Zhang, P., Wu, Y., Shen, D., Yap, P.-T.: Collaborative non-local means denoising of magnetic resonance images. In: International Symposium on Biomedical Imaging (ISBI), pp. 564–567 (2015)
9. Moisan, L.: Periodic plus smooth image decomposition. J. Math. Imaging Vis. **39**(2), 161–179 (2011)

A Combined Approach for *k*-Seed Selection Using Modified Independent Cascade Model

Debasis Mohapatra, Ashutosh Panda, Debasish Gouda and Sumit Sourav Sahu

Abstract Seed set selection for Information Maximization Problem (IMP) is an interesting NP-hard problem. It plays an important role in understanding the spread of information. K-shell decomposition is a graph degeneracy approach to find core as a seed set. But if the cascade of information is controlled by a hop limit and the number of nonoverlapping neighbors between the nodes is less, then the information is unable to spread quickly in the network. Here, our objective is to investigate the fast influence maximization. Some of the information is of no use after some time, in such cases, limited hop diffusion is a better model of information diffusion. Also in most of the diffusion model, the probability of diffusion is kept as a constant. But in the real-world scenario, this may be different. Hence, we have proposed a Modified Independent Cascade (MIC) model that diffuses information with different probabilities of influences of the shells and with a limiting propagation. The proposed method selects the nodes of highest degree from each shell. The proposed approach outperforms the random seed set selection in all cases. But it outperforms the k-core in case of minimum nonoverlapping neighbors. The proposed method is investigated upon the MIC model.

Keywords Graph · k-shell · Max degree · Limited hop · MIC model

D. Mohapatra (✉) · A. Panda · D. Gouda · S. S. Sahu
Department of Computer Science & Engineering, PMEC, Berhampur 761003, India
e-mail: devdisha@gmail.com

A. Panda
e-mail: ashutoshpanda7896@gmail.com

D. Gouda
e-mail: debasishgouda071@gmail.com

S. S. Sahu
e-mail: sumitprabhu73@gmail.com

A. K. Das et al. (eds.), *Computational Intelligence in Pattern Recognition*, Advances in Intelligent Systems and Computing 999,
https://doi.org/10.1007/978-981-13-9042-5_67

1 Introduction

Social networking services like Facebook, Twitter, WeChat, etc. provide a platform to share information. Social network analysis is a wide range research domain that performs variety of analysis on different types of data available in such platforms. Nowadays, these social networks are playing an important role in viral marketing due to its enormous diffusion power. The study of influence spread is used in different fields like popularity of new products and services, limit the propagation of rumor controlling the spread of virus, etc.

In the problem of information maximization, the intention is to find a set of k nodes called seed set that maximizes the diffusion of information. Kempe et al. [4] have shown that the IM problem is NP-hard. They proposed a *SimpleGreedy* approach for approximation. But after that, ample of methods have been developed to provide a better approximation. Two most popular information diffusion models are Linear Threshold Model (LTM) and Independent Cascade Model (ICM). In this paper, we have used ICM for information diffusion modeling that uses Monte Carlo (MC) simulation.

Most of the existing models consider a continuous propagation which is quite unreal in most of the situations. Most of information has its impact or utility up to a certain time limit, beyond which the information is futile. Also, most of the works consider that the probability of influence is same for all nodes, which is quite unacceptable as different persons may have different influence capacities. We incorporate the hop limit with different influence probabilities of nodes to the standard IC model and proposed a Modified IC (MIC) model.

To select k-seed node for information maximization in MIC model, we propose a combination of k-shell decomposition [5, 9] with node degree. K is the number of shells and also the number of seeds in the seed set.

In this paper, our vital contributions are

I. We have proposed MIC model by considering both hop limit and non-uniform influence probability.
II. We select k different highest degree nodes from k different shells as seed set, one from each shell.
III. The proposed method always performs better than random seed selection (k-rand) and in some cases better than k-core [5] where k-core seed has minimum nonoverlapping neighbors.

The rest of the paper is organized as follows. Section 2 discusses related work. Section 3 covers the proposed model. In Sect. 4, we discuss our proposed method. Section 5 explains about results and discussion. At last, Sect. 6 concludes the paper.

2 Related Work

Chen and Taylor [1] have proposed a mathematical model to characterize the spread of information in a network. They proposed the Markov model to spread the influence of information. Lv and Pan [8] have proposed Independent Cascade Model with Limited Propagation Distance (ICLPD) where the influence of seed nodes can only propagate to limited hops and the transmission capacities of the seed nodes are different. Here, Local Influence Discount Heuristic (LIDH) is proposed to speed up the greedy algorithm and observed that LIDH significantly outperforms other compared heuristics in terms of effectiveness in the ICLPD. Erlandsson et al. [3] proposed an idea of information spreading in the domain of online social media. The dataset used in this study is a subset of public Facebook pages. The data from these pages were parsed and for each post, the corresponding likes and comments were extracted. Independent cascade model was used for modeling information spread, and it requires a set of activated nodes at beginning and spread in a limited set. They used four methods for information spreading, that is, degree centrality, k-Shell, ARL, and Vote Node and plotted a graph against time consumption for different seed selection methods and concluded with degree centrality as the best method for this process. Zhao et al. [10] proposed that a greedy algorithm is computationally expensive to find optimal result, i.e., k-node seed set. Hence, a k-shell decomposition algorithm for influence maximization is proposed under the linear threshold model; here, the network is decomposed using a *K*-shell decomposition method to calculate the potential influence of nodes. The experiment done by them shows that KDA can achieve both high accuracy and high efficiency. D'Angelo et al. [2] worked on maximizing the influencing speed in the independent cascade model by adding some new edges incident on some given number of initial active nodes. After analyzing the properties of influencing function, which is monotonically increasing, they have proposed a greedy approximation algorithm to generate a set of edges to be added to the graph to maximize the speed of influencing and number of nodes influenced. Ko et al. [6] worked on generating a hybrid solution for efficient and effective influence maximization in social networks. They proposed a hybrid-IM that curbs the issues in both micro and macro levels. At micro-level, it implements Path-Based Influence Maximization (PB-IM) to minimize the cost of running MC simulations. While at macro-level, it implements Community-Based Influence Maximization (CB-IM) to decrease the computational overhead generated by re-evaluating the influence spread of every node for every seed node. Hybrid-IM has a performance lead of up to 43 times. Lu et al. [7] have designed a cascade discount algorithm to minimize the high computational overhead suffered by the greedy algorithm. The experimental results demonstrate that the influence spread is close to CELF and performs better than degree discount IC and two stages, and cascade discount runs two orders of magnitude faster than CELF over real-world dataset.

3 Proposed Model

The first problem with the standard IC model for influence spread is that it assume all the nodes have the same influential power. It considers that the probability of influencing the neighbors is same for all the nodes. Secondly, it considers the spread to be cascaded recursively to all nodes that may not happen in real scenario. Because most of the information are useful only for a limited time after that the information losses its utility.

We have proposed a modified IC (MIC) model with two modifications in the standard cascade model to deal with the above two problems.

I. We use *k*-shell decomposition of the nodes that divides the whole network into core–periphery structure. The influence probability of a node in $\underline{i}$th shell is P_i. It handles the first problem. The probability is assigned according to shell number, i.e., $(P_1 < P_2 < \cdots < P_k)$.
II. To handle the second problem, we incorporate hopcontrol in IC evaluated model. This model allows the information cascade only up to t hops of the seed nodes.

4 Proposed Method

In this section, we explain our proposed methodology for seed selection in IMP. The seeds are selected using the following steps:

I. Perform the *k*-shell decomposition of network.
II. Select highest degree node from each shell that constitutes a seed set S of size k.
III. The cascade capacity of the set S is evaluated under MIC model.

4.1 Perform k-Shell Decomposition of a Network

The objective of *k*-shell decomposition is to split the whole network into k different shells. The *k*-shell decomposition algorithm can be implemented using the following steps.

I. Compute the degree sequence of the whole network and sort it in ascending order say S is the sorted sequence.
II. The smallest shell number t is equal to the smallest degree t in S.
III. Delete all the nodes with degree t from the graph and put it inside the t-shell. By deleting the nodes if the degree of the neighbor of any node goes down to t or less, then delete those nodes and insert it to t-shell. Set $t \leftarrow t + 1$.
IV. Continue step III until all nodes are placed in shells.

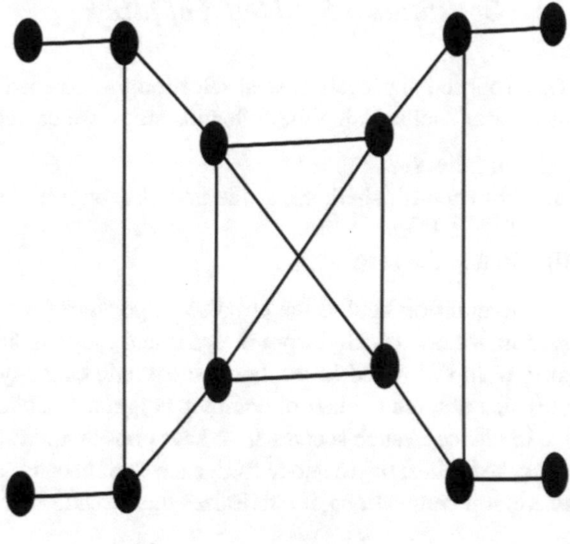

Fig. 1 Sample graph G

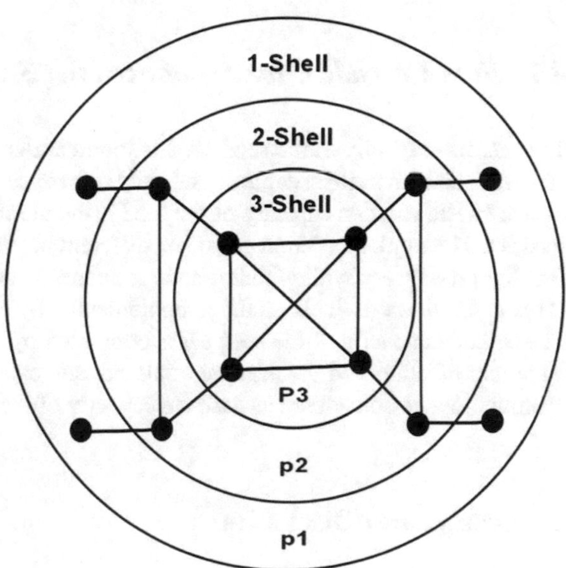

Fig. 2 k-shell decomposition of G

Figure 1 shows a sample graph G, and Fig. 2 shows its k-shell decomposition structure. In Fig. 2, the whole network is divided into three different shells. The innermost shell, i.e., shell number 3, is called as core of the graph G. Each ith shell is assigned with an influence probability of P_i. The influence probability is the probability with which a node influences its neighbors. The influence probabilities are assigned according to shell number, i.e., the node with highest shell number is assigned with highest influence probability ($P_1 < P_2 < \cdots < P_k$). In Fig. 2, probabilities are assigned as $P_1 < P_2 < P_3$.

4.2 Constitute a Seed Set S of Size k

The proposed approach of seed selection finds a seed set S of size k, where k is the number of shells. Using the following steps, we can construct a seed set S.

I. Initially $S \leftarrow \varnothing$
II. For each ith shell, select the node hn_i with highest degree and append it to S $S \leftarrow S \cup hn_i$
III. Return the set S

The intuition behind the proposed algorithm for seed set selection is that if the seed nodes are selected from the core and they are allowed to propagate the information up to limited hops, then the cascade capacity may not increase due to the presence of huge number of common neighbors between them. Hence, to maximize the influence in such scenario, we have chosen nodes from different shells. We consider the highest degree node form each shell because as the highest degree node has maximum connections, it maximizes the cascade capacity.

4.3 Find Cascade Capacity of Seed Set S Using MIC Model

The cascade capacity of the seed set S is found under the MIC model. As discussed earlier, node with highest degree is selected from each shell and constitutes a seed set of size k. The cascade capacity of the seed is the number of nodes influenced by the seed set. The implementation considers different probabilities P_i for different shells. It is the probability with which a node influences another node. Also, it limits the cascade up to t hops. This model is implemented by Monte Carlo (MC) simulation. The cascade capacity of the seed set is computed by executing MC simulations for n number of times and the average of all cascade capacities of the seed set over the n simulations is considered as cascade capacity of the seed set.

5 Results and Discussion

System Specification: A system having 2.40 GHz Intel(R) Core(TM) i7-4770 processor with a memory support of 4 GB and Microsoft Windows 8.1 Professional Operating System is used to run experiments. The implementations are done in python with the help of NetworkX tool.
Data sets: We have created five different synthetic random networks. The random networks are generated using Erdos–Renyi (ER) model. There are two variants of ER model found in the literature, they are $G(n, m)$ and $G(n, p)$ models. In $G(n, m)$ model, n is the number of vertices and m is the number of edges. This model returns a graph uniformly chosen from all the graphs with n nodes and m edges. In $G(n, p)$

Table 1 Cascade capacity measurement

Graph	No. of shells	Hop limit	k random seed set (CC)	k-core nodes as seed set (CC)	Proposed approach (CC)
ER1 ($n = 200$, $m = 800$)	3	2	114	159	168
ER2 ($n = 400$, $m = 1600$)	4	2	151	241	271
ER3 ($n = 600$, $m = 2400$)	4	3	170	277	325
ER4 ($n = 800$, $m = 3200$)	5	3	174	340	371
ER5 ($n = 1000$, $m = 4000$)	5	2	178	313	385

model, n represents the number of nodes and p is the probability of an edge being included in the graph that is independent of other edges of the graph. We have used $G(n, m)$ model to create all five random networks.

We have applied three algorithms, i.e., k-rand, k-core [5] and proposed algorithm on the datasets under *MIC* model. In k-rand algorithm, the k-seed nodes are selected at random, whereas in k-core the seed set of size k is taken from the core of the network.

Table 1 contains the information about the five *ER* graphs those are named as *ER*1, *ER*2, …, *ER*5. The number of shells generated by k-shell decomposition is represented in the second column of Table 1. We have considered k different influence probabilities for k different shells. We have considered hop limit for different graphs as shown in the third column of Table 1. Table 1 also shows the cascade capacity of k-seed set selected by three different methods. Figure 3 shows the comparison between the three algorithms, and the proposed algorithm incurs more cascade capacity than the other two algorithms.

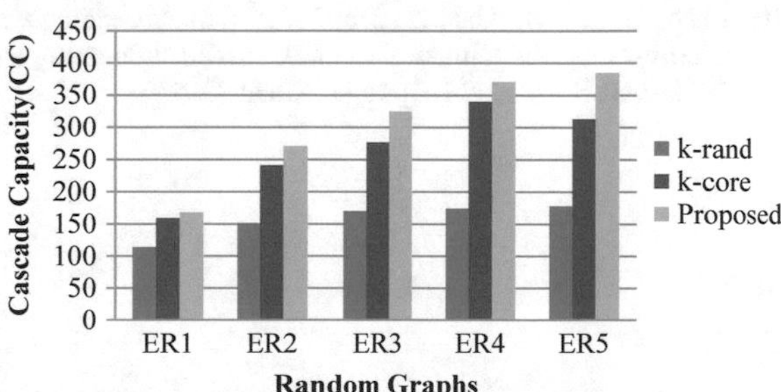

Fig. 3 Comparison between three algorithms

6 Conclusions

In this paper, we have proposed modified IC model. We add non-uniform probability and hop limit to the standard ICM. We have investigated the influence maximization by considering three algorithms: k-rand, k-core, and proposed. The proposed algorithm outperforms the rest two algorithms as it has the highest cascade capacity. In future, we shall explore the use of heuristics and metaheuristics approaches to solve IMP with different modelling parameters.

References

1. Chen, Z., Taylor, K.: Modeling the spread of influence for independent cascade diffusion process in social networks. In: ICDCSW, pp. 151–156 (2017)
2. D'Angelo, G., Severini, L., Velaj, Y.: Influence maximization in the independent cascade model. In: Proceedings of ICTCS, pp. 269–274 (2016)
3. Erlandsson, F., Bródka, P., Borg, A.: Seed selection for information cascade in multilayer networks. In: Cherifi, C., et al. (eds.) Complex Networks & Their Applications VI. COMPLEX NETWORKS 2017. Studies in Computational Intelligence, vol. 689. Springer, Cham (2018)
4. Kempe, D., Kleinberg, J., Tardos, E.: Maximizing the spread of influence through a social network. In: Proceedings of the Ninth ACM SIGKDD International Conference on Knowledge Discovery and Data Mining, KDD '03, pp. 137–146 (2003)
5. Kitsak, M., Gallos, L.K., Havlin, S., Liljeros, F., Muchnik, L., Stanley, H.E., Makse, H.A.: Identification of influential spreaders in complex networks. Nat. Phys. **6**(11), 888–893 (2010)
6. Ko, Y.-Y., Cho, K.-J., Kim, S.-W.: Efficient and effective influence maximization in social networks: a hybrid-approach. Inf. Sci. **465**(2018), 144–161 (2018)
7. Lu, F., Zhang, W., Shao, L., Jiang, X., Xu, P., Jin, H.: Scalable influence maximization under independent cascade model. J. Netw. Comput. Appl. In: KDD '10 Proceedings of the 16th ACM SIGKDD International Conference on Knowledge Discovery and Data Mining, pp. 1029–1038 (2010)
8. Lv, S., Pan, L.: Influence maximization in independent cascade model with limited propagation distance. In: Han, W., et al. (eds.) Web Technologies and Applications. APWeb 2014. LNCS, vol. 8710. Springer, Cham (2014)
9. Sariyuce, A.E., Gedik, B., Jacques-Silva, G., Wu, K.-L., Catalyurek, U.V.: Algorithms for k-core decomposition. Proc. VLDB Endow. **6**(6), 433–444 (2013)
10. Zhao, Q., Lu, H., Gan, Z., Ma, X.: A *K*-shell decomposition based algorithm for influence maximization. In: Cimiano, P., et al. (eds.) Engineering the Web in the Big Data Era. ICWE 2015. LNCS, vol. 9114. Springer, Cham (2015)

Fast Adaptive Decision-Based Mean Filter for Removing Salt-and-Pepper Noise in Images

Amiya Halder, Supravo Sengupta, Pritam Bhattacharya and Apurba Sarkar

Abstract This paper presents a very simple and extremely fast approach to remove salt-and-pepper noise from digital images. Initially, the pixels are roughly divided into two classes, noisy pixels and non-noisy pixels based on the intensity values. Then, the second stage is to eliminate the noise from the images. In this stage, only the noise pixels are processed and the noise-free pixels remain unchanged. This method takes four non-noisy pixels in four directions around noisy pixel and predicts the intensity value of the noisy candidate. This proposed method is suitable to be implemented in consumer electronics products such as digital televisions or digital cameras. Experiments and comparisons demonstrate that our proposed filter has very low detection error rate and high restoration quality.

Keywords Salt-and-pepper noise · Adaptive mean filter · Median filter · Structure similarity index

1 Introduction

Digital signals like digital images are frequently corrupted by different types of noises. One of the very commonly occurring noises is salt-and-pepper or impulse noise. Impulse noise can change the appearance of the image drastically, even with a low noise rate. This is because, the impulse noise, which is a set of random pixels,

---

A. Halder (✉) · S. Sengupta · P. Bhattacharya
St. Thomas College of Engineering and Technology, Kolkata 700023, India
e-mail: amiya.halder77@gamil.com

S. Sengupta
e-mail: supravosengupta1@gmail.com

P. Bhattacharya
e-mail: prb2794@gmail.com

A. Sarkar
IIEST, Shibpur, Howrah 711103, India
e-mail: as.besu@gmail.com

A. K. Das et al. (eds.), *Computational Intelligence in Pattern Recognition*, Advances in Intelligent Systems and Computing 999,
https://doi.org/10.1007/978-981-13-9042-5_68

normally has a very high variation to its backgrounds. Degraded sensors in camera or faulty transmission channels of the image are some of the common reasons for the presence of the impulse noises in images. The strength of the intensity deviation is comparatively very high rather than the strength of the original signal. As a result when the signals are digitized into L intensity levels, the corrupted pixels are generally digitized into either of the two extreme values, which are the lowest or highest values in the dynamic range of the intensity value (i.e., 0 or L − 1). For this phenomenon, the impulse noise present in a digital image normally gives the impression as arrangement of random completely white and completely black dots fragmented over the image.

Over decades median filters are very famous for impulse noise removal from distorted digital images. For example, the Standard Median Filter (SMF) is the most common nonlinear filter. It has been an important technique for good performance impulse noise reduction [1] in low-density noisy images. There are various different approaches that have suggested to modify the standard median filtering technique [2–7]. Some of them are decision-based median filters, in which the pixels are classified into noisy and non-noisy pixels and only the noisy candidates are processed. This approach performs a lot better than SMF because decision-based filters eliminate a lot of blurring effects from the restored image. But this technique cannot perform well for varying noise rate as it considers prefixed sized window. Another approach is to adaptively select the window size based on the noise rate to find the median value. This method performs better than simple decision-based techniques as it takes variable size window which elements the limitation of the decision-based filters. But to find the proper window size and to calculate the median value, this method takes a longer processing time which makes them inefficient for implementation on digital electronic products. Others have suggested a fuzzy switching median filter which is another modification of the decision-based filter and intellectually classifies noisy and non-noisy candidates using fuzzy logic. This approach struggles to preserve the fine details in the image but its result gives a blurred image. Also, since its window size is fixed, some of the windows contain all noisy pixels. In these cases, all noise pixels are not removed from the images and visual impression is decreased significantly. Also, a large number of methods [8–16] have been proposed to remove salt-and-pepper noise.

In this paper, we are presenting adaptive decision-based mean filtering technique to remove salt-and-pepper noise from the digital images efficiently and with a very low computational processing overhead. This method is decision based; the mean of non-noisy neighbors can predict the intensity value of the noisy candidate more accurately and with much less computational complexity. The simplicity and efficiency of the mean-based filter make the method suitable for digital electronic products. The experimental results show that the proposed method performs better than many other well-known existing techniques.

2 Noise Model

The two-sided fixed value impulse noise model is also called the "salt-and-pepper noise." When impulse noise has highest intensity (i.e., L − 1), it is known as the "salt" noise and if the noise is having lowest intensity (i.e., 0) it is known as the "pepper" noise. If "κ1" denotes the probability of occurrence of very low-valued impulse noise in a signal component and that "κ2" gives the probability of occurrence of a very high-valued impulse noise in the same signal, then the total probability of occurrence of "salt-and-pepper noise" in the signal under consideration is given by (κ1 + κ2). If 256 (i.e., 8 bit/pixel) is the number of gray levels used per pixel, and then we assume that 255 represents the "salt" noise, while 0 represents the "pepper" noise. The probability model of an image corrupted by impulse noise is depicted in Eq. 1.

$$\varphi_{i.j} = \begin{cases} 0 & With\ probability\ \kappa 1 \\ L-1 & With\ probability\ \kappa 2 \\ \psi_{i,j} & With\ probability\ 1-(\kappa 1+\kappa 2) \end{cases} \tag{1}$$

where $\psi_{i,j}$ and $\varphi_{i,j}$ are pixels at location (i, j) of clean image and corrupted image, respectively, (κ1 + κ2) is the noise percentage.

3 Proposed Method

To make the restoration process very fast we are proposing very simple mean filtering technique to suppress salt-and-pepper noise. The proposed method is very fast and efficient rather than other median filtering techniques. This method is decision based which helps to consider only the noisy candidates. Taking the mean ensures smooth reconstruction for lower noise rates. Considering the mean value not only decreases the processing overhead but also predicts the intercity value with less deviation from the actual intensity value, which increases the quality of the reconstructed image than the previously proposed median-based filters. The pixels are classified into two groups of noisy and non-noisy candidates based on Eq. 2. That is, if the intensity value is 0 or L − 1 then they are labeled as noisy candidates, otherwise non-noisy candidates assuming that the two intensities that present the impulse noise are the maximum and the minimum values of the image's dynamic range (i.e., 0 and L − 1). Non-noisy candidates are left unchanged and noisy candidates are replaced by the mean of four non-noisy candidates of the north, south, east, and west directions.

$$\alpha(i,j) = \begin{cases} 1, & f(i,j)=0 \\ 1, & f(i,j)=L-1 \\ 0, & otherwise \end{cases} \tag{2}$$

where the value of $\alpha(i, j) = 1$ signifies that the pixel is a "noisy pixel" and $\alpha(i, j) =$ 0 signifies that the pixel is a "non-noisy pixel". The block diagram of the propose algorithm is shown in Fig. 1.

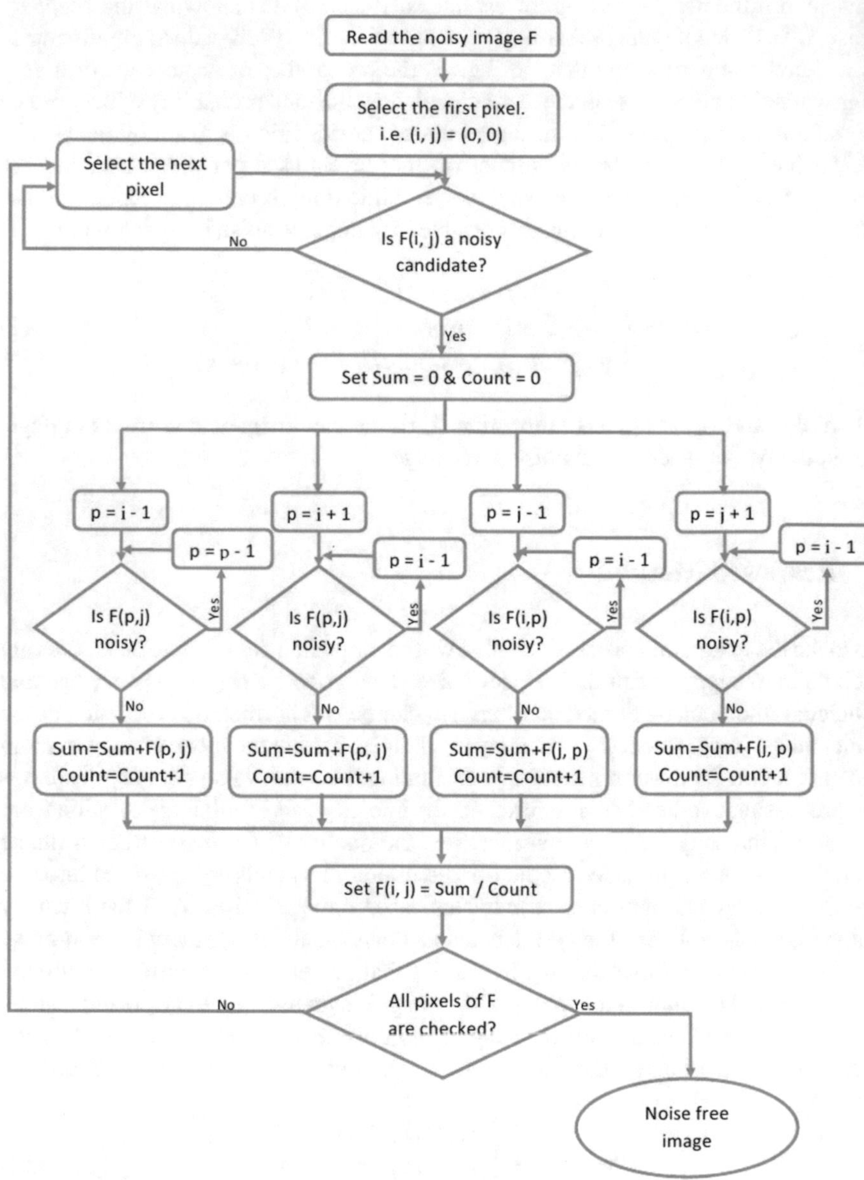

Fig. 1 Block diagram of the proposed algorithm

3.1 Algorithm

1. Read the noisy image F(i, j).
2. Classify all pixels into noisy and non-noisy pixels.
3. For every noisy pixel in F, do.

   3.1. Go toward north direction until a non-noisy pixel is found.
   3.2. Go toward south direction until a non-noisy pixel is found.
   3.3. Go toward east direction until a non-noisy pixel is found.
   3.4. Go toward west direction until a non-noisy pixel is found.
   3.5. Count the number of non-noisy pixels found in these four directions.
   3.6. If count > 0, then $\mathrm{F(i, j)} = \frac{\text{Sum of the values}}{\text{count}}$.

4 Experimental Results

In experimental tests, different existing methods are used to compare the quality of the restored images with different levels of impulse noise in *Lena*, *Cameraman*, *Goldhill*, *Airplane,* and more images. The proposed filter is compared with seven existing methods such as Standard median filter (SMF) [1], Efficient Removal of Impulse Noise (ERIN) [2], Adaptive Median Filter (AMF) [3], Adaptive Fuzzy Switching Median Filter (AFSMF) [4], Unsymmetric Trimmed Median Filter (UTMF) [5], Selective Adaptive Median Filter (SAMF) [6], and Modified Median Filter (MMF) [7]. We use Peak Signal-to-Noise Ratio (PSNR), Image Enhancement Factor (IEF), and Structural Similarity Index (SSIM) to evaluate the performance of different methods as shown in Eqs. 3, 4, 5, and 6, respectively.

$$MSE = \frac{1}{MN} \sum_{i,j} \left(\psi_{i,j} - \varphi_{i,j}\right)^2 \tag{3}$$

$$PSNR = 10 \log_{10}\left(\frac{255^2}{MSE}\right) \tag{4}$$

$$IEF = \frac{\left(\sum_i \sum_j n_{i,j} - \varphi_{i,j}\right)^2}{\left(\sum_i \sum_j \psi_{i,j} - \varphi_{i,j}\right)^2} \tag{5}$$

$$SSIM = \frac{\left(2\mu_\psi \mu_\varphi + C_1\right)\left(2\sigma_{\psi\varphi} + C_2\right)}{\left(\mu_\psi^2 + \mu_\varphi^2 + C_1\right)\left(\sigma_\psi^2 + \sigma_\varphi^2 + C_2\right)} \tag{6}$$

where μ_ψ is the average of ψ, μ_φ is the average of φ, σ_ψ is the standard deviation of ψ, μ_φ is the standard deviation of φ,$C_1 = (\rho_1\ \mathrm{L})^2$, $C_2 = (\rho_2\ \mathrm{L})^2$ are two variables to stabilize the division with weak denominator, L is the dynamic range of the pixel values (for an 8 bit image it takes from 0 to 255), $\rho_1 = 0.01$ and $\rho_2 = 0.03$ by default.

Table 1 PSNR measure for various algorithms at noise densities 10–90% of *Airplane* image

Method	Noise density (%)								
	10	20	30	40	50	60	70	80	90
SMF	31.98	28.51	22.94	18.36	14.66	11.73	9.4	7.568	6.057
ERIN	38.82	34.92	31.19	28.19	24.28	20.89	17.04	13.56	9.515
AMF	35.12	31.49	29.57	27.99	26.62	25.41	23.81	22.38	20.57
AFSMF	38.44	34.82	32.28	30.40	28.56	26.55	24.25	22.27	19.19
UTMF	38.82	35.13	31.79	29.76	27.42	25.08	22.37	18.32	13.07
SAMF	38.61	34.83	31.74	29.78	27.95	26.35	24.57	22.72	20.31
MMF	36.00	32.71	30.70	29.16	27.44	25.27	21.94	17.92	13.73
Proposed	41.96	37.78	35.08	32.69	30.49	28.57	26.34	24.10	21.17

Table 2 PSNR measure for various algorithms at noise densities 10–90% of *Gold Hill* image

Method	Noise density (%)								
	10	20	30	40	50	60	70	80	90
SMF	29.99	27.52	23.25	18.85	15.22	12.34	9.93	8.07	6.59
ERIN	38.65	34.47	31.54	28.55	25.31	21.67	18.03	14.26	10.14
AMF	36.31	32.88	30.96	29.45	28.22	27.01	25.73	24.46	22.83
AFSMF	39.02	35.36	33.09	31.26	29.53	27.81	26.1	24.23	20.99
UTMF	38.66	34.81	32.54	30.35	28.5	26.52	23.73	19.56	14.31
SAMF	38.37	34.5	32.16	30.25	28.75	27.49	25.88	24.37	22.11
MMF	36.34	33.06	31.04	29.52	28.16	26.28	23.1	19.51	15.54
Proposed	40.34	36.77	34.56	32.62	31.08	29.5	27.8	25.86	23.41

Tables 1, 2, and 3 give the PSNR values for *Airplane, Goldhill,* and *Lena* images that are corrupted with 10–90% noise density using the above-existing methods, respectively. Also, Tables 4, 5, and 6 show the SSIM and Tables 7, 8, and 9 show the IEF values for *Airplane, Goldhill,* and *Lena* images. From these tables, it is clearly shown that proposed method provides the better results than other existing methods. In Figs. 2, 3, and 4, compare the restored *Lena, Goldhill,* and *Airplane* images of proposed method with above-existing methods. From these figures, it is simple to say that proposed algorithm achieves appreciably better than existing methods. From these experimental results, it is noticed that our method outperforms all other existing methods for all the tested images.

able 3 PSNR measure for various algorithms at noise densities 10–90% of *Lena* image

Method	Noise density (%)								
	10	20	30	40	50	60	70	80	90
SMF	32.54	29.2	23.68	18.98	15.2	12.25	9.85	7.97	6.51
ERIN	40.59	36.59	33.38	29.99	25.95	21.93	18.05	14.48	10.21
AMF	37.96	34.39	32.41	30.81	29.43	28.27	26.93	25.66	23.95
AFSMF	41.29	37.63	35.21	33.01	31.29	29.23	27.26	25.29	21.73
UTMF	40.72	36.91	34.3	32.16	30.01	27.56	24.5	20.33	14.4
SAMF	40.51	36.54	33.89	31.95	30.34	28.61	27.05	25.36	23.04
MMF	38.06	34.58	32.7	31	29.52	27.01	23.59	19.48	15.32
Proposed	41.99	38.23	35.57	33.57	31.68	29.8	27.9	26	23.11

Table 4 SSIM measure for various algorithms at noise densities 10–90% of *Airplane* image

Method	Noise density (%)								
	10	20	30	40	50	60	70	80	90
SMF	0.94	0.897	0.744	0.495	0.267	0.126	0.062	0.03	0.014
ERIN	0.991	0.979	0.958	0.922	0.842	0.722	0.523	0.307	0.109
AMF	0.982	0.96	0.937	0.911	0.88	0.844	0.786	0.719	0.628
AFSMF	0.991	0.981	0.966	0.948	0.92	0.875	0.797	0.703	0.52
UTMF	0.991	0.98	0.962	0.94	0.907	0.855	0.778	0.615	0.292
SAMF	0.991	0.978	0.96	0.939	0.91	0.875	0.827	0.761	0.655
MMF	0.984	0.967	0.947	0.925	0.879	0.776	0.576	0.32	0.142
Proposed	0.994	0.986	0.973	0.955	0.928	0.89	0.829	0.741	0.603

Table 5 SSIM measure for various algorithms at noise densities 10–90% of *Gold Hill* image

Method	Noise density (%)								
	10	20	30	40	50	60	70	80	90
SMF	0.786	0.741	0.621	0.423	0.231	0.111	0.05	0.024	0.011
ERIN	0.976	0.945	0.907	0.852	0.771	0.642	0.47	0.264	0.088
AMF	0.966	0.926	0.884	0.838	0.787	0.727	0.65	0.558	0.443
AFSMF	0.977	0.95	0.919	0.882	0.837	0.779	0.707	0.611	0.456
UTMF	0.976	0.946	0.912	0.868	0.815	0.746	0.645	0.479	0.224
SAMF	0.975	0.943	0.906	0.862	0.812	0.757	0.683	0.593	0.463
MMF	0.965	0.926	0.886	0.841	0.788	0.707	0.558	0.361	0.17
Proposed	0.982	0.96	0.936	0.903	0.864	0.812	0.74	0.639	0.492

Table 6 SSIM measure for various algorithms at noise densities 10–90% of *Lena* image

Method	Noise density (%)								
	10	20	30	40	50	60	70	80	90
SMF	0.864	0.817	0.665	0.433	0.215	0.096	0.043	0.022	0.01
ERIN	0.984	0.962	0.934	0.888	0.808	0.678	0.49	0.284	0.091
AMF	0.976	0.948	0.918	0.884	0.847	0.804	0.753	0.69	0.599
AFSMF	0.985	0.966	0.945	0.917	0.882	0.831	0.767	0.679	0.529
UTMF	0.984	0.964	0.94	0.909	0.869	0.814	0.727	0.57	0.27
SAMF	0.983	0.961	0.935	0.902	0.866	0.819	0.763	0.694	0.591
MMF	0.975	0.946	0.917	0.882	0.837	0.745	0.575	0.34	0.143
Proposed	0.988	0.972	0.952	0.926	0.893	0.846	0.779	0.687	0.512

Table 7 IEF measure for various algorithms at noise densities 10–90% of *Airplane* image

Method	Noise density (%)								
	10	20	30	40	50	60	70	80	90
SMF	51.2	46	19.2	8.9	4.7	2.9	2	1.5	1.2
ERIN	247.7	201.6	128	85.4	43.5	23.9	11.5	5.9	2.6
AMF	105.7	91.4	88.2	81.6	74.5	67.8	54.8	45	33.4
AFSMF	226.8	197	164.7	142.2	116.4	88	60.7	43.9	24.3
UTMF	247.7	211.3	146.9	122.6	89.6	62.8	39.3	17.7	5.9
SAMF	235.6	197.4	145.3	123.4	101.3	84.2	65.3	48.7	31.5
MMF	129.4	121.2	114.5	106.8	90	65.6	35.7	16.1	6.9
Proposed	510.2	388.9	313.2	241	181.6	140.3	98.1	66.9	38.4

Table 8 IEF measure for various algorithms at noise densities 10–90% of *Gold Hill* image

Method	Noise density (%)								
	10	20	30	40	50	60	70	80	90
SMF	28.7	33.1	18.4	9	4.8	3	2	1.5	1.2
ERIN	210.9	164.2	124.4	83.6	49.3	25.6	12.9	6.2	2.7
AMF	122.8	113.7	108.7	102.8	96.4	87.4	76.3	65	50.1
AFSMF	229.5	201.3	177.3	155.8	130.2	105	83	61.6	32.8
UTMF	211.4	177.4	156.6	126.4	102.8	78	48	21	7.1
SAMF	197.3	165.1	143.4	123.5	108.9	97.6	78.9	63.6	42.5
MMF	123.7	118.6	110.7	104.5	95	73.9	41.5	20.8	9.4
Proposed	310.8	278.7	249.3	213.2	186.4	155	122.7	89.7	57.3

able 9 IEF measure for various algorithms at noise densities 10–90% of *Lena* image

Method	Noise density (%)								
	10	20	30	40	50	60	70	80	90
SMF	53.1	49.4	20.8	9.3	4.9	3	2	1.5	1.2
ERIN	338.7	270.7	193.7	117.6	58	27.6	13.2	6.6	2.8
AMF	185	163	154.9	142.1	129.3	118.7	102.3	87	65.8
AFSMF	397.7	343.4	295.2	236	198.4	148.1	110.5	79.9	39.5
UTMF	349.4	291.3	239.3	193.9	147.7	100.8	58.4	25.5	7.3
SAMF	332.4	267.4	217.8	184.7	159.3	128.3	105.1	81.2	53.4
MMF	189.4	170.2	165.5	148.7	131.9	88.7	47.5	21	9
Proposed	468.2	394.4	320.7	268.3	217	168.7	127.9	94	54.2

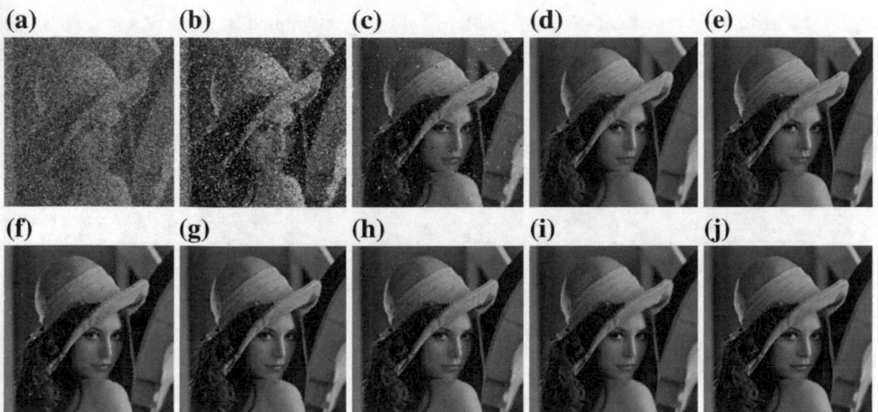

Fig. 2 Experimental results of different methods. **a** image corrupted by 60% of salt-and-pepper noise **b** SMF **c** ERIN **d** AMF **e** AFSMF **f** UTMF **g** SAMF **h** MMF **i** the proposed method **j** the original image of *Lena*

5 Conclusion

In this paper, a novel fast adaptive decision-based mean filter is proposed for removing the salt-and-pepper noise from corrupted signal with low- to high-density noise ratio in images. To demonstrate the better performance of proposed algorithm, different experimental results are conducted on a set of standard images to compare this method with other well-known existing methods.

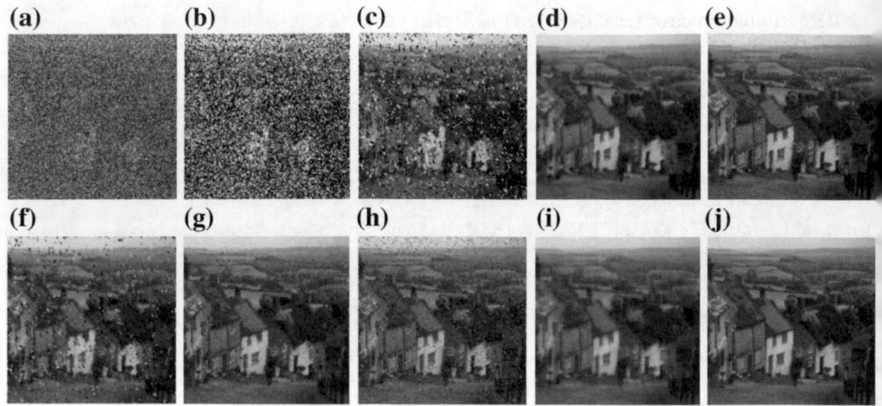

Fig. 3 Experimental results of different methods. **a** image corrupted by 80% of salt-and-pepper noise **b** SMF **c** ERIN **d** AMF **e** AFSMF **f** UTMF **g** SAMF **h** MMF **i** the proposed method **j** the original image of *Gold hill*

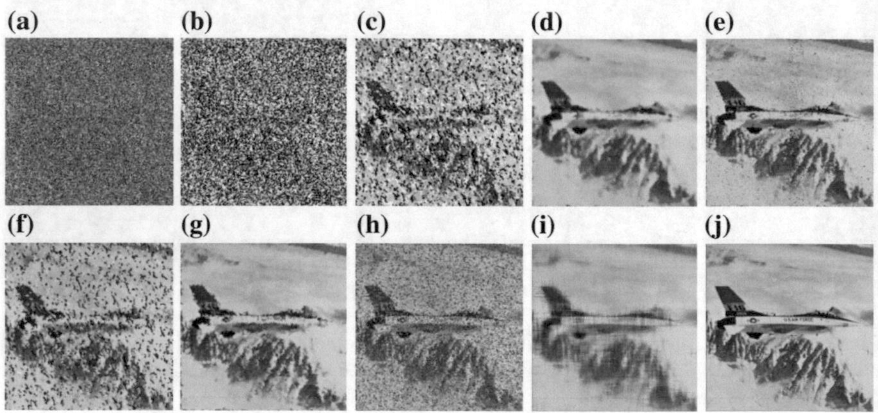

Fig. 4 Experimental results of different methods. **a** image corrupted by 90% of salt-and-pepper noise **b** SMF **c** ERIN **d** AMF **e** AFSMF **f** UTMF **g** SAMF **h** MMF **i** the proposed method **j** the original image of *Airplane*

References

1. Gonzalez, R.C., Woods, R.E.: Digital Image Processing, 2nd edn. Prentice Hall (2002)
2. Luo, W.: Efficient removal of impulse noise from digital images. IEEE Trans. Consum. Electr. **52**(2), 523–527 (2006)
3. Ibrahim H., Kong N., NgF.: Simple adaptive median filter for the removal of impulse noise from highly corrupted images. IEEE Trans. Consum. Electr. **54**(4), 1956–1961 (2008)
4. Toh, K., Isa, N.: Noise adaptive fuzzy switching median filter for salt-and-pepper noise reduction. IEEE Signal Process. Lett. **17**(3), 281–284 (2010)
5. Esakkirajan, S., Veerakumar, T., Subramanyam, A., PremChand, C.H.: Removal of high density salt and pepper noise through modified decision based unsymmetric trimmed median filter.

IEEE Signal Process. Lett. **18**(5), 287–290 (2011)

6. Das, J., Das, B., Saikia, J., Nirmala, S.R.: Removal of Salt and Pepper Noise Using Selective Adaptive Median Filter (2016)
7. Soni, A., Shrivastava, R.: Removal of high density salt and pepper noise removal by modified median filter. In: International Conference on Inventive Communication and Computational Technologies (ICICCT 2017) (2017)
8. Srinivasan, K.S., Ebenezer, D.: A new fast and efficient decision—based algorithm for removal of high-density impulse noises. IEEE Signal Process. Lett. **14**(3), 189–192 (2007)
9. Karthik, B., Kiran Kumar, T.V.U., Dorairangaswamy, M.A., Logashanmugam, E.: Removal of high density salt and pepper noise through modified cascaded filter. Middle-East J. Sci. Res. **20**(10), 1222–1228 (2014)
10. Vijaykumar, V.R., Santhanamari, G.: New decision-based trimmed median filter for high-density salt-and-pepper noise removal in images. J. Electron. Imaging **23** (2014)
11. Wang, Y., Wang, J., Song, X., Han, L.: An efficient adaptive fuzzy switching weighted mean filter for salt-and-pepper noise removal. IEEE Signal Process. Lett. **23**(11), 1582–1586 (2016)
12. Bhateja, V., Verma, A., Rastogi, K., Malhotra, C., Satapathy, S.C.: Performance improvement of decision median filter for suppression of salt and pepper noise. Adv. Intell. Syst. Comput. **264**, 287–297 (2014)
13. Srinivasan, K.S., Ebenezer, D.: A new fast and efficient decision based algorithm for removal of high density impulse noise. IEEE Signal Process. Lett. **14**, 189–192 (2007)
14. Ng, P.E., Ma, K.K.: A switching median filter with boundary discriminative noise detection for extremely corrupted images. IEEE Trans. Image Process. **15**, 1506–1516 (2006)
15. Ng, P.E., Ma, K.K.: Efficient improvements on the BDND filtering for removal of high density impulse noise. IEEE Trans. Image Process. **22**, 1223–1232 (2013)
16. Halder, A., Halder, S., Chakraborty, S.: A novel iterative salt-and-pepper noise removal algorithm. In: Proceedings of the 4th International Conference on Frontiers in Intelligent Computing: Theory and Applications (FICTA), pp. 635–643 (2015)

An Appearance Model and Feature Selection in Kernelized Correlation Filter Based Visual Object Tracking

Vijay K. Sharma, Bibhudendra Acharya and K. K. Mahapatra

Abstract In this chapter, we propose an eigenbasis vector based appearance model in kernelized correlation filter based visual object tracking by detection. The proposed appearance is learned using reconstructed targets in successive video frames. This learned appearance is used in KCF-based tracker in addition to raw pixel based appearance, i.e., without subspace reconstruction. The 2c number of HOG features are extracted from both the appearance models. Using a feature selection strategy, the number of feature channels is reduced to c. The kernel correlation for tracking by detection is computed for only c number of channels. To further improve the tracking accuracy, a discriminative parameter vector is learned in each video frame. This parameter vector is used to compute the scores of two intermediate tracked instances. The proposed method performs better than KCF-based tracker in a number of challenging video sequences.

Keywords Object tracking · Appearance model · Subspace model · Ridge regression · Correlation filter · Discriminative parameter

1 Introduction

In visual object tracking, position (for trajectory) and size of an object is estimated in each successive frame of a video. Object tracking has several important applications including human–computer interaction (HCI), visual surveillance, vehicle navigation

V. K. Sharma · B. Acharya (✉)
Department of Electronics and Telecommunication Engineering, National Institute of Technology, Raipur 492010, India
e-mail: bacharya.etc@nitrr.ac.in

V. K. Sharma
e-mail: vijay4247@gmail.com

K. K. Mahapatra
Department of Electronics and Communication Engineering, National Institute of Technology, Rourkela 769008, India
e-mail: kkm@nitrkl.ac.in

A. K. Das et al. (eds.), *Computational Intelligence in Pattern Recognition*, Advances in Intelligent Systems and Computing 999,
https://doi.org/10.1007/978-981-13-9042-5_69

(obstacle avoidance), traffic monitoring, etc. In defence, missiles are tracked in real time from captured video frames to estimate its trajectory. Object tracking is also used for automatic analysis of sports videos.

Online appearance model construction and target candidates generation are the two basic components of object tracking. Learning an online appearance model is a very challenging work as the target in each successive frame undergoes various changes, e.g., scale change, pose change, deformation, etc. Low to very high target occlusion, brightness variation, and presence of another object similar to the target are the other challenges involved in visual object tracking.

For the construction of mathematical model of target appearance, generative and discriminative are the two statistical learning techniques employed. In generative approach, tracked targets from the previous frames are used.

Some of the generative trackers are kernel based in [1], incremental visual tracker (IVT) in [2], IVT with temporal weights in [3], structural constraint based on [4], and sparse representation based on [5–8].

The background surrounding the target can be used in the creation of appearance model. The tracker of this kind is called discriminative tracker. The aim of the discriminative tracker is to create a binary classifier to discriminate the target against the background. Multiple instance learning (MIL)-based tracker in [9–11], support vector machine (SVM)-based tracker in [12–15], tracker based on tracking learning and detection in [16], and kernelized correlation filters (KCF)-based tracker in [17, 18] use background information to construct a discriminative classifier. In spatiotemporal context (STC)-based tracker [19], spatiotemporal relationships between the target and its local context has been used. Features from various deep layers of convolution neural network (CNN) are also used for discriminative tracking [20].

Although CNN-based tracker performs better, its computational complexity is very high due to large number of filters in pretrained model. KCF-based tracker also performs better in a number of video sequences. One of the good features of KCF-based tracker is its speed which is hundreds of frames per second (FPS). The tracking performances of KCF-based tracker in some of the video frames are poor. It is because of non-robust appearance model.

Generally, principal component analysis (PCA) is used for the reduction of feature dimension. Histogram of oriented gradients (HOG) features have been used with incremental principal component analysis (PCA-HOG) for visual hand tracking in [21]. In [22], PCA has been used for the selection of a subset of HOG features. Fuzzy histogram of optical flow orientation (FHOFO) features, with reduced dimension, is used as descriptor for the recognition of micro-facial expression in [23]. Our feature selection method explores the selection of features along different channels (maps).

In this chapter, we use subspace-based model as an additional appearance model in KCF. The subspace is represented by incrementally learned eigenbasis vectors. In each frame, tracked target is reconstructed in subspace and an appearance model is learned using these reconstructed samples. The learned appearance model is appended by background pixels around the tracked sample to contain the background information. HOG features are extracted from the learned sample with background as well as the raw image sample around the tracked target containing target as well as background pixels. Then we use a feature selection method to select important feature channels out of all feature channels. The kernelized correlation using selected feature channels is performed to detect the target object. A discriminative parameter vector is also learned in each video frame to compute the scores of two different intermediate tracked instances. The proposed method performs better in various challenging sequences.

The remaining part of this chapter is organized as follows. Regularized least squares regression is explained in brief in Sect. 2. A short introduction to KCF technique for visual object tracking is presented in Sect. 3. Section 4 contains the proposed appearance model and feature selection in KCF-based tracker. The experimental results are given in Sect. 5. Section 6 concludes the chapter.

2 Regularized Least Squares Regression

The advantage of KCF-based tracker is that its training sample contains all the patches in the specified region. This is possible by using cyclic shift property to obtain all possible training sets. The computational complexity due to large training set is avoided using DFT.

Consider the n training samples $(\mathbf{x}_1, \mathbf{y}_1), (\mathbf{x}_2, \mathbf{y}_2), \ldots, (\mathbf{x}_n, \mathbf{y}_n)$, where $\mathbf{y}_i$ is the label associated with vector $\mathbf{x}_i$. In ridge regression, we try to find a parameter vector $\mathbf{w}$ that minimizes the function given in Eq. (1) as

$$f_e(\mathbf{w}) = \sum_{i=1}^{n} (\mathbf{y}_i - \mathbf{w}^\top \mathbf{x}_i)^2 + \lambda ||\mathbf{w}||_2^2 \tag{1}$$

where λ is a regularizer. The prediction function is expressed as $f_p(\mathbf{x}) = \mathbf{w}^\top \mathbf{x}$, where $\mathbf{x}$ is an unknown vector. The estimate of vector $\mathbf{w}$ is given in Eq. (2) [18, 24] as

$$\mathbf{w} = (X^\top X + \lambda I)^{-1} X^\top \mathbf{y}. \tag{2}$$

where the matrix X contains all training samples in rows, while the vector $\mathbf{y}$ contains the corresponding labels.

If the input vector $\mathbf{x}$ is mapped to high-dimensional feature space $\varphi(\mathbf{x})$, the estimate vector $\mathbf{w}$ can be expressed using Eq. (3) as

$$\mathbf{w} = \sum_{i=1}^{n} \alpha_i \varphi(\mathbf{x}_i) \tag{3}$$

where α_i is called dual space variable. The prediction function is now expressed in Eq. (4) as

$$f_p(\mathbf{x}) = \sum_{i=1}^{n} \alpha_i k(\mathbf{x}_i, \mathbf{x}) \quad (4)$$

where the kernel function $k(\mathbf{x}_i, \mathbf{x}_j)$ computes dot product $\varphi(\mathbf{x}_i)^\top \varphi(\mathbf{x}_j)$ in high dimensional feature space. The solution to kernelized ridge regression is expressed in Eq. (5) as

$$\alpha = (K + \lambda I)^{-1}\mathbf{y} \quad (5)$$

3 Kernelized Correlation Filters (KCF) Technique for Object Tracking

In this section, KCF-based tracking in [18] is described briefly. For a circulant kernel matrix, solution in Eq. (5) is written in another form in Eq. (6) as

$$\alpha = \mathcal{F}^{-1}\left(\frac{\mathcal{F}(\mathbf{y})}{\mathcal{F}(\mathbf{k}^{\mathbf{xx}}) + \lambda}\right) \quad (6)$$

where the symbols $\mathcal{F}(\mathbf{x})$ and $\mathcal{F}^{-1}(\mathbf{x})$ represent DFT and inverse DFT of $\mathbf{x}$, respectively. The symbol $\mathbf{k}^{\mathbf{xz}}$ represents the kernel correlation between vectors $\mathbf{x}$ and $\mathbf{z}$. It is computed in faster way using Eq. (7) as

$$\mathbf{k}^{\mathbf{xz}} = g(\mathcal{F}^{-1}(\mathcal{F}(\mathbf{x}) \odot \mathcal{F}(\mathbf{x})^*)) \quad (7)$$

where $k(\mathbf{x}, \mathbf{z}) = g(\mathbf{x}^\top \mathbf{z})$ represents dot-product kernel for some function g, (e.g., for polynomial kernel, $g(\mathbf{x}^\top \mathbf{z}) = (\mathbf{x}^\top \mathbf{z} + a)^b$), $\mathbf{x} \odot \mathbf{z}$ represents the element-wise product between vectors $\mathbf{x}$ and $\mathbf{z}$ while $\mathbf{x}^*$ represents the complex conjugate of $\mathbf{x}$.

In object tracking, a patch $\mathbf{z}$ is cropped around the tracked location in the previous frame. The detection response for this patch is given in Eq. (8) as

$$f_{rdt}(\mathbf{z}) = \mathcal{F}^{-1}(\mathcal{F}(\mathbf{k}^{\mathbf{xz}})^* \odot \mathcal{F}(\alpha)) \quad (8)$$

The response $f_{rdt}(\mathbf{z})$ contains the response for all samples which are cyclic shifts of $\mathbf{z}$. The target appearance $\mathcal{F}(\mathbf{x})$ and the model $\mathcal{F}(\alpha)$ in Eq. (8) are learned in successive frames. Let the symbol $\mathbf{x}_f$ denotes the DFT of $\mathbf{x}$ (i.e., $\mathbf{x}_f = \mathcal{F}(\mathbf{x})$ and symbol α_f denotes the DFT of α (i.e., $\alpha_f = \mathcal{F}(\alpha)$). Then the appearance is learned in successive video frames given in Eq. (9) as

$$\mathbf{x}_f^t \leftarrow (1 - R_x)\mathbf{x}_f^{t-1} + R_x \mathbf{x}_f^t \quad (9)$$

where $\mathbf{x}_f^{t-1}$ represents the learned appearance in (t–1)th frame and R_x is the learning rate. Similarly, the model is learned which is given in Eq. (10) as

$$\alpha_f^t \leftarrow (1 - R_m)\alpha_f^{t-1} + R_m \alpha_f^t \tag{10}$$

where α_f^{t-1} in Eq. (10) represents the learned model in (t–1)th frame and R_m is the learning rate.

4 Proposed Appearance Model and Feature Selection in KCF-Based Tracker

In proposed method, we construct a subspace represented by eigenbasis vectors $\mathbf{U}$. The vectors $\mathbf{U}$ are obtained by singular value decomposition (SVD) of the tracked target samples. The basis vectors are incrementally learned to reduce the complexity as in [2, 25] using a batch size of five samples. The maximum number of basis vectors is limited to 16. Starting from fifth frame (when the subspace of minimum size five is available), we reconstruct the tracked target as follows.

Let $\mathbf{T}_{gt} \in \mathbb{R}^{w \times h}$ be the tracked target, where w and h are the width and height of the target sample, respectively. We stack the columns of $\mathbf{T}_{gt}$ to form a 1-D array $\mathbf{T}_{gt1} \in \mathbb{R}^{wh}$. The vector $\mathbf{T}_{gt1}$ is reconstructed in subspace $\mathbf{U}$ by using the relation $\mathbf{T}_r = \mathbf{U}\mathbf{U}^\top \mathbf{T}_{gt1}$, where $\mathbf{T}_r$ represents the reconstructed target in subspace. The reconstructed patch $\mathbf{T}_r$ is rearranged back to 2-D form and augmented in all four sides with the background pixels around the target in that frame. Let the background augmented sample is represented by $\mathbf{z}_r^{t-1}$ in (t–1)th frame. The sample $\mathbf{z}_r$ is updated in each frame using Eq. (11), which is given by

$$\mathbf{z}_r^t \leftarrow (1 - R_r)\mathbf{z}_r^{t-1} + R_r \mathbf{z}_r^t \tag{11}$$

We now have two target appearances $\mathbf{x}_{f1}$ and $\mathbf{x}_{f2}$ with $\mathbf{x}_{f1} = \mathcal{F}(\mathbf{z}_r^t)$ and $\mathbf{x}_{f2} = \mathcal{F}(\mathbf{x})$, where $\mathbf{x}$ represents the raw image pixels. For improved tracking accuracy, c channel HOG features are extracted from $\mathbf{z}_r^t$ and $\mathbf{x}$.

4.1 Feature Selection

The total number of feature channels now is $2c$. We select only half (i.e., c) channels for computing the final response. The proposed feature selection is performed as follows:

(1) Find the response $f_{rdt}(\mathbf{z})$ using Eq. (8) along each channel, separately.
(2) Multiply the response with a Gaussian-shaped 2-D kernel so as to reduce the response values away from the center.
(3) Choose the half feature channels (equal to c) which have the highest response among all.

The subspace-based appearance does not work well when the target is changed significantly from the target in the first video frame. To avoid this problem, we compute 1-D array correlation between the initial target and the target in the current frame. The correlation is computed after extracting HOG features from both the targets and rearranging all the channels separately in 1-D form. The correlation for all the channels is added. For c channel HOG features, the computed correlation is divided by c. If the final correlation ($Corr$) is below 0.15, then the appearance $\mathbf{x}_f$ (which is appearance without subspace) is used to find the response according to Eq. (8). The learning for the model update according to Eqs. (9) and (10) is modified from R_x to $Corr \times R_x$ and from R_m to $Corr \times R_m$ when the $Corr$ drops below 0.1. This is to ensure that the model is not updated by a corrupted or non-accurate target.

4.2 Proposed Discriminative Parameter Learning to Improve Tracking Accuracy

To further improve the tracking accuracy, we use a discriminative parameter vector ($\mathbf{wd} \in \mathbb{R}^{wh}$) learned using LIBSVM [27] in the first video frame. This parameter is updated in successive video frames by the following update strategy. Let $\mathbf{T}_g^t \in \mathbb{R}^{wh}$ be the tracked target in tth video frame. We crop some negative samples around this tracked location. Then a negative example ($\mathbf{T}_{ng}^t \in \mathbb{R}^{wh}$) is selected whose score, computed using $(\mathbf{wd}^t)^\top \mathbf{T}_i^t$, is highest. The update of $\mathbf{wd}$ is given in Eq. (12) as

$$\mathbf{wd}^t \leftarrow \mathbf{wd}^{t-1} + \alpha \times (\mathbf{T}_g^t - \mathbf{T}_{ng}^t) \tag{12}$$

In this chapter, we choose $\alpha = 0.008$. There are two intermediate tracked instances. First one ($\mathbf{X}_1^t$) is obtained using the method as discussed (Proposed 1). Another intermediate tracked instance ($\mathbf{X}_2^t$) is obtained using KCF(HOG) feature set only as in [18]. Scores ($(\mathbf{wd}^t)^\top \mathbf{X}_i^t$) for both the instances are computed. The final tracked instance is the one which has the higher $(\mathbf{wd}^t)^\top \mathbf{X}_i^t$ score. The intermediate instance cropped using basic KCF (HOG) is given a slightly higher weight (1, for instance, cropped using Proposed 1 method and 1.05, for instance, cropped by basic KCF (HOG) method) while comparing the score.

5 Experimental Results

For the execution of tracking codes, the system configuration is a notebook computer with core-i7, 8 GB of RAM and Windows 10 OS. The software used for the implementation is MATLAB 2016a. The values of learning rate parameters used are $R_x = 0.02$, $R_m = 0.02$, and $R_r = 0.015$. All other parameters are set according to paper [18]. The dataset for tracking evaluation is taken from CVPR 2013 paper [26] and it is online available at web.[1]

Two important metrics for tracking accuracy evaluation are average center location error (CE) and average overlap rate (OR) [26]. The center error is the Euclidean distance between tracked target location and the location of ground truth. The overlap rate is used to measure the overlap between tracked target and the ground truth. It is expressed as $OR = \frac{area(B_T \bigcap B_G)}{area(B_T \bigcup B_G)}$, where B_T and B_G, respectively, are the bounding box of the tracked object and the bounding box of the ground truth. Table 1 shows the average CE (where average is computed across the frames) in modified KCF (Proposed 1), basic KCF as well as some other trackers for different video sequences. The value NR in STC tracker indicates that the tracked location till last video frame could not be evaluated. The modified method shows better performance in a number of video sequences. The average OR for proposed (Proposed 1) as well as other trackers is given in Table 2. For the simulation, only one parameter set is used in all video sequences.

The visual tracked results in some of the video frames are shown in Fig. 1. The modified KCF is able to track visual objects even after the object is heavily occluded in Coke11 and Girl video sequences. In shaking video sequence, the object undergoes occlusion, pose change, and rotation. Also, there is very high clutter and brightness variation. The modified KCF is able to detect and hence track the target despite all these variations. In Car4 video sequence, the object undergoes a significant scale change along with brightness variation of the scene.

The improved results after using the proposed discriminative technique, as discussed in Sect. 4.2, are given in Proposed 2 columns of Tables 1 and 2. The Proposed 2 method has better tracking accuracy in a number of video sequences. Our discriminative parameter learning technique is simple as it needs to test only two instances for the higher score. These instances are already obtained using robust tracking methods.

6 Conclusion

This chapter presents a subspace-based appearance as well as feature selection technique in KCF-based visual object tracking. The subspace-based appearance is learned using reconstructed tracked samples in successive video frames. For this additional appearance in KCF tracker, the number of HOG feature channels is doubled. Using a feature selection technique, which is based on comparing the individual responses for all channels, the number of feature channels is reduced by half. The tracking accuracy of the modified KCF method is better than the basic KCF-based tracker in different video sequences. A discriminative parameter vector is also learned using positive

[1] https://sites.google.com/site/wuyi2018/benchmark.

Table 1 Average center error (CE) of trackers

Video	STC (in [19])	DLSSVM (in [15])	KCF (HOG) (in [18])	Proposed 1	Proposed 2
Basketball	75.2	18.6	*7.8*	35.3	**11.0**
Bolt	139.8	*4.1*	**6.3**	8.3	18.1
Boy	25.9	*2.8*	*2.8*	6.4	**3.2**
CarDark	2.8	*1.4*	6.0	3.0	**2.4**
Car4	10.6	18.6	9.8	**3.7**	*3.5*
Coke11	74.3	*8.7*	18.6	**11.9**	**11.9**
Crossing	34.0	*2.0*	2.2	**2.1**	**2.1**
David	12.1	**7.9**	8.0	*7.8*	*7.8*
David2	5.5	*2.0*	*2.0*	**2.9**	*2.0*
David3	6.3	8.2	*4.3*	**4.4**	**4.4**
Deer	400.0	*4.3*	21.1	9.9	**9.5**
Dog1	21.1	**5.1**	*4.2*	10.1	5.9
Doll	NR	**5.7**	8.3	*5.4*	7.0
Dudek	25.5	12.6	**12.0**	12.4	*11.3*
Faceocc	25.0	16.0	**13.9**	*13.4*	**13.9**
Faceocc2	10.1	*7.5*	**7.6**	8.6	8.5
Freeman1	NR	**85.0**	94.8	94.2	*8.0*
Fish	*3.9*	4.6	**4.0**	4.9	5.6
FleetFace	85.0	*23.4*	25.5	**23.9**	24.2
Football1	72.5	5.1	5.4	*4.5*	**4.6**
Girl	21.8	**3.6**	11.9	*3.2*	*3.2*
Lemming	96.2	152.0	77.8	**16.5**	*6.9*
Mhyang	4.5	*2.7*	3.9	4.8	**3.7**
MountainBike	*7.0*	**7.4**	7.6	8.1	8.2
Shaking	9.6	*7.5*	112.5	19.0	**9.0**
Singer1	*5.7*	15.4	**10.7**	18.4	18.4
Singer2	52.4	179.1	**10.2**	*8.4*	*8.4*
Skating1	66.4	75.2	7.6	*6.2*	**6.3**
Subway	NR	3.1	**2.9**	3.2	*2.8*
Suv	51.4	16.0	*3.4*	**6.3**	**6.3**
Sylv	**9.4**	*5.8*	12.9	13.9	19.0
Tiger1	63.5	10.7	**8.0**	12.6	*7.1*
Tiger2	57.4	*16.6*	47.4	**32.3**	36.7
Trellis	33.7	7.7	7.7	*7.3*	**7.6**
Walking	7.1	6.1	**3.9**	*3.0*	*3.0*
Walking2	13.8	18.9	28.9	*9.4*	**10.5**
Woman	NR	*2.9*	10.0	**9.9**	10.0

Table 2 Average overlap rate (OR) of trackers

Video	STC (in [19])	DLSSVM (in [15])	KCF (HOG) (in [18])	Proposed 1	Proposed 2
Basketball	0.287	0.668	**0.674**	0.360	*0.692*
Bolt	0.036	*0.763*	**0.678**	0.635	0.467
Boy	0.543	*0.779*	**0.777**	0.714	0.763
CarDark	0.750	*0.858*	0.614	0.746	**0.784**
Car4	0.351	0.467	**0.483**	*0.493*	*0.493*
Coke11	0.104	*0.729*	0.549	**0.604**	0.602
Crossing	0.248	*0.731*	0.710	**0.711**	**0.711**
David	0.522	*0.544*	0.538	**0.542**	0.541
David2	0.588	*0.835*	0.827	0.779	**0.828**
David3	0.432	0.717	**0.772**	0.764	*0.773*
Deer	0.040	*0.752*	0.623	0.679	**0.684**
Dog1	0.506	**0.550**	*0.551*	0.505	**0.550**
Doll	NR	*0.568*	0.534	0.560	**0.561**
Dudek	0.587	*0.728*	**0.727**	0.726	**0.727**
Faceocc	0.186	0.756	0.774	*0.784*	**0.780**
Faceocc2	0.689	*0.761*	**0.751**	0.736	0.738
Freeman1	NR	**0.217**	0.214	0.205	*0.402*
Fish	0.510	0.824	*0.839*	**0.825**	0.803
FleetFace	0.419	0.609	0.589	*0.613*	**0.611**
Football1	0.354	0.702	0.710	*0.757*	**0.756**
Girl	0.334	0.728	0.545	**0.734**	*0.783*
Lemming	0.231	0.227	0.384	**0.634**	*0.698*
Mhyang	0.688	*0.815*	0.796	0.782	**0.798**
MountainBike	0.583	*0.727*	**0.711**	0.706	0.706
Shaking	0.623	*0.721*	0.039	0.540	**0.690**
Singer1	*0.531*	0.350	**0.355**	**0.355**	**0.355**
Singer2	0.409	0.043	**0.732**	*0.774*	*0.774*
Skating1	0.351	0.395	0.489	*0.493*	**0.490**
Subway	NR	0.713	**0.754**	0.727	*0.766*
Suv	0.512	0.734	*0.883*	**0.752**	0.750
Sylv	0.525	*0.734*	**0.643**	0.641	0.624
Tiger1	0.261	0.726	**0.785**	0.728	*0.800*
Tiger2	0.110	*0.599*	0.354	0.483	**0.492**
Trellis	0.470	**0.621**	*0.631*	*0.631*	*0.631*
Walking	*0.598*	**0.543**	0.530	0.534	**0.543**
Walking2	*0.518*	0.476	0.395	**0.492**	0.490
Woman	NR	*0.773*	**0.705**	0.680	0.693

Fig. 1 Visual tracked results in (from top to bottom) Car4, Coke11, David, Girl, MountainBike, Shaking, Skating1, and Trellis video sequences

..nd negative samples to improve the tracking accuracy. The learned parameter vec..or is used to compute the scores of two different intermediate tracked instances. One ..f them is obtained using basic KCF tracker while the other is obtained using the ..nodified KCF method.

..eferences

1. Comaniciu, D., Ramesh, V., Meer, P.: Kernel-based object tracking. IEEE Trans. Pattern Anal. Mach. Intell. **25**(5), 564–577 (2003)
2. Ross, D.A., Lim, J., Lin, R.-S., Yang, M.-H.: Incremental learning for robust visual tracking. Int. J. Comput. Vis. **77**(1–3), 125–141 (2008)
3. Cruz-Mota, J., Bierlaire, M., Thiran, J.: Sample and pixel weighting strategies for robust incremental visual tracking. IEEE Trans. Circuits Syst. Video Technol. **23**(5), 898–911 (2013)
4. Bouachir, W., Bilodeau, G.-A.: Exploiting structural constraints for visual object tracking. Image Vis. Comput. **43**, 39–49 (2015)
5. Mei, X., Ling, H.: Robust visual tracking using l1 minimization. In: 2009 IEEE 12th International Conference on Computer Vision, pp. 1436–1443. IEEE (2009)
6. Mei, X., Ling, H., Wu, Y., Blasch, E., Bai, L.: Minimum error bounded efficient l1 tracker with occlusion detection. In: 2011 IEEE Conference on Computer Vision and Pattern Recognition (CVPR), pp. 1257–1264. IEEE (2011)
7. Bai, T., Li, Y.F.: Robust visual tracking with structured sparse representation appearance model. Pattern Recognit. **45**(6), 2390–2404 (2012)
8. Zhang, S., Yao, H., Zhou, H., Sun, X., Liu, S.: Robust visual tracking based on online learning sparse representation. Neurocomputing **100**, 31–40 (2013)
9. Sharma, V.K., Mahapatra, K.K.: Mil based visual object tracking with kernel and scale adaptation. Signal Process.: Image Commun. **53**, 51–64 (2017)
10. Babenko, B., Yang, M.-H., Belongie, S.: Robust object tracking with online multiple instance learning. IEEE Trans. Pattern Anal. Mach. Intell. **33**(8), 1619–1632 (2011)
11. Zhang, K., Song, H.: Real-time visual tracking via online weighted multiple instance learning. Pattern Recognit. **46**(1), 397–411 (2013)
12. Sharma, V.K., Mahapatra, K.: Visual object tracking based on sequential learning of svm parameter. Digital Signal Process. **79**, 102–115 (2018)
13. Avidan, S.: Support vector tracking. IEEE Trans. Pattern Anal. Mach. Intell. **26**(8), 1064–1072 (2004)
14. Hare, S., Saffari, A., Torr, P.H.: Struck: structured output tracking with kernels. In: 2011 IEEE International Conference on Computer Vision (ICCV), pp. 263–270. IEEE (2011)
15. Ning, J., Yang, J., Jiang, S., Zhang, L., Yang, M.-H.: Object tracking via dual linear structured svm and explicit feature map. In: Proceedings of the IEEE Conference on Computer Vision and Pattern Recognition, pp. 4266–4274 (2016)
16. Kalal, Z., Mikolajczyk, K., Matas, J.: Tracking-learning-detection. IEEE Trans. Pattern Anal. Mach. Intell. **34**(7), 1409–1422 (2012)
17. Lukezic, A., Vojir, T., Zajc, L.C., Matas, J., Kristan, M.: Discriminative correlation filter with channel and spatial reliability. In: CVPR, vol. 1, Issue 2, p. 3 (2017)
18. Henriques, J., Caseiro, R., Martins, P., Batista, J.: High-speed tracking with kernelized correlation filters. IEEE Trans. Pattern Anal. Mach. Intell. **37**(3), 583–596 (2015)
19. Zhang, K., Zhang, L., Liu, Q., Zhang, D., Yang, M.-H.: Fast Visual Tracking via Dense Spatio-temporal Context Learning, pp. 127–141. Springer International Publishing, Cham (2014)
20. Ma, C., Huang, J.-B., Yang, X., Yang, M.-H.: Hierarchical convolutional features for visual tracking. In: Proceedings of the IEEE International Conference on Computer Vision, pp. 3074–3082 (2015)

21. Yang, H., Song, Z., Chen, R.: An incremental pca-hog descriptor for robust visual hand trackin
In: International Symposium on Visual Computing, pp. 687–695. Springer (2010)
22. Kobayashi, T., Hidaka, A., Kurita, T.: Selection of histograms of oriented gradients featur for pedestrian detection. In: International Conference on Neural Information Processing, p 598–607. Springer (2007)
23. Happy, S., Routray, A.: Recognizing subtle micro-facial expressions using fuzzy histogra of optical flow orientations and feature selection methods. In: Computational Intelligence f Pattern Recognition, pp. 341–368. Springer (2018)
24. Rifkin, R., Yeo, G., Poggio, T. et al.: Regularized Least-Squares Classification. Nato Scienc Series Sub Series III (2003)
25. Levey, A., Lindenbaum, M.: Sequential karhunen-loeve basis extraction and its application t images. IEEE Trans. Image Process. **9**(8), 1371–1374 (2000)
26. Wu, Y., Lim, J., Yang, M.-H.: Online object tracking: a benchmark. In: 2013 IEEE Conferenc on Computer Vision and Pattern Recognition (CVPR), pp. 2411–2418. IEEE (2013)
27. Chang, C.-C., Lin, C.-J.: LIBSVM: a library for support vector machines. ACM Trans. Intell Syst. Technol. (TIST) **2**(3), 27 (2011)

Role of Soft Computing Techniques in Software Effort Estimation: An Analytical Study

P. Suresh Kumar and H. S. Behera

Abstract In developing software, software effort estimation plays an important role in the success of a software project. Inaccurate, inconsistent, and unreliable estimation of a software leads to failure. Because of various special specifications and changes in the requirements, accurate Software Effort Estimation (SEE) for developing software is a difficult task. This software effort estimation must be calculated effectively to avoid unforeseen results. At early development stages, these inabilities to maintain certainty, inaccurate, unreliability are the limitations of expert judgment and algorithmic effort estimation models. After that, attention was turned to machine learning and soft computing methods. Soft computing is an association with the methodologies centering on fuzzy logic, artificial neural networks, and evolutionary computation. These methods will provide flexible information processing capability for handling real-life situations. The main aim of the study is to provide an in-depth review of software effort estimation from the initial stages that included expert judgment-based SEE to the latest techniques of soft computing.

Keywords Software effort estimation · Expert judgment · Soft computing · Artificial neural network · Fuzzy logic · Particle swarm optimization · Evolution computation

1 Introduction

Software effort estimation is a complex task that plays an important role in software development. Software engineering is an organized, disciplined, and quantitative way to develop, run, and maintain software. Software engineering management is defined as the management that follows some activities like planning, measuring,

P. Suresh Kumar (✉) · H. S. Behera
Department of Information Technology, Veer Surendra Sai University of Technology, Burla 768018, India
e-mail: reshu.suri@gmail.com

H. S. Behera
e-mail: hsbehera_india@yahoo.com

A. K. Das et al. (eds.), *Computational Intelligence in Pattern Recognition*, Advances in Intelligent Systems and Computing 999,
https://doi.org/10.1007/978-981-13-9042-5_70

monitoring, and reporting to ensure that the software product delivered efficientl and in time. It is in this context of planning and measurement of software engineerin that the study of software effort estimation came into prominence [1].

1.1 *Software Effort Estimation (SEE)*

Software effort estimation is a procedure that will forecast most realistic quantity o effort which is used to develop and maintain a software from incomplete, inconsistent, uncertain and noisy data input. Usually, effort is measured in terms of money an person-hour, that means number of persons per hour spent in developing a software Accurate estimates are very important to plan properly in developing a software Both overestimation and underestimation will cause serious problems for the company as well as it will lead to failure of a software project [2]. Robert N. Charette [3] explained about common factors that are caused for software project failure, they are unrealistic project goals, indefinite system requirements, incorrect status report of the project status, less communication among customers and developers, unmanaged risks, use of immature technology, inefficient in handling complexity of the project, poor project management, stakeholder politics, commercial pressures. However, success of software project purely depends on how accurate the effort estimation was done. Software effort estimation is to be optimized because right estimation is required for both developers as well as client. While SEE helps the developer in effective development and monitoring of software, it enables the client to carry out contract negotiations, schedule project completion dates and prototype releasing dates, etc. Though many software effort estimation methods exist, finding accurate, consistent effort estimation is a challenging task for researchers.

1.2 *Importance of Soft Computing in Software Effort Estimation*

Software effort estimation is mainly categorized into two models such as algorithmic and non-algorithmic. Some of the most familiar algorithmic software effort estimation models are COCOMO [4], Putnam's SLIM [5], Albrecht's Function Points [6] among others. To estimate effort of software these models need input and some other attributes such as lines of code, complexity; which are very difficult to find before early stage of the software project development. Software project management requires relationships between various characteristics of a software project to be represented. The fact that algorithmic models fail to adequately represent them has been studied in [7]. Studies indicating the need to adjust these characteristics in the context of the existing local circumstances are available in [8].

Disadvantages of algorithmic models will lead to explore non-algorithmic models which are based on soft computing. Soft computing which is a non-algorithmic model was introduced in late 1960s for estimation of software. The ability to understand computational concepts from a human perspective is the hallmark of soft computing. The lack of information with visibility of soft computing is due to some of the problems had no solutions through hard computing techniques, as those are having lack of information or else, the system not defined clearly. Soft computing has the ability to solve a wide array of problems including nonlinear problems, optimization problems, problems involving intelligent control and decision making. Soft computing is intrinsic in nature and due to this, it has the capability to handle inexactness and uncertainty. Fuzzy logic has linguistic nature that can be used to not only modify imprecision on inputs and outputs, but also helps to build Fuzzy logic-based models that rely on inputs from expert judgment. A neural network has computing power through the ability to learn and generalize from data set. Soft computing has the potential to build intelligent systems and automated systems. In summary, soft computing is the amalgamation of several methods which can handle ambiguous conditions like imprecision, inconsistency, and uncertainty. The main goal of using soft computing is to obtain robust solution at a feasible cost.

This paper deals with extensive reviews on various soft computing techniques applied to solve software effort estimation problems. The remaining paper is categorized into 4 sections. In Sect. 2, preliminaries are explained, which contains soft computing techniques such as neural networks, fuzzy logic, particle swarm optimization, and evolutionary computation. Section 3 describes various applications of soft computing techniques. In Sect. 4, critical analysis has been done which includes different datasets and evaluation factors with various intelligent methods. Finally Sect. 5 describes conclusion of work with some further developments.

2 Preliminaries

Soft computing alone is not a single method that can handle complex problems but alternatively it is a mixture of different methods like fuzzy logic, neural network, evolutionary computation, particle swarm optimization, etc. These components are not competitive, but they work together for any solution of a problem [9]. This integration permits complementary reasoning and empirical data used to develop adjustable computing tools that will solve complex problems.

2.1 Artificial Neural Networks (ANN)

Human brains have an amazing power of logical thinking and computational power. ANN is the emulation of human brain. Human brain has two fundamental elements, the first being neurons which are represented as nodes in ANN and the second being synapses which are used to connect these neurons represented using weighted edges in ANN [10]. ANN is one of the oldest (yet latest) computational tools that is invented to simulate the human brain and it analyzes and processes the data. ANN is the collection of artificial neurons which imitate human neurons. ANN is designed to be used on nonlinear function that is used to map from input to output. ANN takes weighted inputs and generates output if the sum exceeds threshold. This output may be excitatory or inhibitory input to the remaining neurons in the network. This process will be continued till it generates one or more outputs [11]. Artificial neural network has successfully solved many complex problems which are considered very difficult to be solved by humans. ANN has the capability of self-learning which will allow to produce good results.

Simple form of ANN is shown in the following figure. It has three layers: they are input layer, hidden layer, and output layer. The number of neurons depends upon the number of input parameters.

The input layer accepts the input that will later be passed onto the hidden layers for processing before finally being sent to the output layer which provides the output. In case of software effort estimation, the input is the set of measures used to represent the size of the software, which is usually represented in terms of functional points and lines of code and some other factors such as experience, database, and complexity of a software.

The hidden layer is connected fully or partially to the input layer and it is fully connected to the output layer. The number of output neurons depends upon the model used for the output layer. In the case of effort estimation output layer has only one neuron and the result is in man-hours [12].

The overall architecture of artificial neural network is shown in Fig. 1.

2.2 *Fuzzy Logic*

Zadeh introduced the concept of fuzzy logic (FL) in the year 1965 [13]. Fuzzy logic is a mathematical system evolved to model the human brain's ability of handling and choosing words [14]. Fuzzy Logic is defined in two ways. In one way FL is a logical system, that is an extension of multivalued logic. In another way, FL can be called as a synonym of theory of Fuzzy sets. Fuzzy logic has the capability of being tolerant on imprecise data. It will change the imprecision into measurement process. Fuzzy Logic is an important tool that will solve very complex problems which are very difficult to solve by some mathematical models. By using this, we can reduce the complexity of the solved problems and at the same time, increase the accessibility of control theory. FL offers a suitable way to create mapping between unknown statistical inputs to scalar statistical output data [15].

Fuzzy logic can be used for software effort estimation by developing a precise mathematical model of the domain and thereby produce real complexity estimation [16]. Some fuzzy logic types are pure fuzzy logic systems, Takagi and Sugeno's fuzzy system [17] and Mamdani's fuzzy logic system [18] with fuzzifier and defuzzifier.

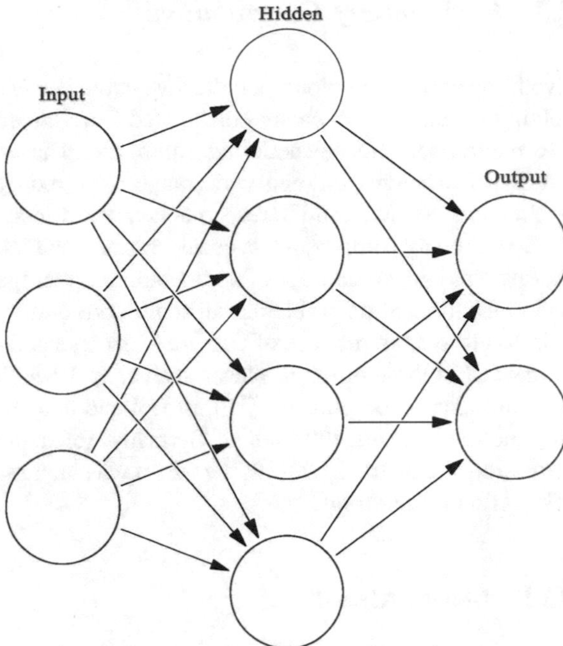

ig. 1 Architecture of ANN

The configuration of fuzzifier and defuzzifier is shown in Fig. 2. From thcsc, Fuzzy logic system with fuzzifier and defuzzifier is the most widely used fuzzy logic system nowadays. This fuzzy logic with fuzzifier and defuzzifier is applied to many industries process and consumer products as seen in the following figure:

- Conversion of crisp input into fuzzy set is done by fuzzifier and narrate the scenarios graphically.
- "If-then" rules used by fuzzy rule base.
- Function aggregation and composition is done by fuzzy inference engine.
- Conversion of fuzzy output into crisp output is done by defuzzifier.

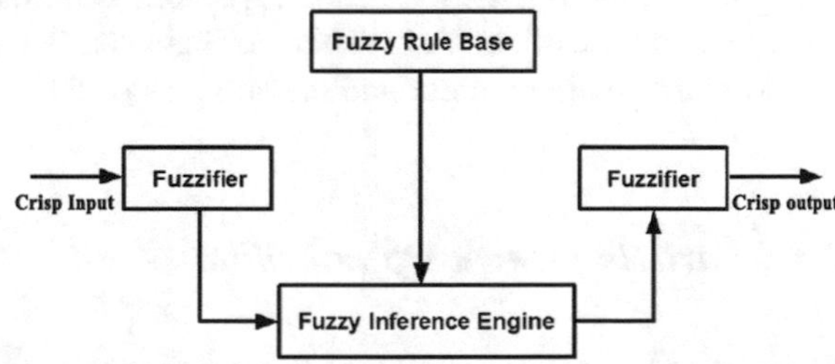

Fig. 2 Basic configuration of fuzzifier and defuzzifier

2.3 Evolutionary Computation

Evolutionary computation is a collective name because of its capability of problem solving techniques which are constructed from the concept of biological evolution like natural selection, genetic algorithm, etc. These applications are applied in a variety of problems through wide range from industry persons using in practical applications to the scientific researchers in the research.

Evolutionary computation is considered as a subfield of Artificial Intelligence and is somewhat linked concept of computational intelligence. It involves lot of combinational optimization problems and continuous optimization. In 1950s this approach is introduced from the idea of Darwin's Principles. After that in 1960 this idea was developed in three different places such as in USA Fogel introduced and named it as evolutionary computation [19], in Holland and Goldberg called their algorithm as genetic algorithm [20] and in Germany, Rechenberg and Schwefel introduced Evolution strategies [21 22]. In the 90s, it was integrated into one technology called "Evolution Computing".

2.3.1 Genetic Algorithm

Genetic algorithms are used for high quality solutions for large space search problems and optimization. Genetic algorithm depends on bio-inspired operators such as crossover, mutation, and selection. One of the characteristics of the genetic algorithm is that, it does not need prior knowledge and expertise. However, the outcome generated is a good approximation to the optimum.

2.3.2 Differential Evolution

For nonlinear function, differential evolution is the promising evolutionary process [23]. It uses the idea of cell mutation and mating to obtain optimal solution. It has a limited number of parameters, due to this differential evolution is fast to intersect and implementation is very easy. Genetic algorithm and differential algorithm are similar because both are population-based approaches and these are also using crossover, mutation, and selection. Main difference between GA and DE is, in GA crossover is considered as main operator and mutation is considered as background operation.

2.4 Particle Swarm Optimization

Eberhart and Kennedy introduced Particle Swarm optimization [24]. This algorithm is derived from the research of social characteristics of bird flock. In this scenario, for example, one bird knows about food place and doesn't know about the distance, est solution for finding the food is to follow the nearest bird to the food. Likewise, ne particle follows another particle and there will be the update in both position and elocity. The population includes particles where each particle acted as a solution or optimized problem. These particles are initialized randomly. This optimization s done through the collaboration and competition among the particles. The main dvantage of PSO is its fast convergence ability compared to other optimization lgorithms. PSO and evolutionary computation have so many similarities; but no volution operators like cross over and mutation in PSO. Main concept of PSO is hat, the particles move like a swarm in problem space and evaluates the position ısing fitness function [25]. In PSO, optimization based on score function which is ıttempted to find weights is in Eq. 1.

$$s = w1^{*}score1 + w2^{*}score2 \tag{1}$$

The main plan behind score fusion is to produce maximal score for genuine one and in the same way minimal score for correct one. In swarm simulation for every iteration the particle keeps track on three parameters such as “present[], that gives the present position”. “pbest[], that denotes the best value found by the fitness function”, “gbest[] best global value got by particle” so far. These three values are denoted in the following Eqs. 2 and 3 below.

$$\begin{aligned} v[] = inertia^{*}v[] + c1^{*}rand()^{*}(pbest[] - present[]) \\ + c2^{*}rand()^{*}(gbest[] - present[]) \end{aligned} \tag{2}$$

$$present[] = prenet[] + v[] \tag{3}$$

where,

v[] particle velocity directed toward the solution.
c1 and c2 learning factors.
rand() generates random numbers ranging between (0,1) in problem-dependent bounded space

Lin and Tzeng [26] used K-means clustering algorithm to project clustering algorithm and they used particle swarm optimization that will take MMRE as fitness merit and N-1 test methods to optimize COCOMO parameters. By this approach, effort estimation is more accurate. To calculate function count (effort), Kaur and Sumeet [27] used swarm intelligence technique which is particle swarm optimization using the parameter value adjusted factor. On the basis of percentage of MARE and RMSE, calculated effort is compared with existing effort models and proved that their proposed model gave better results. Murillo-Morera et al. [28] investigated artificial neural network predictive model which is incorporated with constructive cost model and then it will be improved by applying particle swarm optimization (PSO-ANN-COCOMO II). This model increases the speed of artificial neural network along with the learning ability and handled the error rate. Wu et al. [29] examined the potentiality of software effort estimation by integrating the particle swarm optimization with case-based reasoning. In this, PSO has been used to optimize the weights in the CBR. The results are compared with other old results in respective of the measures such as MMRE, PRED(0.25), and MdMRE and concluded that the weighted CBR is better than unweight CBR.

3 Applications of Various Soft Computing Techniques

Soft computing introduced a new era in artificial intelligence, which has high machine intelligence quotient and human-like information processing capability. This is used for nonlinear problems as well as mathematical systems those are yet to be modeled. The cost of implementing soft computing techniques is considerably low as analyzed by [30]. The widespread use of soft computing is evident in the fact that it has been used in consumer appliances starting from the late 1980s as presented in [31]. Fuzzy logic was introduced later on a pioneering methodology that has wide implications to a large number of fields [32]. Later on, a collaborative methodology based on both neural networks and fuzzy logic was developed called neuro-fuzzy logic. In recent times, evolution computation, particle swarm optimization have shown future in this heterogeneous application field of soft computing.

3.1 Artificial Neural Networks

Capretz and Ho [33] proposed a model that will give promising results and predict the effort of the software through use case diagrams using a cascade-correlation neural network. This approach is evaluated from mean of magnitude of error relative to the estimate and prediction level by using 214 industrial and 26 educational projects toward multiple linear regression model and use case point model. They also investigate the model [34] and again used artificial neural network to predict software effort by using use case diagrams on the use case points. They gave inputs such as software size, productivity and complexity. They evaluated ANN model using MMER and PERD criteria against regression model and found that the CCNN model performs better than the multiple regression model and the Use case Point Model based on the MMER and PRED criteria. Attarzadeh et al. [35] used artificial neural network prediction model that integrates constructive cost model and named that model as ANN-COCOMO II. This model gave accurate results because of the learning ability of artificial intelligence and the excellence of COCOMO.

Popular approach to find the effort of a software in mathematical model is called story point approach. Satapathy et al. [36] considered total number of story points and velocity of the project and they used different types of neural networks such as general regression neural networks, probabilistic neural network, group method of data handling, polynomial neural network, and cascade-correlation neural network.

hese are evaluated and compared with a number of affecting factors. Iwata et al. 37] investigated the effect of classification that involved in estimating amount of ffort of a software development projects. They applied classifications and then reated effort estimation models for every class using linear regression, artificial eural networks and some form of support vector regression. Evaluated these models vith and without classification compared the estimation accuracy and the results have ignificant differences among certain pair of models. Martin [38] considered adjusted unction points and investigated radial basis function neural network to software levelopment effort prediction. He compared the accuracy of RBFNN with simple inear regression, multilayer perception, and general regression neural network. By his comparison, he found some weakness over these methods.

Normally, machine learning especially artificial neural network has adjusted the complex set of bond among independent and dependent variables. Rijwani and Jain [39] used ANN-based model which has used multilayered feed forward neural network that is trained by back propagation method. They have used COCOMO data set and shown accurate estimation for this model. Nassif et al. [40] investigated four neural network models those are multilayer perception, radial basis function neural network, general regression neural network, and cascade-correlation neural network. They compared these models based on accuracy prediction which is based on mean absolute error criteria, whether these models tend to underestimate or overestimate, how every model classified by the importance of their inputs. They used five datasets which are extracted from ISBSG data set and validated these four models and shown pros and cons. Tanveer et al. [41] proposed deployment of practical machine learning and maintenance approaches by industry and research best practices. This was attained by applying ISBSG dataset, smart data preparation, cross-validation and an ensemble averaging of three machine learning algorithms (support vector machines, neural networks, and generalized linear models). By these models software effort and duration has been estimated. Pospieszny et al. [42] proposed deployment of practical machine learning and maintenance approaches by industry and research best practices. This was attained by applying ISBSG dataset smart data preparation, cross-validation and an ensemble averaging of three machine learning algorithms (support vector machines, neural networks, and generalized linear models). By these models, software effort and duration will be estimated.

Some of the intelligent methods using artificial neural network with other information such as datasets used, evaluation factors obtained, and that article published year, author(s), and references are shown in Table 1.

3.2 *Fuzzy Logic*

Wang [65] compared nonlinear mapping obtained by fuzzy systems and neural networks and proved fuzzy systems are better. Flexible learning capability is the main advantage of neural network and by this, it is easy to develop nonlinear problem using input–output data. Added to this, fuzzy systems had the capability of knowledge

Table 1 Various Artificial Neural Network (ANN) methods and evaluation factors

S.No	Year	Author(s)	Intelligent method	Datasets	Evaluation factors	References
1	2002	Idri et al.	Multilayer perceptron	COCOMO 81 Kemereand (15), IBM DPS (24)	MMRE, PRED (0.25)	[43]
2	2004	Idri et al.	Radial Basis Function Network	COCOMO 81, ISBSG	MMRE	[44]
3	2005	Tadayon N	Adaptive Neural Network	COCOMO 81	MRE	[45]
4	2006	Idri et al.	Radial Basis Function with C-means or APC-III	COCOMO 81 (63)	MMRE, PRED (0.25)	[46]
5	2007	deBarcelos Tronto et al.	MLP and Regression model	ISBSG, COCOMO (63)	MMRE	[47]
6	2007	Abrahamsson et al.	MLP Radial Basis Functions (RBF) neural networks	Project data (PROM)	MRE	[48]
7	2008	El-Sebakhy and E. A	Multilayer Perceptron	Kemerer and IBM, COCOMO	MAPE	[49]
8	2009	Li et al.	Multilayer Perceptron	Albrecht, Desharnais, Maxwell, ISBSG	MMRE, PRED (0.25)	[50]
9	2010	Kalichanin-Balich et al.	Feed forward neural network	132 Short-scale projects	MMRE	[51]
10	2010	Jodpimai et al.	Artificial neural network, Feed forward neural network	COCOMO81, IBMDPS, CF, Desharnais, NASA	MMRE, PRED (0.25)	[52]
11	2010	Attarzadeh et al.	Feed forward neural network	COCOMO 63, ARTIFICIAL (100)	MMRE, PRED (0.25)	[53]

(continued)

able 1 (continued)

S.No	Year	Author(s)	Intelligent method	Datasets	Evaluation factors	References
12	2010	Ajitha et al.	Back propagation neural network	UCPWEIGHTS	AAE	[54]
13	2011	Dave et al.	RBNN, FFNN, regression analysis	NASA (60)	MMRE, RSD	[55]
14	2011	Ghose et al.	FFBPNN, CASCADE FFBPNN, ELMAN BPNN	Project data	MMRE, BRE, PRED	[56]
15	2012	Kaushik et al.	Back propagation neural network	COCOMO (63) and NASA (93)	MMRE, PRED	[57]
16	2012	Attarzadeh et al.	FFNN, ANN-COCOMO II	COCOMO I, NASA (93)	MMRE	[34]
17	2012	Benala et al.	DBSCAN-FLANN and UKW-FLANN (Un supervised k-windows)	COCOMO 81, NASA2, and Desharnais	MMRE, PRED (0.25), MdMRE	[58]
18	2012	Nassif et al.	Correlation cascade neural network model	240 projects	MMER, PRED	[32]
19	2013	Sarac et al.	ANN	COCOMO 81 (63)	MRE	[59]
20	2014	Madheswaran M and Sivakumar, D	Multilayer feed forward Neural network	COCOMO (63)	MMRE	[60]
21	2015	Aditi Panda et al.	GRNN, PNN, CCNN, GMDHPNN	Project data	MSE, MMRE, PRED	[61]
22	2015	Lopez-Martin	MLP, GRNN, RBFNN	Project data	MAR, MdAr	[62]

(continued)

Table 1 (continued)

S.No	Year	Author(s)	Intelligent method	Datasets	Evaluation factors	References
23	2016	Azzeh and Ali	RBNN	Project data	MAE, MBRE, MIBRE	[63]
24	2018	Arora and Mishra	Multilayer feed forward Neural network	COCOMO	MRE	[64]

representation. Fuzzy neural network models solved many problems in traditional analytic methods which is also called hard computing [66]. To improve the accuracy of effort estimation Khatibi et al. [67] proposed hybrid estimation method fuzzy clustering with the combination of analogy-based estimation and artificial neural networks. By this fuzzy clustering, the effect of irrelevant and inconsistent data is decreased. C-means clustering is used in fuzzy clustering [68]. In this framework training and testing of old methods (ABE, ANN) are used to cluster and combined to construct hybrid estimation method. Wei et al. [69] utilized the framework which combines neuro-fuzzy technique with system evaluation and estimation of resource-software estimation model. This framework demonstrates capability of improving effort estimation accuracy as well as reduced the mean relative error.

Fatima and Alain [70] previously proposed an approach called fuzzy analogy which combines fuzzy logic and analogy-based reasoning. Extended to their previous work they introduced k-modes algorithm with two initializations. They used ISBSG data set and compared with the old methodologies and shown that the accuracy is improved. Zare et al. [71] considered three levels of Bayesian network as well as 15 components of COCOMO and software size to estimate the effort of a software. They considered fuzzy numbers as intervals of network nodes. From genetic algorithm and particle swarm intelligence, they used the concept of optimal control on genetic algorithm and particle swarm intelligence obtained optimal updating coefficient of effort estimation. They showed that, the accuracy is increased by this proposed method. Ezghari and Zahi [72] proposed an enhancement to fuzzy analogy-based software effort estimation to overcome the drawbacks and they called the model as consistent fuzzy analogy-based software effort estimation. In the introduced CFASEE, it has two capabilities: first one using fuzzy sets they enabled attribute representation to fit the attribute in the software effort by maintaining consistency in it and second is composed of possibility distribution and the confidence possibility distribution. They validated CFASEE with thirteen software project data sets and compared with various methods of software effort estimation models.

Some of the intelligent methods of fuzzy logic along with other details such as datasets used, evaluation factors obtained and that article published year, author(s) and references are shown in the Table 2.

able 2 Various Fuzzy Logic (FL) methods and evaluation factors

S.No	Year	Author(s)	Intelligent method	Datasets	Evaluation factors	References
1	2000	Idri et al.	Fuzzy logic	COCOMO 81	PRED(20), MRE, SDRE	[73]
2	2001	Idri et al.	Fuzzy logic	COCOMO 81,	PRED(20) MRE	[74]
3	2004	Braz et al.	Fuzzy logic	Real projects	FUSP	[75]
4	2010	Verma and Sharma	Enhanced fuzzy logic model	ARTIFICIAL, live Project DATA from COCOMO	MMRE PRED	[76]
5	2010	Hari et al.	Fuzzy logic	NASA	VAF, MARE, VARE	[77]
6	2011	Attarzadeh and Ow	Adaptive fuzzy logic	COCOMO I (63), NASA (93), PROJECT DATA	MMRE, PRED (25)	[78]
7	2011	Malathi and Sridhar	FLS	NASA (93), NASA 60, Desharnais	Not Specified	[79]
8	2011	Sadiq et al.	FL, FFP (Fuzzy function point)	Artificial data	MMRE	[80]
9	2012	Martin et al.	Fuzzy logic	COCOMO 81	MMRE, PRED (20)	[16]
10	2012	Malathi and Sridhar	FA (Fuzzy analogy), LV (linguistic variables)	NASA 93, COCOMO 81	PRED (25)	[81]
11	2013	Azzeh and Nassif	Fuzzy tree model	UCP	MMRE, MdMRE, PRED (0.25), and −0.5	[82]
12	2013	Idri and Zahi	Fuzzy logic	Web data set	MMRE, PRED (0.25)	[83]
13	2014	Kushwaha	Fuzzy logic	NASA 93, COCOMO81	MRE	[84]
14	2016	Lopez-Martin et al.	FLM	Project data	MRE, MdRE	[85]

(continued)

Table 2 (continued)

S.No	Year	Author(s)	Intelligent method	Datasets	Evaluation factors	References
15	2016	Sree and SNSVSC	FLM	NASA 93	VAF, MARE, VARE, MBRE, MMRE, PRED (30%)	[86]
16	2017	Langsari and Sarno	Fuzzy logic	NASA 93	MRE, MMRE	[87]
17	2018	Frank Vijay	Fuzzy logic	Project data	MMRE, VAF	[88]

3.3 Evolutionary Computation

Shim et al. [89] used the application of evolutionary computation to obtain accurate, fast on-site training on neural networks, so that it will control the temperature of the air conditioners. In analogy-based estimation, there are many feature weight optimization techniques for similarity function, but there is no consensus in these methods. Tirumala and Mall [90] investigated the effectiveness of differential evolution algorithm (DE) by applying several alternative strategies to optimize feature weights of similarity function of ABE and named it as DE in analogy-based software development effort estimation. They used PROMISE repository to test and found significant improvements of DABE than ABE.

To predict the best effective effort estimation, Morera et al. [91] used genetic framework. This framework selects data processing, learning algorithms from data set characteristics, attribute selection techniques and then it increases the performance of the predictions and optimizing the processing time. They made a comparison to this framework and done sensitive analysis with old frameworks and reported the proposed framework gave best results. In ANN and GA, the problem of selection of recent projects is not solved fully. The disadvantage in this is similarity measure among the projects and feature selection. Thamarai and Murugavalli [92] used differential evolution to solve feature selection and similarity measures among projects, which is population-based search strategy. Using this method, similarities between projects like key attributes and features are compared key by doing this they achieved optimal projects which are used for effort estimation.

Some of the intelligent methods of evolutionary computation along with datasets used, evaluation factors obtained and that article published year, author(s), and references are shown in Table 3.

able 3 Various Evolutionary Computation (EC) methods and evaluation factors

S.No	Year	Author(s)	Intelligent method	Datasets	Evaluation factors	References
1	2001	Burges and Lefley	GP	DESHARNAIS	MMRE, PRED (0.25)	[93]
2	2002	Shan et al.	GGGP	ISBSG	MMRE, PRED (0.25), and (50)	[94]
3	2002	Sheta	GA	NASA (18)	VAF	[95]
4	2008	Ahmed et al.	GA	ISBSG	MMRE	[96]
5	2008	Alaa and Afeef	GP	NASA (18)	MMRE, VAF	[97]
6	2010	Ferrucci et al.	GP, GGGP	DESHARNAIS	MMRE, MdMRE, PRED (25), MEMRE, MdEMRE	[98]
7	2011	Chavoya et al.	GP	132 SHORT-SCALE PROJECTS	MMRE	[99]
8	2012	Galinina et al.	GA	63 PROJECT DATA	Not Specified	[100]
9	2013	Manikavelan and Ponnusamy	GA, DE	ISBSG	Not Specified	[101]
10	2015	Algabri et al.	GA	NASA, COCOMO	MMRE	[102]
11	2016	Sachan et al.	GA	NASA	MMRE, VAF, RMS	[103]
12	2017	Kumari and Pushkar	GA, MOGA	COCOMO81, COCOMONASA	MRE, ARE, MMRE	[104]
13	2018	Saurabh et al.	Chaotic modified genetic algorithm	COCOMO, Kemerer, Albrecht, Desharnais, and KotenGray	MMRE, PRED	[105]

4 Critical Analysis

After understanding the role that software effort estimation plays in the success or failure of a project, the researcher set out to identify the various approaches to software effort estimation. Methods based on expert judgment, algorithmic models, machine learning-based models, and soft computing-based methods were identified as possible candidates. Realizing the rapid adoption of soft computing techniques in various fields of computer science, the researcher then ventured to understand the existing soft computing techniques used for the purpose of software effort estimation from a representative sample of around 105 research articles which seemed highly relevant by limiting the scope to only three methods including artificial neural networks, fuzzy logic and evolutionary computation. For this purpose, the required research articles were gathered from repositories such as Elsevier, IEEE Explore, Research gate, Springer Link and Inderscience by the keyword search using the key-

Table 4 Frequency of Datasets used versus soft computing technique

Dataset	Soft computing technique			Total
	ANN	FL	EC	
COCOMO	14	8	3	25
NASA	5	6	2	13
ISBSG	3	0	3	6
COCOMO NASA 2	1	0	0	1
Desharnais	3	1	3	7
IBM DSP	3	0	0	3
Kemerer	1	0	1	2
Kemererand	1	0	0	1
Maxwell	1	0	0	1
Albrecht	1	0	1	2
Project data	7	5	2	14
KotenGray	0	0	1	1
Others	3	2	0	5

word "Software effort estimation using soft computing". The resulting articles were then categorized into three broad categories based on the soft computing method including artificial neural network, fuzzy logic, and evolutionary computation.

Table 1 represents the relationship between the artificial neural network, the data set upon which it is trained and tested as well as the evaluation factor based on which the training is done. Similar approach is used to represent the relationships between the intelligent method, the dataset and the evaluation factors for fuzzy logic-based methods, evolutionary computing in Tables 2 and 3, respectively.

Table 4 shows the importance of various datasets in terms of their use in various soft computing techniques used in the various research articles considered in the study which have been individually represented in Tables 1, 2 and 3 for ANN, FL, and evolutionary computation, respectively. This table clearly indicates that the most commonly used dataset among all the soft computing techniques is COCOMO (Fig. 3).

Table 5 represents the frequency of various evaluation factors used by various soft computing techniques among the various research articles considered for the study, where it was found that the most frequently used evaluation factors used are MMRE followed by PRED (Fig. 4).

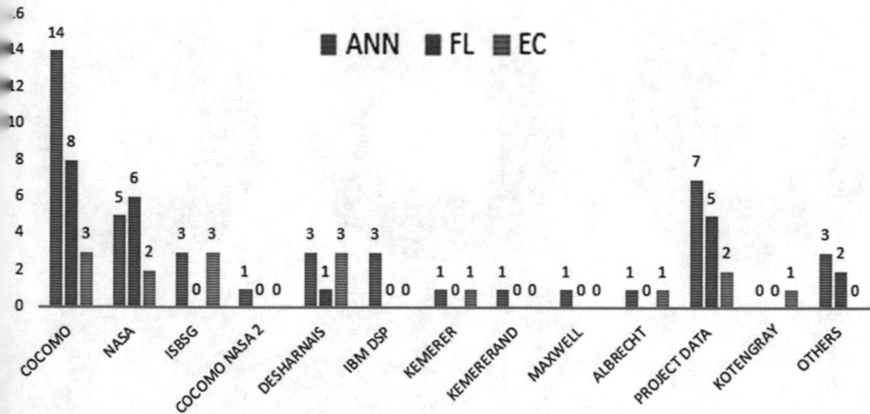

Fig. 3 Datasets verses intelligent methods

Table 5 Frequency of evaluation factor used versus soft computing technique

Evaluation factors	Soft computing technique			Total
	ANN	FL	EC	
MRE	4	5	1	10
MMRE	15	10	10	35
PRED	10	9	4	23
MdMRE	1	0	0	1
MBRE	1	1	0	2
MAPE	1	0	0	1
MARE	0	2	0	2
MAR	1	0	0	1
MAE	1	0	0	1
MIBRE	1	0	0	1
MdAR	1	0	0	1
MSE	1	0	0	1
MEMRE	0	0	1	1
VAF	0	3	3	6
VARE	0	2	0	2
AAE	1	0	0	1
Not specified	0	3	2	5

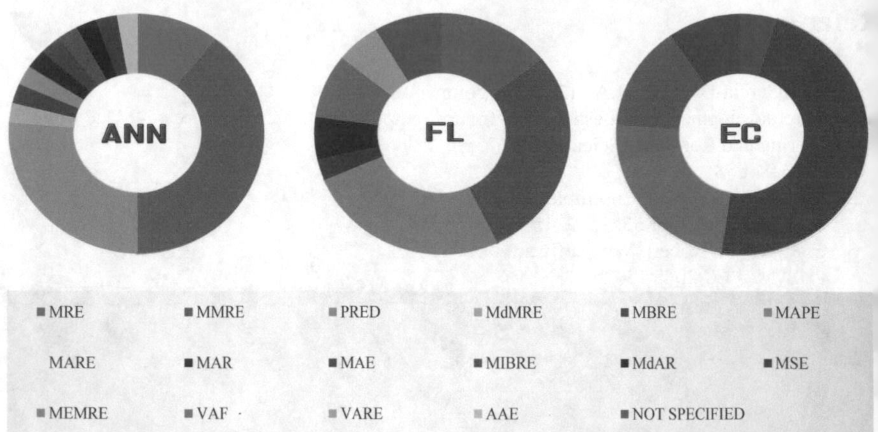

Fig. 4 Evaluation factors and intelligent methods

5 Conclusion

Software effort estimation is a problem for which a "one-size-fits-all" approach does not work. The better the estimate, the better can be the resource, effort, and manpower allocation that can be carried out. The farther the estimate moves away from the actual software development process, the higher is the susceptibility of significant project overheads in terms of cost and delayed delivery. The various soft computing techniques that have been considered in turn are based on different approaches toward the problem and hence the results those are obtained are purely subjective and a comparison of the results obtained by different methods need not provide the required insights as to which method is better or which combination of methods may provide a better estimate. Taking this into consideration, an appropriate method would be to identify the most commonly used methods along with their corresponding data sets and evaluation factors, although reasons for their choice need not have been verified. Understanding why a particular method provided a particular estimate and then interpreting that estimate in the light of assumptions and peculiarities of the method, the chosen data set as well as the evaluation factor has to be ascertained. While the present study does not include these justifications or interpretations, it limits itself to identifying the frequently used methods, data sets, and their evaluation factors. It was observed that COCOMO dataset was used more number of times when compared to other datasets in all intelligent methods such as FL, EC. The next frequently used data sets include NASA, ISBSG, and DESHARNAIS datasets in order. Most performance evolution factors studied were MMRE, MRE, and PRED. Artificial neural network used mostly when compared to other soft computing techniques such as FL, EC which is observed in the proposed study. This study is very much useful for the researchers in the field of software effort estimation.

References

1. NoelGarcia-Diaz, C.-M.A.: (2013) A comparative study of two fuzzy logic models for software development effort estimation. In: 3rd Iberoamerican Conference on Electronics Engineering and Computer Science, vol. 7, pp. 305–314 (2013). https://doi.org/10.1016/j.protcy.2013.04.038
2. Minku, L.L., Yao, X.: Ensembles and locality: Insight on improving software effort estimation. Inf. Softw. Technol. **55**, 1512–1528 (2013). https://doi.org/10.1016/j.infsof.2012.09.012
3. Charette, N.: Robert, Why software fails. Spectr. IEEE **42**, 42–49 (2005). https://doi.org/10.1109/MSPEC.2005.1502528
4. Boehm, B.W.: Software engineering economics. IEEE Trans. Softw. Eng. **SE-10**(1), 4–21 (1984). https://doi.org/10.1109/tse.1984.5010193
5. Putnam, L.H.: A general empirical solution to the macro software sizing and estimating problem. IEEE Trans. Softw. Eng. **SE-4**(4), 345–361 (1978). https://doi.org/10.1109/tse.1978.23152
6. Albrecht, A.J., Gaffney, J.E.: Software function, source lines of code, and development effort prediction: a software science validation. IEEE Trans. Softw. Eng. **SE-9**(6), 639–648 (1983). https://doi.org/10.1109/tse.1983.235271
7. Elish, M.O., Helmy, T., Hussain, M.I.: Empirical study of homogeneous and heterogeneous ensemble models for software development effort estimation. In: Mathematical Problems in Engineering, vol. 2013, 21 p. Article ID 312067 (2013). http://dx.doi.org/10.1155/2013/312067
8. Idri, A., Abran, A., Khoshgofaar, T.: Estimating software project effort by analogy based on linguistic values. In: Proceedings of the 8th IEEE Symposium on Software Metrics, pp. 21–30 (2002). https://doi.org/10.1109/metric.2002.1011322
9. Ibrahim, D.: An overview of soft computing. Proc. Comput. Sci. **102**, 34–38 (2016). https://doi.org/10.1016/j.procs.2016.09.366
10. Haykin, S., Nie, J., Currie, B.: Neural network-based receiver for wireless communications. Electron. Lett. **35**(3), 203–205 (1999). https://doi.org/10.1049/el:19990177
11. Karunanithi, N., Whitley, D., Malaiya, Y.K.: Using neural networks in reliability prediction. IEEE Softw. **9**(4), 53–59 (1992). https://doi.org/10.1109/52.143107
12. Dave, V., Dutta, K.: Neural network based models for software effort estimation: a review. Artif. Intell. Rev. **42**, 295–307 (2014). https://doi.org/10.1007/s10462-012-9339-x
13. Zadeh, L.A.: Fuzzy sets. In: Information and Control, vol. 8, Issue 3, pp. 338–353 (1965)
14. Zadeh, L.A.: Similarity relations and fuzzy orderings. Inf. Sci. **3**, 177–200 (1971). https://doi.org/10.1016/S0020-0255(71)80005-1
15. Zadeh, L.A.: From computing with numbers to computing with words. From manipulation of measurements to manipulation of perceptions. IEEE Trans. Circuits Syst. I: Fundam. Theory Appl. **46**(1), 105–119 (1999). https://doi.org/10.1109/81.739259
16. Martin, C.L., Pasquier, J.L., Yanez, C.M., Tornes, A.G.: Software development effort estimation using fuzzy logic: a case study. In: Sixth Mexican International Conference on Computer Science (ENC'05), Puebla, Mexico, pp. 113–120 (2005). https://doi.org/10.1109/enc.2005.47
17. Scherer, R.: Takagi–Sugeno Fuzzy Systems Multiple Fuzzy Classification Systems. Studies in Fuzziness and Soft Computing, vol. 288, pp. 73–79 (2012). Springer, Berlin. https://doi.org/10.1007/978-3-642-30604
18. Mamdani, E.H., Assilian, S.: An experiment in linguistic synthesis with a fuzzy logic controller. Int. J. Man-Mach. Stud. **7**(1), 1–13 (1975). https://doi.org/10.1016/S0020-7373(75)80002-2

19. Fogel, D.: Evolutionary Computation: Toward a New Philosophy of Machine Intelligenc IEEE (1995). ISBN: 978-0-7803-1038-4, https://doi.org/10.1002/0471749214
20. Goldberg, D.E., Holland, J.H.: Genetic algorithms and machine learning. Mach. Learn. (2–3), 95–99. https://doi.org/10.1023/a:1022602019183
21. Rechenberg, I.: Evolutionstrategie, Optimierung Technisher Systeme nach Prinzipien de Biologischen Evolution. Fromman-Hozlboog, Stuttgart (1973). https://doi.org/10.1002/fed 19750860506
22. Schwefel, H.-P.: Numerical Optimization of Computer Models in Numerical Optimization c Computer Models, 2nd edn. Wiley, New York (1981). ISBN:0471099880
23. Storn, R.: On the usage of differential evolution for function optimization. In: Proceeding of North American Fuzzy Information Processing, Berkeley, CA, USA, pp. 519–523 (1996) https://doi.org/10.1109/nafips.1996.534789
24. Kennedy, J., Eberhart, R.: Particle swarm optimization. In: Proceedings of ICNN'95 - International Conference on Neural Networks, Perth, WA, Australia (1995)
25. Martins, C., Paupitz, R., Lúcio Santos, F.: Simplified particle swarm optimization algorithm Acta Sci. Technol. (2012). https://doi.org/10.4025/actascitechnol.v34i1.9679
26. Lin, J., Tzeng, H.: Applying particle swarm optimization to estimate software effort by multiple factors software project clustering. In: 2010 International Computer Symposium (ICS2010), Tainan, pp. 1039–1044 (2010). https://doi.org/10.1109/compsym.2010.5685538
27. Kaur, M., Sehra, S.K.: Particle swarm optimization based effort estimation using Function Point analysis. In: 2014 International Conference on Issues and Challenges in Intelligent Computing Techniques (ICICT), Ghaziabad, pp. 140–145 (2014). https://doi.org/10.1109/icicict.2014.6781267
28. Murillo-Morera, J., Quesada-López, C., Castro-Herrera, C. et al.: A genetic algorithm based framework for software effort prediction. J. Softw. Eng. Res. Dev. (2017). https://doi.org/10.1186/s40411-017-0037-x
29. Wu, D., Li, J., Bao, C.: Case-based reasoning with optimized weight derived by particle swarm optimization for software effort estimation. Soft. Comput. **22**(16), 5299–5310 (2018). https://doi.org/10.1007/s00500-017-2985-9
30. Dote, Y., Ovaska, S.J.: Industrial applications of soft computing: a review. Proc. IEEE **89**(9), 1243–1265 (2001). https://doi.org/10.1109/5.949483
31. Hirota, K.: History and recent trends in soft computing: research and application aspects in Japan. In: Proceedings of IEEE International Conference on Intelligent Engineering Systems, Budapest, Hungary, pp. 31–37 (1997). https://doi.org/10.1109/ines.1997.632390
32. Dutta, S.: Fuzzy logic applications: technological and strategic issues. IEEE Trans. Eng. Manage. **40**(3), 237–254 (1993). https://doi.org/10.1109/17.233185
33. Nassif, A.B., Capretz, L.F., Ho, d D.: Estimating software effort using an ANN model based on use case points. In: 11th International Conference on Machine Learning and Applications, Boca Raton, FL, pp. 42–47 (2012). https://doi.org/10.1109/icmla.2012.138
34. Nassif, A.B. et al.: Software effort estimation in the early stages of the software life cycle using a cascade correlation neural network model. In: 2012 13th ACIS International Conference on Software Engineering, Artificial Intelligence, Networking and Parallel/Distributed Computing, pp. 589–594 (2012). https://doi.org/10.1109/snpd.2012.40
35. Attarzadeh, I. et al.: Proposing an enhanced artificial neural network prediction model to improve the accuracy in software effort estimation. In: 2012 Fourth International Conference on Computational Intelligence, Communication Systems and Networks, pp. 167–172 (2012). https://doi.org/10.1109/cicsyn.2012.39
36. Panda, A., Satapathy, S.M., Rath, S.: Empirical validation of neural network models for agile software effort estimation based on story points. In: Procedia Computer Science (2015). https://doi.org/10.1016/j.procs.2015.07.474

37. Iwata, K. Nakashima, T., Anan, Y., Ishii, N.: Effort estimation for embedded software development projects by combining machine learning with classification. In: 4th International Conference on Applied Computing and Information Technology (ACIT-CSII-BCD), Las Vegas, NV, pp. 265–270 (2016). https://doi.org/10.1109/ACIT-CSII-BCD.2016.058
38. López-Martín, C.: Predictive accuracy comparison between neural networks and statistical regression for development effort of software projects. In: Applied Soft Computing, pp. 434–449 (2015). https://doi.org/10.1016/j.asoc.2014.10.033
39. Poonam Rijwani, S.J: Enhanced software effort estimation using multi layered feed. In: Twelfth International Multi-Conference on Information Processing-2016 (IMCIP-2016), pp. 307–312. Elsevier, Jaipur (2016). https://doi.org/10.1016/j.procs.2016.06.073
40. Nassif, A.B. et al.: Neural network models for software development effort estimation: a comparative study. In: Neural Computing and Applications, pp. 2369–2381 (2016). https://doi.org/10.1007/s00521-015-2127-1
41. Tanveer, B.M., Vollmer, A., Braun, S.: A hybrid methodology for effort estimation in Agile development: an industrial evaluation In: The Proceedings of the International Conference on Software and System Process, pp. 21–30 (2018). https://doi.org/10.1145/3202710.3203152
42. Pospieszny, Przemysław, Czarnacka-Chrobot, Beata, Kobyliński, Andrzej: An effective approach for software project effort and duration estimation with machine learning algorithms. J. Syst. Softw. **137**, 184–196 (2018). https://doi.org/10.1016/j.jss.2017.11.066
43. Idri, A., Khoshgoftaar, T.M., Abran, A.: Can neural networks be easily interpreted in software cost estimation. Proc. IEEE Int. Conf. Fuzzy Syst. **2**, 1162–1167 (2002). https://doi.org/10.1109/FUZZ.2002.1006668
44. Idri, A., Mbarki, S., Abran, A.: Validating and understanding software cost estimation models based on neural networks. In: IEEE International Conference on Information and Communication Technologies: From Theory to Applications, pp. 433–434 (2004). https://doi.org/10.1109/ictta.2004.1307817
45. Tadayon, N.: Neural network approach for software cost estimation. IEEE Int. Conf. Inf. Technol.: Coding Comput. **2**, 815–818 (2005). https://doi.org/10.1109/ITCC.2005.210
46. Idri, A., Abran, A., Mbarki, S.: An experiment on the design of radial basis function neural networks for software cost estimation. IEEE Int. Conf. Inf. Commun. Technol. **1**, 1612–1617 (2006). https://doi.org/10.1109/ICTTA.2006.1684625
47. deBarcelos Tronto, I.F., da Silva, J.D.S., Sant'Anna, N.: Comparison of artificial neural network and regression models in Software Effort Estimation. In: IEEE International Joint Conference on Neural Networks (IJCNN), pp. 771–776 (2007). https://doi.org/10.1109/ijcnn.2007.4371055
48. Abrahamsson, P., Moser, R., Pedrycz, W., Sillitti, A., Succi, G.: Effort prediction in iterative software development processes–Incremental versus global prediction models. In: IEEE First International Symposium on Empirical Software Engineering and Measurement (ESEM), pp. 344–353 (2007) https://doi.org/10.1109/esem.2007.16
49. El-Sebakhy, E.A.: New computational intelligence paradigm for estimating the software project effort. In: IEEE 22nd International Conference on Advanced Information Networking and Applications Workshops (AINAW 2008), pp. 621–627 (2008). https://doi.org/10.1109/waina.2008.257
50. Li, Y.F., Xie, M., Goh, T.N.: A study of the non-linear adjustment for analogy based software cost estimation. Empir. Softw. Eng. **14**(6), 603–643 (2009). https://doi.org/10.1007/s10664-008-9104-6
51. Kalichanin-Balich, I., Lopez-Martin, C.: Applying a Feedforward neural network for predicting software development effort of short-scale projects. In: IEEE Eighth ACIS International Conference on Software Engineering Research, Management and Applications (SERA), pp. 269–275 (2010). https://doi.org/10.1109/sera.2010.41

52. Jodpimai, P., Sophatsathit, P., and Lursinsap, C.: Estimating software effort with minimu features using neural functional approximation. In: IEEE International Conference on Compu tational Science and Its Applications (ICCSA), pp. 266–273 (2010). https://doi.org/10.1109 iccsa.2010.63
53. Attarzadeh, I., Ow, S.H.: Proposing a new software cost estimation model based on artifici neural networks. In: IEEE 2nd International Conference on Computer Engineering and Tech nology (ICCET), vol. 3, pp. 483–487 (2010). https://doi.org/10.1109/iccet.2010.5485840
54. Ajitha, S., Kumar, T.S., Geetha, D.E., Kanth, K.R.: Neural network model for software siz estimation using a use case point approach. In: IEEE International Conference on Industria and Information Systems (ICIIS), pp. 372–376 (2010). https://doi.org/10.1109/iciinfs.2010 5578675
55. Dave, V.S., Dutta, K.: Neural network based software effort estimation & evaluation criterio MMRE. In: IEEE 2nd International Conference on Computer and Communication Technolog (ICCCT), pp. 347–351 (2011). https://doi.org/10.1109/iccct.2011.6075192
56. Ghose, M.K., Bhatnagar, R., Bhattacharjee, V.: Comparing some neural network models fo software development effort prediction. In: IEEE 2nd National Conference on Emerging Trends and Applications in Computer Science (NCETACS), pp. 1–4 (2011). https://doi.org/ 10.1109/ncetacs.2011.5751391
57. Kaushik, A., Soni, A.K., Soni, R.: An adaptive learning approach to software cost estimation. In: National IEEE Conference on Computing and Communication Systems (NCCCS), pp. 1–6 (2012). https://doi.org/10.1109/ncccs.2012.6413029
58. Benala, T.R., Dehuri, S., Mall, R., ChinnaBabu, K.: Software effort prediction using unsupervised learning (clustering) and functional link artificial neural networks. In: IEEE World Congress on Information and Communication Technologies (WICT), pp. 115–120 (2012). https://doi.org/10.1007/978-3-642-32341-6_1
59. Saraç, Ö.F., Duru, N.: A novel method for software effort estimation: Estimating with boundaries. In: IEEE International Symposium on Innovations in Intelligent Systems and Applications (INISTA), pp. 1–5 (2013). https://doi.org/10.1109/inista.2013.6577643
60. Madheswaran, M., Sivakumar, D.: Enhancement of prediction accuracy in COCOMO model for software project using neural network. In: IEEE International Conference on Computing, Communication and Networking Technologies (ICCCNT), pp. 1–5 (2014). https://doi.org/10.1109/icccnt.2014.6963021
61. Panda, A., Satapathy, S.M.: Rath, S.K.: Empirical validation of neural network models for agile software effort estimation based on story points. Proc. Comput. Sci. **57**, 772–781 (2015). ISSN 1877-0509, https://doi.org/10.1016/j.procs.2015.07.474
62. López-Martín, C.: Predictive accuracy comparison between neural networks and statistical regression for development effort of software projects. Appl. Soft Comput. **27**, 434–449 (2015). ISSN 1568-4946, https://doi.org/10.1016/j.asoc.2014.10.033
63. Azzeh, M., Nassif, A.B.: A hybrid model for estimating software project effort from use case points. Appl. Soft Comput. **49**, 981–989 (2016). ISSN 1568-4946, https://doi.org/10.1016/j.asoc.2016.05.008
64. Arora, S., Mishra, N.: Software cost estimation using artificial neural network. In: Pant M., Ray K., Sharma T., Rawat S., Bandyopadhyay A. (eds.) Soft Computing: Theories and Applications. Advances in Intelligent Systems and Computing, vol. 584. Springer, Singapore (2018)
65. San Diego, CA, USA pp. 1163–1170 (1992). https://doi.org/10.1109/fuzzy.1992.258721
66. Marchi, P.A., dos Santos Coelho, L., Coelho, A.A.R.: Comparative study of parametric and structural methodologies in identification of an experimental nonlinear process. In: Proceedings of the 1999 IEEE International Conference on Control Applications (Cat. No.99CH36328), Kohala Coast, HI, USA, vol. 2, pp. 1062–1067 (1999). https://doi.org/10.1109/cca.1999.801057

67. Bardsiri, V.K., Jawawi, D.N.A., Hashim, S.Z.M., Khatibi, E.: Increasing the accuracy of software development effort estimation using projects clustering. IET Softw. **6**(6), 461–473 (2012). https://doi.org/10.1049/iet-sen.2011.0210
68. Bezdek, James C., Ehrlich, Robert, Full, William: FCM: The fuzzy c-means clustering algorithm. Comput. Geosci. **10**(2–3), 191–203 (1984). https://doi.org/10.1016/0098-3004(84)90020-7
69. Du, W.L. et al.: A Neuro-Fuzzy Model with SEER-SEM for Software Effort Estimation (2010). CoRR, arXiv:1508.00032
70. Amazal, F.A., Idri, A., Abran, A.: Improving fuzzy analogy based software development effort estimation. In: 21st Asia-Pacific Software Engineering Conference, Jeju, pp. 247–254 (2014). https://doi.org/10.1109/apsec.2014.46
71. Zare, F. et al.: Software effort estimation based on the optimal Bayesian belief network. In: Appl. Soft Comput. **49**, 968–980 (2016)., https://doi.org/10.1016/j.asoc.2016.08.004
72. Ezghari, S., Zahi, A.: Uncertainty management in software effort estimation using a consistent fuzzy analogy-based method. In: Applied Soft Computing, vol. 67, pp. 540–557 (2018). ISSN 1568-4946, https://doi.org/10.1016/j.asoc.2016.08.004
73. Idri, A., Abran, A., Kjiri, L.: COCOMO cost model using fuzzy logic. In: 7th International Conference on Fuzzy Theory & Techniques, vol. 27 (2000)
74. Idri, A., Abran, A.: A fuzzy logic based set of measures for software project similarity: validation and possible improvements. In: IEEE Proceedings of Seventh International Software Metrics Symposium, pp. 85–96 (2001). https://doi.org/10.1109/metric.2001.915518
75. Braz, M.R., Vergilio, S.R.: Using fuzzy theory for effort estimation of object-oriented software. In: 16th IEEE International Conference on Tools with Artificial Intelligence (ICTAI), pp. 196–201 (2004). https://doi.org/10.1109/ictai.2004.119
76. Verma, H.K., Sharma, V.: Handling imprecision in inputs using fuzzy logic to predict effort in software development. In: IEEE 2nd International Advance Computing Conference (IACC), pp. 436–442 (2010). https://doi.org/10.1109/iadcc.2010.5422889
77. Hari, C.V., PVGD, P.R., Jagadeesh, M., Ganesh, G.S.: Interval type-2 fuzzy logic for software cost estimation using TSFC with mean and standard deviation. In: International Conference on Advances in Recent Technologies in Communication and Computing (ARTCom), pp. 40–44 (2010). https://doi.org/10.1109/artcom.2010.40
78. Attarzadeh, I., Ow, S.H.: Improving the estimation accuracy of the COCOMO II using an adaptive fuzzy logic model. In: IEEE International Conference on Fuzzy Systems, pp. 2458–2464 (2011). https://doi.org/10.1109/fuzzy.2011.6007471
79. Malathi, S., Sridhar, S.: A Classical Fuzzy Approach for Software Effort Estimation on Machine Learning Technique (2011). arXiv:1112.3877
80. Sadiq, M., Mariyam, F., Alil, A., Khan, S., Tripathi, P.: Prediction of software project effort using fuzzy logic. In: IEEE 3rd International Conference on Electronics Computer Technology (ICECT), vol. 4, pp. 353–358 (2011)
81. Malathi, S., Sridhar, S.: Optimization of fuzzy analog for software cost estimation using linguistic variables. In: International conference on modeling, optimization, and computing. In: Procdia Engineering, pp. 1–14 (2012). https://doi.org/10.1016/j.proeng.2012.06.025
82. Azzeh, M., Nassif, A.B.: Fuzzy model tree for early effort estimation. In: IEEE 12th International Conference on Machine Learning and Applications (ICMLA), vol. 2, pp. 117–121 (2013). https://doi.org/10.1109/icmla.2013.115
83. Idri, A., Zahi, A.: Software cost estimation by classical and Fuzzy Analogy for Web Hypermedia Applications: A replicated study. In: Computational Intelligence and Data Mining (CIDM), pp. 207–213 (2013). https://doi.org/10.1109/cidm.2013.6597238
84. Kushwaha, N.: Software cost estimation using the improved fuzzy logic framework. In: IEEE Conference on IT in Business, Industry and Government (CSIBIG), pp. 1–5 (2014). https://doi.org/10.1109/csibig.2014.7056959

85. Lopez-Martin, C., Yanez-Marquez, C., Gutierrez-Tornes, A.: A fuzzy logic model based upc reused and new & changed code for software development effort estimation at personal leve In: IEEE 15th International Conference on Computing, vol. 5, Issue 3, pp. 298–303 (2016 ISSN: 0975-9646
86. Sree, P.R., SNSVSC, R.: Improving efficiency of fuzzy models for effort estimation by cas cading & clustering techniques. Proc. Comput. Sci. **85**, 278–285 (2016). https://doi.org/1(1016/j.procs.2016.05.234
87. Langsari, K., Sarno, R.: Optimizing effort and time parameters of COCOMO II estimatio using fuzzy multi-objective PSO. In: 4th International Conference on Electrical Engineering Computer Science and Informatics (EECSI), Yogyakarta, pp. 1–6 (2017). https://doi.org/1(1109/eecsi.2017.8239157
88. Vijay, F.: Enrichment of accurate software effort estimation using fuzzy-based function poin analysis in business data analytics. In: Neural Computing and Applications, pp. 1–7. https:/ doi.org/10.1007/s00521-018-3565-3
89. Shim, M., Seong, S., Ko, B., So, M.: Application of evolutionary computations at LG Elec tronics. In: FUZZ-IEEE'99, IEEE International Fuzzy Systems. Conference Proceedings (Cat No.99CH36315), Seoul, South Korea, vol. 3, pp. 1802–1806 (1999) https://doi.org/10.1109 fuzzy.1999.790181
90. Benala, T.R., Mall, R.: DABE: Differential evolution in analogy-based software developmen effort estimation. Swarm Evol. Comput. **38**, 158–172 (2018). ISSN 2210-6502, https://doi org/10.1016/j.swevo.2017.07.009
91. Murillo-Morera, J., Quesada-López, C., Castro-Herrera, C. et al.: A genetic algorithm based framework for software effort prediction. J. Softw. Eng. Res. Dev. (2017). https://doi.org/10. 1186/s40411-017-0037-x
92. Thamarai, I., Murugavalli, S.: Using differential evolution in the prediction of software effort. In: Fourth International Conference on Advanced Computing (ICoAC), Chennai, pp. 1–3 (2012) https://doi.org/10.1109/icoac.2012.6416816
93. Burgess, C.J., Lefley, M.: Can genetic programming improve software effort estimation?: A comparative evaluation. Inf. Softw. Technol. **43**(14), 863–873 (2001). https://doi.org/10.1016/ S0950-5849(01)00192-6
94. Shan, Y., McKay, R.I., Lokan, C.J., Essam, D.L.: Software project effort estimation using genetic programming. IEEE Int. Conf. Commun. Circuits Syst. West Sino Expos. **2**, 1108–1112 (2002). https://doi.org/10.1109/ICCCAS.2002.1178979
95. Sheta, A.F.: Estimation of the COCOMO model parameters using genetic algorithms for NASA software projects. J. Comput. Sci. **2**(2), 118–123 (2006). https://doi.org/10.3844/jcssp. 2006.118.123
96. Ahmed, F., Bouktif, S., Serhani, A., Khalil, I.: Integrating function point project information for improving the accuracy of effort estimation. In: IEEE the Second International Conference on Advanced Engineering Computing and Applications in Sciences, pp. 193–198 (2008). https://doi.org/10.1109/advcomp.2008.42
97. Alaa, F.S., Al-Afeef, A.: A GP effort estimation model utilizing a line of code and methodology for NASA software projects. In: IEEE 10th International Conference on, Intelligent Systems Design and Applications (ISDA), pp. 290–295 (2010). https://doi.org/10.1109/isda. 2010.5687251
98. Ferrucci, F., Gravino, C., Oliveto, R., Sarro, F.: Genetic programming for effort estimation: an analysis of the impact of different fitness functions. In: IEEE Second International Symposium on Search Based Software Engineering (SSBSE), pp. 89–98 (2010). https://doi.org/10.1109/ ssbse.2010.20
99. Chavoya, A., Lopez-Martin, C., and Meda-Campa, M.E. (2011) Applying genetic programming for estimating software development effort of short-scale projects. In: IEEE Eighth International Conference on Information Technology: New Generations (ITNG), pp. 174–179 (2011). https://doi.org/10.1109/itng.2011.37

100. Galinina, A., Burceva, O., Parshutin, S.: The optimization of COCOMO model coefficients using genetic algorithms. Inf. Technol. Manag. Sci. **15**(1), 45–51 (2012). https://doi.org/10.2478/v10313-012-0006-7
101. Manikavelan, D., Ponnusamy, R.: To find the accurate software cost estimation using differential evaluation algorithm. In: IEEE International Conference on Computational Intelligence and Computing Research (ICCIC), pp. 1–4 (2013). https://doi.org/10.1109/iccic.2013.6724240
102. Algabri, M., Saeed, F., Mathkour, H., Tagoug, N.: Optimization of soft cost estimation using genetic algorithm for NASA software projects. In: IEEE 5th National Symposium on Information Technology: Towards New Smart World (NSITNSW), pp. 1–4 (2015). https://doi.org/10.1109/nsitnsw.2015.7176416
103. Sachan, R.K., Nigam, A., Singh, A., Singh, S., Choudhary, M., Tiwari, A., Kushwaha, D.S.: Optimizing basic COCOMO model using simplified genetic algorithm. Proc. Comput. Sci. **89**, 492–498 (2016). https://doi.org/10.1016/j.procs.2016.06.107
104. Kumari, S., Pushkar, S.: A framework for analogy-based software cost estimation using multi-objective genetic algorithm. In: Proceedings of the World Congress on Engineering and Computer Science, vol. 1 (2016)
105. Bilgaiyan S., Aditya K., Mishra S., Das M.: Chaos-based modified morphological genetic algorithm for software development cost estimation. In: Pattnaik P., Rautaray S., Das H., Nayak J. (eds) Progress in Computing, Analytics and Networking. Advances in Intelligent Systems and Computing, vol. 710. Springer, Singapore (2018) https://doi.org/10.1007/978-981-10-7871-2_4

A Nonlinear Telegraph Equation for Edge-Preserving Image Restoration

Subit K. Jain, Jyoti Yadav, Manisha Rao, Monika Sharma and Rajendra K. Ray

Abstract Poisson noise suppression is a challenging and important preprocessing stage for higher level image analysis. Therefore, in this paper, a new attempt has been made using telegraph equation and variational theory for Poisson noise suppression. The proposed approach enjoys the benefits of both telegraph-diffusion equation and Hessian edge detector, which is not only robust to noise but also preserves image structural details. The Hessian function has been used to distinguish between edges and noise. However, to the best of author's knowledge, the Hessian edge detector driven telegraph-diffusion scheme has not been used before for Poisson noise suppression. With the proposed model, restoration is carried out on several natural images. The experimental results of proposed model are found better in terms of noise suppression and detail/edge preservation, with respect to the existing approaches.

Keywords Digital image · Poisson noise · Telegraph equation · Total variation · Hessian function · Edge detection

---

S. K. Jain · J. Yadav · M. Rao · M. Sharma
Starex University, Gurugram 122413, India
e-mail: jain.subit@gmail.com

J. Yadav
e-mail: jyotiyadav171996@gmail.com

M. Rao
e-mail: manisharao1109@gmail.com

M. Sharma
e-mail: monas6182@gmail.com

S. K. Jain
National Institute of Technology Hamirpur, Hamirpur 177005, India

R. K. Ray (✉)
Indian Institute of Technology Mandi, Mandi 175005, India
e-mail: rajendra@iitmandi.ac.in

A. K. Das et al. (eds.), *Computational Intelligence in Pattern Recognition*, Advances in Intelligent Systems and Computing 999,
https://doi.org/10.1007/978-981-13-9042-5_71

1 Introduction

Restoration of images corrupted with Poisson's noise is a prevailing issue in various image-based applications. Its appearance in image reduces the utility and the capability to discriminate different distributed objects. Therefore, in last few decades a variety of research work have been done to develop and implement novel image restoration approaches devoted to the problem of filtering Poissonian images [1–6]. Among various filters, the anisotropic partial differential equations (PDEs) [1, 4] and total variation based frameworks [2, 3, 5] have became the important classes of noise removal techniques. These partial differential equation based methods can effectively simulate and preserve slowly varying signals and important feature of images due to their well-studied mathematical properties and approximation processes. A seminal contribution has been made by Perona and Malik [7] by proposing a non-homogeneous smoothing (diffusion) coefficient driven nonlinear diffusion model as follows:

$$I_t = div(c(|\nabla I|)\nabla I) \qquad in\ \Omega \times (0, T)$$

where Ω is the domain of original image I and observed noisy image I_0. Also, ∇ represents the gradient operator and div represents the divergence operator and $c(s) = 1/(1 + (s^2/k^2))$ is an edge-controlled diffusion function which is designed to preserve the important features and smooth the unwanted signals. Followed this work, several diffusion models have been proposed for computer vision task (see [2] for review). Based on these nonlinear frameworks, several models have been modified and successfully employed to remove the signal-dependent Poisson noise from digital images in an effective manner [3, 4, 8, 9]. To the extent of our knowledge, the first total variation regularization based Poisson noise suppression approach is proposed by Le et al. (TVP model) [3]. Similar to total variation model [10], the TVP model derived using a maximum a posterior (MAP) approach is

$$I_t = div\left(\frac{\nabla I}{|\nabla I|}\right) + \frac{1}{\beta I}(I_0 - I) \qquad in\ \Omega \times (0, T), \tag{1a}$$

$$\frac{\partial I}{\partial n} = 0 \qquad in\ \partial\Omega \times (0, T), \tag{1b}$$

$$I(x, y, 0) = I_0(x, y) \qquad in\ \Omega \tag{1c}$$

where the parameter β has been chosen according to suitably modified discrepancy principle. To our surprise, there is no hyperbolic–parabolic smoothing framework for Poisson noise removal with effective edge preservation yet. Thus, it is beneficial to carry out study into the multidimensional manifold theory for noise removal methods [11, 12] along with the powerful total variational framework [3].

Hence, in the present paper, we provide a new approach by leveraging the telegraph-diffusion equation with the total variation framework for Poisson noise removal. In this paper, we put an emphasis on efficient Hessian edge detector driven telegraph-diffusion model. There are two key advantages of this proposed approach.

irst, we use the telegraph-diffusion model [11], which is more able to provide sharp nd true edges during noise removal than other non-telegraph based diffusion algoithms. Second, higher order derivative based Hessian edge detector is incorporated nto the telegraph diffusion due to its effective edge/noise detection ability. Here, in ne proposed model, the total variation approach is used to preserve significant conents during the restoration which can also be replaced with some other contemporary egularization techniques. The performance of our approach is compared to some xisting filters and has shown better results in terms of accuracy and robustness.

The remainder of this paper is organized as follows. Section 2 describes the roposed method for removal of Poisson noise. The advantages of proposed model re discussed in Sect. 3. We conclude the paper in Sect. 4 with an outlook on future vork.

2 Proposed Model

To avoid the issue of edge localization, smearing of corners, and spatial regularization, we employ a Hessian-based edge indicator along with the diffusion function. Here, first we explain the used differential operator based edge indicator function as follows:

$$\alpha(I) = 1 - H(I) \tag{2}$$

where $\alpha(I)$ calculates a pixel-wise edge assessment using the local structure measure function,

$$H(I) = \delta_1 * \delta_2 + |\delta_1 - \delta_2|^2$$

where δ_1 and δ_2 represent the eigenvalues of Hessian matrix of the image I. From above equation, at the nonhomogeneous region, $\alpha(I) \to 0$, the proposed approach is less smooth at the region; at the homogeneous region, $\alpha(I) \to 1$, the proposed approach is more smooth at the region. The proposed edge indicator is capable of classifying the noisy pixel from edges or boundaries based on the higher order derivatives.

To enhance the understandability and visual quality of digital images, we improved the PDE formulation (1a)–(1c) by incorporating the telegraph equation and the proposed edge indicator with diffusion function. Hence, in order to get filtered image I from a given noisy image I_0, we use the following steps:

1. Start with given degraded image I_0 at time $t = 0$,

2. For time $t = 1 : T$,

(a) Compute $\alpha(I)$ on image $I(., t-1)$ using Eq. (2),

(b) Solve the following proposed telegraph-diffusion model and obtain $I(x, t)$ as

$$I_{tt} + \gamma I_t = div\left(\alpha(I)\frac{\nabla I}{|\nabla I|}\right) - \lambda\left(1 - \frac{I_0}{I}\right) \quad in\ \Omega \times (0, T), \tag{3a}$$

$$\frac{\partial I}{\partial n} = 0 \quad in\ \partial\Omega \times (0, T), \tag{3b}$$

$$I(x, y, 0) = I_0(x, y) \quad in\ \Omega, \tag{3c}$$

$$I_t(x, y, 0) = 0 \quad in\ \Omega \tag{3d}$$

(c) Calculate the $RelError$ between two adjacent diffusion steps as

$$RelError(I^t) = \frac{||I^t - I^{t-1}||_2{}^2}{||I^{t-1}||_2{}^2} \tag{4}$$

(d) If $RelError \leq \varepsilon$ then exit otherwise continue $2(a)$.

Here, we used the $\varepsilon \leq 10^{-4}$ as a fixed threshold. Based on this inference, the proposed model applies telegraph-diffusion-based filtering approach to different regions of same image and enable us for edge and detail preservation. Equation (3a) of the proposed model is discretized using an explicit finite difference method [13]. Hence, the proposed Eq. (3a) in its discretized form can be given as follows:

$$\Rightarrow (1 + \gamma\tau)I_{i,j}^{n+1} = (2 + \gamma\tau)I_{i,j}^{n} - I_{i,j}^{n-1} + \frac{\tau^2}{h}\left[\nabla_x\left(\alpha(I_{i,j}^{n})\frac{\nabla_x I_{i,j}^{n}}{\sqrt{(\nabla_x I_{i,j}^{n})^2 + (\nabla_y I_{i,j}^{n})^2}}\right)\right.$$
$$\left. + \nabla_y\left(\alpha(I_{i,j}^{n})\frac{\nabla_y I_{i,j}^{n}}{\sqrt{(\nabla_x I_{i,j}^{n})^2 + (\nabla_y I_{i,j}^{n})^2}}\right)\right] - \tau^2\lambda^n\left(1 - \frac{I_0}{I_{i,j}^{n}}\right) \tag{5}$$

where edge indicator function can be calculated as

$$\alpha(I_{i,j}^{n}) = 1 - H(I_{i,j}^{n}),$$
$$H(I_{i,j}^{n}) = \delta_1{}^n * \delta_2{}^n + |\delta_1{}^n - \delta_2{}^n|^2$$

With the conditions,

$$I_{i,j}^{0} = I_0(ih, jh) \qquad 0 \leq i \leq N, 0 \leq j \leq M$$
$$I_{i,j}^{1} = I_{i,j}^{0} \qquad 0 \leq i \leq N, 0 \leq j \leq M$$

Through the above numerical discretization, we can obtain the solution at time T. Our numerical results depend on three parameters: the scale of diffusion τ, the damping coefficient γ, and the weight coefficient λ.

Experimental Results

This section presents visual and numerical results of our model and compare with TVP model [3] and NCD model [4]. Both these existing models along with the proposed model are numerically discretized using the same explicit finite difference scheme as given in Eq. (5).

3.1 *Setup and Parameters*

All considered diffusion models were implemented in MATLAB R2015a which runs on a laptop with 2.0 GHZ processing speed, Core 2 Duo CPU, and 3 GB RAM. The stopping time for iterations is chosen according to Eq. (4). The performance of each model was measured through qualitative and quantitative analysis. The value of γ, space step (h) size, and time step (τ) size are chosen as 2, 1, and 0.1, respectively.

3.2 *Comparison Results*

Figure 1 demonstrates final denoising results for the mix image degraded by Poisson noise at image peak intensity of 60 (see Fig. 1a, b). From the visual quality of the results (Fig. 1c, f, i), it is easy to perceive that the TVP model produces staircasing artifacts in flat region, NCD model left spurious noise along edges and proposed telegraph-diffusion model gives better results with efficient edge preservation and less artifacts. A closer observation of the filtered results, in terms of $3D$ surface plots and corresponding contour maps (Fig. 1d–k), reveal that some spurious spikes left in the results obtained using TVP model and NCD model, whereas proposed scheme's result avoids such artifacts.

Next in Fig. 2 we filter the noisy house image corrupted by Poisson noise at image peak intensity of 30. The final result obtained using the proposed filter clearly indicates that the significant edges and textures are preserved and there are no blurring and staircase artifacts. This figure (last column)) also shows results in terms of ratio image (house). These ratio images particularly contain removed image details and noise during restoration process. From these images, it is easy to observe that the proposed algorithm is more capable to remove almost all noise particles with effective structure preservation.

Further, to evaluate and compare the quality of restored results, we use the two different quantitative metrics such as peak signal-to-noise ratio (PSNR) and mean structural similarity (MSSIM) measure and [14]. The PSNR is a measure of ratio between signal and noise, whereas MSSIM metric provides the overall structural similarity between the clean and the filtered images. For both metrics, a high value shows superior performance.

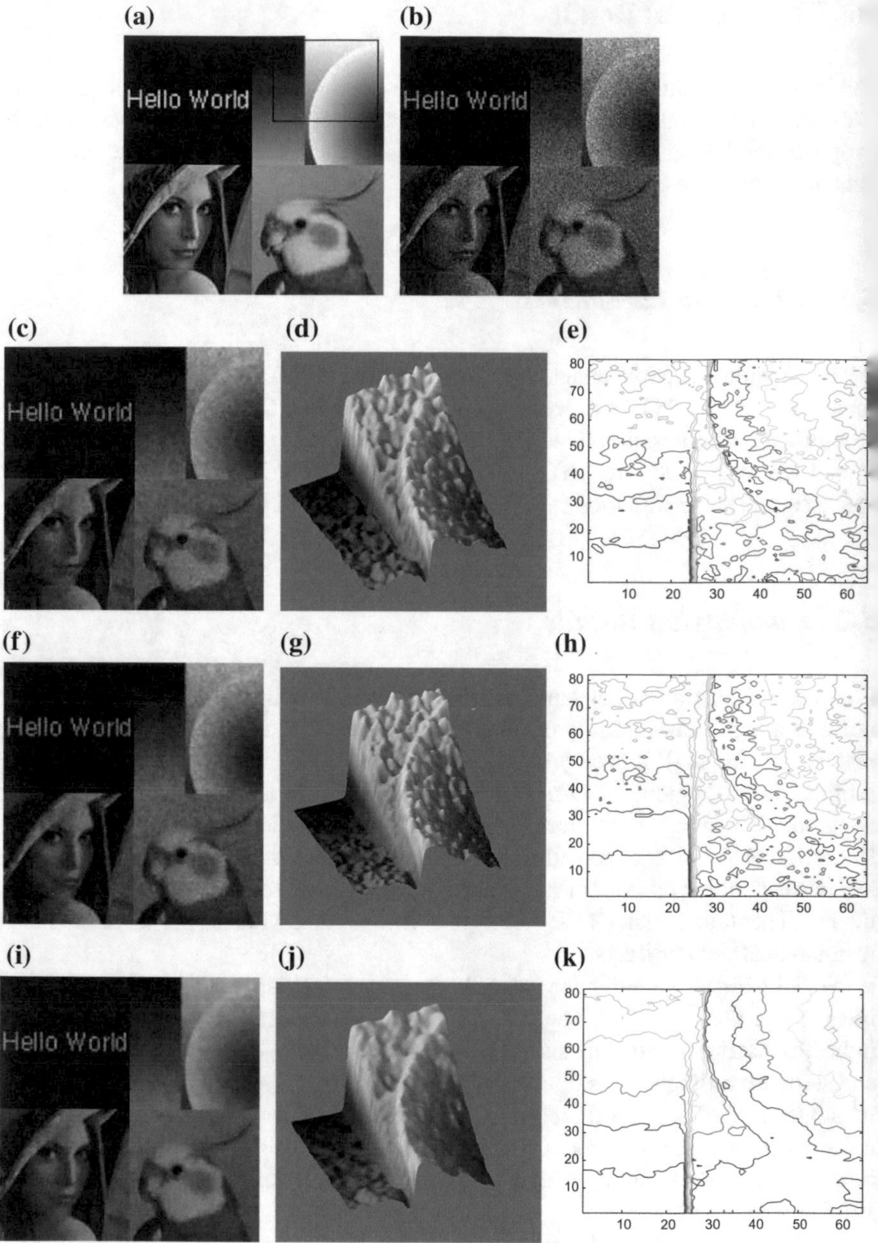

Fig. 1 **a** Original mix image, **b** Noisy images corrupted by a Poisson noise at image peak intensity of 60. Results obtained with **c–e** the TVP model **f–h** the NCD model **i–k** our model. First column: Denoised images. Middle column: 3D surface showing details of a cropped region. Last column: Contours

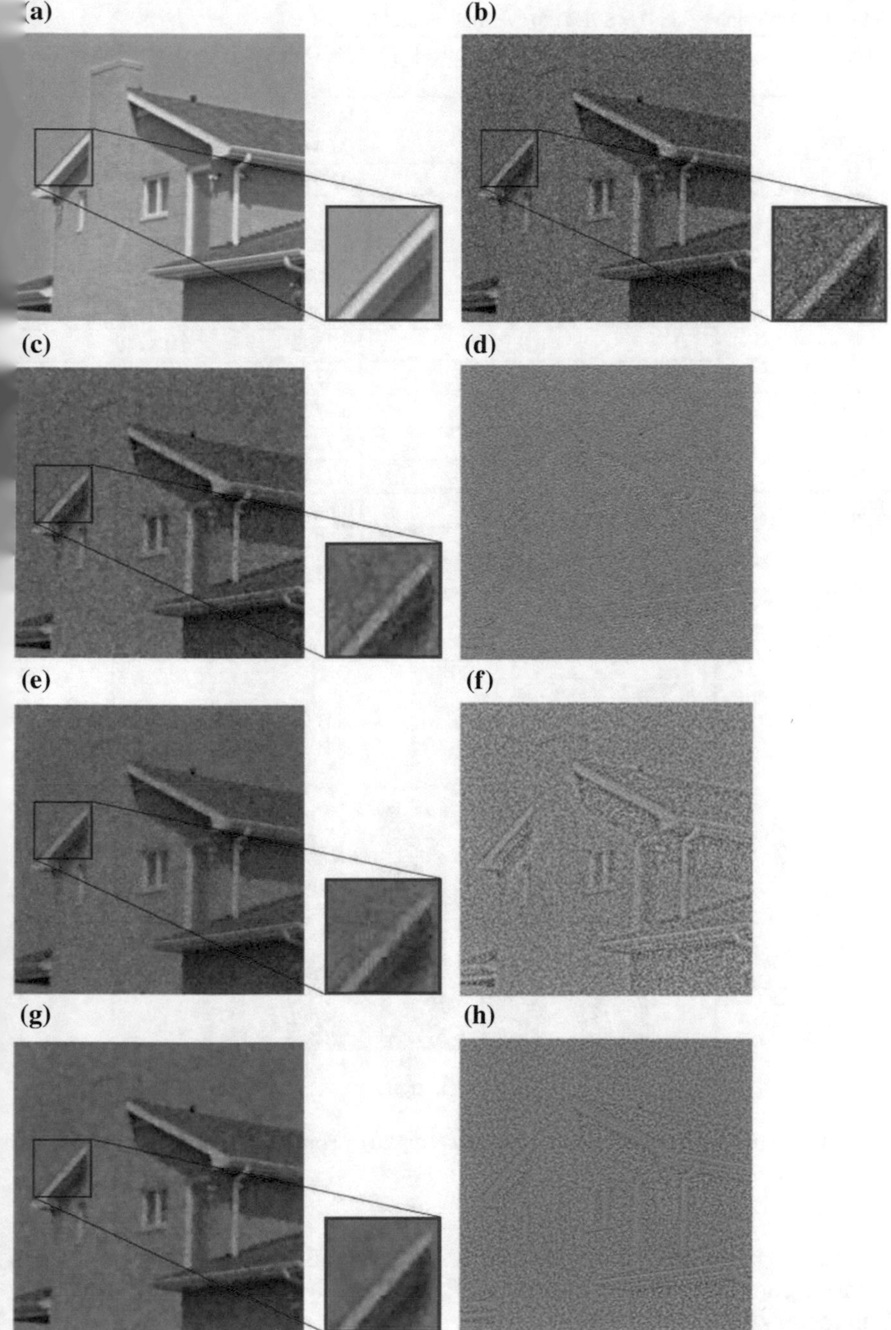

Fig. 2 **a** Original house image, **b** Noisy images corrupted by Poisson noise at image peak intensity of 30. Results obtained with **c–d** the TVP model **e–f** the NCD model **g–h** our model. First column: Denoised images. Last column: Ratio images

Table 1 Denoising results (MSSIM) for various natural images

Image	Poisson noise level with I_{max}	TVP model [3]	NCD model [4]	Proposed model
House	30	0.4949	0.4825	**0.6723**
	60	0.6349	0.6446	**0.7269**
	120	0.7595	0.7517	**0.7967**
Boat	30	0.6751	0.6668	**0.7587**
	60	0.7978	0.7833	**0.8344**
	120	0.8974	0.8676	**0.9025**
Mix	30	0.6132	0.6258	**0.7361**
	60	0.7643	0.7503	**0.8228**
	120	0.8599	0.8256	**0.8648**
Peppers	30	0.5938	0.5788	**0.7087**
	60	0.7442	0.7229	**0.8022**
	120	0.8354	0.7951	**0.8454**
Pirate	30	0.6888	0.6846	**0.7201**
	60	0.8344	0.8218	**0.8506**
	120	0.8842	0.8728	**0.8890**

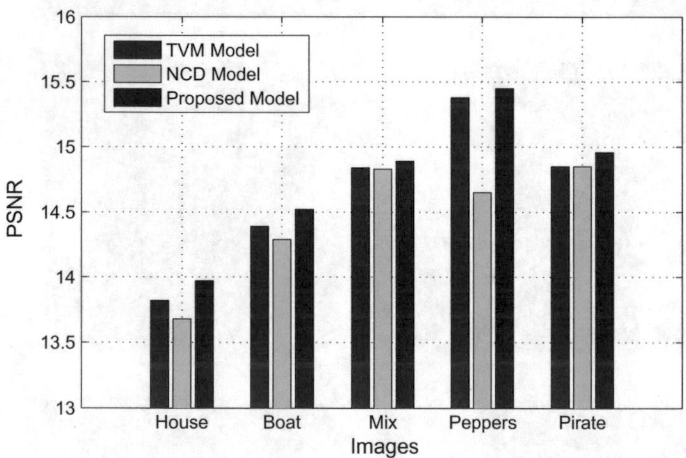

Fig. 3 PSNR (in dB) values for different images corrupted by Poisson noise at image peak intensity of 30

The comparison of MSSIM and PSNR metrics for different test problems as well as varying amounts of noise is shown in Table 1 and Fig. 3, respectively. Comparing both metrics values, the proposed model gives the comparable performance to the existing methods. Overall, our proposed Hessian edge indicator based telegraph-diffusion model performs much better than purely diffusion-based approaches.

Conclusion and Future Scope

In this paper, we propose an efficient pixel-wise adaptive telegraph-diffusion model for signal-dependent Poisson noise removal. The proposed adaptive filtering scheme is a nonlinear hyperbolic–parabolic system which has been solved using standard numerical solvers. By using the proposed scheme, we overcome the limitations of the original diffusion-based models. Also, the performance of the proposed method is quantified through different qualitative measures. The effective performance of the proposed method, previously established for the additive noise removal task, is extended to filter the Poisson noise from natural images. Extensive numerical experiments have been conducted to highlight the efficiency of the proposed model using various natural images. From our numerical experiments, we can consider that the images are well recovered and provide good restoration, without introducing undesired artifacts.

Extending the proposed telegraph-diffusion framework to handle real Poissonian images is an interesting future direction which needs to be explored further.

References

1. Weickert, J.: Anisotropic Diffusion in Image Processing, vol. 1. Teubner, Stuttgart (1998)
2. Aubert, G., Kornprobst, P.: Mathematical Problems in Image Processing: Partial Differential Equations and the Calculus of Variations, vol. 147. Springer, Berlin (2006)
3. Le, T., Chartrand, R., Asaki, T.J.: A variational approach to reconstructing images corrupted by poisson noise. J. Math. Imaging Vis. **27**(3), 257–263 (2007)
4. Srivastava, R., Srivastava, S.: Restoration of poisson noise corrupted digital images with nonlinear pde based filters along with the choice of regularization parameter estimation. Pattern Recognit. Lett. **34**(10), 1175–1185 (2013)
5. Gong, Z., Shen, Z., Toh, K.C.: Image restoration with mixed or unknown noises. Multiscale Model. Simul. **12**(2), 458–487 (2014)
6. Liu, H., Zhang, Z., Xiao, L., Wei, Z.: Poisson noise removal based on nonlocal total variation with eulers elastica pre-processing. J. Shanghai Jiaotong Univ. (Sci.) **22**(5), 609–614 (2017)
7. Perona, P., Malik, J.: Scale-space and edge detection using anisotropic diffusion. IEEE Trans. Pattern Anal. Mach. Intell. **12**(7), 629–639 (1990)
8. Srivastava, R., Gupta, J., Parthasarathy, H.: Enhancement and restoration of microscopic images corrupted with poisson's noise using a nonlinear partial differential equation-based filter. Def. Sci. J. **61**(5) (2011)
9. Zhou, W., Li, Q.: Adaptive total variation regularization based scheme for poisson noise removal. Math. Methods Appl. Sci. **36**(3), 290–299 (2013)
10. Rudin, L.I., Osher, S., Fatemi, E.: Nonlinear total variation based noise removal algorithms. Phys. D: Nonlinear Phenom. **60**(1), 259–268 (1992)
11. Ratner, V., Zeevi, Y.Y.: Image enhancement using elastic manifolds. In: 14th International Conference on Image Analysis and Processing. ICIAP 2007, pp. 769–774. IEEE (2007)
12. Jain, S.K., Ray, R.K.: Edge detectors based telegraph total variational model for image filtering. Information Systems Design and Intelligent Applications, pp. 119–126. Springer, Berlin (2016)
13. Thomas, J.W.: Numerical Partial Differential Equations: Finite Difference Methods, vol. 22. Springer, Berlin (1995)
14. Wang, Z., Bovik, A.C.: Mean squared error: love it or leave it? a new look at signal fidelity measures. IEEE Signal Process. Mag. **26**(1), 98–117 (2009)

A Q-Learning Approach for Sales Prediction in Heterogeneous Information Networks

Sadhana Kodali, Madhavi Dabbiru and B. Thirumala Rao

Abstract In today's world, recommenders have grabbed a major importance to improve the sales where this paper provides the use of machine learning approach which involves machine learning technique like Q-learning that has an evident prediction on improving the sales. The logical network that can be formed for the mobiles and sales can be treated as a heterogeneous information network and traversing this semantic network gives meaningful meta-paths. The reinforcement learning technique, Q-Learning, is applied to predict the sales of a product.

Keywords Machine learning · Meta-path · Q-learning · Heterogeneous information network

1 Introduction

With the evolution of machine learning and innovative algorithms, the approach to reach customer with a good prediction which can improve the sales is the main motto of many companies. With machine learning, we can identify different patterns which have evolved to identify many pattern recognition techniques. To learn from previous models and to predict reliable and better results from previous steps are the main aims of the learning algorithms. A network of objects which interact with each other logically is called heterogeneous information network [1]. These networks can be traversed from one object to the other forming what are called as meta-paths. The

S. Kodali (✉) · B. T. Rao
Department of CSE, Koneru Lakshmaiah Education Foundation, Guntur Dt, Vaddeswaram 522502, Andhra Pradesh, India
e-mail: sadhanalendicse@gmail.com

B. T. Rao
e-mail: drbtrao@kluniversity.in

M. Dabbiru
Department of CSE, Dr. L. Bullayya College of Engineering for Women, Visakhapatnam, Andhra Pradesh, India
e-mail: drlbcse@gmail.com

A. K. Das et al. (eds.), *Computational Intelligence in Pattern Recognition*, Advances in Intelligent Systems and Computing 999,
https://doi.org/10.1007/978-981-13-9042-5_72

main aim of this work is to traverse various meta-paths and iterate over these path to predict the method to improve sales using the Q-Learning algorithm [2].

1.1 Heterogeneous Information Networks

Heterogeneous information networks are the object interaction networks. These objects of various types interact based on semantics and use different object relation to form links between the objects. As different objects are involved in the interaction with different links the network is called a heterogeneous information network. An example of heterogeneous information network can be user–actor–movie–director In the other case, if the same objects are involved with same kind of relationship, i is called a homogeneous information network. The best example can be student–university–student.

A homogeneous information network is a subset of heterogeneous information network. The traversal of different objects with different links can lead to a variety of paths and these are called as meta-paths. User1–(watches)–Movie1–(directed by)–Director1 and so on. Traversing these meta-paths leads to very reasonable and valuable information. On this information, if machine learning strategies are applied they lead to pattern recognition, prediction, classification, and better recommendations also.

2 Related Work

Reinforcement learning (RL) [3] is a machine learning area that helps to decide some agents to take actions that involve better reward. The RL is based upon the environment in which the object acts and the agent which is nothing but the reinforcement learning algorithm. The agents present in RL need to have explicit goals, they should interact with the environment and make decision-making strategies on the environment. An agent upon interaction with the environment can make changes or affect the environment. Examples include making a move in chess, a robot performing an action, hitting the target in a game, etc. The effects of these actions may be unpredicted or unknown.

Reinforcement learning is a nature of learning by observing and interacting with the environment. The reinforcement learning deals with a set of problems that take the feedback from the previous solutions, which mean to say RL deals with closed loop problems. Reinforcement learning helps to learn some control strategies for autonomous agents. Reinforcement learning includes computational approach in order to understand and automate the decision-making. RL assumes to have training information that is available to compute the reward which leads to the next state.

Learning can be mainly categorized as two types [4] as shown in Fig. 1:

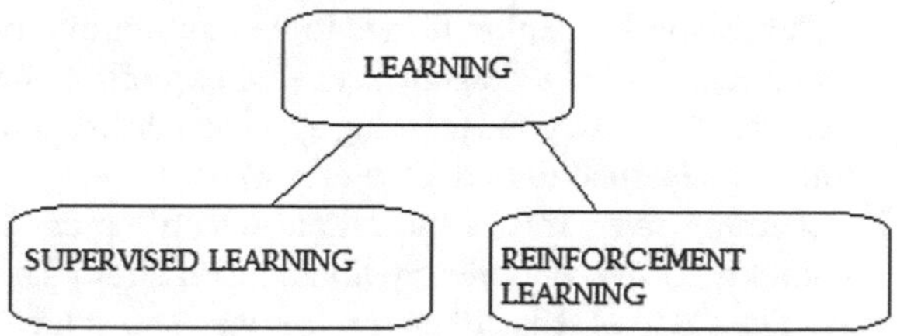

ig. 1 Types of learning

1. Supervised learning: Learning can be done from the training samples of input.
2. Reinforcement learning: Learning can be done with an action that gives a reward which leads to a new state.

The reinforcement learning can further be classified into model-based and model-free approach. In model-based approach, optimal policy is derived by learning from the model. In model-free approach, optimal policy is derived without learning from the model [5]. Reinforcement learning is based upon behavioral pattern of making decision to decide and the kind of action so that the reward is increased. The agent learns the next states in an on-the-go approach without having a prior knowledge of the states. The important notations that are used in the reinforcement learning are the following [6]:

1. Policy: It is a function to calculate the next action for any given state. An optimal policy is a function which maximizes the optimal reward or reinforcement of a particular state.
2. Exploitation: It is used to learn the best action from the current state.
3. Exploration: All the states must be observed and explored to select an action which should be different from the existing best action.
4. Blame attribution problem: The problem which is used to define the occurrence of reward or punishment for a given action. Because an action may lead to a reward or an action can also cause a punishment. The occurrence of a reward can be delayed after a very long time a particular action has occurred or a set of sequence of actions put together can lead to desirable reward.
5. Identification of delayed rewards: The actions which may not seem to be effective now can later lead to great rewards. Later these actions which were ineffective in the beginning are now termed as good actions. The most important aspect of reinforcement learning is that the future rewards must be traced back to previous actions. The backpropagation of actions is simple if the state space is static but if the environment is dynamic the prediction and backpropagation of actions for good reward is very difficult. The complexity of the problem increases if the environment is dynamic.
6. Explore–Exploit dilemma: If a set of actions were identified as good actions from which better rewards were obtained, then the question is whether to choose from these set of actions or is there any probability to have a better action is called the explore exploitation problem. An agent must always try to explore better actions otherwise its policy cannot be improved.

The reason for exploration is to find an optimal policy that leads to an optimal solution. And if the agent keeps on exploring and does not learn from what explored is not a good learning approach the reason for this is that the past actions must be learned to find a better reward.

7. Learning rate: This is the rate at which the agent is either forgetful or learns quickly to override the previous information. The learning rate is indicated as α.The value of α is either zero or one. The value of $\alpha = 0$ if the agent does not learn anything new. If $\alpha = 1$ the agent takes only the latest updated information. (8) Discounted reward: This is an approach to assign weight to the reward so that the previous experiences are relevant to the later ones. The discount factor is given as γ which decreases with time. The value of γ lies between zero and one. If $\gamma = 0$ it means the agent is opportunistic as it considers only current rewards. If γ value is closer to 1 it is an indication of the agent considering the long term rewards.
8. Reward signal: The reward signal is used to identify the good and bad events that affect the agent's rewards. The other special case of a reward signal is the value function which indicates the good events in the long run.

Reinforcement learning is goal-directed learning. Reinforcement learning can also be categorized as model based and model free [7]. Methods or approaches of reinforcement learning that include planning are called model-based methods. Methods which are trial and error based are model-free methods. Reinforcement learning is a good approach for finding structures and patterns from an unreliable environment. Also, the methods are further classified as weak methods and strong methods. A weak reinforcement learning method is based upon search and learning, whereas a strong reinforcement learning is based on specific knowledge.

Applying the reinforcement learning on a heterogeneous information network is a very good example for RL. It contributes to the agent environment interaction may be the environment is an author–paper–author network, a social network, etc. The decision-making goal-oriented steps can be studied with the help of reinforcement learning on the heterogeneous network. RL is an approach which combines the dynamic programming and the neural network when applied over a semantic heterogeneous network can lead to many desirable goals like identifying of the best papers in a paper–author–paper network. Identifying the best sales, finding out more relevant friends in a social network, etc.

3 Methodology

3.1 Q-Learning

Q-Learning is a reinforcement technique used in machine learning. An optimal policy is obtained by an agent by learning from the history of the past actions. The history of events is formed between the agent which is in a particular state upon an action nd reward goes to a next state. This can be indicated as a set of state action and eward.

$$st0 \rightarrow a0 \rightarrow rw1 \rightarrow st1 \rightarrow a1 \rightarrow rw2 \rightarrow st2 \rightarrow a2 \rightarrow rw3\ldots \quad (1)$$

vhere st_0 represents an initial state upon an action, a_0 results in a reward rw_1 that eads to a new state st_1, and this sequence continues. Each sequence is called as an xperience which is indicated as a tuple $\langle sti - ai - rwi + 1 - sti + 1\rangle$.

The Q-Learning algorithm uses a table of Q[S, A] where S represents the set of tates and A represents the set of actions. The recent educated guess of $Q[S, A]$ is epresented as $Q^{*}[S, A]$ which is obtained after calculating a temporal difference vhich is a metric used in Q-Learning.

3.2 *Temporal Difference*

Temporal difference: In order to predict the next grade when a set of previous grades are given by taking an average of the previous grades. Let $g1, g2, g3\ldots gm$ be the set of old grades then the average of these grades is given as the new estimate which can be obtained as $Am = \frac{(g1+g2+g3+g4+\cdots+gm)}{m}$.$Am$ is now the average grade.

$$m^{*}Am = g1 + g2 + g3 + \cdots + gm = (m - 1)(Am - 1) + gm \quad (2)$$

Dividing Eq. 2 by m, we get

$$Am = (1 - 1/m)(Am - 1) + gm/m$$

Let $1/m = \alpha$

$$Am = Am - 1 + \alpha(gm - Am - 1) \quad (3)$$

The temporal difference is $gm - Am - 1$. The value of α lies between “0” and “1”. An experience is a tuple which contains a state, action, reward, new state and is denoted as experience $\langle st - a - rw - st'\rangle$ provides a data point which is a reward plus an estimated future value $Val(s')$. This new data point generated is called a return. Q-learning uses a method called an off-policy method in which it follows an optimal policy in every action it carries out.

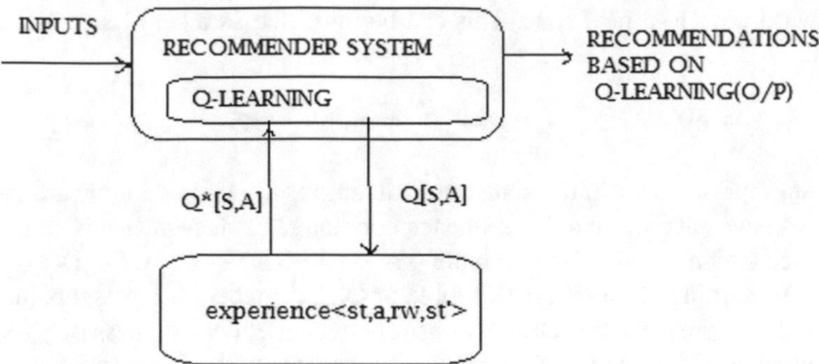

Fig. 2 Workflow in recommendations using Q-learning

Q-Learning algorithm

Input: State St, Action a.
Step 1: Initialize Q[S,A], where S is a set of states and A is a set of actions.
Step 2:for(: :)
{
An initial state st_i upon action a_i gives a reward rw_i which leads to a new state st'_{i+1}
Estimate the temporal difference using $Am = Am - 1 + \alpha(gm - Am - 1)$
$Q[S, A] = Q[S, A] + \alpha\left(rw + \gamma \max Q\left[S', A'\right] - Q[S, A]\right)$
Update $Q[S, A] to Q^*[S, A]$
}

The workflow diagram based on Q-learning for recommenders can be understood from Fig. 2. The inputs are given to the Q-learning-based recommender system and upon the set of model-free learning we get recommendations as required output.

4 Discussion

The Apple iPhone unit sales and revenue data are used to theoretically assess the Q-learning applied on meta-paths. The dataset is available at https://data.world/rflprr/iphone-sales [8]. There are three attributes in these sales data, the first attribute is named as category which gives the quarter number for every year. The first quarter is Q1(October–December), Q2(January–March), Q3(April–June), and Q4(July–September).

The unit sales of iPhone are given in decimal values which are in millions and the third attribute gives year-over-year (yoy) growth. From this dataset, three logical objects quarter-unit of sales-yoy growth form a meta-path and by traversing these meta-paths and applying the Q-learning algorithm, we calculate the new estimated

Table 1 Sample records in the dataset

Category	iPhone	YOY growth
Q2/07	0.141	0
Q3/07	0.489	0
Q4/07	1.036	0
Q1/08	0.817	0
Q2/08	0.483	242.55
Q3/08	4.406	801.02
Q4/08	2.94	183.78
Q1/09	2.427	197.06
Q2/09	3.06	533.54
Q3/09	4.606	4.54
Q4/09	5.578	89.73

points from the existing, from which we can build a best recommender system. The sample dataset values are shown in Table 1.

The first year 2007 is the initial year with initial year-over-year growth equal to zero. The next state that can be reached is Q1 of year 2008. The new estimate from the year 2007 Q4–Q1 in 2008 can be calculated using Eq. 4 with the value of $\alpha = 1/4 = 0.25$

$$A = \frac{(0.141 + 0.489 + 1.036 + 0.817)}{4} = 0.62075 \tag{4}$$

The same can be calculated using previous values given in (2) as follows:

$$A = \frac{(0.141 + 0.489 + 1.036)}{3} + 0.25\frac{0.817 - (0.141 + 0.489 + 1.036)}{3} = 0.62075 \tag{5}$$

The values predicted in Eqs. 4 and 5 give the same result. Equation 5 is computed with the Q-Learning strategy which helps in backtracking to the previous path, i.e., while computing the result of A_4 we can also obtain A_3 which serves as a learning step for A_4.

The yoy growth between Q2 (2007) and Q3 (2008) is 558.47 which is treated as the reward. Figure 3 shows an experience of previous state, action, reward, and the

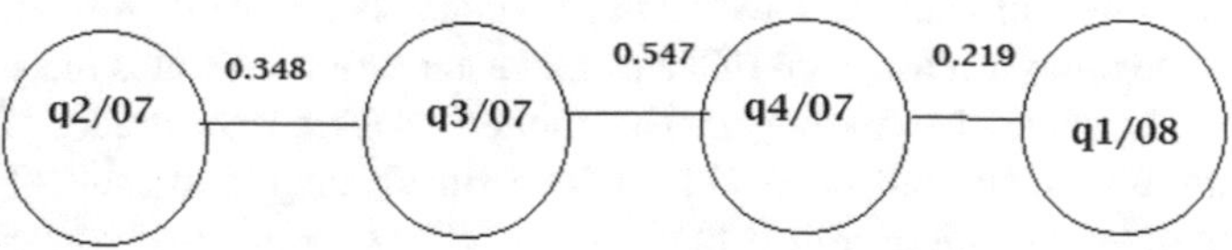

Fig. 3 Experience $\langle st - a - rw - st' \rangle$ for the year 2007–2008

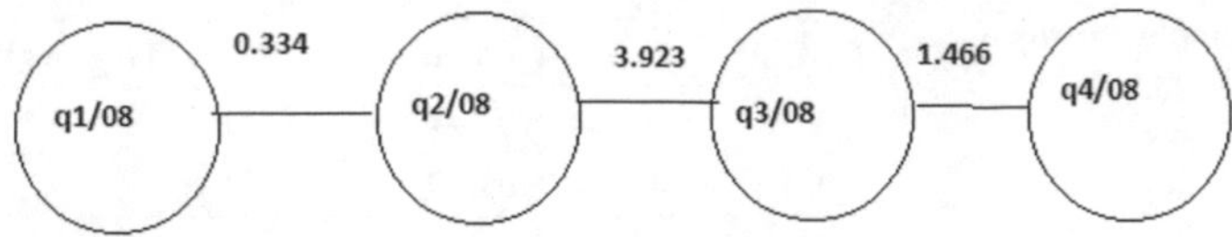

Fig. 4 Experience $\langle st - a - rw - st' \rangle$ for the year 2008

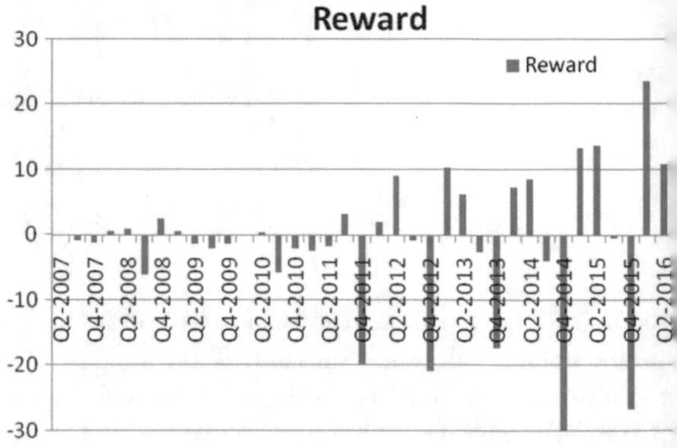

Fig. 5 Graph which depicts the reward for each quarter

next state. From Table 1, the Quarter 2 of year 2007 has made 0.141 million sales and Q3 of 2007 has made 0.489 sales. The reward can be calculated as the difference of 0.141 and 0.489 which is 0.348. With this reward, the old state Q2/7 and an action of sales are moved to Q3/7 and so on. Figure 4 also is a similar example for an experience in the year 2008 that took place between the four quarters.

The sales data reward is calculated and depicted as a graph in Fig. 5.

The other applications of Q-Learning can be in education, medicine, robotics, industrial automation, and finance. In [9], these applications were discussed and a scope for research is also proposed.

5 Conclusion

Machine learning has a wide application in many aspects of our life. This method of machine learning by identifying the various meta-paths in a heterogeneous information network is very useful for effective predictions. The above graph shows the prediction of sales for each year which is calculated as a reward and moving from one year to the other is a traversal of states and each action indicates the sales made in that year. The same methodology can be applied to observe the grades of students, sales of particular items, and so on. One of the main advantages of using Q-Learning which is an off-policy algorithm is that it is an optimal policy and converges to an optimal solution may it be a deterministic or nondeterministic Markov decision process (MDP). The important advantage of choosing reinforcement learning is that it equires very less time to find a good solution. The basic reason is that for every ituation that is encountered a solution is provided of what to do and not how to do.

References

1. Shi, C., Li, Y., Zhang, J., Sun, Y., Yu, P.S.: A survey of heterogeneous information network analysis. J. Latex Class Files **14**(8) (2017)
2. Christopher, J.C.H., Watkins, P. D.: Q-Learning. J. Mach. Learn. **8**(2–4), 279–292. Springer Link
3. Machine Learning for Humans part 5: Reinforcement Learning. https://medium.com/machine-learning-for-humans/
4. Ayodele, T.O.: Types of Machine Learning Algorithms. In: Zhang, Y. (ed.) New Advances in Machine Learning (2010). ISBN: 978-953-307-034-6
5. Quentin, J.M., Huys, A.C., Seriès, P.: Reward based learning, Model based, Model free. Encycl. Comput. Neurosci. (2014). https://doi.org/10.1007/978-1-4614-7320-6_674-1. Springer Science Business Media, New York
6. Volker Sorge: Introduction to AI. http://www.cs.bham.ac.uk/vxs/teaching/ai/slides/lecture12-slides.pdf
7. Sutton, R.S., Barto, A.G.: Reinforcement Learning: An Introduction, 2nd edn. (2014)
8. https://data.world/rflprr/iphone-sales
9. Lorica, B.: Practical Applications of reinforcement learning in industry (2017). Accessed 14 Dec 2017

WNN-EDAS: A Wavelet Neural Network Based Multi-criteria Decision-Making Approach for Cloud Service Selection

O. Gireesha, Nivethitha Somu, M. R. Gauthama Raman, Mandi Sushmanth Reddy, Kannan Kirthivasan and V. Shankar Sriram

Abstract The omnipresence of the cloud-based applications and the exponential growth of the cloud services at different dimensions make the selection of user requirement compliant and trustworthy cloud service provider, a challenging task. Multi-Criteria Decision-Making (MCDM) approaches have their significance in solving cloud service selection problem since they evaluate the alternatives (cloud service providers) based on the intrinsic relationships among the criteria (QoS parameters). However, the assignment of appropriate weights to the criteria has a high impact on the accuracy of the service ranking and the performance of the MCDM methods. Hence, this paper presents a Wavelet Neural Network—Evaluation based on Distance from Average Solution (WNN-EDAS), a novel MCDM approach for the identification of suitable and trustworthy cloud service providers. WNN-EDAS employs WNN to calculate appropriate weights for each criterion and EDAS to rank the cloud service providers. The experiments were carried out on Cloud Armor, a real-world trust feedback dataset to demonstrate the accuracy, robustness, and feasibility

O. Gireesha · M. R. G. Raman · M. S. Reddy · V. S. Sriram (✉)
Centre for Information Super Highway (CISH), School of Computing, SASTRA Deemed to be University, Thanjavur, Tamil Nadu 613401, India
e-mail: sriram@it.sastra.edu

O. Gireesha
e-mail: gireesha@sasta.ac.in

M. R. G. Raman
e-mail: gauthamaraman_mr@sasta.ac.in

M. S. Reddy
e-mail: sushmanthreddy@gmail.com

N. Somu
Smart Energy Informatics Laboratory (SEIL), Indian Institute of Technology-Bombay, Mumbai, Maharashtra 400076, India
e-mail: nivethithasomu@gmail.com

K. Kirthivasan
Discrete Mathematics Research Laboratory (DMRL), Department of Mathematics, SASTRA Deemed to Be University, Thanjavur 613401, Tamil Nadu, India
e-mail: deankannan@sastra.edu

A. K. Das et al. (eds.), *Computational Intelligence in Pattern Recognition*, Advances in Intelligent Systems and Computing 999,
https://doi.org/10.1007/978-981-13-9042-5_73

of WNN-EDAS over the state-of-the-art MCDM approaches in terms of sensitivity analysis and rank reversal.

Keywords Cloud service ranking · EDAS · Rank reversal · Spearman's rank correlation coefficient · Trustworthy cloud service providers · WNN

1 Introduction

"Cloud Computing"– a promising virtualization-based computing technology provides convenient, ubiquitous, and on demand access to the shared pool of platform, infrastructural, and software resources on a "Pay-As-You-Use" fashion over the Internet [1]. The dynamic QoS requirements (functional and nonfunctional) and the competition among various cloud service providers have resulted in the proliferation of functionally similar cloud services at different costs and performance. From the cloud user's perspective, this scenario hardens the process of the cloud service selection, i.e., identification of suitable and trustworthy cloud service providers with respect to the users' desires [2–4]. Further, the nonexpert users who lack the basic knowledge of cloud computing finds the selection of optimal cloud services from a multitude of functionally similar cloud services as a tedious and complex task.

Cloud service selection can be perceived as a Multi-Criteria Decision-Making (MCDM) problem since it involves several alternatives, decision-makers, and criteria to assess the performance of the cloud services [5]. Recent literature reveals the application of MCDM, optimization, logic, description, and trust-based approaches for the evaluation, ranking, and selection of trustworthy cloud services [5]. Among these, several MCDM approaches like TOPSIS [6, 7], PROMETHEE [8], AHP [9], VIKOR [10], BWM [11], etc. have demonstrated their significance in solving cloud service selection problem (Table 1). However, the impact of criteria weights on the performance of the MCDM-based service ranking approaches has not been widely studied. Hence, in this paper, we present Wavelet Neural Network—Evaluation based on Distance from Average Solution (WNN-EDAS), a novel MCDM approach which employs wavelet neural network for criteria weight assignment and EDAS for cloud service ranking. The experimental evaluations were carried out using Cloud Armor, a real-time trust feedback dataset to demonstrate the accuracy, robustness, and feasibility of the WNN-EDAS over the state-of-the-art MCDM approaches in terms of sensitivity analysis and rank reversal.

The rest of the paper is structured as follows: Sect. 2 presents the working of WNN-EDAS, Sect. 3 highlights the performance evaluation of WNN-EDAS in terms of various quality metrics, and Sect. 4 concludes the paper.

Table 1 Related works

Authors	Methodology	Dataset	Validation
Nivethitha Somu et al. [12, 13]	Hypergraph coarsening based robust heteroscedastic probabilistic neural network (HC-RHRPNN)	Quality of web service (QWS) dataset	Classification accuracy, precision, recall, and F-score
	Hypergraph–binary fruit fly optimization based service ranking algorithm (HBFFOA)	WSDream#2	Precision, stability, statistical test, and time complexity analysis
Rajeshkannan Regunathan et al. [14]	Preeminent Service Ranking (PSR)	QWS dataset	TOPSIS, AHP, ANP
Michael Lang et al. [15]	Delphi study	Synthetic dataset	–
Falak Nawaz et al. [11]	Best Worst Method (BWM) + Markov chain	Amazon EC2 cloud services	AHP, convergence
Abdullah Mohammed Al-Faifi et al. [16]	Performance prediction model + Naïve Bayes	Real-world workload dataset from the KSA ministry of finance	Sensitivity, specificity, and accuracy
Chandrashekar Jatoth et al. [17, 18],	Modified Data Envelopment Analysis (MDEA) and Super-efficiency Data Envelopment Analysis (SDEA)	Case study—11 CSPs and7 criteria	Sensitivity analysis, uncertainty, change in alternatives, and adequacy to support group decision-making
	EGTOPSIS	Case study—4 CSPs and 6 QoS parameters	
Rakesh Ranjan et al. [19]	AHP + TOPSIS	Cloud harmony	Sensitivity analysis
Singh and Sidhu [6]	AHP + Improved TOPSIS	Cloud armor	Trustworthiness, untrustworthiness, and uncertainty
Neeraj Yadav, and Major Singh Goraya [20]	AHP	Case Study—3 service providers and 6 criteria.	Sensitivity analysis

2 Proposed Methodology

This section presents an insight into the working of WNN-EDAS, the propose MCDM approach for cloud service selection, i.e., identification of optimal or trus worthy cloud service providers. WNN-EDAS employs WNN to compute appropriat weights for the criteria and EDAS to rank the cloud service providers. The ideolog behind the application of wavelet neural network concepts for weight computatio is to make quick and accurate decisions even in the absence of cognitive levels, i.e. decision-makers for the given cloud service selection problem. Further, the paral lel data processing nature of the neural networks is well suited for solving MCDM problems. Therefore, this work utilizes WNN, a class of neural networks which com bines wavelet function with neural networks, i.e., error backpropagation techniqu to determine the optimal weights of QoS parameters. Further, WNN-based weigh computation technique provides more objective and accurate weights than the othe weight assessment methods like AHP, Delphi method (subjective weight method) Shannon entropy, standard deviation (objective weight method), etc. Further, EDAS is used to rank the cloud service providers based on the criteria weights obtained from the WNN.

In general, WNN is a three-layer architecture with an input layer (QoS parameters), a wavelet layer (Hidden layer), and an output layer (trust value of CSPs-class label) (Fig. 1). Initially, the CSPs (samples with QoS parameter values) are fed as inputs to the input layer. Further, the wavelet or hidden layer computes the optimal weights of the QoS parameters based on the Gaussian wavelet function. Finally, the output layer provides the trust result (class label) of each CSP. The detailed working of the WNN-based EDAS is given in the following steps:

Step-1: Construct the decision matrix with class label. Construct a decision matrix $\left(\mathfrak{D}_{ij}\right)_{m\times n}$ with a set of m alternatives (CSPs), n criteria (QoS parameters), and a class label (trust result).

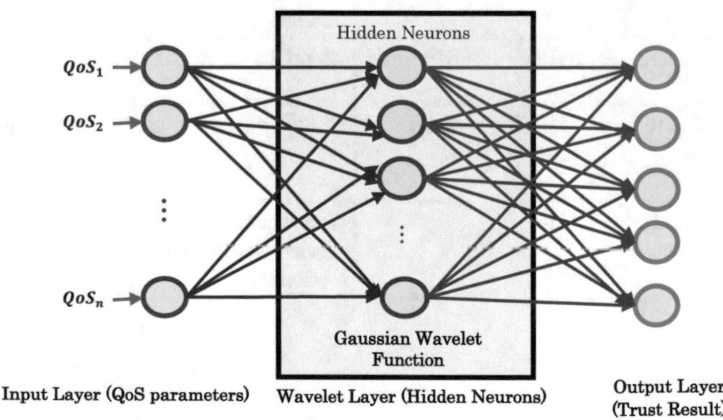

Fig. 1 Wavelet Neural Network (WNN) architecture with $\boldsymbol{m}$ CSPs and $\boldsymbol{n}$ QoS parameters

Step-2: Computation of criteria weights. For a problem with j criteria, a perceptron (error backpropagation) with $j + 1$ neurons is used to construct a WNN. In the first layer, there are n neurons (QoS parameters or criteria's) and the ith neuron has j criteria as input values. Since the different QoS parameters are of different dimensions, normalization technique is applied to the evaluation matrix $(\mathfrak{D}_{ij})_{m\times n}$ to make different QoS parameters comparable. The constructed evaluation matrix $(\mathfrak{D}_{ij})_{m\times n}$ is normalized using vector normalization as in Eq. (1).

$$(\mathfrak{N}_{ij})_{m\times n} = \frac{\mathfrak{D}_{ij}}{\sqrt{\sum_{i=1}^{m}\mathfrak{D}_{ij}^{2}}}, \forall i = 1, 2, \ldots, m; j = 1, 2, \ldots, n \quad (1)$$

Further, the weights of each criterion are computed using WNN with backpropagation technique as discussed below:

Step-2(a): Consider the normalized values (QoS parameters) as the inputs to the neurons. Initialize the weights of the wavelet layer (ω_{hi}) and the output layer (ω_{oj}) randomly with the learning rate (τ).

Step-2(b): For a given input vector, compute the output of the wavelet layer using Eq. (2).

$$f(\mathbb{Q}, \omega) = \psi(\mathbb{Q}.\omega) = \psi(\Sigma_{\mathrm{i}=1}^{\mathrm{n}}(i * \omega_{\mathrm{hi}})) \quad (2)$$

where Q and ω are the variables of input and weight vector of the neuron, respectively, n is the number of inputs fed to the hidden neurons, and Ψ is the activation function (Gaussian activation function) (Eq. (3)).

$$\psi_G(\nu) = \frac{\nu}{\sqrt{2\pi}} exp\left(-\frac{\nu^2}{2}\right) \quad (3)$$

where ν is the matrix multiplication of the Eq. (2).

Step-2(c): Compute the output of the output layer using Softmax activation function (Eq. (4)).

$$y(t) = \varphi\left(\sum_{j=1}^{m}\omega_{oj} * h(j)\right) \quad (4)$$

where $h(j)$ is the output for node j in the hidden layer and φ is the Softmax activation function (Eq. (5)).

$$\varphi_j = \frac{e^{z_j}}{\sum_{j\in group} e^{z_j}} \quad (5)$$

where z is the output nodes and j is the index of the output node.

Step-2(d): Update the weights of the QoS parameters between the layers usin the backpropagation algorithm to reduce the prediction error. The weight updatio procedure is detailed as follows:

1. Compute the error rate (Err_{Rate}) of WNN using Eq. (6).

$$Err_{Rate} = \sum_{t=1}^{s} A(t) - y(t) \quad (6$$

where $A(t)$ is the actual output of the class label (trust result) and $y(t)$ is the predicte value.

2. Update the weights of WNN wavelet function and Softmax parameters w.r.t th error rate using Eq. (7).

$$\omega_{n,t}^{(i+1)} = \omega_{n,t}^{(i)} + \Delta\omega_{n,t}^{(i+1)} \quad (7$$

where $\Delta\omega_{n,t}^{(i+1)}$ is computed by the prediction error.

3. Repeat the steps from (1)–(3) until the error rate is minimum and then obtain the optimal weights of the QoS parameters.
4. Normalize the obtained weights such that $\sum_{j=1}^{n} W_j = 1$.

Step-3: Determine the average solution $(\mathfrak{F}_j)$ w.r.t the criteria. Compute the average solution according to the criteria using Eq. (8).

$$\mathfrak{F}_j = \frac{\sum_{i=1}^{m} \mathfrak{N}_{ij}}{m}, \forall j = 1, 2, \ldots, n \quad (8)$$

Step-4: Calculate the positive and negative distances. Compute the positive distance from average $(\mathcal{P}_{ij})$ and negative distance from average $(\mathcal{N}_{ij})$ based on the type of criteria as defined in Eqs. (9) and (10), respectively.

$$\mathcal{P}_{ij} = \begin{cases} \frac{max(0, (\mathfrak{D}_{ij} - \mathfrak{F}_j))}{\mathfrak{F}_j}, \textit{for benefit criteria} \\ \frac{max(0, (\mathfrak{F}_j - \mathfrak{D}_{ij}))}{\mathfrak{F}_j}, \textit{for cost criteria} \end{cases} \quad (9)$$

$$\mathcal{N}_{ij} = \begin{cases} \frac{max(0, (\mathfrak{F}_j - \mathfrak{D}_{ij}))}{\mathfrak{F}_j}, \textit{for benefit criteria} \\ \frac{max(0, (\mathfrak{D}_{ij} - \mathfrak{F}_j))}{\mathfrak{F}_j}, \textit{for cost criteria} \end{cases} \quad (10)$$

where $\mathcal{P}_{ij}$ and $\mathcal{N}_{ij}$ are the positive and negative distances of performance value of ith cloud service provider (alternative) from the average solution of jth QoS parameters (criteria), respectively.

Step-5: **Determine the weighted sum of the positive distance from the average $(\mathcal{T}_i^+)$ and negative distances from average $(\mathcal{T}_i^-)$**. Compute the $\mathcal{T}_i^+$ and $\mathcal{T}_i^-$ for all CSPs using Eqs. (11) and (12), respectively.

$$\mathcal{T}_i^{+} = \sum_{j=1}^{n} W_j * \mathcal{P}_{ij} \tag{11}$$

$$\mathcal{T}_i^{+} = \sum_{j=1}^{n} W_j * \mathcal{P}_{ij} \tag{12}$$

Step-6: Normalize the values of the weighted sum of $\mathcal{T}_i^{+}$ and $\mathcal{T}_i^{-}$ using Eqs. (13) and (14), respectively.

$$\mu_i^{+} = \frac{\mathcal{T}_i^{+}}{\max_i \mathcal{T}_i^{+}} \tag{13}$$

$$\mu_i^{-} = 1 - \frac{\mathcal{T}_i^{-}}{\max_i \mathcal{T}_i^{-}} \tag{14}$$

Step-7: Compute the appraisal score of each CSP using Eq. (15).

$$\mathcal{U}_i = \frac{1}{2}(\mu_i^{+} + \mu_i^{-}) \tag{15}$$

Step-8: Rank the CSPs in the decreasing order.

3 Experimental Results—Case Study

The implementation of WNN-EDAS, the proposed service ranking approach was carried out using a sample trust feedback dataset (15 CSPs and 9 QoS parameters) from Cloud Armor [21], a real-time trust feedback dataset using Python 3.6 on an Intel core i5 processor at 2.8 GHz system running Mac OS with 4 GB RAM. The trust feedback of QoS parameters ranges from 1 (insignificant feedback score) to 5 (significant feedback score) with a class label of trust result. The case study comprises trust feedback values for 15 CSPs (Backup genie ($\boldsymbol{CSP_1}$), Bluehost ($\boldsymbol{CSP_2}$), Carbonite ($\boldsymbol{CSP_3}$), Elephantdrive ($\boldsymbol{CSP_4}$), Go Daddy ($\boldsymbol{CSP_5}$), ibackup ($\boldsymbol{CSP_6}$), idrive ($\boldsymbol{CSP_7}$), Justcloud ($\boldsymbol{CSP_8}$), Keepit ($\boldsymbol{CSP_9}$), Livedrive ($\boldsymbol{CSP_{10}}$), Mozy ($\boldsymbol{CSP_{11}}$), MyPCBackup ($\boldsymbol{CSP_{12}}$), sos-online-backup ($\boldsymbol{CSP_{13}}$), SugarSync ($\boldsymbol{CSP_{14}}$), and yousendit-online-backup ($\boldsymbol{CSP_{15}}$)) over nine QoS parameters (availability (***A***), response time (***RT***), price (***P***), speed (***S***), storage space (***SS***), features (***f***), ease of use (***EU***), technical support (***TS***), and customer service (***CS***)) by eliminating the missing and redundancy values in the cloud armor dataset. Table 2 presents the 15 CSPs and the trust feedback values for each QoS parameter.

The evaluation matrix (Table 2) was normalized using vector normalization (Step 2, Eq. (1)) to avoid the influence of different dimensions of QoS parameters (criteria) on the accuracy of the service ranking. Further, the relative importance weights of the

Table 2 Evaluation matrix ($\boldsymbol{CSP}_{15\times9}$) of 15 CSPs and 9 QoS parameters

CSPs	Criteria								
	Av	*Rt*	*Pr*	*Sp*	*Ss*	*Fe*	*Eu*	*Ts*	*Cs*
$\boldsymbol{CSP}_1$	5	5	5	3	5	5	5	5	5
$\boldsymbol{CSP}_2$	5	5	5	4	4	5	5	5	5
$\boldsymbol{CSP}_3$	3	3	3	4	5	2	2	2	2
$\boldsymbol{CSP}_4$	5	4	4	4	3	4	5	5	5
$\boldsymbol{CSP}_5$	5	5	5	5	4	5	5	5	5
$\boldsymbol{CSP}_6$	5	5	5	5	5	5	5	5	5
$\boldsymbol{CSP}_7$	4	4	4	4	4	5	5	4	4
$\boldsymbol{CSP}_8$	5	5	5	5	5	5	5	4	4
$\boldsymbol{CSP}_9$	3	4	4	3	4	3	5	4	4
$\boldsymbol{CSP}_{10}$	5	4	4	1	4	4	1	1	1
$\boldsymbol{CSP}_{11}$	2	3	2	3	3	3	3	2	3
$\boldsymbol{CSP}_{12}$	5	4	4	5	5	5	5	4	4
$\boldsymbol{CSP}_{13}$	3	3	3	3	2	3	3	4	3
$\boldsymbol{CSP}_{14}$	5	5	4	5	5	5	5	5	5
$\boldsymbol{CSP}_{15}$	5	5	4	5	4	4	5	5	5

QoS parameters were computed using WNN (Step 2(a)–(d), Eqs. (2)–(7)). Table 3 presents the normalized values of an evaluation matrix and the relative importance (weights) of each QoS parameter.

Finally, the CSPs were ranked in the decreasing order of the appraisal score of EDAS based on the distance of average solution of QoS parameters. The accuracy and validity of WNN-EDAS were compared with the state-of-the-art MCDM techniques like improved TOPSIS [6], VIKOR [22], and PSI [23] in terms of service ranking and Spearman's rank correlation coefficient. Table 4 presents the service ranking obtained by WNN-EDAS and the considered MCDM approaches. We have also utilized the Spearman's rank correlation coefficient ($\boldsymbol{\rho}$) to analyze the correlation among the rankings of WNN-EDAS and the existing MCDM approaches (TOPSIS, VIKOR, and PSI) (Eq. (16)) (Fig. 2).

$$\boldsymbol{\rho = 1 - \frac{6\sum_{i=1}^{m} d_i^2}{m\left(m^2 - 1\right)}} \tag{16}$$

where d_i is the difference in ranks for x and y. From Table 4 and Fig. 2, it is obvious that the service ranking of WNN-EDAS has a high similarity degree with ranking obtained from TOPSIS, VIKOR, and PSI.

Figure 3 shows that the service ranking of WNN-EDAS is consistent with considered MCDM-based service ranking approaches methods. Further, it is clear that WNN-EDAS is highly correlated with TOPSIS (96%), and moderately correlated

able 3 Normalized evaluation matrix ($NCSP_{115\times9}$)

CSPs	Criteria								
	Av	*Rt*	*Pr*	*Sp*	*Ss*	*Fe*	*Eu*	*Ts*	*Cs*
CSP_1	0.2901	0.2931	0.3107	0.1894	0.3054	0.2993	0.2896	0.3077	0.3089
CSP_2	0.2901	0.2931	0.3107	0.2525	0.2443	0.2993	0.2896	0.3077	0.3089
CSP_3	0.1741	0.1759	0.1864	0.2525	0.3054	0.1197	0.1159	0.1231	0.1236
CSP_4	0.2901	0.2345	0.2485	0.2525	0.1833	0.2395	0.2896	0.3077	0.3089
CSP_5	0.2901	0.2931	0.3107	0.3156	0.2443	0.2993	0.2896	0.3077	0.3089
CSP_6	0.2901	0.2931	0.3107	0.3156	0.3054	0.2993	0.2896	0.3077	0.3089
CSP_7	0.2321	0.2345	0.2485	0.2525	0.2443	0.2993	0.2896	0.2462	0.2471
CSP_8	0.2901	0.2931	0.3107	0.3156	0.3054	0.2993	0.2896	0.2462	0.2471
CSP_9	0.1741	0.2345	0.2485	0.1894	0.2443	0.1796	0.2896	0.2462	0.2471
CSP_{10}	0.2901	0.2345	0.2485	0.0631	0.2443	0.2395	0.0579	0.0615	0.0618
CSP_{11}	0.1161	0.1759	0.1243	0.1894	0.1833	0.1796	0.1738	0.1231	0.1853
CSP_{12}	0.2901	0.2931	0.2485	0.3156	0.3054	0.2993	0.2896	0.2462	0.2471
CSP_{13}	0.1741	0.1759	0.1864	0.1894	0.1222	0.1796	0.1738	0.2462	0.1853
CSP_{14}	0.2901	0.2931	0.2485	0.3156	0.3054	0.2993	0.2896	0.3077	0.3089
CSP_{15}	0.2901	0.2931	0.2485	0.3156	0.2443	0.2395	0.2896	0.3077	0.3089
ω_j	**0.1269**	**0.1162**	**0.1087**	**0.1123**	**0.1081**	**0.1071**	**0.1100**	**0.1069**	**0.1038**

Table 4 Service ranking—WNN-EDAS, TOPSIS, VIKOR, and PSI

CSPs	WNN-EDAS	Rank	TOPSIS	Rank	VIKOR	Rank	PSI	Rank
CSP_1	0.8894	4	0.8201	8	0.2295	6	0.9669	3
CSP_2	0.8717	6	0.8700	4	0.1706	4	0.8966	8
CSP_3	0.1082	13	0.4107	13	0.9315	13	0.6489	13
CSP_4	0.6607	9	0.7596	9	0.4187	9	0.8013	10
CSP_5	0.9405	2	0.9108	2	0.1530	3	0.9639	4
CSP_6	1.0000	1	1.0000	1	0.0000	1	1.0000	1
CSP_7	0.5659	10	0.7538	10	0.4278	10	0.8291	9
CSP_8	0.8731	5	0.8656	5	0.1323	2	0.9891	2
CSP_9	0.3913	11	0.6332	11	0.5583	11	0.7284	12
CSP_{10}	0.0652	14	0.3482	15	0.7779	12	0.7435	11
CSP_{11}	0.0000	15	0.3573	14	1.0000	15	0.5326	15
CSP_{12}	0.8104	8	0.8383	7	0.2318	7	0.9473	7
CSP_{13}	0.1163	12	0.4519	12	0.9316	14	0.5666	14
CSP_{14}	0.9374	3	0.9094	3	0.1967	5	0.9617	5
CSP_{15}	0.8216	7	0.8484	6	0.2401	8	0.9508	6

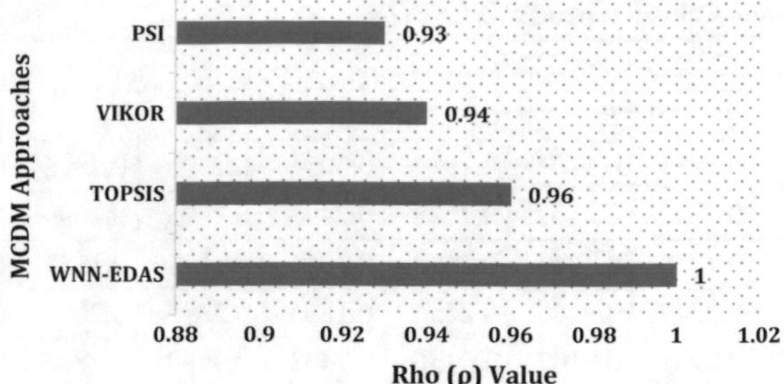

Fig. 2 Spearman's rank correlation inference

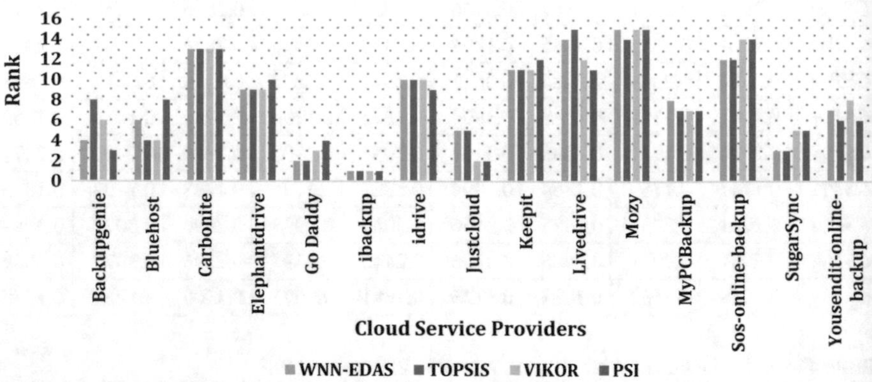

Fig. 3 Service ranking—WNN-EDAS, TOPSIS, VIKOR, and PSI

with the VIKOR (94%) and PSI (93%) which is due to the nature of formulation used by each of the considered MCDM methods.

Further, we have used sensitivity analysis, a significant evaluation factor to reveal the significance of the WNN-EDAS on the accuracy of the service ranking with respect to the change in the weights of the criteria. In simpler terms, sensitivity analysis reflects the stability and robustness of WNN-EDAS with the change in the criteria weights, i.e., WNN-EDAS is robust, if the change in the criteria weights does not affect the ranking order, else it is sensitive. For the same, we have conducted 25 experiments to analyze the impact of weights on the selection of optimal cloud service provider. Figure 4 shows the ranking of each cloud service provider with respect to the change in the criteria weights. The service ranking obtained from the experiments is

$$\begin{aligned}&\boldsymbol{ibackup} > GoDaddy > SugarSync > Backupgenie > Justcloud\\&> \boldsymbol{Bluehost} > Yousendit - online - backup > MyPCBackup\\&> Elephantdrive > idrive > Keepit > Sos - online - backup\\&> Carbonite > Livedrive > Mozy.\end{aligned}$$

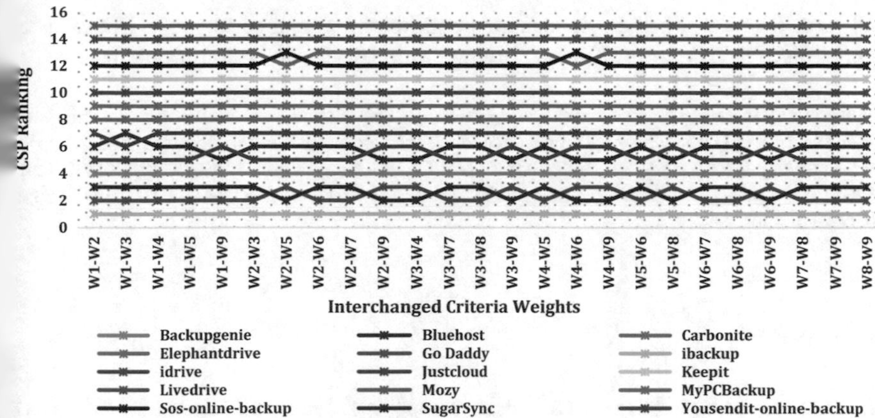

Fig. 4 Robustness of WNN-EDAS using sensitivity analysis

Further, the performance of WNN-EDAS was evaluated in terms of its ability to address the rank reversal problem, i.e., preserve the relative ranking order of the alternatives (cloud service providers) by addition of a new cloud service provider and deletion of the existing alternatives (Fig. 4). The service ranking obtained using WNN-EDAS with the addition of new cloud service provider (Backblaze) and deletion of the existing cloud service provider (Mozy) is $\boldsymbol{Backblaze} \cong \boldsymbol{CSP_6} > \boldsymbol{CSP_5} > \boldsymbol{CSP_{14}} > \boldsymbol{CSP_1} > \boldsymbol{CSP_8} > \boldsymbol{CSP_2} > \boldsymbol{CSP_{15}} > \boldsymbol{CSP_{12}} > \boldsymbol{CSP_4} > \boldsymbol{CSP_7} > \boldsymbol{CSP_9} > \boldsymbol{CSP_{13}} > \boldsymbol{CSP_3} > \boldsymbol{CSP_{10}}$ and $\boldsymbol{CSP_6} > \boldsymbol{CSP_5} > \boldsymbol{CSP_{14}} > \boldsymbol{CSP_1} > \boldsymbol{CSP_8} > \boldsymbol{CSP_2} > \boldsymbol{CSP_{15}} > \boldsymbol{CSP_{12}} > \boldsymbol{CSP_4} > \boldsymbol{CSP_7} > \boldsymbol{CSP_9} > \boldsymbol{CSP_{13}} > \boldsymbol{CSP_3} > \boldsymbol{CSP_{10}}$ respectively.

4 Conclusions

Cloud service selection has been a significant area of research which promotes the growth and rapid adoption of cloud computing technologies for a wide range of business, academic, and personal applications. An efficient solution to the service selection problem involves the expression of intrinsic relation among several QoS parameters (criteria), cloud service providers (alternatives), and decision-makers. Hence, this work presents WNN-EDAS, a novel MCDM approach for the identification of trustworthy cloud services. WNN-EDAS employs Haar wavelet activation function based WNN to determine the optimal weights for the QoS parameters and EDAS, an MCDM approach to rank the cloud services. The efficiency of WNN-EDAS was demonstrated using a case study with 15 CSPs and 9 QoS parameters extracted from the Cloud Armor trust feedback dataset in terms of sensitivity analysis and rank reversal. Further, Spearman's coefficient was used to prove EDAS as a appropriate MCDM approach for solving cloud service selection problem.

Acknowledgements This work was supported by The Council for Scientific and Industri Research, India; The Department of Science and Technology, India; Tata Realty-SASTRA Srin vasa Ramanujan Research Cell, India (Grant No: CSIR - SRF Fellowship/143345/2K17/1, CSI - SRF Fellowship/143404/2K15/1, and SR/FST/ETI- 349/2013, SR/FST/MSI-107/2015) for the financial support.

References

1. Armbrust, M.Z.M., Fox, A., Griffith, R., Joseph, A.D., Katz, R., Konwinski, A., Lee, G. Patterson, D., Rabkin, A., Stoica, I.: A view of cloud computing. Commun. ACM **53**(4), 50–58 (2010)
2. Somu, N., Kirthivasan, K., Sriram, V.S.S.: A rough set-based hypergraph trust measure parameter selection technique for cloud service selection. J. Supercomput. **73**(10), 4535–4559 (2017
3. Somu, N., Kirthivasan, K., VS, S.S.: A computational model for ranking cloud service providers using hypergraph based techniques. Futur. Gener. Comput. Syst. **68**, 14–30 (2017)
4. Nivethitha, S., Raman, M.R.G., Gireesha, O., Kannan, K., Sriram, V.S.S.: An improved rough set approach for optimal trust measure parameter selection in cloud environments. Soft Comput. 1–21 (2019)
5. Sun, L., Dong, H., Hussain, F.K., Hussain, O.K., Chang, E.: Cloud service selection: state-of-the-art and future research directions. J. Netw. Comput. Appl. **45**, 134–150 (2014)
6. Singh, S., Sidhu, J., Compliance-based multi-dimensional trust evaluation system for determining trustworthiness of cloud service providers. **67**, 109–132 (2017)
7. Sidhu, J., Singh, S.: Improved topsis method based trust evaluation framework for determining trustworthiness of cloud service providers. J. Grid Comput. **15**(1), 81–105 (2017)
8. Sidhu, Jagpreet, Singh, Sarbjeet: Design and comparative analysis of MCDM-based multi-dimensional trust evaluation schemes for determining trustworthiness of cloud service providers. J. Grid Comput. **15**(2), 197–218 (2017)
9. Meesariganda, B.R., Ishizaka, A.: Mapping verbal AHP scale to numerical scale for cloud computing strategy selection. Appl. Soft Comput. **53**, 111–118 (2017)
10. Gul, M., Celik, E., Aydin, N., Gumus, A.T., Guneri, A.F.: A state of the art literature review of VIKOR and its fuzzy extensions on applications. Appl. Soft Comput. **46**, 60–89 (2016)
11. Nawaz, F., Asadabadi, M.R., Janjua, N.K., Hussain, O.K., Chang, E., Saberi, M.: An MCDM method for cloud service selection using a Markov chain and the best-worst method. Knowl. Based Syst. **159**, 120–131 (2018)
12. Somu, N., MR, G.R., Kalpana, V., Kirthivasan, K., VS, S.S.: An improved robust heteroscedastic probabilistic neural network based trust prediction approach for cloud service selection. Neural Netw. (2018). https://doi.org/10.1016/j.neunet.2018.08.005
13. Somu, N., MR, G.R., Kirthivasan, K., VS, S.S.: A trust centric optimal service ranking approach for cloud service selection. Futur. Gener. Comput. Syst. **86**, 234–252 (2018)
14. Regunathan, R., Murugaiyan, A., Lavanya, K.: Neural Based QoS aware Mobile Cloud Service and Its Application to Preeminent Service Selection using Back Propagation. Procedia Comput. Sci. **132**, 1113–1122 (2018)
15. Lang, M., Wiesche, M., Krcmar, H.: Criteria for selecting cloud service providers: a delphi study of quality-of-service attributes. Inf. Manag. (2018)
16. Al-Faifi, A.M., Song, B., Hassan, M.M., Alamri, A., Gumaei, A.: Performance prediction model for cloud service selection from smart data. Futur. Gener. Comput. Syst. **85**, 97–106 (2018)

7. Jatoth, C., Gangadharan, G.R., Fiore, U., Buyya, R.: SELCLOUD: a hybrid multi-criteria decision-making model for selection of cloud services. Soft Comput. 1–15 (2018)
8. Jatoth, C., Gangadharan, G.R., Fiore, U.: Evaluating the efficiency of cloud services using modified data envelopment analysis and modified super-efficiency data envelopment analysis. Soft Comput. **21**(23), 7221–7234 (2017)
9. Kumar, R.R., Mishra, S., Kumar, C.: Prioritizing the solution of cloud service selection using integrated MCDM methods under Fuzzy environment. J. Supercomput. **73**(11), 4652–4682 (2017)
0. Yadav, N., Goraya, M.S.: Two-way ranking based service mapping in cloud environment. Futur. Gener. Comput. Syst. **81**, 53–66 (2018)
1. Cloud Armor Project. http://cs.adelaide.edu.au/~cloudarmor/, drafted on 18-08-2018 at 10:00 a.m
2. Alabool, H.M., Mahmood, A.K.: Trust-based service selection in public cloud computing using fuzzy modified VIKOR method. Aust. J. Basic Appl. Sci. **7**(9), 211–220 (2013)
3. Maniya, K., Bhatt, M.G.: A selection of material using a novel type decision-making method: Preference selection index method. Mater. Des. **31**(4), 1785–1789 (2010)

An Improved Feedforward Neural Network Using Salp Swarm Optimization Technique for the Design of Intrusion Detection System for Computer Network

N. Neha, M. R. Gauthama Raman, Nivethitha Somu, R. Senthilnathan and V. Shankar Sriram

Abstract Due to the drastic increase in the rate of cyberattacks, network security has become the highest priority in the recent technological era. As the cyberattacks have become more sophisticated in nature, it emphasizes the need for a second line of defense called the Intrusion Detection System (IDS). In recent years, computational-intelligence-based IDS model has become a major choice of researchers for minimizing the conflicts between the high detection rate and less false alarm rate. In this way, we present a Salp Swarm Optimization based Feedforward Neural Network (SSO-FFN) to build an intrusion detection methodology for computer networks. The performance of SSO-FFN over the state-of-the-art intrusion detection methodologies was validated using NSL-KDD cup intrusion dataset with regards to detection rate and false alarm rate. This paper proposes a technique which has a better accuracy, less false alarm rate, and high detection rate when compared with the existing techniques.

Keywords Intrusion detection system · Network security · Feedforward neural network · Salp swarm optimization

N. Neha · M. R. G. Raman · R. Senthilnathan · V. S. Sriram (✉)
Centre for Information Super Highway (CISH), School of Computing, SASTRA Deemed to Be University, Thanjavur, Tamil Nadu 613401, India
e-mail: sriram@it.sastra.edu

N. Neha
e-mail: nehanageswaran@gmail.com

M. R. G. Raman
e-mail: gauthamaraman_mr@sastra.ac.in

R. Senthilnathan
e-mail: senthilsenthil0028@gmail.com

N. Somu
Smart Energy Informatics Laboratory (SEIL), Indian Institute of Technology-Bombay, Mumbai, Maharashtra 400076, India
e-mail: nivethithasomu@gmail.com

A. K. Das et al. (eds.), *Computational Intelligence in Pattern Recognition*, Advances in Intelligent Systems and Computing 999,
https://doi.org/10.1007/978-981-13-9042-5_74

1 Introduction

With the continuous advancements and increasing popularity of Internet-based appl cations, computer networks have become essential part of our day-to-day moder life through the creation of a universal platform for the users and organizations t store and exchange confidential information. However, the complete dependence o the Internet might lead to security breaches which cause adverse impacts in term of socioeconomic loss to individuals and the nation [1]. Recent security incident like WannaCry Cyber Attack (May 2017) [2], which is also called "the worse ran som ware attack ever", infected an estimate of 300,000 computer systems in just days; Rasputin Attacks (2017) [3], alleged to be a lone hacker, successfully breache databases hosted by dozens of universities and government agencies; Equifax Dat Breach attack, etc. emphasize the need to protect the digital ecosystem from th security threats and intrusions [4].

Among the traditional security mechanisms like firewalls, access control, dat encryption, authentication, etc., Intrusion Detection Systems (IDSs) [5], a second line of defence proves its significance with its ability to detect sophisticated intrusions and security threats. Based on the nature of detection methodology, IDS can be classified into two, namely, signature or misuse detection and anomaly detection. The former detects intrusions based on comparing the current network traffic patterns with the predefined signatures of the known attacks, while the latter detect intrusions based on the deviation of the current observations from the normal or baseline profile [6, 7]. However, either approach has advantages and disadvantages which complements each other, i.e., high detection rate, ability to detect new attacks, and high false alarm rate, in terms of classification detection rate(DR), accuracy, and false alarm rate(FAR).

The perception of intrusion detection as a classification problem has led the researchers to apply several statistical and machine learning techniques [8, 9, 10], (decision tree, K-nearest neighbor (KNN), hidden Markov model, rough set theory [11], Bayesian Network, etc.) for the design of efficient IDS. However, the problem of maintaining the conflict between high detection rate and less false alarm rate makes the design of efficient, adaptive, and robust IDS under continuous development. Hence, in this paper, we present Salp Swarm Optimization [12]–Feedforward Neural Network (SSO-FFNN)-based intrusion detection methodology to achieve high detection rate, less false alarm rate, and time complexity [13]. Here, SSO-FNN uses FFNN, a simple variant of Artificial Neural Network (ANN) [3] which has the potential to handle distorted or incomplete datasets and self-learning capability, thereby addresses the challenges in the conventional learning models. Moreover, they are easy to maintain and also gives a result even if the input is not complete. Further, SSO was introduced to optimize the hyperparameters of FFNN to achieve the major goal of IDS. The experimentations were carried out using the standard NSL-KDD cup intrusion dataset and the implementation of SSO-FFNN was evaluated with regards to classification accuracy, detection rate, and false alarm rate [14] (Table 1).

able 1 Related works

Author	Proposed method	Dataset	Performance metric
Raman et al. [17]	HG AR-PNN(Helly property of hypergraph arithmetic residue-based probabilistic neural network)	KDD CUP 1999	• Classifiers • Stability • Accuracy • Precision • Recall • Detection rate
Raman et al. [18]	Hypergraph-based genetic algorithm and feature selection in support vector machine	NSL-KDD	• Accuracy • Detection rate • False alarm rate
Manzoor and Kumar [13]	Feature ranking and reduction	KDD CUP 1999	• Precision • Recall • Accuracy • TPR (sensitivity) • FPR (specificity)
Papamartzivanos et al. [9]	Genetic trees driven rule induction	KDD CUP 1999 NSL-KDD UNSW-NB15	• Attack accuracy • Attack detection rate • Average accuracy
Kabir et al. [19]	Optimum-allocation-based least square support vector machine (OA-LS-SVM)	KDD CUP 1999	• Probability of Detection (PD) • Classification Accuracy (CA)
Raman et al. [11]	Rough sets (RS), hypergraph (RSHGT)	KDD CUP 1999	• Precession • Recall • Accuracy
Carrasco and Sicilia [5]	Skip-gram models	UNSW-NB 15	• Precision • Accuracy • False positive rate
Yin et al. [8]	RNN	NSL-KDD	• Accuracy • True positive rate • False positive rate
Chiba et al. [20]	BPNN (backpropagation neural network)	KDD CUP 1999	• Precision • True positive rate • True negative rate • False positive rate • False negative rate • Accuracy • F-score • AUC • Average time (to classify a connection instance)

2 SSA-FNN Proposed Methodology

In this section, we briefly discuss the working of proposed improved FFN based on SSA optimization. The workflow of the SSA-FNN as follows. The workflow of the SSA-FNN is as follows.

2.1 *Generation of Initial Population*

The working of SSA-FNN initiates with the initialization of parameters like number of populations, number of iterations, and computation of parameters like c1, c2, and c3. The c2 and c3 parameters depend on the current number of iteration and the total number of iterations. c1 parameter is initialized as given in Eq. (3.2) in [12]. Fitness function is chosen in such a way that it gives an optimized result than the existing techniques. Each population in SSA-FNN was randomly generated in a specified range and is considered as the parameters for FNN.

2.2 *Training and Testing the FFN*

The entire intrusion detection dataset considered for the experimentation is divided into training and testing samples through random sampling without replacement. With the training samples and parameters obtained from each population is used for training the FNN and its performance is evaluated using the fitness function (discussed in step 3) along with the testing samples.

2.3 *Definition of Fitness Function*

The performance of the proposed approach is evaluated using the weighted fitness function discussed in Eq. (1)

$$f_{Fit} = W_1(DR) + W_2(FAR) \tag{1}$$

where DR and FAR are the detection rate and false alarm rate, respectively, W_1 and W_2 are the weights of DR and FAR, respectively.

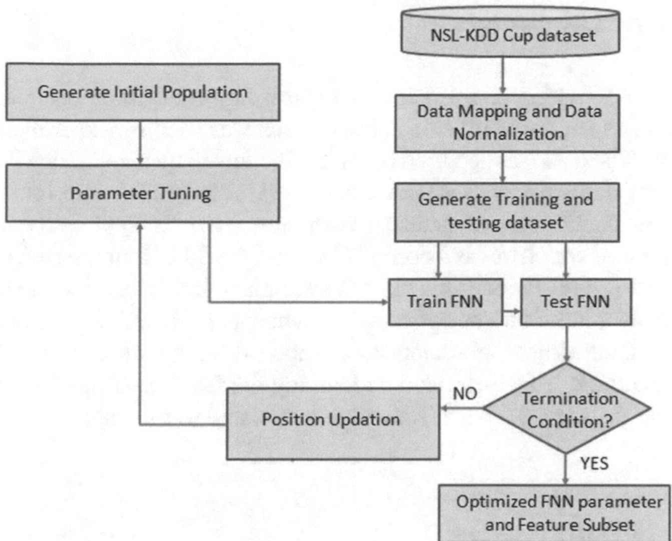

Fig. 1 The proposed flow diagram

2.4 *Termination Condition*

Verify the termination condition, i.e., whether the maximum number of iterations is reached. If the condition is true, then return the optimal number of hidden units else go to step 5

2.5 *Position Updating*

Identify the population with the best fitness value and with its position, update the position of other population using the Eq. (3.1) in [12] (Fig. 1)

3 Results and Discussion

3.1 *Experimental Setup*

SSO-based FFNN was implemented on Python with INTEL® Core™ i3-5005U CPU processor @ 2.00 GHz architecture running Windows 10, and validation process was carried out using WEKA tool.

3.2 Dataset Description

KDD CUP 1999, which is a benchmark dataset in the intrusion detection system is a standardized network intrusion dataset which was used for experimentation and computation. The data was acquired by MIT Lincoln lab by simulating US Air Force LAN and was derived from DARPA 1998. KDD features fall into four categories which are both qualitative and qualitative in nature. Also, KDD-99 dataset consists of five classes out of which one is a normal class and the other four are attack classes—DoS, U2R, R2L, and Probe. This class is redundant and imbalance in nature. DoS attack contains a large number of samples, whereas probe, U2R, etc. have a small sample size. Each time, the attempt to connect to a network can be called either normal or an attack. This is decided by learning the features of the connection such as—the basic features (TCP/IP), content features, and traffic features.

3.3 Data Preprocessing

Data normalization, one of the techniques in data preprocessing is required to improve the accuracy and efficiency of the proposed model (FNN-SSA) and increases detection rate. The main goal of data preprocessing using normalization is to transform the raw data into an appropriate format for further analysis. They have features which are converted from string to numerical values called numerical encoding. Here, the normalization is performed on these numerical values resulting in each value ranging between 0 and 1. One of the methods for normalization is as follows in Eq. (2) [8]:

$$X_{Norm} = \frac{x - x_{min}}{x_{max} - x_{min}} \tag{2}$$

where x: Represents each data point,

x_{min}: Minimal among the data points,
x_{max}: Maximal among the data points and X represents the data point normalized between 0 and 1

The above mentioned Table 2 compares the accuracy of the proposed technique with that of existing techniques. And we can infer that it has achieved 98.63% accuracy.

The above mentioned Figs. 2 and 3 compare the detection rate and false alarm rate of the proposed technique with that of existing techniques. And they are found to be better than that of the existing methodologies.

able 2 Accuracy

Sl. no.	Classification	Accuracy (in %)
1	Naive Bayes	88.7
2	Simple logistics	94.1
3	LWL	93.3
4	Decision stump	9.5
5	REP tree	97.7
6	Decision table	97
7	Random forest	97.1
8	Voted perceptron	65.8
9	One R	95.5
10	Proposed method	98.63

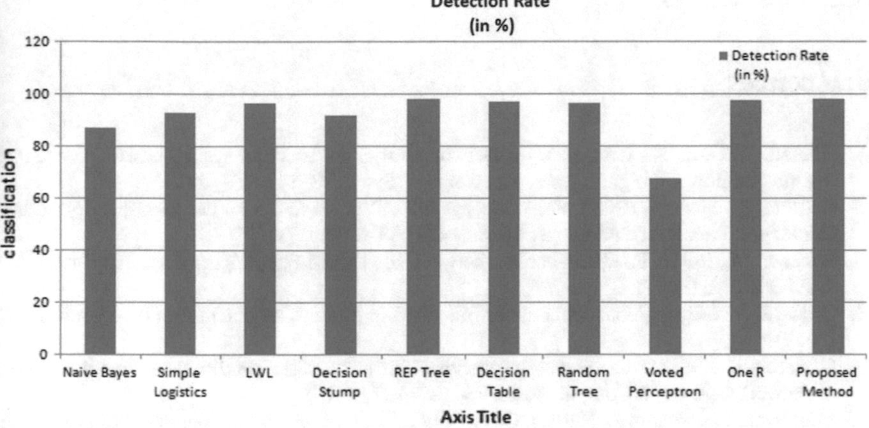

Fig. 2 Detection rate

4 Conclusions

In this paper, an efficient intrusion detection technique that integrates the salp swarm optimization algorithm for optimizing the feedforward neural network parameters is proposed. The NSL-KDD CUP 1999 is a benchmark intrusion dataset which was used for validating the proposed work. Here, the optimization technique is hybridized and outcome of the system shows better results than the existing machine learning based intrusion detection techniques in terms of performance metrics such as detection rate and false alarm rate. This work can be still extended for SCADA systems and the complexity can be minimized by integrating hypergraph property and also be applied for trust prediction in cloud computing [8, 15, 11–16].

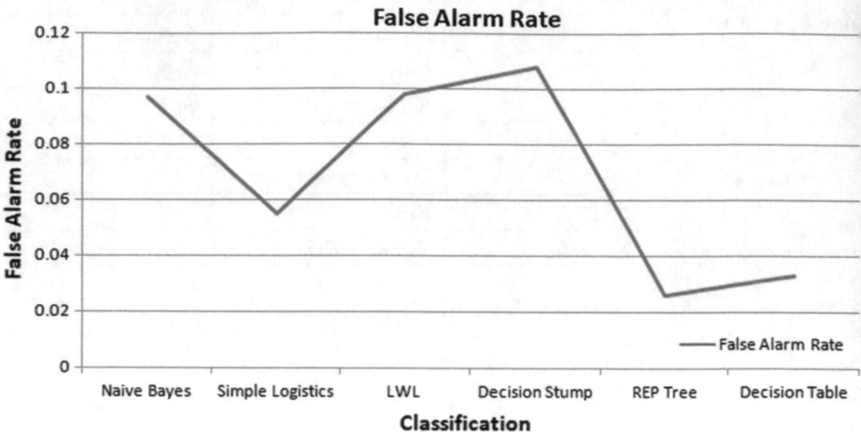

Fig. 3 False alarm rate

References

1. Hamed, T., Dara, R., Kremer, S.C.: Network intrusion detection system based on recursive feature addition and bigram technique. Comput. Secur. **73**, 137–155 (2018)
2. Rathore, S., Sharma, P.K., Loia, V., Jeong, Y.S., Park, J.H.: Social network security: issues, challenges, threats, and solutions. Inf. Sci. **421**, 43–69 (2017)
3. Baezner, M., Robin, P.: Cyber-conflict between the United States of America and Russia, No. 2. ETH Zurich (2017)
4. O'dowd, A.: Major global cyber-attack hits NHS and delays treatment. BMJ: Br. Med. J. **357** (2017)
5. Carrasco, R.S.M., Sicilia, M.A.: Unsupervised intrusion detection through skip-gram models of network behaviour. Comput. Secur. **78**, 187–197 (2018)
6. Muller, S., Lancrenon, J., Harpes, C., Le Traon, Y., Gombault, S., Bonnin, J.M.: A training-resistant anomaly detection system, Comput. Secur. **76**, 1–11, (2018)
7. Koning, R., Buraglio, N., de Laat, C., Grosso, P.: CoreFlow: enriching bro security events using network traffic monitoring data. Futur. Gener. Comput. Syst. **79**, 235–242 (2018)
8. Yin, C., Zhu, Y., Fei, J., He, X.: A deep learning approach for intrusion detection using recurrent neural networks. IEEE Access **5**, 21954–21961 (2017)
9. Papamartzivanos, D., Mármol, F.G., Kambourakis, G.: Dendron: genetic trees driven rule induction for network intrusion detection systems. Futur. Gener. Comput. Syst. **79**, 558–574 (2018)
10. Vijayanand, R., Devaraj, D., Kannapiran, B.: Intrusion detection system for wireless mesh network using multiple support vector machine classifiers with genetic-algorithm-based feature selection. Comput. Secur. **77**, 304–314 (2018)
11. Raman, M.R., Kannan, K., Pal, S.K., Sriram, V.S.: Rough set-hypergraph-based feature selection approach for intrusion detection systems. Def. Sci. J. **66**(6), 1–6 (2016)
12. Mirjalili, S., Gandomi, A.H., Mirjalili, S.Z., Saremi, S., Faris, H., Mirjalili, S.M.: Salp swarm algorithm: A bio-inspired optimizer for engineering design problems. Adv. Eng. Softw. **114**, 163–191 (2017)
13. Manzoor, I., Kumar, N.: A feature reduced intrusion detection system using ANN classifier. Expert Syst. Appl. **88**, 249–257 (2017)
14. Joo, D., Hong, T., Han, I.: The neural network models for IDS based on the asymmetric costs of false negative errors and false positive errors. Expert Syst. Appl. **25**(1), 69–75 (2003)

15. Colom, J.F., Gil, D., Mora, H., Volckaert, B., Jimeno, A.M.: Scheduling framework for distributed intrusion detection systems over heterogeneous network architectures. J. Netw. Comput. Appl. **108**, 76–86 (2018)
16. Somu, N., Raman, M.R.G., Kalpana, V., Kirthivasan, K., Sriram, V.S.S.: An improved robust heteroscedastic probabilistic neural network based trust prediction approach for cloud service selection. Neural Netw. 1–35 (2018)
17. Raman, M.R.G., Somu, N., Kirthivasan, K., Sriram, V.S.S.: A hypergraph and arithmetic residue-based probabilistic neural network for classification in intrusion detection systems. Neural Netw. **92**, 89–97 (2017)
18. Raman, M.R.G, Somu, N., Kirthivasan, K., Liscano, R., Sriram, V.S.S.: An efficient intrusion detection system based on hypergraph-Genetic algorithm for parameter optimization and feature selection in support vector machine. Knowl.-Based Syst. **134**, 1–12 (2017)
19. Kabir, E., Hu, J., Wang, H., Zhuo, G.: A novel statistical technique for intrusion detection systems. Futur. Gener. Comput. Syst. **79**, 303–318 (2018)
20. Chiba, Z., Abghour, N., Moussaid, K., El Omri, A., Rida, M.: A novel architecture combined with optimal parameters for back propagation neural networks applied to anomaly network intrusion detection. Comput. Secur. **75**, 36–58 (2018)
21. Raman, M.R.G., Kirthivasan, K., Sriram, V.S.S.: Development of rough set–hypergraph technique for key feature identification in intrusion detection systems. Comput. Electr. Eng. **59**, 189–200 (2017)
22. Somu, N., Raman, M.R.G., Kirthivasan, K., Sriram, V.S.S.: Hypergraph based feature selection technique for medical diagnosis. J. Med. Syst. **40**(11), 239 (2016)
23. Somu, N., Kirthivasan, K., Sriram, V.S.S.: A rough set-based hypergraph trust measure parameter selection technique for cloud service selection. J. Supercomput. **73**(10), 4535–4559 (2017)
24. Somu, N., Raman, M.R.G, Kirthivasan, K., Sriram, V.S.S.: A trust centric optimal service ranking approach for cloud service selection. Futur. Gener. Comput. Syst. **86**, 234–252 (2018)

An Efficient Intrusion Detection Approach Using Enhanced Random Forest and Moth-Flame Optimization Technique

P. S. Chaithanya, M. R. Gauthama Raman, S. Nivethitha, K. S. Seshan and V. Shankar Sriram

Abstract The recent advancements in the computer networks pave a sophisticated platform to the "Black hat" attackers, which poses a major challenge to network security. Intrusion detection is a significant research problem in network security which motivates the researchers to focus on the development of a robust Intrusion Detection System (IDS). Several research works in network infrastructure reveal the significance of Intrusion Detection Systems (IDS) in protecting the IT infrastructure against the ever-advancing nature of cyber attacks. In a similar manner, machine learning has a pivotal role in enhancing the performance of intrusion methodologies. Hence, this work presents Moth-Flame Optimization Algorithm-based Random Forest (MFOA-RF) to build an efficient intrusion detection methodology for networks, and it enhances the RF technique by tuning the parameter. The experiments were implemented using NSL-KDD Cup dataset and the results were validated in terms of classification accuracy and false alarm rate.

Keywords Intrusion detection system · Network security · Random forest · Moth-flame optimization algorithm

---

P. S. Chaithanya · M. R. Gauthama Raman · K. S. Seshan · V. S. Sriram (✉)
School of Computing, Centre for Information Super Highway (CISH), SASTRA Deemed to Be University, 613401 Thanjavur, Tamil Nadu, India
e-mail: sriram@it.sastra.edu

P. S. Chaithanya
e-mail: chaithanya299@gmail.com

M. R. Gauthama Raman
e-mail: gauthamaraman_mr@sastra.ac.in

K. S. Seshan
e-mail: zeshan.nandan@gmail.com

S. Nivethitha
Smart Energy Informatics Laboratory (SEIL), Indian Institute of Technology-Bombay, 400076 Mumbai, Maharashtra, India
e-mail: nivethithasomu@gmail.com

A. K. Das et al. (eds.), *Computational Intelligence in Pattern Recognition*, Advances in Intelligent Systems and Computing 999,
https://doi.org/10.1007/978-981-13-9042-5_75

1 Introduction

The remarkable increase in the usage of Internet and Internet-based applications has resulted in a global platform for the users to store and share their sensitive and confidential information. However, the complete dependence on the networks leads to several security issues that affect and weaken the security of the private information and economic status of the nation when intrusion or security breach occurs. Such Internet might lead to catastrophic effect on individuals as well as nations. This is evident from recent security incidents like BlueBorne attack, Grabos malware, Uber Data Breach, etc. [1], and their impact on the socioeconomic status of the country. Therefore, cybersecurity has been an active research topic and has attracted the attention of the research community for the past few decades.

Traditional security mechanisms like firewall, authentication, encryption, etc. fail to detect the sophisticated and ever-advancing nature of the intrusions or attacks. Intrusion Detection System (IDS) [2], a second line of defense in depth, proves to a significant and predominant security mechanism to protect the sensitive information system from the adverse impact of the cyber attacks. IDS monitors the system activity by inspecting vulnerabilities in the system, the integrity of files and conducting an analysis of patterns based on the known attacks. Generally, IDS is classified into two, namely, signature detection and anomaly detection with respect to the nature of the detection mechanism. The signature-based IDS detects the intrusion by comparing the incoming traffic patterns with already known patterns, whereas anomaly detection discriminates the network traffic based on the behavior of the normal user, i.e., baseline profile. Either of these detection mechanisms has benefits and limitations with respect to the detection rate and false alarm rate.

The major objective of an IDS is to minimize the tradeoff between high detection rate and less false alarm rate which makes the design of an efficient IDS, an open research challenge. Recent research works reveal the significance of machine learning techniques in the design of an efficient and adaptive IDS. Therefore, several research works were carried out to improve the performance of IDS with machine learning techniques like decision trees, K-nearest neighbor, Support Vector Machines (SVM), rough set theory, genetic algorithm, neural networks, etc. (Table 1). Among them, Random Forest (RF) [3], an ensemble classifier which consists of different regression trees and trained with random variable selection and bagging, was found to be the interesting technique for IDS since it performs well while handling high-dimensional datasets and can also handle binary data, categorical data, and numerical data. However, the identification of the optimal number of regression trees is an NP-hard problem for which several optimization techniques like genetic algorithm, fruit-fly optimization, whale optimization, etc., were employed. In this way, we present a Moth-Flame Optimization Algorithm [4, 5]-Random Forest (MFOA-RF) based intrusion detection methodology to achieve the major objective of IDS. MFOA-RF uses a simple optimization technique named moth-flame optimization algorithm for the identification optimal number of regression tree in order to achieve high detection

able 1 Related works

uthor	Technique proposed	Dataset	Evaluation metrics
Iamed et al. [6]	Recursive Feature Addition (RFA) and bigram technique	ISCX 2012	• Accuracy • F-measure (Detection rate, False alarm rate)
Raman et al. [7]	Hypergraph-based genetic algorithm and feature selection in support vector machine	NSL-KDD	• Accuracy • Detection rate • False alarm rate
Raman et al. [8]	A novel feature selection technique based on RSHGT (Rough Sets—Hypergraph Technique)	KDD cup 1999	• Precision • Recall • Accuracy
Hajisalem et al. [9]	New hybrid classification method called ABC (Artificial Bee Colony) and AFS (Artificial Fish Swarm) algorithms	NSL-KDD, UNSWNB15	• Detection accuracy • Detection rate • False alarm rate • False negative rate • False positive rate
Vijayanand et al. [10]	Genetic algorithm-based feature selection and multiple support vector machine classifiers	ADFA-LD and CICIDS 2017	• Accuracy • Sensitivity • Specificity • Precision • False positive rate • False negative rate
Ambusaidi et al. [2]	Supervised filter-based feature selection algorithm—Flexible Mutual Information, Feature Selection (FMIFS)	KDD Cup 1999, NSL-KDD and Kyoto 2006+	• Accuracy • Detection rate • False positive rate • F-measure • Precision • Recall
Aljawarneh et al. [11]	Hybrid approach and feature selection	NSL-KDD	• True positive rate • True negative rate • False positive rate • False negative rate • Accuracy
Abellan et al. [3]	Credal random forest algorithm	50 different dataset from UCI repository	• Equalized loss of accuracy
Karami et al. [12]	Benign outliers, modified self-organizing map	NSL-KDD, UNSW-NB15, AAGM, and VPN-non-VPN	• Detection rate • False positive rate • Precision • F-measure • Accuracy
Raman et al. [13]	Helly property of Hypergraph (HG) and arithmetic residue-based probabilistic neural network	KDD CUP 1999	• Precision • Recall • Accuracy • Stability (%)

rate. Experimental validations were carried out using the standard NSL-KDD CU [5] dataset in terms of detection rate and false alarm rate.

The rest of the paper is structured as follows: Sect. 2 presents the proposed metho ology of MFOA-RF; Sect. 3 highlights the result and discussion of MFOA-RF; Sect. concludes the paper; and the reference section shows the papers that are reference

2 Proposed Methodology

In this section, the working of the proposed improved random forest technique base on moth-flame optimization algorithm was briefly discussed. The workflow of th MFOA-RF (Fig. 1) is discussed in the following text.

2.1 Generation of Initial Population

The working of MFOA-RF initiates with the initialization of parameters like numbe of moth population, number of flame population, number of iterations, and computations of parameters like flame number, distance, a and t. Each population in MFOA-RF is randomly generated within a specified range and is considered as the parameter for random forest.

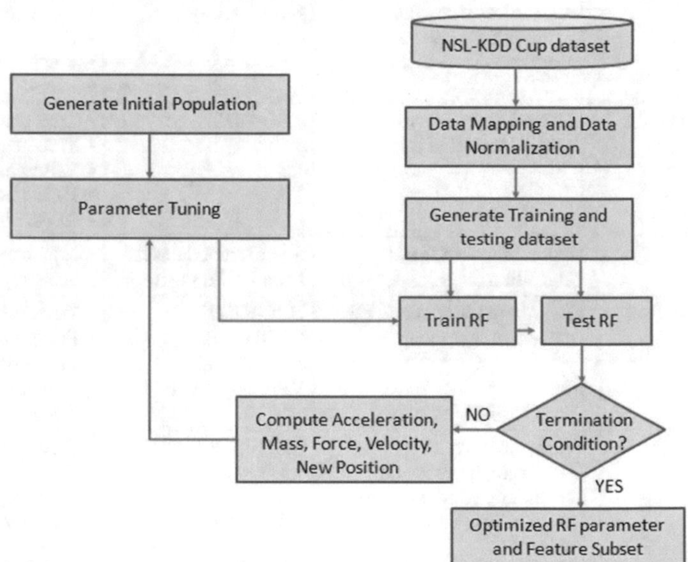

Fig. 1 MFOA-RF architecture with parameter tuning using NSL-KDD dataset

.2 Training and Testing the RF

he testing and training samples are created from the entire intrusion detection dataset y random sampling with replacement. For training the RF, the training samples nd parameters obtained from each population are used, and their performances are valuated using the fitness function (discussed in step 2.3) along with the testing ample.

?.3 Definition of Fitness Function

The weighted fitness function discussed in Eq. (1) is used to compute the performance of the proposed technique

$$f_{Fit} = W_1(DR) + W_2(FAR) \quad (1)$$

where DR and FAR are the detection rate and false alarm rate respectively and W_1 and W_2 are the weights of DR and FAR respectively.

2.4 Termination Condition

The maximum numbers of iterations reached are verified by the termination condition. It returns the optimal parameter only if the condition is true; otherwise, it performs step 5.

2.5 Position Updating

The population is identified with the position and its best fitness value, and then the position of other populations is updated using Eq. (3.12) in [4]

3 Results and Discussions

3.1 Experimental Setup

MFOA-based RF was implemented on Python with INTEL® Core™ i3-5005U CPU processor @ 2.00 GHz architecture running in Windows 8, and the validation process was carried out using WEKA tool.

3.2 Dataset Description

The dataset used is the NSL-KDD Cup dataset, a benchmark intrusion detection dataset which is derived from KDD Cup 1999. This dataset is constructed by MIT Lincoln Laboratory, and it solves the issues of redundancy of records. It consists of 41 features with 3 non-numeric features and 38 numeric features [14]. The training set contains the known attacks while the testing set contains the novel attacks. The major attacks in NSL-KDD Cup dataset are DoS (Denial of Service attacks), R2L (Root to Local attacks), U2R (User to Root attack), and Probe (Probing attacks) [15]. Every attempt to connect to a network can be labeled as either normal or an attack which is decided by studying these high-level features of the connection, classified as basic features (TCP/IP), content features, and traffic features.

3.3 Data Preprocessing

The NSL-KDD Cup dataset is preprocessed by converting the non-numeric features with corresponding numerical values, and this conversion is numerical encoding. After preprocessing the features, these numerical values are normalized by mapping the features with the value ranging from 0 to 1. One of the methods for normalization [14] is as follows:

$$X_{i[0\,to\,1]} = \frac{x_i - x_{min}}{x_{max} - x_{min}} \quad (2)$$

where x_i represents the ith data point, x_{min} represents the minimum among the data points, x_{max} represents the maximum among the data points, and $X_{i[0\,to\,1]}$ represents the normalized data point within the range of 0 and 1. To calculate the efficiency of this algorithm, the following set of data is further processed.

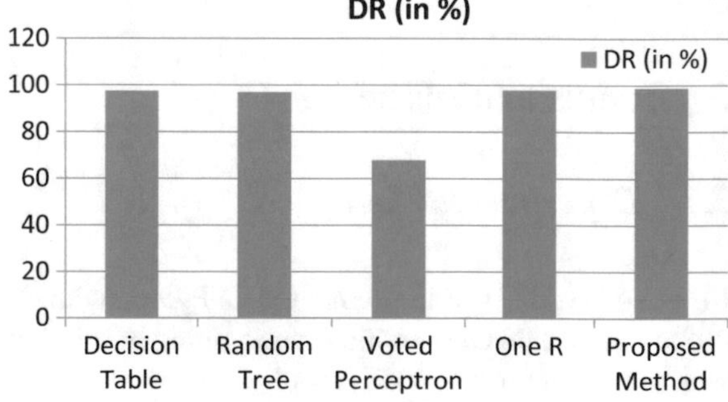

Fig. 2 Detection rate for different classification techniques

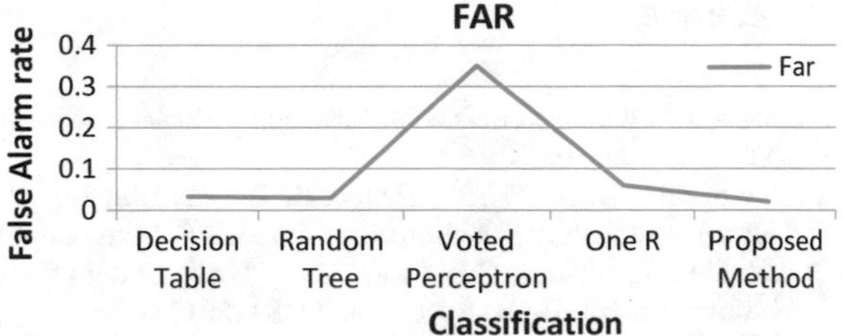

ig. 3 False alarm rate for fferent classification chniques

The detection rate for MFOA-RF approach has high detection rate of 98.7% Fig. 2). It is clear when comparing it with other classification techniques such as lecision table (97.4%), random tree (96.9%), voted perceptron (67.9%), and One R 97.8%).

Figure 3 shows that the proposed method of MFOA-RF approach produces minmum false alarm rate when comparing it with other classification approaches such as decision tree, random tree, voted perceptron, and One R

4 Conclusions

This paper proposed the moth-flame optimization algorithm for optimizing the RF parameters and to obtain the optimal number of regression tree. This MFOA-RF is an efficient intrusion detection technique. This technique was experimented and validated using the benchmark NSL-KDD Cup dataset. Thus, it is concluded by producing high accuracy of 98.7% (Table 2) and also minimize false alarm rate than any other existing machine learning intrusion detection technique. This methodology was more effective and robust even for high-dimensional datasets. However, the performance decreases and also generates overfitting when it is applied to a noisy dataset [16, 17]. Further, the complexity can be reduced by integrating hypergraph property [18–20], and this work can be extended for SCADA system and can also be applied for trust prediction in cloud computing [21, 22].

Table 2 Accuracy for different classification technique

S. No	Classification	Accuracy (%)
1	Decision table	97
2	Random tree	97.1
3	Voted perceptron	65.8
4	One R	95.5
5	Proposed method	98.93

References

1. https://www.exsystemusa.com/singlepost/2018/01/04/TOP-10-Cyber-Attacks-and-Critical-Vulnerabilities-of-2017
2. Ambusaidi, M.A., He, X., Nanda, P., Tan, Z.: Building an intrusion detection system using filter-based feature selection algorithm. IEEE Trans. Comput. **65**(10), 2986–2998 (2016)
3. Abellan, J., Mantas, C.J., Castellano, J.G.: A random forest approach using imprecise probabilities. Knowl. Based Syst. **134**, 72–84 (2017)
4. Mirjalili, S.: Moth-flame optimization algorithm: a novel nature-inspired heuristic paradigm. Knowl. Based Syst. **89**, 228–249 (2015)
5. Zawbaa, H.M., Emary, E., Parv, B., Sharawi, M.: Feature selection approach based on moth-flame optimization algorithm. In: Evolutionary Computation (CEC), pp. 4612–4617 (2016)
6. Hamed, T., Dara, R., Kremer, S.C.: Network intrusion detection system based on recursive feature addition and bigram technique. Comput. Secur. **73**, 137–155 (2018)
7. Raman, M.G., Somu, N., Kirthivasan, K., Liscano, R., Sriram, V.S.: An efficient intrusion detection system based on hypergraph-Genetic algorithm for parameter optimization and feature selection in support vector machine. Knowl. Based Syst. **134**, 1–12 (2017)
8. Raman, M.G., Kirthivasan, K., Sriram, V.S.: Development of rough set–hypergraph technique for key feature identification in intrusion detection systems. Comput. Electr. Eng. **59**, 189–200 (2017)
9. Hajisalem, V., Babaie, S.: A hybrid intrusion detection system based on ABC-AFS algorithm for misuse and anomaly detection. Comput. Netw. **136**, 37–50 (2018)
10. Vijayanand, R., Devaraj, D., Kannapiran, B.: Intrusion detection system for wireless mesh network using multiple support vector machine classifiers with genetic-algorithm-based feature selection. Comput. Secur. **77**, 304–314 (2018)
11. Aljawarneh, S., Aldwairi, M., Yassein, M.B.: Anomaly-based intrusion detection system through feature selection analysis and building hybrid efficient model. J. Comput. Sci. **25**, 152–160 (2018)
12. Karami, A.: An anomaly-based intrusion detection system in presence of benign outliers with visualization capabilities. Expert Syst. Appl. **108**, 36–60 (2018)
13. Raman, M.G., Somu, N., Kirthivasan, K., Sriram, V.S.: A hypergraph and arithmetic residue-based probabilistic neural network for classification in intrusion detection systems. Neural Netw. **92**, 89–97 (2017)
14. Yin, C., Zhu, Y., Fei, J., He, X.: A deep learning approach for intrusion detection using recurrent neural networks. IEEE Access **5**, 21954–21961 (2017)
15. Pervez, M.S., Farid, D.M.: Feature selection and intrusion classification in NSL-KDD cup 99 dataset employing SVMs. In: Software, Knowledge, Information Management and Applications (SKIMA), pp. 1–6 (2014)
16. Joshi, A., Monnier, C., Betke, M., Sclaroff, S.: Comparing random forest approaches to segmenting and classifying gestures. Image Vis. Comput. **58**, 86–95 (2017)
17. Scornet, E., Biau, Gérard, Vert, J.P.: Consistency of random forests. Ann. Stat. **43**(4), 1716–1741 (2015)
18. Raman, M.R., Kannan, K., Pal, S.K., Sriram, V.S.: Rough set-hypergraph-based feature selection approach for intrusion detection systems. Def. Sci. J. **66**(6), 1–6 (2016)
19. Somu, N., Raman, M.G., Kirthivasan, K., Sriram, V.S.: Hypergraph based feature selection technique for medical diagnosis. J. Med. Syst. **40**(11), 239 (2016)
20. Somu, N., Kirthivasan, K., Sriram, V.S.: A rough set-based hypergraph trust measure parameter selection technique for cloud service selection. J. Supercomput. **73**(10), 4535–4559 (2017)
21. Somu, N., Raman, M.G., Kirthivasan, K., Sriram, V.S.: A trust centric optimal service ranking approach for cloud service selection. Futur. Gener. Comput. Syst. **86**, 234–252 (2018)
22. Somu, N., Raman MG, Kalpana, V., Kirthivasan, K., Sriram V.S.: An improved robust heteroscedastic probabilistic neural network based trust prediction approach for cloud service selection. Neural Netw. 1–35 (2018)

Unidentified Input Observer: Comparative Study Through Block Pulse Function and Hybrid Function

Anindita Ganguly, Tathagata Roy Chowdhury, Suman Kumar Ghosh, Aishik Panigrahi, Nilotpal Chakraborty, Anirudhha Ghosh and Neelakshi Ganguly

Abstract In many systems, the inputs cannot be exactly identified. For the implementation of certain desirable control objectives, the inputs and the states of such systems are estimated. In this paper to estimate the desired quantities, observers are designed and linear systems are studied with unidentified inputs. To make the state and input estimation possible, the use of delayed observers is considered here to enlarge the class of systems. The design procedure involves the design of unidentified input observer with appropriate choice of design matrices. The necessary and sufficient conditions proving the existence of such observers are also described in this paper.

Keywords Block pulse functions · Hybrid function · UIO · B.P.F. differential operation · H.F. algorithm · Algebraic approach

---

A. Ganguly
Department of Electrical Engineering, Guru Nanak Institute of Technology, Kolkata 700114, West Bengal, India
e-mail: aninditaganguly80@gmail.com

T. R. Chowdhury · S. K. Ghosh · A. Panigrahi · N. Chakraborty (✉) · N. Ganguly
Department of Electrical Engineering, St. Thomas College of Engineering and Technology, Kolkata 700023, West Bengal, India
e-mail: nilchakroborty100@gmail.com

T. R. Chowdhury
e-mail: tathagataroychowdhury94@gmail.com

S. K. Ghosh
e-mail: ghosh898@gmail.com

A. Panigrahi
e-mail: aishikpanigrahi5@gmail.com

N. Ganguly
e-mail: neelakshi5597@gmail.com

A. Ghosh
Department of Electrical Engineering, Indian Institute of Technology, Dhanbad 826004, Jharkhand, India
e-mail: aniruddhaghosh1993@gmail.com

A. K. Das et al. (eds.), *Computational Intelligence in Pattern Recognition*, Advances in Intelligent Systems and Computing 999,
https://doi.org/10.1007/978-981-13-9042-5_76

1 Introduction

Often, in the process of designing control systems, one has to deal with some amount of uncertainty relating to the plant. Such uncertainties are often considered as unidentified inputs for the purpose of system modelling. The unidentified inputs may represent some input uncertainty, unknown external drivers, or instrument faults. The equation for the unidentified input observer [1] and the derived observer has the differential of the system outputs. To get the differentiated system outputs, an algebraic method based on BLOCK PULSE FUNCTION (B.P.F.) [2] has been used to estimate the unidentified inputs. Over the last few decades, the Unidentified Input Observer (UIO) [1] has gained significant interest. Various approaches have been proposed for the analysis of systems using UIO. The widely popular method is coordinated transformation [3–5] because of its systematic approach. An additional differential operator is needed in this system, which makes the problem more complex. So by an algebraic procedure [1], we are aiming to solve the UIO problem. So the differential of the system output does not need to solve. This is executed by the orthogonal property of the HYBRID FUNCTION (H.F.) [6]. Now, it is possible to re-model the system using simple algebraic manipulations [1] and by the elimination of the unidentified inputs. It can be divided into two interconnected subsystems [3]. Out of which one is dependent on the known inputs. The Luenberger [1, 7] observer is capable of solving any unknown inputs of a system via an algebraic procedure. In this project, we have utilized the hybrid function approximation and its applications to differential operations.

2 Details of Analysis

2.1 Solution by BPF's Differential Operation

2.1.1 A Brief Review of B.P.F.

Block pulse function set $\emptyset(t) = [\emptyset_1(t)\emptyset_2(t)\ldots\ldots\ldots\emptyset_m(t)]^T$ is a set of piecewise constant function and defined in the time interval $\left[0, t_f\right]$ as follows:

$$\emptyset(t) = \begin{cases} 1, \ for \ (i-1)\frac{t_f}{m} \le t < i\frac{t_f}{m} \\ 0, \qquad otherwise \end{cases} \tag{1}$$

where i = 1, 2, …m.

In Eq. (1), m is the B.P.F.'s expansion number. B.P.F. has the following orthogonal and disjoint properties:

$$\text{Orthogonal property}: \quad \int_0^{t_f} \emptyset_i(t)\varphi_j(t)\,dt = \begin{cases} \frac{t_f}{m}, \ i = j \\ 0, \ i \ne j \end{cases} \tag{2}$$

$$\text{Disjoint property : } \quad \emptyset_i(t)\theta_j(t) = \begin{cases} \emptyset_i(t), \ i = j \\ 0, \quad i \neq j \end{cases} \tag{3}$$

If an arbitrary function f(t) is absolutely integrable in the interval $[0, t_f)$, it can e approximated using B.P.F. expansion as

$$f(t) = \sum_{i=1}^{m} F_i \emptyset_i(t) \tag{4}$$

F_i is a coefficient of the ith block pulse function. In Eq. (4), the ith B.P.F'S oefficient F_i of an arbitrary function f(t) is determined as follows:

$$F_i = \frac{m}{t_f} \int_0^{t_f} f(t)\emptyset_i(t)\, dt = \frac{m}{t_f} \int_{(i-1)\frac{t_f}{m}}^{i\frac{t_f}{m}} f(t)dt \frac{1}{2}\left[f\left(i\frac{t_f}{m}\right) + f(i-1)\frac{t_f}{m} \right] \tag{5}$$

where i = 1, 2,…m.

The approximation for the forward integral of B.P.F. is

$$\int_o^t \emptyset_i(\tau)\, d\tau \frac{t_f}{m} \emptyset_i(t) + \frac{t_f}{m} \sum_{j=i+1}^{m} \emptyset_j(t) \tag{6}$$

2.1.2 B.P.F. in Differential Operation

Similar to (4), if a differential of f(t) is absolutely integrable in the interval $\left[0, t_f\right]$, it can be also approximated as (7) by using B.P.F. expansions (9) as

$$f(t) = \sum_{i=1}^{m} F_i \overline{\emptyset_\iota}(t) \tag{7}$$

Many researchers resolved the problem by a calculus of variation as

$$\dot{f}(t) = lim_{\Delta t \to 0} \frac{f(t + \Delta t) - f(t)}{\Delta t} = \frac{m}{t_f} \sum_{i=1}^{m} \left[f(i+1)\frac{t_f}{m} f\left(i\frac{t_f}{m}\right) \right] \tag{8}$$

where $\Delta t = \frac{t_f}{m}$ and i = 1, 2.…m − 1.

But in this paper, a coefficient $\overline{F_\iota}$ is obtained on the basis of B.P.F.'s orthogonal and disjoint properties.

To obtain a recursive algorithm for the B.P.F's coefficients of f˙(t) described by and f(0) which are f(t)'s B.P.F coefficients and their initial values respectively, let us integrate f(t) in the interval [0, t), which gives

$$\int_0^t \dot{f}(\tau)\,d\tau = f(t) - f(0) \tag{9}$$

Equation (9) can be represented as (10) by using B.P.F's expansions as

$$\sum_{i=1}^{m} \overline{F_\iota} \int_0^t \emptyset_i(\tau)\,d\tau = \sum_{i=1}^{m} F_i \emptyset_i(t) - \sum_{i=1}^{m} f(0)\emptyset_i(t) \tag{10}$$

where $\dot{f}(t) = \sum_{i=1}^{m} \overline{F_\iota}\, \emptyset_i(t)$.

Substituting (6) in (10), we can obtain an equation as

$$\sum_{i=1}^{m} \overline{F_\iota} \left(\frac{t_f}{2m} \emptyset_i(t) + \frac{t_f}{m} \sum_{j=i+1}^{m} \emptyset_i(t) \right) \tag{11}$$

Expanding (11) such as (12),

$$\frac{t_f}{m} x \left\{ \overline{\overline{F_1}\left(\frac{1}{2}\emptyset(t) + \cdots \emptyset_m(t)\right) + \overline{F_2}(\frac{1}{2}\emptyset_2(t) + \cdots \emptyset_m(t)) \ldots + \overline{F_m}(\frac{1}{2}\emptyset_m(t)\Big)} \right\} \tag{12}$$

By multiplying $\emptyset_1(t), \emptyset_2(t), \ldots \emptyset_m(t)$ to (12) sequentially and applying B.P.F.'s disjoint property, (12) can be deduced as follows:

$$\frac{t_f}{m}\left(\overline{F_1} + \overline{F_2} + \cdots \overline{F_{\iota-\iota}} + \frac{1}{2}\overline{F_\iota} \right) = F_1 - f(0) \tag{13}$$

Then arranging (13) for $\overline{F_i}$, we can obtain (14)

$$\begin{aligned} \overline{F_1} &= \frac{2m}{t_f}\left[\overline{F_1} - f(0)\right] \quad and \\ \overline{F_{\iota+1}} &= \frac{2m}{t_f}\left[\overline{F_{\iota+1}} - F_1\right] - \overline{F_\iota} \end{aligned} \tag{14}$$

for, i = 1, 2, 3…m − 2.

In (14), the f(t)'s *i*th B.P.F. coefficient $\overline{F_\iota}$ can be obtained by using $\overline{F_\iota}$ and f(0), recursively, and (15) is the generalized form of (14):

$$\overline{F_\iota} = \frac{2m}{t_f}\left[\overline{F_\iota} + (2x(-1)'\sum_{j=1}^{i-1}(-1)^j F_j) + (-1)^i f(0)\right]$$

where

$$\dot{f(t)} = \sum_{i=1}^{m}\overline{F_\iota}\emptyset_i(t) \quad \text{for, i} = 1, 2\ldots\text{m} \tag{15}$$

3 Process of Designing UIO

3.1 *Method of Coordinate Transformation*

Let a dynamic, linear time-invariant (LTI) system be represented as

$$\begin{aligned}\dot{x}(t) &= Ax(t) + Bu(t) + Dd(t)\\ y(t) &= Cx(t)\end{aligned} \tag{16}$$

where $x \in R^n, u \in R^m,$ *and* $d \in R^q$ are the state inputs, outputs, and the unknown input vectors of the system. In (16), ranks of the matrices D and C are p(D) = q p(c) = m, respectively. It is known that if m ≥ q is satisfied, then there exists a similar transformation matrix

T_1 *such that* $T_1^{-1}D = \begin{pmatrix}0\\ I_q\end{pmatrix}$ *and* (16) *can be represented as*

$$\begin{aligned}x_1^*(t) &= A_{11}^*(t)x_1^*(t) + A_{12}^*(t)x_2^*(t) + B_1^*(t)u_1(t)\\ x_2^*(t) &= A_{21}^*(t)x_1^*(t) + A_{22}^*(t)x_2^*(t) + B_2^*(t)u_1(t) + d(t)\\ y(t) &= C_1^*x_1^*(t) + C_2^*x_2^*(t)\end{aligned} \tag{17}$$

where $x = T_1x^* = T_1\begin{pmatrix}x_1^*\\ x_2^*\end{pmatrix}, x_1^* \in R^{(n-q)}, x_2^* \in R^q,$

$$T_1^{-1}AT_1 = \begin{pmatrix}A_{11}^* & A_{12}^*\\ A_{21}^* & A_{22}^*\end{pmatrix},\ T_1^{-1}B = \begin{pmatrix}B_1^*\\ B_2^*\end{pmatrix},\ \textit{and}\ CT_1 = \left(C_1^*C_2^*\right)$$

By defining a variable such as $z(t) = x_2^*(t) - B_2^*u(t) - d(t) - A_{22}^*x_2^*(t)$ in (17), (18) can be obtained

$$\begin{pmatrix}C_1^*C_2^* & \vdots & -1_m0\\ A_{21}^*0 & \vdots & 0-1_q\end{pmatrix}\cdot\begin{pmatrix}x^*(t)\\ y^*(t)\end{pmatrix} = 0 \tag{18}$$

where $y^*(t) = \begin{pmatrix} y(t) \\ z(t) \end{pmatrix}$ *and* $x^*(\mathrm{t}) = A_{21}^* x_1^*(\mathrm{t})$. Thus, the non-singular transformatio $T_2 \in R^{(m+q)+(m+q)}$, which satisfies (19) exists any time:

$$T_2 \begin{pmatrix} C_1^* C_2^* & \vdots & -1_m 0 \\ A_{21}^* 0 & \vdots & 0 - 1_q \end{pmatrix} = \begin{pmatrix} M_1 0 & \vdots & N_1 N_2 \\ M_2 0 & \vdots & N_3 N_4 \end{pmatrix} \tag{19}$$

Using transformation matrix T_2, (18) can be represented as (20), (21)

$$\left(M_1 + N_2 A_{21}^*\right) x_1^*(t) + N_1 y(t) = 0 \tag{20}$$

$$x_2^*(t) = -\left(M_2 + N_4 A_{21}^*\right) x_1^*(t) - N_3 y(t) \tag{21}$$

Let us substitute $x_2^*(t)$ in (17) by (21) and define a variable such as $\bar{y}(t) = -N_1 y(t)$ in (20). As a result of the previous, we can deduce the following (n−q)th order dynamic system (22) which consists of $x_1^*(t)$ only:

$$\begin{aligned} x_1^*(t) &= A^0 x_1^*(t) + B_1 u(t) - A_{12}^* N_3 y(t) \ and \\ \bar{y}(t) &= \left(M_1 + N_2 A_{21}^*\right) x_1^*(t) \end{aligned} \tag{22}$$

where $A^0 = A_{11}^* + A_{12}^* M_2 + A_{12}^* N_4 A_{21}^*$.

3.2 Algebraic UIO Design Procedure

It is obvious that (22) is a convenient form to design a Luenberger type observer. To design an algebraic observer for (22), let us represent (22) by using B.P.F. expansions

$$\sum_{i=1}^{m} \overline{X_{1_i}}^* \emptyset_i(t) = A^0 \sum_{i=1}^{m} X_{i_1}^* \emptyset_i(t) + B_1^* \sum_{i=1}^{m} U_i \emptyset_i(t) - A_{12}^* N_3 \sum_{i=2}^{m} Y_i \emptyset_i(t) \tag{23}$$

$$\sum_{i=1}^{m} \bar{y}^* \emptyset_i(t) = \left(M_1 + N_2 A_{21}^*\right) \sum_{i=1}^{m} X_{1_i}^* \emptyset_i(t) \tag{24}$$

Algebraic Luenberger type observer for (23) can be designed as follows:

$$\begin{aligned} \sum_{\mathrm{i}=1}^{\mathrm{m}} \overline{\mathrm{W}}_{\mathrm{i}} \emptyset_i(t) &= A^0 \sum_{i=1}^{m} W_i \emptyset_i(t) + B_i^* \sum_{i=1}^{m} U_i \emptyset_i(t) - A_{12}^* N_3 \sum_{i=2}^{m} Y_i \emptyset_i(t) \\ &\quad + L\left(\sum_{i=1}^{m} U_i \emptyset_i(t) - \left(M_1 + N_2 A_{21}^*\right) \sum_{i=1}^{m} W_i \emptyset_i(t)\right) \end{aligned} \tag{25}$$

Determine an error function as

$$e(t) = \sum_{i=1}^{m} W_i \emptyset_i(t) - \sum_{i=1}^{m} X_{1_i}^* \emptyset_i(t) \tag{26}$$

From (23) and (25), (27) is derived directly as

$$e(t) = \sum_{i=1}^{m} \overline{W_\iota} \emptyset_i(t) - \sum_{i=1}^{m} X_{1_i}^* \emptyset_i(t) = \left[A^0 - L\left(M_1 + N_2 A_{21}^*\right)\right] e_i(t) \tag{27}$$

The observer gain matrix L in (27) is chosen to have the eigenvalues of the matrix $A^0 - L\left(M_1 + N_2 A_{21}^*\right)\right]$ is negative assignment, proposed algebraic observer can be converged to the actual state of (23) at $t \to \infty$. By using a relation,

$$\sum_{i=1}^{m} \overline{Y_\iota} \emptyset_i(t) = -N_1 \sum_{i=1}^{m} Y \emptyset_i(t),$$

$$\sum_{i=1}^{m} \overline{W_\iota}(\mathrm{t}) \emptyset_i(t) = F \sum_{i=1}^{m} W_i(t) \emptyset_i(t) + G \sum_{i=1}^{m} U_i(t) \varphi_i(t) + H \sum_{i=1}^{m} Y_i(t) \varphi_i(t) \tag{28}$$

where $\mathrm{F} = \left[A^0 - L\left(M_1 + N_2 A_{21}^*\right)\right]$ $\mathrm{G} = B_1^*$ $\mathrm{H} = -\left(A_{12}^* N_3 + L N_1\right)$.

Now the main problem of algebraic observer equation (28) is how to obtain the B.P.F coefficients $W = [W_1, W_2, \ldots\ldots W_m]$. We note the recursive algorithm which is deduced by using a proper B.P.F differential operation summarized as follows:

Step1: Obtain the B.P.F expansion equation from (28)

$$\overline{W_1} = F W_1 + G U_1 + H Y_1$$

Step2: Adopting (14) for $\overline{W_1}$

$$\overline{W_1} = \frac{2m}{t_f}[W_1 - \omega(0)]$$

Step3: The first B.P.F coefficient W_i is obtained from steps 1 and 2 to be equal

$$W_1 = \left(I - \frac{t_f}{2m} F\right)^{-1} \times \left[\omega(0) + \frac{t_f}{2m} G U_1 + \frac{t_f}{2m} \mathrm{H} Y_1\right] \tag{29}$$

Step4: Sum of the ith and (i + 1)th B.P.F coefficients of (28)

$$\overline{W_{\iota+1}} + \overline{W_\iota} = F\left(\overline{W_{\iota+1}} + \overline{W_\iota}\right) + G\left(\overline{U_{\iota+1}} + \overline{U_\iota}\right) + H\left(\overline{Y_{\iota+1}} + \overline{Y_\iota}\right)$$

Step5: Redo like step 2

$$\overline{W_{i+1}} + \overline{W_i} = \frac{2m}{t_f}\left[W_{i+1} + W_i\right]$$

Step6: Redo like step 3 for W_{i+1} by using step 4 and step 5

$$W_{i+1} = \left(I - \frac{t_f}{2m}F\right)^{-1} \times [\left(I + \frac{t_f}{2m}F\right)W_i + \frac{t_f}{2m}G(U_{i+1} + U_i)$$
$$+ \frac{t_f}{2m}H(Y_{i+1} + Y_i)$$
$$\text{For i} = 1, 2, 3\ldots..\text{m} - 1 \tag{30}$$

Therefore, B.P.F coefficients $W = [W_1, W_2 \ldots\ldots W_m]$ can be obtained recursively from (29), (30). Finally from (21), the estimated states of the system (16) can be obtained by state construction such as (31)

$$X(t) = T_1 \times \left[\sum_{i=1}^{m} W_i\varphi_i(t) - \left(M_2 + N_4A_{21}^*\right)\sum_{i=1}^{m} W_i\varphi_i(t) - N_3\sum_{i=1}^{m} Y_i\varphi_i(t)\right] \tag{31}$$

From (17) and (31), an equation for UIO estimation is deduced as follows:

$$d(t) = -\left(M_2 + N_4A_{21}^*\right)(t) - N_3y(t) - \left(A_{21}^* - A_{22}^*\left(M_2 + N_4A_{21}^*\right)\right)(t)$$
$$+ \left(A_{22}^*N_3y(t) - B_2^*u(t)\right) \tag{32}$$

In (32), UIO estimation can be achieved by additional differentiation of system outputs. By using B.P.F expansion and its differential operation (14), it can be estimated as algebraic form (33)

$$d(t) = Q\sum_{i=1}^{m} W_i\varphi_i(t) + R\sum_{i=1}^{m} Y_i\varphi_i(t) + S\sum_{i=1}^{m} U_i\varphi_i(t) - N_3\sum_{i=1}^{m} Y_i^{-}\varphi_i(t)$$
$$\text{where } Q =) = \left[-\left(M_2 + N_4A_{21}^*\right)F - N_3y(t) - \left(A_{21}^* - A_{22}^*\left(M_2 + N_4A_{21}^*\right)\right)\right] \tag{33}$$

$$R = -\left[-\left(M_2 + N_4A_{21}^*\right)H - A_{22}^*N_3\right] \text{ and } S = -\left[\left(M_2 + N_4A_{21}^*\right)G + B_2^*\right]$$

In (33), it is easy to note that the differential of system output can be from (34)

$$\overline{Y_1} = \frac{2m}{t_f}[Y_1 - y(0)]$$
$$\overline{Y_{n+1}} = \frac{2m}{t_f}\left[Y_{i+1} - Y_1\right] - \overline{Y_1} \tag{34}$$

for i = 1, 2, …m−1.

xample 1 Let us consider the following linear time-invariant dynamical systems ith unknown inputs [3, 5]. For the convenience of the simulation, u(t) is omitted ithout loss of generality:

$$\dot{x}(t) = \begin{bmatrix} -2 & -2 & 0 \\ 0 & 0 & 1 \\ 0 & -3 & -4 \end{bmatrix} x(t) + \begin{bmatrix} 1 & 0 \\ 0 & 1 \\ 0 & 0 \end{bmatrix} d(t)$$

$$y(t) = \begin{bmatrix} 1 & 0 & 1 \\ 0 & 1 & 0 \end{bmatrix} x(t) \tag{35}$$

From the previous UIO design procedure, the coordinate transformation matrices 1 and T2 are determined as follows:

$$T_1 = \begin{bmatrix} 0 & 1 & 0 \\ 0 & 0 & 1 \\ 1 & 0 & 0 \end{bmatrix} \ and\ T_2 = \begin{bmatrix} 0 & 0 & 1 & 0 \\ 0 & 0 & 0 & 1 \\ 1 & 0 & 0 & 0 \\ 0 & 1 & 0 & 0 \end{bmatrix} \tag{36}$$

The observer gain is chosen such us $L = \begin{bmatrix} 1 & 1 \end{bmatrix}$, and the derived algebraic observer quation is as follows:

$$\sum_{i=1}^{m} \overline{W}_i \emptyset_i(t) = -4 \sum_{i-1}^{m} W_i \emptyset_i(t) + \begin{bmatrix} 0 & -3 \end{bmatrix} \sum_{i=1}^{m} Y_i \emptyset_i(t) \tag{37}$$

Derived recursive formulas for the algebraic observer equation (37) are as follows:

$$W_1 = \begin{bmatrix} 0 & -0.125 \end{bmatrix} Y_1$$
$$W_{i+1} = 0.667 W_i + \begin{bmatrix} 0 & -0.125 \end{bmatrix} Y_i$$

And an algebraic equation for unknown inputs estimation is as follows:

$$\hat{d}(t) = \begin{bmatrix} 2 \\ -1 \end{bmatrix} \sum_{i=1}^{m} W_i \emptyset_i(t) + \begin{bmatrix} 2 & 5 \\ 0 & 0 \end{bmatrix} \sum_{i=1}^{m} Y_i \emptyset_i(t) + \begin{bmatrix} 1 & 0 \\ 0 & 1 \end{bmatrix} \sum_{i=1}^{m} \overline{Y}_i \emptyset_i(t) \tag{38}$$

In Example 1, we assume that $d(t) = \begin{bmatrix} 5 \\ 3 \end{bmatrix}$ and choose $t_f = 10$ s and $m = 100$.

Now, we have compared the estimated output for $x_1(t), x_2(t), x_3(t)$ in Table 1, thereby providing the graphs in Figs. 1, 2 and 3 to aid the results obtained.

Table 1 Comparison of the estimated output for $x_1(t)$, $x_2(t)$, $x_3(t)$

x1(t) by B.P.F.	x1(t) by R.K.-4	x2(t) by B.P.F.	x2(t) by R.K.-4	x3(t) by B.P.F.	x3(t) by R.K.-4
−0.3399	0	−0.149	0	0.018	0
0.1346	−0.028	0.154	0.298	0.049	−0.039
0.4679	−0.104	0.456	0.590	−0.005	−0.139
0.6877	−0.218	0.750	0.869	−0.117	−0.276
0.8171	−0.361	1.031	1.134	−0.266	−0.435
0.8755	−0.525	1.296	1.382	−0.435	−0.605
0.8793	−0.702	1.543	1.613	−0.614	−0.778
0.8415	−0.887	1.773	1.826	−0.795	−0.949
0.7730	−1.076	1.984	2.023	−0.974	−1.114
0.6826	−1.265	2.178	2.204	−1.145	−1.271

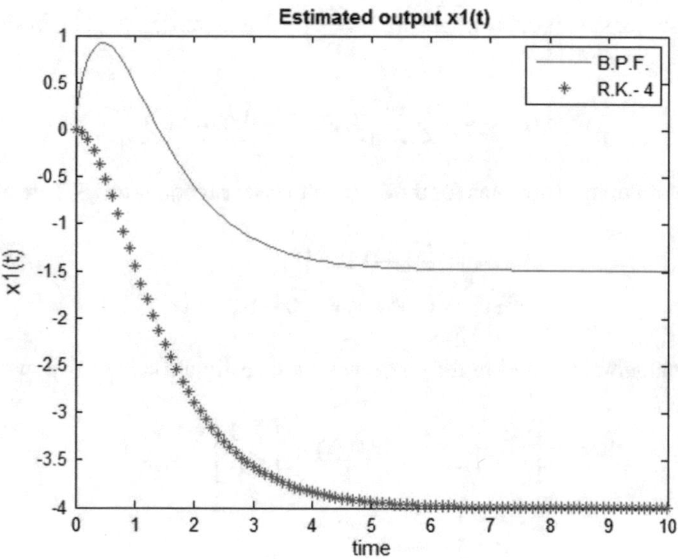

Fig. 1 Comparison between actual and estimated solution via B.P.F. for $x_1(t)$

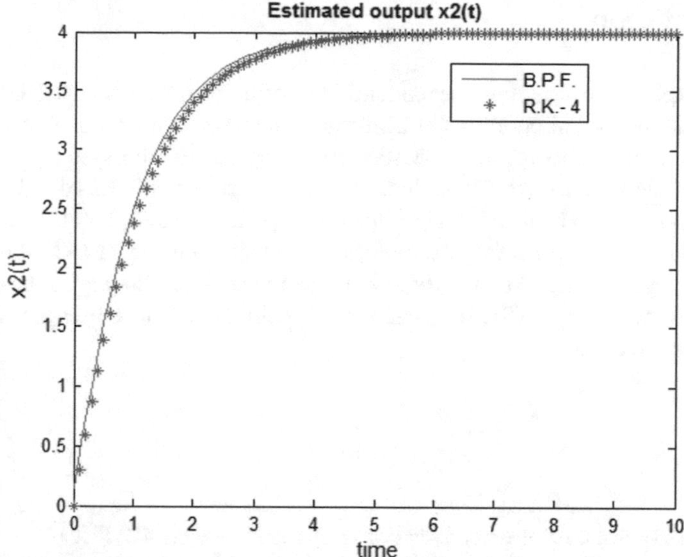

Fig. 2 Comparison between actual and estimated solution via B.P.F. for $x_2(t)$

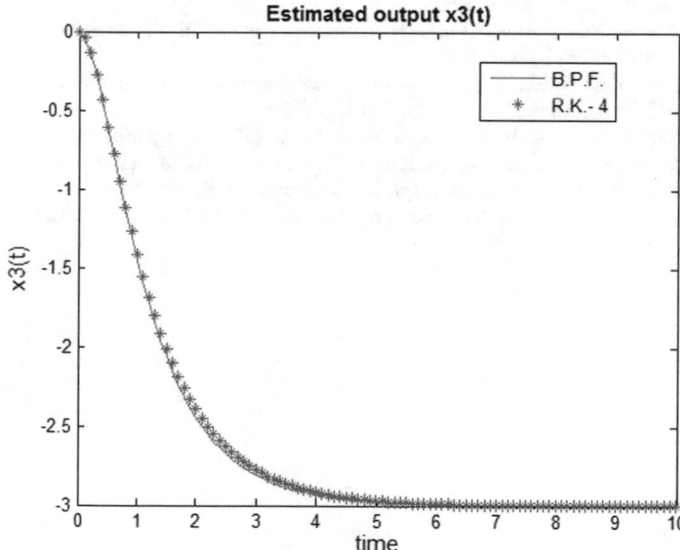

Fig. 3 Comparison between actual and estimated solution via B.P.F. for $x_3(t)$

4 Conclusion

For the purpose of decoupling the unidentified inputs from the systems in UIO desig process, the method of coordinate transformation has been adopted. To estimate th state vectors, a Luenberger type observer is designed for the system from whic the unidentified inputs are decoupled. Here an improved B.P.F. and H.F. differer tial operation is used for the design purposes. To investigate the efficiency of th developed method, an arbitrary example is carried out by using a MATLAB (versio R2015b) programming. All the programming and the simulations with the compara tive study between the B.P.F., H.F., and R.K.-4 methods with its corresponding table and figures are given.

References

1. P. Ahn, Unknown input observer design via block pulse functions. **4**(3) (2002)
2. Z.H. Jiang, W. Schaufelberger, *Block-Pulse Function Andtheir and Their Application in Contro Systems* (Springer, Berlin, 1990)
3. Guan, Y., Saif, M.: A nobel approach to the design of unknown input observer. IEEE Trans Autom. Control **36**, 632–635 (1991)
4. Hou, M., Muller, P.C.: Design of observers for linear systems with unknown inputs. IEEE Trans Autom. Control **37**, 871–875 (1992)
5. Darouach, M., Zasadzinski, M., Xu, S.J.: Full-order Observers for linear systems with unknown inputs. IEEE Trans. Autom. Control **39**, 606–609 (1994)
6. Deb, A., Sarkar, G., Ganguly, A., Biswas, A.: Approximation, integration and differentiation of time functions using a set of orthogonal hybrid functions (HF) and their application to solution of first order differential equations. Appl. Math. Comput. **218**(9), 4731–4759 (2012)
7. Luenberger, D.G.: Observing the state of a linear system. IEEE Trans. Mil. Electron. **MIL-8**, 74–80 (1964)

Data Hiding with Digital Authentication in Spatial Domain Image Steganography

Ayan Chatterjee and Soumen Kumar Pati

Abstract In now a day, Image Steganography is one of the most popular and highly secured wireless communication technique. Digital authentication in image steganography is a special extension in twenty-first century. The approach "steganography" can be categorized into two major wings—spatial domain and frequency domain. Among them, the spatial domain is very common and easy to implement. In this paper, an image steganography scheme is developed in spatial domain with proper digital authentication. The basic idea of integer theory is used to develop this particular data hiding model. More specifically, the concept of triangular number and perfect number is used to develop the system with maintaining all the necessary requirements of information security, i.e., confidentiality, integrity, and availability of data among authorized users. The speciality of this approach is keyless transposition and freeness from different powerful attacks like PRNG (Pseudo Random Number Generator), visual attack, chi-square, and histogram. The efficacy of the proposed method is analyzed with different parameters of data hiding and digital authentication.

Keywords Data hiding · Digital authentication · Image steganography · Spatial domain · Triangular number · Perfect number

---

A. Chatterjee (✉)
Operation Management Group, S P Jain Institute of Management & Research (SPJIMR), 400 058 Mumbai (Andheri West), Maharastra, India
e-mail: fpm18.ayan@spjimr.org

S. K. Pati
Department of Bioinformatics, Maulana Abul Kalam Azad University of Technology, 741249 Haringhata, Nadia, West Bengal, India
e-mail: soumenkrpati@gmail.com

A. K. Das et al. (eds.), *Computational Intelligence in Pattern Recognition*, Advances in Intelligent Systems and Computing 999,
https://doi.org/10.1007/978-981-13-9042-5_77

1 Introduction

Nowadays, the huge improvement of multimedia and dramatic attractiveness of di ital communication is especially notable [1]. The applications of digital communic tion are observed in different sectors of the society like banking, health care, railway governance, etc. [2]. But one of the most important issues regarding digital data com munication is to provide security. More specifically, the security of communicatio should be provided to the authorized users. Information security signifies basicall three major conditions—confidentiality, integrity, and availability of information t the proper authenticated users. These terms are described shortly in the following:
Confidentiality: This feature provides that accessing information is possible only t authorized individuals. In other words, any unauthorized user is not able to acces secret data [3].
Integrity: Validity of secret information is referred to the feature integrity. It ensure that the data can be changed by authenticated users only [3].
Availability: This condition ensures that secret information is fully accessible t the authorized users. More specifically, secret information should not be corrupte before reaching the authorized users [3].

Digital image steganography is an important approach to provide secure commu- nication. Here, the secret information is embedded in a cover image and the embedded image (stego image) is sent to the authorized receiver through internet medium [4, 5]. The uniqueness of the approach steganography is hiding the existence of com- munication from unauthorized users. But the drawback of image steganography is no provision of integrity property. In other words, any criterion is not developed regarding the validity of data in image steganography. So, it is required to set a digital authentication criterion with image steganography. As a result, data hiding with digital authentication makes the communication secure more than only using the steganography approach. For this, some techniques are developed combining the steganography approach with digital authentication.

Digital image steganography can be classified into two major categories—spatial domain and frequency domain. In spatial domain steganography approach, a secret message is embedded directly to the RGB pixels of a digital image. But in the frequency domain, the actual image domain is transformed into another domain using a particular mathematical transformation, and the changed domain is used to insert the secret information. Among these two categories, the spatial domain is very easy to implement. In this paper, a data hiding scheme is proposed with digital authentication in the spatial domain.

Some related works on spatial domain image steganography are described in the following with corresponding uniqueness and drawbacks.

Spatial domain steganography techniques can be categorized into ten different ectors [6]. These are LSB (Least Significant Bit) based, PVD (Pixel Value Differ- nce) based, EMD (Exploiting Modification Direction) based, MBNS (Multi-Base otation System), PPM (Pixel Pair Matching), GLM (Gray Level Modification), VP (Pixel Value Predictor), histogram, edge-based, and mapping-based methods. mong them, LSB-based techniques are popular mostly due to its less difficulty. In is paper, only GLM is considered. The main concept of this technique is that the ecret information cannot be inserted directly in the image domain [7]. The data is mbedded using some proper logic. Recently, a GLM technique [8] is developed ogether with RGB pixel modification. The beauty of this technique is combination f encryption and embedding of secret information. More specifically, the message s encrypted at first and the encrypted message is embedded into the image file. Ulti- nately, it becomes free from steganalysis attack. But the bottleneck of the approach s that no authentication is provided in this particular.

To remove the bottleneck of the approach, the proposed technique is developed. Iere, data is embedded using the approach GLM and authentication is provided with simple idea of integer theory. More crucially, the properties of triangular number and erfect number are used to enhance the integrity level of the proposed scheme.

Remaining of the paper is designed in the following manner. In Sect. 2, the pro- posed scheme is described with embedding and extraction algorithms. In Sect. 3, some experimental results are shown to realize the efficacy of the approach. A con- clusion with corresponding future is given in Sect. 4.

2 Proposed Methodology

In the proposed data hiding model, the basic concept of integer theory is used for embedding and extraction part of steganography. The distinctiveness of embedding is high integrity level with keyless composition. The idea of triangular number and perfect number is used in this scheme. The specific properties of triangular number and perfect number, which are used in this scheme, are noted in the following.

Triangular number: The number of objects that are arranged in an equilateral triangle is called a triangular number. Mathematically, *n*th triangular number can be represented as

$$T_n = \binom{n+1}{2} = \frac{n(n+1)}{2} \tag{1}$$

Perfect number: If the sum of the factors of a number (excluding the number itself) is the number itself, then that number is called perfect number.

The embedding and extraction procedures of the proposed model are described in Sects. 2.1 and 2.2, respectively.

2.1 Embedding Procedure

Embedding phase is the task at the sender side. At first, a cover image (C) is taken into account for embedding the secret message $a_1 a_2 \ldots a_n$. Entropy pixels of the cover image are split into 4×4 blocks, and the message $a_1 a_2 \ldots a_n$ is broken into 4 bits sequential segments. Each of 4 bits is embedded into each 4×4 entropy pixel block (b) using the proposed procedure. First row and third row of each block are replaced with four sequential triangular numbers. More specifically, the positions $b_{11}, b_{12}, b_{13}, b_{14}$ are replaced with four sequential triangular numbers. The same set of triangular numbers is placed in the positions $b_{31}, b_{32}, b_{33}, b_{34}$, respectively. In similar manner, second row and fourth row are replaced with the numbers 33, 55, 03, 36, respectively. In other words, $b_{21} = b_{41} = 33$, $b_{22} = b_{42} = 55$, $b_{23} = b_{43} = 03$ and $b_{24} = b_{44} = 36$. The sequential set of numbers 33, 55, 03, 36 is used to provide the digital authentication at the receiver side. Because—33, 55, 03, 36 is a perfect number set as the factors of 33, 55, 03, 36 (excluding 33, 55, 03, 36) is 33, 55, 03, 36. After that, the 4 bits message segment is converted to corresponding decimal number. This decimal number is added with all the elements of first row and second row of the image block. This operation is performed in each block of the image with each message segment. Combining these blocks sequentially with message segments, the stego image is prepared. This stego image is sent from sender to receiver side through communication channel.

2.2 Extraction Procedure

Extraction phase is basically the task at the receiver side. The main objective in this phase is extraction of secret information from stego image safely. So, the initial phase of this part is the verification of stego image. At first, the stego image (S) is divided into 4×4 entropy pixels. Third and fourth rows of the first block (b') are observed carefully. For the elements of third row, the properties (2) and (3) are checked.

$$b'_{3(j+1)} - b'_{3j} = K_j;\ j = 1, 2, 3 \tag{2}$$

$$K_3 - K_2 = K_2 - K_1 \tag{3}$$

In the case of fourth row, all the elements of that row are taken sequentially and ıerged. If $b'_{41}b'_{42}b'_{43}b'_{44}$ is a perfect number, then the condition of fourth row is erified. If the conditions of third and fourth rows are satisfied in all the stego blocks, ıen the stego image is considered as authentic image. If the conditions violate in ny block, then that particular block is considered as corrupted. After the initial erification step, second verification stage is carried out. The differences between ıe elements of first row and third row with corresponding column positions and the ifference between elements of second and fourth rows with corresponding column ositions are evaluated. If all the differences in a particular block are same, then the tego image is considered as fully verified stego image. This particular difference alue is considered as the decimal value of the message segment. This particular lecimal value is converted to corresponding binary representation. All the binary epresentations from all blocks are merged sequentially and that is taken as secret ıessage.

2.3 *Sender and Receiver Side Algorithms*

The data hiding methodology at the sender side is given below:

Algorithm 1:*Sender Side*
Input:ACover Image *(C)*,
Secret message $(a_1 a_2 \ldots\ldots a_n)$
Output: Stego-image (*S*)

Procedure:
Step 1: Split the cover image (C) into 4×4 entropy pixel blocks (b).
Step 2: Break the secret message $(a_1 a_2 \ldots\ldots a_n)$ into 4 bits sequential segmentations.
Step 3: Take the first message segment $(a_1 a_2 a_3 a_4)$ and the first image pixel block.
Step 4: Replace the 1st and 3rd row of the pixel block (b) with four consecutive triangular numbers i.e. replace $b_{11}, b_{12}, b_{13}, b_{14}$ with four consecutive triangular numbers and$b_{31}, b_{32}, b_{33}, b_{34}$ with same set of triangular numbers.
Step 5: Replace the 2nd and 4th row of the pixel block (b) with the numbers 33,55,03,36 respectively, i.e. replace $b_{21}, b_{22}, b_{23}, b_{24}$ with the numbers 33,55,03,36 respectively and replace $b_{41}, b_{42}, b_{43}, b_{44}$ with same set of numbers.
Step 6: Convert the message segment to corresponding decimal number.
Step 7:Add the decimal number to all the elements of 1st and 2nd rows of the block.
Step 8: Go to next block and next message segment.
If (Next block=NULL)
goto step 9
Else
goto step 4
Step 9: Join the pixel blocks and take the image as stego image.
Step 10: Stop

Algorithm 2: *Receiver Side*
Input: Stego-image (*S*)
Output: Secret message $(a_1 a_2 \ldots\ldots a_n)$

Procedure:
Step 1:Split the stego image (S) into 4×4 entropy pixel blocks (b').
Step 2:Take the first pixel block.
Step 3: Take the elements of 3rd row, i.e. $b'_{31}, b'_{32}, b'_{33}, b'_{34}$ sequentially.
If (conditions (2) and (3) satisfy) then
go to step 4
Else
go to step 9
Step 4: Take the elements of 4th row, i.e. $b'_{41}, b'_{42}, b'_{43}, b'_{44}$ sequentially and merge them.
If ($b'_{41} b'_{42} b'_{43} b'_{44}$ is a perfect number) then
go to step 5
Else
go to step 9
Step 5:for (i=1 to 4)
Evaluate $(b'_{1i} - b'_{3i})$ and $(b'_{2i} - b'_{4i})$
Step 6: If $((b'_{1i} - b'_{3i}) = (b'_{2i} - b'_{4i})\ \forall i)$ then
Take $b'_{11} - b'_{31}$ as message segment and go to step 7.
Else
go to step 9.
Step 7: Go to next block.
If (Next block=NULL)
go to step 8
Else
go to step 3.
Step 8: Convert each message segment to corresponding binary form and merge them sequentially. Take this as message and go to step 10.
Step 9: Print the message to receiver side "The stego image is corrupted during communication".
Step 10: Stop

3 Experiments and Results

The proposed data hiding technique is examined through different JPEG grayscale images [9]. The size of each of the image is 512 × 512. Most common testing parameters are MSE (Mean Squared Error) and PSNR (Peak Signal-to-Noise Ratio). MSE basically determines the distortion between the cover image and stego image. Another parameter PSNR is inversely proportional to the parameter MSE. So, high PSNR and low MSE show the high quality of a data hiding scheme in steganography. Therefore, to show the efficacy of the scheme, it is enough to show the PSNR values only for different images. The working formulae of PSNR and MSE are defined in Eqs. (4) and (5), respectively.

$$PSNR = 20\log_{10}(MAX) - 10\log_{10} MSE \tag{4}$$

$$MSE = \frac{1}{MN}\sum_{i=1}^{M}\sum_{j=1}^{N}|C(i,j) - S(i,j)|^2 \tag{5}$$

g. 1 Cover image 1

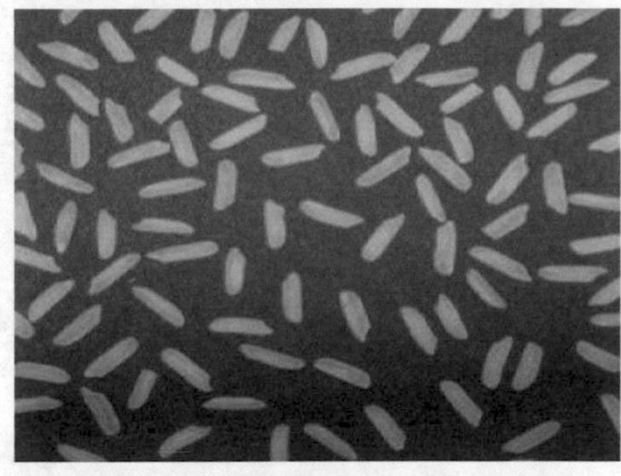

'ig. 2 Cover image 2

Fig. 3 Cover image 3

Here, C and S represent the cover image and stego image, respectively. Also, the symbol "MAX" is equal to $2^4 - 1 = 15$ in this case, since the cover image is split into 4×4 pixel blocks for data embedding.

Here, four images are taken into account for experimentation of the proposed scheme. The cover images and corresponding stego images using the proposed approach are given in Figs. 1, 2, 3 and 4 (cover images) and Figs. 5, 6, 7 and 8 (stego images), respectively.

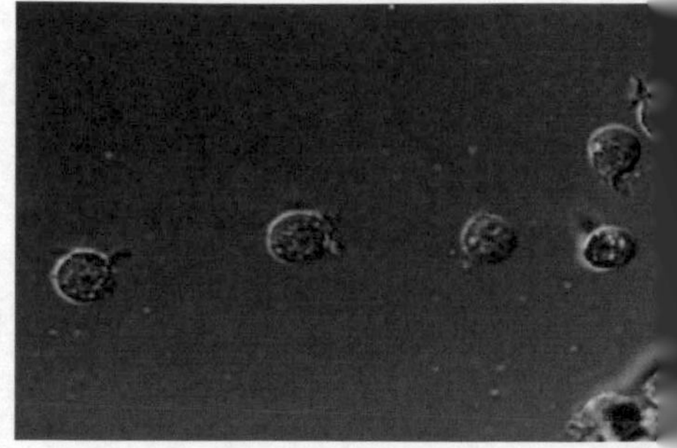

Fig. 4 Cover image 4

Fig. 5 Stego image 1

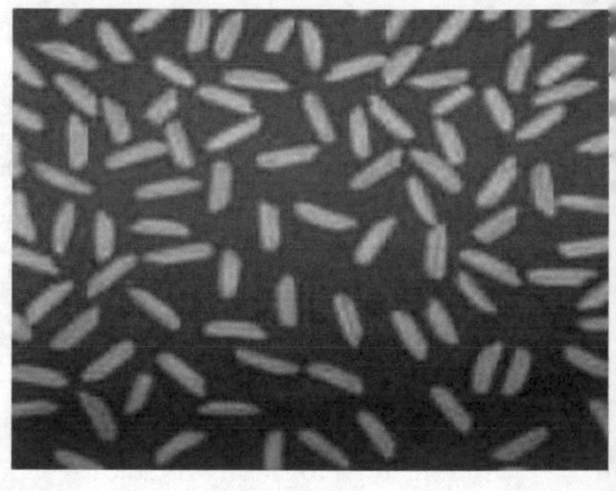

Fig. 6 Stego image 2

PSNR values of the proposed scheme compared with four popular data hiding techniques with four selected images are given in Table 1.

From Table 1, it is observed that PSNR and MSE provide better results over other four popular data hiding models. Now, freeness of the proposed scheme from some popular unintentional attacks is discussed.

PSNR: This is one of the most common unintentional attacks in steganography. This generator randomly selects the least significant bits of pixels of an image and combining the bits extract the secret information. In the proposed scheme, data is

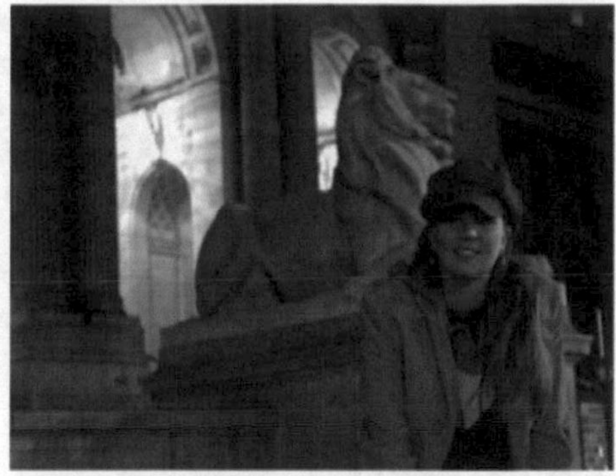

g. 7 Stego image 3

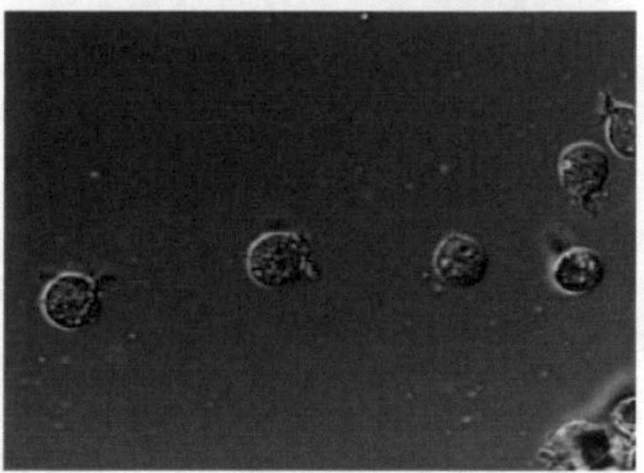

ig. 8 Stego image 4

Table 1 Comparison of PSNR values in different schemes

Images	Size	[8]	[1]	[5]	[11]	Proposed
Image 1	512 × 512	57.82	44.38	43.92	46.32	60.37
Image 2	512 × 512	53.21	40.12	39.44	41.68	57.54
Image 3	512 × 512	59.77	45.96	45.37	48.55	62.98
Image 4	512 × 512	51.32	38.44	37.21	40.63	55.23

not embedded in the LSB positions of the pixel values. So, the scheme is free from most common steganography attack PRNG.

Chi-square attack: This is one of the most powerful attacks in both spatial domain and frequency domain steganography techniques. A particular distribution function is used here to extract the secret information. But this is applicable only in colored images [10]. Actually, intensities of RGB pixel values are used to extract the data from an image. But in the proposed approach, only entropy pixels are used. As a result, it becomes free from chi-square attack.

Histogram attack: This is a powerful attack in steganography. Actually, from the histogram of the stego image, it is tried to extract the message. This is also applicable in LSB-based techniques. From that view, the proposed technique is free from histogram attack.

Visual attack: This is a very common attack in steganography-based communication. The occurrence of the message in an image is realized using this attack by unauthorized user. The loss of originality of an image gives the indication of the existence of secret message. But PSNR is very high in the proposed data hiding scheme. So, it is much secure from visual attack.

Ultimately from different analyses, it is observed that the proposed data hiding scheme is safe and secured from the different unintentional attacks. So, data cannot be extracted easily from the proposed model.

4 Conclusion

A data hiding model in spatial domain image steganography is proposed in this paper. The model of integer theory enhances the protection during both internet and intranet communication. More particularly, the properties of triangular number and perfect number increase all conditions of information security. The uniqueness of this special approach can be measured from two different views. These are data hiding and digital authentication. In the cease of data hiding, the proposed method is unique due to keyless transposition and less complexity of implementation. From another view, digital authentication is provided with double layered checking. Also the freeness of the scheme from different powerful unintentional attacks shows the efficacy with the parameters PSNR. Here, the occurrence of noise effect in stego image can be realized at the receiver end. But the drawback of this technique is that the data cannot be recovered at the receiver side. This particular task can be considered as future task of the proposed work.

References

1. Wedaj, F.T., Kim, S., Kim, H.J., et al.: Improved reversible data hiding in JPEG images based on new coefficient selection strategy. EURASIP J. Image Video Process. **63**, 1–11 (2017)
2. Parah, S.A., Ahad, F., Sheikh, J.A., Bhat, G.M., et al.: Hiding clinical information in medical images: a new high capacity and reversible data hiding technique. J. Biomed. Inform. **66**, 214–230 (2017)
3. Pachghare, V.K.: Cryptography and Information Security. PHI Learning Private Limited, New Delhi (2011)
4. Hussain, M., Wahab, A.W.A., Ho, A.T.S., Javed, N., Jung, K.H., et al.: A data hiding scheme using parity bit pixel value differencing and improved rightmost digit replacement. Sig. Process. Image Commun. **50**, 44–57 (2017)
5. Lin, Y.K.: A data hiding scheme based upon DCT coefficient modification. Comput. Stand. Interfaces **36**, 855–862 (2014)
6. Hussain, M., Wahab, A.W.A., Idris, Y.I.B., Ho, A.T.S., Jung, K.H., et al.: Image steganography in spatial domain: a survey. Sig. Process. Image Commun. **65**, 46–66 (2018)
7. Rashid, A., Rahim, M.K.: Experimental review of "grey level modification" steganography. Int. J. Signal Process. Image Process. Pattern Recognit. **8**(11), 265–272 (2015)

8. Muhammad, K., Ahmad, J., Farman, H., Jan, Z., Sajjad, M., Baik, S.W., et al.: A secure method for color image steganography using gray-level modification and multilevel encryption. KSII Trans. Internet Inform. Syst. (TIIS) **9**, 1938–1962 (2015)
9. Standard test images: http://www.imageprocessingplace.com/root_files_V3/image_databases.htm
10. Westfeld, A., Pfitzmann, A.: Attacks on steganographic systems breaking the steganographic utilities EzStego, Jsteg, Steganos, and S-Tools—and some lessons learned. In: 3rd International Workshop on Information Hiding (2000)
11. Al-Taani, A.T.,AL-Issa, A.M.: A novel steganographic method for gray-levelimages. World Acad. Sci. Eng. Technol. **27**, 613–618 (2009)

mplementation of WOA-Based DOF-FOPID Controller for AGC f Interconnected Power System

ebashis Sitikantha, Binod Kumar Sahu and Pradeep Kumar Mohanty

bstract This write-up presents a controller equipped with a two-degree-of-feedom ractional-order PID (2DOF-FOPID) controller for automatic generation control AGC) of a two-area, four-unit interconnected power system. Controllers employed or AGC must be designed to satisfy the required objectives to get an achievable et of solutions. Design criteria for AGC of the multi-area power system are basi-ally regarded as an optimization problem considering various constraints. Gains of he proposed controller are optimally designed using whale optimization algorithm WOA) taking integral time-multiplied absolute error (ITAE) as the objective func-ion. The superiority and productiveness of proposed 2DOF-FOPID controller are lemonstrated and compared with FOPID and conventional PID controller employing ime domain simulation analysis and statistical analysis. This justifies the potency of he proposed controller for two-area system.

Keywords Automatic generation control (AGC) · Two-degree-of-freedom fractional-order PID controller · Fractional-order PID (FOPID) controller · PID controller · Whale optimization algorithm (WOA)

1 Introduction

The vital role of automatic generation control (AGC) is to match the power generation by generating units with load demand. Any mismatch in generated and demanded power yields variation in the frequency of system from its nominal value [1]. In

D. Sitikantha · B. K. Sahu (✉) · P. K. Mohanty
Institute of Technical Education & Research, Siksha 'O' Anusandhan University, 751030 Bhubaneswar, Odisha, India
e-mail: binoditer@gmail.com

D. Sitikantha
e-mail: debashissitikantha@soa.ac.in

P. K. Mohanty
e-mail: pradipmohanty@soa.ac.in

A. K. Das et al. (eds.), *Computational Intelligence in Pattern Recognition*, Advances in Intelligent Systems and Computing 999,
https://doi.org/10.1007/978-981-13-9042-5_78

electrical power system, every area has its own generating unit/units which is/a accountable for supplying its own load demand and planned interchanges of pow among the neighboring areas. To exchange the contract energy between differe areas and to stipulate inter-area support during abnormal condition, tie lines a used. Any type of unusual conditions and change in loading conditions in eith area or areas results in mismatches in contract power interchanges and frequencie The intense deviation in frequencies may cause partial or entire collapse to the sys tem. Load frequency control (LFC) is put into practice to monitor the output of th generators within a predefined area properly [2]. LFC is taken care regarding th mismatching in system frequency and exchanges of scheduled power. LFC has bee used as part AGC for several years in interconnected power systems [3].

To maintain the desired flow of tie-line power and system frequency implemen tation of control scheme is highly essential. It also yields zero steady-state error an unplanned interchange in power flow. To monitor the big scale power systems, th control algorithm to be used should be capable of dealing with both mechanical a well as electrical nonlinear dynamics. The control algorithm should be operated i every uncertain situation those are instigated by regular variation in load demands The most common employed controller for AGC is conventional PID controller. PID controller is extremely simpler in structure, easily realizable, with low cost, robus in behavior, simple for implementation, and provides relatively better dynamica response.

However, if system complexity increases, its performance decreases. Due to the presence of nonlinear behavior in system and large interconnecting areas, the performance of power system reduces by employing conventional fixed gain PI and PID controllers, even though they ensure zero steady-state error. Fixed gain controllers designed at standard environment also fails to yields better performance over a substantial range of working conditions. Therefore, in [4–7], few classical adaptive controls are stated for LFC.

Despite of getting challenging result by the above controllers, the control algorithms are mostly complex and therefore require a few online model identifications. The widespread models free controllers are based on fuzzy neural networks (FNNs), artificial neural networks (ANNs), fuzzy logic system (FLSs) those are described in [8–12]. The salient feature in fuzzy logic technique is that it provides a model-free depiction in control systems don't require model identification. Controllers designed on the basis of ANN technique again increases the system's performance once it is perfectly trained by employing a worthy quantity of neural network and plant identifications. The FNN considers the benefits of both FLS by handling unpredictable information and ANN is learning the process. A new control strategy cascading fractional-order PI and fractional-order PD (FOPI-FOPD) optimized by SCA is proposed by Tasnin et al. for AGC of interconnected multi-area power system incorporating DSTs and GTPPs [13].

In this article, an FOPID controller is designed using WOA to tackle the AGC issues in a two-area, four-unit interconnected power system. The main contribution of the article are as folows:

- Development of a two-area, four-unit hydrothermal power system in MATLAB SIMULINK environment.
- Implementation of WOA optimization technique in MATLAB.
- Optimal design of 2DOF-PID, FOPID, and conventional PID controller using WOA technique for the proposed power system.
- Validation of robustness of the proposed 2DOF-FOPID controller against random load variations.

2 System Under Investigation

The basic block diagram model of a two-area, four-unit power system comprising of non-reheat thermal units and hydro units is presented in Fig. 1. The power system is furnished with a multivariable and complex structure, which is shown by various blocks. Normally, different structures are equipped with non-minimum phase systems and/or nonlinear time variant systems. Electrical interconnected power system comprise of numerous control areas and are connected by means of tie lines. Generators of one area form a coherent group. Frequency and tie-line power in each area need to be regulated. Figure 1 shows ACE_1 and ACE_2, which represents area control errors, R_1 and R_2 represent governor speed regulations constants, frequency bias factors are B_1 and B_2, time constants of governors T_{g1} and T_{g2}, variation in valve position of governor are ΔP_{g1} and ΔP_{g2}, U_1 and U_2 represent controllers' input to speed governing systems, change in output power of turbine in pu are ΔP_{t1} and ΔP_{t2}, turbine time constants in sec are T_{t1} and T_{t2}, gains of power systems are K_{p1} and K_{p2}, power system time constants in sec are T_{p1} and T_{p2}, ΔP_{d1} and ΔP_{d2} represent change in load demand, system frequency deviations in two areas are Δf_1 and Δf_2 expressed in hertz, Δp_{tie} shows deviation in tie-line power in pu, and T_{12} represents synchronization constants.

Each area consists of its own governing systems, turbines, and generators as presented in Fig. 1. There are three inputs for each area in the interconnected system, i.e., U_1 and U_2 which are the inputs for the speed governing system deviation in demand ΔP_{D1} and ΔP_{D2} and the change in tie-line power ΔP_{tie}. For each area, two outputs are variation in system frequency Δf_1 and Δf_2 and ACE_1 and ACE_2 which are area control errors.

3 Proposed Optimal Controllers

3.1 Conventional PID Controller

PID controller is normally employed to get stability and simultaneously true system output. PID controller comprises of three modes of actions such as P (proportional)

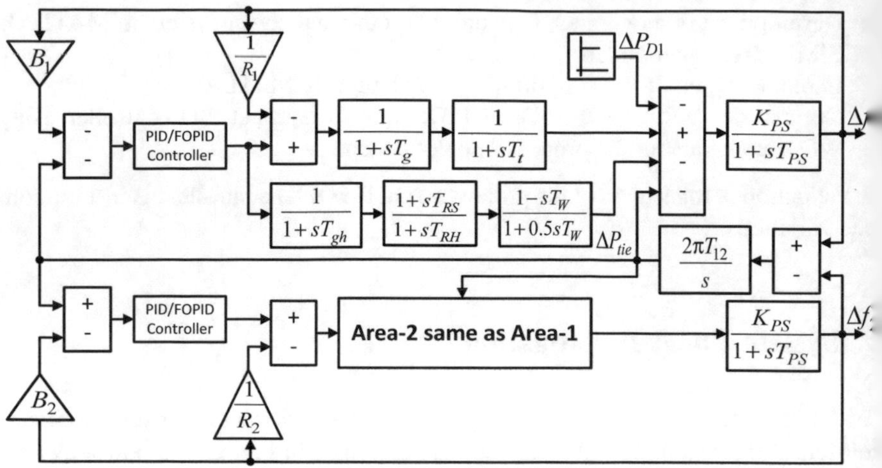

Fig. 1 Transfer function model of the two-area thermal system considered for the study

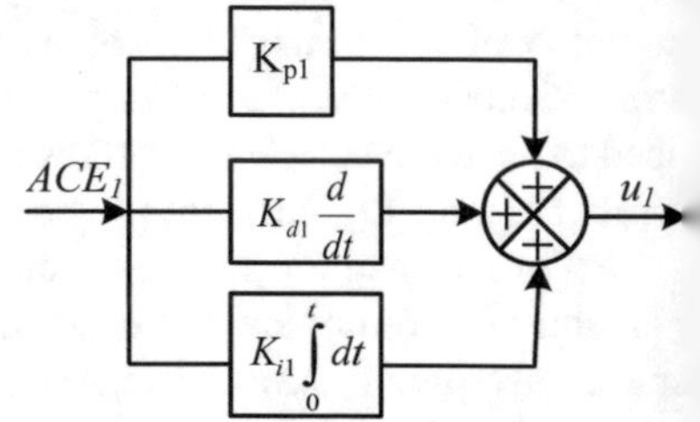

Fig. 2 Structure of PID controller

mode, I (integral) mode, and D (derivative) mode. Difference in reference and actual output is the input to PID controller. Structure of a conventional power system is shown in Fig. 2. Taking appropriate control action, it provides a proper response, which helps to reduce the error in the process output. In time domain, the output expression of conventional PID controller implemented in area-1 is expressed as

$$u_1(t) = K_{p1}e_1(t) + K_{i1}\int\limits_0^t e_1(t)dt + K_{d1}\frac{de_1(t)}{dt} \tag{1}$$

where K_{p1}, K_{i1}, and K_{d1} are the controller's gains of area-1.

3.2 Fractional-Order PID Controller

Massive growth in load demand and the demand for good quality uninterrupted power supply has made the power system complex. Conventional PID controller may not

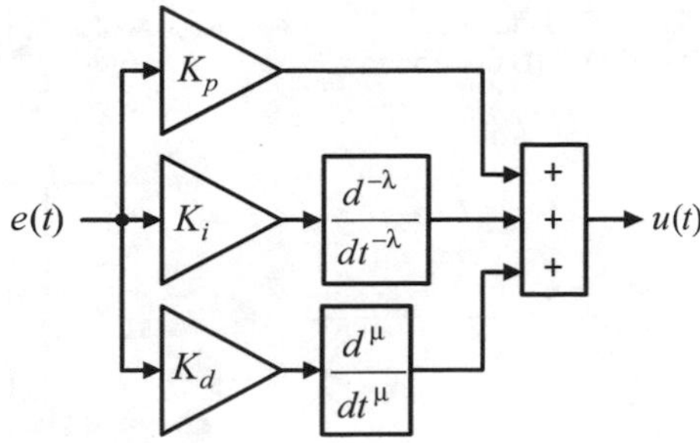

…g. 3 Structure of FOPID …ntroller

…xhibit satisfied performance for the complex power system. Therefore, in this article, …n optimally designed FOPID controller is introduced to tackle the AGC issues with …mproved power quality. In addition to PID controller gains, this controller has two …nore parameters like fractional integral order (λ) and fractional derivative order (μ). …he structure of FOPID controller is shown in Fig. 3. In time domain, the output of …OPID controller of area-1 can be expressed as

$$u_1(t) = K_{p1} \cdot e_1(t) + K_{i1}\frac{d^{-\lambda}}{dt^{-\lambda}}e_1(t) + K_{d1}\frac{d^{\mu}}{dt^{\mu}}e_1(t) \tag{2}$$

In Laplace domain, the transfer function of FOPID controller of area-1 is written …as

$$H(s) = \frac{U(s)}{E(s)} = K_{p1} + \frac{K_{i1}}{s^{\lambda}} + K_{d1}s^{\mu} \tag{3}$$

3.3 Two-Degree-of-Freedom Fractional-Order PID (2DOF-FOPID) Controller

Degree of freedom in control system is the number of closed-loop transfer functions those can be tuned autonomously. A two-degree-of-freedom controller provides an output depending upon the difference between the system output and the reference signal. It computes the biased difference signal for PID controller, depending upon the proportional and integral set point weights. The structure of a 2DOF-FOPID controller is shown in Fig. 4. It comprises of setpoint weights (PW and DW), controllers gains (K_p, K_i, and K_d), and fractional integral and derivative orders (λ and μ). In this article, all these parameters are optimally tuned employing WOA optimization technique.

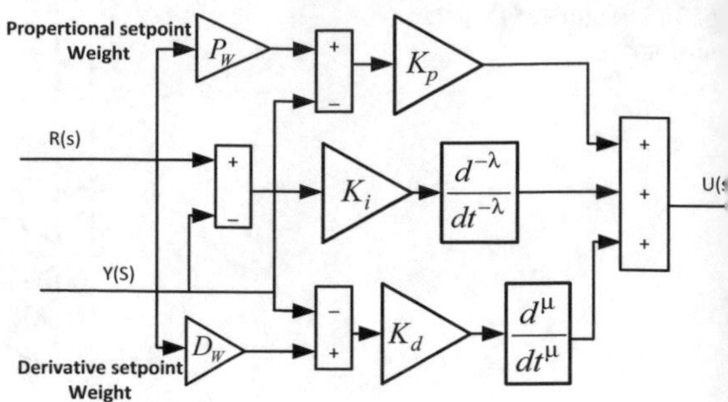

Fig. 4 Structure of 2DOF-FOPID controller

3.4 Whale Optimization Algorithm

Mirjalili and Lewis [14] developed WOA in 2016. This optimization technique resembles the bubble-net hunting approach of humpback whales. The numerous phases o WOA algorithm are as follows:

i. Initialization: This phase generates an initial random population of size [NP × D], where "NP" is number of population and "D" is dimension of the population After accessing the initial random population, the best one is considered as global best search agent.
ii. Encircling prey: This step deals with updation of the initial population toward the leading hunt agent. Mathematical expressions for updating the positions are given by

$$X_{new} = X_{best} - M \times N \tag{4}$$

$$N = |T \times X_{best} - X| \tag{5}$$

where "M" and "T" are coefficient vectors expressed as

$$M = 2pr - p \tag{6}$$

$$T = 2r \tag{7}$$

where "p" decreases linearly from "2" to "0" linearly during iteration process and "r" is a randomly generated number within [0, 1].
iii. Bubble-net attacking strategy: The following approaches given below demonstrates the mathematical modeling of bubble-net behavior of humpback whales

   a. Shrinking encircling mechanism: This strategy is visualized by properly updating the initial randomly generated population with the help of Eqs. (4)–(7).

b. Spiral updating position: This phase updates the population by mimicking helical shaped movement of humpback whales which is mathematically modeled as

$$X_{new} = N' e^{cl} \cos(2\pi l) + X_{best} \tag{8}$$

where $N' = |X_{best} - X|$ constant "c" is taken to express the configuration of spiral movement and "l" is a randomly generated within [0–1].
Humpback whales swim on around the pray taking both spiral and circular movements. This characteristic is precisely modeled by considering 50% possibility for both spiral as well as circular movements and expressed by

$$X_{new} = \begin{cases} X_{best} - M \cdot N & if\ q < 0.5 \\ N' \cdot e^{cl} \cdot \cos(2\pi l) + X_{best} & if\ q \geq 0.5 \end{cases} \tag{9}$$

v. Search for prey: In this step, updation of position is done by randomly selecting hunt agent instead of taking best agent. Mathematical expression of this stage is given by

$$X_{new} = X_{rand} - M \cdot N \tag{10}$$

where

$$N = |T \cdot X_{rand} - X| \tag{11}$$

And $T = 2r$ is given in Eq. (7).

Optimal gains for the proposed controller are determined using WOA algorithm are depicted in Table 1.

4 Simulation Results and Analysis

4.1 Transient Performance Analysis

The simulation is processed in MATLAB SIMULINK environment. A two-area, four-unit interconnected power system is taken for AGC study in which each area has a thermal and a hydro unit. Nominal values of various parameters of the considered power system are indicated in the Appendix. Gain parameters of various controllers are taken in the range [0–5] and fractional order of integration and differentiation is considered in the range of [0–1]. WOA technique is implemented to design the gains of conventional PID, FOPID, and 2DOF-FOPID controller taking ITAE as the objective function. Population size and number of maximum iteration are chosen

Table 1 Optimally designed controllers' gains

Gainrollers									
PID controller									
Area-1					Area-2				
K_{p1}	K_{i1}			K_{d1}	K_{p2}	K_{i2}			K_{d2}
5.0000	4.9967			1.0456	0.0100	3.6984			3.54
FOPID controller									
Area-1					Area-2				
K_{p1}	K_{i1}	K_{d1}	λ_1	μ_1	K_{p2}	K_{i2}	K_{d2}	λ_2	μ_2
0.8407	5.0000	4.3587	0.8644	0.7106	1.7409	0.0100	4.8557	0.5524	0.942
2DOF-FOPID controller									
Area-1					Area-2				
K_{p1}	K_{i1}	K_{d1}	λ_1	μ_1	K_{p2}	K_{i2}	K_{d2}	λ_2	μ_2
1.8536	5.0000	1.4911	0.8004	0.9742	1.3215	3.6021	3.8474	0.1100	1.000
Set point weights					Set point weights				
P_{W1}		D_{W1}			P_{W2}		D_{W2}		
3.5912		4.5897			4.4622		4.1097		

Table 2 Transient performance indicator

Frequency/tie-line power	Transient parameters	Controllers		
		2DOF-FOPID	FOPID	PID
Δf_1	$U_{sh} \times 10^{-3}$ in Hz	−23.1869	−54.4452	−66.3760
	$O_{sh} \times 10^{-3}$ in Hz	1.5285	1.9903	3.7752
Δf_2	$U_{sh} \times 10^{-3}$ in Hz	−3.9465	−15.7466	−20.2520
	$O_{sh} \times 10^{-3}$ in Hz	0.2244	1.2634	0.4579
ΔP_{tie}	$U_{sh} \times 10^{-3}$ in Hz	−2.9350	−6.4507	−8.0938
	$O_{sh} \times 10^{-3}$ in Hz	0.1639	0.4568	0.2218

as 100. Optimized gains of the PID, FOPID, and 2DOF-FOPID controllers and the respective ITAE values are mentioned in Table 1. Step load perturbation of 10% is introduced in area-1 to review the dynamic characteristic of the proposed power system. The result obtained is compared between WOA-based conventional PID, FOPID, and 2DOF-FOPID-controlled AGC system. It is a conventional fact that whenever there is a load variation in one area or both the areas of an interconnected electrical power system, there is always some variation in the frequencies of both the areas and tie-line power. Figures 5, 6, and 7 show the deviation in frequency (Δf_1) of area-1, (Δf_2) of area-2, and tie-line power (ΔP_{tie}), respectively. For PID, FOPID, and 2DOF-FOPID-controlled AGC system, undershoot (U_{sh}), overshoot (O_{sh}), and settling time (T_s) for frequency and tie-line power deviation are depicted in Table 2.

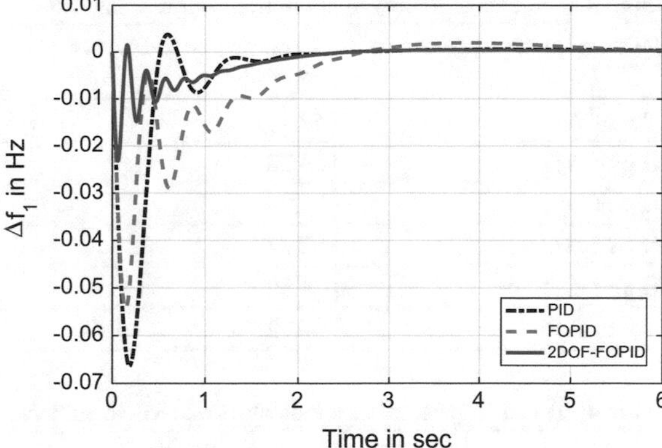

Fig. 5 Change in frequency area-1

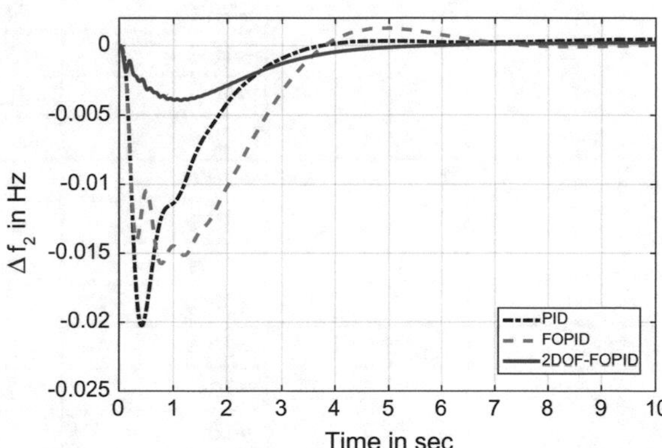

Fig. 6 Change in frequency of area-2

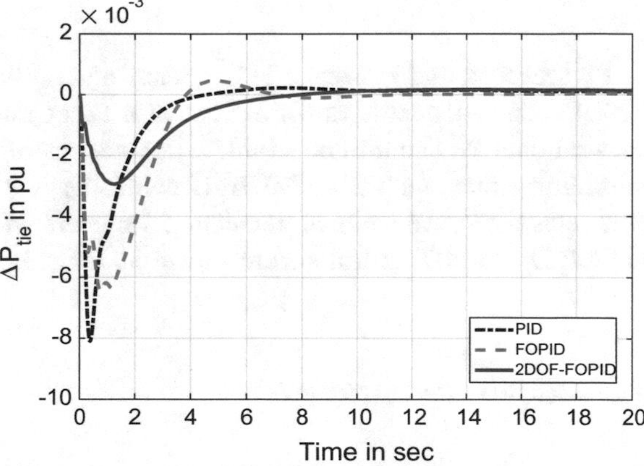

Fig. 7 Change in tie-line power

Table 3 Percentage improvement in transient parameters

Percentage improvement in	As compared to FOPID controller	As compared to PID controll
U_{sh} of Δf_1	57.41	65.07
O_{sh} of Δf_1	23.20	59.51
U_{sh} of Δf_2	74.94	80.51
O_{sh} of Δf_2	82.24	50.99
U_{sh} of ΔP_{tie}	54.50	63.74
O_{sh} of ΔP_{tie}	64.12	26.11

Table 4 Transient performance indicators due to system's parametric variation

	Parameter	% Change	$U_{sh} \times 10^{-3}$ in Hz	$O_{sh} \times 10^{-3}$ in H
Δf_1	R	−20%	−24.8247	0.4520
		−10%	−23.9766	0.9764
		+10%	−22.4802	2.0149
		+20%	−21.8378	2.4769
Δf_2	B	−20%	−3.9048	0.2230
		−10%	−3.9279	0.2240
		+10%	−3.9618	0.2254
		+20%	−3.9746	0.2256
ΔP_{tie}	T_{12}	−20%	−2.7203	0.1589
		−10%	−2.8367	0.1618
		+10%	−3.0218	0.1657
		+20%	−3.1003	0.1671

Figures 5, 6, and 7 reveal that frequency and tie-line power changes with 2DOF-FOPID controller settle down to zero at a faster rate as compared to FOPID and conventional PID controllers. Table 2 depicts that all the transient parameters have minimum values with 2DOF-FOPID controller as compared to other controllers. Percentage improvements in transient parameters with 2DOF-FOPID as compared to FOPID and PID controllers are shown in Table 3.

4.2 Robustness Analysis

Robustness of the proposed PDPID controller with DDF is studied by varying some of the vital systems parameters from −20 to +20% in steps of 10%. Table 4 shows transient parameters of the AGC system with various systems parameters.

From Table 4, it is evident that the suggested 2DOF-FOPID controller is quite bust against parametric variations because all the transient parameters of vary ithin a small range.

Conclusion

n this paper, the role of AGC in a two-area interconnected power system, with each rea having a thermal and a hydro unit is addressed. Each area is employed with DOF-PID, FOPID, and PID controller to enhance the dynamic characteristic of he AGC system. Using WOA technique, the gains of controllers are evaluated. The nost popular objective function ITAE is chosen to optimize the gain parameters of ontrollers. The superiority of 2DOF-FOPID controller is analyzed under a 10% step oad variation in area-1 over other controllers. Time domain analysis is carried out or deviation of frequencies in two areas and tie-line power to claim the superiority of 2DOF-FOPID controller over the PID controller. Finally, robustness analysis is carried out to validate the efficacy of 2DOF-FOPID controller in case of systems' parametric deviation.

References

1. Kundur P.: Power System Stability and Control, 8th reprint. Tata McGraw Hill, New Delhi (2009)
2. Elgerd, O.I.: Electric Energy Systems Theory – An Introduction, 2nd edn. Tata McGraw Hill, New Delhi (2000)
3. Cohn, N.: Some aspects of tie-line bias control on interconnected power systems. Am. Inst. Elect. Eng. Trans. **75**, 1415–1436 (1957)
4. Concordia, C., Kirchmayer, L.K.: Tie line power and frequency control of electric power systems. AIEEE pt. II **72**, 562–572 (1953)
5. Kirchmayer, L.K.: Economic Control of Interconnected Systems. Wiley, New York (1959)
6. Quazza, G.: Non-interacting controls of interconnected electric power systems. IEEE Trans Power Appar. Syst. **85**(7), 727–741 (1996)
7. Aggarwal, R.P., Bergseth, F.R.: Large signal dynamics of load-frequency control systems and their optimization using nonlinear programming: I & II. IEEE Trans Power Appar. Syst. **87**(2), 527–538 (1968)
8. Demiroren, A., Sengor, N.S., Zeynelgil, H.L.: Automatic generation control by using ANN technique. Electr. Power Compon. Syst. **29**(10), 883–896 (2001)
9. Shayeghi, H., Shayanfar, H.A.: Application of ANN technique based on μ-synthesis to load frequency control of interconnected power system. Int. J. Electr. Power Energy Syst. **28**(7), 503–511 (2006)
10. Franoise, B., Magid, Y., Bernard, W.: Application of neural networks to load-frequency control in power systems. Neural Netw. **7**(1), 183–194 (1994)
11. Chaturvedi, D.K., Satsangi, P.S., Kalra, P.K.: Load frequency control: a generalized neural network approach. Int. J. Electr. Power Energy Syst. **21**(6), 405–415 (1999)
12. Sahu, B.K., Pati, S., Mohanty, P.K., Panda, S.: Teaching-learning based optimization algorithm based fuzzy-PID controller for automatic generation control of multi area power system. Appl. Soft Comput. **27**, 240–249 (2015)

13. Tasnin, W., Saikia, L.C.: Maiden application of an sine–cosine algorithm optimised FO casca controller in automatic generation control of multi-area thermal system incorporating dis Stirling solar and geothermal power plants. IET Renew. Power Gener. **12**(5), 585–597 (201
14. Mirjalili, S., Lewis, A.: The whale optimization algorithm. Adv. Eng. Softw. **95**, 51–67 (201

Decision-Making Process of Farmers: A Conceptual Framework

Partha Mukhopadhyay, Madhabendra Sinha and Partha Pratim Sengupta

Abstract With the development trend of agricultural commercialization and due to adverse events, the magnitude of loss is increasing and the farmers committing suicides. This study attempts to underscore the components, which impact the marginal farmers' decision to accept Pradhan Mantri Fasal Bima Yojana (PMFBY). A pilot study reveals that only a few of the determinants is affected most in Burdwan District of West Bengal India. A conceptual model approach links the decision-making with a dynamic and in the light of real-world phenomena so that key factors and drives of decision-making can be related to the point of making better information about future economic and social reports.

Keywords Marginal farmers · Adoption decisions · Crop insurance · Self help groups · Risk attitude · Decision-making · India · Pradhan Mantri Fasal Bima Yojana

1 Introduction

Knowledge of social sciences can help us to improve our societies. With the present and the past, the study of social science allows the human society to take care of future itself. As a science, sociology has two parts, pure (theoretical) and applied (Practical). By searching the phenomena and laws of the society, pure social science

---

P. Mukhopadhyay (✉) · M. Sinha · P. P. Sengupta
Department of Humanities and Social Sciences, Natioanl Institute of Technology, Durgapur, West Bengal 713209, India
e-mail: pmukherjee400@gmail.com

M. Sinha
e-mail: madhabendras@gmail.com

P. P. Sengupta
e-mail: parthapratim.sengupta@hu.nitdgp.ac.in

A. K. Das et al. (eds.), *Computational Intelligence in Pattern Recognition*, Advances in Intelligent Systems and Computing 999,
https://doi.org/10.1007/978-981-13-9042-5_79

established the principles of sciences and also tried to find out which social ph nomena takes place, antecedent conditions, etc., by which the phenomena have be brought into existence as far as the existing social status of human knowledge b psychological, biological and cosmic causes, etc. and applied part points out the actual or possible applications. In our cosmic world, human psychology is a wonde With time, it changes, and is impacted by demography, culture, sex, money, recogn tion and external components, etc. [29]. That psychology is expressed as behaviour factors of any individual. The economic decision plays a standout among the mo fundamental parts throughout our life. Today, researchers explore human psycholog combined with various subjects. Behavioural finance is such a combination of Soc ology, Psychology and Finance. Practical application of behavioural finance implie that decision-making depends upon the behavioural factors of any individual. Sim ilarly, Behavioural Economics is a combination of sociology, psychology and eco nomics. This is a study of impacts of psychological (mass psychology), cognitive social and emotional factors on the economic decision of individuals and institution and the consequences for market price, returns and the asset allocation. Behavioura economics explores why sometimes individuals make irrational decisions and fur thermore, why and how their behaviours does not follow the predictions of economi models. Nobel laureates Herbert Simon (bounded rationality; 1978), Gary Becke (consumer mistakes, motives; 1992), Daniel Kahneman (anchoring bias, illusion o validity; 2002) and George Akerlof (procrastination; 2001) studied in the field o behavioural economics. Decision-making process is influenced by several factors These factors are past experience, belief in personal relevance [2] age and individ- ual differences [33], cognitive biases [43] and also influence what choices people make. Income, education and social class provide also influenced on the quality of our financial decision-making. Understanding the issues that impact the decision-making process is essential to understanding what decisions are made. Often, we have made some mistakes in financial decision-making. We may avoid such mis-takes when decision-making have sufficient buffers. These buffers are describing below as common analysis.

2 Common Analysis of Decision-Making Process

See Table 1.

Table 1 Strategies and criteria for decision-making based on behavioural factors

Sl. No.	Category	Strategy	Criterion
1	Heuristic	Anchoring	Based on bias and their justification
		Representative	Based on common rules of thumb
		Availability	Based on limited information or local source
2	Prospect	Loss aversion	Based on minimum loss
		Regret	Based on minimax of regret
		Mental Accounting	Based on maximizing gain of usability with functionality, reliability, quality, dependability, etc.
3	Market	Price change	Based on interactive action/automata
		Market info	Based on interactive action
		Overreaction of price change	Based on interactive action
		Preference of choice among alternatives in the market	Based on conflict
4	Herding	Buying and selling decision of other investors	Based on common sense and judgments
		Choice of stock to trade of other investors	Based on common sense and judgments
		Volume of stock to trade of other investors	Based on common sense and judgments
		Speed of Herding	Based on common sense and judgments

Source Baker and Nofisinger [8]

3 Decision-Making in General

Decision-making is very crucial for success to everyone. Throughout our life, we make small or big decisions to reduce the burden of the problem or other causes. There are different strategic decision-making theories for farmers and one of the ways to look at the method is using the conceptual model of decision-making process. This method is composed of four stages: (i) Problems identification, (ii) expanding definition, (iii) its analysis and preferences and (iv) implementation. Again, each of 'these phases' has four sub-stages: (i) Planning; (ii) searching and paying attention; (iii) evaluating and choice; (iv) bearing responsibility. Abreast decision-making mechanism is basically of two types [38]. One is analytical and the other is intuitive. A lot of factors act as an important role in the process. This study emphasized the effect of the spontaneous personality's consciousness.

As indicated by Myers, 'a Person's decision-making process depends to a si nificant degree on their cognitive style' [31]. Myers develops a set of four bipol dimensions popularly known as Myers–Briggs type indicator (MBTI). The actu focus on these dimensions are feeling and thinking, introversion and extroversio judgement and perception and intuition and sensing. She asserted that a person decision-making style well relates to how they score on these four measurement For instance, if a person who scored close to the thinking, extroversion, sensin and judgement ends of the dimensions would have a tendency to have an analytica logical, objective, critical and empirical decision-making [31].

The right decision has two definitions [9]:

(1) 'Right decisions are made in the right way and (2) Right decision yields th right results'. The first one denotes the process and process influencing the outcome Subsequently, the two definitions are co-dependent. Here, the key choices (decision are examined which yield results, for instance, quite a while after the choice i rolled out that the environmental change during that time affecting the outcomes. B following Beach (1993), we utilize the first definition [9] (Fig. 1).

The goal of this research is to examine whether a decision maker's knowledg of intuitive personality affects the degree of using analysis versus only intuitio in strategic decision-making. If such effects are available, then deeper research i required. If this is not found, the hypothesis of such a relationship is falsified and nc further research is required.

4 Decision-Making in Critical Condition

This study utilizes Behavioural Finance theory and draws its application to the sociology. Thinking process of any human being does not just work as a calculating machine. In decision-making, information is processing with emotional filters and various biases. These processes influence decision makers in an irrational manner or violate traditional concepts of risk aversion and make predictable errors in their forecasts. These are behavioural factors in which the behavioural scientists is categorized into Heuristic, Prospect, Herding and Market variable. In terms of decision-making, result of the future solely depends upon the decision taken of the present moment by the decision maker. But when the result is uncertain, then the decision maker could not use any mathematical formula. Throughout our life, we have to face a lot of uncertainty. Now, consider the agricultural side of a society. The farmer's invest time, labour and money. But their output is uncertain. When the output is worse to the farmer, sometimes he kills himself. Farmers' suicide is the extreme outburst of crisis in agriculture. Recently, Burdwan district become news headline as in April, 2015, total number of farmer suicides rose to **13** and in West Bengal to **17** [35]. The research would be quite successful in the direction of societal development. Policy makers, farmers and researchers would be getting a clear idea from this research.

Decision Making Structure

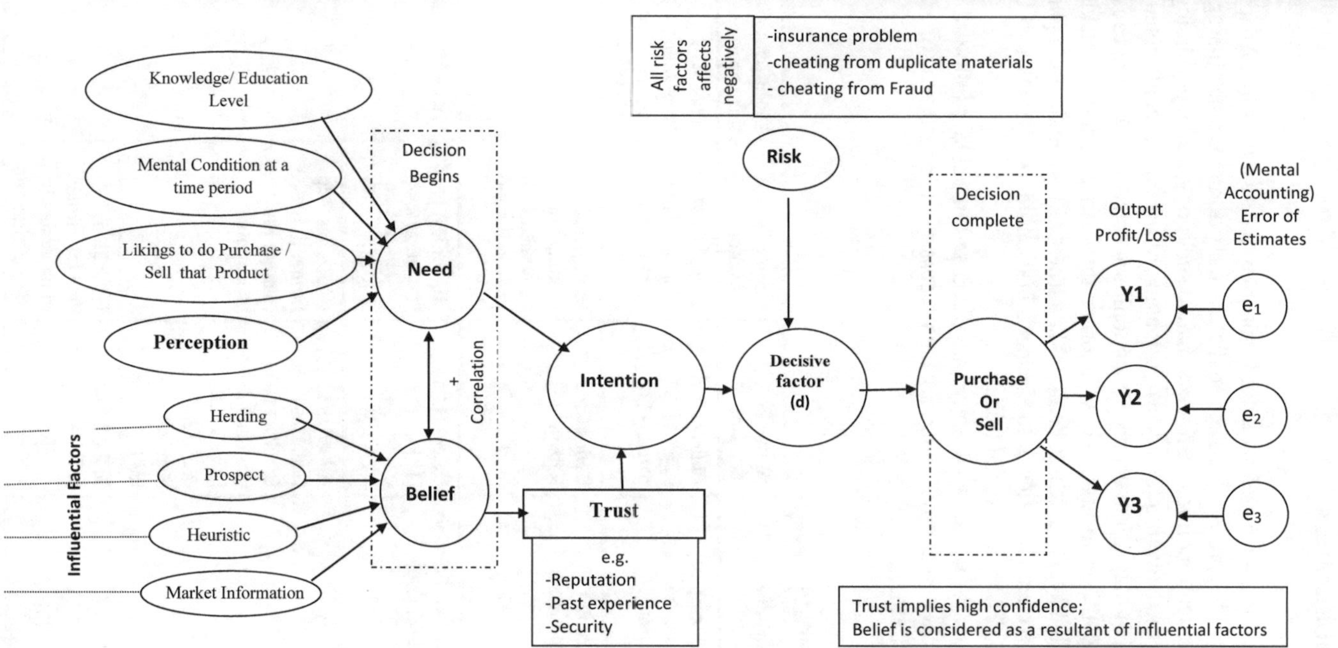

A decision begins together with his Need and belief of a person. Need felt by perception, Mental condition, Knowledge and Likings etc; Belief is influenced by behavioural variables like Herding, Prospect, Heuristic and Market Information. Continuous act of behavioural factors on belief, it turns into trust and Intention arises along with need. Then he thinks for risk. If risks are eliminated, decisive factor become positive and the person purchase or sell. After Purchase or sell there may some error. *(Either he may make more profit or loss). If the error of estimates is large, the person may frustrate.*

Fig. 1 Decision-making structure of a farmer. *Source* Author's own concept

5 Decision-Making Model for Accepting of Crop Insurance by the Indian Farmers

Farmers' decision-making processes are flexible and adaptive over the time. Adapt tion in the decision-making process has been studied in many domains other than agr cultural economics and agronomics. Various domains (sociology, social psycholog cultural research) on the behaviour and decision-making of the farmer contributed t identify the factors that may affect economic, agricultural science and social cause [10]. Jones [17] in an article abridges a few factors in the farmer's decision-makin process as follows: (1) Psychological makeup of the farmer, (2) Socio demographic of the farmer, (3) The characteristics of the farmer's household, (4) Structure of th farmer's business, (5) The wider social milieu.

Some "determinants of adaptive capacity" are discussed in Table 2 followed b a presentation of a conceptual model showing the process of the acceptance of cro insurance by Indian farmers.

Table 2 Determinants of adaptive capacity

Adaptive power rating	Indicators	Source of indicators	Assumption of indicators	Descriptions in this paper
Economic resources	Diversity of source of income	Defiesta and Rapera [12], Armah et al. [5]	There is high adaptability to more income	Income of household as well as individual
Awareness and training	Farming experience	Defiesta and Rapera [12]	More years of farming highly correlated with knowledge and skill	
	Level of literacy (Education)	Deressa et al. [13]	Higher levels of education have increased their potential	Education level
		Adeoti et al. [3], Enete and Igbokwe [19], Gani and Adeoti [23], Martey et al. [28], Randela et al. [37]	Education help farmer to better manage his production	
		Makura et al. [27]	Education enables farmer to understand market situation	

(continued)

…ble 2 (continued)

…daptive power …ating	Indicators	Source of indicators	Assumption of indicators	Descriptions in this paper
	Get Vocational Training	Asian Development Bank [ADB] [6]	Technical and vocational education is illustrated in career in agriculture	Vocational training
		Shankara et al. [40], KrishiVigyan Kendra, Konehally, Tumkur, Karnataka	Training is organized for the farmers to meet the urgent requirements of the peasant community	
Socioeconomic factor	Age	Schultz [39]	Positively correlate with age and adaptive capacity	Age of the farmer
		Abdulai and Delgado [1]	Age has a positive impact on labour supply	
		Potter and Lobley [34]	They have noted farmers have some correlation with age	
		Shem [41]	Individual's age is major factor relating to agricultural decision	
		Attanasio [7]	Personal age (AGE) may be negatively related to savings	
		Jyothi et al. [25], Punia et al. [36]	36–45-year-old women decide on the important matters of small farms and agricultural workers because they earn cash and get control over the resources	

(continued)

Table 2 (continued)

Adaptive power rating	Indicators	Source of indicators	Assumption of indicators	Descriptions in this paper
	Gender	Cunningham et al. [11], Sigei et al. [42]	It is believed that male member of the household have strong communication skills	Gender of the farmer
		Hill and Vigneri [24]	Women have low productive resources and lower participation in the market	
		Dorward et al. [15]	Discrimination observed in agricultural trade	
		Wang'ombe [44]	Women also work but spend more time on home	
	Marital status	Loughrey et al. [26]	Marital status has a significantly negative effect upon off-farm employment in the Italian data while in the Irish data, there is a significant relationship between off-farm employment participation and marriage	
	Private vehicle own	Foxall [21]	Social and economic factor affecting farmer's decision	Have own tractor

(continued)

ble 2 (continued)

daptive power ating	Indicators	Source of indicators	Assumption of indicators	Descriptions in this paper
nfrastructure	Landholding size	Defiesta and Rapera [12]	Large landholding farmers stand a better chance of diversifying their farming practice	Big farmer can farming potato and mustard in one season but smallholders have to adopt opportunity farming
nstitutions	Government subsidy	Defiesta and Rapera [12]	Farmers with access to government subsidies for agriculture input are more resilient	Govt. subsidy
	SHG member	Odebode et al. [32]	Self-Help Groups (SHGs) play effective roles in promoting empowerment through giving of loans to members	SHG = Self help group
		Dwarakanath [16], Elliott [18]	The SHGs play a role as catalysts in farm economic process	
	Loan amount/repayment /restriction	Allen et al. [4]	Farmers faces many challenges to get agricultural loan	Loan amount {1 = Rs. 50,000/-, 2 = Rs. 1 L, 3 = Rs. 2 L, 4 = Rs. 5 L, 5 = 5–6.5 L}

The study presents an empirical report based on a pilot study carried out on randomly selected 52 farmers located in Burdwan district of West Bengal, India. This pilot study has been followed by a complete research study on conducting a primary survey regarding the issues on Pradhan Mantri Fasal Bima Yojana (PMFBY) over 701 farmers located in the same area, as presented by Mukhopadhyay et al. [30] Table 3.

All the variables are framed into 3 components and satisfy 7 determinants out of 13 factors as discussed above. So, the different sections are described in Table 4, which maybecategorized in adaptive power rating independent variables.

Table 3 Pilot test of 52 farmers before adaption of conceptual model

Rotated component matrix[a]			
	Component		
	1	2	3
EDUCATION	0.726		
AGE	0.826		
GENDER	0.848		
MARRIED	0.802		
VOCATIONAL_TRG			
FARMING_EXPERIENCE			
RICE_OUTPUT			
PVT_VEHICLE			
SHG_MEMBER			0.588
LOAN_AMT			
LAND_HOLDING		0.630	
INCOME		0.676	
OTHER_THAN_FARMING_INCOME			

Extraction method: principal component analysis
Rotation method: varimax with Kaiser normalization
[a]Rotation converged in 5 iterations

Table 4 Adaptive power rating of independent variables

Adaptive power rating of independent variables	Indicators considered for model	Codes used in model
Economic resources	Income	A
Awareness and training	Level of literacy (education)	B
Socioeconomic factor	Age	C
	Gender	
	Marital status	
Infrastructure	Landholding size	D
Institutions	SHG member	E

6 Discussion of the Conceptual Model

The conceptual framework of the study focused on linking between the interaction of farmer's socioeconomic characteristics, personal-demographic situation, resource endowments, institutional and policy framework, psychological factor, other supportive factors and their implication in the decision-making process of farmers. The institutional and policy factor also affects the farmer's decision of accepting crop insurance. Other factors like social, economic, psychological factor and asset endowment

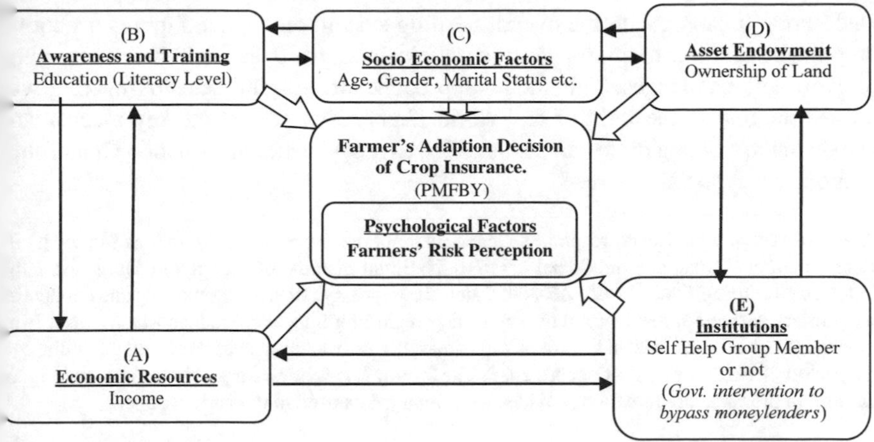

Fig. 2 Conceptual framework of the determinants influencing the farmers accepting crop insurance (PMFBY). *Source* Authors' own concept

factors also determine adoption decision. Such factors affect the farmer's decision on accepting crop insurance positively or negatively. Many cognitive scientists and psychologists have researched the decision-making process for many years from the social perspective. The change within factors may change the complete system as the relationship is non-linear and there is multi-dimension in a self-organization system. Relationships can be fuzzy, in hierarchy or nested, which may limit the content of each element [33] (Fig. 2).

The proposed models will simulate the process of mutual interaction among peasants and institutions; various socio-economic factors like individual economic resources, level of literacy, age, gender, marital status etc., individual farmers' financial flow and management of the family; atmosphere of farmers' family and family sentiment; and crop production in the market with livestock impact etc. As a basis of agricultural activities, we use the concept of a peasant through questionnaire form. The information about the decision-making assets is measured in the concept of capacitances and limitations of farmers. The model might be utilized to comprehend the consequences of the farmers' decision about accepting of crop insurance of an entire agrarian area. Various aspects of crop insurance management, risk perception by the farmers, policy drivers, market impact and worldwide socioeconomic trends are depicted by this model.

7 Conclusion

In this paper, we have discussed about the conceptual model of decision-making. A pilot study reveals that only a few of determinants are affected mostly in the study area. It can explain the overall farming system only when farmers accept t crop insurance. Furthermore, specific policies and practices could help the farme to overcome challenges. This model approaches to link the decision-making wi a dynamic and in the light of real-world phenomena so that the key mechanism and decision-making drive can be identified to create better information about futu economic and social reports.

Acknowledgements The corresponding author is thankful to the Ethics Committee formed by t Department of Humanities and Social Sciences, National Institute of Technology Durgapur, Indi Ref. No. NITD/HSS/PhD/Misc/2017/004, Dated: 16 February 2018, for monitoring and conducti the primary survey for his doctoral research study regarding issues on Pradhan Mantri Fasal Bi Yojana on 701 farmers located at various rural areas of Burdwan district of West Bengal, India, ov the period of February 2017–October 2017. The survey has followed a previous pilot study on 5 farmers in the same areas, which provides the primary data used in the current paper.

References

1. Abdulai, A., Delgado, C.L.: Determinants of nonfarm earnings of farm-based husbands an wives in Northern Ghana. Am. J. Agric. Econ. **81**(1), 117–130 (1999)
2. Acevedo, M., & Krueger, J.I.: Two egocentric sources of the decision to vote: the voter's illusio and the belief in personal relevance. Polit. Psychol. **25**(1), 115–134 (2004)
3. Adeoti, A.I., Oluwatayo, I.B., Soliu, R.O.: Determinants of market participation among maiz producers in Oyo state, Nigeria. J. Econ. Manag. Trade, 1115–1127 (2014)
4. Allen, F., Hryckiewicz, A., Kowalewski, O., Tümer-Alkan, G.: Transmission of financial shocks in loan and deposit markets: role of interbank borrowing and market monitoring. J. Finan. Stab **15**, 112–126 (2014)
5. Armah, F.A., Odoi, J.O., Yengoh, G.T., Obiri, S., Yawson, D.O., Afrifa, E.K.: Food security and climate change in drought-sensitive savanna zones of Ghana. Mitig. Adapt. Strat. Glob Change **16**(3), 291–306 (2011)
6. Asian Development Bank (ADB).: Improving technical education and vocational training strategies for Asia. R. a. SD Department, Manila (2004)
7. Attanasio, O.: Consumption and saving behaviour: modelling recent trends. Fisc. Stud. **18**(1), 23–47 (1997)
8. Baker, H.K., Nofisinger, J.R.: Behavioural Finance: Investor, Corporation and Market. Wiley, New Jersey, USA (2010)
9. Beach, L.R.: Broadening the definition of decision making: the role of prechoice screening of options. Psychol. Sci. **4**, 215–220 (1993)
10. Below, T.B., Mutabazi, K.D., Kirschke, D., Franke, C., Sieber, S., Siebert, R., Tscherning, K.: Can farmers' adaptation to climate change be explained by socioeconomic household-level variables? Glob. Environ. Chang. **22**, 223–235 (2012)
11. Cunningham III, L.T., Brorsen, B.W., Anderson, K.B., Tostão, E.: Gender differences in marketing styles. Agric. Econ. **38**(1), 1–7 (2008)
12. Defiesta, G.D., Rapera, C.L.: Measuring adaptive capacity of farmers to climate change and variability: application of a composite index to an agricultural community in the Philippines. J. Environ. Sci. Manag. **17**(2), (2014)
13. Deressa, T., Hassan, R.M., Alemu, T., Yesuf, M., Ringler, C.: Analyzing the determinants of farmers' choice of adaptation methods and perceptions of climate change in the Nile Basin of Ethiopia. Intl. Food Policy Res. Inst. (2008)
14. Dev, S.M.: Financial inclusion: issues and challenges. Econ. Polit. Wkly. 4310–4313 (2006)

. Dorward, A., Fan, S., Kydd, J., Lofgren, H., Morrison, J., Poulton, C., ... Urey, I.: Rethinking agricultural policies for pro-poor growth. Nat. Resour. Perspect. **94**, 1–4 (2004)

5. Dwarakanath, H.D.: Rural credit and women self-help groups. Kurukshetra, **51**(1), 10 (2002)

7. Edwards-Jones, G.: Modelling farmer decision-making: concepts, progress and challenges. Anim. Sci. **82**, 783–790 (2006). Cambridge University Press

8. Elliott, C.: Some aspects of relations between the North and South in the NGO sector. World Dev. **15**, 57–68 (1987)

9. Enete, A.A., Igbokwe, E.M.: Cassava market participation decisions of producing households in Africa. Tropicultura, **27**(3), 129–136 (2009)

0. Featherstone, A.M., Wilson, C.A., Kastens, T.L., Jones, J.D.: Factors affecting the agricultural loan decision-making process. Agric. Finan. Rev. **67**(1), 13–33 (2007)

1. Foxall, G.R.: Farmers' tractor purchase decisions: a study of interpersonal communication in industrial buying behaviour. Eur. J. Mark. **13**(8), 299–308 (1979)

2. Ganesamurthy, V.S., Krishnan, M.R., Bhuvaneswari, S., Ganesan, A.: A study on thrift and credit utilization pattern of "Self Help Group" (SHG) in Lakshmi Vilas Bank Suriyampalayam Branch, Erode. Indian J. Mark. **34**(1) (2004)

3. Gani, B.S., Adeoti, A.I.: Analysis of market participation and rural poverty among farmers in northern part of Taraba State, Nigeria. J. Econ. **2**(1), 23–36 (2011)

4. Hill, R.V., Vigneri, M.: Mainstreaming gender sensitivity in cash crop market supply chains. In: Gender in Agriculture, pp. 315–341. Springer, Dordrecht (2014)

25. Jyothi, K.S., Gracy, C.P., Suryaprakash, S.: Empowerment and decision-making by rural women-an economic study in Karnataka. Indian J. Agric. Econ. **54**(3), 308 (1999)

26. Loughrey, J., Hennessy, T., Hanrahan, K., Donnellan, T., Raimondi, V., Olper, A.: Determinants of farm labour use: a comparison between Ireland and Italy (No. 545-2016-38731) (2013)

27. Makura, M.N., Kirsten, J., Delgado, C.: Transactions costs and smallholder participation in the maize market in the northern province of South Africa. Integrated approaches to higher maize productivity in the new millennium (No. 338.16 FRI. CIMMYT) (2002)

28. Martey, E., Al-Hassan, R.M., Kuwornu, J.K.: Commercialization of smallholder agriculture in Ghana: a Tobit regression analysis. Afr. J. Agric. Res. **7**(14), 2131–2141 (2012)

29. Mukhopadhyay, P., Sengupta, P.P.: Decision-making process of farmers' in present political economy of agrarian crisis with a study of Burdwan district of West Bengal, India. Int. J. Sustain. Econ. Manag. **6**(1), 96–113 (2017)

30. Mukhopadhyay, P., Sinha, M., Sengupta, P.P.: Determinants of farmers' decision-making to accept crop insurance: a multinomial logit model approach. In: Nayak et al. (eds.) Soft Computing in Data Analytics. Advances in Intelligent Systems and Computing, vol. 758, pp. 267–275. Springer, Berlin (2018)

31. Myers, I.: Introduction to Type: A description of the Theory and Applications of the Myers-Briggs Type Indicator. Consulting Psychologists Press, Palo Alto, CA (1962)

32. Odebode, S.O.: Financing cassava processing among women in rural Nigeria to alleviate poverty: the Place of self-help groups. Bul. J. Agric. Sci. **12**(1), 115 (2006)

33. Parker, D.C., Manson, S.M., Janssen, M.A., Hoffman, M.J., Deadman, P.: Multi-agent systems for the simulation of land-use and land-cover change: a review. Ann. Assoc. Am. Geogr. **93**, 314–337 (2003)

34. Potter, C., Lobley, M.: Ageing and succession on family farms: the impact on decision-making and land use. Sociol. Ruralis **32**(2–3), 317–334 (1992)

35. PTI. West Bengal: another farmer commits suicide in Burdwan district. India Today, April 24 2015. Retrieved From: https://www.indiatoday.in/india/story/west-bengal-potato-farmers-suicide-burdwan-967489–2017-03-24. & https://www.firstpost.com/india/west-bengal-another-farmer-commits-suicide-in-burdwan-district-2211286.html (2015)

36. Punia, R.K., Kaur, S., Punia, D.: Women in agricultural production process in Haryana. Women Agric. **1**, 77–96 (1991)

37. Randela, R., Alemu, Z.G., Groenewald, J.A.: Factors enhancing market participation by small-scale cotton farmers. Agrekon **47**(4), 451–469 (2008)

38. Sadler-Smith, E., Sparrow, P.: Intuition in organizational decision making. In: The Oxfo Handbook of Organizational Decision Making (2008)
39. Schultz, T.W.: Our welfare state and the welfare of farm people. Soc. Serv. Rev. **38**(2), 123–1 (1964)
40. Shankara, M.H., Mamatha, H.S., Reddy, K.M., Desai, N.: An evaluation of training programm conducted by Krishi Vigyan Kendra, Tumkur, Karnataka. Int. J. Farm Sci. **4**(2), 240–248 (201
41. Shem, A.O.: Financial sector dualism: determining attributes for small and micro enterprises urban Kenya; a theoretical and empirical approach based on case studies in Nairobi and Kisu (2002)
42. Sigei, G., Bett, H., Kibet, L.: Determinants of market participation among small-scale pineapp farmers in Kericho County, Kenya (2014)
43. Stanovich, K.E., West, R.F.: On the relative independence of thinking biases and cogniti ability. J. Pers. Soc. Psychol. **94**(4), 672 (2008)
44. Wang'ombe, J.G.: Potato value chain in Kenya and Uganda. Maastritch School of Managemen Netherlands (2008)

n Efficient Approach to Optimize e Learning Rate of Radial Basis 'unction Neural Network for Prediction f Metastatic Carcinoma

rachi Vijayeeta, M. N. Das and B. S. P. Mishra

bstract This paper is based on the optimization of linear weights in a radial basis unction neural network that connects the hidden layer and the output layer. A new ptimization algorithm called dualist algorithm is applied for choosing an optimal earning parameter. A conventional strategy of random selection of radial basis funcion (RBF) centers and the width, as well as the weights, are estimated by gradient escent method and least square methods, respectively. The ideology behind this tudy is to predict the occurrence of metastatic carcinoma in human cells by computational approaches. Our simulation consists of comparing the predictive accuracy of armony search-radial basis function network (RBFN) and dualist-RBFN by optinizing the weight factor. The Wisconsin breast cancer dataset is used as a benchmark o experiment our training pattern. The learning rate (weight factor) is taken as the ptimized parameter to obtain the best possible solution.

Keywords Classification · RBFN · Cancer dataset · Dual-RBFN · *k*-fold validation

1 Introduction

The emergence of machine learning techniques has enabled researchers to predict the occurrence of metastases carcinoma by investigating the samples collected from clinical sources. However, in the present day, medical practitioners have put forward their level best to explore robust computational models for classifying, clustering, and analyzing of certain rare diseases, like metastases carcinoma. Metastases refer to the spread of cancer to other new locations of the body (often by way of lymph

P. Vijayeeta (✉) · M. N. Das · B. S. P. Mishra
School of Computer Engineering, Kalinga Institute of Industrial Technology [Deemed to be University], Bhubaneswar 751024, Odisha, India
e-mail: pvijayeetafca@kiit.ac.in

M. N. Das
e-mail: mndas_prof@kiit.ac.in

B. S. P. Mishra
e-mail: bsmishrafcs@kiit.ac.in

A. K. Das et al. (eds.), *Computational Intelligence in Pattern Recognition*,
Advances in Intelligent Systems and Computing 999,
https://doi.org/10.1007/978-981-13-9042-5_80

system or bloodstreams), where the site of origin is called primary tumor. It is othe wise known as cancer of unknown primary origin (CUP). Several learning methoc such as radial basis function (RBF) neural nets, multi-layer perceptron (MLP) neur nets, Bayesian, decision tree (DT), random forest, hidden Markov models, and de learning, have been experimented on this medical domain of research, thereby deri ing feasible conclusions. Oncologists generally do believe that the pattern of spre may suggest the occurrence of primary site. These patterns, upon being subjecte to computational models in the form of inputs, can help to predict the site of origi of the tumor. In our work, we have applied radial basis function network (RBFN in which the patterns are made as inputs and a Gaussian kernel function is applie in the hidden layer to evaluate the output. This generated output is again compute with relevant weights and summed up to obtain a single output. The main objectiv of our work is to determine the learning parameter (weights) and to optimize it usin dualist algorithm (DA) so that an accurate output is obtained. Furthermore, we hav compared the accuracy with particle swarm optimization in RBF. One of the greates hurdles in medical databases is the high dimensionality of dataset. The basic caus of this dimensionality issue has been identified as the presence of non-contributin features within the dataset. The present research has discovered the hidden pattern and has also suggested an efficient approach to detect CUP.

2 Literature Review and Previous Works

A literature review has showed that there are several studies on cancer prediction problem using statistical learning approaches. According to the Cancer Information Summary report, only 3–5% of the human population is affected by metastases and only 10–15% survival rate has been achieved. In the year 2007, Eccles and Welch [1] identified in their study that the secondary tumor arises by the spread through lymphatic nodes (bloodstream). They had suggested a notable agent targeting oncogenic signaling pathways as a tool for the next generation treatment. Vellido et al. [2] applied support vector machine (SVM) and relevance vector machine to predict the primary site, as well as to successfully colonize the secondary site by optimizing the exclusive contributing features. But the model proposed by them failed when the biomedical data analysis became increasingly multi-variate, multi-scale, and multi-modal. According to a report published by National Center for Biotechnology Information (NCBI) in 2009, the total annual cost of cancer treatment was at around $1.6 trillion, which upon early detection has not only reduced to $1 billion but have also increased the survival rate. To overcome these data management issues, Van Stiphout et al. [3] applied Cox regression followed by proximal SVM (pSVM) to evaluate the effect of several variables at the specified time and also estimated the statistical properties of the relevant inputs. Stojadinovic et al. [4] modeled a Bayesian classification supports system to assess the case-specific oncological outcomes of the patients. Even though it resulted as a good decision support tool but lack of attention was paid to data size and sample-per-feature ratio.

Exarchos et al. [5] formulated a multi-parametric decision support tool that could cceed in integrating heterogeneous data (clinical, image, and genomic-based) to edict the local vs distant vs regional sites of tumor occurrence. Ahmad et al. [6] plemented decision tree, SVM, and artificial neural network to develop predic-ve models. A 10-fold cross-validation was used for measuring unbiased prediction curacy. Chen et al. [7], Wang et al. [8] identified five primary sites of metastases, amely liver, lymph, pancreas, lung, and skin through data analysis using logistic gression, DT, and SVM. They succeeded in generating qualitative and quantitative ataset with a normalized data distribution ratio of training, testing, and validation. li [9] applied deep learning concepts to invade the hidden patterns within the sam-le. Hidden Markov model and Gaussian mixture model [10] were also applied for attern recognitions in metastases carcinoma.

Background of RBFN and Motivation

ne of the major drawbacks of a linear activation function was the occurrence of ingle decision boundary, which subsequently failed in generating a feasible output. o overcome this issue, Broomhead and Lowe in the year 1988 proposed a mathe-natical model termed as radial basis function neural network at Royal Signals and adar Establishment, England. In this model a non-linear activation function $\emptyset_i(\boldsymbol{x})$ s introduced in the hidden layer of the multi-layer perceptron (MLP) to approximate continuous function.

$$y = \sum_{i=1}^{N} W_i \varphi_i(X, \mu_i)$$

This motivation is derived from the techniques for interpolation where the input is modeled as a vector of real numbers $X \in R^n$. The output of the hidden layer is a scalar function of the input vector, $\emptyset : R^n \rightarrow R$ is given by:

$$\varphi_i(x) = \sum_{i=1}^{N} W_i \varphi ||X - \mu_i||$$

where N is the number of samples (or hidden kernels).
$||X - \mu_i||$ is the Euclidean norm.
W_i is the synaptic weights (mostly considered as unity).
φ_i is some radially symmetric function.

But in our work we have taken the help of the most popular function, that is, Gaussian kernel function, which is computed as follows:

$$y_m(x) = \sum_{i=1}^{N} W_{mi} G\left(\left\|x_{ij} - \mu_{ik}\right\|\right) + b_m \quad for\ i = 1, 2, 3..N\ and\ m = 1, 2, 3..M$$

where m is the number of outputs.
W_{mi} is the weight between kernel i and output m.
b_m is the bias on output m.
k is the number of classes.
μ_{ik} is the mean of the expression values of ith gene in kth class.
x_{ij} is the expression values of ith gene in jth sample.

It refers to the determination of the following three basic parameters:

1. Appropriate position of all the kernel centers (μ_i)
2. Significant selection of radius of each kernel (σ)
3. Optimal selection of weights between the kernels and output nodes(W_i).

3.1 Mathematical Formulation of the Problem

A radial basis function is a non-increasing function whose output depends on the distance of the input from a given stored vector. It is represented by local receptors where each point is a stored vector used in one RBF. The hidden processing unit implements a radial basis activation functions for generating a linear combination of output data. The output units implement a weighted sum of hidden units. For a set of "N" different data points such that $\{x_i \in \mathbb{R}^m, i = 1 \ldots N\}$ and a set of N real numbers $\{d_i \in \mathbb{R}^m, i = 1 \ldots N\}$, then a function $\mathrm{F} : \mathbb{R}^m \rightarrow R$ is mapped to satisfy the interpolation condition:

$$F(x_i) = d_i \tag{1}$$

Ideally, the input space contains possibly a large variety of data that are in the form of finite radial centers. These finite data points form cluster which has its own receptive field, and the cluster centre output remains maximal for those corresponding data points. The radial basis function in the hidden layer produces a significant non-zero response only when the output falls within a small localized region of the input space. Each hidden unit has its own receptive field in the input space. The output of the radial centers will be minimum for the data points that are away from the field.

An input vector X_i which lies in the receptive field for the centre C_j *would activate* Φ_j, and by proper choice of weights $\left(w_j\right)$, the target output is obtained. The process of finding appropriate RBF weights is called network training. Therefore, the subsequent expression takes the following form:

$$y = \sum_{j=1}^{h} \Phi_i w_j \tag{2}$$

here $\Phi_{ij} = \Phi\left(\left|\left|X_i - C_j\right|\right|\right)$ is called the radial function.

atrix Notation

we substitute the value of y in Eq. 1, then we can have:

$$\begin{pmatrix} \Phi_1^1 & \Phi_2^1 & \Phi_h^1 \\ .. & .. & .. \\ \Phi_1^N & \Phi_2^N & \Phi_h^N \end{pmatrix} \begin{pmatrix} w_1 \\ : \\ w_N \end{pmatrix} = \begin{pmatrix} d_1 \\ : \\ d_N \end{pmatrix} \Rightarrow \boldsymbol{\Phi w = d}$$

By applying **Micchelli's theorem**, for $\{x_i\}i = 1\,to\,N$ be a set of N distinct oints in $\mathbb{R}^m$, then the N-by-N interpolation matrix $\boldsymbol{\Phi}$ such that the element $\Phi_{ji} =$ $_j - \Phi_i\,is\,non{-}singular.$

.2 *Methods Applied*

.2.1 Architecture of Radial Basis Function Network

The basic architecture of RBFN is a three-layered feed-forward network, which is depicted in Fig 1. Each layer has entirely distinct functionalities to interpolate the est data on the multi-dimensional surface.

1. Input layer: This layer contains the input pattern to be trained.

2. Hidden layer: It consists of hidden nodes that perform a non-linear transformation from the input space to the hidden space, which is of high dimensionality. It is

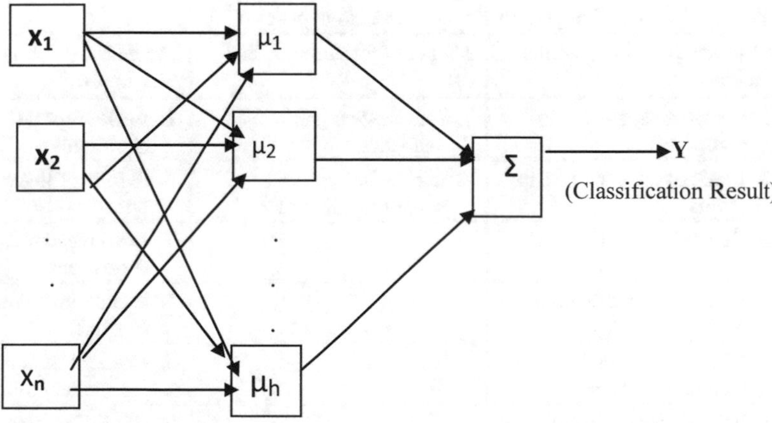

Fig. 1 Architecture of RBFN

further connected with the output layer. Every node of pattern layer is mathematical described by a Gaussian radial basis function as follows:

Let x be the input vector (expression value of a gene in a sample),
P be the maximum number of classes,
μ_i is the radial centre of kernel i,
σ_i is the radius of kernel i,

then we can compute a Gaussian activation function which is a form of Euclide distance between the input and the centre of that unit (also called Euclidean norm

$$\varphi_{ij}(x) = e^{\frac{-\|x-\mu_i\|^2}{2\sigma_i^2}} \quad (3$$

3. Output layer: This layer generates the classification result of the network b computing a linear activation function. Output of the kth neuron in summation laye is given by:

$$O_k(x) = \sum_{i=1}^{P}\sum_{j=1}^{Q_i} W_{ij}^k \emptyset_{ij}, k = 1, 2..P \quad (4$$

where
W_{ij}^k is the weight between the $\emptyset_{ij}$ in the pattern layer and the kth neuron in the summation layer.
O_k is the output of the kth neuron in the summation layer.

In a RBF, even though the input is non-linear the output is linear. Table 1 reveals the possible methods applied to detect the basic three parameters of RBFN.

Table 1 Various techniques applied for training the patterns

Methods for training kernel centers (μ_i)	Methods for training radius (σ)	Methods for training output weights (W_i)
1. Back-propagation algorithm	1. *P*-nearest-neighbor heuristics	1. Back-propagation algorithm
2. Random selection of data points	2. Global first nearest-neighbor, P = 1	2. System of linear equations
3. Clustering algorithm		3. Least mean square regression
4. Pseudo-inverse technique (off-line)		4. Pseudo-inverse technique (off-line)
5. Gradient descent technique (on-line)		5. Gradient descent technique (on-line)
6. Hybrid learning (on-line)		6. Hybrid learning (on-line)
7. Learning vector quantization		

2.2 Working Process Model

this paper we have implemented a population-based human-inspired algorithm lled dualist algorithm (DA) to optimize the weights for obtaining a better global otimal solution. The main aim of the algorithm is to innovate the winner and loser nong the individuals within a given population. In DA, all the individuals in a opulation are defined as dualist who keep on fighting one-by-one to determine the inners or losers or champions by taking into account their strength, skill, luck, nd intellectual capability. The loser learns from the winner and, in the meanwhile, e winner tries to improve their new potential by training or by adopting some new ethods from the opponents. The duel participant with highest fighting skill is called hampion who can train a new dualist. The newly generated dualist shall further articipate in the tournament as a representative of each champion. Combinedly, all e dualists are revaluated and the one possessing the minimal fighting capability ill thereby be rejected from the list.

.2.3 Algorithm for the Model Implemented

Step-1: Input Pattern

Step-2: Computation and Optimization of Ranking Weights for all Features

(i) Calculate the sorted list $C_i=(W_i)^2$.

(ii) Find the features possessing minimum weight:

$f = \arg\min(C_i)$

(iii) Apply *Dualist Optimisation algorithm* to optimize the weights. Call function ***dualist_opti ()***.

Step-3: Output Pattern

Dualist Optimization Algorithm (*dualist_opti()*)

Step-1: Registration of Dual Candidates

Input a n-dimensional binary array for representing the skill set of the Dualist such that:$N_{var} = [1,2,3 \ldots .\, n]$.

Step-2: Pre-Qualification Evaluation

Considering the skillset data, evaluate and test the fighting potential of each candidate within it.

Step-3: Determination of Champion Board

Each champion takes the responsibility to train a new list including himself as for their dual capabilities.

Step-4: Dual List Improvement

Each dualist will fight using the fighting skill and luck to determine the winner and loser follows:-

(a)**Initialization:-** A_Dualist[ ] and B_Dualist[ ] two

Binary array and Luck _coefficient.

(b)**Computation of luck:-**

$$A(\text{luck}) = A(\text{fighting\_skill}) * \left(\text{luck\_coefficient} + (\text{rand}(0-1) * \text{luck\_coefficient})\right) B(\text{luck}) = B(\text{fighting\_skill}) * \left(\text{luck\_coefficient} + (\text{rand}(0-1) * \text{luck\_coefficient})\right)$$

(c)**Winner-Loser Decision making:-**

$$\textbf{if}\left(\left(\mathbf{A}(\mathbf{fighting}_{\mathbf{skill}}) + \mathbf{A}(\mathbf{luck})\right) >= \left(\mathbf{B}(\mathbf{fighting}_{\mathbf{skill}}) + \mathbf{B}(\mathbf{luck})\right)\right)$$

then $\mathbf{A(winner) = 1\ and\ B(winner = 0)}$

else

$\mathbf{A(winner) = 0\ and\ B(winner = 1) End\ if}$

Step-5: Define a Dual schedule between each dualist

The dualist fighting capabilities need to be improved for all the three participants:-

Step-6: Elimination Phase

Elimination of features with low weights.

Step-7: Finish

4 Experiment and Results

We have used Wisconsin breast cancer dataset for our experiment (https://archive.ics.uci.edu/ml/datasets/Breast+Cancer+Wisconsin+(Diagnostic) with the statistical properties depicted in Table 2. The software tool used is Scikit-Learn with python-anaconda. It constitutes 569 samples with 31 attributes and two class labels

Table 2 Statistical property description of dataset

S. no.	Attribute name	Attribute type	Mean		Least value		Computational error		Class label
			Min value	Max value	Min value	Max value	Min value	Max value	
1.	Radius	float64	6.981	28.11	7.93	36.04	0.112	2.873	Malignant
2.	Texture	float64	9.71	39.28	12.02	49.54	0.36	4.885	Benign
3.	Perimeter	float64	43.79	188.5	50.41	251.2	0.757	21.98	
4.	Area	float64	143.5	2501	185.2	4254	6.802	542.2	
5.	Smoothness	float64	0.053	0.163	0.071	0.223	0.002	0.031	
6.	Compactness	float64	0.019	0.345	0.027	1.058	0.002	0.135	
7.	Concavity	float64	0	0.427	0	1.252	0	0.396	
8.	Concave points	float64	0	0.201	0	0.291	0	0.053	
9.	Symmetry	float64	0.106	0.304	0.156	0.664	0.008	0.079	
10.	Fractal dimension	float64	0.05	0.097	0.055	0.208	0.001	0.03	

Table 3 Statistical properties of classifiers

	Statistical properties	Optimized classifiers applied	
		Harmony search-RBF	Dual-RBFN
1	Accuracy	96.04%	97.13%
2	Specificity	0.9523	0.9722
3	Sensitivity	0.9331	0.9708
4	Positive prediction value	0.9210	0.9852
5	Negative prediction value	0.9131	0.9459
6	Prevalence	0.6221	0.6555
7	Detection rate	0.5642	0.6364
8	P-value	0.9622	0.6831
9	Confidence interval	0.8431, 0.9543	0.834, 0.9534
10	Class error	B-0.0254, M-0.0272	B-0.0249, M-0.0236

(benign/malignant). Harmony search-RBFN yields an accuracy of 96.04%, fol lowed by dual-RBFN which yields an accuracy of 97.13%. Other properties of clas sifier are evaluated and stored in Table 3. In our work, we have split our dataset int 80% training and 20% test dataset. The latter one succeeds in generating a bette accuracy of nearly 1.985%, on an average more than the previous method. A 10-folc cross-validation is studied to evaluate the accuracy and reliability of the methods.

5 Performance Measurement Criteria

To improve the predictive performance and accuracy of the learning models, external validation is an absolute necessity. Initially, we need to split the labeled dataset into training and testing subsets using *k*-fold cross-validation (Table 4). In this paper we have applied *k*-fold cross-validation and have tested the accuracy for $k = 10, 5, 3\ folds$. In *k*-fold validation, each sample is executed *k* times for training

Table 4 Confusion matrix representation of classifiers

Confusion matrix for harmony search + RBFN			Confusion matrix for dualist + RBFN		
	Benign	Malignant		Benign	Malignant
Benign	118	2	Benign	133	2
Malignant	4	76	Malignant	4	70

Table 5 Performance estimation using *k*-fold cross-validation

Classifier applied	Average accuracy with 10-fold cross-validation (in %)	Average accuracy with 5-fold cross-validation (in %)	Average accuracy with 3-fold cross-validation (in %)
1. Harmony search + RBFN	93.35	94.93	95.36
2. Dual-RBFN	97.14	97.23	97.16

nd only once for testing. Table 5 is organized to have a glance on the performance stimation of our classifiers applied.

Conclusions and Future Work

From our experiment we have inferred that for different values of *k* the accuracy ate varies and we cannot assure that for increased value of *k*, the accuracy will e higher; rather there can be drastic rise in the computational cost. A line graph as been plotted for demonstrating the *k*-fold validations. In our work we have carried out a comparative study based on the accuracy obtained, by applying harmony search with RBFN and dual-RBFN using the same Wisconsin breast cancer dataset. Ultimately, we reached to a conclusion that dual-RBFN yielded a better accuracy than harmony search with RBFN. From Table 4 we have determined the expected number of malignant and benign samples with the help of confusion matrix. But in our future work we can investigate the effect of these two algorithms on some more datasets, like bone cancer/colon cancer/cervical cancer/lung cancer, and try to predict the primary site of origin. We can even apply many robust swarm optimization techniques or population-based/nature-inspired optimization techniques to obtain an optimal solution in the global search state space.

References

1. Eccles, S.A., Welch, D.A.: Metastasis: recent discoveries and novel treatment strategies. Lancet **369**(9574), 1742–1757 (2007)
2. Vellido, A., Biganzoli, E., Lisboa, P.J.G.: Machine learning in cancer research: implications for personalised medicine. In: ESANN, pp. 55–64 (2008)
3. Van Stiphout, R.G.P.M., Postma, E.O., Valentini, V., Lambin, P.: The contribution of machine learning to predicting cancer outcome. Artif. Intell. **350**, 400 (2010)
4. Stojadinovic, A., Nissan, A., Eberhardt, J., Chua, T.C., Pelz, J.O., Esquivel, J.: Development of a Bayesian belief network model for personalized prognostic risk assessment in colon carcinomatosis. Am. Surg. **77**(2), 221–230 (2011)

5. Exarchos, K.P., Goletsis, Y., Fotiadis, D.I.: Multiparametric decision support system for prediction of oral cancer reoccurrence. IEEE Trans. Inf Technol. Biomed. **16**(6), 1127–11 (2012)
6. Ahmad, L.G., Eshlaghy, A.T., Poorebrahimi, A., Ebrahimi, M., Razavi, A.R.: Using thr machine learning techniques for predicting breast cancer recurrence. J. Health Med. Infor **4**(124), 3 (2013)
7. Chen, Y., Sun, J., Huang, L.C., Xu, H., Zhao, Z.: Classification of cancer primary sites usi machine learning and somatic mutations. BioMed Res. Int. **2015** (2015)
8. Wang, Z., Wen, X., Yaohong, L., Yao, Y., Zhao, H.: Exploiting machine learning for predicti skeletal-related events in cancer patients with bone metastases. Oncotarget **7**(11), 12612 (201
9. Ali, A.-R.: Deep Learning in Oncology–Applications in Fighting Cancer. (2017)
10. Dokduang, K., Chiewchanwattana, S., Sunat, K., Tangvoraphonkchai, V.: A comparati machine learning algorithm to predict the bone metastasis cervical cancer with imbalan data problem. Recent Advances in Information and Communication Technology, pp. 93–10 Springer, Cham (2014)

Voice-Based Gender Identification Using Co-occurrence-Based Features

Arijit Ghosal, Chanda Pathak, Pinki Singh and Suchibrota Dutta

Abstract Automatic detection of gender based on audio is gaining its popularity day-by-day because of its several applications in several domains. But most of the past research works are limited to differentiate male and female voice toward gender detection from audio signal. Very less work has been performed for gender identification considering trans-gender also. As trans-gender has received legal recognition, it is expected that a good gender detection system should be able to identify all the three genders. In this work some judiciously chosen acoustic features have been employed to identify all the three genders. This proposed feature set is of small dimensional to keep computational complexity low. Low-level time-domain acoustic features like short time energy, spectral flux and audio features belonging to frequency domain as well as skewness have been used here. Co-occurrence-based approach has been applied further for precise observation of these features. Neural network, Naïve Bayes and k-nearest neighbor have been used for classification purpose.

Keywords STE · Spectral flux · Gender identification · Co-occurrence matrix · Classification

A. Ghosal (✉) · C. Pathak · P. Singh
Department of Information Technology, St. Thomas' College of Engineering and Technology, 700 023 Kolkata, West Bengal, India
e-mail: ghosal.arijit@yahoo.com

C. Pathak
e-mail: cpathak0096@gmail.com

P. Singh
e-mail: pinki.pinks93@gmail.com

S. Dutta
Department of Information Technology and Mathematics, Royal Thimphu College, 11001 Thimphu, Bhutan
e-mail: suchibrota@gmail.com

A. K. Das et al. (eds.), *Computational Intelligence in Pattern Recognition*, Advances in Intelligent Systems and Computing 999,
https://doi.org/10.1007/978-981-13-9042-5_81

1 Introduction

Speech is the way to express thoughts of a person in spoken words. Speech carrie lot of information related with speaker while working as a communication interfa among people. Linguistic as well as non-linguistic characteristics of speaker g reflected in speech. These characteristics help in identifying gender of speaker. It known that gender detection or identification is gaining its importance because its application in various domains. It has been also observed from the past researc works that gender-dependent systems exhibit better accuracy than systems which a gender-independent. Most of the previous works were limited up to identification male and female voice only. But trans-gender is also legally recognized gender no Unfortunately, only few works have considered trans-gender in their work. So gend identification considering trans-gender too is an open research area.

Audio signal is basically an electric signal. So it has a lot of similarity with electri signal. But as it is an audio, it exhibits some properties or characteristics which ar quite different from ordinary electric signal. It carries features or characteristics i time, frequency and perceptual domains. Acoustic features computed based on thes domains are capable of identifying gender of speaker. The objective of this wor is to propose a small-dimensional acoustic feature set to identify all three gender efficiently.

2 Previous Works

In the respective earlier works, researchers have identified genders as male and female. They have used various types of acoustic features belonging to time, frequency and perceptual domains. Though they have got convincing results their works lack in the point of considering trans-gender. Statistical feature extraction scheme has been proposed by Haralick and Shapiro [1]. He has proposed this concept in the domain of image processing. These statistical features can be extended to be used in other domains too. Using Mel-frequency cepstral coefficients (MFCC), gender classification has been performed by Harb and Chen [2].

Principal component analysis (PCA)-based feature selection scheme has been proposed by Malhi and Gao [3]. Subramanian et al. [4] have performed categorization of different audio sounds by extracting features based on pitch, timbre and rhythm. MIRtoolbox—a MATLAB-based toolbox—has been used by Lartillot et al. [5]. This toolbox contains the efficient features related to time, perceptual and frequency domain. Jabid et al. [6] have proposed an approach of classification by using local directional pattern. Perceptual feature-based speech detection technique has been proposed by Bach et al. [7]. Grosche and Müller [8] have proposed a MATLAB toolbox to extract and analyze tempo-based features. Predominant local pulse from music recording has been identified by Grosche and Müller [9]. Ali et al. [10]

troduced first Fourier transform-dependent methodology for gender categorization om speech.

The novel scheme proposed by Grosche and Müller [8] was again explored based n tempo-based features by Müller and Ewert [11]. Richard et al. [12] have prosed a perceptual feature-based scheme in case of speech detection. Ghosal and utta [13] have used a combination of time-domain feature and frequency-domain ature to identify male and female voices. Srivastava [14] has proposed Weka tool or different pattern detection purposes as well as for classification. Alías et al. [15] ave proposed perceptual feature-based scheme. Kumar et al. [16] have introduced multi-channel gender classification scheme for speech in movie audio. Pahwa and ggarwal [17] have used Mel coefficient feature for gender identification using support vector machine. Ranjan and Hansen [18] have proposed a new framework toward nprovement of performance for gender-independent speaker recognition based on Vector probabilistic linear discriminant analysis. Simpson et al. [19] have investiated acoustic and perceptual correlates of gender in pre-pubertal voice. Alipoor and amadi [20] identified gender of speaker by applying empirical mode decompositionased approach. Doukhan et al. [21] have worked for gender equality for French udiovisual streams by estimating speaking time of female and male. Safavi et al. 22] have identified gender and age group considering speech of children.

3 Proposed Methodology

Based on genders speaker of a speech can be categorized into three—male, female and trans-gender. All these categories differ from each other in point of their respective acoustic properties. While studying the characteristics of male, female and trans-gender it is observed that they mostly differ from each other in frequency-domain as well as in time-domain. The proposed approach depends on low-level time-domain and frequency-domain acoustic features. As it is observed that energy level of male, female and trans-gender varies widely, short time energy (STE) has been considered as time-domain acoustic feature. On the other hand, spectral energy of all three genders also varies widely. Spectral flux is well capable of capturing this characteristic in frequency domain. Hence, spectral flux is used as frequency-domain acoustic feature in the proposed approach. It is also observed that male, female and trans-gender voices maintain certain kind of asymmetry among them. As skewness is well capable of reflecting this asymmetry, it is also used along with STE and spectral flux. Figure 1 reflects the working flow of the suggested scheme.

3.1 Feature Extraction

Previous research works mostly have used MFCC, as well as audio features derived from pitch. Pitch-based features are of long dimensional. To reduce the dimension

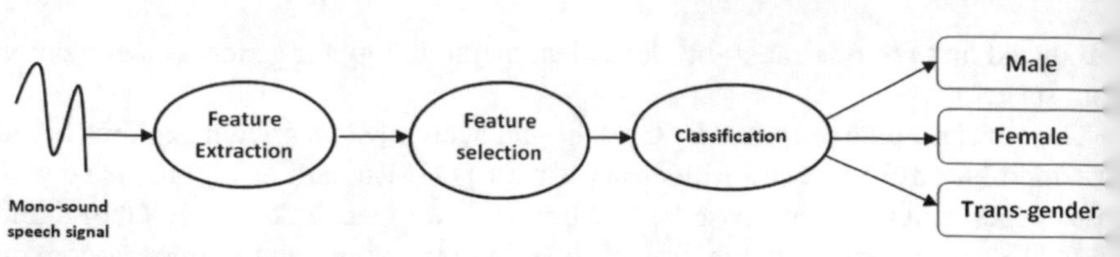

Fig. 1 Working flow of proposed gender identification system

of feature set, PCA is required. This increases the computational complexity and th overall feature extraction time. To reduce feature extraction time and computationa complexity, STE and spectral flux have been introduced along with skewness.

3.1.1 STE-Based Features

Every person speaks differently due to their biological difference. Structure of voca folds varies from person to person. Because of this it has been observed that energ level of speech varies widely from gender to gender. Speech consists of words inter leaved by silence which produces variation in energy. STE is well capable of capturing this phenomenon. This interleaving of words varies widely from gender to gender Motivated by this observation short time energy has been considered. If an audio signal is broken into P frames $\{x_i(m): 1 \leq i \leq P\}$, then short time energy for *ith* frame is defined as:

$$E_i = \frac{1}{n} * \sum_{m=0}^{n-1} [x_i(m)]^2 \quad (1)$$

These frames are of specific duration. Also, all these frames are overlapped with previous frames to keep away from missing of any edge characteristics of any frame. For all frames STE is calculated. Thus, an array of STE, {STEi}, is generated for the corresponding input speech signal.

Generally, standard deviation and mean of all frames are taken into account. But it is also known that mean and standard deviation give only an in-general concept of the distribution. For accurate learning the nature of STE co-occurrence matrix of STE is required to form.

Statistical features [1] are extracted from the co-occurrence matrix and those are used as features. Co-occurrence matrix is capable of capturing the repetitive characteristics of a certain feature. This cannot be measured by just considering mean and standard deviation. Positioning of different STE values surrounded in a region exhibits the characteristics of its related speech signal. As a result, a matrix CO_STE_m having dimension $Q \times Q$ (where $Q = \max\{STE_i\} + 1$) is formed.

A component in the matrix $CO\_STE_m(x, y)$ designates the numeral representation of incidences of STE x along with y in successive time instances. Co-occurrence

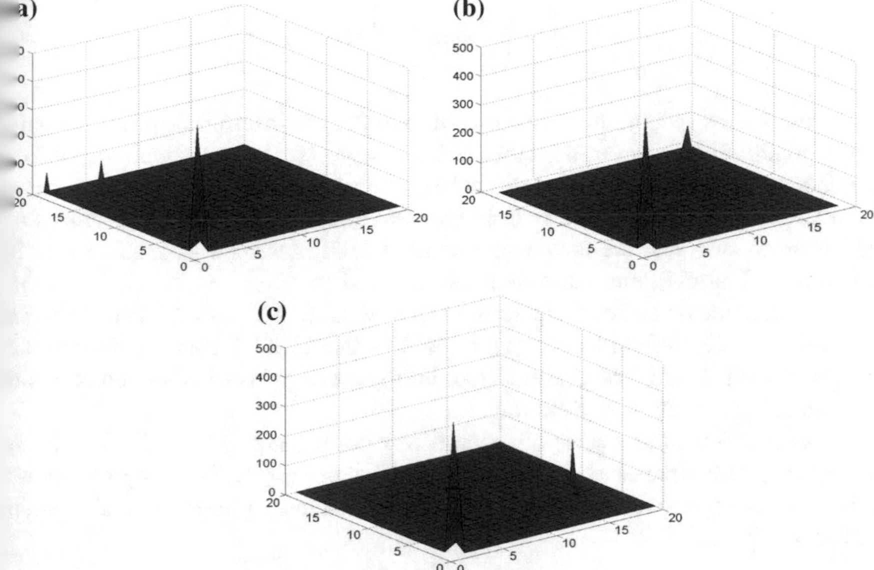

ig. 2 STE co-occurrence matrix of **a** male, **b** female and **c** trans-gender speech

natrix-depended [1] statistical features, like entropy, energy, correlation, contrast nd homogeneity, are then calculated.

Co-occurrence matrices of STE corresponding to male, female and trans-gender re represented in Fig. 2a, b and c, respectively. From these matrices it is clear that ccurrence character of STE is different for male, female and trans-gender.

3.1.2 Spectral Flux-Based Features

Spectral flux is also called as spectral variation. Change of power spectrum for an audio signal can be measured through spectral flux. Spectral flux can be described as the deviation value of spectrum between two successive frames in a very short time span.

It is observed that there exists certain amount of phonetic differences among male, female and trans-gender. These differences cannot be measured by using only time-domain features—rather frequency-domain features are also required to capture these phonetic differences. As spectral flux is capable of measuring the variation in power spectrum, it will also be able to capture the phonetic differences among the genders, because phonetic differences will generate differences in power spectrum in respective speech signal.

Spectral flux is computed as:

$$SF(p) = \sum_{i=0}^{n-1} s(p,i) - s(p-1,i) \quad ($$

SF(p) represents spectral flux corresponding to the *p*th spectrum. s(p,i) is the val corresponding to *i*th bin in the *p*th spectrum; s(p − 1, i) similarly corresponds the spectrum before p. Values collected from each bin of preceding spectrum a subtracted from that of present bin in the current spectrum. These differences a added up to calculate the concluding value, which is termed as spectral flux f spectrum p. For all frames these steps are repeated.

To capture the repetitive characteristics of spectral flux, a co-occurrence matrix spectral flux $CO\_SPEC\text{-}FLUX_m$ is formed, like that of STE. Hence co-occurrenc matrix-based [1] statistical features, like entropy, energy, correlation, contrast an homogeneity, are extracted from it.

Co-occurrence matrix plots of spectral flux corresponding to male, female an trans-gender are represented in Fig. 3a, b and c, respectively. From these plots it i evident that occurrence character of spectral flux is quite different for male, femal and trans-gender.

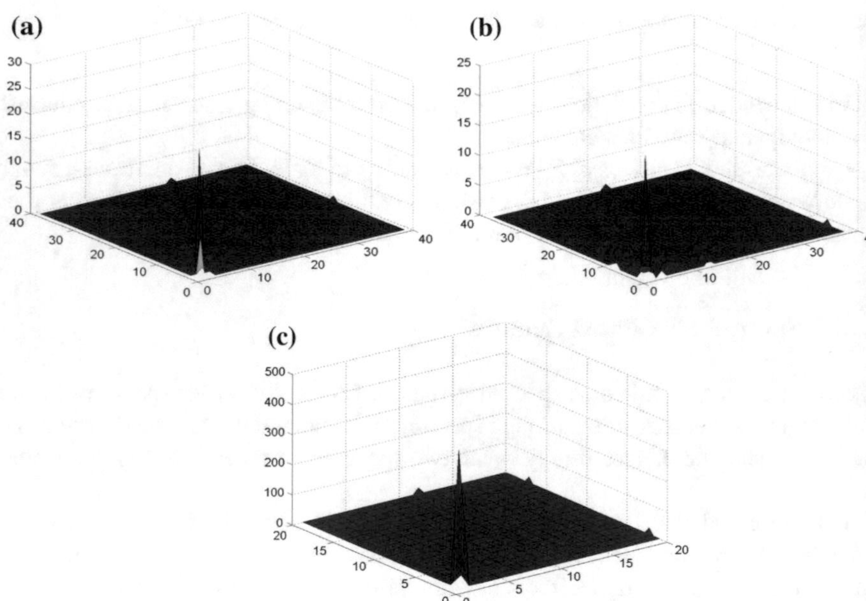

Fig. 3 Spectral flux co-occurrence matrix for **a** male, **b** female and **c** trans-gender speech

1.3 Skewness-Based Feature

is observed that male, female and trans-gender speeches exhibit certain amount ' asymmetry among them. This asymmetry cannot be measured through STE or ɔectral flux, rather a perceptual feature is required which can capture this asymmetry.

Skewness represents the third-order moment of the spectral circulation. It is a ɜrceptual statistical feature. Skewness can measure the irregularity or asymmetry resent for a certain normal distribution related to its mean position. Hence skewness ın measure the asymmetry present among male, female and trans-gender. Skewness *kn* is measured using the following Eq. (3).

$$skn = \frac{E(d-\mu)^3}{\sigma^3} \tag{3}$$

/here μ represents the mean of sampled data d, standard deviation of d is indicated y σ, and the expected value of the quantity p is denoted by $E(p)$.

Finally, an 11-dimensional acoustic feature set is generated (five statistical features rom co-occurrence matrix of STE, five statistical features from co-occurrence matrix f spectral flux and one skewness).

3.2 *Classification*

The foremost purpose of the proposed scheme is to propose a small-dimensional acoustic feature set which will be able to identify male, female and trans-gender very well. To test the discriminating power of the proposed feature set, some standard as well as well-known classifiers like neural network (NN), k-nearest neighbor (k-NN) and Naïve Bayes have been applied. Multi-layer perceptron (MLP) is used to implement neural network (NN). An audio dataset has been generated comprising 300 files. All of these audio files belong to male, female and trans-gender speech voices. Each category contributes equal division in this custom audio dataset. For classification purpose the whole audio dataset was divided into training data set and testing data set. To form training and testing dataset, half of the audio files from each group are considered. To implement MLP type NN, 11 neurons are considered in the input layer, indicating 11 features values; three neurons are considered in the output layer reflecting three categories of audio files—male, female and trans-gender; six neurons are considered in the hidden layer. k-NN classifier is used with k = 5, nearest neighbor rule intended for tie-break and distance metric as “City Block.” Naïve Bayes’ classifier has been used with default configuration.

4 Experimental Results

For classification NN, k-NN and Naïve Bayes have been applied. All of these class fiers are supervised classifiers. The experiment has been conducted with 300 spee files; 100 speech files are considered for male, female and trans-gender. All the speech files are of 90 s duration and of type mono. The dataset has been collecte from various sources, including Internet, recording of live audio performances. T reflect real-life scenario, some of the speech files are considered as noisy too. The speech files are quantized with 16-bit. Weka tool [14] has been used for implement tion of classifiers. 50% of each category of speech files is considered for training an testing purpose. After performing experiment, test and train datasets are reverse Then same classification task has been performed and average of these two stages i considered as final classification result, which is tabulated in Table 1.

4.1 Comparative Analysis

Discriminating strength of proposed feature set has been compared with some pre vious research works. Ali et al. [10] and Pahwa and Aggarwal [17] have alread paid some efforts in this field. Feature sets proposed by Ali et al. [10] and Pahw and Aggarwal [17] have been taken into account for comparative analysis. To mak comparative analysis judicial, the feature sets proposed by them have been applie and extracted from the present dataset and then fed into the same classifiers used i the suggested scheme. The comparative analysis is tabulated in Table 2.

From relative analysis it is clear that the suggested feature set computed on the combination of co-occurrence matrix-based statistical features and skewness-base

Table 1 Classification accuracy of male, female and trans-gender speech

Name of classifier	Male (%)	Female (%)	Trans-gender (%)
Naïve Bayes	95.6	94.6	93.3
Neural network (multi-layer perceptron)	94.3	93.3	92.3
k-nearest neighbor (k-NN)	93.3	92.6	91.6

Table 2 Comparative analysis in terms of classification accuracy

Previous approach	Male (%)	Female (%)	Trans-gender (%)
Ali et al. [10]	92.3	91.6	90.6
Pahwa and Aggarwal [17] (considering MFC1)	92.6	91.3	91.3
Pahwa and Aggarwal [17] (ignoring MFC1)	91.6	91.3	90.6
Proposed work	95.6	94.6	93.3

ature performs better than previous two approaches. So it can be said that the ggested scheme may be adopted as a better approach compared to those previous proaches.

Conclusion

he main goal of this system is to form a good feature set for identification of gender y classifying male, female and trans-gender speeches. As trans-gender has gained gal recognition, they are required to be considered in any gender identification ystem. Experimental result exhibits that the proposed approach can identify all the hree genders with an encouraging result. In this work co-occurrence matrix-based tatistical feature plays the most important contribution in the feature set. In future, ach category of gender may be further sub-divided based on age group.

References

1. Haralick, R.M., Shapiro, L.G.: Computer and Robot Vision, vol. 1 (1992)
2. Harb, H., Chen, L.: Gender identification using a general audio classifier. In: Proceedings of International Conference on Multimedia and Expo, 2003. ICME'03. vol. 2, pp. II-733. IEEE (2003)
3. Malhi, A., Gao, R.X.: PCA-based feature selection scheme for machine defect classification. IEEE Trans. Instrum. Meas. **53**(6), 1517–1525 (2004)
4. Subramanian, H., Rao, P., Roy, S.D.: Audio signal classification. In: EE Dept, IIT Bombay, pp. 1–5 (2004)
5. Lartillot, O., Toiviainen, P., Eerola, T.: A Matlab toolbox for music information retrieval. In: Data analysis, machine learning and applications, pp. 261–268. Springer, Berlin, Heidelberg (2008)
6. Jabid, T., Kabir, Md H., Chae, O.: Gender classification using local directional pattern (LDP). In: 2010 20th International Conference on Pattern Recognition (ICPR), pp. 2162–2165. IEEE (2010)
7. Bach, J.H., Anemüller, J., Kollmeier, B.: Robust speech detection in real acoustic backgrounds with perceptually motivated features. Speech Commun. **53**(5), 690–706 (2011)
8. Grosche, P., Müller, M.: Tempogram toolbox: Matlab implementations for tempo and pulse analysis of music recordings. In: Proceedings of the 12th International Conference on Music Information Retrieval (ISMIR), Miami, FL, USA (2011)
9. Grosche, P., Müller, M.: Extracting predominant local pulse information from music recordings. IEEE Trans. Audio Speech Lang. Process. **19**(6), 1688–1701 (2011)
10. Ali, Md S., Islam, Md S., Hossain, Md A.: Gender recognition system using speech signal. Int. J. Comput. Sci. Eng. Inf. Technol. (IJCSEIT) **2**(1), 1–9 (2012)
11. Müller, M., Ewert, S.: Chroma toolbox: MATLAB implementations for extracting variants of chroma-based audio features. In: Proceedings of the 12th International Conference on Music Information Retrieval (ISMIR) (2012)
12. Richard, G., Sundaram, S., Narayanan, S.: An overview on perceptually motivated audio indexing and classification. Proc. IEEE **101**(9), 1939–1954 (2013)
13. Ghosal, A., Dutta S.: Automatic male-female voice discrimination. In: 2014 International Conference on Issues and Challenges in Intelligent Computing Techniques (ICICT), pp. 731–735. IEEE (2014)

14. Srivastava, S.: Weka: a tool for data preprocessing, classification, ensemble, clustering a association rule mining. Int. J. Comput. Appl. **88**(10) (2014)
15. Alías, F., Socoró, J.C., Sevillano, X.: A review of physical and perceptual feature extracti techniques for speech, music and environmental sounds. Appl. Sci. **6**(5), 143 (2016)
16. Kumar, N., et al.: Robust multichannel gender classification from speech in movie audio. Proceedings of Interspeech 2016, pp. 2233–2237 (2016)
17. Pahwa, A., Aggarwal, G.: Speech feature extraction for gender recognition. Int. J. Image Grap Signal Process. **8**(9), 17–25 (2016)
18. Ranjan, S., Hansen, J.H.: Improved gender independent speaker recognition using conv lutional neural network based bottleneck features. In: Proceedings of Interspeech 201 1009–1013 (2017)
19. Simpson, Adrian P., Funk, Riccarda, Palmer, Frederik: Perceptual and acoustic correlates gender in the prepubertal voice. In: Proceedings of Interspeech 2017, 914–918 (2017)
20. Alipoor, G., Samadi, E.: Robust Gender Identification using EMD-Based Cepstral Feature Asia-Pac. J. Inf. Technol. Multimed. **7**(1) (2018)
21. Doukhan, D., Carrive, J., Vallet, F., Larcher, A., Meignier, S., Le Mans, F.: An open-sour speaker gender detection framework for monitoring gender equality. In: International Confe ence on Acoustics, Speech and Signal Processing (ICASSP) (2018)
22. Safavi, S., Russell, M., Jančovič, P.: Automatic speaker, age-group and gender identificatio from children's speech. Comput. Speech Lang. **50**, 141–156 (2018)

uzzy Posture Matching for Pain .ecovery Using Yoga

hona Ghosh, Sriparna Saha and Amit Konar

.bstract Owing to the present sedentary lifestyle, most of the office employees uffer from pain at different body parts and in turn they need to consult with doctors requently. The doctors prescribe certain exercises (mostly in the form of yoga) to e patients to overcome the stiffness and pain. Yoga is an ancient practice which nvolves specific movements, and breathing exercises used to improve overall health. Vith the rapid advancement in computer vision and video processing technologies, utomatic detection of yoga postures' correctness will help to motivate the patients to o them accurately. Here lies the novelty of our proposed work of automatic detection f yoga postures for pain recovery using type-1 fuzzy set. As care must be taken that he exercises are done appropriately, so a doctor-consulted database of yoga postures as been taken into account. The postures are detected by Kinect sensor first; then the ecognized skeletal forms of the postures are processed to obtain the feature space. Vhile performing the yoga, the exercise pattern may vary for different subjects due o their body structure. So to deal with this uncertainty, fuzzy logic has been used ere. The proposed system is an innovative work in this domain and outperforms all he existing ones in other posture recognition domains also.

Keywords Yoga · Pain recovery · Type-1 fuzzy set · Kinect sensor

---

A. Ghosh (✉) · S. Saha
Department of Computer Science and Engineering, Maulana Abul Kalam Azad University of Technology, Kolkata 741249, West Bengal, India
e-mail: ahonaghosh95@gmail.com

S. Saha
e-mail: sahasriparna@gmail.com

A. Konar
Department of Electronics and Tele-Communication Engineering, Jadavpur University, Kolkata 700032, West Bengal, India
e-mail: konaramit@yahoo.co.in

A. K. Das et al. (eds.), *Computational Intelligence in Pattern Recognition*, Advances in Intelligent Systems and Computing 999,
https://doi.org/10.1007/978-981-13-9042-5_82

1 Introduction

Pain felt at different body parts can interfere with a person's quality of life a general functioning [1]. Acute pain is temporary ache, which is felt generally af some injury or accident. Chronic pain is permanent, which is generally originat from fibromyalgia or arthritis. However, people who suffer from chronic pain ne long-term treatment and therapy to improve the situation. Here, our target patien can belong to both of these two categories, as yoga has been found as an efficie way of exercising to relieve these types of pain [2].

To overcome the limitations of existing related works, we have proposed a sy tem for pain recovery using yoga. Here, the patient performs exercises by watchin exercise-videos containing the yoga postures by trained instructors. These exercis videos are already captured using Kinect sensor [3] and metadata from the exercise is stored. The Kinect sensor along with software development kit (SDK) detec skeleton structure of any human being present in front of the camera. After prepr cessing the acquired skeletons, 14 features in terms of distance and angle betwee different body joints are extracted. As the feature values extracted from same exe cise vary widely due to change in human's body structures as well as habits, thu to deal with this uncertainty Gaussian type-1 membership curves [4, 5] have bee generated for each feature. When the patient is asked to perform a specific exercise then he/she is monitored using Kinect and the same set of features are considered Based on the membership values after matching with stored training dataset, a deci sion has been taken whether the patient is doing the exercise correctly or not. Thu an on-line system is proposed where the patient can perform yoga for pain recover with correct postures while in comfort of his/her home.

The paper is divided into five sections. Some related works are discussed in Sect. 2 Section 3 describes the methodology of our proposed work. Experimental results are shown in Sect. 4. The concluding statements are given in Sect. 5.

2 Literature Survey

Various works have been done in this field. In [6], Reis et al. present a review on rehabilitation of patients by performing exercises using Kinect sensor. Here three types of disabilities have been mostly investigated, including physical, intellectual and visual disability. The current work deals with physical disabilities. In [7], a k-nearest neighbors (k-NN) algorithm has been used to track object. k-NN is a classification process based on a locally approximated function. Li et al. develop a system for human—computer interaction with bare hand which builds a virtual keyboard. In [8], Le et al. explore the ability of Kinect to recognize four postures (lying, sitting, standing and bending) using skeleton information in the context of a health monitoring framework. Support vector machine (SVM) has been selected as classifier because of its ability to work with high-dimensional data. SVM is a supervised learning algorithm which ıssifies based on a hyperplane classifier. In [9], a yoga posture recognition system s been proposed which can recognize the yoga postures presented by the trainee d then retrieve yoga training process from the Internet to make him/her attentive to e posture. The work is limited to only some postures and can be extended to adapt ore exercises. In [10], Nirmal et al. develop a game where the users can interact ith yoga trainer and get audio-based directions based on six standing yoga poses. his work is demonstrated for blind or low-vision people. In [11], a distance-based ɔga learning system has been developed for players to calculate a learning score by atching the distance transformation of the player silhouette with stored standard oga posture. The graphical user interface and the processing speed are not up to the ark here.

Methodology

'his section includes the introduction to the data acquisition device Kinect sensor, long with the description of three stages of posture matching (Fig. 1).

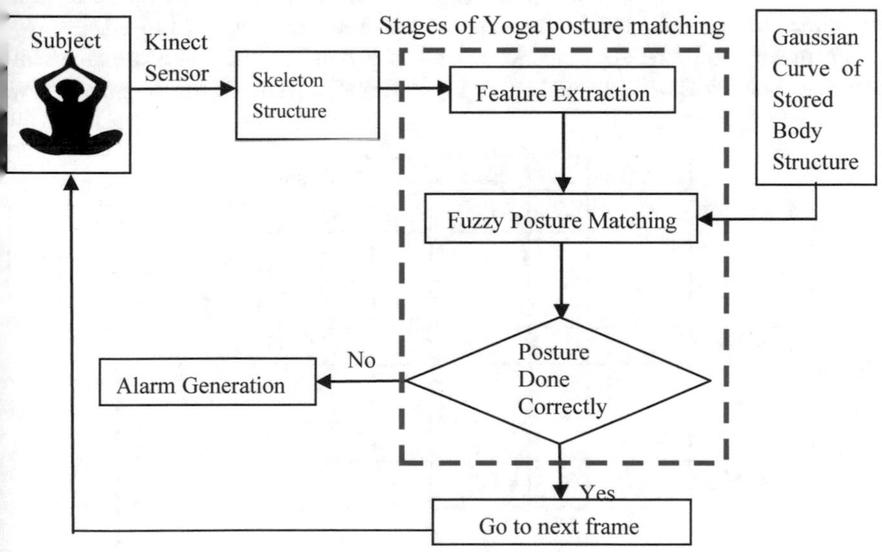

Fig. 1 Block diagram of the overall system

3.1 Kinect Sensor

The Kinect sensor [3] along with SDK tracks motion of human while standing/sitti within 1.2–3.5 m range by detecting his/her 20 body joints in x, y and z coordinat Here, the kinetic sampling rate is taken as 10 frames per second.

3.2 Implementation of Type-1 Fuzzy Set for Pain Recovery Using Yoga

The proposed method involves a training dataset T_e which is represented in Fig. 2 fc any exercise set *e*. In T_e, let a total of *N* number of subjects have participated. Fro each particular subject n $(1 \leq n \leq N)$, *R* number of frames have been taken which ca explicitly describe that exercise *e*. In yoga, a proper form while doing the exercis should be maintained; otherwise there will be adverse effect. Thus each frame c the unknown subject should be matched with the corresponding frame of the yog instructor. Hence, frame-by-frame matching is carried out in the proposed work. Bu we have noticed that while doing the yoga, uncertainties creep during performin the same posture. So to tackle this, type-1 fuzzy set is implemented [4, 5].

From any *r*th frame $(1 \leq r \leq R)$, a total of *F* number of features are taken int account. For *r*th frame of *f*th feature $(1 \leq f \leq F)$, from all the *N* subjects, w

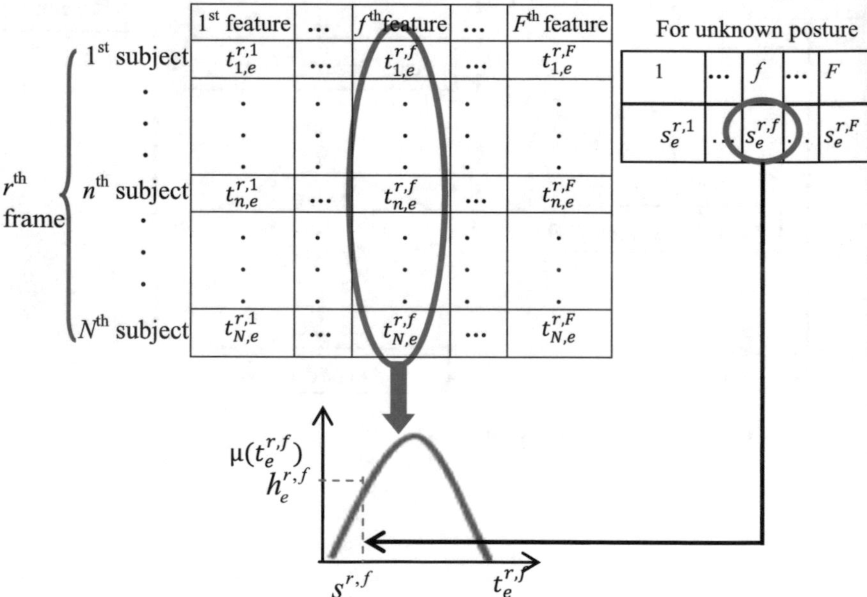

Fig. 2 Calculation of membership value for unknown posture from *r*th frame and *f*th feature

ve a feature space denoted by $\left[t_{e,1}^{r,f},\ldots,t_{e,n}^{r,f},\ldots,t_{e,N}^{r,f}\right]$. From these data points, we ve calculated the mean ($M_e^{r,f}$) and standard deviation ($\sigma_e^{r,f}$) values. The standard viation σ can be denoted as

$$\sigma = \sqrt{\frac{1}{N}\sum_{i=1}^{N}(x_i - M)^2} \quad (1)$$

here N = total number of samples.

Based on these $M_e^{r,f}$ and $\sigma_e^{r,f}$, a fuzzy Gaussian curve is generated with member alue of $\mu\left(t_e^{r,f}\right)$, as pictorially depicted in Fig. 2. The symmetric Gaussian function denoted by

$$f(x;\sigma, M) = e^{\frac{-(x-M)^2}{2\sigma^2}} \quad (2)$$

Now for the unknown posture, the same F number of features is calculated for ach frame. Thus for a particular rth frame, we have a feature set denoted by $s_e^{r,1},\ldots,s_e^{r,f},\ldots,s_e^{r,F}$]. The fth feature of unknown posture is matched with the lready generated fuzzy Gaussian curve $\mu\left(t_e^{r,f}\right)$. The strength of $s_e^{r,f}$ is measured as $h_e^{r,f}$. The same procedure is repeated for all the F number of features.

The strength values obtained from F features are $[h_e^{r,1},\ldots,h_e^{r,f},\ldots,h_e^{r,F}]$. From hese strength values, the minimum value $h_e^{r,\min}$ is taken to consider the contribution of all the features. If this $h_e^{r,\min}$ is greater than an empirically decided threshold (τ), hen it can be stated that the patient is performing the given exercise correctly for rth frame (Fig. 1). The above-stated procedure is repeated for all the R frames.

$$h_e^{r,\min} = \min\left(h_e^{r,1},\ldots,h_e^{r,f},\ldots,h_e^{r,F}\right) \quad (3)$$

4 Experimental Results

The following section deals with different aspects of performance of the proposed work.

4.1 Dataset Preparation

The proposed work is undertaken at Jadavpur University Artificial Intelligence Laboratory, where the research scholars are asked to participate in the data preparation. For the training dataset, 25 male (age 28 ± 7 years) and 9 female (age 31 ± 4 years) students' data are processed and for testing purpose, 13 male (age 27 ± 5 years) a 10 female (age 29 ± 3 years) are requested to perform the exercise sets.

4.2 Description About Exercises

The seven exercises taken into account are widely popular and have been shown Fig. 3 [12]. Spinal twist (Fig. 3a) stretches the back muscles. It lengthens, relaxes an realigns the spine. Hand stretching (Fig. 3b) removes the stiffness in shoulder, elbo wrist and fingers. Surya Pranam or Sun Salutation (Fig. 3c) increases the flexibili of the total body. It serves as an all-purpose yoga tool. Seated child (Fig. 3d) comfor the strain and tiredness by reducing back and neck pain when performed with hea torso support. The sitting triceps stretch (Fig. 3e) focuses on the back of the uppe

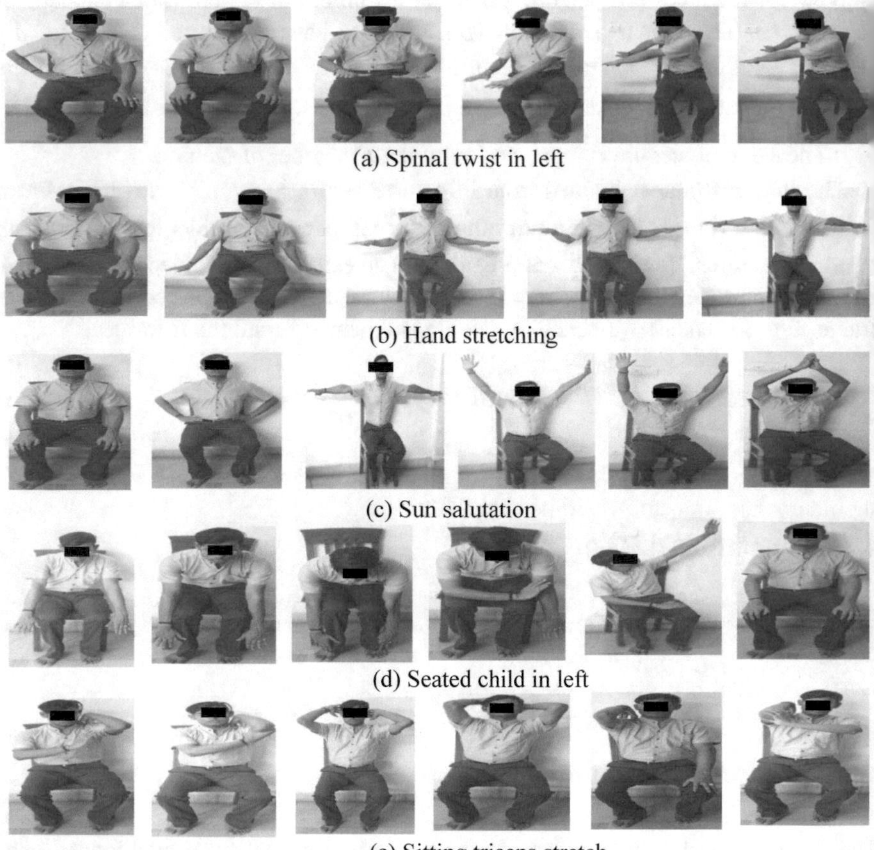

(a) Spinal twist in left

(b) Hand stretching

(c) Sun salutation

(d) Seated child in left

(e) Sitting triceps stretch

Fig. 3 RGB images of the concerned exercise sets for pain recovery using yoga

m and stretches the triceps muscles. The spinal twist and seated child are done in t and right, both the sides similarly.

3 Feature Extraction

total of 14 features are extracted for the proposed algorithm for fuzzy posture atching using Kinect sensor.

1. Distance between spine and elbow left
2. Distance between spine and elbow right
3. Distance between spine and wrist left
4. Distance between spine and wrist right
5. Distance between hand right and hand left
6. Distance between shoulder center and hand left
7. Distance between shoulder center and hand right
8. Distance between wrist left and elbow right
9. Distance between wrist right and elbow left
10. Angle between the line joining shoulder left and elbow left and the line joining elbow left and wrist left

Angle between the line joining shoulder right and elbow right and the line joining lbow right and wrist right

12. Angle between the line joining spine and shoulder left and the line joining shoulder left and elbow left
13. Angle between the line joining spine and shoulder right and the line joining shoulder right and elbow right
14. Angle between the line joining shoulder center and spine and the line joining spine and hip center

The distance features are normalized using the distance between shoulder left and shoulder right joints to overcome the effect of different body structures.

4.4 Recognition of Correctness of Yoga Postures from Unknown Subject

A set of images with corresponding skeletons for an unknown subject while performing a given exercise 'spinal twist in right' is provided in Fig. 4. As it is not feasible to give the strength calculation results for all the frames, thus only for frame number 70 the result is provided in Table 1.

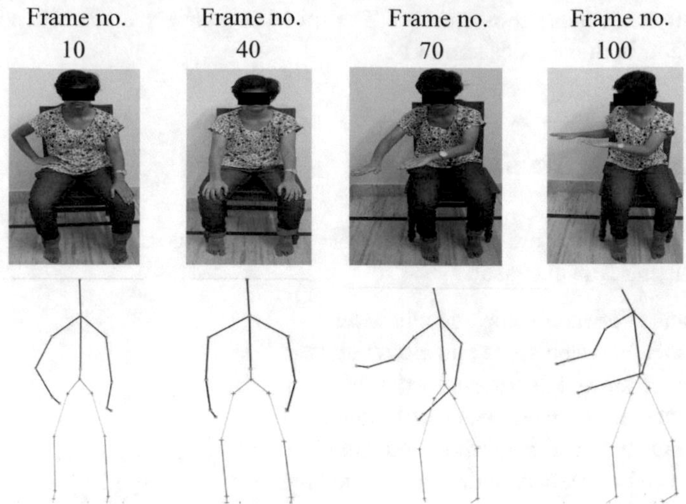

Fig. 4 Postures performed by unknown subject while doing "spinal twist in right"

Table 1 Calculation of strength for frame number 70 from Fig. 4

Feature number	Feature value $\left(s_e^{r,f}\right)$	Strength $\left(h_e^{r,f}\right)$	Minimum strength $\left(h_e^{r,\min}\right)$	Threshold (τ)	Decision
1	0.576	0.6432	0.63	0.6	The posture is performed correctly
2	0.580	0.8718			
3	0.824	0.7777			
4	0.776	0.8595			
5	1.142	0.6900			
6	1.413	0.8032			
7	1.572	0.9341			
8	1.180	0.6341			
9	1.141	0.8912			
10	151.503	0.9991			
11	26.093	0.8029			
12	130.099	0.6963			
13	127.072	0.9929			
14	41.375	0.7612			

5 Performance Analysis

order to evaluate the performance of the proposed algorithm, we have considered four performance metrics, namely recall, precision, accuracy and error rate. om confusion matrix, four parameters, namely true positive (*TP*), true negative N), false positive (*FP*) and false negative (*FN*), can be measured. Based on these arameters, we can calculate the following metrics.

$$\text{Recall} = \frac{TP}{TP + FN} \tag{4}$$

$$\text{Precision} = \frac{TP}{TP + FP} \tag{5}$$

$$\text{Accuracy} = \frac{TP + TN}{TP + TN + FP + FN} \tag{6}$$

$$\text{Average error rate} = \frac{FP + FN}{TP + TN + FP + FN} \tag{7}$$

The results obtained (in percentage) for all these metrics are given in Fig. 5, for ll the seven exercises. From that figure, it can be conferred that the proposed work s well suited for yoga exercises. The proposed work is also compared with the xisting literatures given in [8–11]. In [8], skeletal form of human gestures is used o recognize only four postures, while [9–11] consider depth view of the postures, where intricacies of postures are lost. The comparative results are given in Fig. 6. From the figure, it can be concluded that our proposed work is the best choice for yoga posture recognition.

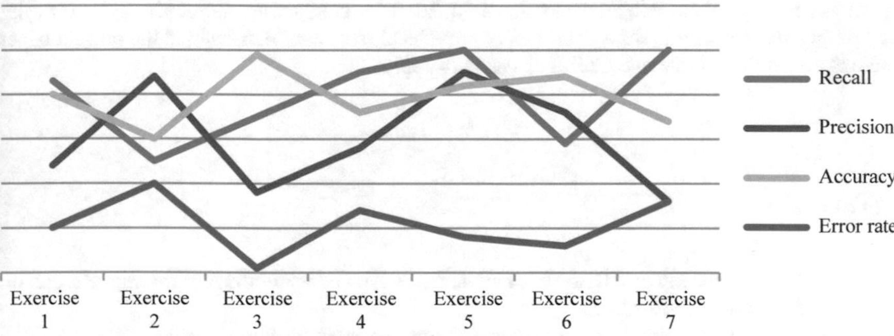

Fig. 5 Comparison of seven different exercises

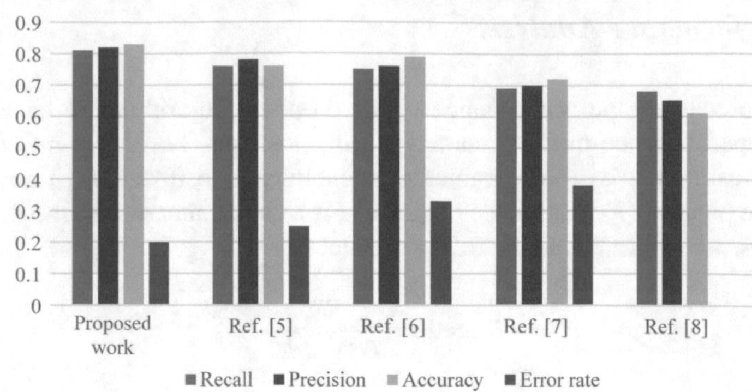

Fig. 6 Comparison of proposed work with existing ones

5 Conclusion

This paper presents a novel approach to posture recognition of yoga by using type-fuzzy set with overall 83% accuracy. We have used Kinect sensor which does nc need refreshing time and can operate throughout the day. As we have taken only th skeletons of the subjects, thus the privacy is also preserved here. The work is wel suited for any age group of patients trying to recover from pain. The system als provides flexibility to the patients, as they do not need to go for frequent visit to th hospitals. The system can be accessed any time of the day in any environment, so i could be very beneficial to the patients.

Acknowledgments The funding of University Grant Commission India (UGC/UPEII/CogSc) i Cognitive Science, Jadavpur University is gratefully acknowledged for the present work. Corresponding author is thankful to the Ethics Committee formed in Maulana Abul Kalam Azad University of Technology, West Bengal dated: April 30, 2019 for conducting, monitoring and certifying the primary survey for this study. The survey provides the primary data used in the present paper. There is no conflict of interest. Useful disclaimers apply.

References

1. https://www.ninds.nih.gov/Disorders/Patient-Caregiver-Education/Hope-Through-Research/Pain-Hope-Through-Research
2. https://yogainternational.com/article/view/restorative-yoga-for-chronic-pain
3. https://en.wikipedia.org/wiki/Kinect
4. Saha, S., Bhattacharya, S., Konar, A.: Comparison between type-1 fuzzy membership functions for sign language applications. In: 2016 International Conference on Microelectronics, Computing and Communications (MicroCom), pp. 1–6. IEEE (2016)
5. Klir, G.J., Yuan, B, Zadeh, L.A.: Fuzzy Sets, Fuzzy Logic, and Fuzzy Systems: Selected Papers. World Scientific Publishing Co., Inc (1996)

. Reis, H., Isotani, S., Gasparini, I.: Rehabilitation using Kinect and an outlook on its educational applications: a review of the state of the art. In: Brazilian Symposium on Computers in Education (Simpósio Brasileiro de Informática na Educação-SBIE), vol. 26(1), p. 802 (2015)

. Li, Y.: Hand gesture recognition using Kinect. In: 2012 IEEE 3rd International Conference on Software Engineering and Service Science (ICSESS), pp. 196–199. IEEE (2012)

. Le, T.L., Nguyen, M.Q.: Human posture recognition using human skeleton provided by Kinect. In: 2013 International Conference on Computing, Management and Telecommunications (ComManTel), pp. 340–345. IEEE (2013)

. Chen, H.T., He, Y.Z., Hsu, C.C., Chou, C.L., Lee, S.Y., Lin, B.S.P.: Yoga posture recognition for self-training. In: International Conference on Multimedia Modeling, pp. 496–505. Springer, Cham (2014)

. Nirmal, M.D., Chandrakant, B.K., Pramod, K.J., Vilas, B.V., Kishor, S.S.: Eye free yoga-an exergame using Kinect. Int. J. Sci. Technol. Res. **4**(4), 13–16 (2015)

1. Hsieh, C.C., Wu, B.S., Lee, C.C.: A distance computer vision assisted yoga learning system. J. Comput. **6**(11), 2382–2388 (2011)

2. https://www.youtube.com/watch?v=KEjiXtb2hRg

.pplication of SCA-Based Two Degrees f Freedom PID Controller for AGC tudy

imai Charan Patel, Binod Kumar Sahu and Manoj Kumar Debnath

bstract The presented article illustrates an effort to integrate a geothermal power lant (GTPP) and a dish-stirling solar-thermal system (DSTS) to conventional eam power plant (SPP) and to investigate the dynamic performance of the sys-m for automatic generation control. Thus, each area of the system has a GTPP,)STS and a SPP. Governor dead band of 0.036 and generation rate constraint GRC) of 3% are taken for the steam power plant. The system dynamics of a two-qual area system is studied using two different controllers independently, that is roportional–integral–derivative (PID) controller and two degrees of freedom PID 2DOF-PID) controller. Particle swarm optimisation and sine cosine algorithm (SCA) .re used to tune the parameters of the controllers and to tune the governor time con-tant and turbine time constant of the GTPP. The study reveals the superiority of ;CA-optimised 2DOF-PID controller than others.

Keywords Automatic generation control · Sine cosine algorithm · Two degrees of reedom PID controller

1 Introduction

Present-day power system is a large network interconnecting various utilities with complex system dynamics. Exchange of power between the utilities takes place through the tie-lines. It is expected that the power system should be able to deliver secured, reliable and uninterrupted power supply to the consumers without any sig-

N. C. Patel
Government College of Engineering, Keonjhar, Odisha 758002, India
e-mail: ncpatel.iter@gmail.com

B. K. Sahu (✉) · M. K. Debnath
Institute of Technical Education and Research, Siksha 'O' Anusandhan University, Bhubaneswar, Odisha 751030, India
e-mail: binoditer@gmail.com

M. K. Debnath
e-mail: mkd.odisha@gmail.com

A. K. Das et al. (eds.), *Computational Intelligence in Pattern Recognition*, Advances in Intelligent Systems and Computing 999,
https://doi.org/10.1007/978-981-13-9042-5_83

nificant deterioration of the power quality. But, to achieve this is not a simple ta due to the complex configuration of the present-day power system. The voltage pr file of the system and the system frequency define the power quality of the pow system, and therefore, they must be maintained within the specified nominal value During power system operation, any sudden disturbance or active load change ma cause the oscillation of the system frequency [1, 2]. In such cases, it is essential keep the frequency oscillations within the limit and at the same time, the frequen oscillations must be damped out within the shortest possible time to retain the pow quality. In order to achieve the above, it is required to balance the active load dema and active power generation. The control technique which ensures the proper balan between the generation and load demand is known as automatic generation contr (AGC). Thus, AGC has the main role for frequency regulation in interconnecte power system.

The idea of AGC was first conceptualised by Cohn in the year 1956 [3]. Late on, many researchers have contributed their efforts in AGC using different contro methodology. Optimum control approach for AGC was recommended by Elgerd an Fosha [4]. Pan and Liaw demonstrated the implementation of an adaptive controlle for AGC of power system [5]. Adaptive fuzzy gain scheduling for AGC was describe by Talaq and Al-Basri [6]. Study of load frequency control (LFC) issues with differen conventional controllers such as PI, integral–derivative (ID), PID and integral double derivative (IDD) controllers have been illustrated in many literatures [7–9]. Thes works mainly emphasise on AGC with conventional energy sources and no effort ha been made for the LFC problem with nonconventional energy sources. Owing to th environmental issues and depletion of conventional energy sources, nonconventiona energy sources will be widely used in near future for power generation. Besides this nonconventional energy sources are abundant and reported in the literature [10] Sharma et al. implemented various classical controllers to study the performance o AGC of a thermal system consisting of three areas in which solar thermal powe plant (STPP) is incorporated in one of the areas [11].

Study of AGC of an unequal two-area thermal system incorporating dish-stirling solar thermal system (DSTS) with various classical controllers such as integral (I), proportional–integral (PI), proportional–integral–derivative (PID) and three degrees of freedom PID (3DOF-PID) have been reported in the literature [12]. Stability analysis of hybrid system with DSTS as one of the energy source has been investigated by genetic algorithm-optimised classical controllers [13]. Hossain et al. [14] described the role of various renewable sources on smart grid. In recent years, geothermal energy has evolved as another possibility for power generation worldwide [15, 16]. A new control strategy cascading fractional-order PI and fractional-order PD optimised by sine cosine algorithm (SCA) is proposed by Tasnin et al. for AGC of interconnected multi-area power system incorporating DSTS and geothermal power plants (GTPPs) [17].

Although geothermal power and solar thermal power are utilised for electric power generation, but so far there has been no extensive research work in AGC for frequency regulation using GTPPs and DSTS. Hence, there is an ample scope of research to investigate the LFC problem incorporating GTPPs and DSTS with earlier existing nrenewable sources. In the present work, an equal two-area interconnected power stem consisting of a steam power plant (SPP), a DSTS and a GTPP in each area considered. Governor dead band (GDB) of 0.036 and generation rate constraint RC) of 3%/min are considered for steam power plant having reheat turbine. The namic response of the system is investigated by separately using PID and 2DOF-D controllers with a step load perturbation (SLP) of 1%. The controller parameters well as time constants of the turbine and GTPP governor are designed by using article swarm optimisation (PSO) and SCA optimisation technique.

System Investigated

igure 1 shows the model of the system under investigation, which consists of two qual areas interconnected together through a tie-line. Power exchange between the reas takes place over the tie-line. Each area is equipped with multiple sources, and e three different sources supplying power in each area are: (i) a GTPP, (ii) a SPP nd (iii) a DSTS. The steam power plant consists of a reheat turbine with GDB of .036 and GRC of 3%/min. Nominal values of different parameters of the power ystem under investigation are depicted in Appendix. Usually, the governor time onstant (Tgg) has the value around 0.1 s, while the time constant of turbine or steam

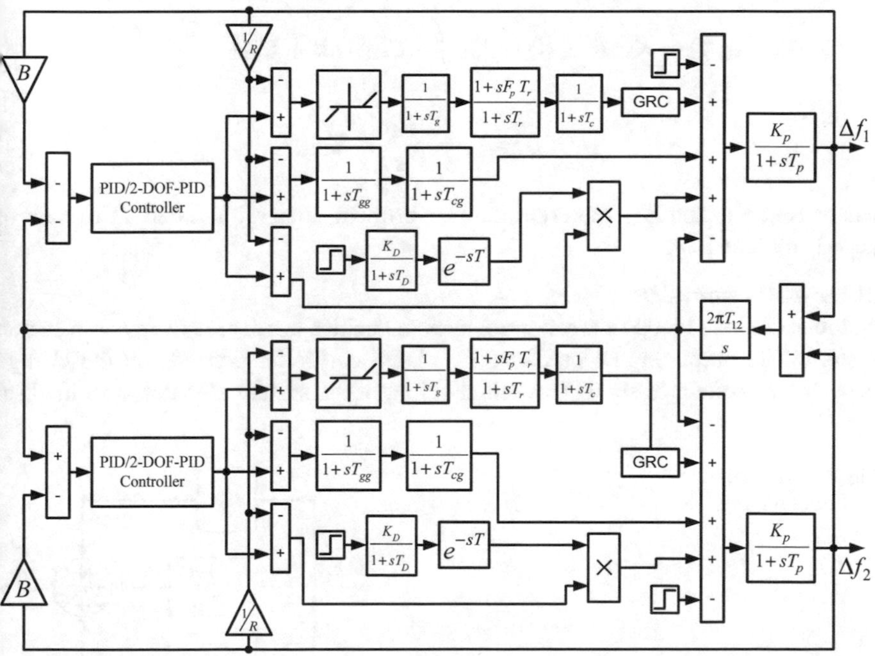

Fig. 1 Transfer function model of the power system under study

chest (Tcg) is in the range of 0.1–0.5 s [17]. The values of these time constants a optimised by using PSO and SCA. Area control errors of each area are the input the controller of the corresponding area and expressed as:

$$ACE_1 = \Delta P_{tie,1-2} + B_1 \Delta\omega_1 \quad (1)$$

$$ACE_2 = \Delta P_{tie,2-1} + B_2 \Delta\omega_2 \quad (2)$$

First, PSO and SCA-optimised PID controller is implemented and then PSO ar SCA-optimised 2DOF-PID controller is implemented to investigate the dynam performance of the system under consideration.

3 Controller Structures and Their Optimal Design

PID Controller:
The structural details of PID controller are described in Fig. 2. The proportiona integral and derivative actions are embedded together in parallel in a PID controlle The output of a PID controller of area-1 in time domain and its transfer function ar expressed in Eqs. (3) and (4), respectively.

$$u_1(t) = K_{p1} ACE_1(t) + K_{i1} \int_0^t ACE_1(t)dt + K_{d1} \frac{dACE_1(t)}{dt} \quad (3)$$

$$G(s) = K_{p1} + \frac{K_{i1}}{s} + K_{d1}s \quad (4)$$

where K_{p1}, K_{i1} and K_{d1} represent the proportional, integral and derivative gain of area-1, respectively.

2DOF-PID Controller:
Number of closed loop in any control system decides the degree of freedom in that system. The output signal produced by 2DOF controller depends on the change between the reference signal and actual output of the system. Separate specified set

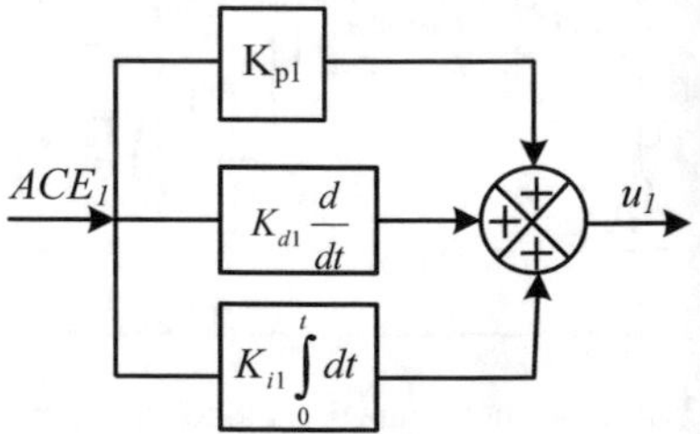

Fig. 2 Structure of PID controller for area-1

g. 3 Structure of OF-PID controller

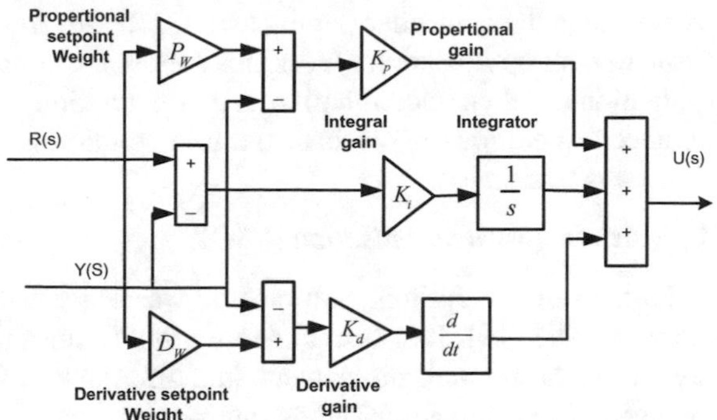

oint weights are assigned in the input to get a weighted difference signal for the erivative and proportional actions. There is no specified set point weights assigned n the input for the integral actions. The output of the controller is computed by umming these control actions, where these control actions are biased according to ne selected gain parameters.

The structure of 2DOF-PID controller is described in Fig. 3. R(s) denotes the eference input, Y(s) denotes the feedback from actual output of the system and J(s) denotes the signal output of the controller. K_p, K_i and K_d are the gains of PID ontroller and P_W and D_W are the proportional and derivative set point weights, espectively. Owing to the two independent control loops, the performance of 2DOF ontrol strategy is indeed superior to the conventional single degree of freedom ontrol strategy [18].

Objective Function:
The performance of any optimisation technique depends on the objective function used in the method. Hence, appropriate selection of objective function is very important in any optimisation technique. Optimisation technique with integral time absolute error (ITAE) as the objective function gives lowest settling time; therefore in this work, ITAE is chosen as the objective function or cost function for tuning the controller gains and the parameters of the GTPP. The mathematical expression of ITAE cost function is given by Eq. (5).

$$ITAE = \int_0^t (|\Delta f_1| + |\Delta f_2| + |\Delta P_{tie}|) \times tdt \quad (5)$$

Optimisation Technique:
The dynamic performance of any system is governed by controller gains and other system parameters. Hence, it is necessary to tune these gains and parameters for optimal system performance. Several optimisation techniques have been illustrated in the literature for tuning the controller gains and other parameters. In this article, Particle Swarm Optimisation (PSO) and Sine Cosine Algorithm (SCA) have been

used to tune the controller parameters, GTPP governor and turbine time constan Basically, all optimisation algorithms follow the common process of initialisatic evaluation and then updation till maximum iteration is reached. When the maximu number of iterations is reached, the best solution is selected and the optimisati process is terminated.

a. *Particle swarm optimisation (PSO)*:

Particle swarm optimisation method was suggested by Kennedy and Eberhart the year 1995 [19]. It is formulated by replicating the social behaviour of varioι organisms during their movements in a bird flock or fish school. The various ste involved in PSO are described as follows:

i. Initialisation: This step deals with random generation of initial population (x and initial velocity (v_i) of size (NPxD), where 'NP and D' are the number c population and dimension of the problem, respectively.
ii. Evaluation: In this step, the power system model is run and the fitness functio $f(x_i)$ of each particle are evaluated and the best fitness function is chosen. Th solution corresponding to best fitness function is set as old global best ($g_{best,old}$ and the initial population (x_i) is set as old local best ($p_{best,old}$).
iii. Updation: In this step, the initial velocity is updated and new particles ($x_{i,new}$ are generated using the following equations:

$$v_{i,new} = w * v_i + C_1 * rand(NP, D) * (p_{best} - x_i) + C_2 * rand(NP, D) * (g_{best\_old} - x_i) \quad (6$$

$$x_{i,new} = x_i + v_{i,new}. \quad (7$$

Thereafter, the fitness function $f(x_i)$ of each newly produced particle is evaluated and the best fitness function is chosen. The solution corresponding to this best fitness function is set as new global best ($g_{best,new}$) and the local best is updated by comparing the performance of $x_{i,new}$ and x_i. Depending upon the performance of $g_{best,new}$, g_{best} is updated and steps (ii) and (iii) are repeated until the maximum iteration is reached and finally the best solution is selected.

b. *Sine cosine algorithm (SCA)*:

SCA algorithm was developed by Mirjalili [20] and it is one of the population-based search algorithm. Like other algorithms SCA follows three steps: initialisation, evaluation and updation. It generates a random solution which oscillates towards or outwards the best solution as per the model of a trigonometric sine and cosine functions. Each candidate solution is evaluated by a fitness function and their positions are updated with the help of the following equations:

$$X_{i,new} = \begin{cases} X_i + r_1 \times \sin(r_2) \times |r_3 P_i - X_i| if\ r_4 < 0.5 \\ X_i + r_1 \times \cos(r_2) \times |r_3 P_i - X_i| if\ r_4 \geq 0.5 \end{cases} \quad (8)$$

here X_i denotes the current position and P_i is the paramount solution of the current osition and is the target point. The four important parameters are r_1, r_2, r_3 and r_4, here r_1 describes the movement direction. It describes the area of the subsequent osition, which may either lie between the solution and target or outside it. Here, $_2$ is between 0 and 2π, which denotes up to what range the movement must be owards or outwards the target and r_3 allots a random load for the target so as to mphasise (r_3>1) or de-emphasise (r_3<1) the effect of target to define the distance. astly, r_4 equally switches amongst the sine and cosine components. In order to have erfect balance amongst exploration and exploitation, and ultimately converge to the ptimum value, the range of sine and cosine in above equation is changed flexibly ith the help of the following equation:

$$r1 = a - t\frac{a}{T} \tag{9}$$

here t and T are the present and the maximum number of iteration, respectively, nd 'a' is a constant.

While the sine and cosine functions are in the range [1, 2] and [− 2, −1], SCA xplores the search space. On the other hand, when the ranges have the interval − 1, 1], exploitation of the search space takes place. The optimisation process n SCA begins with generation of random probable solutions, storing the greatest olutions obtained until now, assigning it as the target point and finally updating ther solutions with reference to the best one. In the meantime, the bounds of sine nd cosine functions are revised to emphasise exploitation of the search space as the number of iteration increases. Upon reaching maximum number of iterations, the optimisation process is ended.

4 Results and Discussion

This article presents comparative performance analysis of PID and two degree of freedom PID (2DOF-PID) controllers optimally designed through PSO and SCA techniques for AGC of a two-area six-unit system. Each area of the power system includes a reheat turbine type: thermal, geothermal and a dish-stirling solar-thermal power system. The power system under study is simulated in MATLAB Simulink environment and PSO and SCA techniques are written in .mfile. Simulink model is called through PSO and SCA programs to optimally design the controllers by minimising the objective function, that is ITAE. A step load disturbance of 10% is enforced in area-1 to investigate the dynamic performance of the system. Optimal controller gains and time constants of geothermal power system are given in Table 1. Frequency deviations in area-1 (Δf_1), area-2 (Δf_2) and tie-line power deviation (ΔP_{tie}) are presented in Figs. 4, 5 and 6, respectively. Undershoot (U_{sh}), overshot (O_{sh}) and settling time (T_s) of frequency and tie-line power fluctuations are depicted in Table 2.

Table 1 Optimal controller gains and time constants of geothermal power plant

Controllers	K_{p1}	K_{i1}		K_{d1}	K_{p2}		K_{i2}	K_{d2}		T_{gg}	T_{cg}
PSO-PID	0.9259	2.0000		0.2826	0.7282		1.7982	1.5070		0.0835	0.1000
SCA-PID	0.9465	2.0000		0.2350	1.3874		0.1590	1.0666		0.0100	0.1053
	K_{p1}	K_{i1}	K_{d1}	P_{w1}	D_{w1}	K_{p2}	K_{i2}	K_{d2}	P_{w2}	D_{w2}	T_{gg}	T_{cg}
PSO-2DOF-PID	0.5410	2.0000	0.1744	3.0880	0.0100	1.6015	0.0100	1.0404	5.0000	5.0000	0.0210	0.1000
SCA-2DOF-PID	**0.6666**	**2.0000**	**0.1883**	**2.5290**	**0.9532**	**1.7465**	**0.0370**	**1.2688**	**5.0000**	**5.0000**	**0.0189**	**0.1000**

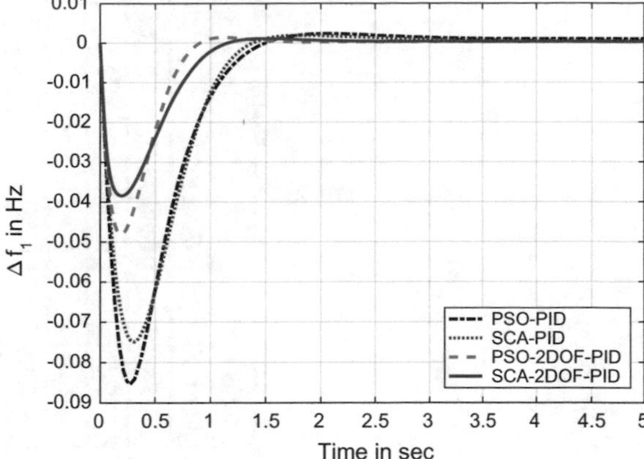

g. 4 Frequency ctuation in area-1

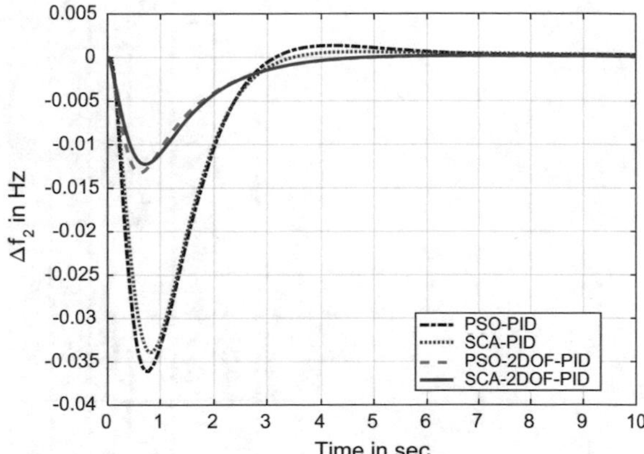

ig. 5 Frequency uctuation in area-2

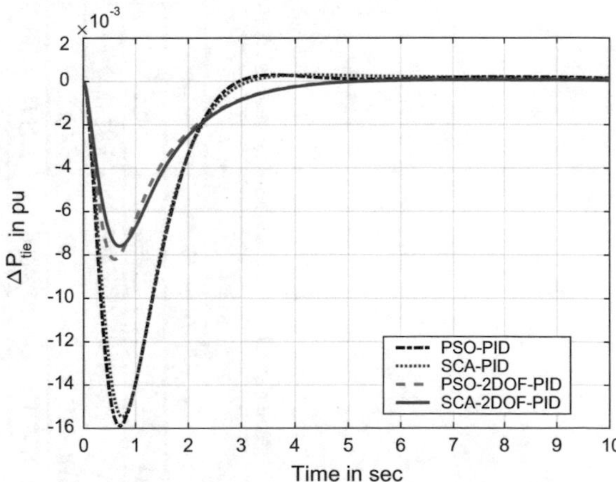

Fig. 6 Tie-line power fluctuation

Table 2 Transient performance analysis of various controllers

Controller	Δf_1			Δf_2			ΔP_{tie}		
	$U_s \times 10^{-3}$	$O_s \times 10^{-3}$	T_s	$U_s \times 10^{-3}$	$O_s \times 10^{-3}$	T_s	$U_s \times 10^{-3}$	$O_s \times 10^{-3}$	T_s
PSO-PID	−85.3498	2.3466	2.37	−36.1189	1.3415	2.68	−15.8997	0.2999	2.18
SCA-PID	−74.9010	1.9161	1.26	−33.9430	0.6200	2.78	−15.4766	0.2950	2.23
PSO-2DOF-PID	−48.1231	1.4252	**0.79**	−13.2158	**0.1772**	**2.69**	−8.2070	0.0891	**2.14**
SCA-2DOF-PID	**−38.4370**	**1.1141**	0.99	**−12.2447**	0.1998	2.70	**−7.6027**	**0.0614**	2.21

From Figs. 4, 5 and 6 and Table 2, it is evident that the performance of 2DOF-D controller is remarkable as compared with other PID controllers. Further, it is und that the overall performance of SCA-optimised 2DOF-PID controller is better an the PSO-optimised 2-DOF-PID controller as most of the transient parameters J_{sh}, O_{sh} and T_s) have minimum values when the power system is equipped with CA-optimised 2DOF-PID controller. Settling time and overshoot of Δf_1 are less ith PSO-optimised 2DOF-PID controller, but are very close to those obtained with CA-optimised 2DOF-PID controller. From Table 2 it is also clear that two degrees f freedom PID controller gives superior dynamic performance in comparison to PID ontroller.

Conclusion

n this work, DSTS and GTPP have been integrated with conventional steam power lant, and PID and 2DOF-PID controllers are selected for frequency regulation when ne system is subjected to an SLP of 10%. PSO and SCA are used to tune the controller arameters as well as governor and turbine time constant of the GTPP. It is demontrated that the 2DOF-PID controller gives superior dynamic performance than the PID controller in terms of various transient response indices, like minimum undershoots, settling time and peak overshoots. Further, it is seen that most of the response ndices have minimum values for SCA-optimised 2DOF-PID controller. Hence, it is concluded that the SCA-optimised 2DOF-PID controller displays superior dynamic performance.

References

1. Kundur, P., Balu, N.J., Lauby, M.G.: Power System Stability and Control, vol. 7. McGraw-Hill, New York (1994)
2. Elgard, O.I.: Electric Energy Systems Theory, pp. 299–362. McGraw-Hill, New York (1982)
3. Cohn, N.: Some aspects of tie-line bias control on interconnected power systems. Trans. Am. Inst. Electr. Engineers. Part III: Power Appar. Syst. **75**(3), 1415–1436 (1956)
4. Elgerd, O.I., Fosha, C.E.: Optimum megawatt-frequency control of multiarea electric energy systems. IEEE Trans. Power Appar. Syst. **4**, 556–563 (1970)
5. Pan, C.T., Liaw, C.M.: An adaptive controller for power system load-frequency control. IEEE Trans. Power Syst. **4**(1), 122–128 (1989)
6. Talaq, J., Al-Basri, F.: Adaptive fuzzy gain scheduling for load frequency control. IEEE Trans. Power Syst. **14**(1), 145–150 (1999)
7. Nanda, J., Mangla, A., Suri, S.: Some new findings on automatic generation control of an interconnected hydrothermal system with conventional controllers. IEEE Trans. Energy Convers. **21**(1), 187–194 (2006)
8. Saikia, L.C., Nanda, J., Mishra, S.: Performance comparison of several classical controllers in AGC for multi-area interconnected thermal system. Int. J. Electr. Power Energy Syst. **33**(3), 394–401 (2011)

9. Mohanty, B., Panda, S., Hota, P.K.: Controller parameters tuning of differential evoluti algorithm and its application to load frequency control of multi-source power system. Int Electr. Power Energy Syst. **54**, 77–85 (2014)
10. Bevrani, H., Ghosh, A., Ledwich, G.: Renewable energy sources and frequency regulatic survey and new perspectives. IET Renew. Power Gener. **4**(5), 438–457 (2010)
11. Sharma, Y., Saikia, L.C.: Automatic generation control of a multi-area ST–Thermal pow system using Grey Wolf Optimizer algorithm based classical controllers. Int. J. Electr. Pow Energy Syst. **73**, 853–862 (2015)
12. Rahman, A., Saikia, L.C., Sinha, N.: AGC of dish-Stirling solar thermal integrated therm system with biogeography based optimised three degree of freedom PID controller. IET Rene Power Gener. **10**(8), 1161–1170 (2016)
13. Das, D.C., Sinha, N., Roy, A.K.: Small signal stability analysis of dish-Stirling solar therm based autonomous hybrid energy system. Int. J. Electr. Power Energy Syst. **63**, 485–498 (201
14. Hossain, M.S., et al.: Role of smart grid in renewable energy: an overview. Renew. Sustai Energy Rev. **60**, 1168–1184 (2016)
15. Hammons, T.J.: Geothermal power generation worldwide. In: 2003 IEEE Bologna Power Tec Conference Proceedings, vol. 1. IEEE (2003)
16. Setel, A., et al.: Use of geothermal energy to produce electricity at average temperatures. I 2015 13th International Conference on Engineering of Modern Electric Systems (EMES). IEE (2015)
17. Tasnin, W., Saikia, L.C.: Maiden application of an sine–cosine algorithm optimised FO cascad controller in automatic generation control of multi-area thermal system incorporating dish Stirling solar and geothermal power plants. IET Renew. Power Gener. **12**(5), 585–597 (2018
18. Sánchez, J., Visioli, A., Dormido, S.: A two-degree-of-freedom PI controller based on events J. Process Control **21**(4), 639–651 (2011)
19. Kennedy, J., Eberhart, R.: Particle swarm optimization. In: Proceeding IEEE Internationa Conference Neural Networks, vol. IV, pp. 1942–1948. Perth, Australia (1995)
20. Mirjalili, S.: SCA: a sine–cosine algorithm for solving optimization problems. Knowl.-Base Syst. **96**, 120–133 (2016)

Classifications of High-Resolution SAR and Optical Images Using Supervised Algorithms

Battula Balnarsaiah, T. S. Prasad, Laxminarayana Parayitam, Balakrishna Penta and Chandrasekhar Patibandla

Abstract Synthetic aperture radar (SAR) images (microwave images) and optical ones have been recognized as important sources to study land use and land cover. The aim of this study is to create land use/cover classification using the maximum likelihood (ML) and support vector machines (SVM) algorithms. Essential geo corrections were applied to the images at the pre-processing stage. To evaluate both the classified images, the metrics of overall accuracy and kappa coefficient were used. The so-evaluated accuracy assessment results demonstrated that the SVM algorithm gave an accuracy of 88.94 and 77.89% in optical and SAR images, respectively, and the kappa coefficients in the same order being 0.87 and 0.75 approximately. The kappa coefficient of the SVM is higher than that of the ML algorithm, both in the case of optical and microwave classified data. Therefore, the SVM algorithm is suggested to be used as an image classifier for both optical and SAR (microwave) high-resolution images.

Keywords Land use and land cover · High resolution · ML · SVM · Optical and SAR image classifications

---

B. Balnarsaiah (✉) · L. Parayitam
Research and Training Unit for Navigational Electronics, Osmania University, Hyderabad, Telangana, India
e-mail: battulabalu@gmail.com

L. Parayitam
e-mail: plaxminarayana@yahoo.com

T. S. Prasad · C. Patibandla
National Remote Sensing Centre, Indian Space Research Organization (ISRO), Hyderabad, Telangana, India
e-mail: shankarprasad_t@nrsc.gov.in

C. Patibandla
e-mail: chandrasekharpatibandla@gmail.com

B. Penta
Department of Geo-Engineering, Andhra University, Visakhapatnam, Andhra Pradesh, India
e-mail: balakrishna.penta@gmail.com

A. K. Das et al. (eds.), *Computational Intelligence in Pattern Recognition*, Advances in Intelligent Systems and Computing 999,
https://doi.org/10.1007/978-981-13-9042-5_84

1 Introduction

1.1 Optical Data

Processing the satellite images obtained using optical data has been recognized as a effective tool for analyzing land cover and usage [1]. Initially, the analysis of land u and land cover by images was started with aerial photography. Optical data has th advantage of being conceptually simpler compared to the microwave images. In th modern era, images obtained by many ways, like synthetic aperture radar, are al in use for analyzing the land use and cover. Vast amounts of optical image data hav been available from earth observation satellites since decades [2]. Advances in th fields of image processing and pattern recognition/machine learning algorithms a the driving forces for development of different techniques for analysis of land usag and coverage. Several external factors play a key role while generating the land cove data by the satellites. One such external factor having an effect on the frequency of th land cover is thematic content [3]. This information is used for many scientific an engineering applications, like military, planning or management, forestry, agricultur purposes and so on. The image processing algorithms play a major role in this contex

1.2 SAR (Microwave) Data

Much of microwave data for image processing/analysis comes from active sensors in particular, from synthetic aperture radar (SAR). Microwave sensors can work in almost all weather conditions [4] and the waves can penetrate into the sub-surface of certain features, viz., soil and moderate vegetation. Thus, the interactions of earth surface features with microwaves are complicated. The partial successes already achieved for soil moisture and snow moisture estimations, and so on [5] from back scatter coefficient data made SAR data attractive for remote sensing applications. SAR (microwave) data offers further advantages since synergy with corresponding optical data is a good possibility [6, 7]. However, the usage of SAR data for land use/cover classification data is complicated partly because of the more textural characteristics present in SAR data than its optical counterpart. Speckle noise is another bottleneck in radar remote sensing data. For this study, dual-polarization SAR DN data (VV and VH) of sentinel-I were used. The time gap between optical and microwave was only five days [8].

3 *Literature Survey*

eed for Research in Land Coverage Using Optical Images

ıe land cover modifications due to human activities affect the climatic and weather ›nditions. The land cover monitoring has to be carried out at regular intervals to udy the changes caused by human activities [9]. This data is helpful to various ›lds like agriculture and watershed management [3, 10]. Research also improved :curacies of land cover classification when other parameters like surface are used to ıpport classification [11]. It is well known that optical data provide important and ıluable data regarding crops and other land covers. However, the satellite images of ıe LISS-IV data or image of the optical sensors experience strong restraints, especially in areas most often under clouds and like obscure covers, saturation. Optical ata are extremely weak regarding penetration aspects [2]. Hence, assessments of arth's land cover images or pictures have to be generated methodically at appro-riate intervals to allow supervising of large duration inclinations as well as inter nnual changeability, to study the human-induced manipulations [12].

! Methodology

)bjectives of Optical and SAR (Microwave) Imagery Data

The present study aims at synergetic utilization of the satellite image of an optical data :o come up with procedures that can be adopted in operational land cover and land use ɔf the image classifications. Accordingly, the following two objectives are the driving forces of the present study. The first objective is to derive land cover information from high-resolution optical satellite imagery using the ML and state-of-the-art SVM. The second objective is to evaluate the quality of classification of image patches with two measures, namely, overall accuracy (OAA) and kappa coefficient (KC). Telangana is one of the states in India, and Hyderabad city is the capital of the state. The study area is at the outskirts of Hyderabad: around Shamshabad International Airport, lakes of Himayath Sagar, Gandipeta (Fig. 2). It is bounded by geo coordinates (longitude and latitude, respectively) [6] (17° 11' 31" N, 78° 24' 31" E) and (17° 26' 13" N, 78° 31' 56" E) and covers an area of 32 sq. km. Settlements, forest and water bodies are the major theme features in the study area [13]. The sentinel image with the dual-polarizations of VV and VH obtained on 22 February 2018 was utilized in this work. The data's slant range dimensions are 5 m × 20 m. Along with SAR data [14], optical data pertaining to IRS Resourcesat-2 LISS-IV sensor dated 17 February 2018 have been used in this study [9]. The following section describes the detailed procedure used to meet the goals of this research. The overview of the methodology adopted for this research has been shown in Fig. 1.

The methodology involves processing of optical LISS-IV multispectral data [9] and microwave data. The methodology followed for microwave intensity data processing involves retrieval of back scatter values, speckle removal, geo coding using

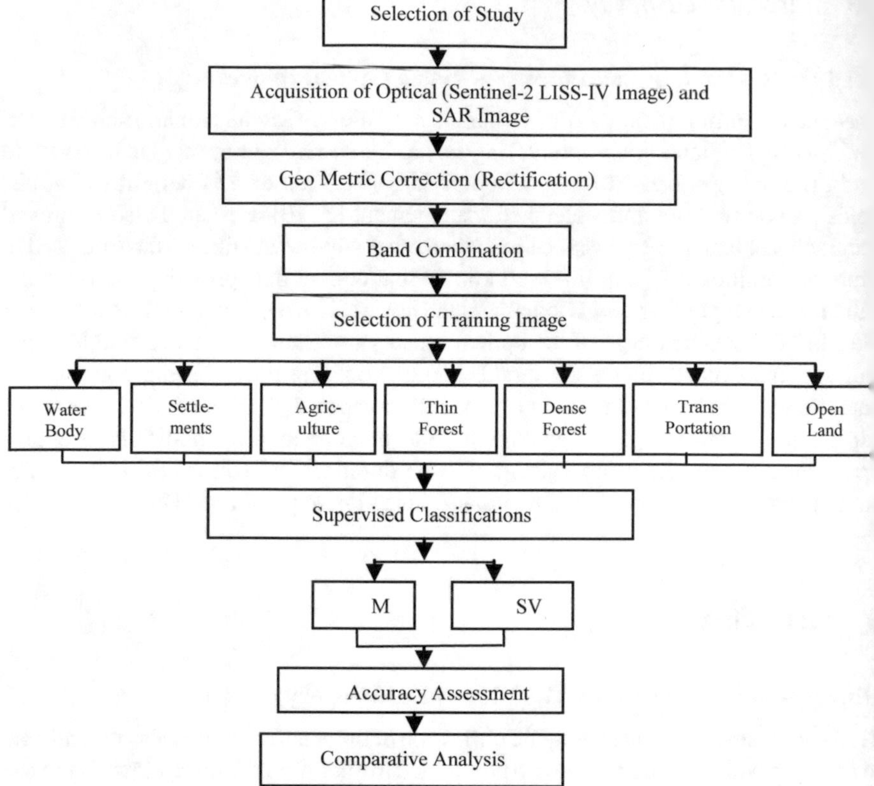

Fig. 1 Overall methodology of SAR and optical images (satellite images)

digital elevation model (DEM), ground truth data collection, supervised classification using ML, SVM techniques and accuracy assessment. The output data of RADARSAT-1 was turnaround and scaled pixel wise. The in-phase (I) and quadrature phase (Q) of the SLC data in the image are represented using a digital number (DN) [14] and is represented in CEOS radiometric data.

3 Data Processing

3.1 Maximum Likelihood Algorithm

The satellite images of both the optical and SAR images were classified using the maximum likelihood algorithm for operational purposes. A probabilistic model is applied to decide whether a pixel belongs to a specific class. All the classes have equal probabilities that the pixel may be classified as one of it, for all training datasets llow the statistical normal distribution. In specific cases, the probabilities of the asses are not always equal. One such example is when Bayesian assessment rule 5] exists. Bayesian classifier equation is as follows:

$$L_k = P\left(\frac{K}{x}\right) = P(k) * P(x/k) \Big/ \sum P(i) * P(x/i) \quad (1)$$

here $P(k)$ is the priori probability of class k, $P(X/k)$ is the conditional probability, is the pixel and k is the class.

$$P(X) = \sum_i p(i) p\left(\frac{x}{i}\right)$$

The probability density function P(K) follows Gaussian distribution [11].

.2 SVM Classification

n any type of the image processing, one of the supervised classification techniques s the support vector machine (SVM) that attempts to generate an optimal plane to naximize the separation between the training data of different classes. The SVM uses a radial basis function (RBF) during the training stage. In this stage, the user provides he training samples on the basis of which the system learns the classifiers [16]. Other kernel functions include linear and polynomial [7, 16]. The SVM is formulated on he principle of minimizing the error in the classifier over the iterations. SVM is capable of generalizing [17] so that it can proceed comfortably in an unseen path [16, 18]. Based on this principle, a systematic method is followed in a linear machine approach to find a linear function with the lowest price of VC (Vapnik Chervonenki's) dimension. Nonlinear data is handled by mapping the input to a high-dimensional space with linear excitation [19].

The classification accuracy of SVM is higher in comparison with other techniques [4].

As discussed above, the nonlinear data classification can be achieved using the underlying equation:

$$f(x) = \operatorname{sgn}\left\{\sum_{i=1}^{m} \alpha_i k_i (x_i, y_i) + b\right\} \quad (2)$$

Here $k(x, y)$ is the kernel function. The kernel plays the key role in the transformation of the input. The available techniques are still lagging in learning the kernel as per the user requirement. Here only common choices of kernel functions in support vector research are the linear, polynomial and Gaussian RBF.

In this study Gaussian RBF kernel was used as follows [7]:

$$k(x, y) = \exp\left\{-\frac{|x - x_i|^2}{\sigma^2}\right\} \quad ($$

In the support vector machine optimization, it is desired to avoid misclassificati of each training example by the use of a regulation parameter that is often referred as the C parameter. This C parameter will generate a smaller margined plane of hyp dimensions if that hyper plane produces an improved result of classifying the pixe exactly. On the contrary, the smaller the value of C, the incorrect the classificatic becomes.

4 Results and Discussion

Accuracy assessment and Kappa Coefficient

Worth of land classification products is indispensable to substantiate the quality c the products for operational applications. Both the LISS-IV data (Fig. 2) and th SAR data were the inputs to ML and SVM classifiers. The results of the classifica tion pertaining to optical and microwave data or image are given in Figs. 3 and 4 respectively.

The accuracy of the classified pixels has to be recorded for the evaluation o the algorithms [13]. Though the researchers in the past did not propose a concrete way to assess the results, in recent years they provided us with methods like PSNR confusion matrix, and so on. These methods accurately compare the methods in term of accuracy and prove the superiority in performance. The confusion matrix display the results in a square matrix with the correct classifications in the diagonal positior and the incorrect classifications in the remaining positions.

A comparison of accuracies given by ML and SVM algorithms for optical and SAR image data are accessible in Table 1. The class identification colours of water

Fig. 2 LISS-IV satellite image

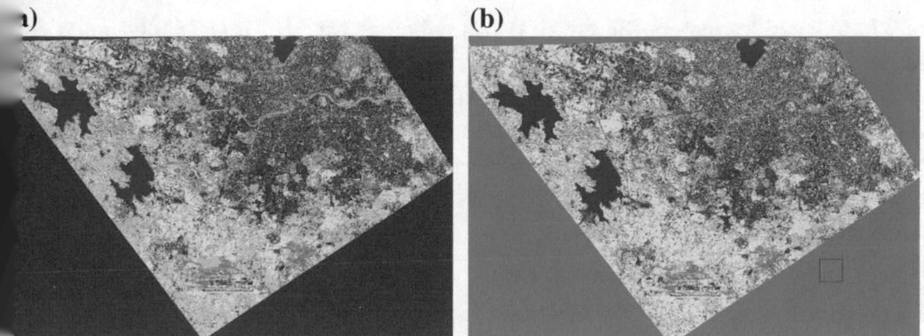

…ig. 3 **a** ML classification of optical images and **b** SVM classification of optical images

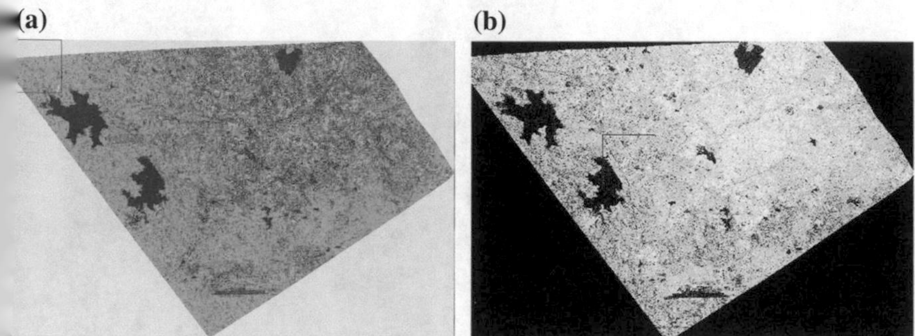

…ig. 4 **a** ML and **b** SVM classification of SAR images

Table 1 Comparison of classifications accuracies of optical, SAR (microwave) data or image

Data	Classification	Overall accuracy (%)	Kappa coefficient
Optical	SVM	88.9472	0.8749
	ML	86.8764	0.8579
SAR (microwave)	SVM	77.8926	0.7683
	ML	75.9864	0.7491

body, settlements, agriculture, thin forest, dense forest, transportation and open land are blue, red, yellow, green, chartreuse green, purple and cyan, respectively in all classified outputs. The present research work analyzed the feasibility of using optical, SAR data outputs in the classification of land cover using SVM and maximum likelihood classifiers. The work reveals that SVM classification gives a slightly better result than ML, having nearly of 88.94% and a kappa coefficient of 0.8749, whereas for ML, it is 86.89% and 0.8579, respectively, for images classified using optical data. Microwave (SAR) data classification has also shown that SVM classification gives a slightly better result than ML, having accuracy of both the images nearly of 77.89% and kappa coefficient of 0.768, whereas for ML, it is 75.98% and 0.74 respectively.

From the aforementioned discussion, kappa coefficient for the SVM classificati is better than ML. Graphical representations of SVM-classified output of optical a SAR data [8] are given in Figs. 5 and 6. Among the two classification techniqu nonparametric SVM classification registered the best results for both optical a SAR datasets.

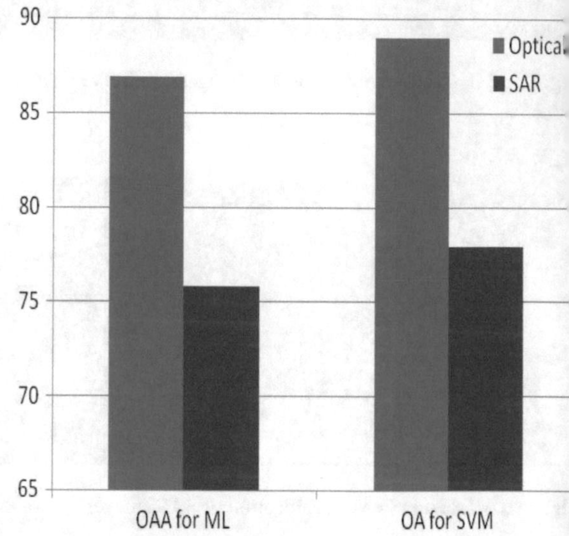

Fig. 5 Overall accuracy (OAA) of optical and SAR images of ML and SVM

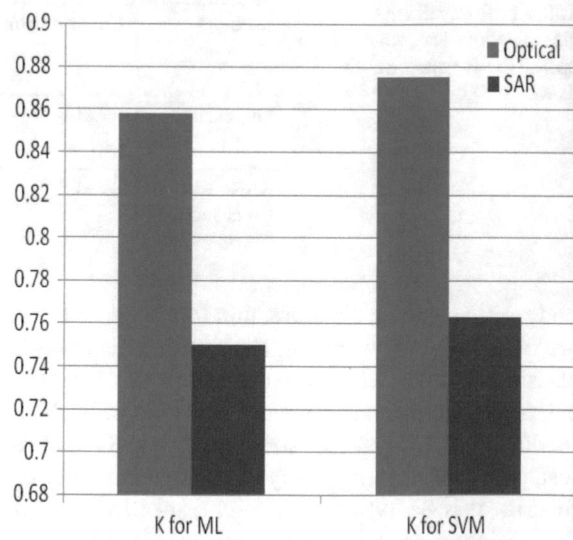

Fig. 6 K: Kappa coefficients of optical and SAR-classified images of ML and SVM

Conclusion and Future Work

ıe optical LISS-IV and SAR image data were classified using supervised techques ML and SVM, and a comparison showed an overall superiority of the support ector machine, for optical data in particular. Application of SVM algorithms to AR imagery also holds an excellent omen. In the case of SAR, accuracy may be nproved if tri-polarization or quad-polarization data are used instead of two. Also, a future work, classification will be tested using neural networks too. Since SAR nages are textural, efforts are being undertaken to introduce texture parameters in assification. Local texture is becoming important in optical imagery too due to vailability of extremely fine spatial resolution, and future study may involve the ıid property in optical imagery too. In order to arrive at better confirmation about ıe results, data covering different areas and land surface features are needed.

cknowledgments The authors would like to thank all the reviewers, Dr. Rajashree Bothale, eneral Manager, Outreach Facility of NRSC and Director of the NRSC (ISRO), Hyderabad for ıeir encouragement. They would also like to extend their sincere thanks to the institute staff for ıe technical support and remarkable suggestions during research work. The authors would like to cknowledge the CSIR fellowship provided by Govt. of India, New Delhi, India.

References

1. Anderson, J. R.: A land use and land cover classification system for use with remote sensor data, vol. 964. US Government Printing Office (1976)
2. Meinel, G., Lippold, R., Netzband, M.: The potential use of new high-resolution satellite data for urban and regional planning. In: IAPRS, vol. 32, Part 4 "GIS—Between Visions and Applications", Stuttgart (1998)
3. Hansen, M.C., DeFries, R.S., Townshend, J.R., Sohlberg, R.: Global land covers classification at 1 km spatial resolution using a classification tree approach. Int. J. Remote Sens. **21**(6–7), 1331–1364 (2000)
4. Ulaby, F.T., Moore, R.K., Fung, A.K.: Microwave Remote Sensing: Active and Passive, vol. 1, pp. 256–337. Reading, MA: Addison-Wesley (1981)
5. Dobson, M.C., Ulaby, F.T.: Active microwave soil moisture research. IEEE Trans. Geosci. Remote. Sens. **1**, 23–36 (1986)
6. Qian, Y., Zhou, W., Yan, J., Li, W., Han, L.: Comparing machine learning classifiers for object-based land cover classification using very high resolution imagery. Remote Sens. **7**(1), 153–168 (2014)
7. Lee, J.S., Wen, J.H., Ainsworth, T.L., Chen, K.S., Chen, A.J.: Improved sigma filter for speckle filtering of SAR imagery. IEEE Trans. Geosci. Remote Sens. **47**(1), 202–213 (2009)
8. Research supported by NASA, NSF, and U. S. Department of Defense. Reading, MA, Addison-Wesley Publishing Co., 1982, 624. Vyjayanthi, P.N.: Synthetic Aperture Radar data analysis for vegetation classification and biomass estimation of tropical forest area (2010)
9. Kumar, P., Prasad, R., Gupta, D.K., Mishra, V.N., Choudhary, A.: Support vector machine for classification of various crop using high resolution LISS-IV imagery. Bull. Environ. Sci. Res. **4**(3), 1–5 (2015)
10. Van Niel, T.G., McVicar, T.R., Datt, B.: On the relationship between training sample size and data dimensionality: Monte Carlo analysis of broadband multi-temporal classification. Remote Sens. Environ. **98**(4), 468–480 (2005)

11. Di Zenzo, S., Bernstein, R., Degloria, S.D., Kolsky, H.G.: Gaussian maximum likelihood a contextual classification algorithms for multi-crop classification. IEEE Trans. Geosci. Remc Sens., **GE-25** (1987)
12. Bayak, H., Yamaguta, Y.: Improved Sub-Space classification method for multispectral Rem Sensing Image Classification. Photogramm. Eng. Remote. Sens. **76**(11), 1239–1251 (2010)
13. Bogoliubova, A., Tymków, P.: Accuracy assessment of automatic image processing for la cover classification of St. Petersburg protected area. Acta Sci. Pol.: Geod. Descr. Terrarum (1–2), 5–22 (2014). ISSN 2083–8662
14. Beaudoin, A., Le Toan, T., Goze, S., Nezry, E., Lopes, A., Mougin, E., Shin, R.T.: Retrieval forest biomass from SAR data. Int. J. Remote Sens. **15**(14), 2777–2796 (1994)
15. Dougherty, G.: Pattern Recognition and Classification. Springer, New York (2013)
16. Pal, M., Mather, P.M.: Support vector machines for classification in remote sensing. Int. Remote Sens. **26**(5), 1007–1011 (2005)
17. Bazi, Y., Melgani, F.: Toward an optimal support vector machine classification system f hyperspectral images. IEEE Trans Geosci. Remote. Sens. **44**(11), 3374–3385 (2006)
18. Pal, M., Mather, P.M.: Support vector classifiers for land cover classification. Int. J. Remo Sens. **29**(10), 3043–3049 (2008)
19. Foody, G.M., Ajay, M.: A relative evaluation of Multi-Class Image Classification by Suppo Vector Machines. IEEE Trans Geosci. Remote. Sens. **42**(6), 1335–1343 (2004)

An Approach for Video Summarization Using Graph-Based Clustering Algorithm

Ghazaala Yasmin, Aditya Chaterjee and Asit Kumar Das

Abstract There has been immense increase in amount of video content over the past few years. This massive growth in the content of video leads to the uncertain outcomes. Processing of these huge chunks of data is demanding plenty of resources, like time, manpower, as well as hardware storage. Video summarization acquires an important remark in this ambiance. It supports in providing efficient storage, fast browsing, and retrieval of huge video data without losing important factors. In this work, video summarization technique has been proposed using graph-based clustering algorithm in three different steps. The presented work has split the video into frames of a predefined time period and computed the similarity between consecutive frames. Based on the similarity values, the frames are grouped into scenes, and the scenes are partitioned separately using video tracks and audio tracks with the help of clustering algorithm. Then the combined clusters of scenes are further analyzed to determine the summary of a video file. This work considers just the video file at hand and attempts to develop a summary file without using any external knowledge from similar videos. As per the authors' knowledge, no previous research work related to video summarization has been conducted in this field. The quality of the summarized video is a measure to express the effectiveness of the proposed methodology.

Keywords Video summarization · Key frames · Information retrieval · Clustering · Audio track · Text summarization

G. Yasmin (✉)
Computer Science and Engineering, St. Thomas' College of Engineering and Technology, Khidderpore, Kolkata 700023, India
e-mail: me.ghazaalayasmin@gmail.com

A. Chaterjee · A. K. Das
Computer Science and Technology, Indian Institute of Engineering Science and Technology, Shibpur, Howrah 711103, India
e-mail: aditianhacker@gmail.com

A. K. Das
e-mail: akdas@cs.iiests.ac.in

A. K. Das et al. (eds.), *Computational Intelligence in Pattern Recognition*, Advances in Intelligent Systems and Computing 999,
https://doi.org/10.1007/978-981-13-9042-5_85

1 Introduction

Video forms an integral part of communication and entertainment in our daily li which is evident by the high-traffic video-streaming sites, like YouTube, Vimeo, a others. Despite the popularity of videos, less research has been conducted on t summarization of videos. Summarization of videos is important as some categori of videos, such as surveillance videos, can be over a yearlong and in other cases, get an idea of the video content before investing time to watch the entire video. T utility of this research work is that it can convert a long video to a shorter duratic and hence can be used by anyone to get an overview of the video. In this researc work, we have developed a novel data mining-based video summarization techniq that provides gist of the video without losing any meaningful information.

1.1 Related Works

A tremendous growth in multimedia domain has been seen in the past few years. Th information has been gathered and elaborated in section, starting with Elkhattab et al. [1], who highlighted the different techniques for summarization of video. Th paper has also provided the comparison along with the advantage of the differen techniques. The above appraisal has also been depicted in the proposal of Ramesh et al. [2]. Freeman and Gregory [3] developed an interactive system for providing an interactive presentation with personalized video based on audio and graphics responses on multiple viewers. Rajendra and Keshaveni [4] did a survey on differen summarization techniques. Sebastian and Jiby [5] presented a survey model on videc summarization. It revealed the application of video summarization technique and its unique specification. He et al. [6] propounded the auto summarization technique using audio–video presentation. Ma et al. [7] prepared a model for summarizing digital video. Smith and Takeo [8] worked on the implementation of video skimming through image and language understanding. Many models have been presented further by Hua et al. [9] and Taskiran et al. [10]. Borji et al. [11] depicted quantitative analysis and compression for visual saliency. Ngo et al. [12] and Jiang et al. [13] propounded the enhanced version of video summarization method through the help of audio and visual analysis. This is one of the efficient and optimal techniques which became more popular among researchers. Earlier research has also been listed with many other works to edit playback and work with multimedia system like digital video and metadata [14, 15].

Proposed Methodology

ıe algorithm is developed in three steps. In the first step, an algorithm is developed easuring similarity between visual scenes (i.e., frames) to generate a summary. the second step, we have developed an algorithm that uses the audio track in a deo to generate a summary. Finally, a video summarization system is proposed that nsiders the previous two approaches to generate a comparatively better summary. basic flowchart of the system is given in Fig. 1.

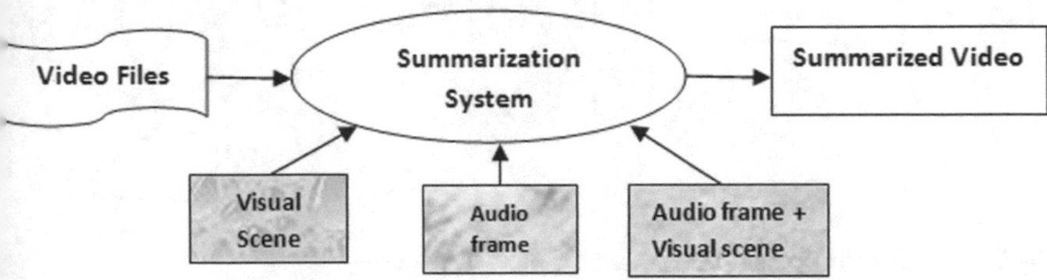

ig. 1 Basic workflow of the proposed system

.1 *Summarization of a Video Using Visual Scene*

n this approach, only the frames in a video are used to generate summary. The audio portion in each frame is not considered during summarization process. The basic idea of this approach is to classify the video into frames of predefined short duration to generate a summary. The broad steps in this approach are: (i) video has been broken into frames of predefined duration and similarities among consecutive frames have been computed; (ii) point of transition of scene (scene is the collection of consecutive frames) has been determined based on a predetermined similarity threshold, and thus the video is divided into a sequence of scenes; (iii) each scene is represented by a set of eigenfaces and a graph is constructed with scenes as nodes and Euclidean distance-based similarity value between a pair of nodes as the weight of the corresponding edge; (iv) use the community detection-based Infomap clustering algorithm [16] to partition the graph into subgraphs of scenes and select the most informative scene from each subgraph; and (v) finally, the selected scenes are arranged according to their initial given order in original video to generate the required video summary.

It is to be mentioned that each scene consists of several frames, and the basic features of each frame are extracted using principal component analysis (PCA) and put together in a new image known as an eigenface. So, the feature vector of an eigenface is same as that of the scene. The workflow of video summarization using visual scene is depicted in Fig. 2. After splitting the video into frames, based on the similarity diagram, as shown in Fig. 2, the scenes are determined. In similarity

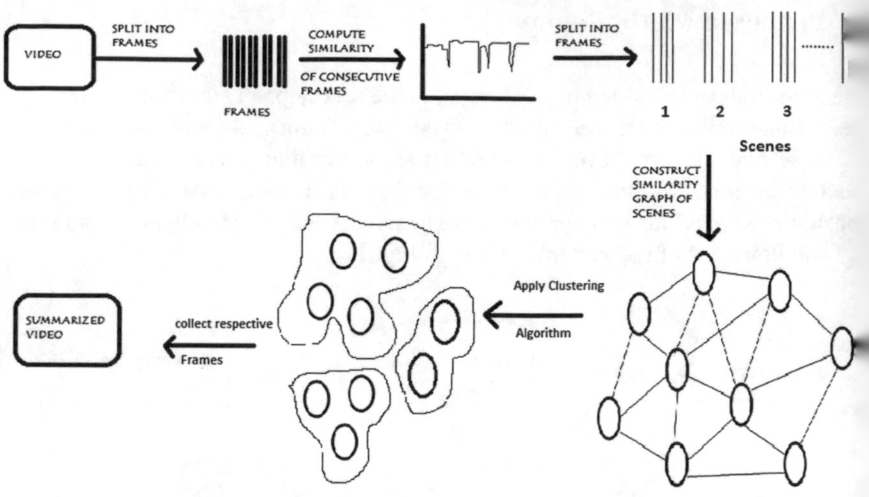

Fig. 2 Flow diagram for summarization through audio frame

graph, until no abrupt change occurs, the associated frames are collected together into a single scene. After clustering, the representative scene, determined by the Infomap clustering algorithm, of each subgraph is collected and arranged according to their original order in the video file, resulting in the summary of video.

2.2 *Summarization of a Video Using Audio Track*

Generally, every video has an audio track. The objective of this method is to generate a summary of a video using the audio tracks. The broad steps involved in this approach are: (i) video has been broken into frames of predefined duration and similarities among consecutive frames have been computed; (ii) point of transition of scene (scene is the collection of consecutive frames) has been determined based on a predetermined similarity threshold, and thus the video is divided into a sequence of scenes; (iii) the audio track of each scene is collected and converted into a text document; (iv) apply text summarization algorithm on collection of text documents. The algorithm selects only few text documents as the summary of all the text documents; (v) find the audio tracks corresponding to the summarized text documents and collect the associated video scenes in the order in which they are appeared in the original video. Figure 3 graphically represents the details of the summarization using audio track.

In this method, the main task is to summarize the text file. Here, text file is considered as the collection of text documents. Each document is obtained from the audio track of a video scene (i.e., collection of video frames). The documents in text file is preprocessed [17] in traditional way using stop word removal and

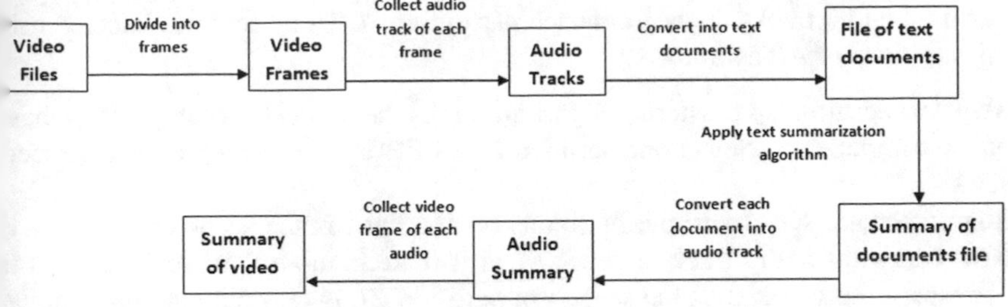

ig. 3 Flow diagram for summarization using audio track

emming techniques, and then term frequency–inverse document frequency (TF-DF) of each remaining word in the file is computed. Thus for each document in the xt file, a vector of TF-IDF value is generated. Then a graph is constructed with text ocuments of associated audio tracks as nodes, and a weighted edge between two odes is drawn. The weight of an edge expresses the strength of similarity between wo associated nodes. Here, the cosine similarity defined in Eq. (1) is used to compute ne similarity S(t_i, t_j) between two nodes t_i and t_j. $\boldsymbol{t_i}$ and $\boldsymbol{t_j}$ are two TF-IDF vectors or text documents t_i and t_j corresponding to the audio tracks of associated video cenes in the video file. Then the graph is partitioned into subgraphs using community letection-based Infomap algorithm [16]. The algorithm provides the most influential ode in each partition. We select these nodes and determine the corresponding text locuments and audio tracks. Finally, the video scenes of corresponding audio tracks re arranged in their original order of appearance in video file and a summary of the given video file is obtained.

$$S(\boldsymbol{ti}, \boldsymbol{tj}) = \frac{\boldsymbol{ti}.\boldsymbol{tj}}{\|\boldsymbol{ti}\|\,\|\boldsymbol{tj}\|} \quad (1)$$

2.3 *Summarization of a Video Combining Visual Scene and Audio Track*

This is the final approach which utilizes both the frames and audio track of a video. This approach tries to overcome the demerits of both the individual methods and gives the importance to both the video and audio tracks of a video during summarization. Video summarization algorithms using visual scenes and audio tracks are named as Algorithm1 and Algorithm 2, respectively, for subsequent references. Both Algorithm 1 and Algorithm 2 give a set of partitions of a video file. Algorithm 1 gives clusters of partitions $EP = \{E_{11}, E_{12},\ldots, E_{1n}\}$ of video scenes based on eigenfaces and Algorithm 2 gives the clusters of partitions $AP = \{A_{11}, A_{12},\ldots, A_{1m}\}$ of scenes based on audio tracks. Now from each partition of *EP*, one scene is selected usi the partitions of *AP* as follows:

Step 1: The Infomap clustering algorithm ranks the scenes of each partition bas on their importance. Select one partition E_{1i} of EP and find its representative scer say s.
Step 2: Search scene s in the partitions of *AP*. Let t is the scene in partition A Now check the audio track of scene t and if there is more than 40% audio in t corresponding scene, then the scene s of partition E_{1i} is selected into the summar otherwise Step 2 is repeated with the next-ranked scene in partition E_{1i}. If no scer of E_{1i} is selected based on this condition, then the scene with the highest average rank and audio is selected.
Step 3: Repeat Step 1 and Step 2 for each partition of *EP*. Thus, a summary of vide scenes is obtained based on both video and audio tracks.

3 Experimental Result

3.1 Dataset Collection

As minimal research has been done in video summarization, there are no globall defined dataset or quality measurement metric which can be used. We have develope and prepared the datasets and measurement metrics in order to measure the effec tiveness of the proposed video summarization algorithm. The main factor that has been considered to collect the data is the quality and content of the video. Majority of the videos that were selected for testing the proposed system are educational and entertainment videos. Educational video consists of videos of subcategories, such as lectures, news, speeches, and national announcements. Entertainment videos consist of videos of subcategories, such as movies, indoor games, and outdoor games' sports commentary.

3.2 Quality Measurement of Summarized Video

The quality measurement strategy uses the generated summary compiled by the experts having the domain knowledge of the video images. Assume that the video on which the method is applied has N scenes. Let the summary generated by the proposed algorithm consists of $N1$ scenes, and that given by the domain experts consists of $N2$ scenes. Then the quality Q of the proposed summarization algorithm is measured using Eq. 2. The value of Q is between 0 and 1.

Table 1 Quality measurement of summaries using three different algorithms

Video category	Algorithm 1	Algorithm 2	Algorithm 3
Lecture	0.34	0.70	**0.76**
Outdoor games	0.26	0.49	**0.52**
Indoor games	0.37	0.54	**0.61**
Movie	0.38	0.68	**0.76**

$$Q = \left(\frac{N1 \cap N2}{N2}\right)^{\left(\frac{N1}{N1 \cap N2}\right)} \tag{2}$$

The base of the equation is the ratio of number of scenes common to both the summaries to the number of scenes in given summary. So, higher the numerator of the base means better the quality of the summary. Similarly, the exponent of the equation is the ratio of number of scenes obtained by the proposed algorithm to the number of scenes common to both the summaries. This ratio is considered because a scene of given summary that does not appear in the obtained summary reduces the quality of the summary. Thus to give higher weightage to this ratio, it is kept as an exponent. It has been observed that the above quality measurement equation works well in denoting the quality of the summary. Higher the value of Q, better is the generated summary. Table 1 summarizes the result for the quality measurement of the three different techniques, namely Algorithms 1, 2, and 3, where Algorithm 3 is the combination of Algorithms 1 and 2. The table also shows that the combined approach provides better quality summary than the individual approach for all four types of videos.

4 Conclusion

The summary serves its purpose of giving viewers a quick overview. In our busy schedule, observing the whole video of any interesting topic is quiet impossible, which motivates the researchers to make a summary of the video. Use of clustering algorithm is very popular in many text summarization purposes. In this paper, we have used the clustering algorithm for video summarization and the experimental result shows quiet impressive quality measurements. The main demerit of the work is that the video dataset is not the benchmark data and is collected from the website. Deep learning and other machine learning techniques may be used for the same purpose to enhance the quality of the video summary.

Acknowledgments This chapter does not contain any studies with human participants or animals performed by any of the authors.

References

1. Elkhattabi, Z., E., Tabii, Y., Benkaddour, A.: Video summarization: techniques and applicatio World Acad. Sci., Eng. Technol., Int. J. Comput., Electr., Autom., Control. Inf. Eng. **9**(928–933 (2015)
2. Ramesh, A., et al.: Video Summarization: an overview of Techniques.
3. Freeman, M.J., Gregory, W.H.: Interactive computer system for providing an interactive prese tation with personalized video, audio and graphics responses for multiple viewers. US Pate 5, 861–881 (1999)
4. Rajendra, S.P., Keshaveni, N.: A survey of automatic video summarization techniques. Int. Electron., Electr. Comput. Syst. **2**(1) (2014)
5. Sebastian, T., Jiby, J.P.: A survey on video summarization techniques. Int. J. Comput. Ap **132**(13), 30–32 (2015)
6. He, L., et al.: Auto-summarization of audio-video presentations. In: Proceedings of the 7 ACM International Conference On Multimedia (Part 1). ACM (1999)
7. Ma, Y.-F., et al.: A user attention model for video summarization. In: Proceedings of the 10 ACM International Conference on Multimedia. ACM (2002)
8. Smith, M.A., Takeo, K.: Video skimming and characterization through the combination image and language understanding. In: IEEE International Workshop on Content-Based Acce of Image and Video Database. IEEE (1998)
9. Hua, X.S., et al.: A generic framework of user attention model and its application in vide summarization. IEEE Trans. Multimed. **7**(5), 907–919 (2005)
10. Taskiran, C.M., et al.: Automated video program summarization using speech transcripts. IEE Trans. Multimed. **8**(4), 775–791 (2006)
11. Borji, A., Dicky, N.S., Itti, L.: Quantitative analysis of human-model agreement in visua saliency modeling: a comparative study. IEEE Trans. Image Process. **22**(1), 55–69 (2013)
12. Ngo, C.W., Ma, Y.F. Zhang, H.J.: Video summarization and scene detection by graph modeling IEEE Trans. Circuits Syst. Video Technol. **15**(2), 296–305 (2005)
13. Jiang, W., Courtenay, C., Alexander, C.L.: Automatic consumer video summarization by audi and visual analysis. In: 2011 IEEE International Conference on Multimedia and Expo (ICME) IEEE (2011)
14. Logan J., et al.: Audio and video program recording, editing and playback systems using metadata. US Patent 10/165,587
15. Sun, X., Kankanhalli, M., Zhu, Y., Wu, J.: Content-based representative frame extraction for digital video. In: International Conference on Multimedia Computing and Systems, pp. 190–193 (1998)
16. Rosvall, M., Bergstrom, C.: Maps of random walks on complex networks reveal community structure. Proc. Natl. Acad. Sci. USA **105**, 1118 (2008)
17. Loper, E., Bird, S.: NLTK: the natural language toolkit. In: Proceedings of the ACL-02 Workshop on Effective Tools and Methodologies for Teaching Natural Language Processing and Computational Linguistics ETMTNLP'02, pp. 63–70 (2002)

arameter Tuning in MSER for Text ocalization in Multi-lingual amera-Captured Scene Text Images

ouvik Panda, Swayak Ash, Neelotpal Chakraborty, yatullah Faruk Mollah, Subhadip Basu and Ram Sarkar

bstract Scene text detection and localization in camera-captured images has ways posed a great challenge to the researchers due to its high complexity in nderstanding the texture and homogeneity of scene text images. The solution to is problem paves way for simplified text extraction and processing, thereby realzing wide range of applications. A very popular method, namely, maximally stable xtremal region or MSER detection is used for localizing text since the text regions re considered to be more stable than other regions in an image. However, it involves nanual selection of several parameters, limiting its usage in many practical applicaions. In this work, the relations among parameters, like image dimension, text size, nd region area, are analyzed by experimenting on an in-house multi-lingual dataset aving 300 images comprising English, Bangla, and Hindi texts and validating the utcome against images of standard datasets, like SVT and MSRA-TD500, which ields encouraging results.

Keywords Scene text · Multi-lingual · Camera captured · Text detection · Text ocalization · MSER

S. Panda · S. Ash
Department of Computer Science and Technology, Indian Institute of Engineering Science and Technology, Shibpur, Howrah 711103, India
e-mail: papupanda421999@gmail.com

S. Ash
e-mail: swayak@gmail.com

N. Chakraborty (✉) · S. Basu · R. Sarkar
Department of Computer Science and Engineering, Jadavpur University, Kolkata 700032, India
e-mail: neelotpal_chakraborty@yahoo.com

S. Basu
e-mail: subhadip@cse.jdvu.ac.in

R. Sarkar
e-mail: raamsarkar@gmail.com

A. F. Mollah
Department of Computer Science and Engineering, Aliah University, Kolkata 700160, India
e-mail: afmollah@aliah.ac.in

A. K. Das et al. (eds.), *Computational Intelligence in Pattern Recognition*, Advances in Intelligent Systems and Computing 999,
https://doi.org/10.1007/978-981-13-9042-5_86

1 Introduction

One of the most important research problems of computer vision in the recent years scene text detection [1] in complex camera-captured images, which is worth explori due to high potential in several application areas, like tourist guide and travel assistance, information retrieval via search engines, unnoticed text detection in t wild, image to text conversions, and so on. However, scene text localization is a ve complex issue because of challenges like varying size, orientation, color, brightnes and contrast of the texts present in the images [2], which are illustrated in Fig. 1.

Several methods have been proposed by the researchers till date to counter the challenges for localizing text regions from complex heterogeneous background Currently, one of the most popular techniques to do so is obtaining MSER [3] maximally stable extremal region, which efficiently estimates blobs of relative stable regions whose pixel values do not vary much within the neighborhood of certain region area. MSER detection is applicable to identify several objects in a image apart from texts, as shown in Fig. 2. But, it depends on tuning of parameters lik percentage for intensity variation and region area selection, in order to specificall get the stable text blobs keeping the number of non-text blobs restricted to a certai limit.

In this work, we intend to determine the relations among parameters, like imag dimension, text size, and region area, by experimenting on an in-house dataset havin images with multi-lingual texts comprising English, Bangla, and Hindi, and validat ing the outcome against the images taken from some standard datasets, like SVT [2 and MSRA-TD500 [4], which provides reasonably good results.

Fig. 1 Scene text images with complexities. **a** Uneven distribution of intensities, **b** warping, **c** blur (absence of well-defined edges), **d** presence of texts with variable size and orientation, **e** Tungsten light effects at night, **f** camera device issues

(a) (b)

Fig. 2 **a** Sample image from an in-house multi-lingual scene text dataset, **b** detected MSER regions

Related Works

Literature survey reveals that till date many methods have been proposed by the researchers around the world to counter the challenges of the said problem.

The work in [1] embeds canny edges with the basic MSER method in order to make it blur-resistant to a certain extent. In another work, as reported in [2], extremal regions are determined by using a sequential classifier to detect text regions. Arbitrary text orientations are managed by the method used in Yao et al. [4] by using a two-level classification scheme to capture intrinsic characteristics of texts. A relatively new concept of two-level binning of gray levels is implemented by the method used in Dutta et al. [5]. The method proposed in [6] attempts to dynamically set the percentage restriction of intensity variation and choosing an optimal region area range for MSER to extract the text regions in an image. In [7, 8], multi-oriented text regions are detected with the help of MSER technique. In [9], MSERs are grouped and perceptually organized to identify the text regions. Some baseline techniques have been highlighted in [10] that demonstrate their efficiency in detecting and recognizing multi-lingual texts from public image datasets.

3 MSER and Its Parameters

The concept of MSER, first proposed by Matas in [3], is based on selecting regions which stay almost the same through a wide threshold range and is used for detecting certain blobs in images, as depicted by color shades in Fig. 3, to extract a number of co-variant regions. In simple terms, an MSER represents a stable connected component of some gray-level sets of the image. Owing to its limitations against blur, later the work in [1] combines MSER with canny edges, which gives better results than earlier.

Although simple and efficient, this popular technique requires manual intervention for tuning certain parameters, like intensity variation threshold and region area range, to specifically get the text regions. In complex scene images, it is often found that a single parametric value set for MSER may be suitable for some but not all the

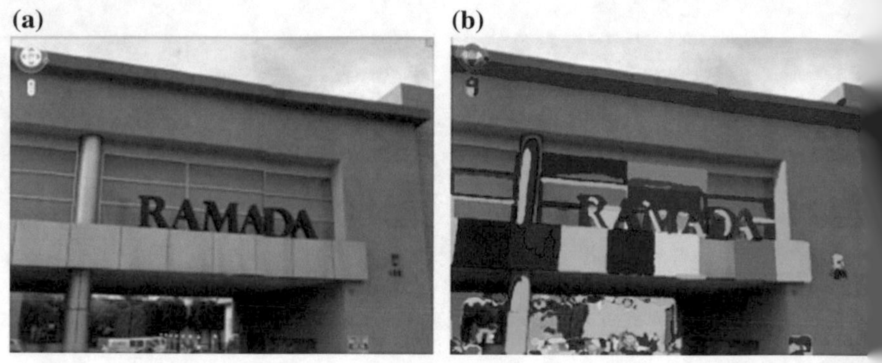

Fig. 3 Original image taken from SVT dataset. **a** No MSERs detected for minimum area (in pixel = 30 and maximum area (in pixels) = 14,000, **b** text MSERs for minimum area = 30 and maximu area = 26,000

images. For instance, as illustrated in Fig. 4, in images with large-sized texts, it i observed that a single MSER region fails to hold the entire desired region, wherea in images with very small-sized texts, many objects other than texts come under single MSER region.

To counter these problems, we conduct several experiments in order to set th optimal values to the parameters, like image dimension, text height, and width, fo

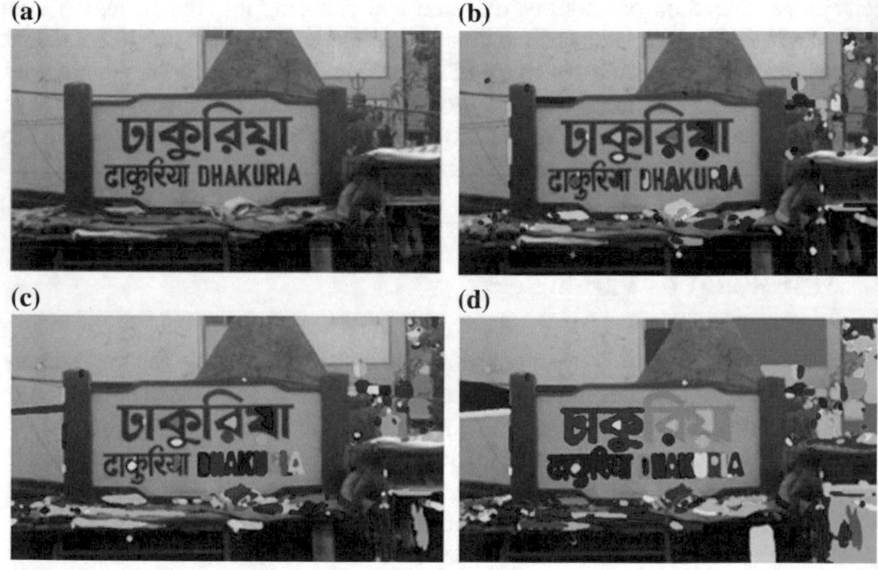

Fig. 4 **a** Original image from an in-house dataset, **b** MSERs detected for minimum area = 30 and maximum area = 14,000, **c** MSERs detected for minimum area = 30 and maximum area = 26,000, **d** MSERs detected for minimum area = 30 and maximum area = 370,000

termining the maximum and minimum area ranges since the region area selection found to play a significant role in determining MSERs.

Proposed Methodology

ext regions, large or small, occupy a certain amount of space in an input image. wing to image-to-image variation in terms of text area or size, there arises a need determine the relation between text region area and MSER region area. Also, this xt region area is found to be dependent on the image dimension. Hence, from this ansitive nature, we can safely claim that region area for MSER depends on the ariation in image dimension too.

To establish these relations, we proceed stepwise following the flowchart as given n Fig. 5, to perform several experiments on standard scene text datasets, like SVT, ASRA-TD500, and an in-house dataset consists of multi-lingual text images.

.1 *Dataset Pooling*

Ve randomly select the images from various datasets and record their individual limension. MSER algorithm is then applied to each image and region area is manually assigned in such a manner that it can cover only the text regions while avoiding the non-text elements as much as possible. After searching an optimal value (manually set) of maximum and minimum region areas for an image, we note these values against image dimension.

4.2 *Image Categorization*

As mentioned earlier, we observe that the images like those depicted in Fig. 6, having different text region areas, show different characteristics of maximum and minimum region areas of MSER. In the case of images with large text region areas, the text region does not fit into one MSER region; even more unnecessary MSER regions

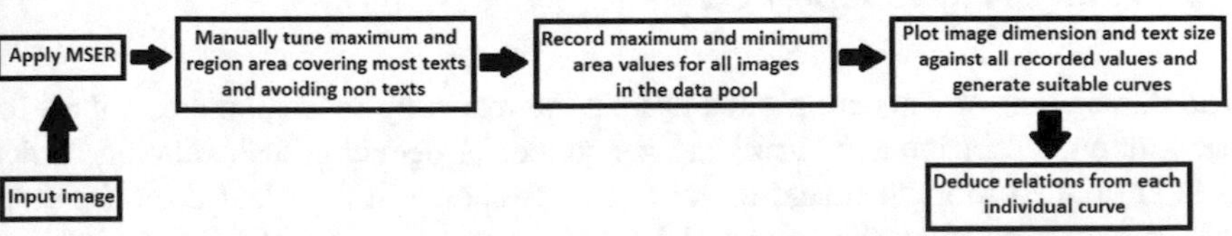

Fig. 5 Flowchart for the proposed methodology

Fig. 6 Effect of minimum and maximum area of sample images from standard multi-lingual datase MSRA-TD500. **a** Image with small-sized text, **b** image with large-sized text, **c** MSER regions fo minimum area = 30 pixels since text is small, **d** MSER regions for maximum area = 14,000 pixel since text is quite larger in size

are formed. Similarly, in the case of the images with small text region areas, the problem which arises is that in a single MSER region, the area covered is more than the desired.

These challenges lead us to formulate the maximum and minimum region areas for MSER. In doing so, we first group the images into two categories: images with large text region areas and images with small text region areas. To have a clear idea, we also plot these values separately for both maximum and minimum region areas for MSER.

4.3 Analysis of Relation Curve

The dimensions of texts are plotted against the manually set region area values for each individual image and curves are generated for deducing their relation, which is later applied to input image to get result, as shown in Fig. 7. It does not seem appropriate to consider the polynomial nature of curves in any of the cases because of its non-monotonous nature and its tendency to force fit all plotted points, which

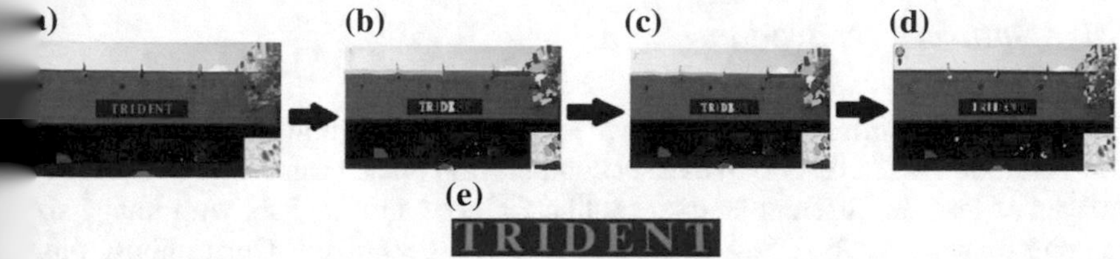

g. **7** Illustration of stepwise text region determination of a sample image from SVT dataset, input image, **b** MSER regions with initial region area ranging from 30 to 14,000 pixels, **c** MSER gions with region area ranging from 1200 to 14,000 pixels, **d** MSER regions with region area nging from 30 to 900 pixels, **e** cropped text region

ads to generating complex relations. Another characteristic observed is that the aximum and minimum region areas of MSER also have a relation with the height f the text region. After searching optimal values of maximum and minimum region reas for an image, we crop the text region.

After observing the graph, we come to a conclusion that minimum area region ollows a polynomial equation of degree 1, that is, it is a straight line with an equation:

$$y_1 = x_1 - c \tag{1}$$

vhere x_1 is the image dimension, which is the number of pixels in the image, y_1 is he minimum region area, c is a positive constant, whose value depends on the text egion height and background complexity.

On the other hand, maximum text region gives us a power curve with the equation:

$$y_2 = a \times x_2 \wedge b + c \tag{2}$$

where y_2 is the maximum region area, x_2 is the total number of pixels of the image and a, b, c are constants, whose values depend on the same parameters, like text region area, text height, and background complexity. We use concepts of regression to optimize the curve for getting a suitable equation and calculate the goodness-of-fit by comparing four parameters—sum of squared error, r-square, adjusted r-square, and root mean square error.

5 Experimental Results and Discussion

We evaluate our proposed idea by executing it on scene text images of standard text datasets, like SVT and MSRA-TD500, along with our in-house multi-lingual dataset. In the following section, we report and analyze the results obtained for each dataset.

5.1 Dataset Preparation

SVT dataset comprises 650 images where texts often exhibit high variability a low resolution. MSRA-TD500 dataset contains 500 pocket camera-captured natu images of texts in different languages, like Chinese and English, with image siz ranging between 1296 × 864 pixels and 1920 × 1280 pixels. Our in-house mul lingual dataset contains 300 images captured with smart phone camera, with text languages like Bangla, Hindi, and English, where the image size varies from 19 × 992 pixels to 4128 × 2322 pixels.

5.2 Curve Plotting

In order to assess the optimal values of minimum and maximum region areas i MSER detection, we analyze over 300 images from image pool we have made of different dimensions having variable size of text regions, and then we plot th threshold values of minimum and maximum region areas against the image size respectively. We observe that for minimum region area the graph shows a linea nature whereas for maximum region area the graph displays the characteristic of power curve, as depicted in Fig. 8.

We further categorize the images depending on the size of the text regions therein Small images have a text region having less number of pixels compared to large images that have text region with greater number of pixels. These regions are man- ually calculated and the minimum and maximum areas are plotted against image dimension respective to each category.

From the curves plotted in Figs. 9 and 10, we observe that apart from the image dimension which varies from 8×10^5 to 2×10^7 pixels, height of the text regions also plays a significant role for deciding the optimal value for minimum and maximum region areas in MSER.

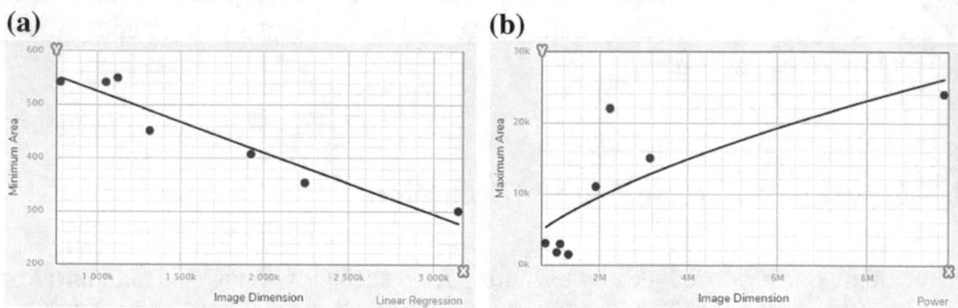

Fig. 8 Variation of image dimension with **a** minimum region area best fit by linear regression, **b** maximum region area best fit by power curve

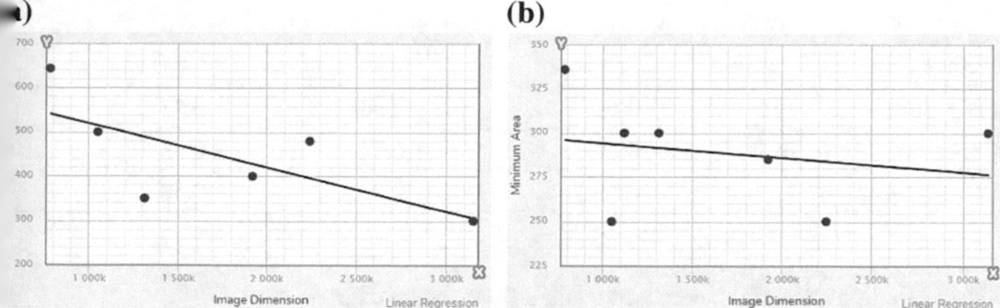

ig. 9 Variation of image dimension with minimum region area based on **a** minimum value of area r large-sized text, **b** minimum value of area for small-sized text

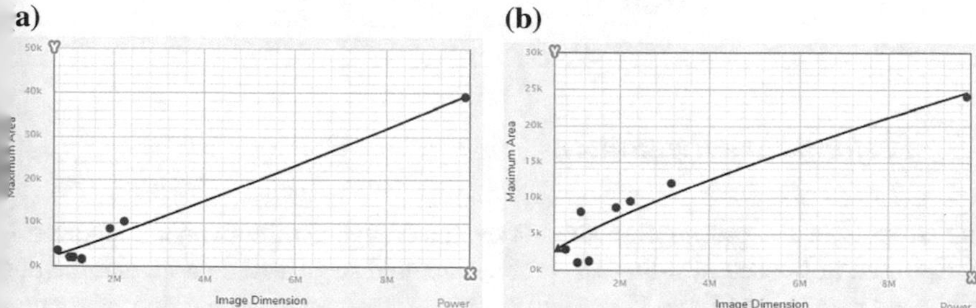

ig. 10 Variation of image dimension with maximum region area based on **a** maximum value of rea for large-sized text, **b** maximum value of area for small-sized text

We determine the height of the text region manually for each image and plot it against the threshold minimum and maximum region area values, thereby giving us he result as shown in Fig. 11.

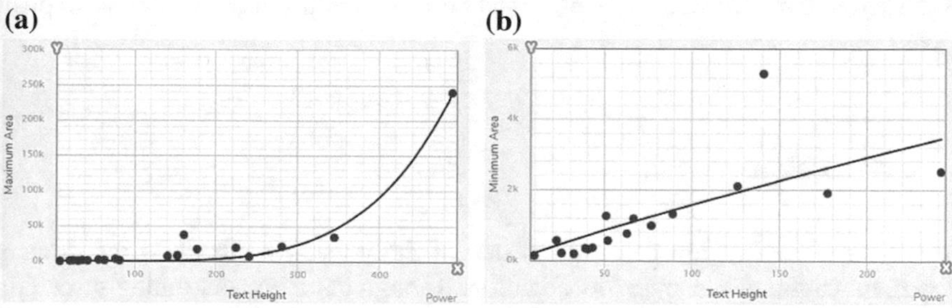

Fig. 11 Variation of area range with text height. **a** Maximum region area versus height of text region, **b** minimum region area versus height of text region

Table 1 Performance of the proposed method for in-house multi-lingual scene text dataset

Method	Precision	Recall	F-measure
Proposed	**0.62**	**0.90**	**0.73**
Chakraborty et al. [6]	0.59	0.84	0.68
Dutta et al. [5]	0.48	0.84	0.58

Table 2 Performance of the proposed method for standard datasets

Method	Dataset	Precision	Recall	F-measure
Proposed	SVT	**0.38**	**0.69**	**0.49**
Chakraborty et al. [6]	SVT	0.19	0.66	0.29
Proposed	MSRA-TD500	**0.67**	**0.74**	**0.70**
Kang et al. [8]	MSRA-TD 500	0.71	0.62	0.66

5.3 Performance Evaluation

The strength of the proposed method is evaluated against the methods reported i [5, 6] for an in-house multi-lingual camera-captured scene text dataset, and the out comes are recorded in Table 1. From this table we can say that the proposed metho outperforms its two recent ancestors in terms of precision, recall, and F-measure.

Some standard scene text datasets, like SVT and MSRA-TD500, are used fo validating the proposed method against the techniques reported in [6, 8]. From Table 2 we notice that the proposed method performs marginally better than the said two methods.

In a nutshell, we can safely claim that the deduced relations between image dimension and MSER region area range give improved results. In the case of MSRA-TD500 dataset, the proposed method achieves better recall and F-measure than the work in [8] although our precision is slightly lower due to significant number of false positives.

6 Conclusion

MSER-based text region detection is one of the popular methods in the domain of camera-captured scene text localization. Though complex, but tuning of certain parameters in MSER is required in order to get good results. Hence, the proposed method attempts to establish the relations among parameters, like image dimension, text size, and MSER region area range by analyzing the curves that best fits value points and achieves reasonably good recall and F-measure. However, the accuracy of the relation curves needs to be improved to get better insight. This shall be done experimenting on more number of images and studying other parameters which ıy crucial role in determining text regions in camera-captured images.

·knowledgments This work is partially supported by the CMATER research laboratory of the ›mputer Science and Engineering Department, Jadavpur University, India, PURSE-II, and UPE-project. Dr. Basu is partially funded by DBT grant (BT/PR16356/BID/7/596/2016) and UGC ·search Award (F.30-31/2016(SA-II)). Dr. Sarkar, Dr. Basu, and Dr. Mollah are partially funded DST grant (EMR/2016/007213).

·eferences

1. Chen, H., Tsai, S.S., Schroth, G., Chen, D.M., Grzeszczuk, R., Girod, B.: Robust text detection in natural images with edge-enhanced maximally stable extremal regions. In: 2011 18th IEEE International Conference on Image Processing (ICIP), pp. 2609–2612. IEEE (2011)
2. Neumann, L., Matas, J.: Real-time scene text localization and recognition. In: 2012 IEEE Conference on Computer Vision and Pattern Recognition (CVPR), pp. 3538–3545. IEEE (2012)
3. Matas, J., Chum, O., Urban, M., Pajdla, T.: Robust wide-baseline stereo from maximally stable extremal regions. Image Vis. Comput. **22**(10), 761–767 (2004)
4. Yao, C., Bai, X., Liu, W., Ma, Y., Tu, Z.: Detecting texts of arbitrary orientations in natural images. In: 2012 IEEE Conference on Computer Vision and Pattern Recognition, pp. 1083–1090. IEEE (2012)
5. Dutta, I.N., Chakraborty, N., Mollah, A.F., Basu, S., Sarkar, R.: Multi-lingual text localization from camera captured images based on foreground homogeneity analysis. In: Proceedings of the 2nd International Conference on Computing and Communications (IC3). Springer (2018). (In Press)
6. Chakraborty, N., Biswas, S., Mollah, A.F., Basu, S., Sarkar, R.: Multi-lingual scene text detection by local histogram analysis and selection of optimal area for MSER. In: Proceedings of the 2nd International Conference on Computational intelligence, Communications and Business Analytics (CICBA). Springer (2018). (In Press)
7. Gonzalez, A., Bergasa, L.M., Yebes, J.J., Bronte, S.: Text location in complex images. In: 2012 21st International Conference on Pattern Recognition (ICPR), pp. 617–620. IEEE (2012)
8. Kang, L., Li, Y., Doermann, D.: Orientation robust text line detection in natural images. In: Proceedings of the IEEE Conference on Computer Vision and Pattern Recognition, pp. 4034–4041 (2014)
9. Gomez, L., Karatzas, D.: Multi-script text extraction from natural scenes. In: Proceedings of the ICDAR, pp. 467–471 (2013)
10. Kumar, D., Prasad, M.N., Ramakrishnan, A.G.: Multi-script robust reading competition in ICDAR 2013. In: Proceedings of the 4th International Workshop on Multilingual OCR, p. 14. ACM (2013)

pplications and Advancements f Firefly Algorithm in Classification: n Analytical Perspective

nmenjoy Nayak, Kanithi Vakula, Paidi Dinesh and Bighnaraj Naik

bstract Nature-inspired algorithms, particularly those located on swarm intelli-ence and population-based, have dragged much attention in the last few years. Fire-y algorithm is one of the leading swarm-based algorithms, which came into view bout 10 years ago. It has become progressively an essential appliance of swarm telligence that has been enforced in many vicinities of optimization along with the ngineering practice. Abundant difficulties from distinctive regions have been solved ffectively by using firefly algorithm and its development. In this paper, we conducted brief study on the applications and advancements of firefly algorithm in the area f data classification. Various classification areas, such as image classification, text lassification, neural network-based classification and some other classifications, are aken into consideration for this study. Special attention has been paid toward the sage level and implementation issues of firefly algorithm in different classification lomains. The main aim of this survey is to inspire researchers toward further research f firefly algorithm in several other application areas other than the intended area.

Keywords Nature-inspired algorithms · Firefly algorithm · Classification · Data mining · Swarm intelligence

---

J. Nayak · K. Vakula · P. Dinesh
Department of Computer Science and Engineering, Sri Sivani College of Engineering, Srikakulam 532410, Andhra Pradesh, India
e-mail: mailforjnayak@gmail.com

K. Vakula
e-mail: vakku.bi@gmail.com

P. Dinesh
e-mail: dinesh.pydi98@gmail.com

B. Naik (✉)
Department of Computer Applications, Veer Surendra Sai University of Technology, Burla, Sambalpur 768018, Odisha, India
e-mail: mailtobnaik@gmail.com

A. K. Das et al. (eds.), *Computational Intelligence in Pattern Recognition*, Advances in Intelligent Systems and Computing 999,
https://doi.org/10.1007/978-981-13-9042-5_87

1 Introduction

Finding an alternate method with most effective or preeminent feasible performan under the given limitations by maximizing or minimizing the wanted or unwant factors is called optimization. In order to get the maximum benefit with minimu cost, we are depending on the optimization techniques. Several problems in differe fields are mapped as optimization problems and resolved using different optimizatio algorithms. It would not be an exaggeration to tell that optimization is far and wide. all over the creation, it is quite common to find optimization in engineering desig business and holiday planning, as well as internet routing. In all these activitie efforts are been placed to get conclusion in specific objectives, like time, quality an gain. The main idea to follow this optimization is to maintain the fixed balance wi exploration and exploitation. In our daily life, we easily come to know that time an gain are the resources that will have a less occurrence in real world. So, we need find a solution from real world to use this resource optimally.

In the real-world application, nature has inspired many researchers in several way and this is the reason that nature is known to be a major source of inspiration. Sinc last decade, most of the newly proposed algorithms are nature-inspired. All thes algorithms are proposed by taking an inspiration from nature only, and a major portio of all these nature-inspired algorithms depends on bio-inspired algorithms, whic have been advanced by illustrating motivation from swarm intelligence. There ar still a number of algorithms based on physical and chemical systems, but bio-inspire and swarm intelligence algorithms became an important topic in advancement of nev algorithms motivated by nature. Many such swarm intelligence and bio-inspire based algorithms are being developed with a very nominal objective toward getting the effective solution. However, algorithms such as ant colony [1], particle swarm optimization [2], cuckoo search [3], bat [4] and firefly [5] are some of the popula swarm-based techniques used for solving many engineering problems.

When we focus on new algorithms based on swarm intelligence, firefly algorithm is on the hot seat and is population-based. If our prominence is dependent on communication of different agents, algorithms can be categorized into attract- or not attract-based algorithms. Firefly algorithm is the best example of attract-based algorithms. It is one of the successful and leading optimization method simulated by nature. Nowadays, firefly algorithm (FA) has turned into a significant element of swarm intelligence, which is applied in many domains of optimization and engineering problems. Firefly algorithm was developed by Yang in 2008 [5]. In FA, heuristic ('lower level') focuses on the production of latest resolutions in a search space, and then selects the optimum resolution for continuity. Diversely, randomization facilitates the search space to divert the solution being caught in a trap into local optima.

The most complicated and demanding managerial process in every day homo-sapiens's life is classification. It is a procedure of classifying objects on a number of observed attributes correlated to that object, which frequently concerned and helps in decision-making activity. There are huge numbers of qualified former classification models [6–13] and have been developed using many conventional neural networks,

.e multilayer perceptron, back-propagation neural network, feed-forward network, d so on, in various specialized classification fields [14]. Supervised learning problems can be clustered into classification and regression problems. Classification is e of the supervised machine learning techniques, which is used to group the accred-d data. In classification, the output value will be 'category'. Many nature-inspired timization algorithms are used for various classifications, as well as for engineer-g and other problems. Some basic classification algorithms are Zero R, One R d Naive Bayesian. Some of the supervised learning algorithms used for classifi-ation problems are decision tree, linear discriminant analysis, logistic regression, -nearest neighbors, artificial neural network and model evaluation. Apart from these eas, classification is used frequently in various ancient and novel domain problems, ch as churn detection, fraud detection, user segmentation, prediction problems, nage classification, document classification, text classification, cancer classifica-on, ECG classification and webpage classification. There is a quick maturity of search in classification enthused by swarm-based, chemistry, physics, biology and volutionary-based algorithms which became popular.

In this paper, a brief review is conducted on the applications of firefly algorithm n solving various data classification problems. A detailed idea on firefly algorithm s illustrated in Sect. 2. Section 3 describes various applications of firefly algorithm vith three different specialized areas of classification, such as neural network, image nalysis and other data classification. Section 4 analyzes the critical aspects of firefly lgorithm in all sorts of classification. The concluding remarks with some important uture directions are mentioned in Sect. 5.

2 Firefly Algorithm

Lampyridae/firefly is a family of insect in the order Coleopteran. They are feathered beetles, commonly called fireflies or lightening bugs, or fire beetle, for their perceptible use of bioluminescence in the period of twilight to attract mates and mug. They chemically produce a cold light from its abdomen that may be pale red, yellow or green. Its wavelengths are from 510 to 670 nm. More than 2100 varieties of fireflies are found in tropical and temperature climates. In many varieties of fireflies, male and female fireflies have strength to fly, but in some categories, females are flightless. Fireflies are soft bodies and brown with front wings and leathery. Some species in shadow areas produce light. Mostly these are bioluminescent in nature. First male fireflies give the signal to flightless female on the ground. In order to give response to their signals, female fireflies produce flashing lights continuously. Both the courtship partners built clear flash signals that are exactly timed to conceal information such as sex and species identity. It is well known that the intensity of flash differs with distance from provenance. Females are attracted toward and prefer brighter males. After the mating process, female firefly lays fertilized eggs on the ground. Later, it will take 3–4 weeks for hatching process and larvae forage until the end of summer. Larvae are known as glowworms. Fireflies sleep through cold weather in the larval stage. Some will do by excavating the underground and others search for below t bark trees. After being as pupate for 1–2 weeks, they will develop into adults. He the pupal stages are controlled by hormones of insects only. The main purpose of t light in adult bug is in mate selection. Fireflies are standard example of an organis which uses bioluminescence for sexual selection. Flashing, steady glows and cher ical signal usage are the different ways to connect or communicate with comparis in wooing (courtship). These flashing lights carry out many physical rules. In th way, firefly mating process will take place and adult species attract their mate due pheromone like in ants.

Firefly algorithm mainly follows three rules [5]:

i. They are attracted toward others by inconsiderate gender.
ii. Attractiveness of fireflies is associated with brightness of fireflies, so that firefl which is less attractive will move in advance to more attractive firefly.
iii. Brightness is based on objective functions.

2.1 Structure of Firefly Algorithm

Yang proposed firefly algorithm [5] based on principle of intensity of light '*I*' i which the origin of light decreases with increase in 'r^2' (square of the distance) Owing to rapid increase in light source, the light turns into unstable state on the basi of retention of light. To ought to be optimizing, this whole principle is integrated with objective function. The two main characteristics attracted the developers and researchers to use firefly algorithm to solve various engineering, classification, as well as other data mining problems, and mostly for solving supervised machine learning problems. One is the variant of light intensity and another one is its attractiveness. However, after the implementation of firefly algorithm, so many advancements and changes were made and it is categorized according to its qualities and behavior, and is classified as follows:

- Rendering of fireflies which are real-valued and binary
- Assessing of fitness function
- Population strategy which are swarm and multiswarm-based
- Establishment of contemporary best solution.

Firefly algorithm was used frequently for most persistent problems of optimization in order to decode aspired commodity (solution). But, in some cases, solutions could not be found for some problems of optimization, which is one of the aspects of no-free lunch theorem [15]. To gain an advantage over this theorem, optimization algorithms resolve granted queries of several optimizations.

In order to entitle the solutions, the concentration of light of firefly '*I*' is correlative to coefficient of the fitness function [5], that is

$I(s) \propto F(s)$. But $I(r)$ differs from the following Eq. (1):

$$I(r) = I_0 e^{-\gamma r^2} \quad (1)$$

ıe deviation at r = 0 in I/r^2 abstained by bringing together by originate the cause inverse square law and Gaussian form's absorption.

Therefore,

$$\beta \propto I(r) \quad (2)$$

here

= attractiveness of firefly

r) = intensity of light

To define β accordingly, Eq. (2) can be rewritten as Eq. (3)

$$\beta = \beta_0 e^{-\gamma r^2} \quad (3)$$

where β is the attractiveness at r = 0.

The intensity of light and brightness are interdependent on each other. According o ancient firefly algorithm, the distance between two fireflies 's_i' and 's_j' in terms f Euclidean distance is given by Eq. (4)

$$r_{ij} = \left\|\mathrm{s_i} - \mathrm{s_j}\right\| = \sqrt{\sum_{k=1}^{k=n} (s_{ik} - s_{jk})^2} \quad (4)$$

The movement of ith firefly is attracted toward other attractive firefly 'j'. As per this condition, the Eq. (5) is derived.

$$\mathrm{s_i} = \mathrm{s_i} + \beta_0 e^{-\gamma r_{ij}^2}\left(\mathrm{s_j} - \mathrm{s_i}\right) + \alpha \varepsilon_i \quad (5)$$

Here ε_i is the arbitrary number exhausted from Gaussian distribution.

3 Applications of Firefly Algorithm

Firefly algorithm is one of the recent SI method introduced by Yang in the year 2008 [5]. It is a type of metaheuristic and nature-inspired algorithm that can be used for resolving NP-hard problems. In this algorithm resides stochastic algorithm. Firefly algorithm helps in finding solution for solving many continuous optimization problems. The main inspiration behind Yang's idea to develop firefly algorithm (FA) is the light intensity of two fireflies that reduces as the distance increases. The behavior of mutual coupling that occurs between both fireflies inspires to get the solution for any real-life problem. Firefly algorithm is specially suited for multimodal optimization and is inspired from the reality that naturally fireflies can be divided into few subgroups due to having a strongest neighboring attraction than long distan attraction.

3.1 Hybrid Firefly Algorithm

In order to solve many problems, various supervised and unsupervised learnin algorithms were integrated with firefly algorithm. Any two normal problem solve are similar while the general performance is correlated beyond all feasible problem Normally, heuristics are engaged to solve a set of problems and improves the outcon of problem solver. Heuristics can also be included into the firefly algorithm. Specif hybrid firefly algorithm advances the outcomes when resolving the given problen Firefly algorithm can also be handled as a study for hybridizing with some othe problems due to their faster convergence and multimodality characteristics.

- Yang stated an eagle strategy which is a new metaheuristic search method th connects the levy flight search with firefly algorithm, and where its mechanisr was applied at local search, which is one of the components in eagle strategy.
- Hassanzab et al. [16] have used firefly algorithm for preparing structure-equivalen fuzzy neural network to speech recognition, where it showed the results that th recognition rate of hybridized speech recognition is higher than other fuzzy neura networks.

Finally, firefly algorithm was superior to exponential particle swarm optimization i both success and efficiency.

The reasons behind efficiency of firefly algorithm are:

- Firefly algorithm can be divided into few subgroups, because of the reality tha neighboring attractiveness is stronger than long-distance attraction.
- Firefly algorithm can deal with multimodal and nonlinear problems efficiently as well as naturally.

Firefly algorithm has been used for resolving classifications and in many optimization problems, such as multiobjective, noisy, combinatorial and dynamic optimizations. In the same way, firefly algorithm has been used in categorization problems using machine learning, neural network and data mining techniques. At last, firefly algorithm focused on image processing, wireless sensors network and some other engineering areas. In this way, firefly algorithm is well appropriate for multimodal optimization. It has faster convergence. The main reason behind the fast convergence is losing the variety of population. To use this algorithm for solving the problems of optimization, balancing of intensification and diversification should be settled.

To regulate the applications of FA in classification as well as other application domains, during the last decennium, this paper reviews the critique survey of various articles up to 2018. A keyword search has been performed on several standard databases (Elsevier, IEEE X-plore, Springer Link and Inderscience online databases). For the period of 2008–2018, various articles based on the keyword search were und. But more number of articles is based on some similar applications. For this ason, in this study, we have considered various classifying features of application eas in various domains using firefly algorithm. The classification of application cinities in this learning has been made in the following way: neural network, nage analysis and others in various application domains. Firefly algorithm has rger domain of applications as compared with other recently used nature-inspired ptimization algorithms. However, in this section, some popular and frequently used assifications using nature-inspired firefly algorithm with its applications are menoned. In addition, some other classifications using firefly algorithm are described the later section.

.2 *Neural Network with Firefly for Classification*

he concept of neural network is another solution to correlative processing of human rain. Neural networks are robustly capable of handling unreliable or inaccurate data nd generating accurate patterns without the use of algorithmic approach. Its other haracteristic is not confined to the problem-solving capability of the system. Neural etworks act as human specialists. They not only solve any particular problem in a ompetence manner but also analyze some complex problems in an efficient way.

Nayak et al. [14] have innovated a modern method for the classification of data with eural networks. They tested the performance of pi-sigma neural network by using irefly algorithm (FA). In view of establishing efficient performance, the authors considered classification accuracy as a performance factor and compared their developed method with some famous nature-inspired optimizations, such as genetic algorithm, particle swarm algorithm. The main aim of their proposed method is to substantiate the enchantment of firefly in data classification and then to examine that their method has desirable performance than other intended methods. The most frequently used learning algorithm for FLANN training is the BP learning algorithm [17]. To overcome the BP learning algorithm's drawback of getting stucked in local optima, a modified bee firefly algorithm has been developed by Hassim and Ghazali [18] for mammographic mass classification task. They used several network models as performance factors on unseen data and later compared their developed method with FLANN back propagation. They claimed that the proposed FLANN architecture generalizes better performance than ordinary FLANN architecture. Behera and Behera [19] innovated a novel ridge polynomial neural network using a famous nature-inspired metaheuristic algorithm named firefly algorithm for classification problem. To measure the performance of their innovative method, they considered various benchmark datasets. Later, they compared their proposed method with firefly-based multilayer perception and found better convergence rate and classification accuracy than the compared method. They claimed that their method will be helpful in certain classification domains to solve various problems of classification: image classification, text classification, and so on. Alweshah and Abdullah [20] developed hybrid firefly algorithm with a probabilistic neural network for solving classification problem. The popular and efficient technique in classification is probabilis neural network. The main work represented in their paper is to develop an efficie method that can find solutions with high accuracy at high speed for the probler of classification. The authors introduced a method to associate the firefly algorith along with simulated annealing, where it is enforced to restrict the randomness the firefly algorithm during developing the probabilistic neural network. This alg rithm was tested on 11 standard datasets. They increased their work by investigatir the efficiency of using levy flight in the firefly algorithm to improve search spa and combining simulated firefly algorithm with levy flight to develop the prob bilistic neural network. Their results show that LSFA is better than SFA and LF LSFA gives better results in classification accuracy. The definite recognition is ve important for nonlinear dynamic part. It is difficult to find out the meticulous mod with less prior knowledge that is highly dynamic and nonlinear, like robotics an autonomous system. Fuzzy neural networks are used to work with classification an real-world applications. Behera et al. [21] introduced a nonlinear dynamic syster identification using Legendre neural network and firefly algorithm, where they clearl mentioned about nonlinear dynamic plant identification. Pseudo-inverse Legendr neural network approaches with chaotic mutation with plant by enlarging input pat tern. For enhanced results, the authors used the firefly technique (which is inspire by the flocking behavior of fireflies) for their intended method. Jinthanasatian [22 has proposed microarray data classification using neuro-fuzzy classifier with firefl algorithm, where neuro-fuzzy is an effective tool in many number of applications The authors recommended neuro-fuzzy along with firefly algorithm and application to microarray classification. They compared their results with seven public datasets such as ovarian, prostate, lung cancer, and so on. Their algorithm provides com- parable results along smaller numbers of the selected features and understandable to humans when compared with other algorithms. Zhang et al. [23] developed a modified firefly algorithm for classifier ensemble reduction. By simulating various datasets, they found that their method is very encouraging for ensemble reduction. They also claimed that their proposed method blunt the premature convergence of primitive firefly model. Adermi et al. [24] introduced a phishing detection system for e-mail by using the combination of firefly algorithm and SVM classifier. They considered firefly algorithm for parameter selection. By using standard dataset con- taining 4000 e-mails, they found higher classification accuracy rate. They claimed that their method has greater performance. Their method elaborates the possibility to increase the robustness of SVM classifier by hybridizing with firefly algorithm. Some of the applications of firefly algorithms with neural network have been illustrated in Table 1.

3.3 Image Classification Using Firefly

Image classification is an important part of digital image analysis. It is a process of selecting information from a raster image. It is known to be assigning pixels in

ble 1 Applications of firefly algorithm in various neural networks

lgorithm used	Type of neural network	Application area	Performance factors	Year	Ref.
A	k-NN	Feature selection	Error rate, classification accuracy	2018	[25]
A	ML NN	Jaw fracture classification	Mean square error rate, precision, recall, accuracy	2018	[26]
inary FA	Medical data classification	Feature selection	Population size, β_0, γ	2018	[27]
A	Sentiment analysis	Feature selection	Classification accuracy	2017	[28]
inary FA	–	Privacy preserving	Precision, recall accuracy, error rate	2017	[29]
MBF	FLNN	Pattern classification	Learning rate, momentum, epoch, minimum error	2017	[30]
FA	ANN	EEG electrode classification	Recall precision, average error rate	2017	[31]
FA	NN	Feature selection	Classification accuracy	2017	[32]
FA	Translucent elastic optical networks	VER classification	Attractiveness parameter, absorption coefficient, maximum number of evaluations, number of fireflies, number of dimensions	2016	[33]
Self adaptive FA	Pattern recognition	Optimal feature selection	Classification accuracy	2016	[34]
FA	BPNN	Power complexity feature selection	Classification accuracy	2015	[35]
FA	Fuzzy min-max NN	Heart disease classification	Sensitivity, specificity, accuracy	2013	[36]

the image to classes or categories of interest. Image classification toolbar has be advanced to support an integrated environment to carryout classifications with too

Darwish [37] introduced a modified multilabel automatic image annotation f classification using the hybridization of firefly algorithm and Bayesian classifier. T main aim of this proposed method is to establish the similarity of image regions a keywords. For this, the author considered image annotation as a problem for t classification of Bayesian framework and compared the author's developed meth with corel dataset. Later, the author found that his way of approach shows higher pe formance and also proved that the developed method is fit for automatic image ann tation. Jothi and Inbarani [38] introduced a hybridization of tolerance rough set a firefly algorithm for the classification of brain tumor images. Later, they consider classification accuracy as performance factor and compared their proposed meth with some standard algorithms, such as cuckoo search and artificial bee colony. The found an improved solution in their method and claimed that their method showe the efficient selection features with high quality. They also claimed that their wa of approach is recommended for medical diagnostics and multimodal data: mam mography, ultrasound, and so on. Keerthana et al. [39] improved band selection wit firefly for the classification of hyperspectral image. They considered band selectio as a problem of optimization. They also considered virtual dimensionality to acquir bands and classification accuracy as performance factor for their proposed metho Later, they compared their method with PSO by using SVM classifier on salina image. They found that their method showed higher classification accuracy rate, an also claimed that their way of approach is so far better than PSO with the accu racy rate of 96.3%. Bartosz and Pawel [40] developed a nuclei detection system fo detailed image examination using integration of firefly, marker-controlled watershe segmentation and one-class decomposition strategy for breast cancer analysis. Fo this, they considered various cytological images. Their experimental results showed desired detection rates for cancer. They claimed that their novel method will be helpful for clinical practices as well as pattern recognition. Su et al. [41] proposed firefly algorithm optimization extreme learning machine for hyperspectral image classification. It is a parameter optimization-based method for extreme learning machine. Gaussian kernel and regularization coefficients are the parameters used by firefly algorithm. The authors have concluded that firefly algorithm-based techniques can show better result for extreme learning machine hyperspectral image classification. It performs some popular algorithms, like particle swarm optimization and genetic evolution algorithm method. Chhikara [42] has proposed an improved discrete firefly and t-test-based algorithm for blind image steganalysis. Feature selection has more importance in data mining, which targets to decrease complexity and increase predictive capability. It represents the combined feature selection that depends on firefly optimization of the parameters, alpha and gamma. t-test filter techniques are used to improve detection of hidden message from blind image steganalysis. The experiments were conducted on different datasets. In the proposed technique, the method is able to identify the sensitive features and decrease the feature set into 67% domain of DCT and 37% of DWT domain. Firefly algorithm inspired optimized extreme learning and band learning for hyperspectral image classification. FA opts

ble 2 Applications of efly algorithm in various age classifications

Type of image	Comparison with other techniques	Year	Ref.
Spatial image	–	2017	[45]
Blind image	DPSO, DFA	2015	[46]
Hyperspectral image	SS-SVM and SS-LSSVM	2015	[47]
Blind image	SVDD classifier	2015	[48]
2D-image	PSO	2015	[49]

or a division of original bands to decrement the difficulty of ELM network. Owing low difficulty of ELM, its accuracy in classification can be used during band selection and parameter optimization as objective function. Su et al. [43] proposed firefly lgorithm-inspired framework with band selection and extreme learning machine for yperspectral image classification. Final results indicated that their method provides etter results when compared to PSO and remaining band selection algorithm. In nodern years, different scientific practices have been fit to the artwork analysis process. In their paper, hyperspectral imaging mixed with classification techniques and ignal processing is recommended as a device to strengthen the procedure for identiication of art forgeries. Polak et al. [44] manage this dual structure and concentrate nitially on testing and system development. However, they explained the problems of data acquisition and its solutions. Finally, this technique increments the number of analyzed artworks and made the total procedure even more efficient. Some of the ntelligent techniques with various types of images can be found in Table 2.

3.4 Other Classification Models Using Firefly

Apart from the above-mentioned classifications, we have some other classifications also by using firefly algorithm, such as ECG-based classification, cervix lesion classification, EEG classification, data classification, ISO-FLANN classifications, bearing fault classification, webpage classification and speech signal classifications.

Saberi et al. [50] developed an efficient clustering-based fuzzy time series algorithm (FEFTS) for regression and classification. FTS methods are normally classified into two types, where one is dependent upon periods of universal set and another depends on clustering algorithms. Clustering-based fuzzy time series algorithm is used due to some problems with the interval-based algorithms, like ideal interval length. Fuzzy logical relationships are normally set up to endure the relationship among the data of input and output in both interval and clustering-based algorithms. In their paper, fast and efficient clustering-based fuzzy time series algorithm is proposed to hold the problems of classification and regression. Audit et al. [51] developed a hybrid approach for micro expression that depends on human emotion recognition. Micro expressions are an automatic indication that people tries to suppress and

Table 3 Applications of firefly algorithm in various other domains of classification

Name of the algorithm	Application area	Integrated with	Year	Ref.
FA	Keystroke dynamics	–	2018	[54]
FA	Webpage classification	–	2013	[55]
FA	Feature selection	–	2018	[56]
FA	Face recognition	–	2017	[57]

obscure. These are used as important devices for good applications, such as enhancir relationships, psychiatry, clinical therapy, national security and lie detection. Mic expressions require classification and extraction. First phase follows firefly algorith and the second one follows improved swarm-optimized functional link artificial ne ral network for classification. From their experimental results, the authors gave th rate of overall recognition is 68.65% and is known for higher performance whe compared with other previous methods. Almonacid et al. [52] introduced solving th manufacturing cell design problem using the modified binary firefly algorithm calle the Egyptian vulture optimization algorithm, which aims to decrease the operation between production cells. This is a NP-hard problem. For the declaration of manu facturing cell design problem (MCDP), the authors engaged firefly algorithm whic is a metaheuristic along real domain. So, binary domain is used here for transfer an discretization. Egyptian vulture optimization algorithm is the second metaheuristi algorithm which is inspired from the Egyptian vulture bird. The authors have foun the best values using firefly algorithm and Egyptian vulture optimization algorithm Preeti and Sowmiya [53] have proposed emotion recognition from EEG signal using ISO FLANN with firefly. Human brains can be analyzed in different ways and its emotions will be measured in EEG signal. From their paper, the authors finalized the six biological tested neural networks that are used to take out from human emotions calculated by EEG. Classification, feature extraction and preprocessor are the three functionalities present in this system. Band pass filter is the first phase. Second phase is expected by the connectivity feature by maximum square coherence estimation. Radial basis and non-negative sparse principal component analysis for the feature selection is the third phase. There are more newly invented results when compared to already existing system. Table 3 describes some applications of firefly algorithm in various domains.

4 Critical Analysis and Discussions

In this paper, a comprehensive study is being conducted on the use and applications of well-known nature-inspired optimization algorithm, named firefly algorithm in classification. Various classification domains with neural network, feature selection other data classification are deemed with firefly algorithm techniques, starting from

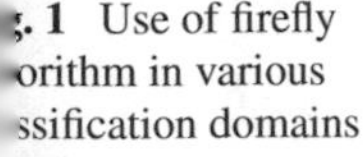

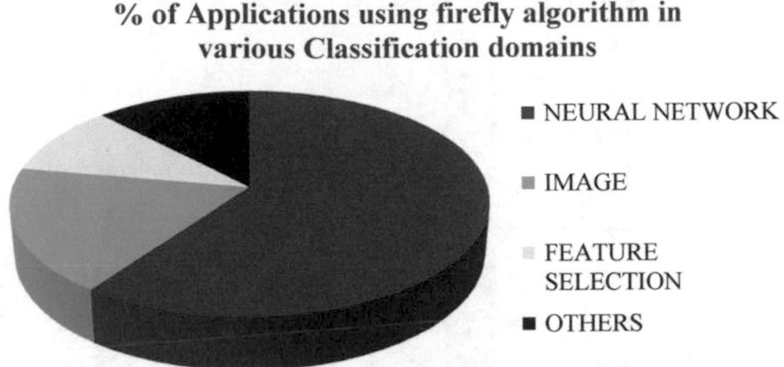

. 1 Use of firefly orithm in various ssification domains

icient firefly algorithm to the integrated firefly algorithms, developed up to 2018. is reasonably noticeable that so many nature-inspired algorithms are used to solve ifferent application areas with many classification algorithm and have successfully roved their effectiveness. However, in this analysis, it is observed that firefly algo-thm is applied to solve various problems of optimization as well as classification. 'arious studies signified that most of the techniques are being developed by firefly lgorithm and its variants. Some important hybridization is also being developed in ntegration with firefly algorithm. The papers from famous online databases, such s IEEE X-plore, Springer, Science direct, based on the keyword search and almost requently used classification domains using firefly algorithm are also covered. A horough analysis has been illustrated in Fig. 1, and it is evident that, as compared vith other classification domains, neural networks with firefly algorithm have been of repeated concern for researchers. In many of the cases, firefly algorithm has been ısed to optimize the weights of the neural network and it has been hybridized with he NN structure also. However, some other classification areas such as webpage classification, key stroke dynamic classifications are also solved using firefly algo-rithm.

It is observed that firefly algorithm is fit for multimodal optimization, combinato-rial optimization, and so on. This is due to its fast convergence. According to Yang [58], in order to remain the use of firefly algorithm in the case of large-scale problems, we should maintain appropriate balance of intensification and diversification. The outcome of firefly algorithm is based on the contemporary finest solutions. There may be a chance in advancement of effectiveness of firefly algorithm by improving the best solution. However, there is a rapid chance of advancements using firefly algorithm for further future research among the researchers.

In real-life applications, firefly algorithms frequently congregate hastily; still, there is no theoretical study how hastily it can really unite. The representation we acquire in Fig. 2 is about the performance of firefly algorithm with an enormous scope in all applications, especially in classifications. Although firefly algorithm is used in huge number of applications, we mainly focus on the use of firefly algorithm in classification domains. Based on the keyword search 'Firefly algorithm', we found 4306, 3819 and 651 papers and based on the keyword search 'Firefly algorithm in classification', we found 143, 1081 and 48 papers from Science direct, Springer Link and

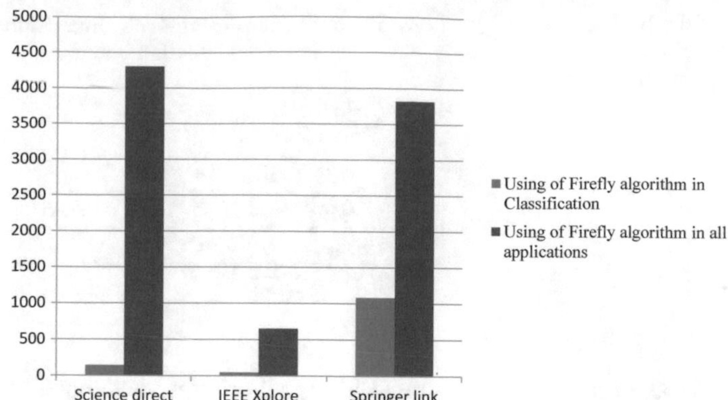

Fig. 2 Analysis of paper publications from standard databases on firefly algorithm

IEEE X-plore standard online databases, respectively. This shows the effectivenes of the algorithm in solving real-life problems.

Sometimes, the solution may not be explored using firefly algorithm in combi natorial optimization problems. For this, further study is required to balance th intensification and diversification in such a case. There are a huge number of ope problems regarding all metaheuristic algorithms [59]. Parameter tuning is one mor important domain that is needed to be studied and considered for all the metaheuris tic algorithms. Also, more surveys recognized that optimal setting to solve severa problems with nominal modification of parameter values is a critical optimizatio problem. From this review, applications of FA are incredibly dissimilar. It migh be more copious to use and apply firefly algorithms to novel domains such as data mining, telecommunications, large-scale real-world applications and bioinformatics [60]. It cannot be overstated to say that more number of applications using firefly algorithms will be developed in the upcoming days.

5 Conclusion

Firefly algorithm extensively prolonged its appliance fields since its initiation in 2008. These days, almost no more area is left there where firefly algorithm had not been applied. In addition, expansion fields of this algorithm are extremely energetic, as novel applications come into view almost every day. This study manifests that firefly algorithm is appropriate for multimodal optimization. It achieves good consequences on functional optimization. FA has fast convergence and is suitable for combinatorial optimization. This article carries a wide-ranging analysis of firefly algorithm (FA) and its classifications. Classification is one of the significant tasks in data mining. Analysis of previous papers hybridizing the firefly algorithm illustrates that there are a few additional algorithms, like genetic algorithm and eagle strategy, which further used as local search for hybridizing these algorithms. Along with other vances, it is worth noting the hybridization of firefly algorithm with neural network. has a better search capability as compared to other nature inspire algorithms. In s paper we have discussed the applications of firefly algorithms in various domains ch as image and neural network. However, there is a quick chance of progressions ing firefly algorithm that can encourage further research in these vicinity in near ospect. It cannot be an exaggeration to say that more number of applications using efly algorithm will be urbanized in the upcoming days.

ppendix

NN	Artificial neural network
PNN	Back-propagation neural network
FA	Discrete firefly algorithm
PSO	Discrete particle swarm optimization
CT	Discrete cosine transformation
WT	Discrete wavelet transformation
CG	Electrocardiogram
EG	Electro encephalogram
LM	Extreme learning machine
A	Firefly algorithm
EFTS	Fast and efficient clustering-based fuzzy time series
LANN	Functional link artificial neural network
FLNN	Functional link neural network
FTS	Fuzzy time series
SO-FLANN	Improved swarm optimized-functional link artificial neural network
KNN	k-nearest neighbor
LFA	Levy flight firefly algorithm
LSFA	Levy flight integrated simulated annealing firefly algorithm
MCDP	Manufacturing cell design problem
MLNN	Multilayered associative neural networks
NN	Neural network
PSO	Particle swarm optimization
SVM	Support vector machine
VER	Virtualized-elastic-regenerator

References

1. Dorigo, M., Stützle, T.: The ant colony optimization metaheuristic: algorithms, applicatio and advances. In: Handbook of Metaheuristics, pp. 250–285. Springer, Boston (2003)
2. Kennedy, J.: Particle swarm optimization. In: Encyclopedia of Machine Learning, pp. 760–7 Springer, Boston, MA (2011)
3. Yang, X.-S., Deb, S.: Cuckoo search: recent advances and applications. Neural Comput. Ap **24**(1), 169–174 (2014)
4. Yang, X.-S.: A new metaheuristic bat-inspired algorithm. In: Nature Inspired Cooperati Strategies for Optimization (NICSO 2010), pp. 65–74. Springer, Berlin (2010)
5. Yang, X.-S.: Firefly algorithm, stochastic test functions and design optimisation. Int. J. Bi Inspired Comput. **2**(2), 78–84 (2010)
6. Macioek, P., Dobrowolski, G.: Using shallow semantic analysis and graph modeling for do ument classification. Int. J. Data Min. Model. Manag. **5**(2), 123–137 (2013)
7. Yin, P., Wang, H., Zheng, L.: Sentiment classification of Chinese online reviews: analyzi and improving supervised machine learning. Int. J. Web Eng. Technol. **7**(4), 381–398 (2012
8. Kanungo, D.P., Naik, B., Nayak, J., Baboo, S., Behera, H.S.: An improved PSO based bac propagation learning-MLP (IPSO-BP-MLP) for classification. In: Computational Intelligenc in Data Mining, vol. 1, pp. 333–344. Springer, India (2015)
9. Uysal, A.K., Gunal, S.: Text classification using genetic algorithm oriented latent semant features. Exp. Syst. Appl. **41**(13), 5938–5947 (2014)
10. Sriramkumar, D., Malmathanraj, R., Mohan, R., Umamaheswari, S.: Mammogram tumou classification using modified segmentation techniques. Int. J. Biomed. Eng. Technol. **13**(3 218–239 (2013)
11. Kianmehr, K., Alshalalfa, M., Alhajj, R.: Fuzzy clustering-based discretization for gene expres sion classification. Knowl. Inf. Syst. **24**(3), 441–465 (2010)
12. Sarkar, B.K., Sana, S.S., Chaudhuri, K.: Accuracy-based learning classification system. Int. J Inf. Decis. Sci. **2**(1), 68–86 (2010)
13. Valavanis, I.K., Spyrouand, G.M., Nikita, K.S.: A comparative study of multi classificatio methods for protein fold recognition. Int. J. Comput. Intell. Bioinf. Syst. Biol. **1**(3), 332–34 (2010)
14. Nayak, J., Naik, B., Behera, H.S.: A novel nature inspired firefly algorithm with higher orde neural network: performance analysis. Eng. Sci. Technol. Int. J. **19**(1), 197–211 (2016)
15. Wolpert, D.H., Macready, W.G.: No free lunch theorems for optimization. IEEE Trans. Evol Comput. **1**(1), 67–82 (1997)
16. Hassanzadeh, T., Faez, K., Seyfi, G.: A speech recognition system based on structure equivalent fuzzy neural network trained by firefly algorithm. In: 2012 International Conference on Biomedical Engineering (ICoBE), pp. 63–67. IEEE (2012)
17. Le Cun, Y.: A theoretical framework for back propagation. In: Touretzky, D., Hinton, G., Sejnowski, T., (eds.) Proceedings of the 1988 Connectionist Models Summer School. June 17–26, pp. 21–28. Morgan Kaufmann, San Mateo, CA (1988)
18. Hassim, Y.M.M., Ghazali, R.: Mammographic mass classification using functional link neural network with modified bee firefly algorithm. In: International Conference in Swarm Intelligence. Springer, Cham (2016)
19. Behera, N.K.S., Behera, H.S.: Firefly based ridge polynomial neural network for classification. In: 2014 International Conference on Advanced Communication Control and Computing Technologies (ICACCCT). IEEE (2014)
20. Alweshah, M., Abdullah, S.: Hybridizing firefly algorithms with a probabilistic neural network for solving classification problems. Appl. Soft Comput. **35**, 513–524 (2015)
21. Behera, S., Sahu, B.: Non linear dynamic system identification using Legendre neural network and firefly algorithm. In: 2016 International Conference on Communication and Signal Processing (ICCSP). IEEE (2016)

Jinthanasatian, P., Auephanwiriyakul, S., Theera-Umpon, N.: Microarray data classification using neuro-fuzzy classifier with firefly algorithm. In: 2017 IEEE Symposium Series on Computational Intelligence (SSCI). IEEE (2017)

Zhang, L., et al.: Classifier ensemble reduction using a modified firefly algorithm: an empirical evaluation. Expert. Syst. Appl. **93,** 395–422 (2018)

Adewumi, O.A., Akinyelu, A.A.: A hybrid firefly and support vector machine classifier for phishing email detection. Kybernetes **45**(6), 977–994 (2016)

Xu, H., et al.: An improved firefly algorithm for feature selection in classification. Wirel. Pers. Commun. 1–12

Hashem, M., Hassanein, A.S.: Jaw fracturc classification using meta heuristic firefly algorithm with multi-layered associative neural networks. Clust. Comput. 1–8 (2018)

Sahmadi, B., et al. A modified firefly algorithm with support vector machine for medical data classification. In: Computational Intelligence and Its Applications: 6th IFIP TC 5 International Conference, CIIA 2018, Oran, Algeria, May 8–10, 2018, Proceedings 6. Springer International Publishing (2018)

8. Kumar, A., Khorwal, R.: Firefly algorithm for feature selection in sentiment analysis. In: Computational Intelligence in Data Mining, pp. 693–703. Springer, Singapore (2017)

9. Kalyani, G., Chandra Sekhara Rao, M.V.P., Janakiramaiah, B.: Privacy-preserving classification rule mining for balancing data utility and knowledge privacy using adapted binary firefly algorithm. Arab. J. Sci. Eng. 1–23 (2017)

0. Hassim, Y.M.M., Ghazali, R., Wahid, N.: Improved functional link neural network learning using modified bee-firefly algorithm for classification task. In: International Conference on Soft Computing and Data Mining. Springer, Cham (2016)

1. Lahiri, R., Rakshit, P., Konar, A.: Evolutionary perspective for optimal selection of EEG electrodes and features. Biomed. Signal Process. Control **36**, 113–137 (2017)

2. Anbu, M., Anandha Mala, G.S.: Feature selection using firefly algorithm in software defect prediction. Clust. Comput. 1–10 (2017)

3. Yamazaki, K., Matsushita, H., Jinno, M.: Virtualized-elastic-regenerator placement by firefly algorithm for translucent elastic optical networks. In: 2016 IEEE Congress on Evolutionary Computation (CEC). IEEE (2016)

34. Zhang, L., Shan, L., Wang, J.: Optimal feature selection using distance-based discrete firefly algorithm with mutual information criterion. Neural Comput. Appl. **28**(9), 2795–2808 (2017)

35. Behnam, M., Pourghassem, H.: Power complexity feature-based seizure prediction using DNN and firefly-BPNN optimization algorithm. In: 2015 22nd Iranian Conference on Biomedical Engineering (ICBME). IEEE (2015)

36. Rajakumar, B.R., George, A.: On hybridizing fuzzy min max neural network and firefly algorithm for automated heart disease diagnosis. In: 2013 Fourth International Conference on Computing, Communications and Networking Technologies (ICCCNT). IEEE (2013)

37. Darwish, S.M.: Combining firefly algorithm and Bayesian classifier: new direction for automatic multilabel image annotation. IET Image Process. **10**(10), 763–772 (2016)

38. Jothi, G.: Hybrid Tolerance Rough Set-Firefly based supervised feature selection for MRI brain tumor image classification. Appl. Soft Comput. **46**, 639–651 (2016)

39. Keerthana, K., Veerasenthilkumar, G., Vasuki, S.: Firefly based band selection for hyperspectral image classification

40. Krawczyk, B., Filipczuk, P.: Cytological image analysis with firefly nuclei detection and hybrid one-class classification decomposition. Eng. Appl. Artif. Intell. **31**, 126–135 (2014)

41. Su, H., Cai, Y.: Firefly algorithm optimized extreme learning machine for hyperspectral image classification. In: 2015 23rd International Conference on Geoinformatics. IEEE (2015)

42. Chhikara, R.R., Singh, L.: An improved discrete firefly and t-test based algorithm for blind image steganalysis. In: 2015 6th International Conference on Intelligent Systems, Modelling and Simulation (ISMS). IEEE (2015)

43. Su, H., Cai, Y., Qian, D.: Firefly-algorithm-inspired framework with band selection and extreme learning machine for hyperspectral image classification. IEEE J. Sel. Top. Appl. Earth Obs. Remote. Sens. **10**(1), 309–320 (2017)

44. Polak, A., et al.: Hyperspectral imaging combined with data classification techniques as an for artwork authentication. J. Cult. Herit. **26,** 1–11 (2017)
45. Ramesh, R., Gomathy, C., Vaishali, D.: Bio inspired optimization for universal spatial im steganalysis. J. Comput. Sci. **21**, 182–188 (2017)
46. Chhikara, R.R., Singh, L.: A hybrid feature selection technique based on improved discr firefly and filter approach for blind image steganalysis. Int. J. Simul. Syst., Sci. Technol. **16**(2–1, (2015)
47. Yang, L., et al.: Coupled compressed sensing inspired sparse spatial-spectral LSSVM for hyp spectral image classification. Knowl.-Based Syst. **79,** 80–89 (2015)
48. Chhikara, R.R., Singh, L.: An improved discrete firefly and t-test based algorithm for bli image steganalysis. In: 2015 6th International Conference on Intelligent Systems, Modelli and Simulation (ISMS). IEEE, 2015
49. Napoli, C., et al.: Toward 2D image classifier based on firefly algorithm with simplified sob filter. In: 2015 Asia-Pacific Conference on Computer Aided System Engineering (APCASE IEEE (2015)
50. Saberi, H., Rahai, A., Hatami, F.: A fast and efficient clustering based fuzzy time series alg rithm (FEFTS) for regression and classification. Appl. Soft Comput. **61**, 1088–1097 (2017)
51. Aadit, M.N.A., Mahin, M.T., Juthi, S.N.: Spontaneous micro-expression recognition usin optimal firefly algorithm coupled with ISO-FLANN classification. In: 2017 IEEE Region 1 Humanitarian Technology Conference (R10-HTC). IEEE (2017)
52. Almonacid, B., et al.: Solving the manufacturing cell design problem using the modified binar firefly algorithm and the Egyptian vulture optimisation algorithm. IET Softw. **11**(3), 105–11 (2016)
53. Preethi, J., Sowmiya, S.: Emotion recognition from EEG signal using ISO-FLANN with fire fly algorithm. In: 2016 International Conference on Communication and Signal Processin (ICCSP), IEEE (2016)
54. Muthuramalingam, A., Gnanamanickam, J., Muhammad, R.: Optimum feature selection usin firefly algorithm for keystroke dynamics. In: International Conference on Intelligent System Design and Applications. Springer, Cham (2017)
55. Saraç, E., Ayşe Özel, S.: Web page classification using firefly optimization. In: 2013 IEEE International Symposium Innovations in Intelligent Systems and Applications (INISTA) (2013
56. Sawhney, R., Mathur, P., Shankar, R.: A firefly algorithm based wrapper-penalty feature selec- tion method for cancer diagnosis. In: International Conference on Computational Science and Its Applications. Springer, Cham (2018)
57. Agarwal, V., Bhanot, S.: Radial basis function neural network-based face recognition using firefly algorithm. Neural Comput. Appl. 1–18 (2017)
58. Yang, X.S.: Review of meta-heuristics and generalized evolutionary walk algorithm. Int. J. Bio-Inspired Comput. **3**(2), 77–84 (2011)
59. Yang, X.S.: Metaheuristic optimization: algorithm analysis and open problems. In: Rebennack, P. (ed.) Experimental Algorithms, Lecture notes in Computer Science, vol. 6630, pp. 21–32. Springer, Berlin (2011)
60. Zamuda, A., Brest, J.: Population reduction differential evolution with multiple mutation strate- gies in real world industry challenges. In: ICAISC (SIDE-EC), pp. 154–161 (2012)

Multi-step-Ahead Fuzzy Time Series Forecasting by Using Hybrid Chemical Reaction Optimization with Pi-Sigma Higher-Order Neural Network

Radha Mohan Pattanayak, H. S. Behera and Sibarama Panigrahi

Abstract Over the years, different researchers have followed one-step-ahead method for forecasting the time series data. However, one-step-ahead forecasting may not assist people for making better decision in problems, like sales data forecast, stock market forecast, weather forecasting, and so on. Therefore, for such type of problems multi-step-ahead forecasting will be a better solution for people to make the correct decision. In this current research, we used chemical reaction optimization (CRO) algorithm hybridized with pi-sigma neural network (PSNN) for multi-step-ahead forecasting of different time series data. In order to measure the accuracy of the proposed CRO-PSNN model, we compare the forecasting results with Jaya-PSNN and TLBO-PSNN using ten time series data sets. The proposed CRO-PSNN provided best result in six time series using RMSE measure and five time series using SMAPE measure. Moreover, in the Nemenyi hypothesis test, the proposed CRO-PSNN model has the best rank, which shows its significance.

Keywords Teaching learning-based algorithm (TLBO) · Pi-sigma (PS) · Higher-order neural network (HONN) · Multi-step-ahead forecasting

---

R. M. Pattanayak (✉) · H. S. Behera
Department of Information Technology, Veer Surendra Sai University of Technology, 768018 Burla, Odisha, India
e-mail: radhamohan.pattanayak@gmail.com

H. S. Behera
e-mail: hsbehera_india@yahoo.com

S. Panigrahi
Department of Computer Science and Engineering, Sambalpur University Institute of Information Technology, Burla 768019, Odisha, India
e-mail: panigrahi.sibarama@gmail.com

R. M. Pattanayak
Department of Computer Science and Engineering, Godavari Institute of Engineering and Technology (Auto), Rajahmundry 533296, Andhra Pradesh, India

A. K. Das et al. (eds.), *Computational Intelligence in Pattern Recognition*, Advances in Intelligent Systems and Computing 999,
https://doi.org/10.1007/978-981-13-9042-5_88

1 Introduction

In forecasting, we will be not only concerned about the accuracy, but also to thi how far the forecaster is able to forecast the data. Owing to this, the multi-ste ahead procedure came into play. Normally, a multi-step-ahead procedure can implemented in two ways: iterative way and direct way, both explained in Sect. 2 In iterative procedure many of the researchers have followed a single neural netwo (NN) where the forecasted output has been fed recursively to the network. Howev in the direct way, the researchers used more than one NN and tuned them separate and forecasted the output in multi-step-ahead way.

In 2017, Yunpeng et al. [1] used LSTM RNN for forecasting the wider range of da pattern; however, they found the normal traditional models still took less number resources and faster forecasting over multi-step-ahead forecasting. In 2010, Mam and Samad [6] compared both iterative and direct methods for multi-step-ahea forecasting by using hybrid RBF neural network. They found that both the procedur reflect good accuracy; however, they revealed that for long-term forecasting the dire method outperformed over iterative method. In 2009, Luna et al. [7] used evolvin fuzzy inference systems for daily inflow forecasting. However, they applied bottom up approach for forecasting daily stream flow and combined the result of more tha one predictor to explore better accuracy for multi-step-ahead forecasting.

Since the late 1980s, more researchers are concentrating their forecasting in one step-ahead only. In the year 1994, Hill et al. [16] and in 2007 Chang et al. [17 implemented the multi-step recursive-type forecasting and found better accurac result. However, after that, in the same year 1994, Zhang [18] implemented multi step-type direct forecasting method and found that the direct method yields bette result than the recursive method.

During the last two decades, researchers used ANN to a large extent for forecasting purpose. Pattanayak and Behera [8] have explained different types of HONN and suggested that due to slow convergence rate and stacked to local minima by ANN, HONN can be a better choice to solve complex time problems. Rao [9, 10] developed Jaya as an optimization model for optimizing the constrained and unconstrained optimization problem to find the best result. Compared to other models, he used less number of independent parameters; hence it is simpler and less complex. In 2010, Lam and Li [5] proposed chemical reaction optimization (CRO) as a population-based metaheuristic optimization technique. Further, in 2017 Panigrahi [22] has developed hybrid model of DE-CRO-PSNN, which has shown significant result in classification accuracy.

In the year 2011, Rao et al. [11, 12] have simulated TLBO as a population-based algorithm based on teaching learning procedure. They developed this algorithm without any algorithm-specific parameter. Again in 2012, Rao et al. [14] have added elitism concept into TLBO to improve the performance of TLBO and used optimization of complex global optimization problem.

Therefore, in our present work, we have followed multi-step forecasting procedure to forecast the sales data by implementing the hybrid CRO-PSNN as higher-order ed-forward neural network. The rest of the paper is organized as follows: Sect. 2 efly explained about the preliminary study, Sect. 3 explains the proposed model chitecture, Sect. 4 explains the result analysis and conclusion, and future scope of e current research is explained in Sect. 5.

Preliminaries

1 *Fuzzy Time Series Forecasting (FTSF)*

efinition 1 (*Singh and Borah* [3, 4]) Let $\mathbf{F}(t)$ be predicted by $(t-1), \mathbf{F}(t-1), \mathbf{F}(t-2)\ldots\ldots\mathbf{F}(t-n)$, then the nth-order fuzzy logical lation will be represented as, $\mathbf{F}(t-n),\ldots\ldots\mathbf{F}(t-1) \rightarrow \mathbf{F}(t)$, where $\mathbf{F}(t)$ is the orecasting component.

efinition 2 (*Zadeh* [2]) Let U be the universe of discourse: $\mathrm{U} = \{\mathrm{u}_1, \mathrm{u}_2, \ldots\ldots \mathrm{u}_\mathrm{k}\}$. fuzzy set 'A' of the universe of discourse U can be defined by its membership unction, $\mu_A : U \rightarrow [0, 1]$ where $\mu_A(u)$ is the degree of membership of the element n the fuzzy set A.

efinition 3 (*Huarng* [21]) Suppose there is more than one fuzzy logical relationship n the same left hand side, that is, $A_i \rightarrow A_{j1} and A_i \rightarrow A_{j2}$. They can be combined nto one fuzzy logical group (FLG) as $A_i \rightarrow A_{j1}, A_{j2}$.

2.2 *Pi-Sigma Higher-Order Neural Network (PS-HONN)*

In order to solve the complex-type problems, Shin and Ghosh [15] developed higher-order neural network as pi-sigma (PS). As shown in Fig. 1, this model has three fixed layers: an input layer, output layer and single hidden layer. Inside the network all the different layers are connected in a feed-forward way toward the output layer. Two different layers are connected with some weight factor. The weight between input and hidden layer is trainable but the weight between hidden and output layer is fixed to one. However, such model has only one layer of trainable weight, so the complexity of the hidden layer reduced by less number of tunable weight. Hence, this type of model is very simple to implement and it shows significant accuracy compared with other models.

The output of the network and the output of the jth hidden unit, that is, $\mathrm{h_j}$ can be calculated as in Eqs. (1) and (2), respectively.

$$Y = f\left(\prod_{j=1}^{k} h_j\right) \tag{1}$$

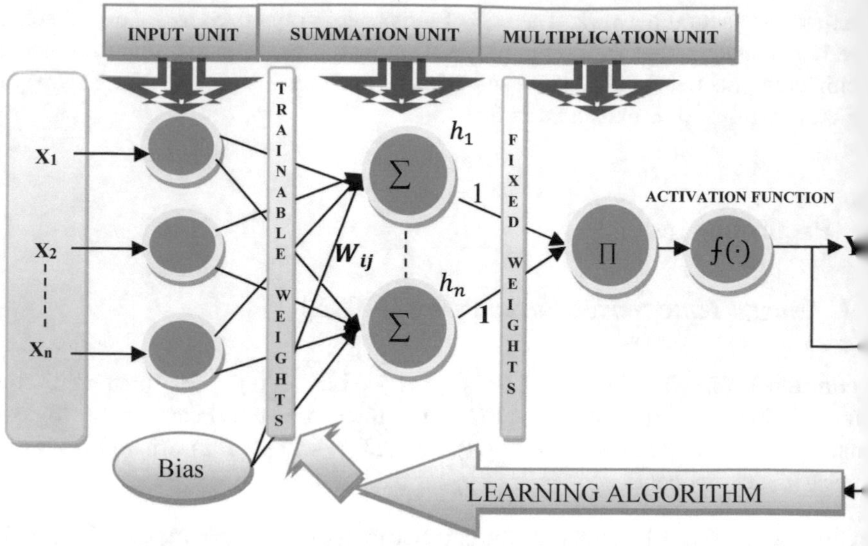

Fig. 1 Pi-sigma higher-order neural network

$$h_j = \theta_j + \sum_{i=1}^{k} W_{ij} X_i \quad (2)$$

where θ_j and W_{ij} are the bias and weight between *i*th input and *j*th hidden unit respectively.

2.3 One-Step-Ahead and Multi-step-Ahead Forecasting

In earlier research most of the researchers have implemented the forecasting in one-step-ahead, which will not be sufficient for the community to take decision from the forecasting result. One-step-ahead forecasting will be represented as in Eq. (3).

$$\left(X_{t-p}, \cdots X_{t-3}, X_{t-2}, X_{t-1}\right) \rightarrow X_t \quad (3)$$

$$\left(X_{t-p}, \cdots\cdots X_{t-3}, X_{t-2}, X_{t-1}\right) \rightarrow (X_t, X_{t+1}, X_{t+2}, \cdots) \quad (4)$$

where X_t is the observation at time t.

But in multi-step-ahead forecasting, it will forecast the next t values for a historical time series data, as represented in Eq. (4). Multi-step-ahead forecasting can be forecasted in two different ways: iterative way and direct way. In the iterative method, it will use the outputs of before time step –1 as an input into the next time step t. The ɔblem faced in this procedure is that the errors may be increased some few steps in ther. But in the direct method, different models will be responsible for forecasting ᵋ time series data. It means that one model can forecast in a one-step forecasting d the other model may be forecasted in two or three-step way. However, the second ocedure will consume more number of resources.

Proposed Model

.1 *A Brief Description of CRO-PSNN*

he total execution process of the model is explained in Sect. 3.2. At first, the given me series data decomposed into some in-sample (i.e. train and validation sample) nd out-of-sample (i.e. test sample). Then for the training pattern, the universe of iscourse U is defined, where the values of D_1 and D_2 are calculated as step-1 in the seudo-code. Then the universe of discourse is partitioned into k intervals of length . Then for each partition, the mid-point has been defined, as shown in step-2 in the seudo-code. Before feeding the time series data into the network, the data has gone hrough the normalization process by following min-max normalization techniques, s explained in Eq. (5). Then the normalized data feed forwarded into the proposed ybrid CRO-PSNN algorithm. The tuning process of the model has been explained in tep-5. Then the resultant forecasted data has gone through de-normalization process, nd the efficiency of the model is measured using Eqs. (6) and (7). To know more bout the CRO optimization process, one can refer to the works [5, 13].

The proposed algorithm is divided into three different phases as initialization phase, iteration phase and the final phase. In the first phase, all the parameters like nitial, termination, length of reactant and reactant number will be initialized. In the iteration phase if the reaction is monomolecular, then it considers decomposition and redox1 but for bimolecular it considers synthesis, displacement and redox2 as three reactions. The type of reaction has been chosen by following the intensification and diversification procedure. In the final phase, best enthalpy reactant will be considered as optimal weight for the network.

3.2 *Pseudo-code for CRO-PSNN*

[Input the data set X. randomly initializes the ReacNum of reactants from a uniform distribution [U, L], Length of reactant, Weight of each reactant in the atom.]

Step 1 : Partition the universe of discourse U into k interval of length w, where the value of D_1 and D_2, as $D_1 = 0.1 * D_{min}$ and $D_2 = 0.1 * D_{max}$

Step 2 : For each partition i, find the mid-point as

$$m_i = [(\text{End of Partition}) - (\text{Start of Partition})]/2$$

Step 3 : Apply min-max normalization to normalize time series data

Step 4 : Apply CRO optimization technique to optimize the weights of PSNN

Step 5 : While the stopping criteria are not satisfied

Generate $rand_1 \in [0,1]$

if $rand_1 \leq 0.7$

Generate $rand_2 \in [0,1]$

if $rand_2 \leq 0.5$

Decomposition R_j to produce a new Reactant R_1,

$W_{1,x} = L + \lambda \times (U - L)$, where λ is a random number

Randomly taken from uniform distribution between [0,1]

else

Apply Redox1 to R_j to produce a new Reactant R_1 as,

$W_{1,x} = L + \lambda \times (U - L)$, where λ is a random number

Randomly taken from cauchyrnd (0.5, 0.1) distribution betwe

[0,1]

[End of if]

else

Generate $rand_3 \in [0,1]$

if $rand_3 \leq 0.33$

Randomly select two best reactants (R_k and R_j where $R_k \neq R_j$)

Apply Synthesis to produce a new Reactant R_1, as

$W_{1,x} = L + \lambda \times (U - L)$,

Where λ is a random number randomly taken from cauchyrnd (M_t, 0.1

Distribution between [0,1.5] and M_t as location parameter

else if $rand_3 \leq 0.66$

Randomly select two best reactants (R_k and R_j where $R_k \neq R_j$)

Apply Displacement to produce a two new Reactants R_1, and R_2

$R_1 = \lambda_t \times R_j + \lambda_t \times (1 - R_k)$ and

$R_2 = \lambda_t \times R_k + \lambda_t \times (1 - R_j)$

Where λ_t is a random number between [0,1] and will update as

$\lambda_{t+1} = 2.3(\lambda_t)^{2\sin(\pi\lambda t)}$Where t is the no of times the reaction occurs.

else

Randomly select a reactants R_k

Apply Redox1 to R_j to produce a new Reactant R_1 as,

$R_1 = R_j + \lambda_t \times (R_k - R_j)$

Where λ_t will calculate as before stage in Synthesis reaction stage.

[End if] …. [End if] …. [End while]

Step 6 : Replace the best solution in place of worst solution in population

Step 7: De-normalize the forecasted data

Step 8: Find accuracy of the model by using Eq. (8) and Eq. (9)

Step 9: End

Experimental Setup and Results

the present study, we implemented the hybrid chemical reaction optimization gorithm and PSNN (CRO-PSNN) for forecasting the different time series data. order to forecast the time series data, in our paper we followed five-step-ahead recasting procedure. To measure the performance of the proposed model, we comred the result of CRO-PSNN with TLBO-PSNN and Jaya-PSNN. This is because the lesser number of algorithmic parameters needed by the TLBO [11] and Jaya] algorithms with improved performance than the other evolutionary algorithms. or CRO, TLBO and Jaya algorithm the ReacNum, number of learners and number individuals are set to 50. The initial reactants in the case of CRO, learner population size in the case of TLBO and number of individuals in the case of Jaya are itialized within the range between 0 and 1. The other algorithmic parameters of RO are as mentioned in the pseudo-code of CRO-PSNN. In Jaya algorithm, the two ontrol parameters are randomly initialized between 0 and 1. In TLBO algorithm, the aching factor is randomly set to 1 or 2. To train the weights of the model, chemical action optimization algorithm has been used. In order to find the number of inputs or the network, we used auto correlation function (ACF). The number of hidden nits ranged from 1 to 20, where the number of output neuron for the network is xed to one.

Ten different time series data shown in Table 1 were used in our experimental tudy. In the experiment, each time series data has been divided into train sample and est sample. Train sample is used to train the weights of model and the test sample s used for measuring the efficiency of the model. For all three models, we divided he pattern into 60% as train set, 20% as validation set and the rest 20% is used as est set for the time series data. The details about total pattern, train, validation and umber of lags for all time series data have been shown in Table 2.

Before processing the data using the model, the data are normalized into a scaled normalized form using min-max normalization, as given in Eq. (5). Twenty average extensive simulation results over 50 executions have been considered for the performance analysis. The best result from 20 simulations for all the different models was collected.

$$y' = \frac{y - min_y}{max_y - min_y} \tag{5}$$

All the models are executed on all data sets to find the root mean square error (RMSE) and symmetric mean absolute percentage error (SMAPE) values, as shown in Eqs. (6) and (7), respectively. Both SMAPE and RMSE over 50 executions are considered for the performance analysis of different network models.

$$RMSE = \sqrt{\frac{1}{n}\sum_{j=1}^{n}\left(y_j - \hat{y}_j\right)^2} \tag{6}$$

Table 1 Description of different time series sales data

Sales data	Description of sales data
Car cells	Monthly car sales in Quebec Jan 1960–Dec 1968
Champagne	Monthly champagne sales (in 1000s) (p. 273: Montgomery: Fore. and T.S.) Ja 1–Dec 8 (1976)
Dietary	Advertising and sales data: 36 consecutive monthly sales and advertising expenditures of a dietary weight control product from Jan 1 to Dec 3, 2013
Gasoline	Monthly gasoline demand, Ontario gallon millions Jan 1960–Dec 1975
Industries	Monthly unit sales, Winnebago Industries, Nov 1966–Feb 1972
Medicine	Annual domestic sales and advertising of Lydia E, Pinkham Medicine, 1907–1960
Paper	Industry sales for printing and writing paper (in thousands of French francs). Ja 1963–Dec 1972
Spare part	Weekly demand of a spare part (p. 271: Montgomery) (Sales, Source: Montgomery and Johnson (1976))
Vehicles	US monthly sales of petroleum and related products. Jan 1971–Dec 1991
Wine	Monthly Australian wine sales: thousands of liters. By wine makers in bottles <= 1 l

Table 2 Description of number of patterns, train, validation and number of lags for each sales da

Data set	Total pattern	Train pattern	Validation pattern	No. of lags
Car cells	108	64	21	12
Champagne	96	57	19	12
Dietary	36	21	7	5
Gasoline	192	115	38	12
Industries	64	38	12	11
Medicine	108	64	21	25
Paper	120	72	24	12
Spare part	60	36	12	6
Vehicles	252	151	50	12
Wine	176	105	35	12

$$SMAPE = \frac{1}{n}\sum_{j=1}^{n}\frac{|y_j - \hat{y}_j|}{(|Y_j| + |T_j|)/2} \tag{7}$$

where y_j and $\hat{y}_j$ are actual and predicted value at jth time, respectively.

[Tab]le 3 Mean RMSE result [D]uncan's multiple range (SPSS V.16.0.1)

Data set	Jaya [9]-PSNN	TLBO [11]-PSNN	Proposed model
	Mean ± SDv	Mean ± SDv	Mean ± SDv
Car cells	3.133 ± 600.756	3.336 ± 330.713	**2.986 ± 561.897**
Champagne	2.182 ± 326.850	2.538 ± 273.146	**1.989 ± 429.971**
Dietary	18.408 ± 2.3771	**17.9204 ± 2.830**	17.932 ± 2.6509
Gasoline	1.940 ± 4430.80	1.921 ± 3546.49	**1.811 ± 3435.53**
Industries	3.7872 ± 80.811	**3.201 ± 24.5112**	3.3819 ± 36.190
Medicine	3.6259 ± 55.539	3.526 ± 46.5391	**3.4326 ± 54.349**
Paper	1.5768 ± 14.460	1.5711 ± 14.525	**1.4350 ± 16.920**
Spare part	3.3472E1 ± 7.88	**2.8467 ± 4.7090**	3.1318 ± 5.5238
Vehicles	3.211 ± 0.35408	**3.0281 ± 0.2728**	3.0794 ± 0.3032
Wine	4.462 ± 524.308	5.0899 ± 261.91	**4.286 ± 649.086**

[4].1 *Performance Analysis by Using RMSE*

[T]he mean RMSE value over 50 iterations has been calculated for the three models and [t]he result is shown in Table 3. From the resultant RMSE value, it has been observed [t]hat the proposed model has outperformed upon six data sets (Car Cells, Champagne, [G]asoline, Medicine, Paper and Wine), and TLBO-PSNN also outperformed on four [d]ata sets (Dietary, Industries, Spare part and Vehicles). However, the forecasting accuracy of Jaya-PSNN has been very poor. Therefore, based on RMSE performance, the proposed model has shown average result, whereas the performances of other traditional hybrid models were not up to the mark.

4.2 *Performance Analysis by Using SMAPE*

The mean SMAPE value over 50 iterations has been calculated for the various models using Eq. (7). The result has been shown in Table 4. From the tabular result it is observed that in five-step-ahead forecasting, the proposed CRO-PSNN as well as TLBO-PSNN has outperformed on five different data sets each. The best result is marked in bold in the table. Therefore, from the SMAPE value it is concluded that both

Table 4 Mean SMAPE result of Duncan's multiple range test (SPSS V.16.0.1)

Data set	Jaya [9]-PSNN	TLBO [11]-PSNN	Proposed model
	Mean ± SDv	Mean ± SDv	Mean ± SD
Car cells	14.825 ± 2.877	15.932 ± 1.808	**14.017 ± 2.423**
Champagne	32.407 ± 5.320	37.913 ± 4.588	**31.371 ± 6.656**
Dietary	73.508 ± 10.66	**69.695 ± 9.754**	72.012 ± 9.859
Gasoline	7.325 ± 1.9287	7.193 ± 1.3740	**6.828 ± 1.4433**
Industries	32.98 ± 9.2599	**26.89 ± 2.9925**	28.558 ± 4.404
Medicine	15.95 ± 3.0057	15.571 ± 2.156	**15.382 ± 2.795**
Paper	15.60 ± 2.6588	**14.154 ± 1.888**	14.87 ± 2.0536
Spare part	6.537 ± 1.8865	**5.371 ± 1.0179**	5.929 ± 1.2444
Vehicles	14.607 ± 2.242	**13.545 ± 1.616**	14.001 ± 1.847
Wine	13.407 ± 1.605	15.08 ± 1.1676	**12.938 ± 2.328**

CRO-PSNN and TLBO-PSNN have shown average accuracy but the performance of Jaya-PSNN is very poor.

4.3 Rank Calculation

With respect to comparison of different models over 12 different time series data sets, Friedman and Yemeni hypothesis test [19] is applied on the SMAPE experimental value. Generally, Friedman's test was required to rank the algorithms by considering each data set individually. However, in the case of distinct ranks, the null hypothesis was rejected, so Nemenyi hypothesis method [20] is applied among all algorithms for making pair-wise comparison. The various experimental results with respect to SMAPE are shown in Table 5. From the tabular result analysis, it is clearly found that the proposed CRO-PSNN model performed better rank over the other two models and the hybrid model Jaya-PSNN has performed worst rank over all other models. A graphical comparison between the forecasted values and the actual values on n various time series data using multi-step-ahead forecasting is plotted and shown in Figs. 2, 3, 4, 5 and 6.

le 5 Ranks of all models g SMAPE value edman and Nemenyi othesis)

Model	Different rank on SMAPE value	Mean rank value
Jaya-PSNN	1	80.15
TLBO-PSNN	3	77.24
CRO-PSNN	2	**69.11**

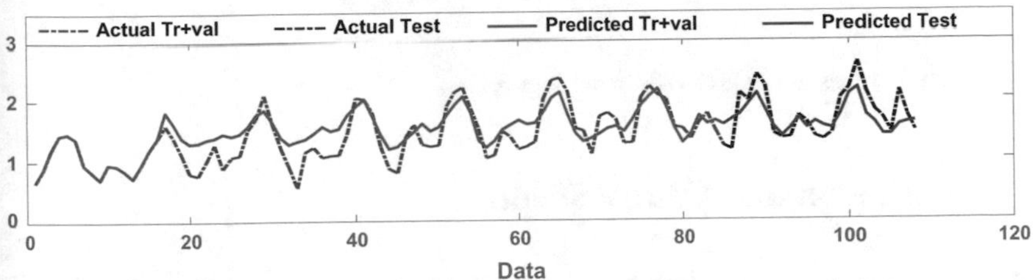

g. 2 The best result of CRO-HONN on Car Cells data

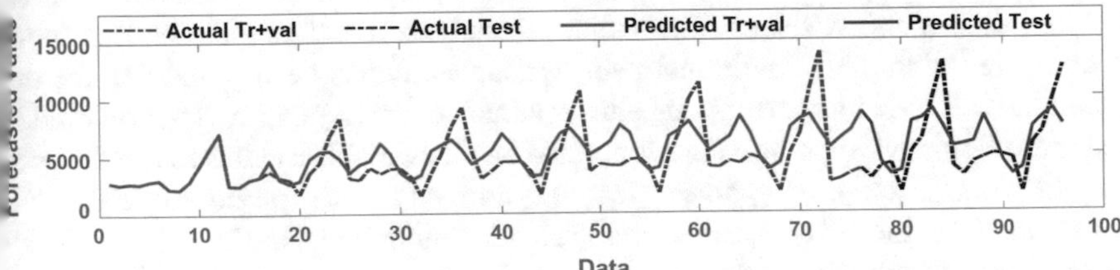

ig. 3 The best result of CRO-HONN on Champagne data

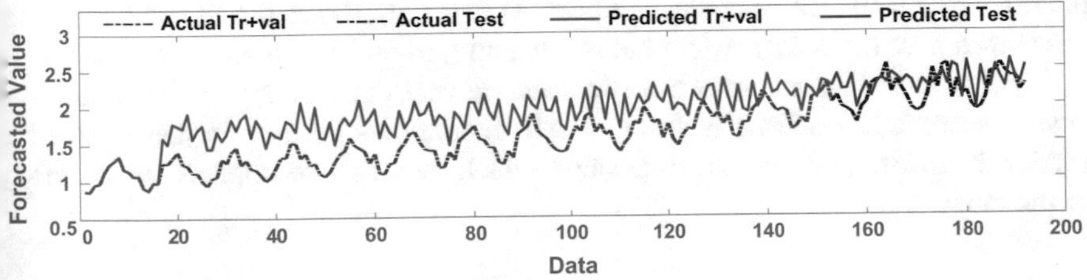

Fig. 4 The best result of CRO-HONN on Gasoline data

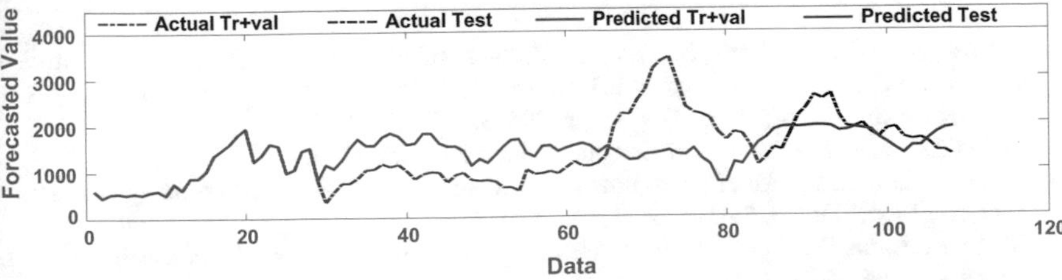

Fig. 5 The best result of CRO-HONN on Medicine data

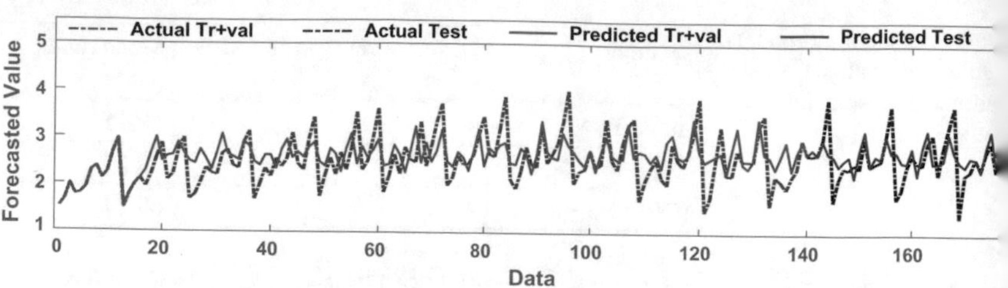

Fig. 6 The best result of CRO-HONN on Wine data

5 Conclusion and Future Scope

In the current research a hybrid, improved CRO combined with PSNN has be implemented for multi-step-ahead forecasting of sales time series data. In ord to test the accuracy, the forecasting result of the proposed model has been pass through descriptive statistics. The statistical analysis result of the proposed mod is compared with other traditional hybrid models and has been found that the pr posed CRO-PSNN has performed better accuracy over the other two hybrid model Later the Nemenyi hypothesis test is applied on all available models to test the acc racy in group. From the hypothesis test, it is noticed that the proposed CRO-PSN hybrid model achieved better rank among all other models and the hybrid Jaya-PSN secured worst rank among the three models. Unlike to one-step-ahead forecasting multi-step-ahead forecasting is a complex task and it takes more complexity, bu in return the result of these models is also not up to the mark. However, from th previous two analyses, it has been observed that the proposed CRO-PSNN mode has shown average accuracy, but not outstanding, in multi-step-ahead forecasting.

So in order to find accuracy in an outstanding way, one can implement either dee neural network or LSTM-RNN model in their research, because in these models i carries the information across time steps, which may help to improve the accurac of the model.

References

1. Yunpeng, L., Di, H., Junpeng, B., Yong, Q.: Multi-step ahead time series forecasting for different data patterns based on LSTM recurrent neural network. In: 2017 14th Web Information Systems and Applications Conference (WISA), pp. 305–310. IEEE (2017)
2. Zadeh, L.A.: Fuzzy sets, information and control, 338–353 (1965)
3. Singh, P., Borah, B.: An efficient time series-forecasting model based on fuzzy time series. Eng. Appl. Artif. Intell. **26**(10), 2443–2457 (2013)
4. Singh, P., Borah, B.: High-order fuzzy-neuro expert system for time series forecasting. Knowl.-Based Syst. **46**, 12–21 (2013)
5. Lam, A.Y., Li, V.O.: Chemical-reaction-inspired meta heuristic for optimization. IEEE Trans. Evol. Comput. **14**(3), 381–399 (2010)

Mamat, M., Samad, S.A.: Comparison of iterative and direct approaches for multi-steps ahead time series forecasting using adaptive Hybrid-RBF neural network. In TENCON 2010–2010 IEEE Region 10 Conference, pp. 2332–2337. IEEE (2010)

Luna, I., Lopes, J.E.G., Ballini, R., Soares, S.: Verifying the use of evolving fuzzy systems for multi-step ahead daily inflow forecasting. In: 15th International Conference on Intelligent System Applications to Power Systems, 2009, ISAP'09, pp. 1–6. IEEE, (2009)

Pattanayak, R.M., Behera, H.S.: Higher order neural network and its applications: a comprehensive survey. In: Progress in Computing Analytics and Networking, pp. 695–709. Springer, Singapore (2018)

Rao, R.: Jaya: A simple and new optimization algorithm for solving constrained and unconstrained optimization problems. Int. J. Ind. Eng. Comput. **7**(1), 19–34 (2016)

Rao, R.V., More, K.C., Taler, J., Ocłoń, P.: Dimensional optimization of a micro-channel heat sink using Jaya algorithm. Appl. Therm. Eng. **103**, 572–582 (2016)

Rao, R.V., Savsani, V.J., Vakharia, D.P.: Teaching–learning-based optimization: an optimization method for continuous non-linear large-scale problems. Inf. Sci. **183**(1), 1–15 (2012)

Rao, R.V., Savsani, V.J., Vakharia, D.P.: Teaching–learning-based optimization: a novel method for constrained mechanical design optimization problems. Comput.-Aided Des. **43**(3), 303–315 (2011)

Sahu, K.K., Panigrahi, S., Behera, H.S.: A novel chemical reaction optimization algorithm for higher order neural network training. J. Theor. Appl. Inf. Technol. **53**(3) (2013)

Rao, R., Patel, V.: An elitist teaching-learning-based optimization algorithm for solving complex constrained optimization problems. Int. J. Ind. Eng. Comput. **3**(4), 535–560 (2012)

Shin, Y., Ghosh, J.: The pi-sigma network: An efficient higher-order neural network for pattern classification and function approximation. In: IJCNN-91-Seattle International Joint Conference on Neural Networks, 1991, vol. 1, pp. 13–18. IEEE, (1991)

5. Hill, T., Marquez, L., O'Connor, M., Remus, W.: Artificial neural network models for forecasting and decision-making. Int. J. Forecast. **10**(1), 5–15 (1994)

7. Chang, F.J., Chiang, Y.M., Chang, L.C.: Multi-step-ahead neural networks for flood forecasting. Hydrol. Sci. J. **52**(1), 114–130 (2007)

8. Zhang, X.: Time series analysis and prediction by neural networks. Optim. Methods Softw. **4**(2), 151–170 (1994)

9. Hollander, M., Wolfe, D.: A Non-Parametric Statistical Methods (1999)

0. Demsar, J.: Statistical comparisons of classifiers over multiple data sets. J. Mach. Learn. Res. **7**, 1–30 (2006)

1. Huarng, K., Yu, T.H.K.: The application of neural networks to forecast fuzzy time series. Phys.: Stat. Mech. Its Appl. **363**(2), 481–491 (2006)

2. Panigrahi, S.: A novel hybrid chemical reaction optimization algorithm with adaptive differential evolution mutation strategies for higher order neural network training. Int. Arab. J. Inf. Technol. **14**(1), 18–25 (2017)

uthor Index

A. K. Das et al. (eds.), *Computational Intelligence in Pattern Recognition*, Advances in Intelligent Systems and Computing 999,
https://doi.org/10.1007/978-981-13-9042-5

Printed in the United States
By Bookmasters